W9-BWM-873

ENCYCLOPEDIA OF HUMAN BIOLOGY

VOLUME 5 Mi–Ph

ENCYCLOPEDIA
of HUMAN
BIOLOGY
VOLUME 5 Mi–Ph

Editor–in–Chief
Renato Dulbecco
The Salk Institute
La Jolla, California

ACADEMIC PRESS, INC. *Harcourt Brace Jovanovich, Publishers*
San Diego   New York   Boston   London   Sydney   Tokyo   Toronto

Copyright © 1991 by ACADEMIC PRESS, INC.

All Rights Reserved.

No part of this publication may be reproduced or transmitted in any form or by any means, electronic or mechanical, including photocopy, recording, or any information storage and retrieval system, without permission in writing from the publisher.

Academic Press, Inc.
San Diego, California 92101

United Kingdom Edition published by
Academic Press Limited
24–28 Oval Road, London NW1 7DX

Library of Congress Cataloging-in-Publication Data

Encyclopedia of human biology / [edited by] Renato Dulbecco.
    p. cm.
   Includes index.
   ISBN 0-12-226751-6 (v. 1). -- ISBN 0-12-226752-4 (v. 2). -- ISBN
0-12-226753-2 (v. 3). -- ISBN 0-12-226754-0 (v. 4). -- ISBN
0-12-226755-9 (v. 5). -- ISBN 0-12-226756-7 (v. 6). -- ISBN
0-12-226757-5 (v. 7). -- ISBN 0-12-226758-3 (v. 8)
    1. Human biology--Encyclopedias.   I. Dulbecco, Renato, 1914-
    [DNLM: 1. Biology--encyclopedias.   2. Physiology--Encyclopedias.
QH 302.5 E56]
QP11.E53   1991
612'.003--dc20
DNLM/DLC
for Library of Congress   91-45538
                                                   CIP

PRINTED IN THE UNITED STATES OF AMERICA
91 92 93 94   9 8 7 6 5 4 3 2 1

CONTENTS OF VOLUME 5

v

HOW TO USE THE ENCYCLOPEDIA

We have organized this encyclopedia in a manner that we believe will be the most useful to you and would like to acquaint you with some of its features.

The volumes are organized alphabetically as you would expect to find them in, for example, magazine articles. Thus, "Food Toxicology" is listed as such and would *not* be found under "Toxicology, Food." If the first words in a title are *not* the primary subject matter contained in an article, the main subject of the title is listed first: (e.g., "Sex Differences, Biocultural," "Sex Differences, Psychological," "Aging, Psychiatric Aspects," "Bone, Embryonic Development.") This is also true if the primary word of a title is too general (e.g., "Coenzymes, Biochemistry.") Here, the word "coenzymes" is listed first as "biochemistry" is a very broad topic. Titles are alphabetized letter-by-letter so that "Gangliosides" is followed by "Gangliosides and Neuronal Differentiation" and then "Ganglioside Transport."

Each article contains a brief introductory Glossary wherein terms that may be unfamiliar to you are defined *in the context of their use in the article*. Thus, a term may appear in another article defined in a slightly different manner or with a subtle pedagogic nuance that is specific to that particular article. For clarity, we have allowed these differences in definition to remain so that the terms are defined relative to the context of each article.

Articles about closely related subjects are identified in the Index of Related Articles at the end of the last volume (Volume 8.) The article titles that are cross-referenced within each article may be found in this index, along with other articles on related topics.

The Subject Index contains specific, detailed information about any subject discussed in the *Encyclopedia*. Entries appear with the source volume number in boldface followed by a colon and the page number in that volume where the information occurs (e.g., "Diuretics, **3:** 93"). Each article is also indexed by its title (or a shortened version thereof) and the page ranges of the article appear in boldface (e.g., "Abortion, **1: 1–10**" means that the primary coverage of the topic of abortion occurs on pages 1–10 of Volume 1).

If a topic is covered primarily under one heading but is occasionally referred to in a slightly different manner or by a related term, it is indexed under the term that is most commonly used and a cross-reference is given to the minor usage. For example, "B Lymphocytes" would contain all page entries where relevant information occurs, followed by "*see also* B Cells." In addition, "B Cells, *see* B Lymphocytes" would lead the reader to the primary usages of the more general term. Similarly, "*see under*" would mean that the subject is covered under a subheading of the more common term.

An additional feature of the Subject Index is the identification of Glossary terms. These appear in the index where the word "defined" (or the words "definition of") follows an entry. As we noted earlier, there may be more than one definition for a particular term and, as when using a dictionary, you will be able to choose among several different usages to find the particular meaning that is specifically of interest to you.

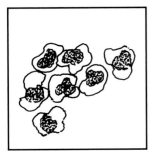

Microbial Protein Toxins

GRAHAM BOULNOIS, *Leicester University*

I. Diversity of Microbial Toxins
II. Toxins with an Intracellular Site of Action
III. Membrane Damaging/Interacting Toxins

Glossary

Endotoxin Structural component of microbe that is toxic when released from cell
Exotoxin Protein toxin secreted from a microbe
Holotoxin Fully assembled, active form of multi-subunit toxin

MANY MICROBIAL COMPONENTS are toxic to plants, animals, and humans. The toxic effects manifest by these components are diverse, often only apparent in special circumstances, and many are of questionable significance in the natural ecology of the microorganism and of unknown importance in terms of human health and disease.

Broadly speaking, toxins can be divided into two classes. The endotoxins, typified by lipopolysaccharide of gram-negative bacteria, are found as important components of microbial cell membranes and cell walls. They exert their toxic effects after release from these structures of which they are an integral and vital part. In contrast, the exotoxins are secreted from the cell by specific or more generalized mechanisms and are typically proteinaceous.

I. Diversity of Microbial Exotoxins

The enormous diversity of microbial life is matched by the production of a large range of microbial products that are toxic to humans. The targets for these toxins (i.e., the tissues and cells of the human body) are equally diverse. In some cases toxicity results from the activity of microbial components such as enzymes such as hyaluronidase and neuraminidase that act in the extracellular milieu of the body. The former depolymerises hyaluronic acid, a component of the extracellular matrix, whereas the latter depolymerizes polysialic acids, important constituents of mammalian cell surfaces. However, the classical toxins are usually considered to act directly on cells either to disrupt membrane function or to modify an intracellular target leading to cell death or deregulation of cellular metabolism. Only in a few cases is the biochemical nature of the "toxic event" known.

II. Toxins with an Intracellular Site of Action

Modification of intracellular targets by this group of toxins requires identification of the target cells via recognition of a specific cell surface receptor, and internalization. The structural complexity of many of the toxins that fall within this general category reflects these steps in activity. This group of toxins are often composed of multiple subunits or are single polypeptides, which are frequently proteolytically cleaved. In the latter case the processed polypeptides are often held together by disulfide bonds. Consequently, this group of toxins are sometimes referred to as the subunit toxins.

The classical subunit toxin is cholera toxin. This toxin is composed of two subunits A and B, and the holotoxin contains five molecules of the B subunit and a single A subunit. The A subunit, although synthesized as a single polypeptide, is cleaved to yield two protein chains termed A_1 and A_2. The heat labile toxin of *Escherichia coli* has a similar molecular organization and a similar mode of action. Another example of a multicomponent toxin is pertussis toxin, produced by the etiological agent of whooping cough, *Bordetella pertussis*. In contrast to cholera toxin, pertussis toxin is composed of five different subunits, and the holotoxin is composed of one molecule of subunits 1, 2, 3, and 5 and two molecules of subunit 4.

Several subunit toxins are synthesized as a single polypeptide precursor and processed after synthesis. Perhaps the classical example is diphtheria toxin, which is made as a single polypeptide (62–63 kDa) containing two disulphide bonds. One of these is formed between two cysteine residues toward the N-terminus, and the intervening region contains three arginine residues, which render the molecule susceptible to cleavage by proteases. Proteolytic nicking generates two polypeptide fragments (A and B) held together by a disulfide bond. This nicking may occur during the internalization process. Although there are differences in detail, pseudomonoas exotoxin A (A and B chains) and tetanus (H and L chains) follow similar principles.

In effect, the subunit structure of the above toxins serves to divorce the toxic component of the molecule, which acts intracellularly, from the components of the toxin that mediate binding of holotoxin to target cells and internalization of the toxic moiety. Receptor recognition determines, at least in part, the cell and tissue tropism of each toxin. In the case of cholera toxin and *E. coli* heat-labile toxin, the B subunits recognize and bind to ganglioside (GM_1) on target cells although *E. coli* heat-labile toxin may also bind galacto-protein. It is generally believed that a single B subunit binds to a single GM_1 molecule. Pertussis toxin, in contrast, has two receptor binding complexes made up of subunits 2 + 4 and 3 + 4, which bind to mannose-containing glycolipids. In some cases, for example, diphtheria toxin, which recognizes a 120–170 kDa glycoprotein, the number of cell surface receptors correlates with the susceptibility of the target cells to the toxin.

Tetanus toxin acts primarily in the spinal cord and is thought to gain access to this site via binding to ganglioside receptors on motor nerve endings in the muscles. Recognition is mediated by a C-terminal fragment of the toxin, which shares amino acid sequence homology with influenza virus hemagglutinin, which also interacts with ganglioside. The toxin is probably taken up into the axon by endocytosis and is transported within vesicles along the nerve trunk to the spinal cord. Here the toxin is released from the motor neuron and interacts with inhibitory neuronal endings.

After binding to target cells, the subunit toxins have evolved several mechanisms to internalize the biologically active moiety of the molecule. In the case of tetanus toxin, insertion into the membrane may be mediated by hydrophobic interactions between the toxin and membrane lipids. The majority of the hydrophobic domains of the toxin lie in the L chain, and one of these may adopt an alpha-helical configuration that may span the membrane. Membrane insertion events may also be a feature of pertussis toxin. In this case, subunits 2 and 3 may insert into membrane. How this facilitates translocation of the active subunit (subunit 2) into the cell remains unclear. Similarly, the mechanism by which cholera toxin is inserted into cells is unclear. It has been postulated that the B subunit inserts into the membrane via hydrophobic interactions to generate a pore that allows passage of the active a subunit through the membrane, events that are also associated with the proteolytic nicking of the A subunit to yield A_1 and A_2. However, B-subunit–receptor interaction may simply serve to bring the A subunit into close apposition with the membrane, which facilitates insertion of the A subunit into the cytoplasm via a membrane vesicle.

In the case of diphtheria toxin, internalization may occur via receptor-mediated endocytosis. After receptor binding by the B fragment, bound toxin molecules may migrate to coated pits and the toxin-receptor complexes may internalize in clathrin-coated endosomes. Loss of the clathrin coat yields smooth endosomes, which are capable of migration within the cytoplasm. A pH drop within the smooth endosome is thought to facilitate membrane insertion of the B domain, which undergoes a conformational change that pulls the A region through the membrane where proteolytic nicking occurs, and the active domain (A fragment) is released by reduction of the disulfide bond. Given the similarity between diphtheria toxin and Pseudomonas exotoxin A, it is tempting to speculate that the latter is internalized by a similar mechanism.

Cholera, *E. coli* heat-labile, diphtheria, and pertussis toxins and Pseudomonas exotoxin A all are toxic as a consequence of their ability to ADP-ribosylate proteins. This occurs via the enzymic transfer of an ADP-ribose moiety from NAD^+. In the case of diphtheria and Pseudomonas exotoxin A, the target for ADP-ribosylation is elongation factor (EF) 2, a component of the protein synthesizing machinery required for polypeptide chain elongation. For diphtheria toxin, a unique amino acid (2[3-carboxy-amido-3(trimethylammonio)propyl]histidine), called *diphthamide* and found only in EF2, is ADP-ribosylated. Diphthamide may be an important target for other cellular regulatory proteins, and its modification may subvert normal regulatory process, resulting in inhibition of protein synthesis and cell death. One molecule of internalized diphtheria toxin can modify many EF2 molecules causing cell death within a few hours. Inhibition of protein synthesis in the human heart may be the cause of death in *Corynebacterium diphtheriae* infections. Pseudomonas exotoxin A may act in a similar fashion.

Although cholera, *E. coli* heat-labile, and pertussis toxins have ADP-ribosylating activity, their target is distinct from EF2. These toxins modify important regulatory components of the cell. Pertussis toxin ADP-ribosylates the G_1 protein and other G proteins (e.g., G_0) of unknown function. Pertussis toxin-mediated modification of G_1, a protein that receives cellular signals for inhibitory receptors (R_1), which in turn receive signals from inhibitory hormones, results in an uncoupling of G_1 from R_1. Because G proteins regulate the activity of adenylate cyclase, modification/inactivation of G_1 renders cells unable to down-regulate adenylate cyclase activity, and thus intracellular levels of cAMP become elevated. Interestingly, G_0 is found in nervous tissue that, when modified by pertussis toxin, may explain the observed neurological lesions associated with infection or vaccination with pertussis toxoid. [*See* G PROTEINS.]

Cholera and *E. coli* heat-labile toxins also exert their toxic effects by mediating an elevation of intracellular cAMP levels. In contrast to pertussis toxin, these toxins achieve this by ADP-ribosylation of G_s, another component of the adenylate cyclase regulatory apparatus. Modification of G_s prevents the hydrolysis of G_s-bound GTP, which results in constitutively high levels of adenylate cyclase. Elevated cAMP levels in the gut mucosa start a train of effects, resulting in the characteristic profuse, watery diarrhea associated with infection by *Vibrio cholerae* and heat-labile toxin-producing *E. coli*.

Tetanus toxin acts to block the release of the neurotransmitter inhibitor (glycine) into the synapse, resulting in constant stimulation of motor neurons. This results in the classical picture of intoxication by *Clostridium tetani* (a generalized muscular contraction ending in respiratory failure). The mechanism by which inhibitor release is blocked is not wholly clear but may involve inhibition of the calcium-dependent ATPase activity of an actomyosin-like protein preventing its contraction. Contraction may normally promote vesicle-synaptic membrane contact required for the exocytosis of vesicle-containing inhibitor.

III. Membrane Interacting/ Damaging Toxins

This is an extremely large and diverse group of toxins, and relatively little is known about their structure, mode of action, and relevance in disease. Gram-positive bacteria, particularly clostridial species, are prolific producers of this type of toxin, although it seems highly likely that many toxins belonging to this group have yet to be described.

One of their best understood membrane-damaging toxins is α-toxin of *Clostridium perfringens*. This toxin has a phospholipase activity, releasing phosphorylcholine and neutral diglyceride from phosphatidyl choline, although other phospholipids are also substrates. The effect of this activity on cell membranes is the release of phosphorylcholine, which renders the membrane unstable, and toxin-treated cells lyse.

In contrast to clostridial α toxin, many membrane-damaging toxins act to disrupt membrane integrity. One example is staphylococcal δ toxin, a heat-labile, hydrophobic peptide of 5,000 kDa, which may act as a surfactant. This toxin may adopt an α-helical, amphipathic rod-shaped membrane-spanning structure, and such molecules may aggregate within membrane to generate pores. The size of the pore may vary with the concentration of toxin applied to the membrane.

The thiol-activated toxins, so-called because they undergo a reversible oxidation-reduction between inactive and active forms, may damage membranes in a manner similar to δ toxin. These toxins represent a large family of similar proteins, all with a single cysteine, which includes Streptolysin O (SLO) from *Streptococcus pyogenes*, pneumolysin from *Streptococcus pneumoniae*, listeriolysin from *Listeria monocytogenes*, and perfringolysin from *Clostridium perfringens*. However, a thiol group is not essential for *in vitro* activity. Although these toxins are produced by disparate gram-positive microorganisms, their primary sequences (and probably their secondary structures) are remarkably similar. These toxins undergo aggregation in membranes, and the aggregates may form pores. All eukaryotic cells that contain cholesterol in their membranes are susceptible, and it has been suggested that the sterol acts as the membrane receptor for these toxins. However, studies on pneumolysin suggest that another membrane component may act as receptor and that sterol–toxin interaction may be important in pore formation. A domain of the toxin containing the unique cysteine residue may mediate sterol interactions. At concentrations of toxin that fail to cause cell lysis, pronounced effects on cellular metabolism become evident. Such effects, particularly on phagocytic cells and lymphocytes, may be important in the disease process. There is convincing evidence for listeriolysin and pneumolysin that they contribute to the virulence of their producing organism.

The α-toxin of *Staphylococcus aureus* probably acts as a channel-forming protein in many mammalian membranes, and channel formation may involve a toxin-aggregation event. The aggregates resemble those produced by the thiol-activated toxin. Rapid binding of toxin to cells is mediated by the C-terminal half of the protein. A specific cell surface receptor may not be involved because the toxin, albeit at higher than normal concentrations, can bind to artificial membranes probably via hydrophobic interactions. A relatively slow onset of membrane damage may reflect a change in conformation of the toxin, perhaps involving hexamerization. Small molecules cross the treated membranes, and there is an influx of water such that cells swell and eventually lyse, a process called *colloid-osmotic lysis*. However, many nucleated cells fail to lyse, but membrane lesion results in aberrant cell metabolism.

Staphylococcal toxic shock syndrome toxin, enterotoxin (of which there are five serogroups A–E), and streptococcal pyrogenic toxin are potent polyclonal mitogens and pyrogens that enhance endotoxic shock. In the case of the type A and B enterotoxins and toxic shock syndrome toxin, it has been demonstrated that the proteins bind to the major histocompatibility complex class II antigen (HLA, DR), and this may be the basis of their mitogenicity. Because such molecules are found on T helper, cytolytic, and suppressor cells, this might explain some of the reported immuno-suppressive and immuno-enhancement effects of some of these toxins.

Bibliography

Bhakdi, S., and Tranum-Jensen, J. (1989). Damage to cell membranes by pore-forming bacterial cytolysins. *Prog. Allergy* **40**, 1–43.

Fraser, J. D. (1989). High affinity binding of staphylococcal enterotoxins A and B to HLA-DR. *Nature* **339**, 221–223.

Hardegree, M. A., and Tu, A. T., eds. (1988). "Handbook of Natural Toxins, vol. 4, Bacterial Toxins." Marcel Dekker, New York, Basel.

Kehoe, M. K., Miller, L., Walker, J. A., and Boulnois, G. J. (1987). Nucleotide sequence of the streptolysin O (SLO) gene: Structural homologies between SLO and other membrane-damaging, thiol-activated toxins. *Infect. Immun.* **55**, 3228–3232.

Saunders, F. K., Mitchell, T. J., Walker, J. A., Andrew, P. W., and Boulnois, G. J. (1989). Pneumolysin, the thiol-activated toxin of *Streptococcus pneumoniae*, does not require a thiol group for *in vitro* activity. *Infect. Immun.* **57**, 2547–2552.

Stephen, J., and Pietrowski, R. A. (1986). "Bacterial Toxins." American Society for Microbiology, Washington, D.C.

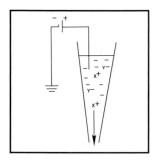

Microiontophoresis and Pressure Ejection

TREVOR W. STONE, *University of Glasgow*

I. Historical Background
II. Applications and Advantages
III. Precautions and Problems of Interpretation

Glossary

Catecholamines Group of phenolic compounds used as neurotransmitters and hormones

End plate Specialized region on the surface of muscle which responds to the transmitter acetylcholine to initiate contraction

Nanoampere One-billionth of an ampere; a measure of electrical current

Neuron Nerve cell consisting of a cell body (i.e., soma) and associated branching processes (e.g., axons and dendrites)

Neurotransmitter Chemical released by one neuron onto another at a synapse, producing excitation (i.e., depolarization) or inhibition (i.e., hyperpolarization)

Synapse Junction between nerve cells

Synaptosomes Isolated nerve terminals obtained by homogenizing and centrifuging a neuronal system

Transport number Proportion of applied current carried by the ion under consideration

MICROIONTOPHORESIS, or microelectrophoresis, literally meaning the movement of small amounts of electrical charge, is a technique for ejecting minute amounts of soluble compounds of interest onto limited regions of a biological tissue by passing appropriate electric currents through the drug solution in a micropipette. For a compound which ionizes in solution to X^+Y^- the passage of outward (i.e., "positive") current from the solution to a distant ground point forces the cation X^+ out of the solution (Fig. 1). Micropipettes are usually multibar-

reled constructions in which three, five, or seven capillary tubes are fused together and then heated and pulled in a microforge to make small openings, usually with a diameter of 1 μm or less.

Compounds ejected by microiontophoresis can therefore be applied to single cells or even restricted areas of the cell surface. The technique is of primary use for comparing the actions of endogenous substances with exogenous molecules and for applying pharmacological antagonists of substances to single cells. Compounds which are not readily ionized in solution could be ejected by the application of pressure pulses to the drug-containing barrels.

I. Historical Background

A. Origins

The beginning of modern microiontophoresis can be attributed to Nastuk, who, in 1953, was seeking a technique to mimic artificially the effects of acetylcholine released as a synaptic transmitter at the neuromuscular junction. He devised what came to be known as the "electrically operated microjet," an apt description for the technique in which he prepared fine glass tubes, micropipettes, by locally heating a region of capillary tubing and then pulling apart the two ends. This yielded two pipettes each, with tip openings of less than 1 μm. The pipettes were then filled with a concentrated solution of acetylcholine chloride. By passing pulses of outward-going (i.e., positive) electrical current through the solution to an earth–ground electrode nearby, the acetylcholine cations could be forced out of the micropipette tip (Fig. 1).

In subsequent developments by del Castillo and Katz, the electrical pulses were made brief, a few milliseconds in duration, by the use of electronic

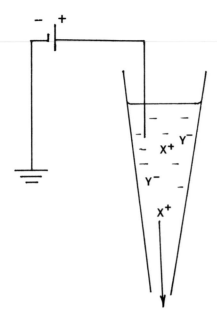

FIGURE 1 The basic principle of microiontophoresis.

switching devices. The depolarization of the muscle endplates induced by this exogenous acetylcholine accurately mimicked the effects of stimulating the motor nerve to the muscle.

B. Developments

The major developments since 1955 have been in the automation of ejection cycling, the sophistication of current pump design to afford greater reliability and reproducibility at the nanoampere current ranges used, and the style of micropipettes. In the central nervous system it was desirable to apply not just one, but several, suspected neurotransmitter compounds and related substances to single cells, and to this end Curtis and colleagues devised the multibarreled micropipette. Several lengths of capillary tubing (usually three, five, or seven) were fused together to form a multibarrel "blank." This was then pulled, as for the single electrodes in a specialized pulling apparatus. After pulling, the tip of the assembly was broken under microscopic control to an overall diameter of about 5 μm. The result was a pair of pipette assemblies in which each individual barrel had a tip diameter of approximately 1 μm. This remains by far the most widely used pipette assembly.

One barrel, usually the center one, of such multibarreled assemblies is normally reserved for recording the responses of cells to the applied chemicals.

This barrel is filled with a highly conducting solution (e.g., 3 M sodium or potassium chloride) and connected electrically to a suitable amplifier. For improving the signal–noise ratio or for intracellular recordings in which the penetrating electrode must have a tip diameter of less than 0.5 μm, independent barrels can be glued along the multibarrel complex or even inserted through the center barrel. In such cases the recording pipette can be arranged to protrude a defined distance beyond the iontophoretic barrels and thus protect the recorded cells from damage.

II. Applications and Advantages

A. Uses

Originally, multibarreled microiontophoresis was used by Curtis and colleagues to examine the effects of a large number of amino acids and analogs on neurons in the spinal cord. This work led to the discovery of the potent excitatory properties of glutamic and aspartic acids, compounds which are now among the most widely used synaptic neurotransmitter substances in the central nervous system. The advantage conferred by the use of multibarreled pipettes, however, is not merely in the possibility of testing a larger number of compounds, but in the opportunity of testing both pharmacological agonists and antagonists simultaneously on any one cell or preparation (Fig. 2).

Thus, for example, the compound from one barrel could block the action of a suspected neurotransmitter applied from a second barrel; then the antagonist substance can be tested for its ability to block the biological response of the cell or the tissue to a physiological stimulus. In this way much of the pharmacological evidence for a neurotransmitter role of acetylcholine, noradrenaline, or amino acids (e.g., glutamate, γ-aminobutyric acid, and glycine) has been obtained.

Microiontophoresis has been applied to an increasingly wide range of biological systems. The early studies mentioned above were performed on anesthetized animals, in which the main advantages of microiontophoresis are readily apparent: By applying substances directly to single cells and recording their immediate responses, it is possible to eliminate the mediating effect of intermediate phenomena (e.g., changes in blood pressure or hormone release). These factors and the attendant al-

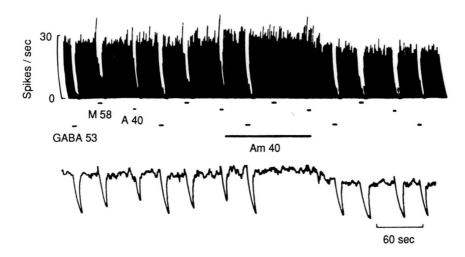

FIGURE 2 An example of records obtained of the firing rate of a neuron in the globus pallidus region of the rat brain, to which compounds were applied by microiontophoresis. The two records show the same sequence, but the upper record was achieved via a resetting integrator, in which the recording pen was reset to zero after counting the activity of the cell for 1-second periods. The lower record was made via an instantaneous rate meter, providing a continuous integrated indication of the firing rate. The bars between the two records indicate the times during which compounds were applied to the cell. In this case the compounds were γ-aminobutyric acid (GABA), applied with a current of 53 nA; morphine (M), applied with a current of 58 nA; and adenosine (A), applied with a current of 40 nA. An application of aminophylline (Am, 40 nA) then causes the blockade of responses to both morphine and adenosine, but not GABA, responses returning after cessation of the aminophylline-ejecting current. These records form part of the evidence that the effects of morphine on the brain neurons might in some situations be mediated by a release of adenosine.

terations of body temperature, blood gases, nutrient supply, etc., present serious problems of interpretation when pharmacological studies are performed by injecting compounds into the whole animal.

Nevertheless, there are still problems with iontophoresis performed *in vivo*. It is never possible to know precisely which region of the cell surface is being studied or whether part of the cell has been damaged by the insertion of the relatively bulky pipette tip into the tissue. It is also a common worry of pharmacological experiments on the central nervous system *in vivo* that the presence of anesthetic could modify neuronal properties or drug sensitivity and thus lead to erroneous conclusions. There has therefore been a general trend to use preparations in which individual cells can be visualized and drugs applied to known areas of the membrane surface. Such preparations include cell cultures and brain slices, in some of which (e.g., the hippocampus) the regular orientation of cell populations facilitates applications to the soma or known regions of the dendritic tree. Cultured cells carry the advantage that, at early stages of development, there might be no intercellular (i.e., synaptic) contacts, so that interfering effects of compounds on synaptic terminals can be excluded. On the other hand, cell cultures are immature and relatively undifferentiated, so great caution should be exercised in extrapolating results to a fully functional intact system.

Microiontophoresis has been used to apply an immense variety of materials to cells ranging from hydrogen ions to a suspension of nerve terminals obtained from a brain homogenate (i.e., synaptosomes). It has been used to apply ADP into vascular capillaries to study platelet aggregation and the formation of thrombi and to apply ATP to activate the flagella of isolated spermatozoa. Its main use, however, continues to be in the study of nerve cells, in which its combination with electrical recording makes it especially useful. [See ADENOSINE TRIPHOSPHATE (ATP); SPERM.]

Microiontophoresis has achieved great value in marking cells. If a micropipette containing a solution of dye is used to penetrate a cell of interest, the dye can then be ejected by microiontophoresis to stain the cell. A subsequent histological preparation of the tissue can then be used to examine the cell and its surroundings in detail under the microscope. Dye marking is particularly valuable for correlating neuronal membrane properties or drug sensitivity with particular structural characteristics (e.g., the presence of certain populations of synaptic terminals or vesicles). Commonly used dyes include Pontamine Sky Blue, Procion Yellow, Lucifer Yellow, and Fast Green. The currents required for dye ejection are usually in the microampere range and therefore generally exceed those used for transmitters and drugs which are between 1 and 100 nA.

B. Variants of Microiontophoresis

Clearly, microiontophoresis, as described above, can only be used for the ejection of substances which ionize in solution, but many compounds of biological interest fail to do so. When an aqueous solution is in contact with glass, however, an electrical "double-layer" is formed, in which negative ions adhere to the glass surface and leave the bulk of the solution carrying a positive charge. The passage of an outward current through a micropipette, therefore, always ejects a minute amount of the fluid it contains. This phenomenon, which results in the ejection of molecules dissolved in the fluid, is known as electroosmosis (since it involves the primary movement of solvent, rather than solute).

C. Micropressure Ejection

By connecting a micropipette or one barrel of a multibarreled assembly to a source of pressure, a small volume of solution can be ejected from the pipette tip. This represents an alternative method for the ejection of un-ionized compounds in solution. Micropressure ejection has an advantage over microiontophoresis, because it is more amenable to quantification. Micropipettes can be filled with a solution of known concentration, and the volume of ejectate at various applied pressures can be calibrated under a microscope. The pressures used normally range to about 30 psi.

III. Precautions and Problems of Interpretation

A. Transport Number

Microiontophoresis is essentially of qualitative value only. By examining the ejection of radiolabeled compounds from micropipettes, it has become apparent that pipette barrels vary enormously in their ability to pass current and, thus, eject dissolved compounds. Even worse, however, there is little correlation between the ability to pass current and the ability to eject any given ion. The factors responsible for this variability are unclear. It is pat-

ently hazardous to compare, for example, the relative potencies of two compounds against this uncertain background. It is only possible to achieve an approximate comparison of potency when the ability of a pipette to eject the substances of interest can be examined *in vitro*. From Faraday's law it is then possible to relate the amount of electric charge passed through a pipette to the quantity of material ejected by the formula

$$Q = It/Fz$$

where Q is the molar flux of compound with valency z released by current I. F is the Faraday constant (96,500 C) and t is known as the transport number for the compound under consideration. t reflects the proportion of applied current which is carried by the ion of interest. Its ideal value in a solution of compound A^+B^- would therefore be 0.5 for each ionic species. In practice, for the reasons noted above, it is frequently much lower than this (Table I).

Recently, Armstrong-James and colleagues have developed a method for directly monitoring the extracellular concentration of catecholamines achieved by microiontophoresis. The technique makes use of electrochemical detection technology, with carbon fiber electrodes associated with the iontophoretic assembly. It obviates the need for assessing transport numbers *in vitro,* but it is a complicated time-consuming technique applicable to a restricted group of chemicals.

Variable ejection is obviously less of a problem with micropressure ejection. The volumes forced out of the pipette tip are linearly related to the applied pressure, and the system can be accurately calibrated by microscopically measuring the diameters of fluid droplets formed in a given period of time under various applied pressures.

TABLE I Transport Numbers of Some Drug Ions Used in Microiontophoresis

Compound	Solution concentration (M)	Ejection polarity	Transport number
Acetylcholine	3	+	0.42
Noradrenaline	0.2	+	0.15
L-Glutamate	0.1	−	0.22
Substance P	0.007	+	0.16
Met-enkephalin	0.01	+	0.07
GABA	1	+	30.0
		−	0.039
ADP	0.01	−	0.007

B. Balance Current

Neurons (i.e., nerve cells) are extremely sensitive to applied electrical current, and it is therefore possible that any changes in cell activity observed in an iontophoretic experiment could be the result of the iontophoretic current per se, rather than the ejected ion. This is a particularly serious problem when current is being passed from the drug-containing barrel to a distant reference or focal electrode, because much of the current will pass through the cell. One way of compensating for this effect is to fill one barrel of a multibarreled assembly with sodium chloride solution and then test through this barrel a current of the same polarity and magnitude as that passed through the drug barrel. A better procedure is to return all of the ejecting iontophoretic channels through the sodium barrel. In this way there should be no net current leakage away to ground; that is, the net current flow at the electrode tip should be zero. The sodium barrel in this situation is known as the balance barrel, and the current it passes is called the balancing current.

C. Retaining Current

However small a micropipette tip is, there will always be a small amount of drug leakage into the tissues or bathing medium. This results partly from simple diffusion of ions into the external medium and partly from the loss of a minute volume of solution, due to the hydrostatic pressure of the column of fluid in the pipette. For some potent compounds (e.g., peptides) this leakage might be sufficient to alter the activity of the cell or to change the sensitivity of its receptors for the compound. It has therefore become standard practice to apply, between the periods of active drug ion ejection, a current opposite that used for ejection, to withdraw the active ion away from the barrel tip.

This current is variously referred to as the retaining, backing, or braking current. It brings with it the complication that, when the ejecting current is applied, a finite time is required (in the range of 1–20 seconds for most situations) to replenish the active ion content in the barrel tip. This phenomenon of drug withdrawal must be taken into account when comparing drug potencies or the time course of action. It is also important to apply compounds in an absolutely strict time cycle, so that the influences of ion withdrawal and replenishment are constant over each cycle.

D. Other Artifacts

In addition to these, several other sources of error are recognized in iontophoretic experiments. These include interference from the drug counterion (e.g., chloride, sulfate, and tartrate) which is, of course, effectively ejected during the relatively long periods of retaining current used between drug ejection. The use of concentrated or dilute solutions of drugs can lead to osmotic artifacts, and if solutions are made at a pH less than 4.0 or greater than 9.0, then resulting changes of extracellular pH around the pipette tip can be produced which can change the cell activity or receptor properties.

Bibliography

Brown, K. T., and Flaming, D. G. (1987). "Advanced Micropipette Techniques for Cell Physiology." Wiley, Chichester, England.

Palmer, M. R. (1982). Micropressure ejection: A complementary technique to microiontophoresis for neuropharmacological studies in the mammalian CNS. *J. Electrophysiol. Tech.* **9,** 123.

Purves, R. D. (1979). The physics of iontophoretic pipettes. *J. Neurosci. Methods* **1,** 165.

Stone, T. W. (1985). "Microiontophoresis and Pressure Ejection." Wiley, Chichester, England.

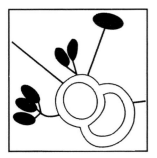

Microtubules

MARGARETA WALLIN, *University of Göteborg, Sweden*

Glossary

Centriole Structure with nine triplets of microtubules, often found close to the nuclear membrane; together with a surrounding amorphous material, two centrioles located perpendicular to each other function as an organizing center for microtubule assembly; when found at the base of a cilium, it is called a basal body

Cilium Tiny, hairlike cellular appendage (also called a flagellum) extending from the surface of a cell; it can move fluid with whiplike movements or propel sperms or flagellated animals through a fluid; inside the projection one finds a complex arrangement of microtubules (nine doublets with dynein arms in the periphery and two single microtubules in the center), which are connected to each other by many different proteins

Microtubule-associated proteins Heterogenous group of proteins that bind to microtubules; several stimulate assembly of microtubules and stabilize formed microtubules, whereas some are seen as projections

Microtubules Long, tubelike structures, with an outer and inner diameter of 25 nm and 15 nm, respectively, composed of tubulin dimers, which make up the backbone of the microtubule

Tubulin Subunit of microtubules; a heterodimer, with a molecular weight of 100,000, composed of one α-tubulin and one β-tubulin monomers, which are encoded by different genes

MICROTUBULES ARE UBIQUITOUS cellular organelles found in all nucleated cells; therefore, they are totally absent in bacteria. Microtubules participate in a wide variety of cellular processes, such as cell division, intracellular transport processes, color changes in pigment cells, cell motility, secretion, and ciliary and flagellar movement. Microtubules also have a role as a cytoskeletal protein. In spite of all these different functions, all microtubules have a generally constant structure: They are long, tubelike structures with an outer and inner diameter of approximately 25 nm and 15 nm, respectively. They differ in length but can be at least 100 μm long. Tubulin dimers form the backbone of microtubules and are aligned longitudinally to protofilaments. Several protofilaments (usually 13) form a microtubule (Fig. 1). A heterogenous group of proteins, the microtubule-associated proteins (MAPs), are bound to the surface of microtubules. Two of these, known as MAP1 and MAP2, are seen as long, extending projections from the surface of microtubules.

How the common microtubule structure can participate in so many different functions is not yet well known. It is tempting to speculate that both a binding of various MAPs and modifications of different tubulins and MAPs after synthesis (post-translational modifications) are involved.

A microtubule is a dynamic structure, which is built up (assembled) and broken down (disassem-

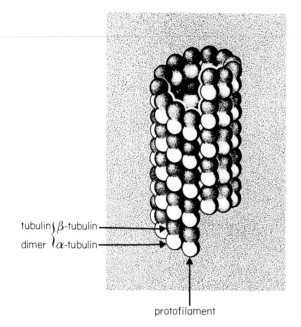

tubulin { β-tubulin
dimer { α-tubulin

protofilament

FIGURE 1 Schematic drawing of a microtubule with 13 protofilaments. Each protofilament consists of tubulin dimers, where α-tubulin (white) and β-tubulin (strippled) alternate. Protofilaments are nearly parallel to the microtubule axis. Tubulin dimers in neighboring protofilaments are displaced by about 0.9 nm, gene-rating a left-hand, three-start helix of the monomers. This arrangement requires a discontinuity in the helix, implying the presence of a ''seam,'' as shown between the two protofilaments in the front. (Reproduced from E.-M. Mandelkow, R. Schultheiss, R. Rapp, M. Müller, and E. Mandelkow, 1986, *J. Cell Biol.* **102,** 1067–1073, with copyright permission of the Rockefeller University Press.)

I. Composition of Microtubules: Tubulin and MAPs

A. Tubulin: A Heterodimer

The functional subunit for microtubule assembly is tubulin, a heterodimer measuring about 8 × 5 nm and consisting of one α- and one β-subunit with approximate molecular weights of 50,000 each (Fig. 3). Tubulin is a typical globular protein, with approximately 25% alpha helix and 47% beta sheet structure. α- and β-tubulins do not have similar amino acid sequences but do have a high degree of homology, about 40%. Tubulin has changed little during evolution, and it seems that some parts are highly conserved throughout evolution. One such example is the guanosine triphosphate (GTP)-binding sites, of which one is located at each monomer. GTP is bound very tightly to α-tubulin and is, therefore, called nonexchangeable. β-tubulin has, in contrast, an exchangeable GTP-binding site relatively close to the amino-terminal. This GTP is hydrolyzed to guanosine diphosphate (GDP) upon assembly of microtubules and, therefore, plays an important role in regulation of the assembly. The highly acidic carboxy-terminals of both α- and β-tubulin are examples of variable regions. They are found on the outside of the assembled microtubule and bind many MAPs.

B. MAPs Confering Different Properties to Microtubules

MAPs are a heterogenous group of proteins that bind to tubulin (Table I). Some MAPs form a helical superlattice on the surface of microtubules. At least three roles have been suggested for the MAPs: to time and localize assembly and disassembly; to link microtubules to microtubules, as well as microtubules to organelles; and in some cases, to force microtubule-dependent movements within the cell. They may also serve as linkers between the other two major cytoskeletal systems: the intermediate filaments and microfilaments. *In vitro* prepared microtubules contain 15–20 associated proteins. It is not yet known whether or not all these proteins are bound to microtubules in the cell and, if so, why so many types of MAPs exist and why their cellular distributions differ. Special MAPs possibly bind to specific tubulin isotypes or posttranslationally modified tubulins to generate microtubule subclasses of as yet undefined functions. Some of the complexity

bled) depending on the needs of the cell. In a nondividing (interphase) cell, microtubules are seen as a network between the area around the nucleus and the cellular membrane (Fig. 2A). This network disassembles upon cell division and reassembles to the mitotic spindle, which segregates the chromatides to the new daughter cells. After a completed cell division, the mitotic spindles disassemble, and new networks assemble in the daughter cells. However, the dynamics differ markedly among different microtubules, ranging from the stable and extremely well-organized microtubules in cilia and flagella to the highly dynamic microtubules of the mitotic spindle in a dividing cell. Between these extremes, one finds the parallel organization of microtubules in the long, thin process (the axon) radiating outward from the nerve cell. These microtubules are more stable than microtubules in the mitotic spindle and the microtubule network seen in the interphase cell.

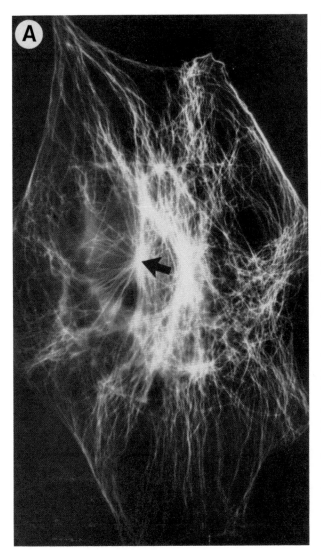

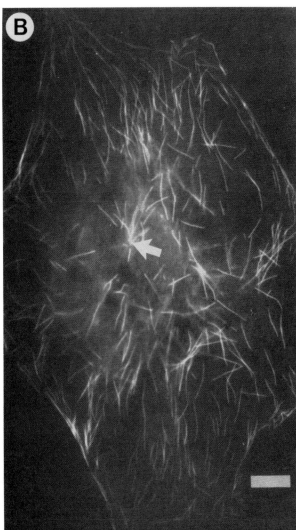

FIGURE 2 A. Immunofluorescence micrograph showing the arrangement of microtubules in a cultured fibroblast cell as revealed by staining with anti-tubulin antibodies. The arrow points to the centrosome. B. The same cell has been microinjected with 5 mg/ml of biotin-labeled tubulin to visualize incorporation of tubulin at the microtubule plus ends. Biotin-labeled tubulin is visualized with rhodamine anti-biotin immunofluorescence. The cell was fixed 66 sec after injection. During this time, biotin-labeled tubulin mainly represents elongation of existing microtubules. The arrow points to the centrosome. Bar, 10 μm. (Reproduced from E. Schulze and M. Kirschner, 1986, *J. Cell Biol.* **102,** 1021–1031, with copyright permission of the Rockefeller University Press.)

in each MAP may arise from posttranslational modifications, but alternative splicing of the MAP gene transcripts are also intriguing possibilities.

MAP1 and MAP2 were some of the first identified and characterized MAPs. They are easily detectable because they extend from the microtubule surface as projections. They are very large proteins, but only a small part binds to tubulin.

MAP1 is a generally abundant protein that stimulates assembly of microtubules. It is composed of two different forms, MAP1A and MAP1B. These forms have molecular weights of approximately 300,000 Da and two smaller peptides of 28,000 and 30,000 Da. The latter are not needed for tubulin-binding.

MAP2 has a somewhat lower molecular weight of 270,000 Da. On sodium dodecyl sulfate (SDS)–polyacryamide gel electrophoresis, it can be seen as two peptides, MAP2A and MAP2B. However, peptide-mapping and reactivity with monoclonal anti-

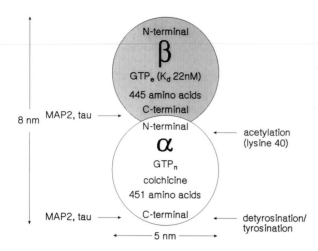

FIGURE 3 Schematic drawing of the tubulin dimer. MAP2 and tau bind preferentially to the β-tubulin carboxy-terminal. (Courtesy of Dr. Martin Billger.)

bodies suggest that these two peptides are essentially identical. MAP2 has a juvenile form, called MAP2C, with a molecular weight of 70,000. MAP2 has only one gene, and the juvenile form is probably formed by a different splicing mechanism. MAP2 is a thermostable protein between 90 and 185 nm long. Only a small part of its total length is microtubule-binding, while the other part extends from the microtubule surface (Fig. 4A). A cyclic adenosine monophosphate (cAMP)-dependent protein kinase is associated with the projection domain of MAP2. Phosphorylation of MAP2 inhibits its ability to stimulate assembly and might be of importance for the cellular control of microtubules. MAP2 is mainly localized to neuronal cells, and in these cells it is absent from the axons. It is, therefore, an example that the localization of a MAP can vary both within and between cells.

Tau is a MAP of lower molecular weight. It extends as short projections on the microtubule surface (Fig. 4B). On SDS–polyacrylamide gels, it can be seen as four to six polypeptides in the molecular mass range of 55,000–62,000 Da. Tau has only one gene, but at least some of the different polypeptides can be explained by a different degree of phosphorylation. There is also a juvenile form, with a molecular weight of 48,000. Tau is heat-stable as MAP2, indicating a low degree of ordered structure. It seems that at least tau, MAP1, and MAP2 bind to the same binding site on the carboxy-terminal of tubulin. Three (or four) repetitive short amino acid sequences (18 amino acids) mediate the binding.

A third, apparently very heterogenous group of MAPs (see Table I) consists of many non-neuronal proteins with molecular weights around 200,000 Da.

TABLE I Microtubule-Associated Proteins

Microtubule-associated protein	Subunit mol. wt. (kDa)	Comments
MAP1A, B	350	Long projections from the microtubule surface stimulate assembly
Associated light chains	28, 30	
MAP1C (dynein)	410 (2), 74 (3), 59, 57, 55, 53 (nine subunits)	ATPase, motor for retrograde axonal transport
MAP2A, B	270	Long projections from the surface of microtubules stimulate assembly; thermostable; mainly present in neuronal cells (absent in axons)
MAP2C	70	Juvenile form of MAP2
MAP3A, B	180	Prominent in astroglia and neurofilament-rich axons
MAP4A, B, C	240, 235, 220	Found only in glial cells
210 kDa HeLa MAP		
205 kDa *Drosophila* MAP	200	Isolated from both neuronal and non-neuronal cultured cells and tissues; thermostable
190 kDa adrenal gland MAP		
190 kDa rat MAP		
Kinesin	110–144 (2), 60–65 (2) (tetramer)	ATPase, motor for anterograde axonal transport
STOPs[a]	145	Slides along microtubules, regulating cold stability of microtubules
Chartins	69–80	Stabilizes microtubules
tau	55–62 (4–5 isoforms)	Short projections from the microtubule surface; stimulates microtubule assembly; thermostable
	48	Juvenile form of tau

[a] Stable tubulin-only peptide.

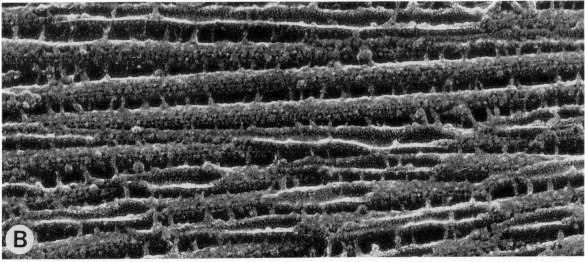

FIGURE 4 A. Electron micrograph of unidirectionally shadowed microtubules, assembled from tubulin and MAP2 and stabilized with taxol. The MAP2 molecules extend outward from the microtubule wall as thin filaments up to 80 or 90 nm in length. (Reproduced from W. A. Voter and H. P. Erickson, 1982, *J. Ultrastruct. Res.* **80,** 374–382, with copyright permission of the Academic Press.) B. Electron micrograph of quick-frozen, deepetched microtubules assembled from tubulin and tau. Tau projections are indicated by arrows. Bar, 100 nm. C. Microtubules assembled from pure tubulin (in the presence of taxol) have a smooth surface and are closely aligned. The samples were made as in B. Bar, 100 nm. (B and C are reproduced from N. Hirokawa, Y. Shiomura, and S. Okabe, 1988, *J. Cell Biol.* **107,** 1449–1459, with copyright permission of the Rockefeller University Press.)

The function(s) of all these different MAPs are not yet known. Examples of other MAPs are STOP (stable tubulin-only peptide) proteins and chartins. MAPs have been defined as proteins that associate and colocalize with microtubules *in vivo*. The two proteins dynein (previously called MAP1C) and kinesin have recently been identified as important motors for the movement of organelles along microtubules. They are treated here as MAPs, although they mainly are attached to transport vesicles and only transiently interact with microtubules. They will be dealt with in detail in the discussion of intracellular transport.

C. Microtubules as Dynamic Structures

Interphase and mitotic microtubules are in a state of dynamic equilibrium with tubulin and MAPs. Most of this knowledge has come from studies of microtubules isolated from neuronal tissue, where they are especially abundant. These microtubules have turned out to be a good model for *in vivo* microtubules. Neuronal microtubules are isolated by repeated temperature-dependent assembly–disassembly steps. Disassembly occurs in the cold and reassembly at higher temperatures. Such a preparation consists of approximately 80% tubulin and 20% MAPs.

Assembly *in vitro* can be divided into three phases: nucleation, elongation, and a steady-state phase. The assembly is very sensitive to many environmental factors. Microtubules assemble *in vitro* in the presence of GTP and Mg^{2+} and when the temperature exceeds 20°C (for mammals). Other requirements include buffers, ionic strength, and pH. Calcium ions are inhibitory and must be removed. The assembly starts from nuclei, the morphology of which is unknown. A few dimers or small sheets have been proposed. Evidence indicates that the amino-terminal domain of α-tubulin interacts with the carboxy-terminal domain of β-tubulin (Fig. 3). During the elongation, assembly is greater than disassembly. Microtubules have an intrinsic polarity, because the α, β-tubulin dimers assemble in a head-to-tail manner in the protofilaments, which all align with the same polarity (Fig. 1).

Several models have been put forward for the assembly process (Fig. 5). The most favored is the dynamic instability model. Tubulin dimers with bound, exchangeable GTP assemble, and after a short time this GTP is dephosphorylated to GDP. The microtubules grow as long as they have GTP-

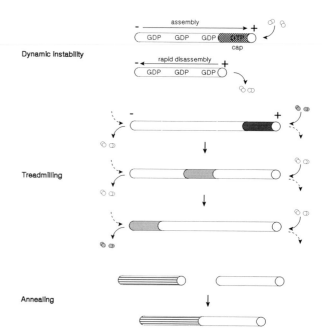

FIGURE 5 Schematic drawing of the major proposed assembly models. In the dynamic instability model, tubulin dimers containing exchangeable GTP are preferentially added at the plus end of a microtubule. It grows continually as long as it has a cap of GTP-containing tubulin dimers. If the GTP cap is lost, microtubules disassemble rapdily but can be rescued before total disassembly. In the treadmilling model, or head-to-tail assembly model, tubulin dimers assemble preferentially at the microtubule plus ends, while they disassemble preferentially at the minus ends (indicated by solid arrows). This leads to a flux of tubulin dimers along a microtubule at steady state. The solid segment represents the same molecules at different times. *In vitro* evidence has also been found for annealing of preexisting microtubules. (Courtesy of Dr. Martin Billger.)

containing tubulin at the tip at the plus end, a so called cap. If the cap is lost, the microtubule disassembles rapidly. Most microtubules are rescued before they are completely disassembled and will reassemble again. The dynamic instability is rapid when microtubules assemble from pure tubulin, but slower in the presence of MAPs, which, of course, is the more physiological situation. With the help of microinjected fluorescent tubulin, microtubules have been found to behave similarly in the living cell (Fig. 2B). Whether this mechanism can account for all the processes in the cell is not yet known; several mechanisms can coexist. A treadmilling of tubulin dimers has also been proposed. Tubulin dimers are preferentially added at the plus end, and lost at the minus end at steady state. This leads to a flux of monomers through the microtubule. Recently, microtubules have also been found to anneal *in vitro*, but whether or not this phenomenon exists in the cell is unclear.

D. Dynamics of Microtubules Under Cellular Regulation

Because microtubules are so dynamic and must fulfill their functions in various ways, an intricate regulatory system that orchestrates microtubule dynamics must exist. Calcium is a strong candidate for the regulation of microtubule assembly and disassembly in the cell. It is inhibitory both *in vitro* and *in vivo*. The effect of calcium is potentiated by calmodulin and S-100, two calcium-binding proteins that, upon activation, inhibit assembly and disassemble preformed microtubules (Fig. 6). Calmodulin has been localized to the mitotic spindle, in the region between the centrosome and the chromatids, a region in which microtubules disassemble upon mitosis. The cell also has a very efficient calcium-sequestering system, providing a reversible pathway in the calcium-mediated regulation of microtubules.

In view of the large quantities of cellular tubulin and MAPs, microtubules must also be regulated by degradation. Calcium-activated proteases (e.g., calpains) mediate an irreversible regulatory pathway, probably of importance for the observed plasticity of nerves during development of the brain. [*See* PLASTICITY, NERVOUS SYSTEM.]

II. Tubulin Synthesized From Many Genes

Tubulin has multiple genes coding both for α- and β-tubulin separately. For each tubulin, 10–20 genes have been found, but only up to 6–7 tubulin gene products are found in the cell, indicating that the remainder are nonfunctional pseudogenes. Species differences exist: in the green unicellular alga, *Chlamydomonas reinhardtii*, only two genes for α-tubulin and two genes for β-tubulin exist. Tubulins from different species are closely related in sequence, and all known tubulins coassemble *in vitro*. The genes of the multigene tubulin families are not closely linked to each other but are located on different chromosomes. In addition to species differences, examples of organ-specific differences are also found in the expression of different tubulins (a testis, a brain, and a chicken red blood cell-specific

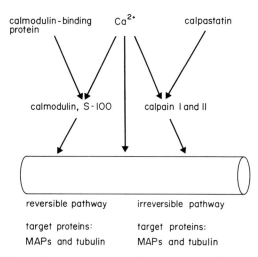

FIGURE 6 Schematic drawing of the calcium-mediated regulation of microtubules.

tubulin). The heterogeneity is most extensive in mammalian brain and increases during development and aging. The existence of several different tubulin isotypes indicate that different tubulins can be involved in different microtubule functions. However, so far very little evidence has been found for such a functional diversity. Neither genetic diversity nor a single mechanism can account for the cellular control over the enormous versatility of microtubule functions. Whether the various tubulin genes are expressed coordinately or selectively during cell differentiation, embryonic development, or the life cycle is not yet known. Tubulin gene expression appears to be under posttransistional autoregulatory control, a result of a cytoplasmic event that modulates stability of the mRNAs bound to the polysomes. If the amount of unassembled tubulin increases, its synthesis decreases, and vice versa.

III. Tubulin and MAPs Modified after Synthesis

Many tubulin isotypes exist in the cell, but not all can be explained by the expression of multiple genes. Tubulin can be modified by different posttranslational modifications, which makes the regulatory system of microtubules very flexible. Three different modifications have been found so far: phosphorylation, detyrosination–tyrosination, and acetylation. Glycosylation, lipid association, and methylation are also indicated.

A. Phosphorylated Tubulin and MAPs

Several different enzymes are involved in the phosphorylation of tubulin, as well as some of the MAPs. The most well known are the cAMP-dependent protein kinase, the Ca^{2+}-calmodulin-dependent protein kinase (kinase II), casein kinases I and II, and the insulin kinase. Of these, the cAMP-dependent protein kinase and a calmodulin-dependent protein kinase mainly phosphorylates MAP1, MAP2, tau, and a 62-kDa MAP of the sea urchin mitotic apparatus. The function of the phosphorylation is not yet known, but seemingly at least the Ca^{2+}-calmodulin-dependent protein kinase reduces the assembly and the MAP binding, and a phosphorylation of MAPs also reduces the assembly competence. Phosphorylation of microtubule proteins can, therefore, be one way to regulate microtubules in the cell. [*See* PHOSPHORYLATION OF MICROTUBULE PROTEIN.]

B. α-tubulin Detyrosinated after Synthesis

α-tubulin is usually synthesized with a tyrosine at its carboxy-terminal. The primary posttranslational modification of α-tubulin is its detyrosination in assembled microtubules, by a tubulin carboxy peptidase, leaving glutamic acid as the carboxy-terminal residue. Tyrosinated α-tubulin has been suggested as a marker of newly formed microtubules, and detyrosinated microtubules may accumulate in "old" microtubules. A tubulin tyrosine ligase may restore the original state of disassembled α-tubulin. No functional differences between tyrosinated and detyrosinated microtubules have yet been found.

C. Stable Microtubules Often Contain Acetylated α-tubulin

When a cell is exposed to antimitotic drugs, such as colchicine and nocodazole, microtubules are broken down; however a small subpopulation of stable microtubules usually remains. These microtubules are, in many cases, acetylated. Acetylated microtubules are also often connected with other stable microtubule structures, such as cilia and flagella microtubules, and the microtubules in the neuronal axon. Only one amino acid, the lysine at position 40 on α-tubulin is acetylated. Recent studies indicate that the acetylation is not responsible for the stability in itself, but it might induce another MAP's composition of these microtubules, which renders them more stable.

IV. Microtubules Constituting a Part of the Cytoskeleton

At least three different filaments are collectively called the cytoskeleton: the microfilaments (actin), the intermediate filaments, and the microtubules. If microtubules are disassembled by any microtubule-binding drug, a differentiated cell will rapidly lose its characteristic form and round up to a more undifferentiated cell form. Therefore, microtubules seem to have an important role as a cellular skeleton. Microtubules may also have a role in the orientation ability of a moving cell. The motive force is generated by microfilaments, but if microtubules disassemble, the cell loses its ability to orient.

In axons, microtubules are extremely well organized in a parallell fashion, conferring stability on the axon. In platelets and nonmammalian red blood cells, a marginal band of microtubules is seen underlying the cellular membrane (Fig. 7). The band gives the blood cells their characteristic discoid shape, which is lost when microtubules are broken down.

Microtubules may also assemble into more complex arrangements. Several unicellular organisms have long, slender extensions, which are used for capturing food particles, for locomotion and for the fixation of the cell by a stalk (Fig. 8A). The cytoskeletal axes of the axopodia, the axonemes, contain various numbers of microtubules (there can be more than 100) (Fig. 8B). These microtubules are often linked to each other in a very specific way. Although they are cross-linked, the axonemes are remarkably labile; disassembly and reassembly of axonemes can take place within minutes or less.

V. Many Transported Organelles Along Microtubules

All cells must have an extremely well-organized transport machinery. In neuronal cells, the transport lengths can be as long as 1 m in some human axons.

Both organelles (e.g., mitochondria and vesicles containing transmittor substance) and soluble proteins are synthesized and transported bidirectionally along microtubules. The types of organelle as well as the transport velocities vary widely. Mitochondria, secretory vesicles, and pigment granules are transported rapidly (50–400 mm/day),

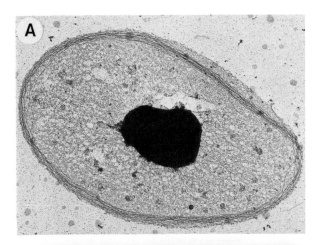

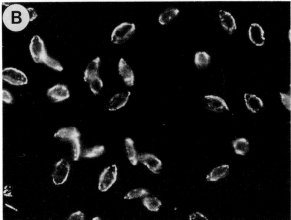

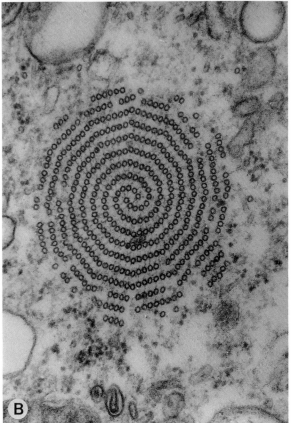

FIGURE 7 A. Electron micrograph of a chicken erythrocyte (red blood cell) showing the marginal band of microtubules as a ring in the periphery of the cell. B. Immunofluorescent images of chicken erythrocytes stained with tubulin antibodies. Microtubules are seen as a bright ring in the periphery. (Reproduced from F. Salmon, 1988, "Intrinsic Determinants of Neuronal Form and Function" (R. J. Lasek and M. M. Black, eds.), pp. 361–370, with copyright permission of Alan R. Liss, Inc.)

FIGURE 8 A. *Echinosphaerium nucleofilium* showing several axopodia radiating from the spherical cell body. The axopodia differ in length but can be as long as 300 μm. The cell body has a diameter of approximately 100 μm. (×200). (Reproduced from L. E. Roth, D. J. Pihlaja, and Y. Shigenaka, 1970, *J. Ultrastruct. Res.* **30,** 7–37, with copyright permission of the Academic Press.) B. Cross section of an axoneme at a level inside the cell body. The microtubules are found in sections. Links are seen between microtubules. (×78,000). (Reproduced by the permission of L. E. Roth, D. J. Pihlaja, and Y. Shigenaka, Biological Photo Service.)

whereas soluble proteins are transported much more slowly (0.2–8 mm/day).

In many instances, mechanochemical enzymes are a prerequisite for motion in the cell. Myosin ATPase has been known for a long time to interact with actin filaments and generate the forces for muscle contraction, and dynein ATPase generate the cilia and flagella bending. Recently, two new motors have been discovered, kinesin and MAP1C.

Kinesin is a motor for anterograde axonal transport (from the nerve cell body to the axonal terminal) (Fig. 9A). It was identified in the cytoplasm of squid giant axons by a new powerful technique, video-enhanced contrast microscopy, which allows for high-resolution light microscopy of unfixed preparations. The cytoplasm was extruded from the squid giant axon, and kinesin moved both vesicles and latex spheres along microtubules. The movement was toward the microtubule plus end at the axonal terminal and with a velocity of about 0.5 μm/sec. Upon further purification, kinesin was also found to be able to move microtubules along a glass surface. Kinesin appears to be a ubiquitous protein motor. It is a heterotetramer with two ATP-binding heavy chains (110,000–140,000 Da) and two 60–65-kDa light chains. Electron microscopy reveals that kinesin has two globular heads at one end of a stalk, where it is connected to a feathery tail (Fig. 9B). The architecture of kinesin is very similar to myosin and dynein. The ATP-hydrolyzing and microtubule-binding regions of kinesin reside at or near the heads, another similarity with myosin. This means that the globular heads interact with microtubules, whereas the feathery tails interact with the organelle to be moved. The ATPase activity of kinesin is coupled to microtubule-binding: assembled microtubules increase the ATPase activity manyfold. Kinesin is strongly inhibited by adenylyl 5'imidodiphosphate (AMP-PNP), the nonhydrolyzable analogue of ATP. In the presence of AMP-PNP, kinesin remains bound to microtubules, and its transport activity ceases. This binding is used for the purification of kinesin and is reversed by the addition of ATP.

In the axon, not one but several types of transport exist. Organelles are moved by fast anterograde transport but also back toward the nerve cell body by retrograde transport. Another motor, first called MAP1C, has been identified as the retrograde motor (Fig. 9A). It moves organelles along microtubules from their plus end to their minus end. MAP1C has many similarities with cilia and flagella dynein and, therefore, is now called dynein. Dyneins are particularly sensitive to inhibition by vanadate and N-ethylmaleimide. In the presence of vanadate and ADP, the dyneins can be cleaved by ultraviolet light in a very characteristic way.

Very little is yet known of the control of the microtubule-dependent motors and how the cell can decide when and where an organelle is to go. It is also unclear whether or not kinesin and dynein are responsible for all microtubule-based movements in the cell; most likely additional microtubule-dependent motors will be discovered. The movements associated with mitosis and the movements between organelles (for instance, between the Golgi apparatus and the endoplasmatic reticulum) are such ex-

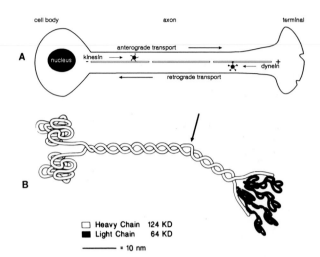

FIGURE 9 A. Schematic drawing of anterograde kinesin-dependent axonal transport and retrograde dynein-dependent axonal transport. (Courtesy of Dr. Martin Billger). B. Schematic drawing of kinesin. Kinesin is composed of two globular head domains, a shaft which is hinged (at arrow) and a tail. (Reproduced from N. Hirokawa, K. K. Pfister, H. Yorofuji, M. C. Wagner, S. T. Brady, and G. S. Bloom, 1988, *Cell* **56**, 867–878, with copyright permission of CELL Press.)

amples. Another example is the transport of pigment granula in pigment cells, chromatophores (Fig. 10). The presence of chromatophores is especially abundant in animals that can change color. The animal has a light color when the pigment granules are concentrated to the cell center, and a dark color when they are dispersed throughout the cell.

VI. Microtubules Transporting Chromosomes During Cell Division

Although the cell division and movement of chromosomes in the mitotic or meiotic spindle have been known for a long time, the molecular basis is still poorly understood. During cell division, the interphase microtubule network disassembles, the centrioles duplicate and move to the poles of the cell, and the mitotic spindle assembles from tubulin and MAPs. In addition to microtubules, the mitotic spindle consists of a spindle matrix, the kinetochores attached to the chromosomes, and the microtubule-organizing centra (MTOCs) which includes the centrioles. The chromosomes move with the microtubules. Three groups of spindle microtubules are recognized: polar microtubules, microtubules attached to kinetochores, and astral microtubules (Fig. 11), the plus ends of these groups are free to assemble and disassemble. The kinetochore

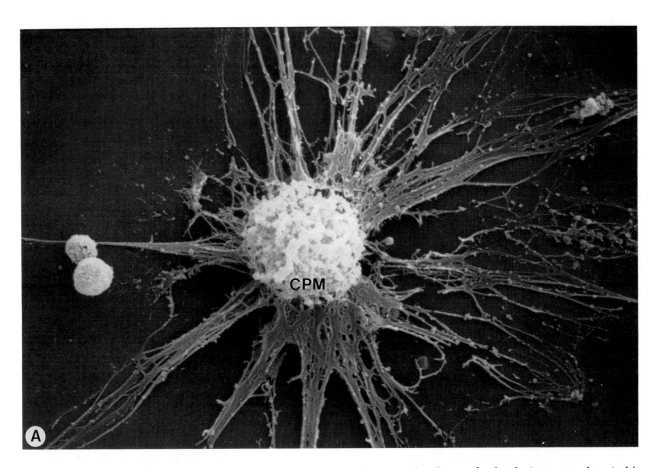

FIGURE 10 Scanning electron microscopy of cultured chromatophores subjected to detergent treatment to remove the plasma membrane. A. Microtubules extend outward from the central pigment mass (CPM) in a nondispersed cell. Some intact plasma membrane is still left. The central pigment mass is composed of thousands of small spherical pigment granules. B. A partially dispersed chromatophore. Plasma membrane (PM) remains between openings where many microtubules (Mts) with pigment granula are seen. C. A higher magnification of a region in B. (Reproduced from M. A. McNiven and J. B. Ward, 1988, *J. Cell Biol.* **106**, 111–125, with copyright permission of the Rockefeller University Press.)

is a specialized region of the chromosome where the plus ends of spindle microtubules (one or several) are attached. It is composed of three platelike layers, 0.3–0.6 μm in diameter. [*See* MITOSIS.]

Polar microtubules from each pole overlap and thus, can be sliding across each other. The cross-bridges can be the motors that generate movement in the overlapping zone. During mitosis, the amount of microtubules in the cell can change from approximately 700 interphase microtubules, with an average length of 20 μm and a mean lifetime of 300 sec, to 3,000 mitotic spindle microtubules, with an average length of 4 μm and a mean lifetime of less than 50 sec.

In an early phase of mitosis (prometaphase), kinetochores capture microtubules from the large number of microtubules that continually grow out from the spindle poles. Opposing forces act on these microtubules, which cause the chromosome to align at the spindle equator. At a later phase (the anaphase), the kinetochore microtubules disassemble at the kinetochore, resulting in a poleward movement of the chromosomes. ATPases have been found associated with the mitotic spindle, and the presence of microtubule sliding based on dynein like motors is, therefore, very probable. During the cleavage of the cell to two daughter cells, another cytoskeletal component is involved, namely the actinomyosin system.

The microtubules in the mitotic spindle are very dynamic. Different microtubules assemble and disassemble asynchronously. Microtubule elongation occurs at a rate of approximately 4 μm/min (about 100 dimers/sec); however the rate of shortening of a microtubule is much faster, >20 μm/min or 500 dimers/sec. Kinetochore microtubules and midbody microtubules are more stable than the other microtubules for unknown reasons, but MAPs have been suggested to be involved.

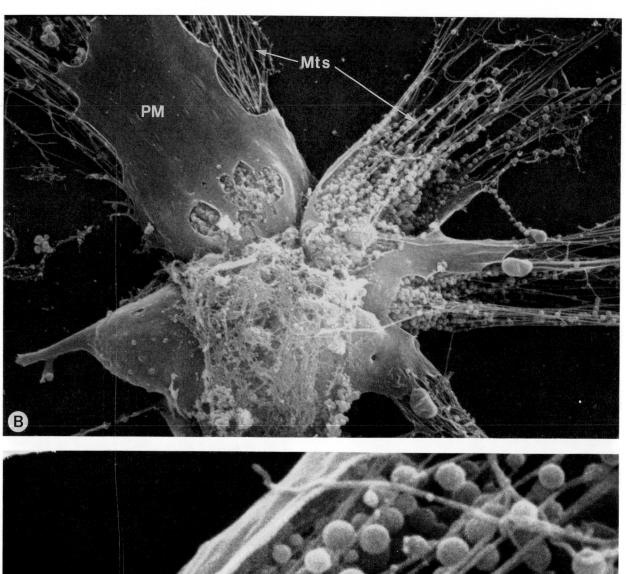

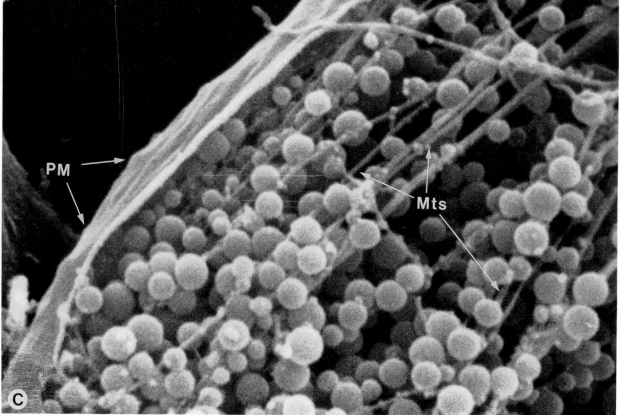

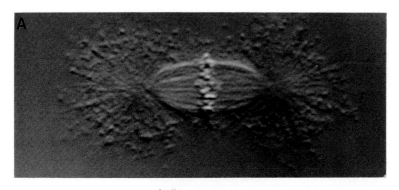

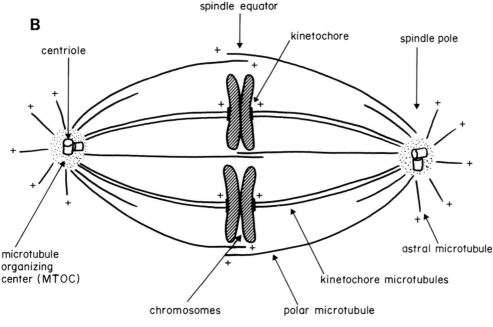

spindle equator

centriole

kinetochore

spindle pole

+

microtubule
organizing
center (MTOC)

astral microtubule

kinetochore microtubules

chromosomes

polar microtubule

half - spindle

FIGURE 11 A. An isolated sea urchin egg spindle at early metaphase viewed with differential interference contrast microscopy. (×1200). (Reproduced from E. D. Salmon and R. R. Segall, 1980, *J. Cell Biol.* **86**, 355–365, with copyright permission of the Rockefeller University Press.) B. A schematic drawing of the metaphase mitotic spindle.

VII. Microtubules Arranged in a Complex Way in Cilia and Centrioles

In the cell, microtubules, closely dependent on each other, are in different and more complex arrangements. In cilia and flagella, nine doublets of microtubules are arranged in a circle around a pair of single microtubules, the "9 + 2" array (Fig. 12). A centriole, a structure composed of 9 triplets (Fig.

15) of microtubules is found at the base of the cilium, then often called a basal body. Apparently the centriole plays two roles in the cell: as a basal body, and as a part of the organizing center for cytoplasmic microtubules during interphase and cell division. Early on much has been learned about these microtubule structures, because they are much more stable than the fragile cytoplasmic microtubules.

A. Ciliary Movement Based on the 9 + 2 Array of Microtubules

Cilia and flagella are tiny, hairlike appendages extending from the surface of many different cells. In mammals, they project from the cell surface of the

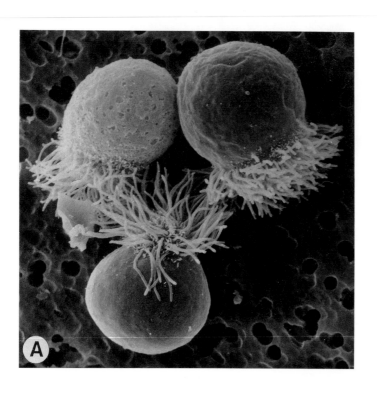

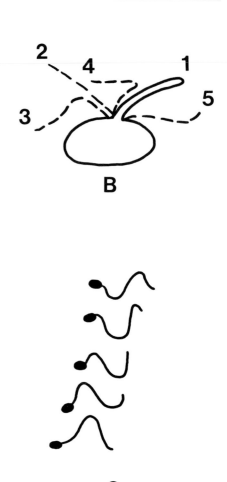

FIGURE 12 Scanning electron micrograph of three ciliated cells isolated from the trachea (×4500) showing the large amounts of cilia on each cell. (Reproduced from E. R. Dirksen and M. Zeira, 1981, *Cell Motility* **1,** 247–260, with copyright permission of Alan R. Liss, Inc.) B. Schematic drawing of the ciliary beat. It starts with a forward power stroke (1 and 2) that propels fluid over the cell. It is followed by a recovery stroke (3–5) by a bending at the base that minimizes viscous drag. Usually each cycle requires 0.1–0.2 sec. C. Schematic drawing of flagella movement. Waves of quasisinusoidal bending originate at the base and move toward the tip, propelling the sperm forward. Waves move continuously from the base to the tip. The waves are generated at a frequency of 30–40 waves/sec.

lung and the oviduct, as well as from the sperm. Their function is to move the mucus with dust and dead cells from the lungs to the mouth, to move the ova along the oviduct, and to propel the sperm. In the lung, fields of cilia bend in coordinated unidirectional waves with a whiplike motion (Fig. 12). The flagella do not make whiplike movements but propagate with quasisinusoidal waves.

Nearly no difference exists between cilia and flagella. Flagella is used for the long and specialized cilia in the sperms and in flagellates. The cilium is often about 0.2 μm in diameter and can be as long as 200 μm in some cells. A ciliated vertebrate cell may contain hundreds of cilia. The microtubule cytoskeleton is surrounded by the cell membrane and is attached to a basal body. Each doublet microtubule containing the A and B microtubules grows from two microtubules in each of the 9 triplets in the basal body, conserving the ninefold symmetry. In the middle usually two single microtubules are evident (Fig. 13). The B tubule shares five protofilaments with the A microtubule and is, therefore, a little wider than the A-microtubule. This gives rise to a slightly different arrangement of dimers, the so-called B-lattice. A dense component is often found close to the wall linking the A and B microtubule. Another nontubulin component is tektin, a protein arranged in 2- to 3-nm-diameter filaments, running longitudinally along the microtubule wall. It seems

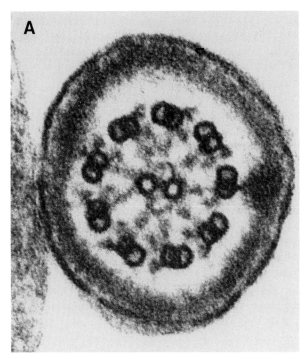

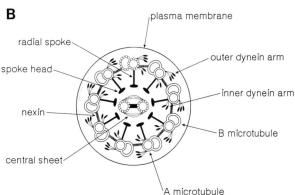

FIGURE 13 A. Electron micrograph of a sperm flaggellum shown in cross section ($\times$320,000). A distinct "9 + 2" arrangement is seen. (Reproduced from B. A. Afzelius and R. Eliasson, 1979, *J. Ultrastruct. Res.* **69**, 43–52, with copyright permission of the Academic Press.) B. Schematic drawing of a cilium in cross section. (Courtesy of Dr. Martin Billger.)

to be involved in formation of the shared wall between the A and B microtubules.

The movement of a cilia is dependent on the presence of a Mg^{2+}-ATPase, the dynein, which is attached along the A microtubule. The two dynein arms are very close to the B microtubule in the next doublet. They act in a very similar way as the actin-myosin sliding in muscle contraction. In the presence of ATP, the arms can bind to the next doublet and slide the doublets along each other. Because the microtubules are bound at their proximal end to the basal body, they can only move a certain distance over each other, resulting in a localized sliding. The sliding is opposed by resistive components that convert it into a transverse bending moment, imparting a swimming motion to the sperm. The inner and outer dynein arms differ (Fig. 14A) and seem to have different functions in generating sliding. The activity of the dynein arms must be regulated to get a local bend propagating from the base to the tip.

Outer dynein arms are very large proteins, with a molecular weight of 1,200,000 to 1,900,000. They contain two or three distinct heavy-chain polypeptide subunits with ATPase activities and molecular weights of approximately 350,000, as well as a varying number of subunits of lesser weight. The dyneins are seen as a set of globular heads joined by flexible stems to a common base (Fig. 14B). Dyneins cannot only generate a bending but, if absorbed to a glass surface, they can translocate microtubules toward their plus ends.

The ciliary structure contains a variety of accessory molecules in addition to the doublet and central pair microtubules and the dynein arms. Peripheral doublets are linked to the central pair with radial spokes. The major functions of the central pair tubules and the radial links seem to maintain a planar form of beat and to provide additional coordination of the sliding. One also finds interdoublet links, the nexins, which probably limit the extent of sliding to approximately 100 nm. One can also see appendages, which are attached to certain individual microtubules, as well as specialized structures in the transition between the basal body and the axoneme. As many as 170 distinct polypeptides are present in the axoneme, of which 17 are associated with radial spokes.

B. Centrioles Play Two Roles in the Cell

Centrioles often exist in pairs, at right angles to each other (Fig. 15). During interphase, they are located close to the nuclear membrane. The centrioles duplicate before mitosis, and during the onset of mitosis the centrioles move to each pole of the mitotic spindle. Most centrioles arise by duplication of preexisting centrioles, always perpendicular to the "mother" centriole. The centrioles can migrate from their normal location to the region from which cilia will form. Each centriole forms numerous so-called satellites from which many basal bodies will

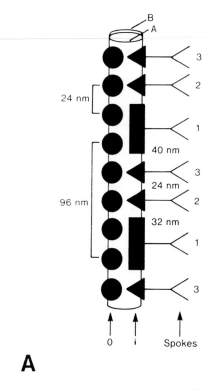

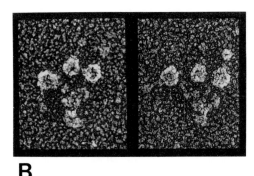

A

B

FIGURE 14 A. Schematic drawing of outer (o) and inner (i) dynein arms on the A-microtubule in relation to the spokes. The outer dynein arms are illustrated as filled circles with a 24-nm longitudinal repeat. The inner arms appear to be of two different forms: one with two globular heads (triangles) and one with three globular heads (rectangles). They seem to originate at or near the base of a spoke. The periodicity of the inner arms coincides with triplets of the spokes. The triplet group of spokes repeat every 96 nm. (Reproduced from W. S. Sale, L. A. Fox, and S. L. Milgram, 1989, "Cell Movement," Vol. 1 (F. D. Warner, P. Satir, and I. R. Gibbons, eds.), pp. 89–102, with copyright permission of Alan R. Liss, Inc.) B. Electron micrograph of quick-freeze, deep-etched outer dynein arms. Three globular heads are connected by a stem to a common base. Arrowheads point to small projections from the three dynein heads. (×350,000). (Reproduced from U. W. Goodenough and J. E. Heuser, 1989, "Cell Movement," Vol. 1 (F. D. Warner, P. Satir, and I. R. Gibbons, eds.), pp. 121–140, with copyright permission of Alan R. Liss, Inc.)

arise. The basal body is anchored to other components in the cells by appendages known as striated rootlets. The centrioles can multiply in ciliated cells to form multiple basal bodies from which the cilia can grow.

Centrioles have a ninefold symmetry. They consist of nine groups of three microtubules and form an elongated cylinder about 400 nm long. The triplets are formed from one complete microtubule with 13 protofilaments (the A microtubule) and two incomplete microtubules each of 10 protofilaments (the B and C microtubules). The B microtubule shares three protofilaments with the A microtubule, and the C microtubule shares three protofilaments with the B microtubule. The C microtubule is somewhat shorter than the A and B microtubules. The triplets are slightly twisted, which makes the centriole a little skew. The microtubules are linked together in several ways: they are connected to the center by radial links, and links also exist between the triplets. Frequently, several dense masses surround the triplets. This whole structure taken together is often called a centrosome. Centrioles are very stable structures, unaffected by most microtubule inhibitory drugs or physical actions. New centrioles can even assemble from a tubulin pool in the presence of such poisons. The reason for this is unclear.

VIII. Inhibition of the Assembly of Microtubules

Microtubule inhibitors are valuable chemical tools in studies of microtubules. A number of drugs bind to microtubule proteins, thereby inhibiting the assembly of microtubules (Table II). Colchicine, an alkaloid from *Colchicum autumnale*, has been known as a dangerous poison since the first century A.D. It causes the disappearance of the mitotic spindle and induces mitotic arrest. A drug with these properties is usually called an antimitotic drug. During the middle of the 1960s, the colchicine-binding protein was identified as tubulin. The tubulin has one high-affinity binding site for colchicine on the alpha-subunit. Colchicine can inhibit assembly at concentrations that are below the concentrations of free tubulin, a so-called "substoichiometric poisoning." Tubulin dimers with bound colchicine can add to a growing microtubule, but their presence prevents further addition

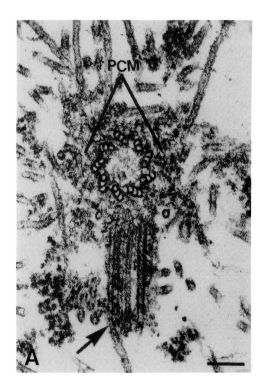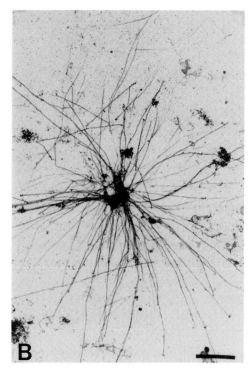

FIGURE 15 A. Section of an isolated centriole duplex showing the nine triplets of microtubules. Microtubules have assembled from the pericentriolar material (PCM) and onto the distal ends of the centrioles (arrows). Bar, 0.1 μm, B. Several microtubules assembled onto an isolated centriole complex. Bar, 1 μm. (Reproduced from B. R. Telzer and J. L. Rosenbaum, 1979, *J. Cell Biol.* **81,** 484–497, with copyright permission of the Rockefeller University Press.)

TABLE II Compounds Affecting Microtubules

Compound	Effect on microtubules	Compounds that bind to the same or overlapping site	Comments
Colchicine	Inhibits assembly	Podophyllotoxin, rotenone, steganazine, nocodazole, benzimidazoles, combrestatin	One high-affinity site for colchicine per tubulin dimer, binds very tight at 37°C, not at 4°C, biphasic binding
Vinblastine	Inhibitory at low concentrations, induces spiralization of protofilaments and paracrystal arrays at higher concentrations	Vincristine, maytansine, phomopsin	Two high-affinity binding sites per tubulin dimer, several low-affinity binding sites
Griseofulvin	Inhibits assembly at 37°C, induces aggregates at 4°C		Mechanism of action relatively unknown; probably binds to both tubulin and MAPs
Ca^{2+}	Inhibits assembly		One high-affinity binding site at the carboxy-terminal of tubulin, several low-affinity sites; inhibition of the calcium-binding proteins, calmodulin and S-100
Sulfhydryl group reagents	Inhibits assembly		Binding to two free sulfhydryl groups per tubulin dimer inhibits assembly
Estamustine phosphate	Inhibits assembly		Binds to the tubulin-binding parts of MAPs, e.g., MAP2 and tau
Heparin	Inhibits assembly		Binds to MAPs
Taxol	Stimulates assembly		One site per tubulin dimer, assembled microtubules are stable to cold, Ca^{2+} and colchicine
Zn^{2+}, Co^{2+}	Induces sheets of protofilaments		Sheets are stable to cold, Ca^{2+}, and colchicine

of tubulin dimers. The exact mechanism of action is, however, unclear.

Many examples of microtubule inhibitors show binding to the same or overlapping site. Podophyllotoxin shares part of its binding site with colchicine. Rotenone is a competitive inhibitor of colchicine, as well as steganazine. All these drugs have similarities in their structure, but benzimidazoles (e.g., nocodazole), which belongs to a different class, also acts on the tubulin colchicine-binding site in a competitive way. Colchicine-binding apparently inhibits lateral tubulin–tubulin association. Because of the very high specificity of colchicine for tubulin, it has been discussed whether or not a cellular factor that binds to the colchicine-binding site exists, perhaps in analogy with morphine and the endorphines. Evidence indicates such a factor, and it might be involved in the regulation of microtubule dynamics.

Vinblastine and its derivatives bind to another site. In the cell, vinblastine treatment induces paracrystalline arrays of hexagonally packed cylinders (280–320 Å in dimater). Tubulin has two high-affinity sites for vinblastine, in addition to several low-affinity sites. A low concentration of vinblastine induces a dimerization of pure tubulin, whereas higher concentrations induce formation of spirals, rings, and paracrystalline structures of microtubule proteins. Addition of vinblastine to assembled microtubules results in the separation of pairs of protofilaments longitudinally, which is followed by coiling of these pairs into spirals. The presence of MAPs is necessary for spiral formation, which indicates their importance for stabilization of longitudinal bonds, whereas vinblastine and derivatives destabilize the lateral bonds.

Several other inhibitors of microtubule assembly exist, as listed in Table II. All these substances inhibit the assembly by binding to tubulin. In contrast, estramustine phosphate is a new and powerful drug, binding to the tubulin-binding domain of MAPs, including at least tau and MAP2, thereby inhibiting the assembly.

A very useful drug has been isolated from *Taxus brevifolia*. It binds to tubulin and lowers the critical concentration for assembly and makes the assembly MAP-independent. The formed microtubules are extremely stable; they are not disassembled by cold or calcium. Taxol has been used for the isolation of microtubule proteins from non-neuronal cells that have a very low microtubule protein content.

IX. Clinical Importance of Microtubules

At least two different forms of disease are linked to microtubules. In the "immotile cilia syndrome," the male patients have immotile sperms. Cilia in the respiratory tract are also immotile, leading to bronchitis and sinusitis. In about half of the cases, the heart and several of the inner asymmetrical organs are located on the wrong side of the body, indicating that cilia are important during embryogenesis. When both defects are found, Kartageners syndrome is diagnosed. The syndrome seems to be caused by the loss of one or both dynein arms, but also the loss of spokes, central microtubules, or some of the doublets results in anomalies. Ciliary abnormalities can also be induced by smoking or infections.

In the United States, approximately 7% of persons over 65 yr of age have Alzheimer's disease, a kind of senile dementia. It is a progressive intellectual deterioration caused by an atrophy of the brain, most marked in hippocampus and in association areas of cerebral cortex. Neurons are lost, and in many neurons neurofibrillary tangles and neuritic plaques are found. The former consists of paired helical filaments, which are not normally found in the cell. They are extremely insoluble and, therefore, have been difficult to analyze. Now however the MAP tau clearly seems to constitute an important part of the filaments. The reason for the altered tau function is not known. [*See* ALZHEIMER'S DISEASE.]

Also, some indirect evidence indicates that microtubules can be involved in aneuploidy, which is the malfunction of the mitotic spindle to transport equal amounts of chromosomes to each sister cell. The relevance of aneuploidy to human health is great since it might induce, for example, spontaneous abortion, early death of newborn infants, and Down's syndrome.

Clinically, antimitotic drugs are mainly used in the treatment of cancers. Although the binding of colchicine to tubulin is rather specific, it has found no use in cancer chemotherapy, mainly because of its high general toxicity. In contrast, vinblastine and derivatives, podophyllotoxin and estramustine phosphate, are used in chemotherapy. Some of the benzimidazoles are used as fungicides and herbicides. These drugs bind preferentially to tubulin from lower organisms, indicating that, although tubulin is an evolutionary, very conserved protein, small difference between species alter the binding and effect of certain drugs.

Bibliography

Bershadsky, A. D., and Vasiliev, J. M. (1988). Cytoskeleton. *In* "Cellular Organelles" (P. Siekevitz, series ed.). Plenum Press, New York.

Dustin, P. (1984). "Microtubules." Springer Verlag, Berlin, Heidelberg.

Katzman, R., and Thal, L. J. (1989). Neurochemistry of Alzheimer's disease. *In* "Basic Neurochemistry; Molecular, Cellular, and Medical Aspects," 4th ed. (G. J. Siegel *et al.*, eds.), Raven Press, New York.

Kirschner, M. W., and Mitchison, T. J. (1986). Beyond self-assembly: From microtubules to morphogenesis. *Cell* **45,** 329–342.

Lacey, E. (1988). The role of the cytoskeletal protein, tubulin, in the mode of action and mechanism of drug resistance to benzimidazoles. *Int. J. Parasitol.* **18,** 885–936.

Maccioni, R. B. (1986). Molecular cytology of microtubules. *In* "Revisiones sobre biologia celular," Vol. 8 (E. Barbera-Guillem, ed.). Servicio editorial universidad del pais vasco. Spain.

Matus, A. (1988). Microtuble-associated proteins: Their potential role in determining neuronal morphology. *Ann. Rev. Neurosci.* **11,** 29–44.

Sullivan, K. F. (1988). Structure and utilization of tubulin isotypes. *Ann. Rev. Cell Biol.* **4,** 687–716.

Warner, F. D., and McIntosh, J. R. (1989). "Cell Movement. Kinesin, Dynein, and Microtubule Dynamics," Vol. 2. Alan R. Liss, Inc., New York.

Warner, F. D., Satir, P., and Gibbons, I. R. (1989). "Cell Movement. The Dynein ATPases," Vol. 1. Alan R. Liss, Inc., New York.

Microvascular Fluid and Solute Exchange

AUBREY E. TAYLOR *University of South Alabama*

Glossary

Capillary absorptive pressure Protein osmotic (oncotic) pressure gradient existing across the capillary wall and is equal to the difference between the protein osmotic pressure of plasma and tissue fluids

Capillary filtration pressure Pressure gradient existing across the capillary wall and is equal to the difference between capillary and tissue fluid hydrostatic pressures

Interstitial edema Build-up of excessive fluid in the interstitium

Interstitial edema safety factor Ability of lymph flow, tissue pressure, and the absorptive pressure to increase to oppose the tendency of the tissues to become dematous

Interstitial fluid Fluid in the spaces between cells

Interstitium Spaces between cells

Net capillary filtration or absorptive pressure Difference between the capillary filtration and absorptive pressures. If this pressure difference is positive, the capillary will filter fluid into the interstitium. If the net pressure difference is 0, the capillary will neither filter nor absorb fluid, and when the net filtration is negative, the capillary will absorb fluid from the interstitium

Plasmalemmal vesicles Vesicles formed in the endothelial cells that are surrounded by a membrane composed of the endothelial cell wall. They appear to be able to fill with certain proteins in plasma and could serve as a specific transendothelial transport system for macromolecules

MICROVASCULAR FLUID and solute exchange refers to the processes by which the capillary system maintains a constant internal environment surrounding the cells. This chapter will present (1) the structural-functional nature of the three basic types of capillary systems in the body; (2) the physics and physiology of solute and solvent exchange in the capillary system; (3) the functional aspects of the lymphatic system; and finally, (4) the effects of altering capillary function on the build-up of excess fluid in the tissues, a condition known as *interstitial edema*.

I. Introduction

Humans have extensive cardiovascular systems that bring needed nutrients to their tissues and remove the unwanted end products of metabolism. The left heart is the major pump of this system, which imparts the energy necessary to propel blood into the aorta where it flows through many small branches of the arterial circulation finally to empty the oxygen-rich arterial blood into a vast microcirculatory system composed of millions of small blood vessels with diameters of 10–12 μm and a length of about 0.5 mm, called *capillaries*. The capillaries course only a short distance through the tissues and then drain into small venules. These venules empty into a larger venous system, which will finally deliver the blood into the right atrium of the heart. The blood then enters the right ventricle and is pumped through the lungs where it is oxygenated and returns to the left atrium where it fills the left ventricle to be pumped again into the aorta. [*See* CARDIOVASCULAR SYSTEM, ANATOMY; CARDIOVASCULAR SYSTEM, PHYSIOLOGY AND BIOCHEMISTRY.]

The capillaries usually are surrounded by a basement membrane, which is immersed in the interstitial fluid compartment that is a complex matrix composed of fluid, collagen, and mucopolysaccharides that surround parenchymal cells. The capillaries are designed to exchange large amounts of solutes and water with the interstitium. Unlike cell membranes, which are almost impermeable to small solutes, the capillary walls are freely permeable to small solutes. A large number of capillaries are present in the tissues and provide a tremendous surface area through which nutrients and other needed metabolites can diffuse out of the capillary into the tissues surrounding the organ's parenchymal cells. These substances can then be used by the cells to produce energy. Conversely, end products of cellular metabolism (e.g., carbon dioxide and urea) can easily diffuse into the flowing blood of the capillaries for removal in the kidneys and lungs.

Certain organs have specialized capillary systems that can obtain nutrients from the environment or secrete unwanted products of metabolism back into the environment. For instance, the capillaries of the lungs exchange oxygen and carbon dioxide with the environment. The O_2 enters the lungs from the inspired air, whereas CO_2, the major end product of metabolism, is expired into the environment. The capillaries of the gastrointestinal tract must absorb nutrients and water from the ingested food stuff to provide metabolic substrates to all tissues of the body. The glomerular capillaries in the kidney filter the entire plasma volume in about 30 min. This filtered plasma passes into a series of kidney tubules that reabsorb the important substances, preventing their loss into the urine. Only 1% of the fluid that enters the renal tubules forms urine, and the remainder is reabsorbed by another extensive capillary system in the kidney, the peritubular capillary system.

The permeability properties of brain capillaries are unique because they are impermeable to all constituents of plasma, with the exception of O_2, CO_2, and fat-soluble compounds. Other water-soluble compounds (e.g., NaCl, glucose) are actively transported by the brain capillary endothelium. This allows the constituents of the fluid surrounding the nervous system to be carefully controlled, ensuring that brain cells can rely on their environment for the necessary ions and energy sources to discharge and relay information repeatedly to other areas of the brain and the rest of the body. Conversely, the capillaries of the liver are permeable and can exchange all components of blood, except the red blood cells with the liver parenchymal cells. This allows the important compounds manufactured by the liver to enter the hepatic circulation easily and allows the end products of metabolism to be processed by the liver cells into compounds that can be used by the body or excreted by the kidney. Many glands require special capillary systems that can exchange necessary fluid and compounds to produce hormones and other biologically important compounds.

Thus, all constituents of plasma are constantly being replenished in the blood by the lung, liver, and gastrointestinal capillaries, whereas muscle, glands, bones, the nervous system, and even large blood vessels use these nutrients and produce end products of metabolism that must be excreted by the lungs, kidneys, and gastrointestinal tract as shown schematically in Fig. 1.

In a living capillary system, the red cells are approximately the same size as capillaries, and they must travel one-by-one from the arterial to the venous end of the capillaries. Substances are always leaving and entering the capillaries, but this process cannot be seen, even under a powerful microscope. The human body must constantly cope with two environments: one from which we glean oxygen, food stuff, and water, and the second, which surrounds the parenchymal cells and is regulated by the vast capillary exchange network of the body. This arrangement allows multicellular organs to obtain necessary nutrients from their external and internal environments and rid their own internal environments of unwanted end products of metabolism.

II. Structural Anatomy of the Microcirculatory Exchange System

A. Types of Exchange Blood Vessels

Figure 2 shows diagrammatically the three different types of capillaries found in body tissues: (1) continuous capillaries, which have a continuous layer of endothelium cells lining their interior surface; (2) fenestrated capillaries, which are also lined by a

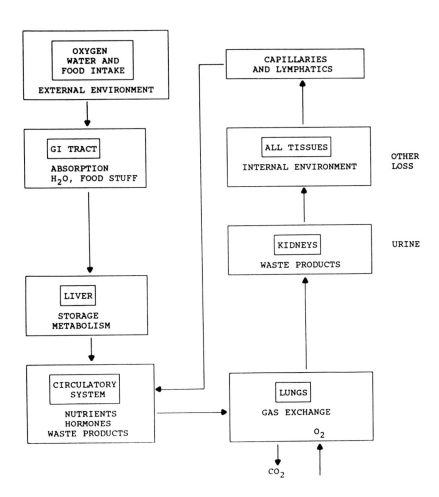

FIGURE 1 Schematic representation of the capillary fluid and solute exchange system in the human body.

continuous endothelium layer, which also contains structures called *fenestrae*; and (3) discontinuous capillaries, which are not lined by a continuous layer of endothelium but have large nonselective gaps. The endothelium that covers the interior surfaces of continuous capillaries is extremely thin; a basement membrane present on the tissue side of the capillary is composed of a dense network of proteoglycans and collagen. The discontinuous capillaries do not have basement membranes but directly communicate with the interstitium. Continuous capillaries are found in lung, brain, skeletal muscle, smooth muscle, and renal peritubular capillaries. Fenestrated capillaries are found in tissues in which large transfers of water and solutes occur such as glomerular, small intestine, stomach, and large intestine capillaries. Discontinuous capillaries are found in the sinuses of the liver.

B. Microvessel Exchange Pathways

The endothelial cell wall is a bileaflet, highly lipid membrane that also has a few small water-filled channels of 0.4 nm radius (called *pores*), which allow the plasma to communicate with the endothelial cell interior. Most biologically important compounds (e.g., NaCl, glucose, hormones) are too big to go through this 0.4-nm pore radius and cannot pass easily into the endothelial cell. Therefore, they must either cross the endothelial cell in the lipid phase of the membrane or pass between adjacent cells through the intercellular clefts to enter the interstitial compartment. Water and small solutes (e.g., NaCl and glucose) can rapidly exchange with their surrounding tissues in these intercellular clefts. The concentration of most compounds in the tissue fluid and the plasma are similar, with the exception of the large plasma proteins, which cannot easily cross the capillary wall. The physiologically important gases, O_2 and CO_2, can diffuse across the

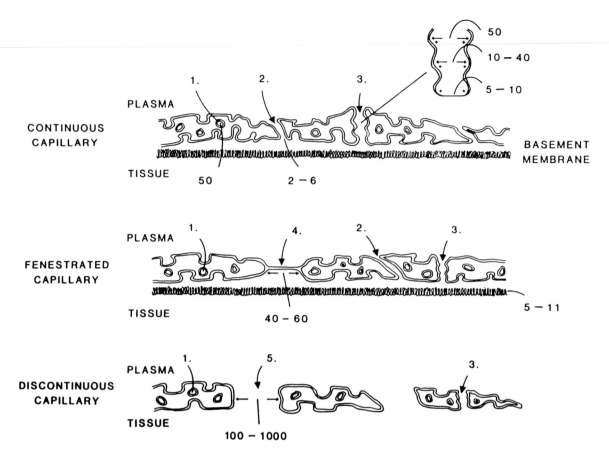

CONTINUOUS CAPILLARY

PLASMA

TISSUE

BASEMENT MEMBRANE

50

10 — 40

5 — 10

50   2 — 6

FENESTRATED CAPILLARY

PLASMA

TISSUE

40 — 60

5 — 11

DISCONTINUOUS CAPILLARY

PLASMA

TISSUE

100 — 1000

entire endothelial cell surface because they are highly fat-soluble. Because the concentration of O_2 (really its partial pressure) is higher in the capillaries than in the surrounding tissues and the concentration of CO_2 is higher in the tissues than in plasma, the necessary diffusion gradients are present in the capillaries and their surrounding tissues to allow O_2 to enter the tissues and CO_2 to enter the capillaries. The brain capillaries are unique in their permeability characteristics in that their walls are almost impermeable to water and totally impermeable to all small water-soluble solutes. This allows the brain endothelial cell to directly regulate the constituents of brain interstitial fluid by active processes, rather than allowing the constituents of plasma to exchange freely with the brain tissue, as occurs in capillaries of other organs.

Figure 2 also shows pathways large molecules (plasma proteins) can use to cross capillary walls and enter the interstitium. It is thought that large plasma proteins are transported across the capillary endothelium by plasmalemmal vesicles because in most capillaries, the intercellular clefts are too small (2.5-nm radius) to allow the passage of the

FIGURE 2 Types of capillaries found in the human body. 1, plasmalemmal vesicles; 2, intracellular clefts; 3, fused plasmalemmal vesicles; 4, fenestrae; 5, nonselective gaps. Diameter of structures is in nm. [Revised from Taylor, A. E., and Granger, D. N. (1984). Exchange of macromolecules across the microcirculation. *In* "Handbook of Physiology, The Cardiovascular System," sect. 2, vol. IV, part 1, chapter 11. (E. M. Renkin and C. C. Michel, eds.) Waverly Press, Bethesda, Maryland, with permission.]

smallest plasma protein, albumin (3.6-nm radius). In contrast, larger intercellular clefts are located in the small venules and are potential pathways for large macromolecules to use to cross the endothelial barrier. Another transcellular pathway, which is formed by three to four fused plasmalemmal vesicles, has recently been discovered in the capillary wall. These transendothelial channels may serve as the major pathway that macromolecules use to enter the tissues. They appear to be the most likely conduit for macromolecule movement across the capillary walls, because some of these channels are covered by diaphragms with pores of 4.5-nm radii, whereas a few have no diaphragms and a pathway equivalent pore radius of 20 nm. These pore sizes are similar to physiological estimates of the pore dimensions in the capillary membrane. However, some plasmalemmal vesicles are known to be selective to certain types of proteins and could provide an additional pathway for these proteins to use when crossing the capillary walls.

Physiological studies indicate that two different size pores must be used to explain the exchange of macromolecules between plasma and the interstitium. The capillary walls must contain a large number of small pores with a radius of 4.5 nm and a small number of large pores with a radius of 20 nm as shown in Table I. Although the anatomical pathways associated with these physiological measures of protein exchange are not presently known, the transendothelial channels formed by the fusion of several endothelial vesicles have the proper pore dimensions and overall characteristics to explain the physiological data.

The effective charge of macromolecules, as well as their molecular radius and configuration, also affects their movement across the capillary wall. The negative charge of plasma proteins causes them to be repelled by the negative charge on the surface of the endothelial cells. It is now known that plasma proteins loaded with fatty acids are more easily incorporated into the plasmalemmal vesicles. Until future studies evaluate the effects of protein charge on transcapillary exchange and the factors that govern the selectivity of endothelial vesicles, no unique structural description of transcapillary macromolecular exchange can be made. Presently, we must be content to describe the macromolecule transport systems in terms of two sets of physiological cylindrical, uncharged, water filled pores of 4.5- and 20.0-nm radii as shown in Table I. [*See* PLASMA LIPOPROTEINS.]

C. Structure of the Interstitium

A complete description of the biochemistry and physics of the interstitium is beyond the scope of this chapter, but it must be realized that the interstitium is not just a bag of water contained within our muscle sheaths and under our skin. The interstitium is a highly organized meshwork that contains long-chained sugar molecules (more than 2 million daltons in size) that are cross-linked to proteins, forming proteoglycans. Proteoglycans serve as the bricks and mortar of the interstitial structural components, and collagen provides the biological equivalent of a steel-like meshwork to provide tensile strength to this interstitial matrix. The proteoglycans are large, negatively charged molecules that combine with positive-charged substances such as sodium in the interstitial fluid. The collagen network forms another tissue space, which contains channels of only 1.5 nm in diameter excluding all plasma proteins. Thus, the interstitium is much like a gel, interspersed with a lattice network of collagen and containing both large and small interstitial channels. [*See* PROTEOGLYCANS.]

Normally, the tissue spaces are compact, which ensures that molecules need diffuse only a short distance to reach the parenchymal cells. In several pathological states, excess fluid enters the tissues, resulting in a condition called *interstitial edema*. In some tissues, the capillary exchange of O_2, CO_2, and nutrients is often affected by the edema fluid because of the longer pathways that these molecules must use to diffuse to the cell surface. But when the lungs or brain become edematous, a life-threatening situation results, and the person will die unless the edema fluid is rapidly removed.

The exact pathway large molecules and water use to move through the tissues to enter the lymphatic system, which drains fluid from the interstitium, is presently not known. Some investigators believe that large preferential pathways exist in the tissues because all plasma proteins ranging in molecular radius from 3.7 to 12.0 nm traverse the interstitium. Others postulate that most solutes and water pass through the proteoglycan and collagen meshwork, with only the larger molecules that cannot enter the collagen fluid phase using larger channels. The fluid and solute that leaks from the capillaries move through the interstitium to finally arrive at the lym-

TABLE I Predicted Pore Sizes for Capillary Walls in Selected Tissues[a]

Tissue	Small pore (nm)	Large pore (nm)	Ratio of large to small pores
Subcutaneous	5.0	20.0	1 : 3,000
Skeletal muscle	6.0	22.0	1 : 3,600
Brain	0.4	—	—
Intestine	4.6	20.0	1 : 6,400
Liver	9.5	33.0	1 : 50
Lung	8.0	20.0	1 : 200

[a] Modified from Taylor, A. E., and Granger, D. N. (1984). Exchange of macromolecules across the microcirculation. *In* "Handbook of Physiology, The Cardiovascular System," section 2, vol. IV, part 1. (E. Renkin and C. Michel, eds.), pp. 467–520. Waverly Press, Bethesda, Maryland, with permission.

phatic capillary, which carries this fluid and solute away from the tissues and back into the circulating blood.

D. Lymphatic Drainage System

Figure 3 shows an idealized capillary interstitial lymphatic system. Most organs contain small lymphatic capillaries (the exception being the brain, spinal cord, and ocular space) that are composed of a discontinuous endothelium with open junctions between the interior of the lymphatic vessels and the tissue fluid. All constituents of the interstitial fluid can pass unhindered into the small lymphatic capillaries. The larger collecting lymphatics contain smooth muscle that contracts vigorously when filled with fluid. Because the larger lymphatics contain valves, these contractions propel fluid from lymphatic segment to lymphatic segment, which moves the tissue fluid away from the capillary environment toward the larger lymphatic vessels that will eventually drain into the circulatory system at the large veins.

Normally, the capillaries leak a small amount of fluid into the tissues, and some plasma proteins also

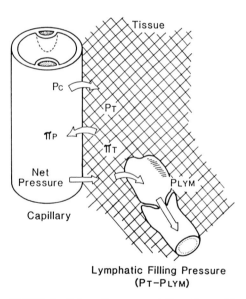

Lymphatic Filling Pressure
(P_T–P_{LYM})

FIGURE 3 Schematic representation of a capillary-tissue-lymphatic system showing capillary (P_C) and tissue fluid (P_T) hydrostatic pressures, and plasma (π_P) and tissue (π_T) protein osmotic pressures. Usually, a small net pressure imbalance exists across the capillary wall [i.e., ($P_C - P_T > \pi_P - \pi_T$)] and fluid enters the tissue and is removed by the initial lymphatic, which has a hydrostatic pressure of P_{LYM}. [Revised from Taylor, A. E., Rehder, Hyatt, R., and Parker, J. C. (1989). "Clinical Respiratory Physiology." Saunders, Philadelphia, with permission.]

enter the interstitial spaces (except in the brain and ocular fluids). The major function of the lymphatic system is to remove this fluid and protein that has entered the tissues. Without the lymphatic removal of solutes and solvent, edema would result because no other mechanisms are present in the body to return the plasma proteins in the tissues back into the circulatory system. Thus, capillaries leak fluid and protein into the interstitium, whereas the lymphatic system removes this fluid and protein. If the lymphatic system were not present, it would be impossible to maintain a normal blood volume without excessive build-up of fluid in the tissues. This protein exchange system is not to be confused with the small molecule exchange system that operates between tissues and the plasma within the capillaries. The exchange of small-molecular-weight substances occurs so rapidly that the plasma and interstitial concentrations are almost identical.

III. Physics of Transvascular Fluid and Solute Exchange

There are many different types of capillaries in the body, and we will not discuss each specific physical and biological process used by them to exchange solutes and water with the interstitium. But we will define some processes that most capillary exchange systems have in common.

A. Physics of Solvent Exchange

Figure 3 demonstrates an idealized capillary system, showing the forces responsible for moving fluid across the capillary wall. These forces are usually referred to as Starling forces, in honor of the great British physiologist who first defined them in 1897. The hydrostatic pressure within the capillary is defined as the *capillary pressure* (P_C), and the hydrostatic pressure in the fluid surrounding the capillary is defined as the *interstitial tissue fluid pressure* (P_T). The differences in these hydrostatic pressures ($P_C - P_T$) is the filtration pressure, which acts to push fluid out of the capillary into the interstitium. The plasma proteins exert considerable osmotic pressure at the capillary wall. This protein osmotic pressure of plasma (π_P) and tissue fluids (π_T) acts to pull fluid into their respective volumes. Their difference across the capillary wall ($\pi_P - \pi_T$) is termed the absorptive pressure. The fluid movement across the capillary wall (J_{VC}) can now be written in terms of the Starling forces as:

$$J_{VC} = K_{FC}[(P_C - P_T) - \sigma_d(\pi_P - \pi_T)] \quad (1)$$

K_{FC} is the filtration coefficient and is a function of both the pore sizes and numbers in the capillary wall. The osmotic reflection coefficient (σ_d) defines how effective the osmotic pressure of a protein solution is in absorbing solvent across a membrane. If the membrane were freely permeable to the plasma protein, $\sigma_d = 0$ and no osmotic pressure would be exerted by the plasma proteins. If the membrane were impermeable to the protein, it would exert its full osmotic pressure ($\sigma_d = 1$) across the capillary wall. Because all proteins leak into the tissues, $\sigma_d \neq 1$ for most capillary walls but is sufficiently close to 1 in most capillary beds, that it can be assumed to be one for practical purposes of describing fluid movement across capillary walls in a general fashion. In humans, the major force that causes fluid to move out of the capillaries into the interstitium is the capillary pressure. P_C is elevated when venous outflow pressure is increased, postcapillary resistance is increased, or precapillary resistance decreased.

Physiology of Solvent Exchange

Figure 4 shows schematically how fluid filters across the capillary wall at the arterial, middle, and venous ends of an idealized capillary. In this example, the capillary pressure is 30 mm Hg at the arterial end of the capillary, $\pi_T = 0$ mm Hg and $\pi_P = 29$ mm Hg and $P_T = 0$ mm Hg. The sum of these forces is $(30 - 0) - (29 - 0) = 1$ mm Hg, which acts in a direction to push fluid out the capillary into the interstitium. The capillary pressure decreases to 20 mm Hg at the midpoint of the capillary because of the capillary resistance, and π_P is still 29 mm Hg. But now P_T is −5 mm Hg and π_T is −5 mm Hg. The sum of forces operating at the midpoint of the capillary is $(20 - (-5)) - (29 - 5)$, still equal to 1, which pushes fluid out the capillaries. At the venous end of the capillary, P_c has now decreased to 10 mm Hg because of the capillary resistance, and P_T is −10 mm Hg, π_T is now 10 mm Hg, and π_P is still 29 mm Hg. The sum of the forces at the venous end of the capillary is $10 - (-10) - (29 - 10)$ or 1 mm Hg. Thus, all segments of the capillary have a small imbalance in forces that filters fluid into the interstitium. However, capillary filtration is a dynamic event, because capillary pressure is not constant

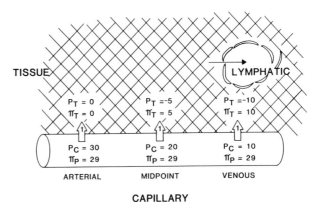

FIGURE 4 Calculation of capillary net filtration pressures at the arterial, midpoint, and venous ends of the capillary as explained in the text. [Revised from Taylor, A. E., and Townsley, M. I. (1987). Evaluation of the Starling fluid flux equation. *NIPS* **2**, 48–57, with permission.]

and because it is the most important determinant of transvascular fluid movement, capillaries may be filtering one moment and absorbing the next moment. In addition, venous pressures are high in dependent regions of the body (e.g., in the feet); therefore, fluid is always leaving the capillaries when we are standing. Without proper lymphatic removal of this fluid, edema forms easily in the ankle regions.

For many years, it has been taught that capillaries filter on their arterial ends and absorb on their venous ends because capillary pressure changes from 30 to 10 mm Hg. But this belief is no longer tenable because P_T and π_T and lymph flow vary in the fluids surrounding the capillary wall, usually causing the capillaries to filter. This force gradient in the interstitium causes fluid to percolate toward the more negative interstitial fluid pressure surrounding the small veins where the initial lymphatics are located. The fluid enters these lymphatics to be restored to the circulating blood; in fact, the total plasma volume exchanges with the tissues every 24 hr, indicating that interstitial fluids are constantly being replenished by the capillary-lymphatic system.

C. Physics of Solute Exchange

Solute flux is more difficult to explain in a mathematical sense, but it can be evaluated using the following equation, which describes the flux of a solute across the capillary wall (J_S) in terms of the osmotic reflection coefficient of the solute (σ_d), and a permeability parameter, PS (the permeability surface area product):

$$\text{Convection} \qquad \text{Diffusion}$$

$$J_S = (1 - \sigma)J_V C_P + \frac{x}{e^x - 1} PS \, \Delta \, C$$

The first term on the right-hand side of the equation is the convective flux term. For instance, if $\sigma_d = 0$, indicating a freely permeable membrane, the solute will move across the membrane in a fluid stream as if there were no capillary wall, similar to water flowing out of a water hose. Small molecules cross capillary walls by convection when capillaries are filtering fluid because their $\sigma_d s$ are equivalent to zero. The second term on the right-hand side of the equation is the diffusional component of transcapillary flux and describes the flux associated with molecules diffusing down their concentration gradients (ΔC). All water-soluble molecules cross capillary walls by both mechanisms. When capillaries are filtering only small amounts of fluid, the solute transport occurs mainly by simple diffusion.

D. Physiology of Solute Exchange

Table II shows PS products, σ_d and K_{FC} for selected organs. Note that PS becomes larger with decreasing molecular size, whereas σ_d becomes smaller. Also note that PS is high for small molecules such as NaCl and glucose but orders of magnitude smaller for albumin. Note that the PS of liver and intestine are two orders of magnitude larger than skeletal muscle. K_{FC} is mainly determined by the surface area of the capillary bed, and a more correct

membrane parameter can be calculated, the membrane volume conductivity, if the surface area for fluid exchange is known.

IV. Normal Physiology of Transvascular Exchange

The data in Table III indicate that the capillaries in most organs produce a slight amount of filtration into the tissues because the sum of the Starling forces is positive ΔP. Normally, the small amount of fluid and protein that leaks out of the capillaries into the interstitium returns to the circulation via the lymphatic system, and the tissues do not either shrink or swell. In some capillaries, however, fluid can be constantly absorbed or filtered. For instance, the glomerular capillaries' filtration pressures are higher than their absorptive pressures, and they continuously filter fluid from the plasma into the renal tubules. Most of this fluid is reabsorbed by the capillaries surrounding the tubules. Reabsorption occurs because the absorptive pressure in these peritubular capillaries exceeds the filtration pressures, so these are continuously absorbing capillaries. Thus, the kidney contains capillaries that are always filtering and others that are always absorbing, as shown in Table III.

The intestinal capillaries normally filter, but when the intestine is absorbing fluid, the tissue protein osmotic pressure decreases to about 5 mm Hg because the transported fluid contains no plasma proteins, and the capillaries now revert to an absorbing

TABLE II Permeability Surface Area Product (PS), Reflection Coefficient (σ_d), and Filtration Coefficients (K_{FC}) for Selected Tissues[a,b]

Tissue	PS (ml/min/100 g)			σ_d		K_{FC} (ml/min/mm Hg/100 g)
	Na	Glucose	Albumin	NaCl	Albumin	
Subcutaneous	15	15	0.100	0.002	0.86	0.020
Skeletal muscle	15	16	0.07	0.002	0.83	0.015
Brain	0.01	0	0	1.0	1.0	0.000001
Intestine	1,400	80	0.09	0.0004	0.92	0.085
Liver	1,500	1,500	0.35	0	0	1.00
Lung	200	150	0.20	0.002	0.50	0.200
Cardiac muscle	100	50	0.08	0.020	0.45	0.70

[a] Modified from Parker, J. C., Perry, M. A., and Taylor, A. E. (1985). Permeability of the microvascular barrier. *In* "Edema" (N. C. Staub and A. E. Taylor, eds.), pp. 143–188. Raven Press, New York, with permission.
[b] The molecular radius of both NaCl and glucose are about 0.5 nm and albumin has a molecular radius of 3.7 nm.

TABLE III Starling Forces for Selected Tissues[a,b]

Tissue	P_C	P_T (mm Hg)	π_P	π_T	LF (ml/min/100 g)	ΔP
Subcutaneous	13	−5	21	4	0.015	+1
Skeletal muscle	9	−3	20	8	0.005	0
Brain	11	7	14	0	—	−10
Intestine (normal)	16	2	23	10	0.08	+1
Intestine (absorbing)	16	3	23	5	0.10	−5
Liver	7	6	22	20	0.10	−1
Lung	7	−5	23	12	0.10	+1
Cardiac muscle	23	15	21	13	0.12	0
Glomerular	50	15[c]	28	0[c]	2.0	+7
Renal peritubular	25	7	32	7	2.0	−7

[a] Modified from Granger, H., *et al.* (1985). Dynamics and control of transmicrovascular fluid exchange. *In* "Edema" (N. C. Staub and A. E. Taylor, eds.), p. 210. Raven Press, New York, with permission.

[b] P_C, P_T, π_P, π_T, and LF denote capillary and tissue hydrostatic pressure, plasma and tissue protein osmotic pressure, and lymph flows, respectively. ΔP is the sum of the forces [$(P_C - P_T) - (\pi_P - \pi_T)$] and represents filtration when positive and absorption when negative. A sum of zero indicates no filtration or absorption.

[c] This represents tubular pressures rather than renal interstitial pressure.

state, because the absorptive pressure becomes greater than the filtration pressure. The brain capillaries show a tendency to reabsorb, but their permeability to water (filtration coefficient) is so small that only minute amounts of fluid can enter the capillaries even for this large gradient in force.

The sum of the Starling forces in the liver is negative, but because the σ_d of the liver capillaries is zero, only the difference between P_C and P_T drive liver transcapillary fluid exchange.

It should be pointed out that the sum of the Starling forces in a capillary system is dynamic and that capillaries can be filtering or absorbing depending on the physiological status of the organ.

CO_2 and O_2 exchange in the tissues is determined by the local blood flow to the tissues, the oxygen consumption of the tissues, and the O_2 extraction and CO_2 removal at the tissue level. The blood flow to an organ will be increased when tissue oxygen is low because the arterioles dilate which delivers more O_2 to the tissues (or carries away more CO_2). If the cells have a higher metabolic state, such as that occurring in exercise, more O_2 must be provided to the muscle to produce the energy necessary for muscle contraction. Normally, three-fourths of the muscle capillaries are not flowing, but when the tissues need more oxygen as occurs in exercise, they all begin to flow, which increases the capillary exchange surface area by fourfold. Providing more blood flow to the tissues is an important means of supplying adequate oxygen supply to the

cells. But the exchange capacity of the capillaries (their total exchanging surface area) is also an important factor determining the amount of oxygen and CO_2 that can be exchanged with the tissues.

The lungs demonstrate this effect of a large exchange surface area for physiological gases in a dramatic fashion. The lung has a enormous capillary exchange surface area of 70–100 m^2, which is about the size of a tennis court. The entire cardiac output must flow through the lungs and even if one-half the lung has been removed, the blood exiting the lung will be fully oxygenated and the blood CO_2 levels will not exceed approximately 40 mm Hg.

V. Pathophysiology of Transvascular Exchange

A. Safety Factors Against Edema

Figure 5 shows how much fluid will enter the tissues as capillary pressure increases. Note that almost no fluid enters the tissues in normal capillaries (solid line) until capillary pressure exceeds 22 mm Hg. Now, if capillary pressure increased, why did fluid not enter the tissues? Obviously, something else must be changing to oppose the tendency for fluid to enter the tissues when capillary pressure is elevated. The insert in Fig. 5 shows what happens to the Starling forces and lymph flow as capillary pressure is increased: the tissue fluid pressure (P_T), the

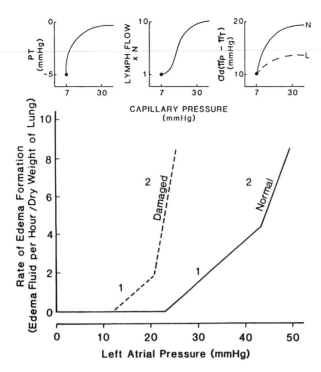

FIGURE 5 Amount of edema fluid formed in a normal (*solid lines*) and damaged (*dashed lines*) capillary system as a function of capillary pressure. The *insert* shows that tissue pressure (P_T), the absorptive pressure [$\sigma_d(\pi_P - \pi_T)$] and lymph flow increases as capillary pressure is increased, which acts to prevent the build-up of excessive tissue fluid at capillary pressure below 25 mm Hg in normal capillaries. When the ability of these factors to change is overwhelmed, edema develops rapidly in the tissues and is accelerated at even higher capillary pressures. When the endothelium has been damaged, the capillary pressure at which edema fluid accumulation occurs is much lower because the absorptive pressure increases only slightly to buffer the increased capillary pressure in leaky capillaries [insert showing change in $\sigma_d(\pi_P - \pi_T)$ as a function of capillary pressure and labelled L]. [Revised from Taylor, A. E. and Townsley, M. I. (1987). *News Physiol. Sci.* **2**, 48–57, and Taylor, A. E., Rehder, K., Hyatt, R., and Parker, J. C. (1989). "Clinical Respiratory Physiology." Saunders, Philadelphia with permission.]

lymph flow, and the absorption pressure [$\sigma_d(\pi_P - \pi_T)$ for the curve marked N] increase by approximately 15 mm Hg. These changes in tissue forces and lymph flow allow the capillary pressure to change by 15 mm Hg before any significant build-up of fluid occurs in the tissues. This is an extremely important phenomenon, because capillary pressure constantly changes in all tissues, and the ability of $\sigma_d(\pi_P - \pi_T)$, lymph flow, and P_T to increase provides what we have termed *tissue edema safety factors* that oppose the tendency for fluid to accumulate in the tissues. Also, note in Fig. 5 that when

capillary pressure exceeds the pressure that can be buffered by the changes in tissue forces and lymph flow, that excessive amounts of fluid begin to fill the tissues. This fluid will not return into the circulation until capillary pressure has returned to normal values. Thus, edema safety factors inhibit edema formation, which would cause deleterious problems in tissue oxygenation and removal of waste products of metabolism. Table IV shows safety factor changes in several different organs when capillary was increased by elevating the venous pressure of the individual organs. The change in absorption pressure is a major force in opposing edema formation in most organs (first column). However, in some organs, π_T is low and cannot change substantially (subcutaneous tissue) as capillary pressure increases, or the capillaries are leaky to proteins and π_T does not significantly decrease to increase the absorptive force (heart and liver). In these organs, other safety factors change to a greater extent, allowing the capillary pressure to increase to levels almost identical to that observed in other organs before edema develops. Cases in point are the edema safety factors in liver and heart, where increased lymph flow and increased tissue pressure become larger safety factors when the absorptive pressure cannot change to any significant extent.

B. Types of Edema Formation

Most edema results because capillary pressures increases although any decrease in plasma proteins, such as that associated with starvation and overzealous fluid therapy in critically ill patients, will

TABLE IV Safety Factors in Various Tissues[a,b]

Tissue	Increased $\sigma_s(\pi_P - \pi_T)$	Increased lymph flow	Increased P_T
Lung	50	17	33
Hindpaw	14	24	62
Small intestine	45	20	35
Colon	52	4	44
Liver	0	42	58
Heart	7	12	81

[a] Modified from Taylor, A. E., and Townsley, M. I. (1987). Evaluation of the Starling fluid flux equation. *News Physiol. Sci.* **2**, 48–57, with permission.

[b] $\sigma_d(\pi_P - \pi_T)$ is the absorptive force and P_T is the tissue fluid pressure. The values are shown as % of the total safety factor measured when capillary pressure was increased 20 mm Hg above control values.

decrease the capillary, absorptive pressure and promote edema formation. Also, many biological compounds such as histamine cause capillary pressure to increase because of venous constriction and venous pressure becomes elevated in heart failure, which increases capillary pressure and results in edema. A curious form of edema occurs when the lymphatic system is blocked or becomes nonfunctional. The most bizarre form of this "lymphatic insufficiency" problem occurs in the natives of the Pacific Islands. A small filaria is present in the bloodstream and tissues of some New Guineans, which damages or obstructs the lymphatic system. When this occurs, grotesque forms of edema develop in areas such as the scrotum, which can sometimes increase to such a great size that a small wheelbarrow must be used to carry the grossly swollen scrotum as the man walks about. Less dramatic forms of edema occur after operations in which the lymphatic system is damaged or removed during surgery. In chronic forms of edema, the tissues are infiltrated with fat and fibrous tissue and the tissues of the swollen areas do not resemble the simple fluid imbibed by the interstitial gel seen with acute forms of interstitial edema.

An unusual form of edema occurs in muscle sheaths, called *compartmental syndrome*. Fluid entering the muscle sheath increases tissue fluid pressure, which compresses the veins and increases capillary pressure to higher values, which causes more fluid to filter, increasing P_C further. Sometimes blood flow will be almost obliterated if this condition persists because the tissue pressure becomes higher than arterial pressure. The muscle will die unless the muscle sheath is opened to liberate the edema fluid. A similar type of edema also occurs in the brain when the cerebral spinal fluid pressure is increased. This usually occurs with bleeding into the cranial vault, causing the cerebral spinal fluid pressure to increase, which compresses the veins and further increases the cerebral spinal fluid volume and pressure. However, it can also occur when too much cerebral spinal fluid is secreted by the choroid plexus or a part of the cerebral spinal fluid circulation system is blocked. Because brain tissue swelling is limited by the cranium, the cerebral spinal fluid pressure can increase to high levels in either of these conditions that severely limit blood flows similar to compartmental syndrome in muscle. When the lungs become edematous, the blood is not properly oxygenated and CO_2 is not properly eliminated by the lungs. When the interstitial fluid volume of the lung increases by 100%, the partial pressure of arterial oxygen decreases to 50 mm Hg.

Finally, Fig. 5 shows how damaged capillaries are more likely to produce edema when capillary pressure is elevated (dashed line). This occurs because (1) the filtration coefficient increases and more fluid will be filtered for a given capillary pressure, and (2) the absorptive pressure $[\sigma_d(\pi_P - \pi_T)$ indicated as the L curve in Fig. 5 insert] will not change greatly because π_T cannot decrease significantly with capillary filtration when capillaries are abnormally leaky causing the edema safety factor to be reduced to 5 mm Hg.

VI. Summary

The tissues must be in a relatively nonedematous state to function properly. The ability of the lymphatics to remove fluid entering the tissues for the absorptive and the tissue fluid pressures to oppose increases in capillary pressure are extremely important biological mechanisms that allow the cells to live in a fluid environment that does not limit O_2 and metabolite exchange or the removal of waste products of their metabolism. The capillary system is the biological system that allows us to function as a multicellular animal. Any disruption of the capillary system will impair cellular metabolism and cause a build-up of unwanted metabolites in the tissues. If proper capillary function cannot be restored, the cells will begin to die and organ failure results with subsequent death of the individual.

Most studies of capillary function reported in this chapter were performed in organs, but recent studies have used isolated endothelial cell membranes to study endothelial transport of fluid and solutes. Although these studies are in the developmental phase, they have begun to yield interesting results relative to transport across endothelial cells grown or filter papers. For instance, albumin flux appears to be greater from the tissue to plasma than from plasma to tissues across these artificial capillary membranes. Although this observation has not been confirmed in organ studies, it does indicate the importance of using these isolated endothelial membranes to design future organ and microscopic microcirculatory studies.

Bibliography

Milici, A. J., Furie, M. B., and Carley, W. W. (1988). The formation of fenestrations and channels by capillary endothelium *in vitro*. *Proc. Nat. Acad. Sci.* **82,** 6181–6185.

Milici, A. J., L'Hernault, N., and Palade G. (1985). Surface densities of diaphragmed fenestrae and transendothelial channels in different musosa capillary beds. *Circ. Res.* **56,** 709–715.

Renkin, E. M., and Michel, C. C. eds. (1984). "Physiology Handbooks of the Microcirculation," vol. IV, parts 1 and 2. American Physiological Society, Waverly Press, Bethesda, Maryland.

Rippe, B., and Haraldsson, B. (1987). How are molecules transported across the capillary walls? *News Physiol. Sci. 2,* 135–139.

Risau, W. (1989). Differentiation of blood-brain barrier endothelium. *News Physiol. Sci.* **4,** 151–153.

Shasby, M. D., and Roberts, R. L. (1987). Transendothelial transfer of macromolecules *in vitro*. *Fed. Proc.* **46,** 2506–2510.

Simionescu, M., Ghitescu, L. Fixman, A., and Simionescu, N. (1987). How plasma macromolecules cross the endothelium. *News Physiol Sci.* **2,** 97–101.

Staub, N. C., and Taylor, A. E., eds. (1985). "Edema." Raven Press, New York.

Taylor, A. E., and Granger, D. N. (1984). Exchange of molecules across the microcirculation. *In* "Handbook of Physiology, The Cardiovascular System." vol. IV, section 2, part 1. (E. M. Renkin and C. C. Michel, eds.), pp. 467–520. Waverly Press, Bethesda, Maryland.

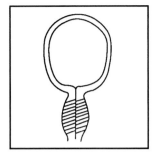

Micturition Control

GERT HOLSTEGE, *Rijksuniversiteit Groningen, The Netherlands*

I. Motoneurons Innervating Bladder and Sphincter
II. Sacral Cord Micturition Reflexes
III. Brainstem–Spinal Pathways Controlling Micturition
IV. Suprapontine Micturition Control

Glossary

Amyotrophic lateral sclerosis (ALS) Disease in which somatic motoneurons die because of reasons still unknown (also called motoneuron disease)
Autonomic motoneurons Sympathetic or parasympathetic motoneurons innervating smooth musculature and glands
Micturition Act of urinating
Perineum Pelvic floor
Somatic motoneurons Innervate striated musculature
Tegmentum Core or central portion of the brainstem—three parts: mesencephalic, pontine, and medullary

MOTONEURONS INNERVATING THE BLADDER and its sphincter are located in the sacral spinal cord, indicating that the sacral cord plays an essential role in micturition control. Nevertheless, the brainstem, via its long descending pathways to the sacral cord, is vital for coordinating muscle activity of bladder and bladder sphincter, during normal micturition. The importance of the brainstem in micturition control is best shown by patients with spinal cord injuries above the sacral level. Such patients have great difficulty emptying the bladder because of uncoordinated actions of the bladder and sphincter (detrusor–sphincter dyssynergia). Such disorders never occur in patients with neurologic

lesions rostral to the pons, which indicates that the coordinating neurons are located in the pontine tegmentum. Barrington showed as early as 1925 that these neurons are probably located in the dorsolateral part of the pontine tegmentum, because bilateral lesions in this area in the cat produced an inability to empty the bladder.

I. Motoneurons Innervating Bladder and Sphincter

The smooth musculature forming the detrusor muscle of the bladder is innervated, via the pelvic nerve, by the preganglionic parasympathetic neurons in the sacral intermediolateral cell group. Sympathetic fibers, running via the pelvic and hypogastric nerves, innervate the bladder also, some directly but others indirectly via connections with the paravesical ganglia of the parasympathetic system. Their preganglionic motoneurons are probably located in the intermediolateral cell group of the upper lumbar cord (L1–L4). The sympathetic fibers have inhibitory effects on the detrusor muscle and excitatory effects on the smooth musculature of the urethra and base of the bladder.

The pudendal nerve innervates the intrinsic external urethral sphincter. This muscle forms part of the pelvic musculature. The motoneurons innervating the pelvic floor muscles are located in the nucleus of Onuf. In 1899 Onuf described a group X in the ventral horn of the human spinal cord, extending from the caudal S1 to the rostal S3 segments. More recent studies demonstrated that motoneurons innervating the urethral sphincter and those innervating the anal sphincter are, respectively, located in the ventrolateral and the dorsomedial part of the nucleus of Onuf. Nucleus of Onuf motoneurons may belong to a distinct class. On the one

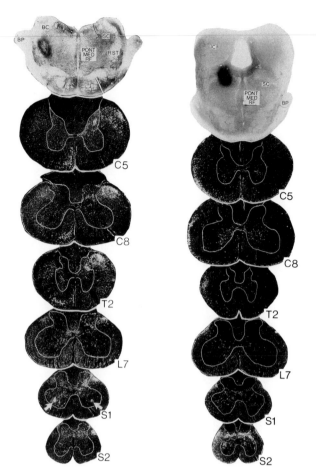

FIGURE 1 Bright-field photomicrographs of autoradiographs showing the tritiated leucine injection areas and dark-field photomicrographs showing the spinal distributions of labeled fibers after an injection in the L-region (on the left) and after an injection in the M-region (on the right) in the cat. Note the pronounced projection to the nucleus of Onuf (arrows in the S1 segment) in the case with an injection in the L-region (left). Note also the dense distribution of labeled fibers to the sacral intermediolateral (parasympathetic motoneurons) and intermediomedial cell groups after an injection in the M-region (S2 segment on the right). (BC, brachium conjectivum; SC, suprachiasmatic nucleus.) From G. Holstege, D. Griffiths, H. De Wall, and E. Dalm. 1986. Anatomical and physiological observations on supraspinal control of bladder and urethral sphincter muscles in the cat. *J. Comp. Neurol.* **250,** 449–461.

hand, they are somatic motoneurons, because they innervate striated muscles and are under voluntary control, but on the other hand they are autonomic motoneurons because (1) cytoarchitectonically they resemble autonomic motoneurons; (2) they have an intimate relationship with sacral parasympathetic motoneurons; (3) they receive direct hypothalamic afferents; and (4) in contrast to the other somatic

motoneurons in the sacral cord, nucleus of Onuf motoneurons are well preserved in the spinal cords of patients who have died from amyotrophic lateral sclerosis (ALS). The sacral autonomic (parasympathetic) motoneurons are also spared in ALS patients, which explains why bladder and sphincter functions remain intact until the latest stages of the disease.

II. Sacral Cord Micturition Reflexes

Behavioral evidence for the existence of sacral cord micturition reflexes was given by De Groat *et al.*, who observed that micturition as well as defecation are elicited in neonatal kittens when the mother licks the perineal region. This is the primary stimulus for micturition, because separation of the kittens from the mother results in urinary retention. The perineal-to-bladder reflex is quite prominent during the first 4 postnatal weeks, after which it becomes less effective and usually disappears by the age of 7–8 weeks, the approximate age of weaning. T1-transections of the spinal cord did not abolish the perineal-to-bladder reflex, indicating that it is a sacral cord reflex. After 7–8 weeks bladder-to-bladder reflexes traveling via supraspinal pathways replace the perineal-bladder reflex. Transection of the spinal cord in older kittens or adult cats causes the reemergence of perineally induced micturition within 1–2 weeks. In humans and animals with intact neuraxes, this reflex system is functionally nonexistent, but it probably plays a role in patients and animals with transection of the pathways between the pons and sacral cord. Thus, it seems that in adult animals and humans there exist pathways within the sacral cord that can produce bladder and sphincter contractions, although these contractions are not necessarily well coordinated (i.e., they are dyssynergic).

III. Brainstem–Spinal Pathways Controlling Micturition

Since the work of Barrington in 1925 it has been known that the supraspinal portion of the micturition reflex is located in the dorsolateral portion of the pontine tegmental field. In most studies this area is referred to as the pontine micturition center. Recent anatomic studies have shown that neurons in the dorsolateral pontine tegmentum, medial to the

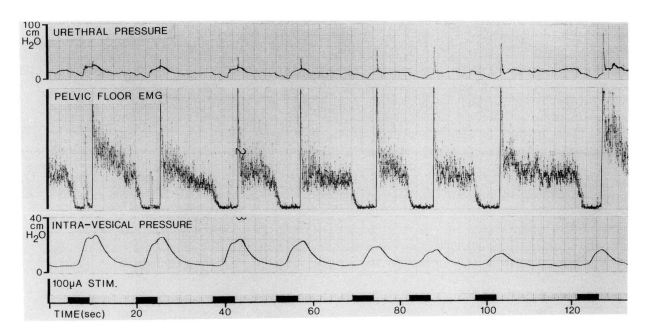

FIGURE 2 Recordings of urethral pressure, pelvic floor EMG, intravesical pressure, and stimulus timing during M-region stimulation in the cat. Note the immediate fall in urethral pressure and pelvic floor EMG after the beginning of the stimulus and the steep rise in the intravesical pressure about 2 sec after the beginning of the stimulus. This pattern mimics complete micturition. From G. Holstege, D. Griffiths, H. De Wall, and E. Dalm. 1986. *Anatomical and physiological observations on supraspinal control of bladder and urethral sphincter muscles in the cat. J. Comp. Neurol.* **250:** 449–461.

locus coeruleus, project directly and specifically to the sacral intermediolateral cell group (parasympathetic motoneurons), as well as to the sacral intermediomedial cell group, but not to the nucleus of Onuf (Fig. 1, right). The nucleus of Onuf receives specific projections from neurons in more lateral parts of the dorsolateral pontine tegmental field; this cell group does not project to the sacral parasympathetic motoneurons (Fig. 1, left). In order to differentiate between them, Holstege *et al.* called them the M- (medial) and L- (lateral) regions. The M-region probably corresponds with Barrington's area. Neither the M- nor the L-region project to the lumbar intermediolateral (sympathetic) cell groups.

Electric stimulation in the M-region produces an immediate and sharp decrease in the urethral pressure and pelvic floor EMG, followed in about 2 sec by a steep rise in the intravesical pressure, mimicking complete micturition (Fig. 2). The decrease in the urethral pressure cannot be caused by a direct M-region projection to the nucleus of Onuf, because such a projection does not exist. Possibly, the M-

region inhibits the L-region, which projects directly to the nucleus of Onuf. Stimulation in the L-region results in strong excitation of the pelvic floor musculature and an increase in the urethral pressure. Bilateral lesions in the M-region result in a long period of urinary retention, during which detrusor activity is depressed and the bladder capacity increases. Bilateral lesions in the L-region give rise to an inability to store urine; bladder capacity is reduced and urine is expelled prematurely by excessive detrusor activity accompanied by urethral relaxation. These observations suggest that during the filling phase the L-region has a continuous excitatory effect on the nucleus of Onuf, which inhibits urethral relaxation, coupled with detrusor contraction. When micturition takes place, the M-region excites, via a direct pathway, the sacral parasympathetic motoneurons, but at the same time the M-region inhibits the L-region, which disinhibits sphincter relaxation so that micturition can take place.

IV. Suprapontine Micturition Control

Micturition is coordinated in the caudal brainstem. Patients with neurological lesions in more rostral portions of the brain never suffer from detrusor-sphincter dyssynergia. However, many of these patients suffer from lack of control over the moment micturition begins. This raises the question of what determines the beginning of the micturition act. Obviously, precise information about the degree of bladder filling is conveyed to supraspinal levels, but specific sacral projections to the pontine micturition center have not been demonstrated. This suggests that other structures, rostral to the pontine micturition center, determine the beginning of the micturition act. Such structures would be expected to project specifically to the M-region of the pontine micturition center. Many clinical studies indicate that cortical structures (the medial frontal gyrus and anterior cingulate lobe) as well as subcortical structures (septum, preoptic region of the hypothalamus, and amygdala) are involved in control of the beginning of micturition. Experimentally, the only structure that has been demonstrated to project specifically to the M-region is the preoptic area in the cat. Stimulation in this area produces micturition-like contractions, but it is not known whether it determines the beginning of micturition. It is possible that regions other than the preoptic area also project to the M-region. Furthermore, the fact that the pelvic floor, including the intrinsic external urethral sphincter, is under voluntary control suggests that direct cortical projections to the nucleus of Onuf may exist. Such projections have not been demonstrated convincingly. Figure 3 gives a schematic overview of the spinal and supraspinal structures involved in micturition control and their role in the neuronal framework of micturition.

Bibliography

Barrington, F. J. F. (1925). The effect of lesions of the hind- and mid-brain on micturition in the cat. *Quart. J. Exp. Physiol. Cogn. Med.* **15,** 81–102.

Blaivas, J. G. (1982). The neurophysiology of micturition: a clinical study of 550 patients. *J. Urol.* **127,** 958–963.

De Groat, W. C., Douglas, J. W., Glass, J., Simonds, W., Weimar, B., and Werner, P. (1975). Changes in somato-vesical reflexes during postnatal development in the kitten. *Brain Res.* **94,** 150–154.

Grossman, R. G., and Wang, S. C. (1956). Diencephalic mechanism of control of the urinary bladder of the cat. *Yale J. Biol. Med.* **28,** 285–297.

Holstege, G. (1987). Some anatomical observations on the projections from the hypothalamus to brainstem and spinal cord: An HRP and autoradiographic tracing study in the cat. *J. Comp. Neurol.* **260,** 98–126.

Holstege, G., Griffiths, D., De Wall, H., and Dalm, E. (1986). Anatomical and physiological observations on

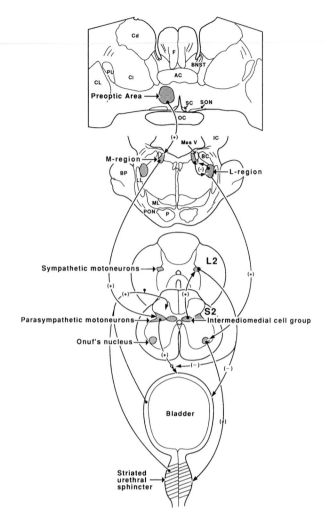

FIGURE 3 Schematic representation of the spinal and supraspinal structures involved in micturition control. Excitatory pathways are indicated by (+), inhibitory projections by (−). AC, anterior commissure; BC, brachium conjunctivum; BNST, bed nucleus of stria terminalis; BP, brachium pontis; CI, capsula interna; CL claustrum; IC, inferior colliculus; F, fornix; LL, lateral lemniscus; MesV, mesencephalic trigeminal tract; ML medial lemniscus; OC, optic chiasm; P, pyramidal tract; PON, pontine nuclei; SC, suprachiasmatic nucleus; SON, supraoptic nucleus.

supraspinal control of bladder and urethral sphincter muscles in the cat. *J. Comp. Neurol.* **250,** 449–461.

Mannen, T., Iwata, M., Toyokura, Y., and Nagashima, K. (1977). Preservation of a certain motoneurone group of the sacral cord in amyotrophic lateral sclerosis: its clinical significance. *J. Neurol. Neurosurg. Psych.* **40,** 464–469.

Onufrowicz, B. (1899). Notes on the arrangement and function of the cell groups in the sacral region of the spinal cord. *J. Nerv. Mental Dis.* **26,** 498–504.

Minerals in Human Life

R. J. P. WILLIAMS, *Oxford University*

Glossary

ATP Adenosine triphosphate is a small organic compound that is used to carry phosphate from site to site.

Bulk diet Sum of all intake of (here) chemical elements into the human body.

Homeostasis Stationary condition of the concentrations of elements (Na, K, Cl) in the body fluids and cells of humans.

Ion In compounds some elements or small molecules carry an electric charge, and the resulting species is called an ion. Where the charge is positive the species is called a cation, e.g., the sodium ion is Na^+. Where the charge is negative the species is called an anion, e.g., the chlorine ion is Cl^-.

Mitochondrion In a cell the major oxidation reactions, burning of fuel, are to be found in a vesicular unit, the mitochondrion.

Osmotic balance Presence of ions or molecules in water causes them to be solvated thus reducing the free water activity per unit volume. Two solutions are in osmotic balance when the total number of molecules or ions in each is such that water does not pass from one to the other to alter the concentration in either.

Oxidative metabolism That part of metabolism which uses oxygen of the air to react with organic molecules such as sugars to produce either energy or new molecules such as are required as components for syntheses.

THE WORD MINERALS in a general sense is used to classify the nature of the nonliving matter of the earth separately from living matter. The distinction is made between the three gross categories minerals, plants, and animals, where plants and animals incorporate all living systems.

The use of the word *mineral* in relationship to human life is somewhat different and refers to the components of the above 'minerals' that are taken in by humans when drinking and eating. Some minerals are present in water supplies as soluble salts, such as sodium chloride and magnesium sulphate, both of which reach very high levels in sea water. All the chemical elements, Table I, are found in simple form in different solid compounds of rocks and soils. Chemists have found it convenient to distinguish these mineral compounds from organic compounds, which are composed overwhelmingly of the elements carbon (C), hydrogen (H), nitrogen (N), oxygen (O), with smaller amounts of sulphur (S), phosphorus (P), chlorine (Cl), and others. These elements are usually associated in complicated molecular forms such as glucose, $C_6H_{12}O_6$, or urea, $CO(NH_2)_2$. All biological systems from bacteria to humans require for their operation elements from the minerals (see below) as well as organic chemicals. The source of minerals (elements) for humans can be either the water or the minerals themselves (such as table salt) or food from a variety of living species including both plants and animals that have obtained their minerals more directly from soils.

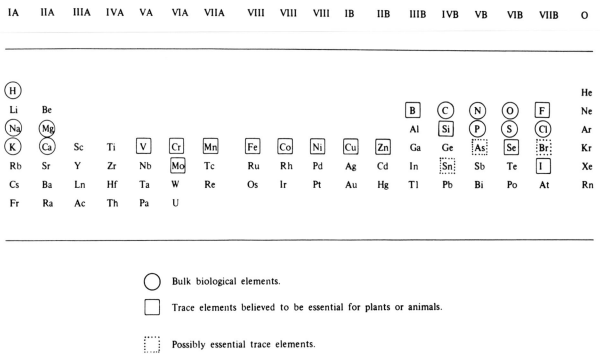

| IA | IIA | IIIA | IVA | VA | VIA | VIIA | VIII | VIII | VIII | IB | IIB | IIIB | IVB | VB | VIB | VIIB | O |

FIGURE 1 The distribution of the elements essential for life in the periodic table.

The simplest starting point for a description of the involvement of minerals in life is with the periodic table of the chemical elements, Fig. 1. In the table are shown the elements that are essential for human life, as far as we know. In the table elements are arranged according to the physico-chemical properties, which can thus be related to their biological functions.

I. Sodium and Potassium

The ions of the elements at the extremes of the periodic table Na^+ and K^+ on one side, and Cl^- on the other, together form soluble salt solutions. They are present in large quantities in the body, both in the extracellular body fluids and inside cells, and are the basis of the maintenance of osmotic pressure in our cells. Through interaction with DNA and proteins they could exert some control functions. Generally however their interactions with organic molecules are very weak. The cell cytoplasm accumulates potassium and rejects both sodium and chloride, a process that requires much of the body's energy utilization. The two positively charged ions, sodium and potassium, are the basis of nerve con-

ductivity and other cellular electrical currents and, therefore of much of the current observed in all organs of the body including the brain. Clearly the requirement for a controlled relatively large amount of these elements as ions is essential for the body's well-being. It is routine practice in medical checking to look at sodium and potassium balances.

II. Calcium and Magnesium

The next chemical elements to be described are called alkaline earths, magnesium (Mg) and calcium (Ca). These elements are present at quite high levels in the body and they also form soluble salts, but the solubility in the human body as elsewhere is limited by reaction with carbonate and phosphate to form precipitates. The most obvious manifestation is the formation of bone in the skeleton and of calcite crystals in the inner ear. Calcium hydroxyphosphate, the major component of bone, is the most abundant structural mineral in the body, while calcite is used for one form of gravity detection. Calcium bound to large organic molecules is important in maintaining the structure of many biopolymer filaments, present within the cells or outside them, as

TABLE I Functional Significance of Trace Elements in the Human Body

Trace element	Significance
Manganese	Catalysis of attachment of sugar to proteins; in some other enzymes.
Iron	Numerous reactions related to catalysis of reactions e.g., of oxygen, peroxides.
Cobalt	In vitamin B_{12} requiring enzymes.
Nickel	In an enzyme urease (rare).
Copper	Many oxidative enzymes usually outside the cell.
Zinc	Many enzymes e.g., alcohol dehydrogenase, peptidases, carbonic anhydrase. Also associated with DNA.
Selenium	In some enzymes using peroxides.
Iodine	In thyroid hormones.
Molybdenum	In enzymes of nucleoside degradation.

in connective tissues. In contrast, magnesium is only a minor component of bone or soft structures. [*See* CALCIUM, BIOCHEMISTRY.]

The free calcium ion has another function in that while it is maintained in extracellular fluids at a high concentration (about 10^{-3} M) it is continually pumped out of the cells, in which it is maintained at very low concentration, generally below 10^{-7} M. The huge difference of the calcium ion concentration (gradient) in resting cells, which costs much energy, is used to alert the cell to change in its surroundings. An initial message following some internal or external disturbance of the resting state usually moves from cell to cell and along cell membranes (nerves) by the sodium/potassium electric current and then a secondary message is transmitted to the interior of some cells by opening the membrane to an influx of calcium ions. On entering the cell the calcium binds to specific proteins and so triggers cellular activity. Many cellular changes of state are activated in this way. Major examples are the transmission of signals between nerve cells at synapses, and muscle contraction elicted by a nerve signal. An activating flux of calcium can also be brought about by chemical messages transmitted by organic chemicals present in the body fluids, such as nerve transmitters and hormones, reacting with their receptors at the cell surface. Activation of receptors frequently opens the way to calcium pulses into cells. Many cellular changes such as fertilization, control of cell-cell contact, and differentiation, all involve calcium acting as the so-called second messenger in cells.

A further important role of free calcium is observed outside cells in body fluids. Here calcium is a major requirement for the proper functioning of many reactions, including food digestion and blood clotting. Following the changes associated with feeding or injury in the body, proteins are released out of the cells from vesicles that are initially inside cells. These proteins are inactive proenzymes. The role of the extracellular calcium is to assist in their activation, by converting them into functional enzymes.

The second alkali earth element, magnesium, is of the greatest importance in the cell. Its ions do not form insoluble salts as readily as calcium ions, but they, together with negatively charged ions, easily form complex ions. Magnesium ions are maintained at a free ion concentration of 10^{-3} M in all cells and in the extracellular body fluids. A considerable amount of magnesium is bound to other molecules. Just as there are many extracellular calcium-dependent enzymes, so there are many intracellular magnesium-dependent enzymes. For example magnesium is deeply involved in phosphate metabolism, and virtually all adenosine triphosphate (ATP), which is crucial to energy exchange in cells, is present in association with magnesium ions as Mg^{2+} ATP^{4-}. The affinity of magnesium for phosphate is also important in controlling RNA and DNA conformations. [*See* ADENOSINE TRIPHOSPHATE (ATP).]

III. Chloride and Other Nonmetals as Anions

Turning to chloride, which forms negatively charged ions (anions), we have noted that it too is present in large amounts in body fluids; in fact they taste salty due to sodium chloride. Chloride helps to maintain the osmotic balance between body fluids and charge neutrality by compensating positive charges. It exchanges with bicarbonate across many membranes, one going in while the other goes out, allowing the removal of carbon dioxide produced by all cells during their function. The next anions of importance (see the periodic table) are sulphate, nitrogen-based anions, phosphates, and carbon-based anions. These elements are all in the upper right hand part of the table. All these anions are metabolized, that is they undergo stable covalent modification in the body and thereby have totally different chemistry from that generated by the ionic interactions, which are readily reversible, associated with the elements Na, K, Mg, Ca, and Cl.

We see that as we move from one extreme side to the other of the periodic table the importance of the mineral elements changes from chemically non-interactive ions, Na^+, K^+, Cl^- to chemically interactive ions, Ca^{2+}, Mg^{2+}, SO_4^{2-} and PO_4^{3-}, and then to covalent nonmetal chemistry. Amongst the elements so far discussed the behavior of sulphur in biology is different from all the others in an outstanding respect, which it shares with carbon and nitrogen especially. All the mineral elements above are formed in the body in the same form or in very similar form to that which they have in the inorganic world. Sulphur, carbon, and nitrogen are transformed, e.g., carbon from carbon dioxide to sugar; nitrogen, from nitrate or ammonia, with carbon to amino acids; and sulphur from sulphate to sulphide. This nonmetal chemistry is at the heart of organic chemical synthesis in biology—of proteins, nucleic acids, fats, and saccharides.

IV. The Special Role of Phosphate

The mineral anion phosphate has already been mentioned in passing as a component of bone and of ATP. Phosphate has an overwhelmingly important role in the metabolism of all living systems. This explains why it is used regularly with potassium, nitrogen, and calcium in agricultural fertilizers. Phosphate enters the body as such or as organic phosphate, which is then broken down to phosphate, and used in three very different ways. In the first it becomes a connecting link in DNA and RNA, which are polymerized forms of nucleotide phosphates. In the second it is combined with a variety of organic molecules—with carbohydrates, especially sugars, it forms the metabolites of such processes as glycolysis. With glycols, it forms the lipids of cell membranes together with fatty acid chains. With polymerized saccharides it gives rise to extracellular glycoproteins. The third use of phosphate is in the transportation of "chemical" energy in all cells. Energy in the human body comes mainly from the burning of sugars with molecular oxygen. Outside the cell this combustion energy is lost as heat. In the cell the energy is trapped in nucleotide polyphosphates so as to make high-energy compounds such as ATP. The separation of one phosphate residue from ATP and similar derivatives is the main driving force of chemical synthesis, of the formation of ion gradients in nerves (as described above) and of muscular activity. This reaction is under numerous controls, such as that of magnesium and calcium ions. The participation of other minerals in other forms of energy exchange is outlined below.

V. The Remaining Light Mineral Elements

The final group of mineral elements in the upper-right of the periodic table includes nitrogen, sulphur, and carbon, which are needed in large amounts by living organisms, and in addition silicon (Si), aluminium (Al), fluorine (F), and boron (B), which though present in large amounts in soils are required in very small amounts (Si), or not at all (Al, B) by the organisms. Fluoride affects the properties of bone and teeth; for this reason it is deliberately added to drinking water in some countries. Nitrogen and carbon present in mineral sources as nitrate and carbonate are utilized by bacteria and plants. The human supply of these elements comes in forms already elaborated by these organisms, e.g., sugars and amino acids. Sulphate is directly used in the body to form many sulphated sugars that are mainly found in intracellular spaces. Some of the elements considered here are deleterious. Nitrate and aluminium are toxic and should be avoided above a certain level in drinking water. This may be also true of borate, though boron is an essential element for plant life.

VI. Homeostasis

Homeostasis, the maintenance of constant levels of minerals in the human body in order to keep all its functions in balance, exists inside cells in their normal active metabolite states, and in the body fluids. As the word implies too much of any element (or compound) is deleterious to human well-being just as too little leads to deficiency symptoms (Fig. 2). Homeostasis is maintained by feedback mechanisms that control the levels of the elements. A controlled mineral diet is usually not necessary for a well-fed person. However sometimes faults develop: for example old age appears to induce problems of calcium metabolism. Lack of balance in

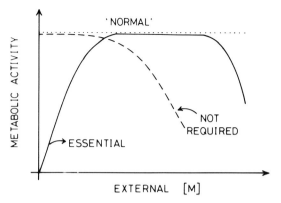

FIGURE 2 The relationship between the effect of a mineral element (response) on the absiscae and the concentration (dose) of the element [M] in diet. This is called a dose/response curve. There are mineral elements that are *essential* for health and they show optimal efficacy over a range of concentration (labeled "NORMAL"). Below this range the body suffers from deficiency: above the range the body suffers from excess and is poisoned to some degree. Other mineral elements labelled "NOT REQUIRED" cause damage to the body over the whole concentration range as shown by the decreased metabolism activities they induce and must only be allowed into the food chain in a very controlled way.

electrolytic message systems can cause hyperactivity, which can be helped by the use of lithium carbonate in relatively large amounts as a drug.

VII. Requirements for Heavier Element Minerals

The elements needed in large amounts are Na, K, Cl, Mg, Ca, P, and S, which have a relatively high free ion concentration in the body, and H, C, N, and O, which constitute most of the essential macromolecules. These elements could be said to be an essential part of *bulk* diet. Their concentration levels are carefully regulated everywhere by special organs such as the kidney. The remaining essential elements are present in much smaller quantities and they are largely trapped in combination with proteins.

It is not known precisely how many heavy chemical elements are absolutely required by man but the following list gives a minimum account: manganese (Mn), iron (Fe), cobalt (Co), nickel (Ni), copper (Cu), zinc (Zn), and molybdenum (Mo) among metallic elements and selenium (Se) and iodine (I) amongst nonmetal elements. To this list some authorities would add vanadium (V), chromium (Cr)

and bromine (Br). There may be further additions. Many other elements listed in Fig. 1 appear to be of little functional value, and some may well be generally deleterious to our well-being if presented in soluble forms. These elements include germanium (Ge), arsenic (As), and cadmium (Cd), as well as many very heavy metal ions such as gold (Au), lead (Pb), and mercury (Hg). It must be noticed that such statements about the value of rare heavier elements are relative, just as they were for light elements because the effect depends on the dose; too little and too much of many elements are equally deleterious. Each required element is held at an optimal level in the body by a homeostatic control mechanism. Even those rare heavy mineral elements that do not appear to have a functional value in the body and are poisonous even in small doses may be useful in medicine. Examples are the use of platinum in $Pt(NH_3)_2Cl_2$ as an anticancer drug and of gold compounds in the treatment of arthritis.

VIII. The Importance of the Trace Elements

The trace elements are usually taken to include all the essential elements required to the right of calcium in the periodic table. They are extremely important in the management of the reactions that are a major source of energy and a major pathway of metabolism. These elements are generally held in tight combination with organic molecules, especially proteins. They may then act merely as structural units or as catalysts. While zinc and the elements magnesium and calcium keep the charge on their bound forms fixed, i.e., Ca^{2+}, Mg^{2+} and Zn^{2+}, it is a characteristic of other trace mineral elements that they can undergo a change of charge, e.g., for iron from Fe^{2+} to Fe^{3+}. This property allows these elements to catalyze oxidation/reduction reactions which consist of transfers of electrons, and therefore of charge, in the body.

Before turning to oxidation/reduction reactions however, we must discuss a trace element that does not undergo changes of valence or oxidation/reduction state, and which has extreme functional value when combined with proteins. This is zinc. The importance of zinc in man's nutrition is not fully understood, but of all trace elements it is present in the body in the largest amounts together with iron. It serves a major role as a catalyst in enzymes. Re-

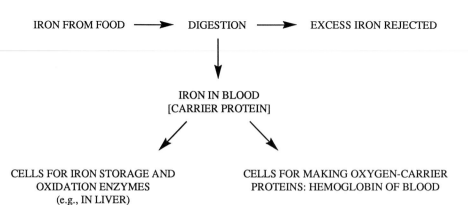

IRON FROM FOOD ⟶ DIGESTION ⟶ EXCESS IRON REJECTED

IRON IN BLOOD
[CARRIER PROTEIN]

CELLS FOR IRON STORAGE AND
OXIDATION ENZYMES
(e.g., IN LIVER)

CELLS FOR MAKING OXYGEN-CARRIER
PROTEINS: HEMOGLOBIN OF BLOOD

cently zinc has been found increasingly associated with DNA-binding proteins and there is suggestive evidence that it is involved additionally in the response of DNA to sterol and similar hormones. There is clear evidence that zinc deprivation in humans causes lack of full development and to skin cell deficiencies. The required free concentration of zinc like that of all other trace elements is very low ($<10^{-10}$ M). [See COPPER, IRON, AND ZINC IN HUMAN METABOLISM.]

An interesting peculiarity of the trace elements is that each is associated with rather special functions. In a manner of speaking they are like vitamins. Examples are the transport of oxygen in the blood, dependent on iron; the clearance of peroxides, which relies on iron in some cases but selenium in others; and the dependence of some synthesis steps, e.g., of porphyrins, on zinc. Virtually all of the cobalt is in special combination with a complex organic molecule forming a genuine vitamin, vitamin B_{12} (see Table I).

Many of the trace elements such as Ni, Mo, and Co are used in small amounts in particular metabolic steps. One or two of the trace elements, however, have dominantly important roles. Iron in two different forms is used to catalyze the oxidation of many organic compounds by molecular oxygen. The two forms of iron are called heme iron and iron sulphur units. The heme iron involves iron locked in an organic ring structure not too unlike that of vitamin B_{12}, while the iron sulphur looks more like a small piece of iron pyrite, the common sulphide ore of iron. Oxidation occurs in the mitochondrial membrane, and is the major source of energy in the body. The process can be likened to the burning of gasoline in an automobile engine. In the mitochondrial membrane, series of these iron enzymes form a type of electronic circuit and are involved in a stepwise process leading to the formation of ATP.

FIGURE 3 The scheme of intake, use, and excretion of a typical mineral, iron. Different cells utilize the element in different functional ways after uptake and distribution. The element is distributed attached to a protein to the cells via the circulating body fluids, e.g., blood. There is a careful balance between intake and excretion so that the body regulates the levels of a mineral as closely as it can to the optimum shown in Fig. 2. The iron that is retained is incorporated into different useful proteins according to the cell type that captures it. Bone marrow produces hemoglobin, which binds iron and ultimately is incorporated into red blood cells. Iron in liver cells which are notably darkly colored due to its presence is incorporated into enzymes.

A second trace element deeply involved in oxidative metabolism is copper, although, unlike iron it is mainly found outside cells. For example, copper is required for the activity of the enzyme in ascorbate oxidase. Together copper and iron assist in the formation of collagen, the filamentous material present throughout the body, for instance in tendons, around muscle, and in the deep layer of the skin. Defects in connective tissue can be traced sometimes to defects of one or the other of these elements in enzymes involved in collagen biosynthesis. Further examples of particular functions of the trace elements are given in Table I.

Finally, as stressed above, all trace elements are kept in controlled concentrations everywhere in the body. Just as bone is a store of calcium and phosphate so there is a small storage mineral deposit of iron in a protein complex called ferritin and a controlled cluster store of zinc in another protein, metallothionine. Excess of the trace elements are usually eliminated by complicated and little understood feedback responses involving special synthesis of chelating proteins, which have the specific functions of building these elements. The transport of the mineral elements to the cells of the body also requires a series of carrier proteins in the blood, such as transferrin (for Fe and Mn), caeruloplasmin (Cu), and albumins (Zn) (See Figure 3).

IX. Summary

It is now known that some 25 of the chemical (mineral) elements are essential for the well-being of the human body. About 10 are required in quite large amounts, while the others are needed only in traces. Their functional significance follows the pattern of their relationships in the periodic table. Human life has adapted itself through evolution to a balanced use of these chemicals. Excess and deficiencies of any one of them cause deleterious conditions. Many of the mineral elements not taken up naturally into the body can become the bases of medicines or can cause serious harm as poisons or pollutants.

Bibliography

Comar, C. L., and Bronner, F., eds. (1960–69). "Mineral Metabolism," vols. 1–3. Academic Press, New York.
Prasad, A. S. (1978). "Trace Elements and Iron in Human Metabolism." J. Wiley, Colchester, England.

Mitochondrial Respiratory Chain

RODERICK A. CAPALDI *University of Oregon*

OxPhos

Glossary

Chemiosmotic energy Potential energy generated by the unidirectional movement of protons across a membrane, contributed by the proton gradient and associated electrical potential.

Coupling site Segment of the respiratory chain at which the electron potential change caused by electron transfer between two prosthetic groups provides sufficient energy to drive the synthesis of ATP

Electron tunneling Passage of electrons (between prosthetic groups of the respiratory chain) via the protein milieu without a defined pathway through side chains or polypeptide backbone

Isoforms Two forms of the same subunit differing in some of the amino acids and in the sequence

Matrix space Interior compartment of the mitochondrion bounded by the inner membrane

mtDNA Circular DNA segment of around 16 kilobases (in mammals) found in the matrix space of mitochondria and coding for several respiratory chain proteins, subunits of the ATP synthase and most of the tRNAs used in mitochondrial protein synthesis

Oxidation–reduction potential Thermodynamic measure of the energy change in an electron transfer reaction

Oxidative phosphorylation Process by which the oxidation of foodstuffs is coupled to the synthesis of ATP

Q cycle Hypothesis to explain the coupling of electron transfer to proton translocation in the ubiquinol cytochrome c oxidoreductase (complex III) segment of the respiratory chain

Respiratory chain complex Complex of proteins, prosthetic groups, and in some cases, lipophilic molecules, forming a unit or building block of the mitochondrial inner membrane; an integrated segment of the electron transfer chain

IN EUKARYOTES, cellular energy is mostly generated by a process called *oxidative phosphorylation* within organelles called *mitochondria*. In this process, NADH and succinate, two key intermediates in the breakdown of foodstuffs, are oxidized by molecular oxygen with the conservation of sufficient energy to drive ATP synthesis. The proteins responsible for the oxidation of these substrates, collectively called *the respiratory chain,* are components of the mitochondrial inner membrane.

The respiratory chain oxidizes NADH and succinate by removing electrons (and protons). The energy available from these high potential electrons is released in a series of discrete steps and used to drive a proton gradient across the mitochondrial inner membrane. Finally, the low potential electrons produced are disposed of by reacting with protons and oxygen to form water. Thus, the respiratory chain components convert chemical energy into electrical energy and electrical energy into potential energy (in the form of a gradient across a membrane, usually called *chemiosmotic energy*), and finally, this potential energy is converted back to chemical energy in the form of ATP (by the ATP synthase). Alternatively, the chemiosmotic energy is used to drive other unidirectional processes such as ion transport. The study of respiratory chain components, along with the closely associated ATP synthase, is commonly called *bioenergetics*.

I. Oxidation-Reduction Properties of the Respiratory Chain

A number of electron transfer steps are involved in the overall oxidation of NADH or succinate by molecular oxygen; these steps are catalyzed by a number of protein complexes with a variety of different prosthetic groups, including flavins, nonheme ion centers, quinone, hemes, and copper atoms. The order of electron transfer through the respiratory chain is shown in Fig. 1, with the various prosthetic groups arranged by oxidation reduction (redox) potential. The reaction

$$NADH + H^+ + 1/2\ O_2 \rightarrow NAD^+ + H_2O$$

involves an electron potential change of 1.14 V, with an overall free energy of oxidation of 52.6 kcal/mole. As indicated in Fig. 1, there are three steps in the overall electron transfer pathway in which the change in reduction potential is large enough to provide the energy needed to drive ATP synthesis. These are the so-called coupling sites of the respiratory chain. Fragmentation studies, using detergents to break up the mitochondrial inner membrane, have established that the respiratory chain components are organized, for the most part, into four large, multisubunit membrane complexes. These are called *complexes I, II, III, and IV* or termed NADH-ubiquinone reductase, succinate ubiquinone reductase, ubiquinol-cytochrome *c* oxidoreductase, and cytochrome *c* oxidase to indicate their functional properties. Complexes I and II include flavin and nonheme iron centers, complex III includes b and c_1 hemes as well as a nonheme iron center (Rieske FeS center), while complex IV con-

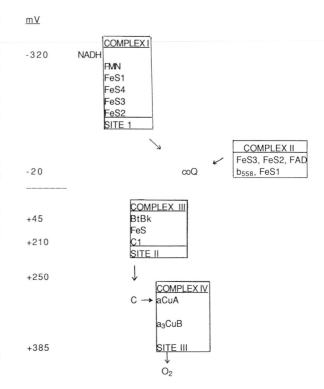

FIGURE 1 Schematic of the mitochondrial respiratory chain showing sequence of electron transfer through prosthetic groups from substrates NADH and succinate to molecular oxygen. FMN, flavin mononucleotide; FAD, flavin adenine dinucleotides; FeS, nonheme iron centers; Bt, Bk, b_{558}, c, c_1, a, a_3, different heme moieties; Cu_A and Cu_B redox active copper atoms.

tains hemes a and a_3 and two copper atoms (Cu_a and Cu_{a3}) as prosthetic groups. From Fig. 1, it can be seen that complexes I, III, and IV are the coupling sites referred to above. There are two important additional respiratory chain components, ubiquinone (or coenzyme Q), a lipophilic molecule embedded in the lipid bilayer, and cytochrome *c*, an extrinsic protein of the inner membrane. Ubiquinone shuttles electrons between complexes I, II, and III, whereas cytochrome *c* is responsible for shuttling electrons between complexes III and IV. [*See* ATP SYNTHESIS BY OXIDATIVE PHOSPHORYLATION IN MITOCHONDRIA.]

II. General Organization of Respiratory Chain Complexes in the Mitochondrial Inner Membrane

The mitochondrial inner membrane is composed of approximately 70% protein and 30% lipid. There are predominantly three types of lipid present: phosphatidylcholine (40%), phosphatidylethanolamine (35%), and cardiolipin (15%). Components involved in oxidative phosphorylation (i.e., the respiratory chain proteins and the ATP synthase) account for as much as 50% of the total protein present in the inner membrane of heart mitochondria. The relative amounts of the different respiratory chain components along with the amounts of other major proteins (e.g., the ATP synthase and ADP-ATP translocase) have been determined, not only from spectral studies, but by protein analysis and by immunological techniques. The ratio of these complexes in heart mitochondria is given in Table I. In liver and some other tissues the ratio of complex III to cytochrome c oxidase is closer to 1:1. Complex II is present as a monomer, but complexes I, III, and cytochrome c oxidase are dimers in the mitochondrial inner membrane. The membrane area per complex III dimer has been calculated at 200,000 Å (in heart mitochondria). Average distances between complex I and complex III dimers is 310 Å and between cytochrome c oxidase and complex III is 225 Å.

The mitochondrial inner membrane is fluid under physiological conditions, with lipid molecules and proteins able to diffuse laterally. The diffusion rates of both ubiquinone and cytochrome c are fast enough that electron transfer between complexes is unlikely to be the rate-limiting step in respiratory chain activity, even if the larger complexes I, III, and IV are hindered from diffusion.

Analysis by electron microscopy and image analysis shows that complexes I, III, and IV are each a transmembrane complex, as shown in Fig. 2, with complex I having most of its mass on the matrix-facing side of the mitochondrial inner membrane and cytochrome c oxidase having most of its mass on the cytoplasmic face of the membrane.

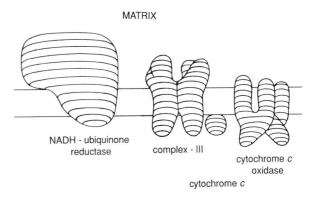

FIGURE 2 Low resolutional structure of three of the four respiratory chain complexes [Reproduced with permission from Capaldi, R. A., *et al.* (1988). *J. Bioenerg. Biomemb.* **20**, 291.]

III. Proteins of Respiratory Chain Complexes and Topology of Subunits

The protein composition of the respiratory chain complexes is different in different organisms, as discussed in a later section, with the most complicated structures being found in higher mammals. There are 25–28 different proteins in mammalian complex I, with an aggregate molecular weight of close to 700,000, 4–5 different proteins in complex II (total Mr 150,000), 11 different proteins in complex III (Mr 250,000), and 13 different proteins in complex IV (Mr 205,000). The different proteins are listed in Tables II–V, with components identified by various terminologies in use and by molecular weights. Also listed are the loci of prosthetic groups to individual subunits, as known, and the site of synthesis of the various proteins. As discussed below, several of the respiratory chain components are coded for by the

TABLE I Ratio of Respiratory Chain Components in Mitochondria[a]

Component	Integer Ratio	Mr of Monomer	Aggregation State
Complex I	1	700,000	Dimer
Complex II	2	150,000	Monomer
Complex III	5	250,000	Dimer
Cytochrome oxidase	10	205,000	Dimer
Cytochrome c	10	12,000	Monomer
ATP synthase	5	550,000	Monomer
ADP-ATP translocase	40	30,000	Dimer
Ubiquinone	70		
Phospholipid	5,000		

[a] Data for beef heart mitochondria.

TABLE II   Composition of Complex I, NADH-Ubiquinone Reductase

Protein Molecular Weight[a]	Prosthetic Group(s)	Site of Synthesis [n(nuclear); m(mitochondial)]	Other remarks
(A) 51,000	FMS.FeS center	n	
23,761[b]	2X FeS centers	n	Missing in some patients with mito- chondrial diseases
9,000	FeS center	n	
(B) 75,000	2X FeS centers	n	
49,000	FeS center	n	
30,000		n	
18,000		n	
15,000		n	
13,000		n	
(C) 66,939[b]		mt	ND5[c]
51,603[b]		mt	ND4, mutant $Arg_{340} \rightarrow His$[d]
38,949[b]		mt	ND2
35,666[b]		mt	ND1, Rotenore binding site
18,689[b]		mt	ND6
13,188[b]		mt	ND3
10,743		mt	ND4L

[a] Proteins listed according to their properties (A) Flavoprotein fraction; (B) iron sulfur–containing fraction; (C) hydrophobic fraction. The hydrophobic fraction contains 10–12 additional proteins not listed in this table.
[b] Molecular weights of the human proteins determined from sequences predicted from cDNAs.
[c] Nomenclature of mitochondrially coded subunits.
[d] Lebers optic atrophy has been found to involve a mutation of ND4.

TABLE III   Composition of Complex II, Succinate-Ubiquinone Reductase

Protein Molecular Weight[a]	Prosthetic Group(s)	Site of Synthesis [n(nuclear); m(mitochondial)]	Other remarks
75,000	FAD.FeS center	n	SD1
28,000	2 FeS centers	n	SD2
15,500	b heme	n	C_{II-3}
10,000[a]		n	C_{II-4}

[a] Subunit C_{II-4} may be two different proteins.

TABLE IV   Composition of Complex III Ubiquinol Cytochrome c Oxidoreductase

Protein Molecular Weight[a]	Prosthetic Group(s)	Site of Synthesis [n(nuclear); m(mitochondial)]	Other remarks
50,000		n	I. Core protein I
46,000		n	II. Core protein II
42,590	2X b hemes	mt	III. Cytochrome b
27,287	c_1 heme	n	IV. Cytochrome c_1
21,406	FeS center	n	V. Rieske FeS protein
13,389		n	VI.
9,507		n	VII.
9,175		n	VIII. Hinge protein[b]
7,998		n	IX. DCCD binding site
7,189		n	X. c_1 associated[b]
6,363		n	XI.

[a] Molecular weights are from sequence data for beef heart subunits except cores I and II, which are based on migration in sodium dodecyl sulfate polyacrylamide gel electrophoresis.
[b] The hinge protein and c_1 associated polypeptide copurify in a water soluble complex with cytochrome c_1.

Matrix

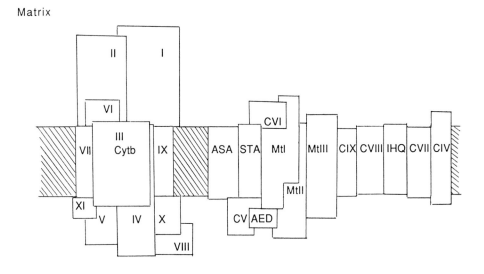

Cytoplasm

FIGURE 3 Arrangement of subunits in complex III and cytochrome c oxidase segments of the respiratory chain. Nomenclature of different subunits of the complexes is provided in Tables III and IV.

mitochondrial DNA (mtDNA) and synthesized on mitochondrial ribosomes.

Figure 3 shows the arrangement of subunits in complex III and cytochrome c oxidase. A majority of the proteins in the respiratory chain complexes are transmembrane.

IV. The Mitochondrial Genome

An important feature of mitochondria is the presence of DNA. The order of genes in human mtDNA is shown in Fig. 4. The DNA is circular and almost all the genes are located on the so-called L strand. The genetic contribution of the organelle to its own assembly is modest as there are only around 16,000 base pairs in the mtDNA of mammals. However, encoded in the organelle are seven subunits of com-

TABLE V Composition of Cytochrome c Oxidase

Protein Molecular Weight[a]	Prosthetic Group(s)	Site of Synthesis [n(nuclear); m(mitochondial)]	Other remarks
56,993	2 heme + Cu	mt	Mt_I (I)[b]
29,918		mt	Mt_{III} (III)
26,049	1 Cu	mt	Mt_{II} (II)
17,153		n	C_{IV}(IV)
12,434		n	C_V (Va)
10,670	Zn	n	C_{VI} (Vb)
10,068		n	AED (VIb)
9,418		n	ASA (VIa) H and L[c]
8,480		n	STA (VIc)
6,258		n	IHQ (VIIb)
6,244		n	C_{VII} (VIIa) H and L
5,541		n	C_{VIII} (VIIc)
4,962		n	C_{IX} (VIII) H and L

[a] Molecular weights are for subunits of the beef heart enzyme.
[b] Two different nomenclatures currently in use for the cytochrome c oxidase subunits.
[c] These subunits each have heart (H) and liver (L) isoforms.

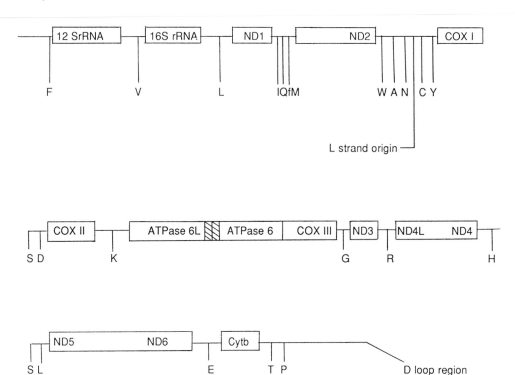

FIGURE 4 Order of genes on mitochondrial DNA. ND1–6 are subunits of complex I, as listed in Table II. CoxI–III are subunits of cytochrome c oxidase. The single letter code denotes tRNAs for the various amino acids.

plex I, one subunit of complex III, three subunits of cytochrome c oxidase, and two subunits of the ATP synthase.

The respiratory chain proteins encoded on mtDNA are all integral membrane proteins with multiple transmembrane domains. In contrast, the nuclear encoded subunits of the inner membrane span the bilayer once via a single transmembrane domain. A good example is cytochrome c oxidase where the three mitochondrially coded subunits have 12, 2, and 7 putative transmembrane segments, whereas seven of the 10 nuclear coded subunits span the membrane only once; the other three nuclear encoded subunits are peripheral membrane proteins.

V. Function of Subunits of Respiratory Chain Complexes—A Comparison of Prokaryotic and Eukaryotic Electron Transfer Chains

There are considerably more subunits of the mammalian respiratory chain complexes than are needed to bind the prosthetic groups, as evident in Tables II–V and Fig. 3. Our present understanding of the roles of this multitude of components is rudimentary. Important clues about those polypeptides critical for electron transfer and proton pumping have come from studies of the respiratory chain of prokaryotes and lower eukaryotes. Prokaryotes do not contain mitochondria, and instead the respiratory chain is located in the plasma membrane. Some aerobic bacteria such as *Paracoccus denitrificans* have a respiratory chain remarkably similar in functional properties to that of higher mammals, but with electron transfer complexes that are structurally much simpler. Complex I of *Paracoccus dentifricans*, for example, contains only 11–13 different polypeptides versus 25–28 in mammals; complex II contains four subunits (vs. 4–5); complex III contains three subunits (i.e., cytochrome b, cytochrome c_1, and the Rieske nonheme iron protein) (vs. 11), and cytochrome c oxidase is made up of three subunits (vs. 13). The proteins of the bacterial respiratory chain (e.g., *Paracoccus*) include all the mitochondrially synthesized subunits of the complexes in eukaryotes. Moreover, there is high sequence homology between the various mitochondrially coded subunits in mammals and their bacterial counterparts, in support of the origin of mitochondia from bacteria.

The comparison of prokaryotic and eukaryotic respiratory chains also helps to establish the position of prosthetic groups in the complex of higher organisms. For example, in cytochrome *c* oxidase the two hemes and copper atoms involved in redox reactions must reside in the three largest and mitochondrially synthesized subunits, as these are the only components in common for the enzyme from the two sources. Moreover, as complex IV in *Paracoccus* couples electron transfer to proton translocation, the mitochondrially encoded subunits of cytochrome *c* oxidase must provide the proton pumping function as well as electron transfer activity. Such comparisons raise the important question of the role of the extra polypeptides found in the respiratory chain of eukaryotes. Possible functions include regulation of electron transfer and/or proton pumping efficiency, control of biogenesis of the various complexes, and ion translocation in association with the redox reactions such as Ca^{2+} pumping by cytochrome *c* oxidase.

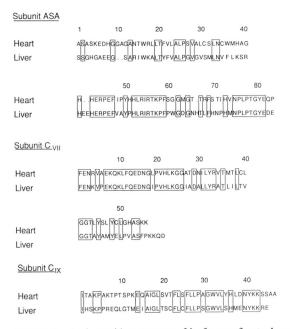

FIGURE 5 Amino acid sequences of isoforms of cytochrome *c* oxidase subunits. Data for C_{VII} and C_{IX} are from beef; data on subunit ASA (see Table IV) are from rat.

VI. Tissue-Specific Expression of Respiratory Chain Complexes

Recent evidence shows that complexes I, III, and IV are present in mammals as several isoforms. Analysis of this tissue specificity is most advanced for cytochrome *c* oxidase. At least three of the nuclearly encoded subunits of this enzymes have two (and possibly more) isoforms. Figure 5 shows the sequences of cytochrome *c* oxidase subunits that occur in different isoforms. Surprisingly, there is considerably more homology of the same isoform in different species than between isoforms of the same subunit in any animal. These proteins may be involved in regulation of enzyme activity or in regulation of biosynthesis of respiratory chain components, but their specific roles have not been defined.

VII. Mechanism of Electron Transfer Between Redox Centers

Electron transfer from one redox center to another in the respiratory chain could be direct, through neighboring atoms; there could be specific electron pathways through the protein or relatively nonspecific transfer via tunneling of electrons through the protein milieu.

Physical measurements of the orientation of prosthetic groups indicate that in both complex III and cytochrome *c* oxidase, the distances between donors and acceptors is too large for direct transfer. Attempts to identify specific electron transfer pathways have been largely unsuccessful, supporting the idea that there is electron tunneling within the respiratory chain complexes.

VIII. Coupling of Electron Transfer to ATP Synthesis: Proton Pumping by Respiratory Chain Complexes

Under physiological conditions electron transfer through complexes I, III, and IV is tightly coupled to proton translocation from the matrix to the cytoplasmic side of the mitochondrial inner membrane. Redox-linked proton translocation has been examined most carefully in complexes III and IV. In complex III, a proton motive Q (ubiquinone) cycle has been proposed to explain the coupling. According to the Q cycle hypothesis, there are two cata-

MATRIX

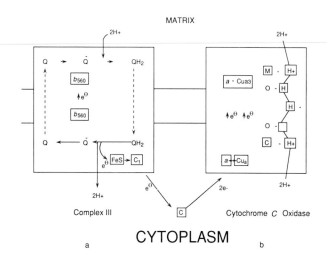

CYTOPLASM

a   b

FIGURE 6 Schematic of mechanisms of proton pumping by complex III and cytochrome c oxidase as described in text.

lytic Q centers: one on the outer (cytoplasmic) side of the inner membrane, involving cytochrome c_1, and the nonheme ion protein; the second on the inner side (matrix), involving cytochrome b. Electrons and protons are cycled across the membrane together between these centers by ubiquinone as it undergoes a two-stage oxidation and reduction via the free radical semiquinone anion. This mechanism is shown schematically in Fig. 6, with two electrons translocated per pair of electrons oxidized. In cytochrome c oxidase, the center involved in redox-linked proton translocation has been variously identified as heme a and more recently as Cu_a. It is thought that electron transfer through one (or both) of these centers is coupled via a conformational change to deprotonation of a closely positioned side chain on the cytoplasmic side of the inner membrane (site C) and protonation of a second amino acid (M) from the matrix side. A proton wire connecting these sites could involve serines, threonines, and peptide backbone as shown schematically in Fig. 6b. Proton pumping in cytochrome c oxidase has been found to involve translocation of two protons per pair of electrons reacted with molecular oxygen.

IX. Human Diseases of the Respiratory Chain

Diseases of mitochondria have been recognized since 1962, when a woman with hypermetabolism caused by defective coupling of oxidation and phosphorylation in muscle mitochondria was described. A large proportion of mitochondrial diseases are due to defects of the respiratory chain; patients with dysfunction of each of the electron transfer complexes, as well as in multiple complexes, have been reported. Most commonly, patients with mitochondrial diseases exhibit disorders of those tissues with the highest demands from oxygen (i.e., brain, skeletal muscle, and heart). The disorders involved can be congenital or late onset and the symptoms varied, with patients sometimes presenting with cardiomyopathies or generalized disease of muscle, liver, and/or kidney. Several well-described syndromes are now known to involve mitochondrial dysfunction. These include Kearns Sayre syndrome, characterized by progressive defects of the eye muscles, pigmentory degeneration of the retina, and sometimes heart block and cerebellar syndrome. Most of the described cases of Kearns Sayre syndrome have proven to be cytochrome c oxidase deficiencies. Many of these patients showed major deletions in the mtDNA. Myoclonus epilepsy with ragged red fibers (MERRF), characterized by ataxia, weakness, and generalized seizures, is another disease caused by cytochrome c oxidase dysfunction, in which there is clear evidence of mitochondrial inheritance of the disease.

Studies of mitochondrial diseases have provided evidence of the tissue-specific expression of complex I, complex III, and cytochrome c oxidase in humans. Cases of myopathy caused by each complex have been reported in which there is severe disease of muscle but no involvement of heart, liver, or kidneys. Also, cardiopathies with complex III and cytochrome c oxidase dysfunction but no involvement of other tissues have been reported. Finally, there are several interesting cases of infants with severe deficiency of cytochrome c oxidase in muscle at birth (floppy baby syndrome), in which functional enzyme appears and respiratory chain activity returns to normal slowly during the first 18 months of life. Such findings are best explained by the existence of fetal and adult forms of cytochrome c oxidase (and possibly other respiratory chain complexes) with a switch from one form to the other in the first few months after birth.

Bibliography

Capaldi, R. A. (1988). Mitochondrial myopathies and respiratory chain proteins. *Trends Biochem. Sci.* **13,** 144.

Capaldi, R. A., Gonzalez-Halphen, D., Zhang, Y-Z., and Yanamura, W. (1988). Complexity and tissue specificity of the mitochondrial respiratory chain. *J. Bioenerg. Biomemb.* **20,** 291.

Hatefi, Y. (1985). The mitochondrial electron transport and oxidative phosphorylation system. *Annu. Rev. Biochem.* **54,** 1015.

Ragan, C. I., Ohnishi, T., Weiss, H., Capaldi, R. A., et al. (1987). Structure, biogenesis and assembly of energy transducing enzyme systems. *Curr. Top. Bioenerg.* **15,** 1.

Wikstrom, M. (1988). How does cytochrome *c* oxidase pump protons. *Ann. NY Acad. Sci.* **550,** 199.

Yang, X., and Trumpower, B. L. (1988). Proton motive Q cycle pathway of electron transfer and energy transduction in the three subunit ubiquinol-cytochrome *c* oxidoreductase complex of *Paracoccus denitrificans*. *J. Biol. Chem.* **263,** 11962.

Mitosis

B. R. BRINKLEY, *University of Alabama at Birmingham*

Glossary

Cell cycle Period of time between the formation of daughter cells from a mother cell and the time when the daughter cells divide. The cell cycle is divided into four phases: G_1 (preparation for DNA synthesis), S (DNA synthesis) G_2 (premitotic phase), and M (mitosis)

Centromere Region of the chromosome located at the primary constriction, which usually contains DNA with highly repeated base sequences. This region organizes the kinetochore and holds sister chromatids together

Chromatid One-half of the replicated chromosomes, also referred to as a sister chromatid

Cytokinesis Division of the cytoplasm during mitosis

Cytoskeleton System of biological polymers, which include microtubules, microfilaments, and intermediate filaments and their regulatory components

Interphase Period between the end of one mitosis and the beginning of the next mitosis, includes G_1, S, and G_2 phases of the cell cycle

Karyokinesis Division of chromosomes in the cell nucleus during mitosis

Kinetochore Specialized structure at the centromere that connects chromosomes to microtubules of the mitotic spindle. Each member of a pair of kinetochores located on sister chromatids is called a *sister kinetochore*

Mitosis M phase of the cell cycle when the chromosomes and nucleus of a parent cell are divided and partitioned equally into two daughter nuclei; accompanied by a division of the cytoplasm to produce two genetically identical daughter cells

Mitotic motor Force-producing molecule(s) in the mitotic spindle, which powers chromosome movement

HUMAN CELLS may be classified into one of two categories: dividing cells such as those of the developing embryo, skin, hair follicles, and bone marrow; and nondividing cells as found in fully developed nerve and muscle tissue. Mitosis is a stage of cell division when the chromosomes and hence the nucleus of a mother cell are divided and partitioned equally into two daughter nuclei. This is accompanied by a division of the cytoplasm producing two genetically identical daughter cells. The process of mitosis is divided into two parts: division of the nucleus (karyokinesis), and division of the cytoplasm (cytokinesis). Dividing cells exist in one of two phases: interphase or mitosis. Both phases are part of a continuous clock-like sequence of events called the *cell cycle*. Before human cells can enter mitosis, they must replicate their complement of DNA contained in 23 pairs of chromosomes. DNA synthesis occurs in a period called *S phase* and is preceded by a period when cells prepare for DNA synthesis called the *G_1 period*. After S phase, cells enter a brief premitotic period called *G_2*. Important biochemical events must occur in the G_2 phase that prepare the cell for mitosis. Mitosis, like the other phases, takes place at a precise time in the cell cycle and is divided into five distinct stages: prophase, prometaphase, metaphase, anaphase, and telophase. Mitosis is usually an error-free process in which the two daughter cells produced are genetically identical copies of the mother cell. When mitotic errors do occur, daughter cells are genetically imbalanced because of the loss or gain of entire chromosomes. Such aberrations lead to serious medical consequences such as birth defects, cancer, or Down's syndrome.

I. The Cell Cycle: Preparing for Mitosis

A. Interphase

Historically, many of the principles of cell division were discovered by 19th century cytologists who identified two main phases in the life cycle of cells: interphase and mitosis. Mitosis was discovered by examining dividing cells in the microscope, and five morphologically distinct stages were identified: prophase, prometaphase, metaphase, anaphase, and telophase (see Color Plates 1 and 2).

At the completion of mitosis, cells were thought to enter a quiescent "resting" period called *interphase,* in which they remained until the subsequent mitosis. The interphase period remained a black box until the middle of the 20th century when DNA was discovered in the cell nucleus, and radiolabeled DNA precursors (tritium-labeled thymidine, a base found specifically in DNA) were used to identify a period in interphase when DNA was synthesized. The interphase period was then divided into the DNA synthesis period or S phase, a period preceding DNA synthesis called *gap-1* or G_1 *phase,* and a period after DNA synthesis but before the next mitosis called *gap-2* or G_2 *phase* (Fig. 1). Cells were grown in the presence of radiolabeled DNA precursors for a brief period of time, followed by growth in nonlabeled precursors, and then analyzed by a detection method called *autoradiography.* These experiments showed that the cell cycle proceeded in a clock-like manner with each of the interphase periods occupying a precise period of time. Moreover, essential biosynthetic events were found to occur in interphase that were required for the onset of mitosis. For example, by inhibiting DNA synthesis or blocking the synthesis of proteins with various drugs, mitosis failed to occur. This led to the conclusion that entry of cells into mitosis was regulated by specific biochemical "triggers" in interphase. [*See* DNA SYNTHESIS.]

B. The Mitotic Trigger

Evidence for a master switch or mitotic trigger was uncovered when it was found that cells in interphase could be induced to enter mitosis prematurely by experimentally fusing them with cells in mitosis. Recently a protein complex known as maturation promoting factor (MPF) has been purified from amphibian eggs and other mitotic cells that regulates mitosis. The discovery of MPF has led to the definition of a mitotic trigger that controls a cascade of biochemical and morphological events, which is depicted in Fig. 2. MPF is an enzyme complex containing a protein kinase that catalyzes the addition of phosphate to a group of proteins causing major morphological changes in cells including chromosome condensation, nuclear envelope breakdown, and the assembly of a mitotic spindle. MPF is composed of several smaller subunits, one of which is known as cdc2. Although present throughout the cell cycle, MPF is active only during mitosis. Another recently discovered protein, cyclin, is also a part of the trigger mechanism. Cyclin is present in mitotic cells but is broken down at the end of mitosis and synthesized anew in the next cell cycle. When cyclin levels reach a critical concentration during interphase, MPF is activated and mitosis begins. Mitosis ends when MPF is inactivated and cyclin is degraded. Therefore, the entire cell division cycle appears to be driven by a small group of regulatory proteins, which themselves are regulated by a set of cell cycle genes encoded in the DNA that control mitosis in all organisms from yeast to humans.

C. Interphase–Mitosis Transition

As described above, cells begin to prepare for mitosis hours before the actual event takes place. In addition, to DNA synthesis and replication of the genome that occurs in S phase, progression into mitosis is also accompanied by replication of spindle poles and major reorganization of cytoplasmic structures, especially the cytoskeleton.

The cytoskeleton, an elaborate system of delicate polymers, is composed of three types of fibers: microtubules, microfilaments, and intermediate filaments. These dynamic polymers are identified by their size and protein composition. Microtubules are hollow tubules, approximately 24 nm in diameter, and are composed of protein subunits known as alpha- and beta-tubulin. Microfilaments are delicate 6-nm filaments composed of a single globular protein, actin. Intermediate filaments are 10-nm in diameter and are composed of several proteins. Collectively, the cytoskeleton plays a variety of roles in interphase cells including motility, maintenance of

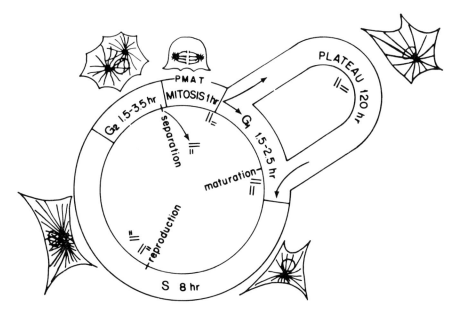

FIGURE 1 Typical cell cycle of a mammalian cell in tissue culture. Mitosis, indicating prophase, metaphase, anaphase, and telophase (PMAT), lasts 1 hr. After division, cells enter a G_1 period of 1.5–2.5 hr, in which they prepare for DNA synthesis. Depending on environmental conditions, cells may exit the cell cycle and enter a period called *plateau* or G_0 in which they either become differentiated into specialized nondividing cells or remain as stem cells capable of reentering the cell cycle at a later time. The DNA synthesis phase, or S phase, requires 8 hr during which the chromosomes become replicated. The G_2 phase shown here requires 1.5–3.5 hr and represents a time when the cell prepares for mitosis. The period required for G_1, S, and G_2 is called the *interphase*. The centriole replication cycle is shown inside the circle, indicating the time in the cell cycle when centrioles reproduce, separate, and mature. Cells on the outside of the circle show the organization of microtubules at different stages of the cell cycle.

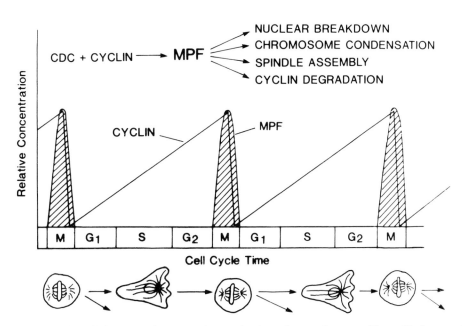

FIGURE 2 Regulation of the cell cycle by maturation promoting factor (MPF) and cyclin. Active MPF consists of a protein complex containing cyclin and cell division cycle protein (CDC), which trigger the mitotic process by inducing a number of biochemical and morphological changes. Cyclin is synthesized during interphase and, on reaching critical concentration, induces cells to enter mitosis. When mitosis is completed, cyclin is degraded and the cells enter the next G_1 phase. This oscillation of biochemical regulators is accompanied by morphological and cytoskeletal changes as shown at the bottom of the diagram.

cell shape, and secretion. Microtubules, which are important in mitosis, form an elaborate network in interphase cells called the *cytoplasmic microtubule complex* (CMTC). As shown in Figs. 3B and 4, the CMTC is organized around a central focus in the cytoplasm known as the centrosome. [*See* MICRO-TUBULES.]

During the G_2 phase, as cells prepare to enter mitosis, the centrosome splits into two parts, and each migrates to an opposite end of the cells. The CMTC then disappears, and a new set of microtubules grow from each of the centrosomes forming the mitotic spindle. Concomitantly, microfilaments and intermediate filaments are also reorganized, and the cells change their shape and become rounded before their entry into mitosis. As diagramed in Fig. 2, these dynamic morphological changes coincide with the activation of mitotic regulator molecules MPF cdc2 and cyclin.

The division of a mother cell into two daughter cells can be described as two related processes: karyokinesis or division of the genetic material, and cytokinesis, division of the cytoplasm.

II. Karyokinesis: Division of the Genome

A. Prophase

As cells enter mitosis, dramatic structural changes take place in the nucleus. The interphase chromatin undergoes condensation, producing a meshwork of coarse fibers. The replicated chromosomes, which are undetectable as discrete units during interphase, appear as paired threads called *sister chromatids* at prophase. As further condensation occurs, a cleft or constriction appears along the paired chromatids, marking the position of the centromere, a specific region on the prophase chromosomes that later plays an important role in the movement of chromosomes.

Other nuclear organelles (e.g., the nucleolus) break down and disappear at prophase. It should be noted, however, that the double membrane surrounding the nucleus, the nuclear envelope, remains intact through prophase.

Major morphological changes also occur in the cytoplasm. The CMTC becomes reorganized into two asters, with microtubules extending radially from the two centrosomes, producing an array of microtubules that will become part of the central spindle. In all dividing human cells, with the excep-

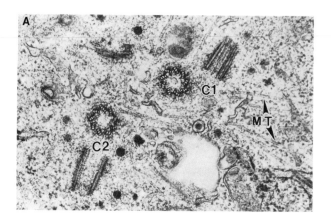

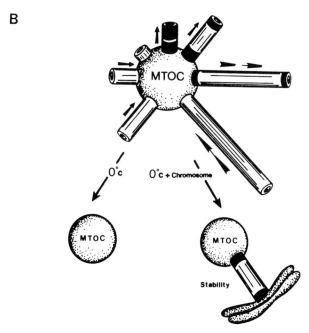

FIGURE 3 Centrosome and microtubule organizing center (MTOC). **A:** Electron micrograph of centriole pairs (C1,C2) present in a region of the cell called the *centrosome*. Microtubules (MT) are seen extending radially from the centrosome. Because there are two pairs of centrioles, the cell is in the G_2 phase of the cell cycle, about to begin mitosis. **B:** Diagram illustrates dynamic growth of microtubules from the CMTC. Microtubules slowly elongate (*outward-pointing arrows*) until their GTP-caps disappear. They then become unstable and shorten rapidly (*inward-pointing arrows*). When experimental cells are cooled to 0°C, microtubules disappear, leaving the MTOC. However, microtubules that are recaptured by the kinetochore become stable and are resistant to cold temperatures. This experiment shows that microtubules become modified when they attach to chromosomes and begin to function in chromosome movement.

tion of fertilized eggs, each spindle pole or centrosome contains a pair of cylindrical organelles known as *centrioles*. The wall of each centriole is composed of nine triplet microtubules, which form

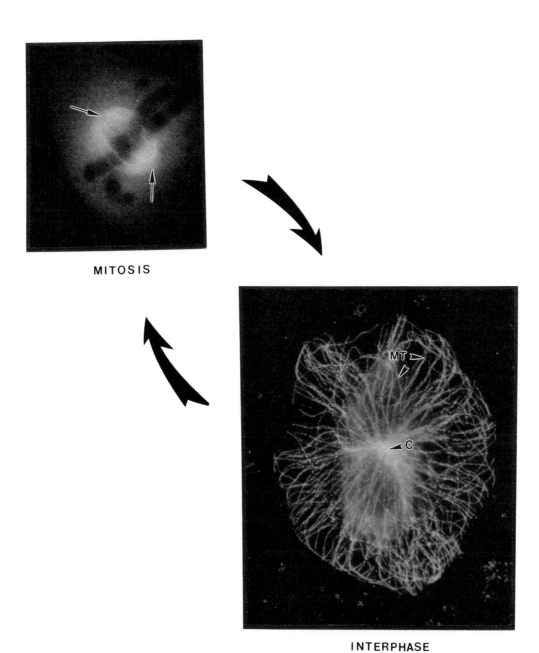

MITOSIS

INTERPHASE

FIGURE 4 The microtubule cytoskeleton undergoes dramatic reorganization as cells progress from interphase to mitosis. During interphase, microtubules (MT) radiate from a central zone, the centrosome (C), and extend outward to the cell periphery. As cells enter mitosis, the cytoplasmic microtubule complex is dissolved and converted into a mitotic spindle (*arrows*), which functions in chromosome partitioning. When mitosis is over, the CMTC reforms in each daughter cell.

a pinwheel design when viewed in cross section (Fig. 3A). Centrioles undergo their own replication cycle, have no known function in mitosis, and are absent from mitotic cells of many organisms. It is

likely that the centrioles, which also form the base (basal body) of the hair-like cilia and flagella, use the mitotic spindle as an apparatus to partition themselves equally into daughter cells.

The centrosome contains centriole pairs that are surrounded and embedded in a dense, amorphous matrix, which is thought to play an important role in generating the growth of microtubules. Microtubules are polarized structures that grow by adding tubulin subunits containing bound guanosine triphosphate (GTP) to one end, the plus or fast-growing end. When it is added, the tubulin-GTP molecules form a cap that stabilizes the growing end of the microtubule. As the tubules elongate, their cap disappears because of the hydrolysis of GTP to GDP. If the cap disappears completely, the walls or lattice of the microtubule become unstable, resulting in rapid shortening caused by depolymerization of microtubules. In living cells, microtubules grow and shrink in a dynamic pattern around the centrosomes. Therefore, the centrosome functions as a microtubule organizing center (MTOC) capable of generating the CMTC of interphase cells as well as the spindle of mitotic cells as shown in Fig. 4.

B. Prometaphase

The onset of this phase of mitosis is signaled by the abrupt disruption of the nuclear envelope and the attachment of chromosomes to the mitotic spindle. Spindle microtubules attach to specialized structures at the centromere called *kinetochores,* and then the chromosomes begin to display erratic patterns of movement.

The dissolution of the nuclear envelope is the pivotal event of prometaphase; furthermore, the molecular control of this process is now beginning to be understood. Special proteins called *lamins* form a thin layer or lamina along the inner surface of the nuclear envelope. At the appropriate moment (near the end of prophase), an enzyme known as a protein kinase is activated and catalyzes the addition of a phosphate to the lamins. At the same time, spindle microtubules grow from the spindle poles and push inward on the nuclear envelope leading to its disruption.

The chromosomes, which were immobile in prophase, undergo active movement the moment the nuclear envelope breaks down. Recent studies have shown that the initial movement of chromosomes at prometaphase coincides precisely with the attachment of spindle microtubules to kinetochores, small disk-like patches located at the centromere and positioned on opposite sides of the paired chromatids.

Spindle microtubules appear to be attracted to kinetochores by a process best described as "capture." That is, once the barrier of the nuclear envelope is removed, microtubules extending from the spindle poles interact with the condensing chromosome. Those microtubules that randomly contact the kinetochore are bound or captured by receptor molecules on its surface. As more microtubules are captured, all the attachment sites become occupied and eventually a stabilized bundle of microtubules

forms between the kinetochore and the spindle pole. This assembly, known as the *kinetochore fiber,* is important in initiating chromosome movement. Initially, the attachment of fibers to the kinetochores is unstable. The hallmark of prometaphase chromosome movement is that of "trial and error," with chromosomes being erratically pushed and pulled along the spindle. This type of chromosome movement is called *congression,* and the forces that are exerted on the kinetochores drive the paired chromatids first to one pole and then to the other until eventually they become positioned midway between the spindle poles. When the forces acting on sister kinetochores are balanced, the chromosomes become aligned at the cell equator midway between the poles, marking the end of prometaphase.

C. Metaphase

At metaphase, sister kinetochores are positioned such that one member of the pair faces one pole and the other faces the opposite pole. As shown in Color Plates 1 and 2, the metaphase spindle consists of two types of microtubules, those that extend from kinetochores to the poles (kinetochore microtubules) and those that extend from the poles and overlap at the cell equator (nonkinetochore microtubules). Also present are asteral microtubules radiating from each spindle pole.

In addition to microtubules, the metaphase spindle contains numerous smooth-surfaced membranes that extend along spindle fibers and aggregate at the poles. These vesicles contain calcium, an ion that may be important in regulating the assembly and disassembly of microtubules during chromosome movement.

Metaphase is significant because it facilitates the proper alignment of paired chromatids at the cell equator so that one complete complement of chromosomes can be partitioned to each spindle pole. It is important that chromatids not split apart prematurely. Although considerable force is placed on each sister kinetochore at metaphase, the chromatids resist being pulled apart by centromere bonds that hold them together.

Throughout metaphase, spindle microtubules remain dynamic. Despite the fact that microtubules are firmly bound to kinetic and spindle poles, tubulin subunits continue to be added to the kinetochore-bound ends. Once incorporated, the tubulin subunits appear to move poleward along the microtubule surface lattice in a treadmill-like manner. Thus, tubulin is incorporated at the kinetochore and comes off at the poles. Although the molecular mechanism controlling the poleward flux of tubulin subunits is unknown, it may be coupled to chromosome movement. The search for a microtubule-based, force-producing molecule or mitotic "motor" to power the movement of chromosomes is a major area of research today.

D. Anaphase

Anaphase is initiated by the splitting of paired chromatids at the centromere and the abrupt movement of sister chromatids toward opposite poles of the spindle. In addition, the spindle itself elongates, pushing the poles further apart. The initial poleward movement of chromosomes is called *anaphase A* and the subsequent elongation of the spindle is called *anaphase B*. At the onset of anaphase A, the balanced forces that serve to align paired chromatids on the cell equator become uncoupled, and chromosomes are pulled poleward by their kinetochores. The interchromatid bonds that held paired strands together during metaphase are broken at anaphase, and the chromatids are split apart, first at the centromere and then along the arms. Although human chromosomes vary considerably in size, they move poleward in unison at approximately the same speed. Neither the net force per chromosome nor the rate of movement is great in comparison with other forms of biological movements such as muscle contraction or the beating of cilia and flagella. However, unlike other types of movement, accuracy and precision are more important in partitioning the genome than speed or expediency. Thus, anaphase chromosome movement is slow but sure.

There are several theories concerning the mechanism of anaphase chromosome movement. The oldest of these is the traction fiber theory, which proposes that the force-producing mechanism or "motor" exists at the poles. According to this theory, chromosomes tethered to the poles by spindle microtubules are pulled by a traction mechanism localized at or near the poles. The kinetochore, according to this model, is a passive device that couples the microtubule to the centromere. Variations in this model place the motor on microtubules, microfilaments, or in a surrounding matrix that extends parallel to the kinetochore fibers, pushing them toward the poles. Recent experiments, however, suggest that the mitotic motor may actually be localized in the kinetochore and functions to pull chromosomes along stable microtubules, like a self-contained locomotive pulls rail cars along a track.

E. Telophase

When the separated chromatids arrive at the spindle poles and kinetochore microtubules disappear, the cell is in telophase. This phase of mitosis is characterized by the further elongation of the spindle and the reformation of the nuclear envelope as a double membrane around the daughter chromosomes. The nuclear envelope forms from vesicles in the surrounding cytoplasm. During prophase, many proteins, including the nuclear lamins and chromatin proteins, were phosphorylated, leading to the breakdown of the nuclear envelope and chromosome condensation. During telophase, these events are reversed with dephosphorylation occurring on the lamins and other proteins, allowing the envelope to reform and chromatin to undergo decondensation. The nucleolus that disappeared at prophase reappears, and a daughter nucleus forms at each end of the elongated mitotic cell. The cell's genome has been successfully partitioned, and karyokinesis has been completed.

III. Cytokinesis: Division of the Cytoplasm

A. Cleavage

Mitosis is incomplete until the cytoplasm of the parent cell is divided into two daughter cells, a process called *cleavage*. In human cells, cleavage begins in late anaphase as a slight furrowing of the cytoplasm at a point midway between the poles of the mitotic spindle. The cleavage furrow continues to constrict around the cell's midsection until only a small bridge of cytoplasm is left connecting the two daughter cells. This bridge or midbody also contains residual overlapping spindle microtubules, which extend from each daughter cell. It is not microtubules, however, that bring about the furrowing process. Cleavage is accomplished by a special cytoskeletal apparatus called a *contractile ring,* made up of a circular bundle of actin microfilaments. The contractile ring is oriented at right angles to the spindle axis, and at the appropriate moment, the actin filaments contract like purse strings, pinching the cytoplasm into two roughly equal parts. The midbody then serves to push the two cells further apart and is then discarded, leaving two complete daughter cells.

B. Variations in Mitosis

The events described thus far are typical of normal mitosis and related cyclic events that function in the growth and renewal of single cells or tissues. Some human cells exhibit variations on the basic theme of mitosis. For example, some cells complete normal karyokinesis but fail to carry out cytokinesis. This type of division results in the formation of multiple nuclei within a single large cytoplasm, or a syncytium. A syncytium may also form after mitosis by the fusion of multiple daughter cells, as in the case of developing muscle tissue. In some cells, the DNA undergoes continuous synthesis without subsequent division of the nucleus or the cytoplasm producing a giant cell (e.g., the megakaryocyte in the bone marrow). Another variation occurs in bone marrow and involves the process of enucleation in developing erythrocytes. In this process the nucleus is extruded, leaving the cytoplasm, which becomes filled with hemoglobin.

IV. Errors in Mitosis

Although the frequency of error in mitosis is low, genetic conditions can lead to aberrations in the mitotic process. Furthermore, drugs and toxic agents in the environment (Table I) can increase the risk of mitotic error.

The most common mitotic error is that of aneu-

ploidy, in which chromosomes are either lost or gained during mitosis. This results in daughter nuclei that contain too few or too many chromosomes. Normally, a somatic cell contains a duplicate set of chromosomes, one member of which is derived from the father and the other from the mother. Such balanced genomes are said to be diploid and contain a 2N number of chromosomes. Human gametes (i.e., sperm and egg) contain only one chromosome from a set and are haploid (N number of chromosomes). Aneuploidy in somatic cells, therefore, results in the formation of nuclei with chromosome numbers of $2N + 1$, $+2$, $+3$, etc., or $2N - 1$, -2, -3, etc. [See CHROMOSOME ANOMALIES.]

Aneuploidy usually results from a process called *nondisjunction*, in which paired chromatids fail to split at anaphase, resulting in the movement of both sister chromatids to one pole. The molecular basis of nondisjunction is unknown, but it is thought to be caused by a defect in the function of one or both members of a pair of sister kinetochores. Down's syndrome in humans results from nondisjunction involving chromosome number 21. [See DOWN'S SYNDROME, MOLECULAR GENETICS.]

Other types of mitotic errors can be caused by aberrations in the formation of the mitotic spindle. The presence of an extra pole determinant (centrosome) can cause the formation of a multipolar spindle, resulting in the abnormal partitioning of chromosomes at anaphase. Multipolar spindles usually appear in cells that contain an abnormal number of centrioles.

A fundamental principle of growth and development is that cells of the human body arise by division of preexisting cells. Mitosis represents but one stage in a clock-like process known as the *cell cycle*. Control of cell proliferation is the hallmark of normal human growth, but errors caused by genetic factors, drugs, and environmental toxins can disrupt the process and lead to serious medical complications. A quotation from a recent editorial by Daniel E. Koshland, Jr., Editor of *Science* magazine serves as an excellent summary:

> In each of our bodies there are choreographers programming a minuet in which chromosomes appear from obscurity, line up with their partners, separate, rejoin and then disperse. That minuet is called the cell cycle, and it must proceed according to certain rules and cadences if we are to lead normal lives. In embryonic cells, the cycle must proceed very rapidly, in some adult cells more slowly and in some neural cells, not at all. If the cycle fails in growing cells, death

TABLE I Some Examples of Potentially Aneuploidy-Producing Compounds That Are of Environmental and Clinical Importance[a]

Fungicides	Benomyl
Herbicides	Trifluralin, atrazine
Organic solvents	Benzene, chloroform, formaldehyde, tricholoroethylene, tetrachloroethylene
Anesthetic	Halothane, chloral hydrate
Anticancer drugs	Diethylstilbestrol, vinblastine, vincristine, taxol, nocodazole
Antianxiety drugs	Diazepam
Antifungal drugs	Griseofulvin

[a] From Liang, J. C., and Brinkley, B. R. (1984). Chemical probes and possible targets for the induction of aneuploidy. *In* "Aneuploidy: Etiology and Mechanisms" (V. L. Dellarco, P. Voytek, and A. Hollaender, eds.), pp. 491–505. Plenum Press, New York.

results. If it grows incorrectly in mature cells, cancer is caused.

Bibliography

Hyams, J. S., and Brinkley, B. R., eds. (1989). "Mitosis: Molecules and Mechanisms." Academic Press, London.

Koshland, D. E., Jr. (1989). Frontiers in cell biology: The cell cycle. *Science* **246,** 545. (See also special articles, pp. 603–640.)

McIntosh, J. R., and Koonce, M. P. (1989). Mitosis. *Science* **246,** 22.

McIntosh, J. R., and McDonald, K. L. (1989). The mitotic spindle. *Sci. Am.* **261,** 48.

Nicklas, R. B. (1988). Chance encounters and precision in mitosis. *J. Cell Sci.* **89,** 283.

Molecular Recognition of Transfer RNAs

PAUL SCHIMMEL, *Massachusetts Institute of Technology*

Glossary

Amber suppression Insertion of an amino acid at a UAG stop codon, done by a transfer RNA whose anticodon has been altered to CUA, so that it can pair with the UAG stop codon, known as an amber suppressor

Amino acids The 20 building blocks for proteins. During protein synthesis, each of the amino acids is joined to a transfer RNA which is specific for that amino acid. The attachment is catalyzed by an aminoacyl transfer RNA synthetase. The amino acids mentioned in this article are alanine, glutamine, leucine, methionine, phenylalanine, serine, tyrosine, tryptophan, and valine

Aminoacylation Attachment of an amino acid to a transfer RNA, catalyzed by aminoacyl transfer RNA synthetases

Aminoacyl transfer RNA synthetase Enzyme that attaches an amino acid to a transfer RNA. Each amino acid is attached to a specific transfer RNA. There is one aminoacyl transfer RNA synthetase for each amino acid and the synthetase matches the amino acid with its cognate transfer RNA

Anticodon Sequence of three bases within a transfer RNA which is complementary to a codon. The complementarity is determined by the rules of base pairing, by which specific bases interact through hydrogen bonds. The rules for complementarity are that A pairs with U and G pairs with C; pairing is antiparallel, so that, for example, the triplet AUG would pair with CAU

Codon Sequence of three bases which specifies a particular amino acid. Because there are four different bases (i.e., A, C, G, and U), there are $4 \times 4 \times 4 = 64$ different codons. For example, AUG specifies the amino acid methionine and UUU specifies phenylalanine. Three of the codons (UAA, UAG, and UGA) specify no amino acid and are interpreted as "stop" codons. Therefore, there are 61 codons which each specify an amino acid. Because there are only 20 amino acids, the code is degenerate, so that more than one codon can specify the same amino acid. For example, in addition to UUU, the codon UUC also specifies phenylalanine

Identity With regard to transfer RNA, it is defined by the amino acid which it accepts. Thus, a transfer RNA which accepts alanine is known as an alanine transfer RNA. The identity is determined by specific bases within the transfer RNA which are recognized by the cognate aminoacyl transfer RNA synthetase

In vitro Refers to experiments performed under laboratory conditions outside of a living organism

In vivo Refers to experiments performed within a living cell

Messenger RNA RNA which codes for a protein. The amino acid sequence of the protein is specified by the consecutive arrangement of codons in the mRNA. During protein synthesis these codons are "read" by transfer RNAs, which insert amino acids in the order specified by the codons

Transfer RNA RNA typically comprising 74–93 bases and having a site for amino acid attachment and an anticodon which can pair with a codon in a messenger RNA

LIVING ORGANISMS depend on an accurate system for decoding the genetic information which is written into the DNA of each cell. Much of the genetic information is a record (in the language of nucleic acids) of the sequences of specific proteins which are required for biochemical functions and transactions. Each protein (e.g., a structural component, enzyme, hormone, or receptor) is comprised of a defined sequence of amino acids. Because there are 20 different amino acids, the number of possible amino acid sequences is virtually unlimited, so that proteins have a great diversity of structures and functions. [*See* PROTEINS.]

Nucleic acids such as DNA comprise just four monomer units, known as bases, which occur in a precise order. These bases are adenine (A), cytidine (C), guanosine (G), and thymine (T). Usually, one gene specifies the instructions for the synthesis of one protein, although in some situations a gene may encode more than one. The decoding apparatus groups together three contiguous bases, known as codons, and interprets them as a single specific amino acid. For example, CCG specifies the amino acid alanine, and ATG specifies methionine. This relationship between codons and amino acids is known as the genetic code. Because there are more base triplets (64) than amino acids (20), there is degeneracy in this code, so that more than one set of triplets can code for a given amino acid. [*See* DNA AND GENE TRANSCRIPTION.]

While the rules of the genetic code have been known for 25 years, the underlying basis for the code has not been understood. Why should a particular base triplet correspond to one amino acid but not another? One speculation was that amino acids might have originally bound directly to base triplets contained in nucleic acids. While there is recent evidence that at least one amino acid (i.e., arginine) can bind selectively to a specific nucleic acid structure, there is no evidence for a general system of amino acid–nucleic acid interactions on which a highly precise coding apparatus could be based.

In this article a summary is given of recent advances in understanding the mechanism that determines the relationship between codons and amino acids. This is accomplished by the molecular recognition of transfer RNAs (tRNAs).

I. Recognition of Transfer RNAs Determines Rules of the Genetic Code

The instructions for proteins which are encoded in DNA are first converted into other nucleic acids, known as messenger RNAs (mRNAs). (The ribonucleic acids are made up of four bases—A, C, G, and U—U being uracil.) Typically, there is one mRNA for each protein. The mRNA are decoded by a family of nucleic acids, designated tRNAs. This occurs by interaction of the tRNAs with an mRNA, where codons in the mRNA pair (through hydrogen bonds) with the complementary sequences (i.e., the anticodons) in the tRNAs. The rules of base pairing specify that A pairs with U and G pairs with C. Each tRNA has one anticodon, but because of degeneracy in the code, there is one or more tRNA for each amino acid. Collectively, these tRNAs can "read" all of the codons contained in an mRNA.

A specific amino acid is attached to each tRNA and corresponds to the triplet specified by the anticodon. Depending on the tRNA, it is typically composed of 74–93 bases (76 being the most common number), which are joined to a sugar–phosphate backbone to make a polynucleotide. This polynucleotide is folded by internal hydrogen bonding between the bases into a cloverleaflike pattern of structure in which the anticodon and amino acid attachment sites are at opposite ends of the molecule (Color Plate 3). There are additional interactions that enable the molecule to adopt an L-shaped three-dimensional structure. Thus, specific amino acids are associated with trinucleotide sequences through the tRNA molecules, although the two components, whose relationship is specified by the code, are spatially separated within the tRNA.

Amino acids are attached to tRNAs by a system of enzymes known as the aminoacyl tRNA synthetases. There is generally one such enzyme for each amino acid. A specific enzyme can attach its cognate amino acid to all of the tRNAs which have anticodons for that amino acid. The enzymes have a high degree of specificity, which prevents them from attaching an amino acid to the wrong tRNA. Thus, the genetic code is based on the molecular recognition of tRNAs by aminoacyl tRNA synthetases.

II. Enzymes That Recognize Transfer RNAs

The aminoacyl tRNA synthetases are diverse proteins, which, unlike tRNAs, have a wide range of sizes. Amino acid sequences are known for over one-half of the enzymes from the bacterium *Escherichia coli*; some of these sequences are unique, having no similarity to those of other synthetases. Through applications of molecular biology and protein engineering, evidence has been obtained for a modular arrangement of functional domains, where subparts of a synthetase are organized into structural and functional units. For example, one segment of alanine tRNA synthetase encodes a domain which recognizes and activates the amino acid alanine so that it can be attached to tRNA. Joined to this segment are sequences which are important for the recognition of tRNA and for assembly of the full structure of the protein.

Three-dimensional crystal structures have been obtained for about one-half of a bacterial tyrosine tRNA synthetase and for most of a bacterial methionine and glutamine tRNA synthetase. These crystals reveal that part of the molecule forms a three-dimensional structural unit, known as a nucleotide fold. This structural unit is important for activation of the amino acid. Molecular modeling has deduced a similar structural unit in isoleucine tRNA synthetase, and it is suggested that at least some synthetases have this design. Thus, even though the enzymes vary considerably in size and sequence, there are some common features to the organization of their three-dimensional structures. Crystals recently have been obtained of a complex between *E. coli* glutamine tRNA synthetase and glutamine tRNA. These have yielded the first high-resolution structure of a complex.

III. Approaches to Molecular Recognition

Early investigations concentrated on defining regions of a tRNA that make contact with bound aminoacyl tRNA synthetase. Photochemical crosslinking and tritium labeling, among other approaches, were used. Based on the results of these investigations, it was proposed that, at least for some complexes, the enzymes make contact along and around the inside of the L-shaped three-dimensional structure of tRNA.

Early research also included the use of genetics to obtain tRNA sequence variants that would have an altered amino acid specificity. These studies took advantage of a special tyrosine tRNA which inserts tyrosine at one of the codons that normally specifies "stop," or termination, of polypeptide synthesis. This tyrosine tRNA has an altered anticodon, so that it recognizes the UAG stop codon and inserts tyrosine. When a stop codon occurs (mistakenly) within the internal part of a polypeptide coding region of an mRNA, a fragment of the protein is made because of premature chain termination. The resulting protein is usually active. When the altered tyrosine tRNA inserts tyrosine at the internal stop codon, however, the full-length protein can be synthesized and is consequently active. This phenomenon is known as suppression, and when UAG is the stop codon that is suppressed, it is called amber suppression. The particular tyrosine tRNA is designated as a tyrosine-inserting amber suppressor.

Mutants of the tyrosine tRNA amber suppressor were obtained that could insert the amino acid glutamine. However, some of the mutations created tRNA sequences that were not present in any glutamine tRNA. The interpretation of these experiments was obscure, especially because it was also found that alteration of the anticodon of a tRNA which normally accepted the amino acid tryptophan also caused this tRNA to accept glutamine.

With modern methods of recombinant DNA technology and molecular biology, it was possible to extend considerably the analysis of tRNA sequence variants. In one study it was noted that there are six different serine tRNAs which are recognized by the same bacterial serine tRNA synthetase. Comparison of the bases which are common to these six tRNAs suggested that a reduced set of 12 might be sufficient to confer the amino acid specificity. These 12 were introduced into a leucine tRNA amber suppressor and this tRNA acquired the ability to insert serine at UAG stop codons. Further work has attempted to identify the nucleotides among these 12 which are essential for the acquisition of amino acid specificity.

The use of amber suppression assays has played an important role for the investigation of tRNA identity *in vivo*. The cellular environment tests for specificity of aminoacylation in the context of all of the tRNAs and synthetases. Because there are 20 synthetases and over 60 tRNAs, more than 1000 pairwise combinations are possible. Alternatively, aminoacylation can be investigated *in vitro,* in which an isolated tRNA species and sequence variants are tested with an individual enzyme. Early work took advantage of the errors that sometimes occur *in vitro* when tRNAs and enzymes are used from heterologous sources. For example, yeast phenylalanine tRNA synthetase can attach phenylalanine to some of the bacterial tRNAs which (in the bacterium) are specific for amino acids other than phenylalanine. By comparing the bases in these "mischarged" *E. coli* tRNAs to each other and to yeast phenylalanine tRNA, sites that might be important for recognition were suggested. In addition, *in vitro* studies with *E. coli* methionine tRNA showed that the anticodon (CAU) triplet was important for recognition by the methionine tRNA synthetase and that change of the anticodon to CUA enabled this tRNA to be aminoacylated with glutamine. Independent studies have shown that sequence alterations to tyrosine, tryptophan, and methionine tRNAs lead to an ability to accept glutamine.) Whether the methionine CAU anticodon was sufficient to confer methionine acceptance on another tRNA was not known.

IV. Role of a Single Base Pair in Establishing the Identity of an Alanine Transfer RNA

Amber suppressor variants have been made for many of the tRNAs. These variants replace the normal anticodon sequence with CUA, which can match by base pairing with the UAG stop codon. (The codon–anticodon pairing is antiparallel, so that the C of CUA pairs with the G of UAG and so forth.) This change in the anticodon does not affect the amino acid specificity *in vivo* of at least 10 different tRNAs. This and other evidence show that the anticodon is not a major determinant for recognition by the associated enzyme.

Alanine tRNA is an example whose anticodon can be changed from GGC or UGC (the two types of alanine anticodons) to CUA without a change in specificity. This tRNA has 76 bases, of which 15 are conserved for the formation of structural features which are common to tRNAs. The remaining 61 bases can be varied within the constraint that base-pairing interactions are preserved for maintenance of the secondary structure. Changes at 36 positions in the sequence of this tRNA were created and the

effect on acceptance of alanine was studied *in vivo* and *in vitro.* From this analysis a single base pair was identified as a major determinant for the identity of the alanine tRNA (Fig. 1).

This base pair is designated G3 : U70 because, the G at the third position (in the numbering system for the bases) is paired with a U at position 70. It is not a standard base pair, because normally G is matched with C and U is usually paired with A. However, if G is changed to an A so as to create an A3 : U70 base pair, or if U is changed to C so as to have a G3 : C70 pair, the resulting tRNA is inactivated for acceptance of alanine, as catalyzed by purified alanine tRNA synthetase. Sequence variations in other parts of the molecule, including the anticodon, do not have this effect.

Because of the striking effect of single nucleotide substitutions at 3 : 70 on recognition by the alanine enzyme, the G3 : U70 base pair was introduced into other tRNAs to determine whether it could confer alanine acceptance on them. The G : U base pair was introduced into position 3 : 70 of cysteine, phenylalanine, and tyrosine tRNAs. These differ from alanine tRNA at 38, 31, and 49 positions, respectively (Fig. 1). When G3 : U70 is introduced

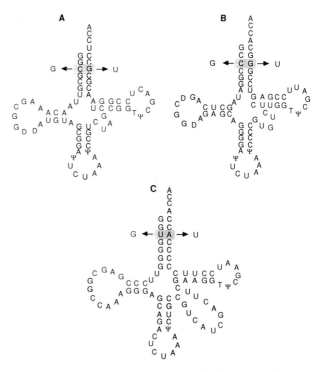

FIGURE 1 Three tRNAs which are specific for (A) cysteine, (B) phenylalanine, and (C) tyrosine. Installation of the G3 : U70 base pair into each of these confers alanine acceptance. (Courtesy of Christopher Franklyn and Ya-Ming Hou.)

into these tRNAs, each acquires the ability to accept alanine. Thus, as little as a single base pair can direct an amino acid to a specific tRNA.

Further investigations have attempted to increase our understanding of the basis for the influence of the G3 : U70 base pair. Evidence has been obtained for a direct interaction between the enzyme and the part of the tRNA structure which encodes this base pair. With an A3 : U70, G3 : C70, or U3 : G70 base pair, high concentrations of the enzyme do not catalyze the attachment of alanine in vitro. Investigations of the A3 : U70 alanine tRNA variant have shown that, at conditions in vitro which are similar to those in vivo, the A3 : U70 is defective for binding to the enzyme. However, additional investigations have demonstrated that, even when bound to the enzyme, this variant does not accept alanine. Thus, the enzyme has a double-barrier (i.e., binding and catalysis) against recognition of the incorrect tRNA. This is of particular significance, because of the strong influence of the G3 : U70 base pair on tRNA identity. Thus, those tRNAs which differ by only one base at the 3 : 70 position (and have, for example, an A : U or G : C pair) are effectively discriminated against by the alanine enzyme, so that there is no interference with the attachment of the correct amino acids to those tRNAs.

V. Attachment of Alanine to Minihelices

The major role for a single base pair inspired an investigation of whether small tRNA pieces could accept alanine. This is possible for the alanine system, because the G3 : U70 base pair recognized by the enzyme is close to the amino acid attachment site, so that a piece that incorporates both the recognition and attachment sites can be conveniently constructed by methods of RNA synthesis.

RNA minihelices were designed that encompass a limited part of the structure (Color Plate 4). These include a minihelix which has 12 base pairs and a microhelix of seven base pairs. When incubated with the alanine tRNA synthetase, each of these can accept alanine. Moreover, the acceptance of alanine depends on the presence of the G3 : U70 base pair: An alteration to G : C inactivates these molecules for efficient aminoacylation by the alanine enzyme. Also, while a minihelix base on the sequence of a tyrosine tRNA is not aminoacylated

with alanine, installation of the G3 : U70 base pair confers alanine acceptance on that minihelix as well.

The alanine minihelix, particularly the microhelix, appears to bind less efficiently to the enzyme than does the intact tRNA. However, an analysis of binding energies suggests that, even with the microhelix, only a minor part of the interaction energy is missing. Moreover, kinetic studies show that, once bound to the enzyme, the rate of amino acid attachment to the mini- and microhelices is comparable to the rate of attachment to the intact alanine tRNA.

VI. A Limited Constellation of Nucleotides Determines the Recognition of Several Transfer RNAs

Recent research has indicated that, in addition to alanine tRNA, for at least some tRNAs a small part of the structure has a major influence on recognition. For arginine, methionine, valine, and isoleucine tRNAs from E. coli, the anticodon has an important role. It is possible to interchange the methionine and valine anticodons, for example, and thereby interchange their amino acid acceptances in vitro. For one of the isoleucine tRNAs, there is evidence that modification of the first base of the anticodon is essential for recognition by isoleucine tRNA synthetase. On the other hand, it appears that yeast phenylalanine tRNA requires parts of the structure in addition to the anticodon. Two additional bases (i.e., G20 and A73) have been implicated and position 20 also plays a role in recognition of E. coli arginine tRNA which has an A. For E. coli glutamine tRNA the crystal structure suggests that the anticodon, and bases in the acceptor helix have an important role.

The anticodon does not play a major role for serine tRNA, and the number of significant bases has been narrowed to eight. There is preliminary evidence that a base (i.e., pair in the acceptor helix) is particularly important for recognition of a histidine tRNA, although much further investigation is required.

These studies show that, while the determinants for identity are idiosyncratic, there are a limited number of bases essential for recognition.

VII. Evolution of Recognition and Concluding Remarks

In general, little is known about the conservation of the determinants for the identities of tRNAs. Among the reported sequences for *E. coli* tRNAs, the G3 : U70 base pair is unique to alanine tRNA. In principle, therefore, it would appear that this tRNA could be distinguished from all others on the basis of the G3 : U70 base pair. In the yeast *Saccharomyces cerevisiae* the most recent compilation of tRNA sequences indicates that G3 : U70 is also unique to alanine tRNA. In addition, G3 : U70 is limited to alanine tRNA in the compilation of reported sequences for mammalian cytoplasmic tRNAs. Thus, in principle, the same determinant for discriminating alanine from other tRNAs in *E. coli* could be use in higher organisms. G3 : U70 has recently been shown to be essential for recognition by alanine enzymes from the insect *B. mori* and human cells.

In addition to the eventual identification of determinants for the identities of all of the tRNAs, the question of context effects and negative determinants requires considerable investigation. It is possible for a major determinant for recognition by a particular enzyme to be obscured by the "context," or surrounding sequences, in which it is placed. For example, certain bases at specific positions might create unfavorable contacts with an enzyme, so that it cannot interact with its major site of recognition. Sites on the tRNA that are covered by the bound alanine tRNA synthetase have been detected, and some of these are in regions not necessary for specific recognition by the enzyme (i.e., they occur outside of the minihelix). These may represent positions where the presence of certain bases would block the interaction with the alanine enzyme. Thus, context effects could assure even more accurate selection of tRNAs by aminoacyl tRNA synthetases.

It can be speculated that a coding system based on aminoacylation of small oligonucleotides could have eventually evolved into the highly elaborated tRNA structures which characterize all present living organisms. For at least one amino acid (i.e., alanine) a small oligonucleotide component of the whole tRNA is sufficient to confer specific amino acid attachment to a site proximal to the major determinant for enzyme recognition. For tRNAs whose determinants for identity are distal to the amino acid attachment site, it is possible that those determinants have been translocated from a position which was proximal at one time. Future research may be able to answer these questions.

Bibliography

Francklyn, C., and Schimmel, P. (1989). Aminoacylation of RNA minihelices with alanine. *Nature (London)*, **337**, 478–481.

Himeno, H., Hasegawa, T., Ueda, T., Watanabe, K., Miura, K., and Shimizu, M. (1989). Role of the extra G-C pair at the end of the acceptor stem of tRNA[His] in aminoacylation. *Nucleic Acids Res.* **17**, 7855–7863.

Hou, Y.-M., and Schimmel, P. (1988). A simple structural feature is a major determinant of the identity of a transfer RNA. *Nature (London)* **333**, 140–145.

Hou, Y.-M. and Schimmel, P. (1989). Evidence that a major determinant for the identity of a transfer RNA is conserved in evolution. *Biochemistry* **28**, 6800–6804.

McClain, W. H., and Foss, K. (1988). Changing the acceptor identity of a transfer RNA by altering nucleotides in a variable pocket. *Science* **241**, 1804–1807.

Muramatsu, T., Nishikawa, K., Nemoto, F., Kuchino, Y., Nishimura, S., Miyazawa, T., and Yokoyama, S. (1988). Codon and amino-acid specificities of a transfer RNA are both converted by a single post-transcriptional modification. *Nature (London)* **326**, 179–181.

Normanly, J., Ogden, R. C., Horvath, S. J., and Abelson, J. (1986). Changing the identity of a transfer RNA. *Nature (London)* **321**, 213–291.

Rogers, M. J., and Soll, D. (1988). Discrimination between glutaminyl-tRNA synthetase and seryl-tRNA synthetase involves nucleotides in the acceptor helix of a tRNA. *Proc. Natl. Acad. Sci. U.S.A.* **85**, 6627–6631.

Rould, M. A., Perona, J. J., Soll, D., and Steitz, T. A. (1989). Structure of the *E. coli* glutaminyl-tRNA synthetase complexed with tRNA[Gln] and ATP at 2.8 angstroms resolution. *Science* **246**, 1135–1142.

Sampson, J. K., DiRenzo, A., Behlen, L., and Uhlenbeck, O. C. (1989). Five nucleotides are required for the identity of yeast phenylalanine transfer RNA. *Science*, **243**, 1363–1366.

Schimmel, P. (1987). Aminoacyl tRNA synthetases: General scheme of structure–function relationships in the polypeptides and recognition of transfer RNA's. *Annu. Rev. Biochem.* **56**, 125–158.

Schimmel, P. (1989). Parameters for the molecular recognition of transfer RNA's. *Biochemistry*, **28**, 2747–2759.

Schulman, L. H., and Pelka, H. (1988). Anticodon switching changes the identity of methionine and valine transfer RNAs. *Science* **242**, 765–768.

Schulman, L. H., and Pelka, H. (1989). *Science* **246**, 1595–1597.

Monoclonal Antibody Technology

YUKIO SUGINO and SUSUMU IWASA, *Takeda Chemical Industries*

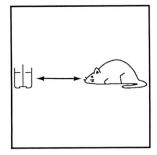

Glossary

Antibody fusion proteins Antibodies with the Fc regions replaced by other bioactive proteins
Bispecific antibodies Immunoglobulin molecules that can react with two distinct antigens
Chimeric antibodies Immunoglobulin molecules comprised of murine variable regions and human constant regions
Epstein–Barr virus transformants Antibody-producing lymphocytes immortalized with Epstein–Barr virus transfection
Hybrid hybridomas Cells constructed by fusing hybridomas with other hybridomas or lymphocytes from immunized animals
Hybridomas Cells constructed by fusing myeloma cells with lymphocytes from immunized animals
Immunotoxins Toxin-coupled antibodies that can kill tumor cells
Myeloma cells Cell lines, derived from tumors of lymphatic cells, used for fusion with antibody-producing lymphocytes
Single-chain antibodies Monovalent antibodies with variable regions of immunoglobulin heavy and light chains joined by a synthetic peptide linker

MONOCLONAL ANTIBODIES (MABs) are produced by lymphocyte clones (i.e., derived from a single cell) and are therefore made up of identical molecules. Because it is not possible to obtain clonal populations from single normal lymphocytes, Köhler and Milstein developed the technique of fusing normal lymphocytes with myeloma cells. The resulting hybrid cells, like lymphocytes, produce antibodies, and, like myeloma cells, can be grown indefinitely. Clonal cultures of such hybrid cells are called hybridomas. Each produces only one kind of antibody, with a single specificity, which is called a monoclonal antibody.

MAB technology is used to construct MAB-secreting hybridomas and to produce and modify MABs. Hybridomas were first produced by fusing a mouse myeloma cell with a spleen lymphocyte from an immunized mouse. This technology has become one of the basic methodologies in biology; mouse MABs are widely used for qualitative and quantitative analyses of antigens, for purification of bioactive substances, and for clinical applications. Some mouse MABs have been found in therapeutic trials to be effective and to have relatively low toxicity, but they are immunogenic in humans, and patients develop antimouse antibodies. Because this complication limits the applicability of mouse MABs in clinical situations, efficient techniques for producing human MABs have been developed. These techniques include the immortalization of human B lymphocytes by Epstein–Barr virus (EBV) and the production of hybridomas, using human B cells. The human MABs thus produced are more suitable for treating tumors as well as viral and bacterial infections.

Recent advances in technologies for hybridoma production, recombinant DNA manipulation, and gene transfection have opened new ways for MAB production, making it possible to obtain novel bioengineered antibodies, called second-generation antibodies. These include bispecific antibodies, chimeric antibodies, single-chain antibodies, and antibody fusion proteins with novel effector functions. Bispecific MABs secreted by hybrid hybridomas can react with and link two distinct antigens. Chimeric MABs, composed of murine variable regions with the desired antigen specificity and human constant regions with the biological functions of human immunoglobulins, are largely human and it is hoped that they will be well tolerated in humans.

Further development of DNA technology has made it possible to produce other chimeric MABs, in which human constant regions are attached to variable regions which are partially human in composition. Such MABs may be more suitable for therapeutic applications. Single-chain antibodies are produced by splicing the genes for variable regions of immunoglobulin heavy and light chains to a DNA segment that encodes a synthetic peptide linker. They are expected to be less immunogenic in humans, and their smaller size might give them the ability to penetrate into body tissues that are normally restrictive to larger molecules.

Another group of genetically engineered antibodies is produced by the ligation of immunoglobulin and bioactive protein genes. Such specific antibody fusion proteins, in which the Fc portion of an antibody is replaced by a toxin protein, are used, for example, as immunotoxins capable of targeting and killing specific cells.

I. Mouse Monoclonal Antibodies

A. Preparation

Cell lines capable of permanently producing a specific antibody directed against a single antigen are constructed by the fusion of myeloma cells and spleen cells from suitably immunized animals. The first step is to isolate myeloma cells that lack the enzyme hypoxanthine guanine phosphoribosyltransferase and are therefore azaguanine resistant, or cells that lack thymidine kinase and are 5-bromodeoxyuridine resistant. Such mutant cells cannot grow in a medium containing hypoxanthine, aminopterin, and thymidine (HAT medium) because, owing to the mutation, they are unable to utilize either hypoxanthine or thymidine to counteract the effect of aminopterin, which inhibits the *de novo* synthesis of DNA.

In contrast, hybridomas deriving from the fusions of such mutant cells and antibody-producing spleen cells can grow in HAT medium and can therefore be selected from the myeloma cells used in the fusion procedure. The normal lymphocytes are lost, due to their limited growth. From among the growing hybridomas, individual clones that secrete the desired antibodies are isolated. Each of the antibodies thus produced is monoclonal, and is called an MAB. The

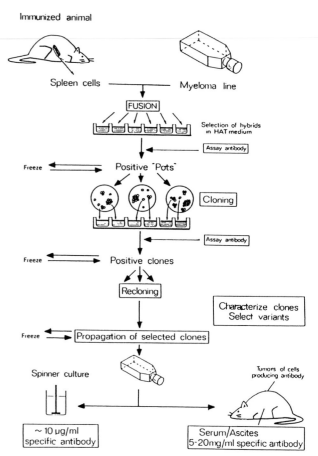

FIGURE 1 Basic protocol for the derivation of monoclonal antibodies from hybrid myelomas. HAT, Hypoxanthine–aminopterin–thymidine.

procedures are depicted in Fig. 1. The MABs are obtained either from supernatants of hybridoma cultures or from ascitic fluids after injecting the hybridoma cells into the mouse peritoneal cavity.

B. Application

1. Affinity Purification

Molecular adsorption to and subsequent elution from special solid adsorbents in granule form are used to produce highly purified compounds which are valuable tools in biology and medicine. Using MABs as the adsorbent confers exquisite selectivity on the technique, allowing the rapid purification of the molecules that have the antigenic determinant recognized by the MAB [e.g., purification of human interferons, human coagulation factor V, human immunoglobulin E (IgE), and mouse histocompatibility antigen].

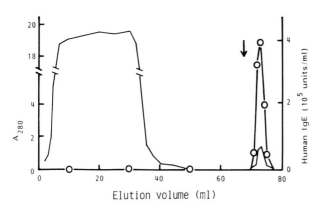

FIGURE 2 Purification of human IgE on an adsorbent containing an IgE-specific mouse MAB. An enriched IgE preparation was applied to the adsorbent preequilibrated with phosphate-buffered saline. Human IgE was eluted from the adsorbent with 0.2 M acetic acid containing 0.15 M NaCl (arrow). ————, Adsorbance at 280 nm (A_{280}); ——O——, human IgE. The IgE has been completely separated from all other proteins.

Human IgE secreted from human myeloma cell line U266 was originally isolated by tedious procedures including several steps. The purification, however, is simplified by using an MAB-coupled adsorbent. The IgE preparation thus obtained represents an approximately 600-fold purification from the original culture fluid (Fig. 2).

Human leukocyte interferon A is also purified from recombinant *Escherichia coli* cells by a procedure using the MAB adsorbent. The preparation obtained has a specific activity of 2.3×10^8 U/mg, and the yield is 95%. [*See* INTERFERONS.]

2. Immunodiagnosis

MABs have been developed for many clinically relevant antigens and used as replacements for animal sera in some of the classical immunoassay procedures (e.g., radioimmunoassays, enzyme immunoassays, immunofluorescence, and immunohistochemical stainings). [*See* RADIOIMMUNOASSAYS.]

Specific MABs have been used to detect carcinoembryonic antigen (CEA), α-fetoprotein, or carbohydrate antigens in cancer diagnosis. As shown in Table I, anti-CEA MABs and antisialosyl Le[a] antibody CA19-9 have been widely used in the diagnosis of gastrointestinal and pancreas cancers. For *in vivo* diagnostic applications (e.g., immunoscintigraphy), MABs are coupled with γ-ray emitting

TABLE I Novel Fucolipids and Fucogangliosides as Human Tumor-Associated Markers Defined by Specific Monoclonal Antibodies

	Association	Structure	Antibody
Lacto series type 1 chain, sialosyl Le	Gastrointestinal/ pancreas cancer	Galβ1 → 3GlcNAcβ1 → 3Galβ1 → R 3 4 ↑ ↑ NeuAcα2 Fucα1	CA19-9
Lacto series type 2 chain	Gastrointestinal/ lung/breast cancer	Galβ1 → 4GlcNAcβ1 → 3Galβ1 → 4GlcNAcβ1 → 3Galβ1 → 4Glcβ1 → 1Cer 3 3 ↑ ↑ Fucα1 Fucα1	FH4
		Galβ1 → 4GlcNAcβ1 → 3Galβ1 → 4GlcNAcβ1 → 3Galβ1 → 4Glcβ1 → 1Cer 3 3 3 ↑ ↑ ↑ NeuAcα2 Fucα1 Fucα1	FH6
		Galβ1 → 4GlcNAcβ1 → 3Galβ1 → R 3 3 ↑ ↑ NeuAcα2 Fucα1	CSLEXI
Globo series	Breast cancer	Galβ1 → 3GalNAcβ1 → 3Galα1 → 4Galβ1 → 4Glcβ1 → 1Cer 2 ↑ Fucα1	MBr1
Ganglio series	Small cell lung carcinoma	Galβ1 → 3GalNAcβ1 → 4Galβ1 → 4Glcβ1 → 1Cer 3 3 ↑ ↑ Fucα1 NeuAcα2	F-12

isotopes, such as iodine-131, iodine-123, indium-111, and technetium-99m.

3. Immunotherapy

The first MAB to come onto the market for therapeutic use was Orthoclone OKT3, (Ortho Pharmaceutical Corp., Raritan, New Jersey) directed against an antigen expressed on all normal peripheral blood T lymphocytes. The MAB can be infused into patients with kidney grafts to abort sudden graft rejection by decreasing the number of circulating cytotoxic T cells, which are responsible for the rejection. This MAB has also been used *in vitro* to remove the T cells that are responsible for acute graft-versus-host disease, in which the cells of the graft attack the cells of the host body.

MABs are now tested extensively for cancer therapy. MABs that bind to tumor-associated antigens could be used either naked or conjugated to cell-killing compounds. MABs specific to each patient's B cell lymphoma have been developed, and some disease remissions have been achieved in clinical trials. The MAB 17-1A, specific for an antigen of gastrointestinal cancer, is now in phase I (concerned with safety) and phase II (concerned with effectiveness) clinical trials. CA125 antibody is currently on the market as a diagnostic tool, but it could also be tested *in vivo* for the treatment of lung and breast cancers. KS$\frac{1}{4}$ antibody, which reacts with an antigen common to human carcinomas, including lung, breast, prostate, and pancreas cancers, has been used in clinical trials.

The antibody has also been used in cancer patients as a chemical conjugate with methotrexate, which is widely used in cancer chemotherapy, to deliver the chemical to the cancer cells. Several other MABs have been chemically coupled with a variety of ribosome-inactivating proteins, including ricin, gelonin, and other related toxins. Linker technology involves the use of bifunctional coupling agents to conjugate MABs to drugs. Agents such as *N*-succinimidyl-3-(2-pyridyldithio)propionate and *N*-(γ-maleimidobutyryloxy)succinimide have made it possible to prepare heterodimeric conjugates (Fig. 3a). The oligosaccharides present on the constant regions of the heavy chains of MABs offer specific sites for coupling, by using periodate oxidation to generate aldehydes. This method of coupling produces homogeneous MAB conjugates with unimpaired antigen-binding activity (Fig. 3b).

MABs linked to the radioisotope yttrium-90 are being considered for tumor therapy. This approach

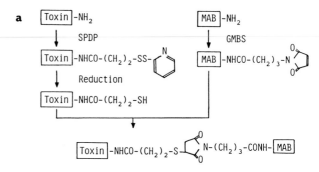

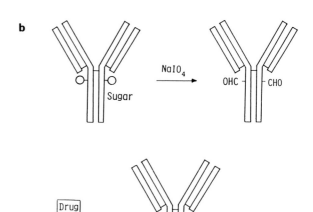

FIGURE 3 Chemical modification of monoclonal antibodies (MABs) using (a) bifunctional agents and (b) oxidation of sugar moieties in the constant regions with sodium periodate. SPDP, *N*-succinimidyl-3-(2-pyridyldithio)-propionate; GMBS, *N*-(γ-maleimidobutyryloxyl)-succinimide.

would overcome the difficulties created by the heterogeneity of tumor cells, a proportion of which contains the antigen recognized by an MAB, because the yttrium–MAB conjugates kill both the antigen-bearing cells and their neighboring nonbearing cells. The conjugates have been used for treating leukemia, lymphomas, and melanomas.

II. Human Monoclonal Antibodies

A. Preparation

Human MABs can be introduced into humans without causing anti-MAB responses. Several approaches have been adopted to produce them. The first approach is the immortalization of antibody-producing human B lymphocytes by EBV transfection. This method, however, is inferior to mouse hybridoma technology in the amount of MAB produced and the clonability and stability of the cells, although a few stable MAB-secreting EBV transformants have been reported. [*See* LYMPHOCYTES.]

The second approach is to fuse human B lymphocytes with mouse myeloma cells, which has led to the production of human MABs to Forssman's antigen, mammary carcinoma, and tetanous toxin. These heterohybridomas constructed by interspecies fusions, however, do not have a high degree of stability, because human chromosomes are preferentially lost during successive passages, often with loss of their ability to produce the MAB.

The development of HAT-sensitive human myeloma cell lines that can be used as a fusion partner for human lymphocytes has provided another approach. Several human myeloma and lymphoblastoid cell lines thus generated are available as a fusion partner for antibody-secreting lymphocytes obtained from peripheral blood, the spleen, and lymph nodes. Human–human hybridomas that produce human MABs directed against dinitrophenol, measles virus, or malignant glioma cells have been reported. However, these fusion procedures have met with only limited success, because fusion efficiency is very low. To overcome this limitation, the techniques of EBV transformation and cell fusion have been combined to produce human MAB-secreting hybridomas (Fig. 4), obtaining high fusion rates and excellent clonability. EBV activates the B lymphocytes to fuse efficiently, resulting in higher yields of human–human hybridomas. Using this

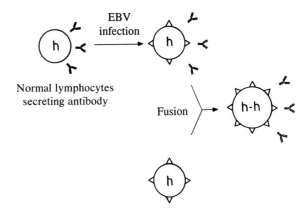

FIGURE 4 Preparation of human–human (h-h) hybridomas using Epstein–Barr virus (EBV) transformation.

technique, human MABs have been produced for a wide variety of antigens, as described in Table II.

B. Application

1. Immunodiagnosis

Considerable expectations are linked to the use of human MABs for the *in vivo* diagnosis and therapy of human cancer. Human MABs conjugated to radionuclides can be used repeatedly in the same patient for the imaging of tumor localization. Such radioimaging could permit the early detection of micrometastatic lesions and the implementation of adequate therapeutic maneuvers. Recent improvements in linker technology have made possible the

TABLE II Human Monoclonal Antibodies Produced by Epstein–Barr Virus Transformation of Human Lymphocytes Followed by Fusion with Immortal Human Partner Cell Lines[a]

Antigen	Lymphoid tissue	Fusion partner cell line	Immunoglobulin isotype secreted	Stability (months)
Tetanus toxoid	PBL	KR-4	M (κ)	>7
	PBL	KR-12	M	>10
	PBL	TAW925	M, G_1 (κ)	>12
Pseudomonas aeruginosa	SP	LTR228	M	—
Exotoxin A	PBL	TAW925	G_2	>6
Mycobacterium leprae	PBL	KR-4	M	>12
Cytomegalovirus	PBL	WI-L2	G (λ)	—
Rubella virus	PBL	Org MHHI	M, G	>2.5–6
Hepatitis B virus	PBL	TAW925	M, G_1	>12
Human leukocyte antigen	PBL	GM1500-6TG-0B	M	—
Lung carcinoma	PBL, LN, ITL	KR-4	M	>5
Breast carcinoma	PBL	KR-4	M	>4
Rheumatoid factor	PBL	KR-4	M, G, A	—

[a] PBL, Peripheral blood lymphocytes; SP, splenic lymphocytes; LN, lymph node lymphocytes; ITL, intratumoral lymphocytes.

binding of MABs to radionuclides without a loss of immunoreactivity. In one approach MABs are modified in the carbohydrate residues, far from the antigen combining site, using derivatives of radionuclide chelators, as shown in Fig. 3b.

2. Immunotherapy

Human MABs that recognize a number of antigens have been produced and can be used for the treatment of human diseases. Passively administered human polyclonal antisera are currently used in several clinical settings, but human MABs probably replace them for safety reasons. In fact, the use of MABs avoids the risk of a possible contamination of pooled globulins derived from human blood with infectious agents. It also diminishes the need for serum donors and improves quality control, with the use of reproducible manufacturing procedures. Human MABs against gram-negative bacterial endotoxins are now in phases II and III trials and will soon be on the market. Antiviral MABs, like antihepatitis B surface antigen (HBsAg) and cytomegalovirus antibodies, will be also supplied for clinical use. Some anti-HBsAg MABs have already been proven capable of preventing infection of a chimpanzee with HB virus.

A number of laboratories have attempted to generate human antitumor MABs. In many cases the human humoral immune response to the tumor is limited because the immunogenicity of tumor cells is weak, and it is thus difficult to obtain human MABs that specifically recognize tumor-associated antigens. In some cases anti-tumor MABs have been generated which bind to and kill human tumor explants in mice. The only MAB that has been used clinically is the antiganglioside GD2 antibody OFA-I-2 secreted by an EBV transformant. The antibody has been successfully applied to the treatment of melanoma patients.

III. Second-Generation Monoclonal Antibodies

Recent advances in the technologies for hybridoma production, recombinant DNA manipulation, and gene transfection have made possible the production of novel antibody reagents. MABs with dual specificities or additional functional moieties can be produced for medical and biological research and for further clinical applications.

A. Bispecific Antibodies

1. Preparation

Naturally occurring IgG antibodies are monospecific and bivalent; that is, they have two binding sites of the same specificity. Bispecific immunoglobulin molecules produced by a number of procedures can react with and link two distinct antigens. Bispecific MABs were first produced by the chemical reassociation of monovalent half-molecules derived from two different MABs (Fig. 5a). An alternative method involves the covalent attachment of whole MABs of different specificities, using a heterobifunctional cross-linking agent (Fig. 5b). These chemical reassociation and aggregation methods have a number of technical disadvantages. The procedures could cause protein denaturation and a subsequent loss of antibody activity or could lead to the formation of oligomeric or polymeric molecules.

A more recent development has been the use of cell fusion techniques used in the production of regular MABs (Fig. 1). Hybrid hybridomas constructed by fusing two different antibody-producing hybridomas express heavy and light chains from both types of antibody. These protein chains recombine in the cell to produce two parental and eight hybrid molecules (Fig. 6). There is a preferential combination of the homologous heavy chain–light chain pairs, so that the yield of the desired bispecific molecule (antibody 5 in Fig. 6) is somewhere between 12.5% and 50% of the total immunoglobulin production.

2. Application

Bispecific MABs containing two different antigen-binding sites could be useful in various fields of

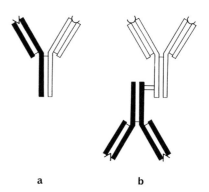

a   b

FIGURE 5 Bispecific antibodies produced by (a) cell fusion or reassociation of monovalent fragments and (b) produced by chemical cross-linking of whole antibodies.

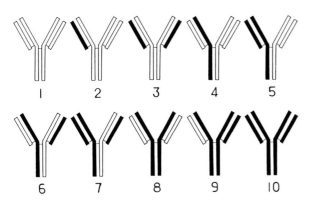

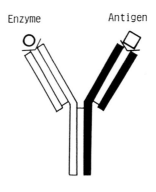

a

Enzyme   Antigen

FIGURE 6 Molecular species in a hybrid hybridoma expected to arise by random association of heavy and light chains. Parental antibodies are 1 and 10, and a bispecific antibody is 5.

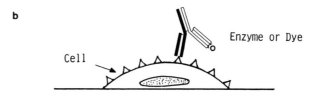

b

Cell

Enzyme or Dye

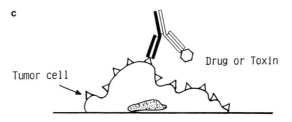

c

Tumor cell

Drug or Toxin

FIGURE 7 Application of bispecific antibodies for (a) immunoassay, (b) immunostaining, and (c) immunotherapy.

biology and medicine for research, diagnosis, and immunotherapy (Fig. 7). For diagnosis antibodies specific for both the antigen of interest and a label (e.g., colorigenic enzymes, fluorescent dyes, chemiluminescent compounds, and heavy metal-chelating compounds) are produced and used for one-step immunoassays, immunocytochemical staining, and *in vivo* radioimaging. In such a diagnosis bispecific MABs provide higher signal–noise ratios and quicker simpler procedures than do regular MABs.

For therapy bispecific MABs are expected to attract immune effector cells or deliver drugs to the specific targets. Cytotoxic T lymphocytes represent an important effector mechanism in the defense against intracellular pathogens, viruses, and possibly tumors. Bispecific MABs that recognize the T cell receptor with one arm and a specific site on target cells with the other can focus the activity of the effector T cells on the specific target site on tumor cells or on virus-infected cells. Bispecific MABs that recognize both a cytostatic or cytotoxic agent and a tumor-associated antigen on cancer cells can direct the agent to the tumor cells (i.e., the so-called "magic bullet therapy"). Toxins (e.g., ricin, abrin, gelonin, and exotoxin) and chemotherapeutic agents (e.g., methotrexate and vindesine) with highly potent cytotoxic effects might be led to tumors by bispecific MABs. Bispecific MABs produced by cell fusion techniques can bind drugs or toxins with one arm without impairment of their antigen-binding activity in the other, and their binding affinities for both antigens are the same as those of the respective parental antibodies with monospecificity. Therefore, they could be more active and

have a longer half-life than antibodies to which drugs or toxins are covalently linked.

B. Chimeric Antibodies

1. Preparation

Some chimeric antibodies are composed of murine variable regions with desired antigen specificities and human constant regions to provide the effector functions of human immunoglobulins (Fig. 8a). These antibodies are introduced by splicing the DNA sequences responsible for the variable regions of the heavy and light chains of an MAB-secreting murine hybridoma to genomic DNA segments encoding the constant regions of human heavy and

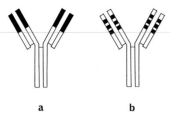

a   b

FIGURE 8 Mouse–human chimeric antibodies. (a) Mouse variable and human constant regions. (b) Mouse hypervariable region (i.e., complementarity-determining region), human variable region framework, and constant region. □ Human; ■ mouse.

light chains. The constructs are inserted into expression vectors, which are then introduced into myeloma cell lines. They could also be introduced into the germ cell line of animals, generating transgenic animals as antibody expression systems. This technology is attractive, because the cell culture of MAB-secreting hybridomas on a large scale is expensive.

The chimeric antibodies depicted in Fig. 8a might still elicit an immune response in humans. A less immunogenic chimeric MAB contains human constant regions attached to variable regions that are partially human in composition (Fig. 8b). The variable domains of an antibody consist of complementarity-determining regions and framework regions. The former (also called hypervariable regions) are mainly responsible for antigen binding. The chimeric MABs have been reshaped by introducing the six hypervariable regions from the heavy- and light-chain variable domains of a murine antibody into human framework regions. Such humanized variable regions retain a binding affinity for the antigen similar to that of the original antibody with a completely murine variable region.

2. Application

Chimeric MABs appear to be best suited as "magic bullets," because they have at least two advantages over murine antibodies. First, the effector functions of human immunoglobulins, including complement fixation, antibody-dependent cell lysis, and phagocytosis, can be selected or tailored as desired. Such effector functions might determine the efficacy of immunotherapy. Second, the use of human constant regions can minimize the antiglobulin immune responses during therapy. The newer types of chimeric antibodies containing humanized variable regions are even less immunogenic and are more useful in the treatment of tumors and micro-

bial infections. This is proven by a number of preclinical tests to determine the efficacy of human–mouse chimeric antibodies against colorectal, lung, and pancreas cancer cells. The chimeric antibody of the 17-1A MAB is now in phase I studies for use in pancreas cancer patients.

C. Single-Chain Antibodies

1. Preparation

The smallest antigen-binding fragment, consisting of a 25-kDa heterodimer of an antibody variable heavy-chain (V_H) and light-chain domain (V_L), was first prepared by peptic digestion of murine IgA, but there have been few reports of the successful isolation of Fv fragments by such proteolytic digestion of intact antibody molecules. Advances in gene manipulation and gene expression techniques now allow the design and synthesis of single-chain antigen-binding proteins. The complete single-chain Fv gene is constructed by linking the V_H and V_L genes through a DNA segment encoding a synthetic peptide linker. The peptide linker must assure the accurate positioning of the individual V_H and V_L domains and the correct folding of the single-chain Fv fragment (Fig. 9). The sequence Glu–Gly–Lys–Ser–Ser–Gly–Ser–Gly–Ser–Glu–Ser–Lys–Ser–Thr or (Gly–Gly–Gly–Gly–Ser)$_3$ is employed as a linker in the semiartificial single-chain antibody designed in this way.

Synthetic sequences encoding the *trp* promoter–operator and the leader peptide are ligated to the single-chain Fv gene, and the expression plasmid, based on the pBR322 vector, is finally constructed. The plasmid is introduced into *E. coli* cells, which synthesize the single-chain antigen-binding protein and accumulate it in insoluble inclusion bodies.

2. Application

Single-chain antibodies prepared so far closely mimic the antigen-binding affinity and specificity of

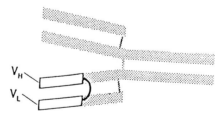

V_H
V_L

FIGURE 9 Single-chain antibody. V_H and V_L regions are joined by a synthetic peptide linker. Stippling indicates the relative size of the whole antibody molecule.

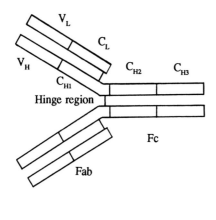

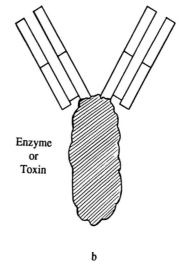

a   b

FIGURE 10 (a) Mouse monoclonal antibody and (b) its genetically engineered fusion protein with novel effector functions. The Fc region (C_{H2} and C_{H3}) of the antibody is replaced by another active protein (e.g., enzyme or toxin).

the parent antibodies. In model experiments antibodies directed against growth hormone, fluorescein, and digoxin were designed and synthesized. Such single-chain antigen-binding proteins are expected to have advantages in clinical applications, because of their small molecular size. These proteins should be cleared from the serum faster than whole immunoglobulin molecules or Fab fragments, and because they lack the Fc portion of an antibody, which binds to cell receptors, they have a lower background in tumor imaging and are less immunogenic. They might also penetrate the microcirculation surrounding solid tumors better than whole antibody molecules. Such single-chain antibodies might be valuable tools for imaging and for the therapy of cancers and cardiovascular or other diseases as well as for the separation and detection of bioactive substances.

D. Antibody Fusion Proteins

1. Preparation

The genes encoding antibodies can be manipulated *in vitro* and introduced into lymphoid cell lines, thus allowing the production of recombinant antibodies that exhibit novel biological functions. A hybrid molecule consisting of an antibody and an enzyme was first constructed by replacing the Fc portion with *Staphylococcus aureus* nuclease. The gene for the variable region was linked to the enzyme-encoding gene through a DNA segment encoding the C_{H1} (part of a constant region) and hinge regions of an immunoglobulin heavy chain (Fig. 10a). In the recombinant antibodies, namely, the C_{H2} and C_{H3} regions were replaced by the enzyme moiety (Fig. 10b). The antibody/enzyme gene was cloned into a vector that functions in mammalian cells and was introduced into mouse myeloma cells. The transfected cells secreted the antibody fusion protein, which has the antigen-binding specificity conferred by the Fab moieties and the DNA or RNA degrading activity of the nuclease moiety. Other recombinant molecules of antibody/DNA polymerase and the antibody/c-myc oncogene product have also been expressed in mammalian cells.

2. Application

Genetically conjugated enzyme-linked antibody can be used in enzyme immunoassays and histochemical stainings. The Fab portion of such an antibody/enzyme fusion protein provides a tag that ensures the export of the enzyme from the cell. These fusion proteins can be applied to the synthesis of specific enzymes and can facilitate the purification of enzymes that cannot be purified easily by the conventional methods.

As for clinical applications analogous antibody fusion proteins could be of use in tumor therapy; the production of immunotoxins might be simplified if immunoglobulin and toxin genes were ligated together by the DNA splicing method and introduced into an expression system. Since success has been achieved using antibodies that were chemically conjugated with toxin molecules, such genetically engineered immunotoxins are expected to provide efficient targeting of tumor cells.

Bibliography

Bird, R. E., Hardman, K. D., Jacobson, J. W., Johnson, S., Kaufman, B. M., Lee, S.-M., Lee, T., Pope, S. H., Riordan, G. S., and Whitlow, M. (1988). Single-chain antigen-binding proteins. *Science* **242,** 423.

Galfre, G., and Milstein, C. (1981). Preparation of monoclonal antibodies: Strategies and procedures. *In* "Methods in Enzymology" (J. J. Langone and H. Van Vunakis, eds.), Vol. 73, p. 3. Academic Press, New York.

Hakomori, S. (1985). Aberrant glycosylation in cancer cell membranes as focused on glycolipids: Overview and perspectives. *Cancer Res.* **45,** 2405.

Harada, K., Ichimori, Y., Sasano, K., Sasai, S., Kitano, K., Iwasa, S., Tsukamoto, K., and Sugino, Y. (1989).

Human–human hybridomas secreting hepatitis B virus-neutralizing antibodies. *Bio/Technology* **7,** 374.

James, K., and Bell, G. T. (1987). Human monoclonal antibody production: Current status and future prospects. *J. Immunol. Methods* **100,** 5.

Milstein, C., and Cuello, A. C. (1984). Hybrid hybridomas and the production of bispecific monoclonal antibodies. *Immunol. Today* **5,** 299.

Neuberger, M. S., Williams, G. T., and Fox, R. O. (1984). Recombinant antibodies possessing novel effector functions. *Nature (London)* **312,** 604.

Samoilovich, S. R., Dugan, C. B., and Macario, A. J. L. (1987). Hybridoma technology: New developments of practical interests. *J. Immunol. Methods* **101,** 153.

Staehelin, T., Hobbs, D. S., Kung, H., Lai, C-Y., and Pestka, S. (1981). Purification and characterization of recombinant human leukocyte interferon (IFLrA) with monoclonal antibodies *J. Biol. Chem.* **256,** 9750.

Steplewski, Z., Sun, L. K., Shearman, C. W., Ghrayeb, J., Daddona, P., and Koprowski, H. (1988). Biological activity of human–mouse IgG1, IgG2, IgG3, and IgG4 chimeric monoclonal antibodies with antitumor specificity. *Proc. Natl. Acad. Sci. U.S.A.* **85,** 4852.

Williams, G. (1988). Novel antibody reagents: Production and potential. *Trends Biotechnol* **6,** 36.

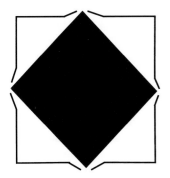

Mood Disorders

CHARLES L. BOWDEN, *The University of Texas*

Glossary

Biogenic amines Naturally occurring chemical substances that appear to be involved in nervous system activity associated with several critical human behaviors, including alertness, concentration, level of interest, speed of thinking, and mood. Several lines of evidence indicate that their function is disturbed in mood disorders. The principal biogenic amines thought to be active in the brain are norepinephrine, epinephrine, serotonin, and dopamine.
Bipolar Describes persons who at different or concurrent times have had both depressive and manic symptomatology. It is synonymous with the earlier term, manic-depressive.
Cognitive psychotherapy Focuses on pessimistic, self-pejorative ideas or misinterpretation of events as key components of depressive disorders which, if corrected, are likely to result in improved mood.
Dysphoria Refers to a relatively enduring feeling state of moodiness, unhappiness, pessimism, and sadness. It is descriptive of the overall mood state seen in depressive disorders.

Electroconvulsive therapy Passage of a small electrical current through the brain by electrodes touching the surface of the head. It has been shown to be an effective treatment in certain forms of depression. The term *convulsive* is added because the current induces a full body seizure, although this is attenuated by the medications now employed.
Interpersonal psychotherapy Focuses on establishing a positive relationship with the patient that allows identification of repetitive, ineffective ways of dealing with characteristic stressors, and suggests more effective means, especially in regard to close relationships.
Monoamine oxidase inhibitors Antidepressant drugs that work by blocking an enzyme involved in the metabolic breakdown of biogenic amines, with evidence that this results, at least temporarily, in increased biogenic amine activity.
Psychotic Refers to disturbed perceptions (such as hallucinations) or disturbed thoughts (such as belief that one is being persecuted) that are generally viewed as both irrational and bizarre, and outside the experience of healthy individuals. These are processes that a lay person would generally view as being abnormal and warranting some outside intervention.

DISORDERS OF MOOD are those in which a predominant and relatively persistent disturbance in mood is a characteristic of the illness. For these purposes, mood is defined as the relatively sustained emotional tone of a person. A depressed mood is evidenced by sadness, blues, or despondency. An elated mood, characteristic of mania, is indicated by elation, expansiveness, or optimism. Two other major dimensions of mood are (1) anger, hostility, and irritability; and (2) blandness, indifference, and underresponsiveness to circumstances that would normally cause sadness or elation. A concept related to mood is affect, which refers to the moment-to-moment equivalent of mood. The term *affective disorder* is sometimes used synonymously for *mood disorder*.

I. Depressive Disorders

Major advances have been made in the diagnosis and treatment of depressive conditions in the past 20 years. Perhaps the primary development is recognition that depression is not one common entity, but rather is made up of several pathophysiological conditions. Because the course and treatment for the disorders differ, accuracy in diagnosis is critical.

II. Major Depressive Disorder

A blue or sad mood state, persistent over most days for a period of weeks, is one of the most common signs of a major depressive disorder. Since sadness is a common and universal experience, it may be useful to summarize some features that differentiate the symptoms seen in major depression. It tends to be less responsive to enjoyable events—a party, time spent with close friends, a good meal. If the person does respond with enjoyment to such events, the response is transient, with a quick lapse to dysphoria when the event is past. It tends to persist in the absence of difficult, stressful life circumstances. Though tearfulness is relatively common, at least as often persons report an inability to experience intense emotions, whether sad or joyful. Symptoms of anxiety are common, and indeed are often as intense as those of depression. Energy is diminished and the person feels prematurely fatigued. Interest in usual activities and social pursuits is diminished, especially as manifested by those characteristics that make the person appear vibrant and interesting. Some persons may be able to muster sufficient effort to present a veneer of normalcy in social situations, yet their inner sense of detachment and dysphoria does not change. [*See* DEPRESSION.]

A set of physical changes is common. The person sleeps less well. The most common characteristic is a pattern of waking during the night and having great difficulty returning to sleep, or being unable to sleep although having awakened well before the intended hour. Often coupled with this is a pattern of

feeling most depressed and pessimistic upon awaking, and experiencing a progressive brightening of spirits as the day goes on. Less often persons may actually sleep more than usual, particularly with excessive daytime napping. Appetite may be disturbed, even favorite foods failing to provide gratification. Weight loss of 5–25 pounds is common. Sexual interest and responsiveness are often impaired. A wide range of general bodily complaints is frequently seen with depression. Unprovoked irritability is common. [*See* APPETITE; SLEEP DISORDERS.]

A substantial percentage of persons with major depression do not subjectively experience their disturbance as depression per se. Usually their principal complaint is a sense of low energy, fatigue, or of nonspecific physical complaints. Such complaints may take the person to an internist, family physician, or other physician, based on the patient's assumption that some physical, non-brain-based disturbance is present. Such "nondepressed" depression may be more common in the elderly.

The untreated course of major depression is often 6 months, a year, or longer. In a majority of untreated cases the illness will recur, often with a pattern of increasing chronicity. Since treatments are effective both for acute episodes as well as reducing the risk of relapse, a systematic, active treatment is warranted.

III. Bipolar Depression

Bipolar depressions are those that occur in individuals who have previously had a manic-type episode. Although the general symptoms are the same as for major depression, some differences may help in differentiating the two conditions. Bipolar depressed patients are less likely to have significant anxiety, restlessness, and physical complaints. Their episodes tend to be shorter, and to have first occurred at an earlier age (the 20s rather than the 40s). A family history is more likely to identify others who have had depressive-type periods. Since several depressive episodes sometimes occur before a first manic episode, definitive diagnosis may not be possible for several years. This is of some importance because of both the different illness course of the two types of depression and the response to medications. Although bipolar depression and major depression are responsive to standard antidepressants and monoamine oxidase inhibitors (MOAIs), there is evidence that drugs of these types may increase the frequency of bipolar episodes in bipolar patients, or induce manic episodes. Since lithium is effective both in the treatment of acute episodes of bipolar depression and in warding off relapse, it is often the initial drug chosen. An additional consideration in beginning with lithium is that the prophylactic effect would be desired in any case, and can be added if the patient does not respond fully to the lithium alone. There is some evidence that the risk of inducing a manic episode is reduced when lithium is being taken concurrently with standard antidepressants. [*See* ANTIDEPRESSANTS.]

IV. Delusional Depression

A lesser percentage of patients with major depression have delusional, or psychotic features. That is, in addition to the earlier catalogued symptoms of depression, they have varying degrees of unrealistic perception of their condition. When mild, these involve an exaggerated sense of their own worthlessness, or guilt about previous misdeeds. When severe, they may entail fixed, false beliefs of cancer or other serious illnesses, or misperception of everyday events as presaging some doomsday occurrence. Depressions of this type do not respond well to psychotherapy or to standard antidepressants. Two forms of treatment appear more effective. The combination of standard antidepressants with an antipsychotic (with the antipsychotic generally continued only during the phase of psychotic symptomatology) is effective, as is electroconvulsive therapy.

V. Dysthymia

Less severe symptoms of depression continuing over a longer period of time (officially, most of the time for at least 2 years) are referred to as dysthymia, or neurotic depression. There is substantial disagreement about milder depressions among authorities. Some view them as mild forms of a chronic, intrinsically biological disturbance. Others view them as learned or, at least, explainable psychological reactions to events in the person's life. Regardless of issues of cause, we have less good data about suitability of and responsiveness to drugs. The studies that have been done indicate that depressed patients selected for mild severity of their symptomatology have quite good responses to standard antidepressants. When drugs are employed, the dosage range needed is the same as for major depression. Patients and some physicians incorrectly equate lesser severity with need for less drug. In fact, the same dosage range appears necessary for any conditions for which standard antidepressants are efficacious.

VI. Double Depression

In recent years, a pattern of depression consisting of occasional episodes of major depression overlaid on an underlying, chronic dysthymic disorder has been identified. These patients appear to respond less well to standard antidepressant treatments. The recognition of this group of patients has been important in underscoring course of illness as an important factor in subtypes of depressive disorder. Increasing attention to differences in chronicity and the degree to which a person fully recovers between episodes may provide as useful a guide to treatment selection as does a particular constellation of symptoms.

VII. Atypical Depression

Several investigative groups have focused on sets of atypical depressive symptoms. It is not clear whether these will prove useful. Some authorities feel that MAOIs may be more effective in atypical depression. The current evidence for atypical depression is summarized here in part because of the confusing ways in which this term is currently used. Atypical has been variously defined to indicate patients who lack characteristic somatic symptoms of depression, or relatedly, have opposite symptoms along some of these dimensions. For example, the patient may have increased appetite and weight gain rather than weight loss, and may feel worse as the day progresses, rather than better. Others have focused on a predominance of anxiety symptoms, hypochondriacal complaints, and emotional lability. Still other investigators have emphasized the patients' chronic dysphoria but the retained ability to respond pleasurably to positive events. The difficulty of getting agreement as to what constitutes "atypicality" is one reason for caution about the applicability of this notion. Further, MAOIs appear to be effective in a wide range of depressive states, including patients with severe major depression with characteristic somatic manifestations.

VIII. Seasonal Affective Disorder

Some individuals regularly become depressed on a seasonal basis. In general the symptoms in such instances are mild ones, principally in the areas of energy, interest, and concentration. The most common, but not sole pattern appears to be depression in the fall and winter, with resolution in the spring. Some of these patients benefit from increased exposure to light with the same spectrum as sunlight. Although of great interest because of theoretical implications regarding the role of light, these conditions do not appear to explain any large percentage of major depressive or bipolar depressive disorder.

IX. Suicidal Concerns

Some depressive states are accompanied by suicidal thoughts. The risk of suicide attempts and suicide is substantially increased by depression. Most of these suicidal thoughts are only mild and transient, without intention of the person to act. At other times the person may be fearful that he or she will not be able to resist the impetus toward self-destructive action. Still other patients who are seriously suicidal may mask their intent on causal inquiry. The complexities of suicidal behavior and its grave consequences dictate that it be treated seriously when it is evident or suspected. As there is good evidence that the impulse to end one's life passes with effective treatment and resolution of depression, it is important to commence such treatment. [See SUICIDE.]

Treatment for the suicidal person is tripartite. If the person has an underlying, associated depressive disorder, treatment will be begun for that disorder. There are no drugs specifically for suicidal ideation. Of equal importance will be the supportive, close, candid, often contractual treatment relationship with the person. Thirdly, appropriate use of a holding environment, such as a psychiatric hospital, to reduce the risk of impulsive or deliberate self-destructive action can also be essential.

Whereas there is conclusive evidence of biological disturbances in major depression, the situation with suicidal behavior is not clear. Suicidal thoughts may plausibly arise as a secondary reac-

tion to sensing oneself as less effective, unable to perform as one would like, etc. Alternatively, there is evidence that patients who are suicidal, especially those who attempt suicide by impulsive, violent means have disturbances in certain neurotransmitter activity. It is possible that additional study could yield biochemical tests to aid in the identification of high-risk suicidal patients.

X. Manic Disorder

Mania is characterized by elevated mood, grandiose, expansive thoughts, accelerated thinking and speech, and reduced need for sleep. Judgment is usually impaired. This often results in unduly risk-taking behavior, and contributes to the frequent bankruptcies, divorces, and legal entanglements of persons with manic disorders. It is associated with difficulty in recognizing that one is ill, thus contributing to resistance to obtaining or continuing with treatment. Many manic patients demonstrate less elation but have prominent anger, irritability, and expressed hostility. The duration of manic episodes is generally shorter than depressive episodes, ranging from a few days to several months.

There appear to be subgroups of manic patients with different symptom expressions and, consequently, differing outcomes and treatment response. These include rapid cycling patients, who have numerous episodes of mania, or mania and depression within a period of time. One convenient guideline is a history of four or more distinct episodes within one year. A second group of patients has concurrent depressive and anxious symptoms with the manic symptoms. A third group has concurrent psychotic or delusional features. All three appear to respond relatively poorly to standard treatment with lithium alone, which constitutes the mainstay of treatment for manic disorders. In the majority of cases, treatment results are quite good.

Recent studies have shown that for a variety of reasons a substantial percentage of patients do not do well with long-term lithium treatment. Factors include the aforementioned different subtypes of manic disorder, poor compliance, and side effects such as shakiness and weight gain. These failures have prompted interest in possible alternative treatments for manic disorder, which is presently focusing on several compounds originally developed for treatment of seizure disorders. These include carbamazepine, valproic acid, and clonazepam, none of which have, as of 1990, had the experimental research conducted that would allow conclusion as to their effectiveness. Counseling aimed at helping the patient recognize the early signs of illness episodes and improving his or her understanding of the impact of the illness on major life functions is generally needed.

XI. Biological Factors in Mood Disorders

Biogenic amine metabolism has been extensively, though not always systematically, studied in mood disorders. Although the evidence is complex, there appear to be abnormalities in the metabolism of norepinephrine in a significant percentage of patients with major depression, bipolar depression, and manic disorder. Additionally, many of the drugs effective in treatment of these disorders alter noradrenergic activity. Additionally, abnormalities of serotonin and dopamine metabolism have been reported, though less consistently. Finally, other neurotransmitter substances that may be linked to or modify action of the above three biogenic amine systems have been reported as abnormal in a significant percentage of patients. These include substances from the hypothalamus (corticotropin releasing factor, thyroid releasing hormone) and adrenal gland (epinephrine, cortisol). The recent availability of analytical techniques sufficiently sensitive to measure the minute concentrations of these substances in body fluids, such as blood, have made this an important area for current research. [See ADRENAL GLAND; HYPOTHALAMUS.]

There is substantial evidence of inherited, genetic factors in mood disorders. These data come largely from familial and especially adopted twin studies, in which the effects of inherited characteristics can be assessed largely apart from learned, life experience characteristics. These data indicate that in bipolar disorder if one identical (monozygotic) twin has bipolar disorder, the other will also have the disorder in over 60% of the cases. If one parent has bipolar disorder, the likelihood that a child will develop bipolar disorder is around 10%. This compares with a likelihood in the general population of around 1%. Although extremely valuable data have emerged from such studies, the studies are difficult to conduct. It is likely that further refinements of our understanding will occur in the future: for example,

there may be both heritable and nonheritable forms of bipolar disorder. [See AFFECTIVE DISORDERS, GENETIC MARKERS.]

XII. Epidemiological Factors in Mood Disorders

A. Age-Related Factors

Depression occurs throughout the life cycle. Although depressive disorders in children and adolescents are less common than in adults, the symptom picture is analogous. Interference with school and developmental activities is often severe. Fortunately, many of the same treatments effective in adults are effective in children.

Bipolar disorder most commonly has its onset in the 20s. Major depressive disorder has its peak incidence in the 40s. Late age onset depression has been little studied. It appears that somatic complaints may be more characteristic of the depressed elderly.

B. Sex-Related Factors

Bipolar disorder is approximately equally common in men and women. Major depressive disorder is approximately twice as frequent in women as men. It is unclear whether this reflects biological differences, greater stressors on women, the fact that men with depressive propensities may be more likely than women to become alcoholic, and thus not be counted as depressed, or some combination thereof.

C. Prevalence and Course

Approximately 5% of persons have major depression at least once in their lifetime. The frequency lifetime prevalence for bipolar disorder is around 1%.

XIII. Treatment

A. Antidepressant Drugs

Standard antidepressants are the mainstay of treatment of major depression. Because of somewhat differing modes of action and side effect profile, a person may do well with one drug from this group but not another. Therefore, the physician will usually wish to try at least two such drugs for adequate periods before considering alternatives. The MAOIs are also effective in major depression. The obligation to adhere to a special diet and the risk of hypertensive reactions tends to result in their consideration as second or third line choice. Current studies may clarify the types of patients for whom MAOIs may likely be effective.

There is usually a lag period of several days to several weeks before patients respond to antidepressant medication. Early indicators of improvement are often apparent to family or professional staff before they are to the patient. In general, the risk of relapse is greater if antidepressants are discontinued sooner than 6–12 months after recovery. A substantial percentage of patients need long-term maintenance treatment with antidepressants. A variety of laboratory and other tests, including drug level determinations, are used to improve the effectiveness and monitor the safety of antidepressant medications.

A number of combination drug therapies are used for patients who do not do well with single drugs. Although the data are not as conclusive as with single-drug therapies, selected patients appear to benefit from combinations of lithium and standard antidepressants; lithium and MAOIs; stimulants and standard antidepressants; the amino acid L-tryptophan and MAOIs; and possibly thyroid enhancing agents and standard antidepressants.

B. Psychotherapy

Especially in the area of depressive disorders, serious investigative efforts in recent years have compared various psychotherapies with antidepressant drug therapy, both alone or in combination with drug therapy. A comprehensive review of these important studies is beyond the scope of this article, but several points are important. Several well-conducted studies have found that in the aggregate, psychotherapy may, when employed concurrently with antidepressant drugs, increase long-term well-being and reduce risk of relapse. It may be more important to think in terms of which patients are improved with combined psychotherapy and pharmacotherapy than the simplistic question of whether drugs are better than psychotherapy, or vice versa. With the individual patient, the decision can often be made over a period of time. If a patient with major depression does not improve fully with medication, or if definitive improvement occurs in some symptoms, yet the patient or the clinician senses that some self-defeating patterns of behavior persist, possibly linked to risk of recurrent depression, a more systematic effort in psychotherapy is warranted. Recent studies suggest possibly greater efficacy for interpersonal, rather than for cognitive psychotherapy when both were employed in a time-limited, systematic fashion by clinicians trained in the principles and methods of each. A further difficulty in this area is that of the gray zone between support, education, and encouragement on the one hand, and formal psychotherapy on the other. No effective drug therapy is conducted without a close, knowledgeable relationship with the patient, which in fact embodies many elements of defined psychotherapy.

Specific psychotherapies appear to be most effective in patients with milder symptomatology. Studies in this area have been difficult to conduct and interpret because spontaneous improvement, independent of specific therapy, occurs frequently. The role of specific psychotherapy or counseling is thus open to different opinions. Perhaps the most common, and certainly a justifiable strategy for these disorders, is to address apparently obvious situational factors, stressors, and psychological factors by psychotherapeutic means. This would hold particularly for depressions of acute onset and short duration. The less apparent the relationship of stressors to onset, the longer the illness course, the greater the case for antidepressant medication. If psychotherapeutic efforts do not largely resolve symptoms within a month or so, reconsideration of drug therapy is indicated.

C. Electroconvulsive Therapy (ECT)

This procedure remains a highly effective treatment for certain depressive disorders. Overall response rates to ECT appear at least as high as to standard antidepressant drugs. It seems particularly indicated in psychotic depression, acutely suicidal patients and patients refractory to other treatments. Despite its efficacy, most authorities consider it a second order treatment for several reasons. Hospitalization, is necessary, with high cost per treatment. There is some risk of memory disruption for the period of treatment. Maintenance treatment is generally required to prevent relapse. Since outpatient maintenance ECT is rarely practicable, antidepressant drugs are indicated. Most psychiatrists thus reason that it is better to determine antidepressant drug response acutely, since it will be needed for continuation treatment.

Bibliography

Keller, M. B., Lavori, P. W., Coryell, W., Andreasen, N. C., Endicott, J., Clayton, P. J., Klerman, G. L., and Hirschfeld, R.M.A. (1986). Differential outcome of pure manic, mixed/cycling, and pure depressive episodes in patients with bipolar illness, *JAMA* **255,** 3138–42.

Klerman, G. L., and Weissman, M. M. (1989). Increasing rates of depression, *JAMA* **261,** 2229–35.

Koslow, S. H., Maas, J. W., Bowden, C. L., Davis, J. M., Hanin, I., and Javaid, J. (1983). Cerebrospinal fluid and urinary biogenic amines and metabolites in depression, mania, and healthy controls: a univariate analysis, *Archives of General Psychiatry* **40,** 999–1010.

Maas, J. W., Koslow, S. H., Davis, J., Katz, M., Frazer, A., Bowden, C., Berman, N., Gibbons, R., Stokes, P., and Landis, D. H. (1987). Catecholamine metabolism and disposition in healthy and depressed subjects, *Archives of General Psychiatry* **44,** 337–44.

Myers, J. K., Weissman, M. M., Tischler, G. L., Holzer, C. E., III, Leaf, P. J., Orvaschel, H., Anthony, J. C., Boyd, J. H., Burke, J. D., Kramer, M., and Stoltzman, R. (1984). Six-month prevalence of psychiatric disorders in three communities: 1980 to 1982, *Archives of General Psychiatry* **41,** 956.

Spitzer, J. C., Endicott, J., and Robins, E. (1978). Research diagnostic criteria: rationale and reliability, *Archives of General Psychiatry* **35,** 773–82.

Mortality

S. JAY OLSHANSKY, *University of Chicago/Argonne National Laboratory*

Glossary

Average lifespan The average genetically-endowed limit to life for a population if free of all exogenous risk factors

Central death rate The number of deaths for a given age, race, or sex group divided by the population at risk of the same age, race, or sex

Immediate cause of death The final disease or condition resulting in death

Life expectancy at age x The average number of years of life remaining for a population of individuals, all of age x, and all subject for the remainder of their lives to the observed age-specific death rates corresponding to the current life table

Life table A statistical method of translating central death rates into the probability of dying and other life-table measures such as the life expectancy at age x

Underlying cause of death Disease or injury that initiated events resulting in death

MORTALITY OR DEATH (used synonymously) is a vital event for which statistics are collected for individuals and populations. Central death rates are most often calculated from death counts observed over a given time frame (usually one year) as a function of estimates of the population alive in the middle of the year (July 1). These central death rates, which may then be converted to probabilities of dying in a life table, are often used as a general gauge of the health status of a population. Mortality rates vary considerably by age, race, sex, and underlying cause of death, and are also known to vary considerably between countries, among subgroups of a population, and across time. The risk of death is high at birth, declines to its lowest levels by the age of 10, increases steadily to the age of 30, and then increases exponentially thereafter. While central death rates throughout most of the world yield an average life expectancy at birth of about 70 years, estimates of the lower limits of mortality (barring major advances in medical technology) indicate that life expectancy could increase to a maximal level of about 85 years.

I. Cause of Death

When a person dies, the single most important factor that contributed to their death is listed on their death certificate. This is referred to as the underlying cause of death. Comparable procedures for recording cause of death on death certificates are followed in most countries throughout the world, although there is some variation. The underlying-cause-of-death data are then compiled for the population, and represent the basis upon which population-level, cause-specific mortality statistics are calculated. While the underlying-cause-of-death information is useful for assessing trends in health within the same population across time, and among subgroups of a population, it is not without its drawbacks. For example, some countries use nonstandardized methods of determining underlying cause of death, in spite of the fact that the World Health Organization has created (and modified ten times) the International Classification of Diseases. This should serve as the standard for classifying cause of death on death certificates. Additionally, the presence of coexisting clinical conditions among the elderly make the determination of their underlying cause of death extremely difficult. Researchers have addressed this problem by analyzing all of the vital information on the death certificate including immediate, contributing, and underlying causes.

Central Death Rates

Central death rates (hereinafter referred to as death rates) are calculated by dividing the number of deaths, taken from the death certificates, by the population at risk (usually a mid-year estimate of the population), and then multiplying by a constant (as per 100,000). Multiplying by a constant yields the number of deaths that may be expected for a population of that size. Death rates may also be calculated by age, race, sex, and for one or more underlying causes of death.

Death rates in the United States from all causes and selected underlying causes are presented in Table 1. These data indicate that while death rates have declined considerably from 1960 to 1985, this has not been the case among all of the underlying causes. Death rates from the number one cause of death, heart disease, declined rapidly during this 25-year period, while death rates from the number two killer, cancer, actually increased. When these death rates are divided further into changes within age–race–sex groups, it shows that summary mortality rates like those presented in Table 1 tend to mask the underlying mortality transitions occurring in the population. The death rates from Table 1 are age adjusted, so that this is what the death rates would look like if the age composition (e.g., distribution of the population across all age groups) in the United States were held constant at a level present during a single year. Adjusting the death rates to a constant age composition eliminates the biasing effect on mortality rates of differences in age composition.

TABLE 1 Age-Adjusted Central Death Rates for the United States (per 100,000)

Cause	1960	1970	1980	1985
All causes	760.9	714.3	585.8	546.1
Heart disease	282.6	253.6	202.0	180.5
Cancer	125.8	129.9	132.8	133.6
Accidents	49.9	53.7	42.3	34.7
Pneumonia and influenza	28.0	22.1	12.9	13.4

Death Rates by Age, Race, Sex, and Underlying Cause

The risk of death is also known to vary by age, race, sex, and underlying cause of death. A relatively high risk of death is present at birth, followed by a declining risk of death to about the age of 10, a gradual increase in the risk of death to the age of 30, and an exponentially increasing risk of death from age 30 on. After the age of 30, the death rate doubles about every 8 years (see Fig. 1). This characteristic mortality curve is common among most animal species and has been referred to as a Gompertz curve, after the scientist who first identified it.

Even within age groups the risk of death varies considerably as a function of race, sex, and underlying cause of death. For example, for those age 40–44 in 1980 in the United States, the death rate from ischemic heart disease for black males was 7.1 times greater than that observed for white females (Table 2). For those aged 60–64 in 1980 in the United States, the death rate from stroke for black males was 3.8 times greater than that observed for white females. However, in the oldest age interval, the death rates for blacks were considerably lower than those observed for whites. This is known as the crossover effect, and is believed to be attributable to either selective survival, in which only the healthiest blacks survive to older ages, or problems of data reliability. Comparable differences in death

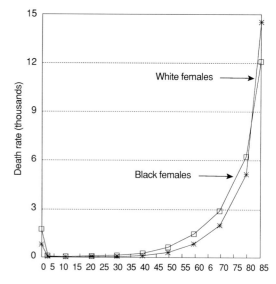

FIGURE 1 Age–race–sex-specific death rates for the United States in 1985 (per 100,000).

TABLE 2 Death Rates from Ischemic Heart Disease and Stroke for the United States for Four Selected Age Groups, by Race and Sex (1980)

	Ischemic heart disease				Stroke			
	WM	WF	BM	BF	WM	WF	BM	BF
20–24	0.7	0.2	1.8	0.3	1.3	0.9	3.5	2.5
40–44	73.8	14.1	100.2	37.3	9.2	8.9	38.8	29.2
60–64	737.3	242.5	707.9	402.9	87.0	63.5	244.7	171.1
85+	6792.2	5471.6	4296.0	3557.2	2260.4	2383.3	1875.4	1898.7

rates from other causes are present among the various age–race–sex groups.

II. International Mortality

Death rates vary not only by age, race, and sex within subgroups of a given population during a single period, but they also show considerable variation between countries. Table 3 shows age-adjusted death rates (per 100,000) for four selected underlying causes of death for five countries. Note that death rates from infectious and parasitic diseases are low in all of the countries except Sri Lanka, where the death rate is from 9 to 14 times greater than that observed for the other countries. Higher death rates from these causes are known to have their greatest impact on infants and children, which explains why life expectancy at birth is much lower in Sri Lanka than in the other countries. It is also apparent that death rates among the longest-lived country, Japan, are quite different from those of other developed nations. For example, note that the relative importance of stroke and ischemic heart disease to overall mortality are reversed for Japan and the United States. Many of these international differences in cause-specific death rates are be-

lieved to be attributable to variation in access to health care and differences in lifestyles (primarily diet and exercise). The presence and use of advanced medical technology may also play a role in these differences.

Another parameter that illustrates international differences in mortality is life expectancy at birth and at age 65. Note in Table 4 that life expectancy at birth ranges from a low of 65.3 in Mexico to a high of 75.5 in Japan. Developing nations have a lower life expectancy at birth than developed countries, primarily as a result of differences in infant and child mortality rates. With the exception of Poland, life expectancy at age 65 is between 14.1 and 16.1 years. The majority of the differences in life expectancy observed at age 65 are attributable to recent declines in old age mortality in the past quarter century in many developed nations.

III. Expectation of Life

Mortality rates may be used to estimate the average number of years of life that remain for a population of any given age. This may be done by calculating

TABLE 3 Age-Standardized Central Death Rates (per 100,000) from Four Causes of Death for Five Selected Countries

Country	Infectious & Parasitic Diseases	Cancer	Ischemic Heart Disease	Stroke
Japan (1986)	4.4	78.5	18.5	56.2
Poland (1986)	5.4	107.0	36.9	42.0
Sri Lanka (1982)	48.4	35.2	12.1	13.8
Sweden (1985)	3.4	99.3	89.6	38.4
United States (1984)	5.7	109.6	92.4	33.2

TABLE 4 Life Expectancy at Birth and at Age 65 for Selected Countries

Country	Life Expectancy at Age	
	0	65
France (1985)	71.8	14.9
Israel (1985)	73.6	15.1
Japan (1986)	75.5	16.1
Mexico (1982)	65.3	14.7
Poland (1986)	66.7	12.3
Sri Lanka (1982)	66.9	14.1
Sweden (1985)	73.8	14.7
United States (1984)	71.3	14.7

from central death rates, what is referred to as a life table. Life tables are most often created from central death rates observed at all ages for a specified cohort (e.g., a country, state, or for subgroups of the population within these larger geographic areas) during a single period. This is called a *period* life table, because it combines into a single metric the mortality experiences of the entire population during a single period (usually a calendar year). It is also possible to calculate a life table based on the lifetime mortality experiences of a single birth cohort, referred to as a cohort life table. However, it would take over 100 years to complete a cohort life table for individuals born today, because it is necessary to wait until the entire cohort has died.

It is important to remember that period life tables are calculated based on the assumption that the mortality rates that prevail at each age during a single period will be experienced by the babies born in that year, for the remainder of their lives. Under conditions of declining mortality, the metric of period life expectancy will underestimate the true expectation of life for a birth cohort. The lifetime mortality risks for those born today are likely to be lower than for those born during the past 100 years.

Figure 2 illustrates that for U.S. females, life expectancy at birth in 1900 was just under 50 years. This increased to over 78 years by 1985, thus representing more than a 30-year increase in life expectancy within just 85 years. The expected remaining years of life for a 65-year-old increased only 3.1 years for males and 6.6 years for females during this same period. Most of the gains in life expectancy in older people occurred after 1968 in the United States, primarily as a result of reductions in death

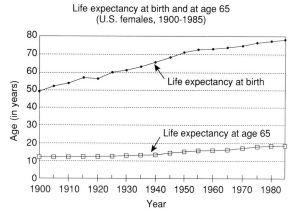

Life expectancy at birth and at age 65
(U.S. females, 1900-1985)

FIGURE 2 Life expectancy at birth and at age 65 (U.S. females, 1900–1985).

rates from the major cardiovascular diseases (e.g., heart disease and stroke). Most of the gains in life expectancy at birth, earlier in this century, are attributable to declines in infant, child, and maternal mortality.

IV. Estimating the Upper Limits to Human Longevity

What are the lower limits to mortality, or conversely, what are the upper limits to human longevity? This question has always interested scientists from a number of disciplines, but it is particularly important to healthcare planners, who must anticipate future demands for healthcare services. For many years it was believed that the longest humans could live was about 70 years. This was based in part on the biblical representation of the upper limits of life as three score and ten years. However, the average life expectancy for all developed nations and some developing countries has already exceeded this limit. In fact, the longest-lived human being was a Japanese man, who was verified to have died at the age of 119.

Since the late 1960s, this question of the upper limits to the average human lifespan has taken on new meaning, because extremely rapid declines in mortality have occurred among the major fatal degenerative diseases (e.g., heart disease, stroke, and some cancers) in most developed countries. This mortality transition surprisingly occurred among the population of middle and older ages. Rapidly declining mortality among the elderly was thought to be nearly impossible just 25 years ago, because it was believed that survivors to older ages were near their biological limits to life. The major fatal degenerative diseases were believed to be age related and therefore somewhat immutable and inevitable.

Research on estimating the upper limits to the average human lifespan has been based on (1) assumptions as to how low mortality rates could go; (2) estimates of optimal improvements in risk factors at the population level; and (3) estimates of the mortality rates that would be required to increase life expectancy beyond 80 years, and the prospects for achieving those levels. The prevailing view is that the maximal average human lifespan, assuming people lived to their biological potential, is somewhere between 85 and 100 years. However, the majority of the current research in this area indicates that the lower end of this range is probably more realistic. This conclusion is based on the assumption that there will not be major advances in medical technology to allow for the alteration of aging at the molecular level. If technology alters the fundamental rate of aging, then it is theoretically possible that the average lifespan could be extended considerably beyond current limits.

V. The Importance of Mortality Statistics

Mortality statistics are a valuable tool for assessing the general health status of the population, and for determining which subgroups of the population should be targeted for specific healthcare programs. For example, racial differences in infant mortality in the United States indicate that further reductions in infant mortality could be achieved if maternal and infant health care was improved for blacks. Gender differences in cause-specific mortality indicate that lung cancer is one of the most rapidly increasing causes of death among women, an indication that more women are becoming smokers when smoking is declining among the rest of the population. Recent trends in age-specific death rates also indicate that the risk of death is being postponed, primarily as a result of declines in the risk from cardiovascular diseases. This has produced an extremely rapid increase in the size of the elderly population, perhaps much greater than anticipated by government sources. These trends have major implications for social programs related to the size and health status of the elderly, such as the Social Security program and Medicare. Finally, mortality statistics are used

(along with data on fertility and migration) to make population forecasts at the local, national, and international level.

Prospects are promising for reducing mortality rates further in both developed and developing nations. In developing nations, infant and child death rates are still quite high, but the means to reduce the risk of death at these ages are readily available. In developed nations, the risk of death is already very low in younger and middle ages, but still quite high in older ages. Finally, while it would appear that the upper limits to human life expectancy have been approached in some parts of the developed world, it is likely that improved lifestyles, medical technology, and breakthroughs in research on aging will reduce the risk of death even further and extend life well beyond its current limits.

Bibliography

Antonovsky, A. (1967). Social class, life expectancy, and overall mortality. *Milbank Memorial Fund Quarterly,* **45**(2), 31–73.

Kitagawa, E. M., Hauser, P. M. (1973). "Differential Mortality in the United States: A Study in Socioeconomic Epidemiology." Vital and Health Statistics Monographs. Harvard University Press, Cambridge, Massachusetts.

Lancaster, H. O. (1990). "Expectations of Life: A Study in the Demography, Statistics, and History of World Mortality." Springer-Verlag Press, New York, N.Y.

Olshansky, S. J., Carnes, B., Cassel, C. (1990). "In Search of Methuselah: Estimating the Upper Limits to Human Longevity." *Science,* **250,** 634–640.

Shryock, H. S., Siegel, J. S. (1975). "The Methods and Materials of Demography." Bureau of the Census, Third Printing (rev.). U.S. Government Printing Office, Washington, D.C.

Motor Control

STEPHAN SWINNEN, *Catholic University of Leuven, Belgium*

Glossary

Central pattern generator Network of neurons responsible for generating properly timed rhythmic output

Closed-loop control Mode of control consisting of a reference mechanism (standard of correctness) against which feedback from the ongoing or completed response is compared to generate error

Coordinative structures Group of muscles constrained to act as a functional unit; related terms are synergies and muscle linkages

Degrees of freedom Number of potential variables that must be controlled to move efficiently

Entrainment Phenomenon that refers to the (mutual) attraction of limb movements or body parts; often used interchangeably with synchronization

Feedforward As opposed to feedback, refers to information used prior to executing a particular act, with the main advantage that the human movement system is prepared in advance of the movement and tuned to processing response-produced information

Motor program Central representation of movement containing abstract movement commands

Open-loop control Mode of control that does not make use of error information but, instead, prepares instructions for actions in advance, without modifications on the basis of feedback

MOTOR CONTROL, a fairly young discipline of scientific inquiry, attempts to describe and explain how movement is accomplished with reference to concomitant postural adjustments. Moreover, it seeks to understand how sensory processes enter into doing. Action emerges as a complex interplay among various forces that characterize the internal (bodily) and external environment, and yet skilled movements are made with great ease and comfort. This is made possible through an ingenious hierarchical control mechanism in which higher levels are responsible for setting overall action goals, whereas lower levels fill in details. The study of motor control has a strong interdisciplinary character. It is an interface among the neurological, biomechanical, kinesiological, and behavioral sciences. The neuroscientific perspective tries to unravel how the brain and spinal cord bring about movement. The biomechanical and kinesiological perspectives seek to understand the physical principles that the human body in action obeys. The behavioral perspective attempts to describe and explain observable movement behavior and the various manipulations that affect it, leading to inferences about the design of the control system. In summary, motor control is an area of study that deals with understanding the neural, physical, and behavioral aspects of movement. Increasingly, these perspectives are merging into a unified approach for understanding human movement behavior. The focus of the present chapter is mainly behavioral.

I. How Does Movement Take Place?

Control of movement can be understood as an interplay with the various forces of nature. On the one hand, movers exploit these forces to produce skillful actions (e.g., ground-reaction forces or elastic forces) generated through the springlike properties of muscles. On the other hand, living beings continuously fight against gravitational forces for their survival. Limbs are like levers, designed to overcome these forces. Standing up against the laws of gravity (which can take as little as a few seconds in some animals and several months in others) is a major accomplishment for many species. Motor control research attempts to explain movement as a complex combination of external and internal forces. On the one hand, mutual interactions between the mover and the (external) environment are a central focus of study. Action leads to changes in the environment, and changes in the environment bring about modifications in action. On the other hand, because the human body consists of a kinematic chain of interconnected body parts, various interactions exist among body parts. Even within a limb, interactions among the joints can be identified that are governed by various types of torque.

For voluntary movement to take place, two elements are required: muscles and a signaling system that makes muscles contract. Many centers in the brain are activated before, during, and after movement execution. Among them are various subcortical and cortical structures working together to control the final outputs from the motor cortex to the spinal cord. An important question concerns how the information is encoded by these high-level systems, which are apparently far removed from the detailed peripheral events in the muscles. Evarts, who pioneered the recording of activity of single cells in the brain of awake animals, has provided a major impetus for understanding the role of the motor cortex. He demonstrated that activity of the nerve cells in this brain structure was primarily related to the amount and pattern of muscular contraction rather than to the displacement that was produced by the contraction.

Neural information from higher levels is relayed to motor neurons whose cell bodies are located in the spinal cord and whose axons terminate on the muscle-fiber membrane. At this site, neural events are translated into mechanical energy by generating a contractile force in the muscles. The properties of these muscles constrain the calculations of the central nervous system: The combination of muscle lengths and tensions are interdependent with the velocity of muscle shortening or lengthening.

To obtain an idea of the relationship between muscle contraction and the resulting physical events, an example of a horizontal flexion–extension–flexion movement, as produced in the right

forearm, is depicted in Figure 1. The subject is requested to flex the elbow toward the body midline, followed by two reversals in direction in between target zones (elbow extension), to finally move toward an endpoint in front of the body (elbow flexion) in the target time of about 500 msec. At the top is shown the angular displacement pattern, followed by velocity and acceleration. This double-reversal movement is accomplished by activity of elbow flexors and extensors (i.e., those muscles that cause the elbow to flex and extend, respectively). Activity in two of them, namely biceps and triceps, as registered by means of surface electromyography, is shown at the bottom. Of course, many more muscles are activated to allow this movement to take place. Among them are those that serve to fixate the limb in reference to the trunk and to stabilize the whole body to ensure postural control. [*See* Muscle Dynamics.]

Before any displacement is evident, the biceps muscle is activated; even peak burst activity is reached before movement onset. A second burst of biceps activity takes place before the second major elbow flexion is initiated. On the other hand, the triceps is marginally active before onset of movement but shows increased activity shortly after movement initiation to oppose elbow flexion and to prepare for elbow extension. A final major burst of triceps activity serves to position the limb at the endpoint, and some degree of biceps activity is evident as well to help achieve that goal. Whenever displacement is recorded, transforming it to time derivatives, velocity and acceleration is useful. After all, acceleration requires force, which is the result of muscle activation. Accordingly, positive acceleration is mainly preceded by biceps activity, and negative acceleration by triceps activity.

Three major points can be made from this example. First, the mechanical effects are lagging behind the electrical changes that can be observed in the muscles. Second, some muscles allow movement to take place (the agonist); others oppose movement or cause movement to proceed in the opposite direction (the antagonist), and the latter need to be timed and organized as well. Third, movement is a complex combination of voluntary and reflex activity, and their rigid distinction is difficult to maintain. Note that biceps activity is low when triceps is activated and vice versa, a phenomenon known as reciprocal inhibition. Reciprocal inhibition is a key principle of reflex organization. Strong reciprocal inhibition allows the limb to be compliant, whereas

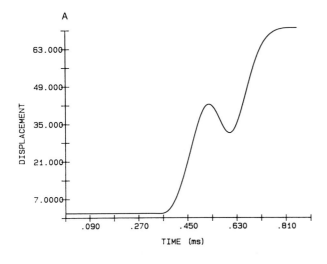

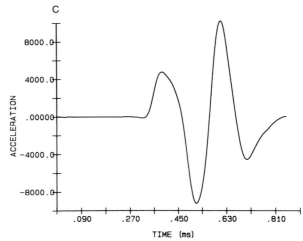

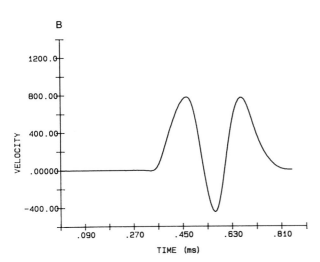

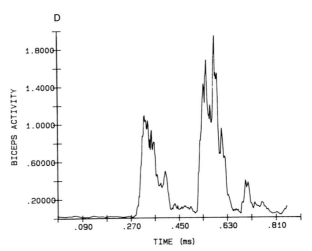

FIGURE 1 Displacement–, velocity–, and acceleration–time trace and rectified and smoothed electromyographic activity in biceps and triceps muscles for production of a horizontal double-reversal movement. Note that when activity is high in one muscle, it is usually low in the other. This is indicative of a key principle of reflex organization known as reciprocal inhibition.

weaker reciprocal inhibition may give way to co-contraction of opposing muscles, thereby stiffening the joint. Cocontraction is scarce in the depicted highly practiced trial, whereas it is usually apparent at initial levels of training. As will be pointed out later, many more reflexes operate during movement production. This intertwining of reflex and voluntary movement leads to a significant simplification of computational burdens for the higher levels of the motor control system.

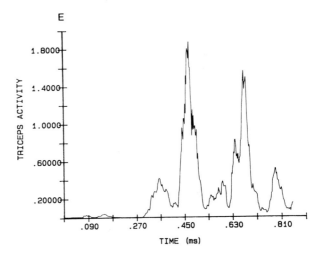

II. General Control Modes

A key issue that has dominated thinking in motor behavior for a long time concerns the locus of movement control. Competing ideas on this matter have formed one of the most persistent controversies throughout this century. On the one hand, the peripheral viewpoint argues that movement is controlled via some combination of feedback from the muscles and joints and vestibular, auditory, and visual systems. The central perspective, on the other hand, assumes that commands are structured in advance and that feedback is not essential to produce patterned movement. These perspectives can be exemplified by two distinct modes of control known as closed- and open-loop systems.

A. Closed-Loop Systems

Closed-loop systems (like home-heating methods) usually consist of three parts. A central feature is the reference mechanism, which represents the goal of the system or something to be achieved (e.g., the thermostat setting at 20°C). Next, information from the environment is gathered to determine the current value that the system seeks to regulate (actual temperature in the house), and this information is relayed to the reference, termed feedback, thereby closing the loop. Comparisons between a desired and actual state give rise to error detection. The executive level, informed about errors, takes action to reduce the error toward zero (e.g., the heater is turned on).

B. Open-Loop Systems

In opposition to closed-loop control, strictly open-loop systems do not contain the feedback cycle and are not directed at nullifying error. Commands are structured in advance and are executed without regard to the effects they have on the environment. The major advantage of such a system is that it can act quickly, as it does not have to process incoming information. Its drawback is that it might lack flexibility in that no adjustments to changing environmental circumstances can be made.

C. Representational versus Nonrepresentational Accounts

In the past decade, the conventional central–peripheral debate has made way for a partly related discussion on the desirability to invoke central representations that are held responsible for the organization of action. On the one hand (and similar to centralists), the so-called motor systems approach, backed by neurophysiology and the information-processing approach, assumes that movements are centrally represented. Movement commands, specified in rather general or abstract terms, are said to be responsible for output organization. Conversely, the action systems approach largely denies the role of representations or central prescriptions in explaining order and regulation manifest in motor behavior. Instead, it is argued that action emerges as an *a posteriori* fact, or as a necessary consequence of the way the system is designed to function. The key question becomes what can be explained in movement organization "for free" before burdening the central nervous system with control problems. This approach is characterized by an emphasis on processes of self-organization, a feature inherent to complex biological and physical systems. It is known as the dynamical account of movement behavior in that the interplay of forces and mutual influences among the components of a system, tending towards equilibrium, is largely held responsible for order in movement.

Although opposite schools of thought have initiated lively debates on major control issues in past decades, this clash of doctrines has created the opportunity to appreciate the benefits of both modes of control, which can be explored by the mover in response to the huge variety of existing task dimensions and environmental constraints. This paves the road for hybrid models of human motor behavior. Next, we discuss some evidence for central and peripheral contributions to motor control. This description goes beyond the formerly made sharp distinction between closed- and open-loop systems.

III. Sensory and Central Elements in Motor Control

A. Sensory Contributions to Motor Control

Two important sources of information play a determining role in movement regulation: proprio- and exteroception. Proprioceptors provide information about body position in space, joint angles, and the length and tension of muscles. It is information about our own (proprio) movements. Exteroceptors inform us about the spatial coordinates of surrounding objects and the environment.

A reciprocal relationship exists between sensory information and movement. Sensory information allows movement to proceed correctly as it tells the mover about the state of the body and environment; conversely, movement allows us to sense and perceive. For example, when holding a pen between your fingers for some time, you will experience a loss of sensation of the pen, and its presence fades from consciousness until you move your fingers again, informing you that the pen is still there. Besides the important role of sensory information in movement, it is also critically involved in the maintenance of posture.

1. Sensory Information and Postural Control

A variety of receptors contribute to the maintenance of posture. The vestibular apparatus in the inner ear contains receptors sensitive to deviations from the vertical and, more general, to the orientation of the body in space. In addition, receptors in the muscles (e.g., muscle spindle, golgi tendon organs) and joints (e.g., Ruffini endings, Pacinian corpuscles) provide information about muscle stretch, degree of muscle tension, and angle positions. Great effort has been spent in studying these ingenious receptor devices; nevertheless, their modes of operation are still a matter of debate. The muscle spindle is particularly interesting for movement control and has, perhaps, been the most extensively studied. From a standing position in a driving bus that suddenly stops, you tend to fall forward, resulting in a stretch of muscles in the lower leg (e.g., gastrocnemius, soleus). This is immediately followed by contraction of these muscles to maintain equilibrium. This response is mediated by the stretch reflex, which originates in the muscle spindle. It causes muscles to increase tension as soon as 20 msec after detection of muscle lengthening and is an example of a fast closed-loop control mode. Another negative-feedback servomechanism originates in the Golgi tendon organ, which senses force rather than muscle elongation, resulting in the reduction of force.

Postural control is not only an end but also a means to an end in that it participates in almost any action. When lifting your arm, leg muscles will be activated before onset of arm activity to secure body equilibrium during this focal act. Thus, posture can be regarded as a background upon which a picture of voluntary movement is "engraved."

In addition to these muscle-specific sensory devices, vision has proven to be a very dominant receptor system in the control of posture and movement. Close your eyes when balancing on one leg, and you will realize that visual information is indeed important for postural equilibration. Vision also provides information about the position and movements of objects in the environment, leading to decisions for action. But, it is a far richer source of information in that it also tells us about our own movements in relation to the environment (also called exproprioception, as distinguished from extero- and proprioception). Such information is mainly derived from the changing pattern of optical arrays, which refers to the particular reflection of light by various objects in the visual field. When the observer moves around in the environment or when an object moves with respect to the observer, changes in the optical array occur, called optical flow. This signifies that visual environmental information (e.g., texture, gradients, surface of objects) flows past us as we move around. The rate at which trees and houses become larger and pass by as we drive a car down a road tells us about our speed and the time at which upcoming objects will be contacted (time-to-contact).

2. Sensory Information and Movement

a. Role of Sensory Information before Movement It is evident that we gather information about the environment before actions are planned and executed. When throwing a ball, information is obtained about the location of the target to be hit, the position of the limbs and body, etc. When enough time is available, this processing of information may occur at a conscious level. In other cases, as in suddenly avoiding obstacles, little or no time is available to pick up relevant information; this requires decisions for action to be taken so fast that time-consuming processing stages are omitted. For example, David N. Lee has argued that information derived from optical flow patterns enables the identification of important external temporal events (the time-to-contact objects), which can be used to control motor activity in humans and animals. This time-to-contact information is derived directly from the rate at which retinal images of an object change in reference to their image size. It allows precise timing of actions without much conscious processing. This is supposedly the way plummeting gannets specify the time at which their wings have to be stretched backward before hitting the water or the way humans specify when to initiate action in hitting an accelerating ball.

b. Role of Sensory Information during Movement

Ongoing movements can also be adjusted on the basis of incoming information that is fed back to a reference, as long as these movements proceed slowly enough. Examples of closed-loop control in human motor behavior are mainly found in continuous tracking tasks such as driving an automobile, flying an airplane, etc. While driving an automobile, for instance, a major goal is to keep the vehicle on the right track without approaching the midline or side of the road too closely. This environmental information is used continuously to guide steering behavior through a series of corrections. The motor system is directed at nullifying error.

c. Role of Sensory Information after Movement

Sensory information can also be processed after movement has taken place. This is especially the case when movements are performed rapidly. For example, after completion of a golf swing, the information produced by this response (response-produced-feedback information) (e.g., the way the ball was hit, the sound produced at contact) is compared to some reference of correctness (or template, a term borrowed from bird-song research), providing information about the degree of correctness of the movement. Any discrepancy between the actual (what is) and expected (what should be) states leads to error-detection and decisions for error-correction that can be of use for the following trial.

This comparator function is hypothesized to reside predominantly in the cerebellum because central motor commands, as well as somatosensory, vestibular, and visual reports from the periphery, converge in this structure. One hypothesis concerning the mode of operation of this comparison process contends that signals are sent to certain neural centers, ahead of the response, to ready the human movement system for upcoming motor commands and for the receipt of feedback information. This advance information is also called corollary discharge, or efference copy, and represents an example of what is called feedforward control. In comparison to feedback, feedforward control refers to sending information ahead of time to prepare the body for upcoming sensory information or for planned motor commands. It has the advantage that incoming information undergoes facilitated process-

ing. Evidence for this viewpoint comes from experiments showing different sensory experiences in the case of passive or active movement production.

B. Central Contributions to Motor Control

As mentioned previously, the central perspective on motor control takes as a dictum that central commands are mainly responsible for organization of and order in movement. To strengthen this argument, various sorts of evidence have been collected to demonstrate that patterned movement can occur in the absence of peripheral feedback. Under conditions of feedback deprivation through deafferentation (severing the sensory nerves that carry information into the nervous system), monkeys have been shown to walk and climb, mice to display grooming patterns, and birds to sing (even though portions of the song are eliminated by denervating one side of the vocal apparatus), etc. This has led centralists to argue that movement is mainly controlled via stored movement commands, often called programs. A motor program can be defined as a central representation of movement or skill. It is a cornerstone of many current motor control theories. Although it is commonly agreed that the program is represented centrally, where exactly it resides remains unclear. Task features may be of concern in resolving this matter. For example, in the case of learned movements, looking for a program in a well-defined location within the central nervous system might not be fruitful as it may consist of several subprograms widely distributed in the nervous system and involving several brain centers. On the other hand, programs for certain inborn rhythmic behaviors, such as locomotion, swimming, and scratching, have a more definite location, taking the form of a network of neurons (called a neural oscillator or central pattern generator), responsible for generating properly timed rhythmic output. Timing of these repetitive movements is then regulated by intrinsic properties of the central nervous system. For example, the central pattern generators (CPGs) for locomotion are located in the spinal cord. CPGs drive the spinal motoneurons (innervating muscles) by rhythmically raising and lowering their membrane potentials, causing them to fire in bursts. The excitability of the CPGs is governed by locomotor centers in the midbrain and brain stem. [*See* LOCOMOTION, NEURAL NETWORK.]

Although CPGs are mainly invoked for the generation of inborn whole-limb cyclical actions, they may be critically involved in, or form the basis of, learned movements. It has been proposed by S. Grillner that whole-limb CPGs are made up of networks of smaller unit generators that control simple muscle synergies. This opens perspectives for parts of the generator network to be used in acquiring new motor acts that are far removed from locomotion. Think, for example, about the leg action in the breaststroke: Leg propulsion is accomplished by extending the hip and knees while dorsally flexing the ankle to maximize grip on the water. This is difficult to accomplish for the beginning swimmer because the natural tendency to flex or extend all joints simultaneously through the concerted action of muscle synergies is strong. It is reminiscent of the tight cooperation among joints that can already be found in the kicking movements of the newborn. With practice, the ankle joint is dorsally flexed while extending the other joints. In other words, a recoordination of parts of the locomotor synergy is accomplished. Thus, we have inherited the neural networks underlying patterned rhythmical movement, as well as the capability to adapt them in a flexible way to meet new environmental demands.

C. Reconciliation of the Peripheral and Central Perspectives

Even though the nervous system is able to issue stored motor commands without reference to peripheral feedback to generate a great repertoire of movements, this should not be taken to imply that feedback is unimportant for goal-directed motor performance and learning in animals and man [*See* MOTOR SKILLS, ACQUISITION.]

As mentioned previously, some have argued that bird song is represented centrally. However, experiments with birds reared from an early nestling stage without experience of their own species-specific song have underscored the learned contributions to development of the normal song template. They point to the important role of auditory experience in species-specific song development. In a similar way, although human actions may be governed by central programs, their particular expression often requires feedback involvement. Evidence pointing to the role of feedback involvement also comes from deafferented humans, who display many skillful activities but seem to have difficulty with performing fine manual skills, such as feeding, writing,

and fastening buttons, and with acquiring certain new skills.

To summarize, while the peripheral and central viewpoints are incompatible at their polar extremes, their reconciliation is possible and desirable. The question can then be asked how central commands and sensory information cooperate to produce skilled action.

D. What is in the Program?

Within a programming perspective, which information is structured in the motor program? The answer involves several viewpoints, two of which have received major attention: the impulse-timing and the mass-spring viewpoints. The first suggests that accelerative and decelerative forces and their timing are programmed, causing the limb to move a certain distance. The second focuses on the programming of end location of a limb movement, without knowledge of starting location. This is hypothesized to occur through the programming of opposing tensions.

1. Impulse-Timing Viewpoint

This viewpoint holds that movements are controlled by specifying the amplitude of forces administered to the muscles in combination with the durations over which these forces act (force × time = impulse). Thus, the program controls the duration and intensity of impulses, i.e., it tells the muscles when to turn on and off, just like the example presented in Fig. 1.

This mode of control results in observable invariances in movement (called invariant features) in the face of changes that can occur in other features (called parameters). This is exemplified in Fig. 2 with a hypothetical movement consisting of three segments. Both sequential movements are performed in different overall movement times (MT = a parameter). Aside from this difference, other things remain unchanged across the two variations. A first aspect is the order of sequences in the action. A second aspect is the temporal structure of the movement (also called relative timing or phasing). One way to evaluate this feature is to determine the ratios of the times of the sequences to overall movement time. These ratios (Fig. 2, in parentheses) are the same for both movements. Thus, absolute timing is hypothesized to be a varying feature and relative timing an invariant feature. The present example only focuses on invariance with respect to

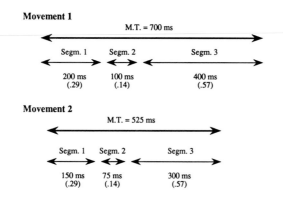

Movement 1

M.T. = 700 ms

Segm. 1 | Segm. 2 | Segm. 3

200 ms (.29) | 100 ms (.14) | 400 ms (.57)

Movement 2

M.T. = 525 ms

Segm. 1 | Segm. 2 | Segm. 3

150 ms (.29) | 75 ms (.14) | 300 ms (.57)

FIGURE 2 Time as a varying and invariant feature for two versions of a hypothetical sequential arm movement. M. T., movement time.

duration of portions of the movement. But movements evolve through the production of forces, and force is generated by muscle contractions. Therefore, invariances can possibly be determined at various levels of movement production. This distinction between parameters and invariant features reflects an interesting property of motor control in that it enables variations in a particular movement while leaving its basic structure intact. (More will be said about this later in this chapter.)

2. Mass-Spring Viewpoint

A remarkable resemblance is evident between a muscle-joint system and a mass-spring system with adjustable equilibrium points. In essence, a mass-spring system is a spring attached at one end to a fixed support and at the other end to a mass. When stretching or compressing the spring, it starts oscillating, eventually coming to rest at a predetermined final position. The system is *attracted* to that state. Consider now the more complex case of two groups of muscles (springs) allowing the elbow joint to flex and extend (see Fig. 3). The model holds that terminal location of a limb is defined as an equilibrium point between the tensions of agonist and antagonist muscle groups and any external forces (e.g., gravity) that may act around it. Regardless of initial conditions or information about the starting position, the limb will move until tensions balance to reach a certain location or endpoint. An appropriate analogy is the saloon door with springs on each side. When opening the door to enter the saloon, one spring shortens, reducing its tension, while the other spring lengthens, thereby increasing its tension. After releasing, the doors will oscillate and finally close again due to the balance of opposing

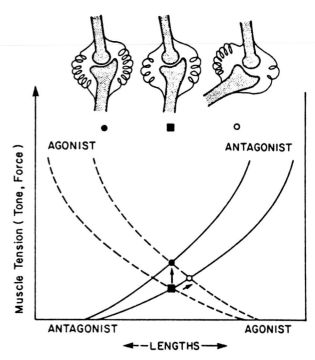

FIGURE 3 Mass-spring control as exemplified for positioning movements in the elbow joint. Comparison of three limb postures that differ in joint angle and/or joint stiffness. Two sets of length–tension curves are shown for muscles that act in opposition to each other (antagonists), as sketched in the diagrams above, where springs represent muscles. Curves for agonists and antagonists, and directions of their length changes, are shown as broken and solid lines. Agonist (broken line) is at greatest length on the left when the antagonist (solid line) is shortest and vice versa. Upward arrow (from filled square to filled circle) indicates increased cocontraction of antagonists, which increases joint stiffness without change of muscle lengths; sloping arrow (to open circle) indicates increase of agonist tone, creating a postural equilibrium at a new position, i.e., at new muscle lengths. Equilibrium at the filled circle can support most weight, at the open circle less, and at the filled square least. (From Brooks, V. B., 1986, "The Neural Basis of Motor Control," Oxford University Press, New York.)

spring tensions and damping effects. In other words, the doors reach a final point of equilibrium, irrespective of conditions that preceded this state (the way the door was opened, the force applied, etc.). Similarly, the position of a limb in space is also considered to depend on the balance of tensions in opposing muscles. In case a new equilibrium point for the doors is needed, a spring with a different stiffness must be attached (in the muscle-joint case, the tension of one muscle group needs to be increased). Evidence for mass-spring control has been gathered in experiments with animals and humans. Studies using arm-positioning movements in monkeys under deprived feedback conditions (deaf-ferented animals with visual information withheld) have shown that a specified movement endpoint can be achieved, regardless of perturbations in the movement, which apparently are not felt by the animal.

The impulse-timing and mass-spring models differ from each other in some essential ways. The impulse-timing model has trouble explaining correct end-positioning when initial location of the limb is suddenly modified. Time, featured heavily in the impulse-timing view, does not receive a primary focus in mass-spring control. Although action theorists disagree with the programmed nature of terminal location, they have extended this mode of control to a more complex dynamical model, consisting of ensembles of nonlinear limit-cycle oscillators or mass-spring systems.

Location programming has some attractive features. Movements can be made largely context-free because variations in starting position and perturbations along the movement path do not appear to affect endpoint control. The mass-spring model seems particularly interesting in view of the general observation that movements are never made twice in exactly the same way. But initial versions of the model do not seem to account well (1) for movements in which the goal is not an endpoint but a well-defined trajectory, (2) for movements consisting of reversals in direction, and (3) for movements with a strong emphasis on timing. Moreover, it is not yet clear how the model operates in multijointed movements. Modified models are now being proposed in an attempt to alleviate the limitations inherent in the basic mass-spring model.

IV. Movement Constraints: Computational Simplification

A. Degrees of Freedom and Control of Movement

Considering all possible movements that can be made in one joint through various muscle combinations and reflecting on the numerous ways to combine the movements of many joints, our ability to control all these muscles for the purpose of goal-directed behavior is rather remarkable. This has come to be known as the "degrees-of-freedom problem." Degrees of freedom refers to the number of independent states that must be controlled at the

same time. How the control and coordination of limb movements is accomplished has become a research matter of major interest in recent years.

When attempting to perform several tasks concurrently, limitations that constrain our capability to do more than one thing at a time often emerge. Experienced drivers easily combine the movements involved in steering a car with listening to the radio and perceptual scanning of the road. But suddenly, information load becomes so heavy due to traffic density that priority is assigned to the crucial tasks while others are ignored (e.g., the radio is turned off). What happened? The driver himself has constrained the amount of information to be processed. In complex motor behavior, natural constraints emerge, which impose limitations on our ability to perform actions concurrently. This often results in the synchronization of movements toward a dominant and unifying pattern of activation. Such constraints are common experience. Do you still remember how much difficulty you had as a preschool kid to pat your head and rub your stomach simultaneously?

Constraints result in preferred modes of interaction between moving limbs. The theoretical and experimental basis for this work was laid down by E. Von Holst some 50 years ago with the investigation of the coordination of fin movements in various anaesthetized fish species. Two important principles of coordination were discovered, which have relevance for other species as well. On the one hand, one fin movement tended to impose its frequency on another, or both fin movements mutually affected each other. This (mutual) attraction was called the "magnet effect," often leading to a 1/1 or 2/1 temporal relationship between fins (Fig. 4a). Besides temporal attraction, fin movements also displayed quantitative effects, and this addition of the level of activity was denoted as "superposition." The latter phenomenon is evident in Fig. 4b as an increased amplitude in those fin movements that temporally coincide with movement in the other fin. Both these phenomena were often found to occur simultaneously, giving rise to various patterns of coordination.

Two important conclusions can be drawn from these interactions. First, they point to the existence of endogenously active (coupled) oscillators as an explanatory principle in the neurobiological control of various cyclical actions, such as locomotion and scratching. Secondly, the fin movement interactions are a foretaste of hierarchical models of con-

FIGURE 4 Interactions among fin movements in anaesthetized fish. Movements of the pectoral (upper trace) and dorsal (lower trace) fin of *Labrus* are shown. The magnet effect (a) results in a strong periodicity between both fins with a frequency relationship of 2/1. Besides the magnet effect, which implies interactions at the frequency domain level, there is also a quantitative effect (b), called superposition, i.e., those dorsal fin movements that coincide with the pectoral fin movements are increased in amplitude. (From Von Holst, E., 1973, The behavioral physiology of animals and man, *in* ''The Collected Papers of Erich Von Holst,'' Vol. 1 [R. Martin, trans.] Methuen, London.

trol whereby various modes of coordination are accomplished by lower centers through the combination of a few elementary principles. This leads to reduction of the potential degrees of freedom and to a decrease of computational complexity.

When humans perform cyclical limb movements simultaneously, similar phenomena can be observed. This does not necessarily imply that the basic modes of operation are the same in both types of species. Nevertheless, noteworthy from an evolutionary perspective, some general principles of coordination arise within the animal world. When making horizontal elbow flexions and extensions in both upper limbs simultaneously, a strong tendency is observed to synchronize the output patterns in a temporally compatible fashion, such as a 1/1 or 2/1 rhythm, similar to the magnet effect (Fig. 5a). Moreover, in the case of a 2/1 rhythm (Fig. 5b), the arm moving at twice the frequency will often display greater amplitudes for those flexion movements that coincide with those in the other limb (the superposition effect). Another preferred mode of interaction is extending one arm while the other flexes but, again, in a fixed temporal order (Fig. 5c). Patterns of activity that do not confer to synchronization or alternation are more difficult to perform and result in degraded consistency.

This phenomenon of interaction, also denoted as entrainment (a term often used interchangeably with synchronization) reflects a fundamental organizational property of movement. It forces limb movements to take on a restricted range of preferred relationships. Similar preferred states of coupling are also evident in the production of rhythms.

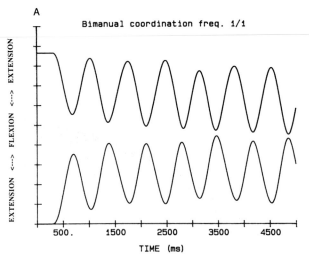

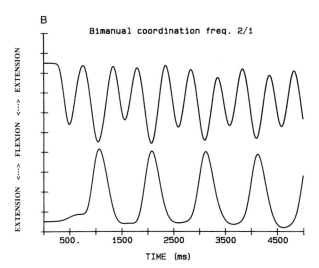

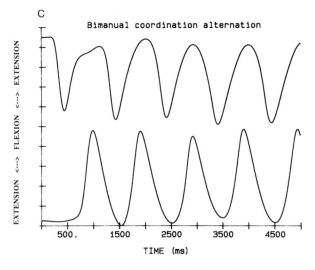

FIGURE 5 Some examples of preferred interactions among upper limb movements in humans.

Tapping 1/1, 2/1, or 3/1 rhythms with the finger of the left and right hands simultaneously is easy; producing polyrhythms (such as a 3/2 or 5/3 tap) is much harder, and many will be unable to do it.

Interlimb interactions not only appear in cyclical movements but are also evident in discrete movement production. In opposition to the former, the latter movements are nonrepetitive and have a distinct beginning and endpoint. Studies using bimanual aiming tasks (pointing to different locations in space) have elegantly demonstrated that the laws governing unimanual performance do not necessarily apply to multilimb action. When aiming at a certain target, more time is needed when the distance to be covered increases or when the target becomes smaller. This can be put in a formal mathematical relationship, known as Fitts' law:

$$\overline{MT} = a + b[\log_2(2A/W)],$$

where $\overline{MT}$ is the average movement time for a series of trials, a and b are empirical constants, A is amplitude of the movement, and W is target width. The distance that the limb moves and the narrowness of the target for which the limb tries to aim provide an indication of the difficulty of the task. If the level of difficulty is increased, the movement will be performed at a slower speed: a speed-accuracy trade-off. This law represents an early attempt to apply mathematical and information-processing principles to the understanding of human movement. It has proven very powerful in the past decades for a variety of aiming tasks and constitutes one of the few lawful relationships known in behavioral motor control. Given the strength of this relationship for unimanual skill, what would you expect to happen when making a short pointing movement to a wide target with one hand together with a long movement to a small target in the other hand? In the case of individualized control of both limb movements, Fitts' law would predict that both movements are completed in different overall times because they differ in degree of difficulty level. This does not, however, appear to be the case. What actually happens is that the slow movement is sped up and the fast movement is slowed down, resulting in a rescaling of the movements toward a common underlying timing basis. As a result, both movements are initiated and terminated virtually simultaneously. Only when they are substantially different in spatiotemporal features can departures from full synchrony be experimentally observed. Studies such as these mark an important turning point in the

history of motor behavior, as they seem to show that the whole is more than or is different from the sum of its parts. In other words, the laws governing multilimb actions are not necessarily linear extrapolations of those applicable to unilimb action. This points to the necessity for investigating movement coordination as a distinct type of control problem with its own lawful modes of operation. In a more general way, global behavior of biological systems cannot always be understood by investigating its elemental units, because interactions among the units are responsible for emergent properties at the macroscopic level that cannot be deduced from knowledge of the individual components. This is a challenge to reductionism!

B. Significant Units of Motor Control

In recent years, systematic observation of coordination phenomena has prompted a lively discussion about the significant units of movement to be controlled by the human motor system. In this respect, the viewpoint has gained acceptance that individual muscles are not controlled but, rather, cooperating groups of muscles, constrained to act as a functional unit (also called muscle collectives or linkages, coordinative structures, and synergies). This has a number of potential advantages. First, it provides an economical solution to the aforementioned degrees-of-freedom problem in that a limited set of preferred modes of movement organization become apparent. Second, computational complexity in higher levels of the control system is reduced, as these muscle synergies not only form the external language of movements but also the internal language of the central nervous system. Although the synergy perspective provides a fruitful and economical way to theorize about the organization of action, it should not be taken to imply that muscles, or even motor units, cannot be controlled individually, or that a release from the constraints imposed by these coordinative structures would be impossible. We will come back to this matter later in this chapter.

What kind of evidence has been advanced in support of the notion that the central nervous system preferably organizes movements by innervating groups of muscles? We have already reported the strong tendency to synchronize patterns of motor output in the limbs for repetitive and nonrepetitive actions. Another line of evidence is provided by the appearance of complete or fractionated reflex patterns in voluntary movement. For example, the tonic neck reflex, which arises in neck proprioceptors, gives rise to body and limb movements through a series of internally triggered reactions. The asymmetric tonic neck reflex, induced by turning the head sideward, is characterized by extension of the limbs on the side of face orientation and flexion on the other side. The symmetric neck reflex causes the upper limbs to flex and the lower limbs to extend when bending the head forward (ventriflexion), whereas the reversed pattern occurs when bending the head backward (dorsiflexion). As head movements also activate the labyrinth receptors, these reflex patterns often operate together with labyrinthine reflexes. Trainers of gymnastic skills have an intuitive knowledge of these triggered reactions. In performing a handstand, bending the head backward (trainers hint: "look at the floor") results in increased extension of upper limbs and trunk, thereby amplifying the support function of the upper limbs. This and many other observations are corroborated by experimental work that has shown actions in accordance with the tonic neck reflexes to confer additional strength. In addition, these reflexes have also been found to emerge during stressful activity to reinforce muscular contractions.

In summary, genetically inherited reflex patterns appear to be part of the behavioral repertoire of the performer who integrates these patterns in the generation of voluntary movement. Such a viewpoint is not incompatible with a programming notion of movement control. Central commands may help orchestrate the possible role of these muscle synergies in movement and may incorporate fractions of it into the overall plan of action. Like synergies or reflexes, learned motor programs impose constraints on action, albeit at higher levels of the central nervous system.

C. Acquiring New Patterns of Coordination

In the previous section, we have emphasized the property of entrainment as a key organizational feature in the control and coordination of movement. This has a pitfall in that entrainment provides a major limitation in terms of the kinds of activities that can be performed together. In other words, entrainment is experienced as an obstacle whenever differing or incompatible movements have to be performed together in the limbs. Remember the difficulty you had in steering a car with one hand and shifting gears with the other when you drove your first miles. Another example is producing

polyrhythms. As discussed above, tapping rhythms with the fingers is not difficult when fingers move in synchrony (1/1) or when tapping in one finger is an integer multiple of the tapping rate in the other (2/1, 3/1). But, when the timing relationships are incompatible (3/2, 5/3), achieving the same degree of consistency in tapping rate is much more difficult. The degree of temporal incompatibility of the movements to be coordinated can then be considered an important criterion for determining task difficulty.

Luckily, the human motor control system has developed the capability to eliminate natural response tendencies (i.e., preexisting synergies, central pattern generators, or other preferred relationships among limb movements) to generate new task-specific forms of coordination. Therefore, a theory of coordination to identify the restricted range of interlimb interactions is as important as obtaining insights into the way patterns of activity can be differentiated within a highly linked neural system, i.e., to dissociate constraints. The latter is not only a matter of selecting the appropriate action patterns but also of inhibiting or repressing unwanted or excessive motor activity. Shaping new forms of coordination requires a release from the constraints imposed by muscle synergies to meet the general principle of minimal expenditure of energy. Recoordination of action patterns sets in at an early age. In the newborn, reaching toward a visual target is initially accomplished by a tight coordination of elbow, wrist, and finger extension. This coordinative pattern evolves into a more differentiated organization of the joint movements in which the fingers can be flexed in anticipation of grasping an object while the arm extends; this is an example of intralimb dissociation. Nevertheless, dissociation of actions performed in different limbs can also be accomplished, as experienced musicians demonstrate. Imagine what would happen if you perform a horizontal elbow flexion movement in one limb with a flexion–extension–flexion movement in the other limb (like that shown in Fig. 1). Figure 6 shows the displacement–time traces of one subject. At the start of bimanual practice (bottom), a strong interlimb synchronization is shown by a tendency to perform a reversal pattern in both limbs. With increasing practice (top), control of both disparate patterns of activation is gradually mastered and interference subsides; in other words, the correct reversal pattern is made in the right limb with the smooth unidirectional movement in the other limb.

The present example illustrates the possibility (at least for some subjects) to overcome natural syn-

subject 3

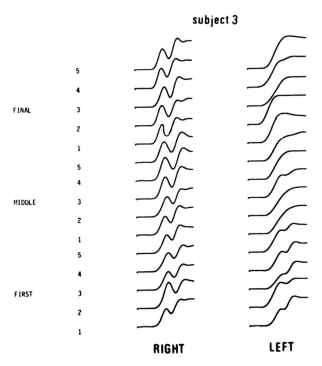

RIGHT   LEFT

FIGURE 6 Displacement–time patterns of simultaneously performed right and left upper limb movements with different spatiotemporal features. The first, middle, and final five trials of bimanual practice (75 trials) are shown to reveal the effects of training. At the start of practice (trials at the bottom), a strong tendency exists for both limb movements to interact, i.e., a discontinuity in the left limb movement can be observed instead of the required smooth flexion movement. At the end of practice (trials at the top), both limb movements are performed correctly and interference in the left limb is no longer apparent. (From Swinnen, S., Walter, C. B., and Shapiro, D. C. 1988, The coordination of limb movements with different kinematic patterns, *Brain and Cognition*, **8**, 326–347.)

chronization effects to meet new task requirements. How the dissociation of action constraints is accomplished has yet to be determined. In this respect, a working hypothesis is that inhibitory neural networks are to be recruited that serve to harness patterns of activation to limit their spread to prevent interference. Inhibitory neurons are of fundamental significance to the operation of the central nervous system. They contribute to shape neural networks and form the basis of action, as much as excitatory pathways do.

V. Hierarchical Control

At various occasions throughout the text, we have referred to the hierarchical nature of the motor control system. This implies that higher levels are mainly responsible for setting overall objectives and

plans. The details of these plans are filled in by progressively lower levels of the central nervous system. This does not imply that lower levels are only following instructions from higher centers. Instead, each level has its own responsibility and, whenever needed, communication can be set up with higher levels to achieve the task goal. At the lowest level, the spinal cord completes the translations from the language of intended movements to that of the muscles. The association cortex, responsible for designing plans and strategies, operates at the highest level of the motor hierarchy. A number of brain structures are located at the intermediate level: the brain stem, motor cortex, cerebellum, and putamen circuit of the basal ganglia. They deal with ways to carry out the intended movement plan. This particular arrangement allows the control system to produce the same effects over and over again but never in exactly the same way. [*See* Brain; Spinal Cord.]

VI. Motor Control Flexibility and Goal Accomplishment

Writing a computer program for analysis of one particular data set is often easy, but the task becomes much more elaborate when the program is to be used for various types of data. Making an adaptive program or any other device to meet general-purpose requirements is indeed a thorny problem. Flexibility and adaptability are interesting properties of the human motor control system, favoring goal-appropriate behavior through various means. At least two types, or levels, of movement plasticity have received some attention in recent years. One type concerns adjustments to environmental circumstances through a rescaling of certain movement variables without affecting the basic structure of movement. A second type of adjustment is more invasive and occurs very fast. The former one may be more preplanned through evaluation of initial conditions, whereas the latter may appear rather suddenly during movement, or whenever attainment of action intentions is in danger. Additional research may prove both to be fascinating levels of flexible goal-directed adjustments.

A. Goal Accomplishment through Parameter Specification

A favorite example to demonstrate that movement can be performed under a variety of circumstances

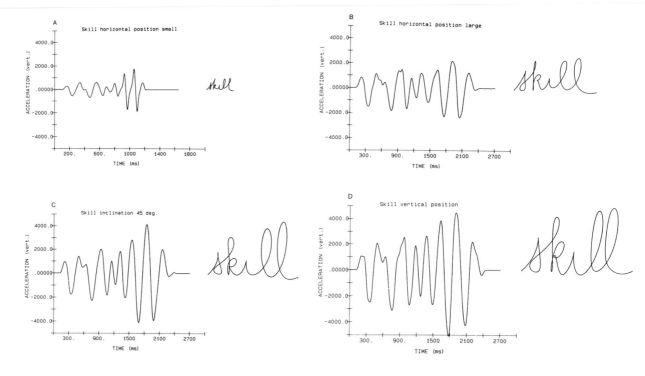

FIGURE 7 Vertical accelerations produced for writing the word "skill" on a digitizer positioned horizontally, with 45° inclination, and vertically. On the left, acceleration patterns are shown, which are roughly proportional to the forces produced in the vertical or upward direction. Note that the time basis of graph A is different from those of graphs B–D. The traces differ from each other in overall size, as shown on the right. Letter sizes in examples C and D are also different from example B, although total duration is similar.

without affecting its basic structure is that of handwriting. Handwriting skill is also a classic case for advocates of programmed control. If you do not believe that movements can run off in a clocklike (programmed) manner once initiated, try this: Write the words "motor control" at normal speed but try to omit the horizontal bar in "t." Although you intend to do so before movement initiation, modifying this particular aspect is difficult unless you slow down considerably. It leaves you with the impression that this action happens beyond voluntary control. But granted these letters seem to be almost rigidly strung together into one unit, what evidence is there in favor of motor output flexibility. Suppose that you write the word "skill" under different circumstances and with different overall sizes and speeds. The basic question arises whether each variation is to be considered as a different movement or as another version of the same movement structure. Current evidence tends to support the second of both possibilities. Figure 7 shows the vertical acceleration patterns generated for writing the word skill on a digitizing board positioned horizontally (small and large size), with 45° inclination, and vertically (as in writing on the blackboard), respectively.

As you can see (Fig. 7), the word skill is written in different overall sizes: The word presented in graph A is two to five times smaller than the word in graphs B–D. The total duration for the smallest word is half that needed for writing the larger words, which require approximately the same overall time. Leaving aside these parameter differences, similarities among these variations are apparent. They are reflective of existing task constraints as well as the individual's particular writing style. Thus, although the absolute values of vertical peak accelerations and decelerations differ among these written words, the general form of these acceleration patterns is similar even though this word has probably never been written before under these varying conditions. The forms are not, however, exactly the same, and the distribution of ratios among chosen landmarks would probably not be as fixed as in the example of Fig. 2. This is probably due to the unique way movement commands unfold on every attempt, whereby unpredictable environmental circumstances have to be met (such as occasional random frictions in the pen–paper medium). It reminds us that each movement is unique and is never made in exactly the same way as before.

Nevertheless, the consistencies or invariances across variations are noteworthy when realizing that the particular set of muscles used is unique and differs on each variation (due to shifts in writing posture, to differential effects of gravity, to the varying positions of the nonwriting hand, etc.). Consequently, the same program can be run off using entirely different muscle groups and with variations in overall size and timing. Whatever is represented in the program for writing this word must be coded in a rather abstract way and does not appear to include information about the particular set of muscles involved; this is the responsibility of lower levels of the nervous system. This elaboration of commands, from general to specific, illustrates the hierarchical nature of the motor control system. It marks once again the essence of the current motor programming perspective: Some features of movement are said to be represented in the motor program, leading to observable output invariances, whereas others can be easily rescaled to produce variations in movement (in this case, overall size and speed). Similar lines of evidence have also been advanced in the field of postural control. Different postural strategies are used for standing on different types of surfaces (e.g., ground, balance beam, board of a sidewalk). However, the proportionate contributions of various muscles in the synergies invoked are argued to be similar for a particular strategy. Again, overall changes in amplitude of postural adjustment are accomplished without modifying their relations one to another. Accordingly, a one-to-one relationship between the program and movement does not exist; rather, programs govern all movements that belong to the same class. R. A. Schmidt therefore uses the term "generalized motor programs."

In spite of the considerable research attention for varying and invariant features in movement, additional experimental evidence is desirable because a number of questions remain to be answered. One of them concerns whether some features of movement are indeed centrally represented. As noted above, motor programming theorists argue that the finding of invariances in the face of changes in countless other variables provides strong suggestions about the information residing in the motor program. Conversely, action theory does not wish to ascribe the resulting invariance and order in movement to higher-level representations; instead, it argues that order in movement emerges spontaneously from the system, in dynamical interaction with its environment. Another issue is that observable invariances at the peripheral level for a particular aspect of movement do not necessarily justify their inclusion in the motor program representation. Often, invariances are found for different but not fully independent variables such as amplitude, time force, etc. These may all result from one and the same higher-order variable that is centrally represented. In any case, the separation of invariant features and adjustable parameters represents an economical solution to human movement production in that movement can be tailored to particular environmental demands without compromising on its basic features. Furthermore, recent evidence points out that particular areas in the brain provide the structural framework for parametrization. Studies with Parkinson's disease patients hint that scaling of movement parameters is attributed to the intact functioning of the basal ganglia.

B. Goal Accomplishment through Response Reorganization: Evidence for Motor Equivalence

Human motor control is flexible in that the same goal can be reached through various ways, a phenomenon known as motor equivalence. For example, when brushing your teeth, you can move your arm or head or you can explore various ways to combine them. The example of writing skill under various board positions, as shown previously, also illustrates this feature. Even when environmental circumstances impede ongoing movement, the intended goal can often be accomplished by means of short-latency but set-dependent responses. Short-latency refers to the rapid character of these responses; set-dependence refers to their dependence on the performer's intention. Speech control research has been particularly productive in illustrating motor equivalence, although it has also been observed in fine-precision grip movements in humans and target-directed arm movements in primates. Mounting evidence indicates that rapid reorganization of movements involved in speech is possible to achieve articulatory goals. For example, when upward movement of the mandible is impeded in the process of saying something that requires a labial occlusion (as in "p" or "b"), the upper lip will increase its downward movement within 50–60 msec to achieve the articulatory goal. It is a case of feedforward open-loop control. These compensatory movements are unique in that they differ from voluntary responses that have much longer latencies; but they are also distinct from reflexes in that they are more adaptive in meeting the subject's goal and intentions. The details of the pathways subserving these adjustments have yet to be uncovered. The premotor cortex, cerebellum, and other structures are proposed to be involved. These short-latency responses indicate that higher-level parts of the control system are relatively unconcerned with the specific contribution of subsystems as long as the intended goal is met.

Recent concerns for the degrees-of-freedom problem on the one hand and for motor equivalence and output variability on the other illustrate that organizational and purposive issues in movement form basic tenets in current theorizing on motor behavior and movement control. Addressing such questions will eventually lead to a better understanding of the ingenious human motor control system.

Bibliography

Brooks, V. B. (1986). "The Neural Basis of Motor Control." Oxford University Press, New York.

Evarts, E. V. (1984). Hierarchies and emergent features in motor control. *In* "Dynamic Aspects of Neocortical Function" (G. M. Edelman, W. E. Gall, and W. M. Cowan, eds.). John Wiley and Sons, New York.

Ghez, C. (1985). Introduction to the motor systems. *In* "Principles of Neural Science," 2nd ed. (E. R. Kandel and J. H. Schwartz, eds.). Elsevier, New York.

Grillner, S. (1985). Neurobiological bases of rhythmic motor acts in vertebrates. *Science* **228,** 143–149.

Keele, S. W. (1986). Motor control. *In* "Handbook of Perception and Human Performance" (K. R. Boff, L. Kaufman, and J. P. Thomas, eds.). John Wiley and Sons, New York.

Kelso, J. A. S. (ed.) (1982). "Human Motor Behavior." Erlbaum, Hillsdale, New Jersey.

Rothwell, J. C. (1987). "Control of Human Voluntary Movement." Croom Helm, London.

Schmidt, R. A. (1988). "Motor Control and Learning. A Behavioral Emphasis," 2nd ed. Human Kinetics, Champaign, Illinois.

Whiting, H. T. A. (ed.) (1984). "Human Motor Actions. Bernstein Reassessed." North-Holland, Amsterdam.

Motor Skills, Acquisition

RICHARD A. SCHMIDT, *University of California, Los Angeles*

Glossary

Acquisition phase Practice phase of an experiment in which some practice variable is manipulated

Generalizability Product of practice enabling performance at task variations other than those actually experienced in acquisition

Generalized motor program Abstract representation for movement whose expression can be varied according to several parameters

Information feedback Augmented information about the outcome of an action with respect to its environmental goal

Motor learning Set of processes associated with practice or experience leading to relatively permanent changes in the capability for motor performance

Motor program Abstract representation that, when initiated, results in the production of a coordinated movement sequence

Retention (or transfer) test Test phase of an experiment with equated treatment conditions by which learning is estimated

THE ACQUISITION OF motor skill, or motor learning, involves the improvement in performance capability for various kins of movement skills, primarily as a function of practice or experience. The emphasis is on skills in which the learner is faced with the problem of how to produce a given movement that requires coordination and motor control (e.g., a novel gymnastics stunt, steering a vehicle), rather than on situations in which the learner must choose among several already learned movements (e.g., playing chess, baking a cake). Thus, motor learning involves the alterations in movement coordination with practice, and much of the emphasis is on the development of motor programs. Learning can also be manifested by modifications in the extent to which sensory information is used in performance, the selection of such information sources, and the involvement of reflexlike processes in motor control. The principles of motor learning are applicable to numerous real-world settings, such as in enhancing training effectiveness in industry, dance, music, or the military, the design of learning environments in educational settings and sport, and the enhancement of recovery from injury or stroke in physical therapy.

I. Fundamental Concepts and Definitions

Generally, when humans engage in practice of almost any movement task, they typically become more proficient in one or more ways, such as performing with less errors or more speed (or both), with greater smoothness and efficiency, with more effective coordination among the participating limbs, with less mental and physical effort, and so on. These gains are assumed to be a product of learning that has resulted in some way from this practice. As studied in the biological and behavioral sciences, motor learning is usually defined as a set of internal processes associated with practice or experience leading to relatively permanent gains in the underlying capability for performance. Therefore, the focus is primarily on the development, with practice, of an internal state (or capability) that provides a basis for skilled performance. However, such a state, as well as the internal processes that contribute to its strength, are generally invisible when examining human subjects, and so alterations in learning must be inferred from changes in performance using the special techniques described below. Fundamental concerns for this area of study involve (1) what processes occur during practice, or how practice "works," (2) how numerous variations in the conditions of practice influence the development of these capabilities (i.e., how practice variations influence learning), and (3) how the fundamental nature of these learned capabilities can be described and understood theoretically. [*See* LEARNING AND MEMORY.]

II. Methods in Examining Learning

A. Learning versus Performance Effects of Practice

It is useful to conceptualize two fundamentally different kinds of changes that occur when people practice some motor skill. First, many changes are more or less temporary in nature, due to momentary fluctuations in motivation, attention, mood states, fatigue, and the like. Many independent variables that can be applied to practice also have this kind of effect, such as motivating instructions, several drugs, fatigue, sleep loss, and others. These effects are termed performance effects, because of their relatively short-term influences on performance. Second, other kinds of changes are clearly more durable (they are relatively permanent) and can survive the imposition of rest periods or slight shifts in the nature of the practice conditions on a later test. A number of independent variables influence performance more permanently, such as feedback about success, and the organization and structure of practice. These latter changes in performance are thought to be due to learning—the development of an underlying capability for responding—and are usually termed learning effects as a result.

B. Transfer and Retention Tests of Learning

One fundamental problem in the study of practice is that learning and performance effects are usually confounded, or mixed, during the acquisition phase; i.e., when people improve performance in practice—or one kind of practice improves performance more than some other—whether (1) such changes are relatively permanent and, therefore,

due to learning, or (2) they are temporary and merely the result of performance effects, is usually not clear. Usually, the most important questions concern those variations that are relatively permanent, thus motivating a search for ways to unravel these two kinds of effects.

One such method of untangling the learning and performance effects is to use a retention or transfer test, two features of which are critical. First, after being treated differentially in an acquisition (or practice) phase, subjects are allowed to rest for a period of sufficient duration so that the temporary effects of the variable, if any, will dissipate. Second, subjects are then tested on the task, but all groups are now tested under the same conditions. Any differences on this test, conducted after the temporary effects of the independent variable have been allowed to dissipate, theoretically leave behind the relatively permanent effects (i.e., the learning effects of that variable). In this method, the performance levels on the acquisition phase are not considered, and the only indicant of relative amount learned between groups is the relative performance levels on the retention or transfer test.

C. Criteria for Learning

The method of transfer tests just described gives information about the effect of some variable in acquisition on a relatively short-term retention test, frequently after a few minutes or a few days. Although immediate retention is an important criterion, or goal, for learning, it is by no means the only one of importance. Another criterion involves the learners' capability to perform after very long retention intervals, and here the retention tests are given after months or even years of no intervening practice. Another involves generalizability, where the goal is to be able to apply what has been learned to some new task version or situation; here, the condition in acquisition judged best according to this criterion will be that which produces the most effective performance in the altered task. Another criterion involves performance under markedly altered environmental contexts, such as extreme stress, sleep loss, the requirement to perform another task simultaneously, and so forth. In all these cases, systematically different kinds of retention or transfer tests will be required to evaluate the effectiveness of the acquisition conditions in meeting these various criteria for learning. Often, those conditions that maximize learning as measured by one criterion will not maximize learning as measured on another. There are even situations where maximizing learning on one criterion reduces learning on another of them (see Section III.D).

III. Principles of Learning and Underlying Theoretical Bases

In the sections that follow, several of the most important variables that have been found to influence the learning process are discussed. In each case, theoretical bases for these effects are discussed.

A. The Role of Practice

Without doubt, the single most important variable that determines learning is practice or experience. This is obvious in many real-world settings—especially in tasks where physical demands of strength and stamina are not critical—where industrial workers (typists, Morse code receivers) and sports–games performers (gymnasts, dancers) achieve high levels of performance only after many years of intensive practice. Connected to this idea is the notion that learning never really ends, as evidence indicates continued improvement in industrial skills after literally millions of practice trials. To be sure, these improvements at high levels of proficiency are quite slow, in contrast to the rather rapid improvements seen in earlier stages of practice.

This phenomenon of rapid improvements early in practice and more gradual ones later is one of the most general principles of motor (and cognitive) learning and has become known as the law of practice. For a task in which the time to complete an action (T) is to be minimized, the principle states that $T = aP^{-b}$, where a and b are constants, and P is some measure of the amount of practice received. This is a power function, where the rate of decrease of T is large for small values of P, with smaller rates of decrease as P increases. Mathematically, this principle implies that a plot of the logarithm of performance (T) against the logarithm of practice (P) will be linear, with a slope of $-b$. Analogous power functions can be derived for tasks in which the performance score increases numerically with practice. Such descriptions have been used in many tasks, and in some cases the performance curves are approximated somewhat more accurately by two- or three-term power functions, or by exponen-

tial equations, with different rate constants for different phases of practice.

In any case, these various mathematical descriptors fit performance curves with practice relatively well, and all describe the common observation that improvements with practice are rapid at first and continue at a slow rate even after years of experience. It is remarkable how many different kinds of tasks—ranging from very cognitive decision-making tasks to movement coordination tasks with extensive muscular involvement—follow this basic principle, and the law of practice is the most fundamental in the field as a result.

Whereas the amount of practice and experience are obviously critical for learning, various ways of modifying practice can make learning more rapid and effective, as well as enhancing one or more of the learning criteria discussed earlier. Some of the most important of these practice effects are discussed next, together with theoretical accounts that seem to explain them most effectively.

B. Ordering of Tasks To Be Practiced

An important determinant of learning is the ordering, or scheduling, of practice experience among several tasks to be learned. Suppose that three different skills (tasks A, B, and C) are to be acquired in a practice session. A traditional practice organization involves what is termed blocked practice (similar to a drill), where all of the experience on task A is completed before shifting to task B, which is completed before moving to task C. Random practice, on the other hand, involves practicing tasks A, B, and C in a jumbled order, perhaps without ever practicing the same task on two consecutive trials. Considerable evidence shows that, compared with blocked practice, random practice degrades performance during acquisition (Fig. 1), with a slower rate of improvement and poorer performance by the end of the practice period; however, when learning is evaluated on retention tests, the subjects that had random practice in the acquisition phase generally perform better than subjects who had blocked practice in acquisition. This benefit is very large when the retention test is under random conditions (circles in Fig. 1) and much smaller when the test is under blocked conditions (squares in Fig. 1). This evidence shows that, relative to blocked practice, random practice degrades performance during practice but aids learning as measured on retention tests.

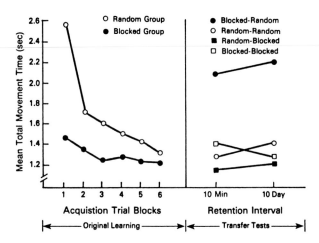

FIGURE 1 Performance on a complex movement speed task under random and blocked conditions in acquisition (left) and for retention tests under random and blocked conditions (right). Regardless of the retention test conditions, the random conditions in acquisition were always more effective for long-term retention. [Adapted from Shea and Morgan, 1979.]

These advantages for random practice have been shown in a variety of laboratory and real-world tasks. However, one apparent exception is that, in the earliest phases of practice, when the subjects are just acquiring the basic idea of the task, blocked practice does appear to have advantages for learning. But soon thereafter, random practice becomes more effective for learning than blocked practice, and it appears to be so at intermediate and advanced practice levels. These effects seem to occur if repititions of a task are separated by other tasks to be learned, or even if separated by other tasks that are intended as distracters (e.g., various mental or motor tasks). The benefits of random practice are *not* associated with the fact that learners might be uncertain about the next task to be performed, because the effect is present even when the subjects can predict the order of practice in advance; rather, the effect is probably due to the fact that learners must change their performance on each trial, with random practice preventing drill-like repetitions. These principles seem analogous to effects seen in experiments on verbal learning discussed under the heading of "spacing effects," where the number of items intervening between a given item to be learned (e.g., a vocabulary word) and its repetition is studied. Here, more spacing of practice, up to a point, enhances learning of simple and more complex verbal materials.

These advantages of random over blocked practice have been understood by two closely related theoretical ideas that have strong implications for how the processes underlying practice are conceptualized. First, a forgetting hypothesis holds that, in blocked practice, repeated attempts at one task allow the learner to use the same "solution" to that movement problem (analogous to a solution to a mental arithmetic problem) on successive trials, which tends to maximize performance in acquisition. In random practice, on the other hand, learners who must switch from one task to the next on successive trials tend to forget the "solution" of a given task to generate the solution for the next, making performance during acquisition less effective. Thus, every time a task is attempted, its solution must be generated again. This generation process is held to be critical for learning, with random practice requiring more generation attempts and blocked practice allowing the learners to avoid generation.

A second view is a depth of processing hypothesis, in which randomized practice forces learners to process information about the tasks with greater elaboration and distinctiveness—so-called "deep" processes in memory. Evidence shows that subjects practicing under random (as opposed to blocked) conditions tend to use more elaboration (connecting the task features to previously learned materials), such as relating the shape of the movement pattern to a known letter or geometric shape. Random subjects also learn the movements' distinctive features (features of the patterns that highlight their similarities and differences), such as noting that one movement is a mirror image of another. Subjects under blocked practice apparently process the task information more superficially because they are not forced to deal with task differences on consecutive trials.

These views are paralleled by the notion of transfer-appropriate processing in more cognitive task situations. This viewpoint holds that, for a given test of retention, a condition in acquisition will be effective to the extent that it stresses the processes that are appropriate for the retention performance. Thus, for performance in most tasks, the learner must generate the solution to the movement problem on demand from the environment, and frequently this performance is required only once, with opportunities for further repetitions or corrections being postponed for several minutes, during which time other actions are being produced. Thus, random practice facilitates this solution-generation process that is appropriate to these kinds of retention tests. This can also be discussed in terms of retrieval practice, where repeated, but spaced, recall attempts are effective for learning because they force practice at retrieving the solution.

Thus, similar viewpoints from several different kinds of learning paradigms have converged more or less independently to suggest that something like solution-generation is a critical feature in skill learning. This thinking has shifted the focus from earlier views of practice that stressed drill and repetition to views that focus more on the processes associated with movement preparation. As we shall see later, this theme runs through several other important areas of research in skill learning.

C. Scheduled Variations in Practice

Another important practice variable concerns the intentional, scheduled variations in a given task to be learned. In this paradigm, a single task is varied along some dimension (e.g., the distance a ball is to be thrown, the speed of an object to be caught, etc.), and the learners experience a variety of task variations during acquisition. (This is distinct from the random-blocked experiments just discussed, where several different *tasks* are experienced; variable practice involves variations among versions of the same task.) In the experiments, learners receive either variable practice of a set of task variants or constant practice with just one of them, and then learning is evaluated on a novel variant of the task in a later test. Thus, these experiments have focused on the capability to apply what is learned to variations of the task that have not received previous practice, and thus have emphasized the generalizability criterion of learning discussed earlier.

The results from many experiments in this area generally show that, as compared with constant practice, learners experiencing variations in practice perform more effectively on a novel retention test. Figure 2 has one such example, where subjects received either a variety of speeds of stimulus display (5, 7, 9, or 11 mph) in a coincident-timing task or just a single one of them, with the novel transfer tests being at 1, 3, 13, and 15 mph—outside the range of prior experience. There was an advantage for the variable practice condition, and this advantage became stronger as the transfer speeds deviated more from the speeds experienced in the acquisition phase. These general effects of variable practice have been shown in several different kinds

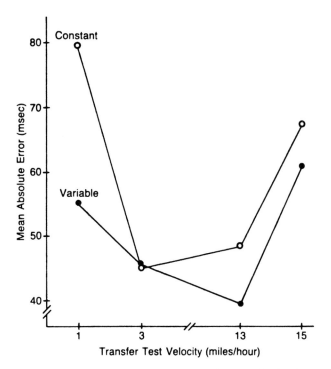

FIGURE 2 Mean performance error on four novel stimulus speeds in a coincident timing task as a function of the variability of practice conditions in acquisition. Variable practice aids generalizability to novel task versions. [Adapted from Catalano and Kleiner, 1984.]

of tasks in adults. Using children as subjects, the effects are even larger in favor of the variable-practice conditions. One investigation using 8-yr-old subjects even shows that variable practice of versions A, B, D, and E is more effective for transfer to an intermediate version C than is practice at version C itself.

These kinds of results have been important in understanding processes involved in practice. First, they tend to argue against so-called specificity viewpoints for learning, which hold that only the particular task versions, or stimulus-response combinations, that are actually experienced in acquisition are learned. These theories cannot explain why novel retention tests can be performed effectively and, more importantly, why variable practice is more effective for these novel retention tests than constant practice. These specificity views are analogous to so-called instance-based theories in visual perception and category-identification tasks, where transfer to novel versions of a display is based on a kind of interpolation between those stimuli that have actually been experienced. While there is a healthy debate about whether or not perceptual-learning and category-identification data can be accounted for by these theories, these notions cannot account for the available evidence on motor learning. Perhaps this is due to the fundamental nature of motor control, where versions of a class of task (such as throwing) seem to lie along relatively simple physical continua (e.g., speed). Analogous continua are not so easily identified in concept identification. [*See* MOTOR CONTROL.]

Rather, evidence favors the notion that people learn classes of actions (e.g., throwing, signing one's signature), these classes being based on abstract memorial structures (termed generalized motor programs) with relatively rigidly defined temporal and structural features. These structures are analogous to central pattern generators that have been identified for locomotion, chewing, and other more genetically defined actions in animals, and are probably based on networks of neurons that operate as a unit to produce the basic elements of the actions. Generalized motor programs can be "run off" at retention using parameters that specify relatively global features of the action, such as the overall amplitude of handwriting, the distance a ball is thrown, or the speed of a spoken word. According to these ideas, subjects learn relationships (called schemas) between the particular parameter values used in practice and the outcomes of the resulting actions. When a novel variant of an action is required, the performer uses the schema to select the parameter that will meet the environmental goal, and then the generalized motor program is executed with that parameter value to produce the novel behavior. Furthermore, variable practice with a variety of parameter values is held to produce more effective rule development than constant practice, so subjects having received variable practice are more accurate in parameter selection, and thus perform more effectively on the novel retention tests as a result.

These results, together with analogous findings from motor control, suggest answers to questions about what is learned during practice. First, learners must acquire the generalized motor programs, or the fundamental aspects of the movement organization. Second, learners must acquire the capabilities to parameterize the program so that the movement output is scaled to the environmental demands. Research has mainly been focused on the second of these problems with respect to schema learning, and little is known about how generalized motor programs for classes of actions are acquired.

Finally, this is not to suggest that these are the only things learned with practice. Much available evidence points to the more effective use of sensory information with practice, more rapid and less attention-demanding decision-making and perceptual analysis, more effective strategies for the selection of movements, and several other processes that result in more effective performance with practice.

D. Information Feedback for Skill Learning

It is well known that the nature and quantity of information about response success—generally termed feedback—is critical for learning. Since early in this century, it has been recognized that if a task was structured so that the feedback inherent in movement production (termed intrinsic feedback, or response-produced feedback) was not sufficient for learners to detect errors on their own (e.g., as in blindfolded limb-positioning tasks), then learning would not occur with practice unless the experimenter provided this information with artificial (extrinsic, or augmented) feedback, usually termed "knowledge of results" in the laboratory. This work suggested that some form of feedback information (either intrinsic or extrinsic) is essential for learning, and this notion remains viable today. For the motor learning that is frequently seen in natural environments, however, the learner can usually detect errors through intrinsic feedback, and this is usually sufficient for effective learning to occur.

For those interested in maximizing learning in various training settings, an important question concerns how feedback can be used artificially by an instructor to speed learning or to facilitate retention or generalization. Over the past few decades, it has been repeatedly shown that, for performance in acquisition, nearly any variation of extrinsic feedback that makes the information more frequent, more precise, more immediate, more informationally "rich," or more "useful" is effective for enhancing performance. Generally, these variations in extrinsic feedback steepen the slopes of performance changes across practice, result in more effective performance at the end of the practice, and apply to a wide variety of movement tasks. This is not surprising: In addition to the motivating and "energizing" effects of feedback presentations, feedback gives information about errors in movement; the learner then corrects these errors on subsequent attempts, which quickly drives the behavior toward the goal state and holds it there throughout practice. However, extrinsic feedback does not always have such potent effects on performance, particularly in tasks for which intrinsic feedback is sufficient to signal errors to the performer. In such cases, intrinsic and extrinsic feedback are redundant, so that the added extrinsic information does not provide much additional benefit over and above the intrinsic information already present.

Because of the marked effects of feedback variations during practice, these lines of investigation erroneously led researchers to the conclusion that all of these feedback manipulations necessarily enhanced learning. This implied that feedback information should be maximized in many real-world training programs, in the design of simulators (e.g., for pilot training), in teaching machines, in computer-aided instruction, and in many other situations; these suspicions have not been found to be correct in several situations. The explanation is that such changes in performance in an acquisition phase are not, as discussed earlier, necessarily due to changes in learning. In fact, recent work suggests that variations of feedback in training that provide less information, and that at the same time retard somewhat the improvement during practice, result in more effective learning. These effects seem particularly strong when measured on long-term retention tests, suggesting that these variations in feedback might act to block forgetting in some way.

A number of feedback variations for practice have this general property. In one, the feedback can be given in a bandwidth format, where errors are signaled only if the learner's movement falls outside some arbitrarily chosen band of correctness. Here, compared with feedback after every trial, the learner receives feedback on only a proportion of practice trials (the incorrect ones). Another variation of feedback is summary or average feedback, where information is withheld for several trials and then provided to the learner either in a graph describing each of the previous trials in the set or as an average value that describes the entire set; or, feedback can be simply withheld on certain trials, as was the case for the data in Figure 3. Here, subjects either received feedback after every trial or after 50% of the trials in acquisition; furthermore, the feedback was "faded" across practice, with more frequent presentations early in practice and less frequent presentations later. Figure 3 shows that, although performance in acquisition was retarded slightly for the 50% group, performance was better on the 10-min and particularly on the 2-day reten-

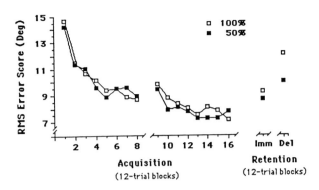

FIGURE 3 Mean root mean squared (RMS) error on a complex limb-patterning task for groups with 100 and 50% feedback in acquisition (left), and for performance on no-feedback retention tests. Reduced feedback frequency in acquisition enhanced long-term retention. [Adapted from Winstein and Schmidt, 1990.] Del, delayed; Imm, immediate.

tion tests; i.e., reducing the amount of feedback presented in this example—and in the other examples mentioned above—enhanced long-term retention of the skills.

This kind of result is contrary to the notion that variations of feedback that enhance performance should necessarily enhance learning and has challenged many earlier viewpoints about how feedback operates. The most important challenge has been to Thorndike's Law of Effect—which holds that trials with feedback are all-important for learning to occur, and trials without feedback are "neutral" and do not promote learning—as well as to the many theories that used Thorndike's views as a starting point.

Instead, the data point to a two-factor view of feedback termed the guidance hypothesis. This view says that frequent feedback has two simultaneous kinds of effects in practice—one positive and several negative ones. The positive effect is the well-known tendency for feedback to guide performance toward the task goal by providing information about errors. However, one negative effect is that the subject may come to rely on this powerful effect of feedback for performance, using it as a kind of "crutch" to maintain performance at a very high level during practice. This can result in the learner failing to process various proprioceptive or task cues that are critical for long-term learning. A second negative effect results from the old finding that giving feedback after a particular trial causes learners to change behavior, whereas withholding it fosters no change in behavior. Performance tends to be more stable in practice when less feedback information is given, generating increased movement consistency. The result is that frequent feedback is not optimal for learning because of the balance between these positive and negative factors.

If this guidance view proves to be a reasonable account of feedback's role in learning, it suggests a number of modifications in the ways that feedback is used in many practical situations. In particular, feedback should probably not be given very frequently, except for early in practice when the learners must acquire the fundamental movement patterning. Feedback should probably also be withdrawn gradually later in practice to prevent the subject from developing a reliance on it. Feedback that summarizes several movement attempts also appears effective, perhaps because it provides more stable measures of performance patterning. Recent data suggest that instantaneous feedback degrades learning—contrary to common-sense perspectives—perhaps by interfering with the subject's capability to detect their own errors; this suggests that feedback should be delayed for a few seconds in practice.

Finally, recent research has begun to ask about kinematic and kinetic feedback—about the subject's movement patterning and force patterns, respectively—rather than feedback only about the movement's outcome in the environment. This feedback is analogous to the instructor-provided information typically seen in training settings. One would suspect that this kind of feedback would be highly effective, but this suspicion was not verified until recently. Many of the principles of feedback discussed in relation to the guidance hypothesis for knowledge of results (extrinsic feedback) seem to hold also for kinematic and kinetic feedback—such as less frequently presented feedback—suggesting that these forms of feedback may operate with similar mechanisms.

E. Several Other Variables in Practice

In this final section, several other variations in practice that have been found to influence the learning process are briefly mentioned. More on these variables, as well on the other topics in this chapter, can be found in the references listed at the end.

The large number of separate ways that skills can be presented to learners prior to practice can have a large role in skill learning. Skills can be demonstrated with films or videotapes, and live demonstrators can be used effectively at all stages of practice. The processes associated with observational learning—or learning through viewing another's performance—are very powerful in early practice, yet they remain relatively poorly understood. Verbal instructions should probably be kept simple, especially with younger learners, due to the difficulties in information processing and the problems associated with describing complex movement patterning in words. Giving learners experience with the stimuli to be experienced in the actual task, termed verbal pretraining, is useful for facilitating early proficiency. Descriptions of the physical principles underlying complex machinery has been found to be useful for learning.

During practice, learners can be directed toward the correct action, or prevented from making errors, by a collection of methods called guidance. Both physical and verbal guidance is useful in preventing costly or dangerous errors. But it can be easily overdone because of the dominance it can have over the learner's natural movement patterns, and guidance procedures often do not contribute very effectively to learning. Mental practice, or imagery, can be used effectively as a companion to physical practice, and the evidence suggests a strong role of this kind of activity, particularly in early learning when the more conceptual elements of the task are being acquired. Procedures in which the learner is encouraged to estimate their own errors prior to receiving feedback can be helpful for learning, probably because they enhance the development of error-detection processes, wherein the learner can evaluate his or her own errors after practice when feedback from an instructor is not present. Finally, even though practice can result in fatigue, these processes are not seriously damaging to learning; careful adjustment of the practice and rest schedules during acquisition, based on the fatigue-producing properties of the tasks to be learned, can be used to optimize the structure of the practice session for learning.

Bibliography

Adams, J. A. (1987). Historical review and appraisal of research on the learning, retention, and transfer of motor skills. *Psychol. Bull.* **101,** 41–74.

Lee, T. D. (1988). Transfer-appropriate processing: A framework for conceptualizing practice effects in motor learning. *In* ''Complex Movement Behaviour: 'The' Motor-Action Controversy'' (O. G. Meijer and K. Roth, eds.). North-Holland, Amsterdam.

Salmoni, A. W., Schmidt, R. A., and Walter, C. B. (1984). Knowledge of results and motor learning. A review and critical reappraisal. *Psychol. Bull.* **95,** 355–386.

Schmidt, R. A. (1975). A schema theory of discrete motor skill learning. *Psychol. Rev.* **82,** 225–260.

Schmidt, R. A. (1988). ''Motor Control and Learning: A Behavioral Emphasis,'' 2nd ed. Human Kinetics Publishers, Champaign, Illinois.

Movement

ANTHONY J. GAUDIN, *California State University, Northridge*

I. Relationship between Skeletal Muscles and Joints
II. Movements of the Skeleton

Glossary

Action Specific movement accomplished by a muscle or group of muscles.
Fulcrum Pivot point around which a lever moves.
Insertion Place of attachment of a muscle that moves when the muscle contracts.
Isotonic contraction Muscle contraction that results in shortening of the muscle fibers.
Lever Solid rod that moves around a pivot point.
Origin Place of attachment of a muscle that does not move when the muscle contracts.

MOVEMENT IS THE ACT of moving, the changing of place, position, or posture. Movement involves a cooperative action between bones and muscles. Most of the bones of the skeleton are connected to one another by articulations or moveable joints. Contraction of the muscles connected to the bones causes movement at these joints. This action provides for movement of the body from one place to another and manipulation of objects in the environment.

I. Relationship between Skeletal Muscles and Joints

A. Isotonic Contractions

Movement of the body and its parts is accomplished by isotonic muscle contractions. During isotonic muscle contraction, chemical energy is released while the longitudinal movement of filaments within the individual muscle fibers causes the muscle to shorten. Practically all skeletal muscles that cause movement are attached to two bones across a movable joint. When the muscle shortens, it moves one of the two bones attached to it. The site of attachment that does not move, or moves least is the *origin* of the muscle, and the place of attachment that moves is the *insertion*. [*See* ARTICULATIONS, JOINTS BETWEEN BONES; SKELETAL MUSCLE.]

B. Levers

Movements of the skeleton may be compared to movements of simple levers. A *lever* is a rigid rod that moves around a pivot point, the *fulcrum*. A force, or *effort,* is required to move the lever, and the weight it must move is the *resistance*. In skeletal movements, a bone acts as a lever moving around a joint that acts as a fulcrum. The strength required to move the bone is provided by the isotonic contraction of a muscle attached to the bone. The resistance in this case is the weight of the bone.

In lifting the arm away from the body (Fig. 1), the lever is the humerus, the fulcrum is the shoulder joint, the effort is provided by contraction of the deltoid muscle, and the resistance is the weight of the whole arm. To increase the resistance, the person could perform the same action while holding an exercise dumbbell. An increased resistance requires a greater effort; this provides the basis for physical fitness programs involving weight lifting.

Three classes of levers are recognized. In a *first-class (class 1) lever,* the fulcrum is located between the point where the force is applied (*lever arm*) and the point of resistance (*resistance arm*) (Fig. 2a). Common examples of first-class levers include a seesaw and a pair of scissors. A first-class lever is used in the body to extend the forearm at the elbow joint. The effort is applied by the triceps muscle, the

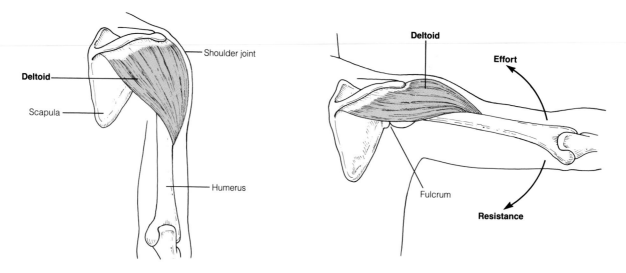

FIGURE 1 Movement of the arm away from the body. When the deltoid muscle contracts, it exerts a force (effort) to move the humerus (the resistance) away from the body at the shoulder joint (the fulcrum). From A. J. Gaudin and K. C. Jones, "Human Anatomy and Physiology," p. 210, Harcourt Brace Jovanovich, San Diego, California, 1989.

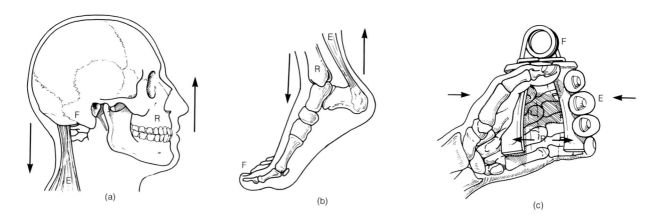

(a) (b) (c)

fulcrum is the elbow joint, and the resistance is the hand or something held in the hand. The actions of throwing a baseball, throwing a forward pass in football, or throwing a javelin are all examples of activities accomplished through the use of a first-class lever.

In a *second-class (class II) lever,* the resistance or weight is located between the fulcrum and the point where the effort is applied (Fig. 2b). Examples of second-class levers include a wheelbarrow and a door hinge. A person uses a second-class lever to stand on his or her tip-toes. In this case, the fulcrum is the ball of the foot, the resistance is body weight,

FIGURE 2 Examples of how levers operate in moving parts of the body. (a) First-class lever; (b) second-class lever; and (c) third-class lever. E = effort, R = resistance, F = fulcrum. From A. J. Gaudin and K. C. Jones, "Human Anatomy and Physiology," p. 211, Harcourt Brace Jovanovich, San Diego, California, 1989.

and the effort is applied by the calf muscles pulling on the heel. Any kind of running, hopping, or jumping activity involves use of leverage gained through this second-class lever.

In a *third-class (class III) lever,* effort is applied between the fulcrum and the resistance (Fig. 2c). A tweezers (forceps) is a common example of a third-class lever. Most of the muscles act as part of third-class levers. For example, flexing the forearm at the elbow joint involves a third-class lever. The elbow joint is the fulcrum, the resistance is the hand, and the effort is applied by the biceps muscle. Performing chin-ups on a horizontal bar also involves a third-class lever.

II. Movements of the Skeleton

Movements of the skeleton are limited to the specific types of movement allowed by the synovial joints in the body. Synovial joints contain a fluid-filled cavity between two bones that facilitates movement by reducing friction. However, the exact nature and extent of the movement of each joint depends on several factors. Ligaments and muscles restrict and direct movement at a joint. For example, raising the arm, as in reaching for a book at the top of a bookshelf, is restricted by muscles from the chest and back that attach to the humerus and limit the ability to reach. Various forms of arthritis, or inflammation of the joints, may inhibit movement. Finally, the mere presence of soft tissues often limits the degree of body movement. For example, muscles of the calf and thigh press against one another when the knee bends, thereby limiting the extent to which a knee joint can be bent. "Muscle-bound" athletes are an example of individuals who may suffer from restricted movement due to the excessive development of certain muscles.

Several specific joint movements are easily recognized and have been designated with descriptive names. These are shown in Fig. 3 and described here.

Flexion is a movement that decreases the angle between two bones. The most common examples of flexion are bending the elbow and knee joints, and closing the fingers in a clenched fist.

Extension, the opposite of flexion, increases the angle between two bones. Straightening out the arm and leg at the elbow and knee joints are examples of extensions. Flexion and extension at the ankle joint have special names. Movement of the foot in a downward direction is *plantar flexion* and movement of the foot upward is *dorsiflexion.*

Hyperextension is overextension of two bones at a joint beyond a 180° angle, as in bending the head

backward. Due to sexual differences in the amount of mineral matrix deposited in the skeleton, females can usually hyperextend their elbow, while males usually cannot.

Abduction is movement of a body part away from the midline. Lifting the arm at the shoulder joint out from the body is abduction of the arm. In the case of the fingers and toes, the midline is in reference to the arm and leg respectively. Consequently, abduction of the fingers and toes is spreading them.

Adduction is the opposite of abduction, that is, movement toward the midline. Lowering an outstretched arm is adduction. Adduction of fingers and toes involves bringing the outstretched digits back together.

Rotation is movement of a bone around its own axis, or around another bone. Turning the head from side to side is rotation of the head and atlas around the odontoid process of the axis. One can rotate the humerus in the glenoid cavity by twisting the entire arm. Rotation of the head of the femur is also possible in the acetabulum. The most common type of rotation involves the forearm where the radius rotates freely around itself and the ulna to turn the hand and wrist. This movement has special names: *supination* is the rotation of the forearm so that the palm faces forward, and *pronation* is turning the palm to the rear.

Circumduction is a complex movement. It involves abduction, adduction, flexion, extension, and some rotation. The movement is usually described as circumscribing a circle in midair. Two children turning a long jump rope for another child jumping in the middle execute this movement with their shoulder joints.

Inversion and *eversion* apply to movements of the foot at the ankle. Inversion turns the foot so the sole faces inward (medially). Eversion is turning the foot so that the sole faces outward (laterally). A person's first experience on ice skates is usually a long and painful series of inversions and eversions of the ankle.

Protraction is moving a body part straight out and away from the midline parallel to the ground. Two examples of protraction are jutting the chin out and sliding the scapulae out away from the vertebral column, as in "rounding" the shoulders forward.

Retraction is the opposite of protraction and involves return of the body part straight back toward the midline.

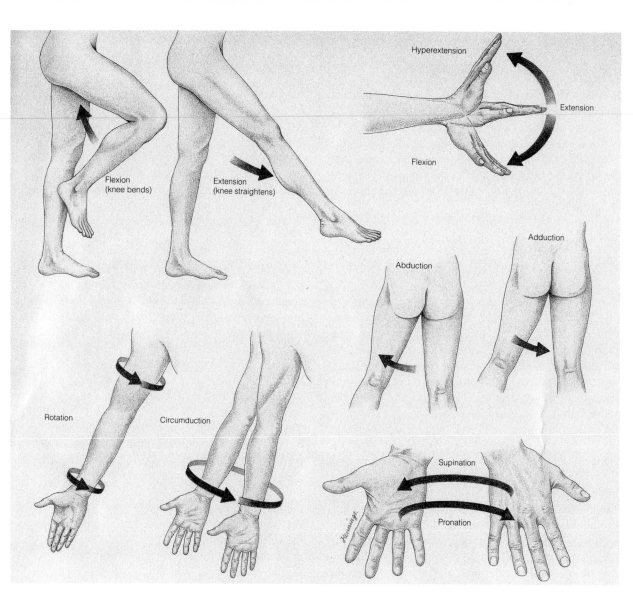

Hyperextension

Extension

Flexion

Flexion
(knee bends)

Extension
(knee straightens)

Abduction

Adduction

Rotation

Circumduction

Supination

Pronation

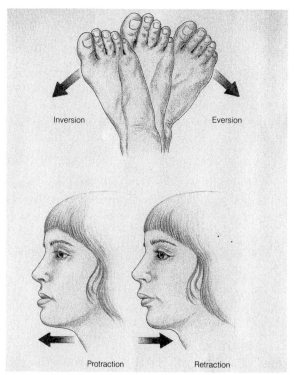

Inversion

Eversion

Protraction

Retraction

FIGURE 3 The movements of the body. From A. J. Gaudin and K. C. Jones, ''Human Anatomy and Physiology,'' p. 172–173, Harcourt Brace Jovanovich, San Diego, California, 1989.

Bibliography

Chapman, C. B., and Mitchell, J. H. (1965). The physiology of exercise, *Scientific American* **212,** 88–96.

McArdle, W. D., Karch, F. I., and Katch, V. L. (1986). ''Exercise Physiology.'' Lea & Febiger, Philadelphia.

Nadel, E. R. (1985). Physiological adaptation to aerobic training, *American Scientist* **73(4),** 334–43.

Rasche, P. J., and Burke, R. K. (1978). ''Kinesiology and Applied Anatomy: The Science of Human Movement,'' 6th ed. Lea & Febiger, Philadelphia.

Rosse, C., and Clawson, D. K. (1980). ''The Musculoskeletal System in Health and Disease.'' Harper & Row, New York.

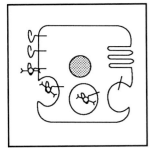

Mucosal Immunology

JERRY R. McGHEE, *University of Alabama at Birmingham*

Glossary

IgA-committed cells B cells in germinal centers of IgA-inductive sites (Peyer's patches), which express the isotype IgA on their membrane (surface IgA-positive; s-IgA$^+$)

Mucosal-associated lymphorecticular tissues Organized lymphoid regions from bronchus-associated and gut-associated lymphoreticular tissues, which take up inhaled or ingested antigens for initial induction of mucosal immune responses

Peyer's patches Major IgA inductive sites in the gut-associated lymphoreticular tissue of mammals where antigen-sensitive B and T cells for subsequent immune responses occur

Secretory IgA Polymeric IgA with J chain is transported across epithelial cells via a specific receptor, termed secretory component, and is released onto the mucosal surface as secretory IgA antibodies

Since its discovery 30 years ago by Joseph Heremans, immunoglobulin A (IgA) and the mucosal immune system has assumed an important position in contemporary immunology. Lars Hanson (1961) showed that IgA is the major isotype in human colostrum and milk, and serum IgA largely consists of a single Ig molecule (monomeric), while external secretions are enriched in polymeric IgA antibodies with distinct features. However, as with much of immunology, we can trace local immunity to Paul Ehrlich, who showed in two papers published in 1891 that guinea pigs orally immunized with the plant toxins abrin and ricin were protected from

conjunctival challenge, whereas toxin-treated control animals developed a severe necrosis of the eye. As briefly summarized below, the plant toxins in Ehrlich's experiments presumably provided protection by stimulating the gut-associated lymphoreticular tissues (GALT) to induce antigen-sensitized cells that subsequently migrated to the lacrymal glands, where they differentiated into plasma cells with synthesis of protective secretory IgA (S-IgA) antitoxin antibodies. This chapter describes the known structures, mechanisms, and functions of the mucosal immune system.

I. The Mucosal Immune System

A. IgA Structure and Distribution

The human mucosal immune system protects more than 400 m^2 of mucosal surface (compared with 1.8 m^2 for skin). Because the major first line of defense against environmental antigens is S-IgA antibodies, vast quantities of these antibodies must be produced. Most IgA is produced at local sites, especially in the gastrointestinal tract. In fact, 65–90% of immunoglobulin (Ig)-producing cells in mucosal tissues are of the IgA isotype, and the human intestine has $>10^{10}$ Ig-producing cells per meter. It has been estimated that approximately 66 mg/kg/day of IgA is made, which is >60% of all Ig produced [*See* IMMUNE SYSTEM.]

Two IgA subclasses occur in humans: IgA1, which predominates in serum (>90%) derived from plasma cells present in bone marrow, spleen, and lymph nodes and is usually monomeric in nature, and IgA2, which is proportionally greater in external secretions and generally comprises 30–50% of the total IgA present. Most s-IgA (>98%) found in external secretions is locally produced by plasma cells in these sites. Monomeric IgA is composed of

IgA Inductive Sites

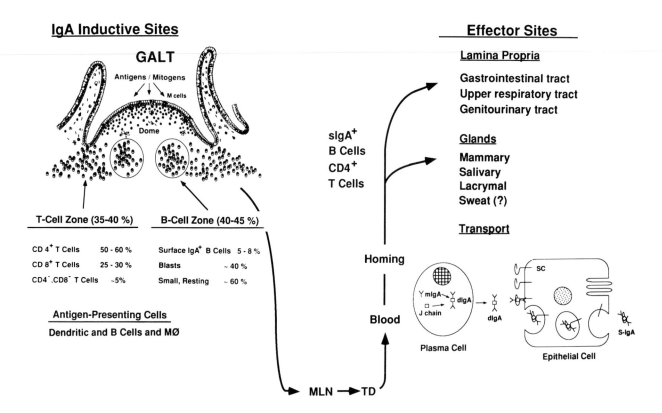

Effector Sites

FIGURE 1 The role of inductive sites for mucosal IgA responses. Antigen uptake by M cells occurs in GALT and BALT (MALT) and results in the initial induction of the immune response. Antigen-sensitized, precursor sIgA+ B cells and CD4+ Th cells in GALT leave via efferent lymphatics and migrate to mesenteric lymph nodes (MLN) and then into the thoracic duct (TD) to reach the bloodstream. These migrating cells enter the IgA effector sites, where terminal differentiation, synthesis, and transport of S-IgA occurs. This induction in MALT and exodus of cells to effector sites is termed the common mucosal immune system. Mϕ, macrophage.

four polypeptide chain subunits; two alpha heavy chains and two light chains (either kappa or lambda), and each alpha heavy chain has one variable region domain and three constant (C) region domains (C_H1, C_H2, and C_H3). Polymers of IgA are formed in plasma cells when a terminal extension of C_H3 with an extra cysteine residue links monomeric subunits together via disulfide bridges. Polymeric IgA usually occurs as dimers; however, tetrameric IgA is present in significant amounts in external secretions. J chain (discovered by Jiri Mestecky, Marian Koshland, and their respective colleagues) is a 15.6-kDa peptide associated with polymeric IgA, which enables it to bind to membrane secretory component (SC), the epithelial cell transport receptor (Fig. 1). The typical 11s IgA found in external secretions contains two monomeric subunits: one J chain and one SC.

B. Transport of Polymeric IgA

Per Brandtzaeg and others have shown that transport of polymeric IgA (and IgM) into secretions is mediated by SC, which is synthesized by and expressed on the surfaces of epithelial cells in the gastrointestinal and respiratory tracts, acinar and ductal cells found in exocrine tissue, and cells lining the

uterus (Fig. 1). In the small intestine, SC is found primarily in the columnar cells of the crypts of Lieberkühn while it is also present on surface epithelial cells in the large intestine. In rats, mice, and rabbits (but not in humans), SC is expressed on the hepatocyte membrane and constitutes an additional receptor and transport mechanism for the secretion of polymeric IgA (present in high amounts in serum of these species), into bile. Polymeric IgA with J chain bound by SC at the cell surface is internalized into vesicles and transported through the cell. After transport to the cell apex, the vesicles fuse with the membrane, and the entire complex is released as S-IgA into the lumen. Because most SC remains bound to the IgA, it is lost from the cell and, thus, has become known as a "sacrificial receptor."

II. Induction of Mucosal Responses

To understand how IgA responses are induced, it is important to first consider how antigens are encountered from the environment. We normally contact antigens by either inhalation or ingestion. Thus, foreign substances are first taken into the bronchus-associated lymphoreticular tissue (BALT) and GALT, respectively. Because BALT is difficult to study, most work has focused on GALT; however, because the two tissues share many features, the findings of the studies with GALT should apply to BALT as well (see below).

A. Mucosal-Associated Lymphoreticular Tissues

In 1964, Gowans and Knight showed that small lymphocytes taken from the thoracic duct of rats and transferred to compatible animals continuously recirculate in the adoptive hosts and enter lymphoid tissues such as mesenteric lymph nodes and Peyer's patches through high endothelial venules. Large blasts (actived lymphocytes) from the thoracic duct migrate to the lamina propria of the intestine, where they divide and differentiate into plasma cells. Craig and Cebra have shown that Peyer's patches contain precursors of IgA plasma cells, which migrate to mesenteric lymph nodes and the thoracic duct, and from the blood populate the lamina propria regions. BALT in the upper respiratory tract also possesses IgA precursors capable of seeding to lamina propria regions of the gastrointestinal tract and upper respiratory tract. This observation led to the proposal by John Bienenstock that BALT and GALT are major IgA inductive sites, which are collectively termed mucosal-associated lymphoreticular tissue (MALT), that provide the host with precursor IgA lymphocytes, which continually supply mucosal sites with IgA plasma cells. [See LYMPHO-CYTES.]

GALT includes the Peyer's patches, which are distinct nodules found mainly in the small intestine (10–12 in mice) and the appendix, which is most pronounced in rabbits and humans. Mammals also contain smaller follicles, termed solitary lymphoid nodules, throughout the small intestine and colon. Solitary lymphoid nodules, which have morphologic similarities to Peyer's patches, may also be more microscopic IgA inductive sites. Although the Peyer's patches, appendix, and solitary lymphoid nodules share similarities, including a dome region containing a specialized epithelium, important differences exist. The appendix, for example, has a less developed parafollicular region (T-cell zone, Fig. 1) than the Peyer's patches, and the greater proportions of B lymphocytes isolated from the appendix differ in numbers of surface IgA (sIgA$^+$), sIgG$^+$, and sIgM$^+$ cells from B-lymphocyte subsets obtained from Peyer's patches. Most studies with GALT cells have been done with Peyer's patches, because they presumably are the major component of GALT. More studies are required to verify that this assumption is correct.

B. Functional Anatomy of Peyer's Patches

Discovered by Peyer in 1677, human Peyer's patches were later shown to develop during gestation, to reach maximum numbers at puberty (>200), and to decline slowly with age. Peyer's patches, found in all experimental animals, have been described in some detail in rabbits and mice and to a lesser degree in rats, swine, calves, dogs, goats, and subhuman primates. Not only is it likely that all mammalian species have Peyer's patches as IgA inductive sites, some studies have also suggested that avian species exhibit Peyer's patchlike structures, which may function as IgA inductive sites.

A key feature of Peyer's patches is the epithelium, which covers the dome region. Because it consists of cuboidal (not columnar) epithelial cells and has few goblet cells, it produces little mucus to impede antigen uptake. Lymphocytes lie next to this epithelium, leading to the term lymphoepithelium. Specialized antigen-sampling cells, termed follicle-associated epithelial (FAE), or microfold (M), cells, are found in this lymphoepithelium and also occur in the appendix, solitary lymphoid nodules, and BALT. FAE cells, which exhibit short microvilli, small cytoplasmic vesicles, and thin extensions around the lymphoid cells in the lymphoepithelium, can actively engulf proteins in droplets (pinocytosis) or particulate antigens (phagocytosis), including viral particles, bacteria, and even small parasites. Uptake of antigen does not result in degradation in lysosomes, so the intact antigen can reach underlying lymphoreticular cells, including antigen-presenting cells (Fig. 1). In some instances, pathogenic viruses and bacteria enter the host by passing through M cells to reach systemic tissues, a process termed translocation. It is unlikely that M cells actually present antigen in association with class II MHC proteins, because they do not express class II determinants; however, they are important for antigen delivery to underlying zones for induction of immune responses. More studies are required to determine the cellular origin of M cells and to understand how their functions are regulated by cytokines and microbial cell products.

Peyer's patches are divisible into three regions: (1) the dome, (2) the follicles with germinal centers (B-cell zone), and (3) the parafollicular, or T cell-dependent, area (T-cell zone) (Fig. 1). Lymphocytes enter Peyer's patches through high endothelial venules in the parafollicular region and all cells exit the tissue via efferent lymphatics, which drain into the mesenteric lymph nodes (Fig. 1). Each Peyer's patch region contains a distinct array of cells. The dome contains B and T lymphocytes with significant numbers of macrophages and small numbers of plasma cells. The B-cell follicles normally exhibit one to two germinal centers, which are enriched in B cells committed to IgA (or sIgA$^+$), and it is there that B cells are continually switching to IgA. The T-cell zone possesses class II$^+$ antigen-presenting cells, most notably dendritic cells. Peyer's patches are less well developed at birth, and T cells from thymus are among the first to populate the tissue. B-cell zones mature slowly and depend on a gut microflora, because germ-free animals exhibit poorly developed Peyer's patches with no germinal centers.

C. Disassociated Cells from Peyer's Patches

Enzymatic procedures have allowed the isolation of all major lymphoreticular cells from murine Peyer's patches and have shown that ~40% of isolated cells are of B lineage and ~35% are mature T cells of various subsets (Fig. 1). Most isolated B cells are sIgM$^+$; however, higher numbers of sIgA$^+$ B cells occur (5–8%) than are seen in other secondary lymphoid tissues. A significant number of B cells are in cell cycle, which is consistent with the prominent germinal centers and continuous influx of environmental antigens into the Peyer's patches. This also explains why Peyer's patches are major IgA inductive sites for response to locally encountered antigen, but it does not explain why the dividing B cells fail to terminally differentiate into IgA (and other isotype)-producing plasma cells in the Peyer's patches themselves.

The IgA response is T cell-dependent, and the Peyer's patches contain significant numbers of functional CD3$^+$, T-cell receptor$^+$ (TCR$^+$) T cells. The majority of isolated CD3$^+$, TCR$^+$ T cells are CD4$^+$ and exhibit the characteristics of T helper (Th) cells. Antigen-specific Th-cell clones can be easily obtained from the Peyer's patch CD3$^+$, CD4$^+$ T-cell subset; however, CD3$^+$, CD8$^+$ T cells that mediate cytotoxic and suppressor functions also occur (Fig. 1). In addition, small numbers of CD3$^+$, CD4$^-$, CD8$^-$ T cells can be isolated from Peyer's patches that contain subsets that are responsible for regulation of immune responses in the presence of suppressor cells (termed contrasuppressor cells). Two major types of CD4$^+$ T cells that affect IgA responses have been isolated and cloned from Peyer's patches. One lineage of CD4$^+$ T cell induces the switch of sIgM$^+$ to sIgA$^+$ B cells, whereas a second CD4$^+$ type with Th-cell characteristics preferentially supports IgA synthesis. The latter Th-cell subset most likely produces interleukins, notably IL-4, IL-5, and IL-6, which support sIgA$^+$ B-cell division and terminal differentiation to IgA production. In addition, cytokines such as transforming growth factor-β (and possibly IL-4) are involved in B-cell switches to IgA. Although studies have shown that antigen and T cells contribute to B-cell switches to IgA, other evidence indicates that commitment to IgA (as with other non-IgM isotypes) is not influenced by antigen. The enzymatic release of Peyer's patch cells also results in significant numbers of class II$^+$ macrophages and dendritic cells. The latter population is particularly supportive of T cell-dependent IgA synthesis in vitro. The Peyer's patches also contain activated B cells with class II expression, which likely serve in antigen presentation to T cells in vivo. The induction of IgA responses in MALT is finely regulated, and the role of Th and dendritic cells and derived cytokines in these responses is receiving much current emphasis. [See CD8 AND CD4: STRUCTURE, FUNCTION, AND MOLECULAR BIOLOGY; CYTOKINES IN THE IMMUNE RESPONSE; T-CELL RECEPTORS.]

III. The Common Mucosal Immune System

A. Mucosal Effector Sites

Local application of antigen onto mucosal tissues or direct injection into exocrine glands can result in the production of S-IgA antibodies at that site. This local immunization often require the use of adjuvants, which draw in inflammatory cells and also trigger the production of antibodies (of IgG and IgM isotypes) in the serum. It is likely that this local stimulation results in an accumulation of clonal B and Th cells (from MALT?) and that cognate interactions result in terminal differentiation of IgA plasma cells and local S-IgA responses. The leakage of antigen adjuvant from local mucosal sites and its entrapment in systemic lymphoid tissues probably account for the serum antibody response. Thus, local immunization is not an optimal means for the selective induction of an S-IgA response in external secretions.

Compelling evidence has been presented over the past 15 years that shows that antigen induction of IgA precursors in MALT results in the dissemination of B and Th cells to remote effector sites such as glands and lamina propria (Fig. 1) for subsequent S-IgA antibody responses. Bienenstock, Cebra Hanson, Lamm, Mestecky, Michalek, Montgomery, and their respective colleagues showed that IgA precursor cells from BALT or GALT populate remote secretory sites, including the mammary, salivary, lacrymal, and uterine cervical glands. Using various oral vaccines, it was further shown that orally administered antigens resulted in the appearance of S-IgA antibodies in these remote site secretions, suggesting that antigen-specific IgA precursors originating in GALT follow a homing cycle to ultimately reach these effector sites (Fig. 1). Adoptive transfer of antigen-specific cells from mesenteric lymph nodes (presumably derived from Peyer's patches) repopulated the lamina propria of the gastrointestinal tract and mammary tissue (of the host into which they were introduced) to become antigen-specific plasma cells. It was also shown that S-IgA responses were identical in individual animals, clearly suggesting a clonal origin in MALT. Peyer's patches as a major IgA inductive site is shown by the observation that immunized intestinal segments containing a Peyer's patch could induce S-IgA antibodies even in adjacent nonimmunized segments, whereas immunized segments not containing a Peyer's patch did not. We conclude that in mammals the MALT, especially the Peyer's patches, are major IgA inductive sites, which lead to S-IgA responses in remote external secretions.

An unresolved issue regards memory in the IgA system and the need for local antigen presence for maximum S-IgA responses. Priming of GALT induces moderate but short-lived plasma cell re-

sponses, whereas local antigen presence greatly increased cell numbers and longevity. However, the antigen used—cholera toxoid—may be unusual in its ability to bind and penetrate epithelial cells, and more studies are required to confirm the universality of this phenomenon.

B. Human Oral Vaccines

Studies in humans suggest the existence of a system where orally administered antigen results in the presence of S-IgA antibodies in remote external secretions. Thus, ingestion of an oral bacterial vaccine induces specific S-IgA antibodies in saliva and tears but not in serum. Furthermore, mitogen stimulation of human peripheral blood results in the synthesis of polymeric IgA (while serum IgA is mostly monomeric) and the induction of both IgA1 and IgA2 plasma cells. In addition, human volunteers ingesting a bacterial vaccine showed that their peripheral blood B cells spontaneously produced specific IgA antibodies, *prior to* the occurrence of S-IgA antibodies in external secretions. Studies with several viral and other bacterial vaccines have clearly shown the subsequent presence of S-IgA antibodies in various human external secretions.

In summary, a common mucosal immune system occurs in mammalian species, where antigen stimulation of BALT and GALT induces an exodus of specific lymphocytes, which home to the various mucosal effector sites (Fig. 1). These responses are finely regulated, and T cells and cytokines are of central importance for ultimate plasma cell differentiation and for production of S-IgA antibodies in our external secretions. The current need for vaccines, including the universal efforts to develop effective immunity to human immunodeficiency virus and acquired immunodeficiency syndrome, compels us to better understand how we can use the common mucosal immune system for eventual prevention of infectious diseases.

Bibliography

Bienenstock, J. (1984). The lung as an immunologic organ. *Annu. Rev. Med.* **35,** 49–62.

Brandtzaeg, P. (1985). Role of J chain and secretory component in receptor-mediated glandular and hepatic transport of immunoglobulins in man. *Scand. J. Immunol.* **22,** 111–146.

Cebra, J. J., Komisar, J. L., and Schweitzer, P.A. (1984). C_H isotype 'switching' during normal B-lymphocyte development. *Annu. Rev. Immunol.* **2,** 493–548.

McGhee, J. R., Mestecky, J., Elson, C. O., and Kiyono, H. (1989). Regulation of IgA synthesis and immune response by T cells and interleukins. *J. Clin. Immunol.* **9,** 175–199.

Mestecky, J., and McGhee, J. R. (1987). Immunoglobulin A (IgA): Molecular and cellular interactions involved in IgA biosynthesis and immune response. *Adv. Immunol.* **40,** 153–245.

Underdown, B. J., and Schiff, J. M. (1986). Immunoglobulin A: Strategic defense initiative at the mucosal surface. *Annu. Rev. Immunol.* **4,** 389–417.

Multiple Sclerosis

DAVID A. HAFLER AND HOWARD L. WEINER, *Harvard Medical School*

Glossary

Adjuvant Substance that enhances immune responses

AMLR Autologous mixed lymphocyte response measures proliferation of T cells to self-MHC

CD4+CD45RA+ T cell T-cell subpopulation that induces suppression in other T cells, particularly CD8 suppressor cells

Con A Mitogen specific for T cells

Cytokine Soluble factor that interacts with immune cells and can affect immune function

MHC Major histocompatibility complex is polymorphic cell surface molecules present on macrophages and B cells that bind and present antigen to T cells

Mitogen Substance that stimulates lymphocytes to divide

Myelin Insulating sheath that covers nerve fibers and that is damaged in multiple sclerosis

T-cell receptor Cell surface molecule on T cells that recognizes antigens presented by MHC gene products

MULTIPLE SCLEROSIS (MS) fits into the category of autoimmune disease, in which immune T cells attack self-organs (in this instance, the white matter of the brain and/or spinal cord). Other autoimmune diseases include juvenile onset diabetes, in which T cells attack the pancreas; rheumatoid arthritis, in which T cells attack the joint; and Hashimoto's thyroiditis, in which T cells attack the thyroid gland.

Although the precise etiology of MS and other autoimmune disease remains unknown, it is believed that a number of factors are involved in the induction of these diseases. One factor involves a nonspecific loss of regulation of the immune system. A second factor involves T cells that are capable of recognizing and attacking self-organs. Although the precise trigger of these events is unknown, clearly most investigators feel viruses or other environmental factors trigger the disease in people whose immune system has certain MHC immune response genes and T-cell receptor genes. [*See* Autoimmune Disease.]

I. Multiple Sclerosis as an Autoimmune Disease

A. Diagnosis of Multiple Sclerosis

As MS results from a T-cell-mediated destruction of white matter, symptoms of the disease can be referable to any myelinated part of the central nervous system. Common symptoms include temporary loss of vision in one eye associated with demyelination and inflammation of the optic nerve (optic neuritis), diffuse numbness and/or tingling, weakness in one or both extremities, and bladder difficulties. Early in the course of the disease, the symptoms tend to occur for discreet periods of time, usually between a few days and weeks, after which the symptoms spontaneously improve. Later in the course of disease, in approximately a third of patients, the symptoms become progressive, usually involving difficulty with ambulation and progressive weakness in the legs. This phase of the disease is known as *chronic progressive multiple sclerosis*. It is not known why some patients go on to the progressive form of the disease and others do not, although most of the defined immune abnormalities in the disease tend to occur in patients of the progressive type of MS. A diagnosis of the disease is now usually accomplished by a history of multiple lesions in time and space; positive lesions on the magnetic resonance imaging (MRI); and lumbar puncture, which reveals inflammation with lymphocytes in the CSF, normal protein glucose, and signs of inflammation in the nervous system with increased immunoglobulin and oligoclonal bands, which represent discrete clonal expansion of B cells in the CSF.

B. Pathology

The term *multiple sclerosis* means "multiple scars" and comes from the hardened plaque-like appearance in sections of the brain. Although MS is thought to be a demyelinating disease, the exact role of demyelination as opposed to inflammation is not entirely clear. That is, in fresh lesions there is infiltration of T lymphocytes accompanied by macrophages, secretion of cytokines, and edema. Many of the acute symptoms observed in the early course of the disease are likely due to swelling in the site of inflammation as opposed to actual loss of myelin. With time, there is loss of myelin associated with the sclerosing plaque. Recent work has indicated that there are certain subpopulations of T cells that are found in MS plaques, and this will be discussed later. The lesions in MS are only found in the central nervous system white matter and almost never found in the peripheral nervous system, although lesions do occur in the dorsal root ganglia. [*See* Myelin and Demyelinating Disorders.]

II. Regulation of Human Immune Responses

A. T-Cell Populations

The immune system can be broadly divided into the following types of cells: (1) T cells, which regulate immune function and can be involved in direct cytotoxicity destruction of other cells; (2) macrophages, which are the phagocytic cell of the immune system and which are involved in the presentation of antigen to T cells; (3) B cells, which may also present antigen to T cells in addition to being responsible for immunoglobulin synthesis. The T-cell population can be broadly divided into T cells that express CD4 and are inducer cells and T cells that express CD8, which tend to have cytotoxic and so-called suppres-

sor function. T cells express a unique T-cell receptor, which recognizes antigen, which is presented in the context of MHC gene products. CD4 cells recognize antigen presented by class II MHC, whereas CD8 cells recognize antigen presented in the context of class I MHC. CD4 cells can be further divided into populations that express different isoforms of the T200 (leuckocyte common antigen) complex. T cells that express the CD45RA (2H4) molecule tend to have suppressor inducer function, whereas T cells that express another isoform of the T200 (CD45R0) have helper inducer function. The CD45 molecule appears to be particularly important in T-cell regulation, as all the tyrosine phosphatase activity in the T cells is immunoprecipitated by anti-CD45 monoclonal antibodies. Phosphatases can up- or down-regulate the cell in terms of its level of activation. It has been recently discovered that both CD4 and CD8 are noncovalently linked to the p^{56}Lyc, which is a tyrosine kinase. Kinases as well are important cell regulation molecules. Thus the bringing together of the T-cell receptor with the various kinases and phosphatases on the T-cell surface are intimately related to the regulation of T-cell function. [*See* CD8 and CD4: Structure, Function and Molecular Biology; Immune System; Lymphocytes; Macrophages; T-Cell Receptors.]

B. Measures of Immune Function

As the mechanisms of suppression have not been clearly defined in immunology, it has been difficult to measure immune function in autoimmune disease states. However, there has been accumulating evidence that measures of immune function that are not antigen specific are altered in a number of autoimmune diseases including MS. In the late 1970s, it was demonstrated that Con A–induced suppression was altered in MS and was later confirmed. That is, the patients had a relative inability to suppress a mitogen-induced suppressor effect as compared with control subjects. Another measure of T-cell function has been the autologous mixed lymphocyte response (AMLR). The AMLR measures proliferation of T cells to self-MHC. As it appears that MHC may have self-antigen bound in its antigen presenting site, it has been postulated that proliferation in the AMLR represents a response to autoantigens. Examination of the AMLR in patients with progressive MS has revealed a decrease in this response as compared with normal controls. Similar decreases have been observed in other autoimmune diseases including systemic lupus erythematosis (SLE), rheumatoid arthritis, and Sjogren's syndrome. At first it might seem paradoxical that proliferation to self-MHC would be decreased in autoimmune disease. However, it has been shown that proliferation in the AMLR is primarily accomplished by T cells that express the CD4 and CD45RA (2H4) molecules which induce a suppressor response. Thus proliferation to self-MHC and possibly self-antigen may result in the generation of a suppressor response, which may be decreased nonspecifically in autoimmunity.

Antigen nonspecific suppressor function in MS has been measured using an AMLR suppressor assay. In this measurement, T cells are cultured for 1 week with self-MHC in an AMLR. The T cells generated in the AMLR are then incubated with whole mononuclear cells and the B-cell mitogen pokeweed, and the ability of the T cells to suppress IgG (a type of antibody) synthesis by the stimulated B cells is measured. As shown in Table I, patients with progressive MS have a decreased ability to suppress this pokeweed Ig response as compared with normal controls or subjects with other neuro-

logic diseases. This is further evidence that antigen nonspecific measures of immune function are altered in MS. Similar alterations are found in patients with SLE.

C. Alterations in Specific T-Cell Populations

When monoclonal antibodies recognizing CD4 and CD8 populations became available in the late 1970s, they were used to examine T-cell populations in MS. (It was originally found there were decreases in CD8 cells in patients with progressive MS as compared with control subjects). This decrease in the so-called CD8 suppressor cells has not been found to correlate with the antigen nonspecific alterations of suppression in MS. The elucidation of cell surface markers that subdivide the CD4 population into helper inducer (CD45R0) and suppressor inducer (CD4, CD45RA) populations has allowed the analysis of the subpopulations in patients with a number of autoimmune diseases including MS and SLE. It has been found that in these autoimmune diseases, there is a decrease in CD4 cells, T cells that have suppressor inducer function (CD45RA). This has been of importance, because decreases of

TABLE I Autologous Mixed Lymphocyte Reaction (AMLR)–Induced Suppression and Suppressor Inducer T Cells in Multiple Sclerosis (MS)[a,b]

| Ratio of AMLR-induced T cells to allogeneic MNCs | Source of AMLR-induced T cells[c] | | | Correlation of percent CD4+2H4+ cells with AMLR-induced suppression in MS patients and its significance[d] | |
	Patients with MS	Normal subjects	Subjects with other neurological diseases	r[e]	p[e]
1:20	81.5 ± 3.5	51.7 ± 5.3[f]	41.1 ± 6.4[f]	0.63	0.0003
1:10	58.8 ± 4.7	25.3 ± 5.3[f]	15.5 ± 3.6[f]	0.73	0.0001
1:4	36.7 ± 4.5	6.9 ± 1.7[f]	4.3 ± 1.4[g]	0.63	0.0002
1:2	17.5 ± 2.4	2.3 ± 0.6[f]	1.0 ± 0.2[g]	0.57	0.0004
1:1	7.4 ± 1.3	0.9 ± 0.2[g]	0.5 ± 0.2[g]	0.59	0.005

[a] From Chofflon, M. M., et al. (1989). *Ann. Neurol.* **25,** 494–499, with permission.
[b] MNC, mononuclear cells; PWM, pokeweed mitogen; Ig, immunoglobulin.
[c] Values represent the percentage of PWM-stimulated Ig secretion of allogeneic MNCs cultured with AMLR-induced T cells as compared with PWM-stimulated Ig secretion of allogeneic MNCs alone (±SEM). Patients with progressive MS ($n = 25$) were compared with age- and sex-matched normal healthy subjects ($n = 11$) and patients with other neurological diseases ($n = 6$). The percent suppression, which was lower in patients with MS, could be calculated.
[d] A linear regression analysis between the percentage of CD4+2H4+ cells in the peripheral blood and the percentage of suppression in MS patients was performed.
[e] Spearman's rank-order coefficient of correlation.
[f] $p < 0.001$ (unpaired t test; MS patients compared with normal subjects or patients with other neurological diseases).
[g] $p \leq 0.002$ (unpaired t test; MS patients compared with normal subjects or patients with other neurological diseases).

CD4+CD45RA+ (suppressor inducer) cells have correlated strongly with decreases in antigen non-specific measures of immune function, both in MS and SLE. These results have united many of the observed immune abnormalities in MS, and to-gether suggest that alterations in subpopulations of CD4 cells and potentially of the CD45 tyrosine phosphatase isoforms that they express may be closely associated with nonspecific alterations in immune function. It can be further postulated that these change may potentially play a major role in the induction of MS.

D. Alterations in T-Cell Receptor Populations

Antigen nonspecific alterations in T-cell function may potentially be associated with alterations of T-cell receptor populations that recognize the antigen in patients with MS. Using a direct single-cell clon-ing assay, T cells can be directly cloned from blood or CSF with high efficiency, which allows a repre-sentative sampling of the original T-cell populations present in these immune compartments. It is possi-ble that all the T cells in the spinal fluid or blood in patients with MS could be using different T-cell re-ceptors. Alternatively, it is possible that all or some of the T-cell clones would have only one (mono-clonal) or a few different (oligoclonal) T-cell popula-tions. The analysis can be performed by a technique in which each T cell gets a type of "fingerprint" by comparing the sizes of the T-cell receptor genes af-ter digestion with different enzymes and Southern blotting. This type of analysis in patients with MS has shown that there are common T-cell clonotypes among T cells in spinal fluid and between spinal fluid and blood in patients with MS, which has not been found in subjects of other neurologic diseases or normal controls. These common T cells have been called *oligoclonal T cells*. The presence of a limited number of T-cell populations in patients with MS is further evidence of an unregulated im-mune system in which T cells are chronically in-duced to divide.

E. Genetics of Multiple Sclerosis

Although environmental factors appear to be a fac-tor in the development of MS and the disease is clearly not an inherited single-gene disorder, there is evidence that genetic factors play some role in the disease, although exactly which genes are involved has not been elucidated. MS appears to involve multiple genes. Identical twins have a strikingly high concordance rate for the disease as compared with nonidentical twins, or siblings. As MS appears to be a T-cell-mediated autoimmune disease, there has been great interest in examining associations between the highly polymorphic genes in the MHC region, which are responsible for presenting the an-tigen, and T-cell receptor genes with expression of the disease. Data have suggested that certain DR 2, DQw1 genes of the MHC region and V_β genes of the T-cell receptor are associated with MS, although other genes may also be involved with the disease.

F. Relation Between Peripheral Blood and Central Nervous System in Multiple Sclerosis

Although MS is a disease of the central nervous system, numerous immune defects have been found in the peripheral blood, indicating that there may be a close relation between the immune abnormalities in the CNS with the peripheral immune system (Fig. 1). The movement of T cells from the peripheral blood into the CSF in patients with progressive MS was found to be rapid. Within 3–4 days, between 60 and 80% of the T cells in the CSF had migrated from the peripheral blood in patients with active disease. This finding, along with the observation that com-mon T-cell receptor clonotypes are found between blood and spinal fluid in the patients, has suggested that there is rapid communication between the sys-temic immune system and the central nervous sys-tem. The systemic (affecting the whole body) nature of MS has been further indicated by the significant effect on the course of chronic progressive MS with total lymph node irradiation (TLI), which is only targeted to non–central nervous system immune structures. [*See* NEURAL-IMMUNE INTERACTIONS.]

G. Experimental Models of Multiple Sclerosis

Much has been learned regarding the ability of T cells to mediate central nervous system inflamma-tion and demyelination from the animal model of MS [experimental allergic encephalomyelitis (EAE)]. EAE can be induced in a number of differ-ent animal species including rodents, rabbits, and primates by the injection of proteins of myelin (e.g., myelin basic protein or proteolipid apoprotein) mixed with adjuvants (e.g., complete Freunds adju-vant). The precise role of the immune adjuvant is unknown; it probably functions to enhance the trig-gering of CD4+ inducer cells, allowing for the

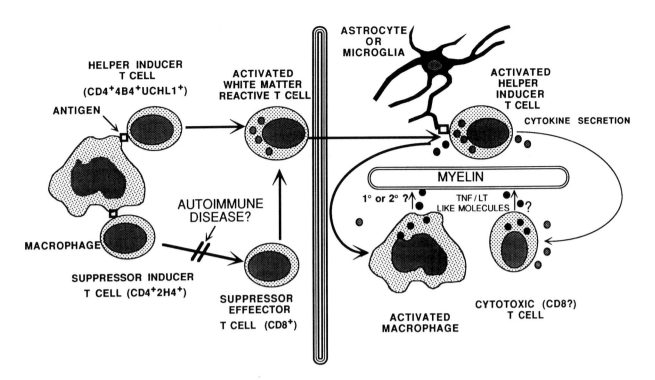

FIGURE 1 Postulated mechanisms involved in the pathogenesis of multiple sclerosis. Defective induction of CD4+ T cells expressing the CD45R molecule (detected by 2H4 mAb) resulting in altered regulation of the immune response is postulated as one of the factors in the pathogenesis of human autoimmune disease. Autoantigen-reactive CD4+ T cells expressing CD29 fibronectin receptor (detected by 4B4 mAb) and UCHL1 molecules, once deregulated, become activated and cross into the central nervous system. Further proliferation of autoreactive T cells may occur by presentation of antigen by astrocytes and/or microglia. Whether antigen-specific cytotoxic T cells or nonspecific cytotoxicity by T cells mediated by CD2-LFA-3 or other adhesion molecules can deliver a lethal hit to membranes of white matter is unknown. Eventually, edema with the release of cytokines may also play a role in the process of demyelination in addition to macrophages, which are known to strip myelin. [From Hafler, D. A., and Weiner, H. L. (1989). *Immunol. Today* **10**, 104–107, with permission.]

proper presentation of the autoantigen to T-cell populations. The injection of brain proteins with adjuvant results in the generation of T cells specific for these proteins. The T cell can then mediate either inflammation or demyelination, depending on the species and route of injection, resulting in a pathologic picture similar to MS. EAE is a CD4+ T-cell-mediated disease in which it has been demonstrated that myelin basic protein reactive T-cell clones can transfer the disease to noninjected animals. This has shown that small numbers of activated autoreactive T-cell clones can mediate an experimental

autoimmune disease against normal brain tissue. Examination of different animal species has indicated that they differ in susceptibility to EAE, in relation, in part, to MHC/T-cell receptor genes, although other genes may also be involved.

H. T-Cell Receptor in EAE

Examination of T-cell receptor sequences from T-cell clones capable of inducing EAE across different species has revealed that common T-cell receptors may be used to recognize different MHC/myelin basic protein fragment combinations. In MS, both normal controls and patients with MS proliferate to myelin basic protein, and myelin basic reactive clones can be isolated from normal subjects. In MS patients, there appears to be reactivity by T cells to specific parts (dominant regions) of the myelin basic protein molecule, which may not be observed as frequently in normal controls of different MHC haplotypes. Thus, T cells with reactivity to certain dominant regions from autoantigens such as myelin basic protein may be analogous to T-cell clones in animals that mediate EAE. Moreover as with EAE, there can be restricted T cell receptor usage in recognition of dominant regions of the myelin basic protein.

III. Immunotherapy for Multiple Sclerosis

A. General

Clinical trials have demonstrated that immune processes play a role in the progression of the disease. Intensive immunosuppression with either a combination of cyclophosphamide and ACTH, cyclosporine-A, or TLI have shown a significant decrease in disease progression in patients with chronic progressive disease. In striking contrast, infusion of the cytokine interferon (IFN)-γ, which can up-regulate class II MHC expression, has caused exacerbations of disease activity. Although the role of immunosuppression in terms of risk/benefit ratio in MS is currently being defined, these clinical trials have indicated that reducing the immune response is associated with slowing or stopping progressive disease, whereas enhancing the immune system results in increased disease activity.

B. Experimental Forms of Immunosuppression

The toxicity and nonspecificity associated with the present forms of immunosuppression have led to attempts in devising specific and safe modalities to treat autoimmune diseases. Trials have begun with antibodies specific for CD4, CD2, and T12 cell surface determinants. Although the anti-T-cell monoclonal antibodies can down-regulate immune space responses, they are associated with the rapid development of human antimouse antibodies, which prevent their further use. Other treatments now undergoing clinical investigation include cytokine manipulation, with the infusion of IFN-α and IFN-β; T-cell vaccination in which autologous T cells isolated from patients with MS are killed by glutaraldehyde, then injected back into the patient the clone was derived from; and the generation of tolerance (absence of immune reactivity) to myelin antigens by the oral route. The goal of immunotherapy is to find a nontoxic, antigen-specific form of therapy that can be administered early in the disease.

Bibliography

Acha-Orbea, H., Mitchell, D. J., Timmermann, L., et al. (1988). Limited heterogeneity of T cell receptors from lymphocytes mediating autoimmune encephalitis allows specific immune intervention. *Cell* **54,** 263–273.

Chofflon, M. M., Weiner, H. L., Morimoto, C., and Hafler, D. A. (1989). Decrease of suppressor inducer (CD4+2H4+) T cells in multiple sclerosis cerebrospinal fluid. *Ann. Neurol.* **25,** 494–499.

Hafler, D. A., Duby, A. D., Lee, S. J., Benjamin, D., Seidman, J. G., and Weiner, H. L. (1988). Oligoclonal T-lymphocytes in the cerebrospinal fluid of patients with inflammatory central nervous system diseases. *J. Exp. Med.* **167,** 1313–1322.

Hafler, D. A., and Weiner, H. L. (1989). MS: A CNS and systemic autoimmune disease. *Immunol. Today* **10,** 104–107.

Ota, K., Matsui, M., Milford, E., Mackin, G., Weiner, H. L., and Hafler, D. A. (1990). T cell recognition of an immunodominant myelin basic protein epitope in multiple sclerosis. *Nature (London)* **346,** 183–187.

Weiner, H. L., and Hafler, D. A. (1988). Immunosuppressive therapy of multiple sclerosis. *Ann. Neurol.* **23,** 211–222.

Wucherpfennig, K. W., Endo, N., Rosenzweig, A., Seidman, J. G., Weiner, H. L., and Hafler, D. A. (1990). Shared T cell receptor V_β gene usage in T cells specific for immunodominant regions of human myelin basic protein. *Science,* **248,** 1016–1019.

Multiple-Task Interactions

HERBERT HEUER, *Philipps-Universität Marburg*

Glossary

Capacity Hypothetical variable with an upper limit tied to a performance measure by a nondecreasing function; interpreted variously as representing energy or structures

Coupling (1) Linkage, generally nonrigid, between two motor-control structures (e.g., oscillators); (2) relation, generally nonrigid, between kinematic or kinetic characteristics of simultaneous movements

Crosstalk Mutual influences between two signals (e.g., coupling, confusions, intrusions)

Performance-operating characteristic Relation between performance scores on two tasks performed simultaneously; typically, performance on one task can be traded against performance on the other

Performance-resource function Describes performance as a function of capacity, or resources

Resources Synonym for capacity with respect to formal characteristics, but less of an energy connotation in its interpretation

MULTIPLE-TASK INTERACTIONS are the cross-task interdependencies between processes involved in the simultaneous performance of two or more tasks. Usually they result in reduced performance on one or both tasks. This performance decrement can be modeled in terms of competition between the tasks for a limited hypothetical entity, variously denoted as capacity or resources. When a more detailed recording of performance is possible (e.g., when two movements must be performed simultaneously), various types of crosstalk can be identified that are beyond the scope of resource-competition models.

I. Introduction

When two tasks are performed simultaneously, performance on one or both of them frequently deteriorates. The multiple-task performance decrement poses a problem for applied research on safety and workplace design. In addition to posing a problem, it has been used as a tool for the assessment of mental workload. Strongly influenced by applied research, models have been developed that account for the performance decrement in terms of a competition between the two tasks for resources or capacity needed to perform them.

From a less applied perspective, the performance decrement may be of less interest than the more fine-grained interactions that are revealed when performance is analyzed in more detail. Such analyses are particularly feasible when signals, which are related to performance on the two tasks, can be recorded more or less continuously. This is only rarely possible for primarily mental tasks, but it is comparatively simple for tasks that require simultaneous movements. Problems of multiple-task performance thus merge with problems of motor coordination.

The broader perspective on multiple-task interactions rather than on multiple-task performance decrements not only calls for other data and other models, but can also deal with occasional improvements on a task that can be achieved by adding a second one. A simple example is mirror writing with the left hand that, in right-handers, is facilitated by simultaneous normal writing with the right hand. Thus, certain constraints that cause performance decrements in some task combinations can enhance performance in others. Multiple-task interactions are no longer viewed only as representing a deficit, although they produce deficits in many combinations of "unnatural" tasks, but they are also seen as serving a function in other situations.

Before turning to models and phenomena of multiple-task interactions, a conceptual weakness of the whole area of research should be mentioned. This is the lack of a concise definition for "task." For example, driving a car and having a conversation would probably be considered as two tasks, whereas breathing and clapping the hands would probably not. Other examples such as holding a bottle with the left hand and opening it with the right might generate more debate. In fact, the distinction between performing one or two tasks is elusive. Intuitively we seem to isolate tasks against each other with reference to conscious goals; however, usually a goal (such as opening a bottle) can be split into subgoals (holding the bottle and manipulating the stopper). Furthermore, what can be two conscious goals at one time (such as operating the clutch and moving the gear shift lever for beginners) can become a unified single goal after some practice. Faced with these difficulties, the only way out seems to be to use the term task in a somewhat arbitrary manner, each task being defined by certain performance measures that are recorded.

II. Competition

The first type of models for multiple-task performance is concerned with the decrement of global performance scores, mainly duration and accuracy measures. The decrement is ascribed to a competition of processes involved in the performance of the two or more tasks for a hypothetical entity denoted as capacity, or resource(s) rather than to direct interference between these processes. Several variants of resource-competition models exist.

A. Capacity and performance

Formally, capacity is defined by its relation to performance scores, i.e., by a performance-resource function (Fig. 1a):

$$x_i(c) = e_i + f_i(c);$$
$$0 \le c \le c_L, f_i(0) = 0, f'_i(c) \ge 0. \quad (1)$$

In this equation, x_i is performance on task i. It is partitioned into the capacity-independent component e_i and the capacity-dependent component $f_i(c)$. When capacity increases from its lower limit of 0 to its upper limit of c_L, performance is assumed not to decline; however, it need not necessarily increase over the full range of c but may remain constant. If that is the case, performance is designated as data-limited rather than resource-limited.

In simultaneous performance, total capacity c_L is assumed to be allocated to the two tasks in a certain proportion that can be voluntarily controlled to some extent. Performance on a second task p are thus given by

$$x_p(c_L - c) = e_p + f_p(c_L - c);$$
$$0 \le c \le c_L, f_p(0) = 0, f'_p(c) \ge 0. \quad (2)$$

The two performance-resource functions specify the relation between the performance scores when the proportion of capacity allocated to the two tasks is varied. An example for such a performance-operating characteristic (POC) is presented in Fig. 1b. Its slope is given by

$$f'_{p \cdot i}(c) = -f'_p(c_L - c)/f'_i(c); \quad 0 \le c \le c_L. \quad (3)$$

This slope is invariant against additional capacity-independent sources of multiple-task interference; capacity-independent sources will modify only the capacity-independent components of performance, e_i and e_p, and thus shift the POC along one or both axes.

The measurement of POCs is a cumbersome task; therefore, empirical POCs are rare. An example is presented in Fig. 2. One task was a simple reaction-time task, which required rapid responses with the left hand to brief tones presented every 4 sec. The second task was two-dimensional pursuit

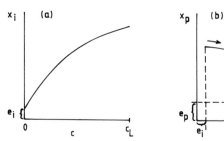

FIGURE 1 Hypothetical performance-resource function (a) and the performance-operating characteristic that can be derived from two such functions (b). Arrow indicates increasing capacity allocated to task i and decreasing capacity allocated to task p.

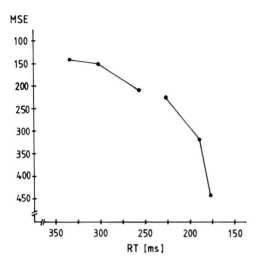

FIGURE 2 Empirical performance-operating characteristic for simple reaction time (RT) and mean squared error (MSE; arbitrary units) on a two-dimensional pursuit-tracking task. The middle two data points were obtained under the instruction "to do best" on both tasks; for the rightmost two data points, RT had to be improved by 20 and 40%, and for the leftmost two data points, tracking performance by the same percentages. Data are plotted such that better performance is up and right on the two axes. ms, msec. [Adapted, with permission, from K.-H. Schmidt, U. Kleinbeck, and W. Brockmann, 1984, Motivational control of motor performances by goal setting in a dual-task situation. *Psychol. Res.* **46**, 129–141.]

tracking. One cursor (target) moved randomly on a screen. Subjects had to follow its motion with a second cursor (follower), the position of which was controlled by a joystick, operated with the right hand. The performance measure was the mean squared error, which is the mean squared distance between the positions of the target and the follower. This measure was computed during reaction-time intervals and the following 1.5 sec. Allocation of capacity was manipulated by setting specific goals for the two tasks. First, subjects were instructed to do "the best" on both tasks (middle two data points in Fig. 2). Then, specific goals were set for one or the other task, which corresponded to 20 and 40% improvement over the "do the best" performance. These goals led to improved performance but at the cost of a decrement on the other task. This kind of tradeoff is typical.

More frequently than POCs, multiple-task performance decrements have been analyzed. Due to the tradeoff between performance scores, different decrements on one task in two conditions are ambiguous. They can indicate a strategy difference so that subjects work on two points of the same POC, or

they can indicate different capacity demands (different performance-resource functions and, thus, different POCs). The former possibility can only be excluded when a larger performance decrement in one task is not accompanied by a smaller decrement in its companion task. In addition to the occasional neglect of the methodological requirement to consider performance decrements on both tasks that have been combined, analyses of performance decrements suffer from the fundamental weakness that they are not suited to separate capacity-based interference from other types; this is because performance decrements depend on capacity-independent factors in addition to resource competition.

B. Variants of resource-competition models

Different models for multiple-task performance decrements do coexist at present. They can be described as variants of the resource-competition notion even when this was not necessarily the conceptual framework of their original presentation. The major differences between them relate to assumptions about capacity allocation and the number of different pools of capacity or resources.

1. Models of Generalized Central Capacity

Models of generalized central capacity share the assumption that there is only one pool of resources, which is accessed by all tasks to varying degrees (including a zero demand of the task). The general opinion appears to be that such models have to be rejected. However, essentially all tests of this type of model involved additional assumptions that are questionable (e.g., the assumption that performance-resource functions for easy tasks have steeper slopes than those for more difficult tasks).

Although generalized-capacity models, which include capacity-independent interference in addition to resource competition, are still tenable, models that posit resource competition as the only source for multiple-task performance decrements have been shown to be wrong. For two tasks, i and j, that are combined with tasks p and q, giving rise to the observed performance scores $x_{i \cdot p}$, $x_{i \cdot q}$, $x_{j \cdot p}$, and $x_{j \cdot q}$, these models predict that the differences $x_{i \cdot q} - x_{i \cdot p}$ and $x_{j \cdot q} - x_{j \cdot p}$ are both either non-negative or non-positive. (The prediction is based on the additional assumption that capacity allocated to tasks p and q is independent of whether they are combined with i or j; however, it is still valid for certain violations of this assumption.) The model frequently failed to pass this test.

One way to make the model fail is to combine tasks that presumably impose particular demands on the same or different cerebral hemispheres. Figure 3 presents an example. Tasks i and j were rapid finger movements with the left and right hand, respectively, between two keys separated by a 6-cm distance; the dependent measure was the standard deviation of key-press intervals. Tasks p and q were running memory tasks. Words or faces were presented in sequence, and the task of the subject was to indicate for each stimulus whether or not it had been presented before. As shown in Fig. 3a, left-hand movements in combination with the verbal memory tasks ($x_{i \cdot p}$) were less variable than in combination with the figural memory task ($x_{i \cdot q}$), whereas right-hand movement were more variable with the verbal ($x_{j \cdot p}$) than with the figural task ($x_{j \cdot p}$).

Performance on the memory tasks reveals that the assumption of task-independent allocation of capacity was not met in this experiment: performance on the memory tasks depended on whether they were combined with left-hand or right-hand movements. This failure, however, does not invalidate the conclusion. Performance on each of the two memory tasks was poorer when combined with movements of that hand, which also exhibited

poorer performance when combined with that memory task. Rather than being attributable to competition for a single type of capacity, these data indicate that performance decrements are larger when the two tasks impose particular demands on the same cerebral hemisphere (left hemisphere: right-hand movements and memory for words; right hemisphere: left-hand movements and memory for faces) than when the functional specializations of both hemispheres are required to perform the task combination.

2. Single-Channel Model

According to the single-channel model, capacity at a particular stage of processing is allocated in an all-or-none fashion to one or the other task at any point in time. Thus, the model holds that only one task can be performed at a time, and that what appears as simultaneous performance of two tasks is really a rapid alternation between them.

Evidently, a single-channel model can be distinguished from a model of generalized central capacity only when a sufficiently high temporal resolution is used in the analysis. With a low temporal resolution, time-sharing (each task is supplied with total capacity for a certain proportion of time) becomes indistinguishable from capacity-sharing (each task is supplied with a certain proportion of capacity at any moment in time) because the amount of capacity allocated to each task during a larger time interval will be graded quantity in both cases. Therefore, the single-channel model has mainly been studied by means of temporally overlapping reaction-time tasks; in fact, one of its initial purposes was to account for the so-called psychological refractory period that can be observed in this task combination.

The psychological refractory period is illustrated in Fig. 4. In this experiment, the first task was to respond as rapidly as possible with one of the four fingers of the right hand to the presentation of a rectangle in one of four positions on a video screen. The second task was similar, but the left hand was used for the response and four tones with different frequencies served as signals. Reaction time for the first task (RT1) showed little variation when the interstimulus interval (ISI) was varied. The same was true for the reaction time (RT2) to the auditory signal, provided that the first response was omitted (control). When it was performed, however, reaction time to the auditory signal (RT2) was delayed at short ISIs. This delay is called the psychological refractory period because of a superficial resem-

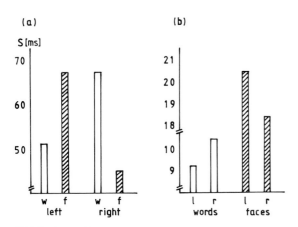

FIGURE 3 Performance on pairs of tasks that impose particular demands on only one or both of the cerebral hemispheres. (a) Variability of rapid left-hand (left) and right-hand (right) finger movements when combined with memorizing words (w) and faces (f). (b) Number of errors in recognizing words or faces when combined with rapid left-hand (l) or right-hand (r) finger movements. ms, msec. [Adapted, with permission, from K. Mc-Farland and R. Ashton, 1978, The lateralized effects of concurrent cognitive and motor performance. *Percep. Psychophys.* **23**, 344–349.]

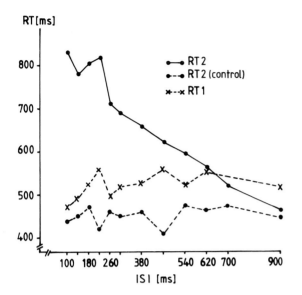

FIGURE 4 The psychological refractory period in a single subject who responded to two signals presented in rapid succession. RT1 is reaction time (RT) to the first signal, and RT2 is reaction time to the second signal that followed the first one after a variable interstimulus interval (ISI). RT2 (control) is from a control condition in which the response to the first signal was omitted. ms, msec. [Adapted, with permission, from H. Heuer, 1981, Über Beanspruchungsänderungen im Verlauf schneller gezielter Bewegungen. *Z. Exp. Angew. Psychol.* **28**, 255–280.]

blance to the refractory period of single nerves (where, however, the response to the second stimulus is omitted or reduced in amplitude rather than delayed).

The single-channel model posits a central mechanism that can deal with only one signal at a time, whereas initial perceptual processes and final motor processes could be parallel. While the central mechanism is dedicated to the processing of one signal, other incoming signals are temporarily held in store for later processing. From this scheme it appears to follow that the slope of the function, which relates reaction time to the second signal (RT2) to ISI, should be −1 as long as the central mechanism is dedicated to processing the first signal when the second signal arrives and zero thereafter. The data of Fig. 4 are typical in that the slope is not sufficiently steep. However, the simple prediction of slopes is valid for a deterministic variant of the model only. Treating the different time intervals for perceptual, "central," and motor processing as random variables alters the expectations dramatically, and slopes considerably less steep than −1 can be predicted. Therefore, although the single-channel

model receives little acceptance today, this is for partly unjustified reasons.

3. Multiple-Resource Models

Multiple-resource models are generalizations that were motivated by the failures of generalized-capacity models to account for the observed patterns of multiple-task performance decrements. Basically, a generalized-capacity model is multiplied by assuming different types of capacities or resource pools, and some complications are added on top of this. The end product is a (convenient?) vocabulary that can be used to describe patterns of multiple-task performance decrements. However, the details of a multiple-resource model have never been formalized, and it is a good guess that the model would turn out to be theoretically empty because it cannot be rejected in principle. This is certain as long as no assumptions are made about the number of resource pools, and rejection might be impossible in practice, even with such assumptions.

According to multiple-resource models, performance on a particular task is a function of several rather than only one type of resource. A multiple-task performance decrement will be observed whenever the demands on at least one type of capacity exceed the limit. Thus, it is an intuitive (but less than stringent) prediction that tasks with similar resource-demand compositions should suffer larger performance decrements when performed simultaneously than tasks with dissimilar resource-demand compositions. This prediction must not be correct because of the additional assumption that a scarce resource can be (partly or completely) replaced by other resources that are available. This assumption paraphrases the fact that many tasks can be performed using different strategies.

A major problem for multiple-resource models is to specify which types of resources do exist. One proposal is that the two cerebral hemispheres correspond to different resource pools. A second proposal distinguishes between separate resource pools for visual and auditory signals, for spatial and verbal material (or manual and vocal responses), and for different stages of information processing (stimulus encoding and central processing, responding). Tests of these proposals have mostly been confined to showing that tasks with presumably common demands on one or the other kind of resource pool are harder to perform simultaneously than tasks with dissimilar demand compositions. Figure 3 gives an example of this kind for hemisphere-specific resources; similarly, combinations of manual-response tasks frequently suffer from larger performance decrements than verbal–manual task combinations. It should be noted, however, that predictions of this kind are not unique for multiple-resource models.

C. The Nature of Capacity–Resources

Although capacity is a hypothetical variable that does not require any definition beyond its formal characteristics, the concept is frequently interpreted by analogies or reference to phenomenal or physiological concepts. Energetical and structural interpretations can be distinguished.

Generalized central capacity in particular has been interpreted as a functional correlate of effort and related to the physiological concept of activation. Given such an interpretation, the relation of capacity to mental workload becomes obvious. However, the effort interpretation raises problems. First, ample evidence indicates that higher motivation (and thus presumable higher effort) can reduce performance on sufficiently difficult tasks as compared with lower levels of motivation. Such findings question the validity of the assumption of nondecreasing performance-resource functions. Second, there appears to be no fixed upper limit for effort as might be inferred from paradoxical improvements observed under adverse conditions. A corresponding flexible upper limit c_L for capacity would complicate tests of the models.

With regard to structures, computer analogies are obvious (e.g., between the central mechanism of the single-channel model and a central processing unit). Resources in particular have been related to functionally defined structures such as capacity-limited memory systems. In general, however, structural referents remain vague.

D. The Scope of Resource-Competition Models

Resource-competition models are deficit models in at least three ways. First, they can account for performance decrements only (except when additional assumptions are made such as a flexible upper limit of capacity). Second, they are limited to global performance measures such as duration and/or accuracy, which characterize the extent to which performance deteriorates and neglect variations in the way a particular performance level is achieved. Third, they are time-independent in that global per-

formance scores can be computed over unrestricted stretches of time. (The single-channel model is an exception.) By virtue of their global dependent variables (overall-performance measures) and their neglect of high temporal resolution, resource-competition models are irrelevant for a variety of interaction phenomena that are revealed by more detailed analyses of multiple-task performance. These phenomena, which give evidence of crosstalk between simultaneously performed tasks, are discussed in the following section.

III. Crosstalk

A second class of models for multiple-task performance focuses on detailed patterns of interaction rather than on gross performance decrements. Although some formal models do exist, the emphasis is still very much on sorting and classifying the varieties of crosstalk that do exist. Therefore, what ties up the phenomena considered in this section is more the general notion of crosstalk than a common characteristic of a set of models such as resource-competition.

A. Varieties of Crosstalk

Crosstalk can be described with reference to a central or peripheral level. For illustration consider coupling, one of the prominent forms of crosstalk in simultaneous movements. Figure 5 presents a mechanical analogue. The device that generates the oscillations of the two pens corresponds to the central level, and the movements of the pens correspond to the peripheral level at which behavioral data are recorded.

If the spring in Fig. 5 were omitted and if the viscous coupling of the weights to the pens were rigid, the driving forces (the weights) would generate two independent oscillations. The coupling between the two oscillators can be described as viscoelastic on the central level. (Coupling appears to be ''soft'' in general with only few exceptions.) On the peripheral level, complex consequences of the central coupling will appear in the relation between frequencies, amplitudes, and phases of the two oscillations. For example, frequencies will become similar; this can be described as frequency coupling. Also, phase will tend to enter a certain relation, describable as phase coupling. These descriptions of the peripheral observations cannot be

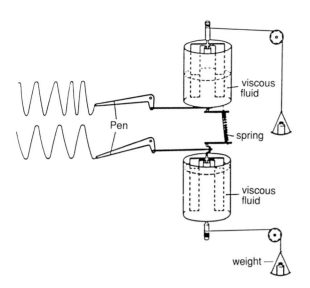

FIGURE 5 A mechanical analogue for a coupling of two oscillators. [Adapted, with permission, from E. von Holst, 1955, Periodisch-rhythmische Vorgänge in der Motorik, pp. 7–15. 5. Conf. Soc. Biol. rhythm, Stockholm.]

mapped on the central coupling in a 1:1 manner; however, they can be used to infer the nature of the central coupling.

Turning from the mechanical analogue to a biological system, apparently a straightforward inference from behavioral indications of crosstalk to the nature of crosstalk between signals on a central level is not possible. So far, only few models of central crosstalk have been formulated; therefore, this section concentrates on its peripheral manifestations.

On the peripheral level, coupling of movement characteristics is only one of the observable phenomena: simultaneous movements tend to become similar, but usually not identical, in certain ways but not in others. In addition to coupling, intrusions and confusions have been observed: one signal can temporarily replace another signal, or two signals can interchange (e.g., in a two-handed tracking task, one hand might do what would be appropriate for the other one). Particularly for movements also superposition of one signal on another and amplitude modulation of one signal with the frequency of the other signal have been found. Overall, the phenomena of crosstalk are such that they withstand simple classifications; thus far no complete classificatory system is available.

The study of crosstalk requires particular types of data. More or less arbitrary performance scores for the tasks combined are not sufficient. Instead, performance must be characterized by compatible descriptions with a sufficiently high temporal resolution. Spatio-temporal descriptions of simultaneous movements fulfill this requirement; therefore, most data on crosstalk are from the motor area. As soon as mental tasks are considered, adequate performance descriptions are harder to obtain. For simultaneous tasks such as driving a car and having a conversation, obtaining compatible measurements might be almost impossible. In such cases, whatever crosstalk may exist on a central level, it will not be identifiable on the behavioral surface as such.

B. Functions and Structural Substrates of Crosstalk

Although crosstalk will impair performance on probably most task combinations, it can be beneficial in others. In principle, it could result from a less-than-perfect design of the brain, but it could also result from a design that is adjusted to particular demands. In the latter case, crosstalk should be expected to support actions that are compatible with the primary purposes of system design, but impede incompatible actions. Several of the phenomena reviewed below suggest that crosstalk might indeed be functional and not only a deficit. Overall crosstalk between simultaneous movements, for example, appears to support "natural" and important behaviors such as keeping balance, locomotion, and cooperation of the two hands.

The conjectural functions of crosstalk give some hints as to its fundamental structural substrate. Locomotion and other basic behavior patterns seem to be supported by specialized structures of the central nervous system. For example, several complex reflex patterns with centers in the brain stem and other lower structures do exist. The increase of behavioral flexibility is achieved through building higher structures on top of lower ones; thus, the activity of lower structures becomes modulated. The lower structures become an integral part of the networks that control voluntary behavior, and they support it to the extent that it is compatible with their original functions and impede it otherwise.

Although this perspective on crosstalk cannot be considered established, it has some supportive evidence. Crosstalk appears to become stronger when the functions of higher brain structures are impaired. For example, section of the anterior commissures and anterior parts of the corpus callosum makes it essentially impossible to rotate two cranks with different frequencies when vision is excluded. Although with vision different frequencies can be produced, they become identical as soon as the eyes are closed. In addition, evidence indicates reflex modulation through superordinate activities and the support of voluntary action through the use of reflex pathways.

Brain structures that cooperate for the production of basic behavior patterns should be located close together. With this design principle not only communication times are minimized, but at the same time unavoidable crosstalk between adjacent locations would be turned from a disadvantage into an advantage. Therefore, crosstalk should be expected primarily to occur between tasks that are subserved by parts of the brain located in close vicinity. It has been proposed that tasks can be mapped on a "functional cerebral space." The distance of tasks in this space would mainly be determined by the strength of linkages between the structures that subserve them. The amount of crosstalk would depend on the distance of tasks in functional cerebral space; whether its effects improve or impair performance would depend on the nature of ongoing activity in the various structures. As far as the effects of task similarity on multiple-task performance are concerned, the notion of functional cerebral space can be used to derive more or less the same prediction as from multiple-resource models, at least to the extent that resource pools are somehow related to brain structures with particular functions. In addition, the concept of functional cerebral space is consistent with enhanced multiple-task performance.

C. Perceptual Crosstalk

Crosstalk can arise in perceptual processing. Perceptions generally serve actions, and depending on which task is performed at the moment, some stimuli are relevant while others are not. The selection of relevant stimuli is a major function of attention. Human actions are frequently directed toward objects in external space; thus, it is functional when attention is directed toward objects in space as well. Irrelevant stimuli can be neglected better when they are spatially remote from relevant stimuli or separated by some other factor that reduces perceptual grouping. Crosstalk will arise whenever attentional selectivity fails to a larger or smaller extent, and whether its effects will be beneficial or detrimental

depends on the exact nature of the stimuli, for example, when colors have to be named and a color word is presented simultaneously in a nearby location (the best-known variant of the task is to name the color of the letters of a color word), naming usually takes less time when the color word corresponds to the color and, which is the more dramatic effect, longer time when the color and the color word are incongruent. Interestingly, crosstalk in this so-called Stroop phenomenon does not result in a higher or lower proportion of errors (pronouncing the color word instead of naming the color), but in longer or shorter latencies of responses. This is probably an indirect effect of crosstalk.

In multiple-task performance, the relevant stimuli for one task must be kept separate from stimuli relevant for another simultaneous task rather than from irrelevant stimuli. Some evidence for perceptual crosstalk can be found in two-handed two-dimensional compensatory tracking in which only the distance between target and follower is shown on a screen. The error can be presented separately for both axes (e.g., using the length of two lines) or in an integrated manner (e.g., using length and direction of a single line). With integrated displays, the movements of the left and right hand, which serve to control the errors on the two axes, tend to become more similar. Thus, with a higher perceptual integration of signals relevant for the left and right hand, each signal affects movements of the wrong hand to a higher degree. This kind of crosstalk, of course, does not necessarily result in performance decrements. To the extent that similar control movements are required by the two hands, it can also enhance performance.

The reduction of perceptual crosstalk in multiple-task performance is accompanied by costs. Crosstalk is reduced whenever stimuli are separated perceptually to facilitate the selectivity of attention. However, when attention is focused on one stimulus, the other one, which is not irrelevant for the task combination, remains unattended. Therefore, attention switching between the stimuli becomes necessary to perform both tasks simultaneously. This additional requirement as well as the temporary attentional neglect of one or the other stimulus could reduce overall performance levels even though reduced crosstalk might serve to improve it. As a consequence, the performance effects of measures taken to reduce perceptual crosstalk are hard to predict; the study of integrated and separated displays in various task combinations has produced rather mixed results.

D. Motor Crosstalk

Crosstalk can arise between processes related to preparation and execution of movements, and it has been studied extensively as compared with other kinds of crosstalk. The following overview focuses on coupling with respect to different movement characteristics. Models of central coupling and only little studied phenomena such as superpositions and intrusions are neglected.

1. Temporal Coupling

a. Discrete Movements Temporal coupling is evident when aimed movements are performed simultaneously. Left-hand and right-hand movements with identical amplitudes share the random variations of movement time; computed across a series of pairs of movements the correlation between durations is high. When aimed movements with different amplitudes are to be performed simultaneously, movement times become almost identical, although they can be quite different when the movements are performed individually.

For movements with more complex trajectories, temporal coupling becomes even more striking than in aimed movements. Simultaneous movements with different geometric forms apparently have never been studied formally, probably because they are almost impossible to perform. The difficulty of the task can easily be demonstrated by asking people to draw a circle and a rectangle simultaneously. Although these patterns differ primarily in their geometric form, they differ in their temporal structure as well because geometric form and temporal structure of a movement are associated. (This association can be demonstrated: draw a circle with variable speed; distortions will be such that curvature is smaller when velocity is higher.) Thus, the problems generated by a request to draw two different geometric forms simultaneously may stem from the inability to produce two different temporal structures simultaneously, except under fairly specific circumstances that are revealed in studies of periodic movements.

b. Periodic Movements When frequencies of individual periodic movements are slightly different, they become more similar when performed si-

multaneously. This has also been observed when one of the movements was involuntary such as breathing. When frequencies do differ more, independence can be found for some task combinations in some subjects. Independence, however, seems to require that at least one of the two periodic movements is highly automated such as walking. For voluntarily controlled movements such as tapping different rhythms with the two hands, temporal coupling appears to be so strong that the task combination can only be performed by way of integrating the two rhythms into a single more complex one. Different lines of evidence point to this conclusion.

Simple rhythms with an integer ratio of frequencies can be integrated in a simple manner: a basic time interval can be defined, which corresponds to the period of the faster rhythm, and can be repeated over and over again. At the end of each repetition a response is performed with one or both hands. This scheme becomes more complicated for polyrhythms with frequency ratios such as $3:2$ or $5:4$. Therefore, polyrhythms should be produced with less accuracy than simple rhythms; this is in fact the case. Figure 6 presents the variability of timing errors in a task in which auditory pacing tones were presented to both ears; subjects had to tap in synchrony with the tones using both hands. Timing consistency was high (variability low) when the periods were identical for the two hands or of an integer ratio; timing consistency was low (variability high) for polyrhythms, and it approached random behavior for rhythms of $5:3$ (400–240 msec) and $4:3$ (400–300 msec). Only the $3:2$ rhythm (600–400 msec) was performed with fairly low variability; however, the subjects in this experiment were musically trained.

If the production of different frequencies with the two hands were in fact controlled by an integrated timing-control structure, integrated pacing signals should enhance performance as compared with separate ones. By presenting pacing tones for the two hands with a small or large difference in pitch, one can elicit the perception of a unitary sequence of tones that have different pitch or the perception of two different streams of tones. With perceptually integrated pacing signals, a $3:2$ rhythm with the two hands is performed more accurately than with "streamed" tones. This finding supports the notion of a single (integrated) timing-control structure. This notion is also supported by an analysis of the variances and covariances between the various time

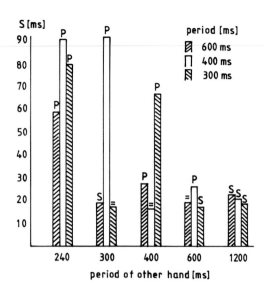

FIGURE 6 Standard deviation of timing errors in the production of intervals of 600, 400, and 300 msec (ms) duration when the other hand produces intervals of 240, 300, 400, 600, 1,200 msec duration. Identical periods in both hands are marked by =; integer ratios (simple rhythms) by S; and noninteger ratios (polyrhythms) by P. [Adapted, with permission, from D. Deutsch, 1983, The generation of two isochronous sequences in parallel. *Percep. Psychophys.* **34,** 331–337.]

intervals in a 3 : 2 rhythm. Thus, temporal coupling appears to be sufficiently strong to enforce an integration of the two tasks into a unitary, but more complex, one, which may become too difficult for certain polyrhythms. It should be noted that the unified timing-control structure is probably not as simple as suggested above (the smallest interval common to the two rhythms is repeated). Whatever its exact nature is, it contains cross-hand intervals, which are triggered by a tap of one hand and end with a tap of the other, and not only within-hand intervals.

Temporal coupling is a rather pervasive phenomenon that cannot only be observed in simultaneous movements, but also when one of the rhythms is perceived and the other one has to be produced; it is hard to dance a waltz to a marching band. Such observations and others do suggest that timing mechanisms in perception and movement production are very much the same.

c. Lateral Asymmetry Lateral asymmetry of temporal coupling can easily be demonstrated. The task is to produce a steady beat with one hand and maximally fast tapping or rhythms of a certain complexity (such as x-xxx-x-) with the other. For right-

handers, this task is easier when the left hand takes the beat. When the right hand takes the beat, performance breaks down already at lower beat frequencies. The performance breakdown in the less favorable condition is characterized by the right hand taking over the rhythm of the left hand. Similar asymmetries can be observed for the feet and hand–foot combinations.

The right–left asymmetry of temporal coupling exhibits a striking relation to the normal cooperation of the two hands in bimanual tasks. The left hand typically has a holding function; it defines a spatial reference with respect to which the right hand performs its manipulations. This assignment would benefit from a coupling of the right hand to the mostly slower movements of the left hand, by which the right hand is supported in following shifts of the spatial reference defined by the left-hand position. Conversely, the left hand should be protected against crosstalk from the faster manipulations of the right hand. Thus, the artificial tapping task seems to reveal a particular kind of asymmetric crosstalk that supports the normal cooperative use of the two hands.

2. Phase Coupling

a. Discrete Movements The phase of a periodic signal characterizes its temporal position relative to some arbitrarily defined zero. For aperiodic signals, the term can be used correspondingly. Simultaneous aperiodic movements tend to become synchronized in time not only with respect to their duration but also with respect to their start. Reaction times for simultaneous movements become more similar as compared with reaction times for the same movements performed individually.

b. Periodic movements Periodic movements of the two hands with identical periods exhibit two stable relative phases. Figure 7 illustrates this for a bimanual tapping task in which taps for the two hands were initiated with the support of pacing signals and thereafter continued without pacing. The period was 1,000 msec and the required phases were 0, 0.1, 0.2, . . . , 1.0. Figure 7a presents the variability of temporal errors. Variability is low at relative phases of 0 (sychronous tapping) and 0.5 (alternating tapping) and higher in-between. The mean errors of Fig. 7b indicate that the intermediate phases are biased toward one or the other of the two stable phases. (The zero errors that are shown for

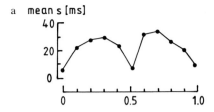

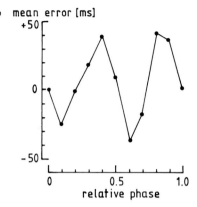

FIGURE 7 Standard deviation of timing errors (a) and mean timing errors (b) in the production of 1,000-msec (ms) intervals with the two hands as a function of relative phase. [Adapted, with permission, from J. Yamanishi, M. Kawato, and R. Suzuki, 1980, Two coupled oscillators as a model of the coordinated finger tapping by both hands. *Biol. Cybernet.* **37**, 219–225.]

two of the intermediate phases are an artefact of averaging across subjects: at these phases some subjects are biased toward the relative phase of 0 whereas others are biased toward the relative phase of 0.5.)

Observations like those of Fig. 7, which indicate two stable relative phases, can also be made when continuous oscillations of fingers or hands are studied rather than discrete tapping responses. For such tasks, it has been found that the relative phase of 0.5, which is stable for low frequencies, becomes unstable when the movements are speeded up. When one oscillates the two hands such that they move in the same spatial direction and gradually increases the frequency, a shift from a relative phase of 0.5 to a relative phase of 0 (symmetric movements) cannot be avoided. Similarly, alternate tapping with the two hands tends to be synchronized (relative phase of 0) when speeded up.

The relative phase of 0, which can hardly or not be avoided with high-frequency movements, is characterized by simultaneous activation of homologous muscle groups of the two sides of the body. The tendency toward this relative phase thus might be better described as a tendency to coactivate ho-

mologous muscles, i.e., a homologous coupling. Evidence for homologous coupling can be found in a variety of tasks in which different muscles on both sides of the body must be activated in rapid succession; however, no evidence of homologous coupling can be seen when there is sufficient time to first select the to-be-activated muscles for each side and only thereafter to activate them (e.g., in slow oscillations of the two hands). Homologous coupling appears to be a transient phenomenon that does exist during selection of to-be-activated muscles, but not thereafter.

Homologous coupling across the body midline is not universal. In swinging the arms back and forth, antagonistic coupling can be seen, which corresponds to the normal use of the arms in walking. It is noteworthy that both symmetric movements in a lateral direction, which are supported by homologous coupling, and simultaneous backward and forward movements support balance in that they keep shifts of the position of the center of gravity small.

In addition to coupling in terms of body-related referents, space-related coupling has been found. Simultaneous up-and-down movements of hand and foot, for example, are easier when both move up or down at the same time than when one moves up while the other moves down. This is true whether the hand is pronated or supinated; by this variation of hand position, the muscle groups that are coactivated are interchanged.

3. Force Coupling

In simultaneous discrete movements of the two arms, forces are essentially independent. In contrast to durations, the random variations of the amplitudes of simultaneously aimed movements are not shared by the two hands. Also, when simultaneously performed movements have different amplitudes, the increased similarity of durations is accompanied by an increased dissimilarity of peak forces. Finally, drawing identical geometric forms with different sizes, using both hands, seems to cause no particular difficulty.

Although force coupling has not been found (or only in a very weak form) in discrete movements, it is apparent in periodic movements. For example, when a single syllable is repeatedly produced in synchrony with finger movements, the instruction to stress certain syllables results in larger-amplitude movements. This effect has been observed, although the subjects were instructed to keep finger movements constant. The complement, stress on syllables, which are temporally coincident with larger-amplitude finger movements, has also been found. The discrepancy between the findings with discrete and periodic movements suggests that force coupling might be a transient phenomenon similar to homologous coupling, which can only be evidenced when forces are specified but no longer thereafter.

E. "Mental" Crosstalk

Crosstalk between primarily mental tasks is hard to identify, as already mentioned; however, some evidence for its existence can be found. This mainly takes the form of intrusions and confusions between responses appropriate for the two tasks. An example is the somewhat esoteric task combination of typing a visually presented text and simultaneously shadowing an auditorily presented text. Remarkably, these two tasks can be performed simultaneously with only negligible multiple-task performance decrement—after a sufficient amount of practice, of course. In this task combination, sometimes words that should be typed are spoken and vice versa. The risk of confusions can be increased when the texts become identical for a while and then different again; this divergence is usually not reflected in the manual and vocal outputs, which tend to remain identical.

IV. Concluding Remarks

Resource-competition models and crosstalk models represent the major approaches to multiple-task interactions; however, they probably do not exhaust the factors that are relevant for the changes of performance on one task when a second simultaneous task is introduced. Attention switching is one of the additional factors that has briefly been touched. Although various factors may affect the final outcome of multiple-task performance, it can be questioned that resource-competition is one of them.

Resource-competition models are less specific than crosstalk models; therefore, one can ask whether or not the typical multiple-task performance decrement is simply a consequence of various kinds of crosstalk and does not require resource-competition as an additional explanatory concept. So far, however, no existing evidence proves that the performance decrement is indeed exclusively caused by crosstalk. Experiments in which both global performance measures and crosstalk phenomena were examined have found performance decrements without accompanying evidence for crosstalk, differences between conditions in the performance decrement without differences in crosstalk, and differences in the amount of crosstalk that were not associated with different performance decrements. Thus, according to the experimental evidence, performance decrements and crosstalk phenomena are only loosely associated.

However, the relation between performance decrements and crosstalk phenomena is more a matter of belief than of experimental evidence. At least two arguments can be raised whenever performance decrements cannot be explained in terms of known crosstalk. The first argument is that the relevant crosstalk has not been identified. The second argument states that crosstalk may have secondary consequences. For example, different rhythms have been found, which, under the instruction to perform them together, were performed in an alternating manner. Generalizing this finding, crosstalk may have the secondary consequence that strategies are employed that serve to avoid it (e.g., a serial rather than parallel organization of operations). The control mechanism of the single-channel model is a device that functions according to this principle. The effect of such devices would be to avoid crosstalk but at the cost of a longer time needed to complete a set of operations (the Stroop phenomenon described above is probably an example for this). With such an argument, crosstalk can cause performance decrements without being itself perceptible.

Bibliography

Damos, D. (ed.) (1991). "Multiple-Task Performance." Taylor & Francis, London.

Heuer, H. (1985). Some points of contact between models of central capacity and factor-analytic models. *Acta Psychol.* **60,** 135–155.

Heuer, H., and Wing, A. M. (1984). Doing two things at once: Process limitations and interactions. *In* "The Psychology of Human Movement" (A. M. Wing and M. M. Smyth, eds.). Academic Press, London.

Neumann, O. (1987). Beyond capacity: A functional view of attention. *In* "Perspectives on Perception and Action" (H. Heuer and A. F. Sanders, eds.). Erlbaum, Hillsdale, Illinois.

Wickens, C. D. (1989). Attention and skilled performance. *In* "Human Skills," 2nd ed. (D. H. Holding, ed.). Wiley, Chichester.

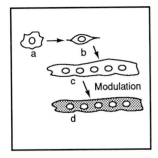

Muscle Development, Molecular and Cellular Control

HENRY F. EPSTEIN *Baylor College of Medicine*

I. Embryonic Muscle Cells
II. Specific Protein Isoforms
III. Gene Regulation
IV. Morphogenesis
V. Extracellular Signals

Glossary

Assembly Organization of specific proteins into biologically active structures

Isoform Structurally distinct form of a functionally defined class of proteins that shows spatial and temporal specificity developmentally

Mesenchyme Population of mesodermal cells that can differentiate into adipose, cartilage, or muscle cells

Myoblast Determined embryonic cell that generates differentiated muscle cells

Myofibril Unit organelle of striated muscle that exhibits contractile structure and function

Myogenesis Overall process of muscle development from determination of embryonic muscle cells to terminal differentiation of specific muscles

Sarcoplasmic reticulum Internal network of membranes that surrounds the myofibril. Specialized proteins of this membrane can pump calcium into the intramembrane space, and others can release calcium to the myofibril on stimulation

THE MOLECULAR AND cellular biology of muscle development is concerned with the identification and characterization of proteins and their genes active in the determination and differentiation of muscle cells. Several genes and their encoded proteins have been identified as necessary for the determination of uncommitted cells to myogenesis. Specific isoforms of actin, myosin, and many other proteins

are synthesized in myoblasts and differentiating muscle cells. The mechanisms of gene activation required for the expression of specific isoforms and the extracellular signals regulating these mechanisms have been characterized. The organization of proteins into the contractile filaments, cytoskeleton, and membranes of differentiated muscle cells is becoming understood in terms of the assembly of specific isoforms. The interactions of specific proteins underlying nerve–muscle transactions are being characterized. The resolution of Duchenne muscular dystrophy in terms of specific defects in DNA, protein, and membrane–cytoskeletal interactions exemplifies the use of the molecular and cellular approach to muscle development.

Embryonic Muscle Cells

A. Mesodermal Origin

Skeletal, cardiac, and smooth muscles in vertebrates arise from the mesodermal layer in post-morula embryos. The exact origin and structure of the mesoderm differs from lower vertebrates such as fish and amphibia, in which true gastrulation occurs, and the higher vertebrates such as birds and mammals, in which the embryo forms in a restricted cell mass, the blastoderm. In the higher forms, a primitive streak marks the start of proliferation and invagination of cells to form the mesoderm. In humans and other higher primates, the intraembryonic mesoderm (i.e., a single row of cells) is derived from the ectoderm and is situated between the ectodermal and endodermal cells. [*See* CARDIAC MUSCLE; SKELETAL MUSCLE; SMOOTH MUSCLE.]

The notochord forms from Hensen's node, a thickening at the anterior of the primitive streak. The intraembryonic mesoderm thickens and forms bilateral masses called *paraxial mesoderm* about the notochord. At the lateral edges of each paraxial mass, the cells thin out to form the lateral plate mesoderm. The paraxial mesoderm will form the somites out of which skeletal muscles arise. The lateral plate will form the splanchnic mesoderm out of which the cardiac and smooth muscles will arise. The possibility that special striated or smooth muscle cells of the head may originate from the head mesenchyme, which in turn arises from the ectodermally derived neural crest, has been much discussed. Current evidence based primarily on the embryology of chick-quail chimeras indicates that nearly all muscle cells are mesodermal in origin. Diffuse groups of mesodermal cells exist that are clearly anterior to the notochord and paraxial masses. These cells are the likely origin of various special head structures such as the extraocular muscles. The iris of the eye and the myoepithelial cells of the sweat glands may be the only "muscle-like" cells of ectodermal origin.

Mesodermal cells give rise to many other cell types besides muscle. These cell types include blood, fat, and connective tissue cells. At the earliest stages, the morphologically undifferentiated mesenchymal cell constitutes the mesoderm. Several major questions remain unanswered about the mesenchymal cell. Are subpopulations of mesenchymal cells determined for generating specific parts of the early mesoderm or of specifically differentiated cells? The latter include fibroblasts, adipocytes, and chondrocytes in addition to myoblasts. Particular controversy exists as to whether some or all fibroblasts are essentially mesenchymal cells. If so, do such fibroblasts remain as potential mesodermal stem cells for the remainder of life? A partial answer may be found in lower vertebrates such as salamanders, which can regenerate limbs. Operationally, stem cells must exist in the blastema of the stump that will generate muscle, bone, and connective tissue.

B. Somitic and Splanchnic Mesoderm

The paraxial masses segment bilaterally into somites. There are usually four occipital, eight cervical, 12 thoracic, five lumbar, five sacral, and 8–10 coccygeal pairs of somites in human embryos. The somite pairs appear temporally in cranial to caudal order. The first occipital pair and seven to eight coccygeal pairs retrogress embryonically and do not give rise to skeletal muscle as the other somites. Similar cell masses develop out of the mesenchyme

of the primitive orbit and about the six pairs of branchial arches. These latter structures will produce striated muscles structurally similar to skeletal muscle but have distinct innervation and are affected differentially by both inherited and acquired muscle disorders.

Each somite differentiates into a myotome, a sclerotome (leading to the axial skeleton), and a dermatome (leading to the integument). The first appearance of spindle-shaped embryonic myoblasts occurs in the myotome. These cells are proliferative and capable of migrating to growing regions of the embryo.

The splanchnic or visceral mesoderm is the origin of cardiac muscle and the smooth muscles of the vascular system and visceral organs.

Paired endothelial tubes form within the splanchnic mesoderm and fuse into a cardiac vesicle. This primitive structure differentiates into a tubular heart (as in fish), which in higher vertebrates undergoes a complex set of morphogenetic processes to form more complex structures. The muscle cells of the artria and ventricles of the four-chambered mammalian heart, the papillary muscles associated with the valves, and the specialized Purkinje fibers of the medial septal conducting system appear to originate in the early cardiac myoblasts. The blood vessels and their smooth muscles begin to form coordinately with the heart. In other regions of the lateral mesoderm, the smooth muscles associated with specific visceral organs begin to form.

It is clear that myoblasts become determined and begin differentiation in many parts of the mesoderm and embryo. This fact alone would suggest that the origin of muscle or even of any type of muscle is polyclonal (i.e., from many different mesenchymal cells).

C. Primary and Secondary Myoblasts

Myogenic cells from embryonic limb buds and developing limbs may be cloned *in vitro*. These cells proliferate in tissue culture to generate colonies of myoblasts that fuse into primitive muscle fibers or myotubes. The clones of myoblasts are already determined to exhibit distinct muscle phenotypes. A subset of clones will proliferate first to form the short primary myotubes of the earliest muscle. Most clones proliferate later in development and form the longer, more permanent secondary myotubes. Both primary and secondary populations may be subdivided in chick embryo limb buds on

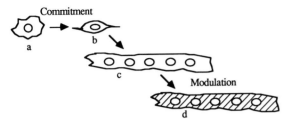

FIGURE 1 Clonal commitment and terminal modulation in skeletal myogenesis: *(a)* uncommitted mesenchymal cell; *(b)* committed skeletal myoblast; *(c)* early developing myotube; *(d)* fully differentiated mature skeletal muscle fiber. [Reprinted with permission from Stockdale, F. E., Miller, J. B., Feldman, J. L., Lamson, G., and Hager, J. (1989). Myogenic cell lineages: Commitment and modulation during differentiation of avian-muscle. *In* "Cellular and Molecular Biology of Muscle Development" L. H. Kedes and F. E. Stockdale, (eds.), pp. 3–13. Alan R. Liss, New York.]

the basis of myosin isoform expression in the myotubes that they generate (Fig. 1). Fast, slow, and mixed (fast/slow) classes of myoblasts may be distinguished by myosin typing. Because many other muscle proteins exhibit multiple isoforms, the possibility exists that still other classes of skeletal myoblasts exist. In mammals, the skeletal muscle fibers appear to be a complex mosaic of expression in terms of various protein isoforms. The fusion of multiple distinct clones of myoblasts in the formation of most mammalian fibers may explain their heterogeneous phenotype.

D. Satellite Cells

In the formation of secondary myotubes, a small number of myoblasts do not fuse nor terminally differentiate and retain their capacity for proliferation. These quiescent myoblasts lie beneath the sarcolemma at the periphery of mature fibers in skeletal muscle, hence the name *satellite cells*. Skeletal muscles can regenerate after a variety of injuries and in certain diseases such as Duchenne muscular dystrophy. The satellite cells proliferate to form new secondary myotubes. These cells may be cultured *in vitro* and will generate new myoblasts and myotubes. Satellite cells cloned from fast or slow chicken muscles will produce myotubes *in vitro* that synthesize only fast or slow myosins, respectively. The maintenance of satellite cells and their clonality permits skeletal muscle to regenerate according to a pattern of differentiation similar to its original development.

E. Cardiac and Smooth Myoblasts

Early in embryonic development, the myoblasts that will form cardiac and smooth muscles are spatially separated and phenotypically distinct from skeletal muscle myoblasts. In contrast to skeletal myoblasts, cardiac myoblasts do not fuse and in mammals lose their proliferative ability with terminal differentiation. Cardiac myoblasts can undergo differentiation *in vitro* into closely adherent sheets, one to two cells thick. The cardiac cells *in vitro* as in the heart are attached to each other by adherent junctions and desmosomes.

Smooth muscle myoblasts also do not fuse but become connected electrically through gap junctions. Differentiated smooth muscles contain cells capable of proliferation. An important example of this capacity is the proliferation of myoblasts in arterial walls after injury or in the presence of high cholesterol levels. This response is believed to be the biological process leading to atherosclerosis. Vascular and uterine muscle cells can be cultured *in vitro*. [*See* ATHEROSCLEROSIS; CELL JUNCTIONS.]

Clearly, cardiac myoblasts form distinct differentiated structures including Purkinje fibers, atrial and ventricular muscle, and the papillary muscles. Similarly, different smooth muscles are highly differentiated including arterial and venous vascular smooth muscle, circular and longitudinal alimentary muscles, and urinary and reproductive tract muscles. Distinct populations of cardiac myocytes that express different myosins have been cultured. Although clonal analysis at the level of the skeletal muscle studies has not been done, the existence of multiple clones of distinct commitment in both the cardiac and smooth myoblast populations must be considered as highly likely.

II. Specific Protein Isoforms

A. Muscle Proteins

Many kinds of proteins are involved in muscle differentiation. In general, specific isoforms of these proteins are associated with one or more types of muscle; other isoforms are produced in other types of muscle cell or nonmuscle cells. Historically, the first examples of muscle-specific isoforms were found in enzymes of intermediary metabolism such as glycogen phosphorylase, hexokinase, lactate dehydrogenase, and creatine kinase.

Contractile proteins such as myosin and actin

were originally considered to be produced only in muscle tissue, as is the case with hemoglobin in erythrocytes or rhodopsin in photoreceptor cells. One of the great triumphs of modern biology has been the recognition that myosin and actin are nearly ubiquitous proteins contributing to the motility and cytoskeleton of most eukaryotic cells. Similarly, the proteins responsible for calcium regulation such as tropomyosin and troponin are homologues of general cell constituents.

The most "muscle-specific" proteins to date appear to be active in the mechanism of excitation–contraction coupling. The calcium-ATPase pump, the ryanodine-blocked calcium release channel of sarcoplasmic reticulum, and the dihydropyridine-blocked channel of the t-system have been characterized in skeletal muscle; the sarcoplasmic reticulum calcium ATPase is also found in cardiac muscle. [*See* MUSCLE, PHYSIOLOGY AND BIOCHEMISTRY.]

B. Temporal and Spatial Specificity

The use of specific isoforms as markers of particular developmental stages and terminal differentiation has been powerful in understanding muscle development. The clonal analysis of skeletal myoblasts depended on the characterization of distinct myosin heavy-chain isoforms of fast and slow muscle. The biochemical and immunological properties of the myosin isoforms served as a tool for demonstrating clonal determination and as a characterization of the final differentiated states of skeletal muscle fibers.

The temporal relation between myoblast fusion and the expression of differentiated characteristics was understood by the study of isoforms. The expression of the M form of creatine kinase rather than the B isoform, α-actin instead of β- and γ-actins, and sarcomere-specific myosin heavy chains in contrast to cytoskeletal myosin are examples of transitions occurring in parallel with myotube fusion.

Skeletal, cardiac, and smooth muscles all produce type-specific isoforms of certain proteins. In particular, specific isoforms of myosin heavy and light chains, α-actin, and tropomyosin are most abundant in each of the major muscle types. At this level, the expression of isoforms may not be exclusive to one type. For example, α-skeletal and α-cardiac actins, isoforms of nearly identical amino acid sequence, are produced in both kinds of muscle but are predominant in one or the other.

In the terminally differentiated state of certain skeletal and cardiac muscles, a second or even third transition of expression occurs. The additional transitions in myosin isoforms have been most intensively studied. The primary myotubes of chicken pectoralis muscle exhibit a sarcomere-specific, fast myosin heavy chain, the embryonic isoform. The secondary myotubes begin to produce another fast myosin heavy chain, the neonatal isoform. Further modulation of expression lends to expression of the adult isoform. During the regeneration process after injury, the three myosin isoforms will also be expressed sequentially.

Another example of modulation in skeletal muscle development of both birds and mammals is the production of a second form of type-one myosin light chain (LC3) in addition to the original LC1. In the cardiac ventricles of small mammals, two isoforms of myosin heavy chain can be expressed. At birth one form is expressed, but postnatally the second isoform becomes increasingly predominant. A variety of physiological perturbations alters the relative expression of the two isoforms.

C. Functional Differences

Two clear examples of functional differences between isoforms in a muscle cell concern the behavior of myosin heavy chains in the body-wall muscle cells of the nematode, *Caenorhabditis elegans*, and in the cardiac ventricles of small mammals. In both cases, two separately encoded myosin heavy chains are synthesized in the same cells. In the nematode muscles, only the homodimers AA and BB occur, whereas in the heart, the heterodimer $\alpha\beta$ and the homodimers $\alpha\alpha$ and $\beta\beta$ are detected. This observation indicates that the interaction between myosin heavy-chain isoforms in the two muscles must differ. It is not known whether the combinational differences are intrinsic to the different amino acid sequences of the heavy-chain isoforms or require the interaction of additional proteins.

In the nematode, the myosin homodimers assemble within the same thick filaments. However, the two myosins segregate to different regions of the filaments. The AA isoform resides in the central zones, and the BB isoform is located in the flanking polar regions (Fig. 2). In this case, the differential assembly of the two myosins requires interaction

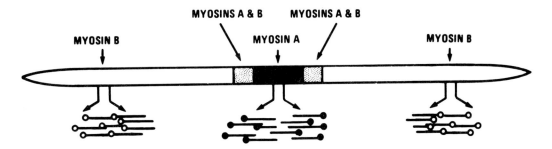

FIGURE 2 Differential location of myosin isoforms in nematode thick filaments. Myosins A and B are the AA and BB isoforms. Solid figures represent myosin A and indicate antiparallel packing of these molecules in the bipolar central zone. Open figures represent myosin B and indicate parallel packing of these molecules in the flanking polar regions. [Reprinted with permission of the Rockefeller University Press from Epstein, H. F., Ortiz, I., and MacKinnon Traeger, L. A. (1986). The alteration of myosin isoform compartmentation in specific mutants of *Caenorhabditis elegans. J. Cell Biol.* **103,** 985–993.

with a third protein, paramyosin. This latter molecule is also separately encoded and is homologous in amino acid sequence to the rod portions of myosin heavy chains. The myosins scramble in mutant thick filaments lacking paramyosin. The paramyosin forms a molecular substratum for myosin in nematode as well as in other nonvertebrate thick filaments. Experiments in rabbit hearts suggest that $\alpha\alpha$ myosin is added preferentially to the polar regions of thick filaments already containing $\beta\beta$ myosin.

The most significant functional differences in the cardiac myosin isoforms are in their enzymatic activities measured *in vitro.* $\alpha\alpha$ myosin exhibits over threefold higher ATPase than $\beta\beta$ myosin. The $\alpha\alpha$-rich epicardial wall shows more rapid excitation and contraction than the $\beta\beta$-rich endocardial wall. The physiological behavior of specific regions of the ventricular muscle correlates with their myosin isoform content.

In contrast to the clear differences in assembly and ATPase of the nematode and cardiac myosin isforms, respectively, many muscle protein isoforms show little or no functional differences. The extreme cases are in the multiple actins of the slime mold *Dictyostelium* and the nematode, which are distinguishable by the nucleotide sequences of their cognate genes but which are identical in amino acid sequence. The functional differences lie not in the proteins themselves but in the differences in regulation of the encoding genes. Certain isoforms appear

to represent distinct regulatory cassettes that are differentially expressed at specific stages and in specific tissues.

III. Gene Regulation

A. Transcriptional Control

The phenomenology of cellular differentiation suggests the activation of expression of certain genes and the suppression of activity of other genes. The specific phenomenon of muscle-specific protein isoforms whose expression may be restricted to particular stages and cell types focused the issue. What mechanisms underlie the observed patterns of isoform diversity, switching between isoforms, and induction of high levels of synthesis of terminal isoforms of certain proteins?

Modern nucleic acid biochemistry and recombinant DNA technology have revolutionized the study of muscle development. The ability to detect mRNAs for specific muscle proteins, to synthesize cDNAs for the specific mRNAs, and to clone and sequence the cognate genes including both coding and regulatory elements now permit the direct analysis of gene activity in chemical terms. This analysis has been most intensive for the expression of actin and myosin isoforms but has been extended to include the spectrum of muscle proteins including tropomyosin and the troponins of the actomyosin regulatory system, the calcium ATPase, calcium release channel and excitatory protein of the sarcoplasmic reticulum and t-system, and the acetylcholine receptor, sodium-potassium ATPase and sodium channel of the muscle fiber membrane.

Transcriptional activation in terms of induction of mRNA synthesis for specific proteins is a central feature of muscle differentiation. In higher vertebrates, two closely related forms of actin, α-cardiac and α-skeletal, have evolved from the single α-actin

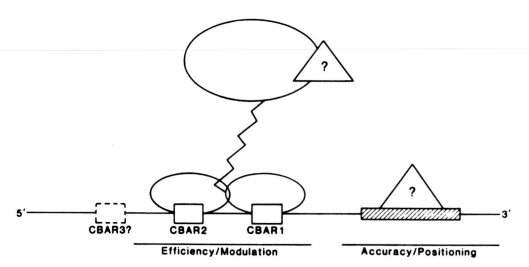

Efficiency/Modulation   Accuracy/Positioning

of lower vertebrates. The two actins are coexpressed in both cardiac and skeletal muscles during development and homeostasis. The α-cardiac mRNA is expressed first in both kinds of muscle, but the α-skeletal mRNA and actin predominates in skeletal myotubes. The tissue specificity of this expression resides within the DNA region flanking the 5' end of the gene and in at least two separable proteins that bind to sites within this DNA region. The DNA sites have been characterized as promoters for the initiation of transcription and as enhancers of transcription (Fig. 3). The combination of specific DNA sites and active protein factors appears necessary for proper expression not only of the actin genes but also in other muscle-related gene families. [See DNA and Gene Transcription.]

B. Alternative RNA Splicing

In several muscle genes, more than one mRNA and polypeptide are encoded by a single gene. The synthesis of fast myosin light chains 1 and 3 is the classic example of alternative splicing in skeletal muscle development (Fig. 4). Amino acid sequence analysis indicated that the two proteins were identical over most of their structure but diverged in their amino terminal regions. Distinct mRNAs were detected by cDNA synthesis; however, only one genomic DNA sequence was found. Two distinct promoter regions exist in close apposition to exons coding for the alternative amino terminals. Remarkably, more than 10 kb of DNA separates the two promoter regions. The most 5' promoter initiates light-chain expression with a loop containing the light-chain 3 promoter and specific exons are removed by the RNA splicing mechanism. The closer

FIGURE 3 Model of promoter region of skeletal α-actin gene. *Upper ellipsoid triangle* reflects combinatorial interactions of higher order *trans*-acting elements. This process produces a transcription activating protein, which recognized the CBAR1 and two sequences. Exact geometry of the CBAR elements with respect to one another (*ellipsoids*) as well as their DNA base sequences (*boxes*) is necessary for tissue-specific activation. A third element (Bar 3) and the overall positioning of the CBAR elements to the starting point of transcription is critical. [Reprinted with permission from Schwartz, R. J., Grichnik, J. M., Chow, K-L., DeMayo, F., and French, B. A. (1989). Identification of *cis*-acting regulatory elements of the chicken skeletal α-actin gene promoter. *In* "Cellular and Molecular Biology of Muscle Development" L. H. Kedes and F. E. Stockdale, (eds.), pp. 653–667. Alan R. Liss, New York.]

promoter initiates light-chain 3 mRNA. When the type-3 promoter is turned on, mRNA for the 5' flanking light-chain exons is synthesized, and a small loop containing a small proximal type-1 exon is removed by splicing (Fig. 4.) [See DNA Synthesis.]

Other examples of alternative processing to produce multiple mRNAs from the same gene include the α-tropomyosins and troponin T isoforms of vertebrate muscles and the myosin heavy chains and tropomyosins in *Drosophila*.

C. Posttranscriptional Modulation

Although transcriptional control of gene expression is the predominant regulatory mechanism during development, there is clear evidence for additional process that affect the stability and translatability of mRNA and the turnover of specific proteins in muscle. These mechanisms are particularly evident in developmental transitions that involve isoform

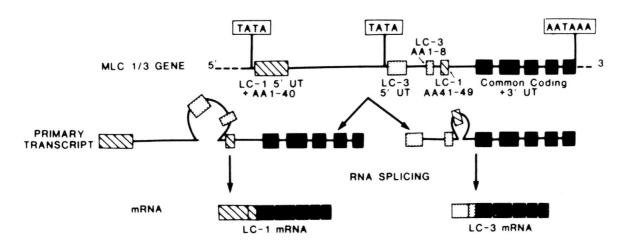

FIGURE 4 Alternative transcription and splicing of myosin light-chain gene. The 5' end of gene is depicted. Separate promoters are activated for LC-1 and LC-3. Resulting transcripts are alternatively spliced to yield mRNAs with different sequences. Alternative mRNAs are translated into related but distinct proteins. *Solid bars* indicate common coding regions. *Dotted bars* are LC-3 unique sequences. *Hatched bars* are LC-1 unique sequences. [Reprinted with permission from Nadal-Ginard, B., Breitbart, R. E., Strehler, E. E., Ruiz-Opazo, N., Periasamy, M., and Mahdavi, V. (1986). Alternative splicing: A common mechanism for the generation of contractile protein diversity from single genes. *In:* "Molecular Biology of Muscle Development" C. Emerson, D. Fischman, B. Nadal-Ginard, and M. A. Q. Siddiqui, (eds.), pp. 387–410. Alan R. Liss, New York.]

switching. The translation of existing mRNA may be reduced, and the degradation of specific mRNA and protein isoforms may be enhanced.

One suggested mechanism of translational regulation involves inhibitor RNA molecules that block initiation of specific mRNA translation by competition for ribosomal binding. Such inhibitors have been isolated from chick and rat embryonic skeletal muscles. The inhibitor RNA molecules appear to be complexed with proteins, but neither the RNA or protein molecules have been characterized in terms of sequence, specificity, or stoichiometry of the complexes.

An important process in skeletal muscle development and homeostasis is the degradation of myofibrillar proteins. As part of isoform switching, the original protein is degraded and disappears. The selectivity of degradation is implied by the distinct half-times of specific myofibrillar proteins. Closely related proteins such as myosin heavy and light chains or actin and tropomyosin show distinct half-times. The differences may be due to characteristic

exchange mechanisms between the polymerized or bound state and the monomeric form of each protein. The actual degradation systems are also capable of discriminating between monomeric proteins composed of very different structural domains and motifs.

D. Gene Regulation is Combinatorial

The regulation of genes in muscle development includes control mechanisms affecting transcription, RNA processing, translation, and protein turnover. Multiple mechanisms also operate at each level of expression. For these reasons, it is unlikely that unique switches control myogenesis or specific myogenic programs. The regulation of commitment and differentiation results from specific combinations of control mechanisms. The regulatory output of these combinations does not appear quantized to two states (i.e., a developmental state is not switched on or off). Instead, the output resembles a complex spectrum of regulatory activity in which thresholds for commitment at early time points and differentiation at later time points may be reached by several combinations. Accordingly, multiple intracellular factors and extracellular signals influence the quantitative and qualitative aspects of development regulation. Each development stage and its spectrum of regulatory activity is linked in time to produce a pathway for molecular differentiation.

Recently, myogenic molecules have been identified. At least three DNA-binding proteins related to known proto-oncogenes of the *myc* family MyoD1, Myf-5, and myogenin, and a fourth, distinct factor *myd* can induce myogenesis in the rat fibroblast cell line 10T½. The cell line ordinarily produces a spectrum of differentiated cell types on treatment with

5-azacytidine, an inhibitor of DNA methylation. The treated cells lead to clonal populations of adipoblasts, chondroblasts, or myoblasts as well as parental fibroblasts. The expression of these DNA-binding proteins permits activation of many muscle-specific genes including specific isoforms of myosin heavy chain, actin and troponin (Fig. 2). The expression of *MyoD1* in a variety of mesodermal and nonmesodermal cells induces both cell fusion and muscle gene expression. The *MyoD1* protein activates its own synthesis and in turn is regulated by phosphorylation.

Two extracellular factors, fibroblast growth factor and tumor growth factor β, play important roles in the commitment of early embryonic cells to mesodermal and myogenic differentiation. Commitment and differentiation may be reversed by the action of oncogenes such as the SV40 large T antigen, by reagents such as ethyl methane sulfonate, a mutagen and carcinogen, and by tetradecanoyl phorbol acetate, a cocarcinogen. Multiple intrinsic and extrinsic signals and switches appear to operate in the myogenic process.

IV. Morphogenesis

A. Myoblast Structure

Myoblasts of skeletal muscle must migrate to proper sites for myogenesis, withdraw from cell division, and be capable of aligning and fusing with other myoblasts to form the striated myotube. Myoblasts may be viewed as a cell with specific structures and functions that will enable terminal differentiation to be achieved.

The overall morphology of skeletal myoblasts is unique compared with other mesenchymal elements, especially fibroblasts. The myoblasts are spindle-shaped because of highly anisotropic cytoskeleton containing both microtubules and actin filaments. The latter stress–fiber-like structures may form a type of nucleation structure for the future thick and thin filaments. Gap junctions form between myoblasts before the onset of fusion. The plasma membranes are rich in N-CAM, which may also play a role in their fusion.

Molecular studies indicate that myoblasts are synthesizing mRNAs for multiple muscle-specific isoforms, particularly the mRNA of α-cardiac actin. Sarcomeric myosin heavy chain is produced in postreplicative myoblasts. Thus, multiple genes appear activated at the myoblast stage before the use of their protein products in the terminally differentiated myotube.

B. Fusion of Myoblasts into Myotubes

The fusion of myoblasts into myotubes is a unique feature of skeletal myogenesis. The process follows withdrawal of the myoblasts from the cell cycle and requires extracellular calcium. The sequence of molecular events underlying myoblast recognition and binding and fusion of plasma membranes has not yet been determined. Current work is proceeding on the identification of membrane proteins active in the fusion process. Morphological evidence suggests that the formation of gap junctions between myoblasts may precede their fusion.

Two features of the mature multinucleated muscle fiber as a syncytium are noteworthy. First, satellite cells capable of further proliferation are associated in close apposition with the membrane of the fiber. The satellite cells permit repair and regeneration of muscle fibers. Second, individual nuclei or groups of nuclei along the fiber appear to be expressing different genes. These nuclear domains may be related to the diverse physiological functions of muscle, including near-muscular junctional transmission, excitation–contraction coupling, contracting and its regulation, and intermediary metabolism and its regulation.

C. Organization of Myofibrils

Myofibrillar structures are directly observed as the definitive striations in skeletal and cardiac muscles. The microscopic striations arise as the result of the assembly of myosin and actin into filaments of narrowly defined lengths, the organization of thick and thin filaments into crystalline arrays, and the alignment of myofibrils both within individual fibers and between separate fibers. Current research work is focused on resolving the roles of accessory proteins in the assembly and organization of the myosin and actin filaments. Interaction between myosin and actin may be necessary for assembly as well as contraction.

In skeletal muscle, myosin is associated with a number of proteins, especially C protein. The C protein is localized at specific sites along the length of the myosin filament and may extend perpendicular to the long axis. The giant protein titin runs parallel to the myosin filament, and C proteins may connect the titin and myosin filaments. These interactions may be important in length regulation or in maintaining orientation of myosin and actin filaments. Several other proteins, most prominently M-band protein and myomesin, cross-link myosin filaments at the M-band.

In an analogous fashion, the large protein nebulin may interact with actin filaments. α-actinin is the major protein involved in cross-linking actin filaments at the Z-disc. Tropomyosin and the three proteins of the troponin regulation system are stoichiometrically assembled with actin. Specific capping proteins can stabilize the lengths of actin filaments.

The alignment of myofibrils with each other and with the sarcoplasmic reticulum and t-system involves specific proteins of the membrane-cytoskeleton system. Spectrin, a membrane-associated actin binding protein, and ankyrin, a protein that connects spectrin and transmembrane proteins, are localized near the Z-disc at the myofibrillar level and at parallel positions near the plasma membrane of the muscle fiber. Cytoskeletal γ-actin and desmin filaments are present at both locations. These two types of filaments can provide the long-range connections required for physical alignment. The spectrin and associated proteins possess the requisite properties for the shorter-range interactions of the sarcoplasmic reticulum and fiber plasma membranes with the cytoskeletal filaments. It is not known which proteins of these muscle membranes anchor the sarcoplasmic and myofibrillar structures.

During development, cytoskeletal isoforms of actin, myosin, and tropomyosin precede the synthesis of sarcomeric forms and the organization of myofibrils. Early myofibrillar structures contain both cytoskeletal and sarcomeric isoforms. Nascent myofibrils form from stress–fiber-like assemblages of early myotubes. The sarcomere-specific protein forms then segregate from the cytoskeletal isoforms. However, myosin and actin filaments can form independently of sarcomere organization. The M-band proteins and cross-linking of myosin filaments occur next. The appearance of membrane-cytoskeletal proteins and myofibrillar alignment are temporally the last organizational events.

D. Assembly of Contractile Proteins

Myosin and actin can polymerize *in vitro*. The myosin and actin filaments possess many of the structural and biochemical properties of native muscle filaments. Genetic dissection of the relations between the amino acid sequences of myosin and actin and their abilities to polymerize *in vitro* or in recombinant cells is now attainable with modern techniques of molecular biology.

An alternative approach is to study genomic mutations affecting these proteins in an experimental organism. At the present, the most productive studies of this kind have been conducted in the nematode *Caenorhabditis elegans* and the insect *Drosophila melanogaster*. Both of these organisms develop striated muscles, which share many general properties with vertebrate muscles. Chemical mutagenesis in both organisms produces multiple specific mutant strains with alterations in myosin heavy chains, actin, and associated myofibrillar proteins.

In the nematode, the most-detailed genetic analysis of assembly has concerned the two myosin heavy chains and the myosin-associated protein paramyosin in the striated body wall myosins. Mutations can lead to either absence or alteration of the affected protein. Absence of myosin heavy chain A located in the bipolar central region of the thick filament leads to a total block of thick filament assembly. Absence of myosin heavy chain B leads to a marked decrement of thick filament number. The assembled thick filaments contain myosin heavy chain A along their entire length, replacing the B chain. Absence of paramyosin leads to abnormal thick filament-like structures with myosin heavy chains A and B scrambled in the central zone and hollow tubes containing some B as the remaining polar structures.

Alterations of myosin heavy chain B, especially dominant mutations, interfere with normal thick filament assembly. Surprisingly, the most mutagenic targets for defective assembly are in the ATP and actin binding sites of the myosin head. Alterations in paramyosin are due to mutations in the gene coding directly for the protein or in gene coding for modification functions such as phosphorylation.

The overall conclusions of the nematode genetic studies are that myosin does not self-assemble independently of other proteins. The two myosins must interact with paramyosin and the products of several other genes to assemble properly. Actin appears to assemble normally into thin filaments in myosin mutants. The paramyosin must be properly modified to interact effectively. Other functions have been genetically specified to play roles in myosin assembly, but their biochemical actions have not been determined. Specific mutants in the myosin, paramyosin, and paramyosin kinase genes accumulate multifilament assemblages, large assemblages generating multiple nascent thick filaments. These structures may be normally transient intermediates in thick filament assembly.

In *Drosophila*, the genetic dissection of actin in the indirect flight muscles has been the most intensive study. There are six genes encoding actin in the fruitfly. The actin of these flight muscles is encoded by only one of these genes; this gene is expressed only in these muscles. Absence of actin leads to no thin filaments or Z-discs within the muscles. Thick filament assembly appears unimpaired. Various alterations of actin differ in the severity of blocking assembly. The most severe appear grossly to disrupt myofibrillar architecture as a whole, whereas milder mutations permit assembly but show improper alignment of thick and thin filaments. Molecularly engineered mutations have been introduced by genomic transformation. Changes equivalent to the differences in amino acid sequence between tissue-specific actin isoforms have been made. The most profound finding is the inability of some chimeric actins to interact with Z-discs for proper alignment. Thus, the different isoforms of actin may be related to interaction with cross-linking proteins and other organizing proteins that are necessary for generating the crystalline arrangements of myofibrils.

E. Cardiac and Smooth Muscle Organization

Cardiac muscle cells originate in morphologically distinct myoblasts, as do skeletal muscle fibers. The cardiac cells are polygonal; they express a variety of proteins shared with other striated muscle, including sarcomeric myosin heavy chain, desmin, vinculin, and α-actin. In the mature heart, the predominant form of α-actin is the cardiac type; skeletal α-actin is always a more minor species.

Cardiac myoblasts can divide even after myofibrils have formed. They elongate into multiple processes that may adhere and form special adherence junctions and intercalated discs with like cells. Internally, the organization of myosin, actin, and associated proteins follows patterns similar to those of skeletal muscle. However, cardiac cells do not fuse. They form a branching network of cells through their elongated processes and by their special intercellular junctions. The overall organization of cardiac muscle is striated with alignment of

myofibrils and cells similar to but less precise than that within skeletal muscle. This decreased order is related to differences at the cellular level and the gross organization of the muscles.

Smooth muscle cells differ from striated muscles in terms of myofilament organization (no striations), expression of distinct isoforms of α-actin, myosin, and other proteins, and their cellular dynamics. The myosin of smooth muscle has now been demonstrated to form thick filaments, although the interactions of myosin are quite different than in striated muscle in vertebrates. The actin filaments form bundles resembling the stress fibers of nonmuscle cells. The major structural difference is the lack of well-defined alignment and cross-linking of myosin and actin filaments.

Smooth muscle cells neither fuse nor form a tight network as skeletal and cardiac cells do. Vascular muscle cells retain the ability to proliferate after differentiation. Such proliferation and the resulting hyperplasia play significant roles in atherosclerosis. All muscle, striated and smooth, shows the ability to hypertrophy in response to various stimuli.

V. Extracellular Signals

The development of all muscle involves the interaction of extracellular factors, hormones, and specific nerves at some level. The discussion here will consider only a limited number of examples that have been best defined in molecular and cellular terms.

A. Induction of Myogenesis

Experiments with amphibian embryos indicate that the commitment of mesoderm to differentiate into skeletal muscle is induced by fibroblast growth factor. This protein, homologous to epidermal and nerve growth factors, in turn stimulates the production of specific proteins that activate the transcription of muscle-specific genes such as α-actin. Other soluble proteins such as tumor growth factor β and the formation of extracellular matrix containing type IV collagen may be significant in early myogenesis. The interaction of mesenchymal elements such as early myoblasts and fibroblasts in the formation of the extracellular matrix is important not only for myogenesis but for the induction of other cell types including nerve, cartilage, bone, and vascular endothelium.

B. Control of Gene Activation

The most detailed studies of the control of gene activation by an extracellular signal have been on the action of thyroid hormone on the expression of myosin heavy chains in both cardiac and skeletal muscles. All vertebrate striated muscles show some differential expression of specific myosin heavy-chain isoforms, which are thyroxine or T_3-mediated. The most dramatic effects are seen in the cardiac ventricles of small mammals in which the β-myosin mRNA is expressed only in the hypothyroid state and the α-myosin mRNA is not expressed in that state. Conversely, α-mRNA is expressed in the euthyroid and hyperthyroid state, and β-mRNA is not expressed in these states. The result of these differences in transcription is a switch in functionally distinct myosins.

The molecular basis of thyroid regulation appears to be combinatorial (i.e., several distinct DNA sites flanking the 58 end of the α-myosin heavy-chain gene appear to be necessary). An implication of this result is that more than one transcription-activating protein is also necessary. Two such proteins have been identified. One is related to the c-*erb*-A proto-oncogene, and second is a specific α-myosin heavy-chain gene binding factor. In addition, a protein complex having high affinity for T_3 has been isolated from cardiac nuclei. This T_3-receptor complex also binds to a specific sequence in the 5' promoter region. [*See* DNA BINDING SITES.]

It is now clear that multiple hormones and other chemical signals have specific effects on muscle development and gene expression. Some of these effects will be summarized below.

C. Nerve–Muscle Interactions

The interaction of nerve and muscle both in physiological homeostasis and development has been a subject of intense study because of both its scientific and medical importance. Mature skeletal muscle atrophies (i.e., loss of fiber thickness without cell death) in the chronic absence of direct innervation. Even short-term denervation produces alterations in multiple systems. Postnatally, switching of fast and slow motor nerves leads to a switching of the speed of contraction of myosin isoforms in their skeletal muscle targets.

The question of the roles of nerve in muscle development is thus compelling. Innervation does not appear necessary for the commitment of early mesenchymal cells to myogenesis. Rather, innervation, first by multiple axons on a skeletal fiber, then by single axons forming a functional neuromuscular junction with the specialized motor end-plate of the muscle fiber, modulates the terminally differentiated state.

Motor neurons and skeletal muscle fibers retain the capability after axonal degeneration or fiber necrosis of reforming neuromuscular junctions at their original sites. The extracellular basal lamina of the junction is necessary and sufficient for this reformation. Although the junctional lamina contains multiple muscle-specific proteins including isoforms of collagen, N-CAM, and acetylcholinesterase, agrin is the critical protein in the recognition process. Agrin is synthesized and secreted by the neurons but interacts with extracellular matrix proteins and membrane proteins produced by the muscle fiber.

The interaction of specific proteins in the extracellular matrix clearly provides specific developmental information to both neurons and muscle fibers. It is now clear that multiple types of extracellular matrix proteins interact with each other and specific cell surface receptors during nerve and muscle development. The molecular recognition of these proteins provides regulatory information in multiple phases in development.

D. Hormonal Regulation of Differentiation

Hormones or soluble extracellular factors regulate the behavior of myoblasts and muscle cells throughout differentiation. An individual factor may act at one or more stages of myogenesis. The spectrum of chemical messengers includes amino acids, carbohydrates, fatty acids, metals, proteins, and steroids. Norepinephrine, thyroxine, and T_3 are derived from the amino acid tyrosine. Prostaglandins are modified fatty acids. Heparin is a complex carbohydrate. Calcium, iron, and potassium are metals. Fibroblast growth factor, growth hormone, insulin, and transferring are proteins. A variety of androgenic, estrogenic, glucocorticoid, and mineralocorticoid steroid hormones are active on muscle. The regulation of muscle development by this wide variety of chemical structures again suggests the combinatorial nature of the control mechanisms as well as the molecular diversity of receptors and pathways for transduction in the control mechanisms. The diversity of regulatory mechanisms is also exemplified by observations that the same hormone mechanisms can have different effects on distinct muscles in the same organism or homologous muscles in distinct organisms.

The action of thyroid hormones in regulating the differential expression of myosin heavy-chain isoforms at the transcriptional level in heart muscle has been discussed. The actions of these hormones serve as a model for the analysis of the actions of other chemical signals. [*See* THYROID GLAND AND ITS HORMONES.]

The withdrawal of myoblasts from the cycle of division and proliferation and their fusion into multinucleated myotubes are unique and critical features of skeletal myogenesis. Several extracellular chemical signals are necessary for this key step in muscle differentiation. The removal of calcium ions from the culture medium prevents myoblast fusion. Basic fibroblast growth factor and other homologous proteins act as mitogens for skeletal myoblasts. Removal or inhibition of factors of this class is required for withdrawal of the myoblasts from proliferation. Insulin has only a small effect on myoblast proliferation but specifically enhances the fusion of myoblasts and the parallel changes in gene expression. Although the molecular mechanisms by which calcium, growth factors, and insulin control the myoblast to myotube transition have not been completely deciphered, the multiplicity of extracellular factors affecting the process and their known actions at multiple sites in other systems suggests that a combination of regulatory interactions is essential.

Bibliography

Emerson, C., Fischman, D., Nadal-Ginard, B., and Siddiqui, M. A. Q., eds. (1986). "Molecular Biology of Muscle Development." Alan R. Liss, New York.

Kedes, L. H., and Stockdale, F. E., eds. (1989). "Cellular and Molecular Biology of Muscle Development." Alan R. Liss, New York.

Pearson, M. L., and Epstein, H. F., eds. (1982). "Muscle Development: Molecular and Cellular Control." Cold Spring Harbor Laboratory, Cold Spring Harbor, New York.

Muscle Dynamics

ROBERT J. GREGOR, *University of California, Los Angeles*

Glossary

Actin Protein molecule found in skeletal muscle that is low in molecular weight and is responsible for forming the thin filament in the sarcomere

EMG Quantification of the electrical signal produced by skeletal muscle as a result of depolarization of the muscle fibers by the alpha motoneuron

Hyperplasia Refers to the possible splitting of muscle fibers resulting in increased muscle size. This phenomenon is not felt to occur under normal conditions. It is generally agreed that muscle increases size by increasing the size of individual fibers and not by gaining more fibers through fiber splitting

L_0 Length at which a muscle produces maximum isometric tension. It represents maximum overlap of the crossbridges within each sarcomere. By definition, P_0 is produced at L_0

Moment of force Mechanically this is the product of force and the perpendicular distance from the line of action of a force and an axis of rotation

Myosin Protein molecule that has several subfragments. The light meromyosin in the tail of the myosin filament comprises the base of the thick filament. The heavy meromyosin is composed of the S2 and S1 regions, where binding takes place with the actin filament

P_0 Maximum isometric tension measured when the muscle is maximally stimulated with maximum overlap of crossbridges

Rate coding Strategy employed by the nervous system to increase muscle tension by increasing the firing rate of motor units already involved in the tension development

Recruitment Strategy employed by the nervous system to increase muscular tension that involves the orderly inclusion of various motor units based on neurophysiological properties of the soma

Size principle Principle that governs the orderly recruitment of motor units based on the size of the soma of the alpha motoneuron innervating the various motor units. Typically, the units recruited first are the ones with the smaller soma, smaller axon, and smaller, more aerobic fibers

Stiffness Mechanically this is the inverse of compliance. If a tissue can be stretched with little force, it is said to be compliant and not very stiff. In skeletal muscle, stiffness is proportional to crossbridge activation. The more crossbridges activated, the more stiff the muscle

Tropomyosin Protein molecule that is bound to the actin filament and is necessary to the control of calcium and the binding of the actin and myosin filaments during excitation–contraction coupling

Troponin Protein molecule that contains three subfragments, all of which are chemically bound to the actin filament. Troponin is responsible for the regulation of calcium and subsequent formation and breaking of the "crossbridge" in the sarcomere

V_{max} Maximum rate of shortening of a sarcomere, fiber, or whole muscle under maximum excitation and zero load. It is a theoretical value representing the maximum rate of crossbridge turnover

THIS ARTICLE will be devoted to a discussion of the mechanical properties of skeletal muscle. Other the mechanical properties of skeletal muscle. Other types of muscle (e.g., cardiac and smooth) will not be included. The text will begin with the fundamental structural unit of skeletal muscle (the sarcomere) and will continue with the arrangement of sarcomeres into fibers, fibers into motor units, and motor units into whole muscle. The chapter will then continue with the structural basis of muscle and how it relates to the mechanical properties regarding length and velocity. Included with length changes will be a discussion on muscle stiffness and the use of elastic strain energy during various movements. Also included will be a discussion of the basic mechanical properties of muscle and issues related to submaximal and maximal activation of the muscle by the nervous system. Finally, the mechanical properties of muscle as estimated during normal movement and the application of these properties to the control of movement will be discussed. Examples related to human movement and the analysis of motor output in normal humans will be the focus of this chapter.

I. General Properties and Structure of Skeletal Muscle

Skeletal muscle is considered to be a motor, a transformer that converts signals from the nervous system into actions in the external environment. It uses and converts chemical energy derived from food into force production. Some general properties of muscle include (1) its irritability, or its ability to respond to a stimulus, (2) its conductivity or ability to propagate a wave of excitation, (3) its contractility or ability to shorten, and (4) its limited ability to grow and potentially, in some situations, to regenerate. Although the muscle fibers possess certain physiological characteristics, these fibers are joined to bone through tendon. Although some properties of muscle are described with respect to the muscle fibers, more realistically the actual performance of

muscle must be considered in light of its interaction with the tendon. It is this muscle–tendon unit that deals with the real world and possesses certain mechanical properties, modulated by impulses from the nervous system, that perform selected functions in the control of movement. [See SKELETAL MUSCLE.]

Muscle contains a great many identifiable elements, and indeed the mechanical aspects of muscle function have been estimated to contain in the order of 10^5 items. It is well beyond the scope of this chapter to discuss all these items because each in itself is worthy of a great deal of research. Consequently, the primary focus of the information presented here will be some of the major aspects of muscle dynamics as they pertain to muscle function in our environment.

When evaluating the gross structure of skeletal muscle, it is apparent that the muscle fibers are linked together by a network of collagenous connective tissue. The endomysium surrounds the individual fibers, whereas the perimysium collects bundles of fibers into what is referred to as *fascicles*. As an outer layer, the epimysium covers the entire muscle. Muscle fibers may vary in length from a few millimeters to more than 400 mm. The effect of fiber length and fiber diameter, which can vary from 10 to 60 μm, will be discussed later in this chapter in relation to speed and force generation. The sarcolemma is referred to as the cell membrane and encircles the myofilaments that comprise a muscle fiber. The sarcolemma provides an active and passive semipermeable membrane and is essential to the excitability property of the membrane. The fluid enclosed within the fiber is referred to as the *sarcoplasm*. Within the sarcoplasm are the fluid sources, organelles, enzymes, and the contractile machinery (bundles of myofilaments arranged into myofibrils) (see Fig. 1).

The fundamental structural unit of the skeletal muscle fiber is the sarcomere. This repeating unit in the muscle fiber represents the zone of a myofibril from one Z band to another. The sarcomere is the basic contractile unit of muscle and comprises an interdigitating set of thick and thin contractile proteins. The sarcomere at rest has a length of approximately 2.2 μm. The exact length, however, is open to debate and can range from 2.2 to 2.5 μm. Each myofibril is composed of bundles of myofilaments (thick and thin contractile proteins). The obvious striations in skeletal muscle are due to the differential refraction of light as it passes through the con-

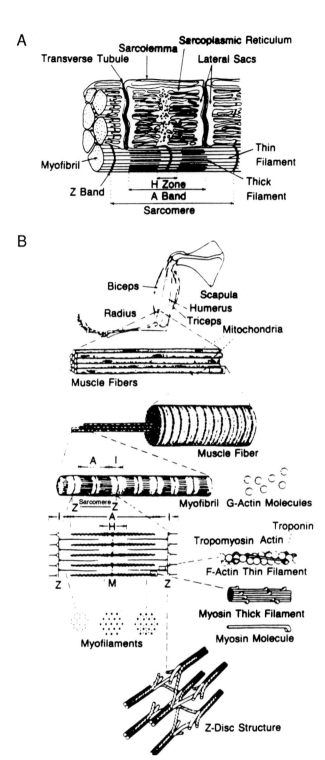

A: Alignment of myofibrils and fundamental structure of the sarcomere. [From W. Bloom, and D. W. Fawcett (1968). "Textbook of Histology," p. 281, W. B. Saunders, Philadelphia, with permission.] B: Organization of skeletal muscle from the whole muscle to the Z-disc structure. [From McMahon, T. A. (1984). "Muscles, Reflexes and Locomotion." Princeton, University Press, Princeton, New Jersey, with permission.]

tractile proteins. The thick filament zone, which includes some interdigitating thin filaments, is doubly refractive and forms the dark or A band (anisotrophic). Within the A band is a zone that contains only thick filaments. Because this zone is clear of the thin filament, it is referred to as the *clear disc* or *H band*. The area between the A bands contains predominantly thin filaments, is singly refractive, and is called the *I band* (isotrophic). Each set of filaments is attached to a central transverse band; the thick filaments attach to the M band located in the middle of the A band, and the thin filaments connect to the Z band, which defines the end of the sarcomere. A cross-sectional view through the A band illustrates the relation between the interdigitating filaments. Six thin filaments surround a single thick filament.

The fundamental elements in the sarcomere that control both force production and velocity are the thick and thin filaments and the several proteins found within each filament. The structure of the thin filament is dominated by the actin molecules. In fact, the thin filament actually contains three proteins that are described to have well-established functions: actin, troponin, and tropomyosin. Each thin filament contains approximately 350 actin monomers and 50 molecules each of troponin and tropomyosin. The major role of the actin filament is to interact with the myosin, whereas tropomyosin and troponin have selected control functions in the sarcomere. The actin filament, which contains both G-actin and F-actin, is a polarized structure. This polarization of the thin filament together with that of the thick filament is responsible for the fact that force acts in a direction toward the center of the sarcomere. Free, unattached muscle will pull its ends toward the center when stimulated.

Each tropomyosin filament is bound to an actin strand and lies in the grooves between the actin strands (see Fig. 2). Each troponin complex consists of troponin-I, troponin-C, and troponin-T. This complex binds to both the tropomyosin and the actin. Troponin-I together with tropomyosin inhibits the activation of myosin ATPase by actin, possibly by physically blocking the myosin-combining site on the actin. Each tropomyosin inhibits seven actin molecules, probably by each of the actins interacting with one of seven rather similar regions on the tropomyosin molecule. Because of its affinity for calcium, troponin sensitizes the thin filament to calcium. When troponin-C is combined with calcium, it is able to remove the inhibitory

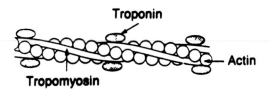

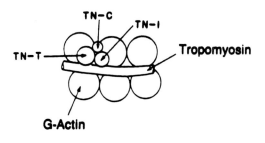

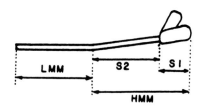

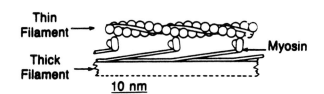

FIGURE 2 Schematic of the basic structure and orientation of selected protein molecules in the actin and myosin filaments. [From Enoka, R. M. (1988). ''Neuromechanical Basis of Kinesiology.'' p. 106, Human Kinetics Publishers, Champaign, Illinois, with permission.]

affects of tropomyosin and contraction takes place.

The main constituent of the thick filament is the protein myosin. Myosin is a large molecule composed of two identical heads connected to a long tail (see Fig. 2). There is a flexible hinge region in the tail approximately 43 nm from the heads. The two halves of the filaments on either side of the M line are of opposite polarity, with the tails of the myosin molecules pointing toward the center of the filament. The heads of the myosin molecules project from the thick filament everywhere except for a region approximately 0.15 μm long in the center where there are only overlapping myosin tails. The exact nature of the packing of the myosin tails in the

thick filaments is uncertain. The myosin heads projecting from the thick filament from a helical array with a periodicity of approximately 43 nm and an axial interval of 14.3 nm. In addition to myosin, the thick filaments contain much smaller amounts of other proteins whose role may be to control and define the structure of the filaments. The specific function of the cross-bridge region, commonly referred to as the *S2 or S1 segment* of the myosin filament, will be discussed later in reference to the mechanical properties of muscle related to length and velocity. Suffice it to say here that the tail of the myosin filament is composed of light meromyosin (LMM), whereas the S2 region and the pair of globular heads in the S1 region are composed of heavy meromyosin (HMM). The structure and function of the S2 and S1 regions as functional contractile elements play an important role in force development in different types of skeletal muscle fibers. The mechanical rotation of these heads and the ability of the globular heads to interact with sites on the actin filament are critical to the generation of force and, at times, storage of strain energy in the muscle fiber. As a force-generating system, the development of force by myofibrils in their act of shortening requires some way of transmitting force from the thick to the thin filament. The cross-bridge hypothesis explains some of the aspects of this force-generating unit. [*See* MUSCLE, PHYSIOLOGY AND BIOCHEMISTRY.]

II. Functional Unit of Muscle: The Motor Unit

Now that the structural elements of striated muscle have been presented, we turn our attention to the functional unit of muscle, which is the motor unit (see Fig. 3). A single motor unit is defined as the cell body and dendrites of a motoneuron, the multiple branches of its axon, and the muscle fibers that it innervates. Each muscle fiber is innervated by a single motoneuron, but each motoneuron will innervate more than one muscle fiber. The number of muscle fibers innervated by a single motoneuron is commonly referred to as the innervation ratio. This ratio can vary from approximately 1 : 2,000 to 1 : 10. This means that for one motoneuron we can see a possible range of 10 to 2,000 innervated muscle fibers. The central nervous system activates the motoneuron and sends action potentials to all muscle fibers in that motor unit. This all-or-none phenome-

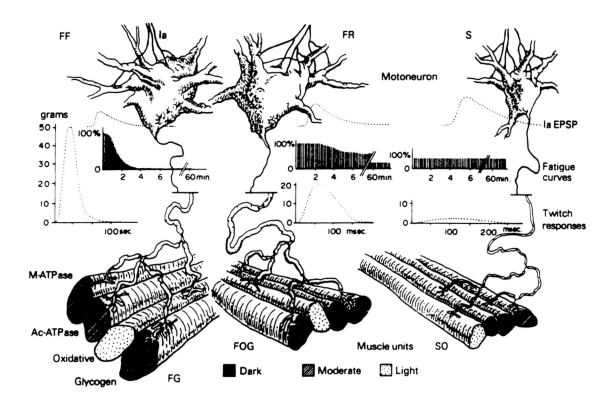

FIGURE 3 Fundamental properties of the FF, FR, and S motor units related to recruitment order, tension development, fatigue resistance, myosin ATPase activity, oxidative capacity, and glycogen content. FF, fast fatigable; FR, fast fatigue resistant; S, slow; FG, fast glycolytic; FOG, fast oxidative glycolytic; SO, slow oxidative; EPSP, excitatory past synaptic potential. [Edington, D. and Edgerton, D. (1976). "The Biology of Physical Activity." Houghton Mifflin Co., Boston, with permission.]

non results in a single depolarization of an alpha motoneuron, resulting in the depolarization of a variety of muscle fibers. We can easily recognize that if the muscle fibers are few in number the motor unit will yield small increments in tension, resulting in finer control. Motor units with such low innervation ratios can be found in the extraocular muscles of the eye. In contrast, motor units in the large muscles of the back and lower extremity may have approximately 2,000 muscle fibers excited by a single alpha motoneuron.

Physiologically, motor units are compared with each other based on a number of properties including (1) characteristics of discharge, (2) speed of contraction, (3) magnitude of force produced, and (4) resistance to fatigue. Two methodologies have emerged to evaluate these parameters. A direct evaluation can be used to measure physiological properties of the motor unit and usually includes measurement of the electrical and mechanical discharge characteristics of that motor unit. These discharge characteristics, or mechanical responses of motor units, are usually observed with reference to different inputs. In contrast, indirect measurements are usually made with regard to the histochemical and biochemical profiles of fibers in the motor unit.

When making direct measurement of the physiological properties of motor units, two different types of input into the motor unit are commonly employed. The first, a single action potential, generates what is referred to as a *twitch,* whereas the second, involving a series of action potentials, produces what is referred to as a *tetanus.* Three measurements are commonly made when evaluating the twitch response. The first involves the time from force onset to peak force and is referred to as the *contraction time.* The second is simply the magnitude of the peak force, and the third is the time from peak force to the time at which the force has declined to one-half of its peak value (i.e., one-half relaxation time). Essentially, the contraction time is used as a measure of the speed of the contractile machinery (i.e., the cross-bridges). If the contraction time is long (i.e., 100 ms), the motor unit is generally referred to as a *slow twitch motor unit.* If the contraction time, however, is short (i.e., 45 msec) the motor unit is referred to as *fast twitch*. Peak force is obviously related to the number of fibers within the motor unit (i.e., the more fibers, the more physiological cross-sectional area and consequently the greater amount of contractile protein involved in generating force). It is currently believed that the peak force exerted in a twitch response is related to and will increase with training.

When a series of stimuli are given to the motor unit, each elicits a single twitch response so that we get a series of twitches. As the stimuli occur closer and closer together, we have what is referred to as a *fused tetanus*. The capability of a motor unit to exert force is usually measured from a fused tetanus rather than the twitch response. The measurement from baseline to the top of the smooth plateau yields a value for maximum force exerted by that motor unit. The ratio between the twitch and the tetanus can vary from 1:1.5 to 1:10 and appears related to the type of muscle being studied. Additionally, force produced in a single tetanus will decline over time if the motor unit is required to produce a series of tetanic contractions. The ability of a motor unit to prevent such a decline is taken as a measure of its resistance to fatigue. A standard fatigue test usually involves a 2–6 min period in which the motor unit is stimulated at a rate of 1 Hz, with each tetanus lasting approximately 330 msec. This particular protocol will yield in fatigue resistance fibers, a relatively continuous output. For fatigable fibers, however, a decline in peak force is observed after 2 min of stimulation.

On the basis of contraction time and fatigue resistance, motor units can be classified into three groups: slow contracting fatigue resistant (S), fast contracting fatigue resistant (FR), and fast contracting fatigable (FF). Type S units typically produce the least amount of force, whereas the FF fibers produce the most. These differences are due not only to variation in the number of muscle fibers within a motor unit (i.e., the innervation ratio for a slow motor unit is typically lower than for a fast motor unit), but also the size of the individual muscle fibers (i.e., the quantity of contractile proteins within the muscle fiber).

Variations in contraction speed among motor units is considered to be a function of variations in enzyme activity (i.e., myosin ATPase). Variation is also due to, in part, the rate at which calcium is released and taken up by the sarcoplasmic reticulum. The fatigue resistant units (i.e., S and FR)

have a fatigue index approximately three times that of the FF fibers. Because the S and FR fibers are much more resistant to fatigue, we might conclude that activation of these motor units is more appropriate for sustained contractions such as those involved in postural stability.

There is a great deal of information on indirect estimates of motor unit output, which will not be discussed in this chapter. These histochemical and biochemical techniques used to measure enzymes (e.g., NADH, SDH, and alpha-GPD) assist in the understanding of the aerobic and anaerobic capacity of the various fibers and motor units. It is generally believed, because fibers in a motor unit have many similar properties and are activated in an all-or-none fashion, that they have similar histochemical and biochemical properties. Recent evidence, however, suggests that individual fibers within a motor unit are not necessarily identical. We can have differences in enzyme profiles, for example, between fibers in the same motor unit as well as at different positions along the membrane of the same fiber. These types of data obviously make the picture more complicated regarding how these fibers are used in the generation of force. For our purposes, however, all fibers within a motor unit are activated together, and consequently, the increments in force produced by a muscle are directly related to the number of fibers within the recruited motor unit. A muscle that has few fibers within a motor unit will be used in activities that require a finer degree of control than a muscle having a large number of fibers within its motor unit. This aspect of control is important, but the need for significant increases in force by calling on additional motor units is also a resource that must be available when needed.

III. Tension Development in Skeletal Muscle

While we are on the topic of motor units, a discussion concerning how motor units are used within the muscle to control movements appears warranted. The nervous system has two options available for varying muscle force production. The force that a muscle exerts is a function of (1) the number of active motor units and (2) the rate at which the actual motor units generate action potentials. These two strategies are typically referred to as motor unit recruitment and rate coding, respectively. It is generally believed that motor unit recruitment follows some orderly sequence, which is essentially based on the size of the alpha motoneuron innervating that motor unit. This phenomenon is referred to as the *size principle* and suggests that the motor unit with the smallest motoneuron be recruited first and the motor unit with the largest motoneuron be recruited last. For example, if we have a pool of motoneurons that experience a common electrical drive, the smaller motoneurons will be recruited first. These typically are motor units in the type S category that produce relatively low maximum tension and have relatively long contraction times. These muscle fibers are used on a daily basis for postural control and stability in common everyday tasks. Studying these motor units in a relatively controlled environment (i.e., a slow ramp contraction), we would see a smooth increase in force as a result of the orderly recruitment of small to large motoneurons and slow to fast motor units. In a ramp contraction with a negative slope, we would see an orderly derecruitment of these same motor units, and this sequence of derecruitment would be opposite that observed during recruitment.

We would be remiss if we did not mention that the threshold of activation for motoneurons is not dependent solely on the size of the motoneuron, but is also influenced by other factors. For example, organization of the input into a motoneuron pool may differ depending on the sensory input to that system. If sensory input comes from sensor A in one case and sensor B in another, the recruitment order of motor units as a result of these two inputs may well be different. Motoneuron size then appears to be a significant but not exclusive factor in determining recruitment order of individual motor units.

Muscle fiber types in athletes have been well-documented in the scientific literature, and it is interesting to note that endurance-type athletes (i.e., marathon runners) have a predominance of type S fibers. Sprinters, however, typically have a predominance of type FF fibers. The majority of other athletes are in the midground 50/50 region. Weightlifters, for example, are generally considered high-power athletes and yet have 50% slow and 50% fast fibers. Their fibers are, of course, much larger than normal because of their high-power output and overload training. It is also interesting to note that this orderly recruitment does not always hold during high-power output activities. In high-power events such as the vertical jump, it is common to see simultaneous activation of both slow and fast motor units.

High-power training can actually force the nervous system to adapt and override the time-consuming orderly recruitment of these units because high power is needed in short period of time. Additionally, it is necessary to emphasize that motor unit types do not exist as discreet populations relative to their excitation threshold. There appears to be considerable overlap between groups, especially between the type S and FR motor units. Consequently, we do not have such a nice uniform sequence in which there are only type S, then only type FR units, then only type FF units that must be recruited in a systematic way. It is possible to activate a muscle such that type S motor units are recruited with type FR units. To make it even more complicated, certain researchers have documented and identified more than 50 types of muscle fiber types. These categories obviously are determined not only by contraction time but by a tremendous array of myosin isoforms and variable enzyme patterns that dictate how these muscle fibers and subsequently motor units are used in the control of movement. For our discussion purposes, we would conclude that there are three types of motor units and that there is considerable overlap between these types of motor units, and although the size principle is robust enough to dictate usage of motor units in many activities, it is conceivable that units from two different types can be recruited at the same time.

Although recruitment is one strategy employed by the nervous system to increase force production in the muscle, the firing rate of that motor unit is a second strategy employed. There have been a variety of studies reported in the literature regarding firing rate as a function of force. Several decades ago it was reported that the upper limit of the motor unit firing rate in humans is approximately 50 pulses/sec. At the lower end of the spectrum, the relation between muscle force and action potential rate (i.e., firing frequency) is not linear. The increase in force caused by the increasing action potential rate from 5 to 10 Hz is not the same as that caused by increases in rate from 20 to 25 Hz. The force-firing rate relation is sigmoidal with the greatest increase in force occurring at lower rates in the range of 3 to 10 Hz. The actual relation depends to some degree on muscle length because the curve shifts to the left for longer muscle lengths and to the right for shorter muscle lengths. Although the relation remains sigmoidal, the frequency for affecting the greatest change in force becomes 3 to 7 Hz for a long muscle and 10 to 20 Hz for shorter muscle lengths.

As is the case with motor unit recruitment, the extent of rate coding appears to be muscle-dependent. For example, if a muscle has the capability to recruit more motor units, that strategy seems to dominate. If motor unit recruitment in a given muscle is completed by 30% of the maximum force, subsequent increases in force must come from increases in firing rate of the active motor units. This carries over to muscle size because larger muscles would most likely have more motor units available for recruitment whereas smaller muscles would have fewer motor units and be, to some degree, dependent on increases in firing rate. An example would be the first dorsal interosseous (FDI) muscle, which has a recruitment rate of about 9 pulses/sec. The peak rate at 40% of maximum voluntary contraction for this muscle is about 25 pulses/sec, whereas the peak rate at 80% maximum voluntary contraction is 42 pulses/sec. In contrast, the larger deltoid muscle has a recruitment rate of about 13 pulses/sec, a peak rate at 40% maximum voluntary contraction of about 26 pulses/sec, and a peak rate at 80% maximum voluntary contraction of only 29 pulses/sec. Consequently, motor units in the smaller FDI are recruited at a lower firing rate than the larger deltoid and have a greater range of firing rates available to increase force from 40 to 80% of maximum voluntary contraction. It should be emphasized, however, that increases in muscle force are not due exclusively to increases in either recruitment or rate coding because the two are undoubtedly going on concurrently.

A final mechanism or strategy employed by the nervous system to increase force in a muscle is related to the temporal relation between the action potentials of different motor units. Normally, we assume the action potentials of different motor units to be discharging at different instances in time (i.e., asynchronous firing). Recent studies, however, report that motor units of a muscle subjected to a strength training program tended to discharge action potentials synchronously. Since these reports, it has been assumed although never proven that motor unit synchronization results in greater muscle force. It also appears that synchronization is more evident when a muscle experiences fatigue. So again we have a strategy that has been employed by the nervous system to modulate muscle force that is affected by not only metabolic properties of those motor units but by sensory feedback to the motor output system of that muscle. It appears then that strategies employed by the nervous system to increase force in the muscle can focus on both recruitment and firing rate. The domain in which each of these is used is open to discussion, but it must remain clear that both strategies are employed.

IV. Skeletal Muscle Length–Tension Curve

Thus far we have discussed the basic structural units of skeletal muscle (i.e., the sarcomere) and the basic functional unit (i.e., the single motor unit). Research in each area is critical to our understanding of the mechanical properties of skeletal muscle and how muscle is used in the control of movement. During the 1930s and 1960s, classic studies were conducted on the effect of length and shortening velocities on skeletal muscle force output. There have been many attempts to dislodge these early theories, and despite a great deal of work, they remain robust. One of the earlier observations made is that there is certain position or length that muscle assumes in which if maximally activated it will produce maximum force. This position is traditionally referred to as L_0. At this position, given maximum excitation of all motor units in the muscle, the muscle will produce what is referred to as its *maximum isometric tension* (i.e., P_0). According to the sliding filament theory, isometric tension is directly proportional to the number of cross-bridges formed between the actin and myosin filaments. This theory has been generally accepted and well-documented in the literature. There are, however, some variations on this theme, and discussion continues regarding the exact nature of the cross-bridge dynamics between the globular heads of the myosin filament and the active sites on the actin filament. It is, however, concluded by almost all scientists that the structural basis for the variation in force seen as a function of changes in muscle fiber length rests in the overlap of the actin and myosin filaments within each sarcomere. As the length of the muscle changes, the actin and myosin filaments slide past one another, changing the number of thin filament binding sites available for cross-bridge attachment. The tension varies as the amount of overlap between actin and myosin filaments varies within each sarcomere and as available binding sites either increase or decrease.

This active length-tension curve and the sarcomere spacing that creates the variation in tension as a function of length is documented in Fig. 4. To be

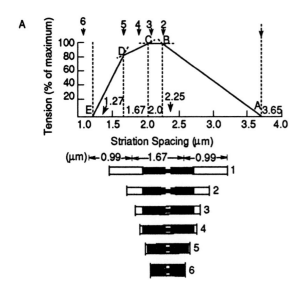

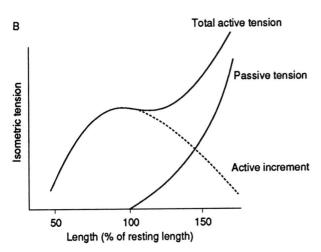

FIGURE 4 A: Relation between filament overlap, sarcomere length (striation spacing) and force production. [From Mc-Mahon, T. A. (1984). "Muscles, Reflexes and Locomotion." Princeton University Press, Princeton, New Jersey, with permission.] **B:** Schematic diagram of active length–tension curve, passive length–tension curve, and total tension curve for the muscle-tendon unit.

The ideal or optimum overlap of cross-bridges (i.e., L_0) seems to be the general location where passive forces begin to appear. Currently, it is believed that this position is different for different muscles. As a muscle is stretched beyond this initial position, the passive contribution causes the force to increase exponentially. The force length relation then becomes a combination of both active and passive compartments and yields what is referred to as a *total tension curve* (Fig. 4B).

V. Skeletal Muscle Force–Velocity Curve

Although cross-bridge overlap and muscle length changes are important to our understanding of muscle function, the mechanical output of this muscle does not respond to isometric loads often during everyday activities. Total body movements involve the acceleration and deceleration of limbs, and these rotary movements about the various joints of the body are controlled by muscle, tendon, ligament, and environmental forces. Consequently, another important parameter is the force–velocity curve. Essentially, the literature tells us that as velocity increases, the force production capacity of skeletal muscle decreases in a nonlinear way. In fact, at V_{max} (the theoretical maximum speed of cross-bridge turnover) zero tension is produced by the muscle.

Under controlled conditions, isolated muscle will yield the curve presented in Fig. 5. These conditions typically involve a quick release or after-loaded method in which the muscle is stimulated maximally and initiates a shortening contraction from L_0 against an external load. The load is incrementally reduced using some form of transducer, and as the load decreases, under maximum stimulation, muscle shortening velocity increases. The experiment is essentially designed to eliminate the elastic elements in the muscle-tendon unit. Consequently the force–velocity curve describes the active cycling of cross-bridges in muscle fibers in the maximally stimulated muscle. This classic force–velocity curve has been the subject of much debate since 1938 when it was first published by A. V. Hill. This curve has been demonstrated for whole muscle and for single muscle fibers.

Although classic experiments have been done to demonstrate the effect of velocity on force during shortening contractions, little work has been done on the effect of changes in lengthening velocities on force production.

more complete, however, activation of the contractile element is not the only process employed by muscle to increase force output. Muscle includes a substantial amount of connective tissue (e.g., sarcolemma, endomysium, perimysium, epimysium, and tendon). These connective tissues behave like stiff elastic bands. When stretched they exert a passive force that combines with the active contribution from the cross-bridges to produce the total muscle force output. The concept of total tension then becomes an additive effect of both the active cross-bridges and the passive connective tissue structure.

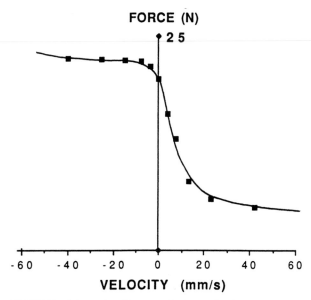

FORCE (N)

2 5

VELOCITY (mm/s)

FIGURE 5 Schematic of force–velocity profile for both lengthening and shortening contractions. This profile represents the force output of maximally active contractile element in skeletal muscle.

The classic response of muscle under varying velocities of lengthening is also presented in Fig. 5. It should be noted that the peak tension increases above P_0, plateaus in a range between 1.2 and 1.3 P_0, and remains constant regardless of increases in lengthening velocity. Data on the lengthening side of the force–velocity curve are not as abundant as those published to describe the shortening side. The significance of this lack of information becomes apparent when we realize that neither isometric nor shortening contractions are performed exclusively in daily activities. We usually employ movement patterns that involve lengthening and shortening cycles. Consequently, the ability to understand force output and skeletal muscle from these isolated muscle experiments becomes attenuated because of lack of information and the ability to extend the results to daily living.

VI. Summary of Factors Affecting Force Production in Skeletal Muscle

Thus far we have discussed the structural and functional units of skeletal muscle and how they affect force production. Actively, muscle produces force using depolarizing currents to initiate cross-bridge coupling. Directly related to cross-bridge kinetics is the amount of active force produced as a function of the amount of cross-bridge overlap within each sarcomere (i.e., length–tension curve). Because sarcomeres are in series it is well known that overall length is distributed among the structural units. It is also well known that not all sarcomeres perform with the same strength and that some will stretch whereas others shorten in the same fiber. X-ray diffraction techniques have been employed to study the differential movement of sarcomeres in series, and it has been reported that not all stretch and shorten to the same degree.

The ability of muscle to produce force at varying velocities also is significant and is a function of a variety of biochemical factors (e.g., type of myosin ATPase). Muscle length, however, is also correlated to velocity. It is generally agreed that the longer the fiber, the higher the potential V_{max}. The semitendinosus muscle in the cat hindlimb, for example, has two compartments separated by a band of connective tissue. If the proximal and distal sections are activated independently, each will arrive at its own V_{max}. If the sections are stimulated simultaneously, the resultant V_{max} will be a summation of the V_{max} obtained for the proximal and distal sections. Also, if a bone is moved at a constant angular velocity, fibers within the muscle that are responsible for the movement and insert at the distal end of the bone are longer than the ones that insert at the proximal end. Structurally, muscle is organized such that fibers responsible for higher velocities tend to be longer. The key here is that we are talking about the structure of the muscle, and we must keep in mind that the total output of a muscle is also regulated by its biochemical and histochemical profile. Experiments showing that length is related to speed are done in muscle with the same histochemical profile. Combining biochemical data with structural data gives us a much more complete picture of total muscle output.

In summary, factors affecting a muscle's ability to produce force are (1) length, (2) velocity, (3) fiber type related to myosin ATPase activity, (4) physiological cross-sectional area, and (5) activation through a combination of recruitment and rate coding. The remainder of this chapter relates to skeletal muscle's ability to enhance its output by using both active and passive components and its ability to do work and generate rotational energy as a member of a group of synergists.

VII. *In vivo* vs *In situ* Properties of Muscle

Although the mechanical properties of skeletal muscle are usually studied in isolation under controlled conditions in which temperature, blood supply, length, and load are regulated, its performance in the real world is influenced by the way it works with its synergists. Another way to look at this is that certain muscles are better suited to specific functions, and the manner in which one muscle responds to its environmental demands will dictate how other muscles in the same group respond. The load sharing that takes place among the various tissues surrounding a particular joint is a research area of much interest in musculoskeletal mechanics. The fact that environmental forces (e.g., the reaction force between the ground and the foot) and inertial forces that are related to the distribution of mass in the limb must be controlled by active muscle forces and passive ligament forces gives us the immediate impression that muscles do not act in isolation. Understanding this interaction with other muscles is a question of highest significance.

Although group dynamics are important, issues related to a muscle's ability to use both active and passive elements within the muscle–tendon unit must be considered first. A. V. Hill described his two-component model (one related to the active contractile element and the other related to the passive series elastic element) and the famous force–velocity relation. Keeping in mind that the force–velocity relation describes properties of the active contractile element muscle does interact, in series, with tendon. It is this interaction that has received a great deal of attention in recent investigations. Before any discussion of this interaction with changes in velocity, we must take one step back and evaluate the muscle–tendon complex as length changes. It is important to know how the length is distributed in the muscle–tendon unit between free tendon, the tendinous component in the muscle, and the sarcomeres in the muscle fibers at submaximal levels of activation.

VIII. Distribution of Length Between the Muscle and Tendon

Structurally, the change in length of the overall muscle–tendon unit can be taken up in (1) the change in length of the sarcomeres and subsequently the fibers, (2) the change in angulation of the fibers with respect to the line of pull of the muscle force, and (3) the change in length of free tendon. In normal movements at varying levels of activation, all these elements contribute simultaneously to the overall length change of the muscle–tendon unit. Although the change in length of the muscle fibers under maximal stimulation has been well-described by the active length–tension curve, length changes under submaximal activation are less understood. Angulation of fibers in skeletal muscle of humans has been presented with data indicating that, at rest, fiber angulation can vary from 0° in the long head of the biceps femoris to approximately 25° in the short head of the same muscle. Research on the cat hindlimb indicates that the soleus muscle has a fiber angulation of only a few degrees, whereas the medial head of the gastrocnemius shows angulations of approximately 20°. Additionally, the medial head of the gastrocnemius in the cat shows a change in angulation from approximately 20° at rest to 45° at maximum shortening. The vast majority of muscles reported in the literature, however, have few degrees of angulation (approximately 5°) at rest, which consequently minimizes the effect fiber angulation changes would have on muscle shortening.

The final piece in the puzzle is the compliance of tendon (i.e., how much tendon stretches under normal loading conditions). Tendon stiffness values (which are the inverse of compliance) have been reported in the literature for certain muscles. Variability is high, however, with some authors reporting stiffness as constant whereas others report it may vary with load. More research is needed here to obtain values that would apply to muscle modeling and provide a better understanding of the role played by tendon and/or the tendinous elements in muscle in the control of movement. As a final note, several reports in the literature present data indicating that the muscle–tendon unit may lengthen whereas the muscle fibers actually shorten at the expense of the tendon. In our understanding of human performance, we typically estimate muscle–tendon length changes using changes in joint angles in the body. If the muscle–tendon unit appears to lengthen, most people feel the actual muscle fibers are lengthening. This type of analysis is especially important to our understanding of muscle injury. Because it is well-documented that as activation increases, more cross-bridges are attached and muscle stiffness increases, the effect of variable activation on the distribution of length between a tendon having a certain stiffness and a muscle that is changing its stiffness as a function of its activation becomes complex. The effect of variable activation on length distribution between muscle and tendon is not well-documented yet basic to our understanding of muscle mechanics and movement control.

IX. Concept of Stretch–Shorten Cycle and Electromechanical Delay

Compliance in the series elastic element of the muscle–tendon unit, which can be in the tendon as well as in the muscle cross-bridges and other tendinous elements in the muscle, also affects the coupling between the electrical signal produced by the muscle (EMG) and subsequent force production by that muscle. This electromechanical delay (EMD) can vary from a few milliseconds to more than 100 msec, depending on the type of contraction. During rapid lengthening of a muscle containing predominantly fast fibers, the delay between onset of EMG and onset of force is approximately 5 msec. During isometric contractions in knee extensors, however, there is a reported delay of 95 msec. Although there is a wide range of values reported for EMD, it is well-accepted that there is a temporal uncoupling between the electrical signal produced by skeletal muscle (EMG) and its associated force output. Again this has clear implications for performance.

As an extension to the two-component model presented thus far, the behavior of muscle, to a first approximation, can currently be represented by a three-component model. In this particular scheme, the characteristics of muscle are represented by three elements: a contractile component (CC), a series elastic component (SEC), and a parallel elastic component (PEC). These three elements now fully explain the behavior of muscle during lengthening, isometric, and shortening contractions. The PEC only comes into effect when the muscle is lengthening and typically when it is lengthened passively. A possible anatomical location for the parallel elastic element is the sarcolemma of individual fibers and connective tissue sheaths of the muscle. The PEC was identified in the late 1940s and described initially in amputee patients being fitted with prostheses. The intent was to see if any properties of the existing muscle could be used by the prosthesis. It has been well-documented that greater force is

exerted for the same metabolic cost during a lengthening contraction when compared with a shortening or isometric contraction. If we were to do a series of idealized experiments, for example, in which we evaluated the force produced at various elbow angles, we would observe a response similar to that illustrated in Fig. 6. We would see the classic length–tension response for each condition, but noticeably higher forces during an eccentric contraction as opposed to a concentric contraction. In reality, what is occurring can be illustrated by lifting and lowering a weight in our hand and observing the electrical output of the biceps brachii muscle. If we put a 20-lb weight in someone's hand and ask them to lower that weight by extending the elbow, we would see lower EMG activity than if we asked the same person to raise the weight through the same range through elbow flexion. The obvious reason is that during the lengthening contraction the biceps brachii uses less metabolic energy because force exerted can come from the active CC as well as the PEC. In the shortening contraction, as one raises the weight, the metabolic cost is higher because the primary demand to perform the task is on the CC. The PEC does not make a contribution during shortening. Consequently, in daily activities, it is easier to lower weights through different distances than it is to raise that weight through the same distance. The application of this phenomenon can be made to any number of settings, ranging from a clinical environment to human factors engineering in the work place.

It becomes evident in the study of human performance that muscles experience regular sequences

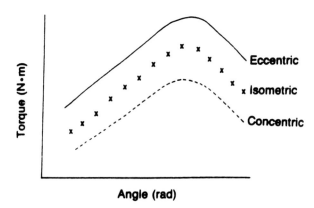

FIGURE 6 Schematic of torque production (joint moment) at different angles during eccentric (lengthening), isometric, and concentric (shortening) contractions. [From Enoka, R. M.(1988). Neuromechanical Basis of Kinesiology, Human Kinetics Publishers, Champaign, Illinois, with permission.]

of lengthening and shortening contractions during daily movements. This phenomenon, referred to as the *stretch–shorten cycle* is designed to use the elastic properties of muscle-tendon to store energy that can then be used during the shortening phase of the muscular contraction. A term commonly used to define a lengthening contraction is *negative work,* whereas the term used to define a shortening contraction is *positive work.* If we define the work done by skeletal muscle on the environment as a product of force and length change (theoretically mechanical work is defined as the integral of force with respect to distance), it has been consistently demonstrated that a muscle can perform a greater amount of positive work during a shortening contraction if that shortening contraction is preceded by active lengthening. It has also been demonstrated that this enhanced force or work output is evident even if the shortening contraction is preceded by an active isometric contraction. From a work-energy standpoint, the increase in the work performed requires an increase in energy expenditure. The question now focuses on the source of this additional energy. Because the SEC is far more important than the PEC in its potential for storage of elastic energy when an active muscle is stretched, an immediate place to look would be the tendon. Essentially, we are back to two-component model, highlighting the SEC. A general scenario may be as follows: An external load stretches the SEC during a lengthening contraction, which can be envisioned as a transfer of energy from the load to the SEC. Essentially, the environment is doing work on the muscle, and energy is stored in the SEC. Once the system is released, the molecular structures in the various tissues return to their original shape and the muscle shortens. The energy received from that stored in the tendon is then used in the shortening contraction.

X. Force Enhancement

The ability to use stored elastic energy is affected by three variables: time, magnitude of stretch, and velocity of stretch. It is currently believed that there should be a minimum time delay between the lengthening and shortening contractions, otherwise most of the stored energy would be lost as heat. The actual magnitude of this delay remains open to discussion, but the shorter the delay, the more the enhancement during shortening. With regard to the magnitude of stretch, if the lengthening contraction is too great, fewer cross-bridges will remain attached after the stretch (i.e., more will be mechanically broken and hence less elastic energy would be stored). Although we feel that tendon plays a major role in this process, some storage may occur in the cross-bridges. If the crossbridges remain attached and the velocity of stretch is high, the percent contribution by the cross-bridges will be higher and the storage of elastic energy will potentially be greater.

Examples of this phenomenon can be cited in everyday activities. If we evaluate, for example, the knee extensor muscles during the early part of the stance phase of locomotion, one of the first responsibilities of this muscle group is to absorb the shock of the ground reaction force. These muscles lengthen early in stance and then shorten during mid and late stance as the knee extends. To maximize the storage and reuse of energy in the SEC, muscles must be active. The quadricep group is active before heel contact and through early stance. Consequently, while shock is being absorbed, the muscles are actively stretched. Immediately after this stretching phase the muscles shorten and have the potential for enhanced contribution to forward body propulsion.

Another example is the vertical jump. It has been widely reported that individuals jump higher if a counter-movement precedes the upward thrust of the body. This counter-movement involves lowering the body by controlled flexion of the hip, knee, and ankle and then raising the body by rapidly extending the same joints. During the counter-movement phase when the body is being lowered, the hip, knee, and ankle extensors are actively lengthening. Elastic energy is stored during this phase and reused, if there is minimal time delay, during subsequent propulsion; positive work is enhanced by the previous active stretch imposed on the muscles. Although other segments of the body assist in propulsion, the active stretch imposed on the lower extremity extensors before shortening undoubtedly enhances their output and results is a significantly higher vertical jump.

Although muscle groups (e.g., lower extremity extensors) can be evaluated in activities such as jumping, recent information has been obtained from animal models that demonstrates similar phenomenon at the level of isolated muscle. Figure 7 displays the force-velocity pattern obtained for the cat soleus muscle *in situ* (i.e., isolated muscle experiment) and during an experiment *in vivo* while the

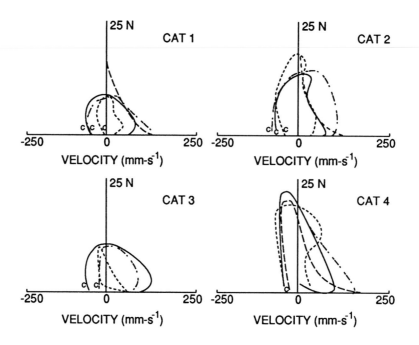

FIGURE 7 Force–velocity patterns within a step cycle for four cats at treadmill speeds of 0.8 (----), 1.3 m/sec (———), and 2.2 (—·—) m/sec. Patterns represent an average of five step cycles beginning at contact (c) with the treadmill. *In situ* force–velocity curves are also presented for cats 1, 2, and 4. [From Gregor, R. J., et al. (1988). *J. Biomech.* **21**, 721–732, with permission.]

cat was walking and running on a motor drive treadmill. The instrumentation used to collect these data were high-speed cinematogrpahy and specialized transducers surgically implanted to measure the force in an individual muscle. The proximal and distal insertions of the soleus muscle were marked and identified, and once position data were collected, muscle length changes were calculated during the entire step cycle. Velocities of lengthening and shortening were then calculated from the original length calculations. Evaluating the four sections of Fig. 7, it becomes apparent that the soleus muscle, during normal locomotion, can produce forces higher than those forces seen during the isolated muscle experiment. To appreciate this diagram fully, we must remember how the data were collected on the isolated muscle. These data were obtained by isolating the muscle from surrounding synergists, maximally stimulating it, and measuring its speed of contraction against variable external loads (a standard force-velocity experiment). Every contraction began at L_0. This classic curve, then, represents the contractile element for the soleus muscle and its ability to produce force under maximal excitation at various velocities of shortening. In contrast, we see the curves obtained during treadmill locomotion begin on the left side of the zero line, progress up the force magnitude scale, cross over to the positive or shortening side of the velocity scale, and decline to zero. These data show that the soleus muscle experiences a lengthening con-

traction immediately after ground contact, with the velocity of stretch on the muscle-tendon unit indicated to the left of zero. As observed, the velocity changes very little as force increases to a peak value around midstance. Another observation is the fact that one of these animals actually produced peak forces during stance that exceeded P_0. As positive velocities increase during the shortening phase (i.e., to the right of the zero line), we see forces produced at certain velocities exceeding the maximal force production capacity of the contractile element of that muscle at the same velocity. Basically, this represents an enhancement of the force production capabilities of the soleus muscle. We say enhancement because the forces are greater than those produced by the contractile element alone. We may then conclude that there is potential for storage and reuse of elastic strain energy by this muscle even at the relatively slow walking speeds used in this experiment.

Although evidence of enhanced force production is well-documented in the literature, the sites at which this energy is stored are more difficult to define. Several investigators argue that energy is stored in the globular heads of the cross-bridge attached to the actin filament and stretched. Others argue that the tendon, as a series elastic element, is a major site for storage. More work is needed to better define sites of storage, the effect of submaximal activation on length distribution between the muscle and tendon, and the whole concept of storage and reuse of elastic strain energy in light of submaximal activation. We must also remember that the lengthening and shortening contractions of the soleus muscle during normal locomotion do not necessarily begin and end at L_0. This makes the experience of the soleus during normal locomotion different from its experience *in situ*. Other evidence indicates that the tendon is actually recoiling during the shortening phase, producing what is referred to as an *enhanced velocity*. If this were the case, the actual velocity seen by the muscle-tendon unit would be dominated by the velocity seen in the tendon. The velocity seen by the contractile element would then be lower than that seen by the whole muscle-tendon unit and would place the contractile element on a different portion of its force–velocity curve than indicated by evaluating the *in vivo* locomotion data. This argument might explain the shift to the right of the *in vivo* force–velocity curve with respect to the force–velocity curve observed in isolated muscle. The basic point of these data are that (1) it appears that forces can be produced at certain velocities that exceed those capabilities of the contractile element and (2) much more information is needed on the response of the total muscle-tendon unit in normal movements.

It has been previously stated that muscles do not act alone in producing movement. When we study the mechanical output of skeletal muscle, we typically take it in isolation to simplify the task of identifying elements in the mechanical system. We have concluded that even when the muscle is isolated from its surrounding musculature and maximally activated, a precise description of the mechanical elements in skeletal muscle is difficult. When we study the same muscle and study its isolated function within a group of muscles during normal locomotion, we began to gain insight into the complexity of how muscles work together. For example, during the vertical jump, the gastrocnemius muscle, a two-joint muscle and typically a fast high-power output muscle, takes on much of the responsibility for upward propulsion. Because it is a synergist to the soleus at the ankle, it is conceivable that the high force produced by the gastrocnemius muscle

could actually unload the soleus during propulsion. The gastrocnemius muscle could move at such high speeds and with such high force that the soleus may be unable to "keep up." Any observation made on the soleus then, as we have seen in the cat treadmill locomotion experiments, must consider the soleus and its involvement with extensor synergists (i.e., the gastrocnemius and plantaris muscles).

XI. Joint Moment, Muscle Force, Work and Power

In a more complete evaluation of human movement, interest continues regarding the effect an individual muscle has on movement at the particular joint. A mechanical term often used to describe output at a joint is the *joint moment*. This *moment* represents the summation of all muscle forces, ligament forces, forces produced by periarticular structures, inertial forces (I_{cg}), and any environmental forces such as the ground reaction force during the stance phase of locomotion. If we were to use the ankle as an example during a cycling task, we would say that the muscles of the triceps surae complex, dorsiflexor muscles such as the tibialis anterior, and the pedal reaction forces contribute most to the joint moment at the ankle. Each of these forces acts some distance from the ankle axis of rotation, and it is the product of this force and the perpendicular distance from the axis of rotation to the line of action of that force that represents the contribution to the joint moment. Many individuals involved in skeletal muscle mechanics are interested in how individual muscles might be used and contribute to the joint moment during normal movements.

The force produced by the triceps surae complex during a normal cycling task is illustrated in Fig. 8. This force is a summation of the forces produced by the gastrocnemius and soleus muscles in attempting to plantarflex the ankle. Synchronous with this force record are EMG records and length changes for the gastrocnemius and soleus muscle-tendon units. Initially, force increases and then decreases as we proceed from top dead center to bottom dead center on the bicycle. This is the phase when the entire lower extremity is extending and providing power to the bike. This force matches the reaction force observed on the pedal. The Achilles tendon force reaches a peak at approximately 115° in the pedaling cycle when the crank of the bicycle is just

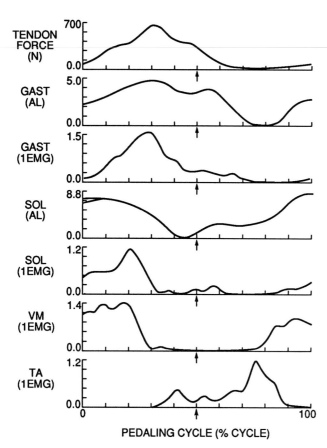

FIGURE 8 Tendon forces from the Achilles tendon in one male subject riding at 270 W on a stationary bicycle. Muscle-tendon length changes for gastrocnemius (GAST) and soleus (SOL) as well as EMG patterns for the gastrocnemius, soleus, vastus medialis (VM), and tibialis anterior (TA) muscles for the same cycle. [From Gregor, R. J., et al. (1987). *Int. J. Sports Med.* **8** (suppl.), 9–14, with permission.]

past a position horizontal to the ground. This force is an average of many cycles and represents a pattern that would commonly be used in the cycling task.

Because length affects force, length change patterns for both gastrocnemius and soleus muscles, were evaluated and are presented in Fig. 8. It appears that the soleus muscle has a greater length change and subsequently higher velocities of shortening and lengthening than those observed in the gastrocnemius muscle. Two joint muscles typically experience lower magnitudes of length change, and this is supported by the comparison between gastrocnemius and soleus muscles. Because the force observed is a summation of both muscles, we may conclude, based on the length change patterns, that

each muscle contributes differently to this total force. Additionally, we must also remember that the soleus is a slow twitch muscle whereas the gastrocnemius a more fast twitch. It is not unreasonable then that these two muscles would contribute differently to the composite force at the tendon.

In Fig. 9, we see a comparison between the mo-

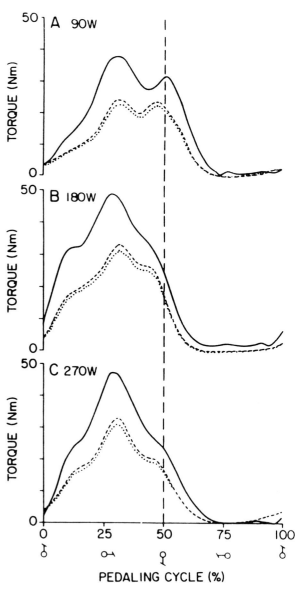

FIGURE 9 Torque (moment) patterns at the ankle for one male subject riding at three separate power outputs on a stationary bicycle. Joint moments (———), and triceps surae moments (--- and ………) are illustrated for each power output. N_m, Newton-meters of torque, W, watts of power. [From Gregor, R. J., et al. (1989). *J. Biomech.* (submitted), with permission.]

ment produced by the triceps surae and what is considered the joint moment at the ankle during cycling. An interesting feature of these joint moment patterns is that the moment produced by the triceps surae complex is temporally in phase with the joint moment. The other interesting feature is that the triceps surae moment consistently represents about 65% of the joint moment at the ankle. This can be interpreted as a contribution of approximately 65% to the total control needed at the ankle. This percentage is relatively constant across three separate power outputs and means that this muscle complex matches the requirements to the bicycle in a proportional way.

This contribution is a result of forces produced by the two muscles at some distance from the ankle axis of rotation. We have already identified the fact that the two muscles are different in their fiber composition and that they experience different length change patterns. It is also conceivable that these two muscles have different force–velocity patterns, and as illustrated in Fig. 8, they have dissimilar EMG patterns. The intent is to dissect and distribute the total control necessary at a joint to the tissues surrounding that joint. It is also interesting to relate these muscle forces and the contributions they make at a particular joint to their mechanical properties. As evidenced in the length change patterns, both muscles experience a lengthening early in the power phase. This active lengthening has the potential to yield enhanced force production during shortening. It is conceivable then that each muscle in a slightly different way uses an essentially similar strategy (i.e., stretch shorten cycle in an attempt to contribute with other synergists to the joint moment).

To summarize, we have evaluated both the structural and functional units within skeletal muscle. We have identified various parameters that affect the muscle's ability to produce force and have associated length changes, velocity changes, activation patterns, fiber type, and previous history to the mechanical output of skeletal muscle. This mechanical output is quantified in units of force, work (i.e., force times distance), power (i.e., the rate of change of work), and joint moments. One item we have not considered in this chapter, other than to discuss it with respect to certain fiber types, is fatigue. The mechanical properties discussed in this chapter are an attempt to understand the factors that affect muscle force production capability from the cellular

to the total joint level. The final section of this chapter will present, to a limited degree, how exercise and disuse affect muscle output.

XII. Exercise and Skeletal Muscle

The effects of exercise on skeletal muscle are specific to the types of exercise employed. For example, if we were to perform an endurance-type activity, we would not see an appreciable increase in the size of the muscle fiber or the whole muscle. We might see an increase in mitochondria and mitochondrial enzymes that enhance aerobic capacity at the cellular level. These types of responses have been documented in the literature, and we need only observe long-distance runners to see the effect of endurance training. The counterpart to this is high-power output training commonly used in football, weightlifting, etc. The high-power output training might selectively recruit the larger fast twitch fibers simultaneously with the slow twitch fibers. Increased fiber size associated with this type of training has been reported in the literature. Although there is continued discussion as to whether fibers may actually split, the fact remains that fibers do get larger with power-type training and consequently increase the size of whole muscle. If we think about force production capability related to physiological cross-section area, we would conclude that the force production capability of the muscle would increase.

In contrast to exercise effects related to increased aerobic capacity or power output, disuse and immobilization produce opposite effects on muscle. For example, if a limb is immobilized for a certain period of time in a cast as a result of a bone fracture, the fibers in the muscle surrounding the immobilized joint will atrophy. It is also documented, although not in the domain of this chapter, that connective tissue properties will degrade. This leaves us with a system, postinjury, that needs rehabilitation structured around the losses realized by the tissues in the system. If we were to embark on a rigorous exercise program and not consider the fact that the muscles have atrophied and connective tissue properties degraded to the extent where injury may occur, we would be remiss in our training regimen design. This latter consideration has resulted in the current practice of immediate mobilization and regulated loading designed to strengthen muscle and minimize stress to the injured tissue. The mechanical properties of muscle obviously change as a result of disuse, and it is imperative to understand the differential loads on normal joints and the differential response of the various tissues surrounding the joint to design an appropriate rehabilitation regimen postinjury.

Bibliography

Cavagna, G. M. (1978). Storage and utilization of elastic energy in skeletal muscle. In "Exercise and Sport Science Reviews" (Robert S. Hutton, ed.) pp. 89–130. Williams and Wilkins, Baltimore, Maryland.

Edgerton, V. R., Roy, R. R., Gregor, R. J., and Rugg, S. (1986). Morphological basis of skeletal muscle power output. "Human Power Output" (N. L. Jones, N. McCartney, and A. J. McComas, eds.) pp. 43–58. Human Kinetics, Champaign, Illinois.

Gordon, A. M., Huxley, A. F., and Julian, F. J. (1966). The variation in isometric tension with sarcomere length in vertebrate muscle fibers. *J. Physiol. Lond.* **184**, 170–192.

Gregor, R. J., Komi, P. V., and Jarvinen, M. (1987). Achilles tendon forces during cycling. *Int. J. Sports Med.* **8** (suppl.), 9–14.

Gregor, R. J., Roy, R. R., Whiting, W. C., Hodgson, J. A., and Edgerton, V. R. (1988). Mechanical output of the cat soleus during treadmill locomotion: *In-vivo* vs *in-situ* characteristics. *J. Biomech.* **21**, 721–732.

Hill, A. V. (1938). The heat of shortening and the dynamic constants of muscle. *Proc. R. Soc. B* **126**, 136–195.

Komi, P. V. (1984). Physiological and biomechanical correlates of muscle function: Effects of muscle structure and stretch shortening cycle on force and speed. "Exercise and Sport Science Reviews" (R. L. Terjung, ed.) pp. 81–122. Williams and Wilkins, Baltimore, Maryland.

Lieber, R. L., and Boakes, J. L. (1988). Sarcomere length and joint kinematics during torque production in frog hindlimb. *Am. J. Physiol.* **254**, C759–C768.

Spector, S. A., Gardiner, P. F., Zernicke, R. F., Roy, R. R., and Edgerton, V. R. (1980). Muscle architecture and force velocity characteristics of cat soleus muscle and medial gastrocnemius: Implications for motor control. *J. Neurophysiol.* **44**, 951–960.

Wickiewicz, T. L., Roy, R. R., Powell, P. L., and Edgerton, V. R. (1983). Muscle architecture of the human lower limb. *Clin. Orthop.* **179**, 275–283.

Muscular Dystrophy, Molecular Genetics

RONALD G. WORTON and ELIZABETH F. GILLARD, *University of Toronto*

Glossary

DMD gene Gene on the human X chromosome that is mutated in boys with Duchenne and Becker muscular dystrophy

Dystrophin Cytoskeletal protein of muscle that is encoded by the DMD gene and is found to be missing or defective in boys with Duchenne and Becker muscular dystrophy

Hind III fragments Fragments of DNA that result from digestion with the restriction enzyme, Hind III

Integral membrane of glycoprotein Protein that passes through the membrane of a cell, with attached carbohydrate side chains on the portion of the protein outside the cell

Myoblasts Muscle precursor cells that fuse together in the process of forming the multinucleate fibers of skeletal muscle

Sarcolemma Outer sheath of a muscle fiber consisting of an inner plasma membrane and an outer basement membrane called the *basal lamina*

Satellite cell Myoblast that remains at the sarcolemma of mature muscle to play a role in the repair of damaged or diseased muscle

DUCHENNE MUSCULAR DYSTROPHY (DMD) is the most common and the most severe of the muscular dystrophies, and Becker muscular dystrophy (BMD) is a milder version of the same disease. DMD is a severe muscle-wasting disorder of young boys that results in early confinement to a wheelchair and death usually before the age of 20. BMD resembles DMD but has a later onset, more benign course, and longer survival. Both DMD and BMD are X-linked genetic diseases caused by mutation in the "DMD gene" that results in a deficiency of the high-molecular-weight cytoskeletal protein dystrophin. This article covers the structure of the gene and the nature of the genetic alterations that cause the disease, as well as the structure and properties of dystrophin, relating these to current ideas about the basic defect.

I. Clinical Description

Boys with DMD are phenotypically unremarkable at birth and remain so for the first year or two of life. They present with muscle weakness at age 3–5 when they begin to have difficulty in climbing stairs and in rising from a sitting position on the floor. Confirmation of the diagnosis is by measurement of serum creatine kinase (CK), muscle histology, and electrical stimulation of muscle or electromyography (EMG). Serum CK is grossly elevated in the preclinical and early clinical stages of the disease as a result of muscle CK release into the serum. Muscle histology characteristic of the disease shows fiber degeneration and regeneration and includes small fibers, variation in fiber size, and invasion by white blood cells that move in to remove the cellular debris resulting from muscle degeneration. In the later stages there is progressive replacement of the muscle by fat and connective tissue. The EMG shows a characteristic abnormal response to electrical stimulation with the evoked action potentials reduced in both duration and amplitude.

A multicenter study has recorded the clinical progression in a large cohort of boys whose symptoms began before 5 years of age. Those with the mildest course would be classified as BMD. In the study, the age of occurrence of a number of "milestones" was recorded, forming a database against which an individual patient could be measured. Typically, affected boys lost the ability to rise from a chair and to climb stairs at around age 9 (range 7–13) and required a wheelchair by age 12 (range 9–16). For a period of about 3 years before wheelchair confinement, leg braces were beneficial in maintaining ambulation.

In later years the major problems were curvature of the spine (scoliosis) and reduction in lung capacity. Although 25% of the patients maintained a relatively straight back, progressive scoliosis was common in the others. Corrective surgery to insert a steel rod along the spine is common practice for boys with curvative of >35 degrees. In many affected boys, degeneration of the diaphragm muscles reduces lung capacity, and multiple bouts of pneumonia are common. The lung volume reduction and the risk of pneumonia were greatest in those who scored lowest in performance. These "weak" patients, with performance ratings below the 50th percentile, died from respiratory failure and pneumonia at age 13–17. The stronger boys who scored above the 50th percentile lived a little longer (age of death 14–21 years), and many died of cardiac failure with respiratory function reasonably well-preserved.

II. Mendelian Inheritance

The incidence of DMD is approximately one in 3,300 male births, with little ethnic variation. An individual case may be either sporadic or familial, and since the DMD gene is carried on the X chromosome familial cases occur in families with a typical X-linked pattern of inheritance (i.e., the disease is inherited by males through carrier females with no male-to-male transmission). Because affected males are unable to transmit the defective gene, these mutations are lost from the population. In a population at equilibrium with respect to disease incidence, the lost mutant genes are "replaced" by the process of new mutation. The calculated mutation rate of 10^{-4} mutations per gamete per genera-

tion is at least an order of magnitude higher than for most other genetic diseases. The assumption of equal mutation rate in males and females leads to the prediction that one-third of affected males (which have a single X chromosome) should result from new mutation, whereas, two-thirds should result from inheritance of the defective gene from a carrier mother. Because females have two X chromosomes, they have twice the chance of having a mutation in one of them, but they remain healthy because the other chromosome performs the necessary function. They transmit the mutation to 50% of their children (their daughters having a 50% risk of being a carrier and their sons having a 50% risk of being affected).

The incidence of BMD is about 10–15% that of DMD. Because about 70% of males with BMD are capable of transmitting the gene to a carrier daughter, a higher proportion of BMD cases (90%) is inherited. The mutation rate for BMD is only 3–5% of that for DMD.

Females who carry a defective gene do not usually express the disease because their second X chromosome produces normal gene product. However, in the somatic cells of females, inactivation of one of the two X chromosomes occurs. This process, described by Mary Lyon (the Lyon hypothesis), is random in nature, with some cells inactivating the paternally derived X chromosome and others inactivating the maternally derived X. Therefore, in females who carry a DMD gene mutation, the proportion of muscle nuclei that have an inactive chromosome carrying the normal DMD gene is randomly variable and leads to mild clinical manifestation in about 8% of carriers. The disease may also be expressed in Turner syndrome females who have a single X chromosome and in females in which one X chromosome has undergone exchange (translocation) with another chromosome, disrupting the DMD gene. The females who carry such a translocation express the disease because the translocation disrupts one DMD gene and the intact X chromosome becomes inactivated in all cells, therefore blocking the action of the second copy of the gene.

III. DMD Gene

The gene responsible for DMD and BMD (i.e., the DMD gene) is located at a position in the middle of the short arm of the X chromosome, at chromo-

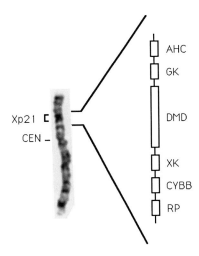

FIGURE 1 Human X chromosome and a schematic of genes mapped in the region of band Xp21. Band Xp21 can be seen to split into two dark bands separated by a light band, in order from the centromere (CEN), Xp21.1, Xp21.2, and Xp21.3. Genes indicated on the map include congenital adrenal hypoplasia (AHC), glycerol kinase (GK), Duchenne muscular dystrophy (DMD), McLeod phenotype (XK), chronic granulomatous disease (CYBB), and retinitis pigmentosa (RP).

some band Xp21 (Fig. 1). The gene itself is the largest yet discovered, covering approximately 2,300 kilobases (kb) of DNA. Because of its enormous size, the entire gene has not yet been isolated, and regions of the gene are identified by certain landmarks. These landmarks are of historical significance only, as they represent the various fragments of the gene that were isolated during the span of about 3 years. To understand the gene, it is necessary to understand a little about the historical developments in the cloning of the gene. This is described briefly below. [*See* GENES.]

A. DMD Gene Localization

Mapping of the DMD gene at Xp21 on the X chromosome was the first step in the cloning of the gene and identification of its protein product, dystrophin. The first evidence came from affected females with translocation exchange points in the X chromosome at band Xp21, suggesting that a gene at this site might be disrupted by the translocation to cause the disease. Subsequent high-resolution chromosome analysis revealed that the exchange points were not precisely the same in all affected females and suggested a target for disruption extending from Xp21.1 to Xp21.3, a region of perhaps 3–4 million

base pairs (bp). This in turn suggested the possibility of a very large gene.

Further evidence came from family studies with DNA probes that detect certain landmarks (genetic markers) in DNA. Several genetic markers from the X chromosome were found to segregate with the disease phenotype in both DMD and BMD families, the strongest co-segregation seen for those markers closest to Xp21. This observation not only mapped the DMD and BMD genes to Xp21, it also provided the first indication that the two diseases might be caused by mutations in the same gene. [*See* DNA MARKERS AS DIAGNOSTIC TOOLS.]

Additional evidence came from a few boys with DMD in combination with one or more additional X-linked diseases. One of these, "BB" was shown to have DMD, retinitis pigmentosa (RP), chronic granulomatous disease (CYBB), and the McLeod cell phenotype (XK), all due to a small deletion of part of band Xp21. Other boys with similar phenotypes, or with DMD plus glycerol kinase (GK) deficiency and congenital adrenal hypoplasia (AHC) have also been shown to have contiguous gene deletions. The most likely gene order is AHC, GK, DMD, XK, CYBB, RP, starting from the end of the chromosome (Fig. 1).

B. DMD Gene Cloning

The DMD gene was one of the first genes to be cloned, based only on the knowledge of its location in the human genome. The approach has been termed *reverse genetics* to distinguish it from the more established procedure of gene cloning from knowledge of the RNA or protein product of the gene. A schematic of the gene is shown in Fig. 2.

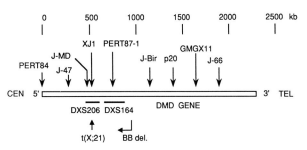

FIGURE 2 Schematic of DMD gene. Gene orientation (5' —> 3') is from centromere (CEN) to telomere (TEL) (i.e., the end of the chromosome). A number of landmarks in the gene are indicated by *arrows*, and the origin or significance of each is described in the text. The edge of the deletion in "BB" is shown, as is the t(X;21) translocation exchange point.

The first recognized fragment of the gene, the "PERT87" clone (Fig. 2), was isolated by enriching for DNA fragments from the region of the deletion in patient "BB." The PERT 87 clone mapped into the BB deletion and also failed to recognize DNA from a subset of boys with DMD, suggesting that these boys had the disease as a consequence of a submicroscopic deletion that removed at least a portion of the DMD gene, including the PERT87 sequence. The cloned region was expanded to give the 220-kb region with the locus designation of "DXS164" (Fig. 2). Another Clone, PERT84 was later recognized to be at the 5' end of the gene (Fig. 2).

The gene fragment "XJ1" was isolated from the junction of a t(X;21) translocation carried by one of the rare translocation females with muscular dystrophy. The translocation involved chromosome 21 in a block of genes encoding ribosomal RNA. Ribosomal gene probes identified a clone containing ribosomal gene at one end and DMD gene at the other. Expansion of the region gave the 140-kb "DXS206" locus (Fig. 2).

Other important landmarks in the DMD gene include clones p20, and GMGX11, isolated as random pieces of the X chromosome, and J-47, J-MD, J-Bir, and J-66, isolated by virtue of their close proximity to PERT87 (Fig. 2).

By analyzing large fragments of the gene, and aligning the pieces the size of the gene was determined to be more than 2,300 kb in size. The entire human genome contains approximately 3×10^9 bp of DNA (3 million kb), and the X chromosome represents about 5% of the genome or about 150,000 kb. The DMD gene, therefore, covers 1.5% of the X chromosome.

C. DMD Gene–Expressed Sequences

The RNA message encoded by the DMD gene is 13.9 kb in size and is transcribed from a minimum of 70 exons (separate coding regions). cDNA clones derived from the RNA message were isolated and are shown schematically in Fig. 3. Each clone represents a portion of the message and each has great value in the detection of deletions and duplications in the DMD gene, as described later.

D. Intron–Exon Structure of the DMD Gene

The DMD gene is so large that it needs to be analyzed in small pieces. One convenient way to do

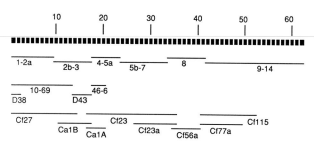

FIGURE 3 Schematic of DMD cDNA. Total length of cDNA is 13.9 kb and for convenience is divided into 62 exon-containing Hind III fragments that are detected by cDNA clones. Fragments are not to scale. Below Hind III fragment lines are cDNA clones in common use for patient analysis. The *line* representing each clone is drawn to indicate Hind III fragments detected by that clone. Clones from Kunkel's laboratory are labeled 1–2a, 2b–3, 4–5a, 5b–7, 8, and 9–14, indicating their position along the 13.9-kb message. Clones from our own laboratory are labeled D-38, 10-69, D-43, and 46-6. Clones from Davies' laboratory are labeled with prefix Cf (fetal) or Ca (adult), depending on the source of muscle cDNA library.

this is to cleave the DNA of the entire chromosome set into fragments by digestion with a restriction enzyme, and then determine which fragments contain coding segments (exons) of the DMD gene. To do this the fragments are separated according to size by electrophoresis through a molecular sieving gel, the smaller fragments migrating farther through the gel than the larger fragments. Those fragments containing one or more exons of the DMD gene have stretches of base sequence identical to the base sequence of the message and therefore identical to the base sequence of the message-derived cDNA. A radioactively labeled cDNA probe will therefore form stable hybrid molecules with the exon-containing fragments (but not with other fragments), and an x-ray film overlaying the gel (or a Southern blot replica of the gel) will turn black whereever exon-containing fragments are located in the gel. The pattern of black DNA "bands" following digestion of DNA with the restriction enzyme, Hind III, is shown in Fig. 4. Each of the 8 lanes across the gel contains DNA from a different individual, some of whom are DMD patients who are missing dark bands as a result of their having a deletion of the corresponding exons from their DNA.

When a complete set of cDNA clones spanning the entire message (Fig. 3) is used as probe on Hind III digested DNA a total of 62 separable fragments are detected. (Seven of these fragments, detected by one cDNA clone are seen in Fig. 4.) Some of these fragments contain more than one exon, the

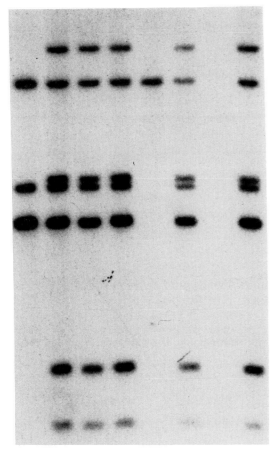

FIGURE 4 Representative autoradiograph of a Southern blot with Hind III–digested DNA from eight affected males. The blot was hybridized with cDNA8. Males 1 and 5 have partial deletions of this region of DMD gene. Male 7 has a complete deletion of this region. The size of Hind III bands is, from top down, 10.0, 7.0, 3.8, 3.7, 3.1, 1.6, 1.25.

TABLE I Size, Order, and 3′ Border Type of Hind III Exon–Containing Fragments[a]

Exon Number	Hind III Fragment No.	Hind III Fragment size	3′ Border Type
1	1	3.20	1
2	2	3.25	3
3	3	4.20	3
4	4	8.50	3
5	5	3.10	3
6	6	8.00	2
7	7	4.60	1
8	8	7.50	3
9			3
10	9	10.50	3
11			2
12	10	3.90	3
13	11	6.60	3
14	12	2.70	3
15			3
16	13	6.00	3
17	14	1.70	2
18	15	12.00	3
19	16	3.00	1
20	17	7.30	3
21	18	11.00	1
22	19	20.00	3
23			3
24			3
25			3
	20	5.20	
	21	12.00	
	22	4.70	
	23	18.00	3
	24	1.80	3
	25	0.45	3
	26	1.30	3
	27	1.50	3
	28	6.00	
	29	6.20	3
	30	4.20	3
	31	11.00	2
	32	4.10	3
	33	0.50	2
	34	1.50	3
	35	10.00	3
	36	1.25 ± 3.8	3
	37	1.60	3
	38	3.70	1
	39	3.10	3
	40	7.00	1
	41	7.8 ± 1.0	3
	42	8.30	2
	43	2.30	3
	44	1.00	
	45	8.80	
	46	6.00	3
	47	3.50	3
	48	6.60	
	(49)	2.80	

(Continued)

total exon number estimated to be greater than 70. The final count will not be known until all the boundaries of the exons to the intervening sequences (introns) are determined. They are now known for more than half of the exons.

Table I provides a list of the 62 Hind III fragments that are known to contain exons of the DMD gene. Where known, the specific exons contained in each fragment are given. For those exons whose intron–exon borders are known, the border type at the 3′ end of the exon is given, and indicates whether the exon ends after the first, second, or third position of a triplet of bases that codes for an amino acid. The significance of the border type will

TABLE I (*continued*)

Exon Number	Hind III Fragment No.	Hind III Fragment size	3' Border Type
	(50)	12.00	
	51	2.40	
	52	2.55	
	53	1.45	
	54	1.50	
	55	6.80	
	56	2.10	
	57	5.20	
	58	1.90	
	59	10.00	
	60	1.80	
	61	5.90	
	62	7.80	

a The size and order of Hind III exon containing fragments is essentially as described by den Dunnen, J. T., et al. (1989), whereas 3' border types are as described by Koenig, M., et al. (1989), with the following exceptions: (1) The order of fragments 24–26 is revised based on exon borders defined by Koenig, M., et al. (1989); (2) Where two fragments share a single exon, we have given them a single number (e.g., 36 and 41), and thus subsequent numbers differ from those of den Dunnen, J. T., et al. (1989), after this point; (3) The order of fragments 48–52 differs from that of den Dunnen, J. T., et al. (1989), and is based on data from a deletion patient and a translocation patient from our laboratory, the order of 49 and 50 being uncertain; (4) 3' border types for exon 22–25 were defined by Karen Bebchuk in our laboratory (*unpublished data*).

become clear in the section describing deletions and duplications in DMD and BMD patients.

The introns in the DMD gene are unusually large, as exemplified by the intron–exon size ratio of $2,300/14 = 164$, far greater than for the average gene. Assuming an exon (and intron) number of approximately 70, then the average intron size is $2,300/70 = 33$ kb. Some introns are much larger than the average. Intron 7, for example, spans most of the DXS206 locus and is 110 kb in size. Introns 1 and 2 are also potentially large as the distance between exon 1 and exon 3 is estimated to be 300–500 kb. Another large intron exists in the region of the gene detected by clone p20. Many of the human genes that have been isolated would easily fit into a single intron of the DMD gene.

E. Conservation of the DMD Gene

Despite the enormous size of the gene, its basic characteristics have been conserved through evolution. The size of the gene in mouse and in other species appears to be similar to that in the human.

One of the mysteries is how a gene with introns of >100 kb in size can be maintained. Because many of the introns are large enough to contain another gene, it suggests the possibility that such genes might be found, thereby accounting for the conservation of large introns during evolution.

IV. Dystrophin, Product of the DMD Gene

The general approach to identify the product of the DMD gene has been to examine the amino acid sequence to identify homologies with other known proteins and to generate antibodies against the protein.

A. Amino Acid Sequence and Proposed Structure of Dystrophin

The complete sequence of the DMD cDNA provided the sequence of the 3,685 amino acids of the protein product. Comparison of this sequence with that of other proteins revealed four distinct domains, three of which show sequence homology and structural similarity to previously characterized cytoskeletal proteins (Fig. 5A). For example, the N-terminal domain (amino acids 14–240) shows sequence homology to α-actinin, a large second domain (amino acids 278–3,080) has a 109-amino-acid repeat pattern similar to the repeats found in both α-actinin and spectrin, (thought to form a tripical helical rod-like segment) and the third domain (amino acids 3,080–3,360) is cysteine-rich and also bears some homology to α-actinin. The C-terminal domain bears no similarity to any known protein. The features are highly suggestive of a cytoskeletal protein, perhaps with a structural role in the cell. The protein was named *dystrophin* to relate it to its role in preventing muscular dystrophy.

B. Localization of Dystrophin in Skeletal Muscle

Further characterization of the protein depended on the use of antidystrophin antibodies. These were raised in animals immunized with either synthetic peptides (with amino acid sequence of part of the dystrophin molecule) conjugated to larger molecules, or "fusion proteins," synthesized by bacteria that had been transfected with a gene containing a

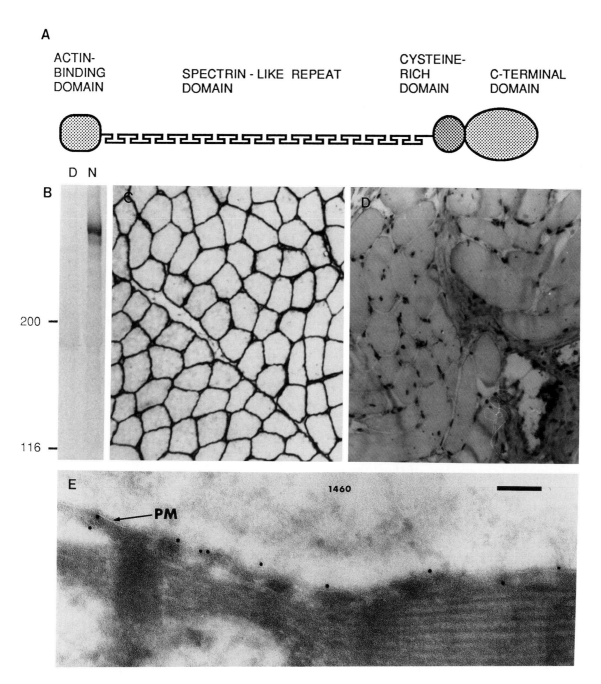

FIGURE 5 Structure and localization of dystrophin. **A**: Schematic of dystrophin showing four domains described in the text. **B**: Western blot of muscle from Duchenne ("D") and normal ("N") muscle. By this technique muscle protein is extracted and separated according to size by gel electrophoresis; dystrophin is identified in the gel by its ability to bind antidystrophin Ab. The major protein so identified in the normal muscle sample (N) but missing from the DMD patient muscle sample (D) is of size 400 kD, as estimated from size markers at 116 and 200 kD. **C**: Section of normal muscle stained with antidystrophin antibody. Note intense staining at periphery of each myofiber. **D**: Section of Duchenne muscle stained with antidystrophin antibody and counterstained with hemotoxylin. No dystrophin is at periphery of myofiber. **E**: Electron micrograph of human muscle fiber (gastrocnemius) labeled with antidystrophin antibody 1460, directed against the last 17 amino acids of dystrophin. Labeled gold particles appear along the plasma membrane (PM). Bar = 100 nm. Figure 5B, C, D courtesy of Dennis Bulman; Fig. 5E courtesy of Dr. Michael Cullen.

portion of the DMD cDNA joined to the 3' end of a highly expressed bacterial gene.

Antidystrophin antibodies recognize a protein of about 400,000 daltons in human and mouse skeletal muscle (Fig. 5B), consistent with the size predicted from the cDNA sequence. Dystrophin is also present in cardiac muscle, with lesser amounts in smooth muscle and small amounts in brain. The protein is missing from muscle of boys with DMD and from *mdx* mice with a mutation in the equivalent mouse gene (Fig. 5B).

Immunofluorescence and immunocytochemistry, using both light and electron microscopy, reveal that the major site of dystrophin in skeletal muscle is at the inner surface of the sarcolemmal membrane (Fig. 5C, D, E), suggesting a possible role for dystrophin in maintenance of the structural integrity of the membrane. Biochemical fractionation has also suggested localization of some dystrophin in skeletal muscle at the junction between the sarcoplasmic reticulum and the sarcolemmal invagination known as the transverse tubule, because immunoreactive dystrophin appeared to copurify with other proteins of this junction. This possibility has not been well-supported by the light microscope and the EM studies. [See SKELETAL MUSCLE.]

C. Evolutionary Conservation of Dystrophin

One of the most conserved regions of the dystrophin molecule is near the C-terminal end of the protein which is highly homologous in chicken and humans. The greatest sequence conservation is in the C-terminal domain of the protein and in the 3' untranslated portion of the gene. This suggests that the C-terminal end of the protein has an important biological function and that the 3' untranslated region must play some as yet undetermined function in the expression of the gene.

D. Dystrophin-Associated Proteins

New insight into the potential function of dystrophin has come from its purification as a complex with an integral membrane glycoprotein. This suggests that the localization of dystrophin at the cytoplasmic face of the sarcolemma results from its tight association with this membrane-spanning glycoprotein and that the complex may serve as a link between the outer membrane and the internal cytoskeleton. This result would be consistent with a role for dystrophin in maintaining the integrity of the

membrane. An alternative suggestion was that dystrophin might maintain a nonuniform distribution of the membrane glycoprotein, perhaps an ion channel or a membrane receptor. Characterization of dystrophin-binding proteins remains a high priority in the continued investigation of the cause of the X-linked muscular dystrophies.

E. Dystrophin-Related Proteins

Further evidence for the functional significance of the C-terminal domain is afforded by the finding that cDNA fragments encoding this part of the protein detect a closely related gene that exhibits sequence similarities with the DMD gene. One such "dystrophin-like" gene encodes a transcript of 13 kb, maps to human chromosome 6, and is a logical candidate for the defective gene in one of the dystrophies inherited with chromosomes other than the X chromosome (autosomes).

An additional dystrophin-related muscle protein has been detected by its ability to react with some, but not all, antidystrophin antibodies. This protein is of a similar size to dystrophin and could, in some situations, be confused with the dystrophin molecule. It is visualized most readily in extracts of muscle protein from patients who are dystrophin deficient. Its relation to the product of the chromosome 6 gene described above is not known.

V. Tissue Specificity and Regulation of the DMD Gene

The availability of DMD cDNA clones and antidystrophin antibodies has prompted studies of the tissue distribution and developmental pattern of DMD gene expression.

A. Tissue and Developmental Specificity

Analysis of the RNA indicated that the DMD transcript is most abundant in skeletal and cardiac muscle, with reduced amounts in smooth muscle and brain, a distribution that is in good agreement with the clinical spectrum of tissue involvement in the disease. Dystrophin transcripts are present in both fetal and adult tissues and in cells capable of differentiation into muscle (myogenic cultures); transcription is initiated as the immature cells (myoblasts) begin to differentiate into multinucleated myotubes. The DMD gene is therefore regulated in a tissue- and a stage-specific manner.

At the protein level, analysis with antidystrophin antibodies has revealed dystrophin in extracts of adult and fetal skeletal, cardiac, and smooth muscle, with an apparently lower molecular weight form in smooth muscle tissue. Lower levels have been detected in brain and neuronal cell cultures. In adult skeletal and cardiac muscle, dystrophin is localized at the muscle cell membrane without preferential distribution to any particular fiber type, and in myogenic cultures dystrophin is observed at the membrane of myotubes after initiation of differentiation. In contrast, dystrophin staining has been reported in fetal muscle at the ends of the myotubes.

B. Dystrophin Isoforms

Some of the tissue and developmental differences might be the result of dystrophin isoforms (different protein forms generated by the same gene), because isoforms of muscle proteins are commonly observed. A brain-specific DMD gene promoter initiates transcription of the gene at a different first exon in brain compared with muscle and is presumed to give rise to a brain-specific isoform. Exon 1 of brain is connected (spliced) to the same second exon as exon 1 of muscle, so that the remainder of the protein is expected to be the same in the two tissues, barring alternative splicing farther along the RNA transcript, which might cause omission of some exons in the RNA messengers.

Alternative splicing is a common mechanism for generating a spectrum of mRNA molecules with different combinations of exons spliced into the message. Alternative splicing at the 3′ end of the DMD gene has been described and presumably gives rise to different dystrophin isoforms in skeletal muscle, smooth muscle, and brain.

C. DMD Gene Regulation

Sequence analysis of several hundred base pairs at the 5′ end of the DMD gene has revealed the presence of several sequence elements that may be involved in the regulation of the gene. These include a "TATA box" and a "GC box," sequence elements usually responsible for a basal level of transcription. Elements thought to be responsible for the enhanced transcription in muscle include a "CArG box" and an "MCAT" concensus sequence. Thus, the DMD gene has a fairly typical muscle-specific

promoter. Confirmation that the DMD gene promoter functions in a muscle-specific fashion has been obtained by joining it to a "reporter gene" (which when expressed generates a characteristic protein). This approach demonstrated enhanced transcription of the reporter gene in fused myotubes in which the reporter gene fused to the DMD promoter was introduced. It has already been mentioned that the brain transcript is initiated from a different promoter located upstream from the muscle-specific promoter.

VI. Mutations in the DMD Gene

Deletion of one or more exons of the DMD gene accounts for more than 60% of mutation associated with DMD or BMD. Tandem duplication accounts for another 6% of mutations in the gene.

A. Deletion and Duplication in the DMD Gene

Many deletions and duplications are detectable with cDNA probes containing sequences from the 70 or more exons of the gene. The cDNA clones most often used for patient analysis are those depicted in Fig. 3. With these probes a broad spectrum of deletions has been detected in DMD and BMD patients. Some of these are depicted in Fig. 4. The largest deletions, several thousand kilobases in size, often remove neighboring genes and result in a more complex disease. Smaller deletions usually remove from one to a few exons, and their distribution in the gene is not random. Two deletion-rich regions are apparent as shown schematically in Fig. 6. Between these two "hot-spots" is a region of the gene relatively free of deletions, and in the last quarter of the gene, deletions are almost never detected.

Duplication of one or more exons constitutes an additional 6% of mutation and are heterogeneous with respect to both size and location. The remaining 25–35% of mutations have yet to be defined.

B. Molecular and Clinical Heterogeneity—The Frame-Shift Hypothesis

An important question is how mutations in the same gene can give rise to the full spectrum of pheotypes, ranging from severe DMD to mild BMD. BMD deletions are not confined to a specific region of the gene, indicating that differences in phenotype are not due to deletion of discrete protein domains with

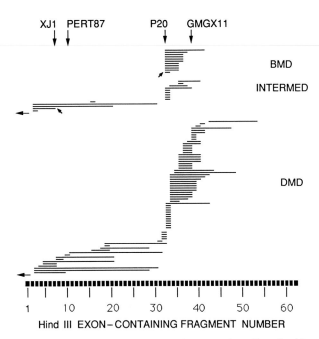

FIGURE 6 Schematic of deletions in 100 males affected with DMD, BMD, or an intermediate phenotype. Each line represents the extent of one deletion, plotted along the map of Hind III exon–containing fragments, numbered as in Table I. *Arrows* at the left indicate deletions extending 5' from exon 1 into the promoter region. *Angled arrows* indicate two exceptions to the frame-shift hypothesis that are discussed in the text.

differing functional significance. Furthermore, phenotypic severity is not simply a function of deletion size, because some deletions associated with BMD are larger than and completely encompass deletions associated with DMD (Fig. 6).

Recent studies have tested the idea that severity of the phenotype might be a direct consequence of the effect of the mutation on the translational reading frame of the mRNA. According to the model, mild disease (BMD) results from a deletion of one or more exons containing an integral number of base triplets, or codons, each coding for a specific amino acid. They would maintain the translational reading frame in the mRNA so that the protein formed by translation beyond the deletion is normal. The dystrophin molecule thus produced would have an internal deletion but intact ends. Severe disease (DMD) would result from a deletion of a set of exons that contain a nonintegral number of codons, resulting in a shifted reading frame in the mRNA sequence after the deletion. In this situation, translation of the message on the 3' side of the deletion inserts incorrect amino acids in the protein until a stop codon is reached, causing the elongation of the protein to stop.

The frame-shift model would apply equally well to duplications, with duplication of an integral number of codons, causing a mild phenotype, and duplication of a nonintegral number of codons, causing a severe phenotype.

Testing of the model required the determination of exon sequence and the precise location of intron–exon borders to predict the effect of deletions or duplications on the translational reading frame of the mRNA. As described above, intron–exon borders have now been defined for more than half of the DMD gene, and exon border types have been designated "1," "2," or "3," depending on whether they occur after the first, second, or third nucleotide of a codon (Table I). Assuming a deletion results in the splicing of the exon immediately preceding the deletion to the exon immediately after the deletion, we can determine the effect of most deletions from the border types listed in Table I. If the splicing joins two borders of the same type, the result is an mRNA that retains the correct reading frame. Conversely, splicing of two exons with different border types will give rise to a message with an altered reading frame.

Several groups have evaluated patients in relation to the frame-shift model and found the hypothesis to hold true for the majority of deletions and duplications. Not surprisingly, patients with an intermediate phenotype (usually those who became wheelchair bound at age 12–16) were found to have deletions of either the frame-shift or the in-frame type. The results indicate that most of the time the reading frame status, when determined for a young patient, will provide a valid prognostic indicator for the severity of the disease.

Exceptions to the frame-shift hypothesis do exist and need further explanation. One of these is seen in Fig. 6 as a deletion of Hind III exon-containing fragment 33 in 11 DMD, five intermediate and one BMD patient (arrow in Fig. 6). This deletion is expected to cause a frame-shift in the message (joins a type 3 border to a type 2 border—Table I); the difficulty is to explain the BMD patient with this deletion. Another exception is a group of patients deleted for exons 3–7. Assuming that exon 2 (3' border type 3) is spliced to exon 8 (5' border type 1), this deletion is predicted to shift the translational reading frame to cause a severe phenotype; yet patients with this deletion include several with BMD, a few with DMD, and some with an intermediate phenotype. (Only one of these, in the intermediate class, is among the 100 patients plotted in Fig. 6.) Explanations that might account for the mild phenotype in the BMD patients included (1) alternative splicing to maintain the reading frame, (2) reinitiation of protein synthesis at a correct triplet from within exon 8, (3) transcription initiation from an unidentified promoter in intron 7, and (4) ribosomal frame-shifting, in which, during protein synthesis, ribosomes switch from one reading frame to another. Finally, there is always the possibility of secondary factors, as yet unknown, that might act to compensate partially for the lack of dystrophin in some mildly affected patients. Such factors, if they exist, might include the dystrophin-like proteins described above.

C. Origin and Mechanism of Mutation in the DMD Gene

The enormous size of the DMD gene and the frequent occurrence of new mutations combine to provide a unique system for the study of mutational origin and mechanism. *A priori*, the chromosomal rearrangements characteristic of the DMD gene may occur in either males or females, in germ cells during meiosis, or in diploid cells during the mitotic cycle, and from either homologous or nonhomologous exchange. These parameters are all amenable to study in DMD families.

There are several indications that some and perhaps many of the chromosomal rearrangements in the DMD gene occur in somatic and/or germ-line *diploid* cells before meiosis. For deletions, the evidence comes from families in which two or more affected boys are found to have a deletion that is not present in the blood lymphocytes of their mother. The deletion must have arisen in a diploid cell of the mother during development, affecting only a proportion of her cells, including germ cells. In terms of parental origin it appears that deletions arise in both males and females with no strong bias toward either one.

Partial gene duplication is rarely described in other genetic disorders, and its study in the DMD gene has provided insight into duplication mechanisms. In a study of the origin of five tandem duplications, (i.e., with segments in the same orientation), four originated by exchange between the two homologous DNA chains in the single X chromosome of the maternal grandfather and the fifth in a single X chromosome of the maternal grandmother. Thus, all five duplications arose by an intrachromosomal event involving an unequal exchange between the two chains. Because the grandfathers were all unaffected, the rearrangements are pre-

sumed to have taken place in the germ line, but at what point in the formation of germ cells is not clear.

Paternal origin of the mutation has been observed for six of six X-autosome translocations ascertained in affected females. This is unlikely to be attributable to increased opportunity for mitotic error in male gametogenesis, because male carriers of X-autosome translocations are often sterile due to faulty germ cell development. These translocations must have arisen at the last stage (i.e., meiosis).

Little information is yet available on the molecular details at the site of the chromosomal rearrangements. To date none of the deletion breakpoints have been sequenced to see if the exchanges are between interspersed repeat units or between unrelated regions of the genome. Sequencing through three duplication junctions has revealed that the exchanges are between unrelated sequences so the exchanges are of the nonhomologous type. Three translocation breakpoints have been sequenced and again no homology was found at the site of chromosomal exchange. In one translocation, the tetranucleotide sequence CGGC, which occurred several times near the exchange site, was implicated as a possible potentiator sequence or recognition sequence for enzymes involved in the translocation process.

VII. Applications to Molecular Medicine

Now that the DMD gene has been cloned and the gene product identified, the diagnosis of DMD and BMD is most readily confirmed by visualizing the defect directly, at the level of either the DMD gene or the protein it encodes.

A. DNA Analysis

Deletions are readily visualized by Southern blot analysis of chromosomal DNA of patients with suitable cDNA probes. Quantitative analysis permits the detection of female deletion carriers who have one copy of the deleted region, and also of males with partial gene duplication who have two copies of the duplicated region.

Deletions may also be detected by the polymerase chain reaction (PCR). Primer pairs have been designed to amplify specific exons and the presence or absence of each exon is scored without the need for radioisotope-based techniques. Primer pairs representing the most commonly deleted exons have been combined into "multiplex" amplification reactions capable of detecting more than 98% of the deletions dectectable by Southern blot analysis.

B. Dystrophin Analysis

Dystrophin analysis requires a muscle biopsy but has the distinct advantage that it allows direct detection of defects in the DMD gene product. However, it is not practical for prenatal diagnosis, and it is not yet reliable for carrier identification. Also, severity of phenotype is likely to be related to both qualitative and quantitative changes in dystrophin, yet neither the levels of dystrophin necessary to prevent muscle weakness nor the essential functional domains of dystrophin have been adequately defined.

Duchenne patients usually exhibit little or no dystrophin, although some antisera directed against the N-terminal end of the molecule may detect significant levels of a truncated protein. Immunostaining of muscle sections using antibodies conjugated to a fluorescent die or to some other detection mechanism also reveals a complete absence of dystrophin in most DMD patients. However, BMD patients have been found to exhibit near normal levels of dystrophin of reduced or increased size with antibodies directed against either the N-terminal or the C-terminal end of the molecule. This is consistent with deletion or duplication that retains the proper reading frame of the message. Immunostaining of muscle sections has been described as "patchy" or "diffuse" in BMD patients. Although immunocytochemical analysis may provide definitive diagnosis in dystrophin negative biopsies, a patchy staining may be difficult to interpret, because this type of staining has been seen in symptomatic carriers as well as in other muscle disorders.

C. Linkage Analysis—Carrier Identification and Prenatal Diagnosis

Before the discovery of genetically linked probes, carrier status was determined by measuring creatine kinase (CK) in the serum, an inadequate test because only 70–75% of obligate carriers have elevated CK levels. For females identified to be at risk, prenatal diagnosis was limited to determining the sex of the fetus by chromosome analysis, with no means of distinguishing an affected from an unaffected male.

This picture has changed completely with the introduction of direct DNA analysis coupled with linked genetic markers for carrier detection and prenatal diagnosis. When a mutation cannot be detected directly, genetic markers within or near the gene may be used to track the defective DMD gene in families. Although the method is highly reliable (approximately 95% accuracy) it is subject to error because of the natural recombination events that act to switch the genetic markers in relation to the disease gene. The accuracy with which carrier status can be determined is greatly enhanced using a pair of flanking markers, the error determined by the frequency of double recombination between the markers. As early as 1986 it was reported that in 75% of cases, flanking markers allowed carrier status to be predicted with 98% accuracy.

Application of molecular techniques to prenatal diagnosis requires fetal DNA derived from either chorionic villi or cultured cells from amniotic fluid. In the 60–70% of families for which the DMD mutation is defined, presence or absence of the mutation in the fetal DNA is scorable directly using DNA analysis. In families for which the DMD mutation is undefined, prenatal diagnosis is dependent on linkage analysis with genetic markers, for which it has been reported that 90% of carrier women have the potential for prenatal diagnosis with 95% or greater accuracy.

VIII. Pathophysiology of Duchenne Muscular Dystrophy

A. Nature of the Basic Defect

Early work on the pathophysiology of DMD suffered from the inherent difficulty in distinguishing the primary defect from the many secondary manifestations of the disease. The possibility of a generalized membrane defect, however, is one outcome that has endured continued support. In brief, the evidence includes (1) the finding of gaps or lesions in the plasma membrane in EM studies of prenecrotic muscle tissue from affected boys, (2) the finding of greatly increased levels of certain muscle enzymes in the serum of young presymptomatic boys suggesting leakage of macromolecules through the muscle membrane, (3) an increased level of Ca^{2+} in muscle fibers possibly caused by increased uptake through a "leaky" membrane, (4) alterations in the binding of sugar-binding proteins (lectins) to glycoproteins on the muscle cell surface, and (5) apparently altered intracellular adhesiveness of skin fibroblasts from DMD patients.

Even with the identification of the dystrophin molecule as the defective protein, in DMD and BMD, the biological role of dystrophin remains speculative and our understanding of the disease is incomplete. An alteration in the muscle membrane is, however, consistent with the identification of dystrophin as a high-molecular-weight cytoskeletal protein, localized at the sacrolemmal membrane. An attractive working model is one in which dystrophin plays a direct and fundamental role in muscle membrane stability. According to the model, the basic defect in DMD is the lack of functional dystrophin, which results in a weakened membrane that is susceptible to contraction-induced tearing. In boys with BMD, the presence of a reduced amount of normal dystrophin or of partially functional dystrophin may give partial stability to the membrane and a milder phenotype.

In both DMD and BMD, localized membrane lesions would be expected to give rise to the segmental necrosis (cell death) that is observed in the early stages of the disease. This is followed by regeneration through the proliferation and differentiation of satellite cells (myoblasts) that move in to repair the damage. In the later stages of the disease the regenerative capacity declines as the finite proliferative potential of the satellite cells is used up and the satellite cells themselves become depleted. This view of the disease progression is consistent with the documented reduction in the growth potential of myoblasts derived from the muscle tissue of affected boys.

In further support of the model is the evidence that dystrophin is bound to one or more integral membrane glycoproteins. A lack of dystrophin could easily result in a deficiency of the glycoprotein in the membrane, and this might explain the lectin-binding alteration in DMD muscle. The presence of an actin-binding domain near the N-terminal end of the dystrophin molecule, if verified, would suggest that the dystrophin–glycoprotein complex might serve to connect the internal cytoskeleton of actin filaments to the basal lamina on the outside of the membrane. It is not difficult to imagine how such a complex might protect the membrane from damage during muscle contraction and relaxation.

B. Animal Models

Models of X-linked muscular dystrophy exist in the mouse and dog. The murine model is the *mdx* mouse. The genetic defect is a point mutation in the mouse DMD gene that changes an amino acid co-

don to a stop codon, resulting in premature termination of the dystrophin molecule. The shortened protein fails to localize at the membrane. In contrast, to the human disease, *mdx* muscle has substantial powers of regeneration, resulting in the restoration of a relatively normal muscle morphology and function by about 4 weeks of age. The lack of a severe phenotype suggests that dystrophin is not required for membrane stability in mouse muscle. One possibility is that a dystrophin-like protein, induced during the first round of degeneration, acts to stabilize the phenotype. One of the two dystrophin-related proteins described above could perhaps fulfill this role.

Canine X-linked muscular dystrophy (CXMD) is a more faithful model of the human disease. The affected animals are dystrophin-negative and deficient in DMD mRNA, although the mutation in the DMD gene has not yet been characterized. The dog model has great potential in the evaluation of new therapeutic strategies.

IX. Transplantation Therapy

One of the major uses of the *mdx* mouse has been in the area of myoblast transplantation. With the finding that the DMD gene product is a large cytoskeletal protein came the realization that treatment of the disease through direct replacement of the protein will be difficult, if not impossible. Therefore, introduction of a new dystrophin gene into the muscle appeared to be the best alternative. Two groups have successfully transplanted normal myoblasts into a single muscle of *mdx* mice and demonstrated that the donor cells fuse with the existing muscle and produce dystrophin. The dystrophin was localized to the membrane, although it had a patchy distribution, and the amount of dystrophin was as high as 30–40% of normal levels. Transplantation also has been done into multiple muscle groups of the *dy/dy* mouse, with an autosomal recessive form of muscular dystrophy, and has resulted in a claim of significant functional recovery after transplantation. This exciting result is promising but the experiment needs to be repeated.

Similar studies have begun in humans, injecting donor myoblasts, usually from the muscle of the patient's father into the muscle of an affected child. Although the few children injected have tolerated the injection well, it is too soon to evaluate the potential benefits of the procedure.

Transplantation studies are in their infancy, and

many unanswered questions remain. These include, for example, the best routes of injection, the number of injection tracts, the number of cells to be injected per tract, and the timing of the injections for the major muscle groups. Also to be determined is the extent to which injected cells migrate through muscle, the proportion of cells that remain as satellite cells, and the level of dystrophin necessary to achieve clinical improvement.

X. Concluding Remarks

Our understanding of DMD and BMD has progressed in remarkable leaps in recent years. The identification of the DMD gene and its product, dystrophin, has given substantial new insight into the basic defect. Although the detailed understanding of the role of dystrophin is not yet in hand, the knowledge gained from the molecular biology approach clearly points the way to future experiments. The discovery of the responsible gene and protein has been referred to as the "end of the beginning." For boys who suffer from the disease there is now hope that the next phase of research will mark the "beginning of the end."

Acknowledgments

This article was abstracted from a more lengthy review by R. G. Worton and E. F. Gillard prepared for a chapter in a neurology text for the series "Molecular Genetics in Clinical Medicine" by Blackwell Scientific Publication. The author is especially grateful to Dr. Elizabeth Gillard for her major contribution to the preparation of both the review chapter and this article, to Drs. Don Love and Kay Davies for information on their cDNA clones for Fig. 3, to Dr. Michael Cullen for the EM picture in Fig. 5, and to Mr. Dennis Bulman for the Western blots and immunocytochemistry pictures in Fig. 5. Many of the ideas expressed in this article originated with the people in my laboratory and in the laboratories of Dr. Peter Ray, Dr. Louis Kunkel, and Dr. George Karpati—to them I am grateful.

Bibliography

den Dunnen, J. T., Bakker, E., Van Ommen, G. J. B., and Pearson, P. L. (1989). The DMD gene analyzed by field inversion gel electrophoresis. *Br. Med. Bull.* **45,** 644–658.

Dubowitz, V. (1989). "Muscle Disorders in Childhood," 1st ed. Year Book Medical Publishers, Chicago.

Emery, A. E. H. (1987). Duchenne muscular dystrophy. Oxford Monographs on Medical Genetics, No. 15. Oxford University Press, Oxford.

Koenig, M., Begg, A. H., Moyer, M., et al. The molecular basis of Duchenne versus Becker muscular dystrophy: Correlation of severity with type of deletion: *Am. J. Hum. Genet.* **45,** 498–506.

Monaco, A. P., and Kunkel, L. M. (1988). Cloning of the Duchenne/Becker muscular dystrophy locus. (Harris, H, and Hirschorn, K. H., eds.) *Adv. Hum. Genet.* **17,** 61–98.

Witkowski, J. A. (1989). Dystrophin-related muscular dystrophies. *J. Child Neurol.* **4,** 251–271.

Worton, R. G., and Thompson, M. W. T. (1988). Genetics of Duchenne muscular dystrophy. *Ann. Rev. Genet.* **22,** 601–629.

Muscle, Molecular Genetics

PREM M. SHARMA, *The Salk Institute for Biological Studies*

Glossary

ATP Nucleotide containing high-energy bonds; provides energy for many biochemical cellular processes by undergoing enzymatic hydrolysis

5′ Cap 7-Methyl-guanosine structure added at the beginning of eukaryotic mRNAs

CAAT consensus sequence Part of a conserved sequence located upstream of the start points of eukaryotic transcription units; recognized by a large group of transcription factors

cDNA Single-stranded DNA complementary to an RNA and synthesized from it by reverse transcription *in vitro*

Cloning Asexual production of a line of cells or organisms or segments of DNA genetically identical to the original

Dalton Unit of molecular weight, 1 dalton equals one-twelfth the mass of carbon-12

Electron microscope (EM) Instrument that uses a focused beam of electrons to produce an enlarged image of an object

Endoplasmic reticulum (ER) System of membrane-enclosed cytoplasmic channels involved in cellular transport processes: Rough ER has ribosomes attached to its outer surface, whereas smooth ER does not

Enhancer Enhancer sequence is a DNA sequence that somehow, without regard to its position or its orientation in the DNA, increases the amount of RNA synthesized from DNA introduced into cells

Exon Portion of a gene that becomes part of the mature mRNA after the intervening sequences (introns) are spliced out

Gene Hereditary unit specifying the production of a distinct protein (e.g., an enzyme) or RNA

Initiation signal AUG codon that specifies the incorporation of *N*-formylmethionine at the 5′ end of a new protein chain

Intron Intervening sequence of DNA, located within a gene, that is not included in the mature mRNA

Isoforms One of several forms in which a protein may exist in various tissues

Myoblasts and myotube Striated skeletal muscles are composed of multinucleate cylindrical fibers, 10–100 μm in diameter and several millimeters or centimeters long. These enormous structures arise in the embryo by the fusion of several primordial cells, the so-called *myoblasts,* which first form a *myotube,* and then a fully differentiated muscle fiber

Poly A Long polyadenylic acid segment added posttranscriptionally to many eukaryotic mRNAs at the 3′ end

Promoter DNA region to which RNA polymerase binds when initiating transcription

Restriction endonuclease Any of several enzymes capable of recognizing and cutting a specific symmetrical nucleotide sequence in DNA

Splicing Precise excision of the intervening sequences from an RNA primary transcript, followed by ligation of the message to produce a functional molecule

TATA box A conserved AT-rich septamer found about 25 bp before the start point of each eukaryotic RNA polymerase II transcription unit; may be involved in positioning an enzyme for correct initiation.

THE STUDY OF molecular biology of muscle is one of the most rewarding examples of the intimate association between structure and function and of the way in which chemical energy is transformed into mechanical work. Muscle cells are adapted to mechanical work by unidirectional contraction. The functional unit is the myofibril, which may be either striated or smooth. In skeletal muscle, myofibrils fill most of the large muscle fiber, leaving small amounts of sarcoplasm, which contains the nuclei, the sarcoplasmic reticulum (SR), and large mitochondria, or sarcosomes. A sarcolemma, with -0.1 V polarization, surrounds the fiber. Myofibrils result from the repetition of sarcomeres. These are limited by Z-line discs and contain the I bands, the A bands, and the H band. The M line may be observed in the middle of the sarcomere. The myofibrils are composed of a repeating assembly of thick and thin myofilaments. Thin filaments are composed mainly of actin, and thick filaments are composed of myosin. The molecular machinery involved in muscle contraction comprises (1) force-generating proteins (myosin and actin), (2) regulatory proteins [tropomyosin (TM) and troponins (Tn)], and (3) structural proteins (α-actinin: the Z-disc protein, M-disc proteins, and the C proteins). Sliding mechanism of muscle contraction postulates that the thin actin filaments are displaced with respect to the thick filaments at each contraction–relaxation cycle.

The role of actin myofilaments is best understood in muscle cells. In striated muscle, the ends of actin filaments are anchored to both sides of the Z discs. These interdigitate with bipolar thick filaments constructed of myosin. Flexible headpieces of the myosin chains are energized by hydrolysis of ATP and pull the adjacent actin filaments toward the center of the myosin fiber. Hydrolysis of ATP, and hence contraction of muscle, is triggered by a rise in cytoplasmic Ca^{2+} concentration. The release of Ca^{2+} stored in the lumen of sarcoplasmic reticulum vesicles is triggered by depolarization of the surface membrane of the muscle cell. In striated muscle, Ca^{2+} induces contraction by binding to troponins, a set of proteins attached to the fibrous protein TM that lies in the groove of the thin filaments; Ca^{2+} binding, in turn, alters the conformation by which TM is bound to thin filaments so that the myosin headpieces can interact with the actin. In vertebrate smooth muscle and invertebrate muscle, Ca^{2+} acts by affecting myosin light chains (MLC) located in the headpieces of myosin.

Isoforms of muscle-specific proteins can confer different regulatory or contractile properties to different types of muscle cells. In some cases, these isoforms are generated by the alternative splicing of an RNA transcript of a single gene. For instance, two MLCs (MLC1 and MLC3) found in striated muscle contain identical C-terminal but different N-terminal sequences and are produced by differential splicing by the use of two different promoters from a single gene. Similarly, two Tn-T isoforms differ by only 14 internal amino acids. These amino acid segments are encoded by two distinct and adjacent small exons of the Tn-T gene; alternative splicing results in the incorporation of one or the other exon into the mature Tn-T mRNA. In general, the generation of protein diversity by alternative splicing is extremely high in the Tn gene, which has the capacity to encode 64 different proteins; is intermediate in TM gene, which has the capacity to produce eight different proteins; and is minimal in MLC genes. The coding capacity of these genes raises important but so far unanswered questions about the physiological roles of the different isoforms produced. It also reveals new, unsuspected, and highly complex aspects of the biochemistry of contractility.

I. Muscle Structure and Function

In vertebrates, there are three classes of muscles: *smooth, cardiac,* and *striated.* Typically, smooth muscles are under involuntary (unconscious) control of the central nervous system. They surround internal organs such as large intestines, gallbladder, and large blood vessels. Contraction and relaxation of smooth muscles control the diameter of blood vessels and also propel food along the gastrointestinal tract. Smooth muscle cells can create and maintain tension for long periods. [*See* Smooth Muscle.]

Muscles under voluntary control have a striated appearance in the light microscope. Striated muscles, which connect the bones in the arms, legs, and spine, are used in more complex coordinated activities (e.g., walking or positioning of the head) and can generate rapid movements by sudden bursts of contraction. Cardiac (heart) muscle resembles striated muscle in many respects, but it is specialized for continuous, involuntary contractions needed in pumping of blood. Study of striated muscle cells, with their regular organization of actin and myosin contractile filaments, has provided important evidence about the mode of contraction in all three types of muscle. Approximately 40% of the body is

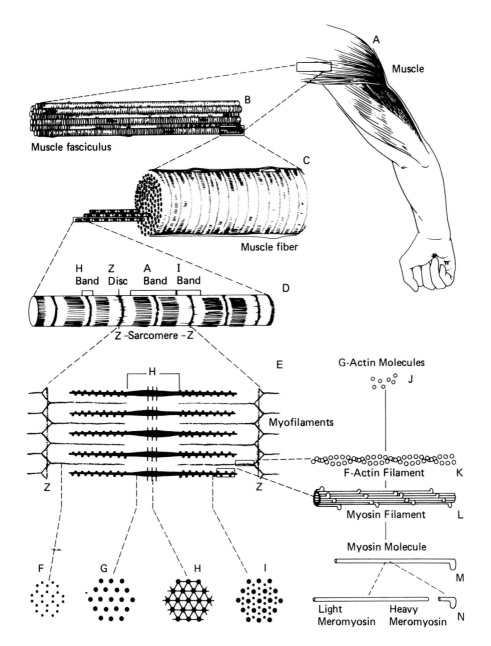

FIGURE 1 Organization of skeletal muscle, from the gross to the molecular level. F, G, H, and I are cross sections at the levels indicated. (Drawing by Sylvia Colard Keene.) [From Bloom and Fawcett (1975). "A Textbook of Histology." W. B. Saunders, Philadelphia, with permission.]

skeletal muscle; another 10% is smooth and cardiac muscles. [*See* CARDIAC MUSCLE; SKELETAL MUSCLE.]

A. Physiologic Anatomy of Skeletal Muscle

Figure 1 illustrates the organization of skeletal muscle, showing that all skeletal muscles are composed of numerous multinucleate cylindrical fibers, ranging between 10 and 100 μm in diameter by several millimeters or centimeters long. Each of these fibers in turn is made up of successively smaller subunits. The entire fiber is surrounded by an electrically polarized membrane with an electrical potential of about -0.1 V, the inner surface of which is negative with respect to the outer surface. This membrane, called the *sarcolemma*, becomes depolarized physiologically each time a nerve impulse that reaches the motor innervation of the muscle (end-plate) activates the membrane. Three cytoplasmic components are highly differentiated in a muscle fiber: myofibrils, SR, and sarcoplasm.

1. Myofibrils

The macromolecular contractile apparatus is made up of actin and myosin filaments. Each muscle fiber contains several hundred to several thousand myofibrils, which are illustrated by the many small open dots in the cross-sectional view of Fig. 1C. Each myofibril (Fig. 1D) in turn has, lying side by side, about 1,500 myosin filaments and 3,000 actin filaments, which are large polymerized protein molecules that are responsible for muscle contraction and are represented diagrammatically in Fig. 1E. The *thick myofilaments* are about 1.5 mm long and 10 nm wide and are separated by a 40-nm space. The *thin myofilaments* are about 1.0 mm long and 5 nm in diameter. Thick myofilaments are made of myosin, and thin myofilaments are composed of a more complex structure containing several proteins (i.e., actin, TM, and Tn) of which actin is the most important. The myosin and actin filaments partially interdigitate and thus cause the myofibrils to have alternate light and dark bands. The *light bands,* which contain only actin filaments, are called *I bands* because they are mainly isotropic (show no directional differences) to polarized light. The *dark bands,* which contain the myosin filaments as well as the end of the actin filaments overlapping the myosin, are called *A bands* because they are anisotropic to polarized light. The two sets of filaments are linked together by a system of cross-bridges. They protrude from the surfaces of the myosin filaments along the entire extent of the filament, except in the very center (Fig. 2B). It is interaction between these cross-bridges and the actin filaments that causes contraction (Section III). The light and dark bands are perpendicular to the long axis of the muscle cell along which the muscle contracts.

The actin filaments are attached to the so-called *Z disc,* and the filaments extend on either side to interdigitate with the myosin filament (Fig. 1E). Localization of the main protein constituents of Z discs shows that α-actinin occupies the central domain of the disc, together with actin and an 85 kDa protein, whereas desmin, vimentin, and synemin are located at the periphery. The Z disc also passes from myofibril to myofibril, attaching the myofibrils to each other all the way across the muscle fiber. Therefore, the entire muscle fiber has light and dark bands, as is also true of the individual myofibrils. These bands give skeletal and cardiac muscle their striated appearance. The portion of a myofibril (or of the whole muscle fiber) that lies between two successive Z discs is called a *sarcomere.* The sarco-

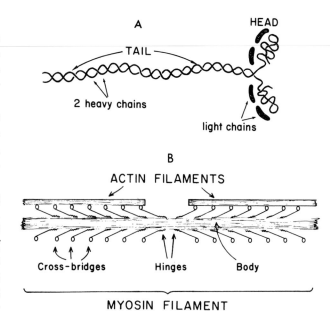

FIGURE 2 (A) the myosin molecule; (B) combination of many myosin molecules to form a myosin filament. Also shown are the cross-bridges and the interaction between the heads of the cross-bridges and adjacent actin filaments. [Reproduced, with permission, from A. Guyton (1986).]

mere is the unit of contraction; movements of actin and myosin within each sarcomere lead to shortening of the sarcomeres and, thus, to contraction of the muscle as a whole. When the muscle fiber is at its normal, fully stretched resting length, the length of the sarcomere is about 2.0 μm. At this length, the actin filaments completely overlap the myosin filaments and are just beginning to overlap with each other. When a muscle fiber is stretched beyond its resting length, as it is in Fig. 1, the ends of the actin filaments pull apart, leaving a light area in the center of the A band, called the *H zone.* In the middle of the H zone, an *M line* can be observed. The H zone rarely occurs in the normally functioning muscle because normal sarcomere contraction occurs when the length of the sarcomere is between 2.0 and 1.6 μm. In this range the ends of the actin filaments not only overlap the myosin filaments but also overlap each other.

2. Sarcoplasm

The myofibrils are suspended inside the muscle fiber in a matrix called *sarcoplasm,* which is composed of large quantities of potassium, magnesium, phosphate, and protein enzymes. Also numerous mitochondria lie between and parallel to the myofibrils, the abundance of which may be related to the constancy with which the muscle contracts: For example, there is a greater number in steadily active muscles (e.g., the heart), indicating the great need of the contracting myofibrils for large amounts of ATP formed by the mitochondria.

3. Sarcoplasmic Reticulum

Also in the sarcoplasm is an extensive endoplasmic reticulum, which in the muscle fiber is called the *sarcoplasmic reticulum*. SR is involved with conduction inside the fiber and with coordination of the contractions of different myofibrils, in addition to being related to the relaxation of the muscle after a contraction.

B. Sliding Mechanism of Muscle Contraction

In a living muscle fiber, changes with contraction can be observed with phase contrast and interference microscope. Figure 3 illustrates the basic mechanism of muscle contraction. In the relaxed state of the sarcomere, the width of the A band of the myosin fibers has been shown to remain constant. The ends of the actin filaments derived from two successive Z discs barely overlap each other while at the same time completely overlapping the myosin filaments. In addition, in the contracted state, these actin filaments have been pulled inward among the myosin filaments so that they now overlap each other to a major extent without any change in the lengths of individual myosin fibers as well as actin filaments. What does change is the width of the I band, the part of the actin myofilament not

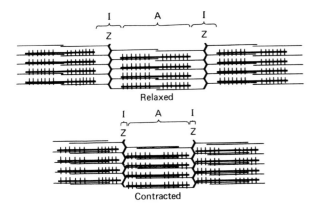

Relaxed

Contracted

FIGURE 3 Relaxed and contracted states of a myofibril, showing sliding of the actin filaments into the channels between the myosin filaments. [Reproduced, with permission, from A. Guyton (1986).]

covered by myosin. Also, the Z discs have been pulled by the actin filaments up to the ends of the myosin filaments. Indeed the actin filaments can be pulled together so tightly that the ends of the myosin filaments actually buckle during intense contraction. Thus, muscle contraction occurs by a sliding filament mechanism. The degree of contraction thus achieved can be measured by determining the length of the sarcomere (i.e., the distance between Z discs) at rest and when it has shortened. Note that insect muscle, in general, shortens only slightly (about 12%), whereas the shortening in vertebrate muscle may be much greater (about 43%). [*See* MUSCLE DYNAMICS.]

C. Macromolecular Organization of Smooth Muscles

The electron microscope has revealed that smooth muscles may have a varied macromolecular organization. In many cases they contain thin and thick myofilaments, as do striated muscles, but the difference lies in the absence of the Z disc and the lack of periodicity. In mollusks and annelids there are muscles with a helical arrangement that have thin and thick myofilaments linked by cross-bridges. In the adductor muscle of the oyster, the so-called paramyosin muscle, each thick filament is surrounded by 12 thin filaments. In smooth muscles, the contraction is slow, but an extreme degree of shortening may be achieved.

What triggers the inwarding sliding of actin filaments toward the myosin filaments? This is caused by mechanical, chemical, or electrostatic forces generated by the interaction of the cross-bridges of the myosin filaments with the actin filaments. In the next few sections we will discuss the details of molecular characterization and genetics of myosin, actin, and various other proteins of the contractile apparatus.

II. Molecular Organization of the Contractile System

The molecular machinery involved in muscle contraction comprises (1) force-generating proteins (myosin and actin), (2) regulatory proteins (TM and Tn), and (3) structural proteins (α-actinin in the Z disc, M-disc proteins, and C proteins). The major protein components of vertebrate skeletal myofibrils are summarized in Table I. Their detailed mo-

TABLE I Major Protein Components of Vertebrate Skeletal Myofibrils[a]

Protein	Percent Total Protein	MW (kDa)	Subunits (kDa)	Function
Force generating proteins				
Myosin	45	500	2×200 (heavy chain)	Major component of thick filaments. Interacts with actin filaments with hydrolysis of ATP to develop mechanical force
Actin	25	42	—	Major component of muscle thin filaments, against which muscle thick filaments slide during muscle contraction
Regulatory proteins				
Tropomyosin	5	64	2×32	Rod-like protein that binds along the length of actin filaments
Troponin	5	78	37–42 (Tn-T) 22–24 (Tn-I) 17–18 (Tn-C)	Complex of three muscle proteins positioned at regular intervals along actin filaments and involved in the Ca^{2+} regulation of muscle contraction
Structural proteins				
Titin	9	about 2,500	—	Large flexible protein that forms an elastic network linking thick filaments to Z discs
Nebulin	3	600	—	Elongated, inextensible protein attached to Z discs, oriented parallel to actin filaments
α-Actinin	1	94–103	2×95	Actin-binding protein that links actin filaments together in the region of the Z disc
M-line protein	1	165	—	Myosin-binding protein present at the central "M line" of the muscle thick filament
C protein	1	140	—	Myosin-binding protein found in distinct stripes on either side of the thick-filament M line

[a] The vertebrate striated myofibril also contains at least 20 other proteins not included in this table.

lecular characteristics and genomic organization follow.

A. Myosin: Its Structure, Function, and Genomic Organization

1. Overview of Myosin Structure and Function

Myosin is an unusual protein—it is both a globular enzyme and a fibrous structural protein that plays a central role in contractile processes of eu-

karyotes. Since its discovery about 50 years ago, biochemical studies have provided a detailed understanding of the structure and organization of myosin in muscle and nonmuscle cells and its structural and enzymatic functions in contractile processes. Myosin from smooth muscle and myosin from striated muscle have slightly different properties, which account for their different contractile regulatory mechanisms. Biological and physiological studies reveal that myosin in muscles of vertebrates and invertebrates has a complex molecular structure that specifies ATPase activity, intramolecular conformational changes, intermolecular interactions with actin during contraction, and assembly into the thick filaments of sarcomeric muscles. In any case, a single monomeric myosin molecule, when isolated by salt extraction from muscle, always contains two identical myosin heavy-chain subunits (MHC) of about 200,000 daltons and two pairs of MLCs of two different types of about 20,000 daltons.

The two heavy chains coil around each other to form a double helix. However, one end of each of these chains is folded into a globular protein mass called the *myosin head* (Fig. 2A). Thus, there are two free heads lying side by side at one end of the double-helix myosin molecule; the other end of the coiled helix is called the *tail*. Bound to each myosin head are two MLCs, one of each type. These MLCs help control the function of the head during the process of muscle contraction. The joints between the heads and the tails of the molecule are flexible. Movements of the head are important in generating the force of contraction in muscle as we shall see.

In muscle, myosin monomers form a specific biopolar aggregate called the *thick filament* containing 300–400 myosin molecules. The central portion of one of these filaments is illustrated in Fig. 2B, showing the tails of the myosin molecules bundled together to form the body of the filament while many heads of the molecules hang outward to the sides of the body. Also, part of the helix portion of each myosin molecule extends to the side along with the head, thus providing an arm that extends the head outward from the body as shown in the figure. The protruding arms and heads together are called the *cross-bridges,* and each of these is believed to be flexible at two points called *hinges,* one where the arm leaves the body of the myosin filament and the other where the two heads attach to the arm. The hinged arms allow the heads to be extended either far outward from the body of the myosin filament or to be brought closer to the body. The heads are believed to participate in the actual contraction process as will be discussed in Section III, A–E.

The total length of the myosin filament is 1.6 μm. However, note that the center of the myosin filament is devoid of the cross-bridge heads for a distance of about 0.2 μm because the hinged arms extend toward both ends of the myosin filament away from the center. Therefore, in the center there are only tails of the myosin molecules and no heads. Also, the myosin filament itself is twisted so that each successive set of cross-bridges is axially dis-

placed from the previous set by 120°, thus ensuring that the cross-bridges extend in all directions around the filament.

a. Actin-Activated ATPase Activity of the Myosin Head Another feature of the myosin head that is essential for muscle contraction is that it can function as an ATPase enzyme. In the absence of actin, this activity is almost undetectable, but whenever pure actin filaments are added, the rate of ATP hydrolysis is increased 200-fold so that each myosin molecule hydrolyzes 5–10 ATP molecules per second, thus providing the ATP's high-energy phosphate bond to energize the contraction process. [*See* ADENOSINE TRIPHOSPHATE (ATP).]

2. Myosin Genes

Numerous isoforms of MLC and MHC proteins are generated from multiple genes or by post-transcriptional mechanisms, resulting in the diversification of the various skeletal and cardiac muscle types. Some of the contractile proteins are expressed at specific developmental stages or at distinct physiological states.

a. Cloning of Myosin Genes and Structural Comparison of Myosin Proteins Within the past decade, genes encoding the MHC subunit and the alkali (alkali MLC) and regulatory (MLC-2) subunits of myosin have been cloned as genomic segments and cDNA clones from humans and several other species. Nonmuscle MHC genes have been cloned from Acanthamoeba and Dictyostelium. Structural data obtained from sequence analyses of these muscle-specific MLC and MHC genes and cDNA clones of their mRNAs have provided an extensive new source of comparative protein and gene structure data for evolutionary analyses. These cloned genes have also been useful as nucleic acid hybridization probes for studies of myosin gene families, the developmental and tissue-specific expression of myosin gene transcripts, and chromosomal locations of myosin gene and myosin gene mutations.

b. Myosin Heavy-Chain Genes Vertebrate (mammalian) sarcomeric MHCs are encoded by conserved, multigene family–encompassing 10–15 genes, each encoding a distinct protein or isozyme. Members of this gene family show both tissue-specific and developmentally regulated expression. During development, fetal, neonatal and adult MHC isozymes are expressed in a sequential program. However, the functional significance of each of these isozymes and the mechanism by which their expression is regulated have yet to be clarified. It may be that changes in the physiological properties of the developing muscle necessitate switches in MHC isozyme expression to accommodate new demands.

Complementary DNA and genomic clones corresponding to several sarcomeric MHC isoforms show strong homology to each other. Those so far isolated range in size from 0.6 to 3.4 kb and represent partial- to full-length protein coding sequences.

In humans, two adult skeletal muscle cDNA clones have been isolated that encode MHCs that are divergent by comparison with each other. The skeletal muscle MHC gene cluster is located on chromosome 7, and the β and α human cardiac MHC genes are located on chromosome 14. When sequences of analogous MHC isoforms from different species are compared, they are even more homologous than intraspecies comparisons, suggesting functional constraints on the divergence of isoform sequences. The 3′ untranslated region (UTR) sequences are also maintained across species when analogous isoforms are compared and appear to be isoform-specific. The evolutionary conservation of MHC isoform sequences implies that they are functionally significant, but to a large degree their distinct roles in sarcomere assembly or contractile function remain to be elucidated. In mouse and human, skeletal MHC genes are localized to a single chromosome.

Expression of MHC isoforms is sequential and controlled by numerous signals including neuronal and hormonal trigger. The fact that some developmental transitions occur in cultured muscle cells suggests that much of the information necessary for sequential MHC expression is intrinsic to the cells. Human muscle cultures may be advantageous for examining regulation of MHC expression because of the longer developmental cycle of primates compared with rodents and chickens. Because of its early appearance in skeletal muscle development and its reappearance during muscle regeneration, the embryonic skeletal MHC isoforms provide a particularly useful marker for charting the progression of myofiber formation and renewal.

A 3.4 kb cDNA representing the human skeletal muscle isoform has been cloned that encodes one of the first MHCs to be transcribed in human muscle development. Expression of this embryonic MHC is a hallmark of muscle regeneration after birth and is a characteristic marker of human muscle dystrophies.

c. Myosin Light-Chain Proteins Amino acid sequence data identifies two classes of MLC, alkali MLC$\frac{1}{3}$ and regulatory MLC-2. Both include multiple isoforms. MLCs are members of the super family of Ca^{2+}-binding proteins that have four EF-hand domains [a] composed of an α-helix E, a divalent cation binding loop, and an α-helix F. Some members of this super family (calmodulin and Tn C) have four functional EF-hand domains, [a] but MLC-2 has only one functional Ca^{2+}-binding EF-hand domain. The other three EF-hand domains of MLC-2 and the four domains of alkali MLC have accumulated deletions and nonconservative amino acid substitutions that inactivate their Ca^{2+}-binding ability.

The globular head of each MHC subunit is associated with one alkali MLC and one MLC-2 subunit (Fig. 2A). Immunoelectronmicroscopy and chemical cross-linking studies show that these MLCs are closely opposed with their N termini extending from the MHC globular head region back into the neck and hinge region. The function of alkali MLC is not understood but may be structural. Vertebrate skeletal MHC retains its ATPase activity after removal of the alkali MLC and regulatory MLC-2.

d. Alkali MLC Genes The structure of alkali MLC (LC1f and LC3f) genes of vertebrate skeletal muscle shows many interesting features.

i. Alkali Light Chains of Fast-Fiber Skeletal Muscles In fast skeletal muscle but not in other muscle types, the alkali MLCs exist in two isoforms of different size, termed *MLC1f* and *MLC3f* (Mr 21,000 and 17,000, respectively). In all vertebrate species

---

[a] EF hand arises from the cystalline structure of carp muscle calcium-binding protein, parvalbumin. The molecule has the approximate shape of a prolate ellipsoid of revolution. The course of the main chain is best visualized in terms of the six helices, A,B,C,D,E, and F, which have been interpreted as being driven from a single gene triplication. Helix C, the CD loop, and helix D are related to the EF region of an approximate 2-fold axis, which roughly coincides with the long axis of the ellipsoid. The over-all configuration of the EF region is remarkably similar to a right hand with thumb and forefinger extended at approximate right angle and the remaining three fingers clenched. The thumb points towards the COOH terminus of helix F. The forefinger points along helix E in the NH2-terminal direction. The clenched fingers trace the course of the EF loop about the calcium ion. The two right hands representing the EF and CD regions are related by a twofold axis.

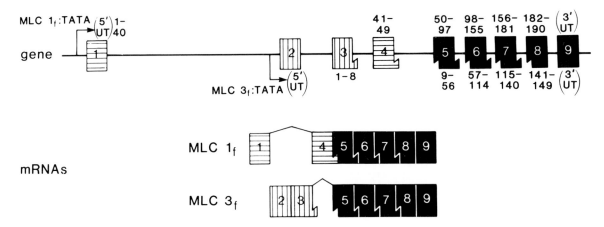

FIGURE 4 Myosin light-chain (MLC) gene organization and two spliced mRNAs (MLC1f and MLC3f). Constitutive (*black*), MLC1f-specific (*horizontal stripes*), and MLC3f (*vertical stripes*) are diagrammed to show their split terminal codon structure and encoded amino acids (*numbered*). The two isoform-specific promoters (TATA) are indicated. Exons 3 and 4 are mutually exclusive cassettes. [From R. E. Breitbart *et al.*, (1987).]

so far analyzed, these two proteins exhibit several interesting features that set them apart from other contractile proteins. MLC1f (190 amino acids long) and MLC3f (149 amino acids long) have complete sequence homology for the last 141 amino acids at the carboxyl terminus. The two proteins differ, however, in the length and amino acid sequence of their amino termini. These two proteins are products of a single gene with an unusual structural organization, containing nine exons (protein-coding segments). The carboxyl termini of the proteins encoding 141 amino acids and the 3'-noncoding sequences of their corresponding mRNAs are identical and are encoded by five common exons. These exons are preceded by four exons of which the first and fourth are specific for MLC1f, and the second and third are specific for MLC3f (Fig. 4). Two different primary transcripts are synthesized from two separate promoters that are more than 10 kb of DNA apart, and both transcripts are spliced alternatively to generate the functional mRNAs. The noncontiguous nature of the coding sequences suggests the function of splicing mechanisms depends on upstream primary and possibly higher order structure in recognition of splice sites. Although use of differential initiation or of differential splicing to different mRNAS are known to occur in eukaryotic systems, the use of two promoter sites combined with alternative splicing to generate two different proteins appears to be novel. This provides a mechanism for

independent regulation of the two MLC mRNAs and thus may account for the observed tissue-specific and developmentally regulated isoform distribution of these MLCs. The promoter regions of the LC1f/LC3f genes from chicken, rat, mouse, and humans contain common sequences (Fig. 5A,B), which may play a role in the regulated expression of these genes.

MLC1f promoter region in humans contains perfect TATA and CCAAT consensus sequences upstream of exon I, whereas upstream of exon II the putative MLC3f promoter contains the deviated TATA box, AAATAAA, and a chloramphenicol acetyltransferase (CAT)-like sequence CAACT. The transcriptional start sites of both putative promoters suggests that the cap site of the MLC1f gene is located 130 nucleotides upstream of the protein start site and for MLC3f mRNA, 71 nucleotides upstream. In the human, rat, mouse, and chicken, the region between the cap site of the MLC1 promoter and the upstream nucleotide −143 exhibits a high degree of sequence similarity, whereas sequences farther upstream are considerably diverged (Fig. 5A,B). In MLC3f promoter, several short regions of sequence homology can be recognized.

The functional activity of the human 5' upstream putative promoter region was determined in transient transfection experiments using the bacterial CAT reporter gene under the control of various 5' deletion mutants of MLC1 and MLC3 promoters. Approximately 100 nucleotides upstream of the two initiation sites of transcription, encompassing the TATA and CCAAT consensus sequences, are sufficient for the promoter activity in cultured cells. The tissue-specific regulation is partially maintained by these promoter constructs as they exhibit preferential but not exclusive activation in myotubes. Similar effects were observed in mice and chickens.

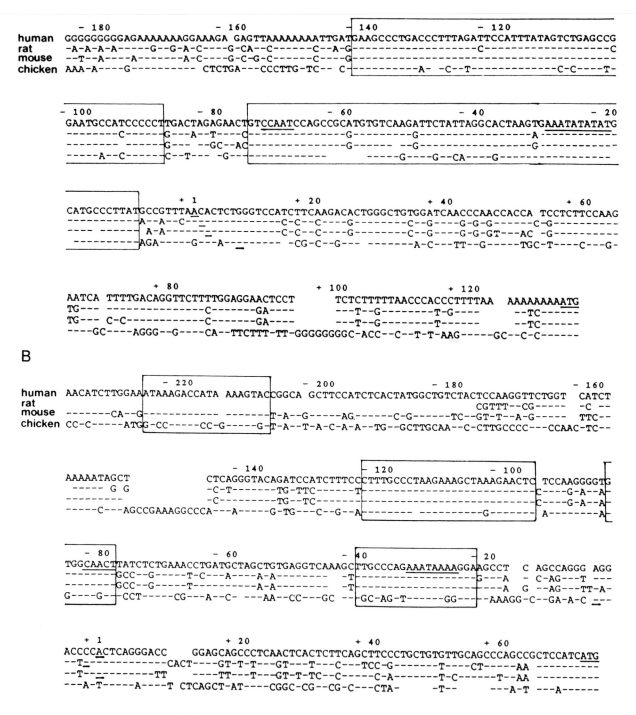

FIGURE 5 (A) Comparison of the MLC1f promoter regions of human, rat, mouse, and chicken. The conserved regions are shown in *boxes*. The numbering refers to the human CAP site as +1. The TATA and CAT box motifs, the ATG start codon, and the respective CAP sites are *underlined*. Identical nucleotides are represented by *dashes; gaps* have been introduced to optimize homology. (B) Sequence comparison of the MLC3f pro- moter regions of human, rat, mouse, and chicken. Details are as described in A. [Reproduced, with permission, from U. Seidel and H. H. Arnold (1989). Identification of the functional promoter regions in the human gene encoding the myosin alkali light chains and MLC3 of fast skeletal muscle. *J. Biol. Chem.* **264,** 16109–16117.]

Recently, a muscle-specific enhancer at the $3'$ end of the rat MLC$_3^1$ gene locus has been identified. This enhancer confers muscle specificity to its own promoters as well as to heterologous promoters in a distance- and orientation-independent way. In contrast, an enhancer-like element found approximately 1 kb upstream of the chicken MLC1 promoter activates only the homologous MLC1 TATA box in the $5'$–$3'$ sense orientation. Both described segments enhance the rather weak basal promoter activity of the MLC1 gene and contribute effectively to the muscle specificity. In humans, an element located downstream of MLC$_3^1$ gene is also capable of enhancing the MLC1 promoter in myotubes. However, the regulatory role of this element *in vivo* needs to be determined.

ii. Alkali Light Chains of Slow-Fiber Skeletal Muscle These MLCs are less well-characterized. In rodents, slow-fiber skeletal muscle predominantly contains a single isoform, MLC1s. It has been suggested that the MLC1s and ventricular MLC (MLC1v) are indistinguishable and are probably encoded by the same gene. In contrast, the slow muscle fibers of large mammals, including humans, contain two distinct alkali MLCs, designated MLC1sa and MLC1sb. The MLC1sb skeletal isoform is also expressed in heart ventricle.

The human skeletal muscle MLC1sb gene is present as a single copy gene and is also expressed in the heart ventricle but not the atria. The cloned cDNA codes for 195 amino acids, has 65 sequences of the $5'$ UTR, and has 238 nucleotides + poly(A) tail of the $3'$ UTR. Human MLC1sb is 79% similar to the chick cardiac isoform and 69% similar to the fast skeletal muscle isoform.

iii. Cardiac Isoforms of MLC Mammalian cardiac muscle contains two alkali MLCs, which represent the major isoforms present in either the ventricular (MLC1v) or atrial (MLCIa) muscle and which are different from the fast skeletal muscle isoforms (MLC1f and MLC3f). The atrial isoform is also expressed in fetal skeletal and fetal ventricular muscle, where it is described as fetal isoform MLC1emb.

The mouse MLC1a and MLC1emb are encoded by a single copy gene, located on chromosome 11. The amino acid sequence deduced from the gene sequence contains 192 amino acids and has a molecular mass of 21,004 daltons. The organization of exons is the same as in the MLC1f gene in the mouse, with the same position of intron–exon junctions. However, the noncoding regions of MLC1f and MLC1a/MLC1emb show no apparent homology and are different in length. For MLC1a/MLC1emb, the $5'$ and $3'$ noncoding sequences are 72 and 150 base pairs (bp), respectively, whereas in MLC1f they are 125 and 278 bp.

Comparison of MLC1a/MLC1emb gene and polypeptide with those of MLC1f and MLC1v (Table II) suggests that MLC1a/MLC1emb and MLC1v were generated from a common ancestral gene. The amino-terminal region of MLC1a/MLC1emb, thought to be involved in the actomyosin interaction, shows conservation with MLC1v but not with MLC1f, suggesting a shared functional domain in these cardiac isoforms. Comparison with the chicken embryonic MLC (L23) suggests that although MLC1a/MLC1emb and L23 show different patterns of expression, both during development and in the adult, they probably represent the homologous gene in these two species (Table II).

A cDNA encoding the embryonic isoform of alkali MLC (MLC1emb) has also been isolated from a human fetal muscle library. The nucleotide sequence analysis of the complete cDNA reveals an open reading frame of 591 amino acids and 56 bp of $5'$ UTR and 186 bp of $5'$ UTR. The derived protein sequence constitutes the first structural information on this myosin isoform of any organism. The embryonic MLC is expressed in fetal ventricle, in adult atria, and faintly in fetal skeletal muscle. Comparison of the human MLC1a/emb sequence with those of other MLCs reveals a strong resemblance to all type-1 essential light chains. This is due to its typical proline + alanine-rich N-terminal part that is not present in MLC3 isoforms.

The cardiac isoforms present in ventricular muscle of rat and human are also called *RVMLC1* and *HVMLC1*, respectively. The RVMLC1 cDNA is 890 bp in length excluding the poly (A)+ tail. The open reading frame encodes a protein that is 200 amino acids in length and of predicted molecular mass 22,150 kDa, and the $5'$ and $3'$ untranslated regions are 45 and 240 bp long, respectively. The HVMLC1 cDNA encodes a protein of 195 amino acids. Its predicted amino acid sequence shows greatest homology (91%) to the human venticular MLC1 sequence, 80% homology to the chicken cardiac MLC1 sequence, and 70% homology to the rat skeletal MLC1-3 sequence. This MLC1 isoform is developmentally regulated. RVMLC1 is expressed at low levels in cardiac tissue during early develop-

TABLE II Homology of Amino Acid and Gene Coding Sequences Between Different Myosin Alkali Light Chains[a]

Exon	Amino Acid Homology			Nucleotide Homology	
	Mouse (MLC1f)	Chicken (MLC1v)	Chicken embryo (L23)	Mouse (MLC1f)	Chicken embryo (L23)
1	55(48)	50	78(58)	60(63)	83(63)
2	56	56	67	54	79
3	72(76)	80	90	76	81
4	78	81	78	67	78
5	77	65	81	82	77
6	89	78	100	86	90
Overall	70(69)	72	82(78)	71(69)	80(76)

[a] From Barton, P. J. R., et al. (1988). *J. Biol. Chem.* **263** (25), 12669–12676, with permission.
[b] Homology of MLC1a/MLC1emb gene (coding regions) and protein sequences compared with those of mouse MLC1f, chicken MLC1v, and chicken embryonic L23. Values represent the number of mismatched residues as a percentage of the number of residues in MLC1a/MLC1emb. Where unmatched residues exist (exon 1 and 5), the overall homology value is indicated in parentheses.

ment but more abundantly after birth and in adult hearts. The two cardiac isoforms, the atrial and venticular MLC1, are products of two separate genes.

The alkali MLC-1 ventricle/slow skeletal and the MLC-1 atrial/embryonic skeletal isoforms each are encoded by a single copy gene that produces a protein with an N-terminal extension homologous to alkali MLC-1 fast skeletal protein. Neither gene has been shown to produce a MLC-3-type protein by alternative splicing, although minor forms of MLC proteins have been detected in some mammals.

e. Regulatory MLC-2 Genes MLC-2 regulates interactions between myosin head and actin through the binding of divalent cations or the phosphorylation of a serine residue. Considerable information has been obtained regarding the cDNA and genomic sequences for light-chain 2 in avian and mammalian striated muscle. The chicken cardiac light-chain 2 gene codes for an mRNA of 700 nucleotides and consists of 4.2 kb of DNA containing 5 introns. The gene is present as a single copy per haploid chicken genome. The gene for rat skeletal fast muscle light-chain 2 also exists in single copy, contains at least five introns, and displays a tissue- and stage-specific expression.

Despite the high level of transcriptional complexity, the light-chain 1 and 3 gene products accumulate in precise stoichiometric relation to the regulatory MLC (LC2) and the associated MHCs.

B. Actin Proteins

Actin is the second major protein and represents about one-quarter of the protein of the myofibril (Table I). The molecular weight of actin is 42,000. The actin myofilament is made up of two helical strands that cross-over every 36–37 nm (Fig. 1K). Each of these cross-over repeats contains between 13 and 14 globular monomers of G-actin about 5 nm in diameter (Fig. IJ). In muscle, actin is mainly present as fibrous or F-actin, which is the polymerized form of G-actin.

Actin is a highly conserved protein that participates in a wide variety of cellular functions in eukaryotes including muscle contraction, ameboid movement, cytokinesis, and mitotic division. Amino acid sequencing data have demonstrated that the present actin isotypes evolved from two major classes of actin, "cytoplasmic" and "muscle." All organisms thus far examined express a cytoplasmic actin form, which is usually used to construct the cellular microfilaments. In simple eukaryotes such as Dictyostelium, Physarum, and Saccharomyces cerevisiae, only one type of cytoplasmic actin is expressed. In mammals, there are two cytoplasmic actins (β and γ), and in amphibians, there are at least three. The second class of actin proteins, the muscle (or "α-like") actins, is found in birds and mammals. It follows that original α-like actin gene evolved from the cytoplasmic type some time before the divergence of birds and mammals. In mammals, four different tissue-specific muscle isotypes have been found: skeletal actin (or α), cardiac actin, and two smooth muscle types, aortic actin and stomach actin.

1. Actin Genes

Considerable information on the structure and organization of actin genes in a wide variety of eukaryotes has been accumulated. The available data reflect the comparative degrees of sequence conservation between species in genes encoding the muscle-specific and the nonsarcomeric isoforms of actin. In contrast to the vertebrate MHC multigene family, only a single member of the actin gene family (which numbers at least 15 in mammals, including inactive pseudogenes) appears to encode α-actin, the adult mammalian skeletal muscle isoform.

This gene is transiently coexpressed with a cardiac isoform in fetal cardiac and skeletal muscle tissues. The tissue-specific isoforms then progressively predominate throughout the development of the corresponding tissue. These striated muscle isoforms and the nonmuscle isoforms are encoded by genes localized on different chromosomes: 3, 5, and 17 in the mouse. This contrasts with the situation in lower organisms in which there is a close physical linkage between several actin genes.

Sequence information on the actin genes from a wide variety of sources indicates several interesting features. The coding sequences are highly conserved: for instance, actins from Physarum and mammalian cytoplasm show about 95% homology. In contrast, the noncoding sequences are widely divergent and vary to an extent that reflects the putative time course of evolutionary divergence. Surprisingly, there is much greater sequence homology within 5′ and 3′ untranslated regions of homologous actins in different mammals than between cytoplasmic β- and skeletal muscle α-actin from a given species. Thus, an identical sequence of 15 nucleotides including sequences such as the CAAT box is present at the 5′ untranslated regions of rat and chick skeletal muscle actin genes, although the adjacent sequences in the two genes are divergent. The location of this conserved sequence strongly suggests that it may have a regulatory function in transcription control. All the six actin genes in *Drosophila melanogester* and at least nine of the 176 actin genes in *Dictyostelium discoideum* are expressed, and multiple actin genes encoding virtually identical actin isoforms are differentially regulated during development. The lack of coordinate control

of the expression of different actin genes in *D. discoideum* is consistent with the differences in their 5′ noncoding sequence. It is possible that each structural gene of this multigene family has been placed under different regulatory "modules" in the genome, presumably being associated with a regulatory "element" that permits its expression in particular cell lineages.

B. Regulatory Proteins

In addition to the force-generating proteins actin and myosin, contraction of muscle requires the coordinated action of the regulatory proteins such as TM and the various Tns.

1. Tropomyosins

Tropomyosins are components of the contractile systems of the skeletal, cardiac, and smooth muscles and the cytoskeleton of nonmuscle cells. Although they are present in all cells, different forms of the protein are characteristic of specific cell types. TM is a rod-like protein composed of two highly α-helical subunits wrapped around each other to form a coiled-coil structure (Fig. 6). Skeletal muscle contains two forms of TM termed α and β, the proportions of which vary with the fiber type. Both of these subunits are 284 amino acids long and differ only slightly in amino acid sequence. In cardiac muscle of small mammals such as rodents, only α-TM is expressed. In striated muscle, TM is localized to the thin filaments, where it is found along both grooves of the actin filament. The function of TM in skeletal and cardiac muscle is in association with the Tn complex (Tn I, T, and C; Section II,B,2) to regulate the calcium-sensitive interaction of actin and myosin.

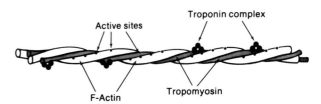

FIGURE 6 Actin filament, composed of two helical strands of F-actin and two tropomyosin strands that lie in the grooves between the actin strands. Attaching the tropomyosin to the actin are several troponin complexes. [Reproduced, with permission, from A. Guyton (1986).]

a. Tropomyosins of Nonmuscle Cells Tropomyosins are found in all smooth muscle and nonmuscle cells. Multiple forms of TM have been detected in many cultures of nonmuscle cells. For example, rat embryonic fibroblasts contain three major TMs termed 1, 2, and 4 (apparent molecular weights of 40,000, 36,500, and 32,400, respectively) and two relatively minor TMs termed 3 and 5 (apparent molecular weights of 35,000 and 32,000, respectively). The multiplicity of TM isoforms in fibroblasts raises questions as to the structure and function of each isoform. In addition, nonmuscle and smooth muscle cells do not contain a Tn complex. In these cell types, the phosphorylation of the MLCs by the enzyme MLC kinase appears to be the major calcium-sensitive regulatory mechanism controlling the interaction of actin and myosin (discussed in Section III,D).

b. Tropomyosin Genes Until recently it had been thought that the nonmuscle TMs were structurally distinct from the smooth and skeletal muscle isoforms. Isoforms of TM from platelets were found to be 247 amino acids in length, whereas isoforms from smooth and skeletal muscle are 284 amino acids long. However, molecular cloning has shown that the heterogeneity of nonmuscle TM is due not only to the expression of different isoforms of the classic 247-amino-acid nonmuscle protein but also to the expression of different isoforms of smooth muscle TMs. Several cDNA and genomic clones representing the TM isoforms from skeletal and smooth muscles and nonmuscle cells (fibroblasts) have been identified and characterized from rodents (rat and mouse) and human species. The cDNA clone (TM3) isolated from human fibroblasts is 1,633 bp long and contains 286 bp of the 5'-UTR, 852 bp that encode a TM sequence, and 495 bp of 3'-UTR. The mouse TM2 cDNA is 1688 bp in length and contains 54 bp of 5'UTR, 852 bp of coding sequence, 769 bp of 3'-UTR, and a poly (A) tail. The encoded protein of TM1, TM2, and TM3 cDNAs is 284 amino acids in length. Comparison of the amino acid sequence with those of other TMs reveals that TM2 is closely related to rat smooth and skeletal muscle α-TMs (Fig. 7). TM2 is identical to rat skeletal muscle α-TM from amino acids 1–257 but differs considerably from amino acids 258–284. It differs from rat smooth muscle α-TM from amino acid 39–80, but other regions are identical. Thus, among the nonmuscle TM isoforms reported so far, TM2 from rat fibroblasts and TM3 from human fibroblasts are closely related to each other (Fig. 7). They differ from amino acids 189–213, but other regions are identical. The TM2, TM3, and smooth muscle α-TM but not skeletal muscle α-TM likely share the common exon encoding the 3'-UTR. The smooth and skeletal muscle α-TMs and the TM3 and TM2 genes appear to be expressed from the same gene via an alternative RNA-splicing mechanism.

The rat α-TM gene generates many different tissue-specific isoforms (Fig. 8). The promoter of the gene is similar to promoters of other housekeeping genes in both its pattern of utilization, being active in most cell types, and its lack of any canonical sequence elements. The gene is 29 kb long and is split into at least 13 exons, seven of which are alternatively spliced in a tissue-specific manner. This gene arrangement, which also includes two different 3' ends, generates a minimum of six different mRNAs, each with a capacity to code for a different protein. These distinct TM isoforms are expressed specifically in nonmuscle and smooth and striated (cardiac and skeletal) muscle cells.

c. hTMnm and hTMa Genes Several contiguous clones covering the human TM (hTMnm) genes are well-characterized. In nonmuscle tissue this gene produces a 2.5-kb mRNA encoding TM30nm, a 248-amino-acid cytoskeleton TM. However, in muscle, alternative splicing of this gene results in the expression of a 1.3-kb mRNA encoding a 285-amino-acid skeletal muscle α-TM. The hTMnm gene spans at least 42 kb of DNA and consists of 13 exons, only five of which are common to both the 2.5-kb and the 1.3-kb transcripts. The boundaries of the exons giving rise to the muscle-specific isoform are identical to those of other genes encoding muscle TMs. The hTMnm gene is unique among Tm genes characterized in that it encodes TM molecules of different sizes from two different promoters. A comparison of the structures of exons encoding the amino-terminal sequences of the muscle and nonmuscle isoforms suggests that the hTMnm gene has evolved by a specific pattern of exon duplication with alternative splicing.

Another cDNA clone of the TM gene family hTMa, which is distinct from the hTMnm, encodes a closely related isoform of skeletal muscle α-TM. In cultured human fibroblasts, the hTMa gene encodes both skeletal muscle- (1.3-kb mRNA) and smooth muscle- (2.0-kb mRNA) type α-Tms by using an alternative mRNA-splicing mechanism. However, the skeletal muscle α-TM expressed in human fibro-

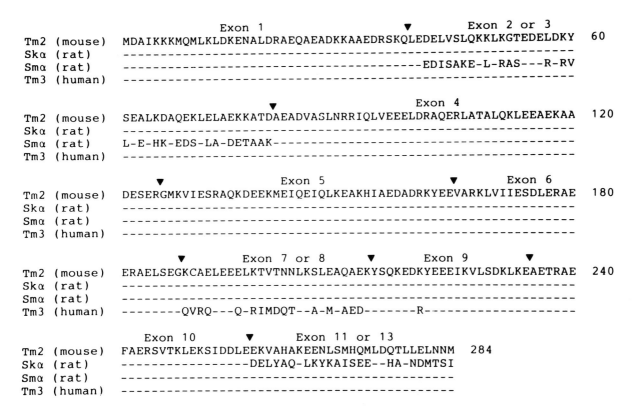

```
                           Exon  1                              ▼            Exon  2 or  3
Tm2 (mouse)  MDAIKKKMQMLKLDKENALDRAEQAEADKKAAEDRSKQLEDELVSLQKKLKGTEDELDKY    60
Skα (rat)    ------------------------------------------------------------
Smα (rat)    --------------------------------------------EDISAKE-L-RAS---R-RV
Tm3 (human)  ------------------------------------------------------------

                                       ▼                     Exon  4
Tm2 (mouse)  SEALKDAQEKLELAEKKATDAEADVASLNRRIQLVEEELDRAQERLATALQKLEEAEKAA   120
Skα (rat)    ------------------------------------------------------------
Smα (rat)    L-E-HK-EDS-LA-DETAAK------------------------------------------
Tm3 (human)  ------------------------------------------------------------

                      ▼             Exon  5                 ▼           Exon  6
Tm2 (mouse)  DESERGMKVIESRAQKDEEKMEIQEIQLKEAKHIAEDADRKYEEVARKLVIIESDLERAE   180
Skα (rat)    ------------------------------------------------------------
Smα (rat)    ------------------------------------------------------------
Tm3 (human)  ------------------------------------------------------------

                      ▼         Exon  7 or  8      ▼         Exon  9          ▼
Tm2 (mouse)  ERAELSEGKCAELEEELKTVTNNLKSLEAQAEKYSQKEDKYEEEIKVLSDKLKEAETRAE   240
Skα (rat)    ------------------------------------------------------------
Smα (rat)    ------------------------------------------------------------
Tm3 (human)  --------QVRQ---Q-RIMDQT--A-M-AED-------R---------------------

                  Exon  10          ▼          Exon  11 or  13
Tm2 (mouse)  FAERSVTKLEKSIDDLEEKVAHAKEENLSMHQMLDQTLLELNNM   284
Skα (rat)    ----------------DELYAQ-LKYKAISEE--HA-NDMTSI
Smα (rat)    -------------------------------------------
Tm3 (human)  -------------------------------------------
```

FIGURE 7 Comparison of the amino acid sequences of mouse TM2 with rat skeletal muscle α-TM (Skα), smooth muscle α-TM (Smα), and human fibroblast TM3 sequences. (–), amino acid identical to that of TM2. The positions of the exon boundaries (*arrowheads*) of the gene encoding rat α-TM and the numbers of exons are from Wieczorek, D. F., et al. (1988). *Mol. Cell. Biol.* **8,** 679–694.

blasts, although similar, is not identical to that expressed in human skeletal muscle.

Expression of rat embryonic fibroblast TM 1 and skeletal muscle β-TM has been demonstrated from a single gene that contains 11 exons and spans 10 kb. The 1.1-kb mRNA expressed in fibroblasts, stomach, uterus, and vas deferens and a 1.3-kb mRNA expressed in skeletal muscle are spliced from a single promoter, a precussor with an identical 5' end. The promoter contains G-C-rich sequences, a TATA-like sequence TTTTA, no identifiable CCAAT box, and two putative SP-1 binding sites.

2. Troponins

Together with TM the Tns are a family of three small proteins that form a complex of about 80,000 daltons (Table I). Both proteins constitute a Ca^{2+}-sensitive switch, within the thin filament, that regulates the contraction. Each of the Tn complexes binds to TM every 40 nm (i.e., every seven active monomers), as can be seen with anti-Tn antibodies. The three components present in equimolar proportions are (1) Tn-C (molecular weight 17,000–18,000), a small protein that specifically binds Ca^{2+}; (2) Tn-I (molecular weight 22,000–24,000), a protein that binds to both actin and Tn-C and inhibits the ATPase of myosin; and (3) Tn-T (molecular weight 37,000–42,000), the largest subunit that binds to TM and also interacts with the other Tns (Fig. 6). Tn has been considered a localized trigger that controls the motion of TM in the thin filament. EM observations have revealed that the Tn complex has a tadpole shape with a globular domain composed mainly of Tn-C and Tn-I, and a long tail that interacts with Tm and corresponds mainly to Tn-I. It is suggested that this lengthwise interaction may be important in the regulation of the switching process.

a. Troponin Proteins and Genes Encoding the Various Isoforms

i. Troponin-C (Tn-C) This is the calcium-binding component of the Tn complex that triggers the contraction of skeletal and heart muscle in response to increasing calcium levels in the sacroplasm.

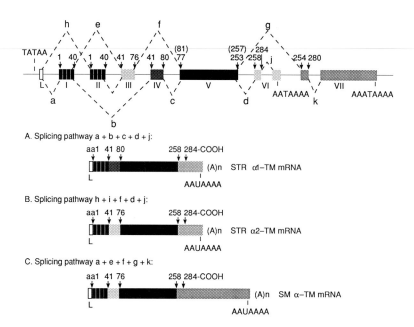

FIGURE 8 Schematic organization and alternative splicing pathways for the origin of three different mRNAs from the α-TM gene. *Boxes* show the regions of the α-TM gene containing leader (L), common (V), and isotype-specific (I–IV, VI, VII) sequences drawn tentatively as single exons in one of several possible 5′ to 3′ arrangements. The *dotted lines* represent proposed splicing events to generate the three different α-TM mRNAs. A, B, and C describe the specific pathways producing the STRα1-TM, STRα2-TM, and SMα-TM mRNAs, respectively. [Reproduced, with permission, from B. Nadal-Ginard, R. E. Breitbart, E. E. Strehler, N. Ruiz-Opazo, M. Periasamy, and V. Mahdavi (1986). Alternative splicing: A common mechanism for the generation contractile protein diversity from single genes. *In* "Molecular Biology of Muscle Development," pp. 387–410, Alan R. Liss, New York.]

There are at least two similar but biochemically and functionally distinct Tn-C isoforms in striated muscles of higher vertebrates. In nonvertebrates, Tn-C has been found in smooth muscle. In adult vertebrates, the relative concentrations of Tn-C isoform proteins is muscle-type-specific. In heart muscle, only one Tn-C isoform protein has been found. For rabbit, this Tn-C protein is found to be identical with one of the two skeletal muscle isoforms, and it has been proposed that the two identical proteins may be encoded by the same gene. The two Tn-C isoform proteins in adult skeletal muscles are found in different ratios, seemingly depending on the fiber composition of individual muscle. Immunocytological methods demonstrate that the levels of Tn-C isoform proteins change during heart and skeletal muscle development in chicken and rabbit.

From cDNAs studies, it is known that human slow skeletal muscle comprises a short 26-bp 5′ UTR, a 486-bp coding region, and a 173-bp 3′ UTR. The fast skeletal muscle troponin C (TC2) contains a 64-bp 5′ UTR, a 483-bp coding region, and a 130-bp 3′ UTR. In both sequences, the canonical polyadenylation signal AATAAA is found 18 bp upstream from the polyadenylic acid tail. The cDNAs of both isoforms contain sequences at the translational start site that are similar to the consensus sequence CCACCATGG; CCACCATGA for the fast isoform and CCAGCATGG for the slow isoform on Tn-C. The coding regions of both cDNAs are about 74% similar. Only the amino-terminal 39 bp of the coding sequences is dissimilar. The amino acid sequence of human cardiac Tn-C is identical with the amino acid sequence of the slow Tn-C protein. Southern blot analyses suggests that both human fast Tn-C and slow Tn-C are encoded by single copy genes. Expression of Tn-C mRNA is developmentally regulated and is tissue-specific. Tn-C is structurally and evolutionarily closely related to a number of calcium-binding proteins referred to as members of Tn-C superfamily.

ii. Troponin-I Troponin is a family of three muscle-specific myofibrillar proteins involved in the calcium regulation of contraction in cardiac and skeletal muscle. Tn-I proteins have multiple functional domains that are distinct and bind with high affinity to actin and Tn-C. The interactions of these domains regulate actomyosin ATPase activity in resting and contracting muscle (Section III,C). Tn-I also interacts functionally with other muscle proteins, including Tn-T.

The complete gene structure for quail fast skeletal muscle Tn-I shows that the 4.5-kb gene has eight exons encoding 830-bp mRNA. The cDNA of the fast skeletal Tn-I gene contains 43 bp of 5′ nontranslated sequence, the entire 546 bp of the protein-coding sequence, and 140 bp of the 3′ nontranslated sequence. The upstream promoter region contains the sequences TTTTATA and TAAA, similar to the TATA consensus sequence. Tn-I proteins have two functional domains: the actin-binding domain and the Tn-C-binding domain. Interestingly, the actin-binding domain of Tn-I is encoded by a single exon, whereas the Tn-C-binding domain is split into at least two exons. The Tn-I gene with several muscle-specific contractile protein genes (e.g., chick α-actin, rat α-actin, chick MLC 3, and rat cardiac MHC genes) showed the presence of homologous sequences in their 5′-flanking regions and large introns of similar size that separate the promoter and the first exon coding for the 5′ nontranslated sequences from the protein-coding segment of the second exon. The presence of these common structural features suggests they may act as regulatory elements in the coordinate control of muscle gene expression during myogenesis.

iii. Troponin-T The structures of rat fast skeletal Tn-T gene and the chicken cardiac Tn-T gene have been defined. Several salient features have been conserved in the rat skeletal and chicken cardiac genes. Each contains 18 exons, although the rat skeletal gene comprises 16.5 kbp, whereas the chicken cardiac gene spans 9 kbp. The first exon in both genes is a spliced leader. All the translated exons that are spliced constitutively in the rat (e.g., 2, 3, 9–15, and 18) and exons 4 and 16 show a high degree of conservation for size and coding sequence. In contrast, the chicken cardiac exons 5–8 are significantly larger and exon 17 considerably smaller than their corresponding rat exons with little sequence homology. All the regulatory protein genes can encode multiple isoforms of the proteins by alternative splicing of the primary transcripts. In particular, among the 18 exons of the rat skeletal Tn-T gene, three types of splicing are exhibited. Each of the exons 4–8 near the 5′ end of the primary transcript may be individually included or excluded from the mature mRNA in a combinatorial fashion to generate as many as 32 different sequences

within the amino-terminal region of the protein (Fig. 9). Of these possible combinations, every one that has been probed has been detected. Exons 16 and 17, in contrast, are alternatively spliced in a mutually exclusive manner to encode different internal α- and β-peptides, respectively, in the carboxy-terminal end. The remaining exons are constitutively spliced, representing domains that are constant among all mRNAs encoded by this gene. The differential incorporation of exons 4–8 with either exon 16 or 17 then may yield as many as 64 distinct but related Tn-T isoforms. Such alternative splicing is regulated in a tissue-specific and developmental manner. This phenomenon reflects the developmental induction of modulatory factors.

For the regulatory proteins of the thin filaments, alternative splicing of primary transcripts of single genes provides a powerful mechanism for modulation of protein isoforms. This is in contrast to the regulated transcription of members of multigene families, as in the case of sarcomeric MHC genes.

C. Structural Proteins

The structural proteins comprise α-actinin, 85K protein, C protein, M-line proteins, and the elastic protein, connectin or titin (Table I).

1. Alpha-Actinin

Alpha-actinin is an actin-binding and cross-linking protein found in both muscle and nonmuscle cells at points where actin is anchored to a variety of intracellular structures. It is a dimer, probably a homodimer with a subunit molecular weight of 94–103 kDa, the subunit being antiparallel in orientation. It is visualized as a long rod-shaped molecule in the electron microscope, 3–4 nm wide by 30–40 nm in length. A number of distinct isoforms of α-actinin have been characterized including the skeletal and smooth muscle isoforms isolated from brain and cultured fibroblasts. The only clear functional difference between these isoforms is that binding of the various nonmuscle α-actinin to actin is calcium-sensitive, whereas binding of the muscle isoforms is calcium-insensitive.

a. Structure of α-Actinin Alpha-actinin can be divided into three domains: an N-terminal actin-binding domain, four internal 122-amino-acid repeats, and a C-terminal region containing two EF-hand calcium-binding motifs (Fig. 10). Isoforms of α-actinin with different calcium-sensitive actin-

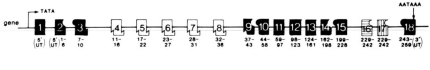

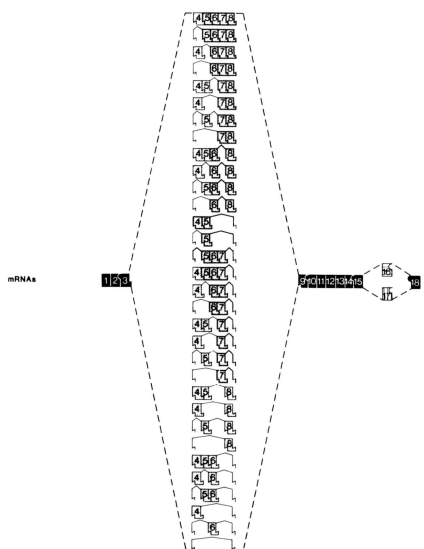

binding characteristics in chicks appear to be generated by alternative splicing. The nucleotide sequence of a chick skeletal muscle α-actinin as well as a near-complete sequence of the chick nonmuscle isoform has clarified the relation between various α-actinin isoforms. Comparison of the sequence of α-actinin with other proteins has revealed extensive homologies with the erythroid and nonerythroid spectrins as well as with dystropin, the protein encoded by the gene at the Duchenne muscular dystrophy locus (see Section IV,B). These observa-

FIGURE 9 Fast skeletal troponin-T gene organization (rat) and 64 possible mRNAs. Exons (*black,* constitutive; *clear,* combinatorial; *striped,* mutually exclusive) are diagrammed to represent their terminal split codons (sawtooth boundaries lie between the first and second nucleotides of the codon, concave/convex between the second and third, and flush boundaries lie between intact codons). The promoter (TATA), polyadenylation signal (AATAAA), untranslated sequences (UT), and encoded amino acids (numbered below exons) are indicated. Each mRNA comprises the constitutive sequences, one of the 32 combinations of exons 4–8, and either exon 16 or 17. [From R. E. Breitbart *et al.,* (1987).]

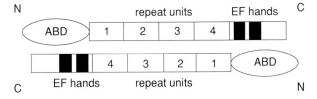

FIGURE 10 Schematic representation of the α-actinin homodimer. (ABD) actin-binding domain. The calcium-binding motifs are shown as *filled boxes* and are labeled *EF* hands. The repeat units are shown to be precisely aligned between the two subunits, although this has not been proven experimentally. [Reproduced, with permission, from A. Blanchard *et al.*, (1989).]

tions suggest that these three proteins are members of the same family of cytoskeletal proteins.

i. Actin-Binding Domain The actin-binding domain of α-actinin is contained within the N-terminal region of the molecule. Because actin is highly conserved across all species investigated, the actin-binding domain of α-actinin would also be expected to be highly conserved across all species. The N-terminal region of α-actinins present in chick smooth muscle, Dictyostelium, and Drosophila shows considerable homology, which extends toward the C-terminal region, where it is markedly reduced. The difference provides some definition of the probable boundary of the actin-binding domain. However, the extreme N-terminal regions are different.

The presumed actin-binding domain in α-actinin is homologous to residues 1–238 of human dystrophin, implying that dystrophin is also an actin-binding protein. Interestingly, the extreme N-terminal region of dystrophin differs between the brain and muscle isoforms and is also different from that found in α-actinin isoforms (Section IV,B). These regions of difference may prove informative in further defining residues involved in actin binding. In addition, the actin-binding domain of α-actinin is homologous to the N-terminal 250 amino acids of other actin-binding proteins.

ii. Central Repeats The central region of α-actinin contains four internal repeats of about 122 amino acids, homologous to the 100 residue repeats found in spectrin and approximately 110 residue repeats found in dystrophin. Repeat 1 displays a greater level of identity (90%) than repeats 2–4 (80%, 76%, and 80%). The level of identity between a given repeat across species is greater than that

between the four repeats within the same species. These repeats in α-actinin appear to be responsible for the formation of the antiparallel dimer.

iii. EF-Hand Calcium-Binding Domain The C-terminal region of α-actinin contains two EF-hand-like calcium-binding motifs. The smooth and fibroblast isoforms of the protein are 98% identical in nucleotide sequence, differing only in the second half of the first EF hand.

Given the close similarity between the smooth and nonmuscle isoforms of the protein, it seems likely that they are encoded by the same gene and arise by alternative splicing of a single primary transcript. In contrast, skeletal isoform is encoded by a separate gene. It is still far from clear, however, how many additional (perhaps tissue-specific) isoforms of skeletal, smooth, and nonmuscle α-actinins exist.

2. 85K Protein
This protein is present in the central region of the Z disc and has a molecular weight of 85,000.

3. C Protein
This protein is present in the A band along the middle portion of the myosin filament. With electron microscopy the anti–C-protein antibody is observed to form between seven and nine transverse stripes in each half of the A band.

4. M-Line Proteins
The center of the H band contains a creatine kinase and a structural protein called *myomesin*. Myomesin has a molecular weight of 165,000 m, can be detected in myoblasts, and has a high affinity for myosin. It acts as a specific marker during the differentiation of myoblasts into myotubes and muscle fibers. Myomesin is present in skeletal and heart muscle but not in smooth muscle and plays a key role in the molecular organization of the thick myofilaments.

5. Elastic Protein of Muscle
In recent years, a high-molecular-weight protein has been isolated from the myofibril and named *connectin* or *titin*. This protein represents 8–12% of the myofibrillar mass. Under the EM this protein appears as an extremely thin and long (more than micrometer) strand. It is extensible and flexible and shows an axial periodicity. This protein interacts

with both myosin and actin and forms nets that are concentrated at the A-I junction of the myofibril. Because of its elastic properties, it is ideally suited to constitute a lattice or scaffold for the thick and thin myofilaments that contribute to the generation of passive tension when the muscle is stretched.

D. Intermediate Filament Proteins

Intermediate filaments (IFs) are unique cytoskeletal structures of 7–12 nm in diameter, which occur in the cytoplasm of most eukaryotic cells. Five subclasses are found in mammalian cells: two complex groups of *acidic* and *basic keratins* of 40–70 kDa; a single protein *desmin* of 52 kDa; a single protein vimentin of 55 kDa; the GFAP of 50 kDa; and three *neurofilament proteins* of 65,100 and 135 kDa. Cytokeratins A and B, which represent the largest and most diverse classes of IF protein, are found in epithelial cells; desmin is synthesized in skeletal visceral and certain vascular smooth cells; the gilial fibrillary acidic protein (GFAP) is found in astroglia and the triplet of neurofilament proteins in neuronal cells; vimentin has a widespread distribution and is found in all mesenchymal derivatives as well as in the progenitors of muscle and neural tissues.

From various structural analyses, it is clear that all IF subunits are built according to a common plan: Each has a central helical rod domain flanked by end domains of variable size and chemical characteristics. The different properties of the IFs are due to their end domains and their structural uniformity to the conservation of the central helical rod domains.

1. Vimentin and Desmin Proteins

The mammalian vimentin and desmin genes are remarkably conserved between various species. In each species, the two genes contain eight introns in identical positions, suggesting that these two genes arose from a common ancestor. In mammalian skeletal muscle, the replicative presumptive myoblasts synthesize predominantly vimentin. However, on fusion, myotubes synthesize high levels of desmin, and as the myotubes mature, vimentin and desmin is found. During the process of maturation of the myotubes, desmin, which is initially homogeneously distributed throughout the myotubes, assembles in the Z bands of the then-forming sarcomeres. In mature muscle fibers, the desmin IFs interconnect myofibrils at the level of the Z bands and attach them to the sarcolemma. In addition, desmin proba-

bly serves to attach actin filaments to the Z bands and to fasciae adherents to the intercalated discs, thus helping to maintain the normal myofibrillar architecture.

The differential expression of desmin and vimentin genes that determines their transcriptional activity during human myogenesis is not clearly understood.

a. Vimentin-Coding Gene Vimentin gene (vim), which is present as a single copy in the hamster and human genome, comprises about 10 kb of DNA and contains more than 80% of intron sequences. It contains nine exons with a total length of 1,398 nucleotides, coding for 466 amino acids giving a molecular weight of 53,000. Comparison of the coding sequences between human and hamster shows 91% homology at the nucleotide level. Vimcoding sequences also have partial homology to several other sequences coding for the IF proteins. At the 3′ nontranslated region of human vim gene there are two canonic poly (A) signals, and at the 5′ end of the gene, there is a consensus promoter sequence. The first AUG start codon is 132 nucleotides downstream of the estimated cap site in hamster vim gene. mRNA from different mammalian sources reveal a single species of 2 kb, and the expression of the gene is growth-regulated.

b. Desmin-Coding Gene It is 8.4 kb long and contains nine exons separated by introns ranging in size from 0.1 to 2.2 kb. The single-copy human desmin (des) gene is located on chromosome 2. The human and hamster des genes show full correspondence in position, size, and sequence of the exons. A predominant feature of the human des gene that is not shared by either the hamster or the chicken is the large number of repeated elements that are found in the introns. The 5′ UTRs of human vim and des genes contain a consensus 16-mer element GTAACGGGACCATGCC and GCTGGGGGCCCTCTC, respectively. The same kind of sequence is found in hamster vim and des genes, GCAAAGG-GACTGTGTC. An 11-bp sequence CAGCTGT-CAGG, with homology to the distal regulatory sequence of human and mouse α-cardiac actin-coding genes, is found in the 5′ flanking region. The activity of the vim and des genes differs in smooth and striated muscles. In human striated muscle, the des gene alone is actively transcribed, whereas the vim gene becomes silent. Only one species of des RNA of 2.2 kb in length is produced. However, in smooth uterine muscle, both genes are coexpressed, whereas in some arterial muscle, vimentin is the predominant protein. The molecular basis for the tissue-restricted expression of des is unknown. However, the 11-bp sequence homologous to α-cardiac actin-coding gene, and a 16-mer element homologous to the hamster des gene could play a role in myogenesis.

In the following section, the molecular mechanism of muscle contraction is discussed.

III. Regulation and Energetics of Muscle Contraction

The detailed ultrastructural and biochemical information available permits an interpretation of the macromolecular mechanisms involved in muscle contraction.

A. ATP Hydrolysis Powers the Contraction of Muscle

Contraction is accompanied by the cyclic formation and dissociation of bridges between the myosin globular heads and the sides of the adjacent actin filaments. Bridge formation pulls the thin filaments toward the center of the A band, and contraction results. ATP hydrolysis is essential to the cyclic formation and dissociation of these actin–myosin bridges. During contraction, the myosin head bound to an actin molecule tilts toward the center of thick filament; this conformational change is the step fueled by energy of hydrolysis of the phosphodiester bond of ATP. Precisely how the change is brought about is uncertain, but it generates the relative movement of the actin and myosin chains with each cycle. As a result of the hydrolysis of one ATP molecule per myosin head, the actin filament is pulled a distance of about 7 nm.

B. Release of Ca^{2+} from the Sarcoplasmic Reticulum Triggers Contraction

The force generating molecular interaction just described takes place only when a signal passes to the skeletal muscle from its motor nerve. The signal triggers an action potential in the muscle plasma membrane and the electrical excitation spreads rapidly into a series of membranous folds, the *transverse tubules* or T tubules, that extends inward

from the plasma membrane around each myofibril. The signal is then somehow relayed to the SR that surrounds each myofibril like a net stocking.

The gap between the T tubules and the SR is only 10–20 nm, but it is unclear how the signal passes between them. After excitation of T tubules, large Ca^{2+} release channels in the SR membrane are opened, allowing Ca^{2+} to escape into the cytosol from the SR where Ca^{2+} is stored in large quantities. The resulting sudden rise in free Ca^{2+} concentration in the cytosol initiates the contraction of each myofibril. Because the signal from the muscle cell plasma membrane is passed within milliseconds (via the T tubules and SR) to every sarcomere in the cell, all the myofibrils in the cell contract at the same time. The increase in Ca^{2+} concentration in the cytosol is transient because the Ca^{2+} is rapidly pumped back into the SR by an abundant Ca^{2+} ATPase in its membrane. Typically, the cytosolic Ca^{2+} concentration is restored to resting levels within 30 msec, causing the myofibrils to relax.

Smooth muscles do not contain a developed SR membrane. Changes in the level of cytosolic Ca^{2+} are much slower than in striated muscle (of the order of seconds to minutes), thereby allowing a slow, steady response in contractile tension.

C. Troponin and Tropomyosin Mediate Ca^{2+} Stimulation of Contraction

Because of the Ca^{2+} ATPase in the SR membrane, the cytosol of resting muscle has a free Ca^{2+} concentration of about 10^{-7} M. An increase in Ca^{2+} concentration to 10^{-5} M initiates contraction. The manner in which Ca^{2+} activates contraction varies for different types of muscle; at least three types of regulation are known. In striated vertebrate muscle, the Ca^{2+} regulation affects the actin thin filaments. In smooth muscles and invertebrate muscle, the regulation affects the myosin headpieces.

The thin filaments in striated muscle contain four proteins involved in the Ca^{2+} regulation of contraction. TM, an elongated protein, lies in each of the two grooves of the actin helix (Fig. 6). Bound to specific sites of each TM molecule are three Tn peptides T, I, and C (for details, see Section II,B). The molar ratio of actin, Tn-T, Tn-I, Tn-C, and TM is 7:1:1:1:1.

A mixture of myosin filaments and purified thin filaments containing only actin hydrolyzes ATP at maximum rate. The presence of Tn and TM on the thin filaments inhibits this ATPase activity by blocking the interaction of myosin head with actin. The Tn-T subunit is responsible for binding the Tn complex to TM, and Tn-I in concert with TM inhibits the binding of actin to myosin and thus the actin-stimulated myosin ATPase activity. Tn-C is the Ca^{2+}-binding subunit with a structure and function similar to that of calmodulin. Occupation by Ca^{2+} of all the Ca^{2+}-binding sites of Tn-C results in a release of the TM—Tn-I inhibition of the actin–myosin ATPase activity and in activation of contraction. How this function of Tn-C is mediated is not clear. A model is proposed that binding of Ca^{2+} to Tn-C induces a conformational change in the TM molecule so that the myosin heads can then interact with actin. This may involve a movement of TM closer to the center of the helical groove of the thin filament.

D. Contraction in Smooth Muscle and Invertebrate Muscle is Regulated Differently by Ca^{2+}

Vertebrate smooth muscle and invertebrate muscle contain TM but not the Tn complex. Ca^{2+} regulation operates differently in these muscles and involves regulatory interactions with the MLCs.

In invertebrates (e.g., mollusks), at low concentrations the interaction of myosin headpieces and actin filaments that allow muscle contraction is inhibited by one of the MLC pairs located in the head region. Binding of Ca^{2+} to one of the MLCs induces a conformational change in the myosin headpiece that allows it to bind to actin; this in turn causes an activation of the myosin ATPase and contraction of the muscle. The so-called regulatory light-chain protein has a high affinity for Ca^{2+}, similar to that of Tn-C of the ubiquitous Ca^{2+}-binding protein calmodulin.

Regulation in vertebrate smooth muscle is similar but more complex. As in mollusks, a MLC attached to the myosin headpiece inhibits the actin-stimulated ATPase activity. In this system, phosphorylation of one of the MLCs relieves this inhibition and thus stimulates contraction.

E. The "Walk-Along" Theory of Contraction

As soon as the actin filament becomes activated by the calcium ions, it is believed that the heads of the cross-bridges from the myosin filaments immediately become attracted to the active sites of the actin filament, and this in some way causes contraction to occur. This interaction appears to cause

contraction according to the "walk-along" theory of contraction.

The walk-along mechanism as illustrated in Fig. 11 shows the heads of two cross-bridges attaching to and disengaging from the active sites of an actin filament. It is postulated that when the head attaches to an active site, this attachment simultaneously causes profound changes in the intramolecular forces in the head and arm of the cross-bridge. The new alignment of forces causes the head to tilt toward the arm and to drag the actin filament along with it. This tilt of the head of the cross-bridge is called the *power stroke*. Then, immediately after tilting, the head automatically breaks away from the active site and returns to its normal perpendicular direction. In this position it combines with an active site farther down along the actin filament; then a similar tilt takes place again to cause a new power stroke, and the actin filament moves another step. Thus, the heads of the cross-bridges bend back and forth and step by step walk along the actin filament, pulling the action toward the center of the myosin filament. Each one of the cross-bridges is believed to operate independently of all others, each attaching and pulling in a continuous but random cycle. Therefore, the greater the number of cross-bridges in contact with the actin filament at any given time, the greater, theoretically, is the force of contraction.

F. ATP as a Source of Energy for Contraction

When a muscle contracts against a load, work is performed and energy is required. The energy for muscle contraction comes from ATP hydrolysis. Yet we detect no major difference in ATP levels between a resting muscle and one that is actively contracting, because a muscle cell has an efficient backup system for regenerating ATP. The concentration of ATP present in the muscle fiber, about 4

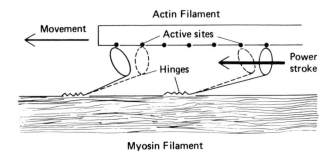

FIGURE 11 "Walk-along" mechanism for contraction of the muscle. [Reproduced, with permission, from A. Guyton (1986).]

mM, is sufficient to maintain full contraction for only a few seconds at most. Fortunately, after the ATP is broken into ADP, the ADP is rephosphorylated to form new ATP within a fraction of a second. There are several sources of the energy for this rephosphorylation.

The first source is phosphocreatine, a compound that carries a high-energy phosphate bond similar to that of ATP. The enzyme phosphocreatine kinase catalyzes a reaction between phosphocreatine and ADP to form creatine and ATP. It is the intracellular level of phosphocreatine that drops after a short burst of muscle activity, even though the contractile machinery itself consumes ATP. The pool of phosphocreatine acts like a battery storing ATP energy and recharging itself from the new ATP generated by cellular oxidations when the muscle is resting.

The next source of energy is released from foodstuffs (i.e., carbohydrates, fats, and proteins). The sarcoplasmic matrix contains the glycolytic enzymes as well as other globular proteins such as myoglobin, salts, and high-phosphate compounds. Glycogen is present in the matrix as small granules or glycosomes observed under EM. Glycogen disappears with contraction through glycolysis, and lactic acid is formed, which can be transformed into pyruvic acid to enter the Kreb's cycle.

Oxidative phosphorylation is the last and most important source of ATP.

IV. Genetics of Muscle-Related Myopathies

Muscle-related myopathies range from the less severe lesions in the genes associated with the muscle subunit of the enzyme(s) of the metabolic pathways [metabolic myopathies: e.g., the phosphofructokinase (PFK) and the glycogen phosphorylase that are the energy source(s) for muscle contraction] to the more severe disease [e.g., the Duchenne muscular dystrophy (DMD)] leading to the progressive degeneration of the muscle. Some of them are discussed below.

A. Metabolic Myopathies

Metabolic myopathies are usually classified according to the area of metabolism affected. The only category relevant to this subject is the disorders of carbohydrate metabolism because a deficiency in the enzyme(s) of this pathway leads to the depletion of energy source for muscle contraction. Two such genetic disorders are discussed.

1. Glycogen Storage Disease Type VII (Tarui's Disease)

Inherited deficiency of PFK is characterized by the coexistence of muscle disease and hemolytic process. Typically, the muscle disease begins in early childhood and consists of easy fatiguability, transient weakness and muscle cramps, and myoglobinuria after vigorous exercise.

Human PFK is under the control of three structural loci that encode muscle (M), liver (L), and platelet (P) subunits residing on chromosome 1, 10, and 21, respectively, and these isozymes are variably expressed in different tissues. Mature skeletal muscle expresses only the M subunit and contains a single isozyme species, the homotetramer M4. Liver and kidney express only L and P subunits and are not affected in muscle PFK deficiency.

Homozygous deficiency of the M subunit in most cases results from a mutation(s) affecting the muscle PFK structure, because catalytically inactive immunoreactive M subunit is present. Genes encoding the human muscle and liver subunit of PFK have been cloned. Tissues expressing the muscle or the liver PFK contain a specific mRNA of 3.0 kb, encoding a protein of about 780 amino acids. Alternative splicing of the transcript encoding the human muscle isoenzyme of PFK has also been observed.

2. Molecular Mechanisms of McArdle's Disease (Muscle Glycogen Phosphorylase Deficiency)

McArdle's disease is a rare form of metabolic myopathy, genetically transmitted as a recessive autosomal trait. Clinically this disease is characterized by asthenia, cramps, stiffness, and attacks of myoglobinuria after physical effort. These manifestations of the disorder originate from a depletion in the ATP stores needed for muscle contraction and relaxation induced by the absence of glycogen phosphorylase enzyme activity. This enzyme catalyzes the first step of the glycogenolytic pathway that provides most of the readily available energy for intensive muscle contraction.

In human tissues and those of other mammals, glycogen phosphorylase exists in three isoenzymatic forms: liver, muscle, and brain. The muscle form is the only isoenzyme expressed in skeletal muscle and is also found in heart and brain along with the cerebral form. This explains why the defect in McArdle's disease is confined to muscle tissue.

Tissues expressing muscle phosphorylase contain a specific mRNA of 3.4 kb, encoding a protein of about 841 amino acids. In four patients with muscle phosphorylase deficiency, the gene was present and apparently not rearranged. However, a clear heterogeneity was found at the mRNA level. Phosphorylase mRNA was undetectable in five patients; in three patients, it was in decreased amounts but with a normal length.

B. Duchenne Muscular Dystrophy

Duchenne muscular dystrophy is one of the most devastating illnesses that can afflict humankind. It is an X-linked recessive disorder, resulting in progressive degeneration of the muscle, and affects about one in 3500 male children. Affected boys are typically diagnosed at about 2 years of age, no longer ambulatory by 12 years of age, and rarely live beyond the second decade in life. In the past few years, remarkable progress has been made toward the isolation and characterization of the 2 million-bp DMD gene, possibly the largest in the human genome, and recently a 14-kb human DMD cDNA corresponding to a complete representation of the fetal skeletal muscle transcript has been cloned. The DMD transcript is formed by at least 60 exons that have been mapped relative to various reference points within Xp21. The first half of the DMD transcript is formed by a minimum of 33 exons spanning nearly 1000 kb, and the remaining portion has at least 27 exons that may spread over a similar distance. [*See* MUSCULAR DYSTROPHY, MOLECULAR GENETICS.]

The gene product, dystrophin, a protein of molecular weight 427,000 appears to be completely absent from DMD muscle. DMD mRNA is most abundant in skeletal and cardiac muscle and less in smooth muscle. DMD selectively affects a subset of skeletal muscle fibers specialized for fast contraction.

Bakers muscular dystrophy (BMD), a less severe and less frequent disease allelic to DMD, is also localized to the same region in the X chromosome. Several other dystrophin-related muscular dystrophies include limb-girdle muscular dystrophy and spinal muscular atrophy.

The techniques of recombinant DNA technology are leading to a revolution in our understanding of the molecular genetics of DMD. This new knowledge has tremendous implications and has led to new methods for diagnosis, prognosis, and genetic counseling.

Acknowledgments

Part of this work was carried out in the laboratory of Dr. Alan Mclachlan at Scripps Clinic and Research Foundation to whom I am indebted for his support and expert scientific guidance. I like to thank my colleague Dr. Roman G. Reddy for his scientific association. I am also indebted to Dr. Sara Sukumar for her enthusiastic support and encouragement in the completion of this work. The expert secretarial support of Ms. Lorna White is gratefully appreciated. This work was supported in part by a grant from the American Cancer Society, #CD402.

Bibliography

Alberts, B., Bray, D., Lewis, J., Raff, M., Roberts, K., and Watson, J. D., eds. (1989). The cytoskeleton. *In* "Molecular Biology of The Cell." Garland Publishing, New York, London.

Blanchard, A., Ohanian, V., and Critchley, D. (1989). The structure and function of α-actinin. *J. Muscle Res. Cell. Motil.* **10**(4), 280–289.

Breitbart, R. E., Andreadis, A., and Nadal-Ginard, B. (1987). Alternative splicing: A ubiquitous mechanism for the generation of multiple protein isoforms from single genes. *Annu. Rev. Biochem.* **56**, 467–495.

Emerson, C. P., Jr., and Bernstein, S. I. (1987). Molecular genetics of myosin. *Annu. Rev. Biochem.* **56**, 695–726.

Guyton, A. (1986). Contraction of skeletal muscle. *In* "Textbook of Medical Physiology" (D. Dreibelbis, ed.). W. B. Saunders, Philadelphia.

Koenig, M., Hoffman, E. P., Bertelson, C. J., Monaco, A. P., Feener, C., and Kunkel, L. M. (1987). Complete cloning of the Duchenne muscular dystrophy (DMD) cDNA and preliminary genomic organization of the DMD gene in normal and affected individuals. *Cell* **50**, 509–517.

Li, Z., Lilienbaum, A., Butler-Browne, G., and Paulin, D. (1989). Human desmin-coding gene: Complete nucleotide sequence, characterization and regulation of expression during myogenesis and development. *Gene* **78**, 243–254.

Sharma, P. M., Reddy, G. R., Babior, B. M., and McLachlan, A. (1990). Alternative splicing of the transcript encoding the human muscle isoenzyme of phosphofructokinase. *J. Biol. Chem.* **265**(16), 9006–9010.

Stedman, H., and Sarkar, S. (1988). Molecular genetics in basic myology: A rapidly evolving perspective. *Muscle Nerve* **11**, 668–682.

Wieczorek, D. F., Smith, C. W. J., and Nadal-Ginard, B. (1988). The rat α-tropomyosin gene generates a minimum of six different mRNAs coding for striated, smooth, and nonmuscle isoforms by alternative splicing. *Mol. Cell. Biol.* **8**, 679–694.

Muscle, Physiology and Biochemistry

F. NORMAN BRIGGS, *Virginia Commonwealth University*

Glossary

ATPase Enzyme that catalyzes the hydrolysis of ATP
Contraction Shortening and/or force development
Isoforms Proteins of homologous, but not identical, structure expressed by isogenes from the same genome
Isozymes Isoforms of enzymes

THE TWO MUSCLE proteins, actin and myosin, which are expressed in large amounts in muscle cells and are responsible for producing muscle contraction, have been highly conserved during evolution. Before the appearance of muscles, these proteins were used to produce protoplasmic streaming and pseudopod motion in nonmuscle cells, a function they still have. The splitting of the terminal phosphate of ATP by the actin-activated myosin ATPase provides the energy to drive both contraction and cellular motion. In muscle cells myosin is organized into thick filaments and actin, into thin filaments. As ATP is hydrolyzed the thick and thin filaments move past each other and the muscle shortens. When muscle is at rest actin is prevented from activating myosin. In vertebrate striated muscles, both skeletal and cardiac, contraction is initi-ated by the release and binding of calcium to a regulatory protein, troponin, on the thin filament. In smooth muscle calcium is bound to a regulatory protein, calmodulin, in the cytoplasm and promotes the phosphorylation of a subunit of myosin which activates the thick filament. Contraction is terminated by the withdrawal of calcium from these regulatory proteins.

The force that muscles generate and the velocity at which they shorten are determined by the ATPase activity of myosin. Myosin ATPase is determined by (1) the gene expressing the myosin, (2) the extent to which thick and thin filaments overlap, and (3) the load on the muscle. The initiation of skeletal muscle contraction is under voluntary control by the central nervous system. The involuntary autonomic nervous system regulates the strength of contraction of cardiac muscle. The regulation of smooth muscle contraction depends on the autonomic nervous system, hormones, and intrinsic nerve activity.

I. Muscle Structure

A. Striated Muscle

Figure 1 shows the structure of a typical skeletal muscle. The muscle is an aggregate of muscle cells, or fibers, bound together by connective tissue into fasciculi, which, in turn, are bound together by connective tissue into a muscle. Skeletal muscle fibers tend to be long and are formed by the fusion of embryonic cells, which make the fibers multinucleated. Part of a muscle fiber is shown at the top of the figure. The external membrane, the sarcolemma, is partially stripped away to show the contents of the fiber. The sarcolemma is the limiting membrane of the cell and generates the action potentials which excite the muscle to contract. The mitochondria

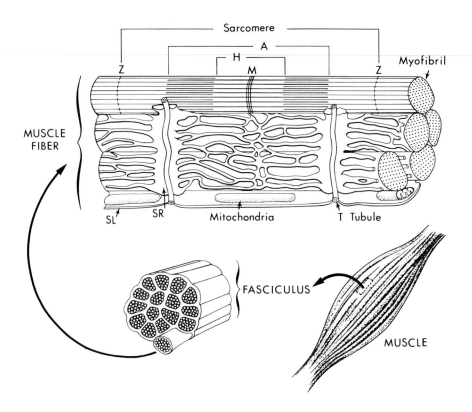

produce ATP by aerobic metabolism. The myofibrils, the contractile apparatus, occupy a major portion of the cell. They are composed of repeating segments, called sarcomeres. Surrounding each myofibril is the sarcoplasmic reticulum, which releases calcium in response to an action potential. This calcium activates contraction. Penetrating the sarcolemma and lying adjacent to the terminal portion of the sarcoplasmic reticulum are T tubules which conduct the action potential into the muscle fiber and to the sarcoplasmic reticulum. The structure of cardiac muscle fibers are similar to the skeletal muscle fibers shown in the top panel of Fig. 1 and in Fig. 2. [*See* CARDIAC MUSCLE; SKELETAL MUSCLE.]

An expanded view of a small segment, one sarcomere long, of a myofibril is shown in Fig. 2. The myofibril is composed of thick and thin filaments (i.e., myofilaments). A sarcomere is bounded by Z bands that hold the thin filaments in proper array. The M line is thought to help hold the thick filaments in array. The I band contains only thin filaments and the Z band. The A band contains thick and thin filaments. In the H band region of the A band the thin filaments do not overlap the thick filaments. The projections from the thick filaments in the A band are crossbridges, which are absent from

FIGURE 1 The structure of skeletal muscle is shown at the level of the whole muscle, the fasciculus, and the muscle fiber. A portion of the sarcolemma (SL) of a muscle fiber has been removed to reveal its contents of myofibrils, mitochondria, and sarcoplasmic reticulum (SR). The T tubule is a specialized invagination of the SL. The fine structure of a portion of a myofibril, the sarcomere, is shown at the top of the figure.

the central portion of the A band. Alternating A and I bands give skeletal and cardiac muscles their striated appearance in the light microscope. Cross-sections at different regions of the myofibril are shown at the bottom of Fig. 2 and show the ordered structure of the thick and thin filaments and how the thin filaments interdigitate with the thick filaments.

Skeletal muscles contain mixtures of muscle fiber types. Type I (slow-twitch) fibers are found in muscles activated for prolonged periods. They contain a slow isozyme of myosin (the myosin splits ATP slowly) and large concentrations of mitochondria and do not fatigue easily. Cardiac muscle fibers resemble type I skeletal fibers. Type II fibers are found in muscles activated for brief periods. They contain a fast myosin isozyme. Type IIB fibers contain few mitochondria, are significantly dependent on anaerobic (i.e., glycolytic) metabolism to sustain activity, and fatigue easily. Type IIA fibers have a

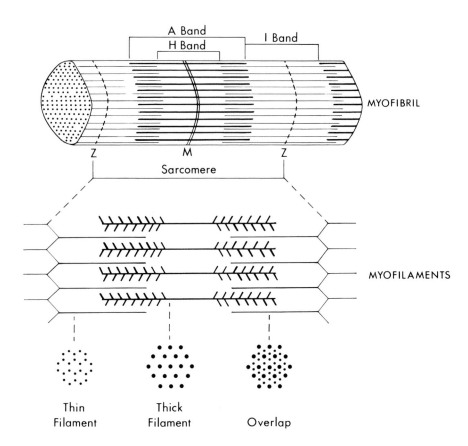

FIGURE 2 Fine structure of a sarcomere. The myofibril identifies A, I, and H bands. A sarcomere is that part of a myofibril between two Z bands. The A band is in the center of the sarcomere. One-half of each I band is on each side of the Z band. The myofilaments are the thin filaments connected to the Z band and the barbed thick filaments. The barbs are crossbridges originating from myosin. Cross-sections through the myofibril, at the bottom of the figure, show the ordered structure of the myofilaments and how they interdigitate at different levels through the sarcomere.

higher concentration of mitochondria than do type IIB fibers and fatigue less easily. They use anaerobic and aerobic metabolism to sustain contractile activity.

It appears that conditioning, if intense enough, leads to the expression of type I myosin isozyme in muscle fibers that otherwise would contain type II myosin isozymes. Other protein isoforms and functions are altered, and the type II muscle can be converted to a type I muscle fiber. Deconditioning causes a reversion to type II fibers. Conversely, denervation leads to the conversion of type I muscle fibers to type II fibers. In humans the proportion of types I and II muscle fibers in different muscles is determined in part genetically and in part can be altered by training. It has been demonstrated that marathon runners have higher than normal percentages of type I fibers in the muscles used for running.

B. Smooth Muscle

Smooth muscles, with few exceptions, surround hollow organs. Contraction causes the pressure within the organ to increase and often propels the contents out of the organ. They do not have the banded structure seen in striated muscle fibers, because their thick and thin filaments are not organized into the sarcomeric structure characteristic of striated muscle myofilaments. The fibers are much smaller than skeletal muscle fibers and have a single nucleus. Sarcoplasmic reticulum is present, but there are no T tubules. [*See* SMOOTH MUSCLE.]

C. Structure of Thick Filaments

The predominant and most important protein in the thick filament is myosin, which is a complex of six polypeptide subunits (Fig. 3A). There are two identical heavy chains of approximately 200 kDa which form a supercoil and two pairs of light chains. The

A

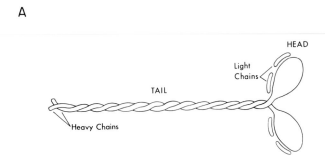

B

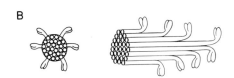

FIGURE 3 Structure of the thick filament. (A) A myosin molecule composed of two heavy chains and two light chains per heavy chain. (B) One end of a thick filament formed by aggregation of myosin molecules. A cross-section through one end of a thick filament is shown on the left.

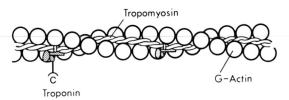

FIGURE 4 A portion of a thin filament composed of polymerized G-actin monomers tropomyosin and troponin. The calcium binding subunit of troponin, troponin C, is labeled.

head portion of each heavy chain contains an ATPase site. The head region of myosin forms the crossbridges seen protruding from the thick filaments in Fig. 2. Each heavy chain binds one alkali light chain, which is essential for ATPase activity, and one phosphorylatable light chain (P-light chain) of about 21 kDa. Two types of alkali light chains, of about 17 and 21 kDa exist. Although a single myosin can have only one type of alkali light chain a single muscle fiber can have both types. The function of the P-light chain in striated muscle is not known, though it is known that it can be phosphorylated. Phosphorylation of the P-light chain in smooth muscle activates smooth muscle contraction (see Section III,B).

Dispersion of myosin into a salt solution produces thick filamentlike structures, in which approximately 500 myosin molecules aggregate along their helical tails. One end of such a thick filament is shown in Fig. 3B, which shows paired heads of the heavy chains protruding from the aggregate. It is interesting that all attempts to discover a function for the pairing of myosin heads have failed.

D. Structure of Thin Filaments

The thin filament of striated muscle is composed of tropomyosin, troponin, and polymerized actin, as shown in Fig. 4. Actin monomers (i.e., G-actin)

have a molecular weight of 42,000. In muscle the actin monomers are polymerized into long coiled structures (F-actin). Two strands of F-actin form the two-stranded twisted structure seen in Fig. 4. Tropomyosin is a rodlike protein composed of two subunits of around 32.7 kDa. The two subunits form a coiled-coil structure which gives the molecule rigidity. X-Ray diffraction studies indicate that tropomyosin lies close to the grooves formed by the F-actin helix. Troponin is composed of three polypeptide subunits and is bound to the end of each tropomyosin. Troponin can bind calcium at four sites.

II. Sliding Filament Model of Muscle Contraction

Muscle contraction, whether in skeletal, cardiac, or smooth muscle, is propelled by the translocation (i.e., sliding) of thin filaments past thick filaments.

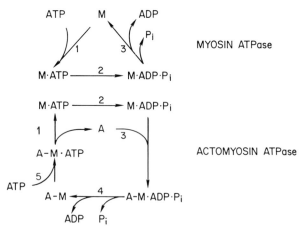

FIGURE 5 Myosin- and actomyosin-catalyzed ATP hydrolysis. (Top) Myosin ATPase reaction scheme. (Bottom) Actomyosin ATPase reaction scheme. A, Actin; M, myosin; A–M, actomyosin; P_i, inorganic phosphate.

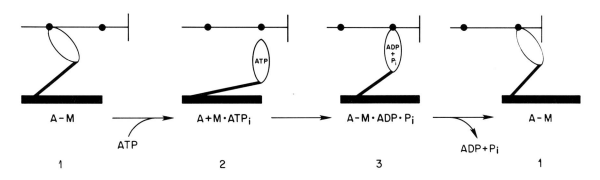

FIGURE 6 The coupling of ATP hydrolysis to the translocation of thick and thin filaments. The thin line with solid circles represents crossbridge attachment sites on a thin filament. The thick line represents the helical part of a thick filament. Coming out from the helical part of the thick filament is a crossbridge containing a myosin ATPase site. A, Actin; M, myosin; A–M, actomyosin; P_i, inorganic phosphate.

The energy for this motion is derived from myosin-catalyzed hydrolysis of ATP. A minimum reaction scheme is shown in Fig. 5. In Figs. 5 and Fig. 12B dots indicate that a chemical is physically bound, while a dash indicates that it is covalently bound. At step 1 of the myosin ATPase reaction scheme, ATP binds to myosin. At step 2 ATP is hydrolyzed and the $ADP \cdot P_i$ (inorganic phosphate) complex is formed. At step 3, the rate-limiting step, ADP and P_i are released. The rate at which myosin hydrolyzes ATP is slow and prevents the muscle from wasting energy when not contracting.

In the actomyosin ATPase reaction scheme actin activates myosin ATPase activity (Fig. 5) 400- to 500-fold by increasing the rate at which ADP and P_i leave the enzyme at step 4. Note that actin enters at step 3 after the myosin $ADP \cdot P_i$ complex is formed. ATP, which enters the cycle at step 5, forms an actomyosin–ATP complex and dissociates actin from actomyosin at step 1. Otherwise, the myosin and actomyosin ATP hydrolytic schemes are similar. [*See* ADENOSINE TRIPHOSPHATE (ATP).]

To produce macroscopic changes in muscle length, thick and thin filaments slide past each other. This translocation is hypothesized to be produced by motion of myosin crossbridges, as shown in Fig. 6. In passing from state 1 to state 2 the crossbridge is detached by ATP and the orientation of the crossbridge changes. Between states 2 and 3 ATP is hydrolyzed to ADP and P_i, and the crossbridge is attached to the thin filament (state 3). Between states 3 and 1 a change in the orientation

of the crossbridge is produced by the energy released from the hydrolysis of ATP. The products of ATP hydrolysis, ADP and P_i, are released before the crossbridge is returned to state 1. The changes in orientation of the crossbridge during the cycle shown in Fig. 6 slide the thin filament past the thick filament. Each cycle accounts for only a small change in muscle length. Macroscopic changes in length require many repetitions of the cycle.

III. Regulation of Crossbridge Cycling

Since muscles contain ATP, thick filaments, and thin filaments one would expect them to contract continuously, but this is not the case. The thin filament proteins tropomyosin and troponin prevent the attachment of thick filaments to thin filaments, unless calcium is bound to troponin.

A. Thin Filament Regulation in Striated Muscle

A specific site on each actin monomer (i.e., G-actin) can bind myosin (thick filament) and activate myosin ATPase. The tropomyosin–troponin complex on the thin filament (Fig. 4) prevents this interaction. Tropomyosin must be moved from its blocking position before thick and thin filaments can interact, and this is accomplished by the binding of calcium to troponin. The binding site is on an 18-kDa subunit with four calcium binding sites, two of which have high affinity for calcium, but do not bind calcium specifically. Two of the sites have a lower affinity for calcium and bind calcium with high specificity. It is the binding of calcium to these low-affinity high-specificity sites which permits the myosin crossbridge to attach to the thin filament. Magnesium is probably bound to the two high-affinity sites in muscle *in situ*. The binding of calcium to troponin produces a conformational change which

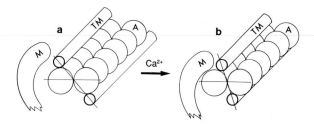

FIGURE 7 Unblocking of an actin–myosin interaction by the movement of tropomyosin. The normal twisted structures of polymerized actin and tropomyosin are not shown for clarity. The binding of calcium to troponin (not shown) rotates tropomyosin from the blocking position (a) to a position which allows the crossbridge from myosin to attach to the thin filament (b). A, Actin; M, myosin; TM, tropomyosin.

is transferred from troponin to tropomyosin. These conformational changes move tropomyosin from its blocking position, as shown in Fig. 7 (the troponin complex is not shown for clarity), and allows the myosin crossbridge to attach to actin.

B. Thick Filament Regulation in Smooth Muscle

Smooth muscle thin filaments do not contain troponin and thus lack thin filament regulation of thick filament–thin filament interaction. They do, however, contain tropomyosin. Calcium activates crossbridge cycling by binding to calmodulin and activating the phosphorylation of the P-light chain of myosin, as shown in Fig. 8. Calmodulin, a protein of 17 kDa, forms a ternary complex with calcium and myosin light-chain kinase (Step I in figure). The ternary complex catalyzes the phosphorylation of myosin P-light chain (Step II). The phosphorylation of P-light chain activates the thick filament–thin filament interaction that catalyzes ATP hydrolysis (Step III) and cycling of the thick filament crossbridges. Contraction is terminated by the removal of calcium and dephosphorylation of myosin P-light chain by a myosin P-light chain phosphatase of 70 kDa.

Although smooth muscle shortening and the development of tension require phosphorylation of the myosin light chain, tension, once developed, can be maintained after a large percentage of the myosin light chains are dephosphorylated. How tension is maintained is unknown, although it clearly reflects attached dephosphorylated crossbridges. Somehow, the attached crossbridge goes into a state, termed the latch state, which is not readily de-

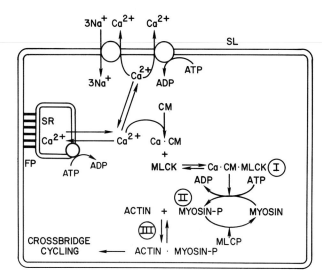

FIGURE 8 Thick filament regulation of crossbridge cycling in smooth muscle. SR, sarcoplasmic reticulum; CM, calmodulin; MLCK, myosin light chain kinase; MLCP, myosin light chain phosphatase; Myosin-P, myosin with phosphorylated P-light chain; FP, foot process.

tached. It is known that little energy is expended by smooth muscle to maintain tension, an advantage in situations when tension may have to be maintained continuously (e.g., by smooth muscle of the vascular system).

IV. Excitation–Contraction Coupling

Calcium must be bound to either troponin (striated muscle) or calmodulin (smooth muscle) to activate contraction. The process which provides calcium and activates contraction, excitation–contraction coupling, has four parts: (1) a signal, a voltage change, or binding of an agonist to receptors in the sarcolemma; (2) a messenger (e.g., calcium); (3) the binding of calcium to troponin or calmodulin; and (4) contraction (i.e., crossbridge cycling). The time courses of voltage, calcium, and force transients for typical skeletal, cardiac, and smooth muscles are shown in Fig. 9. In each case the signal is generated at the sarcolemma before the appearance of the calcium transient. The calcium transient does not persist as long as the tension transient, because the cytoplasmic calcium producing the calcium signal becomes bound to troponin or calmodulin before contraction is initiated.

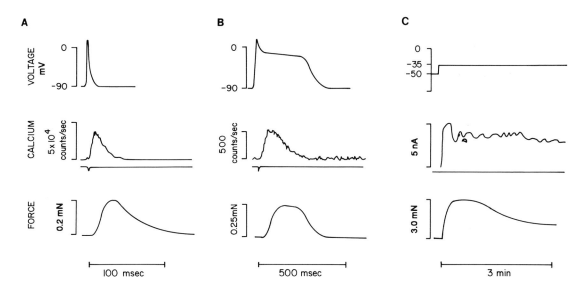

FIGURE 9 Voltage, calcium, and force transients in (A) skeletal, (B) cardiac, and (C) vascular smooth muscle during excitation–contraction coupling. Calcium transients are proportional to free calcium in the cytoplasm.

A. Striated Muscle

Excitation–contraction coupling in striated muscle is shown in Fig. 10. An action potential across the sarcolemma and the T tubule provides the signal to initiate contraction. This action potential, which opens calcium channels in the sarcolemma, is conducted into the T system, where it communicates with the sarcoplasmic reticulum through foot processes, which open calcium channels in the sarcoplasmic reticulum. Calcium outside the muscle cell and stored in the sarcoplasmic reticulum rapidly diffuses into the cytoplasm, binds to troponin on thin filaments, and activates thin filaments. Activation of thin filaments allows the attachment of crossbridges, and crossbridge cycling is initiated. The relative importance of extracellular and sarcoplasmic reticulum calcium for excitation–contraction coupling in skeletal and cardiac muscle differs, being more important in cardiac muscle.

In striated muscle contraction is terminated by the withdrawal of calcium into the sarcoplasmic reticulum or transport out of the cell to extracellular fluid, as shown in Fig. 10 and discussed in detail in Section VI. Cyclic adenosine monophosphate (cAMP) plays a role in modulating excitation–contraction coupling and terminating contraction in cardiac muscle, as explained in Sections V,A and VI,B.

In skeletal muscle only a small percentage of the activating calcium comes from extracellular fluid via the sarcolemmal calcium channel; the majority comes from the sarcoplasmic reticulum. In cardiac muscle a larger fraction of calcium comes from the sarcolemma. In type II skeletal muscle fibers the amount of calcium released is sufficient to saturate the troponin calcium binding sites. In cardiac muscle the amount of calcium released during excitation–contraction coupling is normally sufficient to saturate only about one-half of the sites. All crossbridges do not, therefore, normally contribute to force generation in cardiac muscle. The potential exists, therefore, in the heart to alter the amount of calcium released and thereby to alter force generation.

B. Smooth Muscle

The signal to raise the level of cytoplasmic calcium in smooth muscle can be the binding of an agonist (e.g., norepinephrine) to a receptor or a transient change in voltage (Fig. 11). Contraction is initiated by the binding of calcium to calmodulin, which activates crossbridge cycling, as described in Fig. 8. Opening of the sarcolemmal calcium channels allows calcium from extracellular fluid to diffuse into the cytoplasm. Calcium is released from the sarcoplasmic reticulum either by a voltage-dependent mechanism involving the foot process or by inositol 1,4,5-triphosphate (Ins 1,4,5P$_3$). The mechanisms

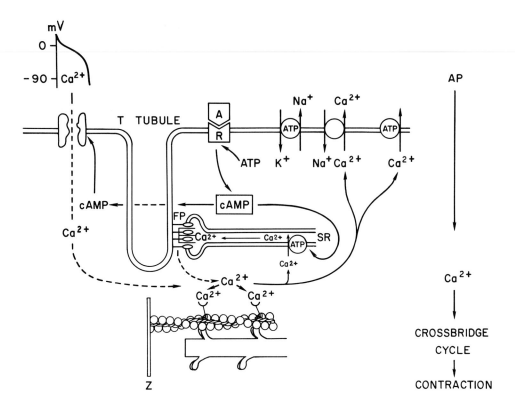

FIGURE 10 Mechanism of excitation–contraction coupling in striated muscle (see text). A, Agonist; R, receptor; AP, action potential; SR, sarcoplasmic reticulum; FP, foot process.

of calcium release are under intense investigation. Some evidence suggests that the binding of an agonist to its receptor activates a GTP-dependent G protein, which in turn activates phospholipase C, which causes a rise in cytoplasmic Ins 1,4,5-P$_3$. Ins 1,4,5-P$_3$ might then open calcium channels in the sarcoplasmic reticulum and release stored calcium. The extent to which cytoplasmic calcium comes from extracellular fluid or from the sarcoplasmic reticulum varies with different types of smooth muscles.

V. Muscle Relaxation

To terminate contraction, calcium is withdrawn from troponin or calmodulin. Calcium withdrawal is achieved by ATP-dependent calcium pumps in the sarcoplasmic reticulum and the sarcolemma and by sodium–calcium exchangers in the sarcolemma.

A. Calcium Pump in the Sarcoplasmic Reticulum

There is a high density of a calcium pump protein of 109 kDa in the sarcoplasmic reticulum that transports calcium into the lumen of that organelle. The enzyme (Fig. 12A) is an integral membrane protein,

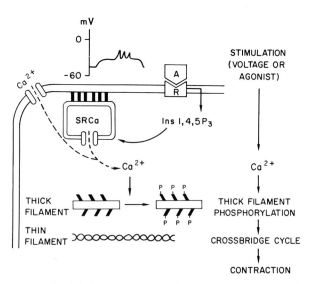

FIGURE 11 Mechanism of excitation–contraction coupling in smooth muscle (see text). A, Agonist; R, receptor; P, phosphate; Ins 1,4,5-P$_3$, inositol 1,4,5-triphosphate; SR, sarcoplasmic reticulum.

A

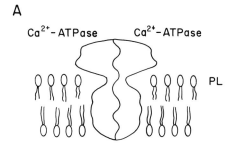

B

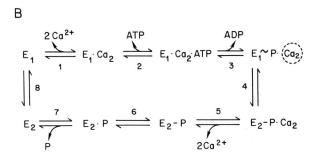

FIGURE 12 Calcium transport mechanism in the sarcoplasmic reticulum. (A) Dimer organization of the Ca^{2+}-ATPase in the membrane of the sarcoplasmic reticulum. Phospholipid (PL) forms the membrane into which the Ca^{2+}-ATPase is immersed. The bulk of the protein, which is outside of the membrane, is on the cytoplasmic side of the sarcoplasmic reticulum. The Ca^{2+}-ATPase transports calcium from the cytoplasmic side to the luminal side of the sarcoplasmic reticulum. (B) Coupling of calcium transport to ATP hydrolysis (see text).

apparently a dimer, that spans the lipid bilayer. The bulk of the enzyme faces the cytoplasm, the side from which calcium is transported. Figure 12B shows the reactions that couple the hydrolysis of ATP to calcium transport. In an ideal cycle one molecule of ATP is hydrolyzed and two molecules of calcium are transported from the cytoplasm into the lumen of the sarcoplasmic reticulum. The unidirectional (i.e., vectorial) transport of calcium is dependent on an alternation in the affinity for calcium. When not phosphorylated, the calcium binding sites of the enzyme face the cytoplasm and have a high affinity for calcium, the E_1 state. During the hydrolysis of ATP the enzyme is phosphorylated (step 3). This induces a conformation change (step 4), which produces the E_2 state. In the E_2 state calcium affinity is three orders of magnitude lower than in the E_1 state. In the E_2 state the calcium binding sites face the lumen of the sarcoplasmic reticulum. The low affinity of the enzyme for calcium in the E_2 state allows calcium to depart from the enzyme and to move into the lumen of the sarcoplasmic reticulum.

After the departure of calcium and phosphate the enzyme undergoes a conformational change (step 8), which restores its high affinity for calcium and its cytoplasmic orientation.

There are at least two Ca^{2+}-ATPase isozymes. Type I isozyme is found in cardiac and type I skeletal muscle fibers. The type II isozyme is found in type II skeletal muscle fibers. The calcium uptake activity of the type I isozyme is modified by the sarcoplasmic reticulum protein, phospholamban. When phospholamban is phosphorylated, the calcium affinity of the Ca^{2+}-ATPase is increased. Phosphorylation of phospholamban is stimulated by norepinephrine, a neurotransmitter from the sympathetic nervous system, by a cAMP-dependent mechanism (Fig. 10). Increasing the calcium affinity of the Ca^{2+}-ATPase of the sarcoplasmic reticulum increases its ability to accumulate calcium and thereby increases the amount of calcium released by an action potential. The contractility of the heart can be increased by this mechanism. The sarcoplasmic reticulum of smooth muscle is found under the sarcolemma. Because this organelle is difficult to free from the sarcolemma, it has been difficult to isolate and study. It appears to transport calcium by a mechanism comparable to that observed in the sarcoplasmic reticulum from skeletal and cardiac muscle.

B. Sarcolemmal Calcium Transport Mechanisms

The sarcolemma of all muscles transports calcium from the cytoplasm to the extracellular fluid. The two calcium transporters are shown in Fig. 10. One is an ATP-dependent calcium pump of 150 kDa which transports calcium from the cytoplasm to the extracellular fluid by a mechanism similar to that used by the sarcoplasmic reticulum. The other transporter is a sodium–calcium exchanger which uses the energy in the sodium gradient across the sarcolemma to transport calcium. The sodium gradient is formed by the sodium pumping action of a Na^+,K^+-ATPase. The sodium–calcium exchanger couples the transsarcolemmal influx of three sodium ions into the cell to the efflux of one calcium ion out of the cell. The sodium–calcium exchanger is the major pathway for the transsarcolemmal ejection of calcium from cardiac muscle cells. In skeletal and smooth muscles ATP-dependent calcium pumps are thought to account for most of the calcium efflux.

VI. Regulation of Muscle Activity

A. Skeletal Muscle

The force that can be developed by skeletal muscle depends on (1) the number of motor units recruited, (2) the frequency at which they are stimulated, (3) sarcomere length, and (4) load.

1. Motor Unit Recruitment

A motor unit is composed of a motor nerve and the muscle fibers it innervates (Fig. 13). The motor nerve originates in the ventral horn of the spinal cord and can be activated reflexly via sensory afferent fibers and by way of higher nerve centers (e.g., the motor cortex). The number of muscle fibers per motor unit varies with the precision required of the motor unit. Small motor units of the extraocular muscles, used to produce very precise motions of the eye, contain only a few muscle fibers. If strong, but less precise, motion is required, large motor units with many hundreds of muscle fibers per motor unit are used. Muscle force is graded by the recruitment of motor units. Small motor units are recruited first. If greater forces are required, larger motor units are recruited. [*See* Motor Control.]

2. Stimulus Frequency

One mechanism for regulating force is modulation of the rate (i.e., frequency) at which motor nerves are activated by the central nervous system. Figure 14 shows a muscle stimulated at increasing

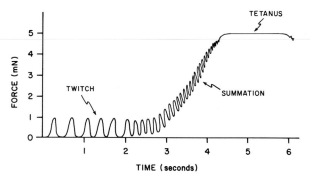

FIGURE 14 Regulation of force by stimulus frequency. As time increases, the interval between stimuli decreases.

frequencies (the interval between stimuli is decreased). Each stimulus to the muscle produces a contraction. If a single action potential is sent down a motor nerve, all of the muscle fibers in that motor unit are activated, and a brief contraction, a twitch, is elicited. Twitches are characterized by complete relaxation between stimuli. Muscles are usually activated by trains of action potentials. Repeated stimuli of low frequency produce summation, while those of high frequency produce tetani. When contraction summates, tension is increased and relaxation is incomplete (i.e., tension does not return to baseline). When the stimulus frequency becomes high, the muscle is tetanized. At tetanus frequency there is no relaxation between stimuli, and the force is maximal. Tetanus tension and summated tensions exceed twitch tension, because repeated excitation of the muscle extends the duration of the calcium transient and allows the myofilament to develop more force.

3. Length–Tension Relationship

In striated and smooth muscle the force developed by a fully activated muscle depends on the length of the muscle. In striated muscle the important length is sarcomere length, which is determined by muscle length. The top of Fig. 15 shows the relationship between sarcomere length and tension in a skeletal muscle. The bottom of the figure shows the relationship between sarcomere length and the overlap of thick and thin filaments. On the ascending limb of the length–tension relationship, an increase in sarcomere length increases the force the muscle develops. On the descending limb an increase in muscle length decreases the force developed. The reduction in tension on the descending limb is due to decreases in the overlap between

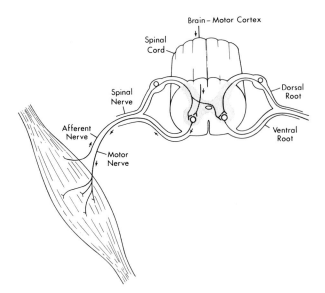

FIGURE 13 Skeletal muscle motor unit.

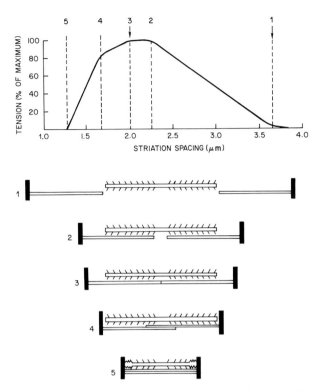

FIGURE 15 Length–tension relationship in skeletal muscle. Sarcomere length is determined by muscle length. Lines 1–5 identify the sarcomere lengths discussed in the text. (Top) The relationship between sarcomere length and relative tension. (Bottom) The overlap between thick and thin filaments at designated sarcomere lengths.

thick and thin filaments. At a sarcomere length of 2.1 μm (line 2), the overlap between crossbridges on thick filaments and a thin filament is maximal. As the muscle is stretched to a sarcomere length of 3.6 μm (line 1), there is no overlap and no tension is produced. Between sarcomere lengths of 2.1 and 3.6 μm, the force is proportional to the overlap.

An explanation for the ascending limb of the length–tension relationship is still being developed. At least three factors contribute. At short sarcomere lengths (e.g., 1.3 μm) (line 5) no force is developed, even though the overlap of thick and thin filaments is complete. The problem is that one of the overlapping thin filaments is from the Z band opposite to the end of the thick filament and cannot produce motion in the correct direction. To develop force, the thick filament must overlap with a thin filament from its own end of the sarcomere. At a sarcomere length of 1.6 μm (line 4), there is less overlap of thin filaments from the opposite side of the sarcomere, and force generation increases. A

second factor affecting force development at short sarcomere lengths is a decrease in the amount of calcium released from the sarcoplasmic reticulum. This decreases the interaction between thick and thin filaments. A third factor is a decrease in the affinity of troponin for calcium. This, too, can lead to a depression in thick filament–thin filament interactions at short sarcomere lengths and, thus, to decreases in force generation.

Skeletal muscles operate at or near optimal muscle length (i.e., sarcomere lengths of 2.0–2.2 μm) and shorten little during contraction. The length–tension relationship is, therefore, of little functional significance to skeletal muscle. The relationship was important, however, in developing the sliding filament model of muscle contraction. The sliding filament theory hypothesizes that force is generated by interaction between the thick and thin filaments. This hypothesis was confirmed by showing, on the descending limb of the length–tension relationship, that the force is directly proportional to the overlap.

4. Load–Velocity Relationship

The shortening velocity of muscle fibers is dependent on the load (Fig. 16). The lighter the load, the faster the muscle can shorten. As the muscle shortens the load–velocity curve on which the muscle fiber operates shifts from curve A to curve B, etc. (i.e., from longer to shorter muscle lengths). The maximal force that can be developed (zero velocity) conforms to the length–tension relationship, as given in Fig. 15. Curve A is for optimal length, while curves B and C are for shorter muscle

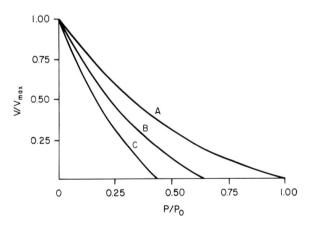

FIGURE 16 Curves A, B, and C show the relationship of the velocity to the load for muscle fibers of decreasing muscle lengths. P/P_0, Load relative to the maximal force the muscle can develop at optimal length (A); V/V_{max}, velocity relative to the maximal velocity the muscle can shorten.

lengths. The maximal velocity that a muscle can develop is independent of muscle length.

B. Cardiac Muscle

The rhythm of the heart is generated by a pacemaker, normally the sinoatrial node, within the heart itself (Fig. 17). The sinoatrial node is composed of specialized muscle fibers, which rhythmically generate action potentials that are conducted throughout the heart and activate all muscle fibers. The conduction of action potentials from one cardiac muscle fiber to the next is via specialized junctions, termed gap junctions, that allow current from one muscle fiber to invade neighboring fibers and thus to excite them. The intrinsic heart rate, established by the sinoatrial node, is modified by the parasympathetic and sympathetic nervous systems, decreasing and increasing the heart rate, respectively.

The heart cannot modify the force of contraction by recruiting muscle fibers, as in skeletal muscle, because all fibers are activated with each beat of the heart and cannot modify force by summation, as in skeletal muscle, because mechanisms exist to ensure that cardiac muscle completely relaxes between each beat. Relaxation is essential to allow the heart to fill with blood between beats. In cardiac muscle sarcomere length (i.e., muscle fiber length) and the sympathetic nervous system determine the force developed at each beat. The lengths of the muscle fibers at the beginning of each heart beat are determined by the volume of blood in the heart when contraction is initiated. The greater the volume of the heart at the beginning of each beat, the greater the force the heart can develop. The heart normally functions on the ascending limb of the length–tension relationship. Activity from the sympathetic nervous system can also increase the contractility of the heart. At normal levels of contractility, about one-half of the thin filament sites are activated by calcium. The sympathetic nervous system can increase the amount of calcium stored in the sarcoplasmic reticulum by increasing cAMP levels (Fig. 10). cAMP activates protein kinases that phosphorylate phospholamban in the sarcoplasmic reticulum and unidentified proteins associated with voltage-dependent calcium channels in the sarcolemma.

Phosphorylation of phospholamban increases the affinity of the sarcoplasmic reticulum Ca^{2+}-ATPase for calcium and allows this organelle to increase its storage of calcium. cAMP-dependent phosphorylation of sarcolemmal proteins (Fig. 10) is thought to increase the amount of calcium coming into the cytoplasm through voltage-dependent calcium channels. The increased influx of calcium through such channels and from the sarcoplasmic reticulum increases the amount of calcium released during excitation–contraction coupling and thereby increases the fraction of thin filament sites that can react with thick filament crossbridges. This increases the force that cardiac muscle can develop. [See PROTEIN PHOSPHORYLATION.]

C. Smooth Muscle

The functions of the gastrointestinal tract, the vascular system, the ureters, the uterus, the lung airways, and the iris of the eye depend on smooth muscle. The smooth muscles in some of these organs contract tonically and are classified as tonic smooth muscles. The other smooth muscles contract phasically and are classified as phasic muscles. Figure 18 shows these two classes of smooth muscles and their types of innervation: multiunit and single unit. It is recognized that this classification is overly simple and that different smooth muscles show characteristics of both classifications to differing degrees.

1. Tonic (Multiunit) Smooth Muscle

Muscle fibers in vascular smooth muscles are arranged circumferentially and contract tonically.

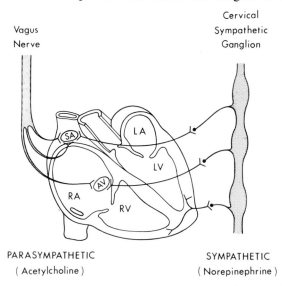

FIGURE 17 Neural innervation of the heart. The transmitter released by each branch of the autonomic nervous system is identified. SA, Sinoatrial node; AV, atrioventricular node; RA, right atrium; LA, left atrium; LV, left ventricle; RV, right ventricle.

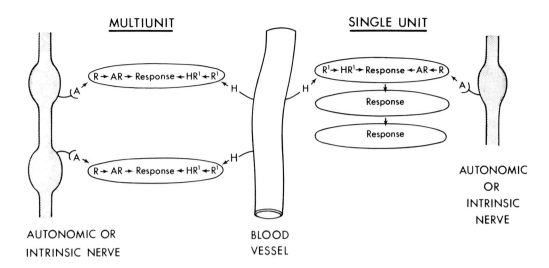

MULTIUNIT

R → AR → Response ← HR¹ ← R¹

R → AR → Response ← HR¹ ← R¹

AUTONOMIC OR
INTRINSIC NERVE

BLOOD
VESSEL

SINGLE UNIT

R¹ → HR¹ → Response ← AR ← R

Response

Response

AUTONOMIC
OR
INTRINSIC
NERVE

FIGURE 18 The regulation of force in smooth muscle. Multiunit muscle fibers are innervated individually. Single-unit fibers are coupled together and can activate each other. Both types of fibers are activated by agonists from nerves or by hormones from the vascular system. A, Agonist; H, hormone, R, receptor for agonist; R¹, receptor for hormone; Response, contraction.

They are classified as "multiunit" because of the extensive innervation of the muscle by the sympathetic nervous system. The transmitters released from the nerves innervating these and most other smooth muscles generate action potentials. In most visceral smooth muscles action potentials are not generated. In these cases transmembrane potential is decreased. Activation of the sympathetic nervous system and the release of transmitters cause contraction or relaxation of vascular and other smooth muscles, depending on the receptor activated.

When the muscle is stimulated to contract, cytoplasmic calcium is increased. This calcium is from both extracellular fluid and sarcoplasmic reticulum. The binding of norepinephrine (e.g., from the sympathetic nervous system) to α_1-adrenergic receptors in the sarcolemma induces the formation of Ins $1,4,5P_3$, which may increase cytoplasmic calcium, as shown in Fig. 11. The increase in cytoplasmic calcium activates actomyosin ATPase by the reactions shown in Fig. 8.

Catecholamines released by the sympathetic nervous system relax some tonic muscles. In these cases norepinephrine binds to a β_2-adrenergic receptor and activates adenylate cyclase or guanylate cyclase in the sarcolemma. This leads to the production of cAMP or cGMP and the activation of a protein kinase which phosphorylates the proteins involved in calcium pumping and sequestration.

2. Phasic (Single-Unit) Smooth Muscle

Phasic muscles (e.g., from the gastrointestinal tract, the ureters, and the uterus) are classified as "single unit" because there is extensive communication between fibers. The electrical field generated by action potentials at the sarcolemma of one muscle cell is transmitted through gap junctions (shown by the arrows in Fig. 18) to other cells. The origin of the action potentials varies. In some cases it is myogenic; that is, the muscle spontaneously develops action potentials (like the pacemaker cells in the heart). In other cases action potentials are generated by sensory fibers within the smooth muscle and reflexly activate effector nerves within the plexus of the smooth muscle. In still other cases the muscle is activated by nerves from the autonomic nervous system. Regardless of the method of activation, there is an increase in intracellular calcium and the initiation of contraction, as described for tonic muscles. The relative amounts of calcium coming from extracellular fluid and from the sarcoplasmic reticulum is not established. It is certain that some comes from the extracellular fluid through voltage-sensitive calcium channels.

VII. Muscle Stiffness

Muscle fibers not only develop force and shorten; they also resist stretch. When a crossbridge is attached to a thin filament, muscle stiffness increases. It follows that the greater the number of attached crossbridges, the stiffer the muscle. Following the death of a muscle cell the concentration of ATP falls, and perhaps all crossbridges become attached to thin filaments. As shown in Figs. 5 and 6, ATP is required to detach crossbridges. A lack of ATP produces rigor mortis. Stiffness in activated muscle is considerably less than that seen in rigor mortis, indicating that when the crossbridges are cycling, some are not attached. As the load on the fiber is decreased and shortening velocity increases, muscle stiffness decreases, indicating that the number of crossbridges attached decreases with increasing shortening velocity. Stiffness reaches a minimum at zero load, when the velocity of shortening is maximal and crossbridge attachment is minimal. From these observations it is concluded that the movement of thin filaments past thick filaments affects the kinetics of crossbridge attachment and detachment and, thus, the number of crossbridges attached at any given shortening velocity.

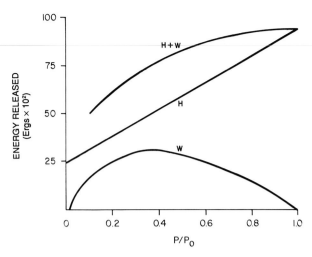

FIGURE 19 Effect of the load on energy utilization by skeletal muscle. P/P_0, Load relative to the maximal force the muscle can develop; H, heat; W, work.

VIII. Muscle Energetics

Part of the energy released from ATP by the crossbridge cycle is used to do work when muscles move loads. The amount of energy used and the partitioning of that energy between work and heat in a twitch is shown in Fig. 19. The amount of energy degraded to heat increases with load (P) and reaches a maximum at P_0, at which there is no muscle shortening. The amount of energy converted to work is maximal at a $P-P_0$ ratio of 0.4. Why the amount of energy converted to work is maximal at this load is not clear. Stiffness measurements indicate that there is a maximal number of crossbridges attached at a $P-P_0$ ratio of 1.0, the load at which energy utilization is maximal. At a load when the $P-P_0$ ratio is 0.4, shortening velocity is not maximal, and stiffness lies between maximal and minimal levels. At this level of crossbridge attachment (i.e., stiffness), an optimal fraction of the energy released by crossbridge turnover produces motion and, thus, work.

Bibliography

Briggs, F. N., and Solaro, R. J. (1976). The role of divalent metals in the contraction of muscle fibers. *In* "Metal Ions in Biological Systems" (H. Sigel, ed.), Vol. 6. Decker, New York. [A brief review, including a historical perspective, of the role of calcium in the regulation of skeletal muscle contraction and the role of the sarcoplasmic reticulum in the accumulation and release of calcium.]

Huxley, A. (1980). "Reflections on Muscle." Princeton Univ. Press, Princeton, New Jersey. [An excellent brief review of the experimental evidence which led Sir Andrew Huxley and others to propose the sliding filament model of muscle contraction.]

Murphy, R. A. (1988). Muscle. *In* "Physiology" (R. M. Berne and M. N. Levy, eds.), 2nd ed. Mosby, St. Louis, Missouri. [An overview of the structure, function, and biochemistry of striated and smooth muscle.]

Needham, D. M. (1971). "Machina Carnis: The Biochemistry of Muscular Contraction in Its Historical Development." Cambridge Univ. Press, London. [A scholarly review of research on muscle going back to earliest recorded history.]

Ruegg, J. C. (1988). "Calcium in Muscle Activation." Springer-Verlag, Heidelberg. [A review of excitation–contraction coupling in skeletal, cardiac, and smooth muscle.]

Siegman, M. J., Somlyo, A. P., and Stephens, N. L. (eds.) (1987). "Regulation and Contraction of Smooth Muscle." Liss, New York. [Proceedings of an International Union of Physiological Sciences satellite conference. Contains sections on structure and crossbridge kinetics, the biochemistry of contractile and regulatory proteins, pharmacology, mechanics, and energetics.]

Squire, J. (1981). "The Structural Basis of Muscular Contraction." Plenum, New York. [A review of the structure and function of the contractile proteins in striated and smooth muscle.]

Swynghedauw, B. (1986). Developmental and functional adaptation of contractile proteins in cardiac and skeletal muscle. *Physiol. Rev.* **66**, 710. [A review of the myofilament isozymes expressed developmentally and during changes in muscle function.]

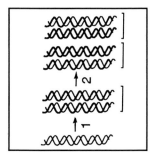

Mutation Rates

H. ELDON SUTTON, *The University of Texas at Austin*

Glossary

Allele   Alternate form of a gene

Chromosome rearrangement   Change in the structure of a chromosome that is microscopically visible

Genomic mutation   Deviation from the normal complement of chromosomes

Genotype   Total complement of specific alleles that are possessed by an individual

Germinal mutation   Mutation in the germ cells that give rise to sperm and ova

Locus   Chromosomal location of any gene or group of genes; often used interchangeably with gene

Mutagen   Substance that enhances the rate of mutation

Mutation   Heritable change in genotype; a change in the nucleotide sequence of DNA

Phenotype   Observable individual, a product of genotype interacting with environment

Point mutation   Mutation that affects a small region of a chromosome and cannot be seen microscopically

Somatic mutation   Mutation in any cells of the body other than germ cells

ALL GENES are subject to structural change in the DNA. Such mutations may occur in germ cells, in which case they may be transmitted to offspring, or they may occur in somatic cells, in which case they are not transmitted except to cells of that individual. Mutations may also involve chromosome rearrangements or abnormal numbers of chromosomes.

The rates at which mutations occur depend on the type of change. Most studies of mutation rates are based on some change in the phenotype. However, not all mutations of a particular gene will cause the variant phenotype. Therefore, mutation rates vary widely depending on the method used to observe them. Phenotypic changes that have been used to measure germinal mutation rates include rare dominant and X-linked recessive diseases, protein variants, chromosomal rearrangements, and genomic changes. Somatic mutations have been measured by the rate of occurrence of cancer and by variant lymphocytes and red blood cells. Exposure to mutagens (rather than mutation) can be detected by increases in sister-chromatid exchange, unscheduled DNA synthesis, and mutagenicity of body fluids (e.g., urine) on microorganisms.

External factors that increase mutation rates are ionizing and ultraviolet radiation, as well as many chemicals. Biological variables that influence mutation rates include variations in DNA repair systems, variations in metabolism, age, and sex.

I. Definition of Mutations

It is imperative that genetic information be stable and that it be copied accurately in biological systems. Were it otherwise, life as we know it, from bacteria to humans, would not exist. Yet DNA is subject to the same physical laws that govern all molecules. These laws predict that chemical changes will occur spontaneously. Furthermore, the concept of an error-free system is hypothetical at best. In the real world, errors are made, however rare they may be.

Changes in the genetic information are called *mutations*. Some mutations provide an expanded repertory of useful genes that are essential for evolutionary changes. Some are functional and provide the genetic variability that characterizes all natural populations. Most mutations are detrimental and are eliminated by natural selection and by chance. These detrimental mutations account for hereditary disease, a substantial part of all human disease, both those diseases inherited in a Mendelian manner and the more complex disorders that may be influenced also by environment.

In this article, the various kinds of mutations will be described, methods of measuring mutation rates will be discussed, observations on human mutation rates will be given, and some factors that influence mutation rates will be identified.

A mutation was traditionally defined as any sudden heritable change. This definition was appropriate for a time when heritability could only be established by testing for segregation among progeny and the traits studied were typically gross phenotypic variations. *Sudden* was defined as one generation. This limitation allowed consideration only of those mutations that occurred in the germ cell line and were transmitted to offspring. Mutations in somatic cells (all cells other than germ cells) could only be studied in a few special experimental circumstances, notably in *Drosophila* and *Delphinium*. The great majority of mutations escaped notice (and still do).

Many new techniques have allowed observation of genetic changes in much greater detail. In many instances, it is possible to look for changes in a single specific nucleotide among the 3 billion in a human haploid complement. The change may or may not have any phenotypic consequence. Changes in DNA can be detected in either germ cells or somatic cells. In the case of somatic cells, this makes it possible to distinguish between changes in the DNA that involve nucleotide sequences—true mutations—as compared with transmissible states (e.g., those associated with differentiation of tissues and that involve stable alterations in DNA function but not structure). Mutations can be detected in any part of the DNA, including regions of the genome thought not to have any function.

These new possibilities for looking at DNA permit us to define mutation more specifically than previously. A contemporary definition would be any change involving the nucleotide sequence of DNA, including organization of DNA into chromosomes. Mutations may consist of substituting one or more nucleotides for the original sequence. They may consist of duplications or, more commonly, deletions of one or more nucleotides. If the addition or deletion is large enough to be visible in a microscopic examination of chromosomes, it is often referred to, along with other types of visible changes, as a *chromosome rearrangement*. If it is not visible microscopically, it is called a *point mutation,* even though it may be a deletion of millions of nucleotides. Such deletions need not be confined to a single gene. Many involve a number of genes.

One particular type of mutation requires special note. This is loss or gain of an entire chromosome, sometimes called a *genomic mutation*. The mechanism for induction of genomic mutations involves abnormal distribution of chromosomes during cell division (nondisjunction) rather than alteration of DNA structure. The only known risk factor is maternal age. Despite this difference from other mutations, genomic mutations do involve a difference in the amount of DNA and therefore should be included in the definition of mutation given earlier.

A note of caution on use of the terms *mutation* and *mutant gene* is in order. Geneticists often use these terms to refer to a variant allele, irrespective of when the mutation occurred. Mutant is a convenient way to distinguish a variant allele from the normal or wild-type allele. In this article, we will be concerned only with the frequency of the events that produce mutant alleles, not the frequencies of those alleles in populations.

II. Germinal Mutation Rates

The rates at which mutations occur have been a matter of great interest, both for better understanding of an interesting biological event as well as for better assessment of the effect on human health of exposure to mutagens (i.e., agents that cause mutations). Mutation rates have traditionally been measured as the frequency of a new phenotype among the offspring of a parent who lacks the potential to produce that phenotype by genetic segregation. Otherwise expressed, the mutation rate is the frequency of newly mutant genes or gene complexes among all gametes (eggs or sperm). Expressing the results per gamete is equivalent to expressing it per generation, an alternate phraseology.

Virtually all observations on mutation rates involve phenotypic or genotypic end points that exclude many possible genetic changes. For example, a number of variant alleles of the X-linked HPRT (hypoxanthine phosphoribosyl transferase) gene have been studied. If activity of the enzyme coded by this gene is zero, a severe condition known as *Lesch-Nyhan syndrome* occurs in males. If the variant allele codes for an enzyme that has some activity, the only consequence is a rare form of gout. Thus a study of the mutation rate for Lesch-Nyhan syndrome only counts those mutations that lead to complete loss of enzyme activity. These might be single nucleotide substitutions or large deletions, either within the coding region of the gene or in noncoding regulatory regions. Many of the same kinds of molecular events would not lead to a recognizable phenotypic change. One's interest may be in the frequency of those mutations that lead to Lesch-Nyhan syndrome. That fraction of mutations is an unknown portion of the total at the HPRT locus. An additional example is provided by the locus that codes for the protein dystrophin. Some mutations produce severe Duchenne muscular dystrophy (DMD). Other mutations produce the milder Becker muscular dystrophy.

Similar considerations apply to other types of mutations. Even the systems that are based on direct analysis of DNA are not designed to pick up all mutations at a locus. This could be done only by determining the nucleotide sequence of a particular gene in vast numbers of persons *and* in both their parents, a project that is not feasible at present. Mutation rates are thus meaningful only with respect to a well-defined phenotypic end point.

A. Sentinel Phenotypes

The traditional study of mutation rates in humans was based on the frequency of a Mendelian trait, either autosomal dominant or X-linked recessive, in families in which there was no prior occurrence of the trait. Traits that are useful for this purpose have sometimes been called *sentinel phenotypes* because of their potential usefulness for monitoring exposure to mutagens. Examples are given in Table I.

Several characteristics are common to these disorders that are readily diagnosed. They are relatively easy to locate through medical, hospital, or public agency records. They are rare because of the low fertility of affected persons. The low frequency is important if a reasonable proportion of all cases are to be new mutations. In the case of autosomal dominant mutations, the condition must not always produce sterility; otherwise, it would not be possible to prove that the trait is inherited.

A typical mutation rate based on sentinel pheno-

TABLE I Mutation Rates Observed for Some Human Genes[a]

Trait	Mutants per million gametes
Autosomal dominant traits	
von Hippel-Lindau syndrome	0.2
Aniridia	2.6–5
Acrocephalosyndactyly	3–4
Retinoblastoma	5–12.3
Achondroplasia	6–13
Neurofibromatosis	44–100
Polycystic disease of the kidneys	65–120
X-linked recessive traits	
Hemophilia B	2–3
Hemophilia A	32–57
Duchenne muscular dystrophy	43–105

[a] Tabulated from Vogel, F., and Rathenberg, R. (1975). *Adv. Hum. Genet.* **5**, 223, which includes citations of the original publications.

types is one mutation per 100,000 gametes. The variation in this figure is large. The rate for Duchenne muscular dystrophy (DMD) is approximately 10 times higher: one new mutation per 10,000 X-chromosome-bearing gametes. This is presumably explained by the large size of the DMD gene, more than 2 million nucleotides as compared with a more typical size of a few thousand nucleotides. As noted earlier, the figure does not include mutations that cause the milder Becker muscular dystrophy or those that do not interfere with normal function of the gene. [*See* MUSCULAR DYSTROPHY, MOLECULAR GENETICS.]

Such mutation rates are subject to several biases. One bias is the assumption that similar phenotypes result from mutations of a single gene. Often it is not known whether this is true. Mutations at any of several genes may produce indistinguishable phenotypes. For some purposes (e.g., monitoring exposure to mutagens), this may not be important. Indeed, it may be advantageous. For other purposes, it would be important to know how many genes contribute to an observed mutation rate and in what proportion.

The argument has also been made that the phenotypes chosen for study of mutation rates are not representative of all genes. We choose them because they are the conditions that we know about, and we know about them because the genes are more mutable than other genes. Based on this argument, a more representative mutation rate would be lower, perhaps one per million gametes per gene locus.

It may also be that most genes are incapable of mutating to a dominantly expressed form. Many genes cause no phenotypic effect, even when completely deleted, as long as there is a functional gene on the homologous chromosome. Therefore, rates of dominant mutations may be biased in favor of a group of genes that are functionally similar but not representative of all genes.

B. Protein Variants

Most genes act by coding for a specific polypeptide chain. Most nucleotide substitutions in a coding region cause substitution of one amino acid for another in the polypeptide. This may or may not alter function of the resulting protein. Approximately one-third of such amino acid substitutions involve a change in the electrical charge of the protein, a variation often detected by its influence on the mobility of the protein in an electrical field (electrophoresis).

The first such electromorphic variant was found in sickle cell hemoglobin (Hb S), a variant that causes sickle cell anemia when a person is homozygous for the sickle cell allele. Heterozygotes, with one allele coding for Hb S and the other coding for the normal Hb A, have no ill effects but produce the two kinds of hemoglobin in each red blood cell. The mutation that produces Hb S is a substitution of the codon GTG, which codes for valine, in place of GAG, which codes for glutamic acid. Hb S has a more positive electrical charge, because glutamic acid is negative whereas valine is neutral. These relations are illustrated in Fig. 1.

The analysis of electromorphic variations pro-

vides an opportunity to detect mutations in specific proteins. As applied to hemoglobin, for example, any variation that is present in a child but in neither of the biological parents must be a new mutation. (It is, of course, necessary to verify the biological relationships by a battery of other genetic markers.) The mutation could be in either of the two types of subunits that are in hemoglobin, each type being coded at a different genetic locus. The mutation could involve virtually any codon in either gene, some of which cause a defect in function but many of which do not. Many mutations (e.g., deletions or other changes within or outside the coding regions that interfere with hemoglobin synthesis) will be missed, but one class of mutations can be identified efficiently and accurately.

A search for electromorphic mutations has been made in two large studies. In Japan, the offspring of atomic bomb survivors have been studied in an attempt to detect effects of the atomic radiation. Because the rate of electromorphic mutations per gene is low, a battery of proteins coded by 33 genes was studied. Several proteins were also included in which reduced activity could be detected. The study population consisted both of offspring of irradiated parents and of nonirradiated control parents. No differences were detected in the two groups. The pooled data yielded six mutations in 1,134,285 loci tested, a rate of 5.3 mutations per million gametes per locus. Similar results were obtained in Germany in a large survey of hemoglobin variants. In view of the fact that only about one-third of nucleotide substitutions cause a change in charge in the protein that is coded, the actual mutation rate for nucleotide substitutions is approximately 16 per million loci per gamete. For a "typical" polypeptide coded by 1,000 nucleotides, this is equivalent to one to two mutations per 100 million nucleotides per gamete.

C. Chromosomal Mutations

1. Rearrangements

Approximately one in each 200 births has a detectable chromosome abnormality, either numerical or structural. Table II summarizes the results of a number of studies of chromosomal abnormalities at various stages of development. Not all these are the result of new events (i.e., mutations). However, only those that are found in liveborn children have the potential for transmission, and only a small frac-

```
         DNA        5'...GTG CAC CTG ACT CCT GAG GAG AAG...3'
Hb A
         Polypeptide        Val-His-Leu-Thr-Pro-Glu-Glu-Lys...
                             ⊕                 ⊖   ⊖   ⊕


         DNA        5'...GTG CAC CTG ACT CCT GTG GAG AAG...3'
Hb S
         Polypeptide        Val-His-Leu-Thr-Pro-Val-Glu-Lys...
                             ⊕                     ⊖   ⊕
```

FIGURE 1 Comparison of the N-terminal region of hemoglobin A (Hb A) with that of sickle cell hemoglobin (Hb S). In Hb A, glutamic acid (Glu) has a negative charge, whereas Hb S has a neutral amino acid (Val, valine) at that position. The two forms of hemoglobin are identical in other respects. When placed in an electric field, Hb A moves faster toward the positive pole than does Hb S. Such a difference is called an electromorphic variation and reflects a change in the gene that codes for the globin β-chain of Hb A.

TABLE II Frequency of Chromosome Abnormalities per Thousand Recognized Conceptions at Various Stages of Human Development[a]

Karyotype	Spontaneous abortions	Stillbirths	Liveborn	Total
Autosomal trisomy	38.7	0.39	1.22	40.3
45, X	12.6	—	0.20	12.8
Sex chromosome trisomies	0.3	0.03+	1.57	1.9
Total sex chromosome aneuploidy	12.9	0.06	1.77	14.7
Triploidy	11.0	0.03	0.02	11.1
Tetraploidy	3.7	0.00	0.00	3.7
Structural anomalies	2.5	0.06	2.12	4.7
Total abnormal karyotypes	69.0	0.56	5.13	74.5
Abnormal karyotypes as percent of pregnancy outcome	46%	5.6%	0.59%	

[a] Modified from Sutton, H. E. (1988). "Introduction to Human Genetics," 4th ed., pp. 154–155. Harcourt Brace Jovanovich, San Diego, with permission. The data are from a number of sources cited in the original.

tion of those is consistent with fertility. Aside from balanced translocations and inversions, most rearrangements are lethal before birth. [*See* Chromosome Anomalies.]

Because of the strong selection against unbalanced chromosomal rearrangements during embryonic and fetal development, it is impossible to be sure of the true rate of formation of structural rearrangements. As shown in Table II, the rate in detectable conceptions is about five per thousand, of which half are aborted. The true rate must be higher, the additional unobserved complements involving unbalanced rearrangements.

2. Genomic Mutations

Genomic mutations account for the great majority of chromosomal abnormalities in recognized conceptions. There are several types. Trisomy involves the presence of an extra chromosome, with trisomy 21 (Down syndrome) being the most common among live births. Among spontaneous abortions, trisomy 16 is the most common, but trisomy 16 fetuses are always aborted.

The mechanism of production of trisomy is nondisjunction (failure of daughter chromosomes to separate into different cells during cell division). Nondisjunction can occur at either division of meiosis and in both males and females. In approximately 95% of cases of trisomy 21, nondisjunction occurs in the formation of ova, primarily in the first division of meiosis. The remaining 5% of cases occur in the first and second meiotic divisions in males.

One clue to the origin and fate of genomic muta-

tions is provided by examination of sperm by means of the hamster fertilization test. In this test, prepared hamster eggs are fertilized *in vitro* by human sperm. The chromosomes from the sperm can be examined at the first cell division for visible aberrations before the cells are eliminated by functional deficiencies. In two studies of sperm karyotypes carried out with the hamster fertilization test, the rate of genomic mutations in normal males varied from 1.6% to 5.1%, about equally divided between sperm with too many and sperm with too few chromosomes. The abnormal complements included some that are not ordinarily seen in spontaneous abortions, suggesting that the genetic complement may be so unbalanced as to lead abortion before a pregnancy is recognized. This is particularly true of sperm that lack a chromosome, because monosomy (presence of one rather than the normal two homologous chromosomes) is not observed in live births and rarely in spontaneous abortions.

The total rate of genomic mutations is about 2–5% in males, based on the above studies. In females, the rate is higher, some 20 times higher in the case of trisomy 21. If we extend these results to other chromosomes, the total rate of conceptions involving genomic mutations would be as high as 25%. Nondisjunction increases strongly with age in females but not in males. It is estimated that most oocytes of women older than 40 years of age are abnormal. Most, of course, do not survive early embryonic development. No environmental agents have been identified that clearly influence the rate of nondisjunction in human beings.

III. Somatic Mutation Rates

Although most studies of mutation have dealt with germinal rates, a few have addressed the issue of somatic rates. Somatic mutations cannot be transmitted to offspring and therefore do not pose a problem for future generations. However, the evidence is now overwhelming that most, perhaps all, cancer includes somatic mutation as a necessary step in its origin. Several other diseases have also been suggested to result in part from somatic mutations.

Somatic mutations have also been studied as indicators of germinal mutation risk. Presumably the same agents act as mutagens in somatic and in germinal cells. Therefore, if it were possible to measure the incidence of somatic mutations in a person or experimental organism, exposures to mutagens could be recognized in the exposed individuals rather than having to delay observation until the next generation. Many fewer individuals would be required for observation, because each somatic cell would be a potential test, with many millions of cells per person. A child is a test of only one cell from each parent. Germinal mutation rates must therefore be based on populations of millions of children.

A. Cancer

There are many surveys on the incidence of cancer. Few of these lend themselves to studies of mutation rates. An exception is found in retinoblastoma, a childhood tumor of the embryonic cells of the retina. It has been demonstrated that mutations of the two homologous *RB* loci on chromosome 13 must occur for development of the tumor. One of these mutations may be transmitted in families, in which case only one additional mutation is necessary in the embryonic cells. Because of the large number of embryonic cells in the retina of a child, even a low mutation rate will likely produce one or more mutations and consequently tumors. Analysis of the frequency of these mutations gives values of about five per million cells, a rate similar to many germinal mutation rates. In this case, the kinds of mutations "counted" are those that cause the locus not to be active, whether point mutations or deletions. [*See* RETINOBLASTOMA, MOLECULAR GENETICS.]

B. Blood Cells

Both red and white blood cells can be viewed as large populations of cells, each with its own chromosome complement and expressing its own genotype. (Mature red cells of mammals do not have nuclei and therefore have no genotype. They express the genotype of the immature precursor cells before the nucleus disintegrates.) Mutation in a cell should lead to a variant phenotype or karyotype. To be useful, a variant phenotype must be reliably detectable, even in the presence of a million-fold ratio of nonmutant to mutant cells, a goal that has been approached but perhaps has not been fully achieved.

1. Chromosomal Variations in Lymphocytes

Chromosome analysis commonly depends on short-term culture of lymphocytes obtained from a sample of blood. It is commonly observed that examination of a substantial number of cells from a person reveals some to have aberrant complements, either in number or because of chromosome rearrangements. The frequency and type depend in part on laboratory variables, and rare variant cells are usually ignored in the interpretation of the karyotype of a person. [*See* LYMPHOCYTES.]

Nevertheless, careful standardization of procedures allows the observation of increased chromosomal variations as a result of exposure to mutagens. If the exposure is recent, many different types of rearrangements are seen. Some (e.g., chromosomes with two centromeres or with large deletions and duplications) would not appear to be compatible with continued cell survival and replication. Indeed, such cells disappear from circulation rather rapidly. Other types of rearrangements are stable and can be seen in increased numbers long after the exposure. Examples would be balanced translocations or inversions, in which there is little or no loss of genetic information and no difficulty in chromosome distribution in mitosis.

An example of the use of karyotype analysis to detect mutational damage is in studies of the survivors of the atomic bomb explosions in Japan. Persons heavily irradiated in 1945 continue to have increased frequencies of aberrant karyotypes in circulating lymphocytes decades after the exposure.

2. HPRT in Lymphocytes

Lymphocytes can be cultured and hence can be tested much as we might test mutant colonies of bacteria by the ability of rare variants to grow under certain conditions of culture. The gene that is ordinarily tested in lymphocytes is the HPRT gene on the X chromosome. One advantage in using this gene is the fact that males have only one X chromosome; hence, a recessive mutant will be expressed. Because of X-chromosome inactivation in females, a recessive mutation that occurs in the active X chromosome will also be expressed. A second advantage is the ability to count mutant cells by their ability to grow on selective media. The enzyme coded by the HPRT locus not only functions in the recovery of normal purines that may be incorporated into DNA, it also causes the incorporation of such toxic analogues as 8-azaguanine (AG) or 6-thioguanine (TG). These synthetic analogues kill cells into which they are incorporated. Cells that lose HPRT activity because of mutation of the HPRT locus are not killed and can grow normally, producing clones (colonies) of mutant cells. Each such clone therefore represents one mutant cell among the original population of cells.

This test has been applied to a variety of situations. The frequency of AG- or TG-resistant lymphocytes in normal blood is about one per million. The frequency increases with exposure to various mutagenic agents. It is difficult to calculate an accurate mutation rate because of the complexity of the process by which stem cells of the bone marrow differentiate into lymphocytes. However, mutation rates of about one per million cells should lead to the accumulation of mutants that are observed.

3. Red Blood Cells

Several red cell variations have been tested as indicators of somatic mutation. The mutation from normal hemoglobin A to sickle cell hemoglobin can be detected on the basis of the antigenic difference. Red cells that are heterozygous for sickle cell hemoglobin occur at a frequency of one per 10 million red cells. This low rate is consistent with the single specific nucleotide substitution that must occur in the hemoglobin A gene to convert it to the sickle cell gene. [*See* SICKLE CELL HEMOGLOBIN.]

Another red cell system that has been used to detect mutations is based on the MN blood groups. In this system, the protein glycophorin A on the surface of red cells exists in two common forms, coded by the two alleles of the MN gene. The two forms differ by two nonadjacent amino acids, indicating at least two nucleotide differences between the alleles. In a heterozygous MN person, any mutation that destroys the antigenicity of the MN region of the molecule will leave the descendent red cells reactive only with the unchanged molecule coded by the locus on the homologous chromosome. In short, an MN person should have rare M and N

cells caused by mutation in the precursor stem cells. Such cells do occur with a frequency of approximately one per 100,000 in persons not exposed to known mutagens. As in the case of TG-resistant lymphocytes, they increase in persons exposed to radiation or chemotherapeutic agents, both potent mutagens.

Similar studies of the ABO blood groups give a somewhat higher frequency of variant cells. In this case, persons who are genetically AO, BO, or AB are tested for loss of A or B antigens. These antigens are produced from the type O precursor by either of two enzymes (A or B). Any mutation that causes substantial loss of enzyme activity will cause the cell to react immunologically as a type O cell. The frequency of O cells in non-O blood is approximately 1–10 per 10,000 cells. The frequency increases several-fold in persons undergoing chemotherapy.

The differing rates observed in the three red cell systems presumably reflect the different spectrum of mutations that are counted. In the sickle cell hemoglobin system, a specific nucleotide substitution must occur. In the MN system, deletions would be counted, as would any point mutation that affects the coding for the MN antigens, but most nucleotide changes would not. In the ABO system, any change that affects the enzymatic activity of the enzymes would be counted, including all deletions and many nucleotide substitutions, including those in regulatory regions outside the coding region. There may also be differences in survival of mutant stem cells in the bone marrow. Because type O blood is the most common, cells with those mutations would appear to have excellent chances of survival. There is no corresponding common mutation known in the MN system in which a glycophorin gene is inactive. When such cells arise in the stem cells, they may not compete well with nonmutant cells.

IV. Indirect Indicators of Mutational Risk

The concern for exposure of humans to environmental mutagens has led to development of a number of test systems designed to detect the presence of mutagenic agents. Such tests do not tell us whether mutations have in fact been induced, but they do provide information on exposure to mutagens, information that is available at the time of the exposure rather than a generation later, as would be the case if we depended on detection of germinal mutations.

The following tests are among those that can be carried out on exposed or potentially exposed persons. Additional tests are available for testing agents in the environment directly.

A. Sister-Chromatid Exchange

Sister chromatids are the two copies of a chromosome that will become the two daughter chromosomes when division is complete. They should have exactly the same information except for rare mutations. In meiosis, they may acquire new information by crossing-over with chromatids of the homologous chromosome, but in mitosis that does not occur or does so rarely. Because sister chromatids of somatic cells have the same genetic information, there is no way of distinguishing them by genetic tests. If breakage occurred followed by reciprocal exchange of distal segments, the product would be genetically identical with the original structure.

Such exchange can be detected, however, by labeling the two chromatids differentially. This can be done if cells are grown for two cycles in a medium that contains bromodeoxyuridine (BrdU). BrdU can substitute for thymidine in the synthesis of DNA. Because it is incorporated only in the newly synthesized strands, the sister chromatids will have different amounts of BrdU after two cycles of synthesis (Fig. 2). The bromine atom of BrdU absorbs light differently than does the methyl group of thy-

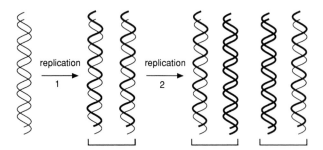

FIGURE 2 When DNA replicates, each of the two strands serves as a template for the synthesis of a new complementary strand. If BrdU is present during replication, it is incorporated into the new strand but not the old. After two rounds of replication, one of the chromatids in each chromosome will have one strand of DNA with BrdU and one strand without BrdU. Both DNA strands in the sister chromatid will have BrdU. When stained with certain fluorescent dyes, the chromatid with two BrdU strands fluoresces less brightly than the sister chromatid. In the drawing, sister chromatids are bracketed together.

mine. If chromosomes are stained with certain fluorescent dyes, chromatids with more BrdU fluoresce less brightly than do those with less BrdU (Fig. 3). If exchange of segments occurs between sister chromatids, that can be noted by the pattern of fluorescence.

It is widely observed that exposure to mutagenic chemicals increases the rate of occurrence of sister-chromatid exchange (SCE). Thus changes in the SCE rate should be useful to monitor exposure to mutagens. SCE itself is not a genetic change in the strict sense and cannot be equated with mutation. But it is an indicator that may be of value in assessing exposure.

B. Unscheduled DNA Synthesis

The normal chemical processes in a cell as well as exposure to many environmental agents result in the constant modification of DNA. Many kinds of changes are possible. Most would cause aberrant function if allowed to remain. Fortunately, there are a number of "DNA repair" systems that detect and correct the great majority of these errors. Many do so by removing one of the DNA strands, replacing it with a newly synthesized strand. Such DNA synthesis is called *unscheduled DNA synthesis,* because it occurs at a part of the cell cycle when normal DNA replication is not scheduled to occur. [See DNA SYNTHESIS.]

Although DNA repair is remarkably accurate, it is not infallible. Errors do occur, and these ultimately are classified as mutations. The number of errors is assumed to be proportional to the amount of DNA repair that occurs. Conversely, the amount of DNA repair can be used as a measure of the induction of mutations, although the individual mutations are not detected. [See REPAIR OF DAMAGED DNA.]

Because thymine occurs only in DNA, the synthesis of DNA can be observed by incorporation of radioactively labeled thymidine, both in normal replication and in unscheduled DNA repair. The nuclei of cultured human cells exposed to mutagens show an increase in thymidine incorporation and therefore in unscheduled DNA synthesis. The same should be true for persons tested during periods of exposure. The test would therefore give results that are in the same time period as the exposure rather than a generation later.

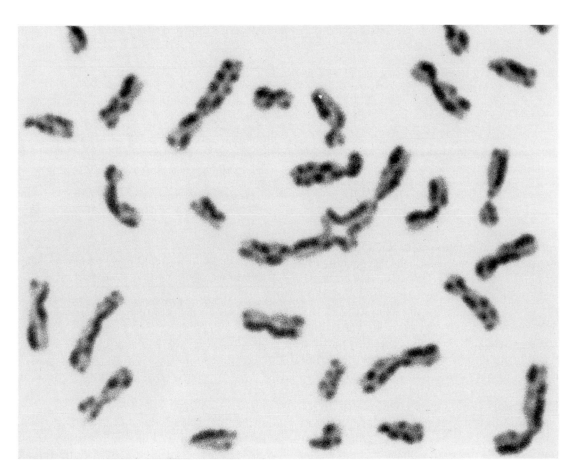

FIGURE 3 Chromosome spread showing sister-chromatid exchanges (SCE) in a patient with Bloom Syndrome. Lymphocytes were cultured through two cell cycles in the presence of BrdU, then spread on a slide and stained with a fluorescent dye. There are dozens of SCEs visible as reciprocal exchanges of light and dark chromatid segments. Bloom syndrome has a chromosome instability that leads to a very high rate of SCE. In a chromosome spread from a person without Bloom syndrome, the number of SCEs is usually <10. [Photograph supplied by James L. German, III, and has been published in Sutton, H. E., and Wagner, R. P. (1985), "Genetics, A Human Concern." Macmillan Publishing Co., New York. Reproduced by permission.]

C. Tests of Body Fluids

If body fluids (e.g., blood plasma) contain mutagens, a reasonable assumption is that some of the mutagens enter cell nuclei and cause mutations. This need not be true, but prudence requires that the potential for risk be accepted until it is proved not to exist.

We can test for mutagens in blood and urine using a variety of microbial tests. The most widely used is the Ames test, which uses special strains of *Salmo-nella* that have been developed to be sensitive to chemical mutagens. The test can be especially useful in monitoring exposure in industrial workers, because it is fast, it is noninvasive when based on urine samples, and it is inexpensive.

V. Factors that Influence Mutation Rates

In 1927, H. J. Muller reported that irradiation of *Drosophila* with X rays increases the mutation rate, an observation supported by the 1928 publication of L. J. Stadler on maize. In 1944, C. Auerback reported that certain chemical substances also are mutagenic. In the years since, many chemicals have been identified as mutagens. There is major concern for the effects of both radiation and chemicals on the genetic health of human populations.

At the DNA level, different mutagens can cause different kinds of changes. However, at the level of genes, mutagens are very nonspecific (i.e., any mutagen is capable of acting on any gene). The basic structure of all genes is similar, although the size of genes varies greatly. Whether a gene is more or less mutable when exposed to a particular mutagen is independent of the function of that gene in the organism.

A. Radiation

Radiation can be classified as electromagnetic (EM) or particulate, although EM radiation is also particulate in a sense. Low-energy EM radiation (e.g., microwaves and visible light) is nonmutagenic. High-energy ionizing radiation (e.g., X rays, gamma rays, and cosmic rays) is capable of disrupting chemical bonds and is mutagenic. Such radiation is also penetrating and readily reaches the gonads. Ultraviolet light (UV) is of too low an energy to cause ionization, but it causes molecules to be chemically reactive, which may also result in mutation. Because UV is absorbed so efficiently by water (and therefore by tissue), it can only penetrate the skin surface and is not of concern with respect to germinal mutations. It does cause mutations in cells near the skin surface.

All living things are exposed continuously to ionizing radiation, a fact that has been true throughout biological evolution. We are exposed to radiation from naturally occurring minerals and rocks and from cosmic radiation. In addition, there are traces of naturally occurring radioisotopes in our bodies. In particular, potassium-40 occurs in all natural sources of potassium, and we cannot exist without potassium. In the past century, we have added radiation from various human sources (e.g., medical X rays and nuclear weapons) to the total radiation exposure. Table III summarizes the principal sources of exposure to ionizing radiation of inhabitants of the United States.

Despite the early demonstration of the mutagenicity of X rays and of the long period of study of radiation, it is still difficult to pinpoint the magnitude of the genetic danger from radiation. It is possible, however, to estimate the upper and lower limits of damage, based primarily on studies of the atomic bomb survivors in Japan and of studies of radiation effects in mice. Table IV presents a recent summary of the genetic effects of one rem of radiation per human generation. This table is concerned only with germinal mutations and does not include somatic effects.

The studies in Japan and on other irradiated groups, primarily patients given high levels of radia-

TABLE III Average per Capita Annual Effective Dose Equivalent of Ionizing Radiation to U.S. Population[a]

Source	Dose equivalent (mSv)[b]
Natural	
Radon[c]	24
Cosmic rays	0.27
Terrestrial (rocks, minerals)	0.28
Internal (potassium-40, etc.)	0.39
Artificial	
Medical: X-ray diagnosis	0.39
Nuclear medicine	0.14
Consumer products	0.10
Occupational	0.009
Nuclear fuel cycle	<0.01
Fallout	<0.01
Miscellaneous	<0.01

[a] Modified from Committee on the Biological Effects of Ionizing Radiation. (1990). "Health Effects of Exposure to Low Levels of Ionizing Radiation (BEIR V)." p. 18. National Academy Press, Washington, D.C.
[b] To soft tissue. One millisievert = 100 millirems.
[c] Dose equivalent to bronchi from radon daughter products. The effective dose equivalent, which corrects for the localized nature of the exposure, is 2.0 mSv.

tion for therapeutic purposes, confirm the somatic effects produced by ionizing radiation. Cancer is a consistent effect of increased radiation. Irradiation during the embryonic and fetal periods may interfere with normal development, particularly of the nervous system.

There has been much debate on the shape of the dose–response curve for cancer and for mutations. In particular, is there a threshold of exposure below which there is no increased risk of mutations or cancer, or is the increased risk linear (or approximately linear) down to the lowest exposure? Theoretical considerations do not provide the answer, although they give no support to the idea of a threshold. Experimentally, it is virtually impossible to carry out experiments at near-zero exposures because of the high background of spontaneous mutations relative to possible rare induced mutations. Most scientists consider it prudent to assume a linear response at low exposures. In other words, even small exposures have the potential to cause a correspondingly small damage. The risk to the individual may be small indeed, and the risk may be far outweighed by the benefits. When the risk is summed up for a large population, the risk may be more significant. It still must be balanced by the benefits of the exposure.

TABLE IV Estimated Genetic Effects of 0.01 mSv per Generation of Chronic Irradiation[a]

Type of disorder	Current incidence per million liveborn offspring	Additional cases per million liveborn Offspring/0.01 mSv/generation	
		First generation	Equilibrium
Autosomal dominant			
Severe	2,500	5–20	25
Mild	7,500	1–15	75
X-linked	400	<1	<5
Autosomal recessive	2,500	<1	Very slow increase
Chromosomal			
Unbalanced translocations	600	<5	Very little increase
Trisomies	3,800	<1	<1
Congenital abnormalities	20,000–30,000	10	10–100
Other disorders of complex etiology			
Heart disease	600,000		
Cancer	300,000	Not estimated	Not estimated
Selected others	300,000		

[a] From Committee on Biological Effects of Ionizing Radiation. (1990). "Health Effects of Exposure to Low Levels of Ionizing Radiation (BEIR V)." p. 70. National Academy Press, Washington, D.C. A number of assumptions and approximations have been made in preparing this table. The original should be consulted for greater detail.

B. Chemical Mutagens

Much less is known about chemical mutagenesis than about radiation mutagenesis. In part this is due to the great diversity of chemicals to which we are exposed, not only manmade chemicals but also the many exotic chemicals that are manufactured by green plants that we consume as food. In part our ignorance is also due to lack of knowledge of the extent to which these chemicals and their metabolic products reach the DNA in various tissues and react with it there. And if damage does occur, how much of it is repaired without any permanent change in the DNA?

It is likely that there will never be much information obtained by direct observation of germinal mutations in humans caused by exposure to chemical mutagens. We must instead rely on other indicators. Perhaps the most significant is the potential of a chemical substance to act as a carcinogen. It has been amply demonstrated that most carcinogens are mutagens, and somatic mutation is an important part of carcinogenic transformation. Therefore, any carcinogen is presumed to be a mutagenic risk until proven otherwise. To be sure, germ cells differ in important ways from somatic cells, but there are likely to be more similarities than differences with respect to risk of mutation. [*See* Carcinogenic Chemicals.]

C. Biological Variations in Mutational Risk

To the intrinsic mutability of all genes and the additional risk from external agents, there must be added the variation in risk caused by biological factors that characterize different individuals in a population. That risk is clearly not evenly distributed, although it is uncertain just how much variation exists. The following are some examples of biological variation in risk.

1. Deficiencies in DNA Repair

The ability to repair DNA defects is the result of the presence of a number of enzymes, which, like all other proteins, are encoded by genes. Deficiencies in such enzymes caused by genetic variation are therefore to be expected.

The best documented case in humans is the deficiency of the repair system found in the autosomal recessive condition known as *xeroderma pigmentosum* (XP). UV light (e.g., that produced by the sun) causes changes in the DNA of cells near the skin surface. Normal persons repair these changes, with only occasional errors that we recognize as mutations. Persons with XP cannot repair the damage from UV light, although they can repair many other types of damage. This defect is demonstrated by the inability of cell cultures from XP patients to carry out unscheduled DNA synthesis after irradiation by ultraviolet light. Inevitably XP patients develop skin changes in areas that are exposed to sun, and death eventually occurs from skin cancers in these areas.

There are other inherited diseases that are thought to involve deficiencies in repair systems. These include Bloom syndrome, ataxia telangiectasia, and Fanconi anemia. The variation in repair efficiency among normal persons has not been studied but may be substantial.

2. Metabolism of Mutagens

Each person inherits an individual metabolism. For the most part, these variations may be of little consequence. However, many chemical mutagens are inactivated by metabolic processes, and other chemicals become mutagenic as a result of enzyme action in the body. It must be assumed, therefore, that some variation in risk occurs because of the variations in metabolic pathways.

3. Age

Studies of irradiated populations demonstrate that fetuses and children who are irradiated are at greater risk of developing cancer than are adults. There are no corresponding studies of chemical mutagens, and there are no data on sensitivity of germ cells rather than somatic cells. It is plausible, however, that young children may have greater risk of mutation than do adults.

Nondisjunction of chromosomes in meiosis is strongly influenced by maternal age but apparently not by paternal age. This is presumably related to the different ages at which meiosis is initiated in males and females. In males, it is continuous in the spermatogonial cells throughout the adult life. In females, it begins in all oogonial cells during the late embryonic to early fetal period.

Point mutations show no such age effect in females, but some genes do show an age effect in males. In brief, the older the male, the more likely he is to transmit a mutant sperm. Other genes do not show this effect. The difference between these two types of genes has not been identified.

With respect to germinal mutations, an additional consideration is the likelihood that a person will have offspring after exposure to some environmental agent. This aspect is concerned with the age of the person, not the physiological damage from the exposure. For example, a 20-year-old person is much more likely to transmit genes at some time in the future than is a 60 year old. These concerns have been incorporated into the concept of a genetically significant dose (i.e., the expected mutational response for a given exposure multiplied by the likelihood that the exposed person will have children subsequent to the exposure).

4. Sex

Males and females do not necessarily have the same likelihood of mutation, either spontaneous mutation or induced. Reference has already been made to the differences in rate of nondisjunction and in rate of certain point mutations, as related to age and sex. There is little information on sex differences in response to mutagenic exposures. Radiation studies in mice do indicate a much lower rate of mutations for female mice than for male mice, related to the length of time for various stages of oogenesis and spermatogenesis. It is likely that differences also exist for other mutagens, but prediction of the differences is difficult because of the unknown interactions of metabolism and repair systems with cell stage.

Bibliography

Committee on the Biological Effects of Ionizing Radiations. (1990). "Health Effects of Exposure to Low Levels of Ionizing Radiation (BEIR V)." National Academy Press, Washington, D.C.

Committee on Chemical Environmental Mutagens. (1982). "Identifying and Estimating the Genetic Impact of Chemical Mutagens." National Academy Press, Washington, D.C.

Crow, J. F., and Denniston, C. (1985). Mutation in human populations. *Adv. Hum. Genet.* **14,** 59–123.

Sutton, H. E. (1988). "Introduction to Human Genetics," 4th ed. Harcourt Brace Jovanovich, San Diego.

Myasthenia Gravis

EDITH G. McGEER, *University of British Columbia*

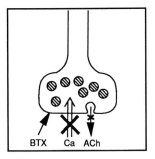

I. General Description
II. Pathology
III. Theories as to the Cause
IV. Animal Models
V. Treatment

Glossary

Anticholinesterase Drug that inhibits acetylcholinesterase and thus prevents the breakdown (metabolism) of acetylcholine

Autoantibodies Antibodies produced to a body's own tissue

***α*-Bungarotoxin** Neurotoxin in the venom of the krait, a snake found in India

Cholinergic Neurons that use acetylcholine as a neurotransmitter

Glycoprotein Compound containing a protein and a carbohydrate (sugar) group

Hyperplasia Abnormal increase in the number of cells in a tissue without abnormality in the cells themselves

Ligand Compound active at a receptor

Plasmapheresis Removal of blood, separation of plasma or plasma components, and reinjection of red blood cells in a suitable suspending medium

Synapse Complex consisting of the nerve ending, synaptic cleft, and postsynaptic receptors

Synaptic cleft Gap between a nerve ending and the tissue it innervates. The neuronal message is carried across this gap by a chemical transmitter released by the nerve ending

Thymus Gland lying just below the neck that plays a major role in the production of antibodies

Transmitter One of a group of specific chemicals used by the nervous system to carry messages from one neuron to another, to muscle, or to other innervated tissue

MYASTHENIA GRAVIS is a relatively rare, chronic disorder in which voluntary muscle activity is weak and easily tired. It appears to be an autoimmune disease (i.e., one in which the body makes antibodies that attack its own tissue). In this case, the tissue attacked is generally the nicotinic acetylcholine receptor (AChR), which receives the nerve input to muscles. If these receptors are blocked or missing, the nerve impulse cannot force the muscle to respond. Antibodies to other components of the neuromuscular system may also be involved in some cases. Treatment strategies aimed at suppressing this disadvantageous immune response are now generally so successful that the victims can lead full, normal lives.

I. General Description

Myasthenia gravis is a chronic neuromuscular disorder characterized by weakness and easy fatigability of voluntary muscles. The pattern of weakness often involves the extraocular muscles but may also involve other facial, bulbar, or limb muscles. The prevalence rate is estimated to be about three per 100,000 population. The age of onset is variable but is often in the third decade in women and somewhat later in men. Many cases show periods of spontaneous remission as well as periods of exacerbation. The evidence is strong that the disease is commonly an autoimmune disorder (see Section II,B). Epidemiological studies have not yielded many instances of the recurrence of myasthenia gravis within families but have provided evidence that relatives often suffer from some other autoimmune condition. This fact and some molecular genetic linkage studies have indicated that an inherited autoimmune susceptibility probably plays a role in many cases. [*See* AUTOIMMUNE DISEASE.]

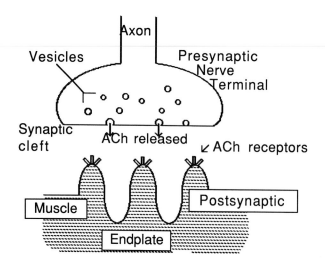

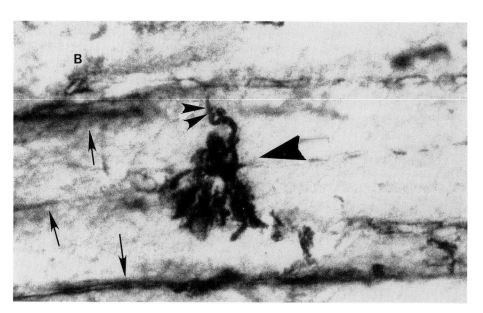

II. Pathology

A. Histological Pathology

The most common findings are isolated patches of necrosis of muscle fibers with infiltration of lymphocytes. Electron microscopic studies have indicated abnormalities of the nerve terminals and synapses. None of these changes, however, correlate clearly with the severity or duration of the disease. There is sometimes a tumor of the thymus, and in more than 50% of the patients, there is hyperplasia of that gland. In many, however, it appears microscopically normal. [*See* THYMUS.]

FIGURE 1 Neuromuscular junction. **A:** Diagram of synapse on muscle. Acetylcholine (ACh) is released by nerve stimulation from vesicles of the nerve ending into the synaptic cleft to act on ACh receptors on the postsynaptic side. ACh is destroyed by acetylcholinesterase (not shown), which occurs in both presynaptic and postsynaptic tissue. **B:** Photomicrograph of guinea pig muscle stained for the specific synthetic enzyme that makes ACh and is unique to cholinergic structures. Axon is stained (*double arrowhead*) as well as the multipronged neuromuscular junction (*single arrowhead*). The black lines (*arrows*) are muscle striations. (Courtesy of H. Kimura, Kinsmen Laboratory, U.B.C.)

B. Chemical Pathology

The similarity of the myasthenic symptoms to those seen in curare poisoning first suggested that the problem was in the neuromuscular junction (Fig. 1). The transmitter of motoneurons is acetylcholine, an identification made by Loewi shortly after he first offered proof of the chemical, rather than electrical, nature of synaptic transmission. Thus, the remarkable improvement seen in many myasthenic patients on treatment with anticholinesterases also indicated that the problem probably lay in the neuromuscular junction. It was not known, however, whether the problem was in the presynaptic cholinergic neuron or in the postsynaptic receptor mechanism. Convincing evidence that the fault involved the postsynaptic receptor was provided by measurements, using α-[^{125}I]bungarotoxin (α-BUTX), of the receptor density in muscle biopsies from myasthenic patients and normal controls. Such measurements are possible because α-BUTX, like many other snake neurotoxins, binds tightly to the neuromuscular junction and kills by blocking cholinergic transmission at that site (Fig. 2A). The myasthenic muscles showed a 70–89% reduction in the number of nicotinic AChRs per neuromuscular junction as compared with muscles from normal controls.

III. Theories as to the Cause

The problem as to what caused the loss in AChRs was clarified by the purification of the AChR from electric eels. Injection of the purified receptor glycoprotein into rabbits in an attempt to raise antibodies for further scientific exploration of the AChR led to the development in the animals of marked muscular weakness closely resembling that seen in the human disease. This serendipitous discovery led to a search for an autoimmune mechanism directed specifically against the cholinergic receptors in the human disease. In the vast majority (80–90%) of cases, circulating antibodies to the AChR itself can be demonstrated, and it is believed that the binding of these antibodies may both block the receptors and lead to accelerated degradation (Fig. 2B). Identification of myasthenia gravis as an autoimmune disease led to treatment by immunosuppression (Section V). However, there are some myasthenic patients in which circulating anti-AChR antibodies cannot be found, and even in those cases where

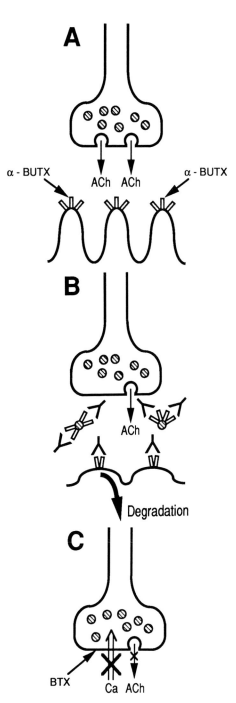

FIGURE 2 Diagrams of neuromuscular junction showing presumed sites of action of various toxins and antibodies. **A:** α-Bungarotoxin (α-BUTX) binds to acetylcholine (ACh) receptors on muscles, thereby blocking interaction of ACh with receptors. **B:** In myasthenia gravis, antibodies bind to ACh receptors on muscles to block action of ACh. Moreover, such binding of antibodies promotes degradation of receptors and membrane lysis, with extrusion of fragments into a widened synaptic cleft, resulting in a decreased number of receptors. **C:** Botulinum toxin (BTX) binds to the nerve ending membrane and blocks calcium-stimulated release of acetylcholine.

they are found, the correlation is poor between the circulating levels and the severity of the disease. This has led to the hypothesis that myasthenia gravis is a heterogeneous disorder in which autoantibodies to other components of the neuromuscular junction may be involved in addition to or instead of the anti-AChR antibodies. A number of such antibodies have been reported, including some to various cycloskeletal proteins found in muscle or to the transmitter, acetylcholine, itself. The exact roles of the various antibodies in the pathogenesis remain to be determined. [*See* AUTOANTIBODIES.]

Mention should be made of the Lambert-Eaton syndrome in which myasthenic symptoms are often seen. This is a rare disorder that is associated, in at least 70% of the cases, with bronchial or oat-cell carcinoma. It has been suggested that such tumors may elicit the production of antibodies that act, like botulinum toxin (Fig. 2C), to inhibit acetylcholine release. Repetitive nerve stimulation has a therapeutic effect on both the Lambert-Eaton syndrome and botulinum poisoning by increasing the quantal release of acetylcholine.

IV. Animal Models

Injection of purified AChR glycoprotein, or some portions thereof, into various animal species produces symptoms highly similar to those seen in myasthenia gravis. This experimental allergic myasthenia gravis (EAMS) is widely used in studies aimed at a further understanding of the pathologic mechanisms or the development of better therapeutic strategies.

V. Treatment

The treatment of myasthenia gravis has progressed to a point where most patients are able to lead a full, productive life. Most treatment strategies presently employed are aimed at suppression of the autoimmune response.

Treatment of myasthenia gravis with anticholinesterases gives only symptomatic relief and is short-acting. It is moreover difficult to titrate the dose for individual patients; the desirable dose changes with stress, exacerbations, or other phenomena, and the possibility of adverse reactions is high. For these reasons, this therapeutic approach is now rarely used except in some mild cases.

Thymectomies were done more than 50 years ago in a few cases of myasthenia gravis in which there was a tumor of the thymus. The beneficial effects reported, and the increasing evidence that this was an autoimmune disease in which the role of thymus in antibody production might be critical, led to increased numbers of such operations. The operative risk being essentially nil, thymectomy is now generally recommended in adult cases with progressive or generalized myasthenia. The advantage lies in the possibility of permanent remission (in 35–45% of cases) or substantial improvement (in another 35–45%). However, such improvement is often slow in developing and may take from just a few months to more than 10 years. For this reason, many patients continue to require immunosuppressive medication for some time after thymectomy.

The most generally used immunosuppressive drugs are the corticosteroids, which generally give rapid improvement. Other immunosuppressive agents used include azathioprine, cyclophosphamide, methotrexate, and cyclosporin. Plasma exchange or plasmapheresis is sometimes used alone or with an immunosuppressive drug to cause a rapid decrease in the level of circulating antibodies.

More selective ways of modulating the immune system are under study in animals with EAMS. One promising approach is to use anti-idiotope antibodies, produced against anti-AChR antibodies, to suppress the immune response to purified AChR protein. When or whether such approaches will prove practical in humans remains to be determined.

Bibliography

Drachman, D. B., ed. (1987). Myasthenia gravis: Biology and treatment. *Ann. N.Y. Acad. Sci.* **505,** 1–914.

Drachman, D. B. (1978). Myasthenia gravis. *N. Engl. J. Med.* **298,** 136–142, 186–193.

Mycotoxins

A. WALLACE HAYES, *Duke University Medical Center*

I. Introduction
II. Human Diseases

Glossary

Agranulocytosis Acute condition characterized by pronounced leukopenia with a reduction in the number of polymorphonuclear leukocytes (white blood cells)

Ataxia Incoordination; an inability to coordinate the muscles in the execution of voluntary movement

Atrophy Wasting of tissues, organs, or the entire body, as from death and reabsorption of cells, diminished cellular growth, pressure, ischemia, malnutrition, decreased function, or hormonal changes

Centromere Nonstaining primary constriction of a chromosome that is the point of attachment of the spindle fiber and is concerned with chromosome movement during cell division

Chromosome One of the bodies in the cell nucleus that is the bearer of genes

Dyspena Shortness of breath, a subjective difficulty or distress in breathing, usually associated with serious disease of the heart or lungs, and occurring in health during intense physical exertion or at high altitude

Emesis Vomiting

Leukopenia Any situation in which the total number of leukocytes in the circulating blood is less than normal

Mendelian disorder Disorder based on heredity as related to or in accord with Mendel's laws

Mosaic Inlaid; resembling inlaid work; the juxtaposition in an organism or genetically different tissues resulting from somatic mutation (gene mosaicism) and anomaly of chromosome division resulting in two or more types of cells containing different number of chromosomes

Necrotic Pertaining to or affected by necrosis. Necrosis is the pathologic death of one or more cells, or of a portion of tissue or organ

Perinatal period Occurring during, or pertaining to, the periods before, during, or after the time of birth (for a few days, 7 days after delivery in humans)

Polysomes Polyribosomes, conceptually, two or more ribosomes connected by a molecule of messenger RNA. These subcellular organelles are active in protein synthesis

Radiomimetic damage Tissue or cellular damage by a chemical emulating the action of radiation (e.g., nitrogen mustards, which affect cells as high energy radiation does)

Ring chromosome Chromosome with ends joined to form a circular structure; the normal form of the chromosome in certain bacteria

Sickle cell anemia Syndrome characterized by the presence of crescent- or sickle-shaped erythrocytes (red blood cells) in peripheral blood

Single cell protein Protein produced in the form of single cellular organisms (i.e., yeast or algae) from generally nonusable waste material

Thalassemia Any of a group of inherited disorders of hemoglobin metabolism in which there is a decrease in net synthesis of a particular globin chain without change in the structure of that chain; several genetic types exist, and the corresponding clinical picture may vary from barely discernable hematological abnormality to severe and fatal anemia

MYCOTOXIN is a generic term used to describe a metabolite produced by a mold(s) in foodstuffs or feeds that can cause illness or death on ingestion, skin contact, or inhalation. The term is derived from the Greek words *mykes*, meaning fungus, and *toxicum*, meaning poison or toxin. Several compounds currently classified as mycotoxins were initially studied as potential antibiotics in the 1930s and 1940s, only to be discarded as being too toxic for higher life forms to be of value in treating bacterial diseases in human populations. The definition of a mycotoxin, nonetheless, encompasses some antibiotics; the difference is one of degree rather than kind. At that time, although the association of ergotism (St. Anthony's fire) with the producing fungus *Claviceps purpurea* had been known for centuries, the potential health problems that these compounds might pose generally were unrecognized outside the veterinary and agricultural communities.

I. Introduction

Although toxic to higher animals and sometimes to plants and microorganisms, mycotoxins are not infectious. The diseases caused by these natural chemicals are called *mycotoxicoses*. Mycotoxicoses differ from mycoses or fungal infections such as ringworm because mold growth in the host is not necessary. With the knowledge stimulated from antibiotic screening programs in the 1940s and 1950s, it is surprising that the potential public health hazard of these mold metabolites was not recognized sooner. In fact, until as late as 1962, mycotoxicoses were described as "the neglected disease."

Before the discovery of the aflatoxins, the role these toxins played in diseases of livestock and other animals was uncertain. But with the discovery of the aflatoxins as the cause of acute liver disease among poultry in the United Kingdom and the subsequent demonstration of their carcinogenicity in laboratory animals, the next question asked was, "Are mycotoxins involved in human health?"

In modern societies, obviously moldy food, or food that has a bad smell or taste, generally is rejected. However, mycotoxins may be present in foods or beverages that do not present such warning signs. As with animals, the element of hunger can be expected to favor ingestion of contaminated food by humans where no alternative is available. Therefore, in developing countries, food shortages, along with a general lack of knowledge as to how food should be cultivated, harvested, selected for use, properly stored, or held after cooking, can augment the intake of undesirable foods. Times of war with resulting food shortages and improper food handling also can lead to conditions optimal for myco-

toxin production as exemplified during World War II in the Orenbury Province of Russia. A human disease, alimentary toxic aleukia (ATA), was caused by consumption of overwintered moldy grain. The mortality rate was as high as 60% in some instances, with 10% of the population being affected. The use of primitive fermentation procedures also favors formation of microbial toxins. Typically, mycotoxins are produced at the time of maximum fungal growth; however, the apparent absence of fungal growth bares no relation to the presence or absence of the toxin. The toxin may remain in the product for years, long after the mold has died. In one instance, corn stored for 12 years and not obviously moldy still contained a mycotoxin, probably formed shortly after harvest. [*See* FOOD MICROBIOLOGY AND HYGIENE.]

When the mycotoxin problem first occurred is difficult to estimate; however, there are reported human mycotoxicoses dating back at least to 1100 AD. It is obvious, however, that a risk assessment of human susceptibility to mycotoxins is difficult if not impossible because we cannot purposely experiment on humans. Although animal data can be extrapolated to humans, it must be appreciated that species variation often makes such data of limited value. Perhaps the most useful estimate of human's susceptibility to a mycotoxin is to be gained by determining the contamination level of human foods by such compounds and by observing the type and severity of diseases associated with the consumption of such foodstuffs.

Although mycotoxicoses in domestic animals have been reported for several decades, not one case of illness or death in swine attributable to aflatoxicoses had been reported in the United States cornbelt as late as 1980. It was subsequently shown, however, that these mycotoxins were involved in episodes of toxic hepatitis among domestic animals, particularly swine, in the southeastern United States. This observation was extended by reports that gave brief but incomplete descriptions of the disease in turkeys, swine, cattle, chickens, and ducklings. However, until 1978, the causal chemical had not been reported in a tissue or fluid from a single case of aflatoxicosis in a domestic animal. Necropsies of young pigs from a herd in which 30 of 250 died indicated aflatoxicoses. Histopathologic findings were characteristic of experimentally induced aflatoxicosis, and aflatoxin B_1, the responsible mycotoxin, was found in the serum and liver. The mycotoxin also was in the corn and in the feed made from the homegrown corn, which was fed to these animals.

In addition to the aflatoxins, other mycotoxins have been detected as contaminants of foods and animal feeds. It is also possible to demonstrate trace amounts of various mycotoxins and their metabolites in edible tissues, milk, and eggs of farm animals fed rations containing small amounts of mycotoxins, particularly aflatoxin. The exposure to humans by an indirect route, as compared with direct dietary intake of mycotoxins, does occur. Data relating to the presence of several mycotoxins (aflatoxin, ochratoxin A, T-2 toxin, and zearlanone) in milk and tissues of cattle fed naturally occurring contaminated rations, as well as mycotoxins in tissues after oral dosing with whole cultures of various fungi, have been reported. Farm animals can convert a mycotoxin into a toxic metabolite by as much as 10–15%. An alternate route leading to human consumption is from contaminated meat and meat products. In addition, some of the mycotoxin may pass into the animal waste, which then could be used for single-cell protein production, thus allowing the mycotoxin to reenter the human foodweb.

The risk to public health associated with continual, low level exposure to dietary mycotoxins has not been satisfactorily assessed. However, most countries now control the use of contaminated peanut meal and other contaminant grains for the manufacture of compounded feeds, particularly those destined for dairy cattle. This practice is important because, despite its almost invariable contamination with one of the mycotoxins (aflatoxin B1), peanut meal is a valuable source of protein and approximately 1% of the ingested mycotoxin appears in milk as a toxic metabolite.

Although a variety of mycotoxins occurs naturally in agricultural materials (Table I), most mycotoxins are not regulated by governments. Control of mycotoxins in food is a complex and difficult task. Information regarding toxicity, mutagenicity, carcinogenicity, teratogenicity, and the extent of contamination and stability of mycotoxins in foods is lacking for most mycotoxins. Such information is necessary to establish regulatory guidelines, tolerances, and seizure policies. In the absence of tolerances, the U.S. Food and Drug Administration (FDA) has set what it considers to be practical lim-

TABLE I Natural Occurrence of Mycotoxins in Agricultural Products

Mycotoxin(s)	Producing fungi	Occurrence
Aflatoxin	*Aspergillus flavus, A. paraciticus*	Corn, peanuts, cotton seed, rye, barley, etc.
Citrinin	*Penicillium citrinum, P. veridicatum*	Wheat, barley, peanuts
Ochratoxin A	*A. ochraceus, P. veridicatum, P. cyclopium*	Corn, barley, wheat, oats, rice
Sterigmatocystin	*A. versicolor, A. flavus, A. rubber, P. luteum*	Wheat, rice, peanuts
Zearalenone	*Fusarium roseum, F. moniliforme, F. nivale, F. oxysporum*	Corn, sorghum, wheat
Trichothecenes	*F. roseum, F. tricinctum, F. nivale*	Corn, barley
Patulin	*A. clavatus, P. patuluns*	Silage, apples
Penicillic acid	*A. clavatus, P. puberulum*	Corn, beans
Alternariol, alternariol monomethyl ether	*Alternaria tenuis, A. dauci*	Weathered grain, sorghum, pecan pickouts
Tenuazonic acid	*Alternaria tenuis, A. tamarii, Sphaeropsidales sp, Pyricularia oryzae, Phoma sorghina*	Diseased rice, plants
Ergot alkaloids (ergotamine, etc.)	*Claviceps* spp, *Aspergillus* spp, *Penicillium* spp	Ergots, ergot-infected pasture grass
Sporidesmin	*Pithomyces chartarum*	0.1% in spores on dead pasture grass
PR toxin	*Penicillium roqueforti*	Silage
Kojic acid	*A. flavus, A. oryzae*	Moldy corn

its for the several mycotoxins in foods and feeds, based primarily on the limits of detection and measurement of analytical methods and, to some extent, on the ability of agronomic and technological practices to prevent contamination.

II. Human Diseases

The involvement of mycotoxins as etiologic agents of human diseases is difficult to determine. Certainly ergotism and ATA can be attributed to a fungal toxin(s). Evidence also suggests that acute cardiac beriberi, common throughout Asia, may be linked to the so-called yellow rice toxins. Examples of human diseases for which evidence suggests involvement of a mycotoxin are summarized in Table II.

The toxicity of mycotoxins potentially harmful to humans also has been demonstrated in human tissue culture. The mycotoxins from *Penicillium islandicum* inhibited growth of Chang's liver and HeLa cells and, at sufficient concentrations, led to morphological changes and cell death. Similarly, the endotoxin extracted from the mycelium of *Aspergillus fumigatus* affected respiration and produced morphological changes in kidney cells grown in culture. Mycotoxins such as aflatoxin B1 and rubratoxin B inhibited a number of human and animal cell culture systems.

In general, mycotoxins are secondary metabolites that perform minor or no obvious function in the metabolic scheme of the organism and are products of reactions that branch off at a limited number of biosynthetic pathways such as those involving acetate, pyruvate, melonate, mevalonate, shikimate, and amino acids. From the standpoint of human and animal health, molds belonging to the genera *Aspergillus*, *Fusarium*, and *Penicillium* have received the most attention, owing to their frequent occurrence in food and feed commodities. The storage fungi *Aspergillus* sp and *Penicillium* sp typically do not invade intact grain before harvest, whereas *Fusarium* sp and *Alternaria* sp are predominantly field (preharvest) fungi. Unfavorable conditions (e.g., drought and damage of seeds by insects or during mechanical harvesting) can enhance mycotoxin production during both growth and storage. Toxin production can take place over a wide range of moisture (10–33%), relative humidity (>70%), and temperature (4–35°C), depending on the fungal organism involved.

After the discovery of aflatoxins and their potent carcinogenicity, the search for mycotoxins in the past three decades has led to the identification of more than a hundred toxigenic fungal organisms and mycotoxins throughout the world; the public health significance, however, of most of these remains unknown. A summary of experimental and domestic animal studies as well as some current

TABLE II Mycotoxicoses in which Analytic and/or Epidemiologic Data Suggest Human Involvement

Disease	Species	Substrate	Etiologic agent
Alimentary toxic aleukia sporotrichoides (ATA or septic angina)	Human	Cereal grains (toxic bread)	Fusarium
Dendrodochiotoxicosis toxicum	Horse, human	Fodder (skin contact, inhaled fodder particles)	Dendrodochium
Kashin Beck disease, sporotrichiella "Urov disease"	Human	Cereal grains	Fusarium
Stachybotryotoxicosis	Human, horse, other livestock	Hay, cereal grains, fodder (skin contact, inhaled haydust)	*Stachybotrys atra*
Cardiac beriberi	Human	Rice	Fusarium
Ergot	Human	Rye, cereal grains	*Claviceps purpurea*
Balkan nephropathy	Human	Cereal grains	Penicillium
Reye's syndrome	Human	Cereal grains	Aspergillus
Hepatocarcinoma	Human	Cereal grains, peanuts	Aspergillus
Pink rot	Human	Celery	Sclerotinia sclerotiorum
Onyalai	Human	Millet	Phoma sorghina

epidemiologic associations of mycotoxins with human disease is pertinent.

A. Ergots

The history of human mycotoxicoses dates back to the Middle Ages when ergotism (St. Anthony's fire) was the scourge of Central Europe. Ergotism, which is now rare, was first associated with the consumption of scabrous (ergotized) grain in the mid-16th century. Subsequent studies led to the identification of *Claviceps purpurea* as the fungal agent invading rye, oats, wheat, and Kentucky bluegrass and *C. paspali* invading Dallis grass. Lysergic acid derivatives, the amine and amino acid alkaloids of ergot, were identified as the causative agents of the gangrenous (*C. purpurea*) and nervous (*C. paspali*) forms of the disease. Gangrenous ergotism typically is manifested as prickly and intense hot and cold sensations in the limbs, and swollen, inflammed, necrotic, and gangrenous extremities, which eventually sloughed off. Convulsive ergotism was characterized by central nervous system signs, numbness, cramps, severe convulsions, and death. Both syndromes have been documented in the recent literature in domestic animals consuming ergotized grains and in humans treated with ergotamine for migraine headaches.

Ergot alkaloids are smooth muscle stimulants, promoting narrowing of blood vessels (leading to gangrenous ergotism) and inducing uterine contractions (oxytocic effect). Ergot alkaloids antagonize serotonin and block both the stimulatory and inhibitory CNS responses of epinephrine. The U.S. Department of Agriculture (USDA) Grains Division has set a tolerance limit of 0.3% (by weight) of contaminated grain in commercial trade. [*See* ALKALOIDS IN MEDICINE.]

B. Trichothecenes

In the first half of the 20th century, a large human mycotoxicosis was reported in Russia. The disease, termed *alimentary toxic aleukia*, was characterized by total atrophy of the bone marrow, agranulocytosis, necrotic angina, sepsis, hemorrhagic diathesis, and mortality ranging from 2 to 80%. It later was linked to the consumption of overwintered cereal grains and wheat or bread made from them. *Fusarium poae* and *F. sporotrichioides*, now considered synonymous with *F. tricinctum*, grow on these grains and have been shown to produce several trichothecene toxins including T-2 toxin, neosolaniol, HT-2 toxin, and T-2 tetraol. Signs of ATA develop in cats given pure T-2 toxin orally.

Several outbreaks of a seasonal intoxication in horses and cattle caused by consumption of hay contaminated with *Stachybotys atra* (*S. alternans*) were reported from Russia between 1930 and 1960. Two forms, the atypical and typical, of intoxication reflecting acute or chronic exposure are characterized by sudden onset of neurological signs (loss of vision, poor control of movements, and tremors) or signs of dermonecrosis, leukopenia, and gastrointestinal ulceration and hemorrhages, respectively. In these regions, humans may exhibit severe dermatitis after handling of or sleeping on contaminated hay. Inhalation of dust from the infected hay can result in inflammation of the nose, fever, chest pain, and leukopenia. Five trichothecene compounds were isolated, of which three belonged to the group of macrocyclics (roridin-verrucarin) containing a conjugated butadiene system attached to the trichothecene structure.

A group of trichothecene toxins including nivalenol and fusarinon-X is produced by *F. nivale* in the flowering grainhead of wheat, barley, rice, corn, other cereals, and certain forage grasses. The disease in cereal grains called *red-mold disease* (akakabi-byo) or *black spot disease* (kokuten-byo) has been associated with intoxications in humans, horses, and sheep in Japan. Symptoms in humans include headaches, vomiting, and diarrhea, with no fatalities. *F. roseum*, capable of producing deoxynevalenol and its acetylated derivatives on rice and barley, also was isolated, suggesting multiple causation.

F. solanae, known to produce T-2 toxin and neosolaniol, has been isolated from bean hulls incriminated in a disease (bean hull poisoning) in horses characterized by retarded reflexes, decreased heart rate, disturbed respiration, cyclic movements, convulsions, with a death rate of 10–15%.

Other diseases attributable to trichothecenes include dendrodochiotoxicosis in horses, sheep, and pigs in Russia that ingested feedstuffs contaminated with *Myrothecium roridum* (produces roridins and verrucarins); various syndromes reported in the United States and Canada involving corn (moldy corn toxicosis) and cereal grains consumed by farm animals (T-2 toxin and others); and finally the alleged use of trichothecene toxins as chemical warfare agent (yellow rain) in Southeast Asia. Recent information suggests that the yellow rain of Southeast Asia was bee feces and not a mycotoxin.

Although many fungal genera such as *Fusarium*, *Myrothecium*, and *Stachybotrys* can produce these toxins, most trichothecenes of health significance are produced by *Fusarium* spp. Despite the diversity of human and animal diseases associated with this group of toxins, characteristic signs and symptoms of radiomimetic damage (e.g., emesis, feed refusal, irritation and necrosis of skin and mucus membranes, hemorrhage, destruction of thymus and bone marrow, hematologic changes, nervous disturbances, and necrotic angina) are common to all toxic syndromes. Feed refusal and vomiting are common problems in farm animals, especially swine, in the midwestern United States and are predominately associated with the presence of the trichothecene deoxynevalenol (vomitoxin) in wheat and corn. Although T-2 toxin and diacetoxyscirpenol (DAS) also can cause emesis and feed refusal, their role in the swine feed refusal syndrome appears negligible because of their rare presence in food and feed commodities in the United States. Trichothecenes (T-2 toxin) can cause fetal death and abortions along with tail and limb abnormalities in rodent offspring. Although fusarinon-X and T-2 toxin are mutagenic at high doses in bacterial and yeast systems, trichothecenes exhibit no mutagenic effect in most other systems. Carcinogenic effects of trichothecenes still need to be systematically addressed, both in epidemiologic and experimental settings.

Metabolism of trichothecenes occurs rapidly through deacetylation and hydroxylation and subsequent glucuronidation in the liver and kidneys. Trichothecenes generally are recognized as potent inhibitors of protein synthesis in eukaryotic systems (animals and plants), inhibiting initiation, elongation as well as termination of protein synthesis by way of their inhibition of peptidyl transferase activity, and also their ability to cause disaggregation of polysomes. T-2 toxins, DAS, nivalenol, and fusarenon-X inhibit initiation, whereas trichodermin, crotocin, and verrucarol inhibit elongation or termination. Many of the toxic effects of trichothecenes can be explained by this mechanism. Despite the severe toxic effects of trichothecenes, their low frequency of occurrence in nature, their rapid metabolism to apparently nontoxic metabolites in animals, and their low potential for residue transfer to humans reduce the risk of human disease to extremely low levels, if not completely. Farm animals, how-

ever, are at a higher risk of intoxication than humans because of the greater likelihood of consumption of trichothecenes in moldy feeds.

C. Aflatoxins

It was not until the early 1960s that aflatoxins were discovered as the causative agents of *turkey X disease* in England, which resulted in the death of thousands of turkey poults, ducklings, and chicks that were fed diets containing *Aspergillus flavus*–contaminated peanut meal. This outbreak coupled with the reported carcinogenicity of the aflatoxins in experimental animals helped fuel the scientific curiosity surrounding this group of food contaminants.

The aflatoxins are a group of highly substituted coumarins containing a fused dihydrofuran moiety. Four major aflatoxins, designated B1, B2, G1, and G2 (based on blue or green fluorescence under ultraviolet light), are produced in varying quantities in a variety of grains and nuts that have not been adequately dried at harvest and/or stored at relatively high temperatures. Commodities most often shown to contain aflatoxins are peanuts, various other nuts, cottonseed, corn, and figs. In addition to *A. flavus*, *A. parasiticus* and to a minor extent some species of *Penicillium* may be capable of producing aflatoxins. Human exposure can occur from consumption of aflatoxins from these sources and the products derived from them, as well as from tissues and milk (aflatoxin M1) of animals consuming contaminated feeds.

Aflatoxin B1 (AFB_1), the most potent and most commonly occurring aflatoxin, has been shown to be acutely toxic (LD_{50} 0.3–9.0 mg/kg) to all species of animals, birds, and fishes tested. Sheep and mice are the most resistant, whereas cats, dogs, and rabbits are the most sensitive species. Acute effects in animals include death without signs, or signs of anorexia, depression, ataxia, dyspnea, anemia, and hemorrhages from body orifices. In subchronic cases, icterus, hypoprothrombinemia, hematomas, and gastroenteritis are common. Chronic aflatoxicosis is characterized by bile duct proliferation, periportal fibrosis, icterus, and cirrhosis of the liver and is associated with loss of weight and reduced resistance to disease. Dietary levels as low as 0.3 ppm can cause such effects.

Prolonged exposure to low concentrations lead to liver tumor, (e.g., hepatoma, cholangiocarcinoma, or hepatocellular carcinoma). In poultry there is a lipid malabsorption syndrome characterized by decreased pancreatic lipase and fats in the feces. Reductions in serum triglycerides, cholesterol, phospholipids, and carotenoids also were seen after AFB_1 exposure. Mutagenicity of AFB_1 has been demonstrated in human cells in culture as well as bacteria. Metabolic activation of AFB_1 was required for the mutagenic effect. AFB_1 is primarily metabolized by enzyme present in the liver and other organs forming detoxified substances, but also generate mutagenic products.

The combination of AFB_1 with DNA or RNA bases leads to formation of repair-resistant adducts, loss of purines and other changes leading to single-strand breaks, base pair substitution, or frame-shift mutations. Involvement of oncogenes in such interactions may result in oncogene activation. In addition, AFB_1 inhibits the synthesis of DNA, RNA, and protein. Inhibition of protein synthesis may be related to several lesions and signs of aflatoxicosis including fatty liver (failure to mobilize fats from the liver), alteration of blood coagulation (inhibition of prothrombin synthesis), and reduced immune function. [*See* DNA SYNTHESIS.]

In addition to aflatoxin contamination of foods such as peanuts and corn, aflatoxins and their metabolites also can occur in animal tissues. Especially important is the metabolite aflatoxin M1 (AFM_1), a product of AFB_1, present mainly in milk of AFB_1-exposed dairy animals. The average daily per capita consumption of AFB_1 and AFM_1 in human populations in the United States has been calculated as 25 ng/kg body weight (BW) and 0.3 ng/kg BW, respectively. By using the epidemiologic data generated from Asia, Africa, and the United States, United States males are twice as resistant to induction of liver cancer by AFB as the males in Asia and Africa. The carcinogenicity of AFM_1 seems to be two orders of magnitude lower than that of AFB_1, and therefore a negligible risk. The effect of AFM_1 on human infants needs to be evaluated further.

Other less widespread human clinical syndromes in which aflatoxins have been implicated include acute hepatitis (aflatoxicosis) in India, Taiwan, and certain countries in Africa; childhood cirrhosis in India; and possibly Reye's syndrome in many parts of the world. Reye's syndrome is a childhood neurologic disease that resembles viral encephalitis and actually involves a viral syndrome followed by rapid progresses into coma and convulsions leading to death. Characteristic lesions include enlarged, pale, fatty liver and kidneys and severe cerebral

edema. Evidence from Thailand and other countries including the United States associates aflatoxin consumption or high levels of AFB_1 in tissues from patients with Reye's syndrome. A syndrome strikingly similar to Reye's is produced in monkeys given AFB_1 orally. Epidemiologic evidence, however, also suggested a link between the use of aspirin and Reye's syndrome, prompting the U.S. Surgeon General to advise against the use of salicylates in children with chickenpox or influenza.

Widespread concern regarding the toxic effects of aflatoxins in humans and animals and the possible transfer of residues from animal tissues and milk to humans has led to regulatory actions governing the interstate as well as global transport and consumption of aflatoxin-contaminated food and feed commodities. Action levels of aflatoxins in corn and other feed commodities used to feed mature nonlactating animals are 100 ppb, although temporary increases in limits are allowed on a case-by-case basis by the FDA in situations such as drought, in which availability of uncontaminated corn is extremely limited. For commodities destined for human consumption and interstate commerce, the action limit is 20 ppb. For milk, the action level of AFM_1 is 0.5 ppb. Among the many approaches tried to limit the aflatoxin contamination of grain, prevention of stress, the use of fungus-resistant varieties of grains, avoiding mechanical injury to grain during harvesting, drying of grains to contain less than 12% moisture, and strict control of humidity during storage are important to prevent AFB_1 production.

D. Ochratoxins

In 1957 and 1958, up to 75% of the households in several villages located in the valley floor in contiguous areas of Yugoslavia, Romania, and Bulgaria were found to be affected by chronic nephropathy (kidney disease, Balkan or endemic nephropathy). Although genetic factors appear to be partially involved, evidence was presented that a mycotoxin, ochratoxin A, produced in foodstuffs by *Aspergillus ochraceus* and a number of other aspergilli and penicilli, was consumed at higher levels more frequently by people in these endemic areas compared with areas free from nephropathy. In addition, a remarkably similar nephropathy was identified in swine (porcine nephropathy) and bovine fed ochratoxin. Signs include lassitude, fatigue, anorexia, abdominal (epigastric or diffuse) pain, and severe anemia followed by renal damage. Reduced concentrating ability, reduced renal plasma flow, and decreased glomerular filtration occur subsequently, accompanied by gross and microscopic renal changes. Deaths result from intoxication caused by insufficient excretion by the kidney.

The ochratoxins are found in barley, corn, wheat, oats, rye, green coffee beans, and peanuts. In experimental animals, ochratoxin A produces renal lesions and hepatic degeneration. In poultry, ochratoxin A causes reduced weight gains and decreased egg shell quality and egg production in addition to renal effects. Teratogenic effects of ochratoxin A in rodents include malformations of the head, jaws, tail, limbs, and heart. Ochratoxins are not mutagenic in several assay systems tested but appear to induce hepatomas and renal adenomas in mice exposed to 40 ppm in the diet.

Ochratoxin A is broken down by carboxypeptidase A and alpha-chymotrypsin to the less toxic ochratoxin alpha. Absorbed ochratoxin A distributes mainly to the kidneys and liver and is excreted rapidly in the urine and feces. Ochratoxin A inhibits mitochondrial respiration and reduces ATP levels. These effects also are produced by ochratoxin alpha. Depletion of glycogen, inhibition of glucose production, a pathway that accounts for 50–60% of the blood glucose in the starved or diabetic stage, and several key cyclic adenosine monophosphate (cAMP)-activated enzymes in this pathway, including phosphoenolpyruvate carboxykinase by ochratoxin A, have received attention in recent years. Whether any of those enzymatic steps is the critical target in the pathogenesis of ochratoxicosis is unknown.

E. Psoralens

Psoralens are furocoumarin compounds that have been used in repigmenting white skin lesions in an acquired disease called *vitiligo*. Psoralens are in some suntan lotions and in drugs used to treat psoriasis. Abuse of such compounds can result in dermatitis after exposure to sun as well as nausea, vomiting, vertigo, and mental excitation. A phototoxic dermatitis in celery pickers also has been linked to the presence of psoralens in stalks of celery infected with *Sclerotinia sclerotiorum* (pink rot), *S. rolfsii*, *Rhizoctonia solani*, or *Erwinia aroideae*. In addition to celery, fig, parsley, parsnip, lime, and clove also contain psoralens. Unlike other photosensitizing agents, psoralens seem to act by reacting with DNA in the presence of light and to a lesser extent

with RNA. Treatment with 8-methoxypsoralen and ultraviolet light induced squamous cell carcinomas of the ear in mice. Psoralens are rapidly excreted in the urine as fluorescent nontoxic metabolites. One of them, however, 8-methoxypsoralen, appears to undergo change similar to those of aflatoxins and may thus react with DNA in a similar fashion.

The mechanism of psoralen photosensitivity appears to involve its intercalation between the DNA bases followed by cross-linking. Intercalation occurs between two pyrimidines on opposing sides of the helix; then the psoralen forms a chemical complex with one of the pyrimidines after absorption of ultraviolet light. Cross-links are formed by further absorption of ultraviolet, which links the psoralen–pyrimidine complex with the second pyrimidine.

F. Citreoviridin (Yellow Rice Toxin)

Acute cardiac beriberi (Shoshin-Kakke), characterized by palpitation, nausea, vomiting, rapid and difficult breathing, cold and bluish extremities, rapid pulse, abnormal heart sounds, low blood pressure, restlessness, and violet mania leading to respiratory failure and death, was observed in Japan in the late 1800s and early 1900s. Evidence suggests citreoviridin as the etiological agent rather than avitaminosis. In fact, the dark yellow toxic metabolite citreoviridin was isolated from *P. citreoviride*, and production of the neurologic syndrome and respiratory failure can be reproduced in rats given extract of *P. citreoviride*–contaminated rice. Other toxins identified failed to produce signs resembling cardiac beriberi. In 1921, the Japanese government passed the Rice Act to reduce the availability of moldy rice in markets. This Act resulted in a sharp decrease in the disease during the same year, while maintaining rice as a prominent dietary ingredient. This sequence of events provides additional support for the involvement of mycotoxins in the causation of cardiac beriberi.

G. Bishydroxycoumarin (Dicoumarol)

Dicoumarol is a Vitamin K antagonist that inhibits the availability of Vitamin K. Resulting effects lead to bleeding disorders and death from blood loss from undetectable bleeding sites. The molecules of naturally occurring coumarins in sweet clover (*Melilotus* sp) hay join together to form dicoumarol as a result of fungal spoilage during curing. Cattle consuming such hay were poisoned in the 1920s in

North Dakota and Alberta. Most human poisonings result from therapeutic accidents involving coumarin therapy of clotting disorders.

H. Zearalenone

In addition to zearalenone and zearalenol being contaminants in grains such as corn, wheat, sorghum, barley, and oats, zearalanol, a synthetic analogue of zearalenol (Ralgro) is used as an anabolic agent (promoting body growth) in cattle. Under natural conditions, zearalenone and its derivatives are produced by *Fusarium roseum*, mostly in ear corn stored in cribs. Zearalenone, despite its structural dissimilarity with estrogens, induces effects similar to those produced by excessive steroidal as well as synthetic estrogens. Among the domestic animals, swine appear to be the most sensitive, exhibiting signs of hyperestrogenic syndrome (i.e., swollen and edematous vulva, hypertrophic myometrium, vaginal cornification, and prolapse in extreme cases).

Zearalenone acts by interacting with estrogen receptors in the cell cytoplasm. The receptor–zearalenone complex is then translocated to the nucleus, where it combines with receptors in the DNA, caus-ing selective gene transcription causing increased water and lowered lipid content in muscle, increased permeability of uterus to glucose, RNA, and protein precursors. Zearalenone induces biphasic changes in the concentration of luteinizing hormone but not follicle-stimulating hormone in serum. The rapid metabolism of zearalenone and zearalanol to conjugated metabolites, which are then excreted in urine and feces, makes consumption of meat and milk from animals receiving Ralgro an insignificant risk to humans.

Recent evidence from Italy and Puerto Rico, however, suggests that estrogenic substances, especially residues of zearalenone and zearalanol in red meats and poultry, may have caused premature thelarche (development of breast before age 8) and precocious pseudopuberty. Also, zearalenone has been shown to be mutagenic and carcinogenic in animals.

I. Other Mycotoxins

A number of other mycotoxins (Table III) have been identified either as contaminants in foods destined for human consumption or as metabolites of fungi isolated from human foods. Although some of these have been associated with outbreaks of do-

TABLE III Miscellaneous Mycotoxins

Mycotoxin	Major producing organisms	Source of fungi	Principal toxic effects
Alternariol and alternariol methyl ether	*Alternaria* sp	Sorghum, peanuts, wheat	Highly teratogenic to mice; cytotoxic to HeLa cells; lethal to mice
Altenuene, altenuisol	*Alternaria* sp	Peanuts	Cytotoxic to HeLa cells
Altertoxin I	*Alternaria* sp	Sorghum, peanuts, wheat	Cytotoxic to HeLa cells; lethal to mice
Ascladiol	*Aspergillus clavatus*	Wheat flour	Lethal to mice
Austamide and congeners	*Aspergillus ustus*	Stored foodstuffs	Toxic to ducklings
Austdiol	*A. ustus*	Stored foodstuffs	Toxic to ducklings
Austin	*A. ustus*	Peas	Lethal to chicks
Austocystins	*A. ustus*	Stored foodstuffs	Toxic to ducklings; cytotoxic to monkey kidney epithelial cells
Chaetoglobosins	*Penicillium aurantiovirens, Chaetomium globosum*	Pecans	Toxic to chicks, cytotoxic to HeLa cells
Citreoviridin	*Penicillium citreoviride*	Rice	Neurotoxic, producing convulsions in mice
Citrinin	*Penicillium viridicatum, Penicillium citrinum*	Corn, barley	Nephrotoxic to swine
Cyclopiazonic acid	*Penicillium cyclopium*	Ground nuts, meat products	Nephrotoxic, enterotoxic
Cytochalasins	*A. clavatus, Phoma* sp, *Phomopsis* sp *Hormiscium* sp, *Helminthosporium dematioideum, Metarrhizium anisopliae*	Rice, potatoes, Kodo millet, pecans, tomatoes	Cytotoxic to HeLa cells; teratogenic to mice and chickens
Diplodiatoxin	*Diplodia maydis*	Corn	Nephrotoxic and enterotoxic to cattle and sheep
Emodin	*Aspergillus wentii*	Chestnuts	Lethal to chicks

continued

TABLE III (*continued*)

Mycotoxin	Major producing organisms	Source of fungi	Principal toxic effects
Fumigaclavines	*Aspergillus fumigatus*	Silage	Enterotoxic to chicks
Kojic acid	*Aspergillus flavus*	Squash, spices	Lethal to mice
Malformins	*Aspergillus niger*	Onions, rice	Lethal to rats
Maltoryzine	*Aspergillus oryzae*	Malted barley	Hepatotoxic and causes paralysis
Oosporein (Chaetomidin)	*Chaetomium trilaterale*	Peanuts	Lethal to chicks
Paspalamines	*Claviceps paspali*	Dallisgrass	Neurotoxic to cattle and horses, causes paspalum staggers
Patulin	*Penicillium urticae*	Apple juice	Lethal to mice; mutagenic; teratogenic to chicks; pulmonary effects in dogs; carcinogenic to rats
Penicillic acid	*Penicillium* spp	Corn, dried beans	Lethal to mice, mutagenic; carcinogenic to rats
PR toxin	*Penicillium roqueforti*	Mixed grains	Hepatotoxic and nephrotoxic to rats; causes abortion in cattle
Roseotoxin B	*Trichothecium roseum*	Corn	Toxic to mice and ducklings
Rubratoxins	*Penicillium rubrum*	Corn	Causes hemorrhage in animals; hepatotoxic to cattle
Secalonic acids	*Aspergillus aculeatus, Penicillium oxalicum*	Rice, corn	Lethal, cardiotoxic, teratogenic, and causes lung irritation in mice
Slaframine	*Rhizoctonia leguminicola*	Red clover	Causes salivation and lacrimation in horses and cattle
Sporidesmins	*Pithomyces chartarum*	Pasture grasses	Hepatotoxic, causes photosensitization in ruminants
Sterigmatocystin	A. flavus	Mammals	Mutagenic, carcinogenic, and hepatotoxic to mammals
Tenuazonic acid	*Alternaria* sp	Grains, nuts	Lethal to mice
Terphenyllins	*Aspergillus candidus*	Wheat flour	Hepatotoxic to mice; cytotoxic to HeLa cells
Tremorgenic mycotoxins			
Fumitremorgens A and B	A. fumigatus	Rice	Neurotoxic (prolonged tremors and convulsions)
Paxilline	*Penicillium paxilli*	Pecans	Neurotoxic (prolonged tremors and convulsions)
Penitrems A, B, and C	*Penicillium cyclopium*	Peanuts, meat products, cheese	Penitrem A: Neurotoxic (prolonged tremors and convulsions) to cattle, sheep, dogs, and horses
Tryptoquivalines	A. clavatus	Rice	Neurotoxic (prolonged tremors and convulsions)
Verruculogen (TR-1)	*Penicillium verruculosum*	Peanuts	Neurotoxic (prolonged tremors and convulsions)
Unidentified toxin(s)	*Aspergillus terrus, Balansia epichloe, Epichloe typhina, Fusarium tricinctum,* and others	Fescue grass	Gangrene (fescue foot); summer slump syndrome, causes fat necrosis and agalactia in cattle
Xanthoascin	A. candidus	Wheat flour	Hepatotoxic and cardiotoxic to mice

mestic animal diseases, no current link between human consumption and disease has been established. Others have been shown to induce toxic and lethal effects in laboratory animals with no association between consumption of these toxins by animals or humans and a disease syndrome. Several of these (e.g., cytochalasins and secalonic acid D) have been used to expand our understanding of normal as well as abnormal cellular responses to xenobiotics.

Although it is difficult to assess the significance of consumption of mycotoxins in human foods, it is clear that such a task requires extensive research into hundreds of known and potentially at least thousands of as yet unknown mycotoxins. Despite the vast number of toxic metabolites, prevention of mycotoxicoses in humans and animals can be achieved for the most part by avoiding stress or damage to seeds by pests and during harvesting, rapid postharvest drying, and finally by avoiding conditions conducive to mold growth during storage.

Bibliography

Beasley, V. R. (1989). "Trichothecene Mycotoxins: Pathophysiologic Effects," vol. 1 (175), vol. 2 (198). CRC Press, Boca Raton, Florida.

Bechtel, D. H. (1989). Molecular Dosimetry of Hepatic Aflatoxin B_1-DNA Adducts: Linear Correlation with Hepatic Cancer Risk, *Regul. Toxicol. Pharmacol.* **10,** 74–81.

Betina, V. (1984). "Mycotoxins: Production, Isolation, Separation and Purification." Elsevier, Amsterdam.

Dvorackova, I. (1990). "Aflatoxins and Human Health." CRC Press, Boca Raton, Florida.

Hayes, A. W. (1981). "Mycotoxin, Teratogenicity and Mutagenicity." CRC Press, Boca Raton, Florida.

Kuratu, H., and Ueno, Y. (1984). "Toxigenic Fungi—Their Toxins and Health Hazards." Elsevier, Amsterdam.

Marasas, W. F O., and Nelson, P. E. (1987). "Mycotoxicology." The Pennsylvania State University Press, University Park, Pennsylvania.

"Mycotoxins, Economic and Health Risks," task force no. 116. (1989). Council for Agricultural Science and Technology. Ames, Iowa.

Richard, J. L., and Cole, R. J., eds. (1989). "Economic and Health Risks Associated with Mycotoxins." Council for Agricultural Sciences and Technology. Ames, Iowa.

Schiefer, H. B. (1988). Yellow rain. *In* "Comments on Toxicology," pp. 1–62. Gordon and Breach, New York.

Seagrave, S. (1981). "Yellow Rain, A Journey Through the Terror of Clinical Warfare." M. Evans and Company, New York.

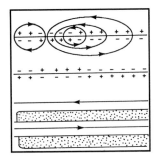

Myelin and Demyelinating Disorders

MARJORIE B. LEES, OSCAR A. BIZZOZERO, *E. K. Shriver*
Center and Harvard Medical School

RICHARD H. QUARLES, *National Institute of Neurological*
Disorders and Stroke, NIH

Glossary

Central nervous system Brain and spinal cord
Glia Non-neuronal cells of the nervous system, including cell types that produce myelin; in the central nervous system, the myelin-producing cells are called oligodendrocytes; in the peripheral nervous system, they are referred to as Schwann cells
Leukodystrophies Group of hereditary metabolic disorders characterized by a widespread, severe deficiency of myelin
Multiple sclerosis Common neurological disease characterized by loss of myelin in the central nervous system and leading to motor and sensory deficits
Myelin-associated enzymes Enzymes present in oligodendrocyte and Schwann cell processes, and associated with limited regions of the myelin sheath
Neuropathy Disease of the peripheral nervous system
Node of Ranvier Interruption in the myelin sheath; space between two myelinated segments
Peripheral nervous system Nerves outside the brain and spinal cord, especially those of the limbs
Saltatory conduction Discontinuous conduction in which the nerve impulse jumps from one node to the next

MYELIN IS A MULTILAMELLAR, membranous sheath that surrounds the nerve axon. In the central nervous system (CNS), it is formed by oligodendrocytes, and in the peripheral nervous system (PNS), by comparable cells called Schwann cells. In contrast to any other cell membrane, 70–80% of the dry weight of myelin is lipid and only 20–30% is protein. However, the major proteins are characteristic of myelin and are not found in other membranes. Myelin is characteristic of vertebrates and appears to be a structural adaptation that facilitates rapid conduction of nerve impulses. It serves as an insulator around the axon and speeds conduction of the nerve impulse with conservation of both space and energy. However, its function appears to be more than that of a passive insulator, and the occurrence of active metabolic processes within the membrane is now recognized. The maintenance of myelin structure and activity is required for normal nervous system function, and myelin pathology leads to serious human disorders such as multiple sclerosis and the leukodystrophies.

I. Overview

A. Evolutionary Appearance

Myelin is a specialization of the vertebrate nervous system, which provides a mechanism whereby the requirement of higher animals for rapid transmission of large numbers of impulses is achieved with an economy of both space and energy. The increase in the speed of conduction along an unmyelinated axon is proportional to the square root of its diameter, whereas that of a myelinated axon is directly proportional to its diameter. Consequently, to conduct at a defined speed, the size of an unmyelinated fiber would have to be dramatically greater than that of a myelinated fiber (Fig. 1). Thus, myelin allows the large number of nerve axons needed to carry out complex activities to be accommodated within a defined space. Furthermore, the periodic interruptions or gaps along the length of the myelin sheath, referred to as nodes of Ranvier, result in saltatory conduction, i.e., the nerve current jumps from one node to the next, and the expenditure of energy is much less than in unmyelinated axons, as it is restricted to the nodes rather than being dissipated along the length of the axon (Fig. 2). The area adjacent to the node is referred to as the paranodal region; the region between two consecutive nodes is the internode.

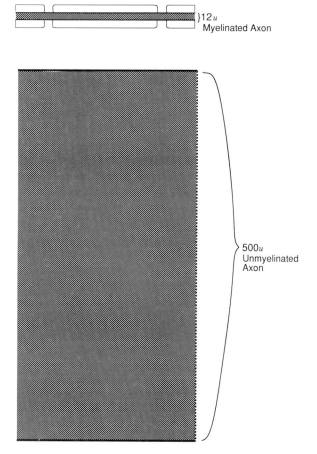

FIGURE 1 Diagrammatic representation of a myelinated nerve fiber (top) and an unmyelinated fiber (bottom) that conduct at the same speed. Many more myelinated fibers can be accommodated in a limited space.

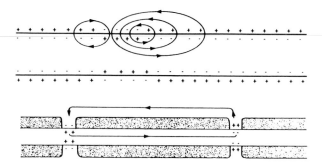

FIGURE 2 Propagation of the action potential in an unmyelinated (top) and a myelinated (bottom) nerve. The arrows indicate the flow of current in local circuits within the active region of the membrane. $(+, -)$ represent the charge stored on the capacitance of the membrane. For unmyelinated fibers, current flows continuously along the membrane, whereas in myelinated fibers, current jumps from one node to the next. (Reprinted from Morell *et al*. 1989, Chapter 6 ("Formation, Structure, and Biochemistry of Myelin") *in* "Basic Neurochemistry," 4th ed. (Siegel *et al*., eds.), Raven Press, New York.)

B. Myelin Function

Myelin is commonly referred to as an insulator around the nerve axon and, indeed, the conservation of energy described above derives from the high electrical resistance due to the large amount of lipid present in the myelin membrane. Stimulation of the nerve axon at the nodal region leads to an inflow of sodium ions and an outflow of potassium ions via specific ion channels in the axolemma (axonal membrane). The insulating properties of the myelin prevent depolarization of the adjacent axolemma and a wave of current jumps from one nodal segment to the next (Fig. 2). The nodal region thus acts as a potassium sink to maintain ion homeostasis. The compact structure of the lamellar membrane and its low water content are additional factors that give rise to the insulating properties of myelin. The stability of the lamellar structure is enhanced by the presence of specialized proteins, which include specific adhesion molecules and proteins capable of binding electrostatically and hydrophobically with lipids and other proteins.

In recent years, it has been recognized that myelin has functions over and above its insulating properties. Because myelin is a part of a living cell, it should come as no surprise that active metabolic processes are occurring. Thus, although myelin is a relatively stable membrane, it undergoes continuing synthesis and degradation, as discussed in Section

V. Further, the presence of specific myelin-associated enzymes, the characteristic lipid and protein composition, the restricted localization of certain myelin components, and the evolutionary conservation of protein sequences all point to an interrelated system with important dynamic properties.

C. Methods of Study of Myelin

The gross anatomical distribution of myelin is evident from its glistening white appearance, and large unmyelinated areas can be identified visually on the basis of their grayish color. Also, light and electron microscopy, combined with appropriate stains, clearly identify the multilamellar structure of myelin and its characteristic staining properties (see Section II). The specificity of these microscopy procedures is enhanced by immunocytochemistry in which antibodies to specific myelin components can be used to follow their developmental appearance. [See ELECTRON MICROSCOPY.]

For structural and metabolic studies, myelin can be isolated on sucrose gradients as a purified membrane on the basis of its low density. Purified myelin collects as a band at a density of 1.08 g/ml (0.65 M sucrose). Myelin can be further separated into subfractions of different densities, presumably corresponding to different regions of the multilamellar structure. Upon isolation, the normal interrelationships between myelin and the cell body and axonal membranes are disrupted, and these may be important for myelin function *in situ*. Brain or nerve tissue can be explanted or dispersed, and oligodendroglial and Schwann cells grown in culture to follow the initial steps of myelination *in vitro*. These cultures express myelin components and significant amounts of membrane sheets are formed, but only minimal amounts of compact myelin. Nevertheless, the timetable of cellular differentiation is comparable to that *in vivo*.

II. Myelin Formation and Structure

Myelination is initiated by a signal between the axon and the myelin-forming Schwann cells or oligodendrocytes, which appear to be activated when the axon reaches a critical size (>1 μm). Myelination in the CNS and PNS are similar in principle but have significant differences in detail. Each Schwann cell directly wraps around and ensheaths a single segment of one axon, whereas a single oligodendrocyte sends out multiple processes that expand into broad, flat sheets as they begin to spiral around different axons and different regions of the same axon. After several turns of the spiraled membrane, the extracellular faces of the sheath become apposed to form the intraperiod line. Concomitantly, cytoplasm is extruded and the intracellular faces of the membranes become closely apposed and compacted. However, a rim of cytoplasm remains at the edges of the compacted membranes, and these appear in longitudinal sections as a series of lateral loops (Figs. 3 and 5). Other areas with remaining cytoplasm are the inner and outer mesaxons and incisures known as Schmidt–Lanterman clefts, which are particularly prevalent in PNS myelin sheaths. Some of the metabolic activity associated with myelin undoubtedly occurs in these cytoplasm-rich regions.

The ultrastructural organization of myelin has been elucidated by X-ray diffraction studies and electron microscopy. The compacted membranes appear in the electron microscope as a series of dark and less dark lines between unstained areas (Fig. 4). The darker line is referred to as the major dense line and corresponds to the apposed cytoplasmic faces of the cell, whereas the less dense line is known as the intraperiod or intermediate dense line and represents the apposition of the extracellular faces of the cell membrane (Fig. 3). The variations in the intensity of staining reflect the differing chemical compositions of the cytoplasmic and extracellular faces of the membrane, particularly the orientation of its proteins.

The ultrastructural organization has also been studied by X-ray diffraction. In CNS myelin, the unit membrane, with a repeat distance of 80 Å, consists of a single lipid bilayer of approximately 50 Å and protein extending approximately 15 Å on either side of the bilayer. Therefore, two fused bilayers have a repeat distance of 160 Å (Fig. 3). These values are slightly larger than the electron microscopic dimensions because electron micrograph specimens are prepared differently from X-ray diffraction samples. In the PNS, the myelin periodicity is 10% higher due to its different protein composition.

Additional differences between the CNS and PNS are seen in the structure of the node of Ranvier (Fig. 5). In the CNS, the node is bare and the axon is exposed directly to the extracellular compartment. By contrast, the node in the PNS is covered

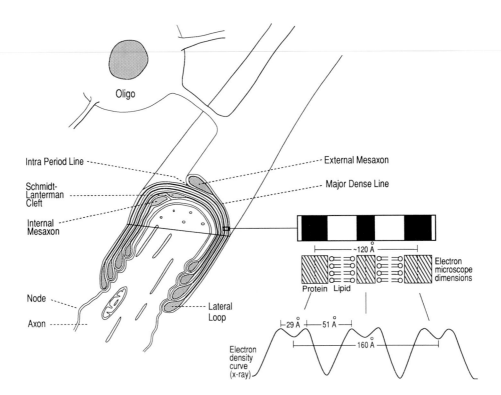

with interdigitated fingers of cytoplasm, and the entire region, including the node, paranode, and internode, is covered with a collagenous basal lamina.

III. Myelin Development

Myelination does not occur simultaneously in all regions but begins in PNS, and then proceeds along the neuraxis. In the human, little or no myelination occurs during the first half of gestation. The dorsal and ventral roots begin to myelinate during the third trimester and are, for the most part, completely myelinated at the time of birth (Fig. 6). Some tracts myelinate rapidly and over a short time, whereas others myelinate more slowly and, consequently, myelinogenesis in the latter extends over a prolonged time. Thus, at any single time point, different stages of myelination are occurring simultaneously. During the first postnatal years, active myelination occurs in most tracts. However, myelination continues in many regions during the first decade of life, and some regions may continue to myelinate up to 30 yr of age. The last areas to be myelinated are the corpus callosum and the intracerebral commisures, regions involved in higher functions such as pattern recognition and reading skills.

FIGURE 3 Diagram of a CNS myelinated internode based on ultrastructural data. Top left shows an oligodendrocyte with several membranous processes, one of which is connected to the myelin sheath. The cut view of the myelin and axon illustrates the 3-dimensional interrelationships between the two structures. The right-hand side of the figure shows the approximate dimensions of one myelin-repeating unit of protein–lipid–protein as seen by electron microscopy of fixed sections (upper) and by X-ray diffraction of fresh nerve (lower). (Modified and redrawn from Morell *et al.*, 1989, Chapter 6 (''Formation, Structure, and Biochemistry of Myelin'') *in* ''Basic Neurochemistry,'' (Siegel *et al.*, eds.), 4th ed., Raven Press, New York.)

IV. Chemical Composition

In both the CNS and PNS, the myelin sheath has a low water content and consequently contains relatively more solids than do other tissue membranes. The solids are mainly lipids, with lesser amounts of protein and only small amounts of carbohydrate.

A. Myelin Lipids

Approximately 70% of the dry weight of myelin is accounted for by lipids, an amount higher than that of other biological membranes. Myelin lipids are not unique to this membrane but, nevertheless, certain

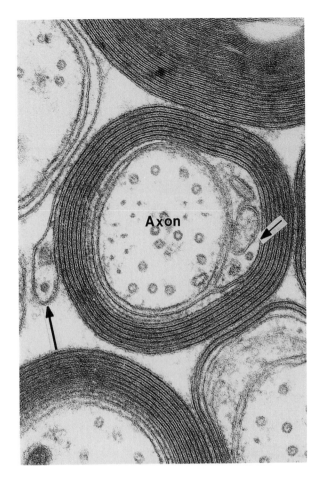

FIGURE 4 Electron micrograph of a transverse section of a myelinated nerve axon in the CNS. The outer (large arrow) and inner (small arrow) mesaxon, containing cellular organelles, the spiral nature of the sheath, and the origins of the intraperiod and major dense lines, are clearly visible (magnification 150,000×). (Courtesy of Dr. Cedric Raine.)

TABLE I Composition of Human CNS Myelin and Brain[a]

Component[b]	Myelin	White matter	Gray matter
Protein	30.0	39.0	55.3
Lipid	70.0	54.9	32.7
Cholesterol	27.7	27.5	22.0
Total galactolipid	27.5	26.4	7.3
Cerebroside	22.7	19.8	5.4
Sulfatide	3.8	5.4	1.7
Total phospholipid	43.1	45.9	69.5
Ethanolamine phosphatides	15.6	14.9	22.7
Lecithin	11.2	12.8	26.7
Sphingomyelin	7.9	7.7	6.9
Phosphatidylerine	4.8	7.9	8.7
Phosphatidylinositol	0.6	0.9	2.7
Plasmalogens	12.3	11.2	8.8

[a] Values from Siegel, G. J., Agranoff, B., Albers, R. W., and Molinoff, P., eds., 1989, "Basic Neurochemistry," 4th ed., Raven Press, New York, p. 121.
[b] Protein and lipid figures are expressed as percent of dry weight; all others are percent of total lipid weight.

sulfatide. The fatty acid composition of galactocerebrosides includes a relatively high proportion of long-chain fatty acids (C20–C24) and of hydroxy fatty acids. The major phospholipids are phosphatidylcholine, phosphatidylserine, sphingomyelin, and ethanolamine phospholipid, with the latter the most abundant. A unique feature of myelin is that 80% of the ethanolamine phospholipid is present in the plasmalogen form (α, β unsaturated ether). In contrast to other myelin lipids, plasmalogens are synthesized in peroxisomes. The high proportion of plasmalogens results in an increase in the overall unsaturation of the membrane and may be impor-

characteristic features can be described. Cholesterol, phospholipids, and galactolipids are the major myelin lipid classes (Table I) and, on a molar basis, they occur in an approximate ratio of 2 : 1.8 : 1. Thus, cholesterol is the most abundant lipid. Essentially all the cholesterol is free (nonesterified), and the presence of significant amounts of esterified cholesterol in adults is indicative of myelin pathology and demyelinating disease. This contrasts with other membranes, in which a specific ratio of free to esterified cholesterol is required to maintain membrane structure. The most typical lipid is galactocerebroside, and the increase in its amount parallels that of myelin during development. A part (15–20%) of the galactocerebroside is sulfated, forming an acidic lipid referred to as cerebroside sulfate, or

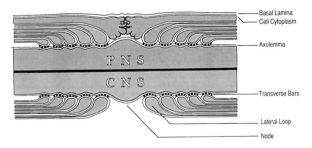

FIGURE 5 Diagrammatic representation comparing the structure of a PNS node with that of a CNS node. Note the interdigitations of Schwann cell processes and basal lamina covering the PNS node. These are absent from the CNS node. Transverse bars connect the lateral loops and the axolemma. (Redrawn from Raine, C. S., 1984, Chapter 1 ("Morphology of Myelin and Myelination.") in "Myelin," 2nd ed. (P. Morrell, ed.), Plenum Press, New York.)

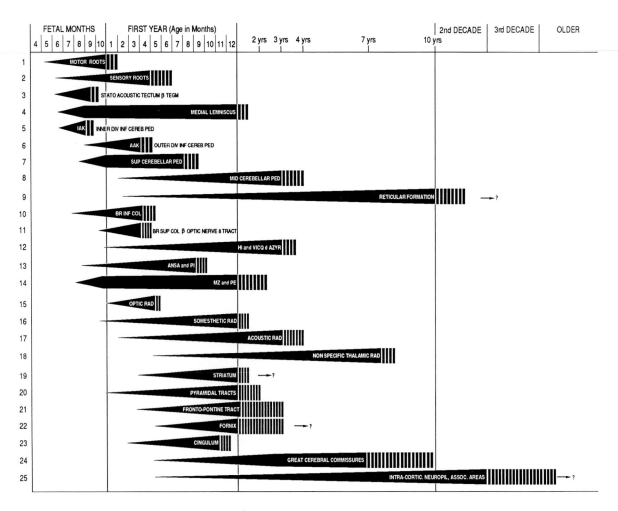

FIGURE 6 Time period of myelinogenesis in different areas of the human brain. The width of the segments denotes the intensity of staining and density of myelinated fibers. Vertical stripes at the end of the bars approximate the age range of the termination of myelination. (Reprinted from Yakovlev, P., and LeCours, A. R., 1967, "The myelinogenetic cycles of regional maturation of the brian" *in* "Regional Development of the Brain in Early Life" (A. Minkowski, ed.), Blackwell Scientific Publications, Oxford, England.)

tant in maintaining its stability. Other lipids (e.g., polyphosphoinositides) are present in only minor quantities, but they are, nevertheless, enriched in myelin and may be further localized to certain regions of the sheath. In other tissues, polyphosphoinositides are involved in signal transduction processes and their presence in myelin may therefore have functional implications. [*See* CHOLESTEROL; LIPIDS.]

B. Myelin Proteins

1. Central Nervous System

In contrast to the lipids, which are ubiquitous components, most of the proteins are unique to myelin. Over 80% of the CNS myelin proteins are accounted for by two protein families, namely proteolipids and myelin basic proteins. These, along with an enzyme called 2'3'-cyclic nucleotide-3'-phosphohydrolase (CNPase), which accounts for an additional 2–3% of the myelin proteins, each have

basic isoelectric points and can potentially interact electrostatically with acidic lipids. The myelin proteins can be separated on the basis of their molecular weights by polyacrylamide gel electrophoresis in a denaturing detergent (Fig. 7) and further identified by staining with specified antibodies. [*See* PROTEINS.]

a. Proteolipid Proteins Myelin proteolipid proteins (PLPs) are a family of hydrophobic integral membrane proteins, which account for more than

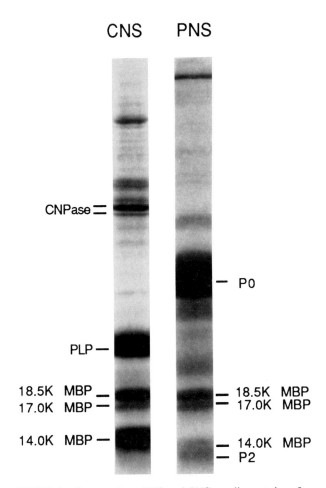

CNS PNS

CNPase ≡

PLP —

18.5K MBP —
17.0K MBP —

14.0K MBP —

— P0

— 18.5K MBP
— 17.0K MBP

— 14.0K MBP
— P2

FIGURE 7 Pattern of rat CNS and PNS myelin proteins after electrophoresis in polyacrylamide gels in the presence of the denaturing detergent sodium dodecyl sulfate. Proteins are visualized by staining with Coomassie Brilliant Blue. In this photograph, the minor PLP, DM 20, is barely evident below the PLP band. The molecular species of the MBPs in human myelin differ from those in the rat.

half of the proteins of CNS myelin. They are restricted to the CNS and first appear during evolution in amphibia. The amino acid sequence is highly conserved with identical sequences found for humans and rodents. The major form of PLP has a molecular mass of 30 kilodaltons (kDa). A second, minor protein, designated DM-20, is formed by alternate splicing of the PLP mRNA. The PLP gene, which is located on the X chromosome, has been isolated and shown to contain seven exons and six introns. The normal protein shows a marked domain structure, i.e., regions of hydrophobic amino acids alternate with positively charged, hydrophilic regions, and the resulting integral membrane protein traverses the membrane several times (Fig. 8).

The precise localization of each of the charged regions in the myelin structure is currently under investigation.

The high proportion of hydrophobic amino acids and the strong domain structure combine to give PLP unusual solubility properties for a protein. PLP is extracted from brain by organic solvents (e.g., chloroform–methanol mixtures), along with the tissue lipids, as a lipid–protein complex. However, after removal of all of the complex lipids, the apoprotein (delipidated protein) retains its solubility in organic solvents. The apoprotein contains 2 moles of covalently bound fatty acid (mainly palmitic, stearic, and oleic acids), and these are attached to cysteine residues. Proteins containing covalently bound fatty acids are referred to as acylated proteins, and acylation is the only known posttranslational modification of PLP. The fatty acid is added, after PLP is inserted into the myelin membrane, by an autoacylation process, i.e., a separate enzyme does not appear to be involved.

b. Myelin Basic Protein Myelin basic proteins (MBPs) are a family of highly positively charged proteins, which have been studied extensively in relation to their role in experimental allergic encephalomyelitis (EAE). At least four isoforms of MBP (21.5, 20, 18.5, and 17 kDa) have been identified in human myelin, and these multiple forms arise from alternate splicing of a single gene, located on the distal end of chromosome 18. Mouse mutants with defective MBP have been identified, but no human counterparts have been found thus far. The major isoform in the human adult is the 18.5-kDa species, consisting of 168 amino acids, 24% of which are basic amino acids (lysine, arginine, histidine). This is in marked contrast to PLP, which contains only 9% basic amino acids. The abundance of lysine and arginine residues provides many sites for cleavage by the proteolytic enzyme trypsin. The MBPs are hydrophilic, extrinsic membrane proteins and have been localized to the major dense line observed in electron micrographs of myelin (Figs. 3 and 4). The MBPs undergo several posttranslational modifications, including N-acetylation, phosphorylation, and methylation. These modifications are assumed to contribute to myelin compaction and also to myelin function.

c. Myelin-Associated Glycoprotein Myelin-associated glycoprotein (MAG) is a 100-kDa protein, which appears to play a role in the formation and

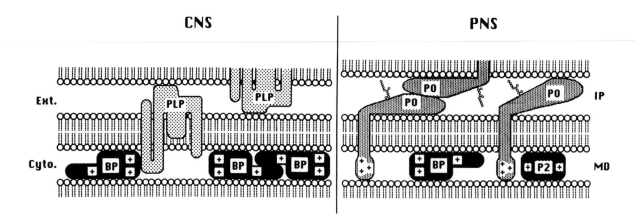

FIGURE 8 Diagrammatic representation of proposed models of the molecular organization of compact CNS and PNS myelin. Apposition of the extracellular (Ext) and cytoplasmic (Cyto) surfaces of the oligodendrocyte or Schwann cell membranes form the intraperiod (IP) and major dense (MD) lines, respectively. Note the greater separation of the intraperiod membrane surface in PNS myelin as a consequence of the large extracellular domain of P0. Positively charged regions (+) of myelin basic protein (BP) and P2 are located at the cytoplasmic face. The correct orientation of the PLP is still unknown. ⋔ represents lipid molecules. ⌇ represents carbohydrate residues. (Reprinted from Morell *et al.*, 1989, Chapter 6 ("Formation, Structure, and Biochemistry of Myelin") *in* "Basic Neurochemistry," 4th ed. (Siegel *et al.*, eds.), Raven Press, New York.)

maintenance of the junction between the myelin-forming glial cell and the axonal surface membrane. It differs from MBP and PLP in that it is absent from multilamellar, compact myelin and is restricted to the inner turn of the oligodendrocyte membrane adjacent to the axon. About one-third of the molecular mass of MAG is accounted for by carbohydrate moieties that are attached by N-glycosyl bonds to asparagine residues in the polypeptide. MAG and other adhesion glycoproteins in the nervous system that function in cell–cell interactions have amino acid sequence similarities with immunoglobulins and other proteins of the immune system. These neural adhesion proteins and related proteins of the immune system make up the so-called immunoglobulin gene superfamily and are believed to have arisen during evolution from a common ancestral protein involved in cell–cell interactions.

d. Myelin-Associated Enzymes Recognition of the existence of a significant number of myelin-associated enzymes has lent support to the concept of myelin as a dynamic membrane. Isolated myelin contains enzymes involved in ion-transport processes, lipid metabolism, and protein modification. Some of these enzymes are ubiquitous (e.g., carbonic anhydrase), whereas others are highly enriched in myelin (e.g., cholesterol esterifying enzymes) or act on components specific to myelin (e.g., certain protein kinases). Each of these enzymes has a characteristic distribution pattern along the neuraxis, probably reflecting different functional needs.

Of particular interest is the enzyme CNPase, which, although present in low concentrations in other tissues, is characteristic of both oligodendrocytes and Schwann cells and their membranous ex-

trusions. CNPase appears early during development and is often used as a myelin marker, although the relationship is not strictly quantitative. The enzyme has been localized to the cytoplasmic face of the myelin membrane and is restricted mainly to the region of the nodal loops and other cytoplasm-rich regions. CNPase exists as two similar monomers with molecular masses of 46 and 48 kDa, and each of the monomers show enzyme activity. The primary sequence is relatively conserved among different species. The human enzyme has been cloned and shown to consist of 400 amino acids, with no membrane-spanning domain. The enzyme has a net positive charge and is posttranslationally modified by phosphorylation of serine and threonine residues. The enzyme contains not only phosphoryl binding sites but also nucleotide binding sites, and these may be important for its function.

2. Peripheral Nervous System

PNS myelin does not contain the hydrophobic PLP found in the CNS but contains a 30-kDa integral membrane protein called the P-zero (P0) glycoprotein (Fig. 7). This protein accounts for over half of the total PNS myelin protein and contains about 5% carbohydrate by weight. Other posttranslational modifications of P0 include acylation, phosphorylation, and sulfation. Like MAG, P0 is a member of the immunoglobulin superfamily, and its extracellular immunoglobulin-like domain is believed to play a role in stabilizing the intraperiod line of compact myelin (Fig. 8). The P0 protein spans the membrane once, and its positively charged C-terminal, cytoplasmic domain is thought to contribute to stabilization of the major dense line of compact myelin. Another PNS myelin-specific protein is a positively charged fatty acid binding protein called P2, which also contributes to myelin stability. PNS myelin also contains the same myelin basic proteins as found in the CNS, but in lesser amounts. As in the CNS, MAG is in the inner Schwann cell membrane of the PNS myelin, where it forms a junction between glia and axons. In addition, MAG appears to mediate interactions between adjacent "semi-compacted" Schwann cell membranes in Schmidt–Lanterman incisures, paranodal loops, and the inner and outer mesaxons. CNPase is associated with uncompacted regions of PNS myelin sheaths but at a much lower concentration than in CNS sheaths.

C. Molecular Organization of Compact CNS and PNS Myelin

A hypothetical model of the molecular organization of myelin proteins is shown in Fig. 8. Integral membrane proteins are defined as having one or more hydrophobic domains which are embedded in the lipid bilayer, whereas extrinsic proteins are attached to only one side of the bilayer by electrostatic interactions. PLP is an integral membrane protein, which passes through the bilayer more than once and may help to stabilize both the intraperiod and major dense lines. In the PNS, the major integral membrane protein is the P0 glycoprotein, which transverses the membrane just once and may also help to stabilize the membrane contacts at both surfaces. The single, large extracellular domain of P0 probably accounts for the well-known, greater separation of the intraperiod membrane surfaces in PNS myelin than in CNS myelin. Both types of myelin contain the extrinsic, positively charged MBPs, which probably help to stabilize the major dense lines of myelin by electrostatic interactions with lipids. P2 protein may function similarly to the MBPs in PNS myelin. MAG and CNPase are not shown in these models of compact myelin, because their lo-

calization is restricted to noncompacted membranes of myelin sheaths, as described above.

V. Myelin Metabolism

Contrary to the earlier concept of myelin as an inert insulator, a variety of active metabolic processes have been shown to occur in this membranous structure. An understanding of how myelin is formed, maintained, and degraded is still far from complete, but most of the current information on assembly and degradation comes from animal experiments using biochemical methods with radioactive precursors, as well as immunocytochemistry.

A. Synthesis

During myelinogenesis, both oligodendrocytes and Schwann cells synthesize large amounts of proteins and lipids and assemble them into extensive sheets of membrane. The synthesis of most myelin proteins and lipids increases rapidly during development, and then declines after the peak period of myelin formation. Increased synthesis, rather than a decrease in catabolism, is mainly responsible for myelin deposition. Intrinsic membrane proteins, such as PLP, P0, and MAG, are synthesized on membrane-bound ribosomes located in the perinuclear region of the glial cells. These proteins are subsequently transported to the Golgi and targeted to the plasma membrane. In contrast, extrinsic proteins (MBP, CNPase) are synthesized on free polysomes and are immediately incorporated into the plasma membrane. From the different time courses of appearance of newly synthesized proteins and lipids, the membrane evidently is not formed as a unit. This concept, as discussed below, also apparently applies to the degradation of myelin. Furthermore, because of the differences in relative synthetic rates of individual components, the composition of myelin changes during development.

The activity of the enzymes responsible for the synthesis of fatty acids and complex lipids increases several-fold during the period of active myelinogenesis and declines thereafter. On the other hand, those enzymes involved in myelin maintenance and turnover also show increased activity during myelin synthesis but remain elevated in the adult. Enzymes responsible for the formation of cholesterol, phospholipids, and galactolipids are located in microsomes, where the bulk of lipid synthesis occurs. However, certain enzymes involved in the recycling of lipids, such as fatty acyl-CoA synthetase and lysophosphatidylcholine–acyltransferase, are present not only in microsomes but are also in myelin, where they may participate in remodelling and maintaining the membrane. Intact phospholipids and their precursors may move from the axon across the axonal membrane to be incorporated into the myelin. The extent to which the latter two mechanisms contribute to myelin deposition and maintenance is unknown.

B. Degradation

Myelin is not degraded as a whole, and even the individual structural components turn over independently. Nevertheless, the turnover of myelin components is relatively slow compared with that of other cell membranes, and half-lives in the order of months have been reported for the major proteins. High molecular weight proteins, some probably representing myelin-associated enzymes, turn over more rapidly. How myelin proteins are physiologically degraded is still unclear, but the presence of neutral proteases in myelin has been documented. Degradation can also be carried out by acid proteases located in lysosomes.

Most of the myelin proteins have functional groups that are added during or after synthesis. Some of these groups are metabolically stable with half-lives comparable to those of the polypeptide backbone (e.g., methyl groups in MBP and sulfate and carbohydrate residues in P0 and MAG) and may have a role in maintaining protein configuration. Other side-chain groups turn over more rapidly and at a rate independent of the protein backbone (e.g., phosphate groups in MBP and fatty acyl groups in PLP) and could, therefore, be actively involved in membrane maintenance and function. For example, the turnover of phosphate groups on MBP is accelerated upon repeated nerve stimulation.

As with the proteins, specific lipids turn over at different rates. Cholesterol, sphingolipids, and cerebrosides are the most stable of the myelin lipids with half-lives on the order of months. Glycerophospholipids, particularly phosphatidylcholine and phosphatidylinositol, seem to turn over more rapidly. Different moieties of the same lipid have also been found to turn over independently. Of particular interest is the rapid turnover of phos-

phate groups in polyphosphoinositides, because these lipids are involved as second messengers in signal transduction mechanisms.

VI. Diseases Involving Myelin

The formation and maintenance of myelin is dependent on the normal functioning of oligodendrocytes or Schwann cells, as well as the viability of the ensheathed axons and their neuronal cell bodies. A deficiency of myelin can result either from an inability to produce the normal complement of myelin during development (hypomyelination) or from its breakdown once formed (demyelination). Demyelinating diseases can be divided into two categories: primary demyelination involving early loss of myelin with relative sparing of axons, and secondary demyelination in which myelin loss occurs as a consequence of damage to neurons or axons. Thus, a myelin deficit can result from a multitude of physiological insults that affect either myelin-forming glial cells, neurons, or the myelin sheaths themselves. Hypomyelination or demyelination can be caused by acquired allergic or infectious factors, genetic disorders, toxic agents, or malnutrition. Many of the biochemical changes in neural tissue are similar regardless of the etiology of the demyelinating disease and include a marked increase in water content and a decrease in myelin lipids and proteins.

A. Multiple Sclerosis

Multiple sclerosis (MS) is the most common demyelinating disease of the CNS in humans, causing severe motor and sensory neurological deficits in about 1 out of 1,000 persons in the United States. It is generally a chronic disease, with typical onset occurring in the second or third decade of life, and is characterized by exacerbations and remissions, which occur over many years. However, it is sometimes slowly progressive from the beginning and, in a small number of patients, occurs in an acute form that progresses to death within a year. Clinically, slowing of nerve conduction can be demonstrated by measurement of auditory or visually evoked potentials. The pathology is characterized by an inflammatory reaction around small veins with lymphocyte infiltration and demyelinated areas called plaques, which can be observed grossly at autopsy

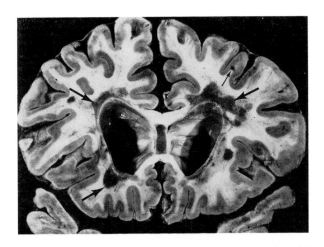

FIGURE 9 Coronal slice of brain from a patient who died with MS. Demyelinated plaques are clearly visible in white matter (large arrows). Small plaques are also observed at the boundaries between gray and white matter (small arrows). (Reproduced from Raine, C. S., 1984, Chapter 8 ("The Neuropathology of Myelin Diseases") in "Myelin," 2nd ed., Plenum Press, New York.)

(Fig. 9). The appearance and disappearance of lesions can be visualized in living patients by magnetic resonance imaging. The plaques are characterized by an absence or severe reduction of myelin and oligodendrocytes, with preservation of relatively normal-appearing axons. Electron microscopy indicates that a major mechanism of myelin destruction is the direct removal of myelin lamellae from the surface of intact sheaths by macrophages. This, along with the presence of apparently healthy oligodendrocytes in areas of active demyelination, suggests that the myelin sheath itself is the primary target of the disease rather than the myelin-forming oligodendrocyte. In early lesions, substantial remyelination is observed, but the newly formed myelin is destroyed and oligodendrocytes are eventually lost from the plaques. Older, chronic MS lesions are sharply defined and contain bare, nonmyelinated axons and many fibrous astrocytes. [*See* MULTIPLE SCLEROSIS.]

Although the cause of MS has not been identified, genetic, immunological, and viral factors are believed to contribute to the pathophysiology. A genetic component is indicated by the increased risk of disease for first-degree relatives of affected individuals and an especially high concordance for the disease in monozygotic twins. Genes coding for histocompatibility molecules are linked to disease susceptibility with a strong association of certain hu-

man leukocyte antigen (HLA) types with the disease. However, these determinants in themselves are neither necessary nor sufficient to confer susceptibility.

An environmental factor, possibly an infectious agent, has been suggested by epidemiological studies. Thus, MS has a high prevalence above latitude 40° north in the United States and Europe and a low prevalence near the equator. In the Faroe Islands, no MS cases existed prior to World War II, but after that time a high incidence was reported, suggesting that an infectious agent related to the arrival of British troops might be important. Further evidence for viral involvement comes from experimental animal models of demyelinating diseases that are caused by viruses. Intensive efforts to identify a causative virus in MS patients have led to reports of at least 12 different infectious agents to date, but none have stood the test of time. A human retrovirus has been the most recent addition to the list.

A number of observations suggest an autoimmune basis for the disease with immunological mechanisms of demyelination. Abnormalities of immunoglobulin synthesis, association with HLA types that are in turn linked to immune response genes, and abnormalities in cellular immune function support this concept. In pathological specimens, perivenular infiltration of inflammatory cells is also observed. Further, administration of pharmacological agents that modify immune function, such as corticosteroids and immunosuppressants, often results in favorable clinical responses. Another factor is the similarity of MS to certain forms of an extensively studied autoimmune disease in animals, called experimental allergic encephalomyelitis (EAE). In EAE, multiple myelin components can elicit disease, but in MS numerous attempts to identify either a brain or viral antigen that is the target of the pathogenic immune response have been unsuccessful. [*See* AUTOIMMUNE DISEASE.]

It should be emphasized that the genetic, viral, and immunological theories for the pathogenesis of MS are not mutually exclusive. Individuals of certain genetic backgrounds may show a characteristic immune response to a specific virus, and this in turn may be responsible for the demyelination. Effective prevention or treatment will probably not be forthcoming until more is known about the cause of the disease. On the other hand, recent demonstrations of active remyelination in acute MS lesions suggest that it may eventually be possible to stimulate a patient's oligodendrocytes to remyelinate.

B. Acquired Demyelinating Neuropathies

Guillain–Barré syndrome is an acute, monophasic, inflammatory, demyelinating neuropathy often preceded by a viral infection. As in MS, cumulative evidence suggests that the disease is mediated by immunological mechanisms. Humoral immunity is suggested by observations that sera from Guillain–Barré syndrome patients cause demyelination in appropriate test systems and that plasmapheresis (serum exchange) is often an effective therapy. However, the relationship of the viral infection to initiation of the disease and the identity of the principal antigen have not been established with certainty.

Neuropathies also occur in association with benign IgM gammopathies, and these diseases are believed to be caused by the reactivity of monoclonal IgM antibodies against neural antigens, including MAG and various complex glycolipids.

C. Other Disorders of Myelin

1. Genetically Determined Disorders

The leukodystrophies are a group of genetically determined disorders of CNS white matter characterized by a diffuse lack of myelin. Most of the leukodystrophies are familial lipid storage disorders involving failure of lipid degradation in cellular organelles called lysosomes. The genetic defect is in specific lysosomal enzymes and, therefore, specific lipids fail to be degraded. Rather, they accumulate and result in a marked but characteristic perturbation of normal cell metabolism. In the lysosomal disease metachromatic leukodystrophy, the enzyme aryl sulfatase A is missing and, consequently, sufatides are not degraded. Because this lipid is present in both CNS and PNS myelin, its accumulation leads to generalized pathology. The most common leukodystrophy is X-linked adrenoleukodystrophy, in which the degradation of very long-chain fatty acids is impaired. The disease is referred to as a peroxisomal disorder, as this metabolic step occurs in cell organelles called peroxisomes.

Other inborn errors of metabolism may also influence developmental processes including myelination. Among them, genetic defects of amino acid metabolism, such as phenylketonuria, are accompanied by a deficiency of CNS myelin.

2. Toxic, Nutritional, and Hormonal Disorders

A variety of biological and chemical toxins can interfere with myelination or result in demyelina-

tion. Diphtheria toxin causes vacuolation and fragmentation of myelin sheaths in the PNS. Lead is a common environmental toxin that produces a multitude of effects on the nervous system, including retardation of CNS myelination as well as demyelination of the PNS. Experimentally, organotins and hexachlorophene cause reversible, edematous changes in myelin characterized by splitting of the intraperiod line. This type of damage is of clinical significance because of the earlier use of hexachlorophene as an antiseptic agent for human infants.

Nutritional deficiencies, including protein malnutrition and vitamin deficiencies, have widespread effects on the nervous system, especially in the early developmental period. During the vulnerable period of myelin formation, undernourishment can lead to impairment of glial cells with consequent depression of the synthesis of myelin-specific components. Hypomyelination can also be observed in essential fatty acid and certain vitamin deficiencies (thiamine, B_6, or B_{12}).

During development, certain hormonal deficiencies can also lead to hypomyelination. Both hypo- and hyperthyroidism can result in a generalized decrease in myelin synthesis. In diabetics, segmental demyelination of peripheral nerves is observed frequently. However, there is no agreement as to whether the latter effects are directly on myelin or secondary to axonal damage. The archetypical example of secondary demyelination is Wallerian degeneration, which occurs following crush or sectioning of a nerve. The proximal nerve segment survives and regenerates, but both the axon and myelin in the distal segment are lost. Myelin and cellular debris are phagocytosed, and eventually the regenerating axons are remyelinated.

VII. Concluding Comments

Over the last 30 years, the view of myelin as only an inert insulator around the nerve axon has changed drastically. Myelin is now considered a metabolically active plasma membrane, which is part of an interrelated functional system. Yet, thus far, we have only a glimmer of understanding of the role of myelin within the overall activity of the nervous system, and much remains to be clarified. The enormous advances in molecular biology and gene cloning in this decade have just begun to unlock the secrets of the control of myelinogenesis. The new technologies of molecular biology are being effectively utilized to increase our understanding of mechanisms whereby the immune system and viruses may cause myelin loss. They have further been used to develop methods for prenatal diagnosis of leukodystrophies. These procedures, along with the possibility of gene transfer, provide potential approaches to prevent and/or reverse devastating demyelinating diseases.

Bibliography

Dyck, P. J., Thomas, P. K., Lambert, E. H., and Bunge, R. (eds.) (1984). "Peripheral Neuropathy." Saunders Co., Philadelphia.

Koetsier, J. C. (ed.) (1985). "Handbook of Clinical Neurology," Vol. 3. Elsevier Science Publishers, B. V. Amsterdam.

Lees, M. B., and Sapirstein, V. S. (1983). Myelin-associated enzymes. *In* "Handbook of Neurochemistry," Vol. 4, 2nd ed. (A. Lajtha, ed.). Plenum Press, New York.

Morell, P. (ed.) (1984). "Myelin," 2nd ed. Plenum Press, New York.

Morell, P., Quarles, R. H., and Norton, W. T. (1989). Formation, structure and biochemistry of myelin. *In* "Basic Neurochemistry" (G. Siegel, B. Agranoff, R. W. Albers, and P. Molinoff, eds.). Raven Press, New York.

Quarles, R. H., Morell, P., and McFarlin, D. (1989). Diseases involving myelin. *In* "Basic Neurochemistry" (G. Siegel, B. Agranoff, R. W. Albers, and P. Molinoff, eds.). Raven Press, New York.

Scriver, C. R., Beaudet, A. L., Sly, W. S., and Valle D. (1989). "The Metabolic Basis of Inherited Disease," Vol. II, Parts 10 (Peroxisomes) and 11 (Lysosomal Enzymes), 6th ed. McGraw-Hill, Inc., New York.

Nasal Tract, Biochemistry

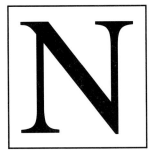

ALAN R. DAHL, *Inhalation Toxicology Research Institute, Lovelace Biomedical and Environmental Research Institute*

Glossary

Cytochrome P-450 General designation for an important family of heme-containing monooxygenases
Dehydrogenases Enzymes that oxidize their substrates by removing the elements of a hydrogen molecule
Monooxygenases Enzymes that oxidize their substrates by ''inserting'' a single oxygen atom
Transferases Enzymes that transfer a chemical moiety from one molecule to another
Xenobiotics Compounds foreign to the body

IN ADDITION TO its roles in warming and humidifying inhaled air, the nose has an important role in removing potentially toxic inhalants before they can reach the lung. Thus unsurprisingly, in addition to the normal complement of enzymes present in all tissues, nasal tissues also contain high levels of the enzymes responsible for metabolism of foreign compounds.

The unique biochemistry of the nasal cavity is related in large part to its capacity to metabolize inhaled materials. Metabolism not only influences the toxicity of inhalants, but, increasingly, it is also thought to influence olfaction. In addition to the capacity to metabolize xenobiotics, a number of biochemical phenomena in the olfactory mucosa appear to be related to the sense of smell. This chapter describes the xenobiotic metabolizing enzymology of the nasal cavity and explores those other areas of nasal cavity biochemistry that are unusual among body tissues.

I. Xenobiotic Metabolizing Enzyme Activities in the Nasal Cavity

A. Enzymes That Catalyze Oxidation or Reduction

Among the enzymes that metabolize xenobiotics by oxidation, the cytochrome P-450 family of enzymes is by far the best studied, and perhaps the most important, in the nasal cavity. Members of this family of enzymes metabolize substrates ranging from lipophilic, polycyclic aromatic hydrocarbons such as benzo(a)pyrene to water-soluble compounds such as ethanol and dimethylnitrosamine. Ordinarily, metabolism leads to production of hydroxylated compounds that are either themselves soluble enough to be excreted or undergo further metabolism by which a water-soluble moiety is attached to the hydroxyl group. Metabolism by these enzymes is particularly interesting, however, when carcinogenic or otherwise toxic metabolites are produced. Metabolism leading to metabolites that are more toxic than the parent molecule is termed activation. Examples of such enzyme-catalyzed increases in toxicity include the activation of acetaminophen, benzo(a)pyrene, and dimethylnitrosamine—all nasal carcinogens in animals.

In addition to their capacity to produce activated metabolites, the nasal cytochromes P-450 influence the scent of an odorant. The difficulty in predicting the quality of a scent from its molecular structure may be a result of chemical transformations in the olfactory tissue that produce chemical species different from those inhaled.

The cytochromes P-450 in their various forms are not distributed uniformly throughout the nasal cavity. Like most xenobiotic metabolizing enzymes, the highest concentrations are found in the olfactory mucosa (Table I). [*See* CYTOCHROME *P*-450.]

The nasal cavity contains high concentrations of flavin-containing monooxygenase, which also occurs in the highest concentration in the olfactory mucosa. It is mostly associated with metabolism of nitrogen-, sulfur-, and phosphorus-containing species to products containing oxygenated heteroatoms.

Aldehyde dehydrogenases, which oxidize aldehydes to the corresponding carboxylic acids, occur at the highest levels in the respiratory mucosa of the anterior nasal cavity (Table II) and are unusual in this regard. The substrate for formaldehyde dehydrogenase is the thiahemiacetal formed by the addition of the tripeptide glutathione to formaldehyde. Glutathione can be depleted in the nasal cavity by high concentrations of inhaled formaldehyde. This depletion may have an important effect on the carcinogenicity of formaldehyde at high inhaled concentrations.

Alcohol dehydrogenase, which oxidizes primary alcohols to aldehydes and secondary alcohols to ketones, also occurs in the nasal cavity. Its distribution has not been extensively studied.

B. Xenobiotic Metabolizing Enzymes That Catalyze Hydrolysis

Epoxide hydrolases constitute a family of enzymes that catalyze the hydrolysis of epoxides to glycols. Glycols are more soluble and less chemically reactive than epoxides and, hence, are less toxic. Epoxide hydrolases are in relatively high concentrations in the nasal cavity. They probably play an important role in protecting the nasal cavity from the carcinogenic effects of inhaled epoxides or of epoxides formed by the cytochrome P-450-mediated oxidation of inhaled, unsaturated hydrocarbons.

Esters of carboxylic acids are common odorants, occurring, for example, in banana oil (amyl acetate) and in solvents such as nail polish remover (ethyl acetate). Carboxyl esterases catalyze the hydrolysis of esters to alcohols and carboxylic acids. High levels of these enzymes are found in the nasal cavity, where they may play an important role in olfaction. They also protect lung tissues from exposure to esters by helping to scrub the esters from the airstream. With the exception of the cyanide-metabolizing enzyme rhodanese, carboxyl esterase activity is the highest xenobiotic metabolizing activity measured in the nasal cavity.

C. Xenobiotic Metabolizing Enzymes That Catalyze Conjugation of Two Molecules

Enzymes that transfer a water-soluble moiety from themselves to a hydrophobic substrate are extremely important in facilitating removal of xenobiotics from tissue. These enzymes occur in the nasal cavity, where they probably are important in detoxication of inhalants. Glucuronyl transferases catalyze the covalent binding of a glucuronic acid moiety, an oxidation product of glucose, to form glycosides from hydroxylated compounds such as alcohols and phenols. Glucuronidation is virtually always a detoxication reaction, because the glycosides are normally water-soluble and easily excreted. This enzyme activity occurs in the nasal cavity at levels nearly the same as those found in liver, on a per gram tissue basis.

The metabolism of reactive carbon species by formation of thio ethers with the tripeptide glutathione is one of the most important detoxication reactions known. These reactions are catalyzed by glu-

TABLE I Distribution of Cytochrome P-450 in Nasal Mucosa

Tissue	P-450 form	Typical substrate	Species
Respiratory epithelium	IIB1	Benzphetamine	Rat, rabbit
	IIIA	Acetaminophen	Rat
	IV	2-Aminofluorene	Rabbit
Olfactory epithelium	IA2	Benzo(a)pyrene	Rabbit
	IIB1	Benzphetamine	Rabbit
	IIE	Dimethylnitrosamine	Rabbit
	IIIA	Acetaminophen	Rat
	IV	2-Aminofluorene	Rabbit
Bowman's gland	IA1	Benzo(a)pyrene	Rat

TABLE II Cellular and Tissue Location of Some Xenobiotic Metabolizing Activities in the Nasal Cavity

Tissue/cell	Activity[a]		
	α-naphthylbutyrate carboxylesterase	Acetaldehyde dehydrogenase	Benzo(a)pyrene hydroxylase
Squamous epithelial cells		+	
Respiratory mucosa			+
Ciliated epithelial cells	+ +	+ + +	
Nonciliated epithelial cells	+	+ +	
Goblet cells	±	+	
Cuboidal cells	+ +	±	
Basal cells	+		
Seromucous glands	±		+
Olfactory mucosa			+ +
Bowman's gland	+ + +	±	+ +
Sustentacular cells	+ +	−	
Basal cells	±	+	
Sensory cells	−	−	

[a] The more plus signs, the higher the activity. ± indicates barely detected activity; − indicates no detected activity.

tathione transferases, a family of enzymes, a number of which occur in the nasal cavity (Tables III and IV). Substrates for glutathione transferases range from halogenated alkanes to epoxides. Typically, these substrates are reactive, electrophilic compounds that can readily bind to proteins or to DNA bases, the latter providing the potential for mutations. Thus, in part, the action of glutathione transferases is to deactivate potential genotoxicants, thereby mitigating their mutagenicity. Although the glutathione transferases are ordinarily associated with detoxication, they can, in certain instances, produce metabolites that are themselves carcinogenic. An example is the formation of an episulfonium ion (a three-atom, positively charged, cyclic sulfur ether) from sym-dibromoethane. This reaction could easily occur in the nasal cavity and may influence the toxicity of this inhalant.

Rhodanese is a sulfur transferase. The best-known activity catalyzed by rhodanese is the transfer of a sulfur atom from thiosulfate to cyanide to produce thiocyanate. Because cyanide is very toxic and thiocyanate is much less so, this is an important detoxication reaction. Rhodanese occurs at very high levels in the nasal cavity: the nasal cavity of rats has almost 10% as much activity as does the entire liver—a remarkable fact considering the differences in mass between the two tissues. This enzyme also occurs in the nasal respiratory epithelium of humans. Human olfactory tissue has not been examined for this activity, but it likely also occurs there at high levels. Hereditary, cyanide-specific anosmia may be related to a surfeit of rhodanese in the nose that prevents detectable concentrations of cyanide from occurring.

TABLE III Relative Activities of Glutathione Transferases from Human Nasal Respiratory Mucosa

Substrate	Relative rate[a]
1-Chloro-2,3-dinitrobenzene	1
Ethacrynic acid	0.10
p-nitrophenylacetate	0.057
Trans-4-phenyl-3-buten-2-one	0.001

[a] As fraction of activity toward 1-chloro-2,3-dinitrobenzene.

TABLE IV Distribution of Glutathione S-Transferases (GSH-T) in Rat Nasal Mucosa[a]

Enzyme	Typical substrates	Respiratory epithelium	Olfactory epithelium
GSH-T 1,1	Δ⁵-androstene-3,17-dione	+ + +	+ +
GSH-T 3,3	1,2-Dichloro-4-nitrobenzenebromosulfophthalin	±	+
GSH-T 5,5	1,2-Epoxy-3-(p-nitrophenoxy)propane, iodomethane	+	+

[a] The more plus signs, the higher the activity. ± indicates barely detected activity.

II. Inhibition and Induction of Nasal Xenobiotic Metabolizing Enzymes

Nasal cytochrome P-450 is readily inhibited *in vitro* by methylenedioxyphenyl (MDP) compounds. MDP compounds are very common and include safrole, which imparts the characteristic odor to sassafras root; heliotropin, used in artificial cherry flavors; and piperonyl butoxide, used as an insecticide synergist in aerosol sprays. The effect of inhibiting liver enzymes on pharmacological and toxicological processes has long been a subject of study, but the effects on olfaction when nasal enzymes are inhibited have only recently been considered to be important. Many of the MDP compounds are themselves potent odorants used in the fragrance industry. Their effect on nasal enzymes opens up an entire area of study relative to the complex interactions between mixtures of inhalants and odorants.

A possible example of inhibition of nasal enzymes is the inhibition of rhodanese by cigarette smoke. After a puff on a cigarette, some smokers are apparently more sensitive to the odor of cyanide than are either nonsmokers or the smokers themselves before they take a puff. The explanation for this phenomenon may be that the enzyme rhodanese is inhibited by the cigarette smoke. As a consequence, the concentration of cyanide in the olfactory mucosa may reach higher levels than it ordinarily would, thereby exceeding the threshold concentration for triggering the olfactory response.

Xenobiotic metabolizing enzyme activity often increases, or is "induced," by exposure to their substrates. Certain isozymes of nasal cytochrome P-450 can be induced. In general, however, the nasal enzymes are not as dramatically inducible as are liver enzymes. Induction of nasal enzymes could have important toxicological implications, however, and may also play a role in olfaction.

III. Effects of Nasal Xenobiotic Metabolizing Enzymes on Inhalants and Pharmaceuticals

The results of *in vitro* experiments indicate that nasal cavity enzymes probably have an important role in the metabolism of inhaled materials. *In vivo* observations verify the *in vitro* results. Over 50% of some inhaled esters are metabolized in the nasal cavity. Such metabolism not only contributes to the protection of the lungs but also influences which tissues are exposed to the hydrolysis products of the esters. Because some hydrolysis products are toxic, the latter effect may have important toxicological consequences. For example, inhaled vinyl acetate would be metabolized in the nasal cavity to vinyl alcohol and acetic acid. Vinyl alcohol rearranges to acetaldehyde. If hydrolysis and rearrangement occur in the nose, the nasal tissue is exposed to the acetaldehyde, not the components of the blood or liver; therefore, naval tissues would be a potential target tissue for the toxic effects of the acetaldehyde.

Inhaled formaldehyde is metabolized to formic acid by aldehyde dehydrogenase in the nasal cavity of rodents, and thereby its toxicity is decreased. The reduction of the toxicity of inhalants is likely to be a common result of nasal metabolism; however, nasal metabolism can also have the opposite effect by increasing the toxicity of inhalants. For example, some nitrosamines are activated to carcinogenic metabolites in the nasal cavity. The soot component benzo(a)pyrene is also metabolized to reactive intermediates in the nasal cavity. Finally, cocaine is metabolized in the nasal cavity, and the resulting formaldehyde may contribute to the toxicity of cocaine to nasal tissue.

Enzymes in nasal tissue metabolize a number of commonly used decongestants. The intranasal route of administration for other pharmaceuticals has resulted in some successes and a large number of failures. A portion of the failures may result from substantial nasal metabolism of the pharmaceutical.

IV. Other Aspects of Nasal Biochemistry

A. General Nasal Biochemistry

The nasal cavity is highly complex, both anatomically and in the distribution of enzymes found in the nasal cavity. Adenosine triphosphatase occurs in the respiratory epithelium, the cilia, the mucus, the olfactory epithelium, and in the ducts of Bowman's glands but not in the nasolacrimal duct, the septal olfactory organ, or the olfactory nerves. Acid phosphatase occurs at its highest levels in the rat olfactory epithelium and in the vomeronasal organ. It also occurs to a lesser extent in other areas of the nose but is absent from the cilia, the mucus, and the ducts of Bowman's glands. Alkaline phosphatase occurs at its highest levels in the vomeronasal organ, the septal olfactory organ, the ducts of Bowman's glands, the olfactory nerves, the olfactory epithelium, the mucus, and the squamous epithelium. It is found to a lesser extent in the other nasal tissues but is not detected in the acini of Bowman's glands. Glucose-6-phosphatase occurs at its highest levels in the acini of the Bowman's glands, the vomeronasal organ, myelin sheaths, and the mucus. It occurs to a lesser extent in the other areas of the nasal cavity but is absent in the nasal lacrimal duct, the ducts of the Bowman's gland, and in the olfactory nerves. Gamma-glutamyl transpeptidase occurs at its highest level in the ducts of the Bowman's glands and at lower levels in most other tissues in the nasal cavity, but it is absent in the myelin sheaths, mucus, the olfactory epithelium, the olfactory nerves, and the septal olfactory organ. Naphthyl butyrate esterase occurs in the highest concentrations in the vomeronasal organ, the acini of the Bowman's glands, the olfactory epithelium, and the respiratory epithelium. It occurs at lower levels in other areas of the nasal cavity but is absent from the myelin sheaths, the cilia, and the olfactory nerves.

Amylase is secreted by nasal glands. Carbonic anhydrase is found at high levels in olfactory nerve cells but is absent in the supporting cells. The function of these enzymes in nasal tissue is not clear. Nicotinamide adenine dinucleotide reduced–cytochrome P-450 reductase occurs in high levels in the olfactory epithelium. This enzyme is necessary to support cytochrome P-450-mediated metabolism.

Nasal tissues contain glycerides, phosphatides, acetylphosphatides, and mucopolysaccharides, mostly associated with the Bowman's glands. In a number of animal species, these glands also contain esterases, lipases, cholinesterases, dopamine-oxidases, and peroxidases.

B. Biochemistry Unique to the Olfactory Tissue

Olfaction is poorly understood at the molecular level. There are some similarities between olfaction and vision. The cilia of the olfactory nerve cells are analogous to the outer segments of rods and cones in that both lack the energy-producing adenosine triphosphatase, dynein. Hence, the olfactory cilia are motile. The dipeptide carnosine (β-alanyl-L-histidine) is highly concentrated in the olfactory bulb and may be a neurotransmitter. Olfactory cilia contain very high levels of adenylate cyclase, at least 10 times higher than the levels found in brain membranes. In the presence of guanosine triphosphate, odorants stimulate the activity of this enzyme. An odorant-binding protein with a molecular mass of around 19,000 daltons has been identified. [*See* OLFACTORY INFORMATION PROCESSING.]

The rat olfactory mucosa also contains a unique cytochrome P-450 form (P-450 IIG). Cytochrome P-450 reductase—a necessary component for cytochrome P-450-dependent monooxygenation—is found at unusually high levels in olfactory tissue. Other enzymes, including ornithine decarboxylase, catalase, and glucose-6-phosphate dehydrogenase, also occur at relatively high levels in olfactory tissue.

Bibliography

Dahl, A. R. (1988). Comparative metabolic basis for the disposition and toxic effects of inhaled materials. *In* "Inhalation Toxicology: The Design and Interpretation of Inhalation Studies and Their Use in Risk Assessment" (U. Mohr, D. Dungworth, G. Kimmerle, J. Lewkowski, R. McClellan, and N. Stöber, eds.). Springer-Verlag, New York.

Dahl, A. R. (1988). The effect of cytochrome P-450-dependent metabolism and other enzyme activities on olfaction. *In* "Molecular Neurobiology of the Olfactory System" (F. L. Margolis and T. V. Getchell. eds.). Plenum Publishing Corporation, New York.

Feron, U. J., and Bosland, M. C. (eds.) (1989). "Nasal Carcinogenesis in Rodents: Relevance to Human Health Risk." Pudoc, Wageningen, The Netherlands.

Snyder, S. H., Sklar. P. B., and Pevsner. J. (1988). Molecular mechanisms of olfaction. *J. Biol. Chem.* **263**, 13971.

Natural Killer and Other Effector Cells

OSIAS STUTMAN, *Memorial Sloan-Kettering Cancer Center*

I. Killer Cells
II. General Properties of NK and Other Effector Cells
III. Function
IV. Recapitulation

Glossary

Interleukin Cytokine (usually a lymphokine or monokine) that usually mediates various immunological effects on lymphocytes and other cells, although it may have nonimmunological effects

Ligand Membrane structure(s) recognized by a receptor

Lymphokine Biologically active product released by a lymphocyte; cytokine is the generic term for products released by cells, and monokine for products released by monocytes or macrophages

NATURAL KILLER (NK) CELLS belong to a family of effector cells in the lymphoid and hemopoietic tissues of most vertebrate species, including man, which can directly kill certain tumor target cells *in vitro*, without prior immunization. NK cells use killing strategies that are different from those of antigen-specific cytotoxic T cells and from activated nonspecific killer macrophages and similar cells.

I. Killer Cells

A. Cell-Mediated Cytotoxicity

Cell-mediated cytotoxicity (CMC) is the lysis of target cells by effector killer cells of lymphoid and myeloid origin. [*See* LYMPHOCYTE-MEDIATED CYTOTOXICITY.]

1. Basic Model of CMC

Lysis requires intimate contact between killers and targets in three steps: (1) recognition (conjugate formation between killers and targets); (2) programming for lysis (includes triggering of synthesis and/or release of lytic molecules by the effector and the unidirectional production of an irreversible lesion in the membrane of the target); and (3) killer-independent lysis (the killers can be dissociated, and the target with its lesion will eventually die; the released effector can recycle and kill another target cell).

2. Targets

Target cells are usually established cell lines. In most cases, they are derived from human or murine leukemias or other tumors, although a number of nucleated cell types have been used as targets. They range from mitogen-induced blasts to fibroblasts infected by viruses, or even protozoan, plasmodial, or metazoan parasites at various stages of their life cycle. The prototypic targets with which most NK work has been done are the K562 erythroleukemia for the human studies and the murine leukemia virus-induced mouse lymphoma YAC-1 for the rodent studies.

3. Assays

The most common method used to measure cell lysis is the release of ^{51}chromium (^{51}Cr) from prelabeled target cells in suspension cultures. The ^{51}Cr binds randomly to internal cellular proteins and is released as a consequence of membrane damage by the lytic event. The maximal lysis produced by the effector cells is compared to the maximal release obtained after treating the labeled target cells with a chaotropic detergent; maximal release is about 80% of the intracellular ^{51}Cr. The results are usually expressed as "percent lysis," derived from equations comparing maximal release and release resulting from the effector–target interaction, corrected for spontaneous release by the target cells alone. Another way of expressing the results is as "lytic units," which are calculated from the linear part of the lysis curve and are expressed as the number of effectors required to produce an arbitrary "percent lysis," such as 20 or 25%. The maximal lysis induced by the killer cells at various effector–target ratios is reached at about 4 hr, although it may require longer assays (i.e., 18 hr), depending on the capacity of the target to repair the lytic damage, the type of effector cell used, and the type of lytic molecules produced by the effectors.

4. Mechanism of Lysis

Depending on the type of effector and target cells, target-killing can be mediated by three nonexclusive mechanisms. These mechanisms include granule exocytosis (the granules contain a variety of lytic molecules), pore formation on the cell membrane (actual holes punched by the lytic molecules in a manner similar to the lysis of cells mediated by complement), and the contact-induced, stimulated nuclear disintegration (a complex event, which has nuclear fragmentation as its end point and is probably mediated by activation of endonucleases). The problem with nuclear disintegration as such is in determining whether it represents the cause or the consequence of the lytic event that programs the cell for death (or suicide). (See Section II,D for more on lytic molecules.)

B. Major Histocompatibility Complex

Major histocompatibility complex (MHC) restriction is required to recognize gene products of the MHC of a given species by some effector cells. Cytotoxic cells can be divided into two categories depending on MHC restriction requirements: (1) T cells that kill in a specific MHC-restricted fashion (i.e., T cells from MHC type A donors immunized to a virus such as influenza will kill only MHC type A-infected targets and not MHC type B-infected targets), and (2) killer cells that do not require MHC recognition for target lysis. T cells have clonally distributed specific receptors for antigen (TcR), which allow recognition of viral and other antigens in the context of self-MHC. The diversity within the repertoire is generated by combinatorial rearrangements, and the interaction with antigen produces a progeny with immunological memory. The TcR structure, with its associated CD3 complex of pro-

teins, binds its specific ligands and transduces signals that trigger cell activation. Activation includes synthesis of a variety of lymphokines, some of them with lytic properties, and cell division. [See T-CELL RECEPTORS.]

C. Non-MHC Restricted Killers

Cytotoxic effector cells, which are independent of recognition of MHC gene products on the targets, include almost any lymphoid or myeloid cell types, such as monocytes, macrophages, B cells, some types of T cells, granulocytes, and even platelets, depending on the type of target cells and/or the stimuli for lytic activation used. Macrophages can kill a wide variety of tumor targets after activation. Many of these effectors will kill preferentially malignant or infected, but not "normal," targets. The nature of such selectivity is still undefined. Resting cells of different types, including T cells, can be made cytotoxic by simply "gluing" them to the target with adhesion lectins or antibodies, thus bypassing the need for the recognition step. Antibody-dependent, cell-mediated cytotoxicity (ADCC) is mediated by cells with receptors for the Fc portion (FcR) of immunoglobulin G (IgG), initially considered a special subset called K cells; however, it was later shown that most of the ADCC effectors belonged to the NK category. NK cells express the surface protein CD16, which is the low-affinity Fc receptor for IgG (FcRIII). In ADCC, antibodies, which are specific for a surface determinant on the target, will bind to the target, and the ADCC effector cells will bind to the antibody via its FcR. Alternatively, FcR+ cells could be "armed" by the specific antibody bound through its Fc portion, leaving the specific binding site available (see chapter on Immunoglobulins). Thus, the specific antibody will provide the "glue" and the FcR+ cell will provide the killer of the target. [See MACROPHAGES.]

NK and other types of effector cells do not require MHC recognition for target lysis, but neither do monocytes and macrophages. As the main property of the NK category, cytotoxic cells are found in fresh blood samples obtained directly from normal donors, and the killing is spontaneous, without any known triggering for activation (thus the "natural" appelative).

D. NK and Other Effector Cells

Terminology in NK studies is not simple. Initially, natural cell-mediated cytotoxicity (NCMC) included a variety of non-MHC-restricted cytotoxic cells obtained directly from blood (in most human studies) and from spleen (in most rodent studies) of normal healthy individuals who were able to kill *in vitro* some selected tumor targets. These targets included NKs, natural cytotoxic (NC) cells, NK-like cells capable of killing virus-infected cells, and a variety of killer cells, which were generated after *in vitro* culture or activation. Almost any cell that could kill the prototypic K562 and YAC-1 target cells was defined as being NK (including macrophages, promonocytes, basophilic granulocyte cell lines, etc.). During 1983–1988, attempts to establish a minimal terminology were made at workshops by researchers in the NK field. The 1983–1986 terminology included two main categories. (1) *NK*—freshly obtained cells with spontaneous cytotoxicity against some prototypic targets, which are lymphoid in appearance, are different from granulocyte–monocyte–macrophage–T cytotoxic cells, and show no MHC-restriction. In addition to the canonical NK (which kill YAC-1 or K562 leukemia targets in mice and humans), this group included some NK subsets with broader target specificity and the NC cells (see Section II.E). (2) *Activated killer (AK)*—includes all the non-MHC-restricted killers, which are produced after short- or long-term *in vitro* culture with or without added lymphokines such as interleukin-2 (IL-2). Although not stated, the AK category should also include the various types of established cell lines with NK, NK-like, or NC activities. The equation of a cloned cell line with NK activity with the prototypic NK cell obtained from a fresh sample has also created some disorder in the field. Because this classification could not exclude nonconventional T cells with NK-like activity from the first category, it generated strong argumentation about NK lineage and confusion regarding the surface-marker phenotype of the effector cells. The 1988 terminology is more restrictive and perhaps more precise. It includes three major types of non-MHC restricted effectors. (1) *NK*—large granular lymphocytes (LGL) that do not express the TcR-related CD3 surface-antigen and do not productively rearrange their beta, gamma, or delta TcR genes. Based on these criteria these cells do not belong to the T-lineage. (2) *NK-like T or non-MHC-restricted cytotoxic T cells*—a fraction of CD3+ T-lymphocytes, which express either the alpha-beta or the gamma-delta TcR, have LGL morphology, and usually kill the same prototypic targets as the NK LGL in the first group. The

NK and NK-like T terminology will be used in this text. (3) *Lymphokine-activated killer (LAK) cells:* activated *in vitro* with IL-2 for a few days, which can fall in either the NK and NK-like T non-MHC-restricted killers. The relative contribution of the T and NK components to LAK depends on the source of cells and the conditions for *in vitro* activation. [*See* LYMPHOCYTES.]

Stricto sensu, only the first category of cells would qualify for the NK appelative. The second category of cells belong to a defined lineage (T) and simply express the NK type of reactivity. LAK cells would fall within the AK category of the 1983 classification. LAK cells became popular in the nonscientific press around 1986 as a potential new treatment modality for advanced human cancer.

This article focuses on a functional aspect of cellular activity (i.e., the capacity to produce lytic molecules which can kill other cells in close proximity) which is also a property of the cells discussed above. The possible noncytotoxic functions of these cells will be mentioned below, with the proviso that in some cases (like the so-called "immunomodulatory" functions of NK cells, it is not clear whether the effects are direct or indirect; see also Section I,G).

E. A Comment on Lymphokines

Since NK cells are lymphokine producers as well as targets for lymphokines produced by other cells, it is worth including a brief comment on lymphokine behavior. The cloning and production of recombinant lymphokines demonstrated without a doubt that a single nonglycosilated recombinant homogeneous protein could have multiple effects on multiple target cells. The basic rules that emerged from studies on the defined recombinant lymphokines and cytokines are: (1) all lymphokine have pleiotropic activities (i.e., a variety of activities that are apparently unrelated to each other); (2) responses usually trigger a cascade of several mediators with synergy (and/or interference) between mediators; (3) targets for each lymphokine are widely distributed among lymphoid, hemopoietic, and other sites (including centran nervous system); (4) lymphokines have a high degree of functional redundancy (e.g., IL-1 and TNF share most functional activities although have limited structural homologies; IL-2, IL-4, IL-6 and IL-7 are growth factors for T cells; IL-1, TNF, IL-2, the interferons, IL-4, IL-5, IL-6 and IL-7 can affect B cell proliferation and function;

IL-1, IL-3, IL-4, and IL-7 affect hematopoietic progenitors); (5) in spite of extensive homologies there are some species restrictions for lymphokine activity (i.e., human IL-1 but not human TNF can act as co-stimulators for mouse thymocytes); (6) T cells produce several lymphokines after stimulation. Two views prevail: (i) The murine TH1/TH2 concept based on the combination of lymphokines produced (TH1: IL-2, interferon gamma, and lymphotoxin; TH2: IL-4, IL-5 and IL-6; with TNF, GM-CSF and IL-3 produced by both types) and (ii) Production of several lymphokines but no clear clustering seen especially with cloned human T cells. Furthermore, the lymphokine production profile of a given T line (especially with human T cells) may not be stable but is convertible by different stimuli; (7) correlation of *in vitro* optimized models may not correlate *in vivo* function; attempts at defining *in vivo* lymphokine biology using transgenic mice (with the lymphokine gene controlled by constitutive mouse promoters) have given unexpected results. For example, the double transgenic for human IL-2/p55 IL-2R showed reduced T and expanded NK functions, rather than the predicted expanded T; the high affinity IL-2 receptor is formed by combinations of two chains, p55 and p70, but both chains can be expressed separately and bind IL-2 albeit with low affinities). [*See* INTERFERONS; INTERLEUKIN-2 AND THE IL-2 RECEPTOR.]

F. History

The NK acronym was coined in 1975 for the spontaneous killer cell activity found in normal mouse spleen. Prior to the 1970s, NCMC was viewed as an unexpected nuisance during attempts to detect tumor-specific, cell-mediated immunity *in vitro* using lymphocytes from cancer patients, because cells from the normal controls could also kill the tumor targets. Although initially considered an artifact, the unexpected reactivity could not be dismissed by technicalities and became a subject of study. As a measure of acceptance of the NK phenomenon by the immunological community, NK appeared in the subject index of the *Journal of Immunology* in 1981 (lymphocyte appeared for the first time in 1948). Because the expected cytotoxicity in cancer patients had to be directed against tumor-associated antigens, as part of the prevalent views on tumor immunology, the first methodical NK studies in normal mice suggested that killing was directed against antigens coded by endogenous leukemia vi-

ruses. However, in 1976 the response clearly was not directed against murine leukemia virus, and the killing showed no MHC restriction. Human NK cells became better characterized, and in 1978 NK activity could be augmented by *in vitro* treatment with interferon in both humans and mice. NC cells were described in 1978, the augmenting effect of IL-2 on NK cells was described in 1981, and LAK cells were described around 1982. Based on the complexities of defining surface phenotype and lineage, as well as on variations on target preference for killing, the notion of activities mediated by an heterogeneous group of effector cells was proposed. Agreement was reached mainly on two NK properties: (1) NK activity was detected in fresh cell samples obtained from normal individuals in the absence of any prior sensitization, and (2) in addition to the independence from MHC restriction, NK cells did not show immunological memory nor clear evidence of clonality in response to a given target, all of which are properties of cytotoxic T cells. Around 1983, the ideas developed that NK and similar cells belonged to the innate or natural type of immunity and that its functions went beyond the potential anti-tumor activities. In spite of the large amount of information generated (the 1989 review article in the bibliography has 1,182 references), questions concerning lineage, receptors, antigens, and functions are still open. All the books cited in the bibliography are multiauthored containing from 12 to 219 individual chapters or papers and are good examples of the different opinions that still predominate in the field.

G. A Problem

Before describing the properties of classical NK cells, a conceptual problem must be discussed. The potential problem is to consider a given *activity* (such as lysis of certain tumor cells) as the *function* of a specialized cellular system of which the NK cell is the prototype, especially if we keep in mind that the lytic capacity can be triggered under appropriate conditions in most nucleated cells and even nonnucleated platelets. For example, if protein synthesis measured by the classic Lowry method could be used as a functional endpoint, the differentiation between B cells (which produce immunoglobulin), beta cells of the pancreas (which produce insulin), and hepatocytes (which produce an array of proteins) would be impossible. Based on that activity, one may conclude that the natural protein synthesis cell is heterogeneous and difficult to place within a given lineage. A somewhat similar situation has evolved regarding NK cells. Two views are prevalent. (1) *A unique cell subset*—NK represent a defined category of cells in the myeloid lineage (i.e., of bone marrow origin) with still undefined overall functions, which depend on their killer capacity (this being the more popular view). And (2) *a functional system*—the NK capacity to kill certain tumor targets is an activity expressed by many cells as part of their developmental or functional programs, rather than the function of a particular cell category. Two variations of both views are as follows. (1) *More than just killers*—the functions of the NK subset or system are related to other activities, such as production of γ-interferon (γ-IFN) or other cytokines. And (2) *complex functions*—the functions of the NK subset or system are complex and include, in addition to the original anti-tumor control of primary and metastatic tumor growth, other functions such as control of some infections (microbial, viral, parasitic, etc., which also include direct killing of certain bacteria), immunoregulatory functions (usually acting as down-regulators or natural suppressors), control of hematopoietic stem-cell growth (and differentiation), and some involvement in rejection of allografts, especially bone marrow grafts.

H. Defense Mechanisms

One of the unique properties of NK and similar cells is that they seem to be naturally activated and, thus, can act instantaneously in response to a foreign stimulus. All of the specific components of the immune system, such as antibody production by B cells or the cell-mediated responses of T cells (including MHC-restricted killing), require time-consuming priming, usually a few days. Priming is followed by clonal expansion and the specific responses against the invading agent as well as important amplification of the specific response by recruitment of a variety of nonspecific effector cells, such as macrophages, monocytes, etc. However, even the nonspecific effectors, such as macrophages, require activation to become lytic, which is usually provided in a functional cascade by the specific priming of the T-cell system. Thus, NK and related cells may act as early defense mechanisms, which may allow the development of the specific defense response, perhaps as part of a more primitive immune system with broad multispecific receptors for recognition. A variety of vertebrate species tested, from catfish to birds, have NK-like cells.

II. General Properties of NK and Other Effector Cells

The properties of the classical NK cells will be discussed in Sections II.A–D, and those of the other effectors in Section II.E. Unless otherwise stated, the descriptions apply to human cells. The phenomenological and mechanistic studies on NK and other NCMC effector cells in experimental animals and humans are derived, in most cases, from the analysis of at least three variables: (1) the donor providing the effector cells, 2) the type of target utilized, and 3) the type of *in vitro* assay employed. A fourth and often forgotten variable is the type of procedure used for the separation of the effector cells, which in some cases can activate them. The first variable is probably the most difficult to control, because at any given time a variety of stimuli may augment (or depress) NCMC effector activity. This intrinsic variation is well exemplified in a human study where the same group of volunteers were tested periodically for blood NK during a lapse of several years, showing almost all the possible permutations in NK activity levels.

A. Phenotypic Properties of NK

1. LGL Morphology

The identification of large granular lymphocytes (LGL) as a population enriched for NK activity in humans and rats, resulted in the tendency to equate NK to LGL. LGL are detected in peripheral blood using discontinuous density-gradient centrifugation on Percoll and are characterized by size and presence of azurophilic cytoplasmic granules. LGL are nonphagocytic, are usually nonadherent, and represent about 5% of the human peripheral blood lymphocytes. Two caveats: (1) although usually most of the NK activity is detected in the LGL fraction, not all LGL have killer activity; and (2) cytotoxic T cells obtained after *in vivo* stimulation or after *in vitro* culture, including specific MHC-restricted T-cell lines, may also have LGL morphology. Thus, the LGL morphology may represent a state of functional activation, poised toward the lytic process or secretion rather than the morphological counterpart of a defined cell subset, such as NK. Using single-cell cytotoxicity assays, with resting or activated blood LGL and a panel of NK susceptible targets, 50–85% of the LGL are capable of killing at least one type of target. The single-cell cytotoxicity assay uses effector–target conjugates transferred to semisolid media and the killing of individual target cells monitored by supravital staining (the dye will penetrate into the damaged cell, whereas healthy cells exclude it). One interpretation of the nonkilling LGL binders is that such cells are at a stage in which they cannot fully express their lytic capacity. Binding is necessary but not sufficient for target lysis, and some targets may lack the capacity to trigger the delivery of the lytic signal by the effector. Another view assumes that the limited number of targets used does not cover the whole repertoire of the NK–LGL cell, which explains why some LGL may not kill any targets.

2. Surface Antigens

The association of a surface marker and a cellular function, such as NK activity, is shown by either negative or positive selection. Negative selection is the elimination of NK activity after treatment of the cells with a given monoclonal antibody in the presence of complement; the positive cells that bind antibody will be killed by the complement. Another type of negative selection is the depletion by immunoaffinity procedures, such as panning, or by fluorescence-activated cell-sorting (FACS) of the fraction, which is positive for a given marker. Positive selection is the enrichment of NK activity in the cell fraction, which is positive for the marker using FACS, panning to antibody-coated plastic dishes or similar immunoaffinity procedures. Table I summarizes the available information on the expression of surface antigens in human blood lymphocytes with NK activity. There is no single surface marker which is unique for NK cells, and the phenotype is defined by a combination of some markers, such as CD16, CD56, and CD57, in cells that do not express CD3. Blood LGL can be divided into CD3+ and CD3− (i.e., T and non-T cells); most of the CD3− LGL express CD16 (the low-affinity receptor for the IgG Fc or FcRIII). Minor subsets of CD3+ T cells can also express CD16. The CD56 reagent, which detects 8–20% of positive cells in normal blood lymphocytes, defines three subsets that kill NK-susceptible targets: CD16− CD56+ CD3− (NK non-T cell, perhaps a precursor of mature NK), the major NK population of CD16+ CD56+ CD3− cells, and a CD16− CD56+ CD3+ NK-like T cell (about 25% of CD56+ are CD3+ and can also express CD8). With CD16 and CD57, four subsets

can be defined in human blood: CD16+ CD57− with high NK, CD16− CD57+ with detectable but low NK, CD16+ CD57+ with wide individual variation of NK, and CD16− CD57− with no NK activity. A fraction of all these subsets also can express CD3 and/or CD8 (i.e., T cells). CD2 was one of the first markers used to define NK subsets using the formation of rosettes with sheep erythrocytes (it was termed the E receptor [ER]); about 50% of human LGL were ER+. The CD2 cluster is the receptor for the LFA-3-antigen (CD58), which is expressed on many cell types and is involved in cell adhesion. CD2 is expressed on most NK and also on thymocytes, T-lymphocytes, and activated T cells. Most of the CD56+ cells, including those expressing CD3, are CD2+. Negative selection with antibodies recognizing CD11b (such as OKM1), CD16, and CD56 usually results in the depletion of most of the NK activity. However, as Table I shows, all of those markers are also expressed on non-NK cell types, including T and non-T lineages. The phenotype of mouse NK is discussed in Section II.E.

One important point to accent here is that many (probably all) of the surface structures listed in Table I, regardless of the cell type on which they are expressed, are not simple "markers" but molecules with transmembrane and cytoplasmic domains and other attributes which can participate in signal transduction leading to activation, including induction or augmentation of cytotoxicity and/or lymphokine production, when properly engaged by ligands or by specific antibodies directed against such structures, especially if cross-linked (i.e., CD2 with its internal ligand CD58/LFA-3 or anti-CD2 antibodies; CD11a/CD18 which form the LFA-1 heterodimer with its internal ligands related to ICAM-1; CD4 with its internal ligands related to Class II of MHC and CD8 with Class I of MHC; etc). The general term of "adhesion" or "accessory" molecules to these types of surface structures denotes their functional importance in signal transduction, cell adhesion, and cell locomotion and recirculation. These are areas which are beginning to be defined, and will be of importance in defining some of the rules for *in vivo* behavior of T, NK, and other cells.

B. NK Lineage

This issue can be viewed from two different angles. (1) If NK are actually a defined cell subset, the issue of lineage may have importance (see Section I.D); and (2) if NK are defined as an activity (killing of

TABLE I Surface-Marker Phenotype of Human NK Cells[a]

Marker	NK	T	B	MO/M	G
CD2 (LAF-3/CD58 R)[b]	+	+	−	−	−
CD3	−	+	−	−	−
CD4	−	+ S	−	+ S	−
CD7 (Leu 9)	+	+	−	−	−
CD8	+ S, L	+ S	−	−	−
CD11a (LFA-1 alpha-chain)[c]	+	+	+	+	+
CD11b (C3R alpha-chain)[c]	+ S, L	+ S, L	−	+	+
CD11c (P150/95 alpha-chain)[c]	+	−	−	+	−
CD16 (FcRIII; Leu 11)[d]	+	+ S?	−	+ A	+
CD18 (beta chain of CD11)	+	+	+	+	+
CD25 (IL-2R alpha gp55)	+ A	+ A	+ A	+ A	−
CD32 (FcRII)[d]	−	−	+	+	+
CD38 (Leu 17)	+ A	+ A	−	−	−
CD56 (Leu 19, NKH-1)[e]	+	+ A	−	+ L	−
CD57 (Leu 7, HNK-1)[f]	+ S	+ S	+ S, L	+ L	−
CD64 (FcRI)[d]	−	−	−	+	−
CD71 (transferrin R)	+ A	+ A	+ A	+	+
4F2[g]	+ A	+ A	+ A	+ A	+

[a] MO/M, macrophages–monocytes; G, granulocytes. + or − indicates positive or negative, respectively, for a given marker. R, receptor. S, (subset) indicates that only a subset is +. L (low) indicates that + cells express low amounts of antigen and are dull by immunofluorescence. A (activated) indicates markers present only on activated cells. CD stands for "cluster of differentiation," as agreed upon by the four International Workshops on Human Leukocyte Differentiation Antigens (Paris, Boston, Oxford, and Vienna, 1985–1989). At present, 79 CDs are categorized by preferential although not usually exclusive expression into T, activation antigens, B, myeloid, NK and nonlineage specific, and platelet. Leu, NKH, and HNK name various individual monoclonals, which in most cases defined the prototypic antigen in the CD and are still used in the older as well as some of the present literature. A CD is defined by more than one antibody reacting to the antigen. CD3, CD4, and CD8 are included as part of the T and non-T controversy discussed in the text. All are related to T-cell function. CD4 is also the receptor for HIV and is expressed on macrophages in humans and rats, but not mice.
[b] CD2 defines the receptor for the LFA-3 (CD58) antigen expressed on many cell types and involved in cell adhesion. Originally, it was characterized as the receptor involved in rosette formation with sheep erythrocytes (E receptor).
[c] CD11a, CD11b, and CD11c define the alpha-chain of molecules, which are always expressed on the cell surface in association with a common beta-chain (CD18).
[d] CD16, CD32, and CD64 define the three types of receptors for the Fc portion of IgG, which have different molecular weights and tissue distributions. Only CD16 (the low-affinity FcR, FcRIII) is detected on NK cells. In the 1989 Vienna Workshop, CD16 showed polymorphism with two possible alleles (NA1 and NA2), which until now have been studied only in granulocytes.
[e] CD56 is similar to the 140-kDa isoform of the N-CAM (neural cellular adhesion molecule) and is also expressed on neuroectodermal cells.
[f] CD57 defines a 110-kDa glycoprotein associated to myelin, and the monoclonal antibodies that define it are against the carbohydrate moiety of the molecule.
[g] 4F2 is a well-characterized activation marker present in many cell types and has not been clustered because no new specific monoclonal antibodies have been submitted to the Workshops.

prototype targets) mediated by a variety of T- and non-T-effector cells, the lineage argumentation becomes less interesting. For example, in both patients and mice with primary immunodeficiencies, such as severe combined immunodeficiency (scid), NK activity is intact, arguing against a lymphoid lineage because scid produces a developmental arrest of both T and B lineages. However, this may also imply that in the absence of the NK-like T activity, only the NK non-T activity is detected. The same interpretation could be applied to the intact NK activity in nude mice and rats (the nude genetic defect produces abnormal differentiation of the thymus and a profound deficiency of classical T cells). Conversely, most of the mouse strains with low NK activity (see Section II.E) have normal T-cell function. In both mice and humans, NK activity is dependent on an intact bone marrow function and can be considered myeloid by definition. Both experimentally induced and genetically determined bone marrow dysfunctions are usually accompanied by low NK function. In humans and rodents, bone marrow contains inactive precursors, which can be induced *in vitro* to express NK activity after cultures with IL-2 (but not with IFN). Such marrow precursors are still not well defined. Another example of the lineage argumentation relates to lytic mechanisms. Around 1982, some studies described that populations enriched for NK cells showed production of superoxide radicals, which are characteristic lytic mechanisms of the professional macrophages (granulocytes and mononuclear phagocytes), prompting editorials with titles suggesting that NK cells were macrophages "in lymphocyte's clothing." However, it was later shown that the detected oxidative bursts were not due to the NK cells but to minor populations of contaminating macrophages.

C. Regulation of NK Activity

1. Augmentation

Short pulses (24 hr or less) of cell populations containing NK cells with the various types of interferons (α-, β-, and γ-IFN) augments lytic activity. Augmentation results from the combination of increased killing efficiency per cell, increased number of cells that bind and kill, and an increased efficiency in the recycling of the killers. NK activity can be also augmented after *in vivo* treatment with either IFN or IFN inducers such as poly I : C. The capacity of IFN to augment NK activity is variable depending on IFN type, because some forms of α-IFN are inactive. Short pulses with interleukin 2 (IL-2) also augments NK activity, and synergy between IL-2 and IFN can be shown. IL-2 is also a growth factor for both NK and NK-like T cells, allowing the long-term growth of cell lines with NK activity. Although recombinant IL-2 has activity, suggesting a direct effect, an indirect role via induction of γ-IFN by the NK cells themselves may also be of importance in the augmentation. From these studies, a theoretical four-stage linear model for development of NK activity has developed: (1) an undefined progenitor (in marrow), which does not bind nor kill and may or may not have a distinctive surface phenotype (there is controversy about CD57 being such a putative marker); (2) a recognizable

precursor, or pre-NK, which binds but does not kill and may bear some distinctive surface markers; (3) the NK effector cell, which binds and kills, may have LGL morphology and the distinctive combination of surface markers; and (4) the activated NK (also known as augmented or boosted) discussed above.

2. Inhibition

This is a less-defined area, and the mechanism of inhibition remains to be defined. Prostaglandins (PG) are potent down-regulators of both spontaneous and IFN-augmented NK activity (especially PGE$_2$, PGA$_1$, and PGA$_2$, with no effect by the PGB and PGF series). A variety of *in vivo* procedures in mice seems to generate cells that can suppress NK activity. The procedures include treatments with various immunoadjuvants (*Corynebacterium parvum*, bacillus Calmette-Gerin or [BCG], etc.) and with some anticancer treatment modalities (whole body irradiation, corticosteroids, adryamycin). The NK suppressor cells are not well characterized but include macrophagelike (perhaps producing PG, as seen in *C. parvum* or steroid models) and nonadherent "null" cells (as seen in the irradiation and BCG models and, by some laboratories, in the *C. parvum* model). The human studies are more limited, but evidence indicates pulmonary alveolar macrophages can produce a dose-dependent inhibition of blood NK activity in *in vitro* mixing experiments.

D. Mechanisms of NK Activity

1. Recognition, Receptors, and Target Antigens

a. Conjugates LGL-NK cells can form stable conjugates after short-term incubation with prototypic targets such as K562 in humans and YAC-1 in mice, and a variable fraction of the targets will actually be killed as a consequence of such interaction. A conjugate is one target cell with at least one bound effector cell, usually conjugates include two to five effectors.

b. Receptors The receptors utilized by NK and NK-like T cells as well as the ligands recognized in the targets are still undefined in humans and rodents. Even the CD3+ NK-like T cells do not seem to use their TcR for mediating their NK activity, based on the lack of inhibition of lysis by anti-CD3 antibodies. Using cloned NK (and NK-like T) cell lines, which in essence represent expanded unique effectors, a clear functional heterogeneity has been found when tested against panels of tumor targets and/or virus-infected targets, with no clear pattern of reactivity being discerned. Thus, one common, recent view is that NK cells may interact with targets not through a single type of receptor but using several surface molecules and their corresponding ligands on the targets. The role of the adhesion molecules discussed in Section II.A.2 as part of the "receptor" system used by NK and NK-like T cells is beginning to be defined.

c. Antigens Concerning the target antigens, the only agreement is that the MHC gene products are not required for successful NK lysis to take place. An inverse relationship between expression of some MHC antigens (Class I) on target cells and sensitivity to lysis has been described, although direct proof is still lacking (i.e., the transfection of MHC into NK-susceptible targets such as K562 or YAC-1 did not change susceptibility to NK lysis). The hypothesis that NK cells can detect absence of MHC as an alternative recognition from the MHC-based recognition of the T-system is still not proven. Although some target-membrane products or antibodies against targets, which can partially block killing, have been described, no clear candidates for target antigens have been characterized. A rather large variety of structures from the cell membranes of NK-susceptible targets have been described in humans and rodents, which could be potential candidates for target antigens; however, none are characterized biochemically, and most of the effects cannot be generalized.

2. Lytic Molecules

The basic mechanisms for cell lysis were mentioned in Section I.A and include granule exocytosis, pore formation, and contact-mediated nuclear fragmentation. A soluble mediator from rodent and human NK cells, termed natural killer cytotoxic factor (NKCF), was described in 1981 and studied in several laboratories but is still not characterized biochemically. On the other hand, perforin (or cytolysin) is a protein of 55–75 kDa contained in granules of MHC-restricted cytotoxic T cells, NK and LAK cells which can form cytolytic porelike lesions on the target cells. Perforin has been cloned in mice and humans and shows some regions of homology with the pore-forming C9 protein of complement.

Although perforin seems to be the main lytic molecules of NK cells, the participation of other still undefined molecules cannot be ruled out. A variety of serine proteases have been described in the granules of cytolytic T cells (known collectively as granzymes), but their actual role in cell-killing is still undefined, and they may be accessory molecules.

E. NC and Other Effectors

1. NC Cells

During studies of specific CMC to anchorage-dependent nonlymphoid fibrosarcomas in mice, lymphoid cells from normal mice were found to kill syngeneic tumors, reminiscent of NK-killing. With inbred mouse strains, syngeneic means genetic identity in members of the same strain; allogeneic means genetic disparity between members of two different inbred strains. The natural effector cells against the fibrosarcomas had some properties different from NK (see below) but shared the same major properties of the NK categories discussed above (no MHC restriction, etc.) and were designated natural cytotoxic, or NC cells, in 1978. The prototypic target line in the early NC studies was Meth-A derived from a chemically induced sarcoma, which grew as an adherent monolayer and was replaced in 1980 by the WEHI-164 cell line, also derived from a chemically induced sarcoma but which grows in suspension and allows the use of standard ^{51}Cr-release assays. The breast cancer cell line BT-20 is the prototype for testing human blood cells with NC activity in ^{51}Cr release assays.

The properties of murine NK and NC show a number of differences, which will be further discussed below. (1) Phenotype: NK appear different from NC; (2) Regulation: NK are regulated mainly by IFN and IL-2; NC are not affected by IFN and regulated by IL-2 and IL-3; and (3) Lytic mechanism: NK cells kill targets by producing a pore-forming perforin; NC cells kill targets by production of tumor necrosis factor (TNF). Only points (2) and (3) have been analyzed in human NK–NC comparisons.

a. Kinetics for Lysis With Meth-A or WEHI-164 targets, it was apparent that the kinetics for lysis by NC was different from NK (and from MHC-restricted or unrestricted T cells): whereas NK and T produce a linear increase beginning at time 0, reaching maximal lysis at 4 hr, NC lysis showed a 4–8-hr lag followed by a linear increase reaching maximal lysis at 18 hr. One explanation for the time gap is the capacity of the targets to repair the early lytic lesion. Such counterlysis is dependent on protein synthesis, can be reversed by protein-synthesis inhibitors, and is the usual basis of all the long-term killing assays, regardless of the actual effector cell (some NK-susceptible targets need long-term assays for killing by NK). Because NC cells kill targets by production of TNF in mice and humans, any TNF producing effector cell, such as monocytes from mice and men, will kill WEHI-164 in 12–18 hr, as will recombinant human or murine TNF. However, the lytic assays can be reduced to 4 hr by using WEHI-164 targets pretreated with protein-synthesis inhibitors.

b. Target Type
The initial studies with murine NK were done using lymphoma or leukemia targets, whereas the identification of NC used nonlymphoid tumor targets, mostly fibrosarcomas. Although never intended as such, the resultant NK–lymphoid and NC–nonlymphoid association was propagated in the literature. It became so established that for a while investigators defined the type of natural effector cells as NK or NC based on the morphology or anchorage-dependency of the target cells, or conversely, they described "NK cells that also kill nonlymphoid targets" as if they represented a special category of effectors. However, neither morphology nor anchorage-dependency determine the type of effector cell that may kill it. Concerning morphology and tumor type, when a number of murine lymphoid and nonlymphoid tumor targets were screened, they usually fell within three main categories: (1) resistant to killing by fresh cells, (2) sensitive to killing by AK cells of either NK or NC type (a small minority are resistant to AK cells), and (3) sensitive to killing by fresh NK or NC (this being a minority and representing mainly the prototypic tumors). Most of the nonlymphoid tumors tested fall within the first two categories (resistant to fresh NK or NC and killed by AK). This statement applies to established cell lines; targets prepared from fresh tumors are usually resistant to fresh cells and only a fraction of them can be killed by AK cells. A similar picture has been seen with human tumor targets. In mice, the *in vivo* passage of NK- or NC-susceptible tumors through a syngeneic host produces variants that become NK- or NC-resistant when tested again *in vitro*. Resistance can be reversible (target cells recover susceptibility after a few weeks in culture) or permanent (the resistant phenotype remains regardless of time in culture). These types of fluctuations suggest *in vivo* selection toward NK or NC resistance which may include modulation of the susceptible phenotype (the cells become resistant but recover susceptibility *in vitro*) or deletion of the susceptible cells (producing the irreversible resistant variants).

c. Recognition, Receptors, and Conjugates
As is the case with NK, the target ligands being recognized by NC are still undefined. Mouse NC form conjugates with their targets and usually kill about 50% of the targets in the conjugates, and TNF mRNA can be detected by *in situ* hybridization on the effector cells.

d. Ontogeny and Maintenance
Murine NK and NC cells differ in their ontogeny and in their persistence in adult and aged animals. Whereas NK activity in spleen develops at approximately 3–4 weeks of age and declines to low levels in adult mice, splenic NC is present since birth and remains constant for the whole life span of the animal. However, no decline of NK activity is observed in murine peripheral blood lymphocytes. NK in humans is detected in blood since birth and is maintained throughout life, although blood NK levels vary widely, even in serial testings of the same individual within a given time. There are no similar studies done in humans for NC activity.

e. Genetics
In mice, marked differences in strain distribution of activity have been described for NKs and NC cells, although in both cases strong influences from a given region (the D-end) of the murine MHC have been detected. A variety of non-MHC linked genes also seem to affect NKs and NC cells. A number of strains show low NKs with normal NC activity and can be divided into two groups: (1) strains with low NK activities, which are augmentable (although not totally normalized) by IFN and IFN-inducers and (2) mouse strains with low or undetectable NK activities, which are not augmentable by IFN and IFN-inducers. Several strains are the prototype of the augmentable NK deficiency including mice homozygous for the beige mutation, which produces among other effects a generalized lysosomal defect and low NK activity (this is the homologue to the human Chediak-Higashi syndrome, which is also accompanied of low NK activity). Some human examples of nonaugmentable NK deficiencies are beginning to be described.

f. Surface Phenotype One difference between murine NKs and NC cells is that while a variety of cell-surface markers are detectable on NK cells, NC cells were consistently negative for those markers as defined by negative selection with antibodies and complement. Thus, NC cells were "null" cells until 1986, when it was shown that NC activity in murine spleen could be enriched in factions that were positive for a variety of surface markers using positive selection with the FACS. The most used surface markers in murine NK studies are: Thy 1 (expressed on most T cells; the human analog is not expressed on T cells; about 50% of murine resting NK cells are Thy 1+; most of the murine AK cells are Th1 +), Qa 5 (not characterized biochemically, no human equivalent defined; all resting and activated NK are Qa 5+, a fraction of T cells are also Qa 5+), NK 1.1 and NK 2.1 (expressed on NK cells; claimed to be a specific marker for NK; although recent studies have shown the existence of CD3+ NK 1.1+ cells, not defined biochemically and no human equivalent), and the ganglioside asialo GM-1 (aGM1; present on NK and T cells). There are no monoclonal antibodies against murine NK 2.1 and aGM1; thus, questions about specificity of these markers still remain. NK 1.1 detects a small fraction of splenic cells (1–5%) where most of the NK activity is found and is the main argument for a unique subset expressing NK functions. Comparing cloned NKs with cloned MHC-restricted T cells, it is apparent that, in addition to having a similar LGL morphology, both types of clones expressed Thy 1, Qa 5, and NK 1.1 or 2.1, the main difference being that the T clones expressed CD8, whereas the NK clones did not. Thus, as it is the case with the human NK studies, and in spite of the periodic claims of exclusive serological NK markers, the picture remains complex. By FACS analysis, it was shown that NC cells in murine spleen were enriched in the fractions that were positive for Thy 1 (some activity in the Thy 1−), Qa 5, and CD4 and the fractions that were negative for CD8 and surface Ig (sIg), and that both the NK 1.1+ and NK 1.1− fractions expressed NC cells. Thus, by single parameter FACS analysis the murine splenic NK phenotype is Thy 1+ or 1−, Qa 5+, NK 1.1 or 2.1+, CD4−, CD8−, sIg−, and aGM1+, whereas the NC phenotype is Thy 1+ (or 1−), Qa 5+, NK 1.1+ or 1.1−, CD4+, CD8−, and sIg− (aGM1 not tested). Both NKs and NC cells were enriched in the larger size fractions as defined by forward light scatter. The conclusion of the NC studies was that NC activity was mediated by cells from various lineages including T and non-T. Two main differences regarding surface phenotype between NKs and NC cells are as follows. (1) Although expressing surface antigens, the cells with NC activity appear refractory to negative selection with antibody and complement; and (2) NC activity is detected in a CD4+ T-cell subset not observed in NK (in mice CD4 is expressed only on T cells; in humans and rats it is expressed also on monocytes and macrophages). Although TNF was initially shown as a monokyne, recent studies showed that in both humans and mice T cells (especially CD4+) could produce TNF under appropriate stimulation, supporting the finding of NC activity in the CD4+ T cells in mice. As was the case for human NK cells, the mouse NK 1.1+ cells cultured in IL-2 do not rearrange the beta or gamma genes of the TcR (i.e., are not T cells), whereas the non-MHC restricted T cell clones with NK activity show rearranged beta genes of TcR. No such studies have been performed yet with cells with NC activity.

g. Regulation Cells with NC activity are augmented by short pulses with IL-2 and interleukin-3 (IL-3) and not by any type of IFN in both mice and humans, whereas NK cells are augmented by short pulses with IFN and IL-2 (see Section II.C).

h. Lytic Mechanisms NC cells in mice and humans kill targets by production of TNF, a well-defined cloned cytokine that is different from the perforin utilized by NK and T cells (see Section II.D). In addition to the cytotoxic–cytostatic activities, TNF has a wide range of other activities in inflammation and immunological regulation (in this latter category, TNF functions overlap with those of interleukin-1).

In summary, NC cells differ from NKs in phenotype, regulation, targets for recognition, and lytic mechanisms, although they share with NKs the properties of being independent of MHC restriction and detected in fresh samples of blood or spleen.

2. NK Cells that Kill Virus-Infected Targets
Human herpes simplex virus (HSV)-infected fibroblasts become sensitive to lysis by a CD16+ NK-like effector, which shows some minor differences from the K562 killers although shares most other properties with classical NKs. The differences include separate variation of activity (low NK-K562 and normal NK-HSV, or the reverse situ-

ation in patients with the Wiscott–Aldrich syndrome) and the requirement for accessory cells for the lysis of the HSV-infected targets, not required by the NK-K562 (the requirement of accessory cells expressing HLA-DR of the MHC has also been described for the lysis of cytomegalovirus-infected fibroblasts by CD16+ NK cells). A somewhat similar system has been described in the mouse against HSV-infected fibroblasts. With other virus-infected targets, such as murine hepatitis virus, natural killing can be mediated by non-NK cells, including B cells, which kill by direct contact. [*See* HERPES-VIRUSES.]

3. AK Cells, Including LAK

As mentioned in Section I.D, a variety of NK-like effector cells have been produced after *in vitro* culture or activation and have been grouped under the AK category. The common AK characteristic, regardless of the procedure used for its generation, is that it usually kills a broader spectrum of targets than the resting NK cells, in some cases including tumor cells obtained from fresh explants in addition to tumor lines. AK cells have been produced by a variety of procedures, including short-term cultures in the presence of heterologous serum (such as fetal calf serum), allogeneic irradiated cells (as in the mixed lymphocyte response), or by addition of crude or purified lymphokines such as IL-2. The earlier studies generated a complex terminology with words such as "anomalous" and "promiscuous" for the killer cells detected in culture. Lymphokine activated killers (LAK) are among the better-studied AK cells, due to their potential use in cancer therapy. LAK are generated from blood lymphocytes after 3–7 days in culture with high dosages of recombinant IL-2 and the phenotype of the progenitors (i.e., the cells at the time of the initiation of the culture) and the effector LAK generated after the culture is identical: mainly associated with LGL, mainly CD16- and CD56-positive, and mostly CD3 negative. It is agreed that LAK is a phenomenon rather than a distinct effector lineage and that in both humans and rodents most of the LAK activity in blood (or spleen in mice) is attributable to IL-2 activated and expanded NK cells. The contribution of NK-like T cells to the LAK compartment is variable and depends on the type of cells used to initiate the LAK cultures. For example, culture of human thymocytes for 5–8 days in recombinant IL-2 would generate LAK cells that will express CD56 but lack CD16 and CD57 and are

mostly CD3−, suggesting derivation from the more immature CD3− thymocyte populations. In cultures of peripheral blood cells with IL-2 and anti-CD3 for up to 20 days, the cells with LAK activity were shown as either CD3+ CD16− NK-like T cells or CD3− CD16+ NKs; these studies also showed that the conventional CD3+ CD4 or CD8+ T cells do not seem to contribute to LAK formation, suggesting the T-like LAK progenitor in the CD3+ CD4− CD8− minor subset of blood T cells. The murine studies also support the view that the T-like LAK activity is derived mainly from the CD4− CD8− subset. IL-4, a different lymphokine with various affects on B and T cells can also induce LAK cells from mouse spleen and synergize with IL-2, and TNF can synergize with IL-2 in LAK generation. Such interactions could be predicted from lymphokine studies in nonlytic systems.

III. Function

Functions of NK and other effector cells have been deduced based on the effects of induced or genetic deficiencies of NK activity *in vivo* and in some *in vitro* tests where NK cells have been shown to modify some type of response by another cell type. Most of the studies have been done in experimental animals and at present writing the functions can only be inferred for NK.

The following functional properties of NK can be considered: (1) *Anti-tumor effects*—good experimental evidence for a role of classical NK in the control of blood borne metastases, suggestive evidence that NK may influence local growth of primary tumors, and no clear evidence that NK may act as an immunological surveillance mechanism preventing tumor development (i.e., NK deficiencies do not show increased risk for tumor development). (2) *Control of infections*—in addition to the direct lysis of some types of bacteria, NK cells have been implied in resistance to some viral infections (cytomegalovirus, herpes simplex, murine hepatitis, etc.), some parasitic infections (by *Plasmodia, Babesia, Toxoplasma, Giardia,* etc.), some fungal infections (*Candida*), and some bacteria (such as *Brucella* and *Listeria*). It is worth noting that resistance to infections is multifactorial, rarely attributable to a single defense mechanism, and probably are but a component in the response. (3) *Immunoregulatory functions*—of two types: (i) Lymphokine production: The production of IL-1, γ-IFN and IL-4 by NK cells is well documented; and there are reports of production of IL-2, IL-3, alpha-interferon, CM-CSF and TNF by CD3- LGL, which depend on the triggering stimulus used. The production of TNF by NC cells, combined with the pleiotropic nonlytic immunological effects of TNF could be another example of possible immunoregulation via lymphokine production. And (ii) "direct" effects by cells: The regulation of immune responses by NK cells is supported by some examples where NK-like cells affect antigen presentation or simply down-regulate a response; in the latter case the regulatory cells have been termed "natural suppressors". (4) *Control of hematopoietic stem-cell growth and differentiation*—NK cells have been shown to directly affect the differentiation of lineage-committed stem cells *in vitro,* such as colony-forming units of the granulocyte–macrophage lineage). NK cells can also influence hematopoiesis by production of cytokines. (5) *Transplant rejection*—NK-like cells are involved in the rejection of bone marrow transplants in mice and probably in humans. In mice, NK cells appear to mediate the phenomenon of hybrid resistance. NKs seem also involved in the development of graft-versus-host disease in mismatched marrow transplantation. The detection of cells with NK markers in organ allograft rejections has also suggested a possible role in organ transplant rejection. (6) *Disease*—the evidence of a possible role in hematological and other diseases is only circumstantial and is mentioned only for the sake of completeness. [*See* IMMUNE SURVEILLANCE.]

IV. Recapitulation

These final comments apply to the NK and other effector cells described in Section II.E.

What are NK cells? For the classical NK, the definition is reduced to what appears to be a subset that shares phenotypic properties with both T and myeloid (monocyte–macrophage and granulocyte) lineages. Two unresolved views are NK as a unique subset versus NK as an activity shared by different cell types, including the NK-like non-MHC-restricted T cells. The activated killers (AK) generated in culture, such as LAK, show the same unresolved issue concerning lineage.

What do NKs recognize? What is recognized on the targets is still undefined; what makes a target susceptible or resistant to NK lysis is still unde-

fined. The receptor(s) on the NK cells are also still undefined.

How do NKs work? Concerning the lytic activity, NKs, NK-like T, and probably LAK cells kill the targets by production of perforin, a pore-forming lytic molecule. NC lytic activity is mediated by TNF, a lytic molecule that produces cell death by undefined mechanisms, which include nuclear fragmentation. The other potential functions are just beginning to be defined.

What do NKs really do? Section III shows a listing of normal and pathological situations where NKs and NK-like mechanisms may have an *in vivo* role. Of all the potential functions described, the most physiological one seems to be that related to cytokine production, which may be involved in regulation of hematopoiesis and immune reactions. On the other hand, as a defense system, NKs had unique qualities (discussed in Section I.E) to act as the first line, or defense, against circulating tumor cells and infectious invaders. Finally, NKs and especially AK cells of the LAK type may be useful as therapeutic agents in advanced cancer and perhaps other pathological conditions.

Bibliography

Herberman, R. B. (ed.) (1980). "Natural Cell-Mediated Immunity Against Tumors." 1309 pp. Academic Press, New York.

Herberman, R. B. (ed.) (1982). "NK Cells and Other Natural Effector Cells." 1546 pp. Academic Press, New York.

Herberman, R. B., and Callewaert, D. M. (eds.) (1985). "Mechanisms of Cytotoxicity by NK Cells." 664 pp. Academic Press, Orlando.

Lattime, E. C., Stoppacciaro, A., Khan, A., and Stutman, O. (1988). Human natural cytotoxic activity mediated by tumor necrosis factor: Regulation by interleukin-2. *J. Natl. Cancer Inst.* **80,** 1035.

Le, J., and Vilcek, J. (1987). Biology of disease. Tumor necrosis factor and interleukin 1: Cytokines with multiple overlaping biological activities. *Lab. Invest.* **56,** 234.

Nelson, D. S. (ed.) (1989). "Natural Immunity." Academic Press, Sydney, pp. 12–828.

Podack, E. R. (ed.) (1989). "Current Topics in Microbiology and Immunology: Cytotoxic Effector Mechanisms." 135 pp. Springer-Verlag, Berlin.

Reynolds, C. W., and Wiltrout, R. H. (eds.) (1989). "Functions of the Natural Immune System." 479 pp. Plenum Press, New York.

Shimizu, A., Kinashi, Y., and Honjo, T. (1989). Structure and function of lymphokines and their receptors. *In* ''Progress in Immunology VII'' (F. Melchers and 17 other, eds.), p. 601. Springer-Verlag, Berlin.

Stutman, O., and Lattime, E. C. (1986). Natural killer and other effector cells. *In* ''Immunology and Cancer,''

M. D. Anderson Symposium on Fundamental Cancer Research, Vol. 38 (M. L. Kripke and P. Frost, eds.), p. 221. University of Texas Press, Austin.

Trinchieri, G. (1989). Biology of natural killer cells. *Adv. Immunol.* **47,** 187.

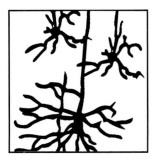

Neocortex

BARBARA L. FINLAY, *Cornell University*

Glossary

Cortical column Fundamental cellular grouping of the neocortex, a collection of cells extending perpendicular to the cortical surface, spanning all cortical layers, receiving and operating upon a common input

Cytoarchitecture Description of the organization of a structure according to the types, sizes, and arrangements of its cells

Lateralization Property in which certain sensory, cognitive, and motor functions are represented preferentially on one side of the neocortex

Modularity Property of the neocortex in which the circuitry for particular sensory, motor, and cognitive functions is kept physically and computationally separate from other sensory, motor, and cognitive functions

THE NEOCORTEX is a layered sheet of cells covering the surface of the forebrain. Homologues of the cells that make up the neocortex can be found in the forebrain of all extant vertebrates, including fish, amphibians, reptiles, and birds, but only in mammals are these cells arranged in the six-layered structure of repeating subunits, termed the neocortex. The neocortex's six layers consist of specialized zones for input, for communication with other areas of the neocortex both locally and at a distance, and for output to noncortical structures. A column of cells arranged perpendicular to the corti-cal surface tends to perform a standardized computation on its input; the input can vary greatly. Through this relatively uniform structure information passes for functions as diverse as recognizing faces, conversing, playing the piano, planning for the future, and adjusting emotional displays to the social context. Components of these capacities are represented in particular parts of the cortical surface, a property termed modularity, and thus local damage will cause disruptions of particular skills and spare others entirely. The human neocortex is lateralized for some computations: In the great majority of individuals, the circuitry for language is located on the left side of the brain.

I. Evolution of the Neocortex

A. Basic Questions

The neocortex is the single largest structure in the human brain and, because of its prominence and importance for many capacities viewed as distinctly human, such as language and pronounced manual dexterity, there has been much inquiry into its origin. Three questions have been asked about the evolution of the neocortex. First, what is the origin of the cell types that make up the neocortex? Second, how do these cells come to have their characteristic layered organization? Finally, what might explain the pronounced enlargement in volume of the neocortex seen most notably in primates, but which has occurred independently in a number of mammalian radiations?

B. Origin and Organization of Cell Types

In many classical textbooks, the neocortex is presented as a structure that is found *de novo* in the mammalian brain. More modern anatomical explo-

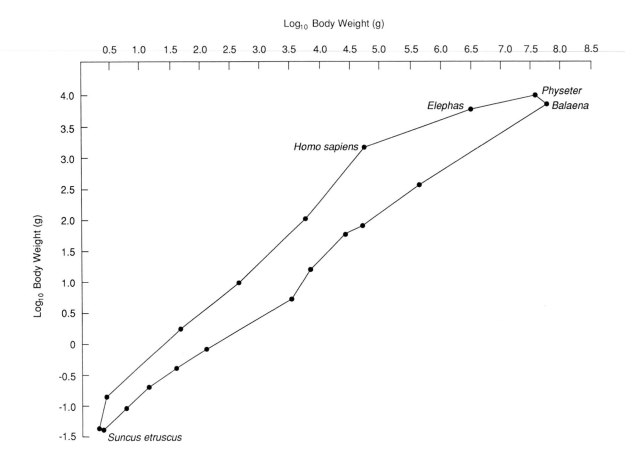

FIGURE 1 Convex polygon bounding the set of mammalian brain–body weight values. Logarithmic scale. Individual species identified represent shrews, humans, elephants, and two whales. (Reprinted, with permission, from Eisenberg, J. F., 1981, "The Mammalian Radiations," University of Chicago Press, p. 227.)

rations of vertebrate brains have demonstrated that this is not the case. All vertebrates possess a number of organized cell masses, termed nuclei, in their forebrains. Even in extant vertebrates believed to have a brain organization resembling the most primitive vertebrate conditions, the lampreys and hagfishes (class Agnatha), cell groups can be found that have, in part, the approximate location, types of input, patterns of connectivity, neurotransmitters, and neuromodulators that are characteristic of the mammalian forebrain and, specifically, the neocortex. The characteristic mammalian pattern of layering is absent, and many aspects of the connections and organization vary. The neocortex of mammals and homologous forebrain structures in vertebrates should be viewed in much the same way the relationship between a human arm and bat wing might be viewed. The major bone and muscle masses have similar embryonic origins, similar topologic relationships and attachments, and similar gross functions. However, the absolute and relative size of the components vary considerably; particular bone and muscle groups may be added,

deleted, or combined, and details of the attachments of bone and muscle may change.

In reptiles, the structure thought to be homologous to the neocortex is called the dorsal ventricular ridge; it receives and integrates sensory input through the thalamus and distributes this information to the midbrain and hindbrain, as does the neocortex. In birds, the laminarly arranged cells of the dorsal telencephalon, and also various divisions of cell masses of the forebrain area, called the corpus striatum, are thought to be homologous to the neocortex. Because these areas receive and integrate the same sort of information that the mammalian neocortex does, functional homology is found as well as anatomical homology. For example, bird song, a complex communicative system that involves auditory learning, elaborate sensory integration, and skilled motor performance, depends in part on the areas in its brain homologous to the neocortex in mammals and humans.

The difference in the developmental programs that produce a layered arrangement of cells rather than a collection of nuclear masses is not yet known, but a difference in the pattern of cell migration during development is likely to be involved, which will be discussed below.

C. Change in Volume of the Neocortex across Mammalian Radiations

The entire brain shows striking differences in volume across mammalian radiations and is most pronounced in primates (Fig. 1). Some of this change in volume can be attributed to change in body size: not surprisingly, bigger bodies are associated with bigger brains. However, at any particular body size, there is a residual variation of at least 10-fold in the relation of whole brain size to body size. The size of the neocortex, in turn, bears an exceedingly regular relationship to whole brain size (Fig. 2): The volume of the neocortex increases exponentially

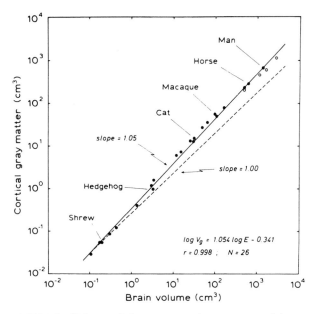

FIGURE 2 Volume of the neocortex (gray matter only) as a function of whole brain volume for a number of mammals. Logarithmic scale. The dashed line represents scaling of cortical volume by the first-power geometric similarity and shows that the amounts of neocortex increases at a greater rate than that expected by geometric scaling alone. Dolphins and whales are indicated by open circles. (Reprinted, with permission, from Hofman, M. A., 1989, On the evolution and geometry of the brain in mammals, *Prog. Neurobiol.* **32**, 137–158.)

with whole brain size, so that larger brains are composed of an increasingly greater percentage of neocortex. The human brain does not differ from all mammalian vertebrates in this respect. Humans have the largest brain–body size ratio of any vertebrate, and a great deal of this hypertrophy is accounted for by the neocortex. However, the amount of neocortex in humans is the amount lawfully predicted from whole brain size.

II. Fundamental Structure and Physiology of the Neocortex

A. Cortical Layers and Columns

When the anatomical organization of the neocortex was first described in the latter half of the nineteenth century, many different schemes for the naming and ordering of the layers of the neocortex were proposed. The scheme that has persisted is the six-layered scheme, laid out by Brodman in his publications on cortical cytoarchitectonics in the period 1903–1920. The neocortex consists of a number of distinct cell types and fiber bands, arranged in strata. Variations in local areas of neocortex are described as having condensations, omissions, or subdivisions of Brodman's fundamental six strata (Fig. 3). These layers will be discussed not in their numerical order but in the order information passes through them (Fig. 4).

The principal input to the neocortex comes from the thalamus, a collection of nuclei in the diencephalon that receives information in turn from various sensory domains, from various areas from the midbrain to forebrain, and from the neocortex itself. The input from the thalamus distributes to layer IV and, to a lesser extent, to the upper part of Layer VI. Layer IV is termed a granular layer, in that it is composed of small, relatively symmetric cells with radial dendrites, called stellate cells (Fig. 5). Several types of stellate cells can be further distinguished, as well as multipolar neurons associated with a variety of neuromodulators in this layer. The processes of cells from layers V and VI also extend through this layer. [*See* THALAMUS.]

This information is then relayed up and down, to layers II and III, and to layers V and VI. The bulk of local interactions in the neocortex are restricted to a column, several cell diameters wide, that extends perpendicularly from layer IV to the cortical surface and down to layer VI. This local interac-

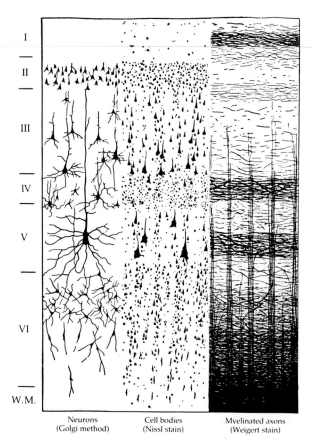

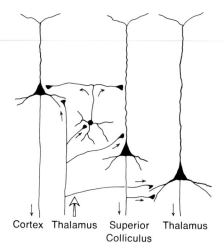

FIGURE 4 Diagram of information flow through the neocoretex. Principal input is from the thalamus to the stellate cells of Layer IV (see also Fig. 3). Cortical output distributes to the rest of the neocortex, to the thalamus, and to many subcortical structures of which one, the superior colliculus, is shown here. (Reprinted, with permission, from Shepherd, G. M., 1988, "Neurobiology," Oxford University Press, New York.)

Cortex   Thalamus   Superior   Thalamus
                    Colliculus

Neurons   Cell bodies   Myelinated axons
(Golgi method)   (Nissl stain)   (Weigert stain)

FIGURE 3 Layers of the neocortex, as described by Brodman. Cortical neurons and axons are arranged in six principal layers, designated by roman numerals. Three types of stains are represented: the Golgi method, which stains whole cells and all their processes; the Nissl stain, which shows only cell bodies; and the Weigert stain, which stains axons. (Reprinted, with permission, from Angevine, J. B., and Cotman, C. W., 1981, "Principles of Neuroanatomy," Oxford University Press, New York.)

tion and distribution of information gives rise to the anatomical and physiological unit of the "cortical column." Layer I, the cell-free outermost fiber layer, is composed of the axons and dendrites of these cells engaging in local interactions. It should be emphasized, however, that longer-range cellular interactions are also important in producing many of the characteristic features of cortical information processing.

Layers II and III are composed of pyramidal cells (Fig. 5), asymmetric cells with a profusion of dendrites at their base, and a long apical dendrite, all of which receive synaptic input. Multipolar neuromodulatory cells are also found in layers II and III. Output from layers II and III is long range and principally intracortical. Axons from these areas dis-

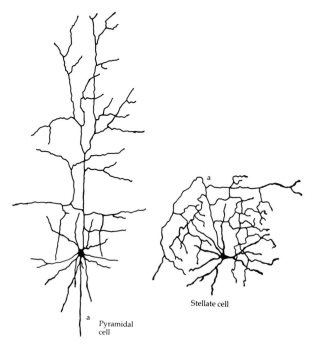

Stellate cell

a
Pyramidal
cell

FIGURE 5 Drawings of a typical pyramidal cell and stellate cell. Pyramidal cells are especially abundant in layers II, III, and V, and stellate cells are found in Layer IV. (a) Designates the axon of each cell. (Reprinted, with permission, from Angevine, J. B., and Cotman, C. W., 1981, "Principles of Neuroanatomy," Oxford University Press, New York.)

tribute to local cortical areas (e.g., from primary to secondary visual neocortex), to distant cortical areas (e.g., from secondary visual neocortex to visuomotor fields in frontal neocortex), and across the corpus callosum to the neocortex on the other side of the brain. The connections across the corpus callosum can link both corresponding and noncorresponding cortical areas. These intracortical connections distribute principally to layers II and III, and Layer V. [See HEMISPHERIC INTERACTIONS.]

Layer V also consists principally of pyramidal cells. These receive input from layers II, III, and IV and distribute their axons subcortically. These subcortical areas are quite diverse and include the basal ganglia and amygdala, the prominent nuclear masses of the forebrain; the superior coliculus, a midbrain structure concerned with attention and eye movements; and a variety of brain stem sensory and motor nuclei. Some giant neurons from the motor cortex, named Betz cells, send their axons to as far as the spinal cord to terminate directly on motor neurons and associated interneurons.

Layer VI contains cells of a variety of morphologies. A principal output connection of Layer VI is a reciprocal connection to the area of the thalamus that innervates the same cortex.

This same organization of layers and repeating columns is found throughout the neocortex. Local subareas of neocortex (discussed below) are modified in such a way that their specialization is reflected. Primary visual cortex, which receives a massive thalamic input of visual information, has a large number of cells in Layer IV. In motor cortex, Layer IV is almost absent, and the cells of Layer V, the subcortical output layer, are unusually large and prominent. These local differences can be employed to divide the neocortex into cytoarchitectonic areas, as Brodman did with his original maps (Fig. 6), the nomenclature of which is still in use today. These cytoarchitectonic divisions, based solely on the visualizable detail of the cellular organization of the neocortex, have proven to typically correspond to functional divisions in the neocortex as well.

B. Cortical Physiology

Electrophysiological recording from single neurons in the neocortex, first undertaken by Mountcastle in the somatosensory cortex and by Hubel and Wiesel in the primary visual cortex, amplified and extended the neuroanatomical picture of columns and layers

in the neocortex. First, the best stimulus to excite single cortical cells was typically a complex transformation of the sorts of stimuli that best excite thalamic cells. In the case of the primary visual cortex, the best stimulus for a typical cortical cell would typically be a bar or edge in a particular orientation, in a specific location in the visual field (Fig. 7). Often, cells would respond only to spatially congruent information with the right and left eye stimulated together. The thalamic input to this area, however, is not binocular, and the visual fields of thalamic neurons are spatially symmetric, not elongate. In the somatosensory cortex, a preferred stimulus would typically be a submodality of touch, like light touch or hot or cold, and cells would often be selective for a particular direction of stimulus movement. All of the neurons in a column perpendicular to the cortical surface had similar selectivity for the appropriate stimulus properties, such as orientation, binocular integration, touch submodality, or location of the receptive field on the skin surface or visual field. Neurons in the different cortical layers vary systematically in some aspects of receptive field structure (e.g., the degree of specificity for location in the visual field), but overall, all neurons in a column process the same type of input. [See SOMATOSENSORY SYSTEM; VISUAL SYSTEM.]

In areas of neocortex that do not receive direct sensory input, the properties of single neurons are often complex combinatorial properties of neurons. These combinations can occur both within and between sensory modalities and can involve aspects of both motor behavior and prior learning cognition. For example, in the neocortex of monkeys and sheep, neurons have been described whose optimal stimulus is the face (or aspects of the face) of the animal's own species. In the area of bat neocortex that processes information relating to echolocation, neurons will respond optimally to an auditory stimulus that is the bat's own call and, with a particular delay, the echo of that call, which thus specifies a target range. In the parietal cortex of monkeys, which is located between the visual and somatosensory–motor areas of the cortex and which receives input from both, neurons can be found that will fire only when a monkey is looking at and reaching for an object of interest. In the motor cortex, the response patterns of neurons can best be related to movements or limb positions that involve a number of muscle groups, and not a single muscle's contraction. In the frontal cortex, neurons have been described that fire only when the animal is attending to

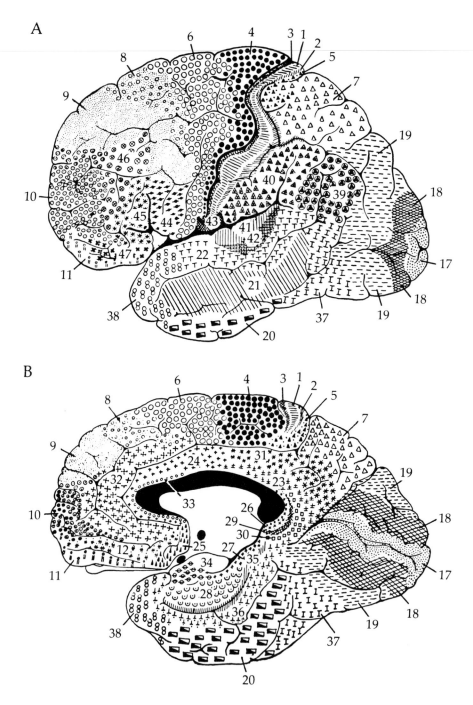

A

B

FIGURE 6 Cytoarchitectonic areas of the neocortex according to Brodman (1909). Lateral (A) and medial (B) views of the neocortex are shown. The numbered divisions, based on the thickness, density, and cell size of the cortical layers, also correspond to functional specializations within the neocortex. (Reprinted, with permission, from Angevine, J. B., and Cotman, C. W., 1981, ''Principles of Neuroanatomy,'' Oxford University Press, New York.)

a stimulus that has previously been associated with reward, but which will not fire if the same stimulus is motivationally neutral.

C. Cortical Maps and Functional Modularity

1. Mapping

The dimensions of single neuron response properties described above are not found randomly dis-

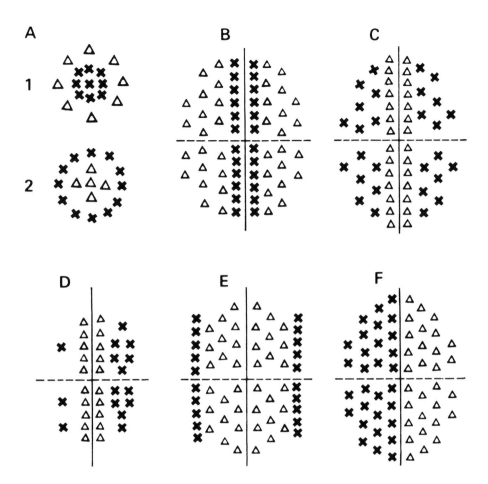

FIGURE 7 Comparisons of receptive fields of neurons in the retina (A1) and lateral geniculate nucleus (A2) with those of neocortical cells in the primary visual cortex (B–F). Whereas the retinal and geniculate cells are symmetric, the cortical cells prefer elongate bars and edges for maximal response. Both classes of neurons will respond to stimuli of light or dark contrast. x = excitatory response in the receptive field; triangle = inhibitory area. (Reprinted, with permission, from Kandel, E. R., and Schwartz, J. H., 1985, "Principles of Neural Science," 2nd ed., Elsevier, New York.)

tributed around the cortical surface but are typically found as orderly dimensional maps laid out across the cortical surface. Neighboring cortical columns typically represent progressive changes in the mapped dimensions. The primary visual cortex is the best described example. In this cortex, the dimensions of location in visual space, preferred eye of activation, and preferred stimulus orientation are all laid out in an orderly way, and these maps are superimposed. Location in visual space is represented once over the full extent in primary visual cortex, with the center of gaze represented at the occipital pole, and the visual periphery buried in the medial cortex. For each mapped location in visual space, a full range of preferred eye activation (ocular dominance) is represented from left-eye dominating to right-eye dominating, changing in orderly sequence. For the same location in visual space, all possible stimulus orientations are also represented, also changing in an orderly way from a preferred angle of 0° to 180° across the cortical surface. These last two dimensional maps are arranged roughly perpendicularly to each other, such that for every location in visual space, every possible value of ocular dominance and preferred stimulus orientation is represented. A block of neocortex that contains one full cycle of all the mapped dimensions has been termed a hypercolumn (Fig. 8). Finally, interposed in this regular map are "color blobs," islands interrupting the regular progression of the other dimensions, where calculations involved in color perception are carried out.

Similar maps can be found in other sensory and motor dimensions, such as the map of the body surface in somatosensory cortex, or changing values of best echo delay seen in the part of the bat auditory

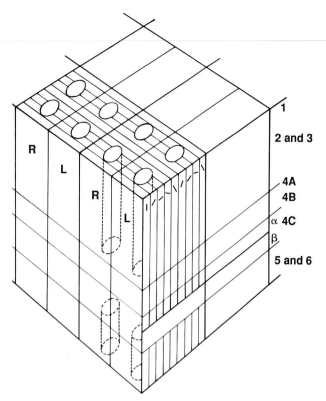

FIGURE 8 Hypercolumn in the visual cortex, representing one location in the visual field. On the left-hand base axis, orderly alternation of preferred activation by the right or left eye is shown. On the right-hand base axis, orderly rotation in the preferred stimulus orientation is shown. On the vertical axis, the Brodman layers are numbered, with Layer IV broken into three subdivisions. Interposed in this regular array are cylinders, called blobs, that are devoted to color processing. This processing unit is repeated over and over throughout the entire neocortex. (Reprinted, with permission, from Livingstone, M. S., and Hubel, D. H., 1984, Anatomy and physiology of a color system in the primate visual cortex, *J. Neurosci.* **4**, 309–339.)

2. Relationships between Maps

The mapped stimulus dimensions described above for various stimulus dimensions are often represented multiple times in the neocortex. For example, in both the monkey and human, the body surface appears to be mapped at least three times over, each representation emphasizing the different features of this sensory array. In the monkey, the part of the neocortex that analyzes visual information consists of not just one but many separate representations of the visual field (Fig. 9). Connections between these representations are both serial and parallel. For example, the representation of stimulus movement is kept separate from the representa-

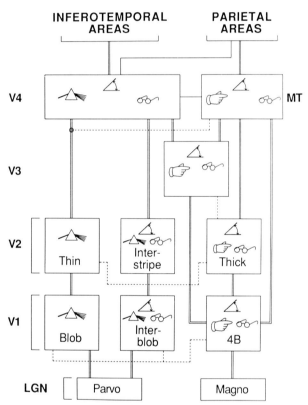

FIGURE 9 Example of serial and parallel connections between the multiple representations of the visual field (V1;V2;V3;V4 and MT) found in the visual cortex. Icons are placed in each compartment to symbolize a high density of neurons showing selectivity for color (prisms), orientation (angle symbols), direction of movement (pointing hands), and binocular depth (spectacles). The input from the thalamus (the lateral geniculate nucleus in this case) is shown with its principal divisions (parvo and magno). The inferotemporal cortex is thought to be involved in recognition and naming, the parietal cortex with pointing, reaching and other operations in space. (Reprinted, with permission, from Shepherd, G. M., 1988, "Neurobiology," Oxford University Press, New York [original figure DeYoe and Van Esen, 1987].)

cortex devoted to echolocation. In many areas of neocortex, the dimension that is mapped has remained elusive but is suggested by the presence of orderly mapping in the neuroanatomical connections between related cortical areas.

Although only the topographic maps of sensory and motor surfaces have been demonstrated directly in the human neocortex, the principle that the neocortex represents complex properties of information as changing dimensions that can be mapped incrementally across the cortical surface appears to be a general one. As such, this insight has guided modeling of the cortical mechanism of such particularly human functions as speech perception, which cannot be directly investigated.

tion of color through several remappings, and a common representation of the output of these two parallel systems is finally found only in the inferotempotal cortex, an area thought to be important for the identification and naming of objects.

3. Modularity

The neocortex keeps stimulus modalities and particular classes of computation physically separate over at least the initial stages of processing. This property is termed modularity. Modularity accounts for the often peculiar effects of localized brain damage, where an individual may have lost the capacity for language but retains the ability to play chess. These dissociations of function can also occur within sensory modalities. For example, the ability to recognize faces can be lost with damage to a particular area of the parietal cortex while sparing the ability to read and navigate visually in the world; or, an individual with damage to a subarea of the language cortex, Broca's area, might lose the capacity to appreciate grammatical relationships in sentences but can fully understand and give the meanings of individual words.

The diversity of the functions that depend on the uniform structure of the neocortex should be underscored. We have already discussed a number of the separable functions dependent on the visual areas of the occipital lobe: the sensory categorization of movement, color, and contour; the dissociability of locating objects for purposes of navigation versus naming those objects; and the specialized social skill of facial recognition and analysis of expression. Also represented in the neocortex are the complex motor patterns for speech; the syntactic and semantic components of language itself, and aspects of the conceptualization, initiation, and execution of complicated learned motor acts, such as piano playing. Represented in the frontal lobe, a number of dissociable complex capacities are often difficult to describe. These include, but are not limited to, interrupting and taking account of one's past performance in planning new action, remembering in what context a particular fact was learned, adjusting emotional expression to the social context, and executing social scripts such as proper eye contact and the maintenance of appropriate interpersonal space. Overall, while excellent progress is being made in understanding the nature of cortical analysis in some sensory domains, the nature of the cortical analysis in cognitive and social domains has hardly begun to be understood.

III. Lateralization of the Human Neocortex

The human neocortex has the relatively unusual property of showing lateralization for functions that bear no intrinsic relationship to the right or left side of the body. For most sensory and motor functions, the brain is lateralized. For example, the left side of the brain receives sensation from the right side of the body and the right half of the visual field and controls the motor functions of the right side of the body; however, in most right-handed people, language is represented on the left side of the brain. Damage to the left temporal cortex can cause complete loss of language (aphasia), whereas damage to the corresponding area on the right side will cause little or no change in language function. Dexterity itself is also lateralized; approximately 90% of individuals prefer the use of their right hand and are more skilled in using it for fine motor functions. The clear lateralization of these two prominent functions led to the characterization of the relationship between the hemispheres as cerebral dominance, meaning the dominance of the left hemisphere for cognitive functions. [See CEREBRAL SPECIALIZATION.]

Later, however, Roger Sperry and his colleagues showed that the right hemisphere was dominant for some other functions, notably aspects of visuospatial understanding, particularly abilities that have a manipulation component such as drawing, arranging blocks, or recognizing something presented visually by feel. These functions are considerably less lateralized than language is in that considerably visuospatial function remains after damage to the right side of the brain. This finding of some functions preferentially lateralized to the right side of the brain had led to a revision of the understanding of the relationship of the cerebral hemispheres, now termed complementary specialization.

The reason for, and genesis of, lateralization is not well understood. In experimental situations, some primates will show evidence of lateralization for particularly tasks, but not in the systematic way seen in the human brain. The most intriguing case of lateralization in animals is bird song, which is also controlled by the left side of the brain. The common feature of complex articulation in the case of bird song and language has led to the hypothesis that the genesis of lateralization might be found in the need for unitary, lateralized motor control of a midline organ, either the various organs of articulation in speech or the syrinx in song.

In right-handed individuals, the language areas in the left hemisphere is larger than in the right hemisphere. This anatomical and behavioral asymmetry is seen in human infants at birth, in the same proportion as right-handed individuals in the population at maturity. Ambidextrous and left-handed individuals show a greater likelihood of their language being bilaterally represented, or represented on the right, and their likelihood of anatomical asymmetry is also less.

IV. Development of the Neocortex

The neocortex is one of the last structures to be formed in the human brain, with the final neurons composing it undergoing their last divisions and migrating into their mature position at about the sixth month postconception. Cells destined for the neocortex are generated at a distant location from their adult position from precursor cells surrounding the lateral ventricles in the forebrain and migrate on supporting glial guides to their position on the brain surface. The first cells generated form the innermost layer (Layer VI) and, later, generated cells bypass these cells and form layers V, IV, III, and II, respectively. Cortical development is further marked by a large transitory population of neurons, which are found on the white matter underlying the neocortex, and in Layer I early in development, and which appear to die after the migration of neurons to the neocortex is completed. The neocortex will continue to increase in size well into early childhood by the growth of neurons and elaboration of their processes, but the full complement of neurons and their fundamental layering is established before birth.

Disorders of early cortical development are, thus, most likely to be caused by disorders of cell generation and migration, which is often visible as unusual cortical thinness, disorders of lamination, or dislocated clusters of cells that have failed to migrate (ectopias).

V. Cortical Pathologies

Focal cortical damage, of the type caused by strokes, space-occupying tumors, and head trauma,

was the first window into the nature of functional cortical organization. Whereas a few deficits appear to be common to all types of focal brain damage (notably, an increased distractibility and inability to screen out irrelevant noise and activity while performing tasks), almost all focal brain damage is followed by inability to perform those particular tasks that correspond to the areas of damage. Particular deficits in sensory systems, language, visuospatial computation, etc. will result, without a generalized decline in all cognitive functions. Therapy for such disfunctions typically involves retraining the individual to substitute intact cognitive functions for the lost ones (e.g., defining maps by verbal landmarks rather than by spatial relations for an individual with a spatial deficit).

Seizure disorders involve abnormalities in the ongoing electrical activity of the brain and often involve the neocortex, particularly the temporal lobe. Normal electrical activity in the neocortex is complex and asynchronous. The neocortex has a number of recurrent and repeating circuits, however, that can produce rhythmic and synchronous activation when desynchronizing activity is removed, as normally occurs in sleep. In seizure disorders, cortical activity becomes massively synchronized and rhythmic. Drugs, or in some cases removal of the area generating the abnormal synchrony, are employed as treatments for seizure disorders. [*See* Seizure Generation, Subcortical Mechanisms.]

Bibliography

Corballis, M. C. (1983). ''Human Laterality.'' Academic Press, New York.
DeYoe, E. A., and Van Essen, D. C. (1988). Concurrent processing streams in monkey visual cortex. *Trends Neurosci.* **11,** 219–226.
Ebbeson, S. O. E. (ed.) (1980). ''Comparative Neurology of the Telencephalon.'' Plenum Press, New York.
Jones, E. G., and Peters, A. (1984). ''Cerebral Cortex. Volume 2. Functional Properties of Cortical Cells.'' Plenum Press, New York.
Peters, A., and Jones, E. G. (1984). ''Cerebral Cortex. Volume 1. Cellular Components of the Cerebral Cortex.'' Plenum Press, New York.
Peters, A., and Jones, E. G. (1988). ''Cerebral Cortex. Volume 7. Development and Maturation of the Cerebral Cortex.'' Plenum Press, New York.

Neoplasms, Biochemistry

THOMAS G. PRETLOW AND THERESA P. PRETLOW, *Case Western Reserve University*

Glossary

Carcinoma   Malignant neoplasm derived from epithelial cells

Karyotype   Chromosomal constitution of the nucleus

Monoclonal   Progeny of a single ancestral cell

Mosaic   Organism whose tissues are composed of two or more kinds of cells with respect to chromosomes or gene expression

Transformation   Conversion of cells in culture to a state in which their growth and morphology are similar to those of some kinds of malignant cells in culture

Translocation   Shifting of part of a chromosome to become attached to and a part of another

Tumor   Literally, a swelling; in common parlance "tumor" is used synonymously with "neoplasm"

A NEOPLASM, or new growth, results from the process of neoplasia, which was defined by Perez-Tamayo in 1961 as "a form of abnormal cellular behavior, the result of many properties manifested by anatomic, functional and biochemical changes in tissues." Others might define neoplasia as a pathological alteration of cells that results in abnormal gene expression and abnormal responses to the factors that normally control growth. Subsumed under the general category of neoplasia, there is a diverse array of thousands of benign and malignant neoplasms in a wide variety of species. In that minority of neoplasms for which the etiology is precisely known, the causes are even more numerous than the varieties of neoplasms.

Biochemically, malignant neoplasms or cancers have been studied much more extensively than have benign neoplasms. Based on the small body of data available, the properties, biochemical and otherwise, of most benign neoplasms are intermediate between those of malignant neoplasms and those of their ancestral precursor types of cells. In a few instances benign neoplasms are thought to be precursors of malignant neoplasms. In many cases the appearance of a malignant neoplasm is associated with a higher than expected frequency of benign neoplasms and "putative preneoplastic" lesions in the same organ. The incomplete state of our understanding of even the definition of neoplasia is exemplified by the fact that our operational definitions of specific neoplasms in animals are based on their gross and microscopic morphological recognition.

I. Introduction

Often there is marked heterogeneity in the behavior and the biochemistry of neoplasms both within and among even the neoplasms that are viewed as being derived from the same type of cell in the same species. This is true even among neoplasms produced experimentally by single chemically pure carcinogens in littermates of the same sex in highly inbred species fed synthetic diets and housed in carefully controlled environments. This heterogeneity markedly limits the breadth of applicability of generalizations that can be made about neoplasms in a description such as that presented here. Each individual neoplasm must be viewed in the context of the category to which it belongs; that is, its properties show enormous variability, but only within the context of the type of cell from which it is derived, its etiology, and the species within which it arises.

As an example of this heterogeneity in a single tissue in humans, there are several dozen kinds of benign and malignant neoplasms of the epidermis (i.e., skin) and its appendages. Among only the neoplasms of the skin that are classified as malignant (i.e., cancers), basal cell carcinomas almost never spread (i.e., metastasize) to grow as tumors (i.e., metastases) at locations distant from their origins. In contrast, malignant melanomas usually metastasize to sites distant from their origins if they are not diagnosed and treated early. The biochemistries of the different cancers of a particular organ are as varied as their behaviors. [*See* METASTASIS.]

In the space available here we should emphasize that we can present only an overview. The biochemical alterations that have been found in neoplastic cells include important changes in the cell membrane, the nucleus, the nucleolus, the mitochondria, and others. It would be difficult to find an enzymatic pathway or organelle that has not been shown to be altered in one or many neoplasms. We refer those who want a more detailed treatment of this subject and an exposition of the history of the biochemistry of neoplasia to the bibliography, particularly to the chapters by Potter and by Busch *et al*.

II. Approaches and Limitations That Affect Our Knowledge

A. Selection of Neoplasms for Study

Our knowledge of neoplasia reflects the variety of neoplasms that have been studied extensively to date. Only a small proportion of our knowledge comes from the investigation of primary or metastatic tumors obtained directly from the hosts in which they originated. A large proportion of our knowledge has been derived from the study of neoplastic cells in culture (*in vitro* models) or animal models of neoplasia. Some of the limitations of culture are discussed below. The two most commonly used animal models are (1) neoplasms produced in animals by carcinogens and (2) transplantable tumors.

Tumors produced by carcinogens in the laboratory are different in many ways from most tumors that occur naturally and, in particular, from tumors that occur naturally in humans. Most tumors known to result from the exposure of humans to carcinogens occur only after years or, more often, decades of exposure to carcinogen. In laboratory animals tumors are generated in months by massive exposure to carcinogens. While the biochemical characterization of neoplasms remains a complex task, the task is simplified somewhat by virtue of the fact that most neoplasms are derived from a single cell (i.e., they are clonal) (see Section VI). [*See* TUMOR CLONALITY.]

In some cases transplantable tumors (e.g., the "minimal deviation" hepatomas investigated by Van Potter, Harold Morris, and their collaborators) are relatively stable, have been studied after specified numbers of passages from one host to another, and have been diligently selected for their suitability for tests of specific theories. In contrast, much of our knowledge of the biochemistry of neoplasia has come from study of tumors (e.g., the Walker mammary carcinoma or the Dunning prostatic tumor) that have been transplanted hundreds of times under poorly defined conditions into hosts for which there are frequently no detailed health records for many decades. These tumors have undoubtedly undergone enormous selection such that any cells that elicited a strong immune response protective for the host have been eliminated and cells that grow slowly have been replaced (i.e., outgrown) by more rapidly growing cells. In the process of transplantation through many different hosts, often in many different institutions, it is likely that these tumors have acquired viruses from successive hosts. While these tumors were appropriate for the state of our knowledge in the eras in which they were developed, some of the most commonly studied such tumors even originated in strains of animals that were different from those in which they are now passaged. One must wonder how applicable is the knowledge gained from these systems to neoplasms as they occur in nature.

B. Direct Biochemical Analysis

Our knowledge of the biochemistry of neoplasms is limited by the methods with which we approach the investigation of this subject. The biochemistry of neoplasia can be explored by the direct chemical analysis of neoplasms. When this approach is taken, the resultant data are limited by virtue of the fact that, on the average, at least half of the cells of most neoplasms are nonneoplastic (i.e., blood vessels, inflammatory and/or immunologically active cells, and fibroblasts). In addition, all neoplasms contain varying proportions of extracellular materials and the normal blood cells that are in the blood vessels that provide the circulation of the tumor. The biochemical data obtained from the direct analysis of neoplasms reflect the properties of both the neoplastic and stromal components of the neoplasms.

C. Analysis of Purified Cells

To circumvent the problem that neoplastic cells are diluted by stromal cells in nature, it is often possible to disaggregate neoplasms into suspensions of single cells that can be separated physically by sedimentation, by reactions with specific (often monoclonal) antibodies attached to insoluble supports, or by many other methods. The biochemical data obtained from the analysis of "purified" neoplastic cells are limited by (1) the degree to which the cells in the initial suspensions of cells from the neoplasm represent the cells that were present *in situ* in the neoplasm before the cells were obtained in suspension, (2) the extent to which they were purified, and (3) the proportion of cells lost during the process of purification. The process of obtaining cells in suspension never yields quantitatively all of the cells available *in situ*, and it is likely that there is usually some selection of subpopulations that might or might not be representative of the neoplasm during the process of obtaining cells in suspension and the purification of cells.

D. Analysis of Cultured Cells

The culture of cloned or purified neoplastic cells can be used to circumvent the problems discussed in Sections II,B and C; however, this is perhaps the most limited approach to the biochemical characterization of neoplasms as they occur in nature. Only a minority of the cells in a neoplasm have the capacity to grow under the conditions for culture that have been developed to date. In most cultures of neoplastic cells from a tumor, there is significant selection of subpopulations of cells that have the capacity to grow in culture and can grow more rap-

idly than the other neoplastic cells from that tumor which are capable of growth in culture. In addition, neoplastic cells often exhibit marked changes (i.e., modulation) in their biochemical properties as a consequence of the foreign environment that they encounter in culture.

III. Generalizations About the Biochemistry of Neoplasms

George Weber has organized much of our thought about the general biochemistry of neoplasia in recent years. In consideration of a large number of hepatocellular carcinomas (i.e., malignant tumors derived from the hepatocyte, the cell that accounts for the largest proportion of the liver) and a small number of other kinds of tumors studied in his and many other laboratories, he has articulated the concepts that the activities and/or concentrations of key enzymes, isoenzymes, and metabolites are "stringently linked" to the process of the transformation of a normal cell to a malignant cell. The concentrations of the same and/or other enzymes, isoenzymes, and metabolites are similarly linked to the progression of less aggressive malignant cells to more aggressive, invasive, and metastatic states. [*See* ISOENZYMES.]

In his masterful reviews of this subject, Weber has pointed out that many dozens of enzymes and metabolites show consistent trends in changes in their concentrations and activities in neoplasms. When one examines successively more aggressive neoplasms of a particular type, it becomes apparent that there are relationships between the increased rates of growth of those particular kinds of neoplasms and the activities of particular enzymes.

In general, the most dangerous malignant neoplasms contain neoplastic cells that divide more rapidly than do normal cells. A few of the enzymes thought to be rate limiting, termed "key" enzymes by Weber, have been studied in detail in many kinds of tumors. For example, glucose-6-phosphate dehydrogenase is the rate-limiting enzyme in one of the two biochemical pathways available to the cell for the production of ribose 5-phosphate, a precursor required for the production of both DNA and RNA. Rapidly growing neoplasms must be able to synthesize DNA and RNA rapidly. This enzyme was shown to be greatly elevated in rapidly growing rat hepatomas as compared to normal liver by Weber, in human breast cancers as compared to normal breast by Russell Hilf, and in human prostatic carcinomas as compared to prostates without carcinoma by our laboratory.

Enzymatic activities that appear to be markedly altered in many different kinds of neoplasms include thymidine kinase, DNA polymerase, lactic dehydrogenase, pyruvate kinase, hexokinase, N-acetyl-β-D-glucosaminidase, phosphoserine phosphatase, ornithine decarboxylase, pyrroline-5-carboxylate reductase, and many others. The selection of enzymes mentioned here is almost arbitrary, since many dozens of enzymes have been shown to exhibit marked changes in activity in neoplasms. These enzymes are found in diverse areas of cellular biochemistry, including nucleic acid, amino acid, and carbohydrate metabolism. In many instances, certain isoenzymes (i.e., variants of the same enzyme) are much more affected than others by the changes associated with neoplasia and the progression of neoplasms. Often, the pattern of isoenzymes in neoplasms represents a recapitulation of the patterns observed during the embryonic development of the cell that is the ancestral precursor of the neoplasm. Most of the biochemical alterations that affect many kinds of tumors can be postulated to involve what Weber has called key metabolic pathways, the alterations of which are tightly linked to the processes of neoplasia and/or progression without reference to any particular kind of cell or organ.

IV. Biochemical Changes in Specific Tumors

In addition to the biochemical alterations common to many tumors, there are numerous alterations known to occur in most neoplasms of a particular kind, but not in most kinds of tumors. An example of this type of change is arginase, which is elevated in most human prostatic carcinomas. Similarly, the expression of leucine aminopeptidase is markedly decreased or absent in most human prostatic carcinomas, but it is present in the epithelial elements of prostates without cancer. While one could speculate about mechanisms that might potentially explain these changes, speculation is useful only as it leads to testable hypotheses; the mechanisms for many of these changes have still not been investigated in detail.

V. Preneoplastic Biochemical Alterations

There are both systemic and focal abnormalities of cells that are associated with a greater propensity for the development of neoplasia. Systemic conditions (i.e., conditions that affect the entire body) include unrepaired chromosomal breaks (e.g., Fanconi's anemia or ataxia–telangiectasia).

Focal or organ-associated conditions associated with the development of neoplasia are numerous, and we shall give only a few examples. As mentioned, in many experimental systems for carcinogenesis *in vivo* in laboratory animals, the production of malignant neoplasms is associated with the production of benign neoplasms in the same organ system. This strongly suggests that similar etiological processes are important in the pathogenesis of both kinds of neoplasms.

In the same vein both humans and experimental animals often present focal changes without the gross appearance of tumors in the same organ system that harbors overt neoplasms. Many of these changes are identified morphologically. Such changes include carcinoma *in situ,* or preinvasive cancer, a change that usually can be identified only microscopically; inborn, often genetic, errors (e.g., familial polyposis); abnormal cell kinetics (e.g., the failure of cells in the colon to mature and exhibit terminal differentiation normally); and many others. We are just beginning to be aware of other focal, probably clonal, changes in certain organ systems that might not always be amenable to detection by the usual histological methods. The most studied of these lesions are "enzyme-altered" foci of the liver and the colon.

Enzyme-altered foci, sometimes termed "putative preneoplastic lesions," are particularly important, because they represent one of the few extensively studied biochemical alterations associated with the process of carcinogenesis. These lesions were first observed in the liver in the 1940s and have been observed more recently in the colon. The hepatic lesions have been characterized in many laboratories and have been shown to exhibit the increased or decreased expression of many enzymatic activities and/or concentrations demonstrated histochemically or immunohistochemically. These lesions might be associated with subtle alterations of the normal histology; however, they could appear histologically normal and could be apparent only when they are demonstrated histochemically or immunohistochemically.

When sectioned serially and investigated for their expression of different phenotypic markers in successive serial sections, hepatic enzyme-altered foci could show abnormal expression of one or many phenotypic markers. Such markers include γ-glutamyl transpeptidase, benzaldehyde dehydrogenase, glucose-6-phosphate dehydrogenase, and many more.

Similar to enzyme-altered foci in the liver, enzyme-altered foci in the colon could appear histologically normal or could show characteristic morphological abnormalities. The most commonly observed markers that are abnormally expressed are N-acetyl-β-D-glucosaminidase, glucose-6-phosphate dehydrogenase, α-naphthyl butyrate esterase, diaminopeptidase IV, succinate dehydrogenase, and β-galactosidase. Despite the small numbers of phenotypic markers that have been examined to date in enzyme-altered foci in the colon, it is interesting that some of them are also abnormally regulated in several other kinds of malignant neoplasms. It appears that the abnormal expression of one or more of these markers precedes the overt dysplastic changes commonly viewed by pathologists as preneoplastic and/or neoplastic.

VI. Other Biochemical Features of Neoplasia

In addition to the biochemical changes already discussed, others have attracted much attention and deserve mention.

A. Antigens in Neoplasia

The genetic alterations present in cancer cells might be expected to be reflected, at least in some cases, in differences in the antigens present in carcinomas compared to normal tissues. In several experimental systems, particularly tumors induced by oncogenic viruses, tumor-associated antigens have been demonstrated, but they appear to be more closely related to the infection by the virus than to the process of neoplasia. With a variety of tests, antigenic differences can be recognized in some human neoplasms. These findings, however, are rare.

Extensive study has concentrated on oncofetal antigens. These antigens are present at relatively high concentrations in some tumors as well as in embryonic tissues during development, but only at low or undetectable levels in adult tissues. These antigens might not cause an immune reaction in the host that carries the neoplasm, but they can be recognized by antibodies raised in other species. Because certain of these antigens are shed from the tumor into the circulation of the patient, their assay in peripheral blood is sometimes useful as a means of monitoring the progress of the tumor.

The antigens are too numerous to list in detail; however, they are well exemplified by carcinoembryonic antigen (CEA). CEA is expressed by the fetal endodermally derived tissues during the first and second trimesters of pregnancy. It is expressed by the colon only at low levels during adult life; however, it can be elevated in the blood of smokers who do not have identified neoplasms. While CEA is not sufficiently specific to be useful for the screening of populations, it is expressed by several epithelially derived tumors (i.e., carcinomas), particularly colon cancers, and has been useful in monitoring their progress and response to therapy. CEA is produced at high levels in a variable proportion of carcinomas of the mammary gland, in neuroblastomas, and in other neoplasms, as well as in some nonneoplastic diseases.

B. Biochemical Demonstration of Clonality

With few exceptions (e.g., neurofibromas found in familial neurofibromatosis) the overwhelming majority of human neoplasms are clonal in origin; that is, they lack the capability of recruiting contiguous cells and grow by the steady proliferation of cells from an original cell that became neoplastic. The fact that most human neoplasms are monoclonal has facilitated their analysis by electrophoretic techniques that allow one to visualize genetic mutations, since a mixture of genetically different cells would introduce much more heterogeneity into the electrophoretic bands obtained.

Two lines of evidence lead one to the conclusion that most human tumors are monoclonal. The first of these is the fact that karyotypic analysis shows distinctively altered marker chromosomes in all cells of many tumors. The second line of evidence is derived from the study of isoenzymes encoded by genes located on the X chromosome. The Lyon hypothesis states that all X chromosomes, in excess of one, in cells are inactivated randomly such that normal tissues from heterozygous females, in which the two X chromosomes differ at recognizable genes, are mosaics of cells, some of which express the paternal X genes, whereas others express the maternal X genes. In contrast, most human tumors express only the maternal or the paternal gene in the malignant cells, whereas stromal cells from the same tumors, which are nonneoplastic, exhibit the same mosaicism as normal tissues.

The early work in this area was done by observing the patterns of glucose-6-phosphate dehydrogenase by electrophoresis of extracted tissues; however, this approach was restricted to the study of a relatively small subpopulation of people who were heterozygous for the enzyme. With the subsequent discovery of new markers on the X chromosome, one can now test for heterozygosity in tissues from most females. These studies have generally confirmed the earlier observations.

VII. Summary

The biochemistry of most neoplasms is characterized by abnormal gene expression and the defective control of growth. In some cases the biochemical alterations are correlated to the rapid growth rate of the cells. Many biochemical changes are shared by several or most neoplasms; others are specific for only one or a small group of them. In some cases they result from specific chromosome alterations.

It is increasingly apparent that most well-studied neoplasms have changes (e.g., translocations, deletions, or duplications) that are not random. Their investigation has become an increasingly important approach to the systematic study of associated biochemical events. Particular changes appear to be characteristic of specific kinds of neoplasm.

Marked differences among different kinds of neoplasms give emphasis to the need for detailed studies of the biochemistry of individual tumors.

Bibliography

Busch, H., Tew, K. D., and Schein, P. S. (1985). Molecular and cell biology of cancer. *In* "Medical Oncology: Basic Principles and Clinical Management of Cancer" (P. Calabresi, P. S. Schein, and S. A. Rosenberg, eds.). Macmillan, New York.

Horowitz, J. M., Yandell, D. W., Park, S.-H., Canning, S., Whyte, P., Buchkovich, K., Harlow, E., Weinberg, R. A., and Dryja, P. (1989). Point mutational inactivation of the retinoblastoma antioncogene. *Science* **243,** 937.

Klein, G. (1988). Oncogenes and tumor suppressor genes. *Acta Oncol.* **27,** 427.

Mitelman, F., and Heim, S. (1988). Consistent involvement of only 71 of 329 chromosomal bands of the human genome in primary neoplasia-associated rearrangements. *Cancer Res.* **48,** 7115.

Perez-Tamayo, R. (1961). "Mechanism of Disease." Saunders, Philadelphia, Pennsylvania.

Potter, V. R. (1982). Biochemistry of cancer. *In* "Cancer Medicine" (J. F. Holland and E. Frei, eds.). Lea & Febiger, Philadelphia, Pennsylvania.

Pretlow, T. G., Harris, B. E., Bradley, E. L., Bueschen, A. J., Lloyd, K. L., and Pretlow, T. P. (1985). Enzyme activities in prostatic carcinoma related to Gleason grades. *Cancer Res.* **45,** 442.

Rowley, J. D. (1988). Chromosome abnormalities in leukemia (Karnofsky Memorial Lecture). *J. Clin. Oncol.* **6,** 194.

Vogelstein, B., Fearon, E. R., Kern, S. E., Hamilton, S. R., Preisinger, A. D., Nakamura, Y., and White, R. (1989). Allelo-type of colorectal carcinomas. *Science* **244,** 207.

Weber, G. (1983). Biochemical strategy of cancer cells and design of chemotherapy: G. H. A. Clowes Memorial Lecture. *Cancer Res.* **43,** 3466.

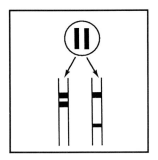

Neoplasms, Etiology

SUSAN J. FRIEDMAN AND PHILIP SKEHAN, *National Cancer Institute*

I. Biology of Neoplastic Transformation
II. Theories of Oncogenesis

Glossary

Autosomal dominant pattern of inheritance In Mendelian genetics the inheritance of a trait carried on an autosome (i.e., non-sex chromosome) by 50% of the offspring of a mating of heterozygotes (i.e., individuals who have alternative forms of the gene (nonidentical alleles) at that locus). At the cellular level a genetic mutation is dominant if it is expressed in the presence of the normal (wild-type) gene

Autosomal recessive pattern of inheritance Inheritance of an autosomal trait by 25% of the offspring of a mating of heterozygotes. At the cellular level a recessive genetic mutation is expressed in the homozygous state, in which both alleles are identical

Cellular oncogene (protooncogene) Gene that normally functions in growth and differentiation, but can contribute to tumorigenesis when mutated or activated by gene rearrangement

Enhancer DNA sequence that increases the transcriptional activity of a promoter

Genotoxic Directly or indirectly damaging to genetic material

Mitotic chromosome nondisjunction Failure of the replicated chromosomes to separate during mitosis

Neoplastic transformation or conversion Accumulation within a cell of multiple individual transformations, which collectively give rise to an expanding population of abnormal or inappropriate cells that progressively destabilize normal tissue structure and function

Promoter Region of a gene that binds RNA polymerase and initiates transcription

Restriction fragment-length polymorphism Normally occurring DNA sequence variation in a population that can be detected by restriction endonuclease digestion and Southern blot analysis

Retroviral oncogene Transforming gene of an RNA tumor virus that is derived from a cellular gene sequence by transduction

Southern blot hybridization Technique that detects specific DNA sequences in the genome by electrophoresis of restriction endonuclease-digested DNA on agarose gels, denaturation, and transfer of the separated fragments by capillary blotting onto nitrocellulose filters, and hybridization to a specific labeled DNA or RNA probe

Stem cells Early embryonic cells with an extensive capacity to reproduce themselves and to give rise to more differentiated daughter cells

Tissue homeostatic mechanisms Regulatory mechanisms that operate collectively to restrict the degree of variation in the size, function, and cytoarchitecture of normal tissues so as to maintain tissue stability

Transduction Process by which a virus transfers genes from one host cell to another

Transformation Long-term deviation of a cellular trait(s) from the normal range of values for that trait(s); changes in culture in the appearance and the behavior of eukaryotic cells that are produced by oncogenic agents

Transgenic mouse Mouse whose somatic cells contain a foreign DNA sequence that was introduced into the germ line

Transposable element (transposon) Mobile genetic element that can transfer genes from one chromosomal region to another; its structure consists of a defined gene sequence flanked by a long terminal repeat of several hundred bases and a short base repeat of genomic DNA at the insertion site

"CANCER" IS A collective term for several hundred diseases characterized by the formation of a tissue mass which is inappropriate for its location, is uncoordinated in its growth and function with surrounding tissues, and persists indefinitely when an evoking stimulus is removed. A unique characteristic of malignant tumors is their ability to invade normal tissues and to metastasize to other sites in the host. The causes of human neoplasms are, for the most part, unknown. It is unlikely that all tumors have a common origin, although, by the process of natural selection during tumor evolution, they might come to acquire common phenotypic features. The most common types of tumors are thought to be nonhereditary, but a subset of these could, nevertheless, share a common etiology with rare inherited tumors.

The mechanisms of tumor development are also uncertain. Cancer is a long-latency multistage process initiated by certain types of viruses, chemicals, radiation, and other tissue-damaging agents. Abnormal cells that arise during tissue injury and adaptation are triggered to begin a series of successive phenotypic changes. They undergo limited self-terminating growth in response to an appropriate signal, accumulate genetic damage by rare mutationlike events, form colonies, and give rise to subpopulations at an increased risk for genetic error.

Neoplastic transformation is a rare event, with an estimated frequency of one in 2×10^{17} mitoses, that occurs when cumulative changes are sufficient to permit growth to a physiologically significant size with disruption of local tissue stability. Changes in the structure and activity of genes involved in the regulation of normal tissue growth and differentiation have been detected in developing tumors and could be necessary for neoplastic conversion. Ultimately, the expression of malignancy is controlled by the specific tissue environment in which these cellular changes have occurred.

I. Biology of Neoplastic Transformation

A. Classic Pattern

The classic pattern of tumor development is observed in chemically induced tumors of skin, liver, bladder, and several other tissues and can be divided into initiation, promotion, and progression stages. The process begins when normal tissue is exposed to and changed by an initiating stimulus. This event occurs in only a small fraction of the population, on the order of 0.01% or less, which varies directly with the strength of the stimulus.

There is evidence from liver carcinogenesis studies that embryonic progenitor cells, rather than mature hepatocytes, could be the relevant targets for initiation. The initiated state is transient and decays, unless it is stabilized within a critical period by cell proliferation. Compensatory proliferation that accompanies cell loss, but not mitogen-induced proliferation, is a highly effective stimulus. Initiated cells deviate minimally from normalcy, are often undetectable, and can remain latent for years, even decades.

The distinguishing characteristic of initiated cells is their ability to undergo limited self-terminating growth in response to a tumor-promoting stimulus that is ineffective for, or actively inhibits, the growth of the surrounding normal tissue. The selective growth advantage of initiated cells partly depends on their acquisition of resistance to the growth-inhibitory and cytotoxic actions of promoting agents. The recently discovered mammalian genetic stress response could allow initiated cells to respond rapidly to the presence of noxious growth-inhibitory agents by synthesizing a common set of proteins to counteract the proliferation block and minimize genotoxic damage. Included in this group of stress-induced proteins are the heavy metal-binding protein metallothionein, DNA repair and recombination enzymes, a gene amplification-producing factor, proteins with cell type-specific functions, and proteins with suspected growth-associated functions (e.g., ornithine decarboxylase, c-fos, and p53).

If initiated cells are repeatedly exposed to promoting agents, they form multicellular colonies called nodules, foci, polyps, or papillomas. Most colonies regress when the growth-promoting stimulus disappears. The cells assimilate into the normal tissue structure, where they remain dormant but viable for extended periods of time. A small proportion (i.e., 1–3%) of the colonies persist and can remain stable in size for decades without net growth, while the cells undergo progressive and potentially destabilizing changes. With time the number of altered traits per cell and the risk of neoplastic conversion increase. The changes occur randomly and independently of one another with no apparent order or pattern. Recent studies on genetic alterations in human colorectal cancer development suggest that the number of accumulated changes, not the order in which they occur, has prognostic significance.

B. Nonclassical Pattern

A form of stomach cancer known as intestinal-type gastric carcinoma deviates in several important respects from the classical initiation–promotion model described above. Certain human populations are at high risk for this disease because of a maternally inherited susceptibility to atrophic gastritis, combined with a high level of exposure to dietary carcinogens. Dietary substances and certain medications chronically irritate the gastric mucosa, which results in an increased turnover of gastric epithelium. With time and under the influence of additional irritating substances, the gastric glands atrophy and the epithelium is replaced, first by small intestine epithelium and later by large intestine epithelium. This atypical epithelium becomes increasingly disorganized and populated by cells that express traits characteristic of less mature gastrointestinal precursor cells. Eventually, from within such an abnormal epithelium, a neoplastic growth will arise. [*See* GASTROINTESTINAL CANCER.]

It is not known whether the targets for neoplastic conversion are a small residual population of embryonic cells in the adult gastric epithelium with multiple developmental fates or, alternatively, whether the loss of gastric phenotypic markers and the expression of intestinal properties represent changes in gene expression in the mature gastric epithelium. Throughout much of the preneoplastic period, the development of gastric cancer remains highly susceptible to modulation by dietary influences, both positive and negative. The major apparent differences between this process and classical tumor development are the requirement for specific etiological agents at various stages of preneoplastic and neoplastic progression, and the emergence of neoplastic cells from tissue that no longer expresses the morphological and functional properties of the tissue originally exposed to the ''initiating'' stimulus.

II. Theories of Oncogenesis

Various theories have been proposed to explain the origin of human cancer. A unifying theme that in-

corporates aspects of these theories is the idea that cancer originates from genetic alterations in somatic cells which act in combination to override the regulation of cell proliferation and differentiation by normal tissue homeostatic mechanisms.

A. Somatic Mutation Theory

According to the somatic mutation theory, cancer is a genetic disease that originates from rare mutational events in somatic cells. Consistent with this idea are the observations that (1) many human tumors appear to be monoclonal (i.e., derived from a single cell), (2) tumor development occurs over many decades and in multiple stages, (3) many environmental carcinogens are mutagenic, (4) inherited DNA repair and chromosome instability disorders greatly increase the risk for certain cancers, and (5) dominantly inherited cancer susceptibility is associated with nonrandom recessive mutations.

1. Monoclonality of Human Tumors

Most human tumors of mesenchymal and hematopoietic tissues, and the few of epithelial origin that have been analyzed, are monoclonal in composition. This is usually considered as evidence that the tumor developed from a single neoplastically transformed precursor cell, although the same result would be obtained by the selective outgrowth of a single clone from an initially polyclonal tumor. The monoclonality of many human tumors is consistent with the idea that the cellular changes involved in neoplastic conversion are rare somatically inherited events.

The clonal composition of tumors can be analyzed by several methods. Tumors of B lymphocytes (i.e., antibody-producing cells) are usually examined for the expression of a common surface immunoglobulin molecule or the corresponding DNA sequence; T cell tumors are analyzed for a common arrangement of antigen receptor DNA; and other tumors, for the presence of unique chromosomal markers or for the expression of one of two allelic forms of an X chromosome-encoded gene product in heterozygous females. The enzyme glucose-6-phosphate dehydrogenase has traditionally been used as a marker for clonal analysis.

Recently, an approach based on restriction fragment-length polymorphisms has been introduced (Fig. 1). This method relies on the fact that, early in embryogenesis, one of the two X chromosomes of a somatic cell is functionally inactivated in a random

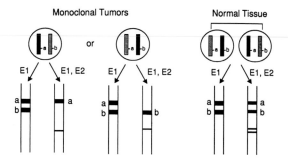

FIGURE 1 Differences in the gel electrophoretic patterns of DNA digested with two restriction endonucleases, E1 and E2, distinguish monoclonal tumors from normal tissues in females heterozygous for an X-linked marker. One of the two X chromosomes in each somatic cell is inactivated at random during embryogenesis (shown as a stippled bar within a circle). All cells of a monclonal tumor contain the same active allele, either *a* or *b*, whereas in normal tissue half of the cells contain an active *a* and half contain an active *b* allele. E1 generates different length fragments for each allele. Inactive (stippled bars) and active (solid bars) X chromosomes differ in the degree of methylation at the E2 restriction site. E2 digests the restriction fragment containing the inactive allele and removes it from its original position in the gel.

and permanent manner (denoted by the stippled bars in Fig. 1). The same inactive X chromosome is inherited by all descendants of that cell. Tumors arising monoclonally in women who are heterozygous for an X-linked gene with allelic forms *a* and *b* express one allele, whereas normal tissue or tumors that originate from multiple cells have active forms of both alleles.

To distinguish between these possibilities, DNA from normal and tumor tissues is separately digested with two restriction endonucleases (i.e., enzymes that recognize specific short base sequences in DNA and cleave DNA at these sites to produce restriction fragments). The first enzyme (E1) generates different-sized restriction fragments for *a* and *b* alleles, which results in a difference in their electrophoretic mobilities on agarose gels. The second enzyme (E2) cleaves DNA at a specific restriction site containing the unmethylated form of the base, deoxycytosine. The rationale for this is that genes in active and inactive X chromosomes differ in the extent of methylation of this base and will therefore differ in their restriction fragment patterns produced by E2 treatment.

Following agarose gel electrophoresis the DNA fragments are transferred to a nitrocellulose or nylon filter and are reacted with a radioactively labeled DNA probe that recognizes a sequence common to both *a* and *b* alleles (i.e., Southern blot

hybridization). The pattern of hybridized bands can be visualized by exposing X-ray film to the filter for an appropriate period (i.e., autoradiography).

Restriction fragments of both alleles are present in the DNA of normal tissue and multiclonal tumors both before and after digestion with E2, although the amount of DNA in each band is reduced after digestion, because a portion of the *a* and *b* alleles is derived from X chromosomes containing an unmethylated E2 restriction site. By contrast, in DNA from monoclonal tumors, either all *a* or all *b* fragments are sensitive to E2 digestion; thus, one of the restriction fragments disappears completely from its original position in the gel.

2. Multistep Nature of Human Tumor Development

The observations that human cancers exhibit a long latency period and an age-dependent increase in incidence and arise from foci of disorganized hyperplastic tissues, rather than from normal tissue, suggest that tumor development is a mutationlike process.

3. Most Environmental Carcinogens Are Mutagens

Most environmental carcinogens can mutate genes and structurally damage chromosomes. However, there are several significant differences between mutagens and carcinogens that suggest that mutagenesis alone might not be sufficient for tumor induction. The most common mutagenic agents—ultraviolet light and ionizing radiation—account for only a small proportion of total cancers; some potent mutagens lack carcinogenic activity (e.g., nitrogen mustards and some nononcogenic viruses). Mutagens also act directly and efficiently to damage their cellular targets, whereas carcinogens damage many more cells than they neoplastically transform and act indirectly to trigger cellular changes that become self-sustaining in the absence of the carcinogen.

4. Disorders of DNA Repair and Chromosome Instability Associated with an Increased Cancer Risk

Rare recessively inherited syndromes that alter chromosome stability and the repair or removal of damaged DNA in somatic cells strongly predispose afflicted individuals to certain types of tumors. These syndromes include xeroderma pigmentosum, ataxia telangiectasia, Bloom's syndrome, Cockayne's syndrome, and Fanconi's anemia. Bloom's syndrome and ataxia telangiectasia are also associated with impaired immune function. Xeroderma pigmentosum, the best studied of these disorders, is a family of diseases that impairs an important cellular mechanism that repairs DNA damage by excising pyrimidine dimers caused by ultraviolet light. Patients with xeroderma pigmentosum are unusually sensitive to sunlight and develop skin tumors at 10,000 times the normal rate.

Diseases that affect the stability of chromosomes and the repair of DNA damage do not increase the risk for all cancers. Rather, for unknown reasons, each disorder is associated with only one or a few types of cancer. Tissue-specific differences in exposure to particular types of DNA-damaging agents, in the kinetics of cell renewal, in the efficiency of repair of DNA damage, in the accessibility of relevant genomic targets to damage, and in the functions of the cellular processes rendered defective by the mutations could contribute to the observed patterns of tumor development.

5. Dominantly Inherited Cancer Susceptibility Diseases

Epidemiological and cytogenetic studies of familial patterns of cancer have uncovered as many as 50 forms of cancer with an autosomal dominant pattern of inheritance. That is, family pedigrees reveal that cancer susceptibility is transmitted to 50% of the progeny of a mating between two heterozygotes. These diseases include tumors of the gastrointestinal tract (e.g., familial polyposis and colon cancer), nervous system (e.g., retinoblastoma and neurofibromatosis), endocrine system (e.g., multiple endocrine neoplasia syndromes), and kidney (e.g., Wilms' tumor). In retinoblastoma, a rare childhood eye tumor that occurs in hereditary and nonhereditary forms, the genetic locus for susceptibility has been mapped to band 14q of chromosome 13, and the putative gene has been cloned. [See Retinoblastoma, Molecular Genetics.]

In the hereditary form of this disease, one of the wild-type alleles is inactivated in the germ cells of the parent and the mutation is therefore present in all somatic cells of the embryo. Inheritance of the disease follows a dominant pattern, although at the cellular level the retinoblastoma susceptibility gene acts recessively. Only after the second allele has become inactivated or lost in a developing retinal cell (i.e., retinoblast) does cancer develop. The reasons that tumors develop exclusively in retinal tissue are unknown. Nonrandom duplications and amplifications of other chromosomes occur frequently in retinoblastoma cells and may be involved in tumor formation. In nonhereditary retinoblastoma the inactivation of both alleles occurs after birth in a single retinoblast by two independent events; thus, the tumor is even more rare.

Fifty percent of the patients who receive radiation treatment for hereditary retinoblastoma develop the rare bone tumor osteogenic sarcoma 6–8 years later. In these tumor cells both copies of the retinoblastoma gene are functionally inactivated. This suggests either that the wild-type retinoblastoma gene serves an important regulatory function in both tissues or, alternatively, that the mechanisms of inactivation of the wild-type retinoblastoma genes can affect nearby genes which predispose to osteosarcoma. In both the hereditary and nonhereditary forms of retinoblastoma, these mechanisms most frequently involve the loss of the wild-type allele in one of the daughter cells produced at mitosis, because of the failure of replicated chromosomes to separate (i.e., chromosome nondisjunction) or because of an exchange of DNA segments between strands of a chromosome pair (i.e., mitotic recombination). Other rare childhood tumors, including neuroblastoma and Wilms' tumor, are thought to originate by similar mechanisms involving different chromosomal loci.

In contrast to the two-step model proposed for retinoblastoma, the deletion of multiple alleles is necessary for the induction of cancer of the colon and the rectum, which likewise occurs in hereditary and nonhereditary forms. In familial polyposis, a dominantly inherited disorder characterized by the development of multiple colon polyps (i.e., adenomas) and a greatly increased risk of colorectal cancer in young adulthood, deletions of the presumptive susceptibility locus (the *fap* locus on chromosome 5) are not found in early adenomatous lesions. Multiple nonrandom deletions of chromosomal segments occur in both hereditary and nonhereditary tumors, most commonly involving putative tumor-suppressor gene loci on chromosomes 17 and 18, but no single deletion is uniquely associated with tumorigenesis.

B. Recombination Theory

In experimental carcinogenesis systems the frequency of cells with altered characteristics is often in the range of 10^{-2} or higher, whereas the spontaneous mutation frequency is estimated to be 10^{-6}

per locus per generation. This discrepancy poses difficulties for the somatic mutation hypothesis. It suggests that tumorigenesis might not be initiated by localized mutations in DNA, but by chromosomal rearrangements that cause large-scale changes in gene expression and destabilize the genome, making possible the appearance of cells with altered characteristics. Consistent with this hypothesis is the observed karyotypic (i.e., chromosomal) instability of most solid tumors and the presence of specific chromosomal translocations in hematopoietic tumors that retain a normal chromosome composition.

Two known mechanisms could contribute to the large-scale genetic rearrangements and changes in gene expression that occur during tumorigenesis; chromosomal translocation and genetic transposition.

1. Chromosomal Translocation

Nonrandom chromosomal translocations, in which parts of two different chromosomes are exchanged, are consistently present in certain tumors of B and T lymphocytes and could contribute to the etiology of these diseases. Chromosomal translocations to 14q32, a region on the long arm of chromosome 14 which contains the immunoglobulin heavy-chain locus, are common in a variety of leukemias and lymphomas. One of the best studied of these translocations has an exchange between this locus and a locus on chromosome 8 that contains the cellular oncogene c-*myc*. This exchange occurs in 80% of a rapidly growing tumor. Burkitt's lymphoma, and deregulates the expression of c-*myc*, presumably by bringing it under the control of the regulatory elements of the immunoglobulin heavy-chain locus. The translocation event occurs in proliferating B lymphocytes that have undergone the final maturation event (i.e., heavy-chain switching) and would normally enter a resting state [*See* LYMPHOMA.]

In the endemic African form of Burkitt's lymphoma, the combined effects of chronic infection of cells with Epstein–Barr virus and malaria could predispose cells to the translocation event. The virus stimulates B lymphocyte proliferation and expands the population at risk for the translocation, while the malaria parasites ''swamp'' the immune system, preventing the virally infected cells from being attacked by the immune defense mechanisms, as they normally would be.

The precise contribution of the translocation to the tumorigenic process is currently unknown. The translocated c-*myc* is no longer responsive to endogenous signals that suppress transcription of the nontranslocated gene. Experiments in transgenic mice have shown, however, that the deregulated expression of c-*myc* is not sufficient to cause tumors. Under the control of the enhancer of the gene for the immunoglobulin heavy chain, c-*myc* is constitutively expressed (i.e., not regulated) by the B cell lineage lymphocytes. This results in a fourfold increase in the total pre-B lymphocytes (an early stage of development in this lineage) and an increase in the level of proliferating cells.

One consequence of the expansion of the pre-B cell population is an increased risk of recombinational errors during the assembly of immunoglobulin genes from their precursor segments. In pre-B cells of non-Burkitt's lymphomas, translocations of chromosomes 11 and 18 to the immunoglobulin heavy-chain locus occur during the process of assembly, at the stage of V–D–J[1] joining. The analysis of the points on these chromosomes at which the interchromosomal exchange takes place suggests that the recombinase enzyme that mediates this exchange may erroneously join a heavy-chain J segment to a cellular oncogene sequence.

In addition to altering the activity of cellular genes, translocations that interrupt genes can cause the production of abnormal gene products. In the early stages of chronic myelogenous leukemia, the distal end of the long arm of chromosome 9 translocates to chromosome 22 to form an abnormal chromosome (the Philadelphia chromosome) and a fusion gene (c-*abl/bcr*) whose product is a mutant protein kinase, an enzyme that is thought to play a significant role in the regulation of gene expression.

2. Genetic Transposition

Transposition is a genetically controlled process in normal tissues that can create new gene combinations by moving a segment of DNA from its normal location on a chromosome to another location on the same or a different chromosome. The segment is inserted at the new location by a recombinational process that does not require extensive homology between donor and recipient DNA. This phenomenon has been studied extensively in bacteria, yeast, and the fruit fly (*Drosophila melanogaster*), which contain special sequences (i.e., transposons) capa-

---

1. These are gene domains that are recombined during immunoglobulin gene formation.

ble of performing these exchanges. By associating structural genes with new regulatory elements, transposition might produce changes in the expression of multiple genetic loci in human cells as well.

If this process were to occur in an uncontrolled random manner, the recombination of unrelated genetic sequences would destabilize the genome and conceivably initiate tumorigenesis. Retroviral sequences with a transposonlike structure are present in many places in the human genome and might serve as transpositional mediators. These sequences insert randomly into DNA, causing gene inactivation, deletion, inversion, and duplications.

Although transposition is a potentially important mechanism of tumorigenesis, its actual role in cancer remains to be demonstrated.

C. Developmental Theory

The developmental theory of oncogenesis proposes that neoplastic transformation is an adaptive response of cells to an abnormal tissue environment that involves nonmutational changes in gene expression. The tumor cell retains the normal differentiation program of its tissue of origin and can be restored to normalcy in an appropriate tissue environment.

Certain highly malignant tumors of embryonal origin spontaneously differentiate into benign growths, and *in vitro* tumor cell lines can be induced to differentiate with appropriate stimuli. However, in none of these instances has it been demonstrated that normalcy has been restored—only that malignant behavior has been suppressed. So far, teratocarcinoma provides the most convincing evidence that certain cancers might be induced epigenetically and are potentially reversible.

Teratocarcinoma is a germ cell tumor that occurs in mice and can be experimentally induced by transplanting normal early embryos to an inappropriate location (e.g., the peritoneal cavity or a subcutaneous site). The resulting tumor arises from the neoplastic transformation of multipotential embryonic cells. The tumor contains stem cells, which form highly malignant rapidly growing tumors, as well as nonmalignant tissues derived from the differentiation of the stem cells. If a stem cell is introduced into a normal early mouse embryo, which is then allowed to develop *in utero,* its malignancy is suppressed and it can contribute to normal embryonic development.

The offspring of mice whose germ cells carry genetic markers derived from the teratocarcinoma cells develop normally and do not develop teratocarcinomas or other tumors at higher than normal rates. Teratocarcinoma is thought to be nonmutational in origin, because it shows a much higher frequency of induction than can be accounted for by spontaneous somatic mutation, and the loss of malignancy that accompanies cell differentiation is permanent.

As tumors develop they become increasingly disorganized and begin to express phenotypes that are inappropriate for their tissue of origin or stage of development. These changes have been attributed to a disturbance of normal differentiation, in which cells become arrested in their development, revert to an immature highly proliferative developmental stage (retrodifferentiate), convert to a different type of differentiated cell not normally found in the tissue (redifferentiate), lose their differentiated characteristics (dedifferentiate), or express genes that are either foreign to that tissue or are not expressed in adult tissue (abnormally differentiate).

The use of these terms to describe the phenotypic characteristics of a tumor has the unfortunate consequence of implying mechanisms that either have not been or cannot be demonstrated. Normal development is a highly ordered process involving a fixed sequence of progressive changes in gene expression that culminates in the creation of a specific set of precise characteristics for each cell lineage. Tumor progression, by contrast, is random and unpredictable. No two cells undergo the same set of changes, nor is there a characteristic end state.

The characteristics of the tumor are determined by several factors: the developmental stage of the target cell(s) at the time of neoplastic conversion, the effect of neoplastic transformation on the process of differentiation, and the rate of intraclonal diversification during tumor progression. This is illustrated by the natural history of chronic granulocytic leukemia, a proliferative disorder of blood cells, which is characterized in its early stages by the overproduction of immature granulocytes. The initial neoplastic transforming event most likely occurs in a pluripotential precursor of the blood cell lineages. The clonal descendants of this cell can be identified by the presence of a unique cytogenetic marker, the Philadelphia chromosome, which, as mentioned in Section II,B,1, results from a translocation involving chromosomes 9 and 22.

In the early chronic stages of this disease, neoplastic transformation does not arrest differentiation nor does it alter the balance between proliferation and differentiation in cell lineages other than granulocytes. The reasons for the selective change in the population dynamics of the granulocyte lineage are not understood. It is only in the later stage of the disease, the blast crisis, that the balance between proliferating and differentiating cells is grossly altered in other lineages and the tumor is overpopulated by immature cells (so-called "blast" cells, from their appearance).

D. Oncogenes and Antioncogenes

The contemporary views are that cancer is caused by mutationlike alterations of critical DNA sequences that regulate normal cell growth and differentiation and that tumorigenesis is controlled by both positive and negative mechanisms.

According to the cellular oncogene hypothesis, the positive control over tumorigenesis is exerted by transforming genes (i.e., oncogenes), first found in retroviruses, which derive from protooncogenes after activation by mutation or recombination. [See ONCOGENE AMPLIFICATION IN HUMAN CANCER.]

Approximately 50 cellular oncogenes have been identified, in part by their presence in retroviruses, in part by their ability to cause a morphological transformation of the NIH/3T3 cell culture line, and in part by their increased expression in tumor cells. The gene products of protooncogenes and their derived oncogenes include components of cell signaling pathways (e.g., growth factors and their receptors, protein kinases and G proteins) and nuclear DNA-binding proteins.

The original version of the cellular oncogene hypothesis viewed cancer as a one-step process caused by the activation of a single protooncogene. Although this was inconsistent with evidence from epidemiological and genetic studies that cancer is a multistep process with a long latency period, the hypothesis was widely accepted after it was known that the transforming sequences of retroviruses able to induce cancer in animals were derived from cellular genes. However, in many cases single activated oncogenes cannot render primary cultures of human or rodent cells *in vitro* tumorigenic or produce tumors *in vivo*. In tumors that consistently expressed a particular oncogene (e.g., colorectal cancer and the mutated *ras* gene), the evidence suggested that oncogene activation was not an obligatory early step in tumorigenesis.

The current hypothesis that cancer results from the synergistic actions of multiple oncogenes conforms to the traditional view of tumor development as a polygenic process. The tumorigenic potential of single oncogenes and oncogene combinations has been tested in transgenic mice and in organ reconstitution systems. Both techniques utilize hybrid genes composed of a truncated viral or cellular oncogene(s) linked to a strong promoter or enhancer. The transgene experiments use regulatory elements (e.g., the mouse mammary tumor virus promoter or the immunoglobulin heavy-chain enhancer) that can be controlled in a tissue-specific manner.

Transgenic mice are produced by microinjecting the gene construct into fertilized eggs. The eggs are then reimplanted into the oviduct of a pseudopregnant mouse and allowed to develop to term. The offspring are then tested for the presence of the transgene by Southern blot hybridization analysis of tail DNA (the technique is basically the same as that illustrated in Fig. 1). Transgenic mice carrying each of the two putative oncogenes that are to be tested in combination are mated to produce offspring whose cells carry both genes. At appropriate times mouse tissues are examined histologically for preneoplastic changes, neoplasms, and oncogene expression.

The organ reconstitution approach uses the fetal mouse urogenital sinus, which, in males, develops into the prostate gland. The sinus can be dissociated, reconstituted with a small number of cells that have been retrovirally transfected with one or more activated oncogenes, and transplanted into the renal capsules of adult isogenic hosts, where it will continue to develop.

An important difference between the two techniques is that, in transgenic mice, all cells contain the transgene, but only certain tissues express it at high levels, whereas in the reconstituted urogenital sinus the proportion of cells in the tissue that constitutively express the gene can be experimentally varied to study the influence of cell–cell interactions on the process of tumorigenesis.

Several conclusions about oncogene cooperativity have emerged from studies on *myc–ras* gene interactions in these experimental models. v-*myc* is the transforming gene of several avian sarcoma viruses and codes for a DNA-binding protein. *ras* is the transforming gene of Harvey (H) and Kirsten murine sarcoma viruses. Its gene product is a GTP-binding protein with GTPase activity.

In transgenic mouse experiments activation of the cellular *myc* gene (c-*myc*) alone is not sufficient to produce tumors, but it accelerates tumor formation with v-Ha-*ras*. Transgenic animals show an age-dependent incidence for cancer, as do humans. Furthermore, the tumors develop from single foci, even though most of the cells in a tumor-forming tissue express the transgene. In lymphoid tissue, which is amenable to clonal analysis, the tumors that form are monoclonal.

Thus, in addition to oncogene cooperativity, other events are apparently required for tumor induction. The only known instance in which a single activated cellular oncogene is able to induce tumors autonomously and synchronously in all cells of a tissue is the induction of mammary tumors by c-*neu* linked to the mouse mammary tumor virus promoter. c-*neu*, a cellular oncogene derived from a chemically induced rat neuroblastoma, codes for a protein related to the epidermal growth factor receptor and other tyrosine kinases. A homologous cellular oncogene present in human cells, c-*erb*-B2 is amplified and expressed in human breast cancers.

Oncogene experiments with reconstituted urogenital sinus have shown that the prostate gland develops normally when only a small percentage (i.e., 0.01%) of cells express v-Ha-*ras*. However, as the percentage of cells expressing the oncogene is experimentally increased, tissue organization is progressively disturbed. Mesenchymal cells overproliferate, there is an increased formation of new blood vessels by endothelial cells (angiogenesis), and epithelial cell proliferation is inhibited. The expression of v-*myc* causes a hyperproliferation of epithelial cells in the periphery of the gland, but does not otherwise interfere with tissue organization or cell differentiation. When the sinus is reconstituted with small numbers of cells carrying both oncogenes, epithelial tumors form. The tumors appear to be monoclonal, chromosomally abnormal, undifferentiated, and invasive. These observations suggest that, in addition to oncogene cooperation, other genetic changes occur during tumor outgrowth.

The basis for the synergistic interaction of oncogenes is not yet understood. A possible explanation is that the selective effects of different oncogenes on multiple cell types within a tissue can severely disturb the cell interactions required for tissue homeostasis. Alternatively, the combined expression of oncogenes with different molecular mechanisms of action within a single cell type could produce local changes in cell interactions that would affect the behavior of neighboring normal cells.

Several mechanisms are thought to contribute to the preferential development of tumors in specific tissues (i.e., tissue selectivity) in oncogene experiments.

In transgenic mice tissue selectivity is partially determined by the type of promoter–enhancer element that controls the transforming gene. The simian virus 40 T antigen, for example, is able to induce tumors in many types of cells in mice. However, when its expression is controlled by the murine promoter–enhancer of a gene expressed normally in the lens of the eye (α-crystallin), it induces lens tumors. Under the control of the insulin-regulatory region, it induces β pancreatic islet cell tumors (the cells that normally produce insulin). With the elastase promoter it induces tumors of the exocrine pancreas, in which the promoter is functional.

In addition, the transforming activity of an oncogene could be restricted to a particular portion of the differentiation program of a cell. Experiments with temperature-sensitive viruses, which express the potential to transform cells at a low, but not high, temperature, show that, as cells progressively become more differentiated at the nonpermissive (i.e., higher) temperature, they could lose their ability to express the characteristics of transformed cells when shifted to permissive (i.e., lower) temperatures.

Certain types of cells must enter into a predifferentiation growth arrest state to begin the differentiation process. The overexpression of certain oncogenes might prevent growth arrest and thereby cause cells to become refractory to normal differentiation signals. For example, it is speculated that the failure of maturing Burkitt's lymphoma B cells carrying the *myc*/Ig translocation to become proliferatively quiescent could be caused by the failure of the translocated *myc* gene to respond to an endogenous signal that turns off the normal allele. In some experimental systems cells can be induced to differentiate by treatment with growth-inhibiting chemicals or by altering culture conditions to limit their proliferative activity.

Intrinsic differences in the proliferative capacity of specific cell types within a tissue might determine the cell type of origin of the tumor. In studies of v-Ha-*ras*/v-*myc* cooperativity in the urogenital sinus, it was found that prostate tumors arose exclusively in the epithelium, even though the mesenchymal cells appeared to be preferentially transformed by the retroviral oncogene carrier. A possible explanation for the formation of epithelial tumors is that, at the time of viral infection, the tissue was at a developmental stage in which epithelial cells had a significant growth advantage over other cell types.

Tumor suppressor activity could also play a decisive role in tissue-selective patterns of tumor formation. The existence of tumor-suppressing genes was originally deduced from somatic cell hybridization experiments, which showed that when normal and tumor cells were fused, the resulting hybrid cells were no longer tumorigenic. The loss of particular chromosomes occurred with prolonged passage of the hybrids and was associated with the reexpression of malignancy. It was hypothesized that these chromosomes contained dominantly acting tumor suppressor genes.

The inactivation or deletion of both copies of specific chromosomal loci is associated with the development of naturally occurring tumors as well. Studies in *D. melanogaster* flies and *Xiphophorine helleri* and *X. maculatus* fishes suggest that the involved loci encode gene products that regulate developmental processes and act to suppress tumor formation. *Drosophila* flies with homozygous recessive mutations (i.e., mutations expressed when both gene copies are inactive or missing) at the lethal *giant larvae* locus become arrested at the larval stage of development and develop neuroblastoma-like tumors. They can be rescued from both defects by introducing into them the normal gene.

Crosses between two species of *Xiphophorine* fish, *X. helleri* (swordtail) and *X. maculatus* (platyfish), give rise to hybrid fish with an increased incidence of melanomas. The melanomas originate from platyfish-derived pigment cells. The ability of the pigment cells to become neoplastically transformed in these crosses maps genetically to a single gene, *Tu*. *Tu* is regulated by two other classes of genes, *R* and *Co*. *R* restricts the expression of *Tu* to specific tissues, while *Co* genes suppress *Tu* in specific regions of the body. Mutation of *R* or its separation from *Tu* by chromosome rearrangement leads to neoplasia. Mutation of a *Co* gene predisposes a fish to malignant transformation of a pigment cell in the part of the body which that *Co* gene regulates.

The loss or inactivation of tumor suppressor genes in human tissues is thought to be a significant factor in human tumor development. A tumor of developing retinal tissue, retinoblastoma, originates from cells in which both copies of the retinoblastoma susceptibility gene have been inactivated or lost (see Section II,A,5). The gene has been cloned and its product identified as a 105-kD nuclear protein. The activity of the protein might be regulated

during the cell cycle by changes in phosphorylation. When transfected into retinoblastoma cells, the wild-type gene inhibits cell growth and tumorigenicity. The gene is expressed by most normal tissues and is dysfunctional or lost in various tumors that are clinically unrelated to retinoblastomas, including breast and small cell lung carcinomas. [*See* TUMOR SUPPRESSOR GENES.]

A second tumor suppressor gene, *p53,* is located on human chromosome 17p13. It encodes a 53-kD nuclear phosphoprotein which is proposed to act as a negative growth regulator. The *p53* gene was initially misidentified as an oncogene based on its transforming activity in primary rat fibroblasts. Subsequently it was found that only the mutated gene possessed transforming activity, which was attributed to the ability of the mutant *p53* protein to complex with and presumably inactivate the wild-type protein. *p53* is considered to be a tumor suppressor gene based on its ability to inhibit transformation of primary rat fibroblasts by mutant *p53* and activated *ras* genes, the expression of *p53* by normal cells, and the inactivation/deletion of both *p53* alleles in colorectal and lung cancers.

Progress is being made in the identification and cloning of human tumor suppressor genes, but it is still too early to know what the functions of these genes might be. Potentially important mechanisms of tumor suppression include (1) interference with neoplastic transformation by directly antagonizing the expression of a transforming gene, or by inducing cell maturation to a stage at which the cells are no longer at risk for transformation; (2) the inhibition of neoplastic cell growth by normal cells, mediated by the cell surface, extracellular matrix, or diffusible molecules; and (3) the prevention of tumor angiogenesis (i.e., the formation of new blood vessels and their penetration into the tumor), which is required for the growth of neoplastically transformed cells to a physiologically significant size.

Bibliography

Bishop, J. M. (1988). The molecular genetics of cancer. *Leukemia* **2,** 199.
Correa, P. (1988). A human model of gastric carcinogenesis. *Cancer Res.* **48,** 3554.
Duesberg, P. H. (1987). Cancer genes: Rare recombinants instead of activated oncogenes. *Proc. Nat. Acad. Sci. U.S.A.* **84,** 2117.
Farber, E. (1984). Cellular biochemistry of the stepwise

development of cancer with chemicals. *Cancer Res.* **44,** 5463.

Hansen, M. F., and Cavenee, W. K. (1987). Genetics of cancer predisposition. *Cancer Res.* **47,** 5518.

Herrlich, P., Angel, P., Rahmsdorf, H. J., Mallick, U., Poting, A., and Hieber, L., Lucke-Huhle, C., and Schorpp, M. (1986). The mammalian genetic stress response. *Adv. Enzyme Regul.* **25,** 485.

Klein, G., and Klein, E. (1986). Conditioned tumorigenicity of activated oncogenes. *Cancer Res.* **46,** 3211.

Rubin, H. (1985). Cancer as a dynamic developmental disorder. *Cancer Res.* **45,** 2935.

Sager, R. (1986). Genetic suppression of tumor formation: A new frontier in cancer research. *Cancer Res.* **46,** 1573.

Sinn, E., Muller, W., Pattengale, P., Tepler, I., Wallace, R., and Leder, P. (1987). Coexpression of MMTV/ v-Ha-ras and MMTV/c-myc genes in transgenic mice: Synergistic action of oncogenes in vivo. *Cell (Cambridge, Mass.)* **49,** 465.

Thompson, T. C., Southgate, J., Kitchener, G., and Land, H. (1989). Multistage carcinogenesis induced by ras and myc oncogenes in a reconstituted organ. *Cell (Cambridge, Mass.)* **56,** 917.

Vogelstein, B., Fearon, E. R., Hamilton, S. R., and Feinberg, A. P. (1985). Use of restriction segment length and polymorphisms to determine the clonal origin of human tumors. *Science* **227,** 642.

Vogelstein, B. Fearon, E. R., Hamilton, S. R., Kern, S. E., Preisinger, A. C., Leppert, M., Nakamura, Y., White, R., Smits, A. M. M., and Bos, J. L. (1988). Genetic alterations during coloreactal tumor development. *N Engl. J. Med.* **319,** 525.

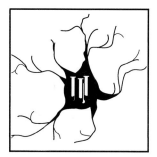

Nerve Growth Factor

SILVIO VARON, THEO HAGG, MARSTON MANTHORPE,
University of California in San Diego

I. Developmental Neurobiological Aspects
II. The NGF Protein
III. NGF Receptors
IV. Molecular Mechanisms of NGF Action
V. NGF and the Adult CNS
VI. NGF and Axonal Regeneration
VII. Summary and Outlook

Glossary

Neural crest Transient structure during early development of the nervous system, which gives rise to most neurons and glial cells of the peripheral nervous system, among others

Neurite Nerve cell process, i.e., a thin extension from a nerve cell body, such as axons and dendrites

Retrograde axonal transport Process by which substances accumulated or taken in at the axonal ending are transported backward along the axon itself toward the cell body

Schwann cell Glial cell of the peripheral nervous system, typically associated with peripheral axons and providing a myelin sheath to many of them

Trophic Characterized processes or substances (other than metabolites or ions) that provide support to life-sustaining activities of a cell

NERVE GROWTH FACTOR (NGF) is a special protein that controls the maintenance, size, extension of processes and transmitter synthesis in selected neurons of the peripheral nervous system (PNS) and central nervous system (CNS). The discovery of NGF, some 40 years ago, was an epoch-making event in modern neurobiology, which provided the first evidence that nerve cells depend on specific, extrinsic factors for their survival and function and, conversely, that developmental or experimental

modulations of such factors can dictate neuronal performances and/or deficits. More recently, recognition of NGF roles in the adult mammalian CNS has raised the possibility that deficits of endogenous NGF (or other neuronotrophic factors) may underlie or aggravate certain human neurodegenerative disorders, as well as the apparent inability of injured adult CNS neurons to regenerate. The NGF field remains a dynamic one: information, interpretations, and conceptual frameworks are likely to continue evolving in the coming years.

I. Developmental Neurobiological Aspects

A. Developmental Neuronal Death and the Concept of Neuronotrophic Factors

In both PNS and CNS, a massive numerical reduction of a given neuronal population takes place when the growing axons reach and connect with their target innervation territory. The extent of this naturally occurring "developmental neuronal death" can be altered by changing the size of the target territory. Such events reveal (1) a timed, target-related appearance of a neuronal vulnerability, which remains to be understood, and (2) a target-dependent protection mechanism, which selectively helps some of the neurons to survive. In the latter context, innervation territories were hypothesized to produce and selectively deliver specific agents, designated as neuronotrophic factors, which would bind to the axon terminals, be retrogradely transported along the axon to the cell body, and promote at the cell body survival-supportive cell machineries.

With increasing age, the rescued neurons appear to reduce their target dependence for survival, perhaps by acquiring additional trophic sources such as

333

their associated glial cells. Trophic support, however, continues to be needed for other neuronal performances. Thus, disconnection from the target territory by axotomy may lead to reductions in cell body size and/or neuron-transmitter synthesis, as well as other neuronal properties.

B. NGF as a Neuronotrophic Factor Prototype

Early ablation of a limb bud in the chick embryo leads to an abnormally high neuronal death in the ventral spinal cord (primary motor neurons) and the dorsal root ganglia (DRG) (primary sensory neurons), the main sources of limb innervation. Conversely, implantation of certain connective tissue tumors reduced the normal developmental death in the DRGs (but not the ventral cord) and also in peripheral sympathetic ganglia. Extracts from these tumors were shown to be responsible for this improved survival as well as for the neuronal hypertrophy and the massive increase in nerve fiber outgrowth that accompanied it, leading to the demonstration of a soluble "nerve growth factor" that addresses the DRG and sympathetic neurons. A similar NGF activity was found to occur in much higher amounts in the adult male mouse submaxillary gland, providing an exceptionally abundant source for purification and characterization of the mouse NGF protein.

Administration of purified, exogenous NGF to neonatal rodents causes massive hypertrophy of sympathetic ganglia and profuse outgrowth of sympathetic fibers. Conversely, administration of antibodies against NGF to newborn animals leads to a dramatic immunosympathectomy. Both effects decrease with increasing postnatal ages; the major effects of such treatments on adult sympathetic ganglia are corresponding changes in the content of certain functional proteins such as tyrosine hydroxylase, the critical enzyme for the synthesis of the noradrenaline transmitter. Similar responses to exogenous NGF or its antibodies can be seen in DRG neurons if the treatments are applied at appropriate embryonic stages. There appears to be little species-related specificity in the biological action of NGF: Mouse NGF is effective not only on mouse but also on chick and rat, as well as (*in vitro*) amphibian and human, ganglionic neurons.

C. *In Vitro* Neuronal Cultures and NGF

Neuronal cultures have been a major tool for NGF investigations from the very beginning. Explant cultures of DRG or sympathetic ganglia (usually from embryonic chicks) respond to NGF with a dose-dependent outgrowth of neurites (i.e., neuronal processes or fibers), which has long served as a semiquantitative measurement of NGF concentrations. Dissociated neuronal cultures from the same ganglia permit quantitative determinations of NGF concentrations in terms of either neuronal survival or neuritic outgrowth responses. By presenting dissociated or explant ganglionic cultures with serial dilutions of NGF, one can define units of NGF activity as the minimal amount of NGF in 1 ml of culture medium that is needed for maximal response (biological units) or for half-maximal responses (trophic units).

Ganglionic cultures have been used to recognize the presence of NGF activity in several source tissues (including both peripheral and central glial cells) and to monitor the progressive purification of their NGF protein. Conversely, presentation of NGF to neuronal cultures from different embryonic origins has extended the spectrum of NGF-responsive neurons to several other ganglionic cells from neural crest origin (but not from placodal origin) as well as to the chromaffin cells of the adrenal medulla (neural crest-derived, adrenaline-producing relatives of sympathetic neurons). A clonal cell line derived from a rat pheochromocytoma (adrenal medulla tumor), designated as PC12, was shown to respond to NGF with the expression of sympathetic neuronal features such as extension of neurites, acquisition of electrical excitability, and enhanced content of transmitter-synthesizing enzymes.

In vitro studies of embryonic ganglionic or PC12 cell cultures have led to the recognition (substantiated, in many cases, by subsequent *in vivo* studies) of most features currently known about biological actions of NGF, among them the occurrence of specific and saturable NGF-binding sites (receptors) on the target cell surface, the internalization and retrograde axonal transport of the receptor-bound NGF, and the alteration of specific cell components involved in or resulting from the molecular transduction processes initiated by the NGF–receptor encounter. Culture studies have also led to the recognition that neurite extension requires the local availability of NGF to the growth cone (the dynamic structure at the tip of an elongating neurite), and that the direction of neuritic elongation can be controlled by the location of an NGF source. Following the example of NGF-responsive systems, other neuronal cell cultures are used to detect and investigate new neuronotrophic factors, such as a brain-derived neurotrophic factor (BDNF) and a ciliary neuronotrophic factor (CNTF); knowledge of these cell cultures lags considerably behind that of NGF.

II. The NGF Protein

In mouse submaxillary gland extracts, NGF occurs in a high molecular weight complex of about 130 kDa (7S NGF), comprising different protein subunits (2 alphas, 1 beta, and 2 gammas) as well as Zn^{2+}. The 7S complex can be isolated in pure form at neutral pH and dissociates reversibly into component subunits at both acid and alkaline pHs. The NGF activity resides exclusively with the beta subunit (beta NGF, also known as 2.5S NGF), which is a basic (pI $\approx$ 10), 26-kDa dimer protein, consisting of two identical chains. The monomeric chain of beta NGF (118 amino acids, with three intrachain disulfide bridges) may undergo proteolytic cleavages during its purification (unless protected in the 7S complex), losing its C-terminal arginine or an N-terminal octapeptide, or both, without apparent effects on its ability to dimerize or express NGF activity. Definitive information on the active domain(s) of beta NGF is still unavailable, and several efforts are directed to the isolation of NGF-derived oligopeptides capable of either mimicking or competing with the intact protein. Substantial amino acid sequence homologies have been detected between beta NGF and the pro-insulin, relaxin, BDNF, and insulinlike growth factor polypeptides, suggesting evolutionary relationships between these hormones and the NGF neuronotrophic factor. NGF proteins have been also purified from other sources, such as snake venoms, bovine seminal vesicles, and human placenta.

NGF genes (cDNA clones) isolated from mouse, rat, bovine, chick, and snake, as well as human sources, have confirmed and extended their very high interspecies homologies. Two major messenger RNAs (mRNA) have been recognized to direct the synthesis of two primary translation products (24 and 32 kDa, respectively). Processing of such precursors to yield the active NGF polypeptide may be carried out by a variety of cells. Regulation of synthesis and/or release of NGF in the producing cells may depend on their innervation by NGF-responsive neurons as well as on other extracellular signals yet to be identified. NGF mRNA measure-

ments have been used to identify NGF-producing tissues and establish that, for example, (1) NGF production in an innervation territory may start only at about the arrival time of innervating axons, and (2) NGF production correlates with innervation density in several sympathetic target territories. Cells in which human NGF gene was artificially introduced are currently investigated as enhanced NGF sources and for potential implantation *in vivo* (see Section V).

III. NGF Receptors

NGF is unable to cross the cell membrane and must operate by binding to cell-surface, ligand-specific, saturable receptors. Binding studies with ^{125}I-NGF have shown two sets of such putative receptors on NGF-responsive cells: a high-affinity (10^{-11} *M*), slow-dissociating set (Type I) and a lower-affinity (10^{-9} *M*), faster-dissociating one (Type II). The Type I receptor is presumed to be the one that actually mediates the NGF action, because its affinity matches the concentration range at which NGF is biologically active and because it has been recognized in responsive, but not in nonresponsive, cells. NGF receptors have been solubilized, apparently yielding two groups of proteins on sodium dodecyl sulfate gels in the 80 and 150 kDa ranges, respectively.

Relationships between Type I and Type II NGF receptors are still an open issue. Their close relationship is encouraged by findings that transferring the Type II gene to cells lacking either type of receptor can lead to the appearance of both sets in the host-cell membrane. They may serve different functions. For example, the Type II receptor may store up NGF (when available at higher concentrations) and relinquish it to the Type I receptor for action. Alternatively, the Type II receptor may convert to Type I upon occupation by the NGF ligand. One speculation is that the truncated Type II receptor may couple with another membrane protein to serve as the NGF receiver for activation of the effector partner in a resulting Type I receptor.

A mouse monoclonal antibody, 192 IgG, raised against a rat PC12 Type II receptor, has been extensively utilized to recognize Type II-positive (and, thus, possibly NGF-susceptible) cells. One finding pertinent to the role of the Type II NGF-receptor role is the display of Type II-immunostaining on Schwann cells of peripheral nerve following axonal degeneration (see Section VI). These peripheral glial cells are known to be able to produce NGF, and their Type II membrane proteins may be involved in the externalization of the NGF product rather than performing a true receptor function, i.e., the elicitation of Schwann cell responses to extracellularly presented NGF.

IV. Molecular Mechanisms of NGF Action

The recognition of NGF by its receptors, as is the case of other ligands with their receptors, is presumed to trigger a series of transduction steps involving the generation of intracellular second messengers, followed by a downstream modulation of various cell machineries, causing the biological effects of NGF. Several investigators have reported molecular changes elicited by NGF in embryonic ganglionic neurons, PC12 cells, or certain NGF-sensitive neuroblastoma cell cultures. Nevertheless, the sequential events underlying either the transduction process or its downstream secondary effects are not yet understood. Among the latter are changes in cell adhesion, cell membrane dynamics, cytoskeletal elements such as microtubule-associated proteins or intermediate filaments, a rise in ornithine decarboxylase (a polyamine-producing enzyme presumed to regulate nucleic acid performances), increased intake of nutrients and ions, activation or maintenance of the Na^+,K^+-pump, and phosphorylation of selected nuclear or cytoplasmic proteins.

Several second messengers known to operate in other ligand-receptor systems have also been reported to be involved in the NGF mode of action. NGF may cause an elevation of intracellular cyclic adenosine monophosphate (AMP), presumably leading to the activation of cyclic AMP-dependent protein kinases and the phosphorylation by them of selected protein substrates. Phosphatidylinositide turnover, which measures the action of ligand-activated phospholipase C on phosphatidylinositide diphosphate, may be increased by NGF in sympathetic ganglia. The enzymatic hydrolysis of phosphatidylinositide diphosphate in turn activates intracellular regulatory controls. An NGF effect on protein kinase C through this system is also supported by the observation that phorbol esters, which activate the kinase, are able to mimic the NGF action on survival and neurite expression of cultured DRG and sympathetic neurons. Less well documented is the ability of NGF to modulate plasma membrane Ca^{2+} channels and, thus, elicit an increase in cytosolic Ca^{2+} and the activation of Ca^{2+}, calmodulin-dependent protein kinases. Finally, NGF-induced increases of membrane lipids or protein transmethylation, another proposed model for ligand-receptor transduction, have been reported.

NGF may also activate cellular regulatory genes (proto-oncogenes) that, when altered, participate in neoplastic transformation (oncogenes). Thus, the protein products of the c-*ras* proto-oncogene is increased by NGF in PC12 cells, and, conversely, the injection of these gene products into either PC12 cells or ganglionic neurons mimics the effects of NGF presentation. Proto-oncogene c-*fos* is also rapidly and transiently activated in PC12 cells by either NGF or phorbol esters. Further progress in this direction is expected in view of the considerable interest in the field. [*See* ONCOGENE AMPLIFICATION IN HUMAN CANCER.]

V. NGF and the Adult CNS

The discovery of NGF and most of its subsequent investigations have centered on peripheral neurons (and related cells) as the targets for NGF action, and much of that action has been studied with perinatal (i.e., developing) tissue; however, no *a priori* reasons exist to restrict the concept of neuronotrophic factors to developing PNS neurons. A CNS neuronotrophic hypothesis has been put forward in recent years, which postulates that adult CNS neurons continue to depend in many critical ways on an adequate supply of endogenous trophic factors. Two major corollaries follow from such a postulate: (1) a relative deficit of trophic support (reduced supply or increased demand) should lead to recognizable neuronal damage or dysfunction, and (2) administration of exogenous factor should compensate for the endogenous deficit, hence prevent or reduce the neuronal damage. Attempts to test these concepts require the selection of a reasonably defined anatomical system and the availability of a neuronotrophic factor active on this test system. Such a model became available with the recent discovery that NGF can affect cholinergic neurons in the basal forebrain, neostriatum and nucleus accumbens (Fig. 1).

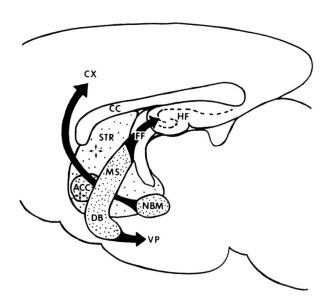

FIGURE 1 Cholinergic neurons of the basal forebrain responsive to NGF include those of the medial septum (MS) and diagonal band (DB), which project to the hippocampal formation (HF) through the fimbria–fornix (FF) and ventral pathway (VP), as well as those of the nucleus basalis of Meynert (NBM), which project to the cerebral cortex (CX). Also responsive to NGF are the cholinergic interneurons (arrow crosses) of the neostriatum (STR) and nucleus accumbens (ACC). CC, corpus callosum.

A. CNS Cholinergic Neurons and NGF

In the basal forebrain, three sets of cholinergic neurons (using acetylcholine as a neurotransmitter), symmetrically located on each side of the brain, are susceptible to NGF action. Messenger RNA for NGF and its protein product are present in the innervation territories, showing that they are NGF sources for the neurons. Radiolabeled NGF injected into the areas of innervation is taken up and retrogradely transported to the corresponding cholinergic neuronal cell bodies. These cholinergic neurons display specific NGF receptors and respond to exogenous NGF (both *in vitro* and *in vivo*) with an increase of their neurotransmitter-synthesizing enzyme (choline acetyltransferase [ChAT]). [*See* NEUROTRANSMITTER AND NEUROPEPTIDE RECEPTORS IN THE BRAIN.]

B. The Septo–Hippocampal Test Model

Cholinergic neurons can be specifically and individually distinguished from other local neuronal populations by their cholinergic markers, e.g., the acetylcholine-synthesizing enzyme ChAT and the acetylcholine-degrading enzyme, acetylcholinester-ase (AChE). Moreover, cholinergic neurons of the basal forebrain display a third marker, the low-affinity NGF receptor (Type II).

A complete fimbria–fornix transection (FFT) on either side of the adult rat brain interrupts the axons going from the medial septum and vertical diagonal band neurons to the hippocampus, presumably interrupting the main delivery route of endogenous hippocampal NGF. Two-thirds of the cholinergic neurons, recognizable by their marker in the medial septum on the same side of the lesion, disappear within 2 wk. Such a disappearance is fully prevented by continuous or intermittent administration of exogenous NGF into the lateral ventricle on the lesion side. Recently, similar results have also been obtained in the monkey. [*See* HIPPOCAMPAL FORMATION.]

The disappearance of the neurons could reflect either their death or a depletion of the markers (by which they are recognized). Marker loss rather than cell death appears to be largely the case, because NGF administration *after* the neurons have disappeared restores the detectability of most of them. Thus, NGF controls several properties of these adult CNS neurons but not necessarily their survival, as already noted with adult PNS target cells. NGF-controlled features include (but are not limited to) the cholinergic neurotransmitter function (AChE and ChAT) as well as the expression of Type II NGF receptors. Furthermore, NGF controls the neuronal cell size, a truly trophic performance.

C. The Aged Rat Model

The cholinergic neurons of the basal forebrain and the neostriatum play a crucial role in the cognitive activities (learning and memory) of the brain. Cognitive behavioral deficits follow cholinergic deficits induced by (1) experimental lesions in the young adult rat, (2) the still poorly understood involutive process of brain aging, and most dramatically (3) the degenerative processes underlying Alzheimer's disease and related senile dementia syndromes. NGF treatments reportedly improve behavioral performances in fimbria-lesioned young adult rats and also affect behaviors in other experimental rats. Such results have further encouraged the speculation that chronic cognitive deficits might also benefit from exogenous NGF administrations. [*See* COGNITIVE REPRESENTATION IN THE BRAIN; LEARNING AND MEMORY.]

In more recent studies, aged (>2 yr old) rats were screened for spatial memory impairment, infused intraventricularly with NGF (or with NGF-free vehicle as controls), and retested on the same behavioral task after 2 and 4 wk of treatment. NGF administration for 4 wk resulted in a reversal of the cognitive impairment, accompanied by a partial reversal of the age-induced size reduction of AChE-positive neurons in nucleus basalis and striatum. These results confirm that NGF can address and help neurons that are already impaired by a chronic insult, as previously seen with acutely damaged ones. Whether NGF improves cognitive behavior by an induced increase in transmitter function, a restoration of old synapses, and/or a sprouting of cholinergic terminals with formation of new synapses is yet to be determined.

D. Unlesioned Young Adult Rat Models

Endogenous sources could provide enough NGF to the responsive neurons to regulate their performances to their maximal capability. Conversely, a variable supply of endogenous NGF might serve to modulate the functional performance of adult neurons. Two recent studies favor the latter view.

A continuous intraventricular infusion of exogenous NGF, which does not directly damage the neighboring striatal tissue, intensifies striatal ChAT immunostaining and increases the size of ChAT-positive neuronal cell bodies. Direct infusion of NGF into the striatum itself elicits similar responses and, in addition, causes the appearance of NGF receptors (Type II) in the striatal cholinergic neurons.

Intact, young adult rats receiving intraventricular NGF infusions also respond with an increase of messenger RNA for NGF receptor in the septum and a corresponding increase of ChAT activity in the hippocampal tissue, the innervation territory of medial septum cholinergic neurons. Thus, like the striatal cholinergic neurons, selected properties of medial septum neurons can be enhanced by exogenous NGF supplementation of their endogenous NGF supply. Conversely, a reduction of neuronal ChAT in the medial septum has been induced by intraventricular administration of antibodies against NGF.

VI. NGF and Axonal Regeneration

A. Peripheral Nerves

Peripheral nerve, such as the sciatic nerve of the adult rat, contain axons from spinal cord motoneurons, dorsal root ganglion sensory neurons, and sympathetic ganglionic neurons, the latter two groups being acknowledged NGF targets. NGF is available to these neurons in their peripheral innervation territories but can also be produced locally by the Schwann cells of the nerve itself. After transection of the nerve, the neuronal cell bodies, the axons of which have been cut, undergo structural and biosynthetic changes that can be prevented by NGF. Recent information suggests that exogenous NGF can promote regeneration of transected axons, a process that under favorable circumstances can occur spontaneously and lead to restoration of sensory and motor functions.

In the distal segment of the transected nerve, the severed axons degenerate and axonal and myelin debris is removed by phagocytosis, but the Schwann cells and the basal lamina surrounding them will persist, thereby creating empty tubular spaces through which regenerating axons can elongate. These Schwann cells, which in the intact nerve do not possess Type II NGF receptors, become receptor-positive after removal of the axon segments and return to a receptor-negative state upon acquiring new contacts with the regenerating axons. In parallel, NGF production by the Schwann cells increases upon loss of axonal contact and recedes when new contacts are established.

The NGF receptors appearing on the Schwann cells may serve a nonreceptor function, externalizing the NGF produced by the Schwann cells themselves so as to make it available to oncoming growth cones in the NGF-susceptible regenerating axons. These growth cones could use the presented NGF for trophic support and also gain directional guidance by turning off NGF receptors on the Schwann cells they have already contacted. The display of NGF receptors in the distal segment of the transected sciatic nerve occurs in all its Schwann cells, and not only those previously associated with sensory or sympathetic axons. This finding raises the possibility of parallel, though yet unrecognized, changes in the Schwann cell production of other neuronotrophic factors and receptors aimed, for example, to the spinal cord motoneurons.

B. Axonal Regeneration in the CNS

In the CNS of adult mammals, axonal regeneration is, at best, a limited, abortive process. This is not necessarily a defect of the adult CNS neurons, because axons from different regions of adult brain or spinal cord can grow through adult peripheral nerve for distances even greater than their original length. At the end of such nerve bridges, however, the regenerating axons only re-enter the CNS for about 1 mm, revealing the resistance to reinnervation of the adult CNS tissue. The nature of such a resistance might reflect (1) the lack of a glial cell and extracellular matrix neurite-promoting organization (contrasting with that of peripheral nerve), (2) the presence of inhibitors (e.g., from CNS myelin and the oligodendroglia that produces it), and/or (3) and inadequate local supply of endogenous neuronotrophic factors. CNS adult tissue can be successfully invaded by axons emerging from fetal CNS grafts, a presumptive source of trophic factors.

The recognition of NGF as a trophic factor for adult CNS cholinergic neurons has suggested the use of the septo–hippocampal system as a model for axonal regeneration into adult CNS (Fig. 2). An aspirative transection of the fimbria–fornix establishes a lesion cavity, into which a segment of sciatic nerve can be implanted to serve as a potential bridge between axotomized septal neurons and their hippocampal territory. Regrowing cholinergic (AChE-positive) axons appear in the bridge and its hippocampal end after 1 wk, and their number increases to a maximal level by the end of 1 mo. Penetration of the hippocampal tissue is much more restricted. Of the cholinergic fibers available at the bridge end, only some 20% have already entered the first 1 mm of CNS tissue at 1 mo, whereas the others do so progressively over the next 5 mo, and deeper hippocampal penetration (2 mm) develops even more slowly. Nevertheless, by 6 mo, reinnervation in the first 2 mm of the dorsal hippocampus displays a nearly normal pattern and density of cholinergic fibers.

Continuous infusion of exogenous NGF directly into the dorsal hippocampus for 1 mo, combined with the nerve bridge, dramatically enhances its reinnervation by the cholinergic axons. Cholinergic fiber numbers increase three- to fivefold in the first 1 mm and 20-fold at 2 mm inside the hippocampal tissue, indicating both an expanded recruitment of the axons available at bridge end and a facilitation of their further ingrowth. Thus, increasing the local availability of NGF can overcome the adult CNS resistance to penetration by adult CNS cholinergic axons, thereby strengthening the view that such a resistance reflects inadequate availability of neuronotrophic factors.

Peripheral nerve whose living cells have been destroyed (e.g., by freeze-thawing) loses its competence as a CNS regeneration bridge. Much of its

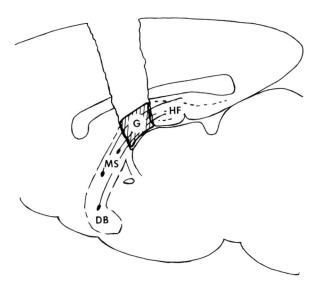

FIGURE 2 In a septo–hippocampal regeneration model, peripheral nerve grafts (G) can serve as a bridge for axons to regenerate from the medial septum (MS) and diagonal band (DB) to the hippocampal formation (HF).

performance in the septo–hippocampal model, however, is regained if a cell-free nerve segment is impregnated with NGF *in vitro* before being implanted. These observations confirm the view that fresh peripheral nerve is itself a source of NGF (beside its likely role as a longitudinally organized scaffolding), which is stopped when its producer cells are eliminated but can be replaced by an exogenous NGF. In the nerve bridge, NGF may be needed only at the septal end to recruit cholinergic fibers into the bridge, or it may also be required over the whole length of the bridge to promote axonal elongation along the basal lamina-lined scaffold spaces. In either case, it is increasingly evident that NGF plays a crucial role in the regenerative capabilities of adult CNS cholinergic neurons at all locations concerned: the cell body level (e.g., septum), the intervening tissue spaces (e.g., nerve bridge), and the reinnervation territory (e.g., hippocampal formation).

VII. Summary and Outlook

The history of NGF is also a history of evolving concepts about neuronal development, functional maintenance, and repair. Much of the NGF history is yet to be written. Current information, and that being acquired at an increasingly rapid pace, will have to be sorted out, revised, and integrated into new cellular and neurobiological frameworks with regard to both the mechanism of action of NGF and the expanding scope of its physiological functions. The newly recognized ability of NGF to improve cognitive deficits in the aged rat is already inspiring attempts to use it in the clinical context of Alzheimer's disease. The NGF importance for CNS neurons is further stimulating the search for other trophic factors with similar capabilities for different sets of CNS neurons and different pathologic situations. And the emerging role of NGF in cholinergic axonal regrowth is redefining the parameters of the more general problem of regeneration by and in the adult CNS, including visual system and spinal cord, two major areas of human trauma.

Bibliography

Barde, Y. A. (1989). Trophic factors and neuronal survival. *Neuron* **2,** 1525–1534.

Hagg, T., Manthorpe, M., Vahlsing, H. L., and Varon, S. (1991). Nerve Growth Factor roles for cholinergic axonal regeneration in the adult mammalian central nervous system. *Comm. Dev. Neurobiol.,* in press.

Levi-Montalcini, R. (1987). The nerve growth factor 35 years later. *Science* **237,** 1154–1162.

Varon, S., Hagg, T., and Manthorpe, M. (1989). Neuronal growth factors. *In* "Neural Regeneration and Transplantation." (F. J. Seil, ed.). *Frontiers of Clinical Neuroscience,* Alan R. Liss, New York, **6,** 101–121.

Varon, S., Manthorpe, M., Davis, G. E., Williams, L. R., and Skaper, S. D. (1988). Growth factors. *In* "Functional Recovery in Neurological Disease," (S. G. Waxman, ed.). *Advances in Neurology,* Raven Press, New York, **47,** 493–521.

Varon, S., Manthorpe, M., and Williams, L. R. (1984). Neuronotrophic and neurite-promoting factors and their clinical potentials. *Dev. Neurosci.* **6(2),** 73–100.

Whittemore, S. R., and Seiger, Å. (1987). The expression, localization and functional significance of β-nerve growth factor in the central nervous system. *Brain Res. Rev.* **12,** 439–464.

Nerve Regeneration

BERNARD W. AGRANOFF AND ANNE M. HEACOCK,
University of Michigan

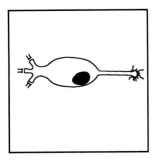

I. Introduction
II. Properties of Regenerating Nerves
III. Extraneural Aspects
IV. Limited Regeneration in the CNS

Glossary

Axon Cylindrical extension of the neuron, which carries electrical impulses (called *action potentials*) away from the cell body of the neuron and toward the synapse where the axon tip interacts with another neuron, a muscle cell, or a gland cell

Dendrites Extension of nerve cells, often forming tree-like arborizations that carry electrical impulses toward the cell body of the neuron

Extracellular matrix Web-like network of fibrous proteins and mucopolysaccharides in the extracellular space

Glia Nonneuronal cells of the brain (astrocytes and oligodendroglia) or peripheral nervous system (Schwann cells)

Growth cone Motile bulbous structure located at the tip of an elongating extension of the neuron (*axon* or dendrite)

Nerve A bundle of nerve fibers

Nerve fiber Axon or dendrite lined by folds of a glial cell, with oligodendrocytes lining axons in the brain and Schwann cells lining axons in the peripheral nervous system

Neuron Nerve cells consisting of cell bodies, dendrites (afferent extensions), and axons (efferent extensions)

Neuronotrophic factors Substances that neurons recognize in their environment and that are required for their growth and maintenance.

INJURY TO A NERVE is followed by degeneration of the nerve fibers distal to the injury. Nerve regeneration describes the process by which this injury is repaired and functional reconnection is attained. Successful nerve regeneration requires both a reprogramming of the biosynthetic capacity of the neuronal cell body and the presence of favorable environmental factors. If these requirements are met, new axons can grow out from the cut ends of the injured nerve, elongating sometimes over long distances, to establish the appropriate specific interaction with the original target tissue.

I. Introduction

The term "regeneration" generally connotes the ability of an organism to replace lost tissue. A starfish or a salamander can regenerate a severed limb. Regeneration in more advanced life forms is more limited but can occur. For example, some mammals have the ability to regenerate lost liver tissue. Thus, after surgical removal of a hepatic lobe, functioning new liver tissue is produced. In contrast, the nervous system usually does not form new nerve cells (neurons) after injury and therefore does not replace lost tissue. Recovery of lost function in the nervous system, when it occurs, is mediated by a limited regenerative process. After section of a bundle of nerve fibers, axons may regrow from surviving neuronal cell bodies, and eventually appropriate connections with other neurons are reformed (Fig. 1). Given the limitation that we live our lives out with those neurons in our brains and spinal cords that we are born with, remarkable recoveries of function can be seen in favorable instances. Suc-

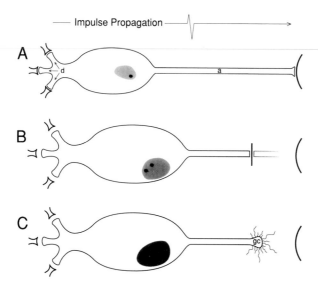

Impulse Propagation

A

B

C

FIGURE 1 Diagrammatic representation of neuronal regeneration. *A:* Intact nerve. Nerve impulses from dendritic processes (d) are conducted as a depolarization wave toward the cell body, which contains the nucleus, and continue away from the cell body, along the axon (a). This electrically mediated impulse is transmitted to a neighboring neuron or peripheral target tissue, depicted on the extreme right, by a chemical rather than electrical signaling mechanism, termed *neurotransmission. B:* Transected nerve. After trauma to a nerve, there is degeneration distal to the lesion, the result of isolation of the axon from the cell body. Characteristic histological (microanatomical) changes in the cell body in response to the injury reflect an increase in RNA and in protein synthesis. *C:* The regenerative response. If successful regeneration is to take place, the proximal stump (the end still connected to the cell body) of the severed axon develops a growth cone (gc) from which new growth occurs. When the growth cone retraces its way to its original site of connection, it is transformed into a presynaptic surface. The growth cone structure is lost, having served its purpose. If the nerve injury is in the CNS, the initial histological changes depicted in (B) may be seen, but there is no significant growth at the site of injury. Death of the damaged neuron often ensues.

cessful regeneration is generally encountered if a nerve bundle is severed in the peripheral nervous system (PNS). The PNS refers to all nervous tissue lying outside the central nervous system (CNS; the brain and spinal cord). Thus if a major nerve bundle of the leg—the sciatic nerve—is damaged, there is a good possibility of recovery of function, because the sciatic nerve lies within the PNS, even though some of the cell bodies that give rise to its axons are in the ventral spinal cord, a part of the CNS. The damaged fibers will eventually regrow, to restore both sensation (sensory) and muscular (motor) function. This is in sharp contrast with the course of events after lesions within the CNS. On damage to

the brain or spinal cord of warm-blooded vertebrates (birds and mammals, including humans), recovery of function is extremely limited. At the anatomical level, there is no evidence of axonal regrowth, such as seen in the lesioned sciatic nerve in the PNS. Apparent functional recovery after CNS injury seen in young individuals is attributed to the "plasticity" of the young CNS (i.e., its ability to remodel its structure in response to environmental stimuli). The concept of plasticity has also been invoked to provide a theoretical framework to explain adaptive changes in the CNS related to the storage of experiential memory. [*See* PLASTICITY, NERVOUS SYSTEM.]

Why lesions in the PNS can induce successful regeneration while those in the CNS cannot be profound theoretical as well as practical ramifications. An important difference between the local cellular environments of nerve fibers in the PNS and those in the CNS may lie in the nonneuronal supporting cells (glial cells) that are in contact with the nerve cell body and its processes in each milieu. In the CNS, the glial cells (astrocytes and oligodendroglia) provide a structural and metabolic scaffolding for the neurons, whereas in the PNS, this role is fulfilled by distinctly different cells, the Schwann cells. It could thus be that Schwann cells provide a nurturing neuronotrophic factor that CNS glia do not or, conversely, that CNS glial cells in some way inhibit the regenerative response, whereas Schwann cells permit it. Still other hypotheses suggest that the ability of a nerve to regenerate lies within a genetic programming potential within each neuronal cell nucleus. Many experimental approaches are directed at answering this question. Convenient model systems for studies of CNS regeneration are to be found in cold-blooded vertebrates, such as fishes and amphibians. For example, surgical crush of the optic nerve of the goldfish produces blindness, but this is followed by complete recovery of vision within a few weeks. This means not only that the 500,000 or so neuronal axons that constitute the optic nerve of a fish eye regenerate and reconnect, but also that the reconnection is very precise, because the retinal map projected on the brain is restored to its original distribution, each point on the optic tectum of the fish brain representing a specific locus on the retina of the eye. The optic nerve lies within the CNS, so that in humans, as in other warm-blooded vertebrates, blindness produced as a result of such damage to the optic nerve is permanent. [*See* ASTROCYTES.]

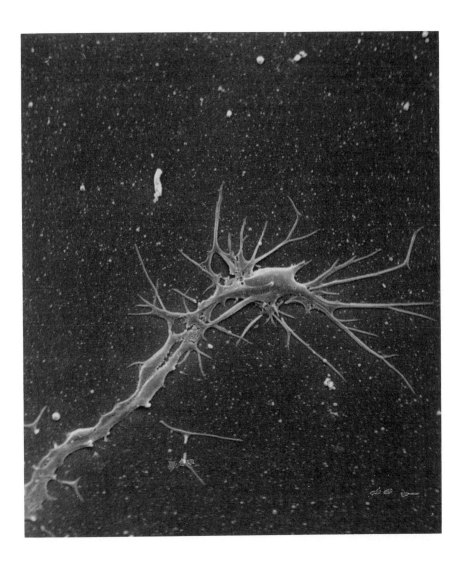

FIGURE 2 A growth cone. A scanning electron micrograph of an outgrowing neuritic process in an explant culture of a goldfish retina (from the authors' laboratory). The nerve fiber represents several neurites that have fasciculated into a single bundle. (Magnification: ×4,000)

II. Properties of Regenerating Nerves

A. Growth Cones

Growth cones are characteristic organelles that are invariably present on the growing tips of nerve processes, whether such processes conduct impulses into the neuron (dendrites) or out of the neuron (axons). The two kinds of processes are collectively termed "neurites." Seen in still or kinetic images of cell cultures, the growth cones are sites of projections possessing vigorous activity: these are microspikes, also termed *filopodia*, which appear to be randomly searching, or even flailing, like so many fingers, while the growing neurite advances in a relatively straight path, elongating by adding new membrane at the "wrist" (Fig. 2). In the nervous system, neurites seldom travel alone. Instead, multiple neurites fasciculate (i.e., form bundles and travel together) sometimes along a path established by "pioneer" fibers.

B. Pathfinding and Specificity

How growth cones find their way to their specific targets is an important, if poorly understood, process. Although the degenerating stump can appear to redirect growing fibers, it can be demonstrated experimentally that the presence of the distal nerve stump is not necessary. In experimental studies, if a nerve segment is removed and the proximal stump misdirected, the nerve will nevertheless regrow appropriately. Eventual regrowth and reconnection with appropriate neuronal sites appear to involve both fiber–fiber and growth cone–target cell recognition. These recognition processes are hypothesized to involve cell surface glycoprotein molecules.

C. Axonal Transport

Neurons have the distinctive property (not shared with any other cell types) that they may possess extremely long processes, often thousands of times as long as the cell body. Because the cell bodies and dendrites are the only parts of the neuron that can synthesize proteins, a means must be present to move newly synthesized proteins to the axonal extremities. This is accomplished by the process of axonal flow (also referred to as *axoplasmic transport*). Not only are new neuronal membranes transported to the growth cone, but cytoskeletal elements, such as tubulin and neurofilament proteins, are supplied via anterograde transport (i.e., from the cell body outward). The neuronal nucleus also must receive information from its extremities (e.g., after injury), and this is accomplished by retrograde transport (i.e., movement toward the cell body).

D. Neuronal Cell Body Response: Growth-Associated Proteins

Axonal damage is often followed by characteristic morphological changes in the neuronal cell body. These include cellular hypertrophy, nuclear eccentricity, and increased numbers of nucleoli. A successful regenerative response is accompanied by a redirection of the biosynthetic apparatus of the neuron. Synthesis of proteins involved in neurotransmitter function is shut down, whereas that of cytoskeletal elements, structural components of the axon, may be greatly intensified. A number of proteins have been identified that are specifically expressed during CNS or PNS regeneration, as well as during development of the nervous system. These have been termed *growth-associated proteins* (GAPs). Of these, the most well-characterized is GAP 43, a membrane protein that is localized to the growth cone. Synthesis of GAPs is specifically induced only in those neurons that exhibit a successful regenerative response.

III. Extraneural Aspects

Successful nerve regeneration requires not only that the neuronal cell body be capable of mounting the appropriate biosynthetic response but also that the environment surrounding the site of the nerve injury be capable of supporting the elongation and eventual functional reconnection of the axon. Thus, it has long been recognized that differences in extraneuronal environment between injured PNS and CNS axons contribute to the success or failure of nerve regeneration. The validity of this concept receives direct support from experiments in which mammalian CNS axons were transected and a segment of peripheral nerve (removed from the leg of a donor animal) was apposed, within the CNS, to the cut ends. Under these conditions, extensive axonal regrowth occurred throughout the length of the PNS graft. Molecular studies of PNS and CNS nerves are elucidating the nature of the environmental components that facilitate axonal regrowth. Available evidence suggests that two components are of primary importance: the presence of neuronotrophic factors and of a supportive substratum.

A. Trophic Factors

Neuronotrophic factors (NTFs) are substances that are required for survival and maintenance of neurons and their processes. Loss of a supply of NTF may result in the death of the neuron. The most well-known example is nerve growth factor (NGF), a well-characterized protein that exerts trophic effects on some populations of PNS and CNS neurons. Trophic factors may be produced by the target tissue that is innervated by the nerve fibers. Delivery of the NTF to the neuronal cell body is accomplished after uptake at the nerve ending and subsequent retrograde axonal transport. Alternatively, the supporting nonneuronal cells may be a source of NTFs, as has been demonstrated for Schwann cells, which, in response to injury, produce not only NGF, but also cell-surface receptors for NGF. It is postulated that NGF is released by the Schwann cells, whereupon it binds to NGF receptor sites on the Schwann cell surface, thereby serving to attract growing axons to elongate along this surface. Such a mechanism illustrates that trophic factors may also have tropic effects (i.e., attracting and guiding growing axons in a specific direction). Thus, in some circumstances, NTFs may at least partially function as supportive substrata. However, the latter most commonly refers to web-like components of the extracellular matrix (e.g., collagen, fibronectin, or laminin). [*See* NERVE GROWTH FACTOR.]

B. Extracellular Matrix

Injury to PNS and CNS nerves of lower vertebrates is often accompanied by an increased amount of laminin in the extracellular environment surrounding the nerve. In the PNS, Schwann cells are the likely source, whereas in the CNS, astrocytes appear to be capable of synthesizing laminin in response to injury. Collagen and fibronectin, but especially laminin, have been found to promote the growth of neurites from cultured neurons or neuronal explants *in vitro*. These extracellular matrix proteins seem to provide an adhesive substratum, which facilitates the forward movement of the growth cone and elongation of the axon. The importance of a supportive extracellular matrix is reinforced by experiments in which a severed nerve tract in the rat brain was induced to grow across an implant of human placenta membrane, a rich source of laminin. [*See* EXTRACELLULAR MATRIX; LAMININ IN NEURONAL DEVELOPMENT.]

Other components of the neuronal environment that may contribute to a successful regenerative response are a family of surface membrane neural cell-adhesion molecules (NCAMs), which are found on neurons. NCAMs are thought to play a role in nerve fasciculation. There is also evidence that extracellular degradative enzymes such as proteases, elaborated by the growth cone or by supporting cells, may facilitate the passage of the growing axon through the surrounding tissue. Somewhat paradoxically, endogenous protease inhibitors (nexins) have been found to exert neuronotrophic effects *in vitro*.

C. Transplantation

Transplantation of neural tissue into injured brain has also been attempted as a means to facilitate directed axonal regrowth. Experiments with fetal neural transplants indicate some success in preventing the neuronal death that often results when a CNS neuron is disconnected from its target tissue. This may reflect the elaboration of trophic factors by the transplant. Although axonal sprouting through the transplant is seen to occur, it should be stressed that recovery of function requires regrowth to appropriate specific neuronal sites. Further ef-

forts are aimed at exploiting advances in genetic engineering to construct implants of cultures that have been previously transfected with genes for trophic factors or other molecules, which may enhance the regenerative response.

D. Inhibitory Factors

In addition to the facilitory effects of trophic factors and extracellular matrix components on nerve regeneration, there are also indications that factors elaborated by CNS white matter are inhibitory to axonal outgrowth. Two proteins present in oligodendrocyte membranes and CNS (but not PNS) myelin have been identified as mediators of this inhibitory effect. Application of specific antibodies to these myelin-associated proteins neutralizes their axonal outgrowth-inhibiting activity and permits extensive neuritic elongation across a CNS myelin substratum *in vitro*.

Injury to the brain and spinal cord is often accompanied by a glial response to the concomitant axonal degeneration, which results in the formation of scar tissue. The latter has been thought to present a physical barrier to nerve growth. Such molecular studies as those outlined above have contributed to the realization that astrocytic scar formation may not be a primary cause of the lack of successful CNS regeneration. Rather it is likely that the presence of permissive substrata and the appropriate neuronotrophic factors in the extraneuronal environment are of greater relevance. Reactive astrocytes have been found to be potential sources of both laminin and neuronotrophic factors, so they may in fact have a major role to play in successful regeneration.

IV. Limited Regeneration in the CNS

A. Collateral Sprouting

Whereas nerve regeneration, as defined by outgrowth from the transected end of an axon leading to functional reconnection with the appropriate target tissue, does not occur in the CNS of higher vertebrates, there is nevertheless a response to injury. Axons from noninjured neurons, either adjacent to the injury or even at some distance from it, may sprout new axons (termed *collateral sprouting*) that will grow into and innervate the regions that had been denervated after the initial trauma. There are abundant examples of such apparently futile CNS responses that are thought to be an injury-induced exaggeration of the normal process of synaptic remodeling, which occurs during learning and memory formation. Whereas lesion-induced collateral sprouting and synaptogenesis have been amply demonstrated, the functional relevance of these processes remains an open question. It is possible that the phenomenon of collateral sprouting may actually compete with the process of terminal sprouting and thus contributes to an abortive regenerative response and lack of functional recovery. These observations of lesion-induced sprouting dramatize the challenge to neuroscientists to discover conditions under which these abortive responses can be converted to purposeful regrowth and recovery of function.

Bibliography

Cotman, C. W., and Anderson, K. J. (1989). Neural plasticity and regeneration. *In* ''Basic Neurochemistry,'' 4th ed. (G. J. Siegel, B. Agranoff, R. W. Albers, and P. B. Molinoff, eds.). Raven Press, New York.

Hammerschlag, R., and Brady, S. T. (1989). Axonal transport and the neuronal cytoskeleton. *In* ''Basic Neurochemistry,'' 4th ed. (G. J. Siegel, B. Agranoff, R. W. Albers, and P. B. Molinoff, eds.). Raven Press, New York.

Seil, F. J., ed. (1989). ''Neural Regeneration and Transplantation.'' Alan R. Liss, New York.

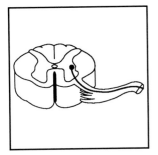

Nervous System, Anatomy

ANTHONY J. GAUDIN, *California State University, Northridge*

Glossary

Autonomic nervous system Portion of the nervous system that controls smooth muscle, cardiac muscle, and glands

Axon Extension of a neuron that carries an impulse away from the cell body

Cerebellum Portion of the hindbrain responsible for coordinating movement

Cerebrum Largest part of the brain, consisting of left and right hemispheres; receives conscious sensation and controls voluntary motor activity

Dendrites Processes that emanate from the body of a neuron that carry impulses in the direction of the body

Diencephalon Portion of the brain lying between the mesencephalon and the telencephalon, connecting the cerebral hemispheres and the midbrain

Distal Farther, or farthest, from the origin of a structure or the midline of the body

Ganglion Mass of nerve cell bodies localized in structures that lie outside the central nervous system

Motor neuron Nerve cell that carries signals away from the central nervous system to a muscle or a gland

Myelin Fatty material produced in sheaths that wrap around the axons of neurons

Nucleus Group of neuronal cell bodies in the central nervous system, with processes that extend into neighboring nervous tissue

Reflex An involuntary response to a stimulus involving a neural pathway in a muscle or a gland

Sensory neuron Nerve cell that carries signals from sensory cells and organs to the central nervous system

Somatic nervous system Portion of the nervous system other than the brain and the spinal cord

Tract Bundle of nerve fibers that form a pathway in the central nervous system

THE NERVOUS SYSTEM is an elaborate communication system, composed of a network of cells that extends throughout the body, receiving information about the internal and external environments, assessing that information, and then sending signals to organs that cause an appropriate response. As such, it is one of two major systems used to regulate body processes. In this system stimuli received by specialized sensory cells (e.g., heat-sensitive cells in the skin or light-sensitive cells in the eye) are transmitted as electrochemical impulses to the spinal column and the brain. When a response is called for, signals are sent to effector organs, glands, and muscles that respond by secreting a hormone or by contracting.

I. Organization of the Nervous System

The nervous system consists of two major portions: the central nervous system (CNS) and the peripheral nervous system (PNS). The CNS includes the brain and the spinal cord, and the PNS consists of all of the nervous tissue outside the CNS (Fig. 1).

The brain consists of the nervous tissue contained within the skull. It is in the brain that the evaluation of impulses from sensory organs is performed; consciousness, personality, and emotion

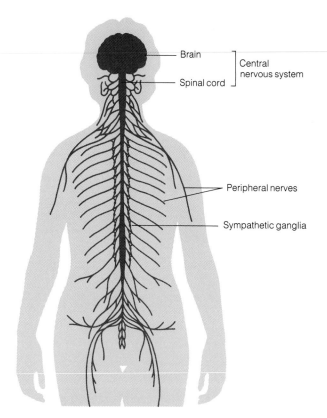

FIGURE 1 Organization of the nervous system. (From A. J. Gaudin and K. C. Jones, "Human Anatomy and Physiology." Harcourt Brace Jovanovich, San Diego, California, 1989.)

originate; and the overall general control of other systems occurs. [*See* BRAIN.]

The spinal cord extends from the brain through the column of vertebrae that compose the backbone, reaching from the lower portion of the brain to the lower spine. Within the spinal cord nervous tissue is arranged in bundles of cells that provide circuits for the transmission of signals. Such bundles are referred to as tracts. [*See* SPINAL CORD.]

The PNS is organized into nerves, bundles of nervous tissue that emanate, cablelike, from the CNS and extend throughout the body, where they provide pathways for signals traveling to and from the CNS. Functionally, the PNS consists of two portions, called the afferent and efferent divisions. The afferent division carries signals to the CNS from sensory organs, while the efferent division carries signals from the CNS to effector organs. The efferent division is further subdivided into the somatic nervous system, which carries signals specifically to skeletal muscles, and the autonomic nervous system, which carries signals to glands, smooth mus-

cle, and the heart. Signals carried by the somatic nervous system are often produced by voluntary conscious activity of the brain, whereas signals produced by involuntary subconscious activity of the brain are usually carried by the autonomic nervous system. [*See* AUTONOMIC NERVOUS SYSTEM.]

II. Anatomy of the Adult Brain

The brain is organized into four major areas (see Color Plate 5). The most anterior portion is the cerebrum, subdivided into two cerebral hemispheres, which fill most of the skull's cranial cavity. Posterior to these are the cerebellar hemispheres tucked under the rear of the cerebrum. Between the cerebellum and the cerebrum lies the diencephalon, composed of the thalamus and the hypothalamus. The remainder of the brain is organized into a rather compact brain stem, which sits at the top of the spinal cord. [*See* HYPOTHALAMUS; THALAMUS.]

A. Cerebrum

The cerebrum consists primarily of two large hemispheres, each composed of several subdivisions (see Color Plate 6). The cerebral cortex forms the outer cerebral layer and is composed of neurons that lack the myelin sheath produced by Schwann's cells. These nonmyelinated cells are dark grayish, so the cortex is frequently referred to as the gray matter of the brain. The neurons in the cortex make numerous interconnections with each other and with fibers in the underlying regions. [*See* CENTRAL GRAY AREA, BRAIN; CORTEX.]

The interior of the cerebral hemispheres is composed mostly of myelinated nerve fibers (i.e., neurons whose fibers are wrapped with a myelin sheath), organized into discrete bundles, or tracts. The corpus callosum in Color Plate 6 is an example of a large tract of the white matter. These tracts connect different areas of the cortex on the same side of the brain and complementary areas on opposite sides of the brain. These myelinated fibers form the so-called white matter of the brain, because they appear white in brain sections. Embedded in the white matter are additional isolated masses of gray matter, the basal ganglia, or basal nuclei (see Color Plate 7). The basal ganglia are important in the control and coordination of voluntary muscular movements. A horn-shaped cavity occupies a small space within each cerebral hemisphere, forming two lateral ventricles. These cavities, along with other cavities of the brain and the central canal of the spinal cord, contain cerebrospinal fluid, the special extracellular fluid of the CNS.

The cerebral surface is highly convoluted, consisting of numerous depressions (i.e., sulci) separated by equally numerous ridges (i.e., gyril). Several sulci, noticeably deeper than others, are referred to as fissures. Color Plate 5 illustrates the fissures of the cerebrum and the major areas (i.e., lobes) they define. In addition to these externally visible lobes, named for the cranial bones they underlie, a large internal fold or cortex, the insula, lies deep to the lateral fissure.

The anatomical significance of the gyri and the sulci is that they increase the volume of the cerebral cortex. Functionally, this increased volume increases the number of cells, cellular interconnections (called synapses), and pathways available in the cerebral cortex, thereby increasing the efficiency of the brain in analyzing incoming information and in generating a complex variety of motor impulses responding to sensory input. [*See* SYNAPTIC PHYSIOLOGY OF THE BRAIN.]

B. Diencephalon

The diencephalon is almost totally surrounded by the enlarged cerebral hemispheres and is composed of three major regions: the thalamus, epithalamus, and hypothalamus (Color Plate 5).

The thalamus is shaped roughly like a dumbbell in cross-section (see Color Plate 8). Two oval masses of nonmyelinated cell bodies form the major portions of the thalamus. These two masses of tissue consist of several nuclei and are joined medially by a short rod-shaped massa intermedia (Fig. 2). The thalamus functions as a relay center for both sensory and motor impulses traveling between the cerebral cortex and other neural areas.

The hypothalamus, a small mass of nerve tissue, is continuous with the inferior end of the thalamus (Color Plate 5 and Fig. 2). It consists of numerous hypothalamic nuclei, which regulate a variety of homeostatic activities, including the control of the autonomic nervous system, the endocrine system through the pituitary gland, hunger and thirst, body temperature, and wakefulness or sleepiness, as well as emotional feelings (e.g., anger, rage, aggression, and stress). [*See* PITUITARY.]

Prominent external landmarks associated with the hypothalamus are the optic chiasma, infundibu-

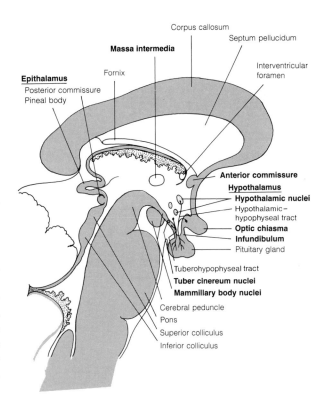

FIGURE 2 Longitudinal section through the brain stem. (From A. J. Gaudin and K. C. Jones, ''Human Anatomy and Physiology.'' Harcourt Brace Jovanovich, San Diego, California, 1989.)

lum, tuber cinereum, and the mamillary bodies (Fig. 2). The optic chiasma is an X-shaped area formed by the crossed-over fibers of the optic nerves coming from the eyes. Posterior to this chiasma is the infundibulum, a stalk of neurons which connects several hypothalamic nuclei with the posterior lobe of the pituitary gland (i.e., hypophysis). The tuber cinereum and mamillary bodies project from the posterior hypothalamic surface.

The most posterior portion of the diencephalon consists of two discrete structures, the pineal body (or gland) and the posterior commissure, collectively referred to as the epithalamus (Fig. 2). The pineal body is an endocrine gland, while the posterior commissure contains fibers that connect the cerebrum and the midbrain. [*See* PINEAL BODY.]

C. Brain Stem

The brain stem is composed of four anatomical subdivisions: the mesencephalon, cerebellum, pons, and medulla oblongata.

The mesencephalon, or midbrain, is a relatively

small structure lying approximately in the middle of the brain (Figs. 2 and Color Plate 9). A narrow canal called the cerebral aqueduct passes longitudinally through the midbrain. This aqueduct contains cerebrospinal fluid and connects two other fluid-filled cavities: the third ventricle above it and the fourth ventricle below.

A longitudinal section through the midbrain (Color Plate 9) reveals the cerebral aqueduct, dividing the midbrain into posterior and anterior regions. The posterior region (i.e., tectum) contains four rounded masses of nerve tissue (the corpora quadrigemina), and the anterior region (cerebral peduncles) includes many fibers and nuclei. The corpora quadrigemina are organized into two pairs of rounded protrusions (Color Plate 9). The upper one is called the superior colliculus, while the lower bulge is the inferior colliculus.

The cerebral peduncles are paired bulges of nerve fibers that project anteriorly from the brain stem. They are composed primarily of fiber tracts that connect the cerebrum with other parts of the nervous system.

The cerebellum bears a superficial resemblance to the cerebrum, being divided into two cerebellar hemispheres. They are shaped like the shells of an open clam and are joined in the middle by the narrow vermis. The nerve tissue of the cerebellum is organized much like that of the cerebrum. A thin cortex composed of gray matter surrounds an interior composed mostly of white matter. The cerebellar cortex is folded into numerous horizontally oriented ridges, called folia cerebelli. Several prominent fissures subdivide the folia into a number of lobules. A longitudinal section through the vermis reveals a unique branching of the white matter, the arbor vitae, or "tree of life" (Color Plate 9).

The cerebellum is connected to other parts of the brain through three paired bundles, or peduncles, of fibers (Color Plate 9). The superior cerebellar peduncles connect with the mesencephalon; the middle cerebellar peduncles, with the pons; and the inferior cerebellar peduncles, with the medulla oblongata.

The pons (pons varolii, meaning "bridge") lies just anterior to the cerebellum. It is approximately the same length as the midbrain (i.e., 2.5 cm) and is separated from the cerebellum by a triangular fluid-filled cavity, the fourth ventricle (Color Plate 9). The pons contains several reflex centers and serves as a link between parts of the brain and between the brain and the spinal cord.

The medulla oblongata (or just medulla) lies at the base of the brain stem (Color Plate 9). It is about 3.5 cm in length and forms the connection between the spinal cord and the brain; thus, all sensory and motor tracts connecting the brain and the spinal cord must cross it. Each lateral surface of the medulla contains a prominent swelling, called the olive, which connects the medulla to the cerebellum. The posterior surface of the medulla forms the floor of the fourth ventricle, which continues into the spinal cord as the central canal. The white matter in the interior of the medulla contains fiber tracts, which transmit motor impulses to voluntary muscles, and several reflex centers.

III. Ventricles and Meninges of the Central Nervous System

In the embryo the development of the brain involves the differentiation of a hollow neural tube that enlarges anteriorly into the adult brain structures. The internal cavity of the tube enlarges simultaneously with the surrounding structures and persists in the adult brain as a series of cavities filled with cerebrospinal fluid. This fluid is formed through a filtration process from three clumps of highly branched blood vessels, each called a choroid plexus, and the specialized cells that cover these vessels. Each choroid plexus is associated with a specific part of the ventricular system (Color Plate 9). The constant addition of cerebrospinal fluid from the choroid plexuses to the ventricular system produces a regular circulation of fluid through the ventricles and the spinal cord. Eventually, the fluid makes its way to the exterior surface of the brain, where it is reabsorbed by the tiny fingerlike projections, called arachnoid villi, that cover the brain (Color Plate 9). Once resorbed, the cerebrospinal fluid returns to the blood from which it was formed. Cerebrospinal fluid functions as a physical "shock absorber," diminishing the effects of any sudden blow or movement to delicate tissues of the brain. In spite of the presence of this fluid, the brain is still susceptible to concussion, especially by violent blows.

A. Ventricles of the Brain

The series of cavities in the brain form four distinct regions, called ventricles (Color Plate 10). The lateral ventricles are U-shaped cavities that occupy the medial portions of the two cerebral hemispheres. The septum pellucidum (Color Plate 9), a vertical membrane, separates the lateral ventricles and the cerebral hemispheres. A small opening, the interventricular foramen, or foramen of Monro, connects each lateral ventricle and the adjoining third ventricle, allowing cerebrospinal fluid to circulate among these ventricles.

The third ventricle, a thin cavity, separates the thalamic halves. It connects the lateral ventricles via the interventricular foramen and surrounds the massa intermedia of the thalamus. Cerebrospinal fluid enters the third ventricle from above and leaves through the cerebral aqueduct, a narrow canal that connects to the cavities within the pons, cerebellum, medulla oblongata, and spinal cord.

The fourth ventricle is triangular and is located between the cerebellum and the pons, and the medulla oblongata (Color Plate 10). Located in its roof, inferior to the cerebellum, is one choroid plexus responsible for the production of cerebrospinal fluid. The fourth ventricle has three prominent foramina (i.e., holes) through which cerebrospinal fluid leaves the fourth ventricle and enters the subarachnoid space. Two of these (called the foramina of Luschka) lie in the superior lateral walls of the fourth ventricle. The third foramen (called the foramen of Magendie) lies in its posterior inferior wall. The fourth ventricle is continuous inferiorly with the central canal of the spinal cord.

B. Meninges of the Brain and the Spinal Cord

The brain and the spinal cord are covered by three membranous tissue layers, called meninges, that channel the flow of cerebrospinal fluid around the brain and the spinal cord and provide additional physical protection to soft nerve tissues. The three meninges are the dura mater, pia mater, and arachnoid (Color Plate 11 [See MENINGES.]

The dura mater is the outermost of the three meninges (Color Plate 11). It consists of a double layer of tough ("dura") fibrous connective tissue that covers the brain and attaches to the cranial bones. The external sublayer terminates at the foramen magnum, the large opening at the base of the skull, but the internal sublayer extends through this opening and continues onto the spinal cord. The two dural layers are fused over most of the brain, forming a single tough membrane whose outer surface is the inner lining of the skull bones. In several places these two layers separate, forming dural sinuses that collect venous blood and cerebrospinal fluid from the brain and return them to large veins that drain blood from this area.

The internal layer of dura mater extends into the fissures of the brain, forming prominent partitions which separate the cerebral hemispheres from one another and from the cerebellum.

The pia mater, the innermost meninx (singular of "meninges"), is a thin vascularized connective tissue membrane that adheres closely to the outer surface of the brain, dipping into sulci and fissures (Color Plate 11). The pia mater extends through the foramen magnum and covers the entire spinal cord medial to the dura mater. At the inferior end of the spinal cord (about the level of the second lumbar vertebra), the pia mater continues as a narrow threadlike extension, the filum terminale, that extends caudally and attaches to the second coccygeal vertebra.

The arachnoid occupies the space between the other two meninges, although it does not attach equally to both. Externally, the arachnoid adheres rather closely to the dura mater, except for the presence of a narrow subdural space between them containing a thin film of cerebrospinal fluid. Internally, the arachnoid is loosely attached to the pia mater by delicate strands of fibrous connective tissue (i.e., trabeculae) that span a larger subarachnoid space filled with cerebrospinal fluid (Color Plate 11). The arachnoid layer extends into the large longitudinal and transverse cerebral fissures and bridges the sulci and other brain depressions.

IV. Anatomy of the Spinal Cord

At the lower end of the medulla oblongata, where the brain stem terminates, columns of gray and white matter in the medulla become organized into the spinal cord, which serves two major functions. First, it contains neurons that connect sensory and motor areas of the brain with other parts of the body. These neurons provide pathways for conducting impulses in either direction—from sensory receptors to the brain then back along motor neurons to the muscles and the glands. Second, the spinal cord directly connects sensory neurons with appropriate motor neurons that produce responses independent of brain influences (i.e., spinal reflexes).

The spinal cord is a cylinder of nerve tissue somewhat flattened anteriorly and posteriorly. It is contained within the vertebral canal of the vertebrae and begins at the foramen magnum, passing through the vertebral foramina. In an embryo the spinal cord and the vertebral column are approximately the same length. However, during late embryonic and postnatal growth, the spinal cord elongates at a slower rate than the vertebral column, so that the adult spinal cord does not extend the full length of the vertebral canal. Instead, it terminates in a cone-shaped conus medullaris between the first and second lumbar vertebrae (Fig. 3). The spinal nerves that exit the lower lumbar and sacral regions are long and they angle inferiorly, until they reach their point of exit from the vertebral canal. Early anatomists thought this group of spinal nerves resembled a horse's tail, so they gave it the name cauda equina.

The anterior surface of the spinal cord has a deep vertical groove, called the anterior medial fissure (Fig. 3), and the posterior side bears a narrower, yet deeper, slitlike groove, the posterior median sulcus. The lateral sides of the cord show two shallow sulci: the anterior and posterior lateral sulci. An additional groove, the posterior intermediate sulcus, lies between the posterior median sulcus and posterior lateral sulcus in the cervical and upper thoracic regions.

As in the brain, nerve tissue in the spinal cord is divided into gray and white matter. In the spinal cord, however, the locations of the two are reversed, the white matter surrounding the gray matter.

Gray matter of the spinal cord consists mainly of neurons, unmyelinated axons, dendrites, and neuroglia (i.e., specialized supportive cells of the nervous system). In a transverse section of the spinal cord, gray matter forms an "H" in the center of the cord (Fig. 3). The central canal is a remnant of the channel formed during the embryonic development of the neural tube and is continuous with the fourth ventricle of the brain.

The horizontal "bar" of the spinal gray matter is called the gray commissure, further subdivided by the central canal into anterior and posterior portions. Two anterior columns of gray matter (composed of the cell bodies of motor neurons) extend anteriorly from the horizontal bar. Two narrower posterior columns of gray matter (composed of sensory neurons) extend from the horizontal bar.

The spinal white matter is mainly axons of myelinated nerve fibers embedded in neuroglia. Practically all the fibers are organized into tracts that run lengthwise, except for the white commis-

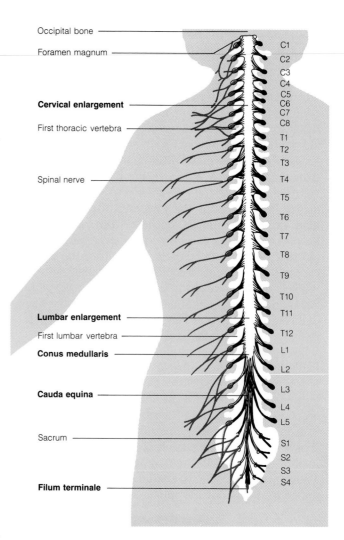

FIGURE 3 Posterior view of the spinal cord and the vertebral column. (From A. J. Gaudin and K. C. Jones, "Human Anatomy and Physiology." Harcourt Brace Jovanovich, San Diego, California, 1989.)

sure, anterior to the gray commissure. It contains horizontal tracts that cross the spinal cord. The white matter of the cord conducts sensory impulses up the cord and motor impulses down the cord.

V. Anatomy of the Peripheral Nervous System

The PNS consists of the cranial nerves, spinal nerves, and autonomic nervous system.

A. Cranial Nerves

A total of 12 pairs of cranial nerves extend from the base of the brain. All but the first pair arise from the brain stem (see Color Plate 12). They are numbered as they emerge from the anterior to the posterior brain: I, olfactory; II, optic; III, oculomotor; IV, trochlear; V, trigeminal; VI, abducens; VII, facial; VIII, vestibulocochlear; IX, glossopharyngeal; X, vagus; XI, accessory; and XII, hypoglossal.

Some cranial nerves include primarily sensory or motor neuron fibers and are called sensory, or motor, nerves. Others are known as mixed nerves, because they include both sensory and motor fibers. Only the axons and the dendrites form the nerves, while the cell bodies of the motor neurons form nuclei within the brain tissue, and those of sensory neurons are grouped in ganglia outside, but adjacent to, the CNS.

Generally, cranial nerves innervate structures in the head and neck region. However, one exception is the vagus (meaning ''wandering'') nerve, a mixed nerve, with innervates the palate, neck, thorax, and abdominal cavity.

The 12 cranial nerves and their sites of origin and modes of action are summarized in Table I.

B. Spinal Nerves

Spinal nerves originate in the spinal cord. These nerves leave through foramina of the vertebral column going to the skin, muscles, bones, and joints of the posterior head region, trunk, and appendages (Fig. 3). In all, 31 pairs of spinal nerves are normal: eight cervical, 12 thoracic, five lumbar, five sacral, and one (occasionally more) coccygeal.

C. Anatomy of Adult Spinal Nerves

Each spinal nerve originates as two extensions from the spinal cord: an anterior root and a posterior root (Fig. 4). The former includes both voluntary and involuntary motor neurons; the latter, only sensory neurons.

Voluntary motor neurons in the anterior root are called somatic motor (or somatomotor) neurons. Each extends its axon from the spinal cord gray matter to a skeletal muscle. Some neurons, such as those innervating the foot, have long axons. These neurons are responsible for voluntary movements produced by contractions of the skeletal muscles attached to bones and certain superficial skin structures in the head and the neck (e.g., the lips and the eyelids).

In addition to these voluntary nerve cells, the anterior roots contain fibers of the autonomic nervous system that control unconscious involuntary activities. The exact positions of these autonomic neurons are discussed in Section V,E.

Posterior roots of spinal nerves have only sensory neurons. Dendrites of these neurons are situ-

TABLE I The Cranial Nerves

Nerve	Origin	Location of skull exit	Functions
Olfactory (I)	Cerebral hemispheres	Cribriform plate of the ethmoid bone	Sensory nerve; smell
Optic (II)	Diencephalon	Optic canal	Sensory nerve; vision
Oculomotor (III)	Midbrain	Superior orbital fissure	Mixed nerve; to the eye muscles
Trochlear (IV)	Midbrain	Superior orbital fissure	Mixed nerve; to the superior oblique eye muscle
Trigeminal (V)			
Ophthalmic	Pons	Superior orbital fissure	Sensory nerve; from the head
Maxillary	Pons	Foramen rotundum	Sensory nerve; from the face
Mandibular	Pons	Foramen ovale	Mixed nerve; facial sensations and chewing motions
Abducens (VI)	Pons	Superior orbital fissure	Mixed nerve; to the lateral rectus eye muscle
Facial (VII)	Pons	Stylomastoid foramen	Mixed nerve; taste sensations and facial muscles
Vestibulocochlear (VIII)	Medulla	Internal acoustic meatus	Sensory nerve; hearing and equilibrium
Glossopharyngeal (IX)	Medulla	Jugular foramen	Mixed nerve; taste and pharyngeal muscles
Vagus (X)	Medulla	Jugular foramen	Mixed nerve; to the head, thorax, and abdomen
Accessory (XI)	Medulla	Jugular foramen	Motor nerve; to the neck muscles
Hypoglossal (XII)	Medulla	Hypoglossal canal	Motor nerve; to the tongue

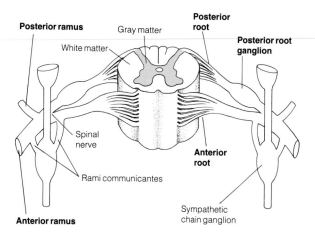

FIGURE 4 The spinal cord and the spinal nerves. (From A. J. Gaudin and K. C. Jones, ''Human Anatomy and Physiology.'' Harcourt Brace Jovanovich, San Diego, California, 1989.)

ated in tissues and organs, and their cell bodies are grouped in a posterior root ganglion (Fig. 4).

Just distal to the posterior root ganglion, the posterior and anterior roots fuse and form a spinal nerve. Because it includes sensory and motor fibers, each spinal nerve is mixed. Cervical, thoracic, and lumbar spinal nerves extend laterally and exit the neural canal through intervertebral foramina. Immediately distal to these foramina, each spinal nerve divides into an anterior ramus and a posterior ramus (Fig. 4). Sacral nerves are slightly different from other spinal nerves in that sacral nerves split within the sacral canal, and anterior and posterior rami exit separately through corresponding sacral foramina.

Compared to its anterior counterpart, the posterior ramus of each spinal nerve is relatively short. Posterior rami innervate only restricted areas of the skin and muscles along the back of the head and the sides of the vertebral column. Anterior rami of the spinal nerves innervate the remainder of the trunk and the limbs.

D. Peripheral Regions Innervated by Spinal Nerves

The pattern of spinal nerve innervation of the skin can be mapped as seen in Color Plate 12. A regular sequential pattern is most prominent on the trunk and slightly modified in the limbs. The nerves that innervate the skin, called cutaneous nerves, contain only sensory and autonomic motor neurons and lack somatic motor neurons.

Spinal nerve innervation of skeletal muscles follows a similar sequential pattern; that is, spinal nerves in the cervical and thoracic region innervate skeletal muscles in the neck, arm, and chest, while nerves from the lumbar and sacral regions innervate abdominal and leg muscles.

Many areas of both the skin and the muscles are innervated by nerves of mixed spinal origin. Such nerves are produced by a fusion and exchange of neurons from the anterior rami of certain neighboring spinal nerves. The resulting mixed nerves form a plexus (discussed in Section V,D,1).

Nerves supplying the skeletal muscles follow one of two pathways. The first pathway involves the thoracic nerves. These 12 nerves leave the intervertebral foramina and pass between the ribs as intercostal nerves. In general, intercostal nerves T1–T6 innervate the skin and the skeletal muscles in the thoracic region. Fibers from their posterior rami are confined to a narrow area bordering the neural spines of the vertebrae, while fibers of the anterior ramus run to the remaining anterior and lateral regions. Intercostal nerves T7–T12 stimulate corresponding intercostal muscles, as well as skin and voluntary abdominal muscles. Nerves T2 also extend into the arms and supply the skin of the axillary (i.e., underarm) region and the surface of the back of the arm.

The second pathway involves anterior rami of the remaining spinal nerves in collections of mixed nerves, or plexuses.

1. Cervical Plexus

The cervical plexus is formed by a mixing of cervical nerves 1–4, with contributions from C5. The pattern of fusion and branching is illustrated in Fig. 5, and the area innervated by each peripheral nerve is summarized in Table II.

2. Brachial Plexus

The brachial plexus is the peripheral nerve supply for the upper extremities. It is formed by the exchange of fibers between the anterior rami of cervical nerves 5–8 and the first thoracic nerve (Fig. 6). Table III summarizes the destinations of the major nerves of the brachial plexus.

3. Lumbosacral Plexus

The lumbosacral plexus combines three groups of spinal nerves: the lumbar plexus, nerves L1–L4; the sacral plexus, nerves L4–L5 and S1–S3; and the pudendal plexus, nerves S2–S4. In a few cases the

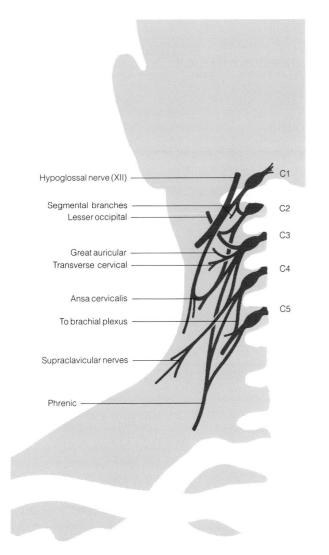

cardiac muscle, certain glands, and smooth muscles in blood vessels and organs of the thoracic and abdominal cavities (Fig. 8). In the past physiologists thought that activities that were clearly controlled at a subconscious level were regulated by a functionally separate, hence, "autonomous," part of the nervous system. We now know that the autonomic nervous system is not truly autonomous. Indeed, it is regulated by other parts of the CNS, specifically, centers in the cerebral cortex, hypothalamus, and medulla oblongata.

In general, the autonomic nervous system has two distinct anatomic and functional subdivisions: sympathetic and parasympathetic. The sympathetic system, also called the thoracolumbar system, because its neurons emerge from thoracic and lumbar regions of the spine, innervates smooth muscles of the arteries. Just as arteries penetrate all parts of the body, so do sympathetic fibers. Sympathetic fibers also innervate several abdominal organs (Fig. 9). The general effect of the sympathetic nervous system is to prepare the body for action in stressful situations.

The other division, the parasympathetic, or cra-

FIGURE 5 The cervical plexus. (From A. J. Gaudin and K. C. Jones, "Human Anatomy and Physiology." Harcourt Brace Jovanovich, San Diego, California, 1989.)

most inferior thoracic nerve (T12) contributes to the lumbar plexus through nerve L1. The pattern of combinations is shown in Fig. 7; Table IV summarizes the destinations of the major nerves of the lumbosacral plexus.

E. Autonomic Nervous System

The autonomic nervous system is that part of the peripheral nervous system that regulates unconscious involuntary activities, such as the control of the heart beat, movements of the digestive system, or glandular activities. It consists primarily of visceral efferent neurons that carry motor impulses to

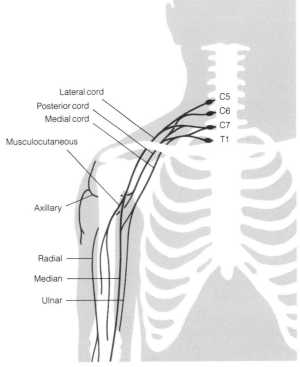

FIGURE 6 The brachial plexus. (From A. J. Gaudin and K. C. Jones, "Human Anatomy and Physiology." Harcourt Brace Jovanovich, San Diego, California, 1989.)

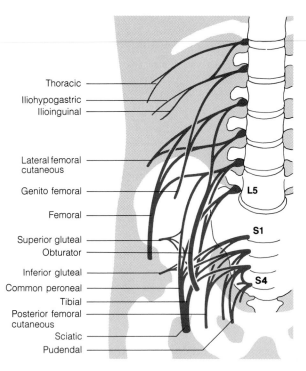

Thoracic
Iliohypogastric
Ilioinguinal

Lateral femoral
cutaneous

Genito femoral

Femoral

Superior gluteal
Obturator

Inferior gluteal

Common peroneal
Tibial
Posterior femoral
cutaneous
Sciatic
Pudendal

L5

S1

S4

FIGURE 7 The lumbosacral plexus. (From A. J. Gaudin and K. C. Jones, "Human Anatomy and Physiology." Harcourt Brace Jovanovich, San Diego, California, 1989.)

niosacral, system, functions as the principal nerve supply to certain structures in the head, digestive organs, and other viscera (e.g., the lungs and the heart). Effects of the parasympathetic system are in some ways opposite those of the sympathetic system. The parasympathetic system stimulates activities of the digestive organs and glands and slows the heart beat and the respiratory rate. It tends to calm the body after a stress-producing experience, and it promotes activities that maintain life-support systems.

In general, motor neurons of the autonomic nervous system are organized into functional units of two neurons each. Each unit includes a pregangli-

TABLE II   The Cervical Plexus

Nerve	Spinal nerve origin	Distribution
Lesser occipital	C2–C3	Scalp behind the ear
Greater auricular	C2–C3	Skin surrounding the ear
Transverse cervical	C2–C3	Anterior skin of the neck
Supraclaviculars	C3–C4	Skin of the upper thorax
Ansa cervicalis	C1–C2	Pharyngeal muscles
Phrenic	C3–C5	Diaphragm

Table III   The Brachial Plexus

Nerve	Spinal nerve origin	Distribution
Posterior cord		
Axillary	C5–C6	Skin and muscles of the shoulder region
Radial	C5–C8, T1	Skin and muscles of the posterolateral region of the arm, forearm, and hand
Lateral cord,		
Musculocutaneous	C5–C7	Skin and muscles of the lateral region of the forearm
Medial cord		
Ulnar	C8–T1	Skin of the medial portion of the hand, muscles of the front of the forearm and the hand
Median	C5–T1	Skin of the lateral portion of the hand, muscles of the lateral region of the forearm and the hand

onic, or presynaptic, neuron with a cell body that lies within the CNS (Fig. 10). The axon of this cell synapses with a second, postganglionic (or postsynaptic), neuron. These nerves synapse either within a ganglion that lies close to the spinal cord or within one of several ganglia located in the thoracic and abdominal cavities. Axons from postganglionic neurons terminate in a specific predetermined organ or tissue. Impulses carried along these neurons either stimulate or inhibit the metabolic activities of these organs.

1. Sympathetic System

Preganglionic neurons of the sympathetic division are anchored in the lateral columns of the spinal cord and their axons emerge from the cord through anterior roots of all thoracic spinal nerves and through the first two lumbar spinal nerves (Fig. 8). These axons remain within a spinal nerve for only a short distance, exiting through one of two communicating rami that connect with a chain of ganglia paralleling the vertebral column. These ganglia are the paravertebral ganglia, which compose the sympathetic trunk, or chain. Since preganglionic sympathetic fibers are myelinated, the rami they run through to enter the sympathetic trunk are called white communicating rami. Figure 10 shows

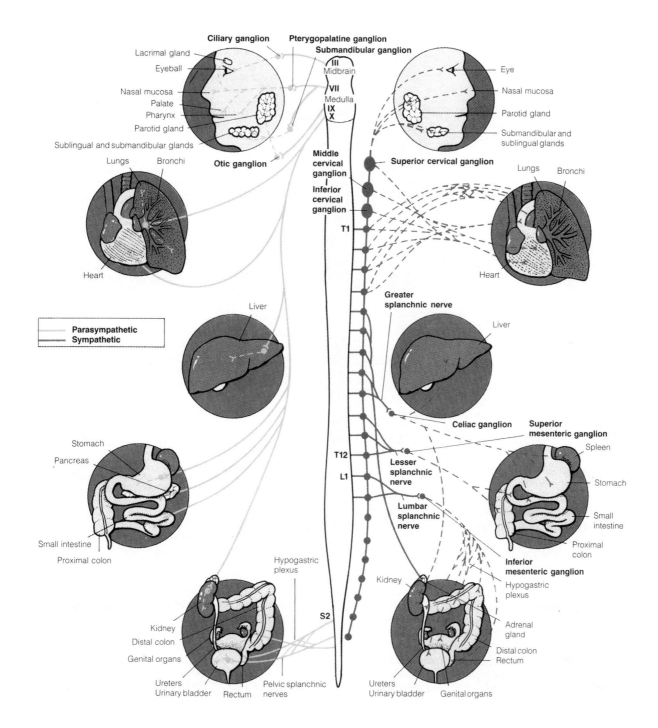

FIGURE 8 Efferent fibers of the parasympathetic (left) and sympathetic (right) autonomic nervous system. (From A. J. Gaudin and K. C. Jones, ''Human Anatomy and Physiology.'' Harcourt Brace Jovanovich, San Diego, California, 1989.)

pathways available to these motor neurons entering the sympathetic chain.

An axon may synapse within the ganglion it en-ters, or pass through and synapse in a ganglion at a different level in the chain (Fig. 10). Other axons run through the sympathetic ganglia and do not synapse until they join neurons in ganglia in or near visceral organs. The nerves formed by these latter preganglionic neurons are called splanchnic nerves.

Impulses in sympathetic neurons travel two different paths to target organs, depending on which

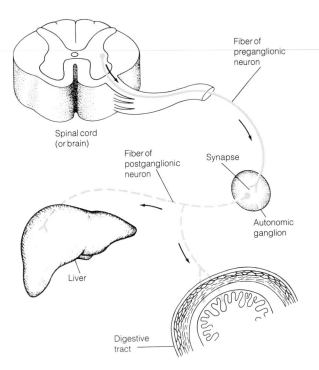

FIGURE 9 Preganglionic and postganglionic fibers of the autonomic nervous system. (From A. J. Gaudin and K. C. Jones, "Human Anatomy and Physiology." Harcourt Brace Jovanovich, San Diego, California, 1989.)

TABLE IV The Lumbosacral Plexus

Nerve	Spinal nerve origin	Distribution
Lumbar plexus		
Iliohypogastric	T12–L1	Skin and muscles of the lower trunk
Ilioinguinal	L1	Skin of the upper thigh and the external genitals in males, muscles of the lower trunk
Genitofemoral	L1–L2	Skin of the external genitals in both sexes and the internal genitals in females, the front of the thigh
Lateral femoral cutaneous	L2–L3	Skin covering the lateral cutaneous region of the thigh
Femoral	L2–L4	Skin of the medial thigh, leg, and foot; certain muscles of the thigh
Obturator	L2–L4	Skin of the medial thigh, certain muscles of the thigh
Sacral plexus		
Superior gluteal	L4–S1	Certain muscles of the thigh
Inferior gluteal	L5–S2	Gluteus maximus muscle
Posterior femoral cutaneous	S1–S3	Skin of the external genitals in both sexes and the internal genitals in females, the posterior surface of the thigh
Sciatic	L4–S3	The tibial and common peroneal nerves
Tibial	L4–S3	Skin of the posterior leg and foot, several muscles of the thigh, leg, and foot
Common peroneal	L4–S2	Skin of the anterior leg and foot, several muscles of the leg
Pudendal plexus		
Pudendal	S2–S4	Skin of the perineum and the genitals

synaptic pattern occurs. In the first pattern impulses are generated in postganglionic fibers that leave the sympathetic trunk ganglia and pass out through spinal nerves. Postganglionic sympathetic fibers are nonmyelinated and pass through gray communicating rami (Fig. 10). Because this happens along the entire length of the cord, all spinal nerves contain at least some postsynaptic sympathetic fibers. Some of these postganglionic fibers leave the sympathetic trunk ganglia in the cervical region and accompany major arteries in the head until they reach their target organs. Within the head they follow the external carotid, internal carotid, and vertebral arteries. The postganglionic fibers are called plexuses and are referred to in terms of the artery involved. Thus, these fibers are distributed into three plexuses: the external carotid plexus, internal carotid plexus, and vertebral plexus.

A second pattern is formed by preganglionic neurons that run through splanchnic nerves and synapse with their postganglionic partners within abdominal, or collateral, ganglia. Collateral ganglia form masses of cell bodies of postganglionic neurons. Here, synapses occur with preganglionic ax-

ons from the spinal cord. Three collateral ganglia (Fig. 10)—celiac, superior mesenteric, and inferior mesenteric—lie near the large arteries for which they are named. Smaller collateral ganglia lie in the pelvic cavity.

Postganglionic axons exit the collateral ganglia to form a series of extrinsic autonomic plexuses. After reaching target organs these axons penetrate the or-

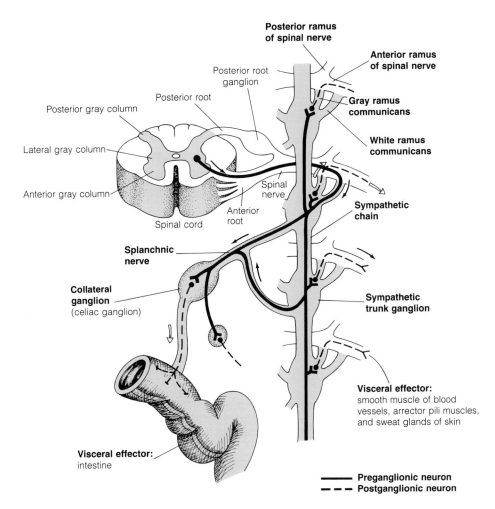

Posterior ramus
of spinal nerve

Anterior ramus
of spinal nerve

Posterior root
ganglion

Gray ramus
communicans

Posterior root

Posterior gray column

White ramus
communicans

Lateral gray column

Anterior gray column

Spinal
nerve

Sympathetic
chain

Spinal cord

Anterior
root

Splanchnic
nerve

Collateral
ganglion
(celiac ganglion)

Sympathetic
trunk ganglion

Visceral effector:
smooth muscle of blood
vessels, arrector pili muscles,
and sweat glands of skin

Visceral effector:
intestine

——— Preganglionic neuron
– – – Postganglionic neuron

FIGURE 10 Motor routes of the sympathetic nervous system. (From A. J. Gaudin and K. C. Jones, "Human Anatomy and Physiology." Harcourt Brace Jovanovich, San Diego, California, 1989.)

gan and are distributed as intrinsic autonomic plexuses within the organ.

One exception to the two-neuron plan in the sympathetic system is known. It involves the adrenal gland, attached to the top of the kidney. Preganglionic sympathetic nerves that innervate cells of the inner portion (i.e., medulla) of the gland do not synapse with postganglionic neurons. Instead, preganglionic cells stimulate secretory cells in the adrenal gland. [*See* ADRENAL GLAND.]

2. Parasympathetic System

The second part of the autonomic nervous system is the parasympathetic system, or craniosacral system. Its neurons arise from opposite ends (i.e., the cranial and sacral ends) of the spinal cord (Fig. 9). It

is the principal nerve supply for certain structures in the eye, glands in the head, heart, reproductive organs, smooth muscles, and glands of the digestive system. The fibers of this system accompany four of the cranial nerves and the anterior roots of spinal nerves S2–S4. Spinal nerves proper do not include parasympathetic motor fibers.

The preganglionic parasympathetic fibers that emerge from the head have their cell bodies in nuclei of the brain stem. They extend to ganglia located in the head, where they synapse with postganglionic neurons which pass to several target organs and tissues (Fig. 8). Preganglionic fibers from the vagus nerve, cranial nerve X, do not pass to a single ganglion. Instead, they innervate the heart, respiratory system, intestinal tract, and their associated glands.

Sacral parasympathetic neurons have preganglionic fibers which arise in the lateral columns of the gray matter in the sacral spinal cord. These axons leave with the anterior roots of sacral nerves 2–4.

They form the pelvic splanchnic nerves and travel to several plexuses in the lower abdominal and pelvic regions. From these plexuses preganglionic axons extend to intrinsic plexuses within the target organs. The primary organs innervated by the sacral parasympathetic fibers are the descending and sigmoid colons, the rectum and the anus, the urinary bladder, and the external genital organs.

Bibliography

Angevine, J. G., Jr., and Cotman, C. W. (1981). ''Principles of Neuroanatomy.'' Oxford Univ. Press, New York.

Barr, M. L., and Kernan, J. A. (1983). ''The Human Nervous System,'' 4th ed. Harper & Row, New York.

Gaudin, A. J., and Jones, K. C. (1989). ''Human Anatomy and Physiology.'' Harcourt Brace Jovanovich, San Diego, California.

McMinn, R. M. H., and Hutchings, R. T. (1988). ''Color Atlas of Human Anatomy,'' 2nd ed. New York Med. Publ., Chicago.

Nauta, W. J. H., and Feirtag, M. (1985). ''Fundamental Neuroanatomy.'' Freeman, New York.

Thompson, R. F. (1985). ''The Brain: An Introduction to Neuroscience.'' Freeman, New York.

Neural Basis of Oral and Facial Function

BARRY J. SESSLE, *University of Toronto*

I. Introduction
II. Sensory Functions
III. Neuromuscular Functions
IV. Autonomic Regulation

Glossary

Periodontal tissues (or periodontium) Supporting tissues of the teeth, which surround the root of each tooth and serve to attach the root of the tooth to its socket in the enveloping alveolar bone; they also contain nerve fibers that control the periodontal blood vessels and, thus, the blood supply of the periodontal tissues, as well as sense organs (receptors) and their associated nerve fibers, which provide the peripheral basis for pain and touch from these tissues

Temporomandibular joint The "jaw joint" immediately in front of the ear; it also has a nerve supply, particularly on the posterolateral aspect of its capsule, that supplies receptors involved in pain, jaw position sense, etc.

Tooth pulp Soft tissues inside each tooth; sometimes referred to as the "nerve" of the tooth because it contains a profuse nerve supply; some of these nerve fibers provide the peripheral basis for pain from the tooth, and some control the blood vessels and blood supply of the pulp

Trigeminal nerve Fifth cranial nerve that has three major branches: the ophthalmic, maxillary, and mandibular; these three branches provide most of the sensory innervation of the face and oral cavity, and the mandibular branch also contains motor axons that supply several muscles of the orofacial region, principally those moving the jaw (mandible)

THIS CHAPTER provides an outline of the sensory and motor neural mechanisms of the face and mouth and, in a more limited sense, of the pharynx and larynx. Few details are provided of some important functions of the face and mouth (e.g., smell, taste, speech) because these topics are covered in other chapters. This chapter also emphasizes the profuse nerve supplies of the oral–facial tissues and their representation in the brain and the exquisite sensory capabilities provided by this rich innervation and extensive central neural representation. It considers the neural basis of oral–facial touch, temperature, and pain and gives particular emphasis to the latter, because pain commonly occurs in the skin, teeth, muscles, joint, and other tissues of the oral–facial region and humans can have long-term suffering from several pain states or syndromes in the face and mouth. Particular attention is also given to the neural processes underlying the many reflex and other motor functions manifested in the oral–facial region, especially those related to mastication (chewing), swallowing, and associated neuromuscular functions. In addition, this chapter briefly considers the role of the autonomic nervous system in regulating oral–facial blood flow as well as the limited understanding we presently have of its role in regulating the sensory and neuromuscular functions of the face and mouth.

I. Introduction

Before reviewing the neural mechanisms underlying the functions of the face and mouth, it first should be noted that the oral–facial region is remarkable in its high level of sensory discriminability and sensitivity. This is probably a reflection of its great innervation density and the large amount of brain tissue devoted to the representation of the oral cavity and surrounding areas. In addition, specialized receptor systems are associated with the periodontal supporting tissues of the teeth and, in lower animals, with the facial whiskers (vibrissae). These receptors provide an added dimension of sensory experience, and together with the tongue and lips, are most important for exploration of the environment and controlling movement and behavior. In addition, some of the most common pains occur in this region (e.g., toothache, headache), and some sensory functions are unique to the region (e.g., taste). Likewise, the oral–facial region is remarkable in the vast array of simple and complex motor activities that are manifested within it. These activities range from relatively simple reflexes, such as the jaw-opening reflex, to the very complex motor activities associated with speech, mastication, and swallowing, which involve the coordinated neuronal activity of many parts of the brain and which provide for social communication and the intake of food and fluid vital for life. It is also important to note that while this chapter will focus on human oral–facial functions, studies in animals have been indispensable and crucial to most of the current knowledge of their mechanisms.

II. Sensory Functions

A. Touch

Our ability to sense touch (i.e., tactile sensibility) is extremely well developed in the orofacial region. Tests to measure touch include tactile threshold, stereognosis (a term referring to the ability to recognize the form of objects), and two-point discrimination (Fig. 1), and they have shown that some oral–facial tissues such as the tongue tip and lips have a greater tactile sensitivity than any other part of the body. These peripheral tissues are densely innervated by primary afferent nerve fibers, each of which terminates peripherally at a sensory organ called the receptor. These receptors "sense" stimuli and changes in the environment and transduce this information into electrochemical energy, which is then carried along the afferent fibers, into the brain, as action potentials. [See SOMATOSENSORY SYSTEM.]

We also have analogous receptor mechanisms for our ability to detect and discriminate the size of small objects placed between the teeth, their hardness and texture, and bite force; these functions have largely been attributed to receptors that are located in the periodontal tissues around the root of each tooth. It is also becoming increasingly appar-

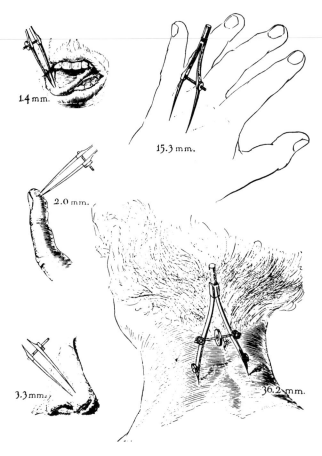

FIGURE 1 Comparison of two-point discrimination thresholds of various parts of the body. In this sensory test, two points are applied simultaneously and with equal force to skin or mucosa; the threshold value is the minimal distance between the points for which they are felt by the subject as two distinct points. [Reproduced, with permission, from L. Langley and E. Cheraskin, 1956, "The Physiological Foundation of Dental Practice," 2nd ed., C. V. Mosby, St. Louis.]

ent that other receptors, such as those in the jaw joint and even in jaw muscles, make an important contribution (Fig. 2). Receptors in this joint (the temporomandibular joint [TMJ]) as well as those in the jaw muscles also largely account for our conscious perception of jaw position.

The receptors in the facial skin, oral mucosa, periodontal tissues, and TMJ that are responsible for the sensibility to mechanical stimuli such as touch or pressure stimuli can be broadly categorized into two types: free nerve endings and corpuscular receptors, of which several anatomically distinct examples exist. Functionally, these so-called mechanoreceptors are primarily associated with large-diameter, fast-conducting afferent nerve fi-

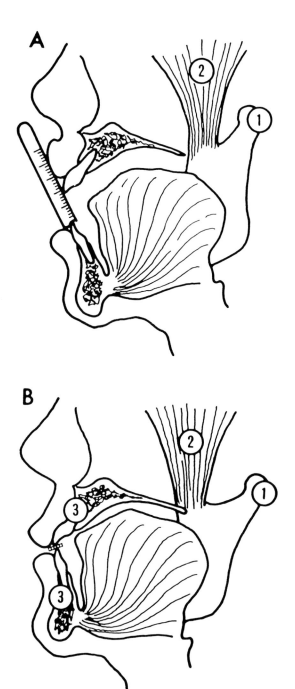

FIGURE 2 Some methods used for determination of mandibular position (A) and interdental size of objects (B), and possible receptor sites involved in these sensory functions. A. Subjects are asked to reproduce certain magnitudes of jaw opening. B. Subjects are asked to discriminate between the size of different objects (e.g., foil, wire, disc) placed between the teeth, or the threshold for detection is obtained. Note that the receptor sites include the temporomandibular joints (1), jaw muscles (2), and periodontium (3). [Reproduced, with permission, from A. Storey, 1968, *J. Can. Dent. Assoc.* **34**, 294–300.]

bers, and they can respond either transiently (so-called velocity detectors) or throughout (static-position detectors) an innocuous mechanical stimulus applied to an oral–facial site. This is reflected, respectively, in a brief or sustained burst of action potentials in their associated nerve fiber, which conducts these action potentials into the brain stem. By these neural signals, the mechanoreceptors collectively can provide the brain with detailed information of, for example, the location, quality, intensity, duration, and rate of movement of an oral–facial tactile stimulus. Many types of mechanoreceptors are exclusively activated by tactile stimuli. This physiological specificity, coupled with the existence of an anatomically recognized receptor structure for some of these mechanoreceptors, and of many neurons in the central relay stations (see below) that respond exclusively to tactile stimulation, strongly supports the concept of a specificity theory. According to this theory, a specific set of receptors and afferent nerve fibers, and nerve cells and relay stations in the brain, respond exclusively to tactile stimuli and provide the cerebral cortex only with neural information related to touch and not, for example, pain or temperature.

The major primary afferent pathway carrying the neural signals from the oral–facial mechanoreceptors is the trigeminal nerve (the fifth cranial nerve). The afferent nerve fibers in this nerve pass via the trigeminal ganglion, where their primary afferent cell bodies are located, and the trigeminal sensory nerve root to the trigeminal brain stem sensory nuclear complex (Fig. 3). The neural signals are transferred to nerve cells (i.e., neurons) at all levels of the brain stem complex. This complex can be subdivided into a main sensory nucleus and a spinal tract nucleus; the latter is subdivided further into the subnuclei oralis, interpolaris, and caudalis. These second-order neurons then project to higher sensorimotor centers as well as to local brain stem regions, including those responsible for activating muscles. Thereby, they serve as so-called interneurons involved in reflexes or more complex sensorimotor behaviors (see Section II,B). A major projection from these trigeminal brain stem neurons, however, is concerned with touch perception and passes primarily to the ventroposterior thalamus of the opposite (i.e., contralateral) side (Fig. 3). After synaptic transmission through third-order neurons, the signals are relayed from here to a particular part of the overlying somatosensory cerebral cortex where the face and mouth are represented and where the corti-

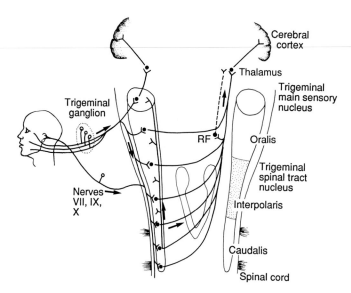

FIGURE 3 Pathways involved in relay of tactile information from the oral–facial region. Mechanoreceptive primary afferent fibers in the three divisions of the trigeminal (V) nerve pass into the brain stem via trigeminal ganglion (where their cell bodies are located). Some afferent fibers from the ear pass via cranial nerves VII, IX, or X; some afferent fibers also pass to the brain stem via cervical nerves (not shown). The information is relayed on neurons located throughout the trigeminal main sensory and spinal tract nuclei. These neurons then send information mainly to the contralateral thalamus, where it is relayed on neurons located particularly in the medial part of the ventroposterior thalamus. Some of the signals reach the thalamus, however, by more circuitous routes, such as via brain stem reticular formation (RF), and some neurons in the trigeminal main sensory nucleus (and possibly in the spinal tract nucleus) project to the ipsilateral thalamus instead of the contralateral thalamus. From the thalamus, information is relayed directly to the somatosensory cerebral cortex. Note that some neurons in the subnucleus caudalis also send an ascending projection to rostral subnuclei (oralis and interpolaris) of the spinal tract nucleus and to the main sensory nucleus; this projection can modulate relay of tactile information through these regions. The horseshoe-shaped structure near the middle of the caudal brain stem represents the solitary tract nucleus. Arrows indicate direction of signals. [Reproduced, with permission, from G. Roth and R. Calmes, 1981, "Oral Biology," C. V. Mosby, St. Louis.]

cal neural processing begins that eventually leads to the perception of a touch stimulus.

The second-, third-, and fourth-order neurons in this pathway to the cortex show many functional properties comparable with those of mechanoreceptive primary afferent fibers. They also retain much of the "specificity" of the tactile primary afferents, thus providing further support for the specificity theory of touch. However, by means of the complex ultrastructure and regulatory mechanisms existing at each of the relay sites, considerable modification of tactile transmission can occur, as a result of other incoming sensory signals and descending influences from higher brain centers. This may explain, for example, how distraction or focusing one's attention on a particular task at hand can depress our awareness of a touch stimulus.

B. Thermal Sensation

Our ability to sense the temperature of an object or substance is particularly well developed in the face and mouth, where thermal changes significantly <1°C can be readily detected and discriminated; however, temperature detection and discrimination can vary depending on the magnitude and rate of the thermal changes, the area of thermal stimulation, whether or not the area has undergone previous thermal changes, and the adapting temperature of the skin or oral mucosa.

The receptors for temperature change (i.e., thermoreceptors) are associated with some of the smaller-diameter, slow-conducting afferent fibers. They are specifically activated by a small thermal

change in either a cooling (cold afferent fibers) or warming (warm afferent fibers) direction, and they provide the brain with precise information on the location, magnitude, and rate of the temperature shift.

The predominant relay site in the brain stem of the afferent signals carried in thermoreceptive primary afferent fibers appears to be the trigeminal subnucleus caudalis. Neurons in this part of the trigeminal brain stem complex are exclusively activated by thermal stimulation of localized parts of the face and mouth, and this thermal information is relayed to the contralateral thalamus and then to the somatosensory cerebral cortex. The specificity theory thereby appears to be consistent with the properties of peripheral and central neural elements underlying our thermal sensibility.

C. Taste

The special sense of taste is covered elsewhere, but three aspects are briefly mentioned here because they relate to some of the other functions discussed in this chapter. First, as with pain (see below), taste has affective, cognitive, and motivational dimensions as well as a sensory-discriminative dimension. For example, we find some tastes pleasurable and are motivated to seek them out, whereas other tastes have the opposite effect. Indeed, humans may have innate as well as acquired taste preferences, and the food industry is well aware of our "sweet tooth," an inborn preference for sweetness. Also, taste sensibility is now known not to be confined to specific areas of the tongue; extralingual (e.g., palatal) taste buds may also make an important contribution to taste. Finally, a number of factors have been reported to modify taste; these may include other sensory experiences (e.g., smell), decreased saliva, wearing of dentures, poor oral hygiene, local anesthetics, plant extracts, and perhaps genetic, metabolic, and endocrine factors and the age of the individual. [*See* TONGUE AND TASTE.]

D. Pain

Pain deserves special emphasis because it is the first orofacial sensory experience that comes to mind for most of us. It also causes great human suffering and represents a major economic burden on society through health care costs, time lost from work, etc. Moreover, orofacial pains are particularly noteworthy because they are very common (e.g., toothache) and are often chronic and disabling. Pain is now conceptualized as a multifactorial experience. It includes a sensory-discriminative component, an aspect that allows us to discriminate the quality, location, duration, and intensity of a noxious stimulus (i.e., a tissue-damaging stimulus), but it also encompasses cognitive, motivational, and affective variables, which can modify a person's response to the stimulus. Thus, the reactions to a noxious stimulus can vary from individual to individual. It can de-

pend not only on the magnitude of the noxious stimulus but also on factors such as the meaning of the situation in which the pain occurs, the person's emotional state and motivation to get rid of the pain, and even racial and cultural background.

Of course, this multifactorial nature of pain can complicate diagnosis and treatment of pain for the clinician and also makes the experimental study of pain exceedingly difficult. Nonetheless, pain studies in humans and experimental animals have used a variety of approaches (e.g., behavioral, electrophysiological, anatomical) to give us some insights into the neural mechanisms underlying pain. [*See* PAIN.]

1. Pain Transmission

The neural basis for orofacial pain and, indeed, pain from anywhere in the body is still only partly understood, but considerable advances have been made in the last few years. These insights into pain and the mechanisms underlying its control have largely come from studies in animals. The classic concept for explaining pain and the other somatic sensations is the specificity theory; while apparently applicable in other sensations such as touch (see above), this theory has been shown to have a number of limitations in trying to explain pain on the basis of a specific peripheral and central system. As a consequence, other theories have been proposed to account for the complexity and multidimensionality of pain. The gate control theory of pain has attracted the most recent interest and research, and although it has its limitations, it does provide a good conceptual framework for considering the multifactorial nature of pain. First, this theory emphasizes the sensory interaction that occurs within the brain between the touch-related neural signals carried into the brain by the low-threshold, large-diameter primary afferent fibers, and those signals conveyed by the small-diameter fibers; the peripheral terminals of many of the latter fibers respond to noxious stimuli. If, as a result of this interaction, the activity in the small so-called nociceptive fibers prevails, central transmission cells are excited (the "gate" opens) and bring into action the central processes related to the perception of and reactions to noxious stimulation. Second, the theory also emphasizes descending central neural controls (i.e., coming from higher brain centers) related to cognitive, affective, and motivational processes that can modulate the gate.

Recent experimental studies have revealed that

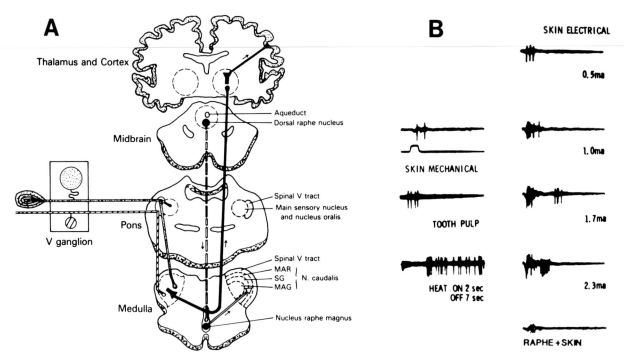

FIGURE 4 A. Diagram of a major ascending trigeminal nociceptive pathway and a descending modulatory pathway that may suppress activity of neurons in the nociceptive pathway. Nociceptive neurons in the subnucleus caudalis receive and relay information from small-diameter primary afferent fibers (crosshatched pathway) only, namely nociceptive-specific neurons, or from small-diameter afferent fibers and from large-diameter primary afferent fibers (stippled pathway) as well, namely wide dynamic range neurons; they predominate in layers I [marginalis, MAR] and V of the subnucleus caudalis. Substantia gelatinosa (SG) and magnocellularis (MAG) are other layers of the caudalis. Responses of both types of neurons to noxious oral–facial stimuli can be suppressed by descending influences from the dorsal raphe nucleus in the periaqueductal gray and nucleus raphe magnus, as shown (bottom right) for the wide dynamic range neuron illustrated in B, when the raphe and skin stimuli were interacted. [Reproduced, with permission, from R. Dubner, B. J. Sessle, and A. T. Storey, 1978, "The Neural Basis of Oral and Facial Function," Plenum Press, New York.] B. This neuron was recorded in the subnucleus caudalis of the anesthetized cat. Its classification as a nociceptive neuron was based on its responses to various types of oral–facial stimuli. Some of these responses are illustrated in the traces. Note, for example, that while the neuron fired with a brief burst of two action potentials when a light mechanical (tactile) stimulus was applied to a localized skin region of the cat's face, the neuron could also be activated when an electrical stimulus was applied to the cat's canine tooth pulp and when noxious radiant heat (and pinch [not shown]) was applied to the same region on the cat's face. Because it responded to innocuous (i.e., tactile) as well as noxious stimuli, this neuron was classified as a wide dynamic range nociceptive neuron. Note on the right that with increasing intensity levels of electrical stimulation of the skin, late as well as early bursts of impulses could be evoked; late discharge probably reflects inputs from nociceptive afferent fibers, and early burst reflects inputs from faster-conducting "tactile" afferent fibers. Time duration of records: 100 msec (except heat record: 10 sec).

some nociceptive primary afferent fibers are specifically sensitive to noxious stimuli, whereas others respond to innocuous stimuli as well. These nociceptive afferents are small diameter and slow conducting, and they terminate in the peripheral tissues as free nerve endings. Those nociceptive primary afferent fibers supplying the face and mouth project via the trigeminal ganglion to part of the trigeminal brain stem sensory nuclear complex. The subnucleus caudalis (see Fig. 3) is especially involved in pain (and temperature) transmission. Many subnucleus caudalis neurons receive the signals from orofacial nociceptive primary afferents and, thus, can respond to noxious stimulation of the face and mouth, TMJ, or jaw and tongue muscles. These neurons are either called wide dynamic range (i.e., they respond to innocuous as well as noxious stimuli) or nociceptive-specific (a wide dynamic range neuron is shown in Figure 4). Also noteworthy, these neurons relay nociceptive information to the contralateral thalamus, from where information is relayed to the overlying cerebral cortex or other thalamic regions. While parts of the thalamus or cortex are involved in the various components of pain behavior (perception, motivation, etc.), the precise function of each region is still uncertain. [*See* CORTEX; THALAMUS.]

The "rostral" parts of the trigeminal brain stem complex are especially concerned with the relay of tactile information (see above), but recent studies also indicate a role for them in orofacial pain mechanisms as well. For example, afferent fibers from the ''nerve'' of the tooth (the tooth pulp), generally assumed to represent a nociceptive input, synapse with neurons present not only in subnucleus caudalis but also at the more rostral levels of the complex.

2. Pain Control

A variety of procedures are available for the control of pain, ranging from pharmacologic measures such as local and general anesthetics and analgesic drugs (e.g., aspirin, morphine) to therapeutic procedures such as acupuncture, transcutaneous electric stimulation, hypnosis, and psychiatric counseling. In extreme cases, neurosurgical methods may be employed. All of these procedures are aimed at blocking pain transmission either at the periphery (e.g., aspirin), before nerve impulses enter the brain (e.g., local anesthesia), or within the brain (e.g., general anesthetics and analgesics). As described above, the trigeminal brain stem sensory nuclei have a complex structural organization by which they can interact in complex ways with many other parts of the nervous system. Interactions involving inhibitory processes that modulate pain transmission have been extensively documented; indeed, inhibitory modulatory processes are widespread in the brain and are involved, as noted above, in touch as well as in reflex activity and more complex behavioral functions (see Section III). Moreover, facilitatory interactions between these various convergent afferent inputs can also occur. These types of interactions are thought, for example, to contribute to the so-called referral of pain, where pain may be felt not at the site of injury or pathology but also, or instead, at other distant sites.

With respect to inhibitory modulation of pain transmission, evidence indicates (1) sensory interaction and (2) descending central control mechanisms, as the gate theory postulates. The output of trigeminal nociceptive neurons, for example, can be markedly suppressed by large fiber afferent inputs (i.e., sensory interaction); however, in some situations, small-fiber nociceptive afferent stimulation may also suppress their activity. They can also be inhibited by stimulation of brain sites such as the midbrain periaqueductal gray matter and the nucleus raphe magnus in the lower brain stem (Fig. 4); stimulation of such descending central controls produces marked analgesic effects in humans and experimental animals. Stimulation of other regions, such as the cerebral cortex, is less effective in suppressing the trigeminal nociceptive neurons, although cortical stimulation does have a profound influence on neurons excited by non-noxious stimuli.

These suppressive effects on nociceptive transmission are most exciting in terms of enhancing our understanding of pain mechanisms and developing better pain control procedures. The effects have been linked in part to the release of endogenous (i.e., naturally occurring) chemicals that may activate the descending control systems or act relatively directly on the pain-transmission neurons. One of these endogenous substances is enkephalin, a peptide that is pharmacologically similar to the opiate drugs such as morphine; other neurochemicals such as 5-hydroxytryptamine (serotonin) also appear to be involved. When injected into certain parts of the brain, enkephalin produces analgesic effects; if applied locally in the vicinity of pain-transmission neurons in the trigeminal subnucleus caudalis or analogous neurons in the spinal cord, the responses of these neurons to noxious stimuli can be suppressed. Thus, pain-suppressing systems appear to occur naturally within the brain, and a number of important therapeutic procedures have been proposed that may exert their analgesic effects by utilizing such systems. The action of narcotic analgesics such as morphine has been linked to such systems, and therapeutic procedures involving skin, muscle, or nerve stimulation (transcutaneous electric stimulation and acupuncture) may exert their analgesic effect in part by exciting pathways to the brain that ultimately lead to activation of endogenous analgesic systems.

3. Oral–Facial Pain States

The oral–facial region is a particular focus of pain, and a brief description and possible mechanisms are given for several of the most common or interesting pain conditions. The first is temporomandibular (or myofascial) pain dysfunction, which appears actually to be a family of disorders and presents a variety of signs and symptoms. The most commonly reported symptoms are pain in the region of the TMJ or jaw muscles or both, limitation of jaw movement, and crackling (crepitus) or clicking in the joint. The pain can sometimes be referred to other structures such as the teeth and muscles of the jaw or neck. Salivation and lacrimation, possibly reflecting involvement of the autonomic nervous system are also frequently associated with the disorder. [See AUTONOMIC NERVOUS SYSTEM.]

The etiology, diagnosis, and treatment of the condition represent some of the most controversial aspects of dentistry. Until recently, occlusal factors related to the faulty interdigitation of upper and lower teeth were considered to be the most important etiologic factors, but recent studies point to the importance of central factors (e.g., stress-related) in its etiology in many patients. Treatment procedures are as varied and numerous as the various theories of its etiology, ranging from balancing the bite (i.e., occlusal equilibration) to the administration of muscle relaxants or anxiety-reducing drugs, to even psychiatric counseling.

The etiology of trigeminal neuralgia (or tic douloureux) is also controversial. This disorder, which rarely occurs in persons <45 yr of age, manifests as paroxysms of excruciating pain that usually last for a few seconds or minutes, with long periods of remission between attacks. It is said to be the most excruciating pain a human can suffer, yet a most interesting and puzzling feature is that the neuralgic attack is usually triggered not by a noxious stimulus but by a light, non-noxious stimulus (e.g., puff of air, wisp of cotton) to certain trigger sites in the perioral region. Theories of its etiology relate to either peripheral or central factors. Some researchers believe that peripheral changes, such as compression of the trigeminal sensory nerve root in the vicinity of the trigeminal ganglion by aberrant vessels or bony outgrowths, are the primary etiologic factor. Despite evidence in favor of a peripheral etiology, the signs and symptoms of the disorder are suggestive of central neural changes, with imbalance in the functional organization of the trigeminal brain stem sensory complex. This view is compatible with the effectiveness of certain anticonvulsant or antiepileptic drugs in depressing the activity of trigeminal brain stem neurons. These drugs are now widely used for the clinical control of the disorder.

Postherpetic neuralgia (shingles) differs from trigeminal neuralgia in that a definitive etiologic factor is known, namely the herpes zoster virus that has gained access to the trigeminal ganglion. The pain usually involves the skin of the ophthalmic division of the trigeminal nerve (i.e., above and around the eye), and a selective loss of large myelinated fibers is apparent. Although this loss could, within the context of the gate control theory, lead to an imbalance of sensory input and an "opening of the gate" (see above), the actual mechanisms responsible for the pain are still unclear.

Another poorly understood condition is so-called atypical facial pain. This pain is diffuse, deep, and dull or throbbing in nature, and can be constant for many days. No trigger zones are associated with the disorder. Its etiology is also uncertain, although a psychogenic mechanism has been suggested for a certain portion of such cases.

Finally, the most common of oral–facial pains is the toothache. This pain is usually associated with trauma or dental decay affecting the "nerve" of the tooth (the pulp) or the overlying hard tissue, the dentine. Much research has centered on the possible peripheral mechanisms of pulp and dentinal sensitivity. Figure 5 illustrates the innervation of the pulp; most of the nerves in the pulp are small-diameter *afferent* fibers associated with sensation, but some are autonomic *efferent* fibers thought to primarily control the blood supply of the pulp. The pulp afferent fibers can be excited by a variety of different types of stimuli (e.g., sugar, hot or cold drinks), as most people can attest to. While it is generally assumed that their excitation is exclusively related to pain, recent studies suggest that

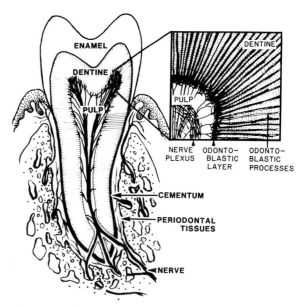

FIGURE 5 The nerves supplying the tooth innervate the periodontal tissues, which are the supporting tissues of the tooth, as well as the tooth pulp. After entering the pulp at the root apex of the tooth, the pulpal nerves arborize extensively, especially in the crown of the tooth. They form a nerve plexus in the periphery of the pulp beneath the layer of odontoblast cells, which have processes extending into the tubules in the overlying dentin (and which are involved in dentin formation). Some individual nerve fibers leave the subodontoblastic plexus; a number enter the dentinal tubules, although many terminate in the odontoblastic layer. [From B. J. Sessle, 1987: modified, with permission, from M. R. Byers, 1984, *Int. Rev. Neurobiol.* **25,** 39–94.]

some pulp afferent fibers and their central connections in the brain may be involved in sensory functions other than pain.

Many of the pulp afferent fibers terminate in close proximity to odontoblasts, but some enter the dentinal tubules, occasionally in close contact with the odontoblastic process (Fig. 5). While these findings are considered by some researchers as evidence supporting a neural theory of dentinal sensitivity, the role of these intradentinal fibers in sensitivity is still unclear. Indeed, it is still conceivable that intradentinal neural processes contribute to the excitation of pulpal nerve fibers, or that intradentinal elements are involved by means of the odontoblast acting as a transducer, or by hydrodynamic processes, or both. More general acceptance exists for the hydrodynamic theory, which suggests that enamel or dentinal stimuli can cause a displacement of dentinal tubule contents; this, in turn, is thought to bring about a mechanically induced excitation of the nerves.

III. Neuromuscular Functions

A. Muscle

The oral–facial region manifests a vast array of both simple and highly complex motor activities, which have important biological significance to the point of being among the most fundamental behaviors required for survival. Movements of the jaw and the surrounding musculature are integrally involved in human behaviors as diverse as mastication (chewing), drinking and suckling, manipulation of objects with the tongue, cheeks and lips, communication through facial expressions, and speech production.

The peripheral motor components of these activities are the muscles of the jaw, face, tongue, pharynx, larynx, and palate. Like muscles elsewhere in the body, they consist basically of a passive elastic component (e.g., tendon, ligaments) and an active contractile component; the latter is composed of numerous individual striated muscle fibers. These so-called extrafusal muscle fibers are connected to the axons of alpha motoneurons present in the brain stem: a single alpha motoneuron plus the muscle fibers that it supplies are known collectively as the motor unit. Impulses from the motoneuron are conducted along its axon (the alpha efferent or motor axon) to the muscle fibers and bring about muscle

contraction through the process of neuromuscular transmission. [See MUSCLE DYNAMICS.]

The peripheral *sensory* components of muscle are receptors. For example, muscle contains free nerve endings and these receptors are associated with muscle pain and possibly responses to stretch. In addition, there are also specialized receptors (e.g., the Golgi tendon organ, which is particularly sensitive to muscle tension and the stretch-sensitive muscle spindle). In addition to a dual afferent supply, the muscle spindle receives a motor innervation from the gamma (fusimotor) efferent fibers of small gamma motoneurons, which modify the sensitivity of the afferent fibers to stretch, thereby indirectly assisting or maintaining muscle contraction. In contrast to muscles in most other places of the body, these specialized receptors have a limited distribution in the craniofacial region.

B. Central Mechanisms

The primary muscle afferent pathways and central connections are poorly documented for most of the oral–facial musculature, except for the trigeminal mesencephalic nucleus, which contains the cell bodies of jaw-closing muscle spindle primary afferent fibers (and some other oral–facial primary afferent fibers). This location of primary afferent cell bodies is the only place in the body where primary sensory cell bodies are located within the central nervous system. Other major distinguishing features of the oral–facial motor systems include the following. (1) Many oral–facial muscles lack muscle spindles and Golgi tendon organs, as noted above. (2) A fusimotor (gamma-efferent) control system is absent due to the lack of muscle spindles in many muscles. (3) Reciprocal innervation, where muscle afferent fibers have reciprocally opposite effects on antagonistic spinal motoneurons, is limited due to the sparsity of muscle spindle and tendon organ afferent fibers. This lack may be compensated for by the powerful regulatory influences afforded by afferent inputs from facial skin, mucosa, TMJ, and teeth (see below). (4) Coordinating pathways and mechanisms exist to allow for the *bilateral* activity of muscles; although activity of a particular limb or trunk muscle can occur on both the left and right sides of the body in some movements, this bilateral activation (or depression) is particularly prominent in orofacial movements (e.g., chewing, swallowing, speech, coughing).

The muscle afferent fibers, along with cutaneous,

joint, and intraoral afferent fibers, make excitatory reflex connections with brain stem motoneurons located within one or more cranial nerve motor nuclei (Fig. 6); these connections are usually indirect and involve interneurons in the trigeminal spinal tract nucleus, the solitary tract nucleus, and the reticular formation. Examples of such "simple" reflexes are the jaw-closing, jaw-opening, and horizontal jaw reflexes; facial muscle reflexes; tongue reflexes; and laryngeal, pharyngeal, and palatal reflexes. In addition to these excitatory reflex effects of various oral–facial stimuli, inhibitory effects are also expressed on the reflexes by sensory stimuli and by descending regulatory influences arising from higher brain centers such as the cerebral cortex (Fig. 6). The excitatory and inhibitory influences are especially involved in protection of the masticatory apparatus (e.g., from biting the tongue during chewing) and in providing much of the neural organization upon which are based more complex motor activities such as the protective reflex synergies of coughing and gagging. An even higher level of complexity of organization is seen with the rhythmic, automated activities of mastication, suckling, and swallowing; speech is another complex sensorimotor behavior utilizing this neural organization.

C. Mastication, Swallowing, and Related Neuromuscular Functions

1. Mastication

Mastication serves to break down foodstuffs for subsequent digestion by means of the masticatory forces generated between the teeth. It is characterized by cyclic jaw movements in three dimensions (vertical, lateral, and anteroposterior) and less rigid facial and tongue motility patterns. These various movements are produced by the coordinated contraction of the jaw, face, and tongue muscles (Fig. 7). Masticatory forces on the teeth are usually in the 5–10-kg range, but can vary depending on such factors as the teeth concerned (molars exert the greatest force, incisors the lowest), practice, toughness of the diet, the wearing of dentures, presence of periodontal disease, tooth–cusp configuration, and the distance that the jaws are separated when the forces are applied. Even greater forces are developed during biting: maximal biting forces on a tooth usually range from approximately 20 to 200 kg (the Guinness world record is over 400 kg!).

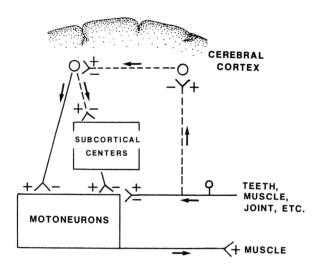

FIGURE 6 Some sensory and higher center controls involved in regulation of brain stem motoneurons supplying muscles of the jaw, face, and tongue. Both alpha and gamma motoneurons can be reflexly influenced (facilitated, +; inhibited, −) by sensory inputs from various receptors. They can also be influenced by descending controls originating from a number of higher brain centers, such as the sensorimotor cerebral cortex. The cortex can exert relatively direct effects on motoneurons or can modulate motoneuron activity indirectly by means of its connections with various subcortical centers (e.g., basal ganglia, brain stem reticular formation, cerebellum).

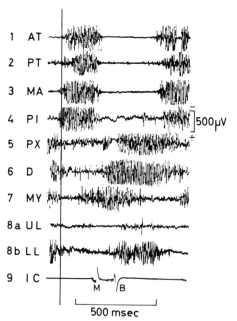

FIGURE 7 Electromyographic activity of various muscles recorded during a natural chewing cycle in a human subject: jaw (1–6); tongue (7); and facial (8). Vertical line indicates onset of activity in right anterior temporalis muscle. M and B indicate, respectively, onset and offset of contact of the opposing incisor teeth. AT, anterior temporalis; PT, posterior temporalis; MA, masseter; PI, medial (internal) pterygoid; PX, lateral (external) pterygoid; D, digastric; MY, mylohyoid; UL, upper lip; LL, lower lip; IC, incisor contact. [Reproduced, with permission, from E. Moller, 1966, *Acta Physiol. Scand.* **69**, 1 (suppl. 280).]

Although mastication was originally thought to be based on alternating simple jaw reflexes of jaw opening and closing, the current concept is that the cyclic, patterned nature of chewing depends on a subcortical center (Fig. 6) comprising a central neural pattern generator, the brain stem "chewing center." This generator is sensitive to descending regulatory influences from higher brain centers and to sensory inputs from oral–facial receptors; the sensory inputs may be particularly critical in the learning of mastication and acquisition of masticatory skills, in actively (i.e., reflexly) guiding masticatory movements, and in guiding the position of the jaw when the teeth come into occlusion and when the jaw is at "rest." The clinical significance of the sensory influences is further exemplified by their probable involvement, along with central influences, in the etiology of pathophysiologic conditions such as temporomandibular/myofascial pain dysfunction and bruxism (tooth grinding).

2. Swallowing

In contrast to mastication, swallowing is an innate, reflexly triggered, all-or-none motor activity that is relatively insensitive to sensory or central control. In addition to its obvious alimentary function, swallowing also serves as a protective reflex of the upper airway because it reflexly interrupts respiration and prevents the intake of food or fluid into the airway. Swallowing consists of a rigid, temporal pattern of muscle activities that appears to depend on a brain stem pattern generator, the "swallow center," for its expression. The coordinated muscle activities provide the means for the propulsion of a food or liquid bolus from the oral cavity to the stomach (Fig. 8) and also afford mechanisms for protection of the airway and prevention of reflux (regurgitation). Some of the muscles are "obligate" swallow muscles (i.e., they always are active in swallowing), whereas others show a variable participation in swallowing (the facultative muscles). The latter muscles especially may be sensitive to alterations in the oral environment and to maturational changes, and thus their participation can vary depending, for example, on the volume or consistency of a foodstuff, or whether the subject is an infant or adult.

3. Other Functions

Mastication, suckling, and swallowing are themselves components of even more complex behaviors. They are associated with feeding and drinking,

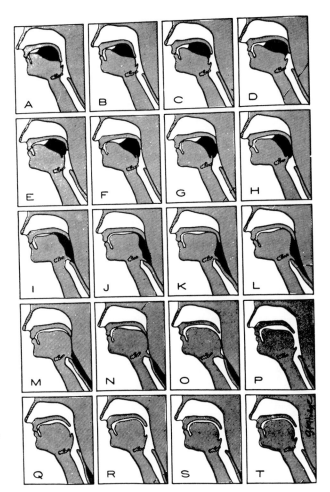

FIGURE 8 The orchestrated sequence of events characterizing a single swallow. At the start of the swallow, the soft palate acts as a partition up to the base of the tongue (A, B) and elevates to engage the posterior pharyngeal wall and close off the nasopharynx as the food bolus (black) moves backward over the tongue surface (C–E). The tongue acts as a piston to squeeze the bolus into the pharynx, and the bolus is conveyed down the pharynx by pharyngeal muscle contractions (F–J); note that the epiglottis is tilted backward (e.g., to protect the entrance into the laryngeal airway [white]); the glottis (not shown) also serves to close off the entrance into the airway. Then, the bolus is slightly delayed at the upper esophageal sphincter (K), which then relaxes to allow the bolus into the upper esophagus (L, M) before closing again to prevent reflux (regurgitation). As the bolus moves down the esophagus (N–T), the soft palate relaxes and the epiglottis resumes its position and the airway is reopened. [Reproduced, with permission, from R. M. Bradley, 1984, "Basic Oral Physiology," Year Book Medical, Chicago; adapted from R. F. Rushner and J. A. Hendron, 1951, *J. Appl. Physiol.* **3**, 622–630.]

which are particularly dependent on the function of higher centers of the brain such as the hypothalamus. Sensory feedback, however, is also utilized for the initiation, maintenance, and cessation of these ingestive behaviors (e.g., a "full stomach" stretches gastrointestinal receptors, which, through their central connections, can inhibit feeding). Many of these higher brain centers are also concerned with other complex functions involving the oral cavity, including oral aggression (e.g., biting), facial expression, and speech.

IV. Autonomic Regulation

The autonomic nervous system also has regulatory effects on a variety of orofacial functions, but with few exceptions, our knowledge of its role in the orofacial region is quite scant; however, it is well known that smooth muscle (e.g., in the gastrointestinal tract, abdominal and pelvic viscera, walls of blood vessels), cardiac muscle, and glandular structures such as the lacrimal and salivary glands receive autonomic efferent innervation. The two major divisions of the autonomic nervous system, the sympathetic and parasympathetic, have important roles in regulating the contractibility, secretion, or other functions of these various tissues.

As far as autonomic regulation in the orofacial region is concerned, most is known about autonomic control of blood flow and salivary secretion. The following will briefly outline autonomic regulation of orofacial blood flow. The oral mucosa indeed has served as a model system for studying autonomic control of the vasculature. Sympathetic vasoconstrictor regulation of the blood supplies of the mucosa, periodontal, and tooth pulp tissues has been well documented, but whether or not a parasympathetic control also occurs in each of these tissues is still unclear. Sympathetic vasoconstrictor tone also takes place in the facial skin and, as elsewhere in the body, is involved in thermal regulatory function (e.g., to prevent heat loss from the body surface). Surprisingly however, in the facial skin, vasoconstrictor tone is primarily restricted to the nose, lips, and ears. Thus, heat loss from the head can represent a considerable proportion of total body heat loss, and these three parts of the orofacial region are also very susceptible to cold, a susceptibility that is well reflected in frostbite. Vasodilation can occur through a release of this vasoconstrictor

tone or through "active" vasodilation, which is especially expressed in the cheek, forehead, chin, and neck. These four areas correspond to those showing increased blood flow in emotional states, such as the flushing of the face that may occur in an embarrassing situation. The extent to which the parasympathetic nerves are involved in this active orofacial vasodilator tone is still unclear. The parasympathetic nerves also do not appear to have a major function, if any, in regulation of the facial sweat glands. The sympathetic efferents again appear to have a major role here and are involved in the common response of these glands to thermal or emotional stimuli.

In addition, it should be mentioned that the autonomic nervous system may also regulate skeletal muscle function as well as the sensitivity of several receptor systems such as the muscle spindle and some cutaneous receptors. Also, certain pain syndromes may involve dysfunction of the autonomic nervous system, but the mechanisms by which the autonomic efferents modulate pain sensibility are still uncertain. Finally, evidence is accumulating to indicate that the autonomic nervous system can exert so-called neurotrophic influences on several sensory-motor functions. In the orofacial region, these include the taste bud and salivary gland functions.

Bibliography

Dubner, R., Sessle, B. J., and Storey, A. T. (eds.) (1978). "The Neural Basis of Oral and Facial Function." Plenum, New York.

Fromm, G. H., and Sessle, B. J. (eds.) (1991). "Trigeminal Neuralgia: Current Concepts Regarding Pathogenesis and Treatment." Butterworths, Stoneham, Massachusetts.

Luschei, E. S., and Goldberg, L. J. (1981). Neural mechanisms of mandibular control: Mastication and voluntary biting. In "Handbook of Physiology. The Nervous System. Motor Control," Vol. II, pt. 2, pp. 1237–1274. American Physiological Society, Bethesda, Maryland.

Sessle, B. J. (1987). The neurobiology of facial and dental pain: Present knowledge, future directions. J. Dent. Res. 66, 962–981.

Wall, P. D., and Melzack, R. (eds.) (1989). "The Textbook of Pain," 2nd ed. Churchill Livingstone, Edinburgh.

Neural-Immune Interactions

DAVID L. FELTEN, SUZANNE Y. FELTEN, *University of Rochester School of Medicine*

I. Evidence for Neural Influences on Immune Responses
II. Endocrine Effects on the Immune System
III. Direct Neural Connections with the Immune System
IV. Functional Roles of Neurotransmitters in the Immune System
V. Cytokine Interactions with Neurons

Glossary

Cytokine Chemical mediator released from a monocyte–macrophage (monokine) or a lymphocyte (lymphokine) that acts on cells of the immune system, providing intracellular signaling to the target cells

Hormone Chemical mediator released from cells in an endocrine gland or from a neuron (neuroendocrine transducer cell) that exerts its effect on target cells via circulation; response of the target depends on the presence of receptors specific for that hormone in or on the target, not on the proximity of the cell releasing the hormone to the target

Mitogen response Stimulated proliferation in a lymphocyte resulting from stimulation by a nonspecific agent, usually a plant lectin; this response generally is viewed as indicative of lymphocyte potential to proliferate but should not be equated with the overall responsiveness of the immune system *in vivo*, particularly in ability to respond to a specific antigen

Neuromodulation Process by which a neurally released mediator produces little or no direct functional effect on a target cell (neural, immune, or autonomic effector cell) but significantly alters the responsiveness of that cell to another signal

Neurotransmitter Chemical mediator released from a neuron that acts on receptors on an adjacent target neuron or effector cell, resulting in an intracellular effect in that cell

Norepinephrine Catecholamine present in the periphery and in the central nervous system, used as a neurotransmitter (central neurons, postganglionic sympathetic neurons) or as a hormone (adrenal medullary chromaffin cells); an important neural mediator for communication with autonomic target tissues

Sympathetic nervous system Subdivision of the autonomic nervous system, supplying innervation to smooth muscle, cardiac muscle, secretory glands, some cells related to metabolism, and cells of the immune system, generally involved in an arousal (fight or flight) response

NEURAL-IMMUNE INTERACTIONS consist of communication channels and mediators that permit bidirectional signaling between the nervous system and the immune system. The nervous system, in response to external and internal cues and inputs, can communicate with cells of the immune system via neuroendocrine secretion and release of neurotransmitters from nerve fibers that innervate lymphoid organs, or tissues where an inflammatory response or immune response is ongoing. Cells of the immune system can secrete cytokines that signal the brain and change its electrical and chemical activity. These bidirectional channels of communication appear to be linked circuitry that permits feedback and feedforward regulation. In addition to the systemic communication channels permitting brain and immune system to signal each other, local signaling within the microenvironment of lymphoid organs permits short-loop communication between cells secreting neurotransmitters and cytokines. These two major classes of mediators likely interact with each other to synergize or counter each other's actions; their actions in turn are regulated by the other mediators present in the microenvironment such as hormones. The consequences of bidirectional neural-immune interactions is a superimposed neural modulation of immune responses and the possibility that psychosocial factors, environmental inputs, and internal cognitive or affective factors may influence the functioning of the immune system. In turn, the brain and its behavior may be responsive to immunologically derived signals that act as ''molecular sensory signals.'' Clearly, the immune system can no longer be considered an autonomous, self-regulating system that operates outside of the context of brain and behavior.

I. Evidence for Neural Influences on Immune Responses

Evidence from many fields points to the ability of the nervous system to modulate immune responses. Ader, Cohen and colleagues have demonstrated that immune responses can be classically conditioned and that such conditioning can alter the course of a genetic autoimmune disease in rodents. Other investigators have shown that numerous cellular activities and responses can be conditioned in the direction of either enhancement or suppression. Numerous stressors, including psychological stressors, can induce alterations in immune responsiveness, although the direction and magnitude of response is highly dependent on the nature, timing, and duration of the stressor. Studies in humans suggest that the stress of examinations in medical students, marital separation, or other aversive circumstances may be accompanied by diminished immune responsiveness. Some investigations of depressed or bereaving individuals suggest that a subset of these people may demonstrate diminished immune responsiveness and increasing incidence of illness, although these two phenomena have not yet been causally linked. A recent study of women with disseminated breast cancer suggests that psychotherapeutic support as an adjunct to standard medical treatment increased the survival of these patients and may permit a better adjustment to their disease. Experimental brain lesions in animals have been shown to cause altered immune responses, with the directionality and duration of the effect dependent upon the site and extent of the lesion.

These many lines of investigation provide evidence that external factors that impinge on the brain, or direct manipulation of neural circuitry of the brain, can result in altered immune responses, sometimes of sufficient magnitude to alter the out-

come of a genetic or acquired disorder. However, in no instance has the entire circuitry from the stimulus to specific brain sites, to specific central pathways using known neurotransmitters, to specific outflow channels from the brain to organs of the immune system, to specific cellular mechanisms that induce specific intracellular changes that can explain the altered immune response been worked out. The task of unraveling the myriads of interactive signaling among the nervous system, endocrine system, and immune system, all capable of changing with the animal's behavioral state, brings together the interpretive complexities of systemic biology magnified at least threefold because of the interactions.

However, we do know that the brain has two major routes by which it can send mediators to alter or modulate activity in the periphery: neuroendocrine secretion and direct neural connections from the autonomic nervous system to the viscera. For many years, glucocorticoids have been known to influence immune responses and have been used clinically as immunosuppressive drugs. Since the 1970s, lymphocytes and monocytes–macrophages have been known to express receptors for neurotransmitters on their surface, but the physiological role for these receptors was unknown. The past 5 years has seen an explosion of information about such hormonal and neurotransmitter signaling of cells of the immune system. The present chapter outlines basic evidence for these mediators playing a modulatory role in immunoregulation. Obviously, we are just beginning to identify the cast of participants in neural-immune signaling; the emerging picture indicates that these mediators can exert a complex array of influences on different cell types of the immune system at different stages of activation and also can interact with cytokine signals from the immune system itself. These neurally derived mediators will likely exert just as complex a regulatory influence over reactivity of lymphoid cells as the cytokines themselves do. [See LYMPHOCYTES; MACROPHAGES.]

II. Endocrine Effects on the Immune System

A. Pituitary

Early studies of hypophysectomized mice and pituitary-deficient dwarf mice reported a resultant diminution in both cellular and humoral immune responses, and involution or hypocellularity of primary and secondary lymphoid organs. The administration of anterior pituitary hormones, particularly growth hormone (GH) and prolactin, led to some degree of restoration of immune function in these animals, suggesting that pituitary hormones could exert a modulatory role over immune responses. [See PITUITARY.]

B. Growth Hormone

GH deficiencies have been associated with diminished T-cell functions, antibody responses, natural killer (NK) cell activity, and decreased cellularity in bone marrow and thymus; these abnormalities are restored to some extent by administration of GH. GH administration enhances mitogen responsitivity of transformed and normal lymphocytes, restores mitogen responsivity in aged animals with depressed responsivity, and increases the activity of alloantigen-specific cytotoxic T lymphocytes. GH also activates macrophages to produce superoxide anions that nonspecifically kill ingested bacteria. Thymocytes, lymphocytes, and monocytes possess high-affinity receptors for GH.

C. Prolactin

Prolactin inhibition or deficiency results in suppressed antibody responses, suppressed delayed hypersensitivity responses, depressed mitogen responses in B and T lymphocytes that are independent of interleukin-2 (IL-2) receptor expression or IL-2 secretion, supressed T lymphocyte-dependent activation of macrophages, and suppressed response to bacterial challenge and to its induction of interferon-gamma production. These responses are reversed or restored, for the most part, by administration of exogenous prolactin. The pituitary secretion of both GH and prolactin is increased by the action of thymosin fraction V, a thymic hormone, suggesting a functional thymic–pituitary axis. Interleukin-1 (IL-1) can inhibit release of prolactin from the pituitary, as can glucocorticoids, whereas estrogen can enhance release. Prolactin may act to counter some of the inhibitory effects of the glucocorticoids and may be active in the early phases of a stress response.

D. Adrenal Corticotropin Hormone

Adrenal corticotropin hormone (ACTH), an anterior pituitary hormone, can exert a direct suppressive effect on antibody production, interferon-gamma secretion, and macrophage tumoricidal activity. ACTH is stimulated by corticotropin-releasing factor (CRF) from the hypothalamus and, in turn, stimulates glucocorticoid secretion from the adrenal cortex, thus establishing the classical CRF–ACTH–glucocorticoid axis. This axis is immunosuppressive to many cellular and humoral immune responses.

E. Opioid Peptides—The Endorphins

The endorphins, also produced in the pituitary, have been reported to modulate immune responses, lymphocyte proliferation, NK cell activity, production of interferon-gamma, and phagocyte chemotaxis. The magnitude and directionality of change of these responses varies considerably, depending on the binding capabilities of the opioid peptide under investigation. In addition, studies of opioid receptors indicate that only some of the effects of endorphins can be attributed to direct receptor-mediated interactions, and nonopioid receptor interactions probably mediate some of the endorphin effects on immune responsivity. Beta-endorphin may modulate the expression of the T-cell receptor, the IL-2 receptor, or other receptors on lymphocytes. Alpha-endorphin can bind preferentially to some MHC class I antigens. Thus, endorphin responses may act through numerous mechanisms and many change the responsiveness of the target cells to other signals.

F. Opioid Peptides—The Enkephalins

Met- and leu-enkephalins also have been reported to modulate antibody responses, antibody-dependent cell-mediated cytotoxicity, lymphocyte numbers and proliferation, resistance to viral and tumor challenges, lymphocyte migration, NK cell activity, expression of IL-2 receptors, and production of IL-2. Again, the magnitude and directionality of the responses are highly variable and depend on the dose, route of administration, timing of the mediator with respect to the antigen, and extent of catabolism following administration. Some of these opioid peptides may be present during some forms of intermittent stress, suggesting that the nature and extent of a stressor may determine which hormonal mediators are secreted. In addition, some nerve fibers in the spleen demonstrate met-enkephalin immunoreactivity, suggesting that the enkephalins also may be utilized as neurotransmitters in lymphoid organs, as they are in the brain.

G. Glucocorticoids

Glucocorticoids were shown in early studies to exert anti-inflammatory effects and immunosuppressive effects, resulting in their clinical use. Subsequent studies showed suppressive effects on lymphocyte proliferation, antibody responses, generation of cytotoxic effector cells, mixed leukocyte responses, generation of T helper cells, NK cell activity, and prolongation of allograft rejection and of tolerance. Moncyte and macrophage functions are suppressed, including numbers of circulating monocytes, expression of MHC class II antigen expression, phagocytosis, cytokine production and secretion, chemotaxis, and synthesis of complement components. However, not all effects of glucocorticoids are immunosuppressive; physiological levels of glucocorticoids can suppress nascent lymphocyte proliferation while enhancing proliferation of activated lymphocytes, suggesting that glucocorticoids may enhance signal-to-noise activity and may regulate overproduction of antibody. Also, conditioned immunosuppression, stress-induced lymphocyte hyporesponsivity, and central effects of CRF administration and IL-1 administration can apparently influence immune responsivenesss independent of the glucocorticoids. Therefore, not all neural influences on the immune system are mediated by glucocorticoids. Unfortunately, many *in vitro* studies of glucocorticoid effects and many *in vivo* studies using exogenous administration of glucocorticoids have utilized pharmacologic rather than physiologic doses.

The immune system is capable of regulating glucocorticoid secretion; IL-1 can elevate CRF secretion from the paraventricular nucleus of the hypothalamus into the hypophyseal-portal circulation to the anterior pituitary, and also possibly ACTH secretion from the anterior pituitary directly. This apparently occurs at the peak of an antibody response, suggesting that the immune system may be capable of turning on and off the secretion of glucocorticoids as key regulators of the immune response. Some stressors may interact at the level of secretion of CRF in the hypothalamus and dysregulate this feedback circuitry. An additional role played by the glucocorticoids is the modulation of receptors for neurotransmitters on lymphoid cells. Glucocorticoids can upregulate the expression of beta-adrenoceptors on many cell types, including lymphocytes, thereby enhancing the responsiveness of those cells to a subsequent interaction with catecholamines. These influences may modulate both the magnitude and timing of neurotransmitter signaling.

H. Hormonal Signaling of Lymphoid Cells

The brief discussion of hormonal effects on immune responsiveness has clearly shown that hormones can interact with each other, hormones can exert principal effects on immunologic receptors or mediators in a true modulatory fashion, hormones can influence the secretion of neurotransmitters or the responsiveness of lymphoid cells to them, and hormones can exert direct or indirect effects through hormone receptors expressed by cells of the immune system. The timing and extent of hormonal secretion, the availability and catabolism of the hormone in different compartments of the immune system, and the presence and interactions of many cell types of the immune system may determine the final effect of a given hormone on a specific response from a specific subset of cells at a specific site at a specific time. However, from this complexity the beginning framework has emerged for linked circuitry by which feedback loops can regulate hormonal secretion, similar to how neuroendocrine feedback regulation occurs. To understand how these hormonal mediators actually are utilized by behaving organisms to modulate responsivity of the immune system, we must add to this complexity the superimposed regulation of hormonal secretion by cortical, limbic, and other central neural circuitry that exerts its effect on the releasing-factor neurons, inhibitory-factor neurons, and other neuroendocrine transducer cells of the brain. There is little doubt that such influences exist and are utilized for modulation of immune responses, as demonstrated by many of the behavioral studies; however, our knowledge of the actual circuitry and mechanisms of influence, particularly with the added factor of cytokine feedback, is at a very elementary stage.

III. Direct Neural Connections with the Immune System

A. Chemically Specific Innervation of Lymphoid Organs

Both primary lymphoid organs (bone marrow) and secondary lymphoid organs (spleen, lymph nodes, gut-associated lymphoid tissue [GALT]) are innervated directly by noradrenergic (NA) postganglionic sympathetic nerves (Figs. 1–4) and by pepti-

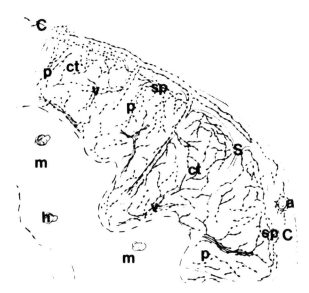

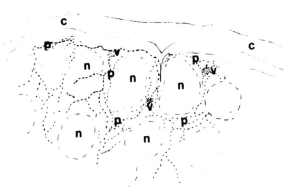

FIGURE 1 Schematic drawing of the noradrenergic innervation of the thymus. Noradrenergic fibers enter the capsule (C) around arterial plexuses (a), travel with vascular plexuses (v) or subcapsular plexuses (sp), and branch into the parenchyma (p) of the thymic cortex. ct, cortex; h, Hassall's corpuscles; m, medulla; S, septa. (From D. L. Felten *et al.*, 1985, *J. Immunol.* **135,** 755s–765s, with permission.)

FIGURE 3 Schematic drawing of the noradrenergic innervation of a lymph node. Nerve fibers enter the lymph node in the hilar region (not shown) and travel through the medullary cords with the vasculature (v) and along the sinuses. The fibers terminate mainly in the paracortical regions (p) around the follicles or nodules (n). Some fibers from the subcapsular region distribute into the adjacent cortical zone as well. Noradrenergic fibers distribute with the vessels and also travel into the surrounding parenchyma, where they end among lymphocytes and other cells. c, capsule. (From D. L. Felten *et al.*, 1985, *J. Immunol.* **135,** 755s–765s, with permission.)

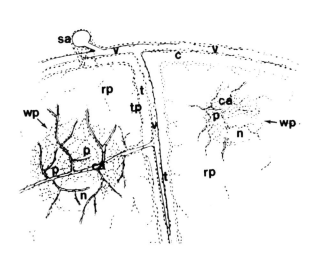

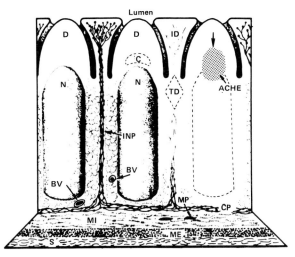

FIGURE 2 Schematic drawing of the noradrenergic innervation of the spleen. Noradrenergic fibers enter the spleen around the splenic artery (sa) as part of the splenic nerve, travel with the vasculature in plexuses (v), and continue into the spleen along the trabeculae (t) in trabecular plexuses (tp). Fibers from both the vascular and trabecular plexuses enter the white pulp (wp), continuing mainly along the central artery (ca) and its branches. Noradrenergic nerve terminals radiate from these plexuses into the periarteriolar lymphatic sheath (p) but mainly avoid the follicular or nodular (n) areas. These parenchymal nerve fibers end among fields of lymphocytes and other cell types. C, capsule; rp, red pulp. (From D. L. Felten *et al.*, 1985, *J. Immunol.* **135,** 755s–765s, with permission.)

FIGURE 4 Schematic drawing of the noradrenergic innervation of the rabbit appendix as a representative of gut-associated lymphoid tissue. Noradrenergic fibers enter along the outer or serosal (S) surface, form a catecholamine nerve fiber plexus (CP) along the muscular layers (MI and ME) adjacent to Meisner's plexus (MP), and then turn radially to run between the lymph nodules (N) in internodular plexuses (INP). Fibers in this plexus travel directly through the T-dependent zone (TD) and enter the lamina propria, or interdomal region (ID), and then branch profusely in this zone among lymphocytes, enterochromaffin cells, and subepithelial plasma cells. Acetylcholinesterase (ACHE) staining is most abundant (arrows) in the region of the crown (C) in the domes (D) and generally is not associated with nerve fibers here. BV, blood vessels. (From D. L. Felten *et al.*, 1985, *J. Immunol.* **135,** 755s–765s, with permission.)

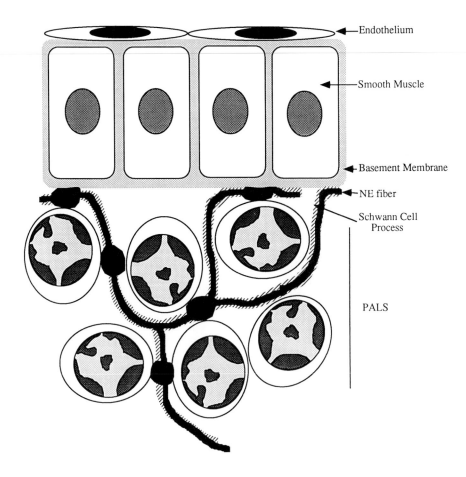

FIGURE 5 Model of the periarteriolar lymphatic sheath (PALS) of the splenic white pulp, demonstrating the presence of noradrenergic nerve terminals along the basement membrane of the smooth muscle of a central arteriole, also in direct contact with lymphocytes, both adjacent and deep to this vessel. These nerve terminals give rise to paracrinelike secretion, with diffusion of the neurotransmitter from the terminal, permitting interaction with receptors on cells that possess them. NE, norepinephrine. (From S. Y. and D. L. Felten *in* "Frontiers of Stress Research" (H. Weiner *et al.*, eds.), pp. 56–71, Hans Huber Publishers, Toronto.

dergic nerves whose origin has not been fully elucidated. Early literature showed that nerve fibers follow the vasculature into a variety of organs, including lymphoid organs, and provide regulatory control over blood flow. The first fluorescence histochemical studies of NA innervation of lymphoid organs reported primarily vascular innervation. However, use of more sensitive fluorescence methods and immunocytochemical methods demonstrated extensive parenchymal innervation, with nerve terminals extending among lymphocytes and macrophages (Figs. 5–7).

In addition, numerous neuropeptide-containing nerve fibers have been found in thymus, spleen, and lymph nodes (Figs. 8–10). Some neuropeptides, such as neuropeptide Y (NPY), may be colocalized with norepinephrine (NE) in NA postganglionic sympathetic nerve fibers. Other peptides, such as substance P and calcitonin gene-related peptide (CGRP), may be found in primary sensory fibers, although experimental confirmation of this origin is not yet available. Other peptidergic nerves, such as vasoactive intestinal peptide (VIP) or met-enkepha-

lin-containing fibers derive from cell bodies whose origin is unknown. Although acetylcholinesterase-positive nerve fibers exist in lymphoid organs, enzymatic studies suggest that they may not be cholinergic nerves; experimental evidence generally is not supportive for cholinergic innervation of lymphoid organs. Tracing studies utilizing retrograde transport of horseradish-peroxidase or other markers have been very difficult to interpret and have produced conflicting results, partly because of the great tendency for these tracers to gradually diffuse from lymphoid organs and give false-positive labeling.

NEURAL-IMMUNE INTERACTIONS

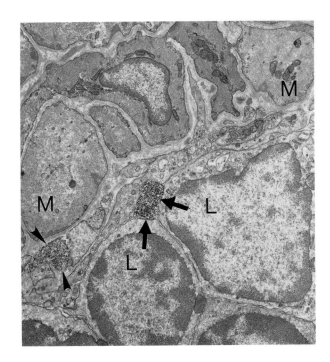

FIGURE 6 Two tyrosine hydroxylase-positive (noradrenergic) nerve terminals at the junction of arteriolar smooth muscle and lymphocytes in the periarteriolar lymphatic sheath of a rat. One terminal (arrowhead) ends adjacent to a smooth muscle cell (M), separated from it by a basement membrane. The second nerve terminal (arrow) ends in direct contact with two lymphocytes (L) in the periarteriolar lymphatic sheath. The contacts are smooth appositions of approximately 6-nm distance. Electron microscopic immunocytochemistry for tyrosine hydroxylase. ×14,000.

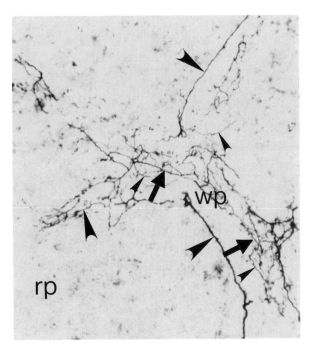

FIGURE 7 Tyrosine hydroxylase-positive (noradrenergic) nerve fibers in the white pulp of the spleen of a rat. These fibers are present along the central artery (arrows), along the marginal sinus (large arrowheads), and within the parenchyma (small arrowheads), where T lymphocytes reside. Immunocytochemistry for tyrosine hydroxylase. rp, red pulp; wp, white pulp. ×160.

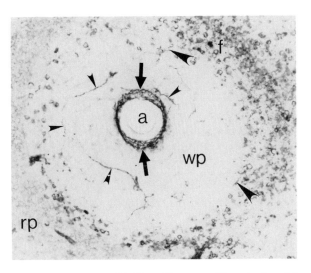

FIGURE 8 Neuropeptide Y-positive nerve fibers in the white pulp of the spleen in a rat. These peptidergic fibers are present around the central artery (arrows), along the marginal sinus (large arrowheads) and within the parenchyma (small arrowheads). This section is counterstained for IgM positive cells (B lymphocytes), which are present in the follicles and in the marginal zone. a, central artery; f, follicle; rp, red pulp; wp, white pulp. Immunocytochemistry for neuropeptide Y. ×250.

Such tracing studies from peripheral organs should also be accompanied by anterograde studies and by experimental manipulations such as ablation studies with neurochemical analysis. These have been done extensively only for the NA nerves; the origin of peptidergic innervation of lymphoid organs awaits more detailed and careful study.

In order for a chemical mediator found in nerves to be considered as a true neurotransmitter, it must fulfill several minimal criteria, including: (1) presence and synthesis in neurons whose fibers end in apposition with, or adjacent to, the target cells; (2) release from the nerves and availability for interaction with target cells; (3) presence of receptors for the mediator on target cells, and second messenger responses or intracellular processes that are affected by the receptor-ligand interaction; and (4) pharmacologically predictable functional consequences of interaction of the mediator with its re-

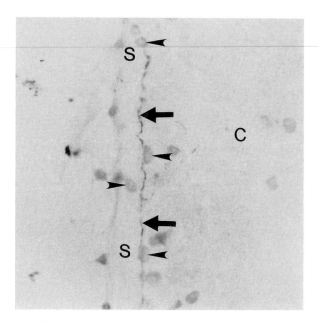

FIGURE 9 Substance P-positive nerve fiber (arrows) running along the margin of a septum in the rat thymus. This fiber is adjacent to mast cells (arrowheads) that stain in this section. C, thymic cortex; S, septum. Immunocytochemistry for substance P. ×250.

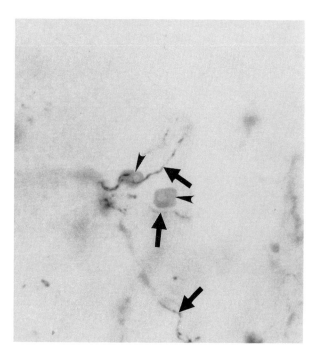

FIGURE 10 Vasoactive intestinal peptide-positive nerve fibers (arrows) adjacent to mast cells (arrowheads) in the outer thymic cortex of a rat. Immunocytochemistry for vasoactive intestinal peptide. ×250.

sponsive target cell. These criteria have been met for NE as a neurotransmitter in several organs, such as spleen and thymus, and appear to be mainly fulfilled for some neuropeptides, particularly substance P and VIP. Intensive efforts are now under way in many laboratories working on various aspects of these criteria. The first step in establishing neurotransmission for a substance with cells of the immune system as targets is to demonstrate the presence of nerves in the appropriate compartments of lymphoid organs, thereby providing a possible source of ligand for interaction with receptors on target cells. The following descriptions are drawn mainly from reports from the laboratories of D. Felten and S. Felten.

B. Primary Lymphoid Organs—Bone Marrow and Thymus

Primary lymphoid organs are innervated by NA nerve fibers that travel along the vasculature and also branch into the parenchyma. In the bone marrow, such fibers end among hemopoietic elements. Early studies suggested that sympathetic stimulation released cells into the circulation from the marrow, and beta-adrenoceptor stimulation of these cells triggered stem cells into cycle or shortened the cell cycle. [*See* HEMOPOIETIC SYSTEM.]

The thymus receives NA innervation from the superior cervical ganglion and from upper-chain ganglia. NA fibers (Fig. 1) distribute with the vasculature into the cortex, forming the densest plexuses along the vasculature at the cortico-medullary junction, and also branch into the cortex among thymocytes. The medulla and associated epithelial zones are innervated only sparsely, mainly along the vasculature. Thymocytes also possess beta-adrenoceptors and respond to catecholamine stimulation by inhibiting proliferation and enhancing the expression of differentiation antigens. Denervation of the NA nerves from the thymus in newborn rats by chemical sympathectomy results in enhanced proliferative responses of thymocytes at 10 days of age. Thus, key functions in primary lymphoid organs, such as proliferation, differentiation, and migration, may be modulated by NE released from sympathetic NA nerve fibers.

Several groups of peptidergic nerve fibers have been found in the thymus, including NPY (appears to be colocalized with NE), substance P (Fig. 9) and CGRP (perhaps localized in primary sensory fibers), and VIP (Fig. 10) (of unknown origin). Thymocytes and lymphocytes possess receptors for numerous neuropeptides, linked with classical second messenger systems, suggesting that neurotransmitters released from these peptidergic nerves into the microenvironment of lymphoid cells may be available for interaction with these cells.

C. The Spleen

NA innervation of the spleen arises from cells in the superior mesenteric–coeliac ganglion, which in the rodent is supplied by the T6–T12 spinal cord intermediolateral cell column of preganglionic neurons. These fibers travel with the splenic artery and distribute with two major systems. The first is the smooth muscle of the capsule and the trabecular–venous sinus system, probably instrumental in the contractile capabilities of this organ. The second system supplied by NA fibers is the white pulp (Fig. 2). NA fibers travel with the central artery of the white pulp and its branches and distribute among lymphocytes in the periarteriolar lymphatic sheath (PALS) where T helper and T cytotoxic/suppressor cells are found, along the marginal sinus and in the marginal zone, where macrophages reside, and in the parafollicular zones along the outside of these B lymphocyte-containing follicles (Fig. 7). NA nerve terminals directly contact T lymphocytes via 6-nm junctional appositions in the PALS (Figs. 5, 6) and along the marginal sinus and also contact macrophages and other accessory cells, demonstrated with electron microscopic immunocytochemistry. This is a closer apposition than found between NA sympathetic nerve endings and other target cells in the periphery.

Substance P fibers also are present in the spleen, around the large venous sinuses and some of the arterioles, and along the trabeculae, with scattered fiber profiles extending into the white pulp and red pulp; CGRP-containing fibers are present in the same compartments and appear to be colocalized with substance P. These substance P and CGRP fibers appear to be compartmentalized in zones distinct from the NA fibers. In addition, NPY (Fig. 8) appears to be colocalized in NA fibers; both types of staining disappear following ganglionectomy or chemical sympathectomy. Other putative neurotransmitters reported to be present in nerve profiles in the spleen include somatostatin, cholecystokinin, neurotensin, met-enkephalin, VIP, and surpris-

ingly, IL-1. The presence of adrenoceptors and receptors for most of these peptides on lymphocytes and monocytes/macrophages raises the specter of highly complex interactions of many mediators derived from nerves and present in specific compartments of the spleen in varying concentrations.

D. Lymph Nodes

Lymph nodes are innervated by sympathetic ganglion cells of the chain or collateral ganglia, whose fibers enter the nodes with the hilar vasculature, distribute through the medullary cords among lymphocytes and macrophages, extend past the cortico–medullary junctions into the paracortical zones, and branch among T lymphocytes (Fig. 3). Additional fibers travel in subcapsular plexuses and arborize among lymphocytes in the cortical zones. NA fibers appear not to innervate the follicles, which contain B lymphocytes. Thus, both the spleen and lymph nodes share common patterns of NA innervation, including innervation of sites of lymphocyte entry, sites of antigen capture, sites of antigen presentation and lymphocyte activation, and sites of lymphocyte egress.

E. Gut-Associated Lymphoid Tissue

GALT is innervated by NA and peptidergic nerves. The NA fibers derive from collateral ganglia, enter the gut with the vasculature, traverse the muscular layers, travel radially toward the lumen, distribute through the T lymphocyte-dependent zones, and arborize in the lamina propria among lymphocytes, mast cells, enterochromaffin cells, plasma cells, and other types (Fig. 4). The NA fibers end along the subepithelial zone alongside the plasma cells. Some peptidergic nerve fibers are localized near immunologically important sites. A plexus of VIP fibers was found by Ottaway adjacent to the postcapillary venules, the site of entry of T lymphocytes bearing VIP receptors into the gut; these fibers probably play a roll in the ingress and retention of T lymphocytes in GALT. Stead and colleagues found substance P and CGRP-containing nerve fibers adjacent to mast cells in the gut, with preferential terminations on these cells. Physiological evidence suggests that these neuropeptides can modulate the release of mediators from mast cells, and mast cell-derived mediators can stimulate the primary afferents electrically.

IV. Functional Roles of Neurotransmitters in the Immune System

A. *In Vitro* Studies of Norepinephrine and Catecholamines

NE and various NA agonists and antagonists have been incorporated into a number of immunological assay systems in an attempt to determine the role of NE in the immune system. The results from these studies are often difficult to interpret due to the wide variability in experimental design, the frequent use of pharmacological rather than physiological doses of drugs, and the unclear relevance of such assays to the living organism in which the lymphoid cell is exposed to a complex local microenvironment. However, a few patterns of action have emerged from these studies.

In general, studies of mitogen-induced proliferation of both T and B lymphocytes have demonstrated that NE, epinephrine, and isoproterenol are catecholamine agonists that act on beta-adrenoceptors and reduce proliferation, an effect that can be blocked by beta-adrenergic antagonists. However, low concentrations of NE in the 10^{-9}–10^{-7} M range can act on alpha-adrenoceptors (blockable by phentolamine) and increase proliferation, particularly if the beta-adrenoceptor effect is blocked. The beta-adrenoceptor decrease in proliferation is mediated through stimulatory G proteins in lymphocytes that increase intracellular cyclic adenosine monophosphate (cAMP). Increases in cAMP in lymphocytes can decrease the synthesis and secretion of IL-2, suggesting a possible mechanism for diminished proliferation. Other possible mechanisms include decreased expression of IL-2 receptors and decreased ability of the effector cell to respond to IL-2.

Effects of NE on the ability of cytotoxic cells to lyse target cells has been reported to increase or decrease following incubation with NE. Incubation of NK cells with NE prior to the addition of target cells results in increased killing activity, whereas the presence of NE throughout the entire incubation period results in a dose-dependent decrease in NK activity.

Epinephrine or isoproterenol can decrease phagocytosis by macrophages and can decrease killing of viral-infected cells and tumor cells by interferon-gamma-activated macrophages. These effects probably are mediated through cAMP. Secre-

tion of IL-1 by macrophages has been reported to either increase or decrease after the addition of adrenergic agents, depending on the source of the macrophages.

The conflicting results from these *in vitro* studies make it difficult to predict what effect adrenergic agents should have on more complex responses, such as T-dependent antibody responses. Based on these *in vitro* data, one might be inclined to predict that NE would decrease primary antibody responses, because it appears to decrease many of the processes involved in the response, such as phagocytosis, IL-2 secretion, and proliferation. However, in a series of *in vitro* studies by Sanders and Munson, NE was reported to enhance the plaque-forming cell response to sheep red blood cells (RBC) if NE was present during the first 6 hr of culture. This increase was blocked by the beta-antagonist propranolol, but not the alpha-antagonist phentolamine. The effects of NE on this response were additionally complicated by an alpha-2-receptor-mediated (clonidine) decrease in the response seen on day 5 in culture, and an alpha-1-receptor-mediated (methoxamine) increase in the response unmasked on day 4 if propanolol blockade of the beta-adrenoceptor effect occurred. Adoptive transfer studies in mice showed the same directional results, with increases in the plaque-forming cell response to sheep RBC if cells were removed from a mouse treated with epinephrine or incubated with epinephrine for 1 hr before transfer into syngeneic irradiated recipients that were then immunized with sheep RBC. These more complex *in vitro* studies are in general agreement with our results obtained from denervation of the spleen and lymph nodes by chemical sympathectomy (reported below). However, what clearly emerges from these studies is the notion that adrenergic effects on immune responses are not simple unidirectional changes that are inhibitory in nature. Even though mitogen-induced proliferation is diminished by NE, primary antibody responses are generally enhanced.

B. *In Vivo* Studies of NA Neural Influences on Immune Responses

The most careful and detailed *in vivo* studies have utilized chemical sympathectomy with 6-hydroxydopamine (6-OHDA), an agent that is taken up into NA nerve terminals by the high-affinity uptake carrier and destroys the terminals. Not only is NE removed via this sympathectomy, but so are co-localized neuropeptides. The alternate approach, surgical ganglionectomy, removes NA innervation and colocalized neuropeptides, other peptidergic neurons, and nerve fibers, perhaps of sensory origin, that are merely passing through the ganglion on the way to their target.

Denervation studies using 6-OHDA for chemical sympathectomy of secondary lymphoid organs can be classified into two general categories: (1) denervation of rodents at birth with 6-OHDA, or denervation with surgical ganglionectomy in adults, generally results in augmented primary and secondary antibody responses; and (2) acute chemical sympathectomy in adult rodents results in diminished primary and secondary antibody responses. In adult mice, this diminution of primary antibody responses is 80% for splenic node response following systemic challenge with antigen and 97% for popliteal lymph node response following footpad challenge. This effect can be prevented by preincubation with desmethylimipramine, a tricyclic inhibitor that prevents uptake of 6-OHDA and protects the NA terminals, suggesting that the effect is not due to toxic action of 6-OHDA and requires uptake of the agent into NA nerve terminals to produce the alteration. As an additional control, propranolol treatment was carried out for several hours immediately following administration to block the surge of release of NE that inevitably occurs when NA terminals are damaged and release their neurotransmitter; this treatment had no effect on the altered response, suggesting that the diminished antibody responses were not due to acute stimulation by NE following excess release.

Further studies from our laboratories have shown that chemical sympathectomy in adult rodents results in suppression of delayed-type hypersensitivity responses to contact sensitizing agents (by approximately 50%), diminished cytotoxic T-lymphocyte responses to alloantigen accompanied by decreased IL-2 production (by approximately 50%), and enhanced NK cell activity *in vivo* and *in vitro*. B-lymphocyte proliferative responses were augmented from denervated inguinal lymph nodes, consistent with an inhibitory role for NE on proliferation, but T-lymphocyte responses in these nodes and other sites, and B-lymphocyte proliferative responses in spleen and other lymph nodes, were highly complex and sometimes were diminished or showed no response. Removal of some of the other colocalized neuropeptides possibly complicated the

response compared with blockade of beta-adrenoceptors in culture.

These findings suggest that acute removal of sympathetic nerves causes dysregulation of immune function, but not in a simple fashion. For example, chemical sympathectomy leads to an apparent disinhibition of B-lymphocyte proliferation from lymph nodes that one might hypothesize would lead to increased antibody production following immunization. Instead, the primary immune response is virtually abrogated by chemical sympathectomy, probably as a result of effects at an early step in the immune response, such as antigen recognition, processing, or presentation. These observations, and the sometimes conflicting results in the literature, lead to the suspicion that many processes and cell types are affected by NE in different ways, depending on the concentration of NE available to those cells, the other mediators present in the specific microenvironment at that time, and the state of activation of the cells at the time of investigation. In addition to these complications, timing of the exposure to signal molecules may be important. For example, agents that increase intracellular cAMP in B lymphocytes *in vitro* generally diminish antibody production by decreasing the synthesis and secretion of IL-2; however, a transient increase in cAMP is important in the induction of antibody production.

Finally, an interesting connection exists between the integrity of NA nerves in lymphoid organs and the onset of autoimmune disease. 6-OHDA-induced sympathectomy in neonatal rats has been demonstrated to increase the severity of T lymphocyte-mediated experimental allergic encephalomyelitis. Administration of catecholamine agonists can reduce the severity of symptoms in this same disease. Beta-adrenoceptor antagonist administration to humans for anti-hypertensive therapy coincides with the development of systemic lupus erythematosis, an autoimmune disease, in a small subset of patients. In mouse models of autoimmune disease studied in our laboratory, decreased innervation in lymph nodes has been found prior to the expression of the autoimmune disease (in NZB mice and (NZB × NZW)F1 mice compared with NZW control mice). We also have found that MRL mice homozygous for the lpr gene, resulting in fatal lymphoproliferation, have substantially less splenic NE than age-matched heterozygous littermates. Because adrenergic agents are readily available and can be administered with minimal adverse responses, perhaps a neurotransmitter-based approach to symptoms of autoimmune disorders in humans may be possible in the future. [*See* Autoimmune Disease.]

C. Effects of Neuropeptides on Immune Responses

As noted above, peptidergic nerves (NPY, VIP, substance P, CGRP, somatostatin, etc.) have been identified in the thymus, spleen, lymph nodes, and areas of GALT. Lymphocytes, or subsets of lymphocytes, possess receptors for most of these neuropeptides and can generate intracellular second-messenger responses following stimulation; these intracellular second messengers frequently are utilized by cytokines to achieve an effect on lymphocytes, suggesting the possibility of dual signaling into the same intracellular systems. In keeping with neuropeptides acting as neurotransmitters with lymphocytes as targets, several of the more thoroughly studied neuropeptides have been shown to have functional effects on the immune system.

Substance P has actions in several target cells that directly or indirectly stimulate inflammation. Substance P enhances vascular permeability and increases local vasodilation, both of which contribute to the ability of lymphocytes to migrate into the area. Substance P receptors have been identified on both T (helper and cytotoxic/suppressor subsets) and B lymphocytes, and on macrophages. Substance P is a T-lymphocyte mitogen and can enhance T-cell proliferation to lectins. It also enhances concanavalin A-induced IgA production by lymphocytes from mesenteric lymph nodes, spleen, and Peyer's patches. Substance P also enhances macrophage phagocytosis and chemotaxis of segmented neutrophils. Substance P has been proposed as an important mediator in the expression of rheumatoid arthritis. Adjuvant-induced arthritis can be reduced in severity by removal of substance P with capsaicin, whereas stimulation of these nerves or injection of substance P can exacerbate inflammation and the severity of arthritis. Stress also has been observed to exacerbate rheumatoid arthritis, and NE also may play a role. β-2-adrenergic blockade or NE depletion delay the onset of experimental arthritis and reduce joint injury. Therefore, both substance P and NE may act in concert in this condition.

Somatostatin appears to be inhibitory to many of the actions of substance P. It inhibits the release of substance P from the peripheral terminals of primary afferent nerves. Somatostatin also has a direct receptor-mediated effect on lymphocytes and monocytes; it inhibits PHA-induced human T-lymphocyte mitogenesis, suppresses endotoxin-induced leukocytosis, and suppresses the release of colony-stimulating factor activity by splenic lymphocytes.

VIP also appears to inhibit a variety of immune functions. It is found in peripheral nerves in the thymus, spleen, lymph nodes, and GALT. VIP decreases mitogen-induced proliferation of T lymphocytes but has no effect on B lymphocytes. T lymphocytes possess high-affinity receptors for VIP, related to intracellular cAMP stimulation, generally inhibitory to proliferation in T cells. These VIP receptors on T lymphocytes can be downregulated by occupancy and internalize considerably more rapidly than they dissociate or are reinserted on the membrane. Ottaway incubated T lymphocytes in VIP to downregulate the receptors and then injected these cells back into the host; these T lymphocytes failed to home to Peyer's patches and mesenteric lymph nodes but migrated properly to spleen, intestine, liver, and lungs. The internalization of the VIP receptor altered the interaction of these lymphocytes with the specialized high endothelium on the postcapillary venules where lymphocyte ingress occurs. Thus, T-lymphocyte trafficking into Peyer's patches and mesenteric lymph nodes requires the expression of high-affinity receptors for VIP.

The availability of NE and numerous neuropeptides in specific compartments of lymphoid organs suggests that a lymphoid cell in such a compartment may be exposed to a unique combination of neurally derived mediators. An *in vitro* investigation of lymphokine-activated killer (LAK) cell activity and proliferation suggests that such presence of multiple mediators may be extremely important to optimal functioning of the immune system. LAK cell activity is stimulated by IL-2. The maximal effect of a specific concentration of this cytokine can be augmented considerably by the presence of somatostatin and beta-endorphin, both of which are synergistic with each other as well as with IL-2. This synergistic effect is countered or blocked by prostaglandin E2 and by agents that stimulate intracellular cAMP directly or indirectly. Thus, neuropeptides may augment, and NE may inhibit, maximal IL-2-stimulated LAK cell activity directed toward specific tumor or viral-infected targets. As these cells and IL-2 have been injected into humans experimentally as cancer chemotherapy, such synergistic effects are not merely of academic interest, particularly in view of the severe toxicity of IL-2. We suggest that it might be possible to reduce the amount of IL-2 used to stimulate LAK cell activity if somatostatin and an opioid peptide are used to synergize the response, endomethacin and propranolol are used to block elevations in intracellular cAMP, and glucocorticoids are withheld to prevent upregulation of receptors on the LAK cells, which could be detrimental to the synergistic effects of the neuropeptides and IL-2. It may be possible to exploit the interactions of cytokines and neurotransmitters for therapeutic benefits in the future, thereby reducing the considerable toxicity, including neurotoxicity, that many cytokines exert. [See INTERLEUKIN-2 AND THE IL-2 RECEPTOR.]

V. Cytokine Interactions with Neurons

Studies of the effects of immunization have demonstrated that during an immune response some soluble mediators can alter electric activity in neurons in key hypothalamic sites, such as the dorsomedial and paraventricular nuclei, and can alter NE and serotonin levels and turnover in the hypothalamus and limbic sites. These affected regions are the same sites that regulate neuroendocrine and central autonomic outflow to the immune system, thereby establishing complete loops of communication between the nervous and immune system. [See HYPOTHALAMUS.]

Recent investigations with IL-1 have demonstrated that both peripherally injected and intracerebroventricularly injected IL-1 can produce electrical or neurochemical effects similar to those seen with immunization, suggesting that IL-1 may be one mediator of such effects. In addition to its classical effect on thermogenesis and induction of slow-wave sleep, IL-1 enhances the turnover of NE in the hypothalamus and enhances the secretion of CRF, thereby elevating ACTH and glucocorticoids. This action may explain the elevation of glucocorticoids that was observed by Besedovsky and colleagues at the peak of an immune response. It is not yet clear whether IL-1 also acts at the pituitary to release anterior pituitary hormones in addition to its effect on releasing factor cells in the hypothalamus. It also is not clear whether IL-1 can cross the blood–brain barrier in sufficient concentrations to produce these effects directly, can cross at circum-

ventricular organs and along pial sleeves around the vasculature, or can initiate a secondary response by other mediators that in turn achieve these effects in hypothalamus and other central nervous system (CNS) sites. Extremely low concentrations of IL-1 in the lateral ventricles, in the 10-femtomolar range, have been reported to alter peripheral NK cell activity and immune responses, apparently via autonomic outflow to the spleen and perhaps other organs.

IL-1 has been reported to exist in hypothalamic neurons in humans and may act within the CNS as a neurally derived mediator. IL-1 immunoreactivity also has been reported in sympathetic (presumably NA) nerves in the spleen. Microglia are capable of synthesizing IL-1 when stimulated by such agents as interferon-gamma. Recently, activated T lymphocytes have been shown to cross through the blood–brain barrier and traverse the brain, providing a possible source for such stimulatory or activating cytokines. A breakdown in the blood–brain barrier also would permit access to the brain by such cytokines, thereby permitting the microglia to synthesize IL-1, upregulate MHC antigens, and present antigen to initiate an immune response in the CNS. [See BLOOD–BRAIN BARRIER.]

Other cytokines, such as IL-2, IL-4, IL-6, and tumor necrosis factor, have been reported to influence neural or glial responses. Thus apparently numerous immune-derived mediators may provide molecular sensory signaling to the CNS. Recent therapeutic administration of interferons and interleukins for cancer have resulted in severe central side effects, including depression, confusion, and other cognitive or affective side effects, suggesting that some cytokine signaling might profoundly alter behavior and higher functions of the nervous system. [See CYTOKINES IN THE IMMUNE RESPONSE.]

We have only begun to scratch the surface of interactions of mediators of the immune system with the nervous system and interactions of neural mediators with the immune system. However, an important conceptual understanding has emerged: These two systems are in intimate contact with each other, share mediators with each other, and produce signal molecules that may interact with each other. It is no longer possible to view the immune system as an autonomous self-regulated system, and it is not longer possible to ignore immunological mediators when considering neural responses and behavior. Perhaps this field of neural-immune interactions has opened the door for a better understanding of common principles of signal molecules and will lead to a unified conceptual understanding of their actions on target cells, including immunocytes and neurons.

Bibliography

Ader, R., Felten, D. L., and Cohen, N. (1989). Interactions between the brain and the immune system. *Ann. Rev. Pharm. Tox.* **30,** 561–602.

Ader R., Felten, D. L., and Cohen N. (1990). Psychoneuroimmunology, 2nd Ed. Academic Press, San Diego.

Berczi, I. (1986). "Pituitary Function and Immunity." CRC Press, Boca Raton.

Blalock, J. E. (1989). A molecular basis for bidirectional communication between the immune and neuroendocrine system. *Physiol. Rev.* **69,** 1–32.

Bost, L. K. (1988). Hormone and neuropeptide receptor on mononuclear leukocytes. *Prog. Allergy* **43,** 68–83.

Carlson, S. L., and Felten, D. L. (1989). Involvement of hypothalamic and limbic structures in neural-immune communication. *In* "Neuroimmune Networks: Physiology and Diseases" (E. J. Goetzl and N. H. Spector, eds.), pp. 219–226. Alan R. Liss, New York.

Dinarello, C. A. (1989). Interleukin-1 and its biologically related cytokines. *Adv. Immunol.* **44,** 153–205.

Dunn, A. J. (1989). Psychoneuroimmunology for the psychoneuroendocrinologist: A review of animal studies of nervous system–immune system interactions. *Psychoneuroendocrinology* **14,** 251–274.

Felten, D. L., and Felten, S. Y. (1989). Innervation of the thymus. *In* "Thymus Update," M. D. Kendall, and M. A. Ritter, London. Harwood Academic Publishers, p. 73–88.

Felten, D. L., Felten, S. Y., Bellinger, D. L., Carlson, S. L., Ackerman, K. D., Madden, K. S., Olschowka, J. A., and Livnat, S. (1987). Noradrenergic sympathetic neural interactions with the immune system: Structure and function. *Imm. Rev.* **100,** 225–260.

Felten, S. Y., and Felten, D. L. (1990). The innervation of lymphoid tissue. *In* "Psychoneuroimmunology," 2nd Ed. R. Ader, D. L. Felten, and N. Cohen. Academic Press, San Diego.

Felten, S. Y., Felten, D. L., Bellinger, D. L., Carlson, S. L., Ackerman, K. D., Madden, K. S., Olschowka, J. A., Livnat, S. (1988). Noradrenergic sympathetic innervation of lymphoid organs. *Prog. Allergy* **43,** 14–36.

Fredrickson, R. C. A., Hendric, H. D., Hingtgen, J. N., and Aprison, M. H. (1986). "Neuroregulation of Autonomic, Endocrine, and Immune Systems." Martinus-Nijhof, The Hague.

Goetzl, E. J., Sreedharan, S. P., and Harkonen, W. S. (1988). Pathogenetic roles of neuroimmunologic mediators. *Immunol. Allergy Clin. of North Am.* **8,** 183–200.

Goetzl, E. J. (Ed.) (1985). Supplement on neuromodulation of immunity and hypersensitivity. *J. Immunol.* **135.**

O'Dorisio, M. S., Wood, C. L., and D'Dorisio, T. M. (1985). Vasoactive intestinal peptide and neuropeptide modulation of the immune response. *J. Immunol.* **135,** 792s–796s.

Payan, D. G., McGillis, J. P., Renold, F. K., Mitsuhashi, M., and Goetzl, E. J. (1987). The immunomodulating properties of neuropeptides. *In* "Hormones and Immunity" (I. Berczi and K. Kovacs, eds.), pp. 203–214. M.T.P., Norwell, Massachusetts.

Smith, E. M. (1988). Hormonal activities of lymphokines, monokines, and other cytokines. *Prog. Allergy* **42,** 121–139.

Weigent, D. A., and Blalock, J. E. (1987). Interactions between the neuroendocrine and immune systems: Common hormones and receptors. *Imm. Rev.* **100,** 79–108.

Wybran, J. (1986). Enkephalins as molecules of lymphocyte activation and modifiers of the biological response. *In* "Enkephalins and Endorphins," (N. P. Plotnikoff, R. E. Faith, A. J. Murgo, and R. A. Good, eds.), pp. 253–282. Plenum Press, New York.

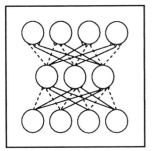

Neural Networks

LEIF H. FINKEL, *University of Pennsylvania*

I. Biologically-Based Networks
II. Applied Neural Networks
III. Conclusion

Glossary

Distributed representation Scheme by which a population of units can represent a stimulus property, such as orientation, velocity, or depth, in which each unit responds over a limited range of stimulation (e.g., to lines oriented between 85° and 95°)

Network System consisting of nodes and connections between the nodes in which the state of a node depends on signals it receives along its connections

Neural unit Individual processing unit in a neural network, modeled with greater or lesser accuracy after physiological properties of real neurons

Synaptic rule Formal rule for altering the efficacy of network connections, thereby changing the probability of one unit activating another unit

A NEURAL NETWORK is a system of interconnected, excitable, neuronlike units. Units can be activated by inputs from other units in the network, by external inputs, or by intrinsic processes. Based on these inputs, each unit generates an output that is transmitted to all units to which it projects. The efficacy of connections in transmitting inputs and outputs can be modified, usually in an activity-dependent manner, and various rules have been proposed to govern these synaptic modifications. The most widely used rule (the Hebb rule) modifies connection efficacies according to the correlation of the excitation in the two connected units. Neural networks can act to generate, transform, or recognize patterns of activity, to store information, and to perform learning. Networks can be simulated on computers or directly implemented in hardware such as passive electrical circuits, VLSI chips, or optical components.

In biologically based networks, the anatomical and physiological properties of individual units are closely modeled after real neurons. These model neurons typically have many compartments, each of which may contain several species of simulated ion channels, transmitter receptors, and associated biochemical machinery. In applied neural networks, the units are greatly simplified and abstracted. For example, the so-called McCulloch–Pitts formal neuron computes a weighted sum of its inputs, subtracts a threshold value from the result, and generates one of two binary output states, 1 or 0, depending on whether the result is positive or negative. Most recent networks employ a continuous sigmoidal output function in place of the McCulloch–Pitts step function. In both biologically based and applied networks, the temporal evolution of network activity occurs in discrete time elements, with each unit in the network evaluated either serially or randomly.

The major characteristics that distinguish current neural network models are (1) the anatomical architecture of the network, (2) the physiological properties of the individual units, and (3) the rules used for modifying the efficacy of the interconnections between units. Biologically based networks have as their primary motivation the understanding of basic principles of brain function. Applied neural networks seek to use some of these principles to perform useful tasks such as categorization, memory storage, or optimization. We will review key examples of both types of network and briefly discuss some of the more intriguing applications that have been recently developed.

I. Biologically-Based Networks

The concept of a neural network can be traced back to the British associationist philosophers and to early work by Nicholas Rashevsky. In the 1950s, Beurle; Cragg and Temperly; and others began to study the dynamical properties of networks of interconnected excitable elements. However, in the last two decades, stimulated by advances in neuroscience and by the advent of high-speed computers, activity in the area of neural network research has exploded. Starting with David Marr's (1969) model of the cerebellum, many neural systems have been modeled including, for example, those responsible for vision and flight control in the housefly, swimming in the lamprey, auditory localization in the barn owl, and eye movements in the monkey. Networks have been used to study the developmental problems of how neural maps are created (e.g., the map of the retina onto the optic tectum), and how the internal structural organization of these maps arises (e.g., the generation of ocular dominance and orientation columns in the visual cortex). However, the greatest concentration of modeling effort has been applied to studying different regions of the cerebral cortex. [*See* CORTEX.]

One approach to cortical modeling has been to study the simplest type of cortex, the archicortex, which is composed of only three layers (instead of the six layers of the neocortex). Both the olfactory cortex (responsible for smell) and the hippocampus (implicated in long-term memory) have been modeled. The difficulty in these studies is that the nature of the sensory inputs that activate this ancient cortex are not well defined. Nonetheless, significant progress has been made in understanding the dynamics of the intrinsic cortical circuitry.

Recently, a detailed network model of a region of mammalian hippocampus has been developed. The network consists of approximately 10,000 neural units, each of which incorporates realistic models of sodium, potassium, and calcium ion channels. Based on recent anatomical information, the units are very sparsely interconnected—each unit contacts less than 3% of the other units in the network. Despite this sparse connectivity, rhythmical firing patterns emerge (the hippocampus displays a prominent firing rhythm known as the theta rhythm at 4–8 firings per second). The interesting aspect of this emergent activity pattern is that no unit fires regularly at this frequency—individual units typically fire much more sporadically. Rather, the rhythm is only displayed by the population of units as a whole; thus, it is a true network phenomenon. This model also predicts how loss of inhibition in the hippocampus can give rise to epilepsy; in fact, the model produces epileptiform activity, which exactly mimics that found in humans and monkeys. [*See* HIPPOCAMPAL FORMATION.]

Recently, efforts have been directed at modeling the piriform cortex of the rat, which is the first cortical processing station for olfactory information. In computer simulations, the waves of rostral-to-caudal activity that are observed *in vivo* have been reproduced. Simulations show that these oscillatory waves arise from two types of inhibition (feedforward and feedback) and from a temporal anisotropy in rostral versus caudal connection velocities. [*See* OLFACTORY INFORMATION PROCESSING.]

Another study has modeled how the piriform cortex can categorize olfactory signals into recognized smells, even when multiple odorants are presented together (e.g., one can smell the brewed coffee in a restaurant kitchen). In this simulation, the network performs a temporally distributed analysis of the "odor-scene," with early responses signifying the general class of the odorant (fruity, salty, aromatic, etc.) and later responses of the same network signaling the individual identity of the odorant (orange, grapefruit, etc.). The theta rhythm, which is also displayed by olfactory cortex (and which is an optimum frequency for inducing long-term potentiation in the hippocampus), is, in fact, the frequency at which rodents sniff the environment. Thus, these models serve as conceptual links between sensory categorization, behavior, and memory.

Work on archicortex is closely related to network studies of associative memory (see below) because the neural architectures are organized in a distributed fashion. The neocortex, in contrast, is predominantly organized into well-defined topographic maps in which adjacent regions of cortex respond to adjacent regions of the environment (or to closely related movements in the motor cortex). The somatosensory cortex, for example, which is responsible for the perception of touch, contains a map of the entire body skin surface. Recent physiological data have shown that this map is, in fact, *dynamic* and undergoes continual change and reorganization due to the ongoing tactile inputs received by the skin. A recent network simulation of monkey somatosensory cortex accounts for the major features of this map plasticity. In this model, the network consists of roughly 1,500 units that receive topographically organized projections from receptor sheets representing the front and the back of the hand. The network contains excitatory and inhibitory units, which are interconnected so that focal cortical stimulation yields a pattern of short-distance excitation and long-distance inhibition (this dichotomy is critical to maintain network stability). Network stability (freedom from oscillation, explosion, or dampening of activity) is also fostered by the use of shunting inhibition (a type of inhibition in which currents are effectively short-circuited at the cell body); similar effects of shunting inhibition have been found in other simulation studies. Excitatory connections in the network are modifiable according to a voltage-dependent synaptic rule. The result of these synaptic modifications is that for a wide range of stimulation protocols, the network organizes itself into neuronal groups—local collections of cells with strong mutual connections and which all share similar functional properties. The network also develops a topographic map of the receptor surfaces. Perturbations of stimulation to these receptors causes the maps to reorganize in patterns that closely correspond to *in vivo* results. The simulations show how representational plasticity can emerge from a biologically based synaptic modification rule embedded in a network with simplified anatomical and physiological properties. [*See* SOMATOSENSORY SYSTEM.]

No area of the nervous system has received more attention from neural network modelers than the visual cortex. Physiological studies have shown that in higher animals the visual cortex is composed of multiple, distinct areas (up to two dozen areas in the rhesus monkey). Each of these visual areas is functionally specialized for particular visual tasks, although there is substantial overlap in the properties of different areas. Thus, for example, area V1 (the *first* visual area) performs a high-resolution analysis of the visual scene, area V4 (the fourth area) is specialized for color and texture vision, and area V5—also called MT, as it is located in the medial temporal cortex—is specialized for discriminating motion. Models have been developed for each of these visual processes. Stereopsis is one such process in which the slight horizontal disparity in the views seen by the two eyes is transformed into a perception of three-dimensional depth. Early network models focused on the problem of determining the exact correspondence between individual points seen by the two eyes. However, recent physiological evidence has shown that disparity-

sensitive cells in the visual cortex are tuned for three broad classes of disparity (which correspond to objects located nearer, farther, and in the plane of visual fixation). A more recent model of the representation of disparity in the cortex is physiologically based and investigates the basis of distributed representations. In this model, each cell only crudely codes for the actual disparity of an object; however, the population of cells, taken together, is able to discriminate extremely fine-depth discontinuities, such as arise in hyperacuity experiments. (The visual system can resolve minute discontinuities that are far smaller than the spacing between cells in the retina.) Such population codes are probably used throughout the nervous system. For example, in the monkey, there appears to be a distributed representation of arm movements in the motor cortex and of eye movements in the superior colliculus. The disparity model exemplifies how relatively simple computational models can elucidate complex principles of brain operation. [*See* VISION (PHYSIOLOGY); VISUAL SYSTEM.]

The problem of motion detection has also motivated a number of network models. Most such models use a mechanism of temporally-delayed inhibition in which excitation at a spatial location (location 1) inhibits activity from an adjacent location (location 2) a short time later. If an object travels from location 1 to location 2 within that time span, it will not result in visual activation due to the inhibition. However, if the object travels in the reverse direction, in some oblique direction, or at a significantly different velocity, it will produce activation. In this manner, a population of cells can represent the direction of motion of an object.

Recently, a model has been proposed of how single cells in the retina (ganglion cells) could perform such a computation. The temporally delayed inhibition comes about through a mechanism of shunting inhibition, which acts to veto any excitatory inputs that synapse more distally on the dendrite. In this way, individual cells will display a preferred direction of motion, provided that the connections to a cell are anatomically arranged so that the geometrical distance relationships in space are preserved by the relative order of inputs to the dendrite. Physiological measurements of cat retinal ganglion cells have been used to make detailed simulations of motion detection by individual cells. [*See* RETINA.]

Given that the visual cortex splits the analysis of the visual scene among multiple areas, the question arises of how the visual "picture" is put back together again in a coherent, unified manner. In one model, this integration takes place dynamically through an ongoing process of signaling along the extensive anatomical connections between the cortical areas. This model has been tested in one of the largest network simulations ever developed (containing nearly $\frac{1}{4}$ million units and over 8 million synaptic connections). The simulation, which is based on three interconnected areas of the macaque visual cortex, shows how network mechanisms allow multiple areas to work together. One network mechanism leads to the resolution of conflicts between areas; because the areas receive inputs conveying different information about the scene, such conflicts are unavoidable. Another mechanism involves recycling the outputs of higher areas back to lower areas, which allows responses to complex objects of scenes to be built up reiteratively. A final integration mechanisms allows the outputs of one area to be used by others for their own operations. For example, information about object motion, which area MT discriminates, can be used by another area to discern structure-from-motion, a cue to depth. In fact, the same area that generates structure-from-motion based on inputs from MT can also generate responses to illusory contours based on inputs from another area (V1), which signal information about line terminations. Thus, an area can perform similar operations on very different inputs and, in doing so, can use different cues to detect a common process.

One of the outstanding problems that remains to be solved is how collections of interacting networks generate behavior that allows the animal to interact with the world (and therefore, to alter the information that it receives from the world). One approach to this problem has been to simulate an entire organism with a nervous system, body form, and behavioral repertoire. An automaton, called Darwin III, has been constructed and has demonstrated an ability to perform some rudimentary categorization and generalization tasks. This automaton is based on a general theory of brain function called "Neural Darwinism" which holds that the nervous system operates as a selective system (analogous to the operation of natural selection in evolution.) Neural Darwinism is a global model of neural function from the molecular level to consciousness. Such theories are necessary to bridge the gap from individual neural processes to an understanding of how psychology emerges from neurobiology.

All these biologically based simulations share several characteristics. They are primarily motivated by a goal of understanding something about how the nervous system works. They are based, with greater or lesser verisimilitude, on anatomical and physiological data. They do not include any mechanisms that are biologically unfeasible, such as requiring long times to converge to a result or only working in rigidly defined anatomical networks. Their results are framed in biological terms, and they generate experimental predictions. However, none of these models is able to handle complex scenes or extended time periods, or is capable of being dramatically scaled-up in numbers of neurons. In addition, many of the models are so complicated that it is sometimes not clear how they do what they do. For these reasons, a significant effort has recently been expended in developing simplified network models that are not biologically based, although they do incorporate certain neurallike features. These networks, to which we now turn, are much easier to analyze and have already accomplished some rather remarkable feats.

II. Applied Neural Networks

Over the last 20 years, a number of different applied neural networks have been used to perform pattern recognition, associative memory, and learning tasks. Networks have been proposed that act as transformation matrices to separate sets of input vectors into orthogonal sets of output vectors. In one of these simulations, a network could "remember" pictures of several hundred human faces. When presented with part of one individual face (e.g., the top half), the network would generate the missing portion of the correct face.

These studies, which used techniques derived from linear algebra, were transformed when it was shown how certain classes of associative networks could be treated with the formalism used in statistical physics to describe spin glasses. The networks considered are rather severely constrained—they must be symmetric (i.e., if neuron A is connected to neuron B, then B must also send a connection to A with the identical synaptic strength). This is a totally unbiological assumption; however, it guarantees a very strong learning result. Namely, if the synaptic strengths are modified according to a correlation rule, such as the Hebb rule, then the network will monotonically converge to a learned state. In the Hopfield formalism, the network is viewed as a dynamical system in a high-dimensional space, and learning consists of creating "attractors," which are activity states of the network that attract other activity states. Thus, if the network receives some arbitrary activation from a set of inputs, the activity will rumble around for a bit but will eventually converge on one of these attractors. One can imagine memory space as a kind of evolving golf course, with little hills and valleys. When you create a memory, you make a new hole, and if the "golf ball" is placed anywhere near this hole and is jostled around a bit, it will eventually fall into the hole.

One problem with this scheme is that the system can get caught in local minima (e.g., the golf ball can get stuck in a little gully, preventing it from reaching the hole). A technique for overcoming this problem, called simulated annealing, has been borrowed from materials science where an analogous problem occurs in making metal alloys. The technique consists of heating the alloy to a high temperature, and then letting it cool very slowly. This would correspond (somewhat forcing the analogy) to giving the golf ball a lot of kinetic energy (having it move fast), and then gradually slowing the ball down. The extra energy allows the system to escape from local minima and to find the global minimum. A network called the Boltzmann machine uses simulated annealing to converge to learned responses. The Boltzmann machine has been used to perform figure-ground separation problems, such as distinguishing the inside of a maze from the outside.

A label that has been applied to much of current neural network research is "connectionism." Connectionism means that the advanced functions of the nervous system, such as pattern recognition, memory, speech, etc., can only be carried out in distributed networks. In other words, mental representations are stored in patterns of activity and any memory or action is distributed over populations of cells. This view is opposed to that of artificial intelligence in which both symbols and rules for manipulating these symbols are assumed to be *directly* represented. One particularly active area of connectionist research is parallel-distributed processing (PDP), in which extremely simple networks are used to perform complex tasks. A typical PDP network is shown in Figure 1 and consists of three layers of cells: an input layer, an output layer, and an intermediate (or "hidden") layer. The network operates in a feedforward manner; each cell is connected to all cells in the next higher layer. A special training procedure, known as back-propagation, is used to adjust the connection weights. In back-propagation, a number of different stimuli are presented to the input units, which activate the hidden

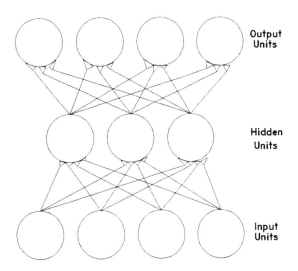

FIGURE 1 A typical PDP network architecture featuring three layers of cell units. Cell units are indicated by large circles, lines between units indicate feedforward connections, and small triangles indicating the synaptic connections. Biological networks typically feature recurrent connections and connections between units at each level (not present in this figure).

units, which in turn, activate the output units. For each input stimulus, the activity of the output units is compared to the desired output activity pattern, and an error signal is computed (error = desired activity − actual activity). This error signal is used to adjust the connection weights of the network such that if a unit is active and the error signal is positive, the weight is increased; otherwise, if the unit is active and the error signal is negative, the weight is decreased. The amount of the change depends on the size of the error (given by an equation known as the delta rule). This rule is a close relative of the Perceptron learning rule introduced by Rosenblatt in the early 1960s. The major difference is that back propagation can be applied to multilevel networks in a step-by-step process. Thus, back-propagation solves the so-called "credit-assignment" problem, which is the determination of the role played by each unit in a multilevel network in arriving at an output.

There is a major difference in how connection strengths are modified in the back-propagation algorithm and in the biologically based networks considered above. Back-propagation and related methods have been called "supervised" learning, and are based on a "teacher" that tells the system whether its response to each stimulus was correct or incorrect. In "unsupervised" learning, no such information is available. For example, in the Hebb rule, synaptic changes occur as a result of correlations in activity patterns *regardless* of the behavioral consequences of these correlations. The Hebb rule and other unsupervised schemes work because the correlations in the system ultimately reflect correlations in the stimulus world. In general, unsupervised schemes deal with lower-level processes (such as early sensory categorization), and implicitly defer any external teaching to higher cognitive tasks. For example, complex behaviors such as talking, reading, or baking a cake require an external teacher. Whereas discriminating the orientation of lines or visually tracking a moving object require exposure to these stimuli, but do not explicitly require an external reward system.

One spectacular example of the power of back-propagation is given by the network NETtalk, which learns to read. The network is presented with written text, such as a book or newspaper, and after training on a number of examples, learns to translate the written letters into phonemes, which a voice synthesizer pronounces out loud. The NETtalk network consists of 309 units (203 input units [29 groups of 7 units], 26 output units, and 80 hidden units, each of which is connected to all the input and output units) and 18,629 connections. At the start of the training procedure, all connection strengths among units in different layers are given random values. As sequences of letters and spaces are presented to the input array, the output units produce patterns of activity that correspond to 1 of 79 possible phonemes. The difference between the correct output pattern and that actually obtained is then fed back to the earlier layers and used to modify the connection strengths of the network via the back-propagation algorithm. After tens of thousands of training cycles, the network achieves excellent accuracy and is able to generalize to read words it has never before seen. Most interestingly, the hidden units develop an organization in their patterns of response. For example, certain hidden units respond only to vowels, others only to consonants, and the response to related vowels (e.g., the "a" in b<u>a</u>ke and the "e" in b<u>e</u>g) is more similar than to unrelated vowels. NETtalk thus illustrates how a network can learn a system of complex and detailed rules, and how these rules can be represented in a distributed fashion.

Applied neural networks have also been used to solve several computationally difficult problems. The dynamical properties of some networks can be used to solve optimization problems such as the traveling salesman problem (a salesman must visit *N* cities and wants to find the shortest circuit that allows him to visit each city once and only once). The network finds a good solution quickly; however, it usually does not find the absolute optimal solution. Moreover, linear programming techniques are still the method of choice for such problems. However, neural networks do offer the hope of being able to rapidly perform difficult computations (based on the possibly fallacious promise that the brain must perform similar calculations). Networks have also recently been used to solve such diverse problems as how to balance a pencil on its point, how to read handwriting on bank checks, how to efficiently manage office systems, and how to teach other neural networks.

III. Conclusion

Neural networks offer powerful insights into how the nervous system may represent and manipulate information. Applied neural networks have recently made great strides in solving difficult computational problems and promise to gain wide application in diverse areas of society. When network techniques are used in conjunction with physiological and psychophysical studies, it is possible to consider systems level questions of how regions of the nervous system function together. The great challenge for the next decade is to integrate computational studies with basic neuroscientific research to uncover the basic principles of brain function.

Bibliography

Anderson, J. A., and Rosenfeld, E. (1988). "Neurocomputing: Foundations of Research." MIT Press, Cambridge, Massachusetts.

Cowan, J. D., and Sharp, D. H. (1987). "Neural Nets." Los Alamos Research Publication LA-UR-87-4098, U.S. Government Printing Office.

Eckmiller, R., and von der Malsberg, C. (1988). "Neural Computers." Springer-Verlag, Berlin.

Edelman, G. M. (1987). "Neural Darwinism." Basic Books, New York.

Hanson, S., and Olsen, C. (1990). "Connectionism in Neuroscience: The Developing Interface." MIT Press, Cambridge, Massachusetts.

Koch, C., and Segev, I. (1989). "Methods in Neuronal Modeling." MIT Press, Cambridge, Massachusetts.

MacGregor, R. J. (1987). "Neural and Brain Modeling." Academic Press, New York.

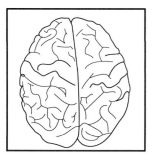

Neuroendocrinology

JOSEPH B. MARTIN, *University of California, San Francisco*

Glossary

ACTH Adrenocorticotropic hormone; a pituitary hormone that stimulates secretion of glucocorticoids from the adrenal cortex

Adenohypophysis Anterior portion of the pituitary that secretes six principal trophic hormones that act on target glands such as the thyroid, adrenal, and the gonads

Amenorrhea Prolonged abnormal cessation of menses

Corticotropin-releasing hormone Hypothalamic hormone that regulates the secretion of ACTH

Cushing's disease Oversecretion of adrenal glucocorticoids caused commonly by an ACTH-secreting pituitary tumor

Galactorrhea Inappropriate secretion of milk from the breast, which can occur in both men and women

Gonadotropin-releasing hormone Hypothalamic hormone that regulates anterior pituitary secretion of luteinizing hormone and follicle-stimulating hormone

Growth hormone–releasing hormone Hypothalamic hormone that stimulates the secretion of growth hormone from the anterior pituitary

Hypothalamus Region of brain below the thalamus that regulates secretion of pituitary hormones

Neurohypophysis Posterior portion of the pituitary, which releases vasopressin and oxytocin

Somatostatin Hypothalamic hormone that inhibits the secretion of thyrotropin and growth hormone

Thyrotropin-releasing hormone Hypothalamic hormone that stimulates the secretion of thyrotropin and prolactin

NEUROENDOCRINOLOGY encompasses study of the relation between the nervous and endocrine systems, which act in concert to regulate many of the metabolic and homeostatic activities of the organism. Neuroendocrinology includes understanding anatomic relations between the hypothalamus and the pituitary gland, the chemistry of the hypothalamic peptides and biogenic amines vital to physiologic regulation, and appreciation of the mechanisms of feedback regulation whereby hormones secreted by glands such as the adrenal, gonads, and thyroid act back on the pituitary and brain.

I. Introduction

Neuroendocrine mechanisms are involved in the regulation of reproduction, growth and differentiation of tissues, water and salt balance, and a number of behaviors. Disorders in neuroendocrine regulation can result in deficits in reproduction and alterations in intellectual function and in exacerbation of other metabolic diseases such as diabetes. Overproduction of hormones by the pituitary can lead to serious clinical disturbances such as Cushing's disease caused by oversecretion of cortisol by the adrenal, excessive growth caused by hypersecretion of growth hormone, and infertility caused by increased secretion of prolactin. Clinical disturbances may also arise from underproduction of hormones caused by destructive lesions (e.g., tumors arising in either the pituitary or the hypothalamus).

II. Hypothalamic–Pituitary Unit

The basic functional unit of the neuroendocrine system is the hormone secretory cell. Secretory cells can be classified into three types: *exocrine*, which secrete into a lumen that connects to the exterior of the body (e.g., sweat glands); *endocrine*, which secrete directly into the blood (e.g., insulin from the pancreas); and *neurosecretory*, which refers to release of a hormone into the circulation from a nerve terminal. Nerve cells in the hypothalmus perform two important neurosecretory functions. In the first case, cells making vasopressin and oxytocin secrete hormones directly into the systemic blood for distribution to distant sites of action. In the second case, hypothalamic peptides are secreted from nerve terminals in the base of the hypothalamus into a specialized portal circulation to be delivered a short distance away to the anterior pituitary. [*See* ENDOCRINE SYSTEM.]

A. Anatomy of the Hypothalamic–Pituitary System

Visualized from below, the midline ventral surface of the forebrain forms a convex bulge termed the *tuber cinereum*. This region of the brain is attached by a stalk to the pituitary gland. Arising from the tuber cinereum in the midline is the *median eminence*, recognized by its intense vascularity. The median eminence forms the floor of the third ventricle.

The boundaries of the *hypothalamus* are somewhat arbitrary. Its anterior limits are defined as the optic chiasm and lamina terminalis. It is continuous here with the preoptic area, the substantia innominata, and the septal region. Posteriorly, it is bounded by an imaginary plane defined by the posterior border of the mammillary bodies ventrally and by the posterior commissure dorsally. Caudally, the hypothalamus merges with the midbrain periaqueductal gray and the reticular formation. The dorsal limit of the hypothalamus is defined by the hypothalamic sulcus on the medial wall of the third ventricle. At this junction, the hypothalamus is continuous with the subthalamus and, above it, the thalamus. Laterally, the hypothalamus is bounded by the internal capsule and the basis pedunculi. [*See* HYPOTHALAMUS.]

B. Pituitary Gland

The pituitary gland, or *hypophysis*, lies in close proximity to the medial basal hypothalamus and is partially enclosed in humans by a bony cavity called the *sella turcica*. The pituitary is divided into two lobes, the anterior, or *adenohypophysis*, and the posterior, or *neurohypophysis*. The adenohypophysis develops from Rathke's pouch, an evagination extending upward toward the brain from the primitive buccal ectoderm, and consists of three parts. The *pars distalis* is the primary source of the anterior pituitary hormones, adrenocorticotropic hormone (ACTH), thyroid-stimulating hormone (TSH), growth hormone (GH), prolactin (PRL), and the two gonadotropins, luteinizing hormone (LH) and follicle-stimulating hormone (FSH). Other peptide hormones and growth factors have been purified from anterior pituitary tissues, but their precise physiological functions remain uncertain. The second part of the anterior pituitary is the *pars intermedia*, which is vestigial in adult humans but is well-defined in the fetus and in lower mammals such as rodents. It secretes other hormones such as melanocyte-stimulating hormone, which has an important function in the regulation of skin pigmentation. The third component of the adenohypophysis is the *pars tuberalis*, which consists of secretory cells similar to those of the pars distalis that envelop the upper pituitary stalk extending over the surface of the median eminence. Its physiological function remains unclear. [*See* PITUITARY.]

The neurohypophysis develops as an inferior extension of the hypothalamus, carrying with it, as it grows into the sella turcica, the nerve terminals of hypothalamic neurons that synthesize and secrete vasopressin and oxytocin. The neuronal cell bodies within the hypothalamus that project to the neural lobe are readily visible in microscopic sections by their large size (*magnocellular neurons*) and by their staining for neurosecretory material. They are clustered in two pairs, one located adjacent to the third ventricle, named the *paraventricular nuclei*, and the second straddling the optic tract, called the *supraoptic nuclei*. The supraoptic- and paraventricular-neurohypophysial nerve fibers are unmyelinated; they course through the median eminence, entering the pituitary stalk to terminate on blood vessels in the posterior lobe.

III. Median Eminence

It should be apparent from these descriptions that the *median eminence* is an important and complex organ that serves several functions. In the first place, it is the principal site of termination of the hypothalamic neurons that regulate the anterior pituitary. These neurons are widely dispersed within the hypothalamus but tend to be particularly concentrated along the margins of the third ventricle. They produce the hypothalamic regulatory peptides that induce secretion (or inhibit it) of hormones in the anterior pituitary. Secondly, the median eminence contains in its deeper layers the axons of nerve cells directed toward the posterior pituitary, which produce the peptides vasopressin and oxytocin. In addition, the median eminence is covered on its ventricular margin by specialized ependymal cells that separate the median eminence from the cerebrospinal fluid. Finally, the median eminence capillaries are fenestrated, permitting access of peripheral blood to the nerve terminals of the hypothalamic-adenohypophysial system.

Some of the cells of the mesial paraventricular nuclei also end in the median eminence; vasopressin can thereby also reach the anterior pituitary, where it has several functions, the most important of which is to stimulate ACTH secretion.

A. Pituitary Portal Blood Supply

The arterial blood supply of the hypothalamus and pituitary are intimately linked. Arterial branches from the internal carotid penetrate the layers of the median eminence to form a capillary bed that lies in direct contact with neurosecretory nerve terminals. The blood is then collected into pituitary portal veins that traverse the pituitary stalk to form a secondary capillary network bathing the secretory cells of the anterior pituitary. This so-called hypotha-lamic-pituitary portal circulation is the mechanism by which hypothalamic regulatory peptides reach the anterior pituitary to either stimulate or inhibit secretion of pituitary hormones. The posterior pituitary receives a second blood supply from other branches of the internal carotid that directly penetrate the posterior lobe. Into these capillaries are secreted the hormones of the posterior lobe, vasopressin and oxytocin. Although the arterial blood supply varies somewhat in its detail in different species, it is similar in all mammals in being derived from branches of the internal carotid.

B. Hypothalamic Releasing Hormones

Functional control of the hypothalamus over hormonal secretion by the anterior pituitary is exerted by hormones (or factors), most of which are small-molecular-weight peptides that reside in nerve fibers in the median eminence. These hypothalamic factors have been structurally characterized and are summarized in Table I. The release of some of the anterior pituitary hormones appears to be regulated by a single hypothalamic factor: thus the gonadotropins depend on gonadotropin-releasing hormone (GnRH) for stimulation of their release. If GnRH is absent during development, sexual functions fail altogether and puberty fails to emerge. In the adult, the development of GnRH deficiency can lead to secondary failure of sexual function with infertility and deficient secretion of sexual steroid hormones (testosterone in the male and estrogen and progesterone in the female). In the case of the other pituitary hormones, a more complex set of regulatory factors is involved. ACTH secretion is stimulated by corticotropin-releasing hormone (CRH) but also by vasopressin. GH secretion is stimulated by growth hormone–releasing hormone (GRH) and is inhibited by somatostatin. Thyrotropin (TSH) is stimulated by thyrotropin-releasing hormone (TRH)

TABLE I Hypothalamic Hormones Active in Anterior Pituitary Regulation

Hypothalamic hormone	Structure	Action
Thyrotropin-releasing hormone	Tripeptide	Stimulates TSH and PRL secretion
Corticotropin-releasing hormone	41-amino acids	Stimulates ACTH secretion
Gonadotropin-releasing hormone	Decapeptide	Stimulates LH and FSH secretion
Growth hormone–releasing hormone	44-amino acids	Stimulates GH secretion
Somatostatin	14-amino acids	Inhibits GH and TSH secretion
Dopamine	Catecholamine	Inhibits PRL secretion
Vasoactive intestinal polypeptide	28-amino acids	Stimulates PRL secretion
Vasopressin	9-amino acids	Stimulates ACTH secretion

and inhibited by both somatostatin and dopamine. PRL secretion is tonically inhibited by the hypothalamus via the actions of dopamine, but stimulation can also occur by effects mediated through TRH and oxytocin.

With the availability of pure synthetic hypothalamic releasing hormones, it has been possible to develop methods to study their precise localization within neuronal subdivisions of the hypothalamus. TRH, CRH, and somatostatin are principally found within neurons of the medial paraventricular nucleus (designated the parvicellular periventricular nucleus). GRH, dopamine, and GnRH are found mainly within neurons situated immediately above the median eminence in the infundibular or arcuate nucleus.

The overall regulatory mechanisms required to achieve homeostasis are based on the ability of the brain to sense the circulating levels of the peripheral hormones. This is achieved primarily by negative feedback on the brain and, in some instances also on the pituitary, by hormones released by target organs. Thus the thyroid hormones (thyroxine and triiodothyronine), the adrenal hormone cortisol, and the gonadal sex steroids each acts to suppress further secretion of the tropic hormone that stimulates its production. As levels of cortisol rise, for example, the hypothalamus secretes less CRH, and the pituitary becomes more resistant to the effects of CRH. As cortisol levels fall, this inhibitory effect lessens and ACTH secretion increases.

A few physiological observations further complicate our understanding of the system regulation of the brain-endocrine-target gland axis. One factor is the finding that all hormonal secretions are pulsatile and episodic rather than steady state. Although the full physiologic rationale for this observation remains conjectural, it can be argued that fine tuning of regulation is achieved more accurately by pulsed signals than by steady-state outputs. Another exceedingly important aspect of neuroendocrine regulation is the influence of circadian rhythms on endocrine function. The most apparent of these daily rhythms is that described for ACTH and cortisol. In humans, levels of ACTH rise during the late night hours to peak at about the time of awakening. They decline throughout the day to their lowest point in the early evening and then rise again. The secretion of hormones in a circadian fashion is regulated by paired nuclei located immediately above the optic chiasm in the anterior hypothalamus called the *suprachiasmatic nuclei*. These nuclei have been shown to direct a host of circadian rhythms in addition to their influence on hormones, including motor activity cycles, feeding and drinking behavior, and body temperature regulation. Disruptions of these structures experimentally or after brain lesions such as tumors result in disturbances of circadian functions—a so-called free-running state.

IV. Neuropharmacology of Pituitary Hormone Secretion

In a conceptual sense, the neurosecretory neurons of the hypothalamus can be likened to the motor neurons of the brainstem and spinal cord, which serve as the final common pathway for initiating motor functions by their connections to the muscles. In the case of the neuroendocrine system, the neurosecretory system is the recipient of incoming information from many brain areas, which integrate responses appropriate to the organism. In terms of the paradigms of "fight" or "flight," for example, it is possible to delineate specifically the hormonal responses accompanying these behavioral reactions. The output system comprising the hypothalamic-pituitary axis transduces these neural responses into hormonal outputs. Other examples of a more chronic nature also serve to illustrate this point. The young woman who first discovers the vulnerability of her menstrual cycle to dysregulation after embarking on the new experience of leaving home for the first time to attend college has experienced the temporary shutdown of the hypothalamic drive essential for reproductive capacity. The psyche in this instance can indicate its perturbation by a dramatic interruption of a previously well-coordinated function. The interruption is almost always temporary but in a few instances may be followed by prolonged abnormalities of menstrual regulation (amenorrhea).

These neural influences are mediated over a number of different pathways, many of which are relayed via systems that use neurotransmitters such as dopamine, norepinephrine, serotonin, acetylcholine, and γ-aminobutyric acid. These effects are mediated over long pathways in the brain because, at least in the case of the biogenic amines (serotonin, epinephrine, dopamine, and norepinephrine), the cell bodies containing them are located in the brainstem. The effects of the biogenic amines can be demonstrated by the disruptions of pituitary hormone secretion occurring with pharmacologic manipulations. For example, administration of drugs that interfere with dopamine synthesis, release, or receptor attachment is sometimes followed by striking rises in PRL secretion, which in turn may be manifest by the inappropriate secretion of milk from the breast (galactorrhea) and by suppression of gonadotropin secretion resulting in amenorrhea and infertility. Fortunately, these effects are almost always reversible, resulting in restoration of the normal reproductive cycle with discontinuation of drug administration. Drugs that can induce this kind of effect include antihypertensives such as alphamethyl DOPA, antipsychotic drugs such as chlorpromazine and haloperidol, and in some instances, the antidepressants such as amitriptyline or imipramine.

The pharmacology of pituitary regulation can be demonstrated in another manner. All pituitary hormones are released in a pulsatile, bursting pattern, which appears to be caused by the phasic release of hypothalamic releasing hormones acting to drive secretion of anterior pituitary hormones. This pulsatile mechanism appears to be generated by mechanisms that in some cases, particularly involving GH and the gonadotropins, reside in systems regulated by the biogenic amines. One important area that continues to receive great attention is the disruption of neuroendocrine regulation that accompanies psychiatric illness (*vide infra*). [*See* NEUROPHARMACOLOGY.]

V. Regulation of Individual Pituitary Hormones

A. Neurohypophysial Hormones

Vasopressin and oxytocin are small peptides containing nine amino acids differing from each other by two residues. Each is synthesized within the cell bodies of different magnocellular neurons as preprohormones. Each is associated with a distinct neurophysin whose function is to transport the peptide within neurosecretory granules to nerve terminals in the posterior pituitary or the median eminence. Nerve terminal depolarization results in release of both the peptide and the associated neurophysin.

The peripheral actions of vasopressin are exerted on the collecting tubules of the kidney to induce reabsorption of water. In the absence of vasopressin, dilute urine is excreted, leading to stimulation of thirst as plasma osmolality falls. The syndrome of polyuria and polydipsia resulting from a deficiency in vasopressin is called *diabetes insipidus,* to be distinguished from the much more common association of polyuria that occurs with increased glucose present in the urine in diabetes mellitus.

Oxytocin acts on smooth muscle cells of the uterus during parturition and upon the mammary ducts to bring milk from the breast glands to the nipple. What function oxytocin serves in men is unclear, although it can be detected in both pituitary and in blood.

B. Regulation of Vasopressin Secretion

Neurons of the anterior hypothalamus are capable of sensing fluctuations in blood osmolality and signaling vasopressin (also referred to as *antidiuretic hormone*)–containing neurons to increase or decrease their firing rates appropriately to either increase or decrease the amounts of vasopressin secreted into the blood. Blood osmolality is zealously guarded over a narrow range around a setpoint of about 280 mOsm/kg. It can be shown that vasopressin release is elicited with a rise in osmolality to about 287 mOsm/kg. The system rapidly responds to these fluctuations, and urinary volume is appropriately increased or decreased to maintain narrow plasma osmolality levels. A common clinical problem in very ill, bedridden patients is the syndrome of inappropriate vasopressin secretion in which small increases in plasma vasopressin levels result in inappropriate water retention and a fall in blood osmolality and in serum sodium concentrations. The result can be seizures, coma, and permanent neurological injury.

Other factors are also important in regulating vasopressin secretion. Hemorrhage or decreased blood volume from any cause, if sufficiently severe, can stimulate massive secretion of vasopressin. Receptors for volume regulation are located in the left atrium and in baroreceptors of the carotid sinus. Although vasopressin also can stimulate rises in blood pressure, the levels required to accomplish this effect are believed to be greater than those ordinarily found in humans. Nevertheless, the possibility has been suggested that vasopressin may have central effects within the brain to regulate blood pressure control systems.

Stress and nausea are also able to stimulate the release of vasopressin. Inputs from the gastrointestinal tract and other viscera relayed via the glossopharyngeal and vagal nerves activate neurons in the nucleus of the tractus solitarius of the medulla oblongata. The signals are then relayed to the hypothalamus to trigger vasopressin release. It is also a well-known experience that stress can inhibit vasopressin secretion, resulting in diuresis. [See PEPTIDE HORMONES OF THE GUT.]

C. Adenohypophysial Hormones

Pituitary secretion of ACTH is both episodic with alterations in hormone secretion occurring over minutes and circadian with levels changing throughout a 24-hr period, resulting in increased levels during the night and falling levels during the daytime. This rhythm is among the most robust of the various circadian rhythms and persists even with change in the daily sleep-wake cycle (as occurs, for instance, in jet travel) for 7–10 days. The disruptions in mental and physical performance that accompany jetlag are believed to be the result primarily of the temporal disruption of the hypothalamic-pituitary ACTH adrenal rhythm (i.e., the rhythm persists in an inappropriate relation to the activity cycle for about 7–10 days).

The hypothalamic regulation of CRH secretion is influenced by a number of central nervous system neurotransmitters including norepinephrine, serotonin, and acetylcholine. These transmitters are speculated to have a role in the "stress-associated" responsiveness of CRH secretion. Indeed it is the activation of this system by both psychological and physical stresses that are most frequently used to define the very essence of "stress" itself. Disorders of the regulatory setpoint of the system are known to occur in about one-half of all patients with severe depression, and abnormalities in suppression of ACTH secretion is tested by the use of dexamethasone, a synthetic highly potent steroid. The dexamethasone suppression test (DST) is widely applied in the investigation of depressed patients as a biologic marker of the depressed state. Abnormalities characterized by failure of ACTH or cortisol suppression usually revert to normal after recovery of the patient (i.e., the abnormality is state-dependent rather than trait-dependent).

Abnormalities of hypersecretion of ACTH also occur in Cushing's disease, usually caused by a small benign tumor of ACTH-secretory cells located in the pituitary. The symptoms include weight gain, particularly of the body, caused by excessive deposition of fat, thinning and easy bruising of the skin, and psychological changes including depression. The diagnosis is confirmed by demonstrating increased secretion of cortisol in the urine and by failure of ACTH to suppress with small doses of dexamethasone. The treatment is surgical, with removal of the tumor resulting in resolution of the symptoms in more than 60% of patients.

D. Regulation of Gonadotropin Secretion

Fundamental to the survival of any species is the ability to propagate its own. This capacity is determined by the coordination of sexual functions defined both by successful and appropriate copulatory activities and by hormonal readiness of the partners. This complex set of behaviors and hormonal regulation is accomplished by the neuroendocrine axis. The hypothalamus develops the capacity at the time of puberty to release GnRH episodically, first at night and then throughout the 24-hr circadian cycle. This awakening of the capacity for reproduction can be detected by sleep-time monitoring of LH and FSH secretion. Gonadal steroids rise in response to the gonadotropins, and secondary sex characteristics begin to develop. Full reproductive capacity appears at about 12–13 years in the female with the appearance of menarche (first menstrual period). In the male, puberty occurs slightly later. The same hormones are made in both sexes both at the hypothalamic level (GnRH) and at the pituitary (LH and FSH). The ovaries in the female produce estrogen and progesterone, which act in a negative feedback loop to reduce further secretion of the gonadotropins. In the male, testosterone production by the testis elicits a similar negative effect on the pituitary. The complexities involved in the hormonal changes of the menstrual cycle are beyond the subject of this review. [See PUBERTY; STEROIDS.]

Disorders of reproductive function can occur at several levels. Precocious development of sexual characteristics (precocious puberty) can result from hypothalamic tumors that alter the normal constraint imposed by the brain on endocrine maturation releasing pathways that prematurely trigger development. However, destructive lesions of the hypothalamus or tumors that destroy functions of the pituitary can prevent the normal appearance of puberty. These conditions occurring before puberty and resulting in failure of reproductive function in the woman are called *primary amenorrhea*. They are also sometimes referred to as *secondary hypogonadism* as opposed to a primary defect located in the gonads themselves. In the adult, pathological processes at either the hypothalamic or pituitary level can result in secondary failure of sexual function (secondary amenorrhea in the female) or secondary hypogonadism in the male. Hormonal replacements with GnRH, which must be administered frequently and by a systemic route, have been successful in restoring sexual competence in both primary and secondary forms of amenorrhea and in male infertility caused by hypothalamic or pituitary failure.

An interesting syndrome of hypogonadism caused by excessive PRL secretion has also been identified. Inappropriate circulating levels of prolactin are able to diminish hypothalamic secretion of GnRH resulting in amenorrhea (often accompanied by galactorrhea caused by PRL effects on stimulating milk production) in women. In men, elevated PRL as, for example, occurring in pituitary tumors that secrete prolactin can also lead to hypogonadism caused by gonadotropin deficiency caused by a similar mechanism of suppression of GnRH secretion. A common accompanying symptom is impotence.

E. Regulation of Prolactin Secretion

The hypothalamic regulation of PRL secretion differs from that of all the other pituitary hormones by its inhibitory nature. The secretion of dopamine from cells of the arcuate nucleus in the base of the hypothalamus acts on dopamine receptors (class D_2) on the pituitary lactotrope cell membrane to suppress both the release and the biosynthesis of PRL. This tonic inhibition persists throughout most of the day but is interrupted during the night when pulsatile bursts of PRL occur, resulting in elevation in PRL before morning awakening. PRL rises during the third trimester of pregnancy to prepare the mammary gland for lactation after parturition. PRL continues to be secreted after delivery with stimulation evoked by mechanical stimulation of the nipple of the breast as the infant suckles.

Abnormal secretion of PRL can occur after administration of therapeutic drugs that interfere with dopamine metabolism. Drugs that have this effect include some antihypertensive drugs like alphamethyl DOPA, and drugs such as the psychotropic agents chlorpromazine and haloperidol that block dopamine receptors. Abnormal PRL secretion also occurs from pituitary adenomas that synthesize abnormal amounts of PRL. These are the most com-

mon of all pituitary tumors. They can arise in either men or women at all ages. In women they are often recognized early because of the associated amenorrhea and galactorrhea. In the case of men they usually grow to a size sufficient to cause visual disturbances or severe headaches before coming to diagnosis. Associated symptoms of impotence and infertility are common manifestations in men of the hyperprolactinemic state.

F. Growth Hormone

Normal somatic growth, which requires the actions of GH, is mediated by growth factors (somatomedins) produced in peripheral tissues such as the liver under the stimulation of the hormone. The secretion of GH requires stimulatory actions by hypothalamic GRH. The output of GH is pulsatile with four to seven pulses of secretion occurring during a 24-hr period, the largest of which is sleep-entrained evident about 2 hr after onset of sleep. GH secretion can be completely inhibited by somatostatin. It is believed based on extensive experimental analysis of pulsatile GH secretion that the coordinate hypothalamic secretion of GRH and somatostatin are required for normal secretion. Somatostatin arising from cell bodies in the anterior hypothalamus seems to determine the rhythm of the pulses, whereas GRH determines their timing and magnitude.

Abnormalities of GH secretion occurring as a result of defective production of GRH during childhood leads to growth failure (dwarfism). This can be the result of an inherited defect in the gene that regulates GH biosynthesis in the anterior pituitary or to destructive lesions of either the hypothalamus or pituitary that disrupt the effects of either GRH or of GH secretion. Treatment of growth failure with human GH obtained from cadaver pituitary glands has been suspended after recognition that the slow virus causing a fatal neurological disease (Creutzfeldt-Jakob disease) could be transmitted to subjects. Fortunately, recombinant DNA technology has led to the *in vitro* production of GH, which can be substituted for the purified material. At the present time, several thousand children are receiving recombinant GH for treatment of GH deficiency.

Excessive pituitary GH secretion, usually resulting from the formation of a pituitary adenoma that secretes GH, causes increased growth (gigantism). In the adult, hypersecretion leads to a clinical syndrome of broadened hands and fingers (acromegaly) and to altered metabolism including diabetes mellitus, hypertension, and in long-standing cases, heart disease. The treatment of pituitary tumors is surgical, accomplished by a procedure called *transsphenoidal partial hypophysectomy,* in which the pituitary tumor is approached from the nasal cavity through the sphenoid sinus. Successful treatment of early cases can be achieved in more than 75% of cases.

G. Thyroid-Stimulating Hormone

Thyroid-stimulating hormone regulation is achieved primarily via the effects of TRH. In its absence, TSH secretion falls to less than 10% of normal. TSH acts on the thyroid to stimulate synthesis and secretion of the two thyroid hormones thyroxine (T_4) and triiodothyronine (T_3). These hormones act on most tissues in the body to maintain appropriate levels of general metabolic activity. Both hormones can act on the hypothalamus and pituitary to inhibit secretion of TRH and TSH, respectively. However, it appears that the principal site of feedback regulatory control is exerted at the level of the pituitary.

Diminished output of TSH can occasionally follow on large destructive lesions of the hypothalamus, particularly those affecting the anterior regions. More commonly, destructive lesions of the pituitary such as nonfunctional large pituitary tumors result in deficiency of TSH. It should be recalled that hypothyroidism is most often the result of primary thyroid disease. Excessive thyroid function (hyperthyroidism) is almost always due to Grave's disease, a defect caused by hypersecretion of thyroid hormone induced by an immunologic disorder in which antibodies mimic the effects of TSH on the thyroid. In this condition, TSH levels are very low or completely undetectable. Pituitary tumors rarely secrete TSH or the gonadotropins, likely because they consist of different molecular structures than ACTH, GH, and PRL. The former are glycoproteins, each comprised of two separate subunits: an alpha subunit common to all three and a beta subunit that is unique for each and that imparts the specific biological effects of TSH, LH, and FSH. [*See* THYROID GLAND AND ITS HORMONES.]

Bibliography

Martin, J. B., and Reichlin, S. (1987). *Clinical Neuroendocrinology.* Contemporary Neurology Series, 1–759. 2nd ed. F. A. Davis Company, Philadelphia.
Hökfelt, T., Johansson, O. and Goldstein, M. (1984). Chemical anatomy of the brain. *Science* **225,** 1326.
Black, P. Mc.L., Zervas, N. T. Ridgway, E. C., and Martin, J. B. (1984). Secretory tumors of the pituitary gland. 1–400. New York, Raven Press.
Müller, E. E., and Nisticö, G. (1989). Brain messengers and the pituitary. Academic Press, New York.

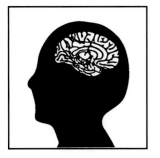

Neurology of Memory

LARRY R. SQUIRE, *University of California San Diego and Veteran Affairs Medical Center*

Glossary

Anterograde amnesia Loss of the ability to learn
Diencephalic midline Area of the diencephalon of the brain surrounding the ventricles, including the medial thalamus and hypothalamus
Magnetic resonance imaging Noninvasive technique for visualizing the structure of living brain
Medial temporal lobe Inner area of the temporal lobe away from the lateral surface, including the hippocampal formation, amygdala, parahippocampal gyrus, entorhinal cortex, and perirhinal cortex
Neocortex Recently evolved outer region of the brain, which occurs as a large thin sheet of highly infolded tissue

LEARNING IS the process of acquiring new information, and memory refers to the persistence of learning in a state that can be revealed at some time after learning is completed. Memory is localized in the brain as physical changes produced by experience. Damage to specific brain regions causes amnesia (i.e., loss of the ability to acquire new information and impaired recall of recently acquired information). A major principle of the organization of memory in the brain is the distinction between conscious and unconscious memory systems. The ability to store and use conscious memories depends on the integrity of the hippocampus and adjacent, anatomically related structures.

I. Memory as a Topic for Neuroscience and Psychology

Before the technological developments of neuroscience, memory was a problem studied primarily by psychologists. In the past few decades, however, the successes of basic neuroscience have made it possible to study the nervous system in increasing detail and to obtain information about cellular and molecular events within single neurons. It has thus become possible to investigate memory at many levels of analysis, from the analysis of synaptic plasticity to the analysis of brain systems and cognition.

The neurology of memory can be advantageously investigated by combining the approaches of neuroscience and psychology. This is because many of the important questions about the neurology of memory are directed at a relatively broad, global level of analysis, in which the subject matter of neuroscience and psychology significantly overlap. What are the learning processes and memory systems whose neurobiological mechanisms we want to understand? Is there just one kind of memory or are there many? How and where are memories represented? What are the brain systems and connections involved in memory, and what jobs do they do?

II. Memory Storage

The brain is highly specialized and differentiated, organized so that separate regions of neocortex simultaneously carry out computations on separate features or dimensions of the external world (e.g., the analysis of visual patterns, location, and movement). Although the evidence is not yet definitive, memory for an event, or even for something as apparently simple as a single visual object, is considered to be stored in component parts, which are linked to the specific processing areas that are ordinarily engaged during learning. That is, memory is stored in the same neural systems that ordinarily participate in perception, analysis, and processing. Stated differently, memories are stored as outcomes of processing operations and in the same cortical regions that are involved in the analysis and perception of the items and events to be remembered. This leads to the idea that memory is *localized* in the sense that particular brain systems represent specific aspects of each event, and it is *distributed* in the sense that many neural systems must participate in representing a whole event. [*See* LEARNING AND MEMORY.]

III. Memory Dysfunction

A. Anterograde Amnesia

Having emphasized this close association between information processing and information storage, it is important to appreciate that there are brain structures and connections that are not themselves permanent repositories of information but that nevertheless play an essential role in the formation and establishment of enduring memory. These structures lie within the medial temporal lobe and the diencephalic midline. Bilateral damage in either of these regions causes an amnesic syndrome (i.e., a global impairment in the ability to acquire new memories regardless of modality, and a loss of some memories, especially recent ones, from the period before amnesia began). The memory deficit occurs in the absence of deficits in perception or other intellectual functions.

The fact that circumscribed amnesia can occur at all indicates directly that memory is not inevitably and inextricably linked to the processing that precedes memory. If it were, amnesia could not occur as it does against a background of intact intellectual functions, intact immediate memory, intact personality and social skills, and intact ability to recall memories acquired long ago. The point is that there are brain structures that when damaged cause amnesia and that are especially important for memory functions.

The deficit in amnesia is easily detected with tests of paired-associate learning or delayed recall. For paired-associate learning, one presents pairs of unrelated words (e.g., army—table) and later asks for recall of the second word of each pair on presentation of the first. For delayed recall, one presents information to be remembered (e.g., a passage of text, a list of words, or geometric design) and, after a delay of several minutes, asks the subject to reproduce the originally presented material. The deficit is present regardless of the sensory modality in which information is presented and regardless whether memory of a prior occurrence is tested by unaided recall, by recognition (e.g., presenting alternatives and asking the subject to choose the previously encountered one), or by cued recall (e.g., asking the subject to recall an item when a hint is provided). Moreover, the memory impairment involves not just difficulty learning about specific episodes and events that occur in a certain time and place, but also difficulty in learning factual information.

B. Preserved Learning

It is now appreciated that memory is not a unitary mental faculty but depends on the operation of many separate systems. Some of the most compelling evidence for this idea is the finding that amnesia spares certain important kinds of learning and memory. Among the kinds of learning that can be accomplished *normally* by amnesic patients are perceptuo-motor skill learning, word priming, adaptation-level effects, and simple classical conditioning. Perceptuo-motor skills refer to hand–eye coordination skills (e.g., learning to trace a figure that has been reversed in a mirror). Word priming refers to a change in the facility for identifying specific words caused by recent exposure to the words. Adaptation-level effects refer to changes in judgments about stimuli (e.g., their heaviness or size) that are caused by recent exposure to stimuli of particular qualities. For example, experience with light-weighted stimuli subsequently causes other stimuli to be judged as heavier than they otherwise would. Simple classical conditioning refers to the development of a conditioned response to a previously neutral stimulus as a result of its having been temporally paired with an unconditioned stimulus. For example, after temporal pairing of a tone and an airpuff, the tone itself will come to elicit an airpuff. Amnesic patients can learn to read mirror-reversed text at a normal rate, although they do not learn the words that they are reading. In other words, they

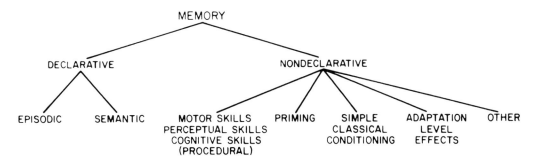

FIGURE 1 Tentative taxonomy of memory. Declarative memory includes episodic and semantic memory and can be retrieved explicitly as a proposition or image. Declarative memory depends on the integrity of brain structures damaged in amnesia. Nondeclarative memory includes skills, priming effects, simple classical conditioning, adaptation-level effects, habituation, sensitization, and perceptual aftereffects. In these cases, learning is nonconscious and can be expressed only through performance. Some forms of nondeclarative memory have sometimes been termed *procedural memory* to capture the idea that learning in these cases occurs as the modification of existing procedures (cognitive operations) or as the creation of new ones.

learn the skill but not the material that was used in the learning of the skill.

Amnesic patients also exhibit intact word priming (i.e., a facilitation in performance caused by recent exposure to words). For example, if the word *baby* is presented, the probability is more than doubled for both normal subjects and amnesic patients that this word will later be elicited by instructions to free-associate a single response to the word *child*. Similarly, presentation of the word *income* produces a strong tendency (about 50%) to produce that word again if the word stem *inc* is given along with instructions to complete the stem with the first word that comes to mind. (The probability that *income* will be produced when it was not presented in the first place is about 10%). These priming effects occur normally in amnesic patients despite the fact that the patients fail conventional memory tasks that ask them to recall recently presented words or to recognize them as familiar. The priming effects illustrated here decline gradually and disappear in a few hours. One view of priming is that these effects operate at a relatively early stage in the analysis of information.

The kind of memory that is affected in amnesia is explicit and accessible to conscious memory. It has sometimes been termed *declarative memory* (also termed *explicit, cognitive,* or *representational*

memory). It can be declared (i.e., brought to mind consciously as a proposition or an image). Declarative memory thus includes the facts, events, faces, and routes of everyday life that comprise conventional memory experiments. By contrast, *nondeclarative memory* (also termed *implicit memory*) comprises a heterogeneous collection of learning and memory abilities, all of which afford the capacity to acquire information implicitly and nonconsciously. For example, in the case of motor skill tasks and perceptual skill tasks (e.g., mirror reading), the knowledge that is acquired cannot be declared. Thus, we do not know what we have learned when we demonstrate a new tennis backhand, at least not in the same sense that we remember the practice sessions themselves. Knowledge of the skill is embedded in procedures and is expressed in performance engaged by the procedures. Skill learning does not require the integrity of the brain structures damaged in amnesia. In the case of priming, already existing cognitive operations are tuned or biased and for a time thereafter can facilitate behavior in a specific way. Thus, the brain has organized its memory functions around fundamentally different information storage systems, some of which are impaired in amnesia and some of which are not (Fig. 1).

C. Retrograde Amnesia

Additional information about the organization of declarative memory comes from the phenomenon of *retrograde amnesia*, which refers to the inability to remember information that occurred before the onset of amnesia. Retrograde amnesia affects both facts and episodes, particularly those that were encountered close to the time when amnesia began. Retrograde amnesia can be relatively brief or quite extensive, but in all these cases it is usually temporally graded. That is, very old (remote) memory is affected less than recent memory. For example, in a middle-aged adult, retrograde amnesia can affect memories from the two or three decades preceding the onset of amnesia and leave relatively intact the memories of childhood and adolescence. Because very old memories are intact, the brain regions damaged in amnesia cannot be the permanent repository of declarative memory. The critical structures damaged in amnesia perform a time-limited function. For a period of time after learning, the storage of declarative memory and its retrieval depend on an interaction between the neural systems damaged in amnesia and memory storage sites located elsewhere. After sufficient time has passed, memories no longer require the participation of these structures. It is possible that memories undergo some lengthy process of reorganization and consolidation whereby they become independent of the structures damaged in amnesia.

IV. Anatomy of Memory

Information about which neural connections and structures belong to the functional system damaged in amnesia comes from two sources: well-studied cases of human amnesia and recent successes at developing an animal model of amnesia in the monkey. New techniques for imaging living brain have also made it possible to detect some kinds of pathological change directly. For example, with magnetic resonance imaging it is possible to detect abnormalities in the hippocampal formation of some amnesic patients. In one carefully studied case of amnesia (R.B.), in which both behavioral and neuropathological data were available, the only significant damage was a bilateral lesion confined to the CA1 field of hippocampus. Thus damage limited to the hippocampus itself is sufficient to cause amnesia. These findings show that the hippocampus is an essential component of the neural system necessary for establishing declarative memory. Cumulative work with animal models suggests that the full medial temporal memory system consists of the hippocampus and adjacent, anatomically related structures including entorhinal cortex, parahippocampal gyrus, and perirhinal cortex (Fig. 2). When several of these structures are damaged together, the severity of amnesia is greater than when only the hippocampus itself is damaged. [See HIPPOCAMPAL FORMATION.]

The critical regions in the diencephalon that when

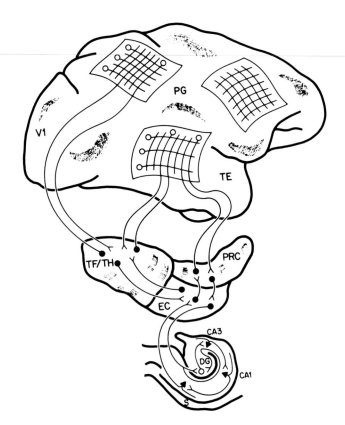

FIGURE 2 Schematic drawing of neocortex together with structures and connections in the medial temporal region believed to be important in transformation of perceptions into memories. Area V1 is the earliest cortical stage for the analysis of visual input. Networks in cortex show putative representations concerning visual object quality (in area TE) and visual object location (in area PG). If this disparate neural activity is to cohere into a stable long-term memory, convergent activity must occur along projections from these regions to the medial temporal lobe. Projections from neocortex arrive initially at the parahippocampal gyrus (TF/TH) and perirhinal cortex (PRC) and then to entorhinal cortex (EC), gateway to the hippocampus. Further processing of information occurs in several stages of hippocampus, first in dentate gyrus (DG) and then in CA3 and CA1 regions. Fully processed input eventually exits this circuit by way of the subiculum (S) and the entorhinal cortex, where widespread efferent projections return to neocortex. Hippocampus and adjacent structures may index or otherwise support development of representations in neocortex, so that, subsequently, memory for a whole event (e.g., representations in both TE and PG) can be revivified even from a partial cue. Damage to this medial temporal system causes anterograde and retrograde amnesia. Severity of the deficit increases as damage involves more components of the system. (This diagram is a simplification and does not show, for example, diencephalic structures involved in memory functions.)

damaged produce amnesia have not yet been identified with certainty. Most likely, the important structures include the mediodorsal thalamic nucleus, the internal medullary lamina, the mammillothalamic tract, and the mammillary nuclei. Damage to the mammillary nuclei alone seems not to be sufficient to produce severe and long-lasting amnesia. Because diencephalic amnesia resembles medial temporal amnesia in many ways, we suppose that these two regions together form an anatomically linked, functional system. It is also possible that the two regions contribute in different ways to the functional system.

Information is still accumulating about how memory is organized and what structures and connections are involved. The disciplines of both psychology and neuroscience continue to contribute to this enterprise. Better understanding of the neurology of memory can be expected to result eventually in quantitative approaches that can provide computer-based models and tests of specific mechanisms. In addition, information obtained at this level of analysis should be directly relevant to neurobiological studies of learning and memory that seek cellular and molecular information. For example, information about the brain systems involved in memory can suggest where to look for cellular and molecular changes and can indicate the functional importance of these changes. Finally, a fundamental understanding of the neurological foundations of memory may lead to better diagnosis, treatments, and prevention of neurological diseases that affect memory.

Bibliography

Squire, L. R. (1987). "Memory and Brain." Oxford University Press, New York.
Squire, L. R., Shimamura, A. P., and Amaral, D. G. (1989). Memory and the hippocampus. In "Neural Models of Plasticity" (J. Byrne and W. Berry, eds.), pp. 208–239. Academic Press, New York.

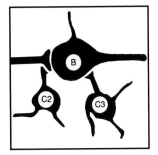

Neuropharmacology

ANTHONY DICKENSON, *University College London*

I. Chemical Transmission
II. Autonomic Nervous System
III. Neuromuscular Function
IV. Transmitters in the Central Nervous System
V. The Future

Glossary

Agonist Chemical that binds to a receptor, initiating a biological response
Antagonist Chemical that binds to a receptor and does not elict a response, but blocks the effects of an agonist
Neurotransmitter Naturally occurring chemical which is released from the terminal into the synaptic gap and acts as a receptor agonist
Receptor Specific binding site on a cell onto which drugs attach to evoke a response (i.e., agonists) or to prevent access of the agonist (i.e., antagonists)
Synapse Gap between a terminal of a neuron and the postsynaptic cell

NEUROPHARMACOLOGY is the study of the actions of drugs on nervous tissue, whether the nervous tissue is the brain or the nerves within the body. The drugs can be the synthetic compounds used for medical or recreational purposes or the natural chemicals that influence nervous tissue as part of the normal functioning of the body. Many of the synthetic drugs given to patients are part of therapies attempting to counter disorders of the natural chemicals. It is obvious, then, that two major challenges of neuropharmacology arise from these dual aspects: (1) Increasing our knowledge of neuropharmacology will inevitably lead to a better understanding of how the nervous system functions in health and disease and (2) therapies or improve-

ments on current therapies will surely follow. Perhaps it is best to put aside the paradox that we are using the neuropharmacological systems within the brain in studying these very systems and to consider instead that chemicals are released from cells in the nervous system, and, by diffusing to other cells, they change activity in parts of the brain; the integrated whole of these events somehow results in human consciousness.

I. Chemical Transmission

It is worth considering first the ways in which nerve cells communicate with each other. The electrical signals, or action potentials, are generated in nerve cells (i.e., neurons) by changes in ionic balance across the membrane of the cell. These action potentials travel from the cell body along the extended process of the cell (i.e., the axon) until they arrive at the specialized end of the axon: the terminal (Fig. 1). The terminal contacts another cell by coming into close proximity to it. It is estimated that a cell can be contacted by several thousand others, as the brain contains several billion neurons. There is, however, no direct contact, and a gap, (the synaptic cleft, which is about 20 nm across) lies between the presynaptic terminal and the postsynaptic cell. Synapses can also be made on other parts of the neurons besides the cell body (i.e., the dendrites, terminals, and axons of neurons). In certain lowly invertebrates a cell communicates with the next neuron in the line by direct electrical events. In higher animals the action potential is transferred via release across the synaptic cleft of a chemical, a process known as chemical transmission. The arrival of the action potential in the terminal depolarizes the membrane, calcium ions flow into the cell, and then a chemical, the neurotransmitter, is released from the presynaptic terminals into the synapse. Neurotransmitters are enclosed in vesicles, and fusion of the vesicle with the membrane of the terminal releases the transmitter. The transmitter then diffuses across the synapse and binds to a specialized part of the postsynaptic membrane: the receptor. This binding process initiates a chain of events in the postsynaptic cell which has been likened to a lock-and-key concept, and the message passes on in the form of another action potential.

The neurotransmitter is synthesized in the neurons from either a simple dietary precursor or a large precursor produced in the cell body in the case of the peptide transmitters. The synthesis is enzymatic and usually consists of a few steps. The active transmitter so produced is then available for release (Fig. 1).

There are many different chemicals acting as transmitters and a greater number of receptors. A transmitter can produce different effects, depending on the receptor type it diffuses to. There is, however, a high degree of specificity, so a transmitter chemical can only activate a limited number of receptor types, in many cases only one or two. The transmitter is then rapidly removed from the synapse or broken down, so turning off the message. The various transmitters used by the nervous system are synthesized in the neuron, from which they are released. Although it was once thought that one neuron contained one transmitter, more than one transmitter per cell now appears to be the rule, a sort of dual-key approach. It is this chemical transmission that allows us to experience the outside world and to respond to our environment in terms of movement, memory, or introspective thought and that forms the basis for nervous system functions ranging from the involuntary control of heart rate to abstract thought. However, it is obvious that higher mental functions have a complexity far beyond our present knowledge, a case in which the relationship between the sum and the parts is hazy. Nevertheless, our knowledge of neuropharmacology provides a good basis for the understanding of neuronal function and a better rationale for therapy.

The receptor for the neurotransmitter is a key target site for synthetic drugs used in clinical medicine. The neurotransmitters bind to their receptor and act as agonists (e.g., they produce an effect by activation of the receptor). This effect, however, can be excitatory (the production of action potentials in the postsynaptic cell), inhibitory (the reduced firing of action potentials), or something between these two extremes (a gentle shift in the

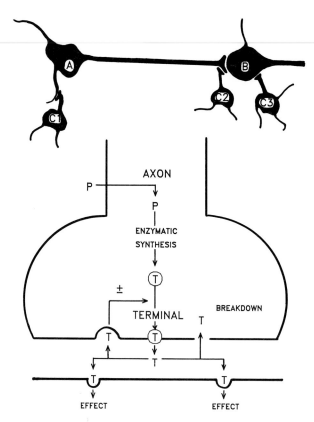

FIGURE 1 The processes of synaptic transmission. The upper portion shows a neuron A contacting the cell body of neuron B via a synapse made by the terminal of the axon of neuron A. C1 is another neuron making a synapse on the dendrites of cell A, C2 synapses onto the terminal of neuron A, and C3 is yet another cell contacting neuron B. Thus, chemical transmission can occur at all of these sites. The lower portion shows an expanded scheme of what happens at a synapse. The precursor (P) is taken up by the cell and converted enzymatically to the transmitter (T), which is then released into the synapse. Effects are produced by binding to receptors in the postsynaptic cells. Autoreceptors on the presynaptic terminal can increase or, more commonly, decrease (±) the transmitter release mechanism. Finally, the transmitter is often taken back up by the terminal prior to breakdown or recycling.

excitability of the neuron, making it more or less susceptible to other inputs). These subtle effects clearly allow flexibility of function within the brain. [*See* NEUROTRANSMITTER AND NEUROPEPTIDE RECEPTORS IN THE BRAIN.]

In addition to these postsynaptic receptors, there are receptors in the presynaptic terminals, so that one transmitter can regulate the amount of another transmitter released, as well as autoreceptors either on the terminal or cell body which allow a transmitter to control its own release. Certain clinically used drugs (e.g., morphine) are agonists; that is, they bind and produce a response at receptors, so enhancing the function of the natural neurotransmitter. Other drugs, however (e.g., those used for schizophrenia), are antagonists. Antagonists bind to the receptor, but produce no activation and also prevent an agonist from binding to the receptor. They therefore block the action of a neurotransmitter. Another target for drugs is the process of removal of the transmitter from the synapse. This removal can occur by reuptake by the terminal or by the breakdown of the transmitter by specific enzymes present in the synaptic cleft. Drugs used to treat depression interfere with transmitter removal processes. Reuptake is most important in the central nervous system, and the transmitter once taken up by the terminal is broken down and becomes available for new synthesis (Fig. 1).

What, then, are the transmitters in the nervous system? It is best to consider the nervous system in two parts, although both influence each other: the central nervous system (i.e., the brain and the spinal cord) and the autonomic nervous system, the controlling system for the unconscious workings of the organs of the body. With the exception of the control of muscles for movement, the autonomic nervous system is the major output path of the central nervous system. [See AUTONOMIC NERVOUS SYSTEM.]

II. Autonomic Nervous System

As discussed the central nervous system is balanced by the activity of inhibitory and excitatory neuronal systems and so, too, is the autonomic nervous system. Two branches—the sympathetic and the parasympathetic—exist, which act in concert to control the body. In both cases the cell body of a first neuron is contained within the central nervous system, and this neuron synapses onto a second cell outside the central nervous system: the ganglionic cell. This second cell then sends an axon to the target organ. Hence, the two neurons involved are termed the pre- and postganglionic cells. The ganglionic cells for both systems use acetylcholine as the transmitter, and the released acetylcholine acts on an acetylcholinergic receptor, termed the nicotinic receptor. However, the postganglionic cells of the parasympathetic nervous system use acetylcholine again, but noradrenaline is the transmitter in the sympathetic nervous system. The noradrenaline

acts on a variety of noradrenergic receptors, and the acetylcholine released from the postganglionic cell binds to a different cholinergic receptor: the muscarinic receptor. [See ADRENERGIC AND RELATED G PROTEIN-COUPLED RECEPTORS.]

The parasympathetic and sympathetic nervous systems innervate the organs of the body (e.g., the heart, lungs, gastrointestinal tract, blood vessels, and glands). The aim of this article is to concentrate on the central nervous system, yet from this account of the autonomic system it is easy to see how drugs acting as agonists or antagonists at cholinergic or adrenergic receptors can be used as therapy for cardiovascular, respiratory, gastrointestinal, and other disorders.

III. Neuromuscular Function

Many analogies have been drawn between the neuropharmacologies of the autonomic and central nervous systems. Further parallels exist between the brain and the neuromuscular junction, the synapse between the motor nerves which pass out from the spinal cord and the muscles. Here, the transmitter is acetylcholine, and the receptor is the nicotinic cholinergic receptor. Thus, drugs which act as antagonists at this site are useful agents for producing muscle relaxation during surgery, since the release of acetylcholine from the terminals causes the muscle contraction for movement. This neuromuscular system, even more than the autonomic nervous system, is too simple to be an adequate model for the brain. In the brain many terminals, transmitters, and receptors converge onto single cells, which are themselves part of the massive networks of neurons.

IV. Transmitters in the Central Nervous System

When one considers the enormous complexity of the central nervous system, it is not surprising that there are many transmitters. The criteria for a chemical's being considered as a transmitter are (1) location in appropriate areas; (2) the ability of the candidate transmitter to be released from the terminal by some form of stimulation; (3) application of the candidate transmitter should mimic the physiological events at that synapse; (4) antagonism; (5) processes for breakdown or removal, so that synaptic events can be terminated.

Applying these criteria, the following can be considered as major transmitters in the central nervous system: for the excitatory amino acids, glutamate and aspartate; for the inhibitory amino acids, γ-aminobutyric acid (GABA) and glycine; for the monoamines, noradrenaline, serotonin (i.e., 5-hydroxytryptamine), and dopamine; for acetycholine and for the neuropeptides, enkephalins, substance P, somatostatin, cholecystokinin, and others.

Thus, not only is this list much longer than the two major transmitters in the autonomic nervous system—namely, noradrenaline and acetylcholine—but the two also have major functions in the brain; conversely, others on the list have important roles elsewhere in the body, implying a conservation in these biological systems, so that transmitters act both in the central nervous system and the periphery, but play different roles. This illustrates a major problem in drug therapy, in that at the present time it is next to impossible to target a drug onto a specific site. Many side effects arise from drugs acting on transmitter receptors other than those at the primary therapeutic site.

The central nervous system transmitters fall into two major classes, depending on the speed of onset and the duration of their effects. Thus, the excitatory and inhibitory amino acids have fast actions, turning neurons on or off and the rest of the transmitters have slower, longer, or more subtle effects. They then change the level of excitability of the neurons, so that when a fast excitation or inhibition is evoked, the background state of the recipient cell is altered. Clearly, this type of organization imparts flexibility and adaptability to the functioning of the central nervous system.

Added to this system of fast and slow transmitters is the organization of the central nervous system networks of neurons. Certain neurons send long axons from one area of the brain to another and are involved therefore in the transmission of information from one region to another or the modulation of one zone by another. Here, the transmitters are generally excitatory amino acids or monamines, but some peptides in sensory nerves are also in long axon neurons. Within restricted zones some neurons send only a short axon. These so-called interneurons are involved in the local control of transmission, and the inhibitory amino acids and, again, neuropeptides tend to be found in these types of neurons.

Discussed below are the roles of these transmitters in a variety of disorders, since it is really only when a transmitter system is thrown into disarray

that we can then assign a role to a particular transmitter in the workings of the human brain, based on the actions of clinically effective drugs.

A. Depression

The first hints as to the causal factors in depression came from the use of two drugs for entirely different medical reasons about 40 years ago. One was used to treat elevated blood pressure; although effective, the patients showed depressed mood and suicidal thoughts.

The other, an antituberculotic drug, caused a marked elevation of mood in the patients. These odd findings were reconciled by the finding that both agents could alter the levels of noradrenaline in the brain. From these chance observations a theory was created from the results of studies using behavioral reward in animals, with manipulations of brain levels of noradrenaline. The central tenet was that elevations of noradrenaline are pleasurable, rewarding, or produced enhanced mood, whereas the converse leads to depression. Support for this idea comes from the clinically used present generation of antidepressant drugs and the effects of cocaine. [See ANTIDEPRESSANTS.]

The antidepressants increase levels of noradrenaline in the synapse either by preventing its breakdown or by blocking its reuptake back into the terminal. Cocaine also shares this latter mode of action. The problems with the use of cocaine (and the antidepressants) come from the enhanced levels of noradrenaline in the sympathetic nerve terminal synapses on the heart, lungs, and blood vessels. It is impossible to administer a drug and to target it on the brain, excluding the rest of the body and so circumventing these side effects. However, there are other antidepressants which seem to act on 5-hydroxytryptamine, and since this transmitter appears to balance the rewarding effects of noradrenaline and act within a punishment system, this is clearly a strong possibility as a target for antidepressants too. Long-term therapy with antidepressants may alter the adrenergic receptor characteristics by altering the synaptic levels of the transmitter, and those receptor changes may underlie the need for the therapy to continue before improvement is produced.

B. Anxiety

A sense of anxiety improves human performance when the anxiety is mild, but seriously impairs daily life when the level of anxiety is too great. Anxiety is treated by a class of drugs known as the benzodiazepines (e.g., librium and valium). These drugs, however, can also be used as sleeping pills, as muscle relaxants, and to treat epilepsy. Since the benzodiazepines can elicit a continuum of central nervous system depression—depending on their potency, dose, and duration of effect—particular drugs can be selected for particular problems. These agents, however, all share the same basic neuropharmacological effect, which is unique among all of the drugs used to treat disorders of the brain. GABA is the main inhibitory transmitter in the brain and is found mostly in interneurons controlling the excitability of other cells. GABA can activate two types of receptor: the $GABA_A$ and $GABA_B$ types. The benzodiazepines enhance the action of GABA on the $GABA_A$ site, so increasing inhibitory influences. The means by which they do this is somewhat mysterious; by binding to a site, the benzodiazepine receptor, these drugs make $GABA_A$ receptor activation by GABA itself more effective.

The increased GABA activity causes a greater degree of inhibition, which is important in reducing 5-hydroxytryptamine function in anxiety, relating to the idea that 5-hydroxytryptamine neurons act as a punishment system. The relative safety of the benzodiazepines seems to be due to their enhancement of a preexisting inhibition by GABA. This acts as a ceiling on their effects; that is, they do not produce any effects in the absence of GABA. Does this receptor or site for the benzodiazepines mean that a natural benzodiazepine uses this site, mimicked by the synthetic drug? Several groups are investigating this possibility, but the evidence is equivocal as yet.

C. Schizophrenia

Schizophrenia, a profound disorder of thought processes, mood, and social functioning, has been cruelly described as the epitomy of madness. A triumph of modern neuropharmacology has been the ability to control the symptoms of this disorder by the use of drugs, to the extent that straight jackets, padded cells, and ice baths are now no longer used in psychiatric hospitals. The variety of effective drugs all share a common mechanism; namely, they are antagonists of dopamine. This block of dopamine is able to reduce the schizophrenic symptoms, but does not appear to influence the cause of the disorder, since medication must be continued if

relapse is to be avoided. These drugs have some other pharmacological actions which produce side effects (e.g., antihistaminic and anticholinergic effects), which probably give rise to their sedative effects. It should be pointed out that these effects are not necessary for antischizophrenic action, but that it has not yet been possible to produce drugs that are completely selective for dopamine receptors. [*See* SCHIZOPHRENIC DISORDERS.]

With high doses or long-term treatment some motor problems can arise which relate to the roles of dopamine in movement, which are discussed in the next section. The role of dopamine in affective disorders is likely to reside in the dopamine pathways arising from midbrain areas and innervating the cortex, the motor function from another dopamine system running from the substantia nigra to the motor-integrating areas of the striatum. This is a good example of the different functions of a transmitter in different areas of the brain. Since there is evidence for atrophy of some brain areas in schizophrenics, it could be that the primary lesion removes a control on dopamine, giving rise to the schizophrenic symptoms.

D. Parkinson's Disease

Parkinson's disease is, in most cases, a disease of old age and, with increasing life expectancy of the world population, it will become more and more prevalent. The deficit is, again, amenable to treatment, and the primary cause is a selective loss of cells containing dopamine in the substantia nigra–striatum pathway. The cause for this cell loss is unknown, but the compound 1-methyl-4-phenyl-1,2,3,6-tetrahydropyridine (MPTP), a byproduct of a flawed designer drug, has been shown to exactly mimic the lesion. It is highly unlikely that a environmental toxin such as this produces Parkinson's disease, since rural populations and preindustrial societies still suffer from the disease. Since the loss is of dopamine, it appears that this transmitter exerts a controlling influence on the motor command networks in the striatum; a loss of control causes the rigidity, tremor, and akinesia typical of Parkinsonism. [*See* PARKINSON'S DISEASE.]

Therapy is based on attempting to replenish this transmitter deficit, either by direct dopamine agonists (e.g., bromocriptine), which are moderately successful, or by building up the levels of dopamine in the remaining cells. Dopamine itself cannot be given, because it doesn't gain access to the brain, so a related compound, L-dopa, is used which does get into the brain. L-Dopa is a precursor for dopamine, is taken up into the dopamine neurons, and is converted to dopamine, which is then available for release. This effectively replenishes the deficit, and normal movement is restored. Unfortunately, too much dopamine and schizophreniclike symptoms occur, and as the dopamine levels fluctuate near the end of the dose, so-called "on–off" episodes occur when movement is uncontrolled, then rigidity sets in, and a cycle ensues; this is obviously upsetting for the patient.

In animals the Parkinson's disease-like syndrome can be alleviated by transplanting peripheral or neural tissue containing dopamine, and there have been a few patients in whom this has been attempted. The results remain controversial, and the moral and ethical aspects require debate, but if cells containing dopamine, which are easily grown in culture in the test tube, were shown to be effective, this could be a useful therapy. At present this looks unlikely.

E. Alzheimer's Disease

Alzheimer's disease, or senile dementia, is another disease in which a transmitter deficit seems to cause the symptoms, but, unlike Parkinson's disease, there is no effective replacement therapy. There are several symptoms, but losses of memory, insight, and severe cognitive disorders typify the syndrome. The major lesion seems to be a loss of cholinergic input into the cortex from the basal nucleus, situated just below the cortex itself. It is known that anticholinergic agents can impair memory, so this lesion is consistent with the amnesia symptomatic of Alzheimer's disease.

There is also loss of another transmitter, this time a peptide, somatostatin, which could result from a loss of control of excitatory amino acid mechanisms; this overexcitation might be the cause of death of the somatostatin-containing cells. There is no evidence as to what symptoms result from the loss of these peptide cells. A small number of Alzheimer's disease patients have the familial disease and there is great interest in locating the gene locus, together with attempts to understand the plaques, tangles, and other aberrant neural elements that occur in the disease. Advances in therapy are delayed by our inability to replenish the loss of acetylcholine by drugs acting effectively as cholinergic muscarinic agonists. As in Parkinson's disease, transplants in animals seem to have potential for developing therapy. [*See* ALZHEIMER'S DISEASE.]

F. Epilepsy

The overexcitability of certain areas of the brain leads to the uncontrolled movements, convulsions, and other disturbances underlying epilepsy, which takes certain forms depending on the degree or the severity of the attack. From the points already made in this article, it is obvious that the strategy for therapy will be either decreasing excitation or increasing inhibitory influences. The former approach could become possible with the development of antagonists for the excitatory amino acid receptors, thus reducing the effects of glutamate and/or aspartate. This reduction could be a heavy price to pay, as these chemicals have ubiquitous actions throughout the brain and antagonism might bring the central nervous system to a halt. Most treatments rely on increasing inhibitions, by using either benzodiazepines (see Section IV,B) or other drugs (e.g., barbiturates) which open the channel associated with the GABA receptor. The other approach is to prevent the spread of abnormal activity; some drugs do this, presumably because they can reduce abnormal patterns of activity, leaving normal functioning of the cells elsewhere in the brain. [*See* EPILEPSY.]

G. Pain

Pain is not a disorder of the central nervous system, but is an important signal for protection of the organism from impending or actual tissue damage. However, an excess of pain (e.g., from chronic diseases or injuries) produces suffering and distress, which overwhelm the need for a warning signal. [*See* PAIN.]

The dramatic examples of pain seemingly arising from an area of the body amputated decades beforehand, the transformation of touch stimuli into pain after nerve damage, and the difficulties of treating some types of pain compared to others underlie the complexities of pain transmission and indicate the lack of a specific pain pathway, immutable and hard wired. Nevertheless, we know a reasonable amount regarding both pain transmission and its pharmacological control.

Pain is generated by nociceptors (receptors for chemicals and heat embedded in the skin, viscerae, and muscles), which, in turn, activate peripheral nerve fibers, known as C fibers. Waves of action potentials pass up these nerves and release a variety of transmitters into the spinal cord, the site of the first synapse in the system. These transmitters include glutamate and a number of peptides, including substance P, which excite dorsal horn neurons in the upper half of the spinal cord. These cells, in turn, are modulated, and their responses are integrated and modified prior to transmitting messages to both higher centers of the brain (producing the sensation of pain) and the lower (ventral) part of the spinal cord, where motoneuron activation causes withdrawal from the stimulus.

The activation of the spinal neurons responding to noxious stimulation can be reduced or abolished by the action of drugs such as morphine which act to inhibit the release of the transmitters from the C fibers. This action is via receptors for opioid drugs on the terminals of the sensory fibers and so is presynaptic inhibition. Just over a decade ago a paradox regarding morphine and related drugs was solved: Why should extracts from a plant (i.e., the opium poppy), produce such profound effects on the brain as the abolition of pain? The reason is now clear; natural opioids exist in the central nervous system and the periphery as transmitters. Like any other transmitter system, there are receptors for these chemicals, and morphine, by a chemical similarity, can activate these receptors. There are at least three types of receptor—the μ, δ, and κ subtypes; morphine acts on the μ receptor. Since along with the morphine analgesia are severe unwanted side effects (e.g., respiratory depression, constipation, cough suppression, and dependence), there is great interest in drugs which act on the δ and κ receptors as potential alternatives to morphine with the desirable analgesia, but not the side effects. The so-called endogenous opioids are peptides, and three families exist which correspond to three types of opioid receptors: β endorphins correspond to μ receptors; enkephalins, to δ receptors; and dynorphins, to κ receptors.

There is great interest in the roles of these peptides, since their presence in the brain might relate to a natural painkilling function or even a built-in dependence system. The enkephalins, in particular, are quickly broken down, so drugs which prolong their actions have been produced which reveal that enkephalins are released during the production of a painful stimulus and so could reduce pain but not abolish it. The other roles of opioids remain elusive, but are being pursued, and, in general terms, the study of opioid peptides and nonopioid peptides could provide an important addition to neuropharmacology, since present therapies are all aimed at the nonpeptide transmitters. Peptides such as cholecystokinin might be involved in Parkinson's disease and schizophrenia; angiotensin, in hypertension; somatostatin, in Alzheimer's disease. There is much interest in devising drugs which manipulate peptide transmission.

H. Stroke

A major consequence of disruption of the blood supply to parts of the brain is death of the neurons in the area deprived of their normal blood supply. Stroke can result in marked neurological deficits, including paralysis, amnesia, and sensory loss, and although some recovery can occur (presumably, some neurons are injured, but they do not die), there are long-term effects which cannot be alleviated. There is great interest in the therapeutic use of some antagonists of the excitatory amino acid receptors for glutamate. It might be that the vascular insult and resulting ischemia cause a loss of control of glutamate release. It appears that the massive release of glutamate can provoke overexcitation of neurons, which leads to death of the neurons, because of the resultant huge imbalances in ions within the cells. There is accumulating evidence that blocking one of the glutamate receptors or, indeed, the ion channels associated with the receptor can protect some of the neurons at risk. Remarkably, treatment after the event could be effective. If the potential of these drugs is verified in humans, a treatment which could partially alleviate the consequences of stroke could emerge. This glutamate receptor might also become overactive in epilepsy and, under normal conditions, seems to be involved in memory, learning, and other long-term events. A consequence of therapy via this route could be amnesia, but this might be a worthwhile price to pay for reducing the problems of cerebral ischemia. [See STROKE.]

V. The Future

Given that we now have a reasonable idea of how some of the transmitters in the central nervous system function in health and disease, the future must be determined. It seems to be easier, at present, to block the effects of transmitters by using direct antagonists or by giving inhibitory drugs such as the opiates. The opposite approach, such as the treatment of Alzheimer's disease and Parkinson's disease (in this case after the initial few years of therapy), in which one is trying to increase transmitter levels, is more difficult. The idea of using transplants to augment the transmitter is feasible from animal studies, but the ethical aspects need to be considered apart from the longer-term effects of the transplant within the delicate balanced structures of the brain. Another area of great interest is definition of the roles of the multitude of peptide transmitters found in the brain. These compounds have been found to coexist with other transmitters, yet the present therapies, because of the lack of drugs for the manipulation of peptides, concentrate only on the well-known or classical transmitters, such as those described earlier. More complete and therefore better clinical effects might be achieved by manipulating both peptide and nonpeptide transmitters. Nerve growth factors might provide protection or a survival aid to damaged or lost neurons and could be used in syndromes or diseases in which cell death and damage occur. An area of great interest is the realization that neurotransmitters can influence the genetic apparatus of nerve cells, influencing long-term changes in neuronal function. [*See* NERVE GROWTH FACTOR.]

Another large step forward will be simply improving our knowledge of pharmacology, since at present we have a reasonably broad idea of neuropharmacology, but many areas remain hazy and, in most cases, details are lacking. Thus, we know certain roles of transmitters, such as dopamine in movement and psychoses, opioids in pain states, and noradrenaline in mood, but exactly how, where, and when do these systems function? Answers will hopefully be forthcoming, and when they are this basic, knowledge will surely aid the clinical treatment of central nervous system disorders.

Bibliography

Björklund, A., Lindvall, O., Isacson, O., Brundin, P., Wictorin, K., Strecker, R. E., Clarke, D. J., and Dunnett, S. B. (1987). Mechanisms of action of intracerebral neural implants. *Trends Neurosci.* **10,** 509.

Cooper, J. R., Bloom, F. E., and Roth, R. H. (1986). "The Biochemical Basis of Neuropharmacology," 5th ed. Oxford Univ. Press, New York.

Dickenson, A. H. (1986). Enkephalins. A new approach to pain relief? *Nature (London)* **320,** 681.

Lancet (1982). Neurotransmitters and CNS Disease. Oct. 23–Dec. 18, 1982 (consecutive issues).

Webster, R., and Jordan, C. (eds.) (1989). "Neurotransmitters, Drugs and Disease." Blackwell, Oxford, England.

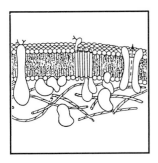

Neurotransmitter and Neuropeptide Receptors in the Human Brain

SIEW YEEN CHAI AND FREDERICK A. O. MENDELSOHN,
University of Melbourne

Glossary

Adenylyl cyclase Enzyme associated with the inner surface of cell membrane, which converts ATP into cyclic AMP. The enzyme is regulated by stimulatory (Gs) and inhibitory (Gi) G proteins coupled to activated receptors

G protein GTP-binding regulatory protein composed of three subunits, α, β, and γ. The α subunit binds and hydrolyzes GTP and the $\beta\gamma$ complex anchors the G protein to the cytoplasmic face of the membrane. In the resting state, the protein exists as a trimer to which GDP is bound. When activated, the G protein binds to the hormone receptor complex, allowing GTP to bind in place of GDP. The GTP-coupled G protein then dissociates from the hormone receptor complex and triggers second-messenger systems

***In vitro* autoradiography** Technique that involves the binding *in vitro* of a specific radioligand to receptors on tissue sections. These are then exposed to photographic emulsion or X-ray film to obtain an image of the binding site distribution and density

Ion channels Channels within the cell membrane that regulate the passage of selected ions into and out of the cell. These channels are regulated by membrane potential, ligand occupancy, or second messengers

Oncogene Genes encoding normal proteins that have become incorporated, often in modified forms, into retroviruses. These genes may induce uncontrolled cell proliferation in the infected cell and thus produce tumors. Protooncogenes are the normal cellular genes from which the oncogene is derived

Phosphoinositide cycle Cascade that is initiated by the binding of a hormone/neurotransmitter to a cell surface receptor, which then activates phospholipase C to hydrolyze phosphoinositide 4,5 diphosphate to inositol 1,4,5 trisphosphate and diacyl glycerol, two separate second-messenger systems, which induce increased intracellular Ca^{2+} and phosphorylation of specific proteins

RECEPTORS ARE cell surface molecules that bind specific neurotransmitters, hormones, or growth factors and then transmit a signal to the cell interior, which causes the cell to respond in an appropriate manner. Some of the genes for these proteins have now been cloned, a number of families of receptors have been recognized, and some insights as to how the receptors work are now becoming available.

I. General

Many receptors were recognized to contain distinct subtypes on the basis of ligand specificity, distribution, and mode of signal transduction. With the remarkable progress in analyzing the genes encoding receptors, two major facts have emerged. First, on the basis of their primary structures and predicted transmembrane topologies, many receptors and ion channels belong to superfamilies. Second, the degree of receptor heterogeneity within a given neurotransmitter receptor type is greater than previously suspected. This diversity of receptor types for one neurotransmitter presumably endows the neural cell with greater information handling ability.

The term *receptor* is also used in neuroscience to denote specialized neural structures involved in transduction of sensory information into neural impulses. This is not the sense in which receptors will be discussed here. However, it is intriguing to note that at least some forms of special sense receptors (i.e., those involved in light transduction and in olfaction) use G proteins in their transduction process, as do some of the cell surface receptors involved in chemical neurotransmission, which are discussed below.

II. Families of Receptors

A summary of some important receptors is given in Table I.

A. Ligand-Gated Ion Channel Receptors

The superfamily of ligand-gated ion channel receptors includes the nicotinic acetylcholine receptor (nACHR), the GABA/benzodiazepine receptor complex, and the glycine receptor. They share the feature of four membrane-spanning domains in their molecular structure and have significant sequence homology.

The nACHR is one of the best characterized receptors because it was isolated early. This was facilitated by an abundant source in the electric organ of an electric fish, Torpedo, and the availability of a snake venom toxin (α-bungaratoxin), which binds tightly and specifically to the receptor. The receptor consists of four homologous but distinct subunits that form a pentameric rosette that spans the membrane to form a central ion channel. In the absence of acetylcholine, the channel is closed, but after binding acetylcholine to the α subunit, the channel opens for a few milliseconds to permit passage of sodium and potassium ions. Elegant studies of single-channel currents of chimeric receptors and of receptors expressed after site-directed mutagenesis are revealing the importance of the subunits and of charged and uncharged residues in function of the nACHR (Fig. 1).

Glycine is the major inhibitory neurotransmitter in the spinal cord and brainstem, whereas GABA is the main inhibitory transmitter in the central nervous system. Binding of both amino acids causes hyperpolarization of the cell by increasing Cl^- conductance. The glycine receptor resembles the nACHR in both amino acid sequence and structural organization. The drug strychnine acts by binding to the glycine receptor and inhibiting Cl^- channel opening.

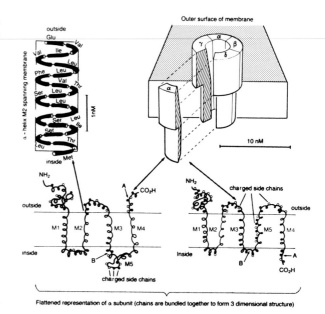

FIGURE 1 A schematic representation of the nicotinic acetylcholine receptor with the four subunits α, β, γ, δ that form a pentameric rosette. Two models of the membrane-spanning, α-helical portions of the channel are proposed: (1) an uncharged pore lined with hydrophilic side chains (left) and (2) a charged pore lined alternately with regions of positive and negative charges (right). [Reproduced with permission from C. F. Stevens (1985). AChR structure: a new twist in the story. *Trends Neurosci.* **8(1)**, 1–2.]

The GABA/benzodiazepine ($GABA_A$) receptor is of particular interest because it contains the binding site for a number of significant drugs that allosterically modify GABA binding or the Cl^- channel. These include anxiolytics (benzodiazepines), anticonvulsants, and convulsants. The amino acid sequences also show homology with the nACHR and identify the $GABA_A$ receptor as part of this superfamily.

The kainate subtype of the glutamate receptor also shares the feature of four transmembrane domains but has little overall sequence homology with the members of the superfamily discussed above and probably therefore belongs to a new family of ligand-gated ion channel receptors.

B. G-Protein–Linked Receptors

The first member of this family identified was the visual pigment rhodopsin, which is embedded in the membrane of discs in rods of the retina. On exposure to light, rhodopsin undergoes a change in its conformation, which enables it to interact with the

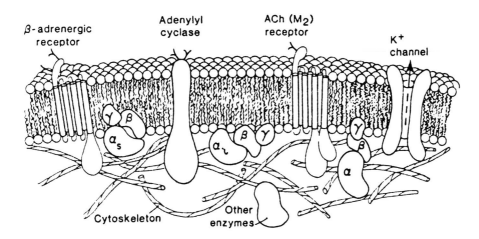

FIGURE 2 A schematic diagram illustrating the organization of G-protein coupled receptors and effector systems in the plasma membrane. The G-proteins are composed of three subunits, α, β, γ, where α_S stimulates adenylate cyclase and α inhibits the enzyme. [Reproduced with permission from E. J. Neer and D. E. Clapham (1988). Roles of G protein subunits in transmembrane signaling' *Nature* **333(12)**, 129–134.]

G protein transducin, thereby triggering cyclic GMP phosphodiesterase, which hydrolyses cGMP, thereby terminating its activation of Na^+ channels to mediate photoreception. [*See* G PROTEINS.]

Other G-protein coupled receptors that have been cloned include the α_1-, α_2-, β_1-, β_2- and β_3-adrenergic receptors, which bind catecholamines such as norepinephrine and epinephrine; the muscarinic M_1–M_5 cholinergic receptors, which bind acetylcholine; the $5\text{-}HT_{1A}$, $5\text{-}HT_{1C}$, and $5\text{-}HT_2$ serotonin receptors; the neurokinin A receptor; and the *mas* gene product, which may correspond to an angiotensin receptor subtype (Fig. 2).

These receptors share many features in parallel. They are composed of single polypeptide chains, 400–500 amino acids long, which bear seven hydrophobic regions separated by alternating extracellular and cytoplasmic loops. The hydrophobic regions of 20–25 amino acids form transmembrane-spanning α helices. The N-terminal extracellular domains show one to three sites that may undergo N-linked glycosylation. This glycosylation is not essential for ligand binding or for coupling to G proteins, but is involved in expression of the receptor on the cell surface and its trafficking through the cell. Surprisingly, the protein coding regions of many of these genes lack introns.

A number of different effector systems may be coupled to these receptors via G proteins (Table II).

Those regulating production of cyclic nucleotides are adenylyl cyclase, which is coupled by both stimulatory (G_S) and inhibitory (G_i) G proteins and regulates the production of cyclic AMP, and guanylyl cyclase, which regulates generation of cyclic GMP. More than 15 different receptors can stimulate adenylyl cyclase through G_S. Cyclic nucleotides can regulate a wide variety of proteins via specific protein kinases and may also directly regulate some ion channels.

Also coupled in this manner by a G protein (G_p) is the phosphoinositide-specific phospholipase C, which is involved in calcium signaling pathways. Activation of this enzyme by an activated receptor liberates from membrane-bound phosphoinositides, the second messengers inositol trisphosphate and diacyl glycerol. These are involved in intracellular signaling by release of Ca^{2+} from intracellular stores and by activation of protein kinase C, respectively.

Phospholipase A2, which cleaves arachidonic acid from the 2 position of phospholipids, is also coupled to some receptors by a G protein. Arachidonic acid is a precursor of a family of signaling molecules (the eicosanoids), which include prostaglandins, thromboxanes, and leukotrienes. Eicosanoids may act as second messengers in some neuronal systems.

Cyclic GMP phosphodiesterase is activated by the G protein transducin, which transmits the signal from activated rhodopsin on exposure of rod cells in the retina to light. As well as these transduction mechanisms, there is also evidence that some ion channels may be directly coupled to G proteins. These include the muscarinic-gated K^+ channel from cardiac atria. Neurotransmitters activate a variety of K^+ and Ca^{2+} channels in neurons by mechanisms involving G proteins, and the abundant G

TABLE I Neurotransmitter and Neuropeptide Receptor Types[a]

Receptor Type	Subtypes	Second Messenger	Agonist	Antagonist	Comments
Aminergic					
α-adrenergic	α_1	PI/Ca^{2+}	methoxamine	prazosin	two subtypes
	α_2	AC, G_i	clonidine, UK 14304	rauwolscine, RX 781094	two subtypes
β-adrenergic	β_1	AC, G_s	R 0363	CGP20712A	
	β_2	AC, G_s	procaterol	ICI-118, 551	
	β_3		BRL 37344		Emorine et al. (1989)
Dopaminergic	D_1	AC, G_s	SKF 38393	SCH 23390	
	D_2	AC, G_i; K^+ channel	quinpirole	(−)sulpiride, YM091512	
Serotonergic	$5\text{-}HT_{1A}$	AC, G_i; K^+ channel	5-HT, 8-OH DPAT	cyanopindolol, metitepine	Not present in human
	$5\text{-}HT_{1B}$	AC, G_i	RU 24969	cyanopindolol	
	$5\text{-}HT_{1C}$	PI/Ca^{2+}		mesulergine, mianserine	
	$5\text{-}HT_{1D}$	AC G_i		metergoline	
	$5\text{-}HT_2$	PI/Ca^{2+}	DOI, DOB	ketanserin, cyproheptadine	
	$5\text{-}HT_3$	?ion channel	2-methyl 5-HT, phenyl-diguanide	ICS 205930, MDL 72222	
Histaminergic	H_1	PI/Ca^{2+}	pyridyl-ethylamine	pyrilamine	
	H_2	AC, G_s	impromidine, dimaprit	cimetidine, tiotidine	
	H_3		R-α-methyl-histamine	thioperamide	Presynaptic inhibition of histamine release
Nicotinic acetylcholine		ion channel $Na^+/K^+/Ca^{2+}$	nicotine	TEA	muscle and neuronal subtypes
Muscarinic acetylcholine	M_1	PI/Ca^{2+}		pirenzepine, (+)-telenzepine	
	M_2	AC, G_i; K^+ channel		AF-DX116, himbacine	Coupled to K^+ channel in heart
	M_3	PI/Ca^{2+}		P-fluorohexahydro-siladifenidol	
	M_4	AC			
	M_5	AC, G_i; PI/Ca^{2+}; AC			
Excitatory amino acids					
	NMDA	int. $Na^+/K^+/Ca^{2+}$	NMDA, L-glutamate, ibotenate	D-AP5, CGS 19755, CPP	
	kainate	int. Na^+/K^+	kainate, domoate	CNQX, DNQX	
	AMPA	int. Na^+/K^+	AMPA	CNQX, DNQX	
	L-AP4	—	L-AP4, L-SOP	—	
	metabotropic	PI/Ca^{2+}	quisqualate	—	

Inhibitory amino acids

Receptor	Effector	Agonists	Antagonists	Comments
GABA/Bzdp GABA$_A$	Cl⁻ channel	muscimol, benzodiazepines	bicuculline, CGS-8216, R015-1788	Major inhibitory receptor. Target of numerous important drugs
GABA$_B$	cAMP↓ K⁺ channel	(−)baclofen	3-APA, phaclofen	May act via a G$_i$ coupled K⁺ channel
Glycine	Cl⁻ channel	glycine, L-serine	strychnine	
Purinergic Adenosine A$_1$	AC, G$_i$ K⁺ channel	cyclopentyl-adenosine	DPCPX	
A$_2$	AC, G$_s$	MECA	triazoloquinazolone	
P	AC, G$_i$	2'5'DDA	5'methylthioadenosine	
Purine P$_2$	K⁺ channel	ATP analogues		x, y, z and t subtypes
Neuropeptidergic Opiates μ	cAMP↓ K⁺ channel	DAMGO, PL017	β-funaltrexamine, naloxonazine	
δ	AC, G$_i$ K⁺ channel	DSBULET	ICI-174864	
κ	Ca²⁺	U69593, dynorphin A$_{1-17}$	nor-binaltorphimine	
Tachykinins NK$_1$	PI/Ca²⁺	substance P SP methylester	L 668169	
NK$_2$	PI/Ca²⁺	NKA	L 659877	
NK$_3$	PI/Ca²⁺	NKB senktide	—	
CCK CCK$_A$	PI/Ca²⁺	CCK-8	lorglumide, MK-329 L-365260	
CCK$_B$		CCK-4		
AVP V$_1$	PI/Ca²⁺	AVP	d(CH$_2$)$_5$Tyr(Et)AVP	
V$_2$	AC, G$_s$	AVP	d(CH$_2$)$_5$-D-Leu², Val⁴AVP	
Angiotensin Type I	PI/Ca²⁺	Ang II	DUP 753	
Type II	?	Ang II	CGP, CGP 42112A, XD 3291	
Bradykinin B$_1$	—	BK$_{1-8}$	[Leu⁸] BK$_{1-8}$	
B$_2$	PI/Ca²⁺	[Phe⁸, ψ(CH$_2$-NH)Arg⁹] BK	Ac-D-Arg[Hyp³, DPhe⁷] BK	Mainly peripheral
ANP ANP$_A$	cGMP			
ANP$_B$	cGMP			
ANP$_C$	—			

ᵃPI, phosphoinositide; Ca²⁺, calcium; AC, adenylyl cyclase; G$_i$, adenylate cyclase inhibitory G protein; G$_s$, adenylate cyclase stimulatory G protein; 8-OH DPAT, 8-hydroxy-2-(di-*n*-propyl-amino) tetralin; DOI, 1-(2,5-dimethoxy-4-iodo-phenyl)-2-aminopropanes; DOB, 1-(2,5-dimethoxy-4-bromo-phneyl)-2-aminopropanes; TEA, tetraethylammonium chloride; NMDA, *N*-methyl-D-aspartate; D-AP5, D-2-amino-5-phosphonovalerate; CPP, 3-3(2-carboxypiperazine-4-yl) propyl-1-phosphate; CNQX, 6-cyano-7-nitro-quinoxaline-2,3-dione; DNQX, 6,7-dinitro-quinoxaline-2,3-dione; AMPA, -amino-3-hydroxy-5-methyl-isoxazole-4-propionate; L-AP4, L-2-amino-4-phosphonobutyrate; L-SOP, L-serine-*O*-phosphate; 3-APA, 3-aminopropyl phosphinic acid; DAMGO, Tyr-D-Ala-Gly-NMe-Phe-Gly-ol; DSBULET, Tyr-D-Ser(OtBu)-Gly-Phe-Leu-Thr; DPDPE, D-Pen²-D-Pen⁵ enkepaline; NKA, neurokinin A; NKB, neurokinin B; CCK, cholecystokinin; AVP, arginine-vasopressin; CPX, 8-cyclopentyl 1,3-dipropylxanthine; MECA, 5'-*N*-methylcarboxamidoadenosine; DDA, dideoxyadenosine.

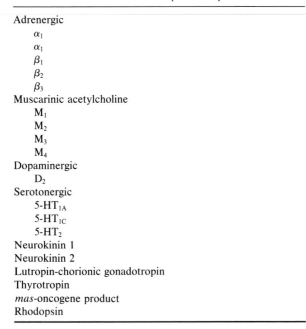

TABLE II Cloned G-Protein–Coupled Receptors

Adrenergic
  α_1
  α_1
  β_1
  β_2
  β_3
Muscarinic acetylcholine
  M_1
  M_2
  M_3
  M_4
Dopaminergic
  D_2
Serotonergic
  5-HT$_{1A}$
  5-HT$_{1C}$
  5-HT$_2$
Neurokinin 1
Neurokinin 2
Lutropin-chorionic gonadotropin
Thyrotropin
mas-oncogene product
Rhodopsin

protein of brain, G_O, can reconstitute this activity. However, in these latter cases it is not known if this is a direct action or whether second messengers are involved.

C. Receptors with Catalytic Activity

Finally, there are a group of receptors that bear intrinsic enzyme catalytic sites that mediate their action. This includes a group of growth factor-related receptors, those for insulin, platelet derived growth factor (PDGF), and epidermal growth factor (EGF). These receptors consist of a single polypeptide chain that traverses the membrane once and bears tyrosine specific protein kinase activity on their cytoplasmic domains.

Some oncogenes encode abnormal receptors that have catalytic domains that are constitutively active in the absence of receptor occupation. These include the verbB oncogene which encodes a truncated version of the EGF receptor, the sis oncogene which encodes an abnormal form of the PDGF receptor, and the neu oncogene, detected in some chemically-induced nervous system tumours, which encodes a tyrosine-specific kinase, but the ligand for the normal receptor is not known.

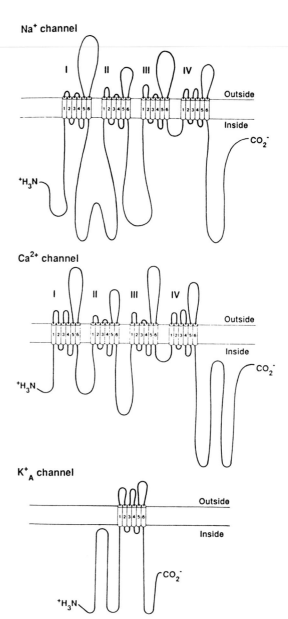

FIGURE 3 Proposed models of the transmembrane arrangements of the principal subunits of the Na^+, Ca^{2+} and the A-current K^+ channels. Both the Na^+ and Ca^{2+} channels have 4 domains with highly conserved amino acid sequences, and each domain contains multiple hydrophobic segments which forms the transmembrane α helices. . [Reproduced with permission from W. A. Catterall (1988). Structure and function of voltage-sensitive ion channels' *Science* **242**, 50–61.]

D. Voltage-Sensitive Ion Channels

These membrane proteins mediate rapid, voltage-gated changes in ion permeability in excitable cells and also modulate membrane potential and thereby excitability. They are not receptors in the usual sense. However, they are included here because they provide important components of the cellular response to receptor-mediated events, and they may function as receptors for drugs and toxins. Three main classes of voltage-sensitive channels have been studied: those for Na^+, Ca^{2+}, and K^+ (Fig. 3).

Voltage-sensitive Na^+ channels mediate rapid depolarization during the initial phase of the action potential. The channel proteins have been isolated with the use of five classes of neurotoxins, which bind tightly to separate sites on the channel and modify its function. The Na^+ channel is a heterotrimeric complex that spans the membrane and shows voltage-dependent changes in ion conductance.

Voltage-sensitive Ca^{2+} channels belong to multiple classes. One class, the L-type channel, is blocked by dihydropyridine drugs and has been isolated from skeletal muscle, although channels in the brain are probably structurally similar. The channel is a heteropentamer of which the α subunits of the Na^+ and Ca^{2+} channels show substantial homology. Ca^{2+} channels are activated by phosphorylation of their α_1 and β subunits by cAMP-dependent protein kinase. This phosphorylation probably underlies Ca^{2+}-channel regulation by β-adrenergic agonists.

Voltage-sensitive K^+ channels terminate the action potential by mediating the outward movement of K^+, which repolarizes the cell. K^+ channels also set the resting membrane potential and thereby modify the excitability of the cell.

This channel has been cloned from Drosophila. Its protein consists of a single polypeptide that has striking homology with the principal subunits of the Na^+ and Ca^{2+} channels. It is therefore likely that the Na^+ and Ca^{2+} channels arose from K^+ channels during evolution.

III. Localization and Properties of Receptors in the Human Brain

Receptors are classically characterized pharmacologically using biological responses to specific agonists and antagonists. On this basis, many subtypes of receptors have been defined. The distribution of receptor subtypes can also be investigated using radiolabeled agonists or antagonists. Early work on receptor mapping involved determination of receptor sites by the binding of radioligand to membrane fractions. A refinement of radioligand binding, *in vitro* autoradiography, involves labeling receptor sites on slide-mounted tissue sections. This provides receptor localization in much greater detail and cellular resolution. More recently, the availability of nucleic acid sequences of many receptors enabled the use of cDNA and oligonucleotide probes to determine the site of synthesis of receptors by *in situ* hybridization histochemistry.

Studies of the localization of receptors in the human brain are not as advanced as those in lower mammals such as the rat. The receptors discussed below have all been studied in the human brain. In general, the distribution of the various receptors show highly characteristic distributions. These patterns give clues to the roles in the overall brain of the neurotransmitters involved, as well as the effects of drugs that interact with these receptors.

A. Cholinergic

The cholinergic muscarinic M_1 receptor is concentrated in the external layers of the cerebral cortex (the highest density in layers I and II and the lowest levels in layer V), in the innermost layers of the primary visual cortex, in the CA fields and dentate gyrus of the hippocampus, and in the basal ganglia. In contrast, the M_2-receptor subtype is more densely distributed in the caudate nucleus, thalamus (particularly the anterior group of nuclei), amygdala, and substantia innominata, with much lower levels in the cortex and hippocampus. [*See* BRAIN.]

The nicotinic cholinergic receptor population is highest in the striatum, nucleus basalis of Meynert, and substantia nigra, with lower levels in the hippocampus, thalamus, and neocortex.

Deficits in the cortical and hippocampal cholinergic nicotinic and muscarinic M_2 receptors have been detected in Alzheimer's dementia, and cholinergic agonists may improve memory in Alzheimer's patients. These observations suggest a role for central cholinergic system in cognitive and behavioral functions. The high concentrations of muscarinic and nicotinic receptors in the human cortex, hippocampus, and substantia innominata support this hypothesis. [*See* ALZHEIMER'S DISEASE.]

B. Adrenergic

The β-adrenergic receptor occurs in high densities in the external cortical layers, hippocampus, striatum, lateral posterior thalamus, and the granular layer of the cerebellum. Lower levels are found in the pallidum, medial thalamic nucleus, substantia nigra, and the red nucleus. The receptors in the cerebellum are predominantly β_2, whereas the frontal cortex and the basal ganglia contain a majority of the β_1-receptor subtype.

The α_2-adrenergic receptor subtype is present in high concentrations in the frontal cortex and in the hypothalamus, with lower levels in the hippocampus and cerebellum. The locus ceruleus and the nucleus of the solitary tract also contain significant densities of α_2 receptors. These brain nuclei, together with certain hypothalamic areas, are thought to be important sites for the central regulation of blood pressure and may be important targets for the centrally acting antihypertensive agents.

Alterations in central adrenergic system have been implicated in the pathogenesis of certain behavioral disorders including schizophrenia and manic depressive disorder. The high levels of adrenergic receptors in the cortex and hippocampus may mediate these effects. [See MOOD DISORDERS; SCHIZOPHRENIC DISORDERS.]

C. Dopaminergic

The actions of dopamine are mediated via either the D_1 receptors (which stimulate adrenylate cyclase) or the D_2 receptors (which inhibit or have no effect on adenylate cyclase). The D_1 and D_2 receptors are both concentrated in the basal ganglia, where the D_1 receptors occur predominantly in the striatum, internal globus pallidus, and substantia nigra and the D_2 in the striatum, nucleus accumbens, and compact part of the substantia nigra. In addition, high densities of D_2 receptors are also present in the olfactory tubercle. Lower levels of D_1 receptors occur in the amygdala, mammillary bodies, and cerebral cortex, and both D_1 and D_2 receptors are present in the hippocampus.

The D_2 receptors are one of the primary targets for drugs used in the treatment of Parkinson's disease and for neuroleptics. The neuroleptic drugs are thought to elicit their inhibitory actions on the dopaminergic mesolimbic system, which includes the nucleus accumbens and olfactory tubercle, brain regions that are rich in D_2 receptors. [See PARKINSON'S DISEASE.]

Recent evidence suggests that at least two forms of D_2-receptor subtype exist as a result of alternative splicing of the mRNA. Both isoforms appear to be present in all brain regions in which the D_2 receptor is found.

D. Serotonergic

5-HT receptors are widely distributed in the human brain and have been pharmacologically characterized into different subtypes; 5-HT_{1A}, 5-HT_{1B}, 5-HT_{1C}, 5-HT_{1D}, 5-HT_2, and 5-HT_3. There are differences in the second-messenger systems; 5-HT_{1A}, 5-HT_{1B}, and 5-HT_{1D} receptors are linked to adenylate cyclase, whereas 5-HT_{1C} and 5-HT_2 are coupled to phosphoinositol turnover.

High concentrations of 5-HT_{1A} receptors occur in the external layers of the cerebral cortex, layers V of the anterior entorhinal cortex, pyramidal and lacunosum moleculare layers of the hippocampus, cortical and granular nuclei of the amygdala, and dorsal, central, and linear raphe nuclei. The high concentrations of 5-HT_{1A} receptors in the limbic system suggest the involvement of these receptors in the emotion and visceral processing. This is supported by the observation of anxiolytic properties of certain 5-HT_{1A} agonists.

5-HT_{1C} receptors are present in high concentrations in the choroid plexus and globus pallidus with lower levels in the striatum, nucleus accumbens, pyramidal layer of the hippocampus, ventromedial nucleus and mammillary bodies of the hypothalamus, and layers III and IV of the cortex.

High densities of 5-HT_{1D} receptors occur in the basal ganglia, with a lower, homogeneous distribution throughout all the layers of the cortex and hippocampus. Although this receptor subtype is the predominant 5-HT_1 class of receptor in the human brain, it only occurs in higher species of mammals and is absent in the rodent.

The highest levels of 5-HT_2 receptors are found in layers III and VA of the isocortex and layer III of the entorhinal cortex. A lower density of receptors occurs in the lateral nucleus of the amygdala, claustrum, striatum, mammillary bodies, substantia innominate, and CA1 of the hippocampus. The presence of 5-HT_2 receptors in the cortex and in components of the limbic system further supports the proposed role of these receptors in anxiety, depression, and senile dementia. These receptors are decreased after chronic antidepressant treatment and in patients with Alzheimer's dementia.

5-HT_3 receptors, which are directly linked to ion channels, are distributed in a distinct pattern because they occur in high concentrations in the dorsal vagal complex, area postrema, and substantia gelatinosa of the medulla oblongata. Lower densities of receptors are present in components of the limbic system. 5-HT$_3$ antagonists are effective antiemetics.

E. Histamine

The histamine H$_1$ receptor occurs in high concentrations in the cerebral cortex and in parts of the limbic system such as the cingulate and orbital cortex, amygdala, and uncus. Moderate levels of the receptors are also present in the striatum, hippocampus, septum, and the hypothalamus.

Some of the most potent inhibitors of the H$_1$ receptors are tricyclic antidepressants. The brain H$_1$ receptors are thought to be involved in arousal and regulation of appetite, and one of the side effects of the H$_1$ antagonist antidepressants is drowsiness.

Recently, histamine H$_3$ receptor–mediated release of histamine was observed from human cortical brain slices, suggesting the presence of H$_3$ receptors in the human central nervous system.

F. GABA-Benzodiazepine Receptor

GABA acts on two receptor subtypes, the bicuculline-sensitive GABA$_A$ and the G-protein–linked, bicuculline-insensitive GABA$_B$ receptor.

The GABA$_A$ receptor is comprised of a multiple subunit, ligand-gated ion channel complex that can be allosterically modulated by the benzodiazepines and barbiturates. The GABA$_A$ receptor occurs in high densities in layer IV of the primary visual cortex and layers III and IV of the cerebral cortex and cerebellum, with lower levels in the basal ganglia and the pyramidal layer and dentate gyrus of the hippocampus.

Because GABA is a major inhibitory neurotransmitter in the human brain and the GABA$_A$ receptor complex includes a binding site for benzodiazepines, a range of clinically important drugs have been found to act via this receptor complex. Alterations in central GABA$_A$ receptor function have been implicated in disorders such as epilepsy, anxiety, and insomnia.

The GABA$_B$ receptors were initially characterized in peripheral sites such as autonomic nerve terminals. It has since been identified in mammalian brains including the human brain, where it occurs in high concentrations in frontal cortex and hippocampus.

G. Glycine

Receptors for glycine occur in high concentrations in the periaqueductal gray and the oculomotor nuclei of the midbrain and in the motor, sensory trigeminal, facial, hypoglossal, cuneate, and gracile nuclei of the medulla oblongata. The concentration of glycine receptors in the brainstem is in agreement with its role as the major inhibitory neurotransmitter in these brain regions. These are possible sites where drugs such as strychnine, which binds to the glycine receptor, may act. Toxic doses of strychnine cause death convulsions.

H. Excitatory Amino Acids

There are at least four different types of excitatory amino acid receptors, characterized on the basis of pharmacological and electrophysiological data: the NMDA, quisqualate, kainate, and L-aminophosphonobutyric acid receptor subtypes.

The NMDA receptor is composed of a number of interacting domains just as for the GABA$_A$ supramolecular complex. In addition to the primary acceptor site for agonists and antagonists, there is an ionophore and sites for glycine, polyamines, magnesium, and zinc. Ligands for the ionophore and the glycine domain generally colocalize with those labeled by agonists or antagonists. The antagonists at all domains seem to protect against the excitotoxic phenomena and delay cell death.

High densities of the NMDA and quisqualate receptors are present in the external layers of the cerebral cortex and in the CA1 and dentate gyrus of the hippocampus. Moreover, high concentrations of NMDA receptors also occur in the primary visual cortex, and amygdala, the mediodorsal nucleus of the thalamus, and the granular layer of the cerebellum, with lower levels in the striatum. Quisqualate receptors are also present in CA3 and CA4 of the hippocampus and the molecular layer of the cerebellum.

In contrast, the kainate receptor subtype is concentrated in the deep and superficial layers of the cortex. It is also present in the CA3 and dentate gyrus of the hippocampus and in the granular layer of the cerebellum. Moderate levels of the kainate receptors are found in the striatum. The kainate receptor has recently been cloned from rat and frog brain cDNA libraries, and the distribution of its mRNA parallels the receptor localization by autoradiography.

The L-aminophosphonobutyric acid receptor is poorly understood, but its physiological roles have been established in the retina and the hippocampus.

Recently, a fifth type of excitatory amino acid receptor, which is linked to the phophoinositide cascade has been suggested. This receptor subtype has a unique pharmacology quite different from those discussed above, and a selective agonist has been identified.

I. Opioid

All the opioid receptor subtypes occur in high densities in the cerebral cortex where the μ and δ sites are concentrated in superficial and the κ in the deep layers of the cortex. The μ and δ receptors are also densely distributed in the visual and auditory cortex.

The most widely distributed receptor subtype, the μ, receptor is also present in high levels in the amygdala, the molecular layer of the cerebellum, and striatum, with lower levels in the pallidum, bed nucleus, nucleus basalis of Meynert, laterodorsal and medial nuclei, and pulvinar of the thalamus.

High densities of δ receptors also occur in the dentate gyrus of the hippocampus, with moderate levels in the basal ganglia, amygdala, claustrum, and thalamus.

The κ receptors are found in high concentrations in the amygdala, claustrum, medial and ventrolateral nucleus of the thalamus, ventromedial and posterior nuclei of the hypothalamus, and the cerebellum.

J. Other Peptides

Many peptide receptors have also been identified in the human brain.

One of the more well-studied peptide receptors, the CCK_B receptor, occurs in high densities in the glomerular and external plexiform layer of the olfactory bulb, layer IV of the cortex, and pontine nuclei, with lower levels in the nucleus accumbens and striatum. CCK_8, the primary form of CCK in the brain, is thought to regulate satiety but exerts many influences on brain peptide and monoamine systems. It has been shown to affect dopamine release and turnover in the striatum, which has high concentrations of the CCK_B receptor.

Somatostatin receptors are highly concentrated in the deep layers of the cortex, in the basal ganglia, and in the hippocampus, amygdala, and habenula nucleus. Lower levels of the peptide receptor occur in the cerebellum, locus ceruleus, spinal trigeminal, and substantia gelatinosa. This discrete distribution of somatostatin suggests that the peptide may be involved in cognitive and sensory processing.

Angiotensin II receptors are found in high densities in the forebrain circumventricular organs, in the median preoptic, paraventricular, and the arcuate nuclei of the hypothalamus, in the striatum and compact part of the substantia nigra, and in the medulla oblongata, in nuclei associated with autonomic control. Centrally, angiotensin II influences fluid balance and hypothalamic and pituitary hormone release and interacts with the autonomic nervous system. The presence of high concentrations of angiotensin II receptors in many regions of the hypothalamus and in the medulla oblongata is consistent with the proposed role of central angiotensin II.

Calcitonin receptors are concentrated in the anterior and posterior hypothalamus, median eminence, inferior colliculus, and substantia nigra, with lower levels in the basal ganglia and hippocampus. Calcitonin is thought to inhibit appetite and to influence gastric acid and pituitary hormone secretion. The dense distribution of its receptors in the hypothalamus may represent sites where calcitonin can elicit these actions.

Calcitonin gene-related peptide (CGRP) occurs from the alternate splicing of the calcitonin gene. Receptors for CGRP are found in high concentrations in the cerebellum, inferior olivary complex, pontine nuclei, the basal ganglia, dorsal vagal complex, and locus ceruleus. CGRP is thought to regulate sympathetic noradrenergic outflow and may influence somatosensory function, actions that may be mediated via these sites.

Neurotensin receptors occur in high concentrations in the substantia nigra and hippocampus where they are concentrated in the presubiculum and entorhinal area.

K. Purinergic

The purinergic receptors are divided into two major subtypes: the P_1, which has a higher affinity for adenosine, and P_2, which preferentially bind ATP. The P_1 receptors can be further divided into A_1 and A_2 sites according to their affinities for different adenosine analogues. Moreover, in most tissues, A_1 inhibits adenylate cyclase whereas A_2 stimulates the enzyme.

Receptor sites for adenosine A_1 have also been localized in the human brain where high concentrations are found in the hippocampus, cerebral cortex, striatum, and anterior and medial nucleus of the thalamus. Moderate levels also occur in the amygdala, olfactory tubercle, and nucleus accumbens.

Adenosine is a potent sedative and has also been implicated as an endogenous anticonvulsant. The high concentration of A_1 in the cortex and hippocampus may represent sites where adenosine can mediate these actions.

IV. Conclusion

The human central nervous system employs multiple classes of chemical transmitters, and in general each interacts with more than one type of cell surface receptor. Many disease processes result from abnormalities of this chemical communication system. Of the large number of drugs that affect brain function, most act by binding to receptors and either stimulating or blocking their function. Although a bewildering number of receptors are already known, the list is almost certainly incomplete—new members of known receptor families are being rapidly discovered by the techniques of molecular biology. Furthermore, there is no reason to believe that the range of known neurotransmitters, particularly neuropeptides, represents the complete list. Extrapolation from studies on molluscan nervous systems suggest that many more neuropeptides may be discovered in mammals. Therefore the field of brain receptors is likely to continue its rapid development and to be a fertile source of understanding brain function and of drugs that modify neurological and psychiatric disorders.

Bibliography

Berridge, M. J., and Irvine, R. F. (1989). Inositol phosphates and cell signaling. *Nature* **341**, 197–205.

Bowery, N. (1989). $GABA_B$ receptors and their significance in mammalian pharmacology. *Trends Pharmacol. Sci.* **10**, 401–407.

Catterall, W. A. (1988). Structure and functions of voltage-sensitive ion channels. *Science* **242**, 50–61.

Emorine, L. J., Marullo, S., Briend-Sutren, M. M., et al. (1989). Molecular characterization of the human β_3-adrenergic receptor. *Science* **245**, 1118–1121.

Hoyer, D. (1988). Molecular pharmacology and biology of $5-HT_{1C}$ receptors. *Trends Pharmacol. Sci.* **9**(3), 89–94.

Monaghan, D. T., Bridges, R. J., and Cotman, C. W. (1989). The excitatory amino acid receptors. *Annu. Rev. Pharmacol. Toxicol.* **29**, 365–402.

Neer, E. J., and Clapham, D. E. (1988). Roles of G protein subunits in transmembrane signaling. *Nature* **333**, 129–134.

Nicoll, R. A. (1988). The coupling of neurotransmitter receptors to ion channels in the brain. *Science* **241**, 545–551.

O'Dowd, B. F., Lefkowitz, R. J., and Caron, M. G. (1989). Structure of the adrenergic and related receptors. *Annu. Rev. Neurosci.* **12**, 67–83.

Pritchett, D. B., Sontheimer, H., Shivers, B. D., et al. (1989). Importance of a novel $GABA_A$ receptor subunit for benzodiazepine pharmacology. *Nature* **338**, 582–584.

Sakmann, B., Methfessel, C., Mishina, M., et al. (1985). Role of acetylcholine receptor subunits in gating of the channel. *Nature* **318**, 538–543.

Schofield, P. R., Shivers, B. D., and Seeburg, P. H. (1990). The role of receptor subtype diversity in the CNS. *Trends Neurosci.* **13**, 8–11.

Stevens, C. F. (1985). AChR structure: A new twist in the story. *Trends Neurosci.* **8**(1), 1–2.

Williams, M. (1987). Purine receptors in mammalian tissues: Pharmacology and functional significance. *Annu. Rev. Pharmacol. Toxicol.* **27**, 315–345.

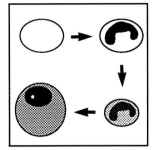

Neutrophils

THOMAS P. STOSSEL, *Harvard Medical School; General Hospital, Boston*

I. Life Cycle of PMNs
II. Functions of PMNs

NEUTROPHILIC POLYMORPHONUCLEAR leuko-cytes (neutrophils; [PMNs]) are bone marrow-derived cells that defend human tissues against bacterial and fungal infections. PMNs circulate briefly in the blood, leave the circulation by active penetration of capillaries, and seek out invading bacteria or damaged host tissue. The accumulation of PMNs at such sites is what leads to pus. PMNs crawl through connective tissues by a kind of amoeboid motion in response to gradients of molecules called chemotactic factors. They ingest bacteria, fungi, and other particulates that they recognize. PMNs kill and digest microorganisms using a variety of hydrolases, reactive oxygen intermediates, and membrane-permeabilizing proteins and polypeptides. The release of these agents into surrounding tissues and the secretion of bioactive lipid mediators cause inflammation. Persons deficient in PMNs, and thus their function, are susceptible to bacterial and fungal infections.

I. Life Cycle of PMNs

A normal adult produces over 100 billion PMNs per day. PMNs mature from a sequence of precursors beginning with the so-called pluripotent hematopoietic stem cell, some progeny of which commit to PMN development, also known as myeloid maturation (Fig. 1). The earliest morphologically recognizable PMN precursor, the myeloblast, derives from the granulocyte–macrophage stem cell, a cell capable of maturing in the direction of the monocyte–macrophage lineage or the PMN lineage. Further maturation of the myeloblast to a mature PMN involves, on the one hand, amplification in numbers

and acquisition of motile and antimicrobial functions and, on the other, loss of the capacity to self-replicate. Hematologists arbitrarily classify PMN progenitors according to their morphology on specially stained films (Wright-Giemsa). Myeloblasts are oval cells with large nuclei containing lacy chromatin and well-demarcated nucleoli. Similar cells containing blue-purple (azurophilic) granules are promyelocytes. These granules become less visible in more numerous myelocytes, which are smaller and have an eccentric nucleus and a prominent Golgi apparatus. These cells are the most mature PMN precursors capable of dividing and represent the main amplification step in PMN maturation. Indentation of the nucleus defines a metamyelocyte, separation of the nucleus into nearly two lobes, a band or stab form, and segmentation of the nucleus into three or more discrete lobes, the mature PMN. The segmentation of the PMN nucleus is what gives the cells their name and may serve to permit the highly deformable PMN to negotiate narrow openings as it crawls through the tissues. [*See* HEMOPOIETIC SYSTEM; MACROPHAGES.]

Regulation of PMN maturation is increasingly well understood. To some extent, the distribution of precursors among lineages may be a stochastic process governed by more or less random intracellular programs. In addition, however, hematopoietic glycoprotein hormones, the colony-stimulating factors (CSFs), drive PMN production in response to demand, such as infection or tissue damage. A variety of peripheral tissues produce these CSFs, which include granulocyte–macrophage CSFs, which are capable of inducing both PMN and mononuclear phagocyte maturation, and granulocyte CSFs, which may be more specific for driving maturation in the direction of the PMN (Fig. 1). CSFs act on specific receptors to stimulate mitosis of early PMN precursors and their maturation along the PMN pathway and also stimulate certain functions of ma-

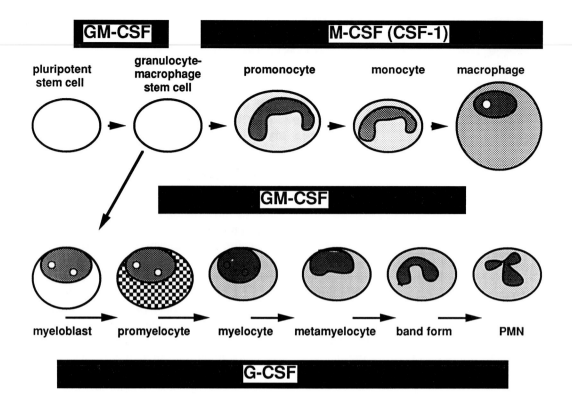

FIGURE 1 Myeloid maturation stages of neutrophils and mono-cyte–macrophages showing approximate places of regulation by cytokines.

ture PMNs. PMN maturation from myeloblast to PMN takes approximately 1 wk. Following release from the bone marrow, PMNs circulate in the peripheral blood for 6–12 hr and then enter the tissues. Once out of the circulation, PMNs do not return to it. In infectious or noninfectious inflammatory states, PMN production rates may increase up to 10-fold.

II. Functions of PMNs

A. Adhesion, Emigration (Diapedesis), Chemotaxis, Phagocytosis, and Degranulation

These related activities of PMNs are responsible for getting PMNs out of the bone marrow and into the tissues and to surround microorganisms and damaged host cells. When a PMN leaves the bone marrow, it acquires a morphology characteristic of motility. The cell becomes elongated and extends a pseudopod in the direction of movement. This pseudopod attaches transiently to the surface, and the cell body is drawn forward. Repetition of these steps results in translational locomotion. Once in the peripheral blood, PMNs become round and

many of them roll along the walls of blood vessels. At places where the vessels have been exposed to inflammatory stimuli, especially in postcapillary venules, the PMNs attach to the vessel endothelium and once again take on the motile shape, penetrate between endothelial cells, and move into the connective tissues (diapedesis). Mediators that promote adhesion of PMNs to endothelium include interleukin-1 (IL-1), thrombin, and tumor-necrosis factor. IL-1 works by inducing the expression of a PMN-specific receptor known as ELAM-1 on the surface of endothelial cells.

Locomotion requires the expression of adhesion receptors at the external surface of the PMN plasma membrane. The best understood receptors are heterodimeric glycoproteins, known as leukocyte cellular adhesion molecules (Leu-CAMs) belonging to the integrin superfamily, which includes receptors for fibronectin, vitronectin, and other connective tissue constituents. PMNs express at least three Leu-CAMs. Each has a common (β) subunit and a larger (α) variable one, which determines the different adhesion molecules known as LFA-1, Mac-1, and Gp150,95. PMN movement also depends on the reversible assembly of the protein actin into linear filaments, the polymerization of which in turn is governed by actin-binding proteins, particularly profilin and gelsolin. Actin-binding protein (PMN filamin) cross-links assembled actin filaments into three-dimensional orthogonal networks that determine the shape of pseudopodia. Myosin causes the movement of some actin filaments and is responsible for pseudopod retraction and the drawing forward of the cell body during locomotion. A complex interplay of ions, especially calcium, and metabolites, such as polyphosphoinositides and diradylglycerols, the production and elimination of which are controlled ultimately by membrane receptor interactions with environmental ligands, serves as second messengers to regulate receptor expression and the function of actin- and myosin-regulatory proteins.

The direction of locomotion of the emigrating PMN is determined by a concentration gradient of molecules known as chemotactic factors. These factors derive from bacteria themselves (N-formyl oligopeptides), denatured proteins, mononuclear phagocytes (interleukin-8), and a variety of mediator systems, including the complement cascade (C5a), the blood coagulation system (fibrinopeptide A), phospholipid (acetylglycerylphosphorylcholine [platelet-activating factor]), and arachidonic acid metabolism (leukotriene B_4). Chemotactic factors bind to specific receptors on the membrane, which transduce signals that result in cell locomotion. It is remarkable that a cell as small as a PMN (15 μ in diameter) can detect a chemical gradient that may be as little as 1% across its surface. Because chemotactic stimuli result in a polarization of the cell's shape, presumably the apparatus for locomotion in a preferred direction becomes established and remains intact as long as the gradient persists in the same direction. PMNs do turn away from the shortest path toward the source of chemotactic molecules, but the closer they come to that source, the less they vary from the straightest trajectory (Fig. 2).

When PMNs arrive at their targets, they recognize them with still different receptors that transduce the movements of phagocytosis in which pseudopodia spread around objects. Many microorganisms and most normal human cells do not natively express ligands that these phagocytosis receptors recognize. Instead, extracellular fluids contain opsonins, which coat microorganisms or, in

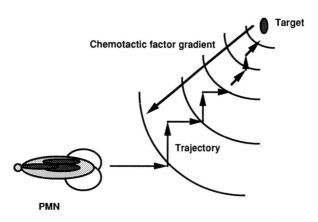

FIGURE 2 Migration of a neutrophil of a chemotactic gradient.

some cases, host cells, that react with PMN receptors. The major opsonins are immunoglobulin G, which binds to PMN Fc receptors, and two fragments derived from the third component of complement (C3). One of these fragments, C3b, ligates a receptor known as CR1; the other, iC3b, binds to the Leu-CAM adhesion molecules described above.

As PMNs ingest particulate objects with outstretched pseudopodia, the azurophilic and specific granules begin to fuse with the plasma membrane at the base of the forming phagocytic vacuole (phagocytic vesicle) and secrete their contents into it directly onto the target. The ingested object now resides within a phagolysosome and the PMN is said to have "degranulated."

B. Microbicidal, Digestive and Inflammatory Functions of PMN

The so-called effector mechanisms of PMNs are usually divided into oxidative and nonoxidative. In the former, perturbation of chemotactic, phagocytic, and certain other receptors results in the activation of an electron transport chain known as the respiratory burst oxidase. The oxidase is a complex enzyme containing a flavoprotein and a two-chain protein that incorporates a heme group and belongs to the cytochrome b family. The activated oxidase is found in the plasma membrane of the PMN, where it catalyzes the one-electron reduction of molecular oxygen (O_2) to the superoxide anion (O_2^-). Superoxide spontaneously "dismutes" to hydrogen peroxide (H_2O_2). Hydrogen peroxide is a substrate of another heme-containing enzyme, myeloperoxidase, which is present in abundance in the azurophilic granules of PMN, where its greenish color is responsible for the characteristic hue of pus. Myeloperoxidase degrades hydrogen peroxide, and this degradation in the presence of chloride, the dominant anion, results in the formation of hypochlorous acid (HOCl). Hypochlorous acid oxidizes amino groups of target proteins to chloramines, relatively stable oxidizing species that can diffuse considerable distances. Another pathway for the disposition of hydrogen peroxide is for it to combine with superoxide, resulting in the production of the highly reactive hydroxyl radical (OH·). This reaction is catalyzed by iron-chelates, which may be presented to this system by ferritin, an iron storage protein that releases iron in the pressure of superoxide, and by another granule-associated constitutent, the iron-binding protein lactoferrin. The large number of different potent oxidants released by stimulated PMNs have considerable antimicrobial and inflammatory potential. The oxidants are capable of cleaving peptide bonds, oxidizing double bonds in lipid fatty acids, degrading DNA, and even altering DNA so as to mutagenize microorganisms and mammalian cells.

Other oxygen-related effector mechanisms of PMNs do not depend on the respiratory burst oxidase. One depends on an arginine-oxidizing enzyme, which metabolizes arginine to produce nitric oxide (NO_2). Another is the arachidonic acid-metabolizing pathway, which converts arachidonate to biologically active prostaglandins and especially to leukotrienes, which have potent vasoconstrictor and permeability-promoting effects. PMNs also synthesize platelet-activating factor. [*See* RESPIRATORY BURST.]

The nonoxidative effectors of PMNs consist primarily of hydrolytic enzymes and of relatively basic (cationic) proteins with antimicrobial activities and reside in the granules of resting cells. The enzymes include proteinases, such as an elastase that is active at neutral pH, collagenase and a gelatinase, phosphatases, lipases, phospholipases, glucosidases, and muramidase (lysozyme). The antimicrobial proteins go under the names bacterial permeability-inducing factor, defensins, and antibiotic peptides. They act by attacking the cell walls of microorganisms rendering them incapable of replication.

C. Antimicrobial Effects of PMNs

Nothing attests more effectively to the importance of PMNs for human defense against microorgan-

isms than clinical states in which the host does not produce adequate numbers of PMNs. This condition is seen when cancer, leukemia, and cancer chemotherapy destroy bone marrow PMN precursors or in certain uncommon congenital or acquired conditions in which PMN production is impaired. In these disorders, when the peripheral blood PMN concentration falls $<500/\mu$, the affected patients suffer from recurrent infections caused by bacteria and fungi, and the lower the PMN count, the greater the incidence of infections. The infections appear particularly on the skin, in the lungs, and in the mouth and lower gastrointestinal tract, places where the patient is exposed most to environmental microorganisms. Accordingly, the constant flow of PMNs from the blood into these tissues is vitally important for human survival. In clinical situations in which the bone marrow is unable to produce PMNs at all, the physician can do little except treat infections with antibiotics until bone marrow function returns or is replaced by transplantation. When some bone marrow capacity to generate PMNs exists, this function may possibly be stimulated by the administration of CSFs, which are now available for such therapy.

Another set of clinical examples that demonstrate the importance of functioning PMNs is where PMNs are present but lack one or more of their normal activities, and the individuals who have such PMNs are subject to recurrent and sometimes life-threatening bacterial and fungal infections. In one autosomal recessively inherited set of genetic diseases, PMN precursors do not synthesize normal quantities of the β-subunit of the Leu-CAM protein, and the PMNs fail to express this important adhesion ligand and opsonin receptor. As a result, these PMNs fail to stick to surfaces and, therefore, cannot locomote properly; they also do not efficiently ingest microorganisms opsonized with iC3b. Another important family of genetic disease affecting PMN function is chronic granulomatous disease, the phenotype of which is PMNs that do not express a normal respiratory burst oxidase and, therefore, lack the major oxidative effector function. Chronic granulomatous disease can be an X-linked or autosomal recessive disease resulting from defective biosynthesis of one or another of the cytochrome *b* oxidase subunits or of the several components involved in its activation. Of particular interest is the fact that a few X-linked (male) patients who lack the *b*-cytochrome show increased cytochrome expression and respiratory burst activity in their PMNs after treatment with the lymphokine interferon-γ.

D. PMN in Inflammation

The contribution of PMNs to inflammation is exemplified by the well-known association of pus with redness, pain, heat, and tissue destruction. The many powerful oxidative and nonoxidative effector mechanisms of PMN explain how these cells could cause these findings of inflammation. In addition, experiments have been performed in which the PMN content of animals was reduced by use of bone marrow toxins or specific antibodies and in which reduced inflammatory responses to immune complexes in the lung, the skin, and the kidneys were demonstrated. Diminished inflammatory reactions are also characteristic for human patients with low PMN numbers or function, although the manifestations of inflammation are never totally absent. Presumably, mononuclear phagocytes (macrophages) are responsible for the residual effects observed.

Inflammation is the price to be paid for effective antimicrobial defense. Vasodilation, which accounts for the redness and some of the swelling observed, brings more PMNs, antibodies, and complement proteins to the invasion site. Tissue destruction by phagocytes may serve a beneficial function resembling surgical debridement of dead tissues, which serves as food for microorganisms. When inflammation becomes chronic, however, the cost exceeds the benefits. One cause of chronic inflammation is the presence of foreign bodies in the tissues. Microorganisms manage to wall themselves off on such objects so that PMN cannot kill them, despite constant bombardment with their microbicidal agents. Other examples of useless inflammation are the autoimmune disorders or idiopathic (unknown cause) inflammatory diseases such as ulcerative colitis. In the latter, chronic pyogenic inflammation of the colon can lead to destruction and scarring of the organ such that its function is seriously impaired. It has even been proposed that toxic oxygen metabolites of PMNs are capable of mutagenizing cells, resulting in neoplastic transformation. This would explain the high incidence of colon cancer in patients suffering from ulcerative colitis.

Unfortunately, few effective treatments for PMN-based inflammation presently exist. Glucocorticosteroids in high doses suppress the emigration of PMNs, and this effect may in part explain the anti-inflammatory effects of this medication. Researchers are studying the possible utility of antibodies directed against Leu-CAMs as a future therapy for reducing PMN functions and inflammatory effects. [*See* INFLAMMATION.]

Bibliography

Babior, B. M., Stossel, T. P. (1991). Functions and nonneoplastic disorders of phagocytic cells. *In* "Blood, Principles and Practice of Hematology" (T. P. Stossel, R. I. Handin, and S. E. Lux, eds.). Lipincott, Philadelphia.

Stossel, T. P. (1987). The molecular biology of phagocytes and the molecular basis on nonneoplastic phagocyte disorders. *In* "The Molecular Basis of Blood Diseases" (G. Stamatoyannopoulos, A. W. Nienhuis, P. Leder, and P. W. Majerus, eds.), pp. 499–533. W. B. Saunders, Philadelphia.

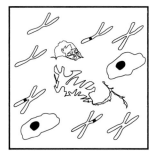

Newborn Screening

LINDA McCABE and EDWARD R. B. McCABE, *Baylor College of Medicine*

Glossary

Autosomal Refers to 22 of the 23 chromosomal pairs which are not the sex chromosomes

Carriers Individuals who are phenotypically normal and have only one mutant gene of a pair for a recessive trait; also known as heterozygotes; expression of the recessive trait requires two mutant genes of the same pair for that trait, one inherited from each carrier parent

Diagnosis In the context of a newborn screening program, this refers to the definitive assignment of a disease classification to an individual after pursuing a positive screening test by taking a history, performing a physical examination, and carrying out the appropriate definitive tests; screening is not the equivalent of diagnosis

Dried blood specimens Samples taken by pricking the skin and allowing blood to soak into a filter paper, where it dries; the blood-impregnated filter paper is then used for newborn screening

False-negative result Newborn screening test result indicating a low risk for the disease, although the individual does have the disease; also referred to as a "missed" patient

False-positive result Newborn screening test result indicating a high risk of the disease, although the individual does not have the disease; also known as a "false alarm"

Negative result Newborn screening test result suggesting that the individual has a low risk for disease

Phenylketonuria Recessive genetic disease due to decreased phenylalanine hydroxylase; treatment consists of a diet restricted in phenylalanine; untreated, this disease can lead to mental retardation

Positive result Newborn screening test result suggesting that the individual has a high risk for disease

Recessive Refers to a genetic trait which requires both genes of a pair to be mutant before expression of the trait is observed; that is, both parents must be carriers

NEWBORN SCREENING involves the application of tests or examinations to newborn infants in order to separate those who are likely to have a disease from those who are not. Every newborn should undergo this testing, as the tests are rapid and inexpensive. Positive results on these tests are not a diagnosis of disease, but, rather, are an indication that the infant needs further testing and evaluation before treatment is initiated. The purpose of newborn testing includes the detection of disease in order to prevent irreparable damage by the early initiation of therapy, the identification of disease-free individuals that carry the gene(s) (i.e., heterozygotes or carriers) for a disease that could appear in their children, and the accumulation of information regarding the genetic composition of the population tested.

I. Introduction

A. Purpose

The goals of newborn screening are to minimize or prevent the effects of untreated or late-treated genetic diseases and other birth defects. The use of automated laboratory techniques facilitate rapid cost-effective identification of populations at significant risk for disorders that can have devastating outcomes, including fatality. These disorders can be treated more effectively when affected newborns are diagnosed early and start appropriate therapy. The diseases are not readily recognized during the newborn period with normal well-child care. Follow-up includes further medical evaluation and treatment when a more extensive examination and testing are diagnostic of the disorder.

B. Principles

To provide screening services efficiently to large populations, the tests must be rapid and reliable, amenable to automation, and inexpensive.

1. The Entire Target Population Should Be Screened

Unless 100% of the newborns are tested, cases will be missed because affected infants are not tested. A child might not be tested in a newborn screening program for various reasons, including lack of information or incorrect information regarding the timing of specimen collection, specific policies of a hospital or practitioner that discourage comprehensive screening, and an individual's disinterest in screening. Populations more likely to escape testing include the homeborn and infants born to transient or homeless families or to parents working or traveling abroad. Other newborns might not be tested because of events during the perinatal period, such as prematurity, early hospital discharge (especially before 24 hours of age), illness, parenteral feeding, and transfer from one hospital to another or from one unit to another within the same hospital.

2. False-Negative Results Should Be Minimized

A child who has the disease, but is classified as being at low risk for the disease by the newborn screening test, will not receive follow-up evaluation and will not be diagnosed during the newborn period. This result is termed "false negative." Certain characteristics influence the number of false-negative results, including the reliability and validity of the test and the quality control of the laboratory running the test. There are also biological explanations for false-negative results, such as an inadequate specimen (e.g., insufficient blood or a mishandled specimen), an inappropriate specimen (cord blood, because the metabolite, such as phenylalanine, has not had time to become elevated; blood collected after a plasma or red blood cell transfusion, because normal protein has been added by the transfusion; exchange transfusion; or dialysis, because the metabolite has been removed), and a specimen obtained from a newborn before physiological changes have occurred to allow the screening test to detect the high risk.

To decrease the possibility of false-negative results, some testing programs use a second newborn screening test, administered when the child is 2 weeks of age. To minimize the frequency of false-negative results, other programs adjust the level of the test result used to distinguish infants at high risk from those at low risk for the disease. This might increase the number of false-positive results and the time and effort in follow-up testing and/or treatment. These decisions are made by the public health authorities responsible for supervising the newborn screening program.

3. False-Positive Results Should Be Minimized

A child who does not have a disease, but is labeled by the screening test as having a high risk for the disease, has received a false-positive result, also known as a "false alarm." This child would receive follow-up evaluation to determine the presence or absence of the disease. A high rate of false-positive results produces unnecessary increases in the cost of follow-up and in the anxiety of the parents of infants receiving such results and diminishes the responsiveness of the health professionals contacted for follow-up evaluations.

4. Adequate Expeditious Follow-up Should Be Available to All Infants with Positive Test Results

Each newborn with a positive newborn screen must receive confirmatory diagnostic follow-up testing without delay. Time could be critical in the prevention of mental retardation, neurological damage, or death. In addition, treatment is necessary for children whose diagnosis is confirmed by the follow-up tests. Unfortunately, many newborn screening programs provide only for the newborn testing, but do not fund personnel, follow-up testing, or required treatment.

5. Benefits of Newborn Screening Should Exceed Costs of the Program

The U.S. General Accounting Office estimated that for each $1 spent on newborn screening, $24 was saved in lifetime care costs for untreated patients. There are additional benefits to newborn screening that cannot be readily assigned a dollar value. These include saving a life, preventing mental retardation, counseling a couple regarding the risk for having other affected children, and improving understanding of the frequency and treatment of these diseases.

C. Outcomes

Newborn screening test results could indicate either a high (positive) or low (negative) risk for having a disease. Infants with a high risk are given follow-up testing. Depending on the disease, the follow-up is a repeat of the newborn screening test or a test that is diagnostic for the disease. Usually, treatment is delayed until these test results are available. However, if the disease is life-threatening and treatment for a brief period of time is safe for unaffected infants, therapy might be instituted immediately upon the initial positive screening result. Once the diagnosis is confirmed and treatment is initiated, most patients remain on treatment for life. This continued therapy is required to prevent the life- or intellect-threatening consequences of the disease.

D. Medical Legal Issues

When a newborn screening test does not identify an infant with the disease, the child has a false-negative result and will not receive the appropriate follow-up to newborn screening. In the absence of follow-up testing and treatment, the infant will develop the undesirable consequences of the disease. If the child is diagnosed as having the disease after developing mental retardation or after death from the disease, the family might seek legal recourse.

II. Phenylketonuria as a Model

A. Development

Phenylketonuria (PKU) is an autosomal-recessive biochemical genetic disease that leads to mental retardation if untreated. The incidence of PKU is one in 10,000 to one in 25,000 in the United States. Individuals with this disease do not metabolize phenylalanine, an amino acid found in protein foods. Treatment involves restricting the amount of phenylalanine consumed. The newborn screening test involves measurement of the level of phenylalanine in the blood. [*See* Phenylketonuria, Molecular Genetics.]

The initial newborn screening test for PKU, which has been abandoned, involved the measurement of phenylpyruvic acid in the urine. Phenylpyruvic acid is produced in the body from phenylalanine, and is therefore an indirect indicator of the level of phenylalanine. This test had a high false-negative rate, classifying one-third to one-half of the infants with the disease as low risk.

The Guthrie test currently used is a bacterial inhibition assay that measures the amount of phenylalanine in the blood; this test is more reliable and valid. The Guthrie test has the additional advantage of using dried blood on a filter paper, which is obtained by pricking the heel of the newborn; this simplifies obtaining the specimen and facilitates specimen transport to regionalized screening laboratories. The Guthrie test is a semiquantitative screening test and is not diagnostic. This test uses a mutant bacteria which requires phenylalanine for growth; the higher the phenylalanine in the dried blood spot, the larger the bacterial growth zone on the test plate.

Another semiquantitative method is paper chromatography, which separates the amino acids as spots on a sheet of paper, one end of which is immersed in a solvent. In addition, there are two tests that provide quantitative diagnostic measures of phenylalanine in the blood: the fluorometric method (based on the fluorescence of phenylalanine with other chemicals) and the column chromatographic method (based on separation and quantitation of each amino acid). Newborn screening is typically administered by each state's Department of Health. Currently, all states in the United States screen for PKU.

B. Treatment Programs

Dietary restriction of phenylalanine with regular monitoring of phenylalanine blood levels is the treatment for PKU. Treatment should begin as early as possible, definitely before 4 weeks of age, to prevent the mental retardation associated with PKU. The initial strategy involved the use of special formulas containing little or no phenylalanine. It was then determined that a small amount of phenylalanine was required for normal growth and development, so a small amount of phenylalanine was added to the diet; for infants this phenylalanine is added in the form of regular infant formula. In the past these practices meant that a breast-fed infant with a positive newborn screening for PKU and positive follow-up diagnostic testing was abruptly weaned from breast milk and switched to the formula diet low in phenylalanine. Subsequent research has demonstrated that phenylalanine supplementation can be provided by human breast milk; hence, breast milk and phenylalanine-free formula are now used to maintain blood levels of phenylalanine at therapeutic levels.

When PKU was initially treated with diet, the expectation was that treatment was only necessary during the first 4–6 years of life, while the nerve cells were undergoing full development (i.e., myelination). After this it was expected that the child could eat a normal diet, without phenylalanine restriction. However, this approach led to poorer school performance and a deterioration in behavioral control. The current recommendation is for a person with PKU to continue on a phenylalanine-restricted diet for his or her lifetime. This, as described below, is particularly true for females with PKU.

C. Areas of Concern

Females of child-bearing age who have PKU need to be on a phenylalanine-restricted diet prior to conception and continuously throughout their pregnancy to protect the fetus. The fetus of a woman on an unrestricted diet might have microcephaly, mental retardation, or congenital heart disease. Currently, there is a widespread search for women with PKU who might not be aware of this potential problem and who are no longer on phenylalanine-restricted diets.

III. Other Diseases for Which Newborn Screening Is Available

A. Congenital Hypothyroidism

Congenital hypothyroidism is caused by inadequate production of thyroid hormone (i.e., thyroxine, or T_4). Patients who are not identified and treated promptly with supplementary thyroid hormone will be mentally retarded, fail to grow, suffer constipation, and have a slow metabolic rate. The incidence of congenital hypothyroidism is one in 3600 to one in 5000 in the United States. In this country routine newborn screening uses dried blood spots, obtained at the same time as those for the PKU test, for immunoassay measurement of T_4 and/or thyroid-

stimulating hormone. All states currently screen for congenital hypothyroidism as well as PKU, although the programs vary in the other diseases screened.

B. Sickle-Cell Disease

Sickle-cell disease, an autosomal-recessive disease, is caused by an abnormality in the β-chains of hemoglobin, which transports oxygen in the blood. Untreated patients could have overwhelming infection, chronic anemia, "crises" involving the blockage of blood vessels, and pooling (or accumulation) of blood in the spleen. The incidence of sickle-cell disease is one in 400 among Afro-Americans. Treatment for these patients includes prophylactic antibiotics, immunization against specific bacterial infections, and rapid access to appropriate medical care. Newborn screening is performed on cord blood or dried blood spots and involves separation of the hemoglobin proteins by electrophoresis or some other method. [*See* SICKLE CELL HEMOGLOBIN.]

C. Galactosemia

Galactosemia, an inherited disorder of galactose metabolism, results in failure to thrive and in vomiting, liver disease, cataracts, and mental retardation. The disease could even be lethal, if untreated, due to overwhelming infection or liver disease in the newborn period. The frequency of galactosemia is one in 60,000 to one in 80,000 in this country. Treatment involves a galactose-free diet, which should be started as early as possible and continued throughout life. Newborn screening involves a test for either elevated blood galactose levels or deficiency of the enzyme that causes galactosemia.

D. Cystic Fibrosis

Cystic fibrosis is an autosomal-recessive disease that results in abnormalities of the lungs, pancreas, and sweat glands. Some affected infants have small-bowel obstruction due to meconium ileus or impaction of the fetal stool with intestinal blockage. Others die from malabsorption and malnutrition. Young adults die from pulmonary disease and infection. The incidence of cystic fibrosis is one in 2000 among whites in the United States. Treatment involves special dietary supplementation, pulmonary management, and antibiotics. The primary benefit of newborn screening appears to be the prevention of early malnutrition; the possibility of benefits to pulmonary function and long-term survival is the subject of current research. The newborn screening test for cystic fibrosis involves an immunoassay measurement of trypsin in dried blood specimens. [*See* CYSTIC FIBROSIS, MOLECULAR GENETICS.]

E. Maple Syrup Urine Disease

Maple syrup urine disease is an autosomal-recessive disorder that results in lethargy, irritability, and vomiting and progresses to coma and death in the first 1–2 months of life if untreated. The incidence of this disease is one in 250,000 to one in 300,000 in this country. Treatment involves the dietary restriction of branched-chain amino acids. Newborn screening for this disorder is similar to that for PKU, but the amino acid leucine is measured by the bacterial inhibition assay.

F. Homocystinuria

Homocystinuria is an autosomal-recessive disorder involving a block in the pathway for breakdown of methionine, an amino acid. It has an incidence of one in 50,000 to one in 150,000 in the United States. Untreated patients develop a marfanoid physique (tall and thin, with long limbs, toes, and fingers), thromboembolism, dislocated lenses, and osteoporosis. Between 65 and 80% of untreated individuals are mentally retarded, and there is also an increased frequency of seizures, psychiatric disturbances, and myopathy. The treatment for homocystinuria depends on the underlying cause of the disease and involves vitamin treatment and/or dietary management. Newborn screening is similar to that for PKU, but methionine is measured by the bacterial inhibition assay.

G. Biotinidase Deficiency

Biotinidase deficiency is an autosomal-recessive disorder due to a block in the recycling of biotin, which, in turn, results in deficiencies of the carboxylase enzymes. Untreated patients can have convulsions, skin rashes, baldness, ataxia, coma, or might even die. Survivors are generally developmentally delayed. The estimate of the incidence of this disorder from limited patient series is one in 70,000. Treatment for this disorder consists of oral biotin. Newborn screening for biotinidase uses a colorimetric enzymatic assay on the dried blood specimen.

H. Congenital Adrenal Hyperplasia

Patients affected by congenital adrenal hyperplasia, an autosomal-recessive metabolic abnormality in steroid hormone synthesis, if untreated, might have life-threatening episodes with dehydration, salt imbalance, and shock in the neonatal period. Females have ambiguous genitalia with masculinization, frequently leading to the incorrect assignment of sex at birth. The incidence is approximately one in 12,000 in the United States, but this number varies substantially among racial and ethnic groups, up to one in 680 in Yupik Eskimos. Medical management involves steroid replacement, but surgical treatment is frequently required for female infants. Approximately 90–95% of the patients with the disease can be detected by newborn screening, which is performed by immunoassay measurement of an abnormally elevated steroid metabolite in the dried blood specimens.

I. Duchenne Muscular Dystrophy

Duchenne muscular dystrophy is an X chromosome-linked recessive disease which primarily affects males and leads to progressive deterioration of muscles, beginning in infancy, with death in the second or third decades. No effective medical therapy is available. Physical therapy and bracing might prolong function and ambulation. Screening involves the measurement of creatine kinase in dried blood specimens and currently requires a specimen drawn after 4 weeks of age. Screening is limited to pilot programs. [*See* MUSCULAR DYSTROPHY, MOLECULAR GENETICS.]

J. Toxoplasmosis

Toxoplasmosis is not a genetic disease, but a congenital parasitic infection contracted by the mother primarily from raw meat and cat litter. Severely affected neonates suffer intrauterine growth retardation, brain calcifications, convulsions, retinal abnormalities, and severe developmental delay. More mildly affected children might have a later onset of developmental abnormalities. This milder form could be amenable to treatment. Newborn screening for this disorder remains investigational, while questions of incidence and benefit from early intervention are being addressed.

K. Acquired Immunodeficiency Syndrome

Acquired immunodeficiency syndrome (AIDS) is a disorder caused by the human immunodeficiency virus, which can be acquired congenitally by transmission from the mother. At this time newborn screening has been completely anonymous and used solely for epidemiological purposes. [See Acquired Immunodeficiency Syndrome (Virology).]

IV. Future Directions

A. New Diseases

Currently, all states and the District of Columbia have newborn screening programs for PKU and congenital hypothyroidism. A 1987 National Institutes of Health consensus conference recommended universal screening for sickle-cell disease. Newborn screening programs continue to evolve, with different states and regions carrying out groups of tests generally from the above list.

B. New Technology

Molecular genetic techniques enable researchers to obtain DNA from dried blood specimens. Using a technique called the polymerase chain reaction, small amounts of DNA are taken from the dried blood spots and amplified to amounts that can be analyzed more easily. The amplified DNA can be used to establish the genotype for many disorders, including sickle-cell disease and PKU. It is expected that this technology will be applicable to cystic fibrosis in the near future.

Bibliography

American Academy of Pediatrics, Committee on Genetics (1982). New issues in newborn screening for phenylketonuria and congenital hypothyroidism. *Pediatrics* **69**, 104.

American Academy of Pediatrics, Committee on Genetics (1987). Newborn screening for congenital hypothyroidism: Recommended guidelines. Pediatrics **80**, 745.

American Academy of Pediatrics, Committee on Genetics (1989). Newborn screening fact sheets. *Pediatrics* **83**, 449.

Andrews, L. B. (ed.) (1985). "Legal Liability and Quality Assurance in Newborn Screening." American Bar Foundation, Chicago, Illinois.

Andrews, L. B. (1985). "State Laws and Regulations Governing Newborn Screening." American Bar Foundation, Chicago, Illinois.

Jinks, D. C., Minter, M., Tarver, D. A., Vanderford, M., Hejtmancik, J. F., and McCabe, E. R. B. (1989). Molecular genetic diagnosis of sickle cell disease using dried blood specimens from newborn screening blotters. *Hum. Genet.* **81**, 363.

McCabe, E. R. B., McCabe, L., Mosher, G. A., Allen, R. J., and Berman, J. L. (1983). Newborn screening for phenylketonuria: Predictive validity as a function of age. *Pediatrics* **72**, 390.

Newborn screening for sickle cell disease and other hemoglobinopathies. (1987). *JAMA, J. Am. Med. Assoc.* **258**, 1205.

Newborn screening for sickle cell disease and other hemoglobinopathies. (1989). *Pediatrics, Suppl.* **83**, 813.

Therrell, B. L., Jr. (ed.) (1987). Advances in Neonatal Screening, *Int. Congr. Ser.—Excerpta Med.* **741**.

Nonnarcotic Drug Use and Abuse

LESTER GRINSPOON, JAMES B. BAKALAR,
Harvard Medical School

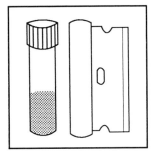

Glossary

Acute effects Caused by a single dose of a drug
Chronic effects Caused by long-term and repeated use of a drug
Cross-tolerance Tolerance to one drug that develops through repeated use of another drug with similar neurophysiological activity
Metabolic tolerance Tolerance that arises because the liver breaks the drug down and eliminates it from the body faster
Pharmacodynamic tolerance Tolerance arising from adaptation of nerve receptors in the brain to a drug
Substance dependence (drug dependence) Psychiatric disorder that involves such symptoms as repeatedly taking a drug while knowing that it does harm but being unable to stop, often taking more of a drug than intended, and developing tolerance or a withdrawal reaction to a drug, or both
Tolerance Loss of sensitivity to a drug that makes it necessary to take more to preserve the original effect
Withdrawal reaction Physical and psychological symptoms that may arise when a dependent user stops taking a drug

SUBSTANCE ABUSE is defined as a maladaptive pattern of substance use indicated by continued use despite the persistent social, occupational, psychological, or physical problems it causes or recurrent use when it is hazardous. Interpretations of drug abuse are vulnerable to cultural prejudice; for the purposes of this article, it means taking drugs at dose levels and in circumstances and settings that significantly augment their potential for harm.

Physical dependence on a drug implies a biochemical or physiological change in the body that makes the continued presence of the drug necessary to avoid a withdrawal reaction. Tolerance refers to a declining effect of the drug on repeated administration of the same dose and the consequent necessity to increase the dose to obtain the original euphoric effect. A widely accepted definition of substance dependence involves at least three of the following: (1) much time spent thinking about the drug or arranging to get it, (2) often taking more than intended, (3) tolerance, (4) withdrawal symptoms, (5) use to avoid or relieve withdrawal symptoms, and (6) persistent or repeated desire or effort to cut down use.

Although neither substance abuse nor substance dependence is now classified as a personality disorder, those who develop a dependency or abuse problem often need a euphoriant to experience pleasure or a respite from pain. People with character disorders, anxiety, depression, feelings of inadequacy, or intolerable life situations are susceptible to abuse of a number of drugs in addition to alcohol and opiates. Just as some adults use alcohol when they feel depressed, anxious, and inadequate, some young people may try to deal with emotional problems through the use of illicit drugs. When drug abuse is symptomatic, the abuser can often make good use of psychotherapy. Once a substance has been abused to the point of compulsive use or physical dependency, treatment becomes more difficult and complicated. In these cases, community resources concerned with drug abuse are an important complement to the traditional psychiatric approach.

I. Amphetamines

Amphetamines and amphetamine congeners are a large group of central stimulant drugs. Among the best known are dextroamphetamine (Dexedrine), metamphetamine (Methedrine), and methylphenidate (Ritalin). Racemic amphetamine sulfate (Benzedrine) was first introduced as a medicine in 1932 and became available in tablet form in 1937. Soon, amphetamine began to receive much sensational publicity, and many physicians came to regard it as a versatile remedy second only to a few other extraordinary drugs such as aspirin. In 1971, the annual legal United States production of all kinds of amphetamines reached more than 10 billion tablets. From the mid-1960s on, considerable growth occurred in illicit laboratory synthesis and black market diversion of legitimately produced drugs.

Since 1970, amphetamine abuse has declined, partly because of legal restrictions. Amphetamines now have accepted therapeutic applications only in treating narcolepsy (a sleep disorder) and attention deficit disorder, and as an occasional adjunct to tricyclic antidepressants in the treatment of depression. Few physicians now prescribe amphetamines for weight loss, because of well-founded doubts concerning efficacy and safety.

Both acute amphetamine intoxication and chronic use have numerous adverse physical effects. Symptoms of acute poisoning include flushing, pallor, cyanosis (bluish skin), fever, high heart rate, nausea, vomiting, difficulty in breathing, tremor, lack of coordination, loss of sensory capacities, elevated blood pressure, hemorrhages, strokes, convulsions, and loss of consciousness. Death from overdose is usually associated with high fever, convulsions, and cardiovascular shock. Intravenous abuse can produce diseases of the liver, lung, heart, and arteries.

Adverse psychological effects include restlessness, excessive talking, insomnia, irritability, hostility, tension, confusion, anxiety, panic, and, quite commonly, psychosis. Clinical experiments using

volunteers show that a psychosis may be induced in essentially normal people by even short-term administration of dextroamphetamine.

The symptoms of amphetamine psychosis strikingly resemble those of paranoid schizophrenia. Restlessness, increased irritability, and heightened perceptual sensitivity often precede delusions of persecution and bizarre visual and auditory hallucinations. Some authorities believe that predominance of visual hallucinations, appropriate mood, clear consciousness, hyperactivity, hypersensitivity, or absence of thought disorder distinguishes amphetamine psychosis from an acute schizophrenic episode. Others consider a strictly clinical differentiation all but impossible. Laboratory tests for amphetamine in urine are technically complex and are negative after 48 hr. Physicians often recognize amphetamine psychosis only in retrospect, by the rapid disappearance of the symptoms—usually within days or, at most, weeks. Suspiciousness and ideas of reference sometimes persist for months after the overt psychosis has ended. Psychosis usually occurs when an abuser who is already taking large doses takes even larger amounts than usual for a period of time.

Chronic amphetamine abusers often find that the drug begins to dominate their lives through a craving severe enough to be called a compulsion. Their irritability and paranoia may cause unprovoked violence and drive their friends away; their preoccupation with the drug has a disastrous effect on their family relationships and work. A high degree of tolerance develops; an abuser may eventually need up to twenty times as much to recover the original euphoric effect.

A letdown or crash often occurs when an amphetamine abuser is forced to stop using the drug for a time because it is producing agitation, paranoia, and malnutrition. A debilitating cycle of runs (heavy use for several days to a week) and crashes is a common pattern of abuse. The physical symptoms of withdrawal include headache, sweating, muscle cramps, stomach cramps, and hunger. The characteristic psychological symptom is a lethargic depression, suicidal at times, which peaks at 48–72 hr after the last dose of amphetamine but may persist at lesser intensity for weeks. Often a vicious cycle develops. Patients who suffer from chronic depression or feelings of inadequacy take amphetamine for relief and become dependent on the drug. When they try to stop using it, they become depressed even further and feel a need to start again.

II. Cocaine

Cocaine is an alkaloid derived from the shrub *Erythroxylum coca,* a plant indigenous to Bolivia and Peru, where peasants chew its leaves for their stimulating effect. Cocaine was isolated in 1860 and became important after 1884 as the first effective local anesthetic (also the only current medical use). In 1914, cocaine was subjected to the same laws as morphine and heroin and since then has been legally classified with narcotics. Since about 1970, cocaine has steadily been gaining popularity and is now the second most popular illicit drug after marijuana.

Street cocaine varies greatly in purity; it is usually adulterated with sugars, procaine, or other substances. Users rarely take cocaine by mouth, because the effect is regarded as too mild to warrant the expense. Three widespread methods of cocaine ingestion are prevalent: snuffing (snorting), subcutaneous or intravenous injection, and freebasing. Freebasers convert cocaine hydrochloride, cocaine alkaloid (the free base) by treating it with a basic solution such as baking soda or ammonia. They then smoke the alkaloid in a water pipe. Freebase cocaine is now being sold on the streets under the name "crack." Freebasing and injection allow the drug to reach the brain faster, providing an ecstatic rush and subsequent letdown that create particularly serious abuse problems.

The adverse effects of cocaine resemble those of amphetamine. One common acute effect is an anxiety reaction with symptoms including high blood pressure, racing heart, anxiety, and paranoia. High doses or repeated use can produce a state resembling mania, with impaired judgment, incessant rambling talk, and hyperactivity. Because cocaine increases energy and confidence and can produce irritability and paranoia, it may also lead to physical aggression and crime. In high doses, especially when smoked or injected, cocaine can produce cardiac symptoms, nausea, headache, cold sweat, tremors, and fast, shallow breathing. People who have high blood pressure or damaged arteries may suffer strokes. High doses can also cause seizures and death from cardiac or respiratory arrest. Deaths from opiates and cocaine injected in combination intravenously (a speedball) occur frequently. The number of deaths from cocaine and emergency room admissions for drug reactions involving cocaine has increased greatly over the past decade.

Cocaine, along with some amphetamines, is the drug most eagerly consumed by experimental animals; they will kill themselves with voluntary injections. Craving can also become a serious problem for human beings who have constant access to cocaine, especially for freebasers and intravenous abusers. The financial as well as physical and psychological costs of compulsive cocaine abuse can be devastating. A damaging habit usually develops over a period of several months to several years. A compulsive user thinks about the drug constantly, cannot turn it down, and may borrow, steal, or deal to pay for it. Cocaine produces substantial acute tolerance and some long-term tolerance. Experimenters report cross-tolerance to amphetamines. On the other hand, sensitization to the seizure-inducing effect may occur.

A common pattern of cocaine abuse consists of binges followed by crashes—the same pattern found in amphetamine abuse and observed in laboratory animals as well as human beings. A week-long binge produces extreme acute tolerance and exhaustion from lack of food and sleep. The user than falls into several days of severe depression, excessive sleep, overeating, and sometimes chills or muscle pains. Craving is likely to continue after this acute phase is over and may reemerge even many months or even years later, especially when certain moods or situations have come to evoke a conditioned response.

Some effects of chronic cocaine abuse are weight loss, insomnia, perceptual disturbances, paranoid thinking, and psychoses. Users often take alcohol, sedatives, or narcotics to calm cocaine-induced agitation, and this practice may create a dual dependency that is particularly hard to treat. Intensive and compulsive users become jittery, irritable, and self-absorbed. They may also become hypersensitive to sound, light, and touch. At first, cocaine may be sexually stimulating, but chronic abusers tend to lose sexual interest. Like amphetamines, cocaine may also cause absorption in apparently meaningless, repetitive activities— scribbling, counting, pacing, sorting and ordering, and taking apart and reassembling mechanical objects.

Each method of ingestion also has its own dangers. Snorting causes noses to become runny or clogged and sometimes inflamed, swollen, or ulcerated; heavy users occasionally develop perforated septa. Freebasing may damage the surface of the lung and reduce its capacity to exchange gases. Injection may lead to infection or embolism. [*See* Cocaine and Stimulant Addiction.]

III. Barbiturates

The barbiturates most commonly used are the long-acting (12–24 hr) phenobarbital, the intermediate-range (6–12 hr) amobarbital (Amytal), and the short-acting (3–6 hr) pentobarbital (Nembutal) and secobarbital (Seconal). Barbiturates are used mainly as sedatives, hypnotics, tranquilizers, anticonvulsants, and anesthetics. The illicit market meets its needs by diverting shipments from legitimate manufacturers and robbing drug warehouses. The barbiturates most common on the black market are secobarbital ("reds," "red devils," "downers"), pentobarbital ("yellow jackets," "nembies"), and a combination of secobarbital and amobarbital ("reds and blues").

But illicit barbiturates are only part of the problem. For many years these drugs were prescribed rather indiscriminately, and many persons were unwittingly maintained in the habit by family doctors. This problem is becoming less common as benzodiazepines replace barbiturates in many medical uses. Pentobarbital, secobarbital, and amobarbital are now under the same federal legal controls as morphine. Both legal and illicit barbiturate use seems to be declining.

A. Patterns of Abuse

Chronic intoxication occurs especially in middle-aged, middle-class people who obtain the drug from their doctors as a prescription for insomnia or anxiety. Now that refills are limited by law, these abusers may visit many physicians and get a prescription from each one. Their drug dependence may go unnoticed for months or years, until their work begins to suffer or they show slurred speech and other physical signs. This pattern is becoming less common as medical use of barbiturates declines.

A second pattern is periodic intoxication. Users in this category are generally adolescents or young adults who take barbiturates for the same reason they use alcohol, to produce a high or a sense of well-being. Personality, set (expectations), and setting determine whether the effect is regarded as a sedative or euphoriant. Barbiturates may become a fairly constant part of their lives, as alcohol is for the social drinker.

The most dangerous pattern is intravenous barbiturate use, mainly practiced by young adults intimately involved in the drug culture. Their drug experience has usually been extensive, including intravenous heroin and amphetamines. They often use barbiturates because the habit is less expensive to maintain then a heroin habit, even at a level of 2,000–3,000 mg/day. The rush is described as a pleasant, warm, drowsy feeling. Like amphetamine abusers (speed freaks), these barbiturate abusers tend to be irresponsible, violent, and disruptive. The physical dangers of injection include cellulitis, vascular complications from accidental injections into an artery, infections, and allergic reactions to contaminants. Users can often be identified by skin abscesses. Barbiturates are also used incidentally to dependence on other drugs—by heroin addicts to boost the effects of weak heroin, by alcoholics to enhance the intoxication or relieve the symptoms of alcohol withdrawal, and by amphetamine and cocaine abusers as a sedative to avoid paranoia and agitation.

B. Effects

Mild barbiturate intoxication (both acute and chronic) resembles alcohol intoxication. Symptoms include sluggishness, difficulty in thinking, poor memory, slowness of speech and comprehension, faulty judgment, narrow range of attention, emotional lability, and exaggeration of basic personality traits. The sluggishness usually wears off after a few hours, but judgment may remain defective, mood distorted, and motor skills impaired for as long as 10–22 hr. Other symptoms are hostility, quarrelsomeness, moroseness, and, occasionally, paranoid ideas and suicidal tendencies. Neurological effects include nystagmus, diplopia, strabismus, ataxic gait, positive Romberg's sign, hypotonia, dysmetria, and decreased superficial reflexes. The diagnosis may be confirmed by a number of laboratory tests.

All patterns of use present dangers to health. Acute intoxication can produce death from suicide, accident, or unintentional overdose. Barbiturates are the drugs most commonly taken with suicidal intent. In many doubtful cases, automatism must be considered: people whose judgment and memory are already impaired by the drug may forget or disregard previous doses and automatically take another dose (overdose) in order to fall asleep, without any conscious intent of committing suicide. The effects of alcohol and barbiturates are additive, and this combination is especially dangerous. Barbiturate-induced death arrives through a sequence of deep coma, respiratory arrest, and cardiovascular failure. The lethal dose varies with the route of administration, excitability of the central nervous system, and acquired tolerance. The ratio of lethal to effective dose can be as low as 3 to 1 or as high as 50 to 1 for the most commonly abused barbiturates.

C. Tolerance and Withdrawal

Like many other drugs, barbiturates produce pharmacodynamic or central nervous system tolerance: nerve cells adapt to the drug and require more of it to respond. By inducing drug-metabolizing microsomal enzymes in the liver, they also produce metabolic tolerance (and incidentally reduce the effectiveness of a number of other drugs, especially anticoagulants and tricyclic antidepressants). Cross-tolerance with alcohol develops.

A withdrawal reaction occurs when the drug has become necessary to maintain the proper functioning of the body. This usually requires 1–2 mo or more at doses well above the recommended therapeutic level. The barbiturate withdrawal reaction ranges from mild symptoms such as anxiety, weakness, sweating, and insomnia to seizures, delirium, and cardiovascular collapse leading to death. At its worst, it is the most severe of the drug abstinence syndromes. Pentobarbital or secobarbital users with 400 mg/day habits experience only mild withdrawal symptoms. Those taking 800 mg/day experience weakness, tremor, and anxiety; about 75% have convulsions. Users on even higher doses may suffer from anxiety, anorexia, and psychoses as well as convulsions resembling grand mal epilepsy. The psychosis is clinically indistinguishable from alcoholic withdrawal delirium (delirium tremens); its main features are confusion, agitation, delusions, and hallucinations that are usually visual but sometimes auditory. Most of the symptoms appear on the first day of abstinence, but seizures generally do not occur until the second or third day, when the symptoms are at their worst. Psychosis, if it does develop, starts in the third to the eighth day. The various symptoms may last as long as 2 wk.

Dependent users often take an average daily dose of 1.5 g of short-acting barbiturate, and some have been reported to take as much as 2.5/g/day for months. Tolerance to the lethal effect does not develop nearly to the same extent as tolerance to the desired psychoactive effect. Thus, withdrawal in a hospital may eventually become necessary to prevent accidental death from overdose.

IV. Benzodiazepines

The benzodiazepines include diazepam (Valium), fluorazepam (Dalmane), oxazepam (Serax), chlordiazepoxide (Librium), alprazolam (Xanax), and about 20 other drugs. They are used mainly in treating anxiety but also serve as sedatives, muscle relaxants, and anesthetics, and in the treatment of alcohol withdrawal. Introduced in the 1960s, benzodiazepines soon became the most popular prescription drug in the United States. A recent substantial decline in medical use suggests that physicians are becoming more cautious about them. Benzodiazepines produce little respiratory depression, and the ratio of lethal to effective dose is very high, estimated at 200 : 1 or more. Very large amounts (>2 g) taken in suicide attempts produce drowsiness, lethargy, confusion, and ataxia but rarely cause permanent damage. However, chlordiazepoxide in high doses has been known to induce coma. The adverse effects of lower doses include drowsiness, unsteadiness, and weakness. Some benzodiazepines have a disinhibiting effect, which may cause hostile or aggressive behavior in susceptible people subject to frustration.

Benzodiazepines produce less euphoria than other tranquilizing drugs, so the risk of dependence and abuse is relatively low. They do not stimulate the liver enzymes that cause metabolic tolerance nearly as much as barbiturates do, but both tolerance and withdrawal symptoms can develop. The withdrawal reaction may occur not only at very high doses but at therapeutic doses, depending on which benzodiazepine is used and for how long. Symptoms include anxiety, numbness in the extremities, dysphoria, intolerance for bright lights and loud noises, nausea, sweating, muscle twitching, and sometimes convulsions. Withdrawal is not usually accompanied by craving for the drug. Because benzodiazepines are eliminated from the body slowly, symptoms may continue to develop for several weeks. To prevent seizures and other problems, withdrawal is accomplished by gradual reduction of the dose.

V. Volatile Solvents and Aerosols

Some abused volatile solvents and aerosols are gasoline, varnish remover, lighter fluid, airplane glue, rubber cement, cleaning fluid, spray paints, deodorants, and hair sprays. The active ingredients include toluene, acetone, benzene, and halogenated hydrocarbons. Because these substances are legal, cheap, and accessible, they are used mostly by the young (ages 6–16 yr) and poor, who inhale them from a tube, can, plastic bag, or rag held over the nose. The intoxication usually lasts 15–30 min. They have a central depressant effect, characterized by euphoria, excitement, a floating sensation, dizziness, slurred speech, ataxia, and a sense of heightened power. Like alcohol, solvents may cause impaired judgment leading to impulsive and aggressive behavior, and amnesia may occur for the period of intoxication. Other acute effects are nausea, anorexia, and, in high doses, stupor and even unconsciousness. Deaths may be caused by central respiratory depression, asphyxiation, or accident.

Substantial tolerance develops after repeated sniffing. A serious risk is irreversible damage to the liver, kidney, and other organs from benzene or halogenated hydrocarbons. Peripheral neuritis has also been reported. Permanent neuromuscular and brain damage must be considered a possibility. Some clinicians report brain atrophy and chronic motor impairment in toluene users.

VI. Marijuana

Marijuana has been known for thousands of years as a medicine and intoxicant, and it was widely used in the nineteenth century as an analgesic, anticonvulsant, and hypnotic. Recently, interest has developed in using it to treat glaucoma and the nausea produced by cancer chemotherapy. It may also have antitumoral and antibiotic properties. One of its nonpsychoactive constituents, cannabidiol, may prove to be useful as an anticonvulsant. But marijuana has been valued throughout history mainly as a euphoriant. Drug preparations from the hemp plant (*Cannabis sativa*) vary widely in quality and potency, depending on the type (possibly three species or, alternatively, various ecotypes), climate, soil, cultivation, and method of preparation. A resin that covers the plant's flowers and leaves contains the active substances, the chief of which is delta-1-tetrahydrocannabinol. The drug can be taken in the form of a drink or in foods, but in the United States it is usually smoked, either in a pipe or in a cigarette called a "joint."

An extensive literature describes the psychological effects of cannabis (a general term for the psychoactive products of the plant). The intoxication varies a great deal, but common symptoms are calmness, euphoria, laughter, rapid flow of ideas, loss of short-term memory, and visual pseudohallucinations, often followed by drowsiness. The intoxication heightens sensitivity to external stimuli, reveals details that would ordinarily be overlooked, makes colors seem brighter and richer, and enhances the appreciation of art and music. Time seems to slow down, and more seems to happen in each moment. Appetite increases. The effects of smoking last for 2–4 hr; the effects of ingestion 5–12 hr.

Curiously, a splitting of consciousness may occur: The smokers, while experiencing the high, may at the same time objectively observe their own intoxication. They may, for example, have paranoid thoughts yet at the same time laugh at them and, in a sense, enjoy them. This ability to retain objectivity may explain why many experienced users manage to behave soberly in public even when highly intoxicated.

Sometimes referred to as a hallucinogen, marijuana can produce some of the same effects as lysergic acid diethylamide (LSD) and LSD-type substances: distorted perception of body parts, spatial and temporal distortion, depersonalization, increased sensitivity to sound, synesthesia, heightened suggestibility, and a sense of deeper awareness. Marijuana can also cause anxiety and paranoid reactions, but the nightmarish reactions that even the experienced LSD user may endure rarely afflict experienced marijuana smokers, who are using a far less potent drug and have more control over its effects. Marijuana lacks the powerful consciousness-altering qualities of LSD, mescaline, and psilocybin. Some investigators question whether the doses normally used in the United States ever produce true hallucinations. Furthermore, cannabis tends to sedate, whereas LSD and the LSD-type drugs often induce wakefulness and even restlessness. Unlike LSD, marijuana does not dilate the pupils or heighten blood pressure, reflexes, and body temperature. Tolerance develops rapidly with LSD-type drugs but much more slowly with marijuana. For these reasons, some clinicians doubt its credentials for inclusion among the hallucinogens.

The significance of tolerance and withdrawal with regular heavy use of cannabis remains uncertain. Some indications of tolerance and a mild withdrawal reaction after frequent use of high doses exist, but little clinical evidence indicates that a withdrawal syndrome or a need to increase the dose presents serious problems to users or causes them to continue using cannabis. Craving or difficulty in stopping use may rarely occur as part of a pattern of regular heavy use.

Cannabis probably tends to suppress inclinations toward violence by inducing serenity or lethargy and reducing aggressiveness. Release of inhibitions is expressed in word, thought, and fantasy rather than in action.

Although many users report that marijuana enhances the enjoyment of sexual intercourse, as well as food, art, and music, little evidence exists that it weakens or stimulates sexual desire or potency.

Chronic heavy use is sometimes said to cause an amotivational syndrome of sloth and apathy. Reports by many investigators, particularly in Egypt and in parts of the Orient, indicate that long-term users of the potent versions of cannabis are passive, unproductive, and lacking in ambition. This suggests that chronic use of the drug in its stronger forms may have debilitating effects, as prolonged heavy drinking does.

There is substantial evidence that moderate use of marijuana does not cause physical or mental deterioration. For example, in the La Guardia study conducted in New York City and published in 1944, an examination of chronic users who had averaged about seven marijuana cigarettes a day over a mean period of 8 yr showed that they had suffered no mental or physical decline as a result. The report of the National Commission on Marijuana and Drug Abuse (1972), although it did much to demythologize cannabis, cautioned that 2% of American marijuana users were at risk as chronic abusers, but it did not make clear exactly what risk was involved. Controlled studies of chronic heavy use have not found personality deterioration, amotivational syndrome, or other psychological harm.

The official diagnostic manual of the American Psychiatric Association refers to a cannabis delusional disorder, but some clinicians question its existence. In the form of ''hemp insanity,'' or cannabis psychosis, this disorder has been reported mainly in India, Egypt, and Morocco, more often in the late nineteenth and early twentieth centuries than today. It is described as a prolonged psychosis caused mainly by chronic heavy use of the drug. This phenomenon has proved to be peculiarly elusive as a clinical entity. The symptoms said to identify it also occur in acute schizophrenia, mania, and

toxic states associated with malnutrition and endemic infection.

One explanation for such psychoses is that persons maintaining a delicate balance of ego-functioning may be overwhelmed and precipitated into a psychotic reaction by a drug that alters their consciousness, however mildly. This concatenation of factors—an ego struggling to manage overwhelming anxiety and to prevent distortion of perception and body image, which is subjected to a drug that, in some persons, promotes these effects—may, indeed, be the last straw in precipitating a psychotic break. Several clinical reports also suggest that cannabis can cause a relapse or worsen symptoms in patients with schizophrenia or mood disorders.

In susceptible people, cannabis may precipitate several other types of mental dysfunction. The most serious is the reaction called toxic psychosis or acute brain syndrome. It is caused by the presence in the brain of substances that interfere with cerebral functions. Delirium usually ends when the toxins disappear from the brain. The symptoms include clouding of consciousness, restlessness, confusion, bewilderment, disorientation, dreamlike thinking, fear, illusions, and hallucinations. Delirium occurs only with a large, ingested dose of cannabis and rarely when it is smoked.

Cannabis users may also suffer short-lived acute anxiety states, sometimes accompanied by paranoid thoughts. The anxiety may become intense enough to be called panic. Although uncommon, this is probably the most frequent adverse reaction to the moderate use of smoked marijuana. Sufferers may believe that body-image distortions mean that they are ill or dying; they may interpret the psychological changes as an indication that they are losing their sanity. Rarely, the panic becomes incapacitating, usually for a relatively short time. Simple reassurance is the best treatment; it is dangerous to mistake this for toxic psychosis and subject the user to physical intervention that heightens the panic. The anxiety reaction is not psychotic, because the ability to test reality remains intact. Any paranoid ideas can ordinarily be dispelled by reassurance.

Anxiety reactions and paranoid thoughts may occur in those taking the drug for the first time or in unpleasant or unfamiliar settings. Such reactions are less common in experienced users. The likelihood varies directly with the dose and inversely with the user's experience; thus, the most vulnerable person is the inexperienced user who inadvertently (often precisely because of unfamiliarity with the drug) takes a large dose that produces unexpected perceptual and somatic changes.

One rather rare reaction to cannabis is the flashback, or spontaneous recurrence, of drug symptoms while not intoxicated. Although some reports suggest that this may occur in marijuana users even without prior use of any other drug, in general it seems to arise only when people who have used more powerful hallucinogenic or psychedelic drugs smoke marijuana at a later time.

Among new users of marijuana, an acute depressive reaction occasionally occurs. It is generally rather mild and transient but sometimes requires psychiatric intervention. This reaction is most likely in a user with underlying depression.

Much recent research on cannabis has been concerned with possible adverse physical effects of chronic use. Researchers have studied cerebral atrophy, seizure susceptibility, chromosome damage and birth defects, impairment of immune response, and effects on testosterone and the menstrual cycle. The results are complicated, but the commonly accepted conclusion is that they are inconclusive. Clinical observations of marijuana users, including recent studies of long-term users in the Caribbean and Greece, show no evidence of disease or organic pathology attributable to any of the causes mentioned above.

The only well-documented adverse effects of chronic marijuana use are produced in the lungs. Mild airway constriction is reported in studies of both animals and human beings. Marijuana smoke also contains many of the same carcinogenic hydrocarbons as tobacco smoke. Although not yet confirmed clinically, chronic respiratory disease and lung cancer must be considered dangers for long-term heavy users. [See MARIJUANA AND CANNABINOIDS.]

VII. Phencyclidine

Phencyclidine (PCP), 1-(1-phenylcyclohexyl)piperidine, was first investigated as a surgical anesthetic and analgesic. Because of disorientation, agitation, and delirium on emergence from anesthesia, it is now medically available only for veterinary use. Illicit PCP may be taken by snuffing, orally, or intravenously, but it is usually smoked in a cigarette, because that is the best means of titrating the dose. The most popular street names are "angel dust," "crystal," and "hog." PCP is occasionally misrep-

resented as mescaline, psilocybin, tetrahydrocannabinol (THC), cannabinol, cocaine, or methaqualone, although this is less common now than in the past.

PCP is relatively cheap and easy to synthesize in garage laboratories. About 30 chemical analogues exist, some of which have appeared on the illicit market. Another related drug is ketamine (Ketalar), a short-acting anesthetic with psychoactive properties similar to those of PCP. PCP is one of the more popular drugs for a new generation of drug users, among blacks and the white working class as well as the middle class; some people have used it daily for as long as 6 yr.

One gram of PCP may be used to make as few as four or as many as several dozen cigarettes. This variability, together with the extreme uncertainty of PCP content in street samples, makes it difficult to predict the effect, which also depends on the setting and the user's previous experience. PCP doses <5 mg are considered low, whereas doses >10 mg are considered high. The effects of 2–3 mg of smoked PCP begin within 5 min and plateau in 30 min. In the early phases, users are often uncommunicative and lost in fantasy. They experience "speedy" feelings, euphoria, body warmth, tingling, peaceful floating sensations, and occasional feelings of depersonalization or isolation. Striking alterations of body image, distortions of space and time perception, and delusions or hallucinations may occur. Thought may become confused and disorganized. The user may be sympathetic, sociable, and talkative at one moment, hostile and negative at another. Anxiety is often the most prominent symptom in an adverse reaction. Head-rolling movement, stroking, grimacing, and repetitive chanting may occur. The intoxication lasts about 4–6 hr, sometimes giving way to a mild irritable depression. Users may find that it takes from 24 to 48 hr to recover completely from the high. Laboratory tests show that PCP may remain in the blood and urine for more than a week.

Mild cases of adverse PCP reaction or overdose usually do not come to medical attention, and when they do they may often be treated as an emergency in the outpatient department. Symptoms at low doses may range from mild euphoria and restlessness to increasing levels of anxiety, fearfulness, confusion, and agitation. Patients may exhibit difficulty in communication, a blank staring appearance, disordered thinking, depression, and occasionally self-destructive or belligerent and irrationally assaultive behavior. Accidents, suicide, and homicide are the main causes of death in PCP use.

Like the other effects of PCP intoxication, neurological and physiological symptoms are dose-related. Among the common symptoms in cases brought to emergency rooms are hypertension and involuntary movements of the eyeballs. At low doses there may be loss of coordination and muscle rigidity, particularly in the face and neck. Heightened deep tendon reflexes and diminished response to pain are commonly observed. Higher doses may lead to high fever, agitated and repetitive movement, jerking of the extremities, and occasionally involuntary assumption of fixed body postures. Patients may become drowsy, stuporous, or even comatose. Vomiting, hypersalivation, and sweating are common. Seizures and death from respiratory arrest may also occur.

Two to three days may elapse before psychiatric help is sought, because friends are trying to deal with the psychosis by providing resources and support; patients who lose consciousness are seen earlier. Although most patients will recover within a day or two, some will remain psychotic for as long as 2 wk. Patients who are first seen in a coma often show disorientation, hallucinations, confusion, and difficulty in communication upon regaining consciousness—symptoms also seen in noncomatose patients. Other symptoms of PCP psychosis are staring, posturing, sleep disturbances, paranoid ideation, and depression. The behavioral disturbance may include inappropriate laughing and crying, public masturbation, urinary incontinence, and violence. Often there is amnesia for the entire period of the psychosis.

The proportion of psychotic reactions to PCP occurring in people with a pre-existing disposition to psychosis is unclear. Many patients who appear at hospitals with PCP psychoses return later with acute psychiatric reactions not related to the drug. Schizophrenics are very sensitive to PCP, and apparently they cannot easily distinguish between its effects and an intensification of their symptoms. Many PCP psychoses possibly involve mainly an aggravation of underlying psychopathology, whereas the physical symptoms are more specifically drug-related. Some observers believe that PCP psychosis is seriously underdiagnosed, because toxic symptoms indicating the presence of a drug are often not obvious and because the most commonly used tests for PCP in blood and urine are unreliable.

The term "crystallized" is sometimes applied to chronic PCP users who seem to suffer from dulled thinking and reflexes, loss of memory and impulse control, depression, lethargy, and difficulty in concentration. No clinical evidence of permanent brain damage exists, but some clinicians report neurological and cognitive dysfunction in chronic users even after 2–3 wk of abstinence. Users may also develop tolerance and a withdrawal reaction consisting of lethargy, depression, and craving.

VIII. Psychedelics

There are many psychedelic or hallucinogenic drugs, some natural and some synthetic. The best known are mescaline, which is derived from the peyote cactus, psilocybin, found in about 100 mushroom species, and the synthetic drug lysergic acid diethylamide (LSD), which is related to psychoactive alkaloids found in morning glory seeds, the lysergic acid amides. Other psychedelic drugs include the natural substances harmine, harmaline, ibogaine, and dimethyltryptamine (DMT) as well as a large number of synthetic drugs with a tryptamine or methoxylated amphetamine structure. A few of these are diethyltryptamine (DET), dipropyltryptamine (DPT), 3,4-methylenedioxyamphetamine (MDA), 3,4-methylenedioxymethamphetamine (MDMA), and 2,5-dimethoxy-4-methylamphetamine (DOM, also known as STP). The average effective dose varies considerably: for example, 75 micrograms for LSD, about 1 mg for lysergic acid amides, 3 mg for DOM, 6 mg for psilocybin, 30 mg for DMT, 100 mg for MDA, and 200 mg for mescaline. Only LSD, psychedelic mushrooms, and to some extent MDA and MDMA (discussed later) are now available in any quantity on the illicit market.

The subjective effects of these drugs differ somewhat in quality and duration, but LSD produces the widest range of effects and can be taken as a prototype. The reaction varies with personality, expectations, and setting even more than the reaction to other psychoactive drugs, but LSD almost always produces profound alterations in perception, mood, and thinking. Perceptions become unusually brilliant and intense: colors and textures seem richer, contours sharpened, music more emotionally profound, smell and taste heightened. Normally unnoticed details capture the attention, and ordinary things are seen with wonder, as if for the first time. Synesthesia ("hearing" colors and "seeing" sounds) is common. Changes in body image and alternations of time and space perception also occur. Intensely vivid, dreamlike, kaleidoscopic imagery appears before closed eyes. True hallucinations are rare, but visual distortions are common. Emotions become unusually intense and may change abruptly and often; two seemingly incompatible feelings may be experienced at once. Suggestibility is greatly heightened, and sensitivity to nonverbal cues is increased. Exaggerated empathy with or detachment from other people may arise. Other features that often appear are seeming awareness of internal body organs, recovery of lost early memories, release of unconscious material in symbolic form, and regression and apparent reliving of past events including birth. A heightened sense of reality and significance suffuses the experience. Introspective reflection and feelings of religious and philosophical insight are common. The sense of self is greatly changed, sometimes to the point of depersonalization, merging with the external world, separation of self from body, or total dissolution of the ego in mystical ecstasy.

People sometimes maintain that a single psychedelic experience or a few such experiences have given them increased creative capacity, new psychological insight, relief from neurotic and psychosomatic symptoms, or a desirable change in personality. For many years, and especially from 1950 to the mid-1960s, psychiatrists showed great interest in LSD and related drugs not only as possible drug models for schizophrenia but also as therapeutic agents in a wide variety of diagnoses. Since 1966, obtaining the drugs legally for therapeutic purposes has become impossible, and professional interest has declined.

The most common adverse effect of LSD and related drugs is the "bad trip," which resembles the acute panic reaction to cannabis but can be more severe and occasionally produces true psychotic symptoms. The bad trip ends when the immediate effect of the drug wears off—in the case of LSD, 8–12 hr. Psychiatric help is usually unnecessary; the best treatment is protection, companionship, and reassurance, although occasionally tranquilizers may be required.

Another common effect of hallucinogenic drugs is the flashback, a spontaneous transitory recurrence of drug-induced experience in a drug-free state. Most flashbacks are episodes of visual distortion, time expansion, physical symptoms, loss of ego boundaries or relived intense emotion lasting usu-

ally a few seconds to a few minutes but sometimes longer. Probably about a quarter of all psychedelic drug users have experienced some form of flashback. As a rule they are mild, sometimes even pleasant, but occasionally they turn into repeated frightening images or thoughts resembling a traumatic neurosis; in that case they may require psychiatric attention. Flashbacks decrease in number and intensity with time. They are most likely to occur under stress or at a time of diminished ego control; thus, they can be induced by fatigue, drunkenness, marijuana intoxication, or severe illness. Marijuana smoking is possibly the most common single source of LSD flashbacks.

Prolonged adverse reactions to LSD present the same variety of symptoms as bad trips and flashbacks. They have been classified as anxiety reactions, depressive reactions, and psychoses; often they resemble prolonged and more or less attenuated bad trips or flashbacks. Most of these adverse reactions end after 24–48 hr, but sometimes they last weeks or even months. Psychedelic drugs are capable of magnifying and bringing into consciousness almost any internal conflict, so there is no typical prolonged adverse reaction to LSD, as there is a typical amphetamine psychosis. Instead, many different affective, neurotic, and psychotic symptoms may appear, depending on individuals forms of vulnerability. This lack of specificity makes it hard to distinguish between LSD reactions and unrelated pathological processes, especially when some time passes between the drug trip and the onset of the disturbance.

The most likely candidates for adverse reactions are schizoid and prepsychotic personalities with a barely stable ego balance and a great deal of anxiety, who cannot cope with the perceptual changes, body image distortions, and symbolic unconscious material produced by the drug trip. People hospitalized for LSD reactions have a high rate of previous mental instability. In the late 1960s, a number of adverse reactions occurred because LSD was being promoted as a self-prescribed psychotherapy for emotional crises in the lives of seriously disturbed people.

Long-term psychedelic drug use is not very common. There is no physical addiction, and psychological dependence is rare because each LSD experience is different and the drug does not produce a reliable euphoria. Tolerance to these drugs develops very quickly but also disappears quickly—in 2 or 3 days. There is no clear evidence of organic brain damage, drastic personality change, or chronic psychosis produced by long-term LSD use, although there is some controversy about this.

The religious and other experiences produced by psychedelic drugs have the capacity, in certain circumstances, to catalyze a transformation in beliefs and ways of life, which may be mistaken for a drug-induced personality or organic change. Drastic changes in cultural and metaphysical identity after taking psychedelic drugs were most common in the 1960s, when the hippie movement promised to build a new society on the use of these drugs.

A persistent issue has been genetic damage and birth defects. The available evidence suggests that LSD produces no serious chromosome damage in reproductive cells; the same is true of other psychedelic drugs to the extent that they have been tested. Nor is there evidence of teratogenicity in human users at normal doses. Nevertheless, all drugs should of course be avoided if possible during pregnancy, especially in the early stages, when the embryo is most vulnerable.

IX. MDMA

A word should be said about MDMA (3,4-methylenedioxyamphetamine), a psychedelic drug that has only recently received widespread publicity. Although structurally related to both amphetamine, a stimulant, and mescaline, a psychedelic drug, MDMA produces neither a stimulant effect nor perceptual distortions, body image alterations, and changes in the sense of self. It is said to evoke a gentler, subtler, and more controllable experience than LSD—one that invites rather than compels self-exploration and the intensification of feelings. In particular, it reduces defensive anxiety. The effects last 2–4 hr. Bad trips, psychotic reactions, and flashbacks of the type produced by LSD and mesca-

line are apparently rare. A few users have reported prolonged adverse reactions such as anxiety or depression for a week or two. Little is known about long-term toxic effects or abuse potential.

Bibliography

American Psychiatric Association (1987). ''Diagnostic and Statistical Manual III,'' revised. American Psychiatric Press, Washington, D.C.

Bromberg, W. (1934). Marihuana intoxication: A clinical study of *Cannabis sativa* intoxication. *Am. J. Psychiatry* **91,** 303.

Grabowsky, J. (ed.) (1984). ''Cocaine: Pharmacology, Effects, and Treatment of Abuse.'' N.I.D.A. Research Monograph 50, National Institute of Drug Abuse, Rockville, Maryland.

Grinspoon, L. (1977). ''Marihuana Reconsidered,'' 2nd ed. Harvard University Press, Cambridge.

Grinspoon, L., and Bakalar, J. B. (1985). ''Cocaine: A Drug and Its Social Evolution,'' 2nd ed. Basic Books, New York.

Grinspoon, L., and Bakalar, J. B. (1979). ''Psychedelic Drugs Reconsidered.'' Basic Books, New York.

Grinspoon, L., and Hedblom, P. (1975). ''The Speed Culture: Amphetamine Use and Abuse in America.'' Harvard University Press, Cambridge.

Institute of Medicine. (1982). ''Marijuana and Health.'' National Academy Press, Washington, D.C.

Petersen, L. C., and Stillman, R. C. (eds.) (1978). ''Phencyclidine (PCP) Abuse: An Appraisal.'' N.I.D.A. Research Monograph 21, U.S. Government Printing Office, Washington, D.C.

Rubin, V., and Comitas, L. (1975). ''Ganja in Jamaica.'' Mouton, The Hague.

Szara, S. I., and Ludford, J. P. (eds.) (1980). ''Benzodiazepines: A Review of Research Results, 1980.'' N.I.D.A. Research Monograph 33, U.S. Government Printing Office, Washington, D.C.

Wesson, D. R., and Smith, D. E. (1977). ''Barbiturates: Their Use, Misuse, and Abuse. Human Sciences Press, New York.

Nuclear Pore, Structure and Function

JOHN A. HANOVER, *NIDDK, National Institutes of Health*

Glossary

Nuclear envelope Double membrane surrounding the nucleus of eukaryotic cells

Nuclear pore complex Organized structure found at the junction of inner and outer nuclear membranes; involved in transport into and out of the nucleus

Nucleoplasm Part of the cytoplasm that is surrounded by the nuclear envelope

Ribonucleoproteins Macromolecular complexes composed of both polypeptide backbone and ribonucleic acid

O-linked   *N*-acetylglucosamine *N*-acetylglucosamine linked via an O-glycosidic linkage to a Ser or Thr residue in a protein; present on glycoproteins of the nuclear pore complex

THE NUCLEAR PORE COMPLEX is present in all eukaryotic cells and is the portal through which transfer of materials between cytoplasm and nucleus occurs. Nuclear pore complexes appear in electron micrographs as an octagonal collection of ribosome-sized particles. The pore is some 1,200 Å in diameter and traverses both bilayers that make up the nuclear envelope, thereby creating a connec-

tion between nucleus and cytoplasm. The pore complex allows bidirectional transport: proteins enter the nucleus from the cytoplasm, and RNA is exported from the nucleus. Molecules approximately <40 Å can diffuse freely across the pore complex. Larger proteins require specific targeting signals before they cross the pore. The mechanisms involved in protein and RNA transport across the pore complex are poorly understood.

I. Introduction

The two major compartments of the eukaryotic cell, the nucleus and cytoplasm, are morphologically distinct and separated by a membrane barrier. This structural compartmentalization also reflects a segregation of function. Transcription of RNA from DNA occurs exclusively in the nucleus; translation of RNA into protein occurs on cytoplasmic ribosomes. Exchange of molecules between the nucleus and cytoplasm is indispensable for gene regulation and normal cell growth and development. RNA is exported from the nucleus into the cytoplasm, where the protein-synthesizing machinery translates the genetic information. The proteins needed for the replication of DNA or factors required for transcription are made in the cytoplasm and must enter the nucleus to perform their functions. Moreover, the nuclear envelope breaks down during mitosis, and the nuclear proteins must also have the ability to reenter the nucleus after the nuclear envelope is reformed. The mature nuclear protein must be capable of nuclear localization. Their transport, therefore, is different from many other intracellular sorting mechanisms, which involve precursor sequences removed during transport. [*See* DNA AND GENE TRANSCRIPTION.]

The nuclear envelope is unlike any other membrane system in the eukaryotic cell. It consists of two lipid bilayers enclosing a cisternal space. The outer nuclear membrane, which is continuous with the endoplasmic reticulum, has attached ribosomes; the inner membrane may form attachment points for chromatin. The nuclear envelope is traversed by nuclear pore complexes, which form a visible link between nucleus and cytoplasm. These pores allow passive diffusion of electrolytes and other small molecules as well as some small macromolecules. However, dextrans with a molecular weight (MW) >22,000 and globular proteins with a MW >65,000 are excluded from the nucleus. Measurements such as these have established that the functional radius of the pore for simple diffusion is approximately 35 Å. Larger proteins require specific stretches of amino acid (nuclear localization sequences) for their proper transport across the nuclear membrane. A number of such sequences have been identified. As for protein transport across the nuclear pore, simple diffusion does not account for the export of RNA-containing particles from the nucleus. RNA typically leaves the nucleus in a mature, fully processed form. The transport event is energy-dependent and involves release of the RNA from some component of the nuclear substructure followed by movement across the nuclear pore. The mechanism of these steps, while under investigation, is poorly understood. Current evidence suggests that one pore complex may be capable of concurrent RNA transport *out of* and protein uptake *into* the nucleus. [*See* CELL MEMBRANE TRANSPORT.]

II. Ultrastructure of the Nuclear Pore Complex

When viewed by electron microscopy, the nuclear pore complex is a prominent feature of the nuclear envelopes of all eukaryotic cells. It forms the junction of the inner and outer nuclear membranes and has an outer diameter of about 1,200 Å. It forms a clear morphological connection between the nucleus and the cytoplasm and is thought to be the route of most nucleo–cytoplasmic traffic. The number of pores per nucleus varies greatly, from 1×10^2 to 5×10^7. The density of pores seems to depend on the metabolic or developmental stage of the cell. Nucleated red blood cells or lymphocytes, for instance, have about 3 pores per micron², rat liver nuclei have 14–16 pores per μm², and the giant nucleus of *Xenopus* oocytes (the germinal vesicle) contains about 50 pores per micron². The degree to

which the morphological characteristics of the nuclear pore are conserved throughout eukaryotic evolution is even more striking. Indeed, by electron microscopy, a pore complex from a human liver sample is very hard to distinguish from that of a yeast or fungus. Figure 1 provides a composite of the images that have been obtained from the pore complex. Various interpretations of the ultrastructure of pore complex have been made, and some of these are summarized in Fig. 2. It must be remembered that the structural information gained from electron micrographs is of low resolution and can be greatly influenced by such factors as tissue fixation, embedding, and staining. Using electron microscopy, the pore is a large, organized, and at least partially symmetrical structure. In the most widely accepted model (Fig. 2, model 16), the pore complex is envisioned to consist of two rings of eight globular subunits, called annular granules. These subunits are symmetrically arranged about a central axis, and a central "plug" may be present. In some cases, small filaments are seen emerging from either the nucleoplasmic or cytoplasmic face of the pore. The transport of large molecules across the pore may also be directly visualized. During mitosis in most eukaryotes, the nuclear membrane breaks down and the pore complexes become dispersed throughout the cytoplasm. When the nuclear envelope reforms after telophase, the nuclear pores return to the nuclear periphery of the daughter cells. [*See* ELECTRON MICROSCOPY; LYMPHOCYTES; MITOSIS.]

Recently, using scanning transmission electron microscopy, attempts have been made to quantify the amount of mass represented by the pore complex. The outer ring contributes about 32 MDa, and the outer ring 20 MDa; the plug and spokes may contribute another 65 MDa. Thus, the entire assembly is massive: 120 million Da.

III. Biochemistry of the Nuclear Pore Complex

Central to the identification of the proteins making up the pore has been the finding that they contain a novel carbohydrate modification consisting of O-

FIGURE 1 Section through the nuclear envelope examined by electron microscopy. The inner and outer nuclear membranes are clearly visible, as are the complexes, indicated by the arrows. (Reproduced, from The Liver: Biology and Pathology, 1988 with permission, from Raven Press, New York.)

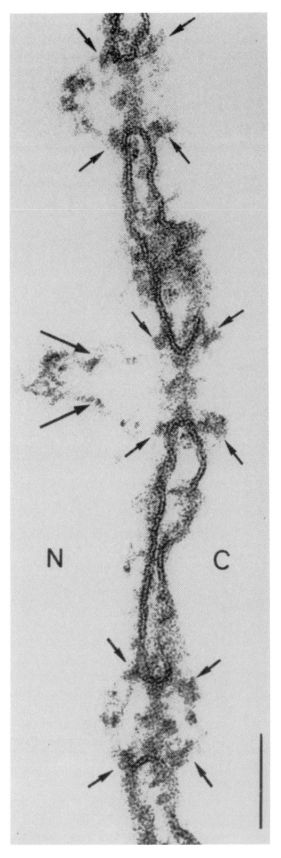

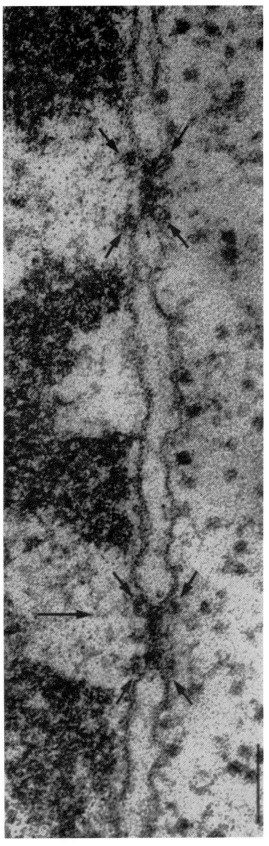

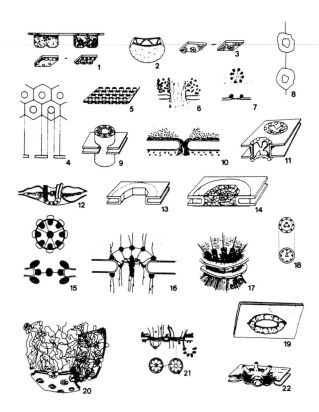

FIGURE 2 Compilation of the various models proposed for the morphological appearance of the nuclear pore. The models (except model 8) are arranged in the same orientation, with the nucleoplasmic side up and the cytoplasmic surface down. Models 1, 3, 12, and 21 should be oriented upside down to obtain the perspective intended by the authors. The models are roughly in chronological order and were proposed from 1950 to 1985. (Reproduced with permission from Dr. G. H. Bahr.)

cytoplasmic surfaces. When examined in light microscopy by immunofluorescence, the antibodies give a highly punctate pattern. This pattern changes dramatically during nuclear envelope breakdown, reflecting the dispersal of pore components. [*See* ANTIBODY-ANTIGEN COMPLEXES: BIOLOGICAL CONSEQUENCES; MONOCLONAL ANTIBODY TECHNOLOGY.]

Another component believed to be part of the pore complex is a high-MW glycoprotein tightly associated with the nuclear membrane. In contrast to the other components of the pore, this glycoprotein bears N-linked oligosaccharides of the high mannose type and is an integral membrane protein, i.e., crosses the membrane. This glycoprotein has been suggested to anchor the other pore components to the nuclear envelope.

IV. Molecular Characterization of Nuclear Pore Proteins

Availability of highly specific antibodies has allowed the molecular cloning of two of the components of the pore complex. The predicted secondary structure and presumptive location of these proteins in the pore complex is shown in Fig. 3.

A major nuclear pore glycoprotein from rat with a MW of 62 kDa (p62) was purified and subjected to microsequencing, obtaining two short stretches of protein sequence. Using their information, a DNA clone corresponding to the mRNA encoding the protein was identified. This peripheral membrane protein, which is extensively modified by O-linked GlcNAc is extremely rich in the amino acids serine and threonine. One of the most striking structural features of this molecule is the presence at the carboxyl terminus of alpha-helical stretches with some sequence similarity to proteins of the cellular cytoskeleton, the keratins, myosin, actin, and tropomyosin. The amino terminus is made up of a number of relatively short repeats similar to collagen. The role of this protein in maintaining the structure of the pore is unknown. The gene encoding p62 in rat has also been isolated. It is a rather short, single-copy gene that lacks intervening sequences. Efforts are underway to understand the regulation of p62 gene expression.

The other component of the pore that has been cloned by a similar approach is an integral membrane protein with a MW of 210 kDa. As shown in

linked *N*-acetylglucosamine (GlcNAc) covalently attached to Ser and Thr residues of the polypeptide. A plant lectin, wheat-germ agglutinin (WGA), binds to these proteins by associating with the O-linked GlcNAc residues. This group of glycoproteins are peripheral membrane proteins, i.e., do not cross the membrane bilayer, because they can be dissociated from the nuclear membrane by high salt extraction. They are exposed only on the cytoplasmic and nucleoplasmic surface of the pore; they are not components of the space between the outer and inner nuclear membrane (cisternae).

Monoclonal antibodies, which react with the nuclear pore proteins, recognize a family of proteins of 45–210 kDa in mass, which all seem to bear O-linked GlcNAc. Immunoelectron microscopy has demonstrated that the antigens recognized by the antibodies are located exclusively within the nuclear pore complex at both the nucleoplasmic and

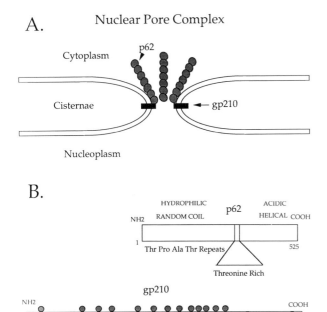

FIGURE 3 Membrane topography and secondary structures of p62 and gp210 membrane glycoproteins. A, the position of p62 and gp210 in the pore. B, a secondary structure prediction based on the primary sequence of p62 and gp210 nuclear pore glycoproteins. The gp210 molecule has N-linked glycosylation sites, indicated by small circles. Several molecules of p62 are shown, which are thought to interact with one another at the carboxy-terminal helical regions. The p62 molecule is in the cytoplasmic space, whereas gp210 is within the cisternal space of the nuclear envelope. The inner and outer nuclear membranes come together at the pore complex; the membranes are indicated by the bilayer-appearing structure.

Fig. 3, this protein is presumed to reside largely in the cisternal space of the nuclear membrane. It is glycosylated with N-linked oligosaccharides in several locations, and only a short segment near the carboxyl terminus may be exposed to the cytoplasmic (or nucleoplasmic) face of the pore. The sequence of this protein shows little similarity to other known proteins, and no obvious features emerge from analysis of the predicted secondary structure.

V. Functions of the Nuclear Pore: Nuclear Protein Transport

The nuclear pore confers upon the nuclear membrane properties of a molecular sieve: Small molecules diffuse across while larger molecules are excluded. Large nuclear proteins have sorting signals

that lead to their proper nuclear compartmentalization. The existence of these signals was established in dissected germinal vesicles from amphibian oocytes of the *X. laevis*. When cytoplasmic and nuclear proteins were injected into the oocyte, cytoplasmic nuclear proteins accumulated in the nucleus; cytoplasmic proteins remained in the cytoplasm. By proteolytic cleavage of nucleoplasmin, the most abundant nuclear protein, the property of nuclear localization resided in a distinct part of the protein molecule, the tail carboxy-terminal fraction, could be shown. The inserted tail fragment itself rapidly accumulated in the nucleus. This approach localized the signal for nuclear entry to a 12,000-Da protein fragment.

To date, the best studied nuclear protein is the large T-antigen of SV40 virus. The T-antigen subunit has a MW of 94,000, and it is too large to enter the nucleus by passive diffusion. Recombinant DNA techniques have been used to identify the nuclear localization signal. The conversion of a Lys within the sequence Pro-Lys-Lys[128]-Lys-Arg-Lys-Val into any amino acid other than Arg completely abolishes nuclear localization. If the sequence is attached to other proteins, it causes their localization to the nucleus. Similar sequences, found in nuclear SV40, contain two independent sequence elements that contribute to its nuclear proteins specified by other viruses, such as polyoma virus and adenovirus.

The primary sequences of many of the known nuclear localization signals are summarized in detail in Table I. With the exception of MATα2, they all consist of several highly basic and hydrophilic amino acids with a structure-perturbing amino acid, such as Gly or Pro on the N- or C-terminal side. Given their diversity, overall shape and charge rather than the exact amino acid sequence could be responsible for nuclear import. In addition, proteins may enter the nucleus by associating with another nuclear protein bearing a localization signal: for instance, antibodies against SV40 large T-antigen are cotransported with the antigen into the nucleus. Therefore, it is unlikely that protein unfolding occurs during transport of proteins into the nucleus.

There is reason to believe that the nuclear targeting sequence has to be exposed on the surface of the molecule to interact with a receptor for nuclear localization sequences. Thus, the SV40 T-antigen sequence can be introduced into several different sites within pyruvate kinase and still function correctly. However, when inserted into a part of the molecule

TABLE I Nuclear Localization Sequences

Protein	Sequence	Amino acid position
SV40 large T-antigen	PKKKRKV	126–132
SV40 VP1 capsid protein	APTKRKGS	1–8
SV40 VP2/3 capsid proteins	PNKKKRK	317–323 (VP2)
Polyoma virus large T-antigen	PKKARED	280–286
	VSRKRPR	189–195
Adenovirus E1a protein	KRPRP	285–289 (C-terminus)
Influenza virus	DRLRR	34–38
NSI protein	WGSSNENGGPPLTPKQKRKMARTARSKVRRDKMAD	203–237
Yeast MATα2 protein	MNKIPIKDLLNPQ	1–13
Yeast ribosomal motif protein L3	PRKR	1–21
Yeast H2B protein	GKKRSKA	28–33
Yeast GAL4 gene product		1–74
Nucleoplasmin	AKKKKLDKEDE	162–172
	(AVK)RPAATKKAGQAKKK(KLD)	(153)156–169(172)
X. laevis oocyte N1 protein	VRKKRKT	531–537
	KDKDAKKSKQE	544–554
Glucocorticoid	YRKCLQAGMNLEARKTKKKIKGIQQATA	497–524
Receptor (rat)		540–795
c-myc protein human	PAAKRVKLD	320–328
Progesterone receptor human	RKTKKKIK	638–642

that is buried according to crystallographic studies, the signal is inactive. Moreover, monoclonal antibodies against the localization signal of SV40 bind to *in vitro* translated SV40 T-antigen and to the T-antigen expressed in cultured cells. The exposure of the nuclear localization signal in proteins may be developmentally regulated. Thus, in *Xenopus,* the "late migrating nuclear antigens" are nuclear in the oocyte, cytoplasmic in the early embryo, and nuclear again in later stages of development. Exposure of a localization signal can also occur in response to external stimuli. For instance, the glucocorticoid receptor may enter the nucleus only when it is occupied by the hormone.

The component of the nuclear membrane that recognizes the nuclear localization signal and the transport machinery is under active investigation. Electron microscopic studies showed that gold particles coated with nuclear proteins readily enter the nucleus regardless of their size (between 50 and 200 Å in diameter). In content, colloidal gold particles coated with cytoplasmic control proteins are excluded from the nucleus and only at very high concentration are associated with the nuclear envelope.

The association of nucleoplasmin-gold particles may occur along fibrillar structures that are associated with the nuclear pore and extend into both the cytoplasm and the nucleoplasm. If gold particles are coated with intact nucleoplasmin, they become associated with the cytoplasmic side of the fibrils and inside the nuclear pores, but not on the nucleoplasmic side of the fibrils. Whether or not these fibrils correspond to the proteins described earlier as components of the pore (for example, p62) is an open question. Nuclear protein transport via localization sequences has a size exclusion limit: Uptake occurs up to a MW of 465,000 (ferritin), but not with IgM (MW 900,000).

Nuclear uptake is time-dependent. Microinjection studies with nucleoplasmin or proteins conjugated with the SV40 large T-antigen nuclear localization sequence have shown that they rapidly bind around the nuclear envelope and that accumulation in the nucleus follows in a time-dependent manner. The number of nuclear localization peptides has an influence on the speed of uptake; the more peptide moieties, the faster the uptake. Nuclear protein transport exhibits saturable kinetics, i.e., reaches a maximum rate with increasing protein concentration, and is temperature-dependent. The optimal temperature for nuclear transport *in vitro* using rat is 30–33°C, probably due to a loss of stability of the nuclear membrane at 37°C. In microinjected living cells, uptake of conjugates bearing a nuclear localization sequence does not occur at 0°C and can be induced by raising the temperature to 22°C, and even more at 37°C. Nuclear uptake is adenosine triphosphate (ATP)-dependent. If ATP pools are lowered using poisons such as sodium azide and deoxyglucose, transport into the nucleus is inhibited. An intact nuclear membrane is also required for nuclear transport. Nuclear uptake can be inhibited by molecules that bind to pore proteins, such as WGA and nuclear pore-specific antibodies. However, electron microscopy has shown that binding of nuclear proteins to the pore is not inhibited by WGA, suggesting that some nuclear pores are specifically involved in protein transport into the nucleus. [See ADENOSINE TRIPHOSPHATE (ATP).]

The available experimental evidence suggests that the nuclear pore is the principal route of nuclear protein import. The pore complex must accommodate large nuclear proteins and selectively exclude cytoplasmic proteins. Uptake requires ATP, yet it is unlikely that this energy is involved in the unfolding of nuclear proteins prior to uptake.

Three models, based on available information showing how nuclear protein transport might occur, are presented diagrammatically in Figure 4. In these models, a binding protein is thought to exist that specifically recognizes nuclear localization signals present on the surface of nuclear proteins. This binding protein would discriminate nuclear proteins from other proteins in the cytoplasm. In model A, the receptor would be associated with the filaments that have been shown to be associated with the pore complex; as mentioned earlier, some morphological evidence supports this idea. Once bound, the receptor might move along the filaments by using the hydrolysis of ATP as an energy source. In model B, occupancy of the receptor by a nuclear protein would trigger pore expansion, allowing selective transport of the nuclear protein–receptor complex into the nucleus. Finally, in model C, the receptor would be attached to filaments that might be tethered to some component of the nuclear substructure. This would allow transport of nuclear proteins associated with the cytoplasmic filaments in a coordinate and efficient manner. Of course, other models may be envisioned; for example, filament translocation and pore expansion may both operate during import.

VI. Functions of the Nuclear Pore: RNA Transport

All known RNA species, mRNA, tRNA, and rRNA, are transcribed from the DNA sequences in

A. Receptor binding; filament translocation

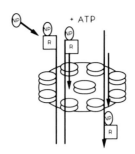

B. Receptor binding; pore expansion

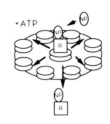

C. Receptor binding; mechanical uptake by tether

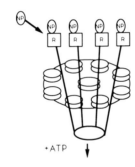

FIGURE 4 Models of nuclear protein import. Nuclear proteins are indicated by the NP within ovals. A binding protein, or receptor, is indicated by the boxed R.

the nucleus. The RNA species undergo a series of posttranscriptional modifications within the nucleus, including processing, splicing, and polyadenylation, and must ultimately be transported to the cytoplasm. Although RNA species are single-stranded, they may have substantial secondary structure arising from folding of the chain into double-stranded regions (hairpins). The extent to which the secondary structure contributes to the recognition, processing, and transport of RNA species is, at present, unknown. The tRNAs are the smallest RnA molecules, containing between 73 and 93 nu-

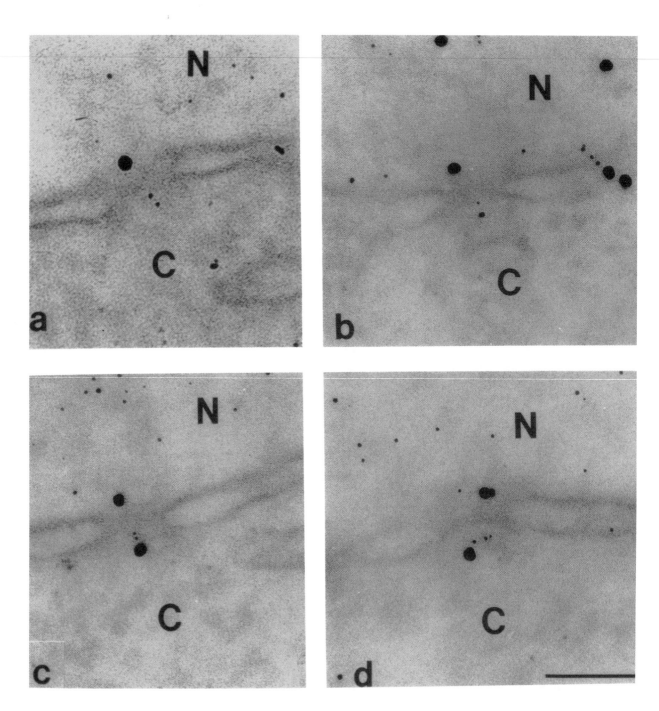

FIGURE 5 Morphological evidence for transport of nuclear proteins and RNA across the nuclear pore. Nucleoplasmin and human RNA species were coated onto colloidal gold and injected into *X. laevis* oocytes. RNA-gold was injected into the nucleus; protein-gold was injected into the cytoplasm. Both gold-labelled species become localized near the nuclear pore after a short incubation period. Panels a–d represent multiple examples of the image observed.

cleotides and having a molecular mass of about 25 kDa. They might be expected to diffuse through the nuclear pore passively; however, the recent identification of transport defective mutants of tRNA species strongly argues against passive diffusion. A single G-to-U substitution at position 57 in the vertebrate $TRNA_{Met}$ molecule was found to reduce its transport rate 95%. This highly conserved region of the tRNA molecule is therefore thought to be critical for recognition by the transport mechanism. This mechanism has been studied in the oocyte of

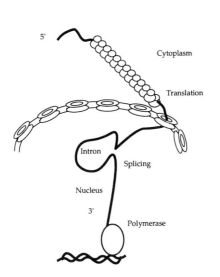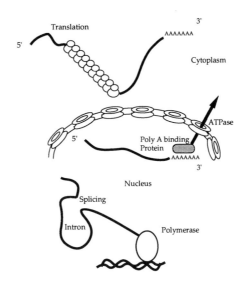

FIGURE 6 Models of mRNA transport. Ribosomes are indicated by the large and small ovals; RNA polymerase is the large oval. The pore complexes are indicated by the oval holes through the membrane. The 5′ and 3′ ends of the RNA are indicated.

X. laevis using nuclear microinjection and microdissection techniques. Transport behaves as a saturable, carrier-mediated translocation process rather than diffusion through a simple pore or channel. It is possible that Ribosome-like components surrounding the nuclear pore function as a tRNA translocation motor. Transport of tRNA species has also been examined morphologically in *Xenopus* oocytes using gold particles of different sizes (Fig. 5). Each pore seems to be capable of bidirectional transport of RNA out and protein in. Whether or not the mechanisms involved in these coordinated transport processes are similar is unknown.

Transport of mRNA into the cytoplasm seems to occur in two steps: (1) mRNA processing and the release of heterogeneous nuclear RNA from the nuclear matrix, and (2) translocation through the nuclear pore. In eukaryotic cells, 90% of the mRNA species is tightly bound to elements of the nuclear substructure. Current evidence suggests that the mRNA is released from this structure and translocated to the nuclear periphery, apparently along defined tracks. The energetics of mRNA transport has been the focus of a large number of investigations. ATP is necessary for transport but not processing. The stoichiometry of ATP hydrolysis appears to be one ATP per nucleotide transported. A nucleoside triphosphatase, present in nuclear membrane, may be activated by polyadenylated RNA; this enzyme

has been implicated in transport. However, despite extensive characterization, the relationship between the ATPase and the RNA transport machinery has not been firmly established. Two models of mRNA transport are depicted in Figure 6. In the Model A, RNA transport would be coupled to translation. In fact, one potential source of energy for the transport of RNA is the movement of ribosomes across the RNA during translation. Support for this model comes from experiments in which translation is arrested by premature termination; the level of cytoplasmic RNA is often reduced under these circumstances. A prediction of this model is that the 5′-terminus of the mRNA would be transported first to allow translation initiation. In model B, the processes of transport and translation would not be coupled. Transport would occur via an energy-dependent mechanism distinct from translation. Efforts are underway to discriminate between these two models.

VII. Summary

Communication between the nucleus and cytoplasm is mediated by the nuclear pore complex. The nuclear pore is an elaborate structure consisting of integral and peripheral membrane glycoproteins, whose primary structure has recently been elucidated. The nuclear pore must be capable of discriminating between proteins that must remain in the cytoplasm and those that must enter the nucleus. This uptake process might be tightly regulated and involves recognition of signals encoded in the structure of nuclear proteins. In addition, the pore appears to play a critical role in the controlled release of RNA species from the nuclear interior to the cytoplasm. Understanding the structure of the pore complex should lead to a more complete understanding of its role in mediating these important cellular processes.

Bibliography

Davis, L. I., and Blobel, G. (1986). Identification and characterization of a nuclear pore complex protein. *Cell* **45**, 699–709.

Dworetzki, S. I., and Feldherr, C. M. (1988). Translocation of RNA-coated gold particles through the nuclear pore of oocytes. *J. Cell Biol.* **106**, 575–584.

Hanover, J. A. (1988). Molecular signals controlling membrane traffic. *In* "The Liver: Biology and Pathobiology," pp. 189–205. Raven Press, Ltd., New York.

Hanover, J. A., Cohen, C. K., Willingham, M. C., and Park, M. K. (1987). O-linked *N*-acetylglucosamine is attached to proteins of the nuclear pore. Evidence for cytoplasmic and nucleoplasmic glycoproteins. *J. Biol. Chem.* **262**, 9887–9894.

Maul, G. G. (1982). Aspects of a hypothetical nucleo–cytoplasmic transport mechanism. *In* "The Nuclear Envelope and the Nuclear Matrix" (G. G. Maul, ed.), pp. 1–11. Alan R. Liss, Inc., New York.

Newport, J. W., and Forbes, D. J. (1987). The nucleus: Structure, function and dynamics. *Annu. Rev. Biochem.* **56**, 535–565.

Park, M. K., D'Onofrio, M., Willingham, M. C., and Hanover, J. A. (1987). A monoclonal antibody against a family of nuclear pore proteins (nucleoporins): O-linked *N*-acetylglucosamine is part of the immunodeterminant. *Proc. Nat. Acad. Sci. USA* **84**, 6462–6466.

Schröder, H. C., Bachman, M., Diehl-Seifert, B., and Müller, W. E. G. (1987). Transport of mRNA from nucleus to cytoplasm. *Prog. Nucleic Acid Res. Mol. Biol.* **34**, 89–142.

Smith, A. E., Kalderon, D., Roberts, B. L., Colledge, W. H., Edge, M., Gillett, P., Markham, A., Paucha, E., and Richardson, W. D. (1985). The nuclear location signal. *Proc. R. Soc. London* **226**, 43–58.

Snow, C. M., Senior, A., and Gerace, L. (1987). Monoclonal antibodies identify a group of nuclear pore complex proteins. *J. Cell Biol.* **104**, 1143–1156.

Wozniak, R. W., Barnik, E., and Blobel, G. (1989). Primary structure analysis of an integral membrane glycoprotein of the nuclear pore. *J. Cell Biol.* **108**, 2083–2092.

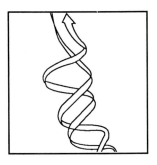

Nucleotide and Nucleic Acid Synthesis, Cellular Organization

CHRISTOPHER K. MATHEWS, *Oregon State University*

Glossary

Alarmone Denotes an intracellular chemical messenger, just as "hormone" denotes an extracellular messenger, which transmits a signal between distant cells; alarmone senses potentially harmful metabolic change and coordinates a metabolic response to that change to minimize deleterious effects on the cell

Compartmentation Maintenance within a cell of two or more distinct pools of a particular metabolite; pools may be physically separated (present in different cell compartments or organelles), or they may be distinguished kinetically

dNMP, dNDP, dNTP Deoxyribonucleoside 5'-mono-, di-, and triphosphate, respectively

Pool Intracellular or intraorganellar content of a metabolite; pool size refers to the amount of a particular metabolite in a cell or organelle

Reconstitution Duplication of part or all of a metabolic process *in vitro,* achieved by mixing separately purified components, including enzymes and other proteins

rNMP, rNDP, rNTP Ribonucleoside 5'-mono-, di-, and triphosphate, respectively

EVEN THOUGH MOST OF THE ENZYMATIC reactions of nucleotide and nucleic acid synthesis are catalyzed by enzymes that are readily soluble proteins, the processes and pathways are organized within cells. Organization is brought about by intracellular compartmentation of the enzymes involved, by the frequent use of multifunctional enzymes that catalyze more than one reaction, by multienzyme complexes that involve noncovalent protein–protein interactions, and by dynamic associations (sequential interactions among proteins, nucleic acids, and membranes, which generate structures essential for initiation of replication and transcription). Because deoxyribonucleoside triphosphates (dNTPs) find virtually their sole functions as DNA precursors, a close coordination exists between dNTP biosynthesis and DNA replication. The importance of this coordination can be seen from the severe genetic consequences of loss of control of one or more of the biosynthetic reactions.

I. General Importance of Metabolic Organization

Until recently, enzymes were characterized as particulate or soluble. Enzymes isolated in particulate form were usually assumed to represent binding of enzyme protein to a membrane in an organelle, whereas soluble enzymes were thought to represent proteins that existed in a less structured form, either as a highly concentrated solution or a gellike phase. Processes such as biological electron transport, fatty acid oxidation, and photosynthesis are catalyzed by enzymes firmly bound to membranes in specific organelles. Thus, the cellular organization of these processes could be described concurrently with the description of the processes themselves. Learning how to solubilize each enzyme of necessity required learning something about the structure within which each enzyme was embedded in the cell.

By contrast, pathways and processes that are catalyzed largely by soluble proteins have resisted attempts to define intracellular structures with which these processes are associated. The ease of isolating an enzyme in soluble form tends to lead an investigator past the question of cellular organization to focus on issues such as mechanisms of catalysis, allosteric regulation, and the quantitative role of an enzyme in a particular metabolic pathway. Glycolysis is an excellent example. All of the enzymes that ferment glucose to ethanol in yeast were crystallized decades ago, and their catalytic mechanisms are well understood. Only recently has it become apparent that these soluble proteins interact with one another *in vitro*, albeit weakly, and so they also form aggregates. At the same time, intracellular concentrations of these enzymes have been determined as very high; in fact, the molar concentrations of glycolytic enzymes are higher in some cases than those of low-molecular-weight intermediates in the same pathway. Such observations point strongly toward mechanisms to assemble and organize the enzymes *in vivo*, for catalytic facilitation of this central energy-generating pathway. However, the evidence supporting such interactions is controversial, in part because of the difficulty of demonstrating such associations *in vitro* at the vastly lower protein concentrations used in standard enzymological investigations.

A similar situation exists with regard to the processes we discuss in this chapter—the biosynthesis of nucleic acids and their precursors, the nucleotides. Like glycolytic enzymes, most of the enzymes involved are readily isolated as soluble proteins; however, they also form complexes. With a few exceptions, protein–protein associations in this group of enzymes are relatively weak. Moreover, some of the complexes that form are dynamic in nature, such as specific DNA-protein structures, which form transiently during the initiation of DNA replication. Nevertheless, information is accumulating about the intracellular organization of nucleic acid metabolism, partly as a result of reconstitution studies with purified proteins, partly from the isolation and characterization of multifunctional enzymes and multienzyme complexes, partly from

new methods for viewing metabolic processes in living cells, and partly from investigations of the biological consequences of disturbances in nucleic acid metabolism.

II. Overview of Nucleotide Biosynthetic Pathways

The pathways of nucleotide biosynthesis are well understood. Figure 1 recapitulates these pathways in summary fashion. Nucleotides can be synthesized either *de novo,* from pentose phosphates, amino acids, and folic acid coenzymes, or by salvage pathways, from nucleobases or nucleosides containing preformed purine or pyrimidine rings. The figure shows only the *de novo* pathways, which are essentially identical among all organisms. [*See* PURINE AND PYRIMIDINE METABOLISM.]

In mammalian cells, the branch point between RNA and DNA precursor biosynthesis is at the level of ribonucleoside diphosphates (rNDPs). All four common rNDPs are substrates for a multifunctional enzyme, ribonucleotide reductase, which reduces adenosine diphosphate (ADP), cytidine diphosphate (CDP), guanosine diphosphate (GDP), and uridine diphosphate (UDP) to the corresponding deoxyribonucleoside diphosphates (dNDPs). Beyond this stage compounds are destined to be used for DNA synthesis. In mammalian cells,

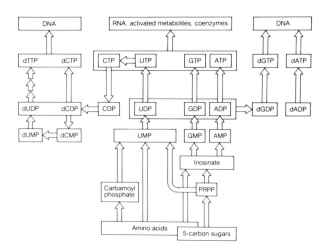

FIGURE 1 Abbreviated scheme showing pathways of ribo- and deoxyribonucleotide biosynthesis. (From *Biochemistry,* by C. K. Mathews and K. E. van Holde. Copyright © 1990, Benjamin/Cummings Publishing Company, Redwood City, California. Reprinted by permission.)

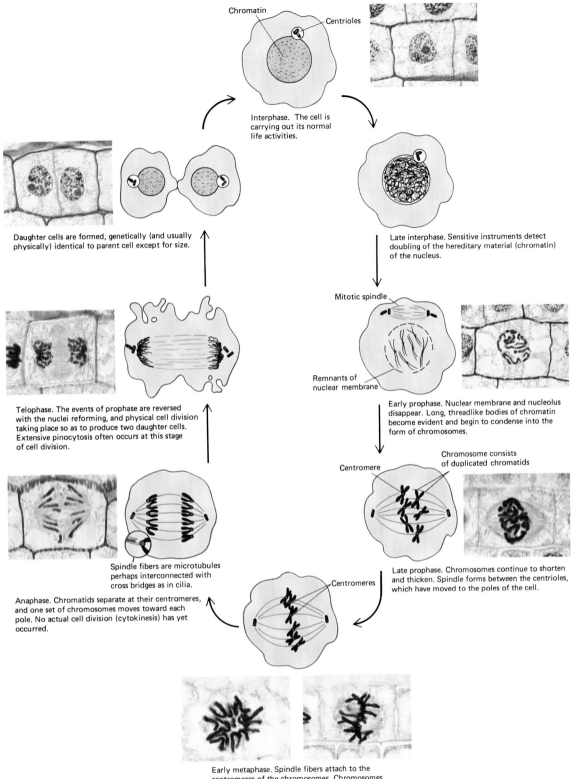

Chromatin

Centrioles

Interphase. The cell is carrying out its normal life activities.

Daughter cells are formed, genetically (and usually physically) identical to parent cell except for size.

Late interphase. Sensitive instruments detect doubling of the hereditary material (chromatin) of the nucleus.

Mitotic spindle

Remnants of nuclear membrane

Telophase. The events of prophase are reversed with the nuclei reforming, and physical cell division taking place so as to produce two daughter cells. Extensive pinocytosis often occurs at this stage of cell division.

Early prophase. Nuclear membrane and nucleolus disappear. Long, threadlike bodies of chromatin become evident and begin to condense into the form of chromosomes.

Chromosome consists of duplicated chromatids

Centromere

Spindle fibers are microtubules perhaps interconnected with cross bridges as in cilia.

Late prophase. Chromosomes continue to shorten and thicken. Spindle forms between the centrioles, which have moved to the poles of the cell.

Anaphase. Chromatids separate at their centromeres, and one set of chromosomes moves toward each pole. No actual cell division (cytokinesis) has yet occurred.

Centromeres

Early metaphase. Spindle fibers attach to the centromeres of the chromosomes. Chromosomes line up along the equatorial plane of the cell.

COLOR PLATE 1   The stages of the cell cycle. Individual steps in the cycle are explained in labels within the figure. The animal cells shown here have a diploid chromosome number of six. Photomicrographs are of plant cells (onion root tip, *Allium cepa*); drawings are of animal cells. [Source: Davis, P. W., and Solomon, E.P. (1986). "The World of Biology." p. 161. Saunders College Publishing, Philadelphia.]

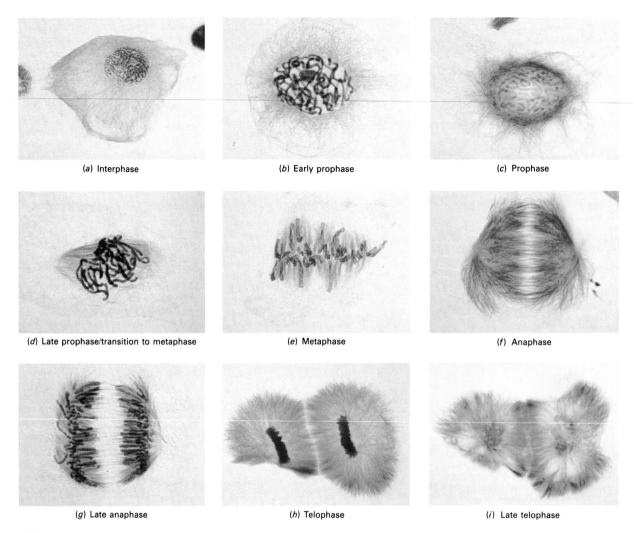

(a) Interphase   (b) Early prophase   (c) Prophase

(d) Late prophase/transition to metaphase   (e) Metaphase   (f) Anaphase

(g) Late anaphase   (h) Telophase   (i) Late telophase

COLOR PLATE 2 Interphase and the stages of mitosis in plant cells *(Haemanthus)* prepared with stains. [Source: Davis, P. W., and Solomon, E.P. (1986). "The World of Biology." p. 162. Saunders College Publishing, Philadelphia.]

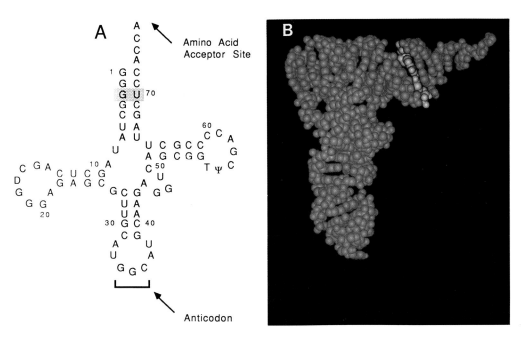

COLOR PLATE 3 (A) The sequence of bases and the cloverleaf structure of an *E. coli* alanine tRNA. The hydrogen bonding between bases is indicated: A pairs with U and G pairs with C. There are a few nonstandard bases (e.g., D and T). The locations of the amino acid attachment site and the anticodon are indicated. Highlighted is the G3:U70 base pair, which is a major determinant of the identity of an alanine tRNA. (B) Three-dimensional space-filling structural representation of the alanine tRNA. The anticodon is at the bottom, and the amino acid attachment site is at the upper right-hand end of the molecule. The G3:U70 base pair is highlighted. (Courtesy of Christopher Franklyn and Ya-Ming Hou.)

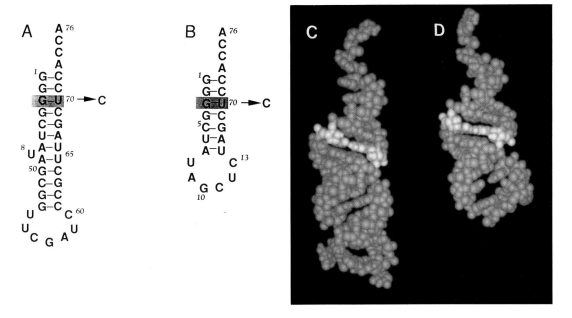

COLOR PLATE 4 (A) An RNA minihelix and (B) a microhelix. Each of these can be charged with alanine. Acceptance of alanine is dependent on the G3:U70 base pair, as demonstrated by the lack of charging of a G3:C70 variant. (C and D) Space-filling models of the RNA minihelix and of the microhelix, which can each be charged with alanine. The vertical orientation of these helices is the same as in the left panel, so that the amino acid acceptor site is at the top. The "bulged" U shown in the double-helix portion of the minihelix (C) has been omitted from the space-filling model (D) because of the uncertainty of its conformation. These molecules can be compared with the whole tRNA shown in Color Plate 3. (Courtesy of Christopher Franklyn and Ya-Ming Hou.)

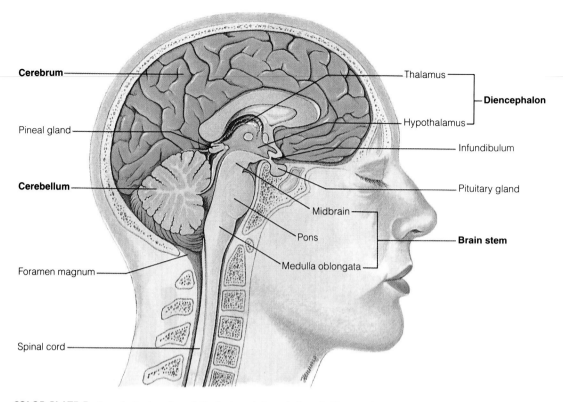

COLOR PLATE 5 Longitudinal section of the brain and the spinal cord. [Source: Gaudin, A. J., and Jones, K. C. (1989). "Human Anatomy and Physiology." Harcourt Brace Jovanovich, San Diego, p. 291. Reproduced with permission.]

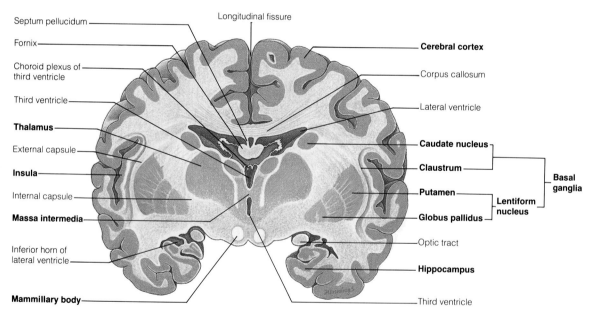

COLOR PLATE 6 Frontal section through the cerebral hemispheres of the brain. [Source: Gaudin, A. J., and Jones, K. C. (1989). "Human Anatomy and Physiology." Harcourt Brace Jovanovich, San Diego, p. 297. Reproduced with permission.]

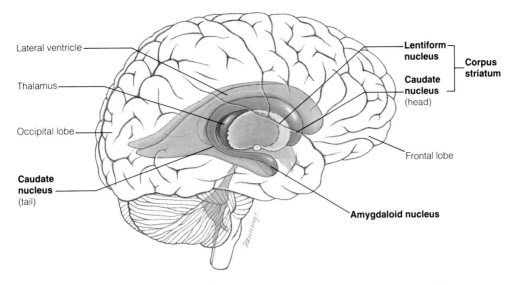

COLOR PLATE 7 The basal ganglia of the brain. [Source: Gaudin, A. J., and Jones, K. C. (1989). "Human Anatomy and Physiology." Harcourt Brace Jovanovich, San Diego, p. 299. Reproduced with permission.]

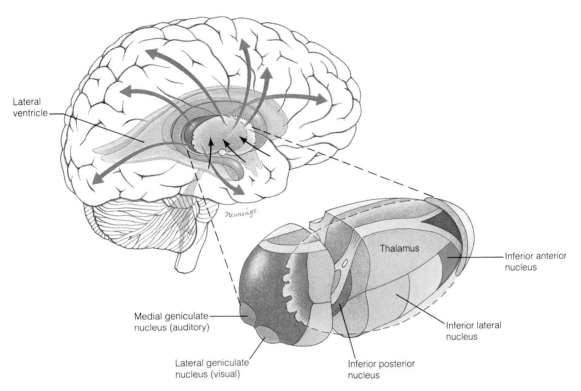

COLOR PLATE 8 The thalamic nuclei. [Source: Gaudin, A. J., and Jones, K. C. (1989). "Human Anatomy and Physiology." Harcourt Brace Jovanovich, San Diego, p. 300. Reproduced with permission.]

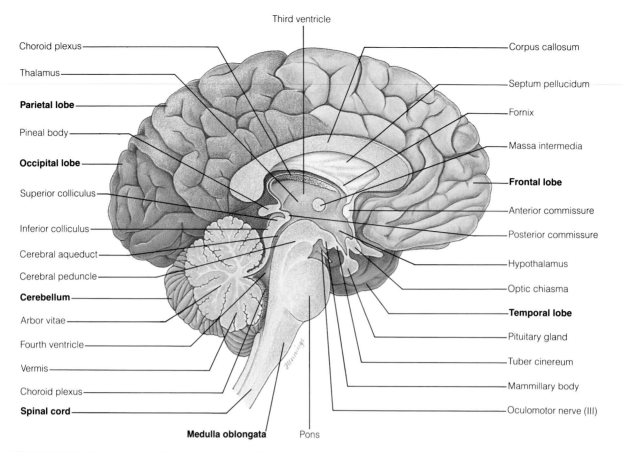

Third ventricle

Choroid plexus

Thalamus

Parietal lobe

Pineal body

Occipital lobe

Superior colliculus

Inferior colliculus

Cerebral aqueduct

Cerebral peduncle

Cerebellum

Arbor vitae

Fourth ventricle

Vermis

Choroid plexus

Spinal cord

Corpus callosum

Septum pellucidum

Fornix

Massa intermedia

Frontal lobe

Anterior commissure

Posterior commissure

Hypothalamus

Optic chiasma

Temporal lobe

Pituitary gland

Tuber cinereum

Mammillary body

Oculomotor nerve (III)

Medulla oblongata   Pons

COLOR PLATE 9   Longitudinal section through the brain. [Source: Gaudin, A. J., and Jones, K. C. (1989). "Human Anatomy and Physiology." Harcourt Brace Jovanovich, San Diego, p. 302. Reproduced with permission.]

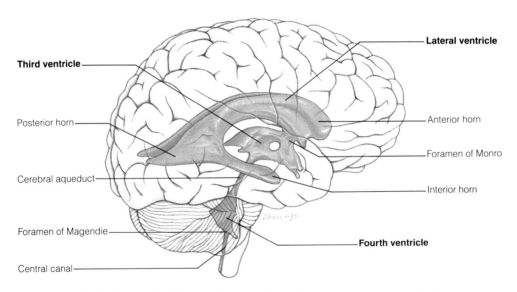

Lateral ventricle

Third ventricle

Posterior horn

Cerebral aqueduct

Foramen of Magendie

Central canal

Anterior horn

Foramen of Monro

Interior horn

Fourth ventricle

COLOR PLATE 10   Ventricles of the brain. [Source: Gaudin, A. J., and Jones, K. C. (1989). "Human Anatomy and Physiology." Harcourt Brace Jovanovich, San Diego, p. 309. Reprinted with permission.]

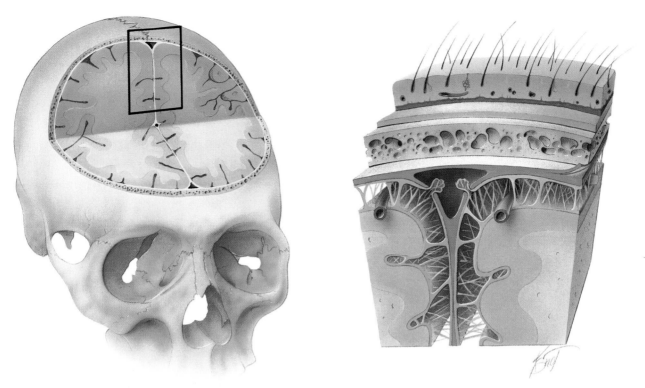

COLOR PLATE 11 Meninges of the brain. [Source: Gaudin, A. J., and Jones, K. C. (1989). "Human Anatomy and Physiology." Harcourt Brace Jovanovich, San Diego, p. 310. Reprinted with permission.]

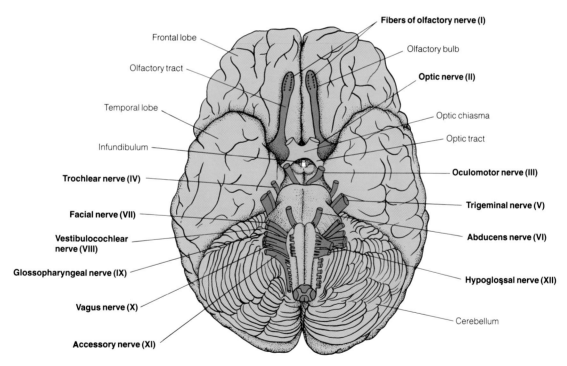

Frontal lobe
Olfactory tract
Temporal lobe
Infundibulum
Trochlear nerve (IV)
Facial nerve (VII)
Vestibulocochlear nerve (VIII)
Glossopharyngeal nerve (IX)
Vagus nerve (X)
Accessory nerve (XI)

Fibers of olfactory nerve (I)
Olfactory bulb
Optic nerve (II)
Optic chiasma
Optic tract
Oculomotor nerve (III)
Trigeminal nerve (V)
Abducens nerve (VI)
Hypoglossal nerve (XII)
Cerebellum

COLOR PLATE 12 The cranial nerves. [Source: Gaudin, A. J., and Jones, K. C. (1989). "Human Anatomy and Physiology." Harcourt Brace Jovanovich, San Diego, p. 323. Reprinted with permission.]

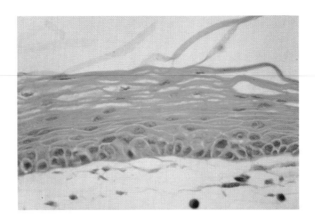

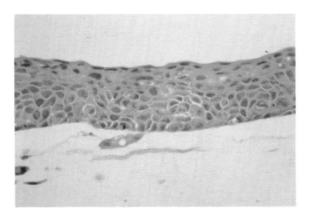

COLOR PLATE 13 Effect of HPV-16 DNA on *in vitro* differentiation of primary human keratinocytes as assayed in the organotypic culture system. *Top panel,* A control that received no HPV DNA and shows a normal pattern of keratinocyte differentiation; *Bottom panel,* The effect of the HPV-16 E6 and E7 genes on keratinocyte differentiation. Dividing cells are observed through the full thickness of the epithelium, and the gradient of differentiation is totally disrupted. Severe cases of cervical dysplasia show similar histological features.

dNTPs have no known roles other than those of DNA precursors and regulators of their own synthesis. dNTPs are the substrates for action of DNA polymerases, which catalyze the template-guided insertion of nucleotide units into polymeric chains. A similar process occurs in RNA synthesis, as detailed elsewhere in this encyclopedia. However, the ribonucleoside triphosphate (rNTP) precursors for RNA synthesis have numerous additional metabolic roles—as energy cofactors (notably adenosine triphosphate [ATP]), as precursors of coenzymes, and in the synthesis of activated metabolic intermediates such as uridine diphosphate glucose and acetyl-coenzyme A. Because of these additional metabolic roles, and because most cells contain five or six times as much RNA as DNA, at least 90% of the metabolic flux through the *de novo* nucleotide synthetic pathways is destined for ribonucleotide, rather than deoxyribonucleotide, synthesis.

III. The Need for Organization of DNA Metabolism

Because of their diverse roles in intermediary metabolism and transcription, ribonucleotides are utilized in all compartments of the cell and at all phases of the cell cycle. By contrast, DNA replication is confined to the nucleus (except for mitochondrial DNA) and takes place during just a fraction of the lifetime of a cell. Therefore, it is appropriate that the cell organize and regulate the synthesis of deoxyribonucleotides so that they are delivered to replication sites on demand. Moreover, because deoxyribonucleotides do not play any additional known roles in mammalian cells, their rates of synthesis must be controlled so that unwanted pool imbalances do not develop; the four dNTPs must be provided at the same relative rates at which they are incorporated into DNA. As described later, deoxyribonucleotide pool imbalances have deleterious biological effects. [*See* RIBONUCLEASES, METABOLIC INVOLVEMENT.]

Another factor that must be considered is the low intracellular concentrations of deoxyribonucleotides. From data on pool sizes and intracellular volumes, one can estimate that rNTP concentrations are in the millimolar range in mammalian cells. ATP, the most abundant ribonucleotide, is present at concentrations ranging from 3 to 10 mM in different cells and organelles. By contrast, dNTP pool

sizes are one to two orders of magnitude lower. These pools must be maintained in the face of vastly increased demand as proliferating cells move from the G1 phase of the cell cycle into the S phase. A typical somatic cell will double its DNA content within 6 hr, using between 1,000 and 10,000 replication forks per cell. Clearly, it is essential that the cell be able to deliver precursors to these sites at high enough rates to saturate replicative DNA polymerases. The demand is even higher in early development, where a cleavage-stage embryo may double its DNA content in <0.5 hr. Thus, DNA precursor biosynthesis must be organized and coordinated with DNA replication. Much of this coordination involves allosteric regulation of the function of enzymes of deoxyribonucleotide biosynthesis by

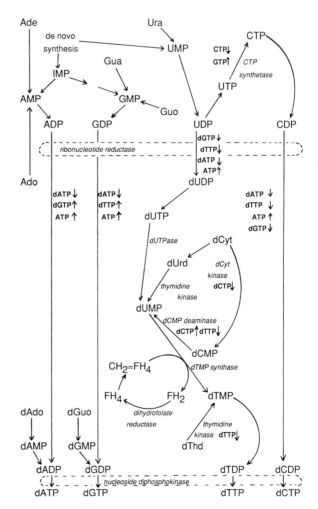

FIGURE 2 Pathways of deoxyribonucleotide biosynthesis in mammalian cells. Key enzymes are named, and allosteric effectors are identified with an upward arrow (for activators) and a downward arrow (for inhibitors).

precursors or intermediates. The relevant pathways, critical enzymes, and control sites are identified in Fig. 2. The key regulated enzyme is ribonucleotide reductase, which catalyzes the first step committed to DNA replication. A single enzyme reduces all four common ribonucleoside diphosphates (ADP, CDP, GDP, and UDP) to their 2'-deoxy derivatives. The enzyme protein, which is an $\alpha_2\beta_2$-tetramer, is controlled by interaction of nucleoside triphosphates with two classes of regulatory sites. One site regulates primarily the catalytic activity of the enzyme, and the other controls its specificity, the efficiency with which it acts upon each of its four substrates. Together, these interactions help to maintain balanced rates of synthesis of the four DNA precursors—dATP, dCTP, dGTP, and dTTP.

As noted earlier, the demand for ribonucleotides is more generalized, and rNTPs must be made available continuously throughout the cell. From the standpoint of intracellular organization, the most remarkable feature of RNA metabolism is the splicing of exons to each other, as introns are excised during posttranscriptional RNA processing. This process involves an intricate structure called the "spliceosome," which forms in the nucleus from small ribonucleoprotein particles. The process is not driven by kinetic factors; it is quite slow, but very precise.

IV. Mechanisms and Processes

Most of the mechanisms known to organize nucleotide synthesis and coordinate it with nucleic acid synthesis are comparable to processes that organize other areas of metabolism—subcellular compartmentation and the use of multienzyme complexes and multifunctional enzymes to channel scarce or labile intermediates. In addition, some nucleotides act directly to regulate processes other than their own biosynthetic pathways.

A. Intracellular Compartmentation

Most of the reactions of purine and pyrimidine nucleotide synthesis take place in the cytosol of mammalian cells. Obviously, both replication and transcription must occur in the nucleus, where the DNA template resides. Because the nuclear membrane contains large pores, transport of nucleotides to their sites of utilization should be straightforward.

However, under some conditions the intranuclear deoxyribonucleotide pools seem to be metabolically distinct from the much larger cytosolic pools. Some of this compartmentation probably results from the utilization of deoxyribonucleoside cytidine triphosphate (dCTP) in a recently discovered pathway unrelated to DNA replication—the synthesis of the deoxyribonucleotide analogs of the phospholipid biosynthetic intermediates cytidine diphosphate (CDP)-choline and CDP-ethanolamine. The levels of these "liponucleotides" are much lower than those of the well-known ribo compounds, and these deoxy analogs probably lay minor or insignificant metabolic roles. Nevertheless, the membrane-associated synthesis of dCDP-containing liponucleotides would tend to create at least two metabolic pools of dCTP.

A significant amount of the cytosolic ribonucleotide output is transported into mitochondria to participate in energy-genrting pathways. The mitochondrial inner membrane contains an adenine nucleotide translocase system, which pumps ADP in concomitantly with the efflux of ATP. In HeLa cells, about 15–20% of the total content of each rNTP is in mitochondria. Mitochondria also contain several deoxyribonucleotide biosynthetic enzymes, and most of the dNTPs destined for mitochondrial DNA are probably synthesized within the organelle, from ribonucleotides and deoxyribonucleosides that are transported inward. [*See* HeLa Cells.]

From the standpoint of cellular organization, one aspect of nucleotide compartmentation seems paradoxical. The *de novo* synthesis of pyrimidine nucleotides involves six reactions:

$$CO_2 + H_2O + ATP \rightarrow \text{carbamoyl phosphate} \rightarrow$$
$$\text{carbamoyl aspartate} \rightarrow \text{dihydroorotate} \rightarrow$$
$$\text{orotate} \rightarrow \text{orotidylate} \rightarrow \text{UMP.}$$

The first three reactions are catalyzed by a trifunctional enzyme in eukaryotic cells. The last two reactions are catalyzed by a bifunctional enzyme (see next section); both of these multifunctional enzymes are cytosolic. However, the fourth enzyme in the pathway, dihydroorotate dehydrogenase, is located on the mitochondrial inner membrane. Therefore, the pathway begins and ends in the cytosol, but intermediates must travel into, and then out of, the mitochondrion. Because the intermediates are not used for other known processes within the mitochondrion, the metabolic logic behind this subcellular organization remains elusive.

B. Multifunctional Proteins

An effective way to organize metabolic pathways is to juxtapose the active sites of enzymes catalyzing sequential reactions so that intermediates are transferred directly from site to site, equilibrating either little or not at all with bulk pools. This phenomenon, called metabolic channeling, can in principle improve cell functions in various ways. First, by preventing the diffusion of intermediates away from their sites of utilization, channeling provides for more efficient utilization of substrates and delivery of end products to sites of ultimate disposition. Second, by keeping average intracellular concentrations of most intermediates low, channeling prevents the accumulation of intermediates to levels where they would adversely affect osmotic pressure in the cell. Finally, a multifunctional protein allows for joint regulation of activities that act in one pathway, rather than requiring several distinct control mechanisms.

Table I lists several multienzyme complexes and multifunctional proteins involved in nucleotide biosynthesis. Although most of these have been observed in mammalian cells, some have not. In fact, enzymes such as the bifunctional dihydrofoalte reductase–thymidylate synthase found in protozoal cells are so different from their counterpart enzymes in mammalian cells that they present attractive targets for chemotherapy of malaria and other protozoal diseases. Similarly, the herpes virus deoxypyrimidine kinase is an actual chemotherapeutic target because, by broad specificity, it permits the intracellular activation of drugs only in infected cells. For example, acyloguanosine (Acyclovir) is activated to the inhibitory nucleotide acycloguanosine triphosphate, through the action of the viral kinase.

Tetrahydrofolate coenzymes participate in purine nucleotide synthesis and thymine nucleotide synthesis. These coenzymes are unstable and also present at very low concentrations in cells. Activities that interconvert these coenzymes are, perhaps not surprisingly, covalently linked, particularly the widely distributed formylmethenylmethylenetetrahydrofoalte synthetase, which contains formyltetrahydrofolate synthetase, methylenetetrahydrofolate dehydrogenase, and methenyltetrahydrofolate cyclohydrolase activities in one protein. Perhaps surprisingly, these juxtaposed active sites do not always create metabolic channeling, as shown by *in vitro* studies of some reaction sequences.

Another class of multifunctional enzymes does not catalyze sequential reactions but, instead, acts upon different substrates in the same active site. An excellent example is ribonucleotide reductase, which, as noted earlier, catalyzes reactions leading to all four DNA precursors. It seems certain that the architecture of this enzyme molecule is designed so it coordinates the four activities, thereby, permitting balanced synthesis of the four dNTPs. Another example is deoxycytidine kinase, which in vertebrate cells has a high affinity for deoxycytidine, but which also acts upon deoxyadenosine or deoxyguanosine. Affinities for the purine substrates are much lower than for deoxycytidine, meaning that the enzyme will preferentially phosphorylate deoxycytidine under most physiological conditions.

C. Multienzyme Complexes

In many cases, enzyme active sites that catalyze sequential reactions are not covalently joined on single protein molecules but reside on separate proteins that interact with one another strongly enough to allow the isolation of a multienzyme complex. Three examples from nucleotide metabolism are listed in Table I. The one that has been best characterized in mammalian cells is a four-protein complex containing the two transformylases of the purine nucleotide biosynthetic pathway and two enzymes of tetrahydrofolate coenzyme interconversion (Fig.3).

Multienzyme complexes that carry out the synthesis of deoxyribonucleotides have been described in both prokaryotic and eukaryotic systems. The only one that has been highly purified and characterized kinetically is a complex from T4 bacteriophage-infected *Escherichia coli*. This complex demonstrates substrate channeling *in vitro*, in both crude and highly purified preparations. Moreover, a number of viral gene mutations that affect individual enzymes of the complex also disrupt the integrity of the complex as isolated *in vitro*. This type of genetic evidence, which goes hand-in-hand with the kinetic studies mentioned earlier, suggests that the complex is formed by specific protein–protein interactions and not by protein aggregations that occur only after cells are disrupted for enzyme isolation. In general, the demonstration that a multiprotein aggregate functions as an organized complex in intact cells requires the gathering of quite diverse lines of evidence—from enzyme purification, enzyme kinetics, molecular genetics, intracellular lo-

TABLE I Multifunctional Enzymes and Multienzyme Complexes in Nucleotide Biosynthesis

Enzyme or complex	Reactions catalyzed	Source
Multifunctional Enzymes		
H_2-folate reductase/dTMP synthase	Synthesis of dTMP from dUMP; reduction of H_2-folate	Protozoa
Deoxypyrimidine/dTMP kinase	Phosphorylation of dTMP and many nucleosides	Herpes virus-infected cells
GART protein	Three reactions of *de novo* purine synthesis	*Drosophila* and other eukaryotes
CAD protein	First three reactions of pyrimidine synthesis	Eukaryotes
UMP synthase	Orotate → orotidylate → uridylate	Eukaryotes
Formylmethenyl-methylenetetrahydrofolate synthetase	Three reactions shown in Figure 3	Eukaryotes
FIGLU transformylase/formimino-H_4-folate cyclodeaminase	Formiminoglutamate → formimino-H_4-folate → 10-formyl-H_4-folate	Bacteria
Multienzyme Complexes		
Purine transformylase complex	Transformylase reactions in purine synthesis (Fig. 3)	Eukaryotes
Carrot thymidylate synthesis complex	dTMP synthase, H_2-folate reductase, serine transhydroxymethylase	Carrot cells
T4 dNTP synthetase complex	rNDPs →→→ dNTPs dNMPs →→→ dNTPs	T4 phage-infected *E. coli*

dTMP, deoxyribosylthymine monophosphate; dUMP, deoxyuridine monophosphate.

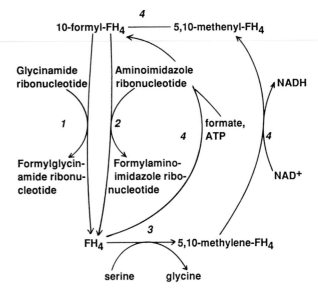

FIGURE 3 Reactions catalyzed by a multienzyme complex for purine nucleotide biosynthesis. The two transformylase enzymes are identified by *1* and *2*; *3* denotes serine transhydroxymethylase, whereas *4* identifies the three reactions catalyzed by the trifunctional formylmethenylmethylenetetrahydrofolate synthetase. (From *Biochemistry,* by C. K. Mathews and K. E. van Holde. Copyright © 1990, Benjamin/Cummings Publishing Company, Redwood City, California. Reprinted by permission.)

calization studies, and reconstitution from purified components. Multienzyme aggregates, such as the T4 dNTP synthetase, have been described in mammalian cells, but to date none has been characterized by the full complement of approaches described above.

Interest in dNTP-synthesizing complexes is focused on the likelihood that they participate as part of a substrate-channeling mechanism for delivery of DNA precursors to replication sites. Evidence for such a model is accumulating in prokaryotic systems, where replicative polymerases are characterized by high K_m values (50–100 μM) and high V_{max} values (500–1,000 nucleotides/sec). Diffusion of dNTPs from remote sites of synthesis probably does not suffice to maintain dNTP levels at or near saturating concentrations at replication sites. In eukaryotic cells, however, direct coupling of dNTP synthesis to DNA synthesis seems impossible, if only because the enzymes of dNTP synthesis are found in the cytosol. In addition, the kinetic properties of replicative DNA polymerases are quite different from those of prokaryotes. V_{max} and K_m values for mammalian DNA polymerases are at least an order of magnitude lower than those of prokaryotes. Recent studies on calf-thymus DNA polymerase-α indicate K_m values for dNTPs in the range of 2–5 μM. At the same time, estimations of intranuclear dNTP concentrations in mammalian cells give values for (dGTP), the *least* abundant dNTP, of about 10 μM. Thus, one need not invoke a channeling mechanism to understand how replicative DNA polymerases can operate at or near substrate saturation in mammalian cells. However, little is known at present about the transfer of dNTPs from

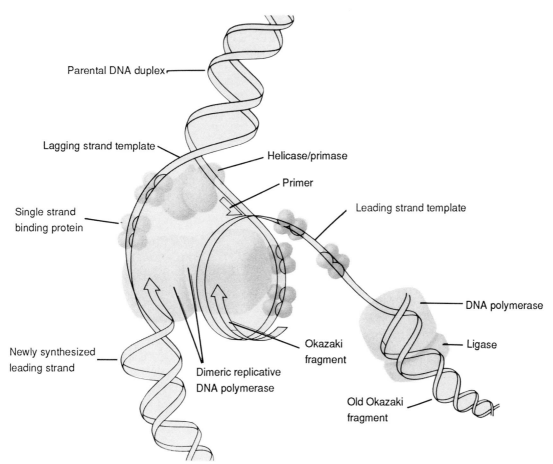

Parental DNA duplex

Lagging strand template

Helicase/primase

Primer

Leading strand template

Single strand
binding protein

DNA polymerase

Ligase

Newly synthesized
leading strand

Dimeric replicative
DNA polymerase

Okazaki
fragment

Old Okazaki
fragment

FIGURE 4 Interactions of proteins at the replication fork during DNA chain elongation. Each arrow denotes a 3' DNA terminus. (From *Biochemistry*, by C. K. Mathews and K. E. van Holde. Copyright © 1990, Benjamin/Cummings Publishing Company, Redwood City, California. Reprinted by permission.)

cytoplasmic sites of synthesis to the interior of the nucleus. In studies where radiolabeled deoxyribonucleosides were microinjected into mammalian cell cytoplasm, radioactivity was rapidly and efficiently incorporated into DNA. Whether movement into the nucleus occurs by simple diffusion or by a facilitated process is not yet known.

D. Dynamic Protein Associations

While DNA replication and transcription have long been known to require enzymes in addition to the polymerases that form internucleotide bonds, isolating multienzyme complexes capable of faithfully carrying out all steps in nucleic acid synthesis (initiation, chain elongation, and termination) has been difficult. A complementary approach has been far more successful—purification of individual proteins

in the process and reconstitution of systems capable of carrying out one or more steps in the process. This is because complexes form transiently and then must dissociate so that the macromolecular synthetic process can proceed, from initiation to elongation, or from elongation to termination. Thus, multiprotein associations exist transiently. We can refer to the protein–protein interactions as dynamic, rather than static, as in those stable multienzyme complexes that can be isolated and purified. [*See* DNA AND GENE TRANSCRIPTION.]

Chain elongation in DNA replication involves the cooperation of about a dozen proteins that interact at the replication fork, as shown in Fig. 4. In this figure, a dimeric replicative DNA polymerase holoenzyme is simultaneously copying both leading and lagging strands. Single-stranded DNA-binding protein serves as a "movable scaffold," holding DNA strands in the optimal conformation to serve as template, to the 3' side of each polymerase molecule. A helicase-primase complex (the "primosome") runs along the lagging DNA strand in a 3'-to-5' direction, unwinding the parental DNA strands and at intervals synthesizing short RNA primers (5'-to-3') on the lagging strand. Another DNA polymerase (polymerase I in bacteria) carries out a repair-type reaction, simultaneously cleaving out RNA primers and replacing them with deoxyribonucleotides. Finally, DNA ligase links the newly synthesized DNA chain to high-molecular-weight DNA on the lagging strand. Whereas the model depicted in the figure is derived primarily from research in prokaryotic systems, the overall enzymatic processes are quite similar in eukaryotic cells. One significant difference is that in bacteria DNA replication occurs in close proximity to the cell membrane; interactions between replication proteins and membrane proteins have been poorly understood and are just now being defined. In eukaryotic cells, considerable evidence favors the concept that DNA replication occurs at a structure called the nuclear matrix. Experimentally, the nuclear matrix is the material left when isolated nuclei are treated sequentially with nucleases, salt, and detergent. The material retains considerable organized structure, and it has been proposed that replicative proteins are anchored to this structure, with loops of DNA being replicated as they move through sites of binding of these proteins.

The initiation of DNA replication also involves transient protein–protein associations, as shown in Fig. 5, which represents DNA initiation in bacteriophage λ. A site-specific DNA-binding protein binds in four copies at a specific replication origin on the DNA molecule. The DNA then bends, forming a tightly wrapped circle, which creates negative superhelical tension on nearby DNA sequences, causing them to unwind. The single-stranded regions that appear become coated with single-stranded DNA-binding protein. This leads to transcription of short single-stranded DNA regions (not shown in the figure), with the resultant short RNA chains next undergoing elongation by insertion of deoxyribonucleotides, as in replicative chain initiation. If the proteins involved in this process interacted strongly with one another, then the initiation complex might never decay and elongation might never begin.

Transcription presents an interesting example of interactions between proteins that may never associate. As shown in Fig. 6, RNA polymerase must unwind duplex DNA to expose a single-stranded template to copy. If the polymerase molecule does not rotate around the duplex (as is apparently the case), overwinding of the DNA duplex ahead of the polymerase and underwinding behind it occurs. If a

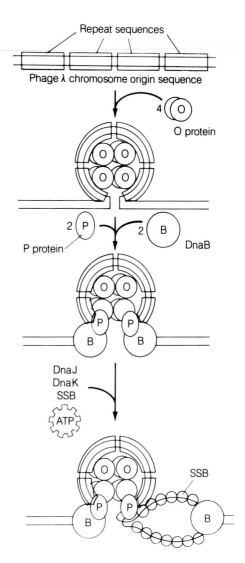

FIGURE 5 Dynamic associations of proteins in the initiation of DNA replication by bacteriophage λ. O and P are virally encoded proteins that form a tightly wrapped nucleoprotein structure by reacting specifically with four repeated sequences at the replication origin (shown by arrows). B represents the proteins of the *E. coli* primosome; S is *E. coli* single-stranded DNA-binding protein; and J and K are additional *E. coli* replication proteins. (From *Biochemistry,* by C. K. Mathews and K.E. van Holde. Copyright © 1990, Benjamin/Cummings Publishing Company, Redwood City, CA. Reprinted by permission.)

circular template is transcribed by just one RNA polymerase molecule, the positive and negative superhelical turns cancel each other out, the linking number remains unchanged. However, if two or more polymerase molecules are transcribing in opposite directions, the circular template develops areas of both over- and underwinding, which cannot spontaneously relax. Under these conditions, a to-

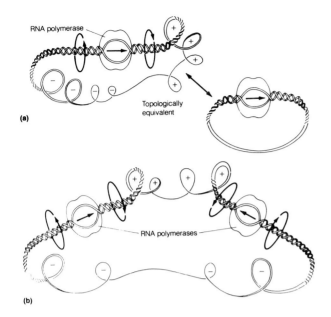

FIGURE 6 Supercoiling of DNA by transcription. If a circular DNA is transcribed in opposite directions by two or more RNA polymerase molecules, the DNA becomes overwound ahead of the transcribing polymerases and underwound behind them. Action of a topoisomerase is necessary to relieve this superhelical tension. (From *Biochemistry*, by C. K. Mathews and K. E. van Holde. Copyright © 1990, Benjamin/Cummings Publishing Company, Redwood City, CA. Reprinted by permission.)

poisomerase can cause the superhelical turns to relax. Therefore, the interaction between RNA polymerase and topoisomerase is essential if transcription is to continue on a circular template, even though the two enzymes may never physically associate with each other.

E. Substrate Cycles

The above examples highlight the importance of dynamic associations among protein molecules, which interact transiently with one another and, thus, do not form an isolatable complex. A different type of interaction is seen in substrate cycling, a mechanism recently proposed to participate in controlling the intracellular concentrations of deoxyribonucleotides. In mammalian cells, a substantial fraction of each dNTP is not incorporated into DNA. Instead, it breaks down, with release of the resulting deoxyribonucleoside to the medium. The rate of this breakdown increases when DNA replication is blocked. When synthesis of deoxyribonucleotides is blocked, it has been observed that the deoxyribonucleosides are efficiently taken back up into cells and converted to dNTPs by salvage pathways. The in-

terconversion of deoxyribonucleoside monophosphates and deoxyribonucleosides has been proposed as the key regulated step, which determines whether substrate flux is primarily in the direction of dNTP synthesis or breakdown (Fig. 7). In turn, this mechanism is believed to help control dNTP pool sizes within the narrow limits compatible with a cell's well-being. A "substrate cycle" of the type depicted presents an efficient way to control a metabolic process; in the example shown, a small change in the activity of either a nucleotidase or a kinase can have a substantial effect upon the direction of metabolic flux—either toward dNTP synthesis or breakdown. While all of this seems quite reasonable, however, it is not yet clear that nucleotidase activities are subject to allosteric regulation, although most deoxyribonucleoside kinases are subject to feedback control by dNTPs. On the other hand, both enzymes could be controlled simply by the intracellular concentrations of their substrates.

F. Nucleotides as Metabolic Regulators

Nucleotide molecules themselves serve as widely distributed metabolic regulators, both controlling

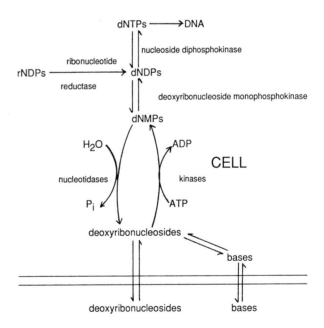

FIGURE 7 A substrate cycle as a possible means for regulation of dNTP pool sizes. Small changes in activity of nucleotidases or nucleoside kinases can have large effects on the rate and direction of flux between deoxyribonucleosides and dNMPs.

their own biosynthesis and coordinating other metabolic processes. As noted earlier, ribonucleotide reductase and other deoxyribonucleotide biosynthetic enzymes are controlled by dNTPs, so that balanced supplies of DNA precursors are produced (see Fig. 2). Other nucleotides play more specific regulatory roles. The best understood examples are 3',5'-cyclic AMP and 3',5'-cyclic GMP, which control numerous processes via the activation of protein kinases, as detailed elsewhere in this encyclopedia. In bacteria, a guanine nucleotide, guanosine 3',5'-tetraphosphate (ppGpp), coordinates a metabolic response to inhibition of protein synthesis by somehow modulating the synthesis of ribosomal and transfer RNAs. Whether or not comparable regulatory mechanisms exist in mammalian cells is not yet known. However, a dinucleotide, diadenosine 5',5'''-P^1,P^4-tetraphosphate (Ap$_4$A), is believed to act as a regulator of mammalian DNA replication. Concentrations of this interesting nucleotide increase in the nucleus as cells move from G1- to S-phase, and Ap$_4$A may somehow trigger the initiation of DNA replication, although the mechanism is still obscure. However, it has been proposed that other dinucleotides, comparable in structure to Ap$_4$A, act as "alarmones"—compounds that somehow sense cellular damage and coordinate a response to the metabolic lesion. It has been proposed that these dinucleotides coordinate a response to oxidative damage, much as ppGpp coordinates a bacterial cellular response to impaired protein synthesis; however, the precise mechanisms involved remain obscure. [See METABOLIC REGULATION.]

V. Biological Consequences of Nucleotide Metabolic Imbalance

As noted earlier, deoxyribonucleotides are used only for DNA replication in mammalian cells. It is important that the four dNTPs be produced at rates equivalent to their relative abundances in DNA. If the normal control of dNTP synthesis is perturbed, for example, by a mutation that inactivates an enzyme control site, pool imbalances develop. Faulty control of ribonucleotide reductase, dCMP deaminase, and CTP synthetase are all known to generate abnormal dNTP pools. For example, inactivation of dCMP deaminase elevates the dCTP pool, with concomitant depletion of dTTP. This type of imbalance has numerous genetic and metabolic consequences,

including enhanced mutagenesis, chromosomal abnormalities, induction of endogenous viruses, enhanced genetic recombination, and cell death. These responses illustrate the importance of maintaining normal control of dNTP pools within rather narrow limits. The mechanism of enhanced mutagenesis has been thoroughly studied, primarily by nucleotide sequence determination of mutant sites. As one might expect, dNTP pool imbalances force replication errors; for example, a dCMP deaminase deficiency stimulates mutagenic AT-to-GC pathways, because the accumulated dCTP competes with the depleted dTTP for insertion sites opposite dAMP in the template. However, a quantitative relationship between pool changes and mutation rates is not predictable. Moreover, different sites within a gene undergo mutation at widely differing rates under the same pool imbalance conditions. These observations suggest that (1) pool imbalance mutagenesis is dependent on events in DNA replication other than the insertion step, possibly exonucleolytic proofreading, and (2) the nucleotide sequences that flank a mutant site somehow influence the mutagenic process. Further analysis of mutagenesis under these conditions should illuminate the mechanisms of other responses to dNTP pool imbalance, which at this writing are much less understood. [See MUTATION RATES.]

Bibliography

Berezney, R. (1985). The nuclear matrix model and DNA replication. *UCLA Symposia Molec. Cell. Biol.* **26**, 99.

Echols, H. (1986). Multiple DNA-protein interactions governing high-precision DNA transactions. *Science* **233**, 1050.

Kornberg, A. (1990). "DNA Replication," 2nd ed. W. H. Freeman & Co., San Francisco.

Mathews, C. K., Moen, L. K., Sargent, R. G., and Wang, Y. (1988). Enzyme interactions in deoxyribonucleotide synthesis. *Trends Biochem. Sci.* **13**, 394.

Mathews, C. K., and van Holde, K. E. (1990). "Biochemistry," Chapters 22, 24, and 26. The Benjamin/Cummings Publishing Co., Redwood City, California.

Meuth, M. (1989). The molecular basis of mutations induced by deoxyribonucleoside triphosphate pool imbalances in mammalian cells. *Exptl. Cell Res.* **181**, 305.

Pruss, G. J., and Drlica, K. (1989). DNA supercoiling and prokaryotic transcription. *Cell* **56**, 521.

Reichard, P. (1988). Interactions between deoxyribonucleotide and DNA synthesis. *Ann. Rev. Biochem.* **57**, 349.

Nutrients, Energy Content

G. B. STOKES, *Murdoch University*

Glossary

Biosphere/biomass Biosphere refers to the total habitat of living things, from the ocean depths to mountain tops; biomass refers to all of the matter associated with living things, including an elephant's leg and the limb of a tree

Calorie Defined as the amount of heat required to raise 1 g of water from 15° to 16°C; the term used in dietetics is the Calorie, which refers to 1,000 cal or 1 kcal and is equivalent to 4.186 kJ (kiloJoules)

Carbohydrate A simple sugar, or complex molecule derived from the polymerization of sugar molecules and their derivatives (such as malt and starch); general empirical formula $(CH_2O)_n$

Digestible energy Difference between the energy of the food consumed and that excreted in the feces; sometimes corrected for metabolic fecal energy, which is the energy excreted during fasting

Heat of combustion Also referred to as caloric value; heat evolved (kJ/mol, kJ/g) when a compound or substance is fully burned

Macronutrients Fat, carbohydrate, and protein, the major dietary compounds; distinct from micronutrients such as vitamins and trace elements

Metabolizable energy Difference between digestible energy and the energy excreted in urine; that portion of dietary energy actually available for supporting metabolism

Metabolism Biochemical reactions of living things, made up of the interaction between synthetic reactions (anabolism) and the reactions of degradation or breakdown (catabolism)

Redox In intermediary metabolism (the set of biochemical reactions concerned with the processing of food molecules for energy), electrons are transferred in certain reactions only (the exchangeable electrons are also called reducing equivalents); the reactions in which the donor molecule contributes electrons to the acceptor molecule are termed redox reactions and are catalyzed by a family of enzymes termed dehydrogenases; in the transfer, the donor is oxidized and the acceptor is reduced; the electrons are passed between donor and acceptor molecules via cofactors, which may be free or, in some cases, are bound to the particular dehydrogenase enzyme

Respiratory quotient Ratio between carbon dioxide breathed out to oxygen breathed in; used to provide an indication of which depot molecules (carbohydrate, fat, or protein) are being catabolized to support the body's energy needs

TABLES THAT LIST THE CALORIC VALUE of foods afford a ready means to estimate the effect on body weight of eating various amounts and combinations of foods. Fat has a higher caloric value (kJ/g) than carbohydrate or protein, so the Calorie-conscious consumer pays particular regard to limiting fat consumption. But why are fats so much higher in Calories? Where are the Calories to be found in foods?

In this article, the characteristics of molecular structure that determine the caloric value of food molecules will be discussed, and arithmetic methods are given for estimating the energy content of macronutrients based on their structure or empirical formula.

The major energy transformations in the biosphere will also be examined in order to appreciate the flow of energy that passes from incident solar radiation, through biomass, to final heat release during biological work.

Oxygen is released into the atmosphere during the photosynthetic reaction and reabsorbed in biological oxidation reactions when fuels are used up. The amount of energy captured during synthesis and released during their utilization (measured as the caloric value or heat of combustion of the nutrient) is directly proportional to the amount of oxygen consumed. The approach presented for understanding the energy content of nutrients also allows a simple explanation of the concept of respiratory quotient.

When nutritionists consider the body's capacity to utilize the digestible energy available from macronutrients, a special case must be made for protein, because about 20% of the energy in protein is excreted in urine. The basis for this energy loss is also discussed.

The biological energy depot molecules of most interest to man, the macro-nutrients: fat, carbohydrate and protein, will also be reviewed.

I. Energy Flow through the Biosphere

Energy enters the biosphere principally as reducing equivalents (electrons) derived from the photolysis of water by incident solar radiation during the process of photosynthesis. The immediate products of photolysis are molecular oxygen, protons, and reducing equivalents. The electrons released during photolysis end up mostly in the food chain as reduction products of carbon dioxide (carbohydrates and fats).

A simplified diagram of the major energy transformations in the biosphere is presented in Fig. 1. Electrons released from water are transferred in complex biochemical pathways via cofactors such as nicotinamide adenine dinucleotide phosphate ($NADP^+$) and reduce oxidized acceptor molecules (CO_2, N_2, NO_3^-, and intermediate products). The ultimate products of these reductive pathways are the food chain fuels.

On subsequent reoxidation of fuels, the electrons end up in an electron sink (usually transferred to molecular oxygen that is reduced back to water). The energy released during this reoxidation process is coupled to the phosphorylation of adenosine diphosphate (ADP) to ATP. ATP serves throughout the plant and animal kingdoms as the main energy currency of the cell. When ATP is hydrolyzed back to ADP and inorganic phosphate (Pi) as part of the coupled biochemical sequences that drive the reactions of the cell, the energy captured initially by photosynthetic cells is finally released from the biosphere as heat. [See ADENOSINE TRIPHOSPHATE (ATP).]

During photosynthesis, plants take up carbon dioxide and release oxygen (described in the left half of Fig. 1). In contrast, during the process of respiration (described in the right half of Fig. 1), animals breathe in oxygen and breathe out carbon dioxide. Similarly, oxygen is consumed and carbon dioxide and water vapor are released when fossil or other fuels are burned in engines or fires. Hence, the total amounts of free and dissolved oxygen and carbon dioxide vary reciprocally and depend on the relative global rates of photosynthesis and combustion, as well as on changes in the amount and composition of organic molecules in the biomass and fossil fuel reserves.

II. Molecular Structure and Heats of Combustion

Table I shows the heats of combustion for a series of simple compounds from propane to carbon dioxide. The compounds are listed in order of decreasing number of reduced bonds, representing oxidation due to the sequential loss of electrons from their bonds. In this article, the bonds that store chemical potential energy (C—C, C—H, N—C and N—H) will be designated as "reduced bonds".

It is apparent from Table I that the energy in a C—C bond is similar to that in a C—H bond (acetic acid 874 kJ/mol, methane 891). Also, only minor changes occur when alkenes are hydrated (propene 2,050, propanol 2,021), thus a double bond represents about twice the potential energy of a single bond. The oxidized bonds (C—O and O—H) contribute no heat on combustion of the parent compound because they are preserved in CO_2 and H_2O, the fully oxidized products of combustion, which cannot be burned further.

The heats of combustion of many organic compounds containing C, H, and O are described by the relationship

$$Q = 110n + C, \qquad (1)$$

where Q is the heat of combustion expressed in kJ (110 kJ the net energy change per mole electron), n

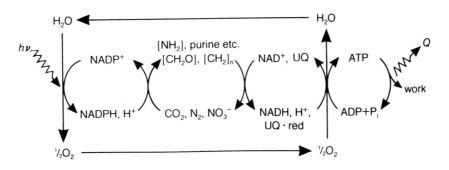

FIGURE 1 Major energy transformations in the biosphere. (Reprinted, with permission, from G. B. Stokes, 1988, *Trends Biochem. Sci.* **13**, 422–424, Elsevier Publications, Cambridge.)

is the number of reducing equivalents, and C is a constant characteristic of the functional group in the compound. The effect of the functional group is apparent from Table I if, for example, the oxidation of an alcohol to an aldehyde is compared with the oxidation of an aldehyde to an acid.

One consequence of the oxygen–water cycle illustrated in Figure 1 is that the amount of oxygen in the atmosphere provides a rough estimate of the energy reserves (reduced bonds) stored in the biosphere. Furthermore, the amount of oxygen taken up in catabolism will be proportional to the number of bonds that are oxidized. Equation (1) also amounts to stating that the heat of combustion of a compound is proportional to the oxygen taken up.

TABLE I Heats of Combustion of Some Simple Compounds

No. of C—C and C—H bonds (k)	Compound	Heats of combustion (kJ/mol) Observed[a]	Predicted (220k)
10	$CH_3—CH_2—CH_3$	2,222	2,200
9	$CH_3—CH=CH_2$	2,050	1,980
9	$CH_3—CH_2—CH_2OH$	2,021	1,980
8	$CH_3—CH_2—CHO$	1,816	1,760
7	$CH_3—CH_2—COOH$	1,527	1,540
7	$CH_3—CH_3$	1,540	1,540
6	$CH_2=CH_2$	1,410	1,320
6	$CH_3—CH_2OH$	1,372	1,320
5	$CH_3—CHO$	1,167	1,100
4	$CH_3—COOH$	874	880
4	CH_4	891	880
3	CH_3OH	728	660
2	CH_2O	569	440
1	$H—COOH$	255	220
0	CO_2	0	0

[a] Data obtained from the CRC Handbook of Chemistry and Physics.

This may be seen from the following equations:

$$H_2O + \begin{array}{c}|\\—C—\\|\end{array}\begin{array}{c}|\\C—\\|\end{array} + \tfrac{1}{2}O_2 \rightarrow \begin{array}{c}|\\—C—OH\\|\end{array} + HO—\begin{array}{c}|\\C—\\|\end{array} \quad (2)$$

$$\begin{array}{c}|\\—C—H\\|\end{array} + \tfrac{1}{2}O_2 \rightarrow \begin{array}{c}|\\—C—OH\\|\end{array} \quad (3)$$

In the reaction shown in equation (2), one C—C bond is oxidized with the uptake of one atom of oxygen, whereas in equation (3), one C—H bond is oxidized, again with the uptake of one oxygen atom.

Equation (1) can be extended to nitrogen-containing compounds. Thus, the change in enthalpy (ΔH) or heat content following combustion, when expressed in kJ/mol, is:

$$\Delta H = 220k + 105l \quad (4)$$

where k is the number of C—C and C—H bonds, and l is the number of C—N and N—H bonds. The constant for the C—N and N—H bonds (i.e., 105 kJ/mol) is close to one-half of that for C—C and C—H bonds (220 kJ/mol). This corresponds with the fact that in the oxidation of compounds to CO_2, H_2O, and N_2, each C—N and N—H bond is oxidized with one-half the oxygen uptake of a C—C or C—H bond, as follows:

$$H_2O + 2[\begin{array}{c}|\\—C—N\\|\end{array}] + \tfrac{1}{2}O_2 \rightarrow 2[\begin{array}{c}|\\—C—OH\\|\end{array}] + 2N \quad (5)$$

$$2[\begin{array}{c}|\\N—H\\|\end{array}] + \tfrac{1}{2}O_2 \rightarrow 2N + H_2O \quad (6)$$

Nitrogen in the products is represented as unbonded, since N_2 contains no bond to another element.

Thus the oxidation of any compound can be represented as the oxidation of its constituent C—C, C—H, C—N and N—H bonds. Hence heats of combustion (kJ/mol) can be calculated either by counting the number of reduced bonds and applying equation 4, or by determining the oxygen uptake from the overall equation of oxidation:

$$C_aH_bN_cO_d + (2a + b/2 - d)/2 \ O_2$$
$$= a \ CO_2 + c/2 \ N_2 + b/2 \ H_2O \quad (7)$$

and multiplying the total oxygen uptake $(2a + b/2 - d)/2$ by $110n$ (according to equation 1), where $n = 4$ for each mol of oxygen.

During oxidative phosphorylation in the mitochondria of cells, fuels are first oxidized by the enzymatic transfer of electrons to the cofactors nicotinamide adenine dinucleotide (NAD^+) and ubiquinone (UQ). The electrons are then passed into the electron transport chain and ultimately reduce oxygen to water. When electron transport is coupled to the phosphorylation of ADP, two or three ATP molecules are formed per electron pair, depending on the nature of the cofactor carrying those electrons. Reduced NAD^+ introduces electrons into the "top" of the chain and three ATP molecules are formed (P/O ratio = 3), whereas UQ introduces electrons into the chain at a lower potential and only two molecules of ATP are formed per pair of electrons (P/O ratio = 2). Therefore, it is interesting to note from Table I that the change in enthalpy for biological oxidation reactions where UQ accepts the pair of electrons via a flavoprotein (alkane–alkene) is substantially less and approximately two-thirds of the value for reactions where NAD^+ accepts the pair of electrons (alcohol–aldehyde; aldehyde–acid). [*See* ATP SYNTHESIS BY OXIDATIVE PHOSPHORYLATION IN MITOCHONDRIA.]

The enthalpy of combustion of nutrients has biological interest merely because it can provide a rough guide to the number of ADP molecules that can be phosphorylated to ATP.

III. Counting Reduced Bonds and Calculating ATP Yield

A useful approximation for students of bioenergetics is that three ADPs are phosphorylated with the energy released from the oxidation of each reduced C—H and C—C bond in substrate molecules. This estimation, based on counting the number of reduced bonds, does not specify the immediate source of the ATP, which may come both from substrate-level phosphorylation and from oxidative phosphorylation using electrons donated by reduced NAD^+ (P/O = 3) and/or reduced UQ (via a flavoprotein) (P/O = 2).

Overall, 12 ATP molecules can be generated from the oxidation of one molecule of acetyl-CoA ($CH_3CO \sim CoA$). The sites of energy storage in the molecule responsible for the 12 ATPs are the single C—C and the three C—H bonds.

Glucose ($C_6H_{12}O_6$) contains 5 C—C bonds and 7 C—H bonds, a total of 12 reduced bonds. We can expect that all 12 reduced bonds will be preserved in the products of glucose fermentation (glycolysis), because no net oxidation occurs in this anaerobic process, which occurs in the absence of oxygen. In muscle, glycolysis produces lactate (CH_3—CHOH—COOH), whereas in yeast the products of glycolysis are ethanol (CH_3—CH_2OH) and CO_2. A count of C—C and C—H bonds shows that lactate contains four C—H and two C—C bonds, whereas ethanol contains five C—H bonds and one C—C bond. Because 2 mol of lactate (or ethanol) are produced per mol of glucose, the 12 reduced bonds of glucose have been preserved in the 6 bonds in each pair of product molecules.

This simple calculation provides a useful check. At each redox step in a metabolic pathway, the number of reduced bonds will change by one, whereas no change will occur at other steps, such as those involving hydration (addition of water across a bond), phosphorylation (formation of a phosphate ester bond), or association or dissociation (gain or loss of a proton). Thus, the oxidized product of a dehydrogenase reaction will always contain one less reduced bond than the substrate (six for glyceraldehyde-3-phosphate and five for 1,3-bisphospho-glycerate; six for lactate and five for pyruvate). Furthermore, we would expect to generate 18 ATPs from the oxidative metabolism of lactate or ethanol (6 bonds × 3 ATP per bond).

The bond count approach can also be used to quickly assess a redox sequence of reactions. For example, the parasitic helminth *Ascaris* metabolizes glucose anaerobically to malate (via PEP carboxy kinase) and the malate is dismutated into succinate (via fumarate reductase) and pyruvate (via malic enzyme). Hence, the 12 bonds of glucose are found in the 7 bonds of succinate plus the 5 bonds of pyruvate. In this way, *Ascaris* is able to recycle its redox cofactors under anaerobic conditions.

This viewpoint may also help one understand why more energy can be stored as depot fat than as an equal weight of glycogen: because these fuels contain different densities of reduced bonds. The point is illustrated by considering the oxidation of the repeating unit of carbohydrate and of fat:

$$\text{Carbohydrate: } -[-\underset{\underset{OH}{|}}{\overset{\overset{H}{|}}{C}}-]- + \quad O_2 \rightarrow CO_2 + H_2O \tag{8}$$

$$\text{Fat: } \qquad -[-\underset{\underset{H}{|}}{\overset{\overset{H}{|}}{C}}-]- + 1\tfrac{1}{2}\,O_2 \rightarrow CO_2 + H_2O \tag{9}$$

The —CHOH— group of carbohydrate has two reduced bonds (one C—H and one C—C, counting half for each of the two C—C bonds shared with its two neighbors) and therefore, requires two oxygen atoms for complete oxidation. It has a molar mass of 30 g/mol, so at 220 kJ/mol for each reduced bond, it yields $220 \times 2 \div 30$ kJ/g, or 14.7 kJ/g. The —CH_2— group of fat has three reduced bonds (requiring three oxygen atoms for complete oxidation) and a molar mass of 14 g/mol, yielding $220 \times 3 \div 14$ kJ/g, or 47 kJ/g. Hence, to a first approximation, fat on a dry weight basis contains about three times as much energy (47/14.7) as carbohydrate.

IV. Understanding the Respiratory Quotient

These same equations (8) and (9) also show the chemical basis of the respiratory quotient (RQ). The RQ has been found particularly useful for estimat-

ing the nature of the body's fuel reserves (e.g., carbohydrate or fat) being used to support metabolism during exercise and rest. The RQ is calculated as the moles ratio of carbon dioxide produced to oxygen consumed. The ratio for pure carbohydrate metabolism is 1.0 (see equation (8)) and for fat metabolism it is 0.7 (see equation (9)). Experimentally, the RQ at the end of a brief sprint is found to be close to 1.0, showing that carbohydrate is the major fuel source. In contrast, the RQ is closer to 0.7 at the end of a marathon run, indicating that metabolism is being supported principally by the oxidation of fat.

The RQ for protein metabolism is 0.8. This ratio is more complicated to explain because the constituent amino acids of protein are more variable in their empirical formula than carbohydrate or fat molecules, and they are incompletely oxidized during catabolism. This is because protein nitrogen is excreted in a reduced form as ammonia or urea. A more detailed explanation is given below, when the difference between digestible and metabolizable energy is discussed.

V. Calculating the Caloric Value of Nutrients

One test of the validity of this approach for understanding the energy content of nutrients would be its capacity to predict the caloric value of macronu-

TABLE II Predicted and Observed Heats of Combustion of Macronutrients

		Number of reduced bonds	Heats of combustion			
			kJ/mol		kJ/g	
Macronutrient	Mr		Predicted	Observed	Predicted	Observed
Fat component (e.g., palmitic acid)						
		C—C = 15	$220 \times 46 = 10{,}120$		$10{,}129 \div 256 = 39.5$	
$CH_3(CH_2)_{14}COOH$	256	C—H = 31		10,031		39.3
		Σ = 46				
Carbohydrate component (e.g., glucose)						
		C—C = 5				
$CHO(CHOH)_4CH_2OH$	180	C—H = 7	$220 \times 12 = 2640$		$2{,}640 \div 180 = 14.7$	
		Σ = 12		2,815		15.7
Protein component (e.g., alanine)						
		C—C = 2				
$NH_2CH(CH_3)COOH$	89	C—H = 4	220×6			
		Σ = 6				
		+	+			
		N—C = 1	$105 \times 3 = 1635$		$1{,}635 \div 89 = 18.4$	
		N—H = 2		1,623		18.2
		Σ = 3				

trients. All that should be required is the empirical formula (for estimating the oxygen uptake) or the structure (for counting the number of reduced bonds), and the molecular weight according to equations (4) and (7).

The heats of combustion for the building blocks of fat, carbohydrate, and protein have been calculated using this approach. Values determined experimentally in a bomb calorimeter (and which appeared in the popular press in units of kcal as Calories) and those predicted by the bond-count approach (using equation (4)) are listed in Table II. The comparisons are remarkably close and justify confidence in the application of this method.

In studies of nutrition, the total energy of the diet is measured, and the digestibility of that energy is calculated after subtracting the energy excreted in the feces. When protein is fed, it is found that approximately 20% of the digestible energy is subsequently lost to the body through urinary excretion. Thus, the metabolizable energy of protein is about 80% of digestible energy. This difference between digestible energy and metabolizable energy is not observed for dietary carbohydrate or fat.

This effect presents another important concept that can be explained using the approach of counting reduced bonds. Carbon and hydrogen in digested nutrients are excreted from the body in a fully oxidized state (CO_2 and H_2O), whereas nitrogen is excreted from the body in a reduced state (as ammonia, urea, or uric acid). Hence, the 105 kcal/mol contributed to digestible energy by each reduced nitrogen bond in the food is not available to support metabolism.

In the example for alanine calculated in Table II, the metabolizable energy should be approximately 81% of digestible energy [$(220 \times 6) \div (220 \times 6 + 105 \times 3)$]. Thus, for a completely digested protein, we would predict a metabolizable energy value of 14.9 kJ/g (18.4×0.81).

VI. Conclusion

The examination of the chemical basis for the energy content of macronutrients leads to the conclusion that the source of energy in the foods that nourish mankind is the reducing equivalents or valence electrons that bond their carbon and hydrogen atoms together. These electrons are initially derived from water and are sequestered during the reactions of photosynthesis. They move through the food chain until their release to sustain metabolism. Their ultimate fate is to reduce molecular oxygen to water, thereby completing the cycle. Energy release is directly proportional to oxygen consumption during combustion, whether the combustion occurs within a cell or an automobile engine.

With this understanding, many of the difficult concepts of bioenergetics can be unravelled. Moreover, two approaches become available for calculating the energy content of macronutrients, based only on a knowledge of the structure or empirical formula and the molecular weight of the material of interest.

In answering the question "Why are fats so much higher in calories than carbohydrate?," one can now reply with confidence: "because fat contains three times as many reduced bonds per unit weight!"

Bibliography

Kharasch, M. (1929). Heats of combustion of organic compounds. *Bur. Standards J. Res.* **2,** 359–430.

Saz, H. J. (1981). Energy metabolism of parasitic helminths: Adaptions to parasitism. *Annu. Rev. Physiol.* **43,** 323–341.

Stokes, G. B. (1988). Estimating the energy content of nutrients. *Trends Biochem. Sci.* **13,** 422–424.

Nutritional Modulation of Aging Processes

EDWARD J. MASORO, *University of Texas Health Science Center at San Antonio*

Glossary

Age-associated diseases Diseases that become more prevalent with advancing age or occur primarily during a specific age range

Gerontology Science of the study of aging and of the special problems of the aged

Life expectancy Mean length of life remaining for a population of a given age

Life span of species Length of life of the longest-lived members of a species

Primary aging processes Intrinsic basic biological processes underlying aging

IT IS GENERALLY HELD that aging is an intrinsic property of most, if not all, living organisms. Although the primary processes underlying aging have yet to be identified, an issue often debated is whether and to what extent aging can be influenced by extrinsic or environmental factors. Of these, nutrition is viewed by many as being a particularly important factor influencing the primary aging processes. The currently available information on the interaction of nutrition with the aging processes is the subject of this article.

I. Criteria Indicating Modulation of the Aging Processes

A. Current State of Knowledge of Aging

What is aging? In basic biological terms we do not have an answer. This statement is hard for many to believe because most of us have an intuitive under-

standing of aging. For example, most observers have no problem in distinguishing between young animals and people and the old members of the species. However, this distinction is based on observable characteristics, of which some do indeed relate to the primary aging processes; others may be due to prolonged exposure to environmental and/or lifestyle factors.

In the absence of knowledge of the primary aging processes, gerontologists have resorted to defining aging in terms of the decreasing ability of an organism to survive with advancing age. Although the probability of dying does increase with increasing adult age, factors other than age can also be, and usually are, involved in the death of an organism. Thus, a definition of aging based on mortality suffers from a lack of specificity. Indeed, the almost universal use of mortality by gerontologists when defining aging underscores our lack of knowledge of the nature of the basic aging processes.

Nevertheless, mortality data, in spite of these conceptual problems, are of great use to gerontology in general and to the area of focus of this article in particular. *Life expectancy* at birth (i.e., the mean length of life of a population of a given species born at a particular moment in time) is the most used in this regard. Unfortunately, life expectancy is of limited value for gerontology because it is markedly influenced by many factors in addition to the aging processes. For example, the increase in life expectancy of the U.S. population during the first half of the twentieth century appears to have resulted primarily from protecting the young from infectious disease and therefore bears little relationship to the aging processes. In contrast, *the maximum life span of a species* (the age at death of the longest-lived members) is felt to strongly relate to the aging processes. For example, the maximum life span of humans is about 110 years, of elephants about 70 years, of dogs about 20 years, of the house mouse about 3 years. These data are interpreted by gerontologists as indicating that humans age the most slowly and mice the most rapidly; i.e., it is felt that the rate of aging of a species inversely relates to the maximum life span of the species. Although the life expectancy of Americans increased markedly during the first half of the twentieth century, the life span of Americans did not. This fact is interpreted as indicating that the medical and technological advances during this time did not influence the aging processes of humans.

Physiological changes with age (most of which are of a deteriorative nature) have long been believed to be an inevitable consequence of the aging processes. However, recent studies suggest that the contribution of intrinsic aging processes to age-associated physiological deterioration may have been exaggerated. Factors such as diseases, lifestyle, and a variety of environmental influences may play major roles in the frequently observed age changes in the physiological systems. Indeed, the current challenge is to design studies that can sort out the contribution of the primary aging processes to the physiological and psychosocial changes observed with advancing age. Until this is accomplished, physiological activities cannot be viewed as valid markers of biological age. The current lack of valid nonchronologic markers of age is unfortunate since their availability would greatly assist in the evaluation of the effects of nutrition on human aging.

Many diseases are age-associated in that they are more prevalent at advanced ages or in a specific age range. Examples are coronary heart disease, cerebrovascular disease, many kinds of tumors, type II diabetes, Alzheimer's disease, and Parkinson's disease. However, the relationship of these diseases to the primary aging processes remains to be defined. Do aging and a particular disease process share the same time frame but relate in no other way? Is the disease process promoted by the aging processes? Is the disease process a part of the aging processes? Currently these questions cannot be answered for any age-associated disease. Research aimed at yielding answers should be of high priority.

B. Longevity as a Criterion

From the above discussion, it is evident that an increase in life expectancy from birth (or in the median length of life) in response to a manipulation does not provide strong evidence that the aging processes have been retarded. Yet the claim is often made that aging has been retarded on the basis of an increased life expectancy. For example, it is claimed that exercise retards the aging processes in humans, apparently because of the emerging evidence that exercise may result in an increased life expectancy. The basis of this increase in life expectancy (if indeed it does occur) appears to be the retarding influence that long-term exercise has on the occurrence of coronary heart disease and other chronic cardiovascular disease processes. Since the relationship of these disease processes to the aging processes has yet to be defined, the claim that exercise or any other manipulation retards the aging processes on the basis of an increase in life expectancy due to retardation of cardiovascular disease seems inappropriate. [See LONGEVITY.]

If exercise were also to be shown to increase the maximum life span of humans, the case for its retarding action on aging processes would be much stronger. However, since the information on life expectancy and exercise has only recently emerged, it will be many years before the effect of exercise on maximum life span will be known. Moreover, even in the case of maximum life span of a species, it is possible that there may be factors other than aging that underlie an increase. Thus, although an increase in the maximum life span provides strong evidence that a manipulation has decelerated the aging processes, it alone cannot be considered to provide unequivocal proof.

C. Age-Associated Physiological Changes as a Criterion

Although many age-associated physiological changes occur, the extent to which a particular change is secondary to primary aging processes is not known. Thus, the ability of a manipulation to retard a particular age-associated physiological change is not strong evidence that the manipulation has influenced the primary aging processes. However, retarding a broad spectrum of age-associated physiological changes by a manipulation does indicate that the primary aging processes have been slowed.

D. Age-Associated Disease Processes as a Criterion

The relationship of a particular age-associated disease process to the aging processes has not been determined for any of the age-associated diseases.

For this reason, retardation by a manipulation of a particular disease process (e.g., atherogenesis) cannot be used as a criterion that the aging processes have been decelerated. However, the retardation of a broad spectrum of age-associated diseases by a manipulation is indicative of the aging processes in general having been affected rather than the pathogenesis of only a particular disease.

E. Difficulties in Determining if Human Aging Has Been Slowed

To determine if a manipulation has slowed the aging processes is particularly difficult in the case of humans. For example, the effect on the maximum life span of humans is almost impossible to learn because of its length, the longest life span of any mammalian species. Unfortunately, an increase in maximum life span is probably the strongest single piece of evidence indicating that a manipulation has decelerated aging. Although providing less compelling evidence, it is necessary with humans to rely on the effects on the physiological and disease processes as the major vehicles for evaluating the effects of an intervention on aging. Moreover, even this information is often difficult to obtain reliably because of the lack of control of, indeed lack of an accurate knowledge of, the lifelong environment and lifestyle of human subjects.

II. Nutrition and the Aging Processes

A. Nutrition and Human Aging

The statement is often made that nutrition influences the aging processes of humans. Unfortunately, there is little hard evidence in support of or against this claim.

There is evidence that several age-associated diseases can be retarded by nutritional means. The progression of atherosclerosis appears to be influenced by dietary lipids and by caloric intake relative to energy expenditure. Inadequate dietary calcium and a lack of Vitamin D have been proposed as contributors to osteoporosis; however, it has been questioned if nutritional factors are major players in this disease process. The incidences of many types of cancer are age-associated and some (not totally convincing) evidence indicates nutrition may play some role in the occurrence. Nutrition plays a role in several other age-associated pathologies, e.g., di-

etary protein in renal failure, dietary sodium and calcium in hypertension, caloric intake in adult-onset diabetes. [See ATHEROSCLEROSIS; LIPIDS.]

If these disease processes are part of or influenced by the primary aging processes, it can be concluded that nutrition influences aging in humans. However, not enough is known at this time about the relationship of the aging processes to these diseases to permit such a conclusion to be drawn. Indeed, the specificity of the nutritional manipulation required to influence these diseases indicates that it is the pathogenesis of the specific disease process rather than aging in general that is being affected.

B. Food Restriction and the Aging of Laboratory Rodents

That nutrition can influence the primary aging processes of mammals is clearly shown by the action of long-term food restriction on laboratory rodents. Restricting the food intake of rats, mice, and hamsters to levels significantly below (30–60% below) that of those fed ad libitum (i.e., animals with continuous access to food) markedly decelerates the aging processes. The evidence supporting this claim includes the effects on longevity, on a wide spectrum of age-changes in the physiological systems, and on a broad range of age-associated disease processes.

Probably the strongest of this evidence is the effect of food restriction on the maximal life span. This is illustrated by the survival curves in Fig. 1 for

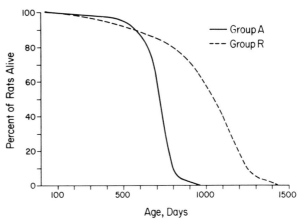

FIGURE 1 Survival curves for a population of 115 ad libitum fed male Fischer 344 rats (group A) and 115 food-restricted (60% of the ad libitum intake) male Fischer 344 rats (group R). [Reproduced with permission from Yu *et al.* (1982), *J. Gerontol.* **37**, 130.]

a group of 115 ad libitum–fed male Fischer 344 rats (Group A) and a group (Group R) of 115 food-restricted male Fischer 344 rats (fed 60% of the mean intake of the Group A rats from 6 weeks of age on). Although the median length of life and the life expectancy of the Group R rats were much greater than for the Group A rats, the finding of most importance in regard to the aging processes is the marked increase in the maximum life span. Indeed, when the last of the 115 Group A rats died, approximately 70% of the Group R rats were still living.

Food restriction also retards most age-associated changes in the physiological systems. Typical of this action are the data in Fig. 2 on the serum cholesterol concentrations of male Fischer 344 rats ad libitum–fed or food-restricted from 6 weeks of age on. Although food restriction did not influence the serum cholesterol levels of young rats (6 months of age) it markedly decreased the increase in serum cholesterol that occurs with advancing age. A broad range of physiological processes are similarly protected from age change. Examples are the loss of gamma crystallins from the lens of the eye, the loss of neurotransmitter receptors from the brain, the loss in immune function, the loss in the ability to learn a complex maze, the loss in the ability of fat

cells to respond to hormones, the loss of reproductive function, and many others. The breadth of these effects provides strong evidence for the view that food restriction is slowing the primary aging processes. [*See* CHOLESTEROL.]

Food restriction has also been found to delay the occurrence and in some cases prevent the clinical expression of many of the age-associated diseases of rodents. Examples are chronic kidney disease, heart disease, cancer, benign tumors, autoimmune diseases, and others. The breadth of these effects points to an action on the primary aging processes rather than on the specific pathogenesis of a particular disease process.

Recent work indicates that these many actions of food restriction are due to the reduction in caloric intake and not to the reduced intake of specific dietary components. It has further been shown that the actions of food restriction are not due to a retardation of growth and development or to a reduction in body fat content or to a reduction in metabolic rate. Current research is being focused on the biochemical and neuroendocrinologic mechanisms by which food restriction retards aging processes in rodents. It is felt that knowledge of these mechanisms will provide insights in regard to the nature of the primary aging processes.

Although other nutritional manipulations in rodents have been found to influence specific age-associated problems (e.g., reducing dietary protein retards the development of kidney failure in rats), none of these manipulations of specific nutrients has the broad influence on longevity and the many other aspects of age changes as food restriction. Thus, of these dietary manipulations, food restriction appears to be unique in its ability to decelerate the primary aging processes.

C. Relevance of Rodent Findings to the Human Situation

If the premise is correct that the primary aging processes are similar in all mammalian species, it is likely that food restriction will decelerate the aging processes in all mammals, including humans. However, food restriction has only been studied adequately in rodent species because of the long life span of most other mammalian species. This long life span makes it difficult to execute the relevant research because of large financial resources that would be required and because of the unwillingness of investigators to invest the large amount of time

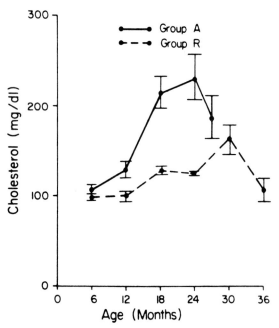

FIGURE 2 Changes in serum cholesterol concentrations with age in ad libitum fed male Fischer 344 rats (group A) and in food-restricted (60% of ad libitum intake) male Fischer 344 rats (group R). [Reproduced with permission from Liepa *et al.* (1980), *Am. J. Physiol.* **238**, E253.]

(much of their scientific life span) to complete such a study.

Since on theoretical grounds it seems likely that food restriction would be effective in retarding aging in humans, should this regimen be employed by people? The answer is no because of the risks of malnutrition. For food restriction regimens in rodents to be successful in retarding the aging processes without any adverse consequences, it is necessary to produce undernutrition without causing malnutrition. Such can be readily accomplished with mice and rats because of our knowledge of their lifelong nutritional needs and because of the extensive experience that has been gained in the experimental use of food restriction in these species. Neither the knowledge nor the experience is available in regard to human use and thus the risk inherent in using any uncharted procedure. [*See* MALNUTRITION.]

Does this mean that the extensive research that has been done on the effects of food restriction on rodent aging is of no value for humans? No, but its value resides in learning the mechanisms by which food restriction slows the aging processes of the rodent models. An understanding of these mechanisms is likely to provide the knowledge base needed for the development of safe and effective interventions for human aging. Moreover, such interventions are likely to be more palatable for humans than the restricting of food intake for most of the life span would be.

III. Conclusions

Whether proper nutrition can slow the aging processes of humans is difficult to establish because of our lack of knowledge of the nature of the primary aging processes. On the basis of current knowledge, the following criteria should be met for it to be concluded that a manipulation has decelerated the aging processes: an increase in life expectancy; an increase in life span; retardation of most age-associated changes in the physiological systems; retardation of most age-associated diseases. Based on these criteria, there is little hard evidence to support the often made claim that nutrition can decelerate the aging processes of humans. The only manipulation that has been shown to slow the aging processes in a mammalian species is that of food restriction (apparently because of caloric restriction) in rodent species. Indeed, the findings on food restriction clearly show that nutrition can markedly influence the aging processes of mammals. Although food restriction is likely to have similar actions in all mammalian species including humans, it has not been directly tested in long-lived species because of the high cost in resources and time of doing so. At this time the use of food restriction by humans should be discouraged because of the possible adverse effects of malnutrition that might be encountered during the lifelong use of an uncharted procedure. Exploration of the mechanisms by which food restriction retards the aging processes of rodents is encouraged because the knowledge from such studies may well lead to the development of safe and palatable interventions of human aging.

Bibliography

Committee on Chemical Toxicity and Aging (1987). "Aging in Today's Environment." National Academy Press, Washington, D.C.

Holehan, A. M., and Merry, B. J. (1986). The experimental manipulation of aging by diet, *Biological Reviews* **61,** 329.

Holliday, R. (1988). Toward a biological understanding of the aging process, *Perspectives in Biology and Medicine* **32,** 109.

Hutchinson, M. L., and Munro, H. N., eds. (1986). "Nutrition and Aging." Academic Press Inc., Orlando, Florida.

Masoro, E. J. (1988). Food restriction in rodents: an evaluation of its role in the study of aging, *Journal of Gerontology: Biological Sciences* **43,** B59.

Rowe, J. W., and Kahn, R. L. (1987). Human aging: usual and successful, *Science* **237,** 143.

Nutritional Quality of Foods and Diets

DENNIS D. MILLER, *Cornell University*

Glossary

Essential mineral elements Elements other than carbon, hydrogen, oxygen, or nitrogen that are constituents of plant or animal tissues and are found in the ash remaining after incineration of the tissue; also necessary for the normal metabolic functioning of the body

Essential nutrients Chemical substances required for metabolism that cannot be synthesized within the body and must be present in the diet; they include essential amino acids and fatty acids, vitamins, and minerals

Kilocalorie Unit used to quantify the energy value of food (1 kilocalorie = 1,000 calories); in common usage, calorie refers to the kilocalorie

Lipids Substances in foods that are insoluble in water and soluble in organic solvents such as chloroform and diethyl ether; include triacylglycerols (esters composed of three fatty acids and one glycerol residue) and cholesterol

Monounsaturated fatty acids Fatty acids with one double bond per molecule

Polyunsaturated fatty acids Fatty acids with two or more double bonds per molecule

Proteins Large molecules composed of amino acids linked together by peptide bonds; contain the elements carbon, hydrogen, oxygen, nitrogen, and sulfur

Saturated fatty acids Fatty acids with no double bonds between carbons

Vitamins Organic substances necessary for the normal metabolic functioning of the body; dietary requirements for vitamins are small, falling in the milligram range or less

THE IMPORTANCE OF DIET in health promotion and disease prevention is widely recognized by consumers, physicians, nutritionists, and policy makers. In the United States and other highly developed, industrialized countries with abundant food supplies, nutritional quality of diets and national food supplies has a larger impact on health than quantity of available food. The relationship between diet and health is complex, and this makes it difficult to define nutritional quality. Clearly, however, nutritional quality is determined by the nutrient composition of the diet. In the first half of this century, nutrition research identified the nutrients essential for the prevention of nutrient deficiency diseases and established recommended daily intakes for many of them. Government policy makers, the food industry, and nutrition educators developed policies and programs based on this research that resulted in the virtual elimination of the deficiency diseases that were so prevalent in the early part of the century. For example, fortification of salt with iodine and milk with vitamin D is credited with preventing goiter and rickets, respectively. Nutrition educators used the concept of the basic four food groups to effectively teach people how to make appropriate food choices. Sadly, in recent years, the alarming increases in poverty, hunger, substance abuse, school drop-out rates, and teen-age pregnancy may cause a return of nutrient deficiency diseases in the United States. Nevertheless, with the conquest of nutrient deficiency diseases and infectious diseases, chronic diseases such as heart disease, cancer, and osteoporosis have become the principle causes of morbidity and mortality in the developed countries. Appropriately, the focus of nutrition research has shifted to questions related to the relationships between diet and chronic disease. As these relationships are being clarified, policies and programs are being developed to guide food industry practices and consumer food choices in the direction of improved nutritional quality of foods and diets. Current dietary guidelines issued by the government and other health agencies stress the need for reduced intakes of total fat, saturated fat, cholesterol, and salt and increased intakes of calcium, iron, and complex carbohydrates and fiber. Nutritional quality of a food or diet, then, is determined by the concentrations of essential nutrients and the relative amounts of carbohydrate, fat, cholesterol, and sodium. Chemical form as well as concentration may have significant effects on nutritional quality.

I. Criteria for Assessing Nutritional Quality

Criteria for assessing nutritional quality may be established for the evaluation of individual foods, diets, or national food supplies. Evaluation of diets is most appropriate, because nutritional status (nutritional well-being) is determined by total nutrient intake. Individual foods contain varying amounts and proportions of nutrients. No single food, with the exception of breast milk for infants <6 mo of age and a few specially formulated liquid formulas, contains all of the essential nutrients in the proportions and amounts necessary to meet nutrient needs. Nevertheless, nutritional comparisons among individual foods are useful and scientifically valid, particularly when evaluating the effects of food processing or when comparing foods within a food group. Monitoring of the nutritional quality of the national food supply is useful because it provides information about the contributions of various food groups to nutrient supplies. Moreover, data describing the nutritional composition of national food supplies have been extremely valuable to epidemiologists who study relationships between diet and health. For example, the positive correlation between the concentration of saturated fat in national food supplies and the prevalence of coronary heart

disease in different countries is strong evidence in support of the hypothesis that diets high in saturated fat increase risk of heart disease.

A. Recommended Dietary Allowances

An adequate diet must contain all of the essential nutrients in the proper amounts and proportions. When intakes are below requirements, signs and symptoms of nutrient deficiency disease develop. On the other hand, excessive intakes of nutrients may also result in adverse health consequences. The relationship between the intake of one class of nutrients (trace minerals) and the concentration of these nutrients in the body has been described (Fig. 1). When intakes are deficient, the concentrations in the body are below levels necessary for optimal physiological function. For example, when intakes or dietary iron are below the requirement for an extended period of time, the supply of iron to the bone marrow becomes insufficient to support hemoglobin synthesis at an adequate rate and iron deficiency anemia develops. On the other hand, when iron intakes are excessively high for an extended period of time, a condition of iron overload may develop. Iron overload is the term used to describe the toxic accumulation of iron in tissues. Iron overload can damage the liver, heart, and other vital organs. Fortunately, the range of safe and adequate intakes for most nutrients is relatively wide. This is because physiological mechanisms operate to allow the body to adapt to low or high intakes of nutrients. To use the iron example again, when iron intakes are low, the body adapts by absorbing a higher percentage of dietary iron from the gastrointestinal tract. Conversely, individuals with adequate or high levels of body iron absorb iron less efficiently. The net result is that concentrations of nutrients in body tissues are maintained within a narrow range even when intakes may vary quite widely. The adaptive processes responsible for maintaining steady-state levels of nutrients are collectively referred to as homeostatic mechanisms. Without them, intakes would have to be held to very narrow ranges. It is important to understand, however, that homeostatic regulation has its limits, and it is possible to overwhelm the system with intakes that are either too high or too low (see Fig. 1).

Nutrient requirements are determined by measuring the amount of a nutrient that must be ingested to meet the nutrient needs of an *individual*. Nutrient

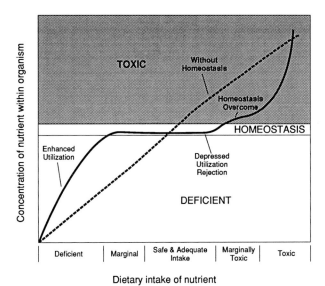

FIGURE 1 Body processes collectively referred to as homeostasis operate to maintain tissue levels of nutrients within an acceptable, often narrow, range, even when dietary intakes vary. High levels of nutrients may be toxic, and low levels result in impaired function. From W. Mertz, Our Most Unique Nutrients. *Nutrition Today*, **18**(2), 6–10, 27–29, and 32–33. © by Williams & Wilkins, (1983).

requirements vary among individuals. Therefore, recommendations for nutrient intakes, which are based on research on individual subjects, are set higher than the average requirement to ensure that the needs of most individuals will be met. The Food and Nutrition Board of the National Research Council, National Academy of Sciences, has established Recommended Dietary Allowances (RDAs) for the United States population (Table I). The Food and Nutrition Board defines RDAs as ". . . the levels of intake of essential nutrients that, on the basis of scientific knowledge, are judged by the Food and Nutrition Board to be adequate to meet the known nutrient needs of practically all healthy persons." This means that individuals with nutrient intakes below the RDA do not necessarily have an inadequate intake, because individual requirements are usually lower than the RDA. However, when intakes are significantly below the RDA, the probability that intakes by a given individual are inadequate is increased. Because of an inadequate research base, RDAs have been established for only about half of the known essential nutrients. In its two latest revisions, the Food and Nutrition Board published "estimated safe and adequate intake" levels for several additional essential nutrients (Ta-

ble II). The Food and Nutrition Board revises the RDAs periodically to reflect new research findings. The most recent edition of the RDAs at the time of this writing was published in 1989.

B. Dietary Guidelines

Criteria based on the RDAs are useful for evaluating the contribution a food or diet will make toward meeting requirements for protein, vitamins, and minerals. Expanding knowledge in the second half of this century has made it clear that criteria based solely on RDAs is insufficient for evaluating the potential impact diet may have on health. As a result, emphasis in dietary advice has shifted from protein, vitamins, and minerals to food components that are associated with chronic diseases such as heart disease, cancer, osteoporosis, and hypertension. These include fat, saturated fat, cholesterol, sodium, calcium, complex carbohydrates, and dietary fiber. Since the 1960s, numerous publications have recommended adjustments in the typical American diet to improve public health. Generally, these recommendations from a variety of government and private health organizations have been to reduce fat, saturated fat, cholesterol, sodium, alcohol, and sugar and to increase dietary fiber and complex carbohydrates. Tables III, IV, and V summarize recommendations published by the United States Department of Agriculture (USDA)–United States Department of Health and Human Services; The Surgeon General of the U.S. Public Health Service; and the Committee on Diet and Health of the Food and Nutrition Board, Commission on Life Sciences, National Academy of Sciences. The recommendations in the latter two cases are based on a comprehensive review of the literature by expert committees. Readers interested in more detail on relationships between diet and health should consult the "Surgeon General's Report on Nutrition and Health" and "Diet and Health, Implications for Reducing Chronic Disease Risk."

II. Problems with the American Diet

A. Prevalences of Diet-Related Diseases

Cause-of-death statistics provide clues about imbalances in national diets. Numbers and causes of deaths in the United States in 1987 are listed in Table VI. While factors associated with deaths in

TABLE I Food and Nutrition Board, National Academy of Sciences—National Research Council Recommended Dietary Allowances,[a] revised 1989; Designed for the Maintenance of Good Nutrition of Practically All Healthy People in the United States

Category	Age (yr) or condition	Weight[b] kg	Weight[b] lb	Height[b] cm	Height[b] in	Protein (g)	Fat-soluble vitamins Vita- min A (μg re[c])	Vita- min D (μg)[d]	Vita- min E (mg α-te[e])	Vita- min K (μg)
Infants	0.0–0.5	6	13	60	24	13	375	7.5	3	5
	0.5–1.0	9	20	71	28	14	375	10	4	10
Children	1–3	13	29	90	35	16	400	10	6	15
	4–6	20	44	112	44	24	500	10	7	20
	7–10	28	62	132	52	28	700	10	7	30
Males	11–14	45	99	157	62	45	1,000	10	10	45
	15–18	66	145	176	69	59	1,000	10	10	65
	19–24	72	160	177	70	58	1,000	10	10	70
	25–50	79	174	176	70	63	1,000	5	10	80
	51+	77	170	173	68	63	1,000	5	10	80
Females	11–14	46	101	157	62	46	800	10	8	45
	15–18	55	120	163	64	44	800	10	8	55
	19–24	58	128	164	65	46	800	10	8	60
	25–50	63	138	163	64	50	800	5	8	65
	51+	65	143	160	63	50	800	5	8	65
Pregnant						60	800	10	10	65
Lactating	1st 6 mo					65	1,300	10	12	65
	2nd 6 mo					62	1,200	10	11	65

Source: Subcommittee on the Tenth Edition of the RDAs, 1989, by the National Academy of Sciences, National Academy Press, Washington, D.C.
[a] The allowances, expressed as average daily intakes over time, are intended to provide for individual variations among most normal persons as they live in the United States under usual environmental stresses. Diets should be based on a variety of common foods to provide other nutrients for which human requirements have been less well defined. See text for detailed discussion of allowances and of nutrients not tabulated.
[b] Weights and heights of reference adults are actual medians for the U.S. population of the designated age, as reported by NHANES II. The median weights and heights of those under 19 yr of age were taken from Hamill et al. (1979, pp. 16–17). The use of these figures does not imply that the height–weight ratios are ideal. NHANES II, National Health And Nutrition Examination Survey II.

TABLE II Estimated Safe and Adequate Daily Dietary Intakes of Selected Vitamins and Minerals[a]

Category	Age (yr)	Vitamins Biotin (μg)	Pantothenic acid (mg)	Trace elements[b] Copper (mg)	Manganese (mg)	Fluoride (mg)	Chromium (μg)	Molybdenum (μg)
Infants	0–0.5	10	2	0.4–0.6	0.3–0.6	0.1–0.5	10–40	15–30
	0.5–1	15	3	0.6–0.7	0.6–1.0	0.2–1.0	20–60	20–40
Children and adolescents	1–3	20	3	0.7–1.0	1.0–1.5	0.5–1.5	20–80	25–50
	4–6	25	3–4	1.0–1.5	1.5–2.0	1.0–2.5	30–120	30–75
	7–10	30	4–5	1.0–2.0	2.0–3.0	1.5–2.5	50–200	50–150
	11+	30–100	4–7	1.5–2.5	2.0–5.0	1.5–2.5	50–200	75–250
Adults		30–100	4–7	1.5–3.0	2.0–5.0	1.5–4.0	50–200	75–250

Source: Subcommittee on the Tenth Edition of the RDAs, 1989, by the National Academy of Sciences, National Academy Press, Washington, D.C.
[a] Because there is less information on which to base allowances, these figures are not given in the main table of RDA and are provided here in the form of ranges of recommended intakes.
[b] Because the toxic levels for many trace elements may be only several times the usual intakes, the upper levels for the trace elements given in this table should not be habitually exceeded.

Water-soluble vitamins							Minerals						
Vita-min C (mg)	Thia-min (mg)	Ribo-flavin (mg)	Niacin (mg ne^f)	Vita-min B$_6$ (mg)	Fo-late (μg)	Vita-min B$_{12}$ (μg)	Cal-cium (mg)	Phos-phorus (mg)	Mag-nesium (mg)	Iron (mg)	Zinc (mg)	Iodine (μg)	Sele-nium (μg)
30	0.3	0.4	5	0.3	25	0.3	400	300	40	6	5	40	10
35	0.4	0.5	6	0.6	35	0.5	600	500	60	10	5	50	15
40	0.7	0.8	9	1.0	50	0.7	800	800	80	10	10	70	20
45	0.9	1.1	12	1.1	75	1.0	800	800	120	10	10	90	20
45	1.0	1.2	13	1.4	100	1.4	800	800	170	10	10	120	30
50	1.3	1.5	17	1.7	150	2.0	1,200	1,200	270	12	15	150	40
60	1.5	1.8	20	2.0	200	2.0	1,200	1,200	400	12	15	150	50
60	1.5	1.7	19	2.0	200	2.0	1,200	1,200	350	10	15	150	70
60	1.5	1.7	19	2.0	200	2.0	800	800	350	10	15	150	70
60	1.2	1.4	15	2.0	200	2.0	800	800	350	10	15	150	70
50	1.1	1.3	15	1.4	150	2.0	1,200	1,200	280	15	12	150	45
60	1.1	1.3	15	1.5	180	2.0	1,200	1,200	300	15	12	150	50
60	1.1	1.3	15	1.6	180	2.0	1,200	1,200	280	15	12	150	55
60	1.1	1.3	15	1.6	180	2.0	800	800	280	15	12	150	55
60	1.0	1.2	13	1.6	180	2.0	800	800	280	10	12	150	55
70	1.5	1.6	17	2.2	400	2.2	1,200	1,200	320	30	15	175	65
95	1.6	1.8	20	2.1	280	2.6	1,200	1,200	355	15	19	200	75
90	1.6	1.7	20	2.1	260	2.6	1,200	1,200	340	15	16	200	75

^{c} Retinol equivalent: 1 re = 1 μg retinol or 6 μg β-carotene.
^{d} As cholecalciferol: 10 μg cholecalciferol = 400 IU of vitamin D.
^{e} α-tocopherol equivalent: 1 mg d-α tocopherol = 1 α-te.
^{f} Niacin equivalent: 1 ne = 1 mg niacin or 60 mg of dietary tryptophan.

most of the categories listed are multiple, strong evidence indicates that diet is a factor in heart disease, cancer, stroke, diabetes mellitus, and atherosclerosis. Furthermore, appropriate dietary changes will likely reduce death rates for these diseases by either helping to prevent them in the first place or by slowing their progression.

Poor diets may also contribute to conditions that, while not directly life threatening, result in a lower quality of life. In the United States, iron deficiency, osteoporosis, and dental disease are other significant public health problems that are diet related.

1. Iron Deficiency

Iron deficiency is a term used to describe a state of impaired iron status. The severity of iron deficiency can range from mild (depletion of tissue iron stores without impairment of metabolic functions) to severe (iron deficiency anemia with resulting deficits in oxygen delivery to tissues and impairment of metabolic reactions dependent on iron-containing enzymes). Iron deficiency may impair immune response, work performance, and neurological function. Although the prevalence of iron deficiency in the United States has been declining in recent years, it remains a problem particularly among children aged 1–2 yr and women aged 15–44 yr. Diets low in iron and iron absorption-enhancing factors (ascorbic acid and meat) and/or high in factors that inhibit iron absorption (phytic acid and dietary fiber) have been associated with iron deficiency.

TABLE III USDA–DHHS Dietary Guidelines for Americans

- Eat a variety of foods
- Maintain desirable weight
- Avoid too much fat, saturated fat, and cholesterol
- Eat foods with adequate starch and fiber
- Avoid too much sugar
- Avoid too much sodium
- If you drink alcoholic beverages, do so in moderation

Source: USDA–DHHS, 1985.

TABLE IV Dietary Recommendations from "The Surgeon General's Report on Nutrition and Health"

Issues for Most People

- *Fats and cholesterol:* Reduce consumption of fat (especially saturated fat) and cholesterol. Choose foods relatively low in these substances, such as vegetables, fruits, whole grain foods, fish, poultry, lean meats, and low-fat dairy products. Use food preparation methods that add little or no fat.
- *Energy and weight control:* Achieve and maintain a desirable body weight. To do so, choose a dietary pattern in which energy (caloric) intake is consistent with energy expenditure. To reduce energy intake, limit consumption of foods relatively high in calories, fats, and sugars, and minimize alcohol consumption. Increase energy expenditure through regular and sustained physical activity.
- *Complex carbohydrates and fiber:* Increase consumption of whole grain foods and cereal products, vegetables (including dried beans and peas), and fruits.
- *Sodium:* Reduce intake of sodium by choosing foods relatively low in sodium and limiting the amount of salt added in food preparation and at the table.
- *Alcohol:* To reduce the risk for chronic disease, take alcohol only in moderation (no more than two drinks a day), if at all. Avoid drinking any alcohol before or while driving, operating machinery, taking medications, or engaging in any other activity requiring judgment. Avoid drinking alcohol while pregnant.

Other Issues for Some People

- *Fluoride:* Community water systems should contain fluoride at optimal levels for prevention of tooth decay. If such water is not available, use other appropriate sources of fluoride.
- *Sugars:* Those who are particularly vulnerable to dental caries (cavities), especially children, should limit their consumption and frequency of use of foods high in sugars.
- *Calcium:* Adolescent girls and adult women should increase consumption of foods high in calcium, including low-fat dairy products.
- *Iron:* Children, adolescents, and women of childbearing age should be sure to consume foods that are good sources of iron, such as lean meats, fish, certain beans, and iron-enriched cereals and whole grain products. This issue is of special concern for low-income families.

Source: DHHS Public Health Service (PHS) Publication No. 88-50210, 1988.

2. Osteoporosis

Osteoporosis is an enormous public health problem affecting over 15 million Americans. Although the relationship between diet and osteoporosis risk is only poorly understood, diets low in calcium and high in protein have been associated with osteoporosis. Other factors associated with the disease include age, female gender, premenopausal oophorectomy, oral corticosteroids, and extreme immobility.

TABLE V Dietary Recommendations of The Committee on Diet and Health, Food and Nutrition Board, Commission on Life Sciences, National Research Council, 1989

- Reduce total fat intake to ≤30% of calories. Reduce saturated fatty acid intake to <10% of calories and the intake of cholesterol to <300 mg daily. The intake of fat and cholesterol can be reduced by substituting fish, poultry without skin, lean meats, and low- or nonfat dairy products for fatty meats and whole-milk dairy products; by choosing more vegetables, fruits, cereals, and legumes; and by limiting oils, fats, egg yolks, and fried and other fatty foods.
- Every day eat five or more servings of a combination of vegetables and fruits, especially green and yellow vegetables and citrus fruits. Also, increase intake of starches and other complex carbohydrates by eating six or more daily servings of a combination of breads, cereals, and legumes.
- Maintain protein intake at moderate levels.
- Balance food intake and physical activity to maintain appropriate body weight.
- The committee does not recommend alcohol consumption. For those who drink alcoholic beverages, the committee recommends limiting consumption to <1 oz. of pure alcohol in a single day. This is the equivalent of two cans of beer, two small glasses of wine, or two average cocktails. Pregnant women should avoid alcoholic beverages.
- Limit total daily intake of salt (sodium chloride) to ≤6 g. Limit the use of salt in cooking and avoid adding it to food at the table. Salty, highly processed salty, salt-preserved, and salt-pickled foods should be consumed sparingly.
- Maintain adequate calcium intake.
- Avoid taking dietary supplements in excess of the RDA in any one day.
- Maintain an optimal intake of fluoride, particularly during the years of primary and secondary tooth formation and growth.

3. Dental Diseases

Costs for dental care in the United States in 1985 have been estimated at $23.3 billion. Diet plays a crucial role in the growth and development of healthy teeth as well as in the cause and prevention of dental caries. It is well established that sugars, especially sucrose, can cause tooth decay. It is also clear the fluoride is effective in preventing dental caries. [*See* ORAL PATHOLOGY.]

B. Problem Nutrients

The nutritional quality of the U.S. diet is monitored periodically by various surveys. The USDA's Human Nutrition Information Service (HNIS) prepares estimates of nutrients available for consumption and contributions of food groups to nutrient supplies. The estimates are based on data compiled by the Economic Research Service and reflect food "consumed" in the economic sense. Food "avail-

TABLE VI Estimated Total Deaths and Percent of Total Deaths for the Ten Leading Causes of Death: United States, 1987

Rank	Cause of death	Number	Percent of total deaths
1[a]	Heart diseases	759,400	35.7
	Coronary heart disease	511,700	24.1
	Other heart disease	247,700	11.6
2[a]	Cancers	476,700	22.4
3[a]	Strokes	148,700	7.0
4[b]	Unintentional injuries	92,500	4.4
	Motor vehicle	46,800	2.2
	All others	45,700	2.2
5	Chronic obstructive lung diseases	78,000	3.7
6	Pneumonia and influenza	68,600	3.2
7[a]	Diabetes mellitus	37,800	1.8
8[b]	Suicide	29,600	1.4
9[b]	Chronic liver disease and cirrhosis	26,000	1.2
10[a]	Atherosclerosis	23,100	1.1
	All causes	2,125,100	100.0

Source: DHHS (PHS) Publication No. 88-50210, 1988.
[a] Cause of death in which diet plays a part.
[b] Cause of death in which excessive alcohol consumption plays a part.

able for consumption'' is calculated as (total production + imports + beginning inventories) − (exports + year-end inventories + nonfood use + military procurement). Per capita consumption is calculated by dividing total consumption by the size of the civilian population. Data on individual nutrients are calculated by applying appropriate food composition data to the food consumption figures. The data are not corrected for losses due to food waste or nutrient destruction during processing, storage, and home preparation. Table VII lists the contribution of major food groups to nutrient supplies for 1985, the most recent data available at the time of this writing.

The Nationwide Food Consumption Survey (NFCS) is conducted at about 10-yr intervals by USDA's HNIS. In this survey, food consumption by individuals is determined from 3-day dietary reports. In the 1977–1978 NFCS, food intakes of 37,785 persons from across the conterminous United States were measured. The data, when compared with RDAs, reveal nutrients that may be low in the diets of some people. While no firmly established guidelines exist for what constitutes a low or deficient intake, intakes below 70% of the RDA may lead to adverse health consequences. The data in Table VIII show that one-third to one-half of the

population consume less than 70% of the RDA for several nutrients.

Food consumption data from these and other surveys, coupled with causes of death statistics and data from surveys designed to assess nutritional status in the population, have been used by the Joint Nutrition Monitoring Evaluation Committee (JNMEC) to identify nutrients that are consumed in either excessive or inadequate amounts by significant numbers of the population. In a 1986 report titled "Nutrition Monitoring in the United States," the JNMEC singled out the following dietary components as being consumed in either excessive or inadequate amounts:

High dietary consumption	Low dietary consumption
Food energy	Vitamin C
Total fat	Calcium
Saturated fatty acids	Iron
Cholesterol	Fluoride
Sodium	
Alcohol	

III. Factors Affecting Nutritional Quality

A. Food Choices

The most important factor affecting the nutritional quality of individual diets is food choice by individuals, families, or institutional food service managers. Because no single food or food group contains all the essential nutrients necessary for an adequate diet, variety is essential. To assist consumers in making appropriate food choices, nutrition educators have grouped foods according to the nutrients they contain. Table IX is the USDA's revised listing of food groups. Food guides suggesting a certain number of servings from each group are designed to make appropriate food selections relatively simple. The USDA's HNIS recently published a revised food guide in *Home and Garden Bulletin,* Number 232 (see Table X). The guide is designed so that healthy persons consuming at least the minimum number of daily servings in each food group should get the required amounts of protein, vitamins, minerals, and fiber. Also, by choosing carefully within groups, too much fat, saturated fat, sodium, and sugar can be avoided.

TABLE VII Contributions of Food Groups to Nutrient Supplies in the United States in 1985[1]

Food group	Food energy	Protein	Fat	Choles- terol	Carbo- hydrate	Minerals Cal- cium	Phos- phorus	Magne- sium	Iron	Zinc
Meat, poultry, and fish	20	43	33	40	*	4	28	15	29	48
Eggs	2	4	2	40	*	2	4	1	4	4
Dairy products, excluding butter	10	21	11	14	6	76	36	20	2	20
Fats and oils, including butter	20	*	46	5	*	*	*	*	*	*
Citrus fruits	1	1	*	0	2	1	1	2	1	*
Noncitrus fruits	2	1	*	0	5	1	1	5	2	1
White potatoes	3	2	*	0	5	1	3	6	4	3
Dark green, deep yellow vegetables	*	*	*	0	1	1	1	2	2	1
Other vegetables	2	2	*	0	3	4	4	7	6	3
Legumes, nuts, and soy	3	5	3	0	2	3	6	12	6	5
Flour and cereal products	20	19	1	0	36	4	14	20	38	13
Sugar and other sweeteners	18	*	0	0	40	*	*	*	1	*
Miscellaneous[2]	1	1	2	0	1	2	2	9	4	2

Source: Marston and Raper (1987).
* Less than 0.5%.

The adequacy of diets based on any food guide depends on the nutrient composition of the individual foods that are chosen. If the nutrient composition of foods within a food group is lower than expected, then diets may be nutritionally inadequate even when food guides are followed. As it turns out, many factors can affect the nutrient composition of foods. Therefore, we must understand these factors

TABLE VIII Distribution of Individuals According to Nutrient Intakes Expressed as a Percentage of the 1980 RDAs

Nutrient	Percentage of individuals receiving: ≥100% RDA	70–99% RDA	<70% RDA
Energy	24	44	32
Protein	88	9	3
Calcium	32	26	42
Iron	43	25	32
Magnesium	25	36	39
Phosphorus	73	19	8
Vitamin A value	50	19	31
Thiamin	55	28	17
Riboflavin	66	22	12
Preformed niacin[a]	67	24	9
Vitamin B$_6$	20	29	51
Vitamin B$_{12}$	66	19	15
Vitamin C	59	15	26

Source: Pao and Mickle, 1981.
[a] Based on RDA values as milligrams of preformed niacin.

and strive to minimize nutritional quality losses in all foods.

B. Agricultural Practices

Many agricultural practices can affect the nutrient composition of foods. For plant foods these include geographic location, soil type, fertilization, moisture availability, sunlight exposure, maturity at harvest, method of harvest (mechanical vs. hand picking), season of planting and harvest, and genetics, and for animal foods feeding practices, age at time of slaughter, season, sex, hormone use, and genetics. Because the chemical composition of both plant and animal tissues is dictated by the genetic code, genetics is probably the one factor that has the largest influence on nutrient composition at the time of harvest or slaughter.

C. Food Storage, Processing, and Preparation

1. Storage

Storage conditions postharvest can have a marked effect on the nutrient composition of foods, particularly fruits and vegetables. Most plant tissues continue to transpire and respire after harvest up until they are processed or cooked. Control of storage conditions can markedly extend the storage life of fresh fruits and vegetables. Improper storage can result in wilting, decay due to bacterial and fun-

TABLE VII Extended

Food group	Vitamins							
	Vitamin A	Vitamin E	Ascorbic acid	Thiamin	Riboflavin	Niacin	Vitamin B_6	Vitamin B_{12}
Meat, poultry, and fish	17	10	2	26	22	47	40	72
Eggs	2	2	0	1	4	*	2	6
Dairy products, excluding butter	10	2	3	9	35	2	11	20
Fats and oils, including butter	6	61	0	*	*	*	*	0
Citrus fruits	1	1	28	3	1	1	2	0
Noncitrus fruits	4	3	15	2	2	2	9	0
White potatoes	0	2	16	4	1	6	12	0
Dark green, deep yellow vegetables	49	3	11	1	1	1	3	0
Other vegetables	8	5	21	5	3	4	7	0
Legumes, nuts, and soy	*	6	*	5	2	5	4	0
Flour and cereal products	*	4	0	43	24	29	10	2
Sugar and sweeteners	0	0	*	*	*	*	*	0
Miscellaneous	3	1	4	*	4	3	*	0

[1] Based on unrounded nutrient data; expressed as the percentage of total available nutrients provided by the respective food groups.
[2] Coffee, chocolate liquor equivalent of cocoa beans, tea, spices, and fortification of products not assigned to a specific group.

gal growth, weight loss, overmaturity, and chilling injury. While the deterioration that results from improper storage adversely affects appearance, flavor, and texture, it also may cause nutrient loss. For example, studies have shown that the vitamin C content of wilted spinach is much lower than in spinach that has been stored to minimize wilting. Approaches to improve storage conditions include rapid cooling after harvest, control of storage temperature and humidity, and manipulation of the storage atmosphere (this usually involves lowering the O_2 and increasing the CO_2 in the air in the storage chamber). For optimum retention of quality, storage conditions must be tailored for each individual product.

2. Food Processing and Preparation

Modern food-processing methods have increased the variety and year-round availability of foods. In addition, many processed foods offer a high level of convenience. Although it is true that in most cases foods fresh from the garden, farm, or sea have superior nutritional quality, few of us can rely directly on these sources for most of our foods. On the other hand, the greater variety and convenience afforded by processed foods would be expected to improve the overall nutritional quality of the diet by increasing the variety of foods that people actually eat.

Food processing methods can be divided into eight basic categories:

1. Heat processing: blanching, pasteurizing, sterilizing
2. Low-temperature processing: refrigerating, freezing
3. Moisture removal: drying, concentrating
4. Chemical processing: addition of salt, sugar, antioxidants, emulsifiers, stabilizers, thickeners, nutrients, etc.
5. Mechanical processing: milling of cereals, removal of peels, etc.
6. Microbiological processing: cheese-making, brewing, etc.
7. Packaging: enclosing in a container
8. Final Preparation: peeling, chopping, thawing, heating (boiling, broiling, microwaving, baking, frying), holding (steam table, etc.).

Virtually all of the processes listed above will alter, to a greater or lesser degree, the nutritional quality of foods. A brief description of some of the more important processing effects follows. The reader is referred to Karmas and Harris (1988) for an in-depth treatment of processing effects on nutritional quality.

a. Nutrient Destruction. Changes in the chemical structure of some nutrients may result in loss of nutritional activity. Rates of these chemical reactions are affected by several parameters including temperature, pH, concentration, oxygen, light, prooxidants (e.g., iron ions), and water activity (the amount of "free water" available in the food). A

TABLE IX Food Groups and Some Foods They Contain

Breads, Cereals, and Other Grain Products

Whole Grain

Brown rice	Oatmeal	Whole wheat bread and rolls	
Buckwheat groats	Popcorn	Whole wheat cereals	
Bulgur	Pumpernickel bread	Whole wheat crackers	
Corn tortillas	Ready-to-eat cereals	Whole wheat pasta	
Graham crackers	Rye crackers		
Granola			

Enriched

Bagels	Farina	Muffins
Biscuits	French bread	Noodles
Corn bread	Grits	Pancakes
Corn muffins	Hamburger rolls	Pasta
Cornmeal	Hot dog buns	Ready-to-eat cereals
Crackers	Italian bread	Rice
English muffins	Macaroni	White bread and rolls

Fruits

Citrus, Melons, Berries

Blueberries	Honeydew melon	Raspberries
Cantaloupe	Kiwifruit	Strawberries
Citrus juices	Lemon	Tangerine
Cranberries	Orange	Watermelon
Grapefruit		

Other Fruits

Apple	Grapes	Pineapple
Apricot	Guava	Plantain
Banana	Mango	Plum
Cherries	Nectarine	Pomegranate
Dates	Papaya	Prune
Figs	Peach	Raisins
Fruit juices	Pear	

Vegetables

Dark Green

Beet greens	Dandelion greens	Romaine lettuce
Broccoli	Endive	Spinach
Chard	Escarole	Turnip greens
Chicory	Kale	Watercress
Collard greens	Mustard greens	

Deep Yellow

Carrots
Pumpkin
Sweet potatoes
Winter squash

Starchy

Breadfruit	Lima beans
Corn	Potatoes
Green peas	Rutabaga
Hominy	Taro

Dry Beans and Peas (Legumes)

Black beans	Lima beans (mature)
Black-eyed peas	Mung beans
Chickpeas (garbanzos)	Navy beans
Kidney beans	Pinto beans
Lentils	Split peas

Other Vegetables

Artichokes	Cabbage	Green beans	Radishes
Asparagus	Cauliflower	Green peppers	Summer squash
Beans and alfalfa sprouts	Celery	Lettuce	Tomatoes
Beets	Chinese cabbage	Mushrooms	Turnips
Brussels sprouts	Cucumbers	Okra	Vegetable juices
	Eggplant	Onions (mature and green)	Zucchini

Meat, Poultry, Fish, and Alternates

Meat, Poultry, and Fish

Beef	Ham	Pork	Veal
Chicken	Lamb	Shellfish	Luncheon meats
Fish	Organ meats	Turkey	Sausage

Alternates

Dry beans and peas (legumes)	Nuts and seeds
	Peanut butter
Eggs	Tofu

Milk, Cheese, and Yogurt

Lowfat Milk Products

Buttermilk	Lowfat plain yogurt
Lowfat milk (1%, 2%)	Skim milk

Other Milk Products with More Fat or Sugar

American cheese	Chocolate milk	Fruit yogurt	Swiss cheese
Cheddar cheese	Flavored yogurt	Processed cheeses	Whole milk

Fats, Sweets, and Alcoholic Beverages

Fats

Bacon, salt pork	Mayonnaise
Butter	Mayonnaise-type salad dressing
Cream (dairy, non-dairy)	Salad dressing
Cream cheese	Shortening
Lard	Sour cream
Margarine	Vegetable oil

Sweets

Candy	Jam	Popsicles and ices
Corn syrup	Jelly	Sherbets
Frosting	Maple syrup	Soft drinks and colas
Fruit drinks, ades	Marmalade	Sugar (white and brown)
Gelatin desserts	Molasses	
Honey		

Alcohol

Beer
Liquor
Wine

TABLE X Suggested Pattern for Daily Food Choices

Food group	Suggested daily servings
Breads, cereals, and other grain products Whole grain Enriched	6–11 (include several servings a day of whole grain products)
Fruits Citrus, melon, berries Other fruits	2–4
Vegetables Dark green leafy Deep yellow Dry beans and peas (legumes) Starchy Other vegetables	2–5 servings (include all types regularly; use dark green leafy vegetables and dry beans and peas several times a week)
Meat, poultry, fish, and alternates (eggs, dry beans and peas, nuts and seeds)	2–3 servings; total 5–7 oz. lean
Milk, cheese, and yogurt	2 servings (3 servings for teens and women who are pregnant or breastfeeding; 4 servings for teens who are pregnant or breast-feeding)
Fats, sweets, and alcoholic beverages	Avoid too many fats and sweets; if you drink alcoholic beverages, do so in moderation

Source: USDA/HNIS, 1986.

TABLE XI Factors That Affect Nutrient Stabilities in Foods

Nutrient	Factors that protect against losses	Factors that accelerate losses
Protein		heat, especially in the presence of sugars; high pH;
Carbohydrates		no nutritionally significant losses occur
Polyunsaturated fatty acids	antioxidants	air, oxygen, light, pro-oxidants such as iron ions; hydrogenation
Ascorbic acid	low pH, sulfite, low moisture	heat, oxygen, light, high pH, iron and copper ions, some plant enzymes
Thiamin	low pH	heat, neutral and high pH, sulfite, thiaminases (enzymes in raw fish)
Riboflavin	low pH	light, especially in liquid foods such as milk
Niacin		stable under most conditions; leachable
Vitamin B6		heat, light (stability varies depending on the form of the vitamin)
Folacin		heat (stability varies depending on form)
Vitamin A	antioxidants	heat, oxygen, iron and copper ions, low pH

primary objective of the food-processing engineer is to design food-processing systems that will meet the objectives of microbiological safety and organoleptic quality and at the same time will result in optimal retention of nutrients.

3. Vitamins

The class of nutrients most susceptible to chemical destruction is the vitamin. Many vitamins are rather fragile organic molecules, which lose their biological activity when even very small changes in their chemical structures occur. Some of the more significant factors affecting vitamin stabilities in foods are listed in Table XI. Fortunately, conditions that result in accelerated rates of destruction are similar for most nutrients. Thus, if processing conditions are chosen to protect one nutrient, most other nutrients will be protected as well.

4. Proteins

Processing may alter protein quality by promoting chemical reactions between essential amino acid residues and other food components. In general,

amino acid residues in intact protein molecules are quite stable to most processing and storage conditions. The exception to this is the amino acid lysine, which contains a free amino group on the epsilon carbon of the side chain. Under some conditions, this amino group can react with reducing sugars (e.g., glucose, lactose) to form brown products. This is the familiar browning that occurs when bread is baked or breakfast cereal flakes are toasted. It is known as nonenzymatic browning, or Maillard browning (after the French chemist who first described the reaction). When this browning occurs, the nutritional activity of the lysine is destroyed and protein quality is diminished. The nutritional impact is the greatest in cereals, which are low in lysine to begin with. Fortunately, processing effects on protein quality have little significance in the United States where protein intakes exceed requirements by substantial margins. Moreover, the class of foods where protein quality losses are greatest (i.e., ready-to-eat toasted breakfast cereals) are normally consumed with milk, which is rich in lysine and therefore compensates for the lost lysine in the cereal.

In many cases, processing improves protein quality by enhancing protein digestibility. Legumes, for example, contain antinutrients called trypsin inhibitors. These substances interfere with the activity of trypsin, a digestive enzyme that breaks down proteins in the small intestine. It is well known that raw soybeans, which are a rich source of a relatively high-quality protein, will not support growth in rats and other animals. When the soybeans are treated with moist heat, growth improves dramatically. This is because the trypsin inhibitor is heat-sensitive and heat destroys it, thereby preventing it from interfering with the digestion of the protein in the soybeans. [See PROTEINS (NUTRITION).]

5. Minerals

Minerals, unlike vitamins and proteins, are virtually indestructible. Losses may result from leaching, milling, and trimming but not from actual chemical destruction. The bioavailability of minerals, however, may be altered by processing. Bioavailability may be defined as the percentage of a nutrient in a food that is potentially available for absorption from the gastrointestinal tract in a physiologically active form. Processing may change the oxidation state of a mineral, and this may affect its bioavailability. For example, iron may be oxidized from the ferrous form (Fe^{+2}) to the ferric form (Fe^{+3}) during processing. Ferrous iron is thought to be more bioavailable than ferric iron. Processing may also affect mineral bioavailability indirectly by altering other nutrients in the food. For example, ascorbic acid is a potent enhancer of iron bioavailability. If a food process destroys ascorbic acid, iron bioavailability from that food and other foods consumed with it in the same meal will be diminished. Conversely, fortifying a food with ascorbic acid will improve iron bioavailability. [See MINERALS IN HUMAN LIFE.]

6. Physical Separation of Nutrients From Foods

In many food processes, nutrients are physically removed from the edible portion of foods. Probably the most dramatic example of this is the nutrient changes that occur when wheat is milled into white flour. The wheat kernel may be divided into three basic parts: (1) the endosperm is the starchy interior of the kernel, making up about 83% of the total weight; (2) the bran includes the outer layers of the kernel, making up about 14.5% of the intact kernel; and (3) the germ is the smallest part, making up only about 2.5% of the total. Whole wheat flour includes all three parts of the kernel. White flour is mostly endosperm. Many of the nutrients in the intact wheat kernel are concentrated in the bran and/or germ layers, so removal of these in milling causes dramatic losses. Four nutrients, iron, thiamin, riboflavin, and niacin, are routinely restored to white flour through fortification (enrichment), but other nutrients such as zinc, magnesium, and vitamin B_6 are not.

Leaching may also separate nutrients from the edible portion of the food. This is mainly a problem with the water-soluble nutrients such as vitamin C and potassium. It occurs when foods are blanched, canned, or cooked in water. Losses are greatest when the food is finely chopped, when a large water to food ratio exists, when the food is processed at high temperatures for long periods, and when processing or cooking water is discarded. An example of how leaching can affect the content of two different minerals in opposite directions is shown in Table XII. Processing causes potassium losses because potassium is a highly soluble mineral nutrient. Losses are greatest with canning because of the relatively large amount of water that is used and because the food remains in contact with the water not only during processing but also during storage of the product. Leaching losses in frozen vegetables

TABLE XII Effects of Processing on Sodium and Potassium in Vegetables

	Content (mg/100 g edible portion)	
	Sodium	Potassium
Carrots, raw	26	306
Carrots, canned	236	120
Corn, raw	trace	280
Corn, frozen	1	202
Corn, canned	236	97
Peas, raw	trace	170
Peas, frozen	129	150
Peas, canned	236	96

Adapted from USDA, 1976–1989.

occur primarily during blanching. In many cases, processing results in an increase in the sodium content of the edible portion of the food. This is because salt is added during processing. Salt is not usually added to frozen vegetables. Peas are an exception, because a brine solution is used to separate peas by size prior to blanching and freezing.

Unfortunately, because of suspected relationships between sodium and potassium intakes and hypertension, both the decreases in potassium and the increases in sodium shown in Table XII are nutritionally undesirable.

7. Nutrient Dilution

Nutrient dilution is a term used to describe the addition of sugars or fats to foods. Food grade sugars include sucrose, high fructose corn syrup, corn syrup, lactose, and honey. These are highly purified carbohydrates that contain only insignificant concentrations of nutrients. Sugars do contain approximately 4 kcal/g dry weight. Food grade fats and oils include butter, margarine, shortening, lard, beef tallow, and vegetable oils. Fats and oils contain varying amounts of the fat soluble vitamins A, D, and E as well as essential fatty acids; however, they are highly concentrated sources of calories (9 kcal/g).

Both sugars and fats are added to a wide variety of foods either during processing and preparation, or at the table. It was estimated that in 1985, fats and oils provided 20% and sugars provided 18% of the available calories in the food supply. The net effect of this nutrient dilution is that more calories have to be consumed to get the quantities of nutrients that would be available in the foods before sugar or fat was added. For some people, choosing a diet adequate in all essential nutrients may be very difficult if the foods contain excessive amounts of added sugar and fat. In addition, evidence indicates that diets high in fat may lead to obesity, heart disease, and/or cancer in some people. [*See* Fats and Oils (Nutrition).]

8. Addition of Nutrients to Foods

Several terms have been used to describe the practice of adding nutrients to foods, including enrichment, fortification, and restoration. Although these terms initially had slightly different meanings, in practice they are frequently used interchangeably. Here, nutrification will be used to indicate any addition of nutrients to foods. Nutrification has a long history in the United States. Iodization of salt began in 1924, vitamin D was first added to milk in

1933, and a program for enrichment of flour and cereal products began in 1941. Originally, nutrification was instituted as a public health measure for reducing the widespread prevalence of nutrient deficiency diseases such as goiter, rickets, and pellagra. The early successes of nutrification efforts are well documented and have led to the nutrification of a much broader spectrum of food products. Today, foods ranging from breakfast cereals to candy and soda pop are being nutrified. While it is generally felt that the addition of nutrients to foods enhances nutritional quality, many nutritionists are beginning to view with concern the widespread nutrification of the U.S. food supply without apparent regard to potentially adverse nutrient–nutrient interactions. Federal regulations do not specify which foods may be fortified, but the FDA has recently issued a fortification policy in the form of a series of guidelines. The FDA urges food manufacturers to adhere to its policy when adding nutrients to foods. The following quote from this policy statement reflects the feelings of many nutritionists and food scientists on this important issue: "The addition of nutrients to specific foods can be an effective way of maintaining and improving the overall nutritional quality of the food supply. However, random fortification of foods could result in over or under fortification in consumer diets and create nutrient imbalances in the food supply. It could also result in deceptive or misleading claims for certain foods. The FDA does not encourage indiscriminate addition of nutrients to foods, nor does it consider it appropriate to fortify fresh produce; meat, poultry, or fish products; sugars; or snack foods such as candies and carbonated beverages."

IV. Nutritional Quality of Major Food Groups

This section lists the major nutrients found in each of the food groups listed in Table IX and makes comparisons among foods within each food group. [*See* Food Groups.]

A. Breads, Cereals, and Other Grain Products Group

Foods in this group are prepared from cereal grains. Important food cereals include wheat, rice, maize (corn), oats, barley, and rye. Cereals have been major dietary staples in most cultures throughout recorded history. Cereal grains are compact, stable, and relatively easy to ship and store. They may be processed into a wide variety of food products.

Whole grain cereal products are good sources of energy (mainly as starch); protein (cereals contain 7–15% protein by weight), trace minerals including iron, zinc, and magnesium; vitamins, especially thiamin, riboflavin, niacin, and folic acid; and dietary fiber. Cereals contain very little fat, calcium, or vitamins A, D, C, and B_{12}.

A large proportion of wheat and rice is milled before it is consumed. Milling involves the removal of bran and germ from the intact kernel leaving the white, starchy endosperm. This process results in a marked reduction of B vitamins, trace minerals, and dietary fiber, which are concentrated in the bran and germ portions of the kernel. Iron, riboflavin, niacin, and thiamin are added back in enriched cereal products. The nutrient compositions of selected cereal foods are listed in Table XIII.

B. Fruits Group

Most fruits are low in calories and are good sources of fiber and some vitamins and minerals. Citrus fruits are particularly rich in ascorbic acid (vitamin C) while deep yellow fruits such as apricots, cantaloupes, and mangos are good sources of vitamin A. Table XIV lists the nutrient composition of some representative fruits.

C. Vegetables Group

The vegetables group encompasses a wide variety of plant species. Nutrient composition varies considerably among the various species. All vegetables are sources of dietary fiber and, as a group, vegetables contain vitamins and minerals that are not found in sufficient concentrations in other food groups. Because of the diversity in the vegetables group, the USDA has subdivided the group into five subgroups (see Table IX).

Dark green vegetables are good sources of vitamins A and C, riboflavin, folic acid, iron, calcium, magnesium, and potassium. Deep yellow vegetables are rich in vitamin A. Dry beans and peas contain thiamin, folic acid, iron, magnesium, phosphorus, zinc, potassium, protein, and starch. Starchy vegetables are high in starch and contain varying levels of niacin, vitamin B_6, zinc, and potassium. The other vegetables subgroup is quite diverse, but individual vegetables are significant sources of vitamins A and C, potassium, and other nutrients. Table XV lists the nutrient composition of some representative vegetables.

D. Dairy Products Group

The most striking nutritional attribute of dairy products is their calcium content. Three-fourths of all the calcium in the American diet comes from dairy foods. Dairy products are also good sources of pro-

TABLE XIII Nutrient Composition of Selected Foods from the Breads and Cereals Group

Nutrient	Enriched white bread (2 slices)	Whole wheat bread (2 slices)	Enriched white rice ($\frac{1}{2}$ cup)	Brown rice ($\frac{1}{2}$ cup)	Corn chips (1 oz.)
Calories	150	140	111	116	157
Protein (g)	4.6	6.0	2.0	2.4	2.0
Fat, total (g)	2.2	2.4	0.1	0.6	9.1
Fat (% cal)	13	15	1	5	53
Fat, saturated (g)	0.7	0.7	—	0.2	1.8
MUFA (g)	0.8	0.8	—	0.2	3.4
PUFA (g)	0.5	0.6	—	0.2	3.7
Cholesterol (mg)	0	0	0	0	0
Thiamin (mg)	0.26	0.2	0.1	0.1	0.04
Riboflavin (mg)	0.17	0.12	0.01	0.02	0.05
Niacin (mg)	2.1	2.2	1.0	1.4	0.4
Iron (mg)	1.6	1.9	1.4	0.6	0.5
Zinc (mg)	0.34	1.0	0.42	0.52	0.44
Sodium (mg)	233	360	0	0	236

Source: USDA Agriculture Handbook 456.

TABLE XIV Nutrient Composition of Selected Foods from the Fruits Group[a]

	Apple, raw, with skin (1 apple = 138 g)	Banana, raw (1 banana = 114 g)	Orange, raw (1 orange = 131 g)	Peaches, canned, heavy syrup (1 half, 1¾ tbsp liquid = 81 g)	Peaches, canned, juice pack (1 half, 1⅔ tbsp liquid = 77 g)
Calories	81	105	62	60	34
Water (g)	115	82	114	64	67
Carbohydrate (g)	21	27	15	16	9
Protein (g)	0.27	1.2	1.2	0.4	0.5
Fat, total (g)	0.49	0.55	0.16	0.08	0.03
Fat (% cal)	5	5	2	1	0.3
Saturated fatty acid (g)	0.08	0.2	0.02		
MUFA (g)	0.08	0.3	0.05		
PUFA (g)	0.15	0.1	0.03		
Cholesterol (mg)	0	0	0		
Calcium (mg)	10	7	52	2	5
Iron (mg)	0.2	0.4	0.13	0.22	0.21
Zinc (mg)	0.1	0.2	0.1	0.07	0.08
Sodium (mg)	0	1	0	5	3
Potassium (mg)	159	451	237	74	98
Ascorbic acid (mg)	8	10	70	2	3
Thiamin (mg)	0.02	0.05	0.11	0.01	0.01
Riboflavin (mg)	0.02	0.11	0.05	0.02	0.01
Niacin (mg)	0.1	0.6	0.37	0.5	0.4
Folacin (mcg)	4	22	40	2.6	
Vitamin A (RE[b])	7	9	27	27	29
Total dietary fiber (g)	2.2	1.6	2.4		

[a] Values are from USDA Agriculture Handbook No. 8-9.
[b] Retinol equivalent.

tein, riboflavin, vitamins A and B_{12}, and thiamin. Virtually all of the fluid milk sold in the United States is fortified with vitamin D. Many dairy products contain significant amounts of fat, saturated fat, cholesterol, and sodium. Table XVI lists the nutrient composition of some representative dairy products.

E. Meat, Poultry, Fish, and Alternates Group

Foods in this group are rich in a wide variety of nutrients. They are particularly good sources of protein, iron, zinc, niacin, and vitamins B_6 and B_{12}. The iron in meat, poultry, and fish has high bioavailability, and these foods enhance the absorption of iron from other foods when consumed together in a meal. The iron in eggs and legumes, on the other hand, is less bioavailable, and eggs and legumes may actually depress iron absorption from other foods consumed along with them. The fat content of foods in this group is highly variable. Lean cuts of meat and poultry with the skin removed are quite low in fat. Hamburger, luncheon meats, and sausages tend to be quite high in fat. In addition, the fat is more saturated than most vegetable fats. Egg yolks are very high in cholesterol and also contain a considerable amount of fat. The cholesterol content of most meat products is similar on a weight basis because the cholesterol content of lean and fat tissues is approximately the same. Dry beans and peas are very low in fat, but nuts, seeds, and peanut butter may be higher in fat than many meat products. Table XVII lists the nutrient composition of selected foods from the meats group.

F. Fats, Sweets, and Alcoholic Beverages Group

The foods in this group are notable for their high caloric density and low levels of nutrients. The USDA recommends that foods from this group be limited and that those who choose to consume alcoholic beverages should do so in moderation.

TABLE XV Nutrient Composition of Selected Foods from the Vegetables Group[a]

	Potatoes, boiled, w/o skin ($\frac{1}{2}$ cup = 78 g)	Potatoes, french-fried in vegetable oil (10 strips = 50 g)	Carrots, raw (1 carrot = 72 g)	Broccoli, boiled, drained ($\frac{1}{2}$ cup = 78 g)	Tomatoes, raw (1 tomato = 123 g)	Lettuce, Iceberg, raw (2 leaves = 40 g)
Calories	67	158	31	23	24	6
Water (g)	60	19	63	70	116	38
Protein (g)	1.3	2	0.7	2.3	1.1	0.4
Lipid (g)	0.08	8	0.14	0.2	0.26	0.08
Carbohydrate (g)	16	20	7.3	4.3	5.3	0.82
Fat (% cal)	1	45	4	8	1	12
Saturated fatty acid (g)	0.03	2.5				
MUFA (g)		1.6				
PUFA (g)	0.04	3.8				
Calcium (mg)	6	10	19	89	8	8
Iron (mg)	0.24	0.38	0.36	0.89	0.59	0.2
Potassium (mg)	256	366	233	127	254	64
Sodium (mg)	4	108	25	8	10	4
Zinc (mg)	0.21	0.2	0.14	0.12	0.13	0.08
Ascorbic acid (mg)	6	5	7	49	22	2
Thiamin (mg)	0.08	0.09	0.07	0.06	0.07	0.02
Riboflavin (mg)	0.02	0.01	0.04	0.16	0.06	0.01
Niacin (mg)	1.0	1.6	0.7	0.6	0.7	0.07
Folacin (mcg)	7	14	10	53	11	22
Vitamin A (RE[b])			2,025	110	139	14
Total dietary fiber (g)		2.1	2.3		1	

[a] Values are from USDA Agriculture Handbook No. 8-11.
[b] Retinol equivalent.

TABLE XVI Composition of Selected Foods from the Dairy Group[a]

	Whole vitamin D milk (8 oz.)	Skim milk, vitamins A and D added (8 oz.)	Yogurt, low-fat, fruited (8 oz.)	Cheese, cheddar (1 oz.)	Ice cream, vanilla (1 cup)
Calories	157	86	231	114	269
Water (g)	213.96		169	10.42	80
Protein (g)	8	8	10	7	5
Lipid (g)	9	0.4	2.5	9	14
Carbohydrate (g)	11.35	11.88	43.24	0.36	31.72
Fat (% cal)	52	4	10	71	47
Saturated fatty acid (g)	5.6	0.1	1.6	6	9
MUFA (g)	2.58	0.116	0.67	2.66	4.14
PUFA (g)	0.3	0.016	0.1	0.3	0.5
Calcium (mg)	290	302	345	204	176
Iron (mg)	0.1	0.1	0.2	0.2	0.1
Potassium (mg)	368	406	442	28	257
Sodium (mg)	119	126	133	176	116
Zinc (mg)	0.93	0.98	1.68	0.88	1.41
Ascorbic acid (mg)	4	2	1.5		1
Thiamin (mg)	0.1	0.1	0.1	0.01	0.05
Riboflavin (mg)	0.4	0.343	0.4	0.1	0.3
Niacin (mg)	0.2	0.2	0.2	0.02	0.1
Folacin (mcg)	12	13	21	5	3
Vitamin A (RE[b])	337	500	104	300	133
Total dietary fiber (g)	0	0	0	0	0

[a] Values from USDA Agriculture Handbook No. 8-1.
[b] Retinol equivalent.

TABLE XVII Composition of Selected Foods from the Meat Group[a]

	Beef, eye of round, separable lean, choice, roasted (3 oz. = 85 g)	Beef, ground, extra lean, pan-fried (3 oz. = 85 g)	Beef, ground, regular, pan-fried (3 oz. = 85 g)	Chicken breast, meat+skin, roasted (½ breast = 98 g)	Chicken breast, meat only, roasted (½ breast = 86 g)
Calories	156	224	243	193	142
Water (g)	53	46	45	61	56
Protein (g)	25	24	23	29	27
Lipid (g)	5.7	13.5	16	7.6	3.1
Carbohydrate (g)					
Fat (% cal)	33	54	59	35	20
Saturated fatty acid (g)	2.2	5.3	6.3	2.2	0.9
MUFA (g)	2.5	5.9	7.0	3.0	1.0
PUFA (g)	0.2	0.5	0.6	1.6	0.7
Calcium (mg)	4	7	11	14	13
Iron (mg)	1.6	2.3	2.3	1.0	0.9
Potassium (mg)	336	306	283	240	220
Sodium (mg)	52	69	79	69	63
Zinc (mg)	4	5.3	4.8	1.0	0.9
Ascorbic acid (mg)					
Thiamin (mg)	0.07	0.06	0.03	0.06	0.06
Riboflavin (mg)	0.14	0.25	0.18	0.12	0.1
Niacin (mg)	3.1	4.6	5.5	12.4	11.8
Folacin (mcg)	6	9	8	3	3
Vitamin A (RE[b])				26	18
Total dietary fiber (g)					

[a] Values are from USDA Agriculture Handbook No. 8-5.
[b] Retinol equivalent.

V. Nutritional Quality of Food Components

As discussed above, different foods, even within a given food group, may have markedly different nutrient compositions and, therefore, selecting the recommended number of servings from each food group does not necessarily assure a balanced diet. Moreover, composition alone does not always give a clear picture of the potential physiological effect a food or diet may have. For example, it is well known that coconut oil will produce increases in serum cholesterol levels when it is consumed in large amounts. Corn oil, on the other hand, appears to have the opposite effect. Both oils are nearly 100% triglyceride (fat), but the fatty acids that make up the triglycerides are different. Therefore, it is necessary to characterize foods not only according to the amount of a given nutrient but, often, by the chemical form of that nutrient as well.

A. Protein

The nutritional quality of a protein is determined by its amino acid composition and digestibility. The amino acid pattern in proteins of high nutritional quality matches the pattern of amino acids required for protein synthesis in the cells of the animal or human consuming the protein. If just one essential amino acid in a food protein is present in suboptimal amounts, the quality of the protein will be lower than that of an ideal protein. Therefore, protein quality is determined by the limiting amino acid in the protein. In general, proteins in animal foods such as meat, milk, and eggs have high quality, whereas proteins from plant sources usually contain suboptimal levels of one or more amino acids. Fortunately, the limiting amino acids in plant proteins vary with the source, so it is generally possible to "complement" proteins by mixing foods with different limiting amino acids. For example, cereal proteins are low in lysine but relatively high in sulfur amino acids, whereas legumes are low in sulfur amino acids but high in lysine. Therefore, mixing a cereal with a legume will result in a combination of relatively high protein quality.

B. Lipids

Lipids are defined as the class of compounds that are soluble in nonpolar organic solvents such as chloroform or diethyl ether. Lipids in foods include triacyl glycerols, phospholipids, cholesterol, and plant sterols. [See LIPIDS.]

Triacyl glycerols (also known as triglycerides) are esters of fatty acids and glycerol. The fatty acids may vary in length from 4 carbons to more than 24 carbons. Food fatty acids may be saturated (no double bonds), monounsaturated (one double bond per molecule), or polyunsaturated (two or more double bonds). Table XVIII lists the fatty acids commonly found in foods. Concentrations of free fatty acids (i.e., not esterified to glycerol) in foods are generally very low. Triacyl glycerols make up the bulk of the lipid fraction in most foods, approaching 100% of the total weight of vegetable oils. Other lipid components make up a much smaller fraction but may still have a significant nutritional impact.

Phospholipids are made up of fatty acids, glycerol, phosphate, and, with few exceptions, a nitrogen-containing group. Lecithin is an example. Phospholipids play a structural role in cell membranes and lipoproteins. [See PHOSPHOLIPID METABOLISM.]

Cholesterol is an alcohol structurally unrelated to triacyl glycerol. It is found only in foods of animal origin. Egg yolks, liver, and brain are especially high in cholesterol. [See CHOLESTEROL.]

There is substantial evidence that fatty acid composition of dietary fat as well as total dietary fat is related to serum cholesterol levels. Saturated fatty acids, particularly C14:0 and C16:0 tend to raise serum cholesterol. Polyunsaturated fatty acids (PUFAs) and monounsaturated fatty acids (MUFAs) have the opposite effect and tend to lower serum cholesterol. In general, animal fats tend to be more saturated and vegetable oils more unsaturated with a few exceptions. Coconut oil, for example, is a highly saturated vegetable oil. While PUFA is effective in lowering serum cholesterol levels, there is concern that high intakes of PUFA may not be desirable. Some studies with animals have shown that PUFA acts as a promotor for some cancers. Recent studies have shown that MUFAs are as effective as PUFAs in lowering serum cholesterol and may be preferable to PUFAs.

A common practice for food manufacturers is to hydrogenate oils. This is the process of adding hydrogen to the oils to reduce the level of unsaturation. Oils are hydrogenated for two reasons: (1) to convert oils to semisolids (e.g., margarine) and (2) to stabilize oils against oxidation. Most hydrogenated oils remain highly unsaturated; the major change is a reduction of PUFA with an increase in MUFA. The fatty acid composition of some fats and oils are listed in Table XIX.

TABLE XVIII Fatty Acids Commonly Found in Foods

Common name	Shorthand representation	Food source
Saturated fatty acids		
Butyric	$C_{4:0}$[a]	butterfat
Caproic	$C_{6:0}$	butterfat
Caprylic	$C_{8:0}$	coconut oil
Capric	$C_{10:0}$	coconut oil
Lauric	$C_{12:0}$	coconut oil
Myristic	$C_{14:0}$	butterfat/coconut oil
Palmitic	$C_{16:0}$	most fats and oils
Stearic	$C_{18:0}$	most fats and oils
Monounsaturated fatty acids		
Palmitoleic	$C_{16:1}$, ω-7	fish oils
Oleic	$C_{18:1}$, ω-9	most fats and oils
Polyunsaturated fatty acids		
Linoleic	$C_{18:2}$, ω-6[b]	most vegetable oils
Linolenic	$C_{18:3}$, ω-3	soybean oil, canola oil
Aracidonic	$C_{20:4}$, ω-6	lard
Eicosapentaenoic	$C_{20:5}$, ω-3	fish oils
Docosahexaenoic	$C_{22:6}$, ω-3	fish oils

[a] Example: $C_{4:0}$ indicates 4 carbon atoms and no double bonds in the fatty acid molecule.

[b] Example: $C_{18:2}$, ω-6 indicates 18 carbons and 2 double bonds, with the first double bond on the number 6 carbon counting from the methyl end of the molecule.

TABLE XIX Fatty Acid[a] and Cholesterol Content of Selected Fats and Oils

Fat/oil	Cholesterol (mg/tbsp)	Saturated fat	Polyunsaturated fat	Monounsaturated fat
Coconut oil	0	92	2	6
Butterfat	33	66	4	30
Beef tallow	14	52	4	44
Lard	12	41	12	47
Chicken fat	11	31	22	47
Peanut oil	0	18	33	49
Soybean oil	0	15	61	24
Corn oil	0	13	62	25
Sunflower oil	0	11	69	20
Canola oil	0	6	32	62
Stick margarine	0	18–24	14–35	45–67
Tub margarine	0	17–21	27–44	36–55

[a] Percentage of total fatty acids.

C. Carbohydrates

Carbohydrates are composed of the elements carbon, hydrogen, and oxygen in the approximate ratio of 1 carbon–2 hydrogens–1 oxygen (CH_2O). They are frequently classified as simple or complex. Simple carbohydrates are sugars and include primarily mono- and disaccharides. Glucose, fructose, and galactose are examples of monosaccharides. Disaccharides are composed of two monosaccharides chemically linked together. They include sucrose (glucose + fructose), maltose (glucose + glucose), and lactose (glucose + fructose). Sugars occur naturally in a wide variety of foods and are also available in highly refined form. Ordinary white table sugar is one of the most chemically pure foods available, consisting of more than 99% sucrose. [*See* CARBOHYDRATES (NUTRITION).]

Complex carbohydrates are composed of monosaccharides chemically linked to form polysaccharides. Complex carbohydrates may be further divided into available (digestible) and unavailable (indigestible) carbohydrates.

Starch is the most common available polysaccharide in foods and contains several hundred to several thousand glucose units in each molecule. Molecular weights of starches range from 50,000 to several million. Starches are hydrolyzed to glucose by digestive enzymes in the gastrointestinal tract.

Unavailable carbohydrates are more commonly known as dietary fiber. Dietary fiber consists of a wide variety of polysaccharides and lignin, which, technically, is not a carbohydrate. Because of its complexity and problems with measurement and characterization, fiber is difficult to define precisely. A useful definition of dietary fiber follows: components of plant material that are resistant to digestion by endogenous enzymes of the mammalian gastrointestinal tract. Dietary fiber is present only in foods of plant origin. The composition of dietary fiber varies with the food. Components include cellulose, noncellulose polysaccharides (hemicellulose, pectins, mucilages, gums), and lignin (a nonpolysaccharide). The physiologic effects of fiber components appears to be related to water solubility. Therefore, another possible classification of fiber is based on water solubility. The water-insoluble fractions (e.g., cellulose, hemicellulose) aid in laxation and increase stool weight. The water-soluble fractions (e.g., pectin, gums) have little effect on stool weight but appear to be effective in reducing serum cholesterol levels. Total dietary fiber is defined as the sum of the soluble and insoluble fractions. [*See* DIETARY FIBER, CHEMISTRY AND PROPERTIES.]

Carbohydrates are the major source of energy in most diets. Sugars and starches provide about 4 kcal/g. While dietary fiber is not digested in the small intestine and, therefore, cannot be absorbed, it is partially degraded by the microflora of the large intestine. The microflora metabolize the products of fiber degradation to short-chain fatty acids (acetic, propionic, and butyric), which are rapidly absorbed and used for energy.

Health implications related to dietary carbohydrates have been hotly debated by professional nutritionists and the general public for decades. Refined sugars have been maligned as contributors to health problems ranging from hyperactivity in children to heart disease in adults. However, little clear evidence links refined sugars to any health problem other than dental caries. On the other hand, the addition of refined sugar to the diet clearly will cause nutrient dilution, because refined sugars provide calories but no nutrients and may displace more nutritious foods from the diet.

Interest in dietary fiber has increased dramatically in the past 2 decades. Many organizations are currently recommending that dietary fiber intakes be increased to a total intake of 25–50 g/day. Putative benefits of dietary fiber range from prevention or relief of constipation to prevention of colon cancer. Recently, several studies have been published showing that soluble dietary fiber, when consumed as part of a low-fat, low-cholesterol diet, reduces serum cholesterol levels. Food sources of soluble dietary fiber include oat bran, legumes, and fruits. Most nutritionists agree that fiber should be obtained from food, not in concentrated forms present in fiber-supplement pills.

VI. Recommendations for Improving Nutritional Quality

The Committee on Diet and Health of the Food and Nutrition Board concluded in 1989 that the evidence on relationships between diet and health is compelling and warrants efforts to promote dietary change for the American public. The Committee recommends two approaches: (1) the public should be informed about risks and benefits of dietary modification and (2) the food supply should be modified to facilitate dietary change. The two approaches must be carried out simultaneously. It will be easier for the public to improve the quality of their diets if appropriate foods are available in the marketplace. At the same time, it must be recognized that the food industry will not introduce new products with improved nutritional quality unless a market exists for the products. Therefore, the following steps will lead to the improvement of the nutritional quality of foods, diets, and the national food supply:

1. Educate the public about nutrition so that they can make informed choices in the marketplace.
2. Improve food-labeling so that consumers have the information necessary to make healthy food choices.
3. Include more nutrition in the curriculum of medi-

cal, nursing, teacher training, elementary, and secondary schools.
4. Strengthen regulations governing the marketing of food products to reduce false and misleading nutrition claims in advertising and on food labels.
5. Strengthen the nutrition monitoring system to allow for improved data on the nutritional composition of the food supply and the nutritional status of the population.
6. Increase funding for nutrition research designed to further our understanding of relationships between diet and health.

Bibliography

Committee on Diet and Health, Food and Nutrition Board, Commission on Life Sciences, National Research Council (1989). "Diet and Health—Implications for Reducing Chronic Disease Risk." National Academy Press, Washington, D.C.

Committee on Technological Options to Improve the Nutritional Attributes of Animal Products (1988). "Designing Foods—Animal Product Options in the Marketplace." National Academy Press, Washington, D.C.

Karmas, E., and Harris, R. S. (eds.) (1988). "Nutritional Evaluation of Food Processing," 3rd ed. Van Nostrand Reinhold Company, New York.

Marston, R., and Raper, N. (1987). Nutrient content of the U.S. food supply. *National Food Rev.* **Winter–Spring,** 18–23.

Pao, E. M., and Mickle, S. J. (1981). Problem nutrients in the United States. *Food Technol.* **35(9),** 58–69, 79.

Subcommittee on the Tenth Edition of the RDAs, Food and Nutrition Board, Commission on Life Sciences, National Research Council (1989). "Recommended Dietary Allowances," 10th ed. National Academy Press, Washington, D.C.

U.S. Department of Agriculture (1986–1989). "Composition of Foods: Raw, Processed, Prepared," Agriculture Handbook 8, revised, Vols. 8-1–8-17, U.S. Department of Agriculture, Washington, D.C.

U.S. Department of Agriculture–Human Nutrition Information Service (1986). "Nutrition and Your Health—Dietary Guidelines for Americans." Home and Garden Bulletin Number 232-1. U.S. Department of Agriculture, Human Nutrition Information Service, Hyattsville, Maryland.

U.S. Department of Health and Human Services (1988). "The Surgeon General's Report on Nutrition and Health." U.S. Government Printing Office, Washington, D.C.

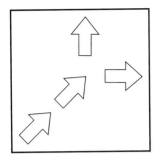

Nutrition and Immunity

GABRIEL FERNANDES, *The University of Texas Health Science Center at San Antonio*

Glossary

Calorie Unit in which energy is measured; food energy is measured in calories

CD4+ Thymic-derived helper T cells

CD8+ Thymic-derived cytotoxic (or suppressor) T cells

Class II antigens Antigens encoded by the major histocompatibility complex (MHC) region are important for recognition of foreign antigens

Cytokines Group of proteins produced by variety of cells including lymphoid cells

Ig+ Immunoglobulin-producing bone marrow–derived cells (B cells)

THE SCIENCES of nutrition and immunity are intimately linked to each other and are currently facilitating the understanding of cellular and molecular mechanisms involved in maintaining a healthy and disease-free lifespan. In recent years, nutrition research, along with immunological studies, has been extremely valuable in understanding human growth and development, maintenance of proper health, disease prevention, and disease control. The nutrients in food are chemical components derived from the diet that provide the following: (1) fuel and energy for regulation and maintenance of physiological processes, and (2) promotion of growth and repair of body tissues. The immune system obtains nutrients from the food consumed to produce and distribute normal, healthy immune cells throughout the body to combat invasive pathogenic microorganisms and to discriminate between self- and nonself-antigens to protect the integrity of the host organism. The immune system is a highly intricate, delicately regulated network of cells that are intimately involved in both cellular and humoral immune responses. The tools of immunology and molecular biology provide a resourceful approach to investigate the role of a variety of dietary components in maintaining the optimal function of immune cells not only to prevent infection but possibly to protect against the occurrence of cardiovascular disease, cancer, autoimmune disorders, and acquired immunodeficiency syndrome (AIDS) in humans. Several aspects of diet and nutrition interactions with immune function and their role in modulating the occurrence of life-shortening diseases will be explored in this article.

I. Interaction of Diet and Immune System

Immunological investigations are of utmost importance in studying the role of nutrition, infection, and age-associated disease processes both in humans and experimental animals. It is well known that several functions of the immune system, particularly those that are thymus-related, consistently decline in efficiency during acute malnutrition or with normal aging. This decline has been regularly linked to increased susceptibility to bacterial, fungal, and viral infections as well as to the development of cancer and autoimmune diseases. It seems reasonable, therefore, to postulate that any diet and nutritional factors that might delay, decrease, or reverse the rate of immunological decline may also delay or modify the development of infection and age-related diseases. Cellular and molecular analyses have revealed that the function of the immunoregulatory network is intimately linked to other important biological systems, including the brain and endocrine system. Prolongation of immune function by dietary manipulation to prevent neuroendocrine stress might delay the development of diseases associated with severe malnutrition and aging process and could contribute toward a full, healthy, and disease-free lifespan for more individuals. Indeed, although most humans are genetically endowed to live a minimum of 100 years, in many ways the survival rate is decreased, mainly because of acute viral and bacterial infection and early appearance of several age-associated diseases, particularly in individuals with genetic susceptibility.

Indeed, the immune system is fully equipped to recognize and to respond constantly not only to viral and other infectious agents but also to a large number of invading foreign antigens. Several new sources of antigens are regularly infused into the body by absorption of daily intake of a variety of common food products, including various sources of liquids. Further, insults from carcinogens, food preservatives (e.g., nitrate and nitrite), or polycyclic aromatic hydrocarbons from vegetables or fruits may initiate the activation of immune responses. The generation of an immune response involves a series of interactions between lymphocytes and mononuclear cells, which includes cell–cell communications, generation of immunoreactive molecules, mitotic division, immunoglobulin (Ig) synthesis and secretion, and expression of several cell surface markers not found on resting lymphocytes. An effective immune response requires balanced functioning of thymic-dependent T lymphocytes, helper (CD4$^+$) and cytotoxic subsets (CD8$^+$), antibody-producing B lymphocytes (Ig$^+$), macrophages (Mϕ), natural killer cells (NK), and enhancing or inhibiting soluble factors or cytokines (Table I). [*See* IMMUNE SYSTEM.]

Nutritional modulation of immune functions along with regular physical activity (exercise) is currently viewed as important in preventing numerous diseases, particularly in inhibiting the progress of age-related diseases, including certain malignancies, and prevention of infectious diseases. It is well-established that long-term chronic nutrition deficiency, particularly severe protein deficiency along with essential vitamins and trace elements, has been associated with increased susceptibility to infectious diseases in humans. In the past several

TABLE I Major Immune Cells and Cytokines Involved in Immune Function

Cells	Cytokines
Macrophages	IL-1, TNFs
T helper (CD4)	IL-2–IL-8
T suppressor (CD8)	IFNs, CSFs
B cells (Ig$^+$)	
NK cells	

years, the deleterious effects of overnutrition or excess intake of one or more specific dietary factors coupled with obesity and less physical activity has also been noted and linked to the impairment of the immune system as well. Either unbalanced or excess dietary factors may also lead to malabsorption of nutrients and may increase enteric infection by altering the environment of the digestive system, thereby increasing the pathogenicity of resident bacteria and viruses, which could rapidly accentuate sepsis and the disease process.

In recent years, analyses of the impact of nutrients, including rapid progress in enteral and parenteral products, on immunity, infection, and development of age-related diseases occurring in both clinical and experimental systems have generated considerable information on interaction of nutrition and immune response. These investigations have shown that protein deficiency and protein-calorie-malnutrition syndrome in humans is accompanied by a high frequency of infection, profound depression of both T-cell and B-cell immune functions, and malfunction of the complement system and effector processes (e.g., inflammation and phagocytosis).

There are various ways in which nutritional factors can influence host immune response to pathogenic challenge. Nutritional and dietary factors (protein, carbohydrates, fats, minerals, and vitamins) that are acting directly on cells of the immune system may change the functional capacity of the immune cells by altering both intracellular and/or the outer membrane components and their receptors to recognize foreign stimulus and thereby to modify cell proliferative responses to several hormonal and growth factors, in general. Initially, failure of the antigen presentation mechanism of Mϕs or their deficient phagocytic capacity may lead to disruption of the immunological circuits, which are necessary for selective potentiation of either cytotoxic T cells and/or inhibition of excess proliferation of lymphocyte subsets, particularly B cells.

Nutrient availability in general may also have an indirect impact on overall immunological responsiveness. The simultaneous interaction of different well-balanced nutrients is most critical for synthesis of Igs, enzymes, and several types of cytokines that participate closely during the immune response.

Earlier studies on nutrition and immunity focused on protein-energy malnutrition, indicating immunological impairment depends on malnutrition as well as presence of infection. Although a single nutrient imbalance is not very common in humans, particularly in many developed countries in recent years, acute deficiencies of protein, along with certain vitamins and trace metals, are still regularly associated with immunological impairments in many underdeveloped countries. However, excessive calorie and fat intake, particularly both saturated and polyunsaturated fatty acids, are presently suspected to be immunosuppressive. Also, the role of certain trace elements in immune functions is receiving considerable attention, although it is a complex issue because it is linked to the soil conditions from which many grains, fruits, and vegetables are produced. Another aspect of concern is increase in the intake of highly processed foods currently consumed by most of the population by which both useful fiber and vitamins are regularly lost. The present review will focus on the role of dietary proteins, fats, vitamins, minerals, and trace elements on immune functions, particularly in relation to T-cell and B-cell functions, cytokine production, and their association with a few life-shortening disease processes in humans.

II. Proteins and Immunity

Dietary protein has a potent effect on the development of humoral immune responses. It exerts its influence by acting through either humoral or cellular mechanisms and affects synthesis and secretion of Igs. Very low levels of dietary protein has long been known to reduce host resistance to a variety of infectious processes, particularly by lowering both systemic and local immune responses against various pathogenic organisms, including tubercle bacillus, primarily by inducing depression in both Mϕ-mediated and T-cell–mediated immunological response *in vivo*. Low protein diets also markedly depress response to several types of common bacterial infections. However, several experimental studies have revealed that moderate protein depri-

vation enhances cell-mediated immune function, resulting in increased resistance to viral and fungal infections. [*See* PROTEINS (NUTRITION).]

Dietary protein may influence growth of lymphoid tissue throughout the body, and thymic tissue is particularly sensitive to protein deficiency. Spleen and mesenteric lymph nodes are generally affected by protein deficiency. Protein deficiency alters the functional activity of nonmigratory T cells, stem cells, and other cells in the reticuloendothelial system due to the loss of key amino acids and metal ions associated with proteins. Protein-depleted animals have diminished titers of specific antibodies when immunized with specific antigens. Protein deprivation leads to short-lived antibody production, resulting in circulation of high levels of IgM consistent with defective function of the T-helper cell population. The depressed responsiveness to antigens such as sheep red blood cells (SRBC) immunization in protein-deficient mice may be primarily caused by impaired macrophage lineage cell function and deficient levels of protein synthesis. Generally, both T-cell and B-cell populations involved in the production of IgG responses are markedly depleted in protein deficiency.

The major influence of acute dietary protein deficiency on antibody production involves loss of the helper T cell functions. However, moderate dietary protein deficiency leads to enhancement of splenic lymphocyte response to T-cell mitogen and decreased response to B-cell mitogens. Most dietary proteins (animal and vegetable sources) may have a substantial effect on immune functions, depending on the stage of life cycle when the deprivation occurs. A newborn's immature immune system is generally more susceptible to protein deprivation compared to the mature system, primarily because the immature immune system does not possess the specificity of humoral or cell-mediated immunity. Both protein excess (animal source) or total deprivation directly affects the clonal expansion of all immune cell populations or specifically affect the clonal expansion of memory cells necessary for the generation and expansion of other critically interacting cells.

Not much is known about the dietary deficiency or excess of individual amino acids, although tryptophan has been studied to some extent. Tryptophan deficiency is known to result in immunodepression. Combined deficiencies of methionine and cysteine may cause depression of humoral immune function. Specific deficiencies of lysine, histidine, or arginine produce reductions in antibody responses. Diets deficient in branched chain amino acids may also impair host resistance, and dietary deficiency of lipotropic amino acids may result in thymic involution and impaired antibody response. Thus, a well-balanced diet with moderate protein consumption can be effective in maintaining an active immune system throughout the lifespan, although, with age, the requirements for protein intake may alter the balance of the ratio between protein, carbohydrates, and lipids as well as altered calorie intake.

III. Calories and Immunity

Total calorie or energy content of a person's diet may influence physiological processes and immune functions. Both calorie excess as well as deficit may affect immunological functions as well as response to infectious challenge in humans and experimental animals. However, moderate starvation in animals may contribute to increased resistance to viral diseases. Food or calorie restriction (40%, undernutrition without malnutrition) has been known to prevent development of renal disease in experimental animals. The spleen cells of these animals respond better to T-cell mitogens and consistently show an increased capacity for thymidine incorporation into DNA *in vitro,* provided the diet is adequate in vitamins and trace elements. Moderate calorie restriction has also been shown to prevent tumor incidence in mice by producing less antimammary tumor virus antibodies. An imbalance in calorie intake and expenditure in genetically prone obese mice leads to lower spleen and thymic weight, decreased response to antigens, fewer Thy1.2-positive cells, reduced NK activity, elevated thymic hormone activity, and lower cell-mediated cytotoxicity compared with lean controls.

In humans, carefully monitored total fasting as well as self-induced anorexia nervosa have an impact on immunocompetence. However, short-term starvation did not alter IgG class antibodies of serum or circulatory T cells or B cells or monocytes, yet resulted in a decrease in DNA synthesis and delayed hypersensitivity response, which may be corrected by adequate nutrient supplements. In humans, obesity is associated with higher incidence of infectious diseases and early death. The mechanisms leading to compromised immune function in obesity remain obscure but require immediate attention, primarily because of increased prevalence of overweight and obesity, which induces several immunological dysfunctions both in children and adults. Disorders of lipoprotein metabolism have been shown to alter the function of lymphoid cells (e.g., lymphokine production and Class II antigen expression). These deficiencies may impair antigen processing and presentation to T cells. In laboratory animals, restriction of protein alone or total calorie restriction causes a marked depression of certain aspects of B-cell (humoral) immunity. This reduction was in direct proportion to the degree of dietary restriction. In striking contrast, many T-cell immune functions (cellular) have regularly been increased by dietary restriction when instituted soon after weaning age. These elevated T-cell functions include (1) accelerated allograft rejection, (2) increased proliferative responses to mitogens (phytohemagglutinin and concanavalin-A), (3) increased cytotoxic activity against xenogeneic, allogeneic, and syngeneic tumors, (4) increased sensitization to simple protein antigens in delayed allergic reactions, (5) increased resistance to certain viral infections, and (6) increased IL-2 production. In contrast, the same degree of protein or protein-calorie restriction can greatly enhance the susceptibility to bacterial infections (e.g., streptococcal infections). Thus, a paradoxical relation exists between cellular and humoral immunodeficiency observed in protein and protein-calorie–deprived humans and in several animal systems, which is currently pursued by many investigators to understand the role of dietary components in balancing the immune system to perform optimally throughout life.

There is ample evidence describing that T cells, after acquiring self-non–self-discriminatory ability in the thymus, carry on immune surveillance by seeking changes in MHC-coded surface antigens in normal as well as in transformed cells. T cells from malnourished or aging animals and humans respond less vigorously than those from young animals and that the defects are intrinsic to aged T cells. Further alteration in T-cell proliferative response is also related to increased CD8+ and adherent cell population. However, based on several sources of evidence generated by nutritional studies, it is possible that both age and diet can dramatically change the fatty acid composition of thymic tissue and also the membrane composition of cells that emerge from the thymus and lymphoid tissues. Increased levels of polyunsaturated fatty acids in membranes and their susceptibility to form free radicals in the absence of antioxidants in the dietary lipids may cause intrinsic immune dysfunction in all immune cells, particularly during the rapid growth stage or during the aging process.

Indeed, an extensive study on long-lived Fischer-344 male rats to compare the effect of *ad libitum* (AL) and 40% food restriction (FR) revealed striking changes in spleen cell fatty acids and higher membrane fluidity in the FR animals (Fig. 1). These changes appeared to facilitate Ca^{2+} entry into the T cells leading to maintain better immune functions (Fig. 2). As improved cellular immune functions were noted in the past in FR animals, these studies suggest that preventing the decline of Ca^{2+} influx by FR maintains prolonged T-cell function during aging. A similar mechanism was also noted in very short-lived autoimmune-prone [(NZBxNZW)F₁ and MRL/lpr] mice. These mice had significantly longer lifespan with a reduced disease state and maintained an active immune function with a calorie- or food-restricted dietary regimen. Food restriction clearly inhibits renal disease and prevents rise in B cells, thereby producing less autoantibodies, which are highly detrimental and accelerate renal disease and lymphoid cell hyperplasia and/or vasculitis in these animals. Some of the lymphocyte proliferation is closely linked to the production of sex hormones, which are markedly lowered by food restriction and increased by diets containing high levels of lipids. Both systemic lupus erythematosis and rheumatoid arthritis are more common in females and are linked to deregulation of sex hormone levels in females. Moderation in food intake along with low fat diet (source of fat is important) may be immensely valuable in lowering autoimmune disease and maintaining a youthful immune function during the aging process.

IV. Lipids and Immunity

In many Western countries, intake of high-saturated fat has risen significantly after the turn of the century and is strongly linked to increased cardiovascular disease incidence in men and women. Excess dietary lipids along with genetic factors primarily serve as precursors for the production of different series of prostaglandins (PGs) either inflammatory or anti-inflammatory types and may modulate immune functions through the increased production of PGs of two series. Generally, the PGs and/or leukotrienes are a group of biologically ac-

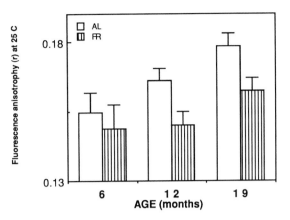

FIGURE 1 Membrane fluidity was measured by steady-state fluorescence depolarization using 1,6-diphenylhexatriene (DPH) probe. Increasing fluorescence anistrophy (r) values were obtained with DPH, which incorporates into the deep hydrophobic core of the membranes. AL, ad libitum; FR, food restriction.

tive molecules derived primarily from oxidative metabolism of polyunsaturated fatty acids that produce lipoxygenase in several forms. Leukotrienes are synthesized from arachidonic acid by lipoxygenase in lymphoid cells (e.g., Mφs, monocytes, neutrophils, and lymphocytes). They possess potent and varied proinflammatory activities and mod-

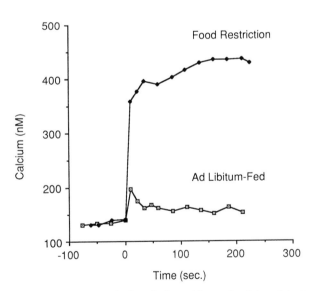

FIGURE 2 Calcium rise in cells from old rats. A calcium-detecting fluorescent probe (Fluo-3) was introduced into immune spleen cells of *ad libitum*–fed or food-restricted rats. At time 0, a suboptimal level of ionomycin was added to Fluo-3–labeled cells to enable calcium to enter the cells, and the rise in their intracellular free calcium concentration was measured by using a FACS IV system.

ulate many physiological and immunological functions in cardiovascular and renal disease as well as other chronic diseases (e.g., atherogenesis, thrombosis, RA, several forms of inflammatory disease, including inflammatory bowel disease, various types of tumors (e.g., Hodgkin's disease, breast cancer, etc.), asthma, psoriasis, septic shock, and Type II diabetes. A few studies have already shown that excess PGE_2 can inhibit T-cell function by down-regulating the IL-2 receptor functions. Other recent studies have also indicated that either restricted calories or fat intake can decrease PGE_2 production and increase functional activity of IL-2 receptors in mice and rats. Dietary lipids also influence Ig levels, cytotoxicity of Mϕs, serum cholesterol, various lipoproteins, and formation of immune complexes. Dietary lipids from saturated, monounsaturated (ω-9), polyunsaturated (ω-6), and marine oils (fish oil, ω-3 series) modify the phospholipid fatty acid profile of the immune cell membranes in mice and thereby alter the fluidity or microviscosity and can thereby modify the functional nature of the immune cells (e.g., antibody synthesis and T-cell–mediated functions, membrane–ligand interactions, permeability, cell-to-cell interaction, and membrane-bound enzyme activities). Membrane fatty acids profile may further affect the availability of eicosanoid precursors to the cyclooxygenase and lipoxygenase systems and thereby can regulate the immune responses differently. Fatty acid profiles and serum lipoproteins, especially low-density lipoproteins, are known to influence receptor binding, signal transduction, and stimulation of lymphocytes to perform their effector functions. [See LIPIDS.]

Generally, excess protein and saturated lipid intake increase serum cholesterol and triglyceride levels and modulate the function of Mϕs much more than other lymphoid cells. Mϕs are major producers of cyclooxygenase and lipoxygenase products and are responsible for PG production, thromboxanes, and leukotrienes. Changes in the production of PGs as well as other cytokines occur primarily because of the levels and source of dietary fat intake, which determine the phospholipid levels in the plasma membranes. The phospholipids in activated lymphocytes are now known to influence the binding of mitogens or various antigens to lymphocyte receptors, which is followed by series of biochemical events (e.g., protein kinase C (PK-C) translocation, inositol phosphate generation, and Ca^{2+} influx), leading to induction of mRNA, protein synthesis,

and cell proliferation. Indeed, recent evidence indicates that proliferation is also closely linked to the activation of several diet lipid-mediated eicosanoids, which are currently implicated in the modulation of IL-1, IL-2, and interferon (IFN) and TNF production by various cells in the body. Both IL-2 concentration and IL-2 receptor density are essential for synchronized proliferation of T cells, which have recently been found to be altered by food intake. Thus, higher levels of dietary fats and excess eicosanoid products may influence cytokines, natural cell-mediated cytotoxicity, IL-2 receptors, and killer T-cell function, which may eventually increase the risk to malignancy (Fig. 3).

A. Lipids, Immunity, and Cancer

The major macronutrient associated with increased cancer risk, particularly colon and breast cancer, is repeatedly alluded to dietary fat, primarily because nearly 40% of total dietary calories in many countries are derived from various sources of fats and oils. The effect of various types of dietary fat on mammary carcinogenesis is an area of intense interest. Several studies have shown that high dietary fat intake, particularly saturated fats and also certain vegetable oils, is related to increased incidences of breast, prostate, and colon cancer though this issue remains very controversial. A number of mechanisms have also been proposed to explain the enhancement of mammary tumorigenesis in experimental animals by feeding diets containing either vegetable ω-6 oils or animal fats. Immune suppression, PG production, free radical formation,

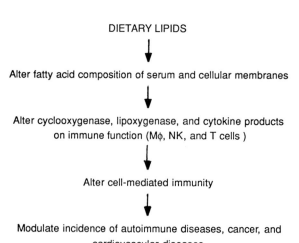

DIETARY LIPIDS

↓

Alter fatty acid composition of serum and cellular membranes

↓

Alter cyclooxygenase, lipoxygenase, and cytokine products on immune function (Mϕ, NK, and T cells)

↓

Alter cell-mediated immunity

↓

Modulate incidence of autoimmune diseases, cancer, and cardiovascular diseases

FIGURE 3 Possible action of dietary lipids on immune function.

membrane fluidity changes, intracellular transport system modulation, increased caloric use, and increased mammotrophic hormone secretion are some of the possible mechanisms through which mammary tumorigenesis may be modulated in animals. Recently the literature on micronutrients and cancer has emphasized the role of dietary micronutrients to serve as key antioxidants in preventing lipid peroxidation and for maintaining adequate immune function by the lymphokine cells. Several studies in mice in the past have repeatedly reported the beneficial effects of calorie/fat restriction including changes on $CD8^+$ T-cell functions and mammary tumor incidence, as well as decreased prolactin and anti-MMTV antibody levels and decreased mammary tumor virus particles in mice with hormone-dependent tumors. However, restriction of total calorie intake of a high-fat diet did not modulate the tumor incidence in mice, indicating high calories derived from high-fat diets may still induce mammary tumors because of the presence of excess immunosuppressive PG products in various cells, which may modify the production and interaction of various growth hormones and lymphokines, resulting in less cytotoxic and/or antiviral action against malignant cells in certain genetically susceptible strains of mice.

However, diets based on fish oil containing high levels of ω-3 fatty acids [e.g., eicosapentanoic acid (EPA, 20:5) and docosahexaenoic acid (DHA, 22:6)] without reduction in food intake appear to have beneficial effects in rodents against several types of cancer compared with vegetable oil diets, which are a rich source of arachidonic acid (20:4), a precursor of PGE_2, and induce tissue inflammation. The ω-3 fatty acids differ from arachidonic acid (ω-6) and inhibit both the cyclooxygenase and lipoxygenase enzyme systems. Diets containing 20% menhaden oil (ω-3) significantly decreased both number of mammary tumors and tumor weight in rats treated with carcinogens. Recent epidemiological studies also favor fish consumption on reduction of breast cancer incidence. Thus, several favorable results induced by increasing the anti-inflammatory ω-3 fatty acids have also been linked to fish consumption, not only on reduced cardiovascular disease but also on many inflammatory autoimmune diseases in humans. Despite a rise in the recognition of favorable health benefits of fish oil, skepticism still prevails because of inconsistencies in clinical or experimental data caused by variation in supplementing adequate levels of antioxidants. A recent study revealed the dramatic effects of menhaden fish oil (ω-3) on reducing the growth of human breast cancer cells in nude mice. Decreased tumor growth was correlated with decreased PGE_2 production and increased NK cell activity in spleen cells and less c-*myc* mRNA expression in tumor cells. Thus, the role of various dietary lipids and their mechanism of action on inducing malignancy especially in animal models has been recently receiving considerable importance. Although model systems in animals are very useful for understanding the pathology of disease, there are profound differences in animals and humans in the response to different kinds of agents.

B. Lipids, Immunity, and Autoimmune Disease

Rheumatoid arthritis affects more than 7 million adults and systemic SLE strikes approximately one in 1,000 individuals of the general population in the United States alone. Patients with these disorders are mostly female. Both are multisystem autoimmune diseases characterized by the production of a variety of autoantibodies and a resulting immune-mediated inflammation. The inflammatory process in SLE may spare the major organs of the body or may involve vital organs such as brain and kidney. There is no known cure for either SLE or RA, but several treatments are directed to relieve symptoms and prevent progression of the disease. However, many of the drugs (e.g., corticosteroid, azathioprine, cyclophosphamide, gold salts) used in treating these diseases have substantial toxicities. Patients with these diseases regularly seek dietary cures as therapies that have no toxicity, but there is no convincing evidence that any specific food or vitamin has any effect on RA or SLE diseases. However, several experimental studies have clearly shown that either dietary calorie or fat restriction has a profound effect on autoimmune SLE in mice. Nonetheless, patients with RA are advised to use nutritionally well-balanced diets and also to exclude certain food items. Stringent caloric restriction after onset of this disease may not have practical therapeutic value in human disease. Although weight reduction in obesity seems to alleviate RA symptoms in many of these patients, no specific dietary factor has yet been described that either improves autoimmune disease or reduces drug dosage and toxicities. [*See* AUTOIMMUNE DISEASE.]

Recent studies have, however, reported significant beneficial effects of calorie-restricted diet and ω-3 fatty acid diet (without calorie reduction) in short-lived autoimmune-prone mice. These mice serve as a suitable model for SLE to study the mechanisms underlying B-cell hyperactivity, and its relation to increased Class II antigen expression or immune response-associated genes (Ia) which is influenced by diet and endocrine hormonal levels.

In the context of an association between Ia antigen expression and autoimmune disease, these mice are particularly interesting because they exhibit an age-related increase in both the proportion and absolute number of Ia$^+$ Mϕs that parallels lymphoid hyperplasia and symptoms of SLE. Recent studies have suggested that both FR and ω-3 diet can decrease Ia antigen expression and B-cell functions. Both these dietary therapies have revealed that only diet high in calories with fish oil can inhibit autoimmune disease, whereas lard and corn oil diets when fed ad libitum accelerated the disease, causing early deaths. In the absence of adequate vitamin E supplementation, fish oil diet, however, decreased NK activity, increased susceptibility to coxsackie virus–induced myocarditis, and increased peroxidation of subcellular membrane lipids. The beneficial effects of fish oil with adequate antioxidant appear to be caused by the following major physiological changes: (1) decreases in plasma estrogen and prolactin levels, (2) prevention of loss in CD8$^+$ cells, (3) preventing rise in Ia$^+$, Ig$^+$, and Ly1/B$^+$ cells (source of autoantibody-producing cells), and (4) decreasing in anti-DNA antibody levels (a cause for renal disease). Further, decreased arachidonic acid (20:4) and increased ω-3 fatty acids (20:5 and 22:6) concentrations in spleen cells of fish oil–fed mice and changes in TNF production were also noted. Clinical studies carried out with fish oil supplement in RA patients as well as in normal subjects revealed changes in PGE$_2$, IL-1, and TNF production in peripheral blood cells. In the case of RA patients, a significant relief in joint pains and swelling by ω-3 fatty acid intake was noted, but the findings were not consistent or dramatic, indicating either more refined ω-3 products or increase in antioxidant supplementation, particularly certain vitamins such as vitamin E, may be necessary for fish oil to act effectively on inflammatory cells to decrease pain and to improve cellular immunity in patients by modulating cytokine and PG production.

V. Nutrition and Cytokines

Recently, cytokine, as a common term, was introduced to a group of protein cell regulators that were previously called lymphokines, monokines, ILs, and IFNs. Produced by several cells in the body, they have a pivotal role in immune response regulation (Table II). This heterogeneous group of well-characterized proteins plays an important role both in many normal physiological responses and also in pathophysiology of several diseases, particularly autoimmune disorders. Cytokines are low-to-medium-molecular-weight (<80 kDs) proteins and interact with specific high-affinity cell surface receptors. Some cells may have multiple cytokine receptors, and/or each cell may produce more than one cytokine. Most cytokines are highly potent and generally act at picomolar concentrations but are produced locally. The role of nutrition, either deficient or excess state, on the production and use of various lymphokines by lymphoid cells is not yet fully understood, mainly because it is a relatively new area and the complexities are too numerous to sort individually, yet some progress is underway. It is now well-established that both lymphokines and monokines influence the behavior of many target cells involved in immunity and inflammation. The balance of immunoregulatory lymphokine-monokine mediators is important for overall immunological function and host defense. [*See* CYTOKINES IN THE IMMUNE RESPONSE.]

Among all lymphokines, IL-2, which is secreted by mitogen-induced T cells that induce proliferation and differentiation of lymphocytes, has been extensively studied. IL-2 stimulates specific immune responses and amplifies certain cell-mediated and humoral immune reactions. The proliferative capacity of target lymphocytes to IL-2 is altered in severe protein deficiency. However, moderate calorie, or FR, and low dietary fat intake decrease IL-1 production and increase IL-2 production as well as its use by other T cells in experimental animals. The mechanism of increased IL-2 production appears to be related to the changes in membrane fatty acid composition and increase in membrane fluidity, which may permit efficient activation of growth factor–associated genes. Other ILs (e.g., IL-3, IL-4, IL-5, IL-6, IL-7, and IL-8) have been characterized, but their role with regard to nutrition and immunity still remains to be elucidated. [*See* INTERLEUKIN-2 AND THE IL-2 RECEPTOR.]

TABLE II Major Cytokines and Their Immune Functions[a,b]

Immune functions	Cytokines						
	TGF-β	IFN-γ	TNF	IL-1	IL-2	IL-4	IL-6
Induce cellular antiviral state		*	*	*			
Mitogenic for various cells	*	*	*	*	*	*	*
Cytostatic for various cells	*	*	*	*		*	
Cytostatic for tumor cells	*	*	*	*			
Activate macrophages	*	*	*	*	*		
Stimulate NK cell activity	+	*	*		*		
Stimulate LAK activity		*	*		*		
Enhance MHC Class I		*	*				
Enhance MHC Class II		*	*			*	
B-cell activation		+				*	
B-cell proliferation	+	*	*	*	*	*	*
B-cell differentiation	+	*	*	*	*	*	*
T-cell activation	+	*		*	*	*	*
T-cell proliferation	+		*	*	*	*	*
T-cell differentiation	+				*		*
Activate endothelial cells		*	*	*			
Induce fever		*	*	*	*		
Induce acute phase proteins		*	*	*		*	
Adjuvanticity		*	*	*	*	*	
Antitumor activity		*	*	*	*		

[a] Summarized from *Immunol. Today* 1989, **10**, 300.
[b] *, stimulates; +, inhibits; TGF, transforming growth factor; IL, interleukins; TNF, tumor necrosis factor; IFN, interferon.

Interferons are proteins that induce new RNA and synthesis of new proteins in cells and inhibit viral replication. IFN-α and IFN-β are antiviral IFNs produced by leukocytes and fibroblasts in response to viral infection and bacterial stimulation. IFN-γ is produced by stimulated T lymphocytes and NK cells in response to alloantigens, mitogens, and soluble antigens. IFN-γ is an immunostimulatory lymphokine that enhances the expression of Class II MHC antigens on cell surfaces, proliferation of T lymphocytes, and metabolic activation of monocytes and Mϕs. IFN titer in marasmic or malnourished children is lower compared with controls, indicating that adequate nutrition is a key factor in its production. [*See* INTERFERONS.]

IL-1 proteins (molecular weight, 13,000–20,000) have multiple effects on lymphocytes. Induction of IL-2 synthesis in induced/helper T-cell subsets leads to IL-2 stimulation of S-phase entry of activated T cells. Altered functional activity and/or depressed lymphocyte response to IL-1 occurs in protein malnutrition. Both IL-1 and PGs produced by Mϕ have important regulatory effects on other cell functions. Excess PGE$_2$ production by polyunsaturated dietary fats inhibits mitogen-stimulated thymidine incorporation in lymphocytes, reduces cytotoxic T-lymphocyte generation, and inhibits NK activity IL-1 and IL-2 production. In some instances, enhanced PGE$_2$ production coupled with reduced production of IL-1 may be responsible for nutritionally induced immunodepression in malnourished children. Generally, febrile and acute phase responses to infection are decreased by several factors, including nutrition, and increasing age and protein malnutrition have specifically been noted frequently as important contributors to this decrease. It is often a diminished acute phase response that increases mortality from infection in elderly people who are undernourished and, particularly, in children in underdeveloped countries. However, carefully planned experimental studies are not being carried out, and uncontrolled, poorly designed studies may often give conflicting results because of the differences in dietary components and existing disease conditions, particularly because of the presence of parasitic infections during the malnourished state.

VI. Vitamins, Trace Elements, and Immunity

A. Vitamins

Most vitamins are enzyme cofactors with specific roles and have been considered to protect tissues and cells against oxidative damage and thus to be protective against premature aging, atherosclerosis, ischemic tissue damage, and prevention of carcinogen toxicity. There are many synthetic and naturally occurring antioxidants in our environment that are known to play a critical role in cellular and plasma components to prevent free radical formation in subcellular membranes and peroxidation of membrane lipids.

1. Vitamin A

Vitamins, especially the antioxidant vitamins, have a favorable influence on cellular and humoral immune functions. Vitamin A and synthetic retinoids and carotenoids appear to have an active role in cancer resistance. Hypovitaminosis-A suppresses phagocytosis, mitogen-induced lymphoproliferation, delayed-type hypersensitivity, and specific antibody responses in mammals. Vitamin A deficiency increases neutrophils and decreases lymphocyte number and changes membrane glycoproteins of lymphocytes that could modify cell activation signaling mechanism. Modification of cellular immune functions by providing adequate vitamin A increases cytotoxic function of immune cells which may be one of the mechanism of action of retinoids in reducing tumors. Retinoids also enhance Mϕ differentiation and their function, which may provide adequate source of IL-1 for T-cell activation to increase the cytotoxic action against tumor cells. [*See* VITAMIN A.]

2. Vitamin E

Vitamin E, vitamin C, and β-carotene are considered as essential antioxidant vitamins. An antioxidant stabilizes highly reactive, potentially harmful, free radicals that are generated either during cellular metabolism or through inhaling or ingesting environmental pollutants during the metabolism of various drugs. The free radicals are known to cause damage to membranes, enzymes, and nuclear mate-

rial of cells. As initiation of immune response is considered to occur at the level of cell membranes, vitamin E plays an important role in preserving the integrity of membranes. Vitamin E deficiency has been associated with increased lipid peroxidation and loss of integrity of membranes. Mixed lymphocyte responses and plaque-forming capacity were found severely depressed in animals fed vitamin E–deficient diets. Furthermore, changes in Mϕ membrane receptor functions, IL-2 production, and PG production were also noted with vitamin E deficiency. A positive correlation exists between splenic cell vitamin E levels and mitogenic activity of splenocytes in mice. Although NK cell ability to lyse tumor cells does not alter with vitamin E status, tumor cells injected in vitamin E–deficient diet–fed mice grew faster. Vitamin E also seems essential for maintaining lymphocyte membrane fluidity, which is essential for cell activation and for mitogenesis. Inflammatory responses were altered in the fluids in the joints of RA patients, indicating vitamin E acts as an anti-inflammatory agent. [*See* VITAMIN E.]

3. Vitamin C

Vitamin C has been found to be a blocking and detoxifying agent essential for neutrophil chemotaxis and has been suggested to have adequate immunostimulatory effects based on animal and human studies. Interestingly, vitamin C may help to maintain vitamin E levels and inhibit free radical reactions. There has been considerable interest in the role of vitamin C in cancer prevention. Vitamin C may be involved in blocking the formation of some carcinogen or tumor promoters, converting them to a less harmful metabolites and enhancing host resistance to progression and dissemination by increasing the functional activity of various immune cells. Leukocytes have high levels of vitamin C, suggesting its important role in maintaining the immunological surveillance mechanism. [*See* ASCORBIC ACID.]

4. β-Carotene

β-Carotene has been known recently to have a role in protection against certain types of human cancers. β-Carotene may be a powerful immunomodulator because of its association with low risk of cancers. The frequency of T lymphocytes bearing the CD4$^+$ or OKT4 marker found on helper/inducer T lymphocytes increase with high levels of oral β-carotene intake in healthy humans. Both T-lymphocyte and Ig$^+$ functions were enhanced by β-carotene supplementation in animals. β-Carotene was found to overcome the inhibition of IFN-induced Mϕ activation by suppressor molecules synthesized by tumors, suggesting β-carotene may be involved in modulating the effects of IFN.

5. B Vitamins

The B complex vitamins may affect immunocompetence, and a severe deficiency of these vitamins may impair immune response in animals and humans. The B complex vitamins are generally involved in cellular metabolism. Both reduced intake or chronic deficiency of B vitamins are associated with alteration of immune functions, and cellular immunity may be severely compromised when there is a decrease in the tissue level of these vitamins. Clinical data indicate that increased thiamine dosage stimulates neutrophil chemotaxis. Furthermore, pyridoxine deficiency has also been associated with decreased size of lymphoid tissues, impaired humoral immune responses, and diminished cell-mediated immune responses. Pyridoxine-free diet in rats markedly decreases the antibody response to *in vitro* immunization. In animals, pyridoxine deficiency is known to depress immune response to tumor cells but does not affect phagocytosis of SRBC by Mϕs or NK cell activities. Rats consuming reduced levels of pyridoxine showed an impaired immune response to infection. Data on animals and humans indicate that high levels of pyridoxine intake did not enhance immunocompetence. Low levels of B_{12} were associated with impaired neutrophil function. *In vitro* studies on normal human lymphocytes indicated that B_{12} was needed for DNA synthesis and B_{12} enhanced antibody function and mitogenic responses. Severely reduced folic acid intake is known to impair the immune response to rotavirus in mice. Clinical studies indicate that adequate folic acid levels are necessary for optimal functioning of the immune system. Biotin deficiency is also associated with abnormal humoral and cell-mediated immune responses. As biotin serves as a cofactor for carboxylases, biotin deficiency may result in multiple carboxylase deficiency and lead to impaired immune responses (e.g., poor lymphocyte-mediated CD8$^+$ activity in response to mitogens). Riboflavin and pantothenic acid deficiencies are reported to alter immunocompetence.

B. Trace Elements

Many of the trace metals are critical for mammalian survival and reproduction, and there is concern that marginal trace element deficiency may be a public health problem. The intake of various essential micronutrients has been repeatedly suggested to decrease infection as well as cancer risk by modifying specific phases of carcinogenesis. Micronutrients may also enhance the functional activities of the immune system and its interacting mechanism of T cells and B cells, Mϕs, and NK cells specifically by enhancing the production of various cytokines to facilitate their phagocytic and cytotoxic action against invading pathogens and/or to destroy emerging premalignant cells in various vital organs. Although trace metal and immunology research is active, still considerable research efforts need to be focused toward understanding solubility properties and bioavailability of micronutrients in different age groups, particularly in the elderly population to prevent occurrence of age-associated immune deficiency, which perhaps could enhance the development of several malignant diseases. [*See* TRACE ELEMENTS (NUTRITION).]

1. Iron

Iron deficiency may have deleterious effects on immune functions. Iron deficiency may have both augmenting and inhibiting effects on host defense mechanisms. Human iron deficiency is associated with frequent infections, and iron administration may reduce the morbidity. Although iron deficiency is associated with immune dysfunction, iron overload may also have immunomodulatory or toxic effects on lymphoid cells.

2. Zinc

Zinc is one of the most widely studied trace elements. It is cofactor for several enzymes. Zinc-deprived animals have impaired immune response to pathogenic challenge. Insufficient zinc intake may cause metabolic and immunological changes in animal species. Patients with low levels of serum zinc had increased susceptibility to a variety of infectious disorders and abnormal immune responses. Immunodeficiency complications of Down's syndrome, obesity, sickle cell disease, infection, burns, etc., are associated with low plasma zinc levels. In mice, even marginal deprivation of zinc, imposed in the early postnatal period, leads to substantial reduction in immunological responsiveness, particularly T-cell–mediated immunity. Zinc deficiency may lead to decreased mitogenic responses to lectins, decreased or delayed hypersensitivity to skin test antigens, decrease in circulating T lymphocytes, thymic involution, depressed plaque-forming cell responses, depressed NK cell activity, hypogammaglobulinemia, and several other health-related abnormalities.

3. Copper

The mechanism by which copper acts in altering immune responses may involve an interaction at the plasma membrane level. Copper deficiency may result in increased T-cell controlled infections, inhibition of T-cell mitogenic responses, decrease in phagocytic cell number, and proinflammatory effects, whereas excess of copper may inhibit complement-mediated hemolysis of sensitized SRBC and thereby can reduce the immunological function *in vivo*.

4. Manganese

The role of manganese in immunological functions is not clear. Rats fed marginally deficient levels of manganese had reduced levels of IgG agglutinins and the 19S fraction of gammaglobulins. Excess of manganese may increase susceptibility to pneumococcal infection, elevated antibody titers, inhibition of chemotaxis, and inhibition of coagulation.

5. Selenium

By 1970, selenium was recognized as nutritionally important when it prevented degeneration of the acinar pancreas in vitamin E–deficient chicks. Selenium has antioxidant functions and may enhance immune responses and may increase resistance to initiation and growth of malignant cells. Selenium is a part of glutathione peroxidase that scavenges free radicals formed during lipid peroxidation and is a catalyst for destruction of hydrogen peroxide. Mice fed selenium-deficient diets for two generations exhibited depressed mitogenesis, decreased thymic size, and depressed specific and nonspecific antibody titers. Selenium may be important as an antioxidant and as a factor in energy metabolism of phagocyte cells, but the mechanism of its modulation of humoral immunity and its role in modulating proinflammatory lipids is not clear.

6. Magnesium

Magnesium deficiency is also related to reduced immune responses caused by thymic hyperplasia.

Some of the trace elements (e.g., zinc, selenium, and magnesium) may interact with one another for maintaining optimum cellular functions. Essential and nonessential mineral elements levels may have a role in neoplasia, and their deficiency as well as excess may lead to malignancy.

VII. Nutrition, Immunity, and Gene Expression

To achieve real progress in the field of nutritional and immunological studies, efforts have begun to elucidate the role of plasma membrane receptor functions, particularly on antigen-presenting cells (self- and non–self-antigens), such as Mϕs, to activate CD4$^+$ T-cell functions. Careful analysis of the relation between several gene functions of major histocompatibility complex (MHC), a system of surface molecules, and occurrence of disease in animal models should provide information regarding the role of nutrition and immunologic abnormalities in the receptor structure in several human autoimmune diseases. During the past decade, several aspects of receptor structure and function have been elucidated. For instance, the murine Class II molecule composing either I-A or I-E regions on the chromosome contains an α and a β chain which are noncovalently associated. A number of mutants of Ia molecules, primarily of I-A molecules, with altered amino acid sequence, have also been characterized. In general, T-cell recognition mechanism is affected by mutation in the α_1 and β_1 extracellular domains, whereas alteration in the cytoplasmic tail can affect levels of expression of a Class II gene product and signal transduction. Studies of naturally occurring Class II variants have shown that just a few amino acid substitutes may be associated with altered immune response and with susceptibility to autoimmune disease. Dietary treatments, both calorie restriction and/or ω-3 feeding, reduce autoimmunity by decreasing the Ia antigen expression. Whether these diet therapies prevent any possible loss in T-cell function on MHC levels has to be investigated to determine the exact role of nutrition in modulating the autoimmune disease process in susceptible animals.

Nevertheless, an indirect clue regarding changes in antigen presentation seen in mice maintained on calorie-restricted diet and immunized with acetylcholine receptor antigen (an autoantigen) demonstrated a marked reduction in T-dependent antigen-specific lymphocyte proliferation to that antigen. Antibody response was also found reduced in these mice when compared with immunized mice fed a normal AL diet. The depressed lymphocyte response seen in calorie-restricted mice was found because of a significant change (reduction) in the Mϕ ability to process and to present antigen as well as reduced T-cell proliferative response. Because FR, on one hand, increases T-cell function and IL-2 production and, on the other hand, is known to decrease Ia expression and antigen presentation by Mϕ, it appears that the marked changes induced by diet on specific receptor expression and functions can, thereby fail to present antigen to T cell.

At present, studies are also underway by many investigators to understand the role of nutrients at molecular level in other tissues such as liver and kidney. Indeed, a major focus of molecular approach is to understand how external growth factor signals are transmitted from the surface of the cell to the nucleus. The changes in the plasma membranes induced by dietary lipids may play a dominant role in changing the membrane-associated proteins and eventually the growth regulation (by altering a normal gene into an oncogene).

Studies undertaken in patients with autoimmune diseases have shown increased expression of several protooncogenes, particularly of c-*myc*, c-*myb*, and c-*ras* in peripheral blood lymphocytes, suggesting that environmental factors, sex hormones, and dietary factors may play a critical role in regulating the expressions of protooncogenes.

Recent studies on autoimmune mice maintained on a calorie-restricted dietary regimen showed decreased protooncogene mRNA expression. These changes appear to be related to the activation of lymphocyte *in vivo*, increased Class II antigen and B-cell numbers, and decreased IL-2 production by T cells both in saturated and polyunsaturated, high-calorie diet–fed animals. In contrast, mice who were food restricted and fed ω-3 lipids showed decreased serum estrogen and prolactin levels and a decreased protooncogene expression and increased IL-2 production, indicating that dietary factors may activate certain genes more than others. Studies in nonautoimmune-prone animals also show a close link between dietary factors involved in gene expression with age in other tissues. For example, FR in aging rats significantly modulated both α_{2u}-globulin mRNA and senescence marker protein 2 (SMP-2) expression, which are regulated by hormones. Indeed, α_{2u}-globulin mRNA in the liver tissue, which declines with age in AL-fed rats, was prevented in FR rats. Similarly, SMP-2 mRNA, which increases with age in male rats, was maintained significantly lower in FR rats. Steady-state expression of cytochrome P-450b mRNA both in AL and FR rat livers was found much higher in FR rats and less in AL-fed aging rats, indicating FR preserved cytochrome P-450 enzyme system much more effectively. The P-450 system is known to play a critical role in metabolism of steroids and drugs as well as detoxification of carcinogens. In the case of β-actin mRNA expression, a rise in AL-fed rats with age was prevented by FR, suggesting that both diet and age may play a critical but, in some instances, opposing role in activating certain genes. For a long time, FR rats and mice have been known to develop fewer numbers of tumors and other diseases. It appears, therefore, that future studies, with respect to nutrition, immunity, and gene expression, should expand considerably, but it should also be targeted to transgenic mice as experimental animal models. Transgenic mice do offer a great opportunity to manipulate them nutritionally by feeding them with known dietary components and study the expression and function of each well-characterized, microinjected gene of interest, particularly cellular and humoral immunity genes of T cells and B cells along with the immune response.

Bibliography

Beisel, W. R. (1982). Single nutrients and immunity. *Am. J. Clin. Nutr.* **35,** 417–468.

Bishop, J. M. (1983). Cellular oncogenes and retroviruses. *Biochemistry* **50,** 301–354.

Chandra, R. K., ed. (1988). "Nutrition and Immunology: Contemporary Issues in Clinical Nutrition." Alan R. Liss, New York.

Faulk, W. P. (1975). Effects of malnutrition on the immune response in humans: A review. *Trop. Dis. Bull.* **72,** 89–103.

Fernandes, G. (1984). Nutritional factors: Modulating effects on immune function and aging. *Pharmacol. Rev.* **36,** 123S–129S.

Fernandes, G. (1987). Influence of nutrition on autoimmune disease. *In* "Aging and the Immune Response. Cellular and Humoral Aspects" (E. A. Goidl, ed.), pp. 225–242. Marcel Dekker, New York.

Fernandes, G. (1989). Effect of dietary fish oil supplement on autoimmune disease: Changes in lymphoid cell subsets, oncogene mRNA expression and neuroendocrine hormones. *In* "Health Effects of Fish and Fish Oils" (R. K. Chandra, ed.), pp. 409–433. Arts Biomedical Publishers, Newfoundland, Canada.

Fernandes, G. (1989). The influence of diet and environment. *Curr. Opinion Immunol.* **2,** 275–281.

Fernandes, G., and Venkatraman, J. (1990). Micronutrients and lipid interactions in cancer. *Ann. N.Y. Acad. Sci.* **587,** 78–91.

Fernandes, G., Venkatraman, J., Khare, A., Horbach, G. J. M. J., and Friedrichs, W. (1990). Modulation of gene expression in autoimmune disease and aging by food restriction and dietary lipids. *Proc. Soc. Exp. Biol. Med.* **193,** 16–22.

Gershwin, M. E., Beach, R. S., and Hurley, L. S., eds. (1985). "Nutrition and Immunity." Academic Press, Orlando, Florida.

Gross, R. L., and Newberne, P. M. (1980). Role of nutrition in immunologic function. *Physiol. Rev.* **60,** 188–302.

Hansen, M. A., Fernandes, G., and Good, R. A. (1982). Nutrition and immunity. *Annu. Rev. Nutr.* **2,** 151–177.

Johnston, P. V. (1985). Dietary fat, eicosanoids, and immunity. *Adv. Lipid Res.* **21,** 103–141.

Mountz, J. D., Steinberg, A. D., Klinman, D. M., and Smith, H. R. (1984). Autoimmunity and increased c-*myb* transcription. *Science* **226,** 1087–1089.

Olson, R. E., ed. (1975). "Protein-Calorie Malnutrition." Academic Press, New York.

Scrimshaw, N. S., Taylor, C. E., and Goodon, J. E. (1988). Interactions of nutrition and infection. *Nutrition* **4,** 13–50.

Suskind, R. R., ed. (1977). "Malnutrition and the Immune Response." Raven Press, New York.

Watson, R. R., and McMurray, D. N. (1979). The effects of malnutrition on secretory and cellular immune processes. *CRC Crit. Rev. Food Sci. Nutr.* **12,** 113–159.

Weindruch, R., and Walford, R. L., eds. (1988). "Retardation of Aging and Disease by Dietary Restriction." Charles C. Thomas, Springfield, Illinois.

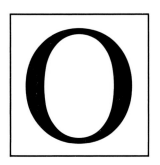

Obesity

GEORGE A. BRAY, *Pennington Biomedical Research Center, Baton Rouge*

Glossary

Bioelectric impedance analysis Method of determining body fat from measurement of impedance

Hypothalamus Region of the brain, located just above the pituitary and below the thalamus, that regulates food intake, body temperature, pituitary function, and other vital functions

^{40}K Radioactive isotope of potassium, which occurs naturally and can be measured to determine the amount of lean tissue, as most of the potassium in the body is in muscle

Norepinephrine Chemical neurotransmitter

Nucleus Group of cells within a region of the brain

Paraventricular nucleus Collection of brain cells located near the ventricle but rostral to the third ventromedial nucleus

Satiety State of feeling satisfied

Serotonin Neurotransmitter also called 5-hydroxytryptamine

Ventromedial nucleus Nucleus in the hypothalamus that is near the ventricle

OBESITY IS AN EXCESS of body fat. Obesity should be distinguished from overweight, which is an excess of weight relative to some standard for height. Overweight and obesity are frequently used interchangeably. When overweight is sufficiently large, it almost certainly implies obesity. Near the upper end of the normal range of body weights, however, some individuals may be obese but not overweight, whereas others are overweight but not obese. For this reason, measurements of total body fat and its distribution are important.

Once obesity and overweight are defined, the prevalence of obesity and overweight can be determined. Underlying the development of obesity is an imbalance between food intake and energy expenditure, which results in expansion of the stores of fat in adipose tissue. These concepts will be dealt with next. Then the types of obesity and the attendant risks associated with obesity will be reviewed. Finally, treatments for obesity will be briefly discussed. For more details on any aspect of this major public health problem, the reader is referred to several recent monographs.

I. Measurement of Body Fat

Accurate measurement of body fat requires sophisticated techniques. These techniques include measurements of body density determined by weighing an individual out of water and after a submersion with correction for the air in the lungs. Fat can also be determined by various isotopic and chemical means, which measure total body fat or water content from which fat is calculated. Measurement of the naturally occurring isotope of potassium (^{40}K) may be used for accessing lean body tissue, as most potassium is in muscle. Fat is determined by subtracting lean body mass from total body weight. Measurement of impedance (bioelectric impedance analysis [BIA]) has recently been introduced and may provide accurate measurements of body fat on most adults. The measurement of height, weight, skinfold thickness, and circumference can also be used to assess body composition and the distribution of body fat.

The term overweight can be expressed in several ways, including relative weight, which is the weight of an individual in relation to their height compared to the median weight for individuals of the same height. It can also be expressed in terms of various ratios of body weight and height. Based on the lowest rates of associated health risks, Tables I and II provide good body weights for most adult men and women according to age.

In addition to total body fat, measuring regional distribution of fat is also important. This can be done by determining the circumference of the waist relative to the circumference of the hips. When this ratio (waist circumference divided by hip circumference) is higher than 1.0 for men or higher than 0.85 for women, fat distribution is unhealthy. Fat distribution can also be estimated from thickness of fat skinfolds measured alone or as the ratio between the trunk and limbs. Finally, intrabdominal fat can be measured using sophisticated scanning techniques known as computed tomography or magnetic resonance imaging. Depending on the definition chosen, various estimates of overweight or obesity can be derived. The National Center for Health Statistics uses the 85th percentile of weight for height of men and women aged 20–29 yr to define the upper limit of normality. Using this criterion, the prevalence for American men and women are shown in Table III. Using a measurement of weight in relation to height known as the body mass index, overweight in North America is more common than in Europe or Australia.

TABLE II Healthy Weights for Men

Height[a]	Age (yr) 19–24	Age (yr) 25–65
60 (5'0")	91–115	96–119
61 (5'1")	94–119	99–124
62 (5'2")	97–123	102–128
63 (5'3")	100–127	106–132
64 (5'4")	104–131	109–136
65 (5'5")	107–135	113–141
66 (5'6")	110–140	116–145
67 (5'7")	114–144	120–150
68 (5'8")	117–148	124–155
69 (5'9")	121–153	127–159
70 (5'10")	125–158	131–164
71 (5'11")	128–162	135–169
72 (6'0")	132–167	139–174
73 (6'1")	136–172	143–179
74 (6'2")	140–177	147–184
75 (6'3")	144–182	151–189
76 (6'4")	148–186	155–194

[a] Inches.

II. Pathogenesis of Obesity

Body fat is stored primarily in adipose tissue. The maintenance of normal body fat stores in relation to total body weight, like many other systems, is regulated. The average adult male eats nearly 1 million kcal/yr and expends essentially the same amount. An error of 1.0% (10,000 kcal/yr more intake than

TABLE I Healthy Weights for Women

Height[b]	Age (range)[a] 20 (19–24)	30 (25–34)	40 (35–44)	50 (45–54)	60 (55–64)	65+
58 (4'10")	91–115	96–119	100–124	105–129	110–134	115–138
59 (4'11")	94–119	99–124	104–128	109–133	114–138	119–143
60 (5'0")	97–123	102–128	107–133	112–138	118–143	123–148
61 (5'1")	100–127	106–132	111–137	116–143	122–148	127–153
62 (5'2")	104–131	109–136	115–142	120–147	126–153	131–158
63 (5'3")	107–135	113–141	118–146	124–152	130–158	135–163
64 (5'4")	110–140	116–145	122–151	128–157	134–163	140–169
65 (5'5")	114–144	120–150	126–156	132–162	138–168	144–174
66 (5'6")	117–148	124–155	130–161	136–167	142–173	148–179
67 (5'7")	121–153	127–159	134–166	140–172	146–178	153–185
68 (5'8")	125–158	131–164	138–171	144–177	151–184	158–190
69 (5'9")	128–162	135–169	142–176	149–182	155–189	162–196
70 (5'10")	132–167	139–174	146–181	153–189	160–195	167–202
71 (5'11")	136–172	143–179	150–186	157–193	166–200	172–208
72 (6'0")	140–177	147–184	154–191	162–199	169–206	177–213

[a] Years.
[b] Inches.

TABLE III Percentage of Overweight and Obese Persons in Several Affluent Countries

	Age (yr)	% Overweight		% Obese	
		Men	Women	Men	Women
United States	20–74	31	24	12	12
Canada	20–69	40	28	9	12
Great Britain	16–65	34	24	6	8
The Netherlands	20+	34	24	4	6
Australia	25–64	34	24	7	7

expenditure) would increase the body weight by 1 kg every year or 10 kg (22 lb.) per decade. Because this obviously doesn't happen to most people, the regulation over time for most individuals must be quite good. This regulatory system is depicted schematically in Figure 1.

Food intake is initiated by a variety of factors including social environment, availability of food,

Control

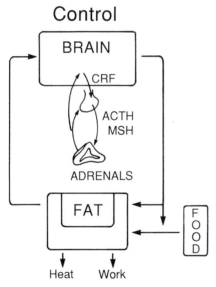

FIGURE 1 Diagram depicting a feedback or regulatory system for regulation of food intake. Food enters the system on the right. Signals from the ingestion of food can act to reduce future short-term food intake by one of several mechanisms, including hormones, nutrients, or neural signals. Information about the status of energy in the body and food intake is integrated in the brain, where signals are generated that lead to food-seeking and ingestion or terminate meals. Other signals activate various systems for storage or release of nutrients from fat tissue. The pituitary gland and adrenal are depicted through their connections with the brain. This is included because development of obesity appears to depend on the presence of adrenal steroids to a much greater extent than any other endocrine system. ACTH, adrenocorticotropin; CRF, corticotropin-releasing factor; MSH, melanocyte-stimulating hormone.

and internal hunger drives, which lead to food seeking. Once initiated, intake of a single food item occurs at a decelerating rate; however, if a second food item is provided, there is usually "room" for it. This phenomenon of sensory-specific satiety related to different foods is an important item in short-term regulation. The decelerating rate of food intake implies an initial stimulation to food intake produced by the smell, taste, and physiologic effects of food followed by inhibitory events collectively called satiety, meaning the events associated with termination of a meal. [*See* SATIATING POWER OF FOODS.]

Inhibition of food intake may be produced by the nutrients themselves as well as hormones released when food enters the gastrointestinal tract. Of these hormones, cholecystokinin has received the most study, but several other important candidates exist. Nutrients themselves may directly initiate feedback signals after transport into the brain. Alternatively, nutrients may trigger signals transmitted over the vagus or sympathetic nerves as the nutrients are absorbed from the gastrointestinal tract or pass through the liver. For example, as glucose is absorbed and passes through the liver the activity of the vagus nerve slows, and the signal is transmitted to the brain.

Several centers in the brain are involved in the feeding process. Afferent messages from the vagus nerve enter the vagal center (nucleus of the tractus solitarius) and are relayed from there to the hypothalamus. The hypothalamic coordinating centers also receive information from other parts of the brain that monitor sight, smell, and taste of food. Internal metabolic signals such as low glucose or high insulin levels may also act directly on areas in the medial or lateral hypothalamus. Destruction of the medial hypothalamus (specifically the paraventricular and ventromedial nucleus) is associated with an increase in food intake and obesity in a wide variety of mammalian species. Alternatively, damage to the more lateral hypothalamic area is associated with reduced food intake and weight loss in these same species. [*See* HYPOTHALAMUS.]

The central integration of messages about feeding involves two primary neurotransmitters (norepinephrine and serotonin), whose affects may be modulated by a variety of small peptides released from nerve endings. In the medial hypothalamus, norepinephrine stimulates feeding, whereas in the more lateral region it can inhibit feeding. Conversely, serotonin (5-hydroxytryptamine) in the medial hypothalamus lowers food intake. Several peptides including the opioids, neuropeptide Y, and gallanin are known to stimulate food intake when injected into appropriate brain areas. Several other peptide hormones are known to depress food intake. These inhibitory and stimulatory systems operate on the motor control of food-seeking and food intake and indirectly on the metabolic processing of food.

The ingestion of food can activate the sympathetic nervous outflow system as it leaves the brain. This alteration in sympathetic activities associated with food intake may also modulate the release of insulin from the pancreas. Insulin is a key circulating hormone involved with storage of nutrients from the diet. Insulin receptors on liver, muscle, and adipose tissue are important in facilitating the response to insulin expressed as an increased storage of fat, glycogen, and protein. [See INSULIN AND GLUCAGON.]

Human beings can store fat by increasing the size of individual cells or by increasing the number of cells. In individuals who are modestly overweight, most of the increase in fat stores occurs by increasing the size of preexisting fat cells. On the other hand, individuals who are markedly overweight almost always have an increase in the total number of fat cells as well as enlargement of the individual cells. With weight loss, fat cells shrink in size but, in most instances, the number of fat cells does not decrease. Thus, individuals who become markedly overweight in childhood with an increase in total number of fat cells find it difficult to lose weight in adult life.

III. Types of Obesity

As noted above, the number of fat cells is one basis on which types of obesity can be classified. Some individuals are fat and have a normal number of cells, which are enlarged (hypertrophic obesity), whereas others have both an increased size and an increase in the number of fat cells (hyperplastic, hypertrophic obesity). Alternatively, the distribution of fat can be used to define individuals as either upper body, or android, obesity (malelike) versus lower body, or gynoid, obesity (femalelike).

The age at which obesity begins is a third way in which individuals can be classified. For some individuals obesity begins in childhood, and for others in adult life.

Several defined causes of obesity have been identified. Both single and polygenic inheritance is involved in the transmission of obesity in human beings. In rare circumstances, specific genetic diseases are associated with obesity. In most individuals, however, an underlying genetic predisposition is essential for development of obesity. Within the same family, some individuals will become fat and others will not. Environmental and nonhereditary factors account for about one-half to two-thirds of this difference in individuals within families, whereas genetic factors account for the remaining 30–50%. The best estimates suggest that genetic factors may be of equal importance to environmental ones in the overall determination of body fat and that genetic factors may be more important than environmental ones in determining fat distribution.

Obesity can be caused by a variety of neuroendocrine diseases. Damage to the medial hypothalamus, as noted earlier, can cause obesity in humans as well as most other mammals. An ovarian disease known as polycystic ovaries is frequently associated with increased body fat. Likewise, hypersecretion of steroids from the adrenal glands can produce obesity. Increased insulin secretion or administration of excess insulin, which may occur during treatment of diabetes, may also lead to obesity.

Diet also plays an important role in the development of obesity. A high-fat diet in animals and probably humans enhances the likelihood of becoming obese. Certain drugs taken for other diseases may increase body fat (e.g., tricyclic antidepressants, glucocorticosteroids).

Finally, social, economic and psychological factors play a role in the development of obesity. In the United States, obesity is more prevalent in the lower social-economic groups than in the higher ones.

IV. Risks of Obesity

Regardless of its cause, obesity may be associated with a variety of risks. Life insurance data and studies of large populations have shown that overweight and obesity are associated with increased risks of premature death. In addition to the direct effects on mortality, obesity increases the risk of a variety of diseases. Several studies suggest that obesity may be a determinant for the risk of developing heart attacks and other cardiovascular diseases. Obesity is also a primary risk factor for the development of noninsulin-dependent diabetes (type 2 diabetes). Risks of gallbladder disease and some types of cancer are also increased in the overweight. In addition to these significant sources of increased risks, evidence indicates that obesity influences a variety of bodily functions including heart, lung, and metabolic systems.

Increased abdominal or truncal, or upper body, fat, like total body fat, also increases the risks of mortality, heart disease, stroke, and diabetes. This increased risk from abdominal fat can occur even in individuals who are essentially of normal weight.

V. Treatments

A variety of treatments have been used for obesity. At one extreme are the high-risk treatments such as surgical intervention with gastric or intestinal operations; at the other extreme are diets, behavior modification, and minor changes in exercise patterns. Except for the individuals who are massively obese, surgery is inappropriate therapy. For most individuals wishing to lose weight, behavioral modification including changes in eating behaviors as well as changes in exercise behavior should be the centerpiece. To the extent that changes in both eating and exercise patterns represent long-term adaptations in life-style, changes in body weight are a realistic expectation. For individuals who want rapid changes in weight loss without changing the behaviors associated with maladaptive eating and exercise patterns, the prognosis for maintenance of weight loss is low.

Bibliography

Bray, G. A. (1989). "Obesity: Basic Aspects and Clinical Applications." *Med. Clin. N.A.* **73,** 1–269.
Bray, G. A. (1989). Overweight and fat distribution—Basic consideration and clinical approaches. *Disease-A-Month.* **35,** 451–537.
Bray. G.A. (1986). Effects of obesity on health and happiness. *In* "Handbook of Eating Disorders" (K. Brownell and J. Foreyt, eds.) pp. 3–44. Basic Books, New York.
Bray, G. A., and Greenway, F. L. (1988). Obesity: Future directions for research. *In* "Fat Distribution and Metabolic Risk Factors during Growth and Later Health Outcomes" (C. Bouchard and F. Johnston, eds.) pp. 333–350. Alan R. Liss, New York.
Garrow, J. S. (1988). "Obesity and Related Diseases." Churchill Livingstone, Edinburgh and London. pp. 1–329.
Gray, D. S., and Bray, G. A. (1988). Evaluation of the obese patient. *In* "Handbook of Eating Disorders, Part 2: Obesity" (G. D. Burrows, P. J. V. Beumont, and R. C. Casper, eds.) pp. 47–59. Amsterdam, Elsevier.

Olfactory Information Processing

DETLEV SCHILD, *University of Göttingen*

I. Anatomy
II. Physiology
III. Behavior
IV. Olfactory Coding

Glossary

Action potential = spike Electrical pulse generated by a nerve cell and traveling on nerve fibers
Ethmoid bone Section of the skull base above the nasal cavity separating it from the forebrain
Stereotype behavior Behavior that is triggered by a particular stimulus and rigidly performed in a schematic way (i.e., always the same sequence of actions)
Sustentecular cell Nonneuronal cell located between neurons and in functional contact with neighboring neurons
Turbinate Folds of mucous membrane in the nasal cavity holding the olfactory mucosa

ALL LIVING CELLS are able to recognize chemical cues from their environment. In this general sense all cells are chemosensitive. Protozoa are attracted by certain chemical sitmuli, lymphocytes react and respond to an enormous variety of antigens, the growth and metabolism of most cells of an organism is influenced or controlled by hormones, nerve cells are stimulated by neurotransmitters, specialized cells that are involved in the control of cardiovascular function can measure the plasma partial pressures of O_2 and CO_2, and finally, the senses of taste and smell endow animals with the capability of sampling and reacting to chemical stimuli they are exposed to.

Chemoreception of protozoa and lymphocytes, the action of hormones and neurotransmitters, the influence of O_2 and CO_2 receptors in cardiovascular control, and the taste system are dealt with in other chapters of this encyclopedia. In this chapter, the processing of olfactory chemosensory information is described: the action of odors on olfactory receptor cells, the subsequent processing of this information by higher brain regions, and its impact on behavior. The basic principles of olfactory coding as compared with the coding in other sensory systems is analyzed in a separate section.

I. Anatomy

A. Receptor Cells in the Olfactory Mucosa

In vertebrates and humans, the olfactory receptor cells are situated in the epithelium in the nasal cavity, each receptor cell tightly surrounded by sustentacular cells. In humans, this epithelium, the olfactory mucosa, is a few square centimeters large. It is located in the upper part of the superior turbinate, the nasal septum, and the roof between. It is composed of three major cell types: olfactory receptor neurons, sustentacular cells, and basal cells. Olfactory receptor cells have two unique properties: first, they are probably the only neurons that communicate directly with the environment of the organism, and second, they have the capability of regeneration throughout life. Basal cells are progenitor cells of olfactory neurons; they divide, differentiate, and replace dying receptor neurons. The sustentacular cells separate olfactory receptor neurons; they have supporting, glia-like, and secretory functions. The lamina propria underlying these cells contains the axons of the olfactory neurons and secretory glands. Secretion of these glands together with secretion of sustentacular cells provide the mucus, which forms a 20–30-μm-thick layer above the olfactory tissue. Environment and mucus are separated from the extracellular compartment by a net of tight-junctions, which surround all mucosal cells at their apical part.

Olfactory receptor cells are bipolar primary neurons with one long dendrite from the top of which, the so-called olfactory knob, several cilia issue. The cilia are relatively long (up to 150 μm) as compared with the thickness of the mucus layer (i.e., they extend mainly horizontally as a dense long-haired carpet in the mucus layer). The cell body (soma) of an olfactory neuron is usually fairly small; it has an oval form with diameters of about 4 μm and 7 F μm. As primary neurons the cells have an axon attached to the soma. The unmyelinated axons form bundles ensheathed by Schwann cells. All bundles together correspond to the first cranial nerve (olfactory nerve), which passes through small openings of the cribriform plate of the ethmoid bone into the brain. Here the axon terminals make synaptic contact in the glomerula of the olfactory bulb. The nerve passage through the cribriform plate is vulnerable: head traumata often lead to injuries of the olfactory nerve with subsequent complete or partial anosmia.

B. Olfactory Bulb

The olfactory bulb is the second stage in the olfactory pathway. In many animal species, it is also the last stage where the entire olfactory information is processed and from where signals diverge to many other parts of the brain. Unfortunately, little is known about the human olfactory bulb; the anatomical and morphological similarities found in other higher vertebrates (e.g., rat, mouse, and rabbit) suggest, however, that the knowledge we have from these species can be applied to a large extent to humans.

The olfactory bulb in higher vertebrates is a neatly stratified structure with distinct layers: the olfactory nerve layer (ONL), the glomerular layer (GL), the external plexiform layer (EPL), the mitral cell layer (MCL), the internal plexiform layer (IPL), and the granule cell layer (GCL). The outermost layer is the ONL, where bundles of receptor cell axons distribute on the way to the glomerula. The next layer is the GL. Glomerula are spherically shaped morphological entities in which the primary receptor fibers make synaptic contact with the secondary neurons, the mitral cells (MC). MCs and many tufted cells are the only cells that send their output to higher brain structures. There is as yet no explanation for the fact that in most vertebrates the synaptic contacts between primary fibers and MCs

form glomerula. However, in some lower vertebrates (e.g., goldfish), there are no typical glomerula, and it seems that glomerula formation begins in synchrony with the appearance of interneurons, which mediate information between glomerula. These interneurons, the periglomerular cells, are excited by primary fibers and act predominantly on MC dendrites, whereby this action is suppressive (inhibitory) in most of the periglomerular cells and excitatory in only about 20% of the periglomerular cells. The information processing of periglomerular cells appears thus to be an important and intermediate step in the information channel receptor neurons → MCs, and it seems to require the transition of a homogeneous, plexus-like GL into separate glomerula.

Mitral cells in higher vertebrates form a narrow and sharp layer, the MCL. These cells have several dendrites, only one of which, the long primary dendrite, reaches one glomerulum. The other secondary dendrites are oriented mainly orthogonally with respect to the primary dendrite. It is important to note the number of receptor neurons (N_r), glomerula (N_g), and MCs (N_m). Although these numbers differ from species to species, there are fairly constant ratios. The relation $N_r \gg N_m > N_g$ seems to hold in all species, whereby the ratio $N_r : N_g : N_m$ is approximately given by $(F \cdot 1{,}000) : 1 : F$, with F being in the range between 10 and 50.

Along the length of the primary MC dendrites, there is a layer between the GL and the MCL. In this layer, the EPL, information is conveyed vertically in the primary MC dendrite, from a glomerulum to a MC soma. The second type of relay neurons of the olfactory bulb, the tufted cells, are distributed in the EPL. Most of the rest of the information transfer in this layer consists in local lateral interactions between secondary dendrites of different MCs, mainly through granule cell dendrites but also through other cell types such as short axon cells.

Following the EPL and the MCL, there is the IPL, consisting mainly of numerous fibers, particularly axon collaterals of mitral cells. The latter can also be observed in the MCL as well as in the following layer, the GCL, where the granule cells are distributed. These GABA-ergic interneurons make reciprocal synapses on MC dendrites, thereby mediating self-inhibition and lateral inhibition between MCs.

The olfactory bulb also receives centrifugal afferent input from several brain regions. The innervation pattern of these centrifugal fibers within the bulb is correlated within their origin. The most conspicuous sites of origin are the ipsilateral olfactory cortex, ipsilateral horizontal limb of the diagonal band, locus ceruleus, raphe nucleus, and the contralateral anterior olfactory nucleus.

C. Central Structures

The output fibers of the olfactory bulb together with the centrifugal afferent fibers, which reach the olfactory bulb from many parts of the brain, form the olfactory tract. The centripetal fibers make direct contact in the anterior olfactory nucleus, the piriform cortex, the olfactory tubercle, parts of the amygdala, and the lateral entorhinal cortex. These areas, which are considered as the primary olfactory cortex, are interconnected by numerous associational fibers. Projections from the primary olfactory cortical areas run to the thalamus, the neocortex, the hypothalamus, the hippocampus, and other nuclei of the amygdala. The thalamus probably does not have the relay function it has in other sensory systems because the direct projections from the primary olfactory cortex to the neocortex show a considerably higher density than the transthalamic pathway. The fact that only a few cells from many primary olfactory cortical areas project to the thalamus indicates further that this pathway presumably does not reflect a typical sensory information transfer through the thalamus. The involvement of thalamic cells might therefore rather serve to compare olfactory information with neocortical activity. The projection to the hypothalamus is probably concerned with odor-induced intergrative control of autonomic functions; especially, the fibers from the anterior cortical nucleus, which (in the rat) project to the preoptic/anterior hypothalamic area as well as to the medial hypothalamus, might provide a direct influence on reproductive and endocrine functions. [*See* HIPPOCAMPAL FORMATION; HYPOTHALAMUS; NEOCORTEX; THALAMUS.]

II. Physiology

A. Receptor Neurons

Electrophysiological studies on single olfactory receptor cells have proven to be extremely difficult because of their small size. Almost everything we know about vertebrate olfactory receptor cells is therefore based on recordings from amphibia species, which have relatively large olfactory receptor cells.

Odorants, which are always air-borne in higher vertebrates, enter the nose through the nostrils and reach the mucus of the olfactory mucosa. The odorous molecules enter the mucus according to the partition coefficient of the molecule type for the air–mucus interface. Once in the mucus, there is a certain probability for the odorous molecules to hit a molecular receptor molecule on a cilium of an olfactory neuron and to bind to it.

The mucus is an osmotic barrier between air and the nerve cell processes; in addition, there seem to be proteins in the mucus that bind odorant molecules and thereby enhance the probability of detection. Such odor-binding proteins would also transport the odor molecules in the mucus flow away from their site of action.

The activity characteristics of olfactory neurons can be summarized as follows:

1. The neurons show a low spontaneous discharge rate that is well below one action potential (or spike) per second. This feature is advantageous because a relatively small change in firing rate indicates a significant response.
2. When maximally stimulated with natural stimuli, the neurons respond with 20–30 spikes/s.
3. When the intervals between spikes are analyzed, they show an interval distribution that suggests that the spikes are distributed randomly in time (i.e., only the mean firing rate of a response is characteristic for the response).
4. Responses to odors show a characteristic and well-reproducible pattern: the spike rate increases within a few seconds to a maximum and then decreases to a fairly constant niveau (phasic-tonic response).
5. A typical vertebrate olfactory receptor neuron responds to a large number of stimuli, whereby every stimulus brings about a different response spike rate. Another olfactory receptor neuron would also respond to a large number of stimuli, maybe partly overlapping with those of the first-mentioned cell, but a particular stimulus would usually cause different responses in the two cells. From the experiments up to now, it is difficult to establish whether there are receptor neurons that respond identically to any sequence of stimuli. Such cells would form receptor classes (comparable, with those responsive, e.g., for blue light in vision), but unfortunately we presently neither know whether receptor classes exist nor exactly what they could be responsive to. Too few of the several million olfactory neurons per mucosa have been recorded in every single animal, and too few

out of the thousands of possible stimuli have been applied to every studied cell.

The events occurring between the binding of stimulus molecules to the cilia and the generation of action potentials (i.e., the details of stimulus transduction) are only partially known. Binding of a stimulus molecule to a ciliar receptor protein seems to trigger one or more intracellular enzyme cascades, which eventually lead to a net inward current, the so-called generator current, into the cilia. There is evidence that various G proteins and cyclo-adenosine monophosphate (cAMP), cyclo-guanosine monophosphate (cGMP), as well as inositoletrisphosphate (IP_3), are involved in the transduction processes. The precise roles of these second-messenger pathways remain to be established in detail, however. The generator current loads the cell's capacitance, thereby depolarizing (i.e., making it less negative inside) the membrane potential to the cell's firing threshold. It has been shown that olfactory receptor cells are extremely sensitive, responding with spikes to current injections of only a few (<10 pA) picoamperes. The generator current is modulated and processed by the voltage-gated channels of the olfactory receptor neurons. There seem to be two voltage-gated inward currents and three different outward currents in all studied species: (I) a fast inward current carried by Na^+ ions and blockable by tetrodotoxin (TTX): this current is responsible for the initiation of action potentials; (2) a small Ca^{2+} current that partially inactivates (i.e., its conductance decreases with time); (3) a Ca^{2+}-dependent outward current carried by K^+ ions; (4) a fast K^+ current that inactivates within a second; and (5) a third K^+ current that inactivates slowly in the range of tens of seconds. The physiological role of the outward currents are partially to return the membrane potential to or below the original level (repolarization and hyperpolarization), but also the adaptive regulation of the cell's sensitivity as a function of its activity state and recent activity history.

Future understanding of the olfactory system will crucially depend on methodological progress, allowing the study of the stimulus selectivity of many cells in the same mucosa.

B. Olfactory Bulb

Four different approaches have until now been undertaken to investigate the physiological behavior of olfactory bulb (OB) neurons on natural stimula-

tion of the olfactory system: (1) extracellular single- or double-unit recording from OB cells, (2) the distribution of 2-deoxyglucose in OB cells after exposure to stimuli, (3) the activity pattern recorded by many EEG electrodes, and (4) optical recording with video or photodiode techniques.

1. Extracellular recordings from OB neurons have mostly been obtained from MC/tufted cells and more rarely from granule cells. Although MC/tufted cells fire regular action potentials, granule cells generate potentials of longer duration and varying amplitude. In lower vertebrates, the temporal response pattern of MCs on stimulation is of the phasic-tonic type and seems to reflect the response behavior of the receptor neurons. In higher vertebrates, the temporal response patterns appear to be more complex and additionally correlated with the respiratory rhythm. A typical MC responds to more or less one-third of all stimuli, which means that a large number of MCs are usually involved in the coding of every stimulus. Different MCs respond in a different manner to the same stimulus. Recordings with two electrodes from two MCs have shown that the cell responses are positively correlated if one cell is in the neighborhood of the other. In a certain range beyond this neighborhood, the correlation is negative. Adjacent MCs or MCs within a certain neighborhood seem thus to be responsive in a similar way to a given stimulus, whereas an MC that is excited by a certain stimulus suppresses other MCs that are located in a ring around it (lateral inhibition). These phenomena can be explained (1) by the fact that adjacent MCs get similar inputs from glomerula and (2) by the lateral inhibition mediated mainly by granule cells.

For the investigation of such lateral interactions between OB cells, intracellular recordings are particularly appropriate and have in fact revealed details such as time constants of excitation and inhibition of MC/tufted cells. However, the fact that a large number of MC/tufted cells are involved in the coding of most stimuli emphasizes also the main drawback of single-unit recordings in the OB: as coding is done by sets of MC activities, little can be concluded from single-cell recordings. Obviously, to achieve this goal, a recording technique with a sufficiently high spatial resolution is necessary.

2. 2-deoxyglucose (DOG) is taken up as glucose by active nerve cells but not metabolically degraded. Labeled 2-deoxyglucose (DOG) is thus stored only in metabolically active cells. Stimulation of the olfactory system with a certain odor

results in a DOG uptake in the active cells, which can be determined in slices and histological determination, determining which regions of the OB were involved in the stimulus response. As a sufficient DOG uptake takes a much longer time than one short stimulus presentation, stimuli must be applied repeatedly. The DOG staining pattern corresponds then to the average activity pattern over all stimulus applications. The stained areas do not exclusively indicate MCs but all kinds of cells including glia. With this method, a number of important findings evolved: Almost every odor is coded at the same time at many areas of the OB. Activation patterns can be altered experimentally by a preceding stimulus deprivation. In any case, the staining pattern for two different odorants usually overlap considerably but are well distinguishable. Odors similar in concentration are mapped to similar DOG activation patterns. In particular, low odorant concentrations lead to only a few stained foci corresponding to single glomerula or small groups of glomerula. When higher or saturation concentrations were used, the densely stained foci were spread over large areas of the GL. The density of labeling within every stained glomerulum tends to be uniform so that the glomerula are currently thought of as functional units. The spatial resolution of the method does not reach the single-cell level so that no cellular activation patterns can be assessed.

3. Experiments have been performed in which 64 electroencephalogram (EEG) electrodes have been placed on the surface of the OB of the rabbit to record change of electrical potential at the surface. Contrary to the DOG method, the time resolution of this method is excellent but the spatial resolution is poorer. Nonetheless, some more interesting features of the signal processing in the OB, which are different from those formerly described, could be found in this way.

Every EEG electrode signal is oscillatory. The oscillation frequency is not characteristic for any odor, whereas the amplitudes are characteristically influenced by stimulus application. These amplitudes are therefore observables of the OB mass activity. They can be plotted in the conventional way as time functions; what is more instructive is to plot them in state space. As a simple example, take the three-dimensional state space: three appropriate EEG signals are chosen as the x, y, and z axis. Then at a particular time these three signals correspond to a point in this state space. If the points for all recorded time samples are connected, a characteristic curve results. This can be a simple point (fixed point), a closed curve standing for a periodic oscillation, or a chaotic attractor (i.e., a bunch of nonperiodic trajectories in state space that form a certain structure with nonvanishing volume to which all trajectories converge). The EEG approach has shown that all three classes of OB activity states exist and that they can change from one into another according to the consciousness states of the animal. Furthermore, different odors are clearly characterized by different OB activity states. One of the nice points of this approach is that time does not play the role of a variable; it is simply a parameter, and the activity state as a whole is time-independent, corresponding to the time-independent characteristics of the applied odor.

4. Recently it has become feasible to stain nervous tissue with voltage-sensitive dyes which, in a way not completely understood, reflect the electrical potential of the plasma membrane by absorbing transmitted light or emitting fluorescent light in a manner that is proportional to the potential across the membrane. These optical signals can be detected by a video camera or by an array of presently up to 500 photodiodes. Time resolution of a video camera is video frequency, while, with fewer photodiodes, a much higher time resolution can be achieved. These methods have thus an excellent time resolution and at the same time a fairly high spatial resolution, although single-cell activities cannot be resolved. In a sense, the optical methods incorporate the advantages of both the DOG and the EEG approach: They reach the same time resolution of the EEG method and the spatial resolution of the DOG method. The results obtained with these methods show clearly how waves of activity spread over the OB when stimuli are applied. It can also be observed how different odor stimuli lead to different activation patterns, although one picture element still corresponds to thousands of cells.

On the whole, the electrophysiological and optical recording techniques show clearly that large parts of the OB are involved in the processing of most stimuli. The activation of MCs can thus be understood as an image of a chemosensory stimulus. However, many important details of this chemosensory map (e.g., how concentration coding is accomplished or whether the olfactory bulb contains an associative memory) (i.e., a memory not addressed by its address, as in computers, but by parts of its contents) remain to be investigated.

The physiology of higher brain structures of the

olfactory pathways leave even more room for speculation because surprisingly little work has been done in this field. There is, however, evidence that the piriform cortex shows the typical features of a associative memory.

III. Behavior

The sense of smell is considered as one of the so-called "lower senses," and in most situations of human adult life, other senses, above all vision and audition, are of greater importance for (fast) behavioral responses. However, this would not exclude that olfaction might influence behavioral responses on a slower time scale and the general motivational state. Olfactory stimuli do not seem to enter consciousness in the way visual or auditory stimuli do. Many of the olfactory stimulus responses might even bypass consciousness. Olfactory behavioral responses are not well controllable by interference with consciousness or conscious memory. From an evolutionary point of view, this makes olfactory-guided behavior stable and also stereotype.

It is remarkable that olfactory cues are most often described by comparisons, analogies, or metaphors (i.e., they are classified almost exclusively by invoking associations). Our knowledge in this field stems from direct studies of human olfactory-guided behavior and also from olfactory dysfunctions in humans. Comparisons with animal behavior especially in vertebrates have proven to be most instructive because they lead to new experiments and have suggested fairly general and widespread mechanisms of olfactory behavior. It must be mentioned in this context that our conceptual framework of behavior and olfaction should not only be based on our present cultural understanding and attitude concerning the existence and effects of olfactory stimuli. During the past centuries (i.e., in the very recent past), olfaction played a remarkably different role in human behavior, so that in comparison, our age seems odorless and largely deodorized.

Olfactory stimuli can lead to a variety of behavioral responses that can be grouped into four categories: genotype recognition, reproduction, homing, and food intake control. The latter is obvious and well known. A host of compounds can be differentiated, and with some training, which enhances the system's resolution considerably, humans are capable of distinguishing between many tens of thousands of odor notes. This ability is particularly developed in wine tasters and perfumers. Olfaction-guided homing is certainly present in some fish such as salmon, which find "their" river arm after a migration cycle by the chemical cues they were exposed to before the first migration. Olfaction-controlled homing might also occur in other less studied species of the animal kingdom (e.g., birds). Amphibia are among the best studied species as far as olfaction is concerned, particularly when morphological or electrophysiological questions were addressed. Amazingly, the mucosa of many frogs undergo characteristic changes when the animals move from one medium to the other (water/air), but nobody really knows what frogs smell in their terrestrial life. An attractive hypothesis is that the sense of smell is important during spawning time; the animals find the pond in which they grew up (a special case of homing). For instance, salamanders displaced from their usual habitat, found "their brook" among many others in the immediate neighborhood, although, for humans, there were no essential differences in vegetation, climate, soil characteristics, etc., between the brooks. It seems to be a common feature of olfaction-guided homing that the animals return always to their reproduction grounds; they seem to be imprinted with odors of this place. Imprinting with odors immediately after birth is also well-established in mammals (e.g., rodents and rabbits appear to find their mothers' nipples by smelling a pheromone (a hormone-like acting substance) that is secreted at the nipples). When the nipples are odorized with a different odor (e.g., a perfume) before the first contact, rabbit pups searching for the nipples approach objects with that perfume odor.

There is a host of examples which show clearly that many types of reproductive behavior are induced or modulated by odors. For example, in many mammals, sex attraction, sex status recognition, and the advertisement of sex status is communicated by odors, and odors can even induce oestrus and/or ovulation. Further, courtship and mating are often under the control of odor stimuli. Mice, for example, show clear odor-induced mating preferences; the animals prefer some individual members of the species over others, and this preference depends on the individual odors of the different animals. Genotypic variations are reflected in characteristic (body) odors that, in principle, allow the distinction between different members of the same species. This fact was suggested long ago as a hypothesis in the context of territory demarcation.

The major histocompatibility complex (MHC) on chromosome 17, which plays a key role in the immune response, is the genetic basis not only for the individuality of the body's cells but also for the fact that gene differences can be identified by scent. Genotype recognition plays thus an important role in reproduction behavior. Another remarkable example of the interference of olfactory genotype recognition in reproduction behavior is the Bruce effect discovered in 1959: In the mouse and other species, pregnant females abort with a high probability if they are exposed to a genetically strange male or its urine. Injection of prolactine counteracts this pregnancy block.

Little is known about the connection between reproduction and olfaction in humans. There is, however, no doubt that in women, some olfactory thresholds are lower around ovulation and during pregnancy. It is also interesting that the sense of smell in humans changes in a characteristic way with age. Human neonates detect odors within hours of birth as observed by consistent variations of salivation and respiration. A 2-week-old infant orients reliably toward his or her mother's breast pad as opposed to that of another woman. There are probably two fairly different age-related phases of odor perception, the first lasting until the age of 5 or 6 years and the second throughout the rest of the life whereby a decline of thresholds occurs in the last decades. A classical study presented children with the odors of amyl acetate (fruity banana-like smell), feces, and sweat. Reliable differentiation between the pleasant and unpleasant odors did not occur until the age of about 5. Most younger children (more than 95%) showed no displeasure at odors that adults normally find unpleasant. After age 6, adults and children show the same pattern of reactions.

Although a normal sense of smell remains almost unnoticed by humans, disorders pose severe problems to those who have them. They reduce the quality of life and can even cause anorexia, stress, or depression. Dysfunctions are described by the following terms: (partial) anosmia, absence of the capability of smelling (some) odors; hyposmia, decreased sensitivity of the sense of smell; and dysosmia, distortion of normal smell.

The most frequently encountered reasons for a loss of odor perception are viral infections, normal aging, and head traumata, which affect the fibers passing through the cribriform plate. However, many other diseases can cause a diminished or even abolished odor perception as well. These include Alzheimer's disease, multiple sclerosis, Parkinson's disease, cirrhosis of the liver, and vitamin B_{12} deficiency. Some typical changes of nerve cells in Alzheimer's disease have been shown to appear also in olfactory receptor neurons. The easier access to these neurons than to other neurons of the central nervous system might facilitate the diagnosis at the cellular level. The sense of smell is further diminished by many drugs, although the precise actions are only rarely established. Many of the diseases mentioned and some drugs might slow the rate of receptor neurons turnover (i.e., the continuously occurring degeneration and regeneration). Other olfactory symptoms in diseases such as olfactory hallucinations as part of the aura of epileptic seizures are probably not related to morphological alterations. [See ALZHEIMER'S DISEASE.]

IV. Olfactory Coding

A. Topological Aspects

Contrary to other sensory systems, the basic coding mechanisms in the olfactory system have withstood all attempts of explanation during the past decades. One reason for this is probably the topology of the map stimulus → neuronal response: In the visual black/white system, for example, the stimuli are images (i.e., intensity functions on two spatial dimensions); the same is true for the sense of touch. In audition, the spectra of sound waves are detected by the hair cells of the inner ear; a one-dimensional cell array codes one parameter, the frequency, and spectral intensities are coded by spike rates. In these sensory systems, the stimuli can be described by one or two ordered parameters (frequency in audition and the two spatial coordinates in vision); the stimulus space, including intensities, is thus at most three-dimensional. In principle, a one-to-one or point-to-point map of the stimuli to the activities of a one-dimensional chain or a two-dimensional grid of neurons is therefore straightforward.

In contrast, if the stimuli of a sensory modality must be described by more than two ordered parameters, the stimulus space is higher than three-dimensional and a single one-to-one map of the stimulus space to the activities of a two-dimensional grid of receptor cells is not possible. This seems to be the situation for olfaction and taste. Accordingly, receptor cells in the olfactory mucosa cannot be placed in an ordered way with respect to stimulus parameters. In fact, no such order has as yet been found. It has rather to be expected that there are classes of receptor cells, each of them being responsive to one or some molecular stimulus properties, and that the responses of these cells reflect the high dimensionality of the stimulus space. When the responses of receptor cells to many stimuli are analyzed by principal component analysis, the number of necessary factors to describe the data adequately is about 10. To obtain this result, a number of orthogonal transforms (rotations of the coordinate system) are applied to the data. This means that the number of receptor cell classes must exceed the number of the 10 factors. In other words, at receptor cell level, the olfactory information is coded in a space of more than 10 dimensions.

B. Glomerula, The Core of Olfactory Coding

The cell layer that follows the receptor neurons are MC/tufted cells. The axons of these primary cells of the OB project directly to higher brain structures. MCs are neatly placed on a two-dimensional surface so that they can be viewed as a two-dimensional grid. At this level of the system, olfactory stimuli are coded as the activities or amplitudes of all MCs (i.e., as a function on a two-dimensional grid). Such a function can also be interpreted as an image whereby activities are coded on a gray scale. Activity images are nothing unusual in the nervous system and may be even the most common form of nervous information. A good comparison to the MC image is probably the activity image of ganglion cells in the retina. The crucial difference between these systems is that the input to the retinal network is a real image in the everyday sense, whereas the input to the olfactory network is chemical substances with no definite spatial ordering.

The most marked morphological structures between olfactory neurons and MCs are the glomerula. This means that synaptic connections in the glomerula (i.e., the direct ones from olfactory axons to MC dendrites and the indirect ones from primary nerves through periglomerular cells to MC axons) accomplish the transformation that is characteristic for the olfactory system. It has to transform vectors in the high-dimensional receptor activity space into an ordered image of MC/tufted cell activities.

The term *order* needs an explanation. MCs can be imagined in two different ways: The first is simply the geometrical one, where every cell occupies a certain place on a two-dimensional grid. It turns out, however, to be convenient to imagine MCs as a function of their inputs. For the sake of simplification we assume that there are only two classes of olfactory receptor cell neurons and that every MC is connected to both in varying proportions. Then an MC can be described by two numbers: the connection strength to class 1 and the connection strength to class 2. In this picture, MCs are described in their *input parameter space*. The example can easily be extended to many dimensions. MCs are adjacent in their input parameter space if they are connected in a similar way to the receptor cell classes. MCs that are adjacent in the input parameter space are generally not adjacent in the geometrical space on the grid. However, in the special case in which MCs are neighbors in the input parameter space *and* in the geometrical space, the map from receptor neurons to MCs and the MC sheet structure itself is called *ordered*. Ordered maps are essential for the nervous systems, because the neurons of the second and higher nerve cell layers behind the sensory cells cannot obtain any information about relations between the stimuli if these relations are not mapped in an ordered way. The experimentally established fact that spatially adjacent MCs respond in a similar way to the same stimulus suggests that the projection to MCs is in fact ordered. It follows that glomerula have the function of connecting the primary fibers to MCs in such a way that an ordered projection from a high dimensional space (the receptor activity space in which odors are represented) onto an activity function on a two-dimensional grid (the activities of MCs) results. Odor images of this form can subsequently enter the olfactory cortex, the amygdala, the hypothalamus, and other central structures to lead to an adequate behavioral response.

Bibliography

Corbin, A. (1982). "Le miasure et la Jonquille. L'odorat et l'imagiuaine social XVIIIᵉ-XIXᵉ sièceles." Aubier Gontainge, Paris.

Freeman, W. J. and Skarda, C. A. (1985). Spatial EEG patterns, non-linear dynamics and perception: The neo-Sherrington view. *Brain Res. Rev.* **10**, 147–175.

Getchell, T. V. (1986). Functional properties of vertebrate olfactory receptor neurons. *Physiol. Rev.* **66**, 772–817.

Hudson, R. and Distel, H. (1986). Olfactory guidance of nipple search behavior in newborn rabbits. *In* "Ontogeny of Olfaction'' (W. Breipohl, ed.), pp. 243–254. Springer, Berlin, Heidelberg, New York.

Mori, K. (1987). Membrane and synaptic properties of identified neurons in the olfactory bulb. *Prog. Neurobiol.* **29,** 275–320.

Schild, D. (1988). Principles of odor coding and a neural network for odor discrimination. *Biophys. J.* **54,** 1001–1011.

Schild, D. (1989). Whole-cell currents in olfactory receptor cells of *Xenopus laevis. Exp. Brain Res.* **78,** 223–232.

Schild, D., ed. (1990). ''Information Processing in Olfactory Systems.'' Springer, Berlin, Heidelberg, New York.

Stoddard, M. D. (1980). ''The Ecology of Vertebrate Olfaction.'' Chapman and Hall, London, New York.

Omega-3 Fatty Acids in Growth and Development

ARTEMIS P. SIMOPOULOS, *The Center for Genetics, Nutrition and Health, American Association for World Health*

Glossary

AA Arachidonic acid (20-carbon)
DHA Docosahexaenoic acid (22-carbon)
EFA Essential fatty acids
EPA Eicosapentaenoic acid (20-carbon)
LA Linoleic acid (18-carbon)
LNA Alpha-linolenic acid (18-carbon)
PUFA Polyunsaturated fatty acid

OMEGA-3 AND OMEGA-6 FATTY ACIDS are the two classes of essential polyunsaturated fatty acids (PUFAs). The omega-3 class is represented by $18:3\omega3$, alpha-linolenic acid (LNA), and the omega-6 class by $18:2\omega6$, linoleic acid (LA) (Table I). Animals and humans do not have the capacity to synthesize either $18:3\omega3$ or $18:2\omega6$; therefore, both LNA and LA must come from the diet.

In 1918, Aron was the first to suggest that fats have nutritional functions in addition to being a good source of energy from food, providing 9 kcal/g. In growth and development, nutritional thinking has been dominated for a long time by concepts concerning protein and body growth. M. A. Crawford in 1981, suggested that it is quite possible that lipids and essential fatty acids, such as LNA and LA, and their long-chain metabolic products, particularly docosahexaenoic acid (DHA), were postulated to be of greater significance to early human development than proteins. This paper presents information on omega-3 fatty acids: their sources and metabolism; elongation and desaturation of LNA; and the role of omega-3 fatty acids in pregnancy, fetal growth, human milk, infant feeding, childhood, and aging; as well as the evolutionary aspects and the omega-3–omega-6 balance and dietary recommendations.

I. Omega-3 Fatty Acid Sources and Metabolism

A. Sources

LNA is the predominant terrestrial omega-3 fatty acid, and eicosapentaenoic acid (EPA) and DHA are the predominant aquatic omega-3 fatty acids. LNA is found in the chloroplast of green leafy plants (Table II) and in a few vegetable oils, specifically linseed, rapeseed, walnut, wheat germ, and soybean (Table III). LA is widely distributed in the vegetable kingdom and is particularly rich in most, but not all, vegetable seeds and in the oils produced from the seeds (with coconut oil, cocoa butter, and palm oil being exceptions).

Although food-selection patterns of land mammals vary and the diet of some herbivorous species

α-LINOLENIC ACID

5,8,11,14,17-EICOSAPENTAENOIC ACID

4,7,10,13,16,19-DOCOSAHEXAENOIC ACID

may include more leaf or more seed material, most mammals will obtain LNA and LA from their food.

Plankton, on which fish feed, is rich in LNA, EPA, and DHA. Thus, fish get EPA and DHA from eating plankton or from metabolizing LNA to EPA and DHA. Although plankton mainly provide dietary omega-3 fatty acids for aquatic animals, it is sometimes eaten by land animals. Both marine and freshwater algae contain LNA, EPA, and DHA. The fatty acids in marine algae vary according to species, season of collection, and nutrient supply. Zooplankton, such as krill (whale food), has very little LNA (1%) but is very rich in EPA and DHA. Its contents of EPA and DHA are 12–22% and 5–11%, respectively. Zooplankton in the North Atlantic and in the Mediterranean contains 7–16% EPA and 16–24% DHA, respectively.

Oily, or fatty, fish from deep, cold water (ocean) are rich sources of omega-3 fatty acids (LNA, EPA, and DHA) (Table IV). Although all fish and shellfish contain omega-3 fatty acids, the amount that is in a single serving of one species may vary significantly from that contained in a single serving of another, due to differences in the total oil content. In general, a single serving of lean fish, because of its lower oil content, provides a lesser amount of omega-3 fatty acid than does a single serving of oily fish. Seafood and their lipids are important to human nutrition. Unfortunately, most white fish that are popular in the U.S. are low in fat content and, therefore, contain minimal amounts of fish oil and, thus, minimal amounts of omega-3 fatty acids. [*See* FATS AND OILS (NUTRITION).]

B. Elongation and Desaturation

With the exception of carnivores such as lions and cats, who obtain EPA and DHA directly from the flesh of other mammals, animals and humans can convert LNA from the diet to EPA and DHA. This process results in the two classes of the omega-3 and omega-6 fatty acids with 20 and 22 carbon atoms, respectively, and four, five, or six double bonds (Table I). The amounts of the elongated polyunsaturated derivatives are dependent on the dietary source. Because most edible vegetable oils contain smaller amounts of LNA, their higher concentration of LA may depress synthesis of EPA and DHA from LNA, except selectively in tissues such as retina, brain, and testis, which are rich in DHA. Humans who consume large amounts of EPA and DHA, which are present in fatty fish and fish oils, have increased levels of these two fatty acids in their plasma and tissue lipids at the expense of LA and arachidonic acid (AA). Alternately, vegetarians, whose intake of LA is high, have more elevated levels of LA and AA and lower levels of EPA and DHA in plasma lipids and in cell membranes than omnivores.

Elongation and desaturation of LNA to EPA and DHA occurs in human leukocytes and in the liver of both humans and rodents (e.g., rats). Omega-3 and omega-6 fatty acids compete for the desaturation enzymes. But both Δ^4 and Δ^6 desaturases (the enzymes involved in desaturation) prefer the omega-3 to the omega-6 fatty acid. Retroconversion of DHA and of AA to shorter-chain fatty acids has been shown to occur in suspended rat hepatocytes by beta-oxidation (Table I).

TABLE I EFA Metabolism Desaturation and Elongation of Omega-3 and Omega-6[a]

Linolenate series	Linoleate series
C18:3ω3 LNA	C18:2ω6 LA
↓ Δ[6] desaturase	↓ Δ[6] desaturase
C18:4ω3	C18:3ω6 gamma-linolenic acid
↓	↓
C20:4ω3	C20:3ω6 dihomo-gamma linolenic acid
↓ Δ[5] desaturase	↓ Δ[5] desaturase
C20:5ω3 EPA	C20:4ω6 AA
↓	↓
C22:5ω3 docosapentaenoic acid	C22:4ω6
↓ Δ[4] desaturase	↓ Δ[4] desaturase
C22:6ω3 DHA	C22:5ω6 docosapentaenoic acid

[a] The first number (18) refers to the number of carbon atoms in the molecule, the number after the colon (3 for 18:3ω3 and 2 for 18:2ω6) refers to the number of double bonds in the molecule, and the ω3 and ω6 refer to the position of the double bond closest to the methyl end (CH_3) of the molecule. The use of the omega system of designation is based on the fact that the characteristics of unsaturated fatty acids from a nutritional standpoint depend on what exists near the methyl end, not the carboxyl end, of the fatty acid. Dr. Ralph Holman called such a designation the omega (ω) system. Omega, the last letter of the Greek alphabet, implies that the counting of the carbon atoms begins at the methyl end of the fatty acid molecule. The omega system nomenclature is now more popular than the n-3 or n-w nomenclature.

EPA and DHA are found in membrane phospholipids of practically all cells of individuals who consume omega-3 fatty acids. DHA is found mostly in phospholipids, whereas LNA is found mostly in triglycerides, cholesteryl esters, and in very small amounts in phospholipids. EPA is found in cholesteryl esters, phospholipids, and triglycerides.

Human and other mammal's cerebral cortex, retina, testis, and sperm are exceptionally rich in DHA. DHA is one of the most abundant components of the brain's structural lipids, phosphatidyl-ethanolamine (PE), phosphatidylcholine (PC), and phosphatidylserine (PS). As indicated earlier, DHA can be obtained directly from the diet by, for example, eating fish or synthesizing it from dietary LNA.

C. Metabolism

EPA and DHA are the precursors of the prostaglandins and thromboxanes of the 3-series and leukotrienes of the 5-series. AA is the precursor of the 2-series of prostaglandins and thromboxanes and of

TABLE II Fatty Acid Content of Plants[a]

Fatty acid	Purslane	Spinach	Buttercrunch lettuce	Red leaf lettuce	Mustard
14:0	0.16	0.03	0.01	0.03	0.02
16:0	0.81	0.16	0.07	0.10	0.13
18:0	0.20	0.01	0.02	0.01	0.02
18:1ω9	0.43	0.04	0.03	0.01	0.01
18:2ω6	0.89	0.14	0.10	0.12	0.12
18:3ω3	4.05	0.89	0.26	0.31	0.48
20:5ω3	0.01	0.00	0.00	0.00	0.00
22:6ω3	0.00	0.00	0.001	0.002	0.001
Other	1.95	0.43	0.11	0.12	0.32
Total fatty acid content	8.50	1.70	0.601	0.702	1.101

Source: A. P. Simopoulos and N. Salem, Jr., 1986, *N. Engl. J. Med.* **315**; 833.
[a] Milligrams per gram of wet weight

TABLE III Sources of Omega-3 Fatty Acids

Oils	18:3[a]
Linseed oil	53.3
Rapeseed oil (canola)	11.1
Walnut oil	10.4
Wheat-germ oil	6.9
Soybean oil	6.8
Tomato seed oil	2.3
Rice bran oil	1.6

Source: Provisional Table on the Content of Omega-3 Fatty Acids and Other Fat Components in Selected Foods, U.S. Department of Agriculture, Washington, D.C., February 1986.
[a] Edible portion, raw, 100 g.

the leukotrienes of the 4-series. Prostaglandins, thromboxanes, and leukotrienes derived from EPA have different biological properties from those derived from AA. Prostaglandins, thromboxanes, and leukotrienes derived from AA are potent platelet aggregators and vasoconstrictors and are powerful inducers of inflammation, whereas those derived from EPA prevent platelet aggregation and vasoconstriction and are weak inducers of inflammation. Competition also exists between EPA and DHA in prostaglandin formation and leukotriene synthesis. Because omega-3 fatty acids are found in all cell membranes, they have a variety of biological functions that influence inflammation, platelet adhesion, plasma lipid levels, blood pressure, and growth and development.

TABLE IV Relative Content of Omega-3 Fatty Acids and Cholesterol in Fish

Species	% Oil in flesh[a] (a)	% Omega-3 fatty acids in oil[a,b]	% Omega-3 fatty acids in flesh[c]	% Cholesterol in flesh[d]
Haddock	00.5	39.6	0.198	0.060
Snapper	01.1	23.0	0.253	0.040
Tuna, canned	01.0	30.0	0.300	0.063
Shrimp	01.1	28.5	0.314	0.180
Cod, Atlantic	00.7	45.9	0.321	0.050
Pollock	00.8	48.4	0.387	0.071
Sole, lemon	01.4	31.0	0.434	0.050
Ocean perch	02.0	22.0	0.440	0.050
Flounder	01.3	35.0	0.455	0.050
Squid	01.0	53.3	0.533	0.241
Mullet	03.0	19.1	0.573	0.021
Halibut	02.0	36.0	0.720	0.050
Shad	02.8	26.1	0.731	0.038
Whiting, Pacific	03.0	33.3	0.999	0.066
Swordfish	04.4	25.7	1.131	0.057
Trout, rainbow	07.0	17.6	1.232	0.050
Tuna, raw	05.1	30.0	1.530	0.046
Whitefish, lake	07.0	22.2	1.554	0.060
Sardine, canned	06.3	26.8	1.688	0.140
Salmon	09.3	23.0	2.139	0.053
Sablefish	10.0	22.9	2.290	0.040
Mackerel, Atlantic	13.0	19.0	2.470	0.065
Herring	15.0	18.4	2.760	0.085
Dogfish shark	14.1	24.5	3.455	0.039

Source: M. Barton and J. A. Emerson (1986).
[a] Data derived from the National Marine Fisheries Service and the scientific literature.
[b] Total percentages of 13:3ω3, 20:5ω3, 22:5ω3, and 22:6ω3.
[c] Percent oil in flesh × percent omega-3 fatty acids in oil.
[d] Cholesterol percentages represent total cholesterol values. Note: the levels presented here are average figures. These levels may vary widely due to seasonality, fish diet, age, size, and the processing methods employed for the various product forms in which these fish species are sold to the public. Therefore, these figures should not be considered as absolute.

Ingestion of omega-3 fatty acids, particularly of increased amounts of EPA and DHA omega-3 fatty acids, replaces the omega-6 fatty acids in cell-membrane phospholipids. The fatty acid composition, the cholesterol content, and the phospholipid class of biomembranes are critical determinants of membrane physical properties and have been shown to influence a wide variety of membrane-dependent functions, such as integral enzyme activity, membrane transport, and receptor function. This ability to alter both the lipid composition and function *in vivo* by diet, even when essential fatty acids are adequately supplied, demonstrates the importance of diet in growth and development and in health and disease. [*See* NUTRITION AND IMMUNITY.]

II. Animal Studies in Growth and Development

A. Animal Studies

The term "essential fatty acids" was first coined by G. O. Burr and M. M. Burr in 1930. Diets deficient in fatty acids are associated with clinical symptoms such as skin rash, retardation of brain growth, incomplete brain-cell division, and a high neonatal mortality in second-generation rat pups. Subsequent work by G. O. Burr and M. M. Burr showed that the addition of fatty acids to the diet, particularly LA, eliminated the symptoms caused by the fat-free diet in the deprived animals. The EFAs discovered by G. O. Burr and M. M. Burr include both LNA and LA, but their effects differ. LA restores healthy skin, successful growth, reproduction, and lactation, whereas LNA supports growth but does not prevent the skin lesions of EFA deficiency or support reproduction.

Research on the role of omega-3 fatty acids during pregnancy and lactation has expanded over the past 30 years. As a result the work of B. L. Walker, it became possible to produce DHA deficiency in the rat for the first time. B. L. Walker published his studies in 1967 and demonstrated that restriction of dietary LNA in the maternal diet of rats is reflected in a lowering of the $22:6\omega3$ in the brain lipids of the pups at birth. Postnatally, even low levels of linolenate in the diet of the lactating dam result in rapid accumulation of DHA in the brain. This remarkable affinity of brain lipids for DHA raised the possibility of a functional requirement for linolenic acid. The work of B. L. Walker is indeed a milestone because

it made possible an animal model—the rat—that was finally deficient in DHA and the deficiency could be restored with even small amounts of dietary omega-3 fatty acids.

C. Galli *et al.* in 1971 and R. Paoletti and C. Galli in 1972 concluded that dietary EFA deficiency was found to affect the *developing* central nervous system of the developing rat; this nutritional stress could be of comparable severity to that obtained by protein and/or calorie malnutrition. In LNA deficiency, $22:5\omega6$ increases and LNA supplementation raises the proportion of $22:6\omega3$ in brain glycerophosphatides. C. Galli *et al.* (1974) proposed to consider the ratio of $22:5\omega6/22:6\omega3$ in tissue lipids as an index of relative LNA deficiency.

In 1973 R. M. Benolken *et al.* discovered that dietary LNA influences the electroretinogram (ERG) response in the rat. In 1975 T. G. Wheeler and R. M. Benolken measured the ERG response in the rat as a function of dietary supplements of purified ethyl esters of LNA, LA, and oleic acids. Dietary LNA affected the ERG amplitudes to a greater extent than LA. The electrical response of photoreceptor cell membranes appears to be a function of the position of the double bonds as well as a function of the total number of double bonds in fatty acid supplements. These findings firmly establish a selective functional role in the visual system for omega-3 fatty acids and suggest that the observed electrical alterations are associated with fatty acid substitutions in the plasma membrane of the photoreceptor cells. Therefore, polyunsaturated fatty acids (PUFAs) derived from LNA and LA appear to be important functional components of photoreceptor cell membranes. The photoreceptor cells of vertebrates consist of an inner and an outer segment. DHA is the dominant PUFA of the phospholipids of vertebrate membranes. Thus, a selective functional role in the visual system for omega-3 fatty acids is established.

The studies of B. L. Walker, R. M. Benolken, T. G. Wheeler, C. Galli, and C. Galli and R. Paoletti were critical for the understanding of the role of omega-3 fatty acids on retina and brain function. By 1975 progress had been made in the development of a model—the DHA deficient rat—by using omega-3 fatty acid deficient diets during pregnancy and postnatally, and the use of the ERG response in the rat had established that DHA deficiency in retina could be reflected by measuring ERG amplitude changes.

M. S. Lamptey and B. L. Walker extended this work further. These investigators carried out a study in which female rats were fed a diet high in linoleic/linolenic acid ratio prior to mating and during pregnancy and lactation. The offsprings were weaned to the same diet and their physical, neuro-motor, and neurophysical development assessed and compared to pups from dams fed a diet containing soybean-oil rich in omega-3 fatty acids. The brain lipids in the two groups of animals were compared also. This was a monumental study. The soybean-fed progeny had higher levels of 22:6ω3 and lower levels of 22:5ω6 in the brain PE, and their performance in the discrimination-learning test was superior to that of progeny fed safflower-oil rich in omega-6 fatty acids, providing further support for the essential role of dietary linolenic acid for the young rat.

While these above studies were going on, R. N. T. W. Fiennes *et al.* carried out the first experimental studies of LNA requirement in primates (capuchins). The capuchins suffered from symptoms that closely resemble those of EFA deficiency, yet their diet contained adequate amounts of LA. When linseed oil (55.7% LNA) was added to this diet, their symptoms improved. Although the capuchins were on a purified diet (24–28 mo) that contained very little LNA, the liver and red blood cell PE still contained about 6% of their fatty acids as LNA metabolites, mostly DHA. Their red blood cell fatty acids had the profile expected from the studies in rats fed a diet high in LA and poor in LNA, specifically elevated levels of 22:4ω6 and 22:5ω6 and low levels of 22:6ω3.

Beginning in 1984, M. Neuringer *et al.* carried out a series of studies in primates. Primates, of course, are closer to human beings in retinal structure and visual function. Studies were carried out in the infant rhesus monkey that had been DHA-deficient during pregnancy and lactation, caused by a LNA-deficient diet. The maternal omega-3 fatty acid deprivation led to abnormally low levels of DHA in the tissues of the near-term fetus and newborn infant. Vulnerability to dietary omega-3 fatty acid deprivation was even greater after birth. The monkeys were followed up to 22 months. In the control monkeys the DHA (22:6ω3) had increased in the cerebral cortex and retina, twice as high as at birth. Whereas, the deficient monkeys, failed to show an increase in both the cerebral cortex and retina by 22 months. As was shown previously in the rat studies, and in the studies with the capuchins, the deficient animal tissues had a compensatory increase of 22:5ω6. Subnormal visual acuity and prolonged re-

covery time of the dark-adapted ERG after a saturating light flash were noted in the deficient monkeys at 4–12 wk of age. M. Neuringer *et al.* pointed out

> The fatty acid composition of the cerebral cortex described here for control newborn and juvenile rhesus monkeys are very similar to those reported by Svennerholm (1968) for human newborns and adolescents, respectively. However, the brain and retina of human infants are less developed at birth than those of rhesus monkeys (Cheek 1975), so that human infants might be even more vulnerable to postnatal dietary deprivation of ω3 fatty acids.

> The authors concluded our findings provide evidence that dietary ω3 fatty acids are essential for normal prenatal and postnatal development of the retina and brain. Further research will be required to determine the relative contributions of prenatal versus postnatal deprivation to the observed functional deficits and to determine the degree to which the biochemical and functional effects of ω3 fatty acid deficiency are reversible.

B. Aging

N. P. Rotstein *et al.* in 1987 studied the effects of aging on the compositional and metabolic aspects of retinal phospholipids in the rat. The levels of DHA and other omega-3 hexaenoic acids were decreased in retinal glycerophospholipids of aging rats, particularly in those containing choline and serine. *In vitro* labeling of retinal lipids with [2-^{3}H] glycerol and [1-^{14}C] DHA in young and aged animals showed that most retinal lipid classes incorporated DHA. The incorporation of DHA was most marked when DHA was decreased and further stimulated by aging. This indicates that the decrease in the DHA content of retinal phospholipids is simply due to the decreased availability of DHA in the retina rather than to an impaired activity of the enzymes involved. The levels of the omega-3 pentaenoic acids in retinal lipids were much less affected by aging than the omega-3 hexaenoic acids, which indicates no defect in either the availability of 18:3ω3 or its metabolic products up to 22:5ω3.

The above findings are important because they indicate a marked difference between the effects of aging and EFA deficiency in the retina. In LA (18:2ω6) deficiency, the AA (20:4ω6) is decreased in tissue lipids and is replaced by 20:3ω9 (from oleic acid). In LNA (18:3ω3) deficiency, the DHA (22:6ω3) decreases in lipids and is replaced by 22:5ω6 (from LA). In the aging retina, none of these compensatory mechanisms were shown. This suggests that an impairment of the Δ_4 desaturase enzyme system is most likely responsible for the decreased levels of $22:6\omega3$ (and $22:5\omega6$) observed in retinal lipids as a consequence of aging. Both $22:5\omega6$ and $22:6\omega3$ fatty acids require Δ_4 desaturase for their synthesis.

A decrease in DHA possibly plays a significant role in visual impairments that accompany old age, because DHA is required for normal function of photoreceptors in rats and primates. If one were to speculate as to the appropriate dietary $\omega3$ fatty acids for the aged, it would be dietary DHA rather than LNA that might maintain and possibly improve visual function in the elderly.

III. Human Studies in Growth and Development

A. Pregnancy and Fetal Growth

Since the Second World War, the role of maternal nutrition on fetal growth and development has been extensively studied in the context of protein calorie malnutrition. The role of omega-3 fatty acids has only recently come into focus, despite the evidence of its importance having been demonstrated long ago in studies involving rats and primates. Lipid nutrition during pregnancy and lactation is of special relevance to human development, because brain development in the human takes place during fetal life and the first two years after birth. DHA is found in large amounts in the gray matter of the brain and in the retinal membranes, where it accounts for 30% or more of the fatty acids in the ethanolamine and serine phospholipid. DHA accumulates in the neurons of the brain between weeks 26 and 40 of gestation in humans.

The EFA requirements for pregnancy have been estimated by M. Crawford *et al.* (1981) to be between 600 and 650 g. in terms of both LA and LNA. This figure is about 1% of the nonpregnant woman's dietary energy; another 0.5% of energy should come from AA and DHA.

Determinations carried out on the fatty acids in PC and PE in maternal plasma, fetal cord blood, fetal liver, and brain from the human fetus at mid-term abortion show a decrease in LA and LNA and a progressive increase of AA and DHA in the PC and PE from maternal liver to cord blood, to fetal liver, and finally to fetal brain. M. Crawford *et al.* termed this sequence a process of biomagnification, and considered it responsible for the high content of AA and DHA originating from LA and linolenic

acid, respectively, in the brain. Thus, there is a preferential transfer of DHA and AA to the fetal side of the placenta. *In vitro* studies with human placenta show simultaneous accumulation of both omega-3 and omega-6 chain elongation products.

During the third trimester of human development, rapid synthesis of brain tissue occurs, in association with increasing neuromotor activity. The increase in cell size, number, and type requires *de novo* synthesis of structural lipids. Both M. Martinez (1974 and 1978) and M. T. Clandinin noted the accumulation of DHA in the brain of the human infant during the last trimester. The levels of LNA acid and LA were consistently low in the brain, whereas marked accretion of long-chain desaturation products, specifically DHA and AA, occurred. Subsequent studies by M. Martinez and A. Ballabriga (1987) confirmed these findings. More recent data indicate that the main developmental changes in the brain seem to be an increase in DHA at the end of gestation and a decrease in oleic acid $(18:1\omega9)$ and AA in PE. Similar changes occurred in the liver. Therefore, a prematurely born infant (prior to 37 wk) has much lower amounts of DHA in the brain and liver and is at risk of becoming deficient in DHA unless DNA is supplied in the diet. In the full-term newborn about half of the DHA accumulates in the brain before birth and the other half after birth.

There is epidemiologic evidence that the birth weights of newborns in the Faroe Islands (where fish intake is high) are higher than those in Denmark, and so is the length of gestation: 40.3 ± 1.7 wk for the Faroese versus 39.7 ± 1.8 wk for the Danish pregnant women. The average birth weight of primiparas was 194 g higher for the Faroe Islands. The higher dietary omega-3 fatty acid intake quite possibly influenced endogenous prostaglandin metabolism. It is hypothesized that the dietary omega-3 fatty acids inhibit the production of the dienoic prostaglandins, especially PGF_{2a} and PGE_2, because they are involved in the mediation of uterine contractions and the ripening of the cervix that lead to labor and delivery. These important observations need to be further investigated, as the prevention of prematurity is one of the most critical issues to be overcome in perinatal medicine.

B. Human Milk and Infant Feeding

A number of studies from around the world indicate that human milk contains both LNA and LA and their long-chain omega-3 and omega-6 fatty acids, whereas cow milk and infant formula do not; therefore, it seems reasonable to recommend their inclusion in infant formulas. M. Crawford *et al.* (1973) analyzed 32 samples of human milk and reported the presence of LNA and LA and the long-chain omega-3 and omega-6 fatty acids. At that time M. Crawford *et al.* recommended that all these fatty acids should be considered essential, because they can be classified as structural lipids in the human brain. Infant formulas still do not contain long-chain fatty acids (Table V).

The long-chain fatty acid composition of red cell membrane phospholipids may reflect the composition of phospholipids in the brain. Therefore, determination of red cell membrane phospholipids has been carried out by many investigators to determine the long-chain PUFA content in breast-fed and bottle-fed infants. As expected, the fatty acids $22:5\omega3$ and $22:6\omega3$ were higher in the erythrocytes from breast-fed infants than in those from bottle-fed babies and the $20:3\omega9$ was lower in the erythrocytes of the breast-fed infants.

Following birth, the amount of red blood cell DHA in premature infants decreases; therefore; the amount of DHA available to the premature infant assumes critical importance. Preterm infants do not have a limited ability to convert LNA to DHA (Table I); therefore, a number of studies have been carried out on the DHA status of the premature infant. Premature babies have decreased amounts of DHA, but human milk contains enough DHA to support normal growth of the premature baby. M. T. Clandinin *et al.* were concerned that infant formula that does not contain DHA is fed to premature babies and did studies in which they developed infant formulas supplemented with DHA from two different sources, fish oil and hen egg yolk oil (Table VI).

When fish oil is incorporated into formula, DHA is absorbed by the infants' gastrointestinal tract, as seen by measuring DHA in plasma PE. A single dose of 71 mg/kg/day of fish oil DHA in a bolus is absorbed, as well as 11 mg/kg/day fish oil DHA dispersed in the formula. Dispersed fish oil appears to be absorbed as much or even more despite the much lower dose. DHA at a dose of 11 mg/kg/day results in 0.2% DHA in the total dietary fatty acids in the formula, which is within the range of 0.1–0.3% found in human milk. This dose of DHA does not decrease plasma phospholipid AA and appears to be a physiologic amount that could prevent de-

TABLE V Fatty Acid Composition of Human Milk and Formulas (molar percent)

Fatty acid[a]	Human milk (n = 11)	Portagen®[b]	Enfamil® Premature	Similac® Special Care
8:0	0.35 ± 0.00	60	24.5	24.1
10:0	1.39 ± 0.14	24	14.1	17.7
12:0	6.99 ± 0.70	0.42	12.2	14.9
14:0	7.96 ± 0.88	trace	4.7	5.8
16:0	19.82 ± 0.37	0.19	7.5	6.8
16:1	3.20 ± 0.21		0.1	0.2
18:0	5.91 ± 0.3	0.47	1.7	2.3
18:1	34.82 ± 1.4	4.1	12.4	10.0
18:2n6	16.00 ± 1.3	8.1	22.4	17.4
18:3n3	0.62 ± 0.04	trace	0.6	0.9
20:1	1.10 ± 0.2		0.3	0.1
20:2n6	0.61 ± 0.1			
20:3n6	0.42 ± 0.04			
20:4n6	0.59 ± 0.04			
20:5n3	0.03 ± 0.00			
22:1	0.10 ± 0.00			
22:4n6	0.21 ± 0.00			
22:5n6	0.22 ± 0.00			
22:5n3	0.09 ± 0.03			
22:6n3	0.19 ± 0.03			

Source: Carlson *et al.*, 1986, *Am. J. Clin. Nutr.* **44,** 798–804.
[a] Values are expressed as mean ± SEM.
[b] Pediatric Products Handbook, 1983 ed., Mead Johnson Nutritional Division, Evansville, Indiana.

clines in membrane phospholipid DHA following preterm delivery.

The amount of omega-3 fatty acids in human milk, particularly DHA, is lower in vegetarians than in omnivores. One can increase the amount of DHA in human milk by giving fish oil rich in DHA to the mother. In one study, 5 g/day of fish oil raised the levels from a baseline of $0.1 \pm 0.06\%$ to $0.5 \pm 0.1\%$ ($P < 0.001$); 10 g/day raised DHA levels to $0.8 \pm 0.1\%$ ($P < 0.001$); and 47 g/day led to a DHA level of 4.8%. Thus, relatively low intakes of dietary DHA can significantly elevate milk DHA content.

The need to supplement infant formula with omega-3 fatty acids and, particularly, DHA for the premature is now recognized. Studies are in progress comparing the growth and development of premature infants who are fed mother's milk with those who are receiving formula supplemented with omega-3 fatty acids and those whose formula is not supplemented.

C. Childhood

The first case of human LNA deficiency was reported in 1982. The LNA deficiency was induced by long-term intravenous hyperalimentation. The pa-

TABLE VI Fatty Acid Composition of Human Milk and Formulas Prepared by Blending Egg Yolk Lipid or Fish-Oil Products with Other Oils

Fatty acid (%w/w)	Human milk	Formula	
		Egg yolk oil	Fish oil
Short chain	2.0	1.00	1.95
Medium chain	11.0		
16:0	20.0	22.70	19.21
16:1	03.2	3.94	2.44
18:0	05.8	11.94	8.08
18:1ω9	38.0	38.97	39.30
18:2ω6	12.0	11.49	12.22
18:3ω3	00.9	0.46	1.68
20:4ω6	00.5	0.25	0.13
20 and 22ω6	01.0	0.50	0.21
20 and 22ω3	0.1–0.7	0.11	3.30

Source: M. T. Clandinin and J. E. Chappell, 1985, *In* "Composition and Physiological Properties of Human Milk" (J. Schaub, ed.), Elsevier Science Publishers, Amsterdam, p. 221.

tient was a 6-yr-old white female who had 266 cm of her small intestine, the ileocecal valve, and 34 cm of her large bowel removed as a result of a 22-caliber rifle wound to her abdomen. For 7 mo she was on total parenteral nutrition (TPN) solution, which was high in LA and low in LNA. She developed neurological symptoms consisting of episodes of distal numbness, paresthesia, weakness, and blurring of vision. Analysis of fatty acids of serum lipids indicated marginal LA deficiency but marked LNA deficiency. Because neurological abnormalities do not occur with omega-6 fatty acid deficiency, omega-3 fatty acid deficiency was suspected and the patient was given an emulsion that contained LNA; the neurological symptoms disappeared. Based on studies of this patient, the requirement for LNA was estimated to be about 0.54% of calories. Recently, another child who was on TPN without LNA supplementation developed neurological symptoms that subsided with LNA supplementation.

Over the past 20 years, data from other studies involving humans on enteral or TPN, which were low or deficient in omega-3 fatty acids, have focused attention on the importance of omega-3 fatty acids in vision and central nervous system function.

D. Aging

K. S. Bjerve in 1987 described four patients with LNA deficiency as a result of long-term gastric tube-feeding that included large amounts of skim milk without LNA supplementation. These patients, who were in nursing homes, developed skin lesions diagnosed as scaly dermatitis, which disappeared with EFA supplementation. A number of other patients were reported to have omega-3 fatty acid deficiency, again patients on long-term gastric tube-feedings or prolonged TPN because of chronic illnesses. K. S. Bjerve estimated the minimal daily requirement of LNA and of long-chain omega-3 fatty acids in diets is equivalent to 0.2–0.3% and 0.1–0.2%, respectively, of total energy intake. If a deficiency of total omega-3 fatty acid intake is suspected, its concentration in plasma should be measured. A decrease in the concentration of 20:5ω3, 22:5ω3, and particularly 22:6ω3 in plasma or erythrocyte lipids indicates that the dietary intake of omega-3 fatty acids has been low. The presence of clinical symptoms, along with the biochemical determinations, provide additional support for the diagnosis. To verify the diagnosis, it is essential that the clinical symptoms disappear upon supplementation of the deficient diet with omega-3 fatty acids.

With the increase in the number of elderly persons in the population, and the proliferation of nursing homes, particular attention must be given to the nutritional requirements of the elderly, especially those who are fed enterally or parenterally. The optimal and minimal dietary requirements of LNA have been estimated to be 860–1,220 mg/day and 290–390 mg/day, respectively. Similarly, the optimal and minimal dietary requirements of long-chain omega-3 fatty acids are estimated to be 350–400 mg/day and 100–200 mg/day, respectively. Long-chain fatty acids appear to be two to three times more effective than LNA in curing and probably preventing clinical symptoms of omega-3 fatty acid deficiency.

IV. Implications for Dietary Recommendations

A. Evolutionary Aspects and the Omega-3–Omega-6 Balance

The development of agricultural food production took place 10,000 years ago and was one of the main steps toward human cultural evolution. Although precise information does not exist about the diet of humans, the development of agriculture occurred independently in the Middle East and other parts of the earth, such as Central and South America and Southeast Asia. Human beings usually ate wild plants, wild animals, and fish from the rivers, lakes, and oceans. Wild animals and birds who feed on wild plants (rich in omega-3 fatty acids) (see Table II) are very lean, with a carcass fat content of only 3.9%, and contain about five times more PUFAs per gram than is found in domestic livestock. Furthermore, 4% of the fat of wild animals contains EPA, whereas domestic beef contains a very small amount, if any, because cattle are fed grains rich in omega-6 fatty acids and poor in omega-3 fatty acids. Similarly, eggs from free-ranging chickens are rich

TABLE VII Fatty Acid Levels in Chicken Egg Yolks[a]

Fatty acid	Greek egg	Supermarket egg
	milligrams of fatty acid[b]	
Saturated fats		
14:0	1.10	0.70
15:0	—	0.07
16:0	77.60	56.66
17:0	0.66	0.34
18:0	21.30	22.88
Total	100.66	80.65
Monounsaturated fats		
16:1n-7	21.70	4.67
18:1	120.50	109.97
20:1n-9	0.58	0.68
22:1n-9	—	—
24:1n-9	—	0.04
Total	142.78	115.36
n-6 Fatty acids		
18:2n-6	16.00	26.14
18:3n-6	—	0.25
20:2n-6	0.17	0.36
20:3n-6	0.46	0.47
20:4n-6	5.40	5.02
22:4n-6	0.70	0.37
22:5n-6	0.29	1.20
Total	23.02	33.81
n-3 Fatty acids		
18:3n-3	6.90	0.52
20:3n-3	0.16	0.03
20:5n-3	1.20	—
22:5n-3	2.80	0.09
22:6n-3	6.60	1.09
Total	17.66	1.73
Ratio of fatty acids to saturated fats	0.4	0.44
Ratio of n-6 to n-3	1.3	19.4

Source: A. P. Simopoulos, and N. Salem, Jr. (1989). n-3 fatty acids in eggs from range-fed Greek chickens (letter to the editor). *N. Engl. J. Med.* **321**, 1412.
[a] The eggs were hard-boiled, and their fatty acid composition and lipid content were assessed as described elsewhere.[4]
[b] Per g of egg yolk.

in omega-3 fatty acids with a ratio of omega-6 to omega-3 of 1.3 whereas the standard USDA egg is poor in omega-3 with a ratio of omega-6 to omega-3 of 19.4 (Table VII).

From all available information, human beings appear to have evolved on a diet rich in omega-3 fatty acids and balanced between omega-3 and omega-6 fatty acids, which is a more physiological state. Over the past 100 years, Western diet has shifted dramatically toward increased amounts of omega-6 fatty acids at the expense of omega-3 fatty acids. This shift has led to a relatively deficient diet in omega-3 fatty acids, with a ratio of omega-3–omega-6 equal to 1 : 14 instead of 1 : 1 as is the case with wild animals and presumably humans, although the optimal ratio of omega-3–omega-6 in the diet is not precisely known.

The competition between omega-3 and omega-6 fatty acids in their elongation and desaturation products; in prostaglandin metabolism; the increase of the AA metabolites in cardiovascular disease, hypertension, and autoimmune disorders and their decrease with ingestion of LNA or EPA and DHA strongly suggest that a balance is needed between omega-3 and omega-6 fatty acids for platelet function, blood viscosity, monocyte function, membrane function, and central nervous system function.

Over the past 20 years, studies in the rat and the rhesus monkey indicate that dietary restriction of omega-3 fatty acids during pregnancy and infancy interfere with normal visual function and may even impair learning ability in the offspring. These findings and the presence of omega-3 fatty acids in human milk provide evidence for the essentiality of omega-3 fatty acids in growth and development and in health and disease.

B. Dietary Recommendations

In making dietary recommendations about fat intake, particular attention should be paid to the relative amounts of omega-3 and omega-6 fatty acids in the diet. The amounts of omega-3 fatty acids for infant nutrition should be consistent with the amounts found in human milk. After infancy the ratio of omega-3–omega-6 should be either 1 : 1 or 1 : 5.

The type of omega-3 fatty acid to be considered varies with the age of the individual. Because premature infants, and possibly the elderly, are limited in their ability to convert LNA to EPA and DHA,

DHA should be included in their diet. Preliminary evidence indicates that some diabetics and hypertensives lack this ability as well. For others, LNA by itself or LNA, EPA, and DHA from vegetables and fish or fish oils (for those who cannot obtain or will not eat fish) should be included in the diet. Americans today consume 15 lb of fish per person per year. To improve the balance of omega-3–omega-6, fish must be eaten a minimum of two times per week, significantly substituting meat consumption. Vegetarians who do not eat fish should increase the amount of omega-3 fatty acid sources in their diet by eating soybeans, walnuts, and omega-3 fatty acid-containing vegetable oils.

Bibliography

Bjerve, K. S., Fougner, K. J., Midthjell, K., and Bonaa, K. (1989). N-3 fatty acids in old age. *J. Int. Med.* **225,** 191–196.

Carlson, S. E., Rhodes, P. G., and Ferguson, M. G. (1986). Docosahexanoic acid status of preterm infants at birth and following feeding with human milk or formula. *Am. J. Clin. Nutr.* **44,** 798–804.

Clandinin, M. T., Chappell, J. E., Leong, S., Heim, T., Swyer, P. R., and Chance, G. W. (1980). Intrauterine fatty acid accretion rates in human brain: Implications for fatty acid requirements. *Early Hum. Develop.* **4,** 121–129.

Crawford, M. A., Sinclair, A. J., Msuya, P. M., Munhambo, A. (1973). Structural lipids and their polyenoic constituents in human milk. *In* "Dietary Lipids and Postnatal Development" (C. Galli, G. Jacini, and A. Pecile, eds.). Raven Press, New York.

Galli, C., and Simopoulos, A. P. (eds.) (1989). "Dietary ω3 and ω6 Fatty Acids: Biological Effects and Nutritional Essentiality." Plenum Publishing, New York.

Holman, R. T., Johnson, S. B., and Hatch, T. F. (1982). A case of human linolenic acid deficiency involving neurological abnormalities. *Am. J. Clin. Nutr.* **35,** 617–623.

Neuringer, M., Connor, W. E., Lin, D. S., Barstad, L., and Luck, S. (1986). Biochemical and functional effects of prenatal and postnatal ω-3 fatty acid deficiency on retina and brain in rhesus monkeys. *Proc. Nat. Acad. Sci. (USA)* **83,** 4021–4025.

Rotstein, N. P., Ilincheta de Boschero, M. G., Giusto, N. M., and Aveldano, M. I. (1987). Effects of aging on the composition and metabolism of docosahexaenoate-containing lipids of retina. *Lipids* **22,** 253–260.

Salem, N., Jr. (1989). Omega-3 fatty acids: Molecular and biochemical aspects. *In* "New Protective Role for Selected Nutrients" (G. Spiller and J. Scala, eds.). Alan R. Liss, Inc., New York.

Simopoulos, A. P., Kifer, R. R., and Martin, R. E. (eds.) (1986). "Health Effects of Polyunsaturated Fatty Acids in Seafoods." Academic Press, Orlando, Florida.

Simopoulos, A. P., Kifer, R. R., Martin, R. E., and Barlow, S. M. (eds.) (1991). Health Effects of Omega 3 Polyunsaturated Fatty Acids in Seafoods. *In* "World Review of Nutrition and Dietetics" Karger, Basel, Switzerland.

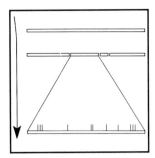

Oncogene Amplification in Human Cancer

PETRI SALVÉN, MANFRED SCHWAB, *German Cancer Research Center*

KARI ALITALO, *University of Helsinki*

Glossary

Oncogenes Genes whose mutations and pathological function influence the steps of tumor development. Mutations within normal protooncogenes, which regulate the growth and differentiation of normal cells, may produce cellular alterations, which give cells a continuous signal to proliferate and block their differentiation.

CANCER IS A disease attributed to complex changes in somatic cells leading to uncontrolled cell proliferation. Increasing evidence points to the fact that acquired genetic abnormalities are the primary causes of these changes. In several instances it has been shown that cancer can be caused by mutations and abnormal expression of certain genes called *oncogenes*.

Regulatory or structural alterations of cellular oncogenes and tumor suppressor genes have been identified in various cancers. Oncogene alteration by point mutations can result in a protein product with a strongly enhanced oncogenic potential. Aberrant expression of cellular oncogenes may also be due to tumor-specific chromosomal translocations that deregulate the normal expression of protooncogenes. Here we summarize data on the third mechanism of oncogene activation in human cancer: oncogene amplification.

I. Chromosomal Basis of Oncogene Amplification

A. Chromosomal Abnormalities Associated with Oncogene Amplification

Gene amplification was initially discovered during the study of chromosomal abnormalities of tumor cells. The amplification of specific genes in somatic cells has been implicated in an increasing variety of adaptive responses of cells to environmental stresses. These amplifications often manifest themselves as chromosomal changes known as double minute chromosomes (DM) and homogeneously staining chromosomal regions (HSR). DMs appear as small, spherical, usually paired chromosome-like structures that lack a centromere and may contain circular DNA. HSRs stain with an intensity intermediate between chromosomal dark and light bands. Because both of these types of chromosomal abnormalities are occasionally found in cells from untreated cancers, the types of genes they contain in cancer cells has been of considerable interest. [*See* CHROMOSOME ANOMALIES.]

Double minute chromosomes associated with DNA amplification were found in tumor cells followed by the discovery of DMs and HSRs in cells selected for drug resistance. Experimental work on drug-resistant cells has shown that in the absence of a selection pressure (drug), DMs and the amplified genes that they contain are lost, whereas amplified DNA in the form of HSRs is retained in the cells. This is explained by the fact that DMs are segregated unevenly in mitosis and are frequently lost from the nucleus because of their lack of centromeres. HSR chromosomes carry centromeres and are therefore divided equally between daughter cells at mitosis. If DMs and HSRs contain amplified genes that encode drug-resistant or growth-stimulating protein products, it would follow that the more stable chromosomal form (i.e., the HSR) confers a greater selective growth advantage for cells. Although DMs and HSRs have been described predominantly in tumor cells selected for resistance to cytotoxic drugs, it is also clear that DMs and HSRs may be present in cancer cells before the initiation of therapy. It was this setting that lead researchers to explore the possible amplification of cellular oncogenes.

B. DMs and HSRs Contain Amplified Oncogenes

The first report of a somatic amplification of a cellular oncogene concerned the c-*myc* oncogene in a promyelocytic leukemia cell line HL-60. The degree of c-*myc* amplification was between 8- and 32-fold both in the HL-60 cell line and in primary leukemic cells from the patient. In a neuroendocrine cell line from a colon carcinoma, COLO 320, approximately 30-fold amplified c-*myc* copies were mapped either to HSRs of a marker chromosome or to DMs, depending on the particular subline studied. Because DMs were present already in the primary tumor cells from this colon carcinoma, it is likely that also c-*myc* had been amplified during *in vivo* growth of the tumor.

An extensive search for changes in other oncogenes and tumor cells has thereafter revealed amplifications that do not show up in cytogenetic analysis. Thus, for example, the c-*myb* oncogene is amplified in a characteristic marker chromosome of a colon carcinoma without evidence of HSRs or DMs and in other tumors, the amplified c-*abl* and c-*myc* oncogene loci map to abnormally banding regions in translocated or resident chromosomal segments. The finding of moderately amplified oncogenes also in chromosomal sites lacking HSRs suggests that (onco)gene amplification may be more common than the structural alterations revealed by chromosome banding and microscopy.

C. Mechanisms of Gene Amplification

The mechanism of gene amplification and the structure of the amplified DNA have been resolved mainly in experimental settings involving selection for drug resistance in cell culture. Although the mechanisms are still incompletely known and may vary in different cases, some general features have emerged. There seems to exist a spontaneous degree of illegitimate DNA replication in normal cells,

so that various segments of DNA are replicated more than once during a single cell cycle. In unselective conditions, this DNA is probably lost because the newly synthesized extra copies of DNA are not covalently linked to chromosomal DNA of mitotic cells. However, if there is a selective pressure to retain an increased gene dosage, a progressive multiplication of gene copy number is obtained. Thus, the generation of DMs in some human tumors and during *in vivo* growth of some experimentally induced tumor cells has been thought to reflect the changes in the copy number of the genes involved in the formation of the malignant phenotype. [*See* GENE AMPLIFICATION.]

One amplification model proposes that multiple cycles of unscheduled DNA replication at a single locus during a single cell cycle result in the formation of superimposed replication bubbles. The ensuing multiple copies of DNA must resolve into a tandem linear array before the next mitosis. It is suggested that this occurs by homologous recombination between any of several repeated sequences within the amplified domain. Part of the recombinations would lead to extrachromosomal circles possessing an origin for replication; these could be the precursors of DMs. Because of the unequal recombinations, the resolved linear structures would consist of tandemly repeated but heterogeneous units. A gradient of amplification may be formed so that centrally located sequences are amplified more than sequences distal to the origin of replication.

Several lines of evidence suggest that gene amplification is particularly common at specific sites of the genome. For example, on introduction of genes into the chromosomes of cells, the chromosomal site of integration of transfected genes significantly affects the frequency and cytogenetic result of their experimentally induced amplification. The amplification frequency in some transfectants has been found to be 100-fold that of the others. This suggests that there are preferred chromosomal positions for amplification of cellular genes and that chromosomal rearrangements (e.g., translocations) may facilitate gene amplification by a favorable positioning of chromosomal sequences.

The incidence of cells bearing amplified genes under conditions of cytotoxic selection can vary by two orders of magnitude and is greatly increased by the presence of mitogenic substances (hormones or tumor promoters) during selection or certain carcinogenic or cytotoxic agents before selection. Mitogenic hormones probably increase DNA overreplication. They may also enhance the colony-forming efficiency of drug-resistant cells in selective conditions. Most of the cytotoxic agents that increase the incidence of gene amplification are inhibitors of DNA synthesis. Aberrant replication is known to take place after transient inhibition of DNA synthesis, and this response can lead to gene amplification. Chemicals known to act as tumor promoters could, in principle, have major effects on the expression and amplification of oncogenes or even possible recessive cancer genes that have suffered carcinogenic insults. However, many carcinogens seem to induce specific DNA amplification in experimental conditions. [*See* DNA SYNTHESIS.]

II. Tumor Specificity of Oncogene Amplification

A. Neuroblastoma: Amplification of The N-*myc* Oncogene

The N-*myc* gene was originally isolated as an amplified DNA fragment with homology to the c-*myc* gene from a neuroblastoma cell line displaying an HSR in its karyotype. Neuroblastoma is one of the most common malignancies of childhood, arising in the neural crest cells most often in adrenal, retroperitoneal and mediastinal tissues. N-*myc* amplification has since been detected in about 40% of advanced neuroblastomas as well as in many cultured neuroblastoma cell lines. For example, the HSR on the abnormal chromosome 1 of IMR-32 neuroblastoma cells is composed of about 3,000 kb of chromosomal DNA, including the N-*myc* locus and sequences from at least three distinct domains of the short arm of chromosome 2. This structure forms a gradient of differentially amplified sequences, so that the centrally located sequences are amplified more than sequences more distal in the amplified DNA.

There is a clear correlation between the aggressiveness of neuroblastoma tumors and their N-*myc* gene copy number, which may range from 3–10 to 100–300. Neuroblastoma can be classified in four different stages according to the degree of spread of the tumors. N-*myc* amplifications have been diagnosed in about 50% of tumors of stage IV, but only in a few tumors of stage I or II. The prognosis of patients at stage III and IV is usually poor, their disease-free 2-year survival is only 10–30% compared with the 75–90% survival of stage I and II

patients. However, it has been shown that the amplification of N-*myc* is associated with a poor prognosis independently of the stage of the tumor: for example, 9 months after diagnosis, no progression of the disease was seen in 61% of stage IV patients having no amplifications of N-*myc*, whereas progression of the disease was seen in more than 90% of the patients having more than 10 copies of the N-*myc* gene.

It is not clear whether the amplification of N-*myc* is involved in the process of tumor progression. *In situ* hybridizations on tissue sections have indicated that even within one tumor, the levels of expression of the N-*myc* gene can vary between individual cells, being highest in undifferentiated neuroblasts. So it is possible that those tumor cells, which have amplified the N-*myc* gene, grow faster than the rest of the tumor cells and are therefore diagnosed at an advanced stage. In fact, two classes of neuroblastoma have been identified that may well correspond to tumors with and without N-*myc* amplification. Although the number of N-*myc* copies in primary untreated neuroblastomas is a clinically important prognostic factor, an even better indicator might be the concentration of the N-*myc* protein in tumor tissue. Immunohistochemical determination of the N-*myc* protein can reveal enhanced expression originating not only from an amplified gene, but also from a deregulated single copy gene.

B. Small Cell Lung Carcinoma: Amplifications of the *myc* Gene Family

The c-*myc*, N-*myc*, and L-*myc* genes have all been found to be amplified in several human tumors. They all encode for nuclear phosphoproteins with the ability to bind to DNA. Amplifications of the c-*myc* gene are most common among the three genes, and they are particularly frequent in lung carcinoma cell lines known as small-cell carcinomas (SCLC) and in breast cancer. It is remarkable that the *myc* amplifications are so frequent in cell lines from SCLC, which represents one-fourth of all human lung tumors. Altogether, 35–40% of established SCLC cell lines have amplified one of the three members of the *myc* gene family, but in no case have two different *myc* genes been shown to be amplified in the same cell line.

Gene amplification in general is associated with highly enhanced RNA expression and abnormally high levels of the corresponding protein product. Amplifications of the *myc* gene family usually lead to enhanced *myc* mRNA and protein expression. Furthermore, amplification of the c-*myc* gene seems to correlate with the phenotypic properties of the corresponding cells. The so-called variant SCLC cell lines grow faster and are more loosely aggregated, and they have an increased cloning efficiency when compared with typical SCLC lines. Also variant cells are less sensitive to irradiation *in vitro* and fail to express certain biochemical markers characteristic for classical SCLC lines. Strikingly, an increased copy number of the c-*myc* gene was detected more commonly in the variant SCLC cell lines but in only rare cell lines with a classic morphology. The variant SCLC lines have been proposed to be derived from SCLC tumors having unusual morphologies resembling those of large cell carcinoma.

Contrary to the results from cell lines, c-*myc* amplifications are not very common in primary SCLC tumors. According to data based on a collection of more than a hundred SCLC patients from different studies, the frequency of amplification of the c-*myc* gene in SCLC patients is only about 2%, whereas it is 7% for N-*myc* and 12% for L-*myc*. In squamous cell carcinomas, c-*myc* amplification was diagnosed in 15% of the cases. At the present, there is no apparent explanation for the observed discrepancy between the *in vitro* and the *in vivo* situation. However, it has been claimed that variant histological features can be seen in less than 10% of SCLC tumors at the time of diagnosis, but in more than 30% of autopsy samples, suggesting a selection of the variant morphology during treatment or the subsequent course of the disease.

Studies on experimental gene amplification suggest that anticancer therapy might not only provide a growth selection for cells, which already have amplified a *myc* gene, but could also promote the generation of gene amplifications. It has been shown that the rate of gene amplification may be enhanced *in vitro* up to 1,000-fold by various treatments that inhibit DNA replication as well as metabolic inhibitors of DNA synthesis, and agents that introduce adducts into DNA (hypoxia). High *myc* expression might allow the cells with an amplified *myc* gene to survive under the poor cell growth conditions prevailing during anticancer therapy. Therefore at relapse, selection might favor clones with extra *myc* copies. Thus far, however, the comparative analyses of *myc* amplifications in primary lung tumors and their metastases have resulted in inconsistent data.

C. Glioblastomas and Squamous Cell Carcinomas: Amplification of c-*erb* B-1, a Growth Factor Receptor Gene

Epidermal growth factor (EGF) is a polypeptide mitogen that induces a pleiotropic response in its target cells. Although the exact physiological functions of EGF remain unknown *in vivo,* it seems to accelerate several developmental processes related to fetal and postpartal growth and maturation. In *in vitro* culture, various cells express receptors for the EGF peptide and respond to EGF stimulation by an intricate sequence of events including tyrosine phosphorylation of the internal part of the transmembrane EGF receptor (EGF-R) protein and culminating in mitogenesis in permissive conditions.

Comparison of the amino acid sequence of purified human EGF-R with several deduced amino acid sequences in a data bank showed that EGF-R is the homologue of the v-*erb* B oncogene of avian erythroblastosis virus. This immediately suggested that molecular lesions of the EGF-R gene, also known as c-*erb* B-1, may be involved in human malignancy as well. Squamous cell carcinomas and brain tumors known as glioblastoma multiforme (GM) were indeed found to contain 6–60-fold amplifications of the c-*erb* B-1 gene and elevated concentrations of the protein product of the gene (EGF-R). GM is a highly anaplastic tumor of relatively undifferentiated neuroglial cells. Overall, GM accounts for about 25–30% of all intracranial tumors and for more than 50% of all primary gliomas. In several GM tumors, rearranged versions of the c-*erb* B-1 gene have been detected. It is also of interest that the c-*erb* B-1 gene is highly expressed in many glioblastomas that do not have amplification of the gene. The c-*erb* B-1 gene is also amplified and highly expressed in several squamous cell carcinomas.

D. Breast Cancer and Some Other Adenocarcinomas: Amplification of *neu,* a Receptor Tyrosine Kinase Gene

The *neu* oncogene was first identified as a transforming gene in ethylnitrosourea (ENU)-induced rat neuro/glioblastomas. The homologous human gene was cloned as an EGF-R related cDNA by two groups and named either HER-2 (human epidermal growth factor receptor) or c-*erb* B-2 according to the designation of the related EGF-R-gene, also known as c-*erb* B-1. The *neu* gene is characteristically activated by an ENU-induced point mutation changing glutamic acid for valine in the transmembrane domain of this receptor protein.

There have been several reports on the frequent amplifications of the *erb* B-2/*neu* gene in human mammary carcinomas and some other adenocarcinomas. Amplification has been reported in about one-fourth of primary breast carcinomas. Statistical evidence has been obtained on the significance of *neu* amplification as a determinant of the prognosis of the breast carcinoma patients. Patients with multiple copies of *neu* in their tumors had sooner relapses and shorter survival times. However, although such an association was obvious in the group of patients with disease spread to lymph nodes, the material was insufficient for estimation of the prognostic value in patients with stage I disease. Several studies have addressed this same question, and despite highly controversial results the general opinion now agrees on the correlation of *neu* amplification and poor prognosis of the disease. A study to support this opinion has been an expanded series of more than 300 patients whose breast carcinomas had metastasized to the lymph nodes. Thus *neu* amplification appears to be among the best prognostic variables after the number of tumor-positive nodes evaluated by histopathological techniques. Amplification means an especially poor prognosis for patients having metastases.

The primary role of *neu* in mammary carcinomas is also supported by studies carried out in transgenic mice. The directed expression of active *neu* oncogene in mammary cells of mice leads to frequent adenocarcinomas of the mammary gland, suggesting that in this experimental model few other oncogenes are needed for carcinogenesis *in vivo*. Besides mammary carcinomas, some other adenocarcinomas (e.g., ovarian carcinomas and stomach carcinomas) have been shown to contain amplifications of *neu*. Other oncogenes have also been found to be amplified in human mammary carcinomas, notably two closely located genes encoding polypeptides of the fibroblast growth factor gene family, *int*-2 and *hst*, but their clinical significance is not known. Despite repeated efforts, expression of these genes has not been detected at the mRNA level, suggesting that the region of chromosome 11, band q13 involved in these amplifications may contain still other genes important in breast carcinoma. [*See* BREAST CANCER BIOLOGY.]

III. Significance of Oncogene Amplification in Multistage Carcinogenesis and Tumor Progression

It is obvious from data summarized in the preceding sections that amplification of certain oncogenes is a common correlate of the progression of some tumors and also occurs as a rare sporadic event affecting various oncogenes in different types of cancer. Amplified oncogenes, whether sporadic or tumor-type-specific, are also expressed at elevated levels, in some cases in cells where their diploid forms are normally silent. Increased dosage of an amplified oncogene may therefore contribute to the multistep progression of at least some cancers.

In a few cases, specific genes are amplified in normal cells during developmental processes; there are examples of prokaryotes, yeast, *Drosphila,* and vertebrates. Amplification can be transient during growth or permanent in terminally differentiated cells. It cannot yet be excluded that amplification of, for example, N-*myc* occurs in rare stem cells of potential future neuroblastomas or transiently during some phases of differentiation of neural cells. However, if this were the case, the finding of amplified oncogenes in cancer cells could just reflect their stage of differentiation in a developmental cell lineage. If they exist, normal cells with amplified oncogenes should be rare, because they have not revealed themselves in analysis of normal cell populations or tissues. Also, we would have to postulate specific mechanisms for ensuring differential and synchronized gene amplification in specific normal cells whenever expression must be faster than can be achieved by transcription from a single copy of the gene. In cancer cells, less specific mechanisms will suffice, because tumors appear to contain ongoing clonal expansions that occur at the cost of phenotypically inferior sibling cell lines.

One possibility is that preexisting mutations of oncogenes could ignite the process of amplification in some tumors by first converting a protooncogene into an active oncogene. Subsequently, a selective growth advantage of the actively dividing initiated cells could lead to amplification of the corresponding gene. Loss of cancer suppressor genes has also been shown in various tumors carrying oncogene amplifications. These results fit to the multistage theory of cancer development and progression. Apparently, cooperating lesions in cellular oncogenes accumulate during tumor growth and clonal selection and increase the malignant potential of the tumor cells. It may be that activated oncogenes have specific roles in the accelerated genomic evolution of tumor cells.

It seems that the molecular diagnosis of oncogene amplification might serve to denote additional genetic lesions associated with a more aggressive disease in small-cell lung cancer and maybe also in neuroblastomas, squamous cell carcinomas, and glioblastomas, and an alternative lesion associated with tumor progression in sporadic cases. However, most leukemias, for example, do not show *myc* amplifications and therefore multiple other somatic lesions such as c-*myc* translocation or point mutations in the c-*ras* genes could suffice for the clonal expansions of these tumor cells. The role of c-*myc* amplification would be to provide increased dosage of a protein that is advantageous for growth during tumor progression, especially in cells that would not otherwise divide easily. The development of solitary mammary carcinomas in a sporadic manner in transgenic pregnant mice that abundantly express c-*myc* in response to corticosteroids in sensitive tissues is also compatible with the view that, besides an elevated dosage of c-*myc*, some other factors are necessary for the emergence of malignant neoplasms.

There may be no obligatory sequence for activation of oncogenes in the genesis of any particular tumor. Amplification of an oncogene could play its part in malignant progression of already initiated cells whenever it happened to occur. Generally, enhanced expression of an oncogene could be a necessary prerequisite for acquisition of a growth advantage by cells having extra copies of the gene. This effect could also be the principal contribution of amplification to tumorigenesis. [*See* MALIGNANT MELANOMA.]

Bibliography

Alitalo, K., and Schwab, M. (1986). Oncogene amplification in tumor cells. *Adv. Cancer Res.* **47,** 235–281.

Alt, F. W., DePinho, R., Zimmerman, K., et al. (1986). The human *myc* gene family. *Cold Spring Harbor Symp. Quant. Biol.* **51,** 931–941.

Biedler, J. L., Meyers, M. B., and Spengler, B. A. (1983). Homogeneously staining regions and double minute chromosomes, prevalent cytogenetic abnormalities of human neuroblastoma cells. *Adv. Cell. Neurobiol.* **4,** 268–301.

Bishop, J. M. (1983). Cellular oncogenes and retroviruses. *Annu. Rev. Biochem.* **52,** 301–354.

Bishop, J. M. (1985). Viral oncogenes. *Cell* **42,** 23–38.

Minna, J. D., Battey, J. F., Brooks, B. J., et al. (1986). Molecular genetic analysis reveals chromosomal deletion, gene amplification, and autocrine growth factor production in the pathogenesis of human lung cancer. *Cold Spring Harbor Symp. Quant. Biol.* **51,** 843–853.

Schimke, R. T., Hoy, C., Rice, G., Sherwood, S. W., and Schumacher, R. I. (1988). Enhancement of gene amplification by perturbation of DNA synthesis is cultured mammalian cells. *In* "Cancer Cells 6. Eukaryotic DNA Replication" (T. Kelly and B. Stillman, eds.), Cold Spring Harbor Laboratory, Cold Spring Harbor, New York. 317–322.

Schwab, M. (1985). Amplification of N-*myc* gene in human neuroblastomas. *Trends Genet.* **1,** 271–275.

Slamon, D. J., Godolphin, W., Jones, L. A., et al. (1989). Studies of the HER-2/*neu* proto-oncogene in human breast and ovarian cancer. *Science* **244,** 707–711.

van de Vijver, M., Peterse, J. L., Moor, W. J., et al. (1988). *Neu*-protein with comedo-type ductal carcinoma *in situ* and limited prognostic value in stage II breast cancer. *N. Engl. J. Med.* **319,** 1239.

Venter, D. J., Kumar, S., Tuzi, N. L., and Gullick, W. J. (1987). Overexpression of the c-*erb* B-2 oncoprotein in human breast carcinomas: Immunohistological assessment correlates with gene amplification. *Lancet* **II,** 69.

Oogenesis

TONI G. PARMER, GEULA GIBORI, *University of Illinois at Chicago*

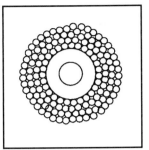

I. Gametogenesis: Primordial Germ Cells and the Formation of Primary and Secondary Oocytes.
II. Growth of the Oocyte
III. Inhibition of Oocyte Maturation
IV. Induction of Oocyte Maturation
V. Regulation of Metabolism during Oocyte Growth

Glossary

Dictyate stage Dipotene stage of the first meiotic division in which the meiotic process of the oocyte has been arrested

Germinal vesicle Nucleus of the primary oocyte

Germinal vesicle breakdown Breakdown or dissolution of the nucleus of the preovulatory dictyate oocyte, initiated when the oocyte resumes meiosis

Meiotic competence Ability of the primary oocyte to resume meiosis *in vitro*

Oogonia Primordial germ cells that have entered the female gonad and continue their mitotic proliferation

Primary oocyte Oogonia that have entered into the first meiotic division

Primordial germ cells Undifferentiated, mitotically dividing stem cells found in both sexes, which are destined to give rise to the gametes

Secondary oocyte Oocyte that has completed the first meiotic division and contains the haploid number of chromosomes

OOGENESIS HAS BEEN DEFINED as "the entire period of female germ cell differentiation, from the oogonial stage to the mature gamete." The following discussion is mainly concerned with mammalian oogenesis with particular reference to humans.

However, when necessary, reference to studies involving lower vertebrates will be made, as knowledge of the processes in mammals is incomplete.

I. Gametogenesis: Primordial Germ Cells and the Formation of Primary and Secondary Oocytes

In vertebrate embryos, cells destined to give rise to the gametes are known as the primordial germ cells. In the mouse embryo, the primordial germ cells are first seen at the posterior end of the primitive streak about 8 days postcoitum. From this location, the cells migrate to the allantois, to the hindgut epithelium, up the dorsal mesentary, and, by 10–11 days postcoitum, they enter the germinal ridges (from which the gonads arise). The migration of these cells is believed to occur by amoeboid movement using lytic enzymes to penetrate cell membranes.

At the time that they can first be recognized, the primordial germ cells number approximately 40–50 in the human embryo. During migration to the germinal ridges, they undergo numerous mitotic divisions and increase to 600–1,300 cells. Once they reach the germinal ridges, they become embedded in the epithelium, where they are known as oogonia. The oogonia continue to proliferate at a high rate markedly increasing in number. In the human, this stage of rapid proliferation begins in the fifth or sixth week of embryonic development. From the 10th week on, an increasing proportion of these oogonia enter meiotic prophase while the remainder continue to proliferate by mitosis.

The epithelium of the germinal ridge forms the surface of the gonad and is known as the cortex. The primitive sex cords, strands of compactly arranged cells in the interior of the gonad, form the medulla. In the female embryo, the medulla be-

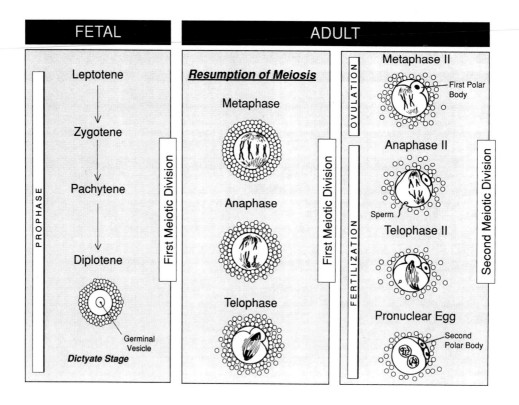

FETAL | ADULT

PROPHASE

Leptotene

Zygotene

Pachytene

Diplotene

Germinal Vesicle

Dictyate Stage

First Meiotic Division

Resumption of Meiosis

Metaphase

Anaphase

Telophase

First Meiotic Division

OVULATION

FERTILIZATION

Metaphase II

First Polar Body

Anaphase II

Sperm

Telophase II

Pronuclear Egg

Second Polar Body

Second Meiotic Division

FIGURE 1 Meiosis in the oocyte. To illustrate, only four pairs of chromosomes have been depicted. During fetal development, the oocyte enters the prophase stage of the first meiotic division. At diplotene, the primary oocyte enters a prolonged resting period and is said to be in the dictyate stage. Shortly before ovulation meiosis is resumed. The oocyte completes the first meiotic division resulting in a secondary oocyte and the first polar body. The second meiotic division begins almost immediately but is arrested at metaphase II. At this stage ovulation occurs. The completion of the second meiotic division takes place only at the time of fertilization.

comes reduced, the primary sex cords are resorbed, and the interior of the gonad becomes filled with loose mesenchyme permeated by blood vessels. The oogonia remain embedded in the cortex, close together in groups. They are connected by intercellular bridges, which are thought to result from incomplete separation of daughter cells during mitotic divisions. These bridges allow for the free exchange of cytoplasmic material between the connected cells and probably play a role in the synchronization of germ-cell differentiation. During the growth of the gonad, the cortical cells on the inner surface of the cortex split up into groups; strands of these cells then surround one or several of the oogonia that have entered meiosis. These oogonia are now called primary oocytes. The strands surrounding them become a single layer of flattened cells called follicular cells. The whole structure is called a primordial follicle. Oocytes at this stage of development can be seen at 4 mo gestation but usually remain few in number until about the fifth or sixth month.

By the fifth month of intrauterine life, the oogonia reach a maximum number of approximately 7 million and consist of both "resting" and dividing oogonia together with primary oocytes at all stages of meiotic prophase up to the diplotene stage. At birth the number of oogonia and oocytes declines to

approximately 2 million. The reason for this loss is not known but is due mainly to a large decrease in the number of small oocytes. Oocytes in the pachytene and diplotene stages of meiosis also degenerate.

Once the primary oocyte reaches the diplotene stage of meiotic prophase, it enters a prolonged resting period until shortly before ovulation (Fig. 1). Before being released from the follicle, the oocyte undergoes its first meiotic division producing a secondary oocyte and the first polar body. Nearly all of the cytoplasm is retained by the secondary oocyte. The second meiotic division begins almost immediately but is arrested at metaphase. At this stage, the secondary oocyte, with its first polar body still contained within the zona pellucida, is released at ovulation. The completion of meiosis takes place only at the time of fertilization. [See MEIOSIS.]

II. Growth of the Oocyte

A. Ultrastructure of the Mammalian Oocyte and Changes in Cellular Organelles during Oocyte Growth

The fine morphology of oogonia is essentially very similar to that of the primordial germ cells. Like all mitotic cells, oogonia exhibit a pair of centrioles at either pole of their mitotic spindles. Shortly after the end of mitotic proliferation, the centrioles disappear, so that from the commencement of the meiotic divisions until the time of fertilization the oocyte is devoid of these structures.

The fine morphology of meiotic oocytes has been studied extensively in the rat, rabbit, cow, and human and has been shown to be essentially identical in these species. During early meiosis, prominent changes occur in the nucleus or germinal vesicle of the oocyte with little change in the cytoplasm. From the preleptotene to pachytene stages of early meiosis, chromatin condenses into isolated chromosomes; later, homologous chromosomes pair with each other. During diplotene, unpaired chromosomes reappear—similar to those seen during leptotene. In the human, this chromosomal pattern persists throughout the length of the resting period. In the rat and rabbit, it disappears and the chromatin assumes a netlike form in which it is distributed uniformly throughout the nucleoplasm.

At the end of meiotic prophase, the intercellular bridges between oocytes disappear, causing isolation of individual oocytes surrounded by several layers of follicle cells.

The life of the oocyte can essentially be divided into three main stages: the quiescent stage, the maturative stage, and the preovulatory stage.

1. Quiescent Stage

The quiescent stage refers to the period of meiotic arrest. This period varies among species of mammals. For example, in the human, it can vary between about 12 yr and several decades. During this period, the oocyte has a spheroidal shape, with a diameter of approximately 50–70 microns, and is still metabolically active; for instance, it continues to synthesize RNA. Close to the nucleus, histological preparations show a crescent-shaped region densely packed with basophilic structures, known as the Balbiani vitelline body, or yolk nucleus complex. This area contains the majority of the cell's organelles: a prominent Golgi apparatus with few mitochondrial and endoplasmic reticulum elements. The rest of the cytoplasm consists almost exclusively of free ribosomes.

The most prominent of the juxtanuclear organelles are the annulate lamellae. These structures have been identified in the rabbit and golden hamster but not in other mammalian species. They consist of stacks of up to 100 parallel paired membranes with flattened cisternae, about 30–50 microns wide. Their number varies considerably, and their function is not known. Some investigators believe that they are produced by a blebbing activity of the nuclear membrane, whereas others believe that they are a specialized form of the endoplasmic reticulum because they are occasionally associated with ribosomes.

2. Maturative Stage

The beginning of the maturative stage is marked by the resumption of the mitotic activity of the cumulus cells surrounding the oocyte, presumably as a result of the actions of follicle-stimulating hormone (FSH) and estrogen. The follicle cells exhibit morphological signs of enhanced secretory activity, such as large numbers of cytoplasmic ribosomes and prominent Golgi complexes. The secretory activity brings about the deposition of liquor folliculi, which in turn causes the formation of a fluid-filled space, the follicular antrum. As the volume of the follicular antrum increases, the cells that immediately surround the oocyte, the corona radiata, and the rest of the surrounding cells, the cumulus oophorus, separate from the remaining follicle cells. As a result, the oocyte and the surrounding cumulus come to occupy an eccentric position in the follicular cavity.

During this stage, the oocyte increases in volume until it reaches a diameter of approximately 70–85 microns. Then its growth ceases, while the follicle continues to grow. This oocyte is said to be meiotically competent (i.e., it has the ability to resume meiosis *in vitro*). Only oocytes that have completed growth and meiotic maturation can be fertilized and undergo normal development.

During the period of active growth, the oocyte develops the zona pellucida, a thick, translucent coat that surrounds the plasma membrane of the oocyte and is made up of three glycoproteins with molecular weights of 83,000, 120,000, and 200,000. These proteins are produced by the oocyte; in fact, in culture, growing mouse oocytes, which have been denuded of follicle cells, still secrete these glycoproteins. The zona pellucida is not formed as a continuous extracellular membrane around the oocyte but, rather, is laid down during oocyte growth in discontinuous portions that eventually fuse.

During the formation of the zona pellucida, the cells of the cumulus oophorus exhibit an increase in surface activity and form cytoplasmic protrusions. These traverse the zona to reach the surface of the oocyte, to which they are firmly anchored by means of intercellular junctions. Also during this time, microvilli develop over the surface of the oocyte.

One of the most noticeable changes during this stage is the resumption and completion of the first meiotic division. It is highlighted by the breakdown of the germinal vesicle, referred to as germinal vesicle breakdown (GVBD). This is actually the condensation of the nucleolus, followed by the disappearance of the nuclear membrane and liberation of the chromosomes in a cytoplasmic area devoid of organelles.

Metaphase I soon follows and the chromosomes become arranged at the equatorial plate of the meiotic spindle. The first meiotic division results in the liberation of the first polar body into the perivitelline space.

The first polar body is morphologically similar to the oocyte: it contains microvilli and cortical granules. The absence of these components in the second polar body provides a means of distinguishing the two. Another difference is that the nucleus of the first polar body consists of isolated chromosomes, without a nuclear membrane, whereas the second polar body contains a nucleus complete with a double membrane.

3. Preovulatory Stage

At the formation of the first polar body, the oocyte enters the preovulatory stage. In this stage, the most prominent change is the relationship between the oocyte and the cumulus cells. The cumulus cells separate from the zona pellucida and from one another.

The organelles of the oocyte also undergo profound structural changes. The mitochondria increase in number and become uniformly distributed throughout the cytoplasm while maintaining a close association with the endoplasmic reticulum. The endoplasmic reticulum develops in an area that is either very elongated or concentrically arranged and become filled with a highly dense material. These changes are thought to be functionally related to the high rate of protein synthesis known to take place during oocyte growth and maturation. The greatest changes occur in the Golgi complex, which after dividing into multiple aggregates, migrates to the periphery of the cell.

Following extrusion of the first polar body the chromosomes of the now secondary oocyte form a compact, crescent-shaped mass that moves to a metaphase II arrangement. In this stage the oocyte is ovulated.

B. Interactions between Oocytes and Follicular Cells

The cumulus cells, which surround the follicle, are believed to regulate oocyte growth by providing nutrients. This role is supported by the observation that oocytes will grow *in vitro* only if cumulus cells are in contact with the oocyte. Gap junctions between cumulus cell projections and the oocyte provide channels for the passage of ions and small molecules (molecular weight <1000 Da). Such metabolic coupling (or metabolic cooperactivity) between the two cells is essential for oocyte function. [*See* CELL JUNCTIONS.]

In vitro studies with oocytes enclosed in cumulus cells, in fact, show that >85% of the metabolites derived from the culture medium and ending up in the oocyte are first taken up by and metabolized in the cumulus cells. They reach the oocyte via these gap junctions. In addition, a positive correlation exists between oocyte growth and the extent of ionic coupling between the oocyte and the cumulus cells.

The gap junctions may be important in maintaining meiotic arrest. The cumulus cells around an immature oocyte are closely apposed to the zona pellucida, which is studded with numerous cytoplasmic projections, whereas around a mature oocyte the cells are loosely organized and few projections are seen in the zona pellucida. Electrophysiological measurements indicate that ionic coupling between cumulus cells and oocyte decreases progressively as the time of ovulation approaches. Thus, the release of the oocyte from meiotic arrest may be the result of metabolic uncoupling of the cumulus cells from the oocyte, possibly induced by the luteinizing hormone (LH) surge. This will be further discussed in Section V.

III. Inhibition of Oocyte Maturation

A. Role of Cyclic Adenosine Monophosphate

A great deal of evidence indicates that cylic adenosine monophosphate (cAMP) plays a fundamental role in the inhibition of oocyte maturation in mammals. Analogs of cAMP, such as dibutyryl cAMP, or inhibitors of cAMP degradation, such as the methylxanthines, have been shown to inhibit oocyte maturation. The resumption of meiosis in follicle-enclosed oocytes is inhibited by both dibutyryl cAMP and isobutylmethylxanthine. These effects on oocyte maturation probably reflect the physiological events, because the levels of cAMP in the follicle significantly decrease shortly before resumption of meiosis, both under *in vitro* and *in vivo* conditions.

Cyclic AMP produces a greater inhibition of maturation in the cumulus-enclosed oocyte than in oocytes denuded of their cumulus cells. Exposure to cholera toxin or forskolin, which activate adenylate cyclase and thus increase levels of cAMP, inhibit spontaneous maturation in rat cumulus-oocyte complexes but not in denuded oocytes. These results led to the suggestion that the cells of the cumulus oophorus are the source of the inhibitory cAMP, which may be transferred to the oocyte via gap junctions. This idea was supported by the observation that during incubation of hamster cumulus-enclosed oocytes in the presence of forskolin, a parallel increase in cAMP was seen in both oocyte and cumulus cells. On the other hand, several investigators were unable to demonstrate the transfer of cAMP from the cumulus cells to the oocyte. Although they were able to demonstrate cumulus cell-dependent inhibition of oocyte maturation by cAMP modulators, they failed to find any difference in cAMP levels between oocytes cultured in the presence of forskolin and controls.

Thus, although compelling evidence indicates that cAMP can be transferred from the cumulus cells to the oocyte and prevent maturation, it cannot be ruled out that, physiologically, the inhibitory effect is mediated by another agent, which requires cAMP for its action.

B. Role of Gonadotropins

FSH has been shown to inhibit the maturation of oocytes of several species including the rat, pig, and mouse. Compounds that increase levels of cAMP also inhibit oocyte maturation in isolated oocyte cumulus complexes, suggesting that cAMP mediates the FSH action. As we will see later, LH also increases the level of cAMP in isolated oocyte cumulus complexes; however, FSH is significantly more active and is probably the major gonadotropin that stimulates cAMP accumulation in the oocyte inhibiting maturation. FSH and cholera toxin, however, do not inhibit oocyte maturation or stimulate cAMP accumulation in the absence of the cumulus cells, whereas forskolin has its effect in the presence or the absence of these cells. Thus, the oocyte appears to have the catalytic subunit of adenylate cyclase but seems to lack the guanosine triphosphate-binding protein, which is necessary for activation of the enzyme by either FSH or cholera toxin.

C. Role of Steroids

The involvement of steroid hormones in the inhibition of oocyte maturation is the subject of much controversy. Some researchers have reported that steroid hormones stimulate the rate of meiosis, although LH can still initiate meiosis in follicles even after the total inhibition of follicular steroidogenesis. Others argue that steroids are involved in the arrest of oocyte maturation. When mouse oocytes were cultured with testosterone, progesterone, or androstenedione, GVBD was partially inhibited, whereas estradiol, estrone, and dihydrotestosterone had no effect. In studies on pig oocytes, however, estradiol reduced the maturation of denuded oocytes, whereas progesterone inhibited this effect. A similar effect of estradiol was seen in denuded, cumulus-enclosed hamster oocytes, in which estradiol enhanced forskolin-dependent meiotic arrest. In other studies, estradiol was found to inhibit spontaneous maturation in denuded mouse oocytes. Thus, the precise role of follicular estrogen on oocyte maturation is obscure. The differences between studies may be due to differences in the nature of the culture media used. Reportedly, estradiol has the capacity to arrest meiosis of either intact or denuded oocytes *in vitro* but only if no exogenous protein(s) are present in the media. Because *in vivo* the oocyte cumulus complex is bathed by a highly proteinaceous fluid, estrogens may not play a physiological role in the maintenance of meiotic arrest. The preovulatory surge of LH induces the resumption of meiosis and a rise in follicular steroidogenesis, including a transient rise in estrogen production. Estrogen remains elevated in some species when GVBD takes place, in other species when the oocyte becomes committed to mature. Even though GVBD takes place later on, these findings argue against the suggested role of estrogen as a follicular inhibitor of the resumption of meiosis. However, estradiol may be involved in the maintenance of the structural integrity of the oocyte cumulus complex. Thus, estrogen has been shown to regulate the appearance and maintenance of gap junctions between cells of the membrane granulosa and of the myometrium, and it may perform a similar role in the maintenance of junctions between cumulus cells or between cumulus cells and the oocyte.

The role of androgens and progestins in the maturation of mammalian oocytes is also unclear. Progesterone has been shown to facilitate the maturation of denuded rabbit and bovine oocytes, but a similar effect has not been seen on the maturation of human and porcine oocytes. Addition of steroids to the culture medium does not induce maturation of bovine, porcine, and rat oocyte-cumulus complexes. Only at high concentrations (100 μM) does progesterone block GVBD and polar-body formation from oocytes in medium- and large-sized follicles. Reportedly, large quantities of progesterone are present in media incubated with oocyte-cumulus complexes, indicating that cumulus cells can secrete steroid hormones. Thus, although the evidence for a primary regulatory role of steroids in oocyte maturation is unconvincing, what role these molecules have in modulating the meiotic process remains to be determined. [*See* STEROIDS.]

D. Oocyte Maturation Inhibitor

In 1935, Pincus and Enzmann showed that when rabbit oocytes are liberated from their follicles they undergo spontaneous maturation *in vitro*, without hormonal stimulation. This was found to be the case for all mammalian species studied, leading to the suggestion that the follicle cells ''supply to the ovum a substance or substances that directly inhibit nuclear maturation.''

Subsequent studies found that resumption of meiosis was inhibited when porcine oocytes were co-cultured with porcine granulosa cells or with medium conditioned by granulosa cells, but not with the theca layer or the ovarian bursa. These results suggested that within the follicle the granulosa cells produce an oocyte maturation inhibitor (OMI), which keeps the oocyte in a state of meiotic arrest.

Follicular fluid from various species, such as the rabbit, bovine, porcine, and human, exerts this inhibitory effect on the oocyte. This effect is not species-specific: human follicular fluid, for example, can inhibit the maturation of rat oocytes. Oocyte maturation inhibitor activity appears to decline as follicular growth increases. Follicular fluid collected from humans indicated that OMI activity was much lower in follicles yielding mature and fertilizable oocytes as compared with follicles yielding immature or atretic oocytes.

Rat or porcine oocytes isolated from either medium or large follicles have been used in culture as a bioassay system to partially characterize and purify porcine OMI. Oocyte maturation inhibitor has been reported to have a molecular weight of less than 2,000 with two major peaks of activity. It is thought to be a peptide, since OMI cannot be extracted by ether or charcoal but is trypsin-sensitive and heat-stable. OMI isolated from porcine granulosa cell extracts has been found to be similar in properties to OMI isolated from porcine follicular fluid, which is similar to OMI from bovine follicular fluid. The characterization and isolation of the putative inhibitor, however, has yet to be accomplished. Problems in the adequacies of the assay systems employed, lack of a standard OMI preparation, and the variability of the experimental approaches have hampered the complete isolation of this molecule.

It has recently been postulated that OMI may be the purine hypoxanthine. This conclusion was based on the findings that the fractions of porcine follicular fluid with the greatest inhibitory effect on oocyte maturation also has an absorption maximum identical to that of hypoxanthine, had a retention time on high-pressure liquid chromatography of pure hypoxanthine, and exerted inhibition of oocyte maturation identical to that exhibited by a commercial preparation of hypoxanthine. Hypoxanthine also augmented the effect of FSH on inhibition of oocyte maturation as well as that of dibutyryl cAMP. At high concentrations hypoxanthine is known to inhibit cAMP phosphodiesterase activity, and this may be the mechanism of its action in the inhibition of oocyte maturation.

Another purine, adenosine, has also been shown to inhibit oocyte maturation at physiological concentrations (0.3–10 μM). Adenosine was also found to increase cAMP and progesterone production in response to FSH and LH in preovulatory oocyte-cumulus complexes, with little response in postovulatory complexes. In addition, adenosine augmented oocyte maturation inhibition by FSH. This response was specific for adenosine: no effect was seen with inosine, hypoxanthine, or guanine. Little effect of adenosine on oocyte maturation is seen in the absence of FSH.

The site of action of adenosine is probably cumulus cell-dependent because adenosine itself has no effect on the rate of spontaneous maturation of denuded oocytes. This conclusion is supported by the fact that FSH action on the oocyte-cumulus complex is the cumulus cell and the action of adenosine is FSH-dependent.

The mechanism of adenosine action is unknown. Adenosine action is associated with an enhancement of the accumulation of cAMP in the presence of FSH; several theories on the biochemical processes involved in this effect have been suggested. For example, adenosine may act on receptors to increase the accumulation of cAMP. Adenosine has been shown to increase adenosine triphosphate (ATP) levels in luteal, granulosa, and cumulus cells, whereas FSH depletes concentrations of ATP in the cumulus cells. Thus, adenosine may buffer the depletion of ATP by FSH, thereby maintaining elevated levels of cAMP, which is then shunted via gap junctions into the oocyte to inhibit maturation. Adenosine may also act on cumulus cells by decreasing intracellular levels of calcium, which has been shown to inhibit FSH-stimulated accumulation of cAMP.

IV. Induction of Oocyte Maturation

A. Role of Gonadotropins

Follicle-enclosed oocytes have been used to study the mechanism of induction of maturation because isolated oocytes resume meiosis spontaneously. The factor responsible for the initiation of meiosis is LH. However, neither the specific target cell(s) for LH nor the underlying biochemical events are known. Inhibition of steroidogenesis does not block the induced maturation by LH, and administration of progesterone, prolactin, or estradiol does not mimic its effect. FSH may be responsible for preparing the follicle for ovulation. Thus, at ovulation the processes that link the oocyte with the cumulus cells retract, and the resulting spaces now present between the cumulus cells are filled with a hyaluronidase-sensitive material. FSH *in vitro* has been shown to specifically stimulate the synthesis of hyaluronic acid by the mouse cumulus; therefore, the cumulus is thought to expand as the result of the gonadotropin-induced deposition of glycosaminoglycans, possibly hyaluronic acid between the cumulus cells. This effect is specific for FSH—it does not occur in response to LH or human chorionic gonadotropin (hCG). The action of FSH is essential for follicular growth prior to the LH surge; the cumulus expansion is seen only *in vitro,* not *in vivo*, suggesting that some follicular component inhibits expansion of the cumulus before the LH surge. An increase in proteinase activities (aminopeptidase, endopeptidase, trypsinlike, and elastaselike activities) in the cumulus cells after the LH surge may play a role in oocyte-cumulus changes.

Cyclic AMP, which appears to inhibit oocyte maturation in the preovulatory follicle, can also induce maturation. In fact, LH elevates cAMP levels in follicular fluid and treatment of follicles with dibutyryl cAMP induces maturation. Therefore, LH may induce oocyte maturation by a cAMP-mediated response. Indeed, forskolin, which enhances cAMP levels, has been shown to mimic the effect of LH on the ovarian follicle.

The seemingly contradictory effects of cAMP can be explained by the changes of the follicle before ovulation, which modulate the cAMP effect. Clearly maturation of the mammalian oocyte in the preovulatory follicle is suppressed by agents present within the follicle and is removed by the LH surge. By changing the metabolism of the oocyte-cumulus complex, LH would induce maturation in a manner similar to the spontaneous maturation that occurs in the isolated oocyte complex. Among the metabolic changes induced by LH, oxygen consumption is decreased with increased lactate production in the preovulatory cumulus at about the time the oocyte increases its respiration. This metabolic shift occurs before any alteration in the morphology of the oocyte-cumulus complex is seen; yet inhibition of the stimulatory effect of LH on follicular lactate production does not prevent its action on the resumption of meiosis.

Another change induced by LH is blocked communication in the oocyte-cumulus complex by causing an uncoupling among cumulus cells and between cumulus cells and the oocyte. This uncoupling reaction is also induced by cAMP and phosphodiesterase inhibitors. Thus, in the preovulatory phase, cAMP generated by the cumulus cells in response to FSH is transferred via the junctional processes to the oocyte to keep it in a state of meiotic arrest, but once the LH surge takes place the interruption of communication between the oocyte and cumulus cells stops the transfer of cAMP, and the oocyte can now resume meiosis.

However, not all laboratories have been able to show that termination of communication between the oocyte and cumulus cells takes place prior to the resumption of meiosis. For rat oocytes, some researchers have suggested that prior to uncoupling and the resumption of meiosis LH-induced maturation may be due to a desensitization of FSH receptor-activation of adenylate cyclase in cumulus cells. Thus, further investigations are required to determine more precisely the role of cAMP in the LH-induced maturation.

B. Role of Peptides and Growth Factors

Gonadotropin-releasing hormone (GnRH) or its agonist analogue, the role of which is to cause FSH and LH release from the pituitary, can induce the resumption of meiosis in the hypophysectomized rat and in follicle-enclosed oocytes *in vitro*. This response appears to be specific for GnRH, because a GnRH antagonist abolished the meiosis-inducing action. The antagonist, however, did not block the effect of LH, showing that GnRH does not appear to mediate the action of LH in the induction of oocyte maturation.

Epidermal growth factor (EGF), transforming growth factor-β (TGFβ), and insulin-like growth factors I and II (IGF-I and IGF-II) have been reported to stimulate oocyte maturation, whereas insulin and platelet-derived growth factor do not. TGFβ, EGF, and the IGFs had minimal effects on cAMP production by oocyte-cumulus complexes, yet their effects were inhibited by cAMP analogues, phosphodiesterase inhibitors, and hypoxanthine. Their mechanisms of action require further study. [*See* TISSUE REPAIR AND GROWTH FACTORS.]

V. Regulation of Metabolism during Oocyte Growth

A. Changes in Enzyme Activity

Knowledge of the enzyme systems that exist in the growing oocyte in mammals is minimal. The activities of several enzymes, however, have been reported to increase during the growth period of the oocyte. Thus, mitochondrial succinate dehydrogenase reaches a peak in preovulatory oocytes, glu-

cose-6-phosphate dehydrogenase (G-6-PD) and lactate dehydrogenase (LDH) reach maximum activity when the oocyte has a diameter of approximately 85 microns. In the mouse, LDH comprises about 5% of the total protein content of the oocyte, whereas the human and the rabbit have much lower levels.

RNA polymerase has also been shown to be present in the oocyte. RNA polymerase I, which is responsible for the synthesis of ribosomal RNA, increases during the initial phase of oocyte growth and then declines. Creatine kinase has been reported in mouse and rat oocytes and may play an important role in maintaining a high ATP–ADP ratio during oogenesis. These and other results indicate that the synthesis of many enzymes is completed in the oocyte before the onset of meiotic maturation.

B. DNA and RNA Synthesis

1. DNA

The DNA content of the growing oocyte has been examined in mammals by Feulgen microspectrophotometry. Oocytes contain more than the diploid amount of DNA; the "extra" DNA is associated with the mitochondria. The fully grown oocyte contains approximately 11 picograms (pg) of genomic and 2–3 pg of mitochondrial DNA. The last synthesis of DNA occurs at meiotic prophase. [*See* DNA SYNTHESIS.]

2. RNA

Most of the knowledge on this subject has been obtained from *in vivo* mouse studies. All the main RNA species have been shown to be present: ribosomal RNA (28S + 18S), 5 S-ribosomal RNA, transfer RNA, messenger RNA (mRNA), and heterogenous nuclear RNA. In the fully grown oocyte, 90–95% of total RNA is ribosomal and transfer RNA, the remainder being mostly mRNA. Total RNA content increases during oocyte growth and reaches a peak in fully grown oocytes. Its synthesis stops with antrum formation. Small, growing oocytes are estimated to contain approximately 0.29 ng of RNA, whereas fully grown oocytes contain about 0.57 ng. RNA of growing oocytes is unusually stable; in the mouse, for example, at least 80% of labeled RNA found 2 days after label administration is retained 10–20 days later.

During meiotic maturation, the fraction of mRNA in total RNA decreases from 19 to 10% and almost 20% of total RNA is degraded or deadenylated (i.e., loses the terminal poly(A) tail). The mechanism(s) involved in the reduction of mRNA and in the degradation–deadenylation processes are unknown, although RNA polymerase activity does decrease in the mature oocyte.

Mammalian oocytes store a large excess of maternal mRNA which is translated in the early postfertilization period. The synthesis of RNA apparently is not necessary for resumption of meiosis, because inhibitors of RNA synthesis do not inhibit nuclear progression.

C. Protein Synthesis

As the oocyte grows, cells volume increases tremendously. In the mouse, for example, its diameter increases from 15–20 microns to approximately 80–85 microns; the total protein content increases linearly with volume, reaching 28–30 ng in the fully grown oocyte. The absolute rate of protein synthesis increases 1.1 pg/hr in a quiescent mouse oocyte to 42 pg/hr in a fully grown oocyte. Ribosomal proteins are synthesized throughout oocyte growth, even though ribosomal RNA synthesis is not detectable. The cytoskeletal proteins, actin and tubulin, are synthesized both during growth and in the fully grown oocyte. Cytoskeletal and ribosomal proteins are the most abundant and most stable proteins in the growing oocyte.

As the oocyte progresses from the dictyate to metaphase II stage of meiosis, the pattern of proteins synthesized change markedly in a stage-specific way. This program is supported by preformed mRNA, because mRNA synthesis at the resumption of meiotic maturation is undetectable at this time. The changes in protein synthesis do not take place in oocytes that fail to undergo GVBD, suggesting that they are triggered by the contact of the oocyte's nucleoplasm and cytoplasm at the time of GVBD.

The precise nature and role of the proteins synthesized during meiotic maturation are not known. In oocytes liberated from follicles, inhibitors of protein synthesis do not prevent GVBD but stop meiosis; they do, however, block LH-induced maturation of follicle-enclosed rat oocytes. Thus, new protein synthesis must take place for nuclear progression to proceed beyond GVBD.

These changes in protein synthesis may be caused by a cytoplasmic factor produced or un-

masked independently of the nucleus, which is also involved in GVBD and nuclear maturation. In fact, nucleate oocyte fragments of mouse oocytes resume protein synthesis and meiosis in vitro, and anucleate fragments undergo some of the changes in protein synthesis. These findings have led to the suggestion that meiotic competence of the oocyte is due to the synthesis of new proteins directed by mRNA already present in the cytoplasm of the oocyte. The nature of the functions of these proteins are unknown.

Bibliography

Behrman, H. R., Preston, S. L., Pellicer, A., and Parmer, T. G. (1988). Oocyte maturation is regulated by modulation of the action of FSH in cumulus cells. In "Meiotic Inhibition: Molecular Control of Meiosis" (F. Haseltine, D. J. Patanelli, and N. First, eds.). Alan R. Liss, Inc., New York.

Brower, P. T., Gizang, E., Boreen, S. M., and Schultz, R. M. (1981). Biochemical studies of mammalian oogenesis: Synthesis and stability of various classes of RNA during growth of the mouse oocyte in vitro. *Develop. Biol.* **86,** 373.

Dekel, N., Lawrence, T., Gilula, N., and Beers, W. (1981). Modulation of cell-to-cell communication in the cumulus-oocyte complex and the regulation of oocyte maturation by LH. *Develop. Biol.* **80,** 356.

Eppig, J. J., and Downs, S. M. (1984). Chemical signals that regulate mammalian oocyte maturation. *Biol. Reprod.* **30,** 1.

Guraya, S. S., ed. (1985). "Biology of Ovarian Follicles in Mammals." Springer-Verlag, Berlin.

Siracusa, G., De Felici, M., and Salustri, A. (1985). The proliferative and meiotic history of mammalian female germ cells. In "Biology of Fertilization" (C. B. Metz and A. Monroy, eds.). Academic Press, New York.

Tsafriri, A. (1988). Local non-steroidal regulators of ovarian function. In "The Physiology of Reproduction, volume I, The Female Reproductive System" (E. Knobil and J. D. Neill, eds.). Raven Press, New York.

Wassarman, P. M. (1988). The mammalian ovum. In "The Physiology of Reproduction, volume I, The Gametes, Fertilization and Early Embryogenesis" (E. Knobil and J. D. Neill, eds.). Raven Press, New York.

Zamboni, L. (1970). Ultrastructure of mammalian oocytes and ova. *Biol. Reprod. Suppl.* **2,** 44.

Zamboni, L., and Thompson, R. S. (1972). Fine morphology of human oocyte maturation in vitro. *Biol. Reprod.* **7,** 425.

Optical Microscopic Methods in Cell Biology

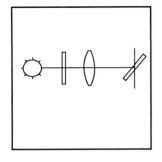

ELLI KOHEN, CAHIDE KOHEN, JOSEPH G. HIRSCHBERG, *University of Miami;*

RENE SANTUS, *Museum National d'Histoire Naturelle, Paris;*

PATRICE MORLIÈRE, *C.H.U. Henri Mondor, Creteil, France*

FREDERICK H. KASTEN, *Louisiana State University Medical Center, New Orleans*

FEROZE N. GHADIALLY, *Canadian Reference Centre for Cancer Pathology, Ottawa, Ontario, Canada*

Glossary

Fluorescence probes Fluorescent compounds, preferably compatible with cell viability (vital stains), which are used to selectively label cell constituents, i.e., organelles, enzymes, membranes, and cations. The intensity, excitation, and emission spectrum of the probe may change as a function of intracellular processes, enzyme activity, membrane potential, or cation levels.

Microscopic resolution A measure of the finest structural detail attainable for the determination of fluorescence lifetimes and recording of time-resolved spectra.

Microspectrofluorometry The topographic temporal and spectral analysis at the microscopic level of fluorescence emitted by cell-surface membranes and intracellular constituents.

Microspectrophotometry The study at the microscopic level of light absorption by intracellular constituents.

Photobleaching Decrease or total destruction of fluorescent groups (fluorophores) due to photochemical alteration of the molecule (which is currently used in a variety of fluorescence approaches).

Contemporary optical microscopic methods in conjunction with spectroscopic instrumentation and low-level light detection have opened new horizons in cell biology by allowing us to probe the internal environment of living cells using nondestructive techniques. The remarkable expansion of fluorescence probes and the rise of new methods, such as one-, two-, and three-dimensional microspectrofluorometry of cell metabolism, organelle interactions, and microarchitecture, have revived the morphological approach aimed at unravelling structure and function relationships in intact cells. The highly sensitive techniques, recently introduced in microscopy, range from multiple spectral parameter imaging to fluorescence photobleaching recovery, fluorescence resonance energy transfer, fluorescence polarization microscopy, polarized fluorescence photobleaching recovery, and combined fluorescence and x-ray microscopy. Further advances on the horizon are pixel-by-pixel deconvolution in living cells of multiexponential fluorescence lifetimes and fluorescence anisotropy decay, fluorescence correlation spectroscopy, and the recording of time-resolved fluorescence spectra, several of which have already been tentatively initiated. The ultimate rewards include an enhanced perspective of *in situ* intracellular processes and a revolution in our understanding of cellular biochemistry, pharmacology, and pathology. There is also the realistic hope, crucial from the standpoint of biotechnology, of comprehending the control mechanisms of biochemical processes within living cells at both qualitative and quantitative levels.

I. Introduction

Contemporary optical microscopic methods have enabled us to probe by nondestructive techniques the intracellular microenvironment of living cells for a better understanding of structure and function. These studies are accomplished in step with the actual speed of life processes. Microscopic approaches are used to monitor *in situ* interactions of cell organelles, intracellular communication between functionally interdependent microdomains (microcompartments), as well as metabolic and functional communication between cells. The cell biologists of today are reaping the benefits of concurrent technical progress in absorption and fluorescence spectroscopy, low-level light detection, and laser and other light sources, as well as extremely fast and efficient data-processing equipment. Quantitative absorption and fluorescence microscopy at the limit of structural resolution, and techniques of one-, two-, and three-dimensional multichannel analyses open the way for pixel-by-pixel or voxel-by-voxel exploration of cell images in terms of physiological activities, drug effects, xenobiotic actions, or pathological changes.

The remarkable expansion of fluorescence probes is reversing a previous trend in biomedical research away from cell imaging, with reliance primarily on solution biochemistry. The rejection of the outmoded morphological approach was due to disenchantment with *cellular autopsy,* i.e., observations based on the use of classically stained tissue sections or semi thick sections for electron microscopy, neither of which was very successful in resolving major questions in biomedicine. Development of sophisticated enzymologic, physical, and spectroscopic techniques, the advent of automated cytology, nuclear energy resonance, gene splicing, and genetic engineering further relegated morphology to the domain of outdated and unfashionable approaches.

A pioneer of fluorescence chemistry, Gregorio Weber, examined the path that leads from solution spectroscopic techniques to their application at the level of cellular spectroscopy. He set a formidable agenda for the cell biologist: "The fluorescence spectroscopist who wishes to carry out a complete analysis of emission must examine the fluorescence excitation and emission spectra, the quantum yield, the fluorescence lifetime, and the state of polarization of the emission. In practice, fluorescence quenching in different intracellular microenvironments must also be given consideration." (pp. 601–609).

The revitalization of the morphological approach rests upon the introduction of new parameters for *in situ* biochemical studies. The cell is thus not a mere bag of enzymes, and the interior of the cell does not represent a thick but homogeneous soup. Therefore, the ultimate value of the morphological approach rests upon the development of appropriate probes for exploring the organization of intracellular compartments and organelles.

Interference or phase-contrast methods are used in combination with fluorescence microscopy to correlate morphological detail and functions probed by endogenous or exogenous fluorochromes. The most striking developments in microscopic techniques are associated with an impressive armamentarium of fluorescence probes to monitor cell enzymes, organelles, cations, and membranes. The synergism between photobiological techniques and microscopic methods keeps expanding. The comprehensive array of microscopic fluorescence techniques now being applied includes biophysical methods relying on determinations of the lifetimes (in billionths of seconds) of excited molecular states within the intracellular microenvironment, as well as the measurement of intermolecular distances (in billionths of a meter) by transfer of energy between molecules.

II. Instrumentation

1. Microspectrophotometers

A series of publications by T. Caspersson from 1936 to 1950 signalled the birth of microspectrophotometry as initially applied to the *in situ* assay of nucleic acids in living cells by UV absorption spectrophotometry. Quantitative cytochemical methods were rapidly applied to the study of nucleocytoplasmic relations in normal and malignant cells. These investigations demonstrated considerable nucleoprotein synthetic activity in the nucleus of normal cells in connection with cell growth and cell function, and a very high concentration of ribonucleoprotein in certain nuclear elements (mitotic chromosomes, nucleoli).

In tumor- and virus-infected cells, the studies revealed considerable disturbance of cytoplasmic protein and RNA metabolism as well as variability of DNA and RNA content.

These studies were based on the observation by Caspersson in agreement with Abbé's sine conditions, that *true absorption measurements can be made on microscopic objects whose sizes are three to four times those of the wavelength used.* A system for parallel studies of nucleic acids and hemoglobin, during red blood cell maturation, used UV light of 260 nm for nucleic acid determinations and light in the visible spectral range (around 450 nm) for hemoglobin (Fig. 1). The advantages of the photographic method used by the Stockholm group over photoelectric intensity measurements were (1) the living bone marrow cells needed to be exposed to the UV for only short periods; (2) the density of the photographic image could be determined within a very small area; and (3) correction was easy for a slight shift in position or change in the form of the living object.

For microspectrography of hemoglobin in the visible spectral range, the source of light was a water-cooled Phillips high-pressure mercury lamp. The light was dispersed by a prism monochromator, and stray light was eliminated with a Schott filter combination. The monochromator exit slit functioned as an illumination field limiter and the microscope condenser diaphragm acted as an aperture stop. This provided illumination with a known aperture at the highest possible spectral purity. At the microscope objective, the image was projected by a prism toward the photocell. The prism holder could be moved with two precision micrometer screws around horizontal and vertical axes. The image was thereby moved over the photocell aperture. Steps cut in the mounting of the micrometer screws rendered possible measurements of definite points in the cell image and of the intensity of the unabsorbed light in the immediate neighborhood of the cell.

A universal microspectrophotometer, built for spot measurements, as well as for integrating measurements in middle- and long-wave ultraviolet and in the visible spectral region, allowed measurements on objects no larger than one to a few wavelengths of measuring light, such as individual chromosomes during cell division, leading to the

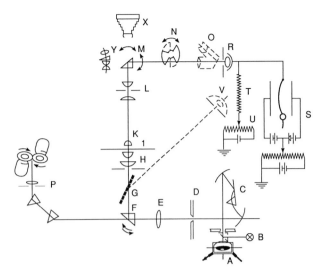

FIGURE 1 Ultraviolet absorption microspectrophotometer. (a) High-pressure mercury arc, (b) auxiliary light source; (c) mirror monochromator (the entrance pupil of the monochromator is filled by light to minimize light loss in the optical system); (d) diaphragm; (e) lens; (f) reflecting prism; (g) beam splitter; (h) Zeiss fused silica condenser (the image of the exit slit of the monochromator is projected onto the object I through the condenser H); (k) Köhler-v Rohr objective for ultraviolet; (l) ocular in molten quartz; (m) quartz glass prism; (n) light-weakening rotating sector (rotating close to the plane of the image); (o) centering device; (t, u, s) electronic circuits; (p) Köhler's device with rotating spark gap [by means of the rotating spark gap as indicated by Köhler (see Naturwiss. 21, 321, 1933) sufficient light intensities can be obtained in the UV region from 2200 Å to 3000 Å. After the collector the spark light is refracted through two 60° quartz prisms filled with water, and then projected on the aperture diaphragm of the condenser (according to Köhler, Z. wiss. Mikros. 10, 433, 1893)]; (r) sodium-noble gas photocell; (y) Köhler's finder device; (x) camera. The image which can be aligned with the help of finder Y is projected through the prism M on the photocell R. The rotating sector can be operated with an accuracy of ¼%. The absorption measurements on the object versus object-free space in the microscope plane. The method of manometrically controlled pneumatic displacement of quartz plates is used to control fine movement of the microscope stage in the range of 0.1 μm. Such accuracy control of stage movement cannot be obtained if quartz plates are replaced by steel plates.

discovery that the DNA patterns are reproducible for identical chromosomes in different metaphase plates but differ distinctly between different chromosomes (1960).

Methods for the x-ray determination of mass allowed an optical resolution below the value of 1 μm. These methods were applicable to single cells as well as cell populations, and were used to study populations from scrapings of tissue (i.e., normal tissue, cancer *in situ,* and invasive cancer; also populations of neoplastic cells in different stages of

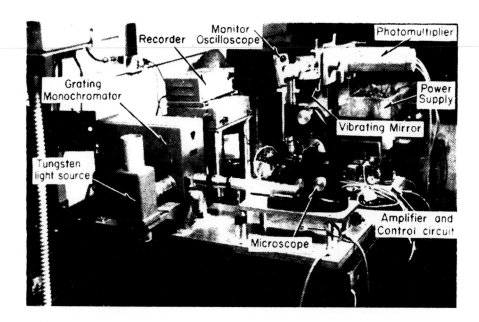

FIGURE 2 Photograph of the highly sensitive recording microspectrophotometer capable of measuring the light absorption of substances present at enzyme concentrations in a single living cell. This photograph of the assembled apparatus shows the optical, electronic, and mechanical components. A tungsten light source and end-on photomultiplier (left) are pictured. (Courtesy of Professor Britton Chance, Department of Biochemistry and Biophysics, University of Pennsylvania School of Medicine, Philadelphia, Pennsylvania.

growth, treated or untreated with chemotherapeutic drugs) and for aspiration biopsy smears.

A sensitive scanning microphotometer led to the discovery of visual pigment in intact rod outer fragments of the frog eye. Further definition of the pigments was achieved using an improved microspectrophotometer (Fig. 2).

Microspectrophotometry (Fig. 2) permitted the analysis of mitochondrial cytochromes in single living cells.

To obtain sharply resolved spectra advantage was taken of the knowledge that respiratory pigments will display the same but intensified spectra at low temperatures. A chamber was constructed on the microspectrophotometer stage for the cooling of cell objects to liquid nitrogen temperature (Fig. 3).

B.2. Microspectrofluorometers.

The first microspectrofluorometer, designed by Borst and Königsdörffer in 1929, basically consists of a fluorescence microscope adapter for visual, microspectrographic, and microspectrophotometric measurements. It was first used to identify individual porphyrins in tissue sections by their fluorescent spectra.

Subsequently Chance and Legallais developed a microfluorimeter giving a differential response to the fluorescence of two areas in the microscope field, built to study the compartmentalization of blue-fluorescing pyridine nucleotides in the mitochondria and cytoplasm. The blue *autofluorescence*

of cells was owing to reduced pyridine nucleotides, based on the kinetics of intracellular reactions in aerobic–anaerobic (or vice versa) transitions.

The rapid-scanning microspectrofluorometer of Olson for the study of chloroplast emission (1960) was built around a Reichert-type metallurgical inverted microscope. Light from a high-pressure mercury arc was filtered through copper sulfate to remove the UV light and through appropriate filters to isolate the 436 nm region of the exciting light. A 1.35 NA apochromatic condenser was used to concentrate the exciting light on the specimen. On the emission side, a filter combination gave a sharp cutoff at about 480 nm to exclude the exciting light. The image of fluorescent cell structures was dispersed by a Zeiss spectometer ocular, modified to focus a spectral image at a spiral scanning slit. The spiral scanning slit was adjustable in width, in order to pick up very weak fluorescence, when necessary, at the expense of spectral resolution. Transient color changes in fluorescence were studied during the photochemical bleaching of chloroplasts.

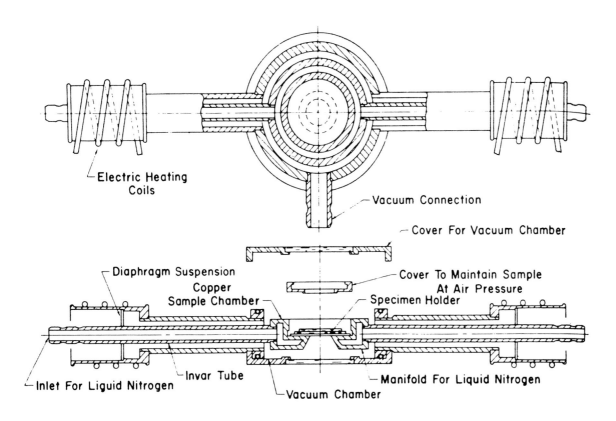

Electric Heating Coils

Vacuum Connection

Cover For Vacuum Chamber

Diaphragm Suspension

Copper Sample Chamber

Cover To Maintain Sample At Air Pressure

Specimen Holder

Inlet For Liquid Nitrogen

Invar Tube

Vacuum Chamber

Manifold For Liquid Nitrogen

FIGURE 3 Diagram of low-temperature attachment to the microspectrophotometer used for study of cytochromes in the single cell at liquid nitrogen temperature. (Courtesy of Britton Chance, *Expt. Cell Res.* **20,** 741–746, 1960).

Subsequent application was to the pattern distribution of chemical substances, such as DNA along the chromosomes which can be used for identification of chromosomes and chromosomal regions using an ultramicrofluorimeter, (Figs. 4–6). For this analysis the fluorescence technique has several advantages over absorbimetric procedures:

1. following the binding of acridine orange to DNA, DNA amounts less than 10^{-15} g can be determined;
2. the resolving power of a fluorescence microscope for particles adjacent to one another is almost twice that of bright-field microscopy in transmitted light;
3. the determination of the amounts of fluorescing substances in an irregularly-shaped object is much simpler than determinations using absorption measurements.

For determinations in the ultrafluorimeter (Fig. 5), selective labeling of bands within metaphase chromosomes with quinacrine mustard (QM) was

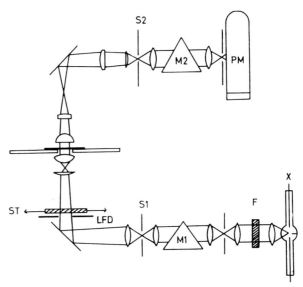

FIGURE 4 Plan of the measuring device (microspectrofluorometer) used for recording the emission spectra of intracellular nucleic acids and nucleoproteins treated with acridine orange. X = Xenon arc lamp; F = Heath protection filter; MI = quartz prism monochromator for selecting the exciting radiation; SI = slit of MI; LFD = light field diaphragm; ST = standard for emission measurement (uranyl glass, Schott GG 17); M2 = monochromator for analysis of the emitted radiation; S2 = slit of M2; PM = photomultiplier. (Courtesy of Professor Rudolf Rigler, Jr., Karolinska Institute, Stockholm, Sweden.)

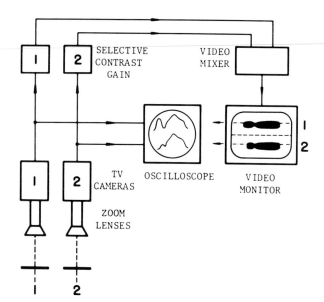

FIGURE 5 Diagram for TV-based chromosome analysis. This apparatus has been developed with a view to the important practical problem of identifying very small chromosome aberrations. This has been accomplished by the introduction of a second TV camera whose picture can be relayed to the same monitor as that of the first camera. In this way, pictures of two chromosomes, e.g., members of one and the same pair, can be juxtaposed with optional enlargement, thus greatly increasing the possibility of identifying small aberrations in the patterns. (Courtesy of Professor T. Caspersson.)

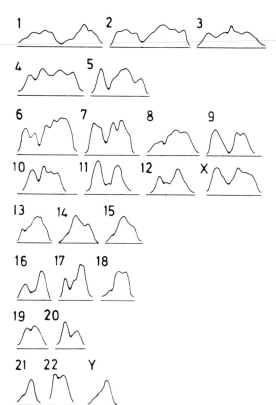

FIGURE 6 The human karyotype, quinacrine mustard preparation: typical fluorescence distribution patterns of chromosomes. To rapidly assemble a large number and range of fluorescence patterns, a very fast fluorometer system was used. It was based on microdensitometry of fluorescence photomicrographs of quinacrine- and quinacrine mustard-stained metaphase plates. In it, both negatives and positives can be used, the latter with the aid of a reflectometer arrangement. Positives can be easily cut up to make a preliminary sorting of the chromosome pictures to be measured. The whole identification system is based upon the fluorimetric measurements. The present cytological technique also permits visual chromosome identification on fluorescence photographs. The above figure shows the typical patterns of the whole human karyotype, and is the accepted norm for chromosome numbering. For each chromosome fluorescence pattern the ordinate is fluorescence intensity and the abscissa is position along the length of the chromosome. The principle followed, as far as possible, in numbering is that the shorter the chromosome, the higher the number (Courtesy of Professor T. Caspersson.)

achieved. Microdensitometry of fluorescence photomicrographs of QM-stained metaphase plates made it possible to assemble rapidly a large number and range of fluorescence patterns. Based on 30,000 fluorescence curves, it was possible to identify all human chromosomes accurately (Fig. 6). An internationally accepted numbering system for definition of human chromosomes is based on these measurements.

Further refinement came with the demonstration by Hirschfeld that individual small molecules could be observed visually and measured photoelectrically by an appropriate fluorescence-tagging and illumination procedure. Between 80 and 100 molecules of fluorescein isothiocyanate were bound to one molecule (20,000 daltons) of polyethylenimine, which in turn was bound to gamma globulin. This reagent was detected by illumination at a light intensity high enough to produce complete photochemical bleaching during the observation period.

Microfluorimetry is suited to intracellular enzyme studies. The subject has two aspects: first, the determination of the amount of enzyme activity present in a given histological or cytological region, and second, the cytochemical determination of kinetic characteristics of an enzyme, such as the Michaelis–Menten constant and its response to activators and inhibitors. The technology was applied to the study of oxidative enzyme systems, glucose-6-phosphate dehydrogenase, hydrolytic enzymes, alkaline phosphatase and acid phosphatase in cryostat sections of frozen living tissue.

A television fluorescence microspectrophotometer developed by West, permitted the determination of an emission spectrum from any area of a cell illuminated with an extremely narrow excitation light produced by a monochromator. The emitted light was analyzed by a second monochromator with an image-intensified orthicon coupled to a picture monitor and a line-selector oscilloscope. The recording of very weak fluorescence spectra was very rapid.

Many of the changes of interest in the living cell take place in the 0.1 to 100 nm range, i.e., the size and range from atoms to macromolecular aggregates. Polarization microscopy has been used from early 1959 to study small structural changes in living cells, especially in the study of cell division. By taking advantage of anisotropic optical properties exhibited by cellular fine structures, polarization microscopy is successful in closing the gap between time-lapse cinematography, pictures with phase contrast optics (which show the continually changing shape and distribution of organelles, but at an insufficient structural resolution), and electron microscopy (which shows fine structural detail, but only in fixed cells). Anisotropic optical properties are

1. birefringence, i.e., the presence of different refractive indices for light waves vibrating in different planes in a medium; and
2. dichroism, i.e., exhibition by a crystalline structure of two different colors depending on the direction of observation or the plane of vibration of polarized light waves traveling through the crystal. This indicates the difference of absorption curves for chromophores oriented in different directions.

Thus, birefringence corresponds to anisotropy in refraction, and dichroism relates to anisotropy in absorption. A microspectropolarimeter can also be used to study optical rotatory dispersion (ORD) from cells to identify a particular intracellular constituent, provide information on its molecular conformation, and study the interaction of intracellular macromolecules. Optical rotation is not to be confused with birefringence. A solution of sugar does not have directionality, but it can twist polarized light in a left-handed or right-handed direction depending on its chemical nature. It is as though the polarizer were turned; in this case, the light can be extinguished by turning the analyzer by the same amount. With birefringence, generally the light can-

not be extinguished by turning the analyzer, but optical rotation does this for monochromatic light. Optical rotation indicates three-dimensional asymmetry of molecules, which themselves can be randomly oriented. Thus, optical activity is a physical property of dyssymmetric substances that results from their ability to refract and absorb left- and right-circularly polarized light to different extents. Optical rotation varies as a function of wavelength, while the variation with wavelength is negligible for birefringence. Optical rotation dependence on wavelength is termed ORD.

Biological macromolecules have the characteristic ability of inducing asymmetry in small molecules, such as fluorescent dyes, with which they form complexes. An interesting application is to acridine orange–mucopolysaccharide complexes present on cell surfaces by ORD spectra in cases of mucopolysaccharide storage diseases (mucopolysaccharidoses). Specific ORD spectra are obtained for acridine orange complexed to different mucopolysaccharides.

Video image processing greatly enhances contrast, quality, and speed in polarization-based microscopy, by correcting for depolarization by rotation of polarized light in the condenser, objective lenses, slides, and coverslips.

A grating microspectrofluorometer for the study of cellular metabolism (Figs. 7, 8) is designed to carry out microassays in single living cells at a time resolution compatible with the real-time changes in metabolism or cell-to-cell transfer of chemicals.

In this apparatus the fluorescence of 40 to 50 individual cell regions (1 to 1.5 μm), localized along the linear-scan axis, are imaged upon individual detector channels for topographical analysis. In addition, for spectral analysis, the fluorescence of larger cell regions (about 3 to 6 μm) are spectrally dispersed, and the various wavelength elements are imaged on a multichannel array, with the attainable spectral resolution in the range of 1 to 2 nm.

The following options are provided:

1. Balzer's dichromatic beam splitter (Filtraflex DC), reflecting wavelengths longer than 600 nm into the path toward the ocular, with transmission of the shorter wavelengths (390–600 nm) toward the detector;
2. special dichromatic filter, reflecting 50% of wavelengths between 440 and 580 nm and nearly 100% of the light at 580 nm and greater; or
3. aluminized mirror, reflecting the entire visible spectrum for fluorescence photography.

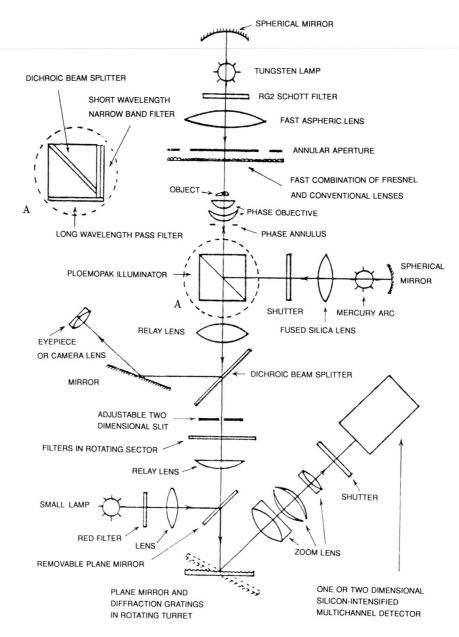

FIGURE 7 Schematic diagram of grating microspectrofluorometer (Miami). The apparatus is operated in two modes: topographic and spectral. For *topographic operation* the sequence in the light path is: mercury arc→Ploemopak illuminator→objective→object→transmission through dichroic mirror of Ploemopak→transmission through dichroic beam splitter separating fluorescence emission from red light used for cell and microinstrument visualization→slit→selected filter in rotating sector→magnifying lens or lens system (usually 10x magnification)→plane mirror→lens (or zoom lens)→OMA detector. For *spectral* operation the mirror–grating turret is rotated to position a selected grating in the light path. The OMA channels (500) are scanned along a single coordinate (unidimensional scan). For two-dimensional scan (see Figs. 13, 14), scanning takes place along 500 channels (x coordinate) and 256 tracks (y coordinate).

For microspectrofluorometry, tissue culture cells are grown on a 0.1 mm-thick coverglass.

The cells to be examined are positioned at the slit in the microscope field. Provision is made in the instrument for injections of metabolites and modifier agents (Fig. 8).

Knowledge of the limits of detection is crucial in assessing the relevance of the method for specific problems in cell biology. The answer in each case is provided by the maximal number of obtainable photons $F(\lambda)$ in the wavelength interval $\Delta\lambda$. Once $F(\lambda)$ is known, the signal-to-noise ratio (S : N) can be calculated from the square root law. A fluorescence

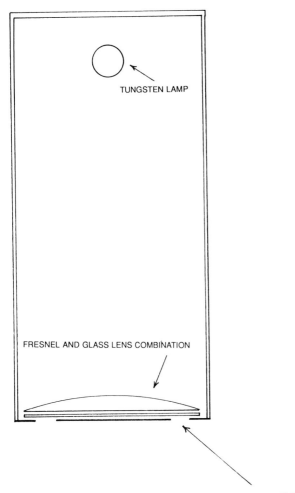

FIGURE 8 Schematic of long–working distance, high-resolution phase condenser used in conjunction with microspectrofluorometer for micromanipulatory procedures.

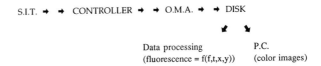

S.I.T. = Silicon intensified target (detector)
O.M.A. = Optical multichannel analyzer
P.C. = Personal computer

FIGURE 9 Diagram of the data-recording and data-processing components of the two-dimensional microspectrofluorometer in Creteil.

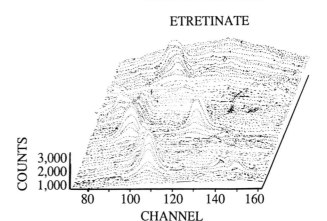

FIGURE 10 Two-dimensional topographic fluorescence scan of human fibroblast incubated with etretinate. Excitation at 365 nm. Etretinate localizes in the lysosomes. The x coordinate corresponds to the channels, y to the frames, the topographic scan being carried along the x and y coordinates. The fluorescence intensity at each site is given by the z coordinate. Each channel views about 0.5 μm along the x axis. Organelle-size resolution is obtained, each etretinate-labelled lysosomal site emerging as a peak over the cytoplasmic background.

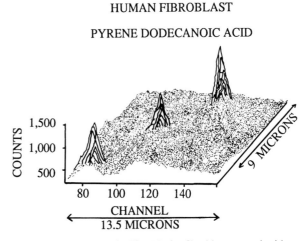

FIGURE 11 Same as in Fig. 14, for fibroblast treated with pyrene dodecanoic acid.

signal with a reasonable S : N requires a cell microdomain containing a minimum of 10^{-20} mol for an ideal quantum efficiency of 1 (10^{-19} and 10^{-18} respectively for quantum efficiencies of 0.1 and 0.01).

This design has been modified by Hirschberg and Nestor, for high-resolution two-dimensional topographic scan of fluorescence (Fig. 9) associated with mitochondria and lysosome-size organelles (Figs. 10, 11), or one-dimensional scan with simultaneous spectral recording from over 100 cell regions.

An apparatus for the measurement of fluorescence lifetimes in single cells was developed. The time behavior of the transient photon flux emitted by a fluorescent sample permits measurement of the interactions of the fluorescent molecule with its environment. For the measurement of fluorescence lifetimes, a nitrogen laser is employed that delivers pulses of 8 nanosec with a repetition frequency of 10 to 500 Hz. The emitted wavelength is 337 nm, which is suitable for excitation of cellular chromophores such as NAD(P)H and also exogenous (cytotoxic) agents such as benzo(a)pyrene. The analysis of the photoelectric signals is carried out by a sampling oscilloscope coupled to a microcomputer. The oscilloscope is triggered by a second photomultiplier, which receives directly a fraction of the exciting beam signal. For each excitation pulse, a voltage is read from the monitor, proportional to the fluorescence intensity. Following repetitive excitation (up to 1000 consecutive excitation pulses), the signal is retrieved from computer memory. The ultimate goal is the pixel-by-pixel determination of the fluorescence lifetime during the intracellular interactions of drugs (or coenzyme activity) in different cell organelles and cytoplasmic compartments.

Fluorescence microscopy in three dimensions was developed by Agard and Sedat, beginning in 1984. The usual high numerical aperture (NA) objectives required for high-resolution spatial analysis and optimal light gathering power have a depth of focus too narrow to image an entire cell clearly, yet too wide to give good optical sectioning. The method of fluorescence microscopy in three dimensions, which uses computational methods for infocus, high-resolution synthetic generation of stereopairs corresponding to interplanar spacing of 0.25 μm, is equivalent to the idealized combination of a high-NA objective lens and an infinite depth of focus. The most straightforward method uses only information from adjacent focal planes to correct a central one. This is accomplished by recuperating the true image i from the observed image o, by correcting for the out-of-focus contamination s (i.e., smearing function of the objective or point spread function, PSF). For calculation purposes, it is often convenient to recast the PSF, which requires a convolution operation (complex summation) in terms of its Fourier transform, called contrast transfer function (CTF). The convolution is now simplified as a multiplication. The microscope's three-dimensional CTF indicates that the resolution is substantially greater in the x–y direction than it is in the z. A significant improvement in imaging can be obtained by collecting several data sets at different tilt angles, from which it should be possible to reconstruct a three-dimensional image with uniform resolution in all directions. In principle the design of an ultraprecise, accurately centered tilt stage is required.

Agard *et al.* have developed the necessary computer programs and successfully merged several three-dimensional data sets.

Currently, real time three-dimensional imaging is limited by the integration time required to form a high quality CCD (charge-coupled device) image. It should now be possible to generate complete stereo pairs at the rate of 1 every 20 seconds, with an expected reduction of the time per stereo pair to about 2 to 5 seconds. This should be sufficiently fast to allow the study of complex dynamic phenomena in living cells. Studies have focused on the three-dimensional organization of chromosomes and other organelles in structural cell biology, i.e., polarization phenomena in epithelial cells and spatial aspects of regulated transport through the Golgi apparatus.

Scanning microfluorimetry has been introduced for the measurement of fluorescence intensities in multiple small regions (e.g., 0.25–0.5 μm) in the microscope preparation. The instruments can be divided into image-plane scanners, flying spot scanners and stage scanners. Laser scanning microscopy (Figs. 12, 13) uses point-by-point illumination to build up the image and is especially useful in the study of weakly fluorescent objects. A three-dimensional image is built up by storing point-by-point measurements, using analog-to-digital conversion to form a matrix in a computer image memory. At present, there are three main types of laser scanning fluorescence microscopes: the ultrafast, using a rotating polygon designed for rapid automation; the fast, using laser beam mirror scanning; and the slower stage scanning type. Laser scanners using confocal microscopy achieve higher resolving power than is feasible with conventional microscopy and can reveal valuable information concerning the three-dimensional structure of cells as revealed by fluorescence probes. Coordinated phase-contrast observation is also very important for living-cell investigation.

The advantages of laser scan fluorescence microscopy are the almost complete elimination of the interface generated by fluorescence of optical parts, high sensitivity, minimal fluorescence fading, and reduced out-of-focus signals.

The idea for confocal imaging in light microscopy was apparently conceived in the late 1970s by Sheppard, who showed theoretically that with confocal scanning microscopy, a fundamental improvement in imaging, as compared to normal light microscopy, could be expected. The confocal principle

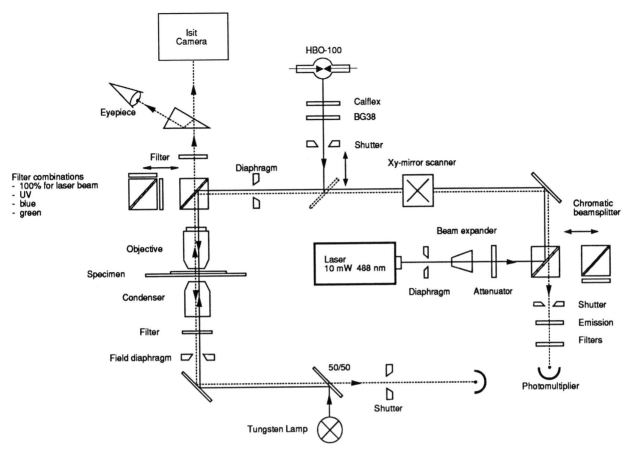

FIGURE 12 Schematic diagram of laser scanning microscope (LSM) including conventional epi-illumination with filter blocks for multiple wavelength excitation and passage of the laser beam. For LSM, frequently HeCd, Ar, and HeNE lasers are used. The laser beam is expanded with a telescope to a size suitable for microscope objectives. The beam can be moved along two axes by a x-y scanner unit consisting of two orthogonal galvanometers, one for the lines and one for the scanned spots. Between the scanners, relay optics are placed, which image the first mirror onto the second one. Both mirrors are imaged onto the pupil of the objective lens. Independent of the scanning angle, the full aperture of the objective is always used. The beam is then focused by the objective to a diffraction-limited spot on the object. This spot is scanned across the object by rotation of the mirrors, whereafter the transmitted rays are collected by the microscope condenser and directed to a photomultiplier tube. The rays generated by reflectance or fluorescence retrace the entire beam path backward (including the scanning mirrors) until the rays are reflected by a beam splitter onto a photomultiplier tube for intensity measurements. The Zeiss laser scan microscope incorporates, besides the laser which can be used for incident as well as transmitted light microscopy, a conventional (tungsten) light source. For transmitted light microscopy, a standard condenser is used to obtain absorbance, phase, and differential interference–contrast images. Reflection and fluorescent contrast images are obtained with incident light. The images of a given microscope field are stored in computer memory and then compared with each other. This allows multiparameter analysis of cells. Additions for conventional epi-illumination fluorescence microscopy include a connection for a mercury lamp (HBO-50 or -100), without interfering with LSM, a mirror to switch the mercury lamp into the light path, a field diaphragm, and filters. (Courtesy of J. S. Ploem, *et al.* Laboratory of Histochemistry and Cytochemistry, Sylvius Laboratoria, University of Leiden, Leiden, The Netherlands).

was applied by several groups of investigators to fluorescence microscopy, resulting in improved imaging and effective suppression of contributions from out-of-focus areas in the specimen.

The basic idea of confocal micrococopy is that one and the same point in the object is optimally illuminated by a point light source as well as imaged on a point detector. In this system, only the radiation generated in the specimen layer imaged reaches the detector efficiently, thus reducing the contribution of out-of-focus areas to the image. Confocal microscopy can directly deliver clear optical sections without the use of the time-consuming image reconstruction algorithms (as in the three-dimensional

FIGURE 13 View of LSM.

fluorescence microscopy where deconvolution techniques are required to eliminate out-of-focus information). However an advantage of that approach is that fully in-focus synthetic mono or stereo images can be obtained with considerably less photobleaching.

The most important property of confocal fluorescence microscopy is the ability to record three-dimensional images (Fig. 14). A point source (laser or other) is imaged into an object utilizing a high-numerical-aperture microscope objective. When applied to fluorescence microscopy, the image of the point source is imaged onto a point detector by the same objective. To achieve confocal microscopy in connection with laser scan fluorescence microscopy, the images of the illumination and detector pinholes are made to coincide at a common point through which the object is scanned mechanically. Unwanted out-of-focus fluorescence signals from structures above and below the plane of focus tend to be eliminated. When using an objective of 100X with a numerical aperture of 1.3, optical sections of 0.7 μm thickness can be obtained. Three-dimensional reconstructions of the object are generated by a computer. If a single objective is used for excitation and emission, source and detection pinholes are optically separated by a dichroic beam splitter. The confocal system produces the image of only a single diffraction-limited light spot in the object, so a complete image must be constructed by scanning the light spot through the object. The confocal scanning system by White *et al.* is based on the principle of epi-illumination and a moving mirror scanning system (Fig. 15).

Biological applications of confocal microscopy have included optical sectioning of chloroplasts and the visualization of the hybridization of a biotinylated probe, labelled with fluoresceinated avidin to chromosomes of human peripheral lymphocytes.

The Iconoscope, developed by Zworykin in 1934 at the RCA Laboratory in Camden, New Jersey, led to the evolution of broadly used television in the early 1950s. Even as Zworykin announced the development of the first practical TV image pickup tube, he was aware of the potential of TV to extend the limits of human vision, including into the microscopic world. Closed-circuit television (CCTV), in conjunction with the light microscope, has provided large-screen display, raised-image contrast, and made visible the images formed by ultraviolet and infrared. Stripped to its bare essentials, a modern video microscope consists of a video camera mounted to a microscope through a light-tight coupler, a monitor for displaying the image, and a shielded cable connecting the two.

The capability and sophistication of computers that can process video images are advancing at a very rapid pace. With modern CCTV devices coupled to digital computers, it is practical to extract an image from a scene that appears to be nothing but noise, capture fluorescence too dim to be seen, visualize structures below the limit of resolution, and record in time-lapsed and high-speed sequences through the light microscope without great difficulty. Video can now capture with different fluorescent stains [e.g., fluorescein isothiocyanate (FITC) for total protein, propidium iodide for DNA] the localization of chemical components in protozoan carriers of Chagas' disease (*Trypanosoma cruzi*), or their movements, previously too fast to be recorded, and the time-lapse sequence of a malaria parasite invading a human blood cell. Video images with *background subtraction* and *contrast enhancement* allow visualization by polarized light microscopy of fine structures in the diatom *Pleurosigma angulatum*. Such observations of a weakly birefringent object could not have been made previously because the planapochromatic objective lens required for such imaging contains birefringent crystalline elements that ordinarily mask those of the specimen.

Video processing tends to be simple and often can be carried out in *real time,* by either analog or digital means. In *analog processing,* the strength of the video signal is a continuously varying function, and the time interval that determines the location of

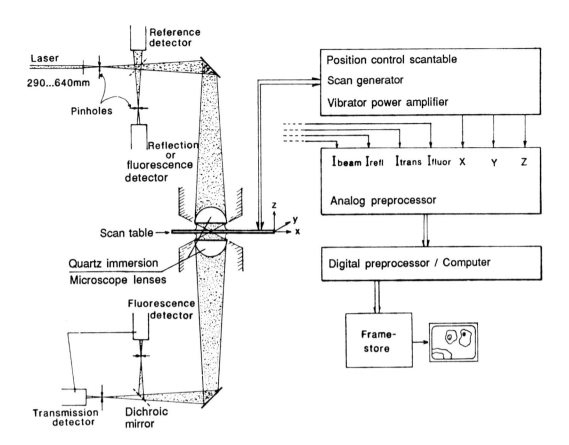

FIGURE 14 Optical system of the Wild-Leitz confocal microscope with laser-beam path.

image points is treated as a continuum. In *digital processing*, the strength of the video signal as well as the location of the image point (the *pixel*) are treated as discrete points, commonly expressed in binary form. By digitizing the pixel location and intensity, low levels of noise or minor fluctuations in pixel locations and signal strength no longer perturb the signal.

Resolution of the image can be classified into two categories: *gray level resolution*, which describes how accurately the digital image represents differences in intensity within the original image, and *spatial resolution*, which describes how well the digital image represents information about the size and position of features within the original image. For display applications in which the purpose is to accentuate subtle differences in image contrast, as many as 256 or more gray values may be required. However, for many analytical applications, one goal of digital processing is to reduce the number of different gray values in the image, in order to accelerate processing and reduce memory storage requirements. A similar consideration applies to the

minimal number of pixels needed to reproduce the original optical image in both the horizontal and vertical dimensions.

Video recording is also applied to low-light–level microscopy as demonstrated by Allen. A video-enhanced microscope with a computer frame memory was used to examine the phenomena of intracellular traffic flow and cell locomotion. The approach was used to detect birefringence in the motile reticulopodial network of *Allogromania laticollaris* and to obtain evidence for the active role of microtubules in cytoplasmic transport by studying the gliding movement along single native microtubules in squid cytoplasm. For video intensification, photon noise and pixel noise are distinct problems at low light fluxes. Therefore, summing and averaging video frames are the most frequently described arithmetic operations.

Based on the parameters evaluated by microspec-

Optical train in MRT 500 LSCM

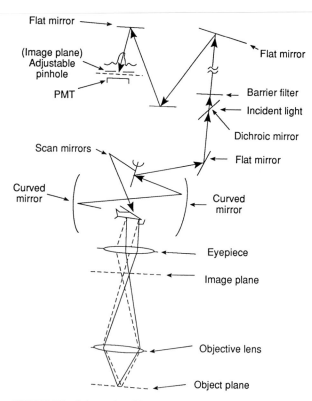

FIGURE 15 Schematic of laser scanning confocal microscope. (Courtesy of W. W. Webb from the *Handbook of Biological Confocal Microscopy,* Pawley, J. (ed.); MR Press, Madison, Wisconsin).

trofluorometry of single living cells, cytofluorometers with more effective and versatile performance can be constructed. Applying this cardinal principle, Thorell (1962) constructed an apparatus (Fig. 16) for simultaneous monitoring of NAD(P)H and flavin redox states in cell populations. The design includes a combination of filters and dichroic mirrors, together with a triple laser arrangement: argon UV (351, 363 nm) for NAD(P)H excitation, argon blue laser (488 nm) for oxidized flavin excitation, and a helium–neon laser for measurements of light scatter and absorption. A high blue-to-green ratio of fluorescence intensity corresponds to the reduced state of cell coenzymes, and a low ratio refers to the oxidized state. The studies of Thorell were focused on liver and bone marrow cells, but the same cytofluorometric approach is capable, with a variety of fluorescent probes, of producing pertinent quantita-

tive cytochemical data for populations of normal and malignant cells.

Flow cytometry was used by Pipeleers *et al.* to prepare functionally homogeneous pancreatic B cells. Ubiquitous autofluorescent molecules such as oxidized flavin adenine dinucleotides (FAD) and reduced nicotinamide adenine dinucleotide [NAD(P)H] are particularly useful as discriminative parameters in a cell sorter, because their intracellular levels vary, not only with the total nucleotide concentration, but also with the metabolic redox state of the cells. Cell sorting resulted in the isolation of B-cell subpopulations heterogeneous in their metabolic responsiveness to glucose and their sensitivity to diabetogenic agents, according to redox state [i.e., NAD(P)H and oxidized flavin ratio].

III. Fluorescence Probes

An extraordinary expansion in the domain of fluorescence probes (Fig. 17) has equipped the cell biologist with a considerable arsenal to explore the intracellular microenvironment in structural and functional terms. Between 1985 and 1989, over 500 new probes were introduced. A recent review and comprehensive listing of fluorescent dyes and probes that includes spectral data and applications has been assembled by Kasten (1989).

Basic studies of metabolic control and intracellular microcompartmentation are carried out with NAD(P)H and flavins that reflect the redox state and metabolic activity of the cell. Many xenobiotics (benzo(a)pyrene, quinacrine) and cancer chemotherapeutic agents (adriamycin, mithramycin) are also fluorescent, which facilitates the monitoring of their intracellular interactions.

Iron-containing respiratory pigments (hemoglobin, myoglobin) fail to fluoresce, whereas their iron-free breakdown derivatives (protoporphyrin, hematoporphyrin, uroporphyrin, coproporphyrin and phylloporphyrin) emit a striking red fluorescence. Policard, Derrien, Bommer, and Figge observed the selective accumulation of porphyrins (i.e., photodynamic photosensitizers) in animal tissues and in tumors, which is at the heart of the modern photodynamic therapy of cancer.

A few of the most popular vital fluorochromes are acridine orange, fluorescein diacetate (FDA), carboxyfluorescein diacetate (CFDA) and 4-acet-

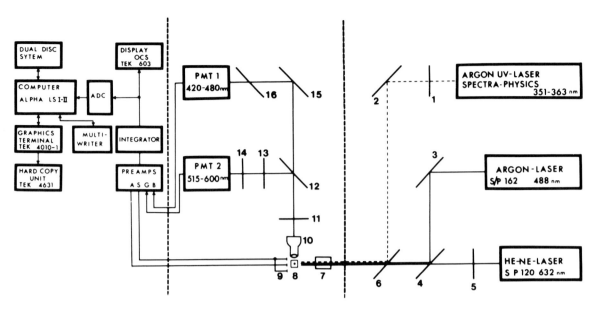

1. EXCITATION FILTER UG11
2. UV MIRROR
3,4. DICHROIC MIRRORS, BIOPHYSICS 45-580
5. NEUTRAL FILTER NG9, T10%
6. DICHROIC MIRROR ZEISS FT 420
7. CYLINDRIC LENS
8. FLOW CHANNEL
9. SCATTER & ABSORPTION SENSORS

10. MICROSCOPE OBJECTIVE
11. LONG PASS FILTER, ZEISS LP 418
12. SHORT WAVE PASS FILTER < 480 nm
    MELLES GRIOT 03 SWP 013
13. 03 SWP 017 < 600 nm
14. OG 515, >515 nm
15. 03 LWP 003, REFLECTING < 480 nm
16. ZEISS KP 500, ADJUSTABLE TILT, ~< 480 nm

FIGURE 16 Basic design of the flow-cytometric instrument for monitoring of intracellular flavins simultaneously with NAD(P)H levels. Combinations of filters and dichroic filters together with a triple laser arrangement allow simultaneous selection of the different fluorescence emissions and also measurement of light scatter and absorption. The filter and mirror combinations for the actual experiments are indicted. The signals are stored and processed by a computer system as shown. Parts from a cytofluorograph FC200 and 4800A [Ortho Instruments, Westwood, MA (see Kamentsky, L. A., Thorell, B., *Acta Cytol* **14**:307, 1970)] were included in the construction. The instrument allows the flow-cytometric analysis of cellular endogenous (blue) fluorescence of NAD(P)H [emission 420–480 nm, excitation by Argon UV-laser (351–363 nm)], green-yellow fluorescence of flavins [emission 515–600 nm, excitation by argon-laser (488 nm)] and light scatter (He–Ne laser 632 nm). (Reproduced by permission of *Cytometry* (**4**:61–65, 1983).)

amido-4′isothiocyanatostilbene-2-2′-disulfonic acid, disodium salt (SITS). The last three are not fluorescent by themselves, but they have the property of penetrating the living cells and liberating fluorescein in the presence of intracellular esterases. Other fluorescent probes are used as vital stains to investigate the functional state of membranes [ani-

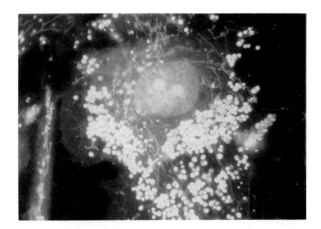

FIGURE 17 Fluorescence micrograph of human fibroblast treated with two fluorochromes: (1) a fluorescent xenobiotic, i.e., quinacrine; and (2) a vital mitochondrial probe, i.e., dimethylaminostyrylmethylpyridinium iodine (DASPMI). Both fluorochromes are excited by the 436-nm line of a mercury arc. Quinacrine accumulates mainly in the lysosomes; DASPMI is bound to mitochondria, which exhibit a filamentous appearance, and also is seen within the nucleus.

linonaphtalene-sulfonic acid (ANS), fluorescamine] and lipid droplets (Nile red). Organelle-specific probes are dimethylaminostyrylpyridiniummethyliodine (DASPMI), dimethylaminostypylpyridiniumethyliodine (DASPEI) and rhodamine 123 for mitochondria, 3-3'-dihexyloxacarbocyanine iodide [$DiOC_6(3)$] for the endoplasmic reticulum, acridine orange, Lucifer yellow and Fluorobora-1 for lysosomes, 6-((N-(7-nitrobenz-2-oxa-1,3-diazol-4-yl)amino)caproyl)sphingosine (NBD ceramide) for the Golgi apparatus and phalloidin-rhodamine or phallacidin for actin filaments. Using hexanoic ceramide-NBD, it was shown that fluorescent sphingomyelin and glucocerebroside are synthesized intracellularly from the fluorescent NBD analogue and subsequently become translocated and processed through the Golgi apparatus and the plasma membrane. The cyanine dye, DiOC6-3, allowed the visualization of the endoplasmic reticulum by fluorescence microscopy showing that it is arranged in a polygonal network in a dynamic state. High-resolution fluorescence micrographs obtained with simultaneous use of vital probes for different organelles allow *in situ* discrimination of activities attributable to these structures, specifically when there is uncertainty about the actual localization of a certain function.

In 1929, Ellinger and Hirt at Heidelberg University developed the first intravital fluorescence or luminescence microscope to examine the microcirculation of exteriorized organs. The apparatus used a water-immersion objective and a standard UV source and filters. Intravital microscopes were used later by various investigators to examine fluorescent dyes (e.g., fluorescein, trypaflavine) in the kidneys and liver of injected frogs and mice, to study the microcirculation of the liver, the conjunctiva, the skin, and the adrenal gland, and to monitor coenzyme fluorescence [NAD(P)H] in exteriorized organs or brains of mammals and birds, and giant neurons of mollusks.

Earlier techniques for measuring cytosolic free Ca^{2+}, such as the luminescent photoprotein aequorin, and the absorbance dye arsenazo III required microinjection and were therefore applied mainly to giant cells, such as *Chironomus* salivary gland cells. The greatest progress since 1980 has been the advent of new fluorescent indicators (Tsien *et al.*) that can be loaded using hydrolyzable esters. These indicators share nearly identical binding sites, which are modeled on the well-known Ca^{2+}-selective chelator, ethylene glycol bis(beta-aminoethyl ether) N, N' tetraacetic acid (EGTA). Currently, four indicators are in use: quin-2; fura-2; indo1; and fluo-3.

Ideally, a Ca^{2+} indicator should be most sensitive to small changes in Ca^{2+} concentration in the physiological range prevailing in most cells (10^{-8} to 10^{-6} mol/liter), and allow discrimination against Mg^{2+} or other cations. The probe affinity for Ca^{2+} should be low in order to minimize Ca^{2+} binding, expansion of the calcium pool of the cytosol, and significant Ca^{2+} buffering that could blunt or distort the changes in Ca^{2+}. The response time of the indicator should be at least one order of magnitude faster than the physiological event to be recorded, with no interference from changes in ionic strength, pH, heavy metals, volume, movement, temperature, autofluorescence, or autoabsorbance. The indicator should not trigger, enhance, or inhibit any cellular function. Its penetration into small mammalian cells should be technically easy, with uniform distribution restricted to the cytosol. Once incorporated, the probe should not leak out of the cell or decay in the cytosol. As much as possible, its fluorescence should decay slowly.

Compared to the calcium probe quin-2, fura-2 has a fivefold higher quantum efficiency and a sixfold greater molar absorption coefficient; therefore its fluorescence is 30 times brighter than quin 2. When fura-2 binds to calcium, the excitation peak shifts from 360 nm to 340 nm. The molar absorption coefficients of free and complexed fura-2 are almost equal, so sequential excitation at the two wavelengths is used to detect calcium by differential recording. The fluorescence emission maximum of indo-1 shifts from 490 nm in Ca^{2+}-free medium to 405 nm when saturated with Ca^{2+}. The intensity at selected wavelengths can be formed into a ratio in which the probe molarity term cancels. Fura-2 is preferred when it is more practical to change the excitation wavelengths, and indo-1 is often used when it is more practical to monitor emission at two wavelengths. The fluorescence of fluo-3 is excited by the argon laser at 490 nm, and the emission maximum is around 520 to 525 nm. Fluo-3 AM is readily loaded into cells as a permeant acetoxymethyl ester and is not fluorescent until hydrolyzed by the cell. Fluo-3 does not undergo significant emission or excitation wavelength shifts with Ca^{2+} binding, but there is an approximately 40-fold enhancement of fluorescence.

On a molar basis, the bioluminescent probe aequorin is by far the most sensitive Ca^{2+} probe available. The relationship between aequorin luminescence and Ca^{2+} concentration spans more than a millionfold range of light intensity. Aequorin can be introduced into small mammalian cells by a variety of methods.

Current pH indicators belong to two fundamental classes: (1) dyes that translocate across membranes in response to pH gradients across those membranes; and (2) dyes that undergo absorption shifts to longer wavelength in basic solution, frequently with considerable fluorescence enhancement. Dyes of class (1) are weak acids or bases that accumulate in alkaline or acid compartments. Accumulation occurs because the neutral form is membrane permeant, but the charged form is not permeant as such, and in continuous local equilibrium with the neutral species. The most commonly used weak bases are 9-aminoacridine and acridine orange (Lee, Simmons). An example of a weak acidic dye is fluorescein, but fluorescein is equally important in class (2) because of the pH sensitivity of its emission. By far the most popular family of pH indicators is fluorescein and its derivatives. Fluorescein fluorescence increases with deprotonation. Derivatives commonly used include: 4′,5′-dimethyl-5(or 6)-carboxyfluorescein [(DMCF) 4′,5′-dimethylfluorescein-dextran [(DMFD) and 2′,7′-bis(carboxyethyl)-5(or 6)-carboxyfluorescein [(BCECF)].

The gentlest means of loading these dyes into cells is by use of membrane-permeant, hydrolyzable esters. Thus the uncharged nonfluorescent fluorescein diacetate (FDA) penetrates into cells, where it is hydrolyzed to the brightly fluorescent fluorescein dianion. The current application is the use of acetoxymethyl (AM) esters; BCECF/AM loading works effectively in a broad variety of cells from cyanobacterial to mammalian.

Major new pH indicators are seminaphtorhodafluor (SNARF) and seminaphtofluorescein (SNAFL) which are most useful for measuring pH changes in the range of 6.3 to 8.6. Calibrations of pH indicators inside cells are usually made with nigericin (which makes the membrane fully permeable) to equilibrate the inside and outside pH. Measurements can be made with one excitation wavelength and two emissions (emission ratio), two excitation wavelengths and one emission (excitation ratio), two excitations and two emissions or one excitation and one emission.

Fluorescent indicators of membrane potentials have been developed over the last 15 years.

Potential sensitive probes fall into four classes: (1) positively charged carbocyanines; (2) negatively charged oxonols; (3) zwitterionic or positively charged styryl dyes; and (4) merocyanines.

The predominant dyes for measurement of relatively slow potential changes are the carbocyanines and merocyanines. Valinomycin is usually used to clamp the membrane potential to a set voltage for calibration of the signal. Potential sensitive translocation of carbocyanine between the inner and outer plasma membrane apparently results in spectral changes. In excitable cells in response to depolarization, there is a potential-sensitive reorientation of the asymmetrically distributed monomer of merocyanine 540 from an orientation in which its long axis is perpendicular to the surface to one in which its axis is parallel to the surface. This results in the concentration-dependent formation of nonfluorescent dimers.

The fast response of styryl dyes is the result of a direct potential sensitive change in the electronic distribution within the dye, which results in spectral shifts. Since the response does not require probe redistribution, spectral changes frequently occur within microseconds. The fast response permits applications in imaging of electrical activity in single neurons of intact brain and cardiac tissues.

Using merocyanine, oxonol, and styryl dyes, absorption and fluorescence signals were measured from Aplysia ganglions imaged on a 12 × 12 diode array and from 124 adjacent regions of rat somatosensory cortex.

Using an oligonucleotide targeted to messenger RNA, complex formation (i.e., base pair formation) is expected to interfere with mRNA processing or translation and consequently inhibit protein synthesis. Increased oligonucleotide-mRNA hybrid stability is obtained by covalent attachment of agents that intercalate between two successive bases, such as 2-methoxy, 6-chloro, o-aminoacridine through a linker pentamethylene arm. Oligodeoxynucleotides covalently linked to intercalating agents have been shown to inhibit mRNA translation both *in vitro* and in microinjected *Xenopus laevis* oocytes. In addition, the linked oligonucleotides can block viral development in cell cultures or kill parasites such as trypanosomes. Because the intercalating agents used are fluorescent, their intracellular fate and their microenvironment can be monitored through their fluorescence excitation spectra. Oligonucleotides with phosphodiester linkages penetrate poorly across cell membranes owing to their high negative charge. Covalently linked intercalating agents improve penetration and appear to drag the oligonucleotide inside living cells.

Photoactive groups covalently attached to 3'- and 5'-ends of oligodeoxynucleotides include proflavine, azidoderivatives, porphyrins, aminoacridines, ellipticine, and diazapyrene derivatives. These oligonucleotide-photosensitizer conjugates bind to their complementary sequences on single-stranded nucleic acids. Upon visible or near-UV irradiation, homopyrimidine oligonucleotides carrying a photoactive group bind to the major groove of duplex DNA. A local triple helix is formed. Oligonucleotide-photosensitizer conjugates open new possibilities to photoinactivate specific messenger RNA or specific genes, and therefore may be used to control gene expression in a highly selective way. They can also be employed to induce site-directed mutations in predetermined regions of the cellular genome.

Fluorogenic substrates of intracellular enzymes are not fluorescent but release a fluorophore when they are cleaved. Fluorogenic substrates used for the *in situ* assay of lysosomal enzymes include 4-methylumbelliferyl-beta-glucopyranoside, fluoresceinyl mono- and bis-beta-glucopyranoside, and BZIPAR[1] (a rhodamine peptide). The most promising are the fluoresceinyl probes, which yield upon cleavage a characteristic increase of emission around 535 nm (436 nm excitation). To obtain *in situ* release of lysosomal enzymes, lysosomal membrane permeabilization is attained by photosensitization (i.e., photofrin-2 or protoporphyrin) or lysosomotropic detergents, i.e., N-dodecylimidazole. Such methods have been applied to the evaluation of beta-glucosidase–deficient Gaucher disease fibroblasts. The techniques are applicable to identification of enzyme defects in other lysosomal storage diseases.

IV. Latest Developments and Emerging New Methods

A. Pixel-by-Pixel Fluorescence Lifetime Determinations

Time-resolved fluorescence spectroscopy is often used for studies of biological macromolecules (Figs. 18, 19). Fluorescence emission usually occurs on the nanosecond time scale, so it is often used to

---

1. Carbobasoxyisoleucine-proline-arginine-rhodamine 110-arginine-proline-isoleucinecarbasoxy . . . . .

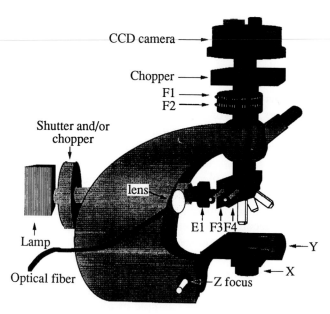

FIGURE 18 CCD camera-microscope system for time-resolved fluorescence spectroscopy. The light from an argon laser is directed to the CCD camera–microscope system via an optical fiber. (Courtesy of Dr. T. Jovin, Max Planck Institute of Biophysical Chemistry, Gottingen, FRG).

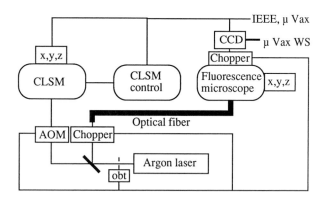

FIGURE 19 Optical and electronic integration of a confocal laser scanning microscope (CLSM) with a CCD camera-microscope system for time-resolved fluorescence spectroscopy (Department of Molecular Biology, Max Planck Institute for Biophysical Chemistry, Gottingen, FRG). A shutter (obt) closes and opens the path of an external laser common to both systems. Acoustico-optic modulators (AOM) or a mechanical chopper can intercept the light directed either to the CLSM or to the other microscope, in the latter case via a UV-transmitting optical fiber. The CLSM is equipped with computerized Z-axis control and galvanometric X, Y scanning. The chopper in the emission path of the fluorescence microscope is phase locked to the excitation to measure delayed luminescence (fluorescence, phosphorescence) with high contrast. System connections are to a local computer and also to a local Vax 2000 through a local network and a standard parallel (IEEE) port (Courtesy of Dr. T. Jovin).

study the dynamic properties of proteins, membranes, and nucleic acids. The sensitivity of fluorescence detection and advances in new detectors have resulted in increased emphasis on the use of fluorescent microscopy to obtain a more detailed understanding of cellular phenomena. Fluorescence spectra recorded from cells are often complex because of spectral overlaps and similarities in the emission of different fluorophores, the simultaneous presence of several fluorophores at a given cell site, or even in macromolecules, plus the intrinsically complex emission of macromolecules. The complex emission from biological samples can sometimes be resolved into their underlying components by measurement of the time-resolved emission. Each fluorophore displays one or more characteristic decay times (monoexponential or multiexponential decay); the time-resolved emission is the sum of the emission from the individual components. Resolution of complex or multiexponential decays requires data with a high signal-to-noise ratio.

Two methods are used to recover the parameters describing the time-resolved decay, time-domain measurements and frequency-domain measurements.

For *time-domain measurements* a pulsed light source (laser or other) is required. The time lag between absorption (excitation) and emission is often on the nanosecond scale, so the excitation pulses must be very brief: e.g., a high-repetition-rate laser that provides pulse rates near 1 MHz and pulse widths near 10 picoseconds. In the time-correlated single-photon technique, fluorescence relaxation microscopy (FRM), when the exciting lamp flashes, a START signal is generated by a photomultiplier for the time-to-amplitude converter (TAC). The first single photon (fluorescence) subsequently detected by a second photomultiplier, located at the detector end of the system, generates a STOP signal. The TAC provides a pulse of amplitude proportional to the time elapsed between the excitation and the detection of the first single fluorescent photon. The pulse is then stored in a multichannel analyzer (MCA). After repeating this cycle a large number of times, a photon-counting histogram of accumulated counts (ordinate) versus time (abscissa) is obtained, related to the rate of decay of intensity.

For *frequency-domain measurements,* the pulsed source is replaced by an intensity-modulated light source. The modulation frequency is varied over the widest possible range, with frequencies comparable to the decay rate of the emission. Because of

the time lag between absorption and emission, the emission is delayed relative to the modulated excitation, and shows a phase shift that increases from 0 to 90° with the increasing frequency. If the modulation frequency is low compared to the decay rate, the modulation will be reproduced less faithfully by the rapid emission; therefore, the finite response time (decay) of the sample results in demodulation, which decreases from 1.0 to 0.0 with increasing frequency. The phase shift and modulation, measured over a wide range of frequencies, constitute the frequency response of the emission, which is analyzed to obtain the fluorescence-decay parameters.

Studies by Lakowicz *et al.* of the frequency response of tyrosine emission from oxytocin, a cyclic nonapeptide with a single tyrosine residue showed a mean decay time for oxytocin of near 0.7 nsec. By use of 2-GHz frequency-domain fluorometry, picosecond resolution of tyrosine intensity decay was achieved. The data are adequate to support a three-exponential decay, with apparent decay times of 80, 359, and 927 psec, probably the first resolution of three decay times all below 1 nsec.

The frequency-domain data provide excellent resolution of complex fluorescence decay processes. Using the chi-square method, typically the values of the goodness-of-fit parameter are compared for the one-, two-, and three-exponential fits. The best-fitting decay model is indicated by the minimal value for the goodness-of-fit parameter. Despite rapid advances in both instrumentation and data analysis, the pixel-by-pixel determination of fluorescence lifetimes from living cells as a probe of intracellular microenvironment, as well as the recording of time-resolved fluorescence spectra from such cells, remains to be achieved.

It is also of interest to recover the fluorescence anisotropy decays for biological molecules. These decays can potentially reveal the size, shape, and segmental mobility of the molecule under investigation. In time-decay fluorometry, the time-resolved decays of the polarized components of the emission are measured. In measurements of frequency-domain fluorescence anisotropy decay, the measured quantities are the phase-angle differences and the amplitude ratio between the perpendicular and parallel components of the modulated emission. Studies on the fluorescence anisotropy decay of melittin, a membrane-disruptive component of bee venom, indicate that the confidence in anisotropy data is increased if the highest modulation frequency (GHz) is attained.

Delayed luminescing immunophosphor-labelled antigens, can be selectively detected by effective suppression of the fast-decaying component (lifetime less than 100 ps) because of autofluorescence. After suppression, delayed luminescence of the immunophosphor is allowed to reach the detector. This technique was applied to immunophosphor-labelled CD4 lymphocytes counterstained for nucleic acids with ethidium bromide (i.e., fast decaying fluorescence). Time-resolved imaging of the field showed an image of the delayed fluorescence-phosphor label only, with suppression of ethidium bromide fluorescence and autofluorescence.

An important aspect of time-resolved phosphorescence microscopy is the fact that the delayed-luminescence images can be visualized with the naked eye directly, without sensitive detection devices such as image intensifier cameras or integrating detectors. Quantification of the delayed fluorescence was carried out with a microscope fluorimeter mounted on the time-resolved fluorescence microscope. Suppression of nonspecific background fluorescence by at least two orders of magnitude was achieved.

B. Polarized Fluorescence Photobleaching Recovery

In biological samples likely to be viewed under a microscope the rotational rates are somewhat too slow to be seen during the fluorescence lifetime. This difficulty can be resolved by switching to phosphorescence, which has a longer lifetime. This is usually achieved by attaching the extrinsic probe eosin thiocyanate to the cell component under study (hundreds-of-microseconds lifetime instead of nanosecond scale). Deoxygenation is required to lengthen the triplet state lifetime and increase the triplet population, which is incompatible with certain biological samples. Thus a brief flash of polarized light can be applied to create an anisotropic distribution of unbleached fluorophore (aligned perpendicularly to the fluorophores excited by the polarized light). The anisotropic distribution of unbleached fluorophores relaxes, with the use of an observation beam parallel to the bleaching (pump) pulse, and a time-dependent fluorescence recovery is observed, allowing deduction of the rotational diffusion coefficient.

C. Chemi- and Bioluminescence Studies in the Microscope Field

The ATP dependence of enzyme-substrate kinetics of the firefly luciferin-luciferase reaction has led to multifaceted applications in different fields. With a range of quantum efficiency that encompasses close to 14 orders of magnitude, interest in chemi- and bioluminescence has stimulated several developments in instrumentation related to photobiology and microspectroscopy.

The present approaches to the measurement of chemiluminescence or bioluminescence include studies with chemical amplifiers such as luminol in phagocytic cells (i.e., human neutrophils). Luminol is used as a probe of the oxidative burst in these cells. Measurable chemiluminescence can be obtained using as few as 10 cells, and it may be possible to do so with a single cell.

D. Fluorescence Redistribution after Photobleaching (FRAP), also Called Fluorescence Photobleaching Recovery (FPR)

The first application of the photobleaching concept for the measurement of translational mobility employed half of a single light-sensitive visual disk from the rod cell of an amphibian retina, which was briefly exposed to illumination in a microspectrophotometer to bleach rhodopsin. The absorbance of rhodopsin measured in both halves as a function of time was shown to increase in the bleached half and to decrease in the unbleached half, indicating that rhodopsin was free to redistribute, and providing an estimate of its mobility. Many other applications followed. As described by Ware, the basic apparatus for fluorescence photobleaching recovery (FPR) generally includes a research microscope equipped with epi-illumination, a laser light source, a fluorescence detection system, and a means for measuring the time dependence of the fluorescence recovery.

Fluorescent analog cytochemistry (FAC) refers to the use of fluorescent analogs of cytoplasmic proteins to study the distribution and dynamics of specific cytoplasmic components. The fluorescent analogue is generally injected into living cells. This approach has been used in combination with FPR, as described above. Fluorescent analogs of actin are among the most common probes employed because they can mimic the function of native cytoplasmic actin. Microfilaments are an essential component of the cytoskeletal system, producing cytoplasmic motility. FAC, when employed in conjunction with the FPR technique, has permitted the diffusion coefficient of actin to be measured in living cells. Also, a fluoresceinated derivative of tubulin has been shown to become associated with cytoplasmic microtubules following injection into cells. The use of FPR techniques showed that the fluorescence redistribution was too rapid to be accounted for by a simple end-dependent exchange of tubulin. The data were more compatible with a mechanism involving rapid exchange of tubulin along the entire lengths of microtubules.

The technique of total internal reflection fluorescence microscopy (TIRF) was developed by Axelrod in 1981 as a means to excite fluorophores very near a glass (or plastic) interface in an aqueous environment. When applied to biological cell cultures, TIRF allows selective visualization of cell–substrate contact regions, especially in samples in which fluorescence from other areas would otherwise obscure the fluorescent pattern in contact regions. According to the theory of TIRF, if an excitation light beam is incident upon the solid–liquid surface to which the cell adheres at an angle large enough for the beam to totally reflect internally, TIR generates a very thin (generally less than 200 nm) electromagnetic field in the liquid (or cell) with the same frequency as the incident light. The intensity of the generated field decays exponentially with distance from the surface. The field, called the *evanescent wave,* is capable of exciting fluorophores in the cell region near the surface while avoiding excitation of a possibly much larger number of fluorophores farther away in the liquid. The interface causes the emission to be anisotropic; therefore, the fluorescence intensity observed depends in a complicated manner on the fluorophore orientation, its distance from the surface, and on the numerical aperture of the objective. However, some of the surface-induced optical phenomena can actually be examined experimentally in systems for selective fluorescence detection, for selective surface fluorescence quenching, and for optimal collection of fluorescence power.

The most straightforward application of TIRF is *long-term viewing* of cells (with time-lapse video) to observe the location and motion of cell–substrate contacts, because the evanescent wave minimizes exposure of the cell's organelles to excitation light. Cells may be labeled by a membrane lipid fluorescent analog (such as 3,3′-dioctadecylindocarbo-

cyanine), specific ligands (rhodamine α-bungarotoxin targeted to acetylcholine receptors), or fluorescent antibodies against particular cell surface and cytoplasmic components. Optical sectioning by TIRF is particularly useful in viewing the submembrane filament structure of cells. Doping of the solution surrounding the cells with a nonabsorbing and nonpermeable fluorescent marker (i.e., fluorescein-labeled dextran) results in negative TIRF. Focal contacts of cell–substrate are dark, whereas cell–substrate gaps appear bright. The intrinsic fluorescence of tryptophan residues on unlabeled proteins and fluorescein-, rhodamine-, NBD- or dansyl-labeled proteins, have been used to study adsorption equilibria of proteins. TIRF has been utilized in combination with FPR.

E. Fluorescence Photoactivation (PAF)

The method of fluorescence photoactivation as described by Ware *et al.*, utilizes labels that are initially nonfluorescent and that can be activated through a photochemical reaction. Such molecules have been called photoactivable fluorophores (PAF).

The fluorescence photoactivation approach could largely replace photobleaching techniques within the next decade. The method would be advantageous with respect to optical contrast and because of its inherent high signal-to-noise ratio. Instead of measuring, as in FPR, a slight reduction in the intensity of fluorescence against a bright background signal, a photoselected fluorescent species would be measured against a dark background.

F. Fluorescence Resonance Energy Transfer (FRET) Microscopy and Digital Imaging of Fluorescence Resonance Energy Transfer

Fluorescence can be used as a spectroscopic ruler, as first enunciated by Stryer and subsequently applied by Jovin, to measure distances between cellular components and gain information about the interactions of these components. This method, known as fluorescence resonance energy transfer (FRET), is founded on the principle that a fluorophore (donor) in an excited state may transfer its excitation energy to a neighboring chromophore (acceptor) nonradiatively through dipole-dipole interactions (Förster-type transfer). This process requires some spectral overlap between the emission spectrum of the donor and the absorption spectrum

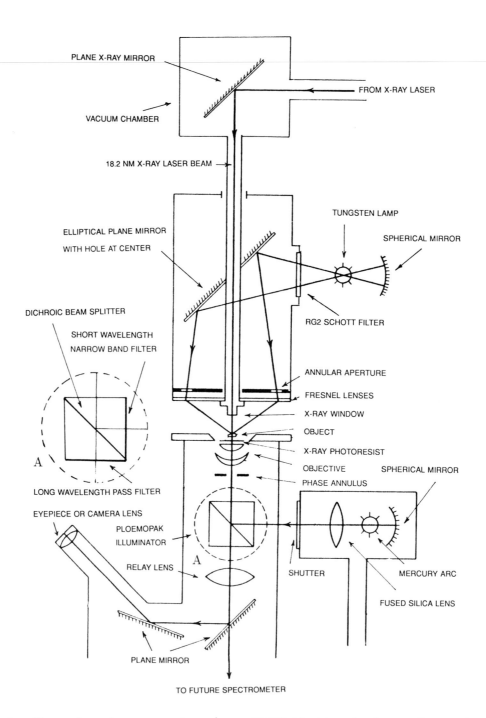

PLANE X-RAY MIRROR

VACUUM CHAMBER

FROM X-RAY LASER

18.2 NM X-RAY LASER BEAM

TUNGSTEN LAMP

SPHERICAL MIRROR

ELLIPTICAL PLANE MIRROR
WITH HOLE AT CENTER

DICHROIC BEAM SPLITTER

SHORT WAVELENGTH
NARROW BAND FILTER

RG2 SCHOTT FILTER

ANNULAR APERTURE

FRESNEL LENSES

X-RAY WINDOW

OBJECT

X-RAY PHOTORESIST

OBJECTIVE

A

LONG WAVELENGTH PASS FILTER

EYEPIECE OR CAMERA LENS

PLOEMOPAK
ILLUMINATOR

RELAY LENS

A

SPHERICAL MIRROR

PHASE ANNULUS

SHUTTER

MERCURY ARC

FUSED SILICA LENS

PLANE MIRROR

TO FUTURE SPECTROMETER

of the acceptor. For a given donor–acceptor pair, the efficiency of the transfer is dependent on the relative orientation of donor–acceptor and on the distance between them.

FRET techniques were used to study endocytosis and intracellular membrane traffic. In this case, N-4-nitrobenzo-2-oxal-1,3,-diazole (NBD) or fluorescein were utilized as the donor and sulforhodamine was the acceptor. Liposomes were examined in a

FIGURE 20 Schematic of composite x-ray laser microscope (COXRALM).

situation in which the geometry of the transfer process was known in great detail. By visually inspecting the specimen for quenching of donor emission or an increase in sensitized acceptor fluorescence, the transfer between the different lipid molecules was detected.

There are limits in practical applications of FRET in microscopy. According to the Förster equation, the probability of energy transfer from donor to acceptor varies with the reciprocal of the sixth power of the intermolecular distance. The range of the finite energy transfer is confined to distances generally shorter than 5 nm, depending on the respective orientation of donor and acceptor electronic dipole moments. If the acceptor molecule is also a fluorescent donor, quenching is accompanied by sensitized emission of the acceptor at a longer wavelength.

FRET depletes the population of excited donors and thus shortens the lifetime. Additional information on the distribution of donors and acceptors can be obtained from analysis of fluorescence spectra taken at intervals during the lifetime of the excited donor population (time-resolved spectra, TRS). Immediately after excitation, the probability of FRET will be very high between close donors and accep-

FIGURE 22 Operation of the phase-fluorescence inverted microscope incorporated in the composite x-ray laser microscope. The large long-working-distance, high-resolution phase condenser is seen showing the working distance (about 6 cm) over the stage of the inverted microscope.

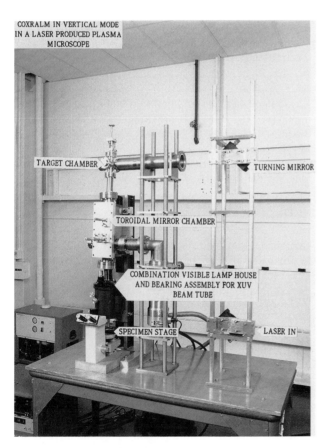

FIGURE 21 View of composite x-ray laser and phase-fluorescence microscope. COXRALM in vertical mode with laser-produced plasma microscope (x-ray laser microscope). The modified inverted Diavert phase and fluorescence microscope is of the same kind as used in the microspectrofluorometer.

tors and will decline rapidly as the remaining donors are separated from acceptors by greater distances. Nanosecond TRS are fluorescence-emission spectra obtained at discrete times during the fluorescence decay, which can be useful to probe the dynamics of FRET.

G. Composite Optical X-ray Laser Microscope (COXRALM)

COXRALM, developed by Suckewer and Hirschberg, is an inverted phase-contrast microscope with the capability of observing UV-induced fluorescence, combined with the option of contact micrograph generation by means of a flash soft x-ray exposure (Figs. 20-22). The source is soft x-ray laser 18.2 nm radiation. A limitation is that penetration of this x-ray is only 0.5–1.0 micrometer. Further structural resolution will be obtained when more penetrating, shorter wavelengths can be used. An alternative approach is the adoption of the reflecting x-ray microscope to the study of cell surface receptors involved in the action of hormones and growth factors. It is conceivable that this method, through the use of heavy metal–tagged antibodies, may be applicable to the study of receptors during cell differentiation and other interactions such as cell aggregations and invasiveness of malignant cells.

The combined x-ray and fluorescence approach should be suitable for ultrastructural and functional studies on the organelles and microarchitecture of the same living cell.

H. Support of Optical Microscopy by Diagnostic Ultrastructural Pathology and Electron Probe X-ray Analysis

The previous discussion has primarily considered techniques based on absorption and fluorescence spectroscopy at the single-cell level. However, diagnostic possibilities are also afforded by transmission electron microscopy and electron probe x-ray analysis in the nanometer domain of structural resolution, which is outside the reach of optical microscopic methods, with the exception of x-ray microscopy and FRET. The diagnostic importance of electron microscopy for a great many malignant or benign tumors and other cell and matrix pathology is well established. An argument can be made that practically every organelle can occasionally serve as a diagnostic marker. As examples, *nuclear pockets* are encountered in lymphomas and leukemias, *nuclear fibrous lamina* are observed in repair tissue but not in malignant tumors, *rod-shaped microtubular bodies* are seen in vasoformative tumors, *massively increased mitochondria* (probably with deficient function) occur in oncocytomas, *myelinosomes* are found in some lysosomal storage diseases and in tumors of pneumocyte type II cells, *mucous and serous granules* are visible in exocrine gland cancers, *APUD granules* are hallmarks in neuroendocrine tumors such as pheochromocytomas and carcinoid tumors, *tubular confronting cisternae* appear in AIDS and in Japanese T-cell leukemia, and *massive hyperplasia of smooth ER* is a feature in hepatocytes of patients taking a variety of drugs (i.e., barbiturates and ethanol).

Electron probe x-ray analysis is based on x-ray emission following irradiation of the specimen with an electron beam. Instances of diagnostic applications are the detection of bismuth in *nuclear inclusions*, magnesium silicate (talc) in *talcosis* associated with *pulmonary fibrosis*, iron in *hemosiderosis*, and iron sulphide in *melanosis duodeni*.

V. Conclusions

New frontiers in cell biology have been revealed by modern optical microscopic methods. Such methods make use of a broad expanse of fluorescence (and potentially bioluminescence) probes and the versatility of new techniques being introduced, i.e., multiparameter studies, laser scan fluorescence microscopy, confocal microscopy, video microscopy, FPR, fluorescence photoactivation, FRET, and the COXRALM. A tremendous flexibility in approach permits the probing of the intracellular microenvironment at both the fine structural and functional levels. It is now possible to probe interactions within microdomains of cell organelles where the delicate interplay between diverse structural and functional compartments can be examined. The ultimate reward may well be an enhanced perspective of *in situ* intracellular processes leading to a revolution in our understanding of cellular biochemistry, pharmacology, and pathology.

Acknowledgments

This work was supported by National Science Foundation Grant DMB 8303691, the National Gaucher Foundation Grant #27, the Biological Stain Commission, and the Institut National de la Santé et Recherche Médicale (Un INSERM U312). The authors acknowledge thankfully the efficient secretarial help of Mrs. Maria Fernanda Reynardus and the art work of Mr. Bill May.

Bibliography

Arndt-Jovin, D. J., Robert-Nicoud, M., Kaufman, S. J., Jovin, T. M. (1985). Fluorescence digital imaging microscopy in cell biology. *Science* **230,** 247–261.

Bessis, M. (1975). Hematology without the microscope. *Blood Cells* **1,** 401–403.

Borst, M., and Königsdörffer, H., Jr. (1929). Untersuchungen über Porphyrie mit Besonderer Berücksichtigung der Porphyria congenita. Hirzel, Leipzig, G.D.R.

Caspersson, T. (1940). Die Eiweissverteilung in den Strukturen des Zellkernen. *Chromosoma* **1,** 562–619.

Caspersson, T., and Lomakka, G. (1970). Recent progress in quantitative cytochemistry: Instrumentation and results. *In:* "Introduction to Quantitative Cytochemistry II." G. L. Wied, and G. F. Bahr, (eds.) Academic Press, New York, N.Y.

Chance, B., and Thorell, B. (1959). Localization and kinetics of reduced pyridine nucleotides in living cells by microspectrofluorometry. *J. Biol. Chem.* **234,** 3044–3050.

Cowden, R. R., and Harrison, F. W. (1985). Advances in Microscopy (Progress in Clinical and Biological Research, Vol. 196). Alan R. Liss, New York, N.Y.

Hopman, A. H. N., Wiegant, J., Raap, A. K., Landegent, J. E., van der Ploeg, M. and van Duijn, P. (1986). Bi-color detection of two target DNAs by nonradioactive *in situ* hybridization. *Histochemistry.* **80,** 1–4.

Inoue, S. (1986). "Video Microscopy." Plenum Press, New York, N.Y.

Kasten, F. H. (1989). The origins of modern fluorescence microscopy and fluorescent probes. *In:* Cell Structure and Function by Microspectrofluormetry. E. Kohen and J. G. Hirschberg, eds., pp. 3–50. Academic Press, San Diego, California.

Kohen, E., and Hirschberg, J. G., eds. (1989). Cell Structure and Function by Microspectrofluorometry." Academic Press, San Diego, California.

Masters, B. R. (1986). Noninvasive corneal redox fluorimetry. *In:* (J. Zadunaisky and H. Dawson, eds). *Curr. Top. Eye Res.* **4,** 139–200.

Tanke, J. H. (1989). Does light microscopy have a future? *J. Microsc.* **155,** 405–418.

Taylor, D. L. and Wang, Yu-Li, eds. (1989). "Fluorescence Microscopy in Living Cells in Culture. Part B. Methods in Cell Biology," Vol. 30. (Leslie Wilson, ed.) Academic Press, New York, N.Y.

Weber, G. (1986). Solution spectroscopy and image spectroscopy. *In:* Applications of Fluorescence in the Biomedical Sciences," pp. 601–615. Alan R. Liss, Inc., New York, N.Y.

Wells, K. S., Sandison, D. R., Stickler, J., and Webb, W. W. (1989). Quantitative fluorescence imaging with laser scanning confocal microscopy. *In:* The Handbook of Biological Confocal Microscopy. (J. Pawley, ed.), pp. 23, 24. IMR Press, Integrated Microscopy Resources for Biomedical Research, University of Wisconsin.

West, S. S. (1970). Optical rotatory dispersion and the microscope. *In:* "Introduction to Quantitative Cytochemistry II." (G. L. Wied, and G. F. Bahr, eds.), pp. 451–475. Academic Press, New York.

Optical Properties of Tissues

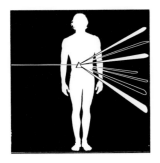

BRIAN C. WILSON, *McMaster University*

Glossary

Bilirubin Yellow bile pigment formed in the breakdown of heme

Chromophore Chemical that absorbs light with a characteristic spectral pattern

Choroid Thin, pigmented vascular coating in the eyeball extending from the anterior margin of the retina to the optic nerve

Fluorescence Property of emitting light of a longer wavelength on absorption of light energy

Light fluence rate Measure of light "intensity," defined in terms of the optical power per unit area

Light scattering Change in direction of propagation of light in a turbid medium caused by reflection and refraction by microscopic internal structures

PROPAGATION OF ULTRAVIOLET (UV), visible, and infrared (IR) radiation ("light") in tissues is governed by the absorption and scattering of light photons. At normal power densities, the absorption is caused by specific molecules ("chromophores"), which have different concentrations in different tissues and may have a strong wavelength dependence. Light energy absorption leads to reversible or irreversible photochemical, thermal, photomechanical, or other changes in the tissue. The scattering of light in tissue, resulting from microscopic variations in the refractive index, does not deposit energy in the tissue, but rather alters the spatial distribution of the light. This and the location of chromophores then determine the absorbed energy

distribution and the resultant photobiologic response. Transmission through the structures of the eye and the skin are important in the proper functioning of these organs. Transmission through or scattering from other organs is increasingly studied as lasers and optical technologies are applied to human biology and medicine.

I. Fundamentals of Light Absorption and Scattering

A. Background

The optical region of the electromagnetic spectrum includes UV (wavelength range 100–400 nm), visible (400–780 nm), and IR (>780 nm) radiation, as shown in Fig. 1. At shorter wavelengths lie X- and γ-rays, in which the photon energy is high enough to cause ionization of atoms in tissue, whereas at longer wavelengths lie the microwave and radiofrequency bands. The visible spectrum is divided into different color bands, the UV into three regions (i.e., UV-A, UV-B, and UV-C), and the IR into IR-A, IR-B, and IR-C. These divisions are related to particular physical and/or photobiological characteristics of the radiation, and other ways of dividing the spectrum are also used. For example, reference will be made to "near" and "far" UV or IR, corresponding to those parts of the spectrum lying closest to or farthest from the visible band, respectively.

A fundamental property of electromagnetic radiation is that the energy of each photon varies inversely with the wavelength. The photon energy in the optical spectrum corresponds roughly to molecular excitation energies, so that these photons can produce specific biophysical and biochemical effects in tissue, and these are strongly energy (i.e., wavelength) dependent.

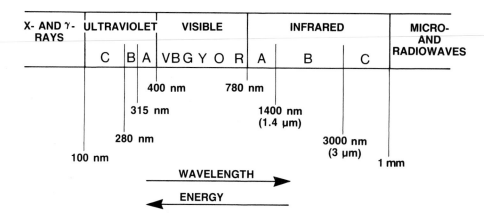

X- AND γ-RAYS	ULTRAVIOLET	VISIBLE	INFRARED	MICRO- AND RADIOWAVES
	C B A	V B G Y O R	A B C	

400 nm   780 nm

315 nm   1400 nm (1.4 μm)

280 nm

3000 nm (3 μm)

100 nm   1 mm

WAVELENGTH →

← ENERGY

FIGURE 1 Electromagnetic spectrum showing division of the optical region of ultraviolet, visible, and infrared radiation. Note nonlinear wavelength scale. Some of the boundaries are variously defined: UVA/UVB may be 315 or 320 nm; UVA/visible, 380–400 nm; visible/IR, 760–780 nm. Normal colors (violet (V)–red (R)) are indicated in the visible region.

The optical region is of particular biological interest. It encompasses the natural spectrum of radiation reaching the Earth's surface from the sun. A minimum exposure to at least parts of the optical spectrum is essential to human health and well-being, whereas overexposure may lead to tissue damage and disease. An example of the former is ricketts resulting from insufficient vitamin D, which is produced in the skin by exposure to UV-B. By contrast, chronic high exposure to UV-B is associated with induction of skin cancers. Treatment of diseases by light has been known since the time of the ancient Egyptians (the word *radiation* itself stems from Aton Ra, the Egyptian sun god). Herodotus, for example, in 425 BC related the strength of the skull to sunlight exposure, whereas sunbathing was prescribed by early physicians for conditions as diverse as obesity, epilepsy, and jaundice. Vision itself depends on the transmission of visible light through the anterior structures of the eye and on the photochemical changes induced in the retina by absorption of the photon energy. Other wavelengths are harmful to the retina or lens and must be absorbed before reaching them.

These examples serve to illustrate fundamental principles in photobiology, namely, that

1. the photon energy must be absorbed by the particular tissue target before any biological effects can occur;
2. the biological effects resulting from energy absorption depend on the target and on the wavelength; and
3. the amount of light reaching a tissue target depends on the tissue through which the light must first travel, and hence on the light absorption and scattering properties of this tissue.

B. Light Absorption in Tissue

At high light power densities, which can be achieved using short-pulse lasers, light absorption in tissue is a complex, nonlinear process, in which molecules or atoms are excited to high energy levels, or even ionized. Although there is great interest in understanding the photophysics and photobiology in this regime, the applications are for the future and will not be discussed further.

At more normal power levels, as found with natural sources, lamps, and most lasers, light absorption is a linear process in which the rate of energy absorption is simply proportional to the incident light fluence rate [i.e., power density (W/cm^2)]. The absorption is due to specific molecules, known as chromophores, each of which has a distinct absorption spectrum. At any wavelength, the absorption coefficient of a particular tissue, which is a measure of the probability that a photon will be absorbed per unit distance traveled through the tissue, is the sum of the absorption coefficients of all the chromophores present. The absorption spectra of some important chromophores in soft tissues are shown in Fig. 2. In the visible region, hemoglobin is a significant absorber in the intravascular component of all tissues. It has a complex spectrum, which changes significantly between the oxygenated and reduced states. The double-peaked structure in the green region of the spectrum at 542/577 nm is particularly

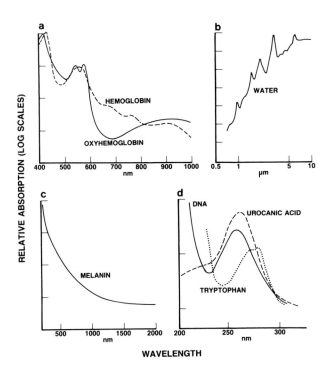

FIGURE 2 Absorption spectra of some tissue chromophores in the UV, visible, and IR. Note different wavelength ranges. Vertical scales are in arbitrary units, but in each case one division represents a factor of 10 in the absorption coefficient.

characteristic for oxyhemoglobin. There are several wavelengths where the absorption coefficient of hemoglobin is independent of the oxygenation ("isobestic points," e.g., ~815 nm). Comparisons of the ratio of absorption at isobestic and nonisobestic wavelengths may be used to determine the oxygenation status of tissues. Although its absorption is less in the near-IR than in the visible, hemoglobin is still a significant chromophore in this region.

The pigment melanin, a protein-polymer complex, is an important visible chromophore in some tissues. It has a relatively simple spectrum, with the absorption decreasing steadily with increasing wavelength. Melanin absorption plays a key role in the photoprotection and photoresponse of skin (see Section II,A). In the eye, the choroid also has a melanin layer, which absorbs light transmitted through the retina, thus preventing light being scattered back to the retina from the sclera. Melanin absorption is negligible above the near-IR range. Other visible pigments are present in different concentrations in various tissues. For example, bilirubin and beta-carotene are also found in skin and, together with hemoglobin and melanin, are responsible for skin color.

In the UV range, absorption in tissue is mainly due to proteins and nucleic acids. The latter have a main absorption peak close to 265 nm because of the pyrimidine structure, whereas the aromatic amino acids are the absorbing sites in protein, with tyrosine peaking at 275 nm and tryptophan at 289 nm (see Fig. 2d). Numerous other small aromatic molecules also absorb in the UV. At wavelengths less than about 240 nm, high absorption results from peptide bonds. Absorption by water also becomes high at less than about 190 nm.

These various absorption characteristics contribute to the form of the "action spectrum" for photobiological effects in tissue (i.e., the wavelength dependence of biological responses to irradiation). This follows the first principle above that energy absorption must occur before biological change can result. The correspondence is not exact, however, both because of the wavelength-dependent photobiological sensitivity of the targets and because of the influence of light scattering and the distribution of the chromophores. Examples are the action spectrum for UV-induced erythema (i.e., reddening of the skin as in sunburn), which peaks between about 240 and 290 nm, whereas skin cancer induction is greatest in the UV-B region between 280 and 325 nm.

As the wavelength increases beyond about 600 nm, and particularly in the IR-B and IR-C regions, the absorption of specific chromophores falls off, and water becomes the dominant IR absorber in soft tissues. The absorption coefficient of any tissue then depends largely on its water content. The IR absorption spectrum of water is complex (Fig. 2b), with numerous narrow peaks. The absolute value of the water absorption can be large. At the highest peak around 2.9 μm, the coefficient is about 13,000/cm, so that, in a tissue with a 70% water content, such radiation is attenuated by 99% in less than 5 μm.

The total absorption in tissue is lowest between about 600 and 1,300 nm, with the extinction of specific pigments having fallen off and that of water still being small. This red/near-IR "optical window" permits the greatest penetration of light into tissue and is, therefore, a region of particular interest in therapeutic and diagnostic applications.

An example of the total absorption spectrum in tissue is shown in Fig. 3. Even in highly pigmented liver, the total absorption is less than about 5/cm (0.5/mm) in the optical window. Thus, on average, such photons travel more than 2 mm before being

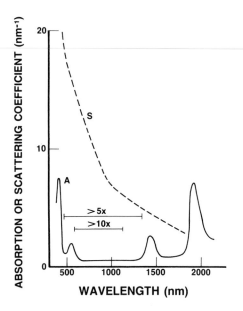

FIGURE 3 Example of the absorption (A) and scattering (S) spectra of tissue (liver *in vitro*). *Bars* show the wavelength ranges where the scattering is greater than 5 times or 10 times the absorption for this tissue, illustrating the concept of the optical window. [From Parsa, P., *et al.* (1989). *Appl. Optics* **28**, 2325–2330, with permission.]

absorbed: however, because absorption and scatter are random processes, the distribution of these interaction distances is exponential, and 1% of the photons will travel more than 1 cm before being absorbed. For lightly pigmented tissues (e.g., brain) the average distance traveled before absorption may be several centimeters. At shorter visible wavelengths, hemoglobin and other pigments cause a marked increase in the total tissue absorption. In the example of Fig. 3, the 555-nm single absorption peak of deoxyhemoglobin is apparent, as is the so-called Soret band just above 400 nm, which corresponds to excitation of the heme ring structure by the light. Strong absorption peaks in the near-IR caused by water are also seen.

C. Light Scattering in Tissue

With the exception of the transparent structures of the eye, tissues are optically turbid, because photons may be scattered as well as absorbed. Scattering is normally elastic, so that the direction of the photons is changed but there is no loss of energy, and the wavelength is unaltered. The scattering is due to microscopic fluctuations in the refractive index of the tissue, corresponding to physical inho-

mogeneities. The angles through which the light is scattered and the scatter coefficient depend on the size and shape of the optical inhomogeneities relative to the wavelength and differences in the refractive index between the inhomogeneities and the surrounding medium. In scattering from molecules or structures whose size is much less than the wavelength (Rayleigh scattering), the scattering is weak, is nearly equal in all directions, and rapidly decreases with wavelength. When the wavelength and inhomogeneities are about the same size, the scattering is stronger, is more forward-directed, and decreases roughly inversely as the wavelength. Scattering from large inhomogeneities is highly forward-directed and only weakly dependent on wavelength (Mie scattering).

The inter- or intracellular structures causing light scattering in tissues have not been clearly identified and may well be different for different tissues. In most soft tissues, scattering is very forward-directed in the visible and near-IR. This suggests that the optical inhomogeneities are of the same order as or larger than the wavelength (i.e., on a micron scale). The scatter spectrum for liver is shown in Fig. 3 and is typical of soft tissues in this spectral range. The scattering decreases approximately linearly with wavelength, with little spectral structure. In the optical window, the scatter coefficient is much greater than the absorption coefficient. Also, over the wavelength range in Fig. 3, the scattering angular distribution is roughly constant, with an average scattering angle of about 20–30° (data not shown). At a given wavelength in the optical window, the scattering coefficients vary between tissues by a factor of about 5–10.

In the dermis of skin, scattering has been associated with collagen fibers, and optical changes seen in aging or photodamage may result from changes in collagen content or structure. In the brain, laying down of myelin with maturation is thought to be important in causing the increased light scattering seen in adult tissue compared with neonatal tissue. [*See* COLLAGEN STRUCTURE AND FUNCTION.]

Although scattering is the result of microscopic variations in refractive index, the average refractive index of tissue determines the speed of light in the tissue and also the regular (specular) reflection at the tissue surface. The latter causes the "shining" appearance of skin or freshly exposed tissue surfaces. In the red part of the spectrum where measurements have been made, the average refractive index of most soft tissues is in the range 1.37–1.45, so that, when a visible light beam strikes an air–tissue interface, typically a few percent is specularly reflected. By contrast, a large fraction (up to ~60%) may be diffusely reflected after multiple scattering within the tissue.

On a microscopic scale, light photons crossing cell membranes between relatively aqueous and lipid environments are subject to both refraction (bending of light waves) and specular reflection, which result in scattering as seen on a macroscopic scale. This can be seen by passing a light beam through a cell suspension. In an aqueous environment, the beam is strongly scattered, but this can be markedly reduced by increasing the refractive index of the suspension medium to match that of the cell membranes.

D. Combined Absorption and Scattering in Tissue

For soft tissues in the optical window, the scattering coefficient is high and also much larger than the absorption coefficient. Typically, the average distance between scattering interactions is only tens of microns. However, many more scattering interactions than absorption events take place, and the scattering is forward-peaked. These properties account for the existence of the optical window: Although the photons interact with the tissue tens or hundreds of times for each millimeter of their travel, in most cases the photons survive with no loss of energy and are deflected only slightly (and randomly) from the initial direction. Thus, a fraction of the photons may penetrate to considerable depth. For example, in near-IR transmission spectroscopy of the neonatal brain to measure the tissue oxygenation, light is applied to one side of the head, and detectors on the other side measure the transmitted fraction. Virtually no photons pass through the head without interacting, but a large transmitted signal can be detected nevertheless. This comes from the scattered light, whose average distance traveled in crossing the head is in the order of meters rather than centimeters.

The forward scattering of tissue effectively reduces the strength of the scattering coefficient. Photons may scatter many times through a small angle and travel an equivalent depth in tissue as they would if the number of scatters was less but each scatter was through a larger angle. The angular dependence of the scatter and the scattering coefficient may be combined in the concept of a "re-duced" or "transport" scattering coefficient, which is typically 10 or 20 times less than the true scattering coefficient. Then, unless the detailed scattering angles are specifically important, the transport coefficient is a convenient measure of effective scattering.

The absorption and scatter in tissue combine to determine how light entering the tissue is distributed. An example is shown in Fig. 4, where different tissue pigmentation and blood content lead to different and wavelength-dependent light distributions. In the extreme case of the absorption being much greater than the scattering (UV and far-IR), a light beam incident on tissue is attenuated exponentially, with little lateral spreading of the light or back-scattering (diffuse reflectance). Conversely, where the tissue is very scattering, there is wide spreading of the beam; the decrease of fluence with depth below the surface is complex and may even show a subsurface peak several times higher than the incident value, and the diffuse reflectance is high.

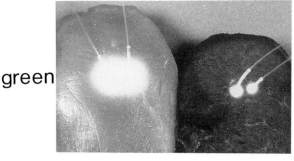

FIGURE 4 Illustration of light absorption and scattering in tissue. Each panel shows two different tissues (lightly pigmented "white" avian muscle on the *left* and "red" bovine muscle on the *right*) with implanted optical fibers delivering either red (630 nm) or green (514.4 nm) light. Photographs show distribution of unabsorbed light reaching the surface of tissues after scattering. The optical power delivered to tissues is the same in all cases. [From Marijnissen, J. A., and Star, W. M. (1987). *Lasers Med. Sci.* **3**, 235–242, with permission.]

An important consequence of high light scattering in tissue is that any simple spectral measurement (e.g., light transmitted through or diffusely reflected from a certain thickness of tissue) does not reveal the true absorption spectrum of the tissue. Rather, the spectrum is distorted by the scattering. Thus, in Fig. 4, the ratio of the intensities of light reaching the surface at the two wavelengths in each tissue is not the same as the inverse ratio of the absorption coefficients at these wavelengths. In general, scattering tends to "flatten out" peaks and valleys in the absorption spectrum.

In many practical situations, it is not necessary to consider separately the contributions of absorption and scattering to the light attenuation. The exponential decrease in fluence as light penetrates through tissue may be described adequately by a single "effective attenuation coefficient." This depends on factors such as the light beam size and shape as well as on the tissue optical properties and so varies from case to case. The effective attenuation coefficient is directly proportional to the tissue optical density (OD = $\log_{10}(I_0/I)$, where I_0/I is the ratio of the incident to transmitted fluence rate of a light beam passing through 1 cm of tissue.

E. Measurement Techniques

Measurement of tissue optical properties is not simple. In addition to the problems caused by light scattering, the tissue handling and preparation are important. Clearly, embedding techniques or histological stains cannot be used. Freezing may change the scattering properties, whereas keeping tissues at room temperature even for a limited period can alter the absorption spectrum and possibly the scattering. Loss of components during excision or sectioning, particularly blood, is an obvious source of error. With solid tissues, the condition of the surface can affect how light is scattered from it. The most direct information has been obtained using optically thin tissue sections (typically <100 μm), but these are most subject to preparation artifacts. Thicker sections ($\sim$mm) may be used, but the determination of the true scattering and absorption values is complicated by the interplay between scattering and absorption in the overall light transmission through the section.

Comparisons of light distributions measured in intact tissues immediately *postmortem* with those in living tissue suggest that the scattering properties are not generally much altered but that the absorption in the visible range can change markedly because of drainage of blood from the tissues and reduced blood oxygenation.

Measurements *in vivo* have generally been done by inserting optical fibers into the tissue. The other end of the fiber is connected to a light detector, so that the distribution of input light in the tissue may be "mapped." The optical properties may be deduced if this mapping is sufficiently detailed. Recently, noninvasive tissue spectroscopy has become feasible, as illustrated by the case of neonatal brain measurements cited above. The general principle is to deduce the optical properties from the (spectrum of) light that is diffusely reflected from or transmitted through the tissue. An area of particularly active study is the use of time-of-flight spectroscopy. In this, short laser pulses are used, and the time spectrum of diffusely reflected or transmitted light is measured using fast detectors. The time taken for a photon to travel between the source and detector depends on the path that it follows through the tissue by multiple scattering. The fraction of the light that survives a particular path length depends on the tissue absorption. Thus, the time spectrum carries information on both the scattering and absorption properties of the tissue. Because the speed of light in tissues is so high (20 cm in 1 nsec), these techniques are at the leading edge of laser and optical detector technology.

F. Data on Human Tissue Optical Properties

The characteristics of tissues discussed above are based primarily on measurements in nonhuman tissues. With the exception of skin and eye, which are discussed below, there have been few systematic studies of human tissues. This situation is rapidly changing with the interest in laser applications in medicine. Although clearly there are differences in the detailed spectral characteristics of given tissues between human and other mammalian species, the general characteristics described above are expected to apply to the human case also, because chromophore content (absorption) and tissue/cellular structure (scattering) are generally comparable, even if not identical. This has been borne out by the few measurements that have been made.

II. Optical Properties of the Skin, Eye, and Blood

The skin and eye are unique organs in that they are naturally and routinely exposed to optical radiation.

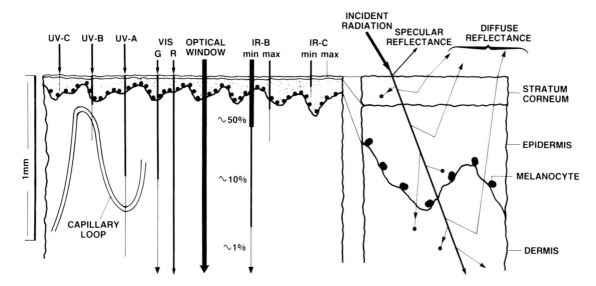

FIGURE 5 Light penetration and interactions in the skin, shown schematically in cross section. Only part of the dermis is indicated: Its total thickness may be as much as about 4 mm, and below it (not shown) lies fatty subcutaneous tissue that acts as a heat insulator and shock absorber. On the *left side,* each line indicates approximate depth of penetration of different fractions of the incident light. For IR-B and IR-C, "max" and "min" represent wavelengths corresponding to local peaks and valleys in the water absorption spectrum. On the *right* of the diagram, the expanded cross section shows absorption and scattering processes in each layer, specular reflection from the stratum corneum, and contributions of light back-scattered from various layers to the total diffuse reflectance of the skin. Most of the diffuse reflection comes from the dermis.

They have evolved to take advantage of parts of the optical spectrum but also to protect themselves from components that are harmful. Unlike most other tissues or organs that are, optically, relatively homogeneous, the skin and eyes are heterogeneous and optically complex. The skin has several distinct layers with widely varying absorption and scattering properties. These properties are also dynamic, changing with alterations in local vasodilation, vascular permeability, blood oxygenation, pigmentation, and thickness of the skin. The anterior structures of the eye are nonscattering, which is unique in itself, but have absorptions and hence transmittances, which are strongly wavelength-dependent in the UV, visible, and IR. There is a large literature on the optics of skin and eye, of which only the major features can be presented.

A. The Skin

In considering the skin, the critical issue is the vastly different penetration of light of different wavelengths into the various layers, namely, the stratum corneum, epidermis, and dermis. This is illustrated in Fig. 5. The stratum corneum is a thin (10–20 μm) proteinaceous layer of flattened, keratinized dead cells. It acts as a specularly reflecting surface with a refractive index of about 1.55, which gives 5–7% specular reflectance over the optical range, independent of pigmentation. It is relatively translucent (i.e., not highly scattering). The absorption is largely due to melanin, except at short UV wavelengths.

The epidermis, which varies in thickness between about 50–150 μm, absorbs most of the UV-C and far-IR striking the skin. The UV absorption is due to aromatic amino acids, urocyanic acid, and nucleic acids (see Fig. 2d). The epidermal absorption in the 250–300-nm range is high, regardless of the melanin content. However, in the UV-A, visible, and near-IR, melanin absorption in the epidermis has a profound influence on the penetration of light to the dermis and the spectrum of light diffusely reflected from the skin, which determines the skin color. Melanin granules are produced in the melanocytes of the basal layer and transferred to keratinocytes, which migrate outward (although in normal skin, the majority of melanin remains located in the basal cell layer). The dermis is also optically highly scattering, being a semisolid mixture of collagen (and reticulum and elastin) fibers embedded in a viscous ground substance of water and polysaccharides. It contains the blood capillaries, sweat and sebaceous glands, nerves, and lymphatics.

Both the production of melanin and its migration can be increased by UV exposure, especially UV-

B, in the process of delayed tanning. (Short-term immediate tanning is the result of photochemically induced changes in preexisting melanin granules.) Once present, the melanin protects the deeper layers. In the absence of this protection or in the case of acute overexposure to UV, damage may be caused to the dermal tissue (e.g., in sunburn), which results in capillary dilation and inflammatory response.

The effect of melanin content (and distribution) on skin color is seen in Fig. 6. In black skin, the melanin in the epidermis markedly reduces the reflectance of visible and near-IV light. At longer wavelengths, above about 1,100 nm, the skin "color" is indistinguishable, because the melanin absorption in stratum corneum and epidermis becomes negligible (see Fig. 2c). High melanin content also reduces or eliminates the characteristic "signature" of other chromophores (e.g., hemoglobin). Thus, although in Caucasian skin the oxyhemoglobin band is clearly seen in Fig. 6, this is lost in Negroid skin because of high absorption by melanin of both the incident visible light and of light that is back-scattered from the dermis.

In the dermis, the main chromophores in the visible region are hemoglobin (oxy- and reduced), beta-carotene, and bilirubin. Hemoglobin is entirely vascular, whereas beta-carotene is sequestered in dermal lipid and subcutaneous fat. Bilirubin may be both intra- and extravascular.

Although the concentration and distribution of chromophores play important roles in the optics of skin, it must be emphasized that the scattering of light, particularly in the dermis, also profoundly influences the spectral distribution of light in the skin. Much theoretical and experimental work has been directed at understanding the complex interplay of absorption and scattering in this multilayered structure. The details are particularly important in therapeutic applications of light in dermatology, where targeting of light energy to the correct structure is critical. Thus, in the laser treatment of portwine stain, the congenitally abnormal blood vessels causing this disfigurement are thermally occluded by pulses of light at 577 nm, corresponding to an oxyhemoglobin absorption peak, with little effect on overlying normal epidermis in which the absorption at this wavelength is low. Use of short pulses also ensures that the effect is produced in the blood vessels before the induced heating has time to spread and damage adjacent structures.

As a result of the complex optics of the skin, the action spectra for particular photobiologic effects depend strongly on the depth of the specific targets and on the absorption and scattering of overlying layer(s). The sensitivity to different spectral regions can also be increased by deliberate or accidental exposure to photosensitizers, of which there are many manmade and natural species. Deliberate medical uses range from psoralen-UVA (PUVA) treatment of psoriasis to the treatment of skin cancers by porphyrins and other red- or near-IR-activated photosensitizers. [*See* SKIN; SKIN, EFFECTS OF ULTRA VIOLET RADIATION.]

B. The Eye

The focus in this section is not on the functioning of the eye as a light-imaging system. Rather, the transmission of optical radiation through the structures of the eye will be considered, as a function of wavelength from UV to IR. The essential characteristics are illustrated in Fig. 7. Visible light is transmitted with little absorption loss to the retina. Near-IR is partially absorbed in the aqueous and vitreous humors. UV and far-IR are strongly absorbed by the anterior elements. UV-C and far-IR are completely blocked by the cornea. UV-B is mainly absorbed by the cornea with some reaching the lens, whereas UV-A passes through the cornea and aqueous with little loss but is completely absorbed by the lens. These various absorption losses are shown quantitatively in Fig. 8 for each component in the eye and for the whole structure through to the retina. Note

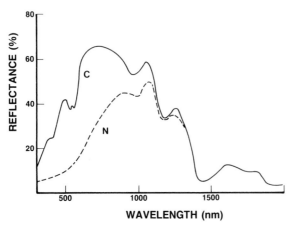

FIGURE 6 Reflectance spectrum of dark Negroid (N) and fair Caucasian (C) skin *in vivo*. Measurements include regular (specular) reflectance of a few percent. [From Anderson and Parrish, J. A. (1982). "The Science of Photomedicine." (J. D. Regan and J. A. Parrish, eds.) Plenum Press, New York, with permission.]

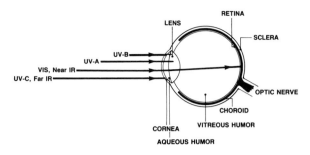

FIGURE 7 Schematic of transmission of different spectral regions through the occular media.

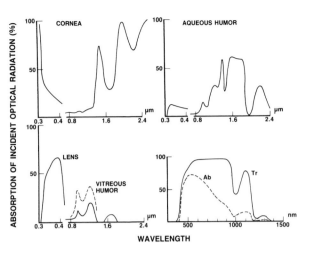

FIGURE 8 Spectral absorption of the ocular media of the human eye (*first three graphs*), expressed as percent of the incident optical radiation absorbed by each structure. Note the break in curves in the visible, where absorption of all structures is low. The *lower right graph* shows percent transmission of the whole ocular structure in the visible and near-IR and percent absorption of light energy in the retina and choroid in this spectral region. Ab, absorbed; Tr, transmitted. [From Sliney, D., and Wolbarsht, M. (1980). "Safety with Lasers and Other Optical Sources." Plenum Press, New York, with permission.]

that the overall transmission is significant only in the visible and near-IR range. All the anterior elements of the normal eye are essentially nonscattering, thus allowing their functioning as high-quality optical elements.

As would be expected, the action spectra for damage to the occular structures are the consequence of the location of the main energy absorptions at each wavelength. Acute exposure to UV-B and UV-C primarily produce effects in the cornea and conjunctiva, which is the thin membrane lining the eyelids and covering the cornea. Examples are conjunctivitis (inflammation of the membrane) and photokeratitis (inflammation of the cornea), as in the effects known as *welders flash* or *snow blindness*. Cataract formation, which is a partial or total loss of the optical transparency of the lens (or lens capsule), is associated with chronic exposure to UV-A. The loss of transparency involves induction of both visible pigmentation and optical scattering.

The sclera is highly scattering for the light that is transmitted through the retina and choroid. The effect of its back-scattering into the eye is greatly reduced by the pigmented epithelial layer between the retina and choroid, as is evident in the case of albino individuals in whom this protective effect does not occur. [*See* EYE, ANATOMY.]

C. Blood

The optical properties of blood are important, not only because the absorption contributes significantly to the absorption of tissues in the visible part of the spectrum, but also because optical techniques are used in a number of *in vitro* and *in vivo* assay techniques to measure quantities such as oxyhemoglobin saturation, hemoglobin concentration, hematocrit (volume percentage of red blood cells), and arteriovenous oxygen difference. The absorp-

tion and scattering of visible and near-IR light by whole blood is determined primarily by those of the red blood cells. The optical coefficients thus depend on the hematocrit value, H: The absorption coefficient is proportional to H, whereas the scattering coefficient increases with H at small values but falls again at high hematocrit.

Because the absorption spectrum is due to hemoglobin, whose spectrum varies with oxygenation (Fig. 2a), the absorption spectrum of blood has a complex shape that depends on the oxygen saturation, S. For example, at 660 nm, the absorption coefficient for H = 50% and S = 0% is about 4/mm, whereas for H = 50% and S = 100% (fully oxygenated), the value is about 0.4/mm (i.e., 10 times smaller). However, the corresponding values at 820 nm are roughly equal at about 1/mm. Thus, the effect of blood oxygenation on the light absorption is wavelength-dependent. The corresponding scattering coefficients do not depend on the oxygenation and change only slowly with wavelength in this range. As with complete tissues, the scattering is much higher than the absorption (for H = 50%, the scattering coefficients at 660 nm and 820 nm are about 700/mm and 500/mm, respectively). In the visible range where measurements have been done, red blood cells are highly forward scattering, with characteristics consistent with their size and refractive properties. [*See* HEMOGLOBIN.]

III. Effects of Light on Tissue

The discussion above has been concerned mainly with how tissue affects light, by absorbing and scattering it. The question of how light affects tissue is of primary concern in photobiology and photomedicine. Only some major aspects of this can be touched on. A central principle has been expounded already (i.e., that the light must be absorbed in the tissue before any biological effects can occur). Further, the spectrum and distribution of light in the tissue determine the chromophores affected and the location of energy deposition. Absorption results initially in raising of molecules to a higher energy state. A variety of mechanisms exists for de-excitation, depending on the chromophore, the degree of excitation, and the physico-chemical environment of the chromophore. De-excitation pathways may be radiative or nonradiative.

In radiative processes, part of the energy may be re-emitted as light. Fluorescence is the commonest mechanism, in which light of a longer wavelength (lower energy) than the incident light is given off, usually on a time scale in the order of nanoseconds. Not all tissues chromophores fluoresce. As with the absorption spectrum, the fluorescence emission spectrum is distinctive for particular fluorophores (fluorescent chromophores). At any given emission wavelength, there are a range of wavelengths that can excite fluorescence. The excitation spectrum is usually similar but not identical to the absorption spectrum. Examples of tissue autofluorescence spectra (i.e., the spectra of naturally occurring fluorophores in the tissue) are shown in Fig. 9. This illustrates the distinctiveness of such spectra and the fact that significant changes may occur with disease, caused by presence or absence of specific molecules. The plots also demonstrate the effect of reabsorption of excitation and emitted light by hemoglobin in the tissue, which accounts for some of the contour valleys (at 420, 540, and 580 nm; compare with Fig. 2a). Such fluorescence spectroscopy holds exciting potential for medical diagnosis. Fluorescence imaging is also widely used in tissue microscopy. Deliberately administered (exogenous) fluorophores may be used, both *in vitro* and *in vivo*.

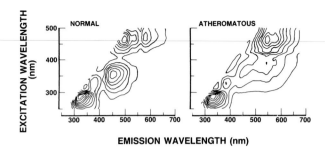

FIGURE 9 Fluorescence contour maps in normal and abnormal human aorta (*in vitro*). Note that the wavelength of emitted fluorescence is always greater than the excitation wavelength. (Data kindly provided by Drs. M. Feld and R. Richards-Kortum, Massachusetts Institute of Technology.)

Although radiative de-excitation is of great practical interest, it normally accounts for only a small fraction of the energy absorbed in tissue. Nonradiative mechanisms lead to the essential biological effects observed. Except in some laser techniques, the main results of energy absorption by tissue are either photochemical changes or heating of tissue. The former may be complex and depend critically on the chromophores present, on the irradiation wavelength, and on the biophysical and biochemical environment. The photochemical events lead to specific biological changes, depending on the chromophores and their location in the cell or tissue. This is described in the action spectrum for the particular photobiological effect.

Tissue heating is usually nonspecific. The biological effects depend on the temperatures produced and the nature of the tissue. This can range from physiologic changes (e.g., altered blood flow) to biochemical changes (e.g., protein or collagen denaturation) to vaporization or even burning of tissues. Deliberate, localized heating is the basis for much of laser surgery. Heating of tissues can be localized to some degree by using wavelengths where particular chromophores absorb strongly (e.g., hemoglobin in blood vessels). Thermal effects may also be confined to some degree by using pulses of (laser) light on a time scale that is short compared with the time for heat to diffuse in the tissue, as illustrated earlier.

In recent years, there have been many reports of rather nonspecific effects of low light levels in tissue (typically ~mW/cm²). Effects have ranged from accelerated wound healing to pain control. The photobiological basis for this "biostimulation," in which the wavelength appears to be critical, is not well understood but is likely related to triggering or stimulation of particular metabolic pathways.

At the other extreme of high optical power densities and short exposure times, photomechanical effects can be produced, which can result in cutting or ablation (physical removal of tissue from an irradiated surface). With intense short pulses of UV radiation, direct photochemical breaking of molecular bonds can occur, again resulting in cutting or ablation of tissue.

Bibliography

Ben-Hur, E., and Rosenthal, I., eds. (1987). "Photomedicine," vols. I–III. CRC Press, Boca Raton, Florida.

Carruth, J. A. S., and McKenzie, A. L. (1986). "Medical Lasers: Science and Clinical Practice." Adam Hilger, Bristol.

Preuss, L. E., and Profio, A. E., eds. (1989). Tissue optics. *Appl. Optics* **28,** 2207–2357 (multiple papers).

Regan, J. D., and Parrish, J. A., eds. (1982). "The Science of Photomedicine." Plenum Press, New York.

Sliney, D., and Wolbarsht, M. (1980). "Safety with Lasers and Other Optical Sources." Plenum Press, New York.

Smith, K. C., ed. (1989). "The Science of Photobiology," 2nd ed. Plenum Press, New York.

Suess, M. J., ed. (1982). "Nonionizing Radiation Protection," chaps. 1–3. Regional Publications, European Series 10, World Health Organization, Copenhagen.

Wilson, B. C., and Jacques, S. L. (1990). Optical reflectance and transmittance of tissues: Principles and applications. *J. Quant. Electr.* (*in press*).

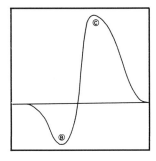

Oral Contraceptives and Disease Risk

ROSS L. PRENTICE AND LUE PING ZHAO, *Fred Hutchinson Cancer Research Center*

Glossary

Combined oral contraceptives Pills containing both an estrogen and a progestin that prevent conception by suppressing ovulation. In each cycle such pills are taken daily for 3 weeks followed by cessation for 1 week.
Meta-analysis Formal means of combining the intrastudy information from two or more reports
Relative risk Ratio of disease rates for ''exposed'' as compared with ''unexposed'' persons

STEROID CONTRACEPTIVES contain an estrogen or a progestin, or a combination thereof. By far the most commonly used steroid contraceptive is the combined oral contraceptive (OC), in which each pill contains both an estrogen and a progestin. Such pills are taken daily for 3 weeks, followed by cessation for 1 week during which withdrawal bleeding occurs, and the cycle is resumed. These preparations have proven effective in preventing pregnancy, primarily by suppressing ovulation. However, intensive study during the past 25 years has identified a corresponding impressive range of risks and benefits. The most important risks include elevations in cardiovascular disease rates including coronary heart disease and stroke, whereas benefits include reduced rates of cancers of the ovary and endometrium. The possibility that combined OCs increase the risk of cancer of the breast among young women continues to be a topic of intensive controversy and study. Increased rates of cervical cancer and liver cancer among OC users have also been reported. Much of the available epidemiologic data involves older OC preparations in use during the 1960 and 1970s. The newer low-dose pills, containing less than 50 μg of estrogen and substantially less progestin than the early pills, now account for the vast majority of usage, particularly in developed countries. Although it is reasonable to anticipate an improved risk-benefit profile and a diminution of cardiovascular disease risk in particular with the low-dose preparations, additional years of experience will be necessary to establish such results.

I. Preparations and Patterns of Use

By far the most commonly used steroid contraceptive is the combined OC in which each pill contains both an estrogen and a progestin. Such pills are to be taken daily for 3 weeks and then use is ceased for 1 week, withdrawal bleeding occurs, and the cycle is resumed. Such preparations suppress ovulation and, in the unlikely event that ovulation occurs, evidently reduce conception rates by affecting ovum transport through the fallopian tubes and by altering cervical mucus so that reduced spermoidal penetration occurs. These products were first marketed in the United States in the early 1960s. Their use rapidly spread throughout the world, and by 1980 approximately 80 million women had used combined oral contraceptives. The original Enovid pill contained 10 mg of norethynodrel (a progestin) and 150 μg of mestranol (an estrogen), but most current usage involves 1 mg or less of progestogen and less than 50 μg of an estrogen. The specific estrogen and progestin used varies among a considerable number of preparations. In the United States, norethindrone is a commonly used progestin and ethinyl estradiol is the most commonly used estrogen.

Sequential oral contraceptives were marketed widely in the United States up to about 1980 and are still in use in some countries. These preparations involve exposure to estrogen alone for approximately 15 days of woman's menstrual cycle, approximately 5 days of an estrogen plus a progestogen, followed by cessation for a week with resultant withdrawal bleeding. Because of the substantial period of "unopposed" estrogen with such sequential preparations, one may gain valuable insights into the hormonal aspects of disease occurrence by comparing the epidemiologic effects of sequential and combined OCs. Such comparison might also lead to hypotheses concerning the effects of other steroid contraceptives, such as the minipill, for which few epidemiologic data yet exist.

The minipill has been sold in the United States since 1973. Such pills contain only a progestin, at a considerably lower dosage than is found in the combined pill. This pill does not suppress ovulation but appears to act by creating a thickened cervical mucus and by inhibiting ovum transport and implantation. This pill is not currently widely used. Some other strongly progestational contraceptives, particularly long-acting injectable preparations [e.g., norethindrone enanthate and, especially, depomedroxyprogesterone acetate (DMPA)], have been used extensively in a number of countries. DMPA, for example, is usually administered every 3 months in a dose of 150 mg. These products are not currently licensed for use as contraceptives in the United States.

The newest combined OCs are the multiphasics. These preparations, which are widely used in some developed countries, vary the doses of the two hormones during the pill cycle. Such variation is designed to keep hormone doses low while preventing ovulation and maintaining menstrual cycle control.

There has been a gradual increase in the use of OCs during the 1980s. In 1988 an estimated 14% of married women of reproductive age in developed countries and 8% of such women in developing countries, for a combined total of more than 60 million married women worldwide, used such preparations. By 1987 pills with estrogen doses less than 50 μg accounted for about 85% of all pharmacy purchases of combined OCs among developed countries.

From the beginning it has been clear that combined OCs offer almost perfect, reversible, contra-

ceptive effectiveness when used properly. However, a range of noncontraceptive side effects have also been identified. Early clinical trials identified headaches, nausea, irregular menstruation, breast tenderness, and weight gain as possible short-term side effects of OC use. By the late 1960s and 1970s, more serious adverse effects including thromboembolism, stroke, and heart disease were identified, stimulating the dosage changes noted above. Since the late 1970s there has been much study of OC use in relation to the risk of several cancers. Combined OCs use has been found to convey noteworthy protection against ovarian and endometrial cancer, while possibly increasing the risk of certain other cancers. The next sections provide additional detail on the health risks and benefits of OC use. Much of the material presented is extracted from a 1987 review article by R. L. Prentice and D. B. Thomas with updating if there has been important subsequent information.

II. Risks and Benefits for Cardiovascular Diseases

A. Background and Methods

In 1968 three substantial cohort studies were initiated to study the relation between oral contraceptive use and subsequent disease occurrence. The Royal College of General Practitioners' (RCGP) Oral Contraception Study enrolled 46,000 women aged 15 or older, about one-half of whom were never-users of OCs. The Oxford Family Planning Association Contraceptive Study recruited 17,032 women aged 25–39, 56% of whom where current OC users. The Walnut Creek Contraceptive Drug Study enrolled 16,638 women aged 18–54. At recruitment 28% of the women were OC users and 33% were former users. About 40% were 40 years of age or older. These cohort studies have provided much valuable information on the risks and benefits of OC use. Case-control studies have also been carried out using a variety of study populations to relate OC use to specific diseases. These studies compare patterns of OC use among the "cases" developing a certain disease to the usage patterns for suitably selected "control" subjects, thereby allowing the dependence of the disease odds ratio on OC use to be estimated. In the present context the odds ratio is always sufficiently close to the relative risk that the two are interchangable and the terminology relative risk is used throughout. The relative risk is simply the ratio of the disease occurrence rate among OC users to that among nonusers. Of course, the magnitude of the relative risk for a given disease can depend on such aspects of OC use as duration of use, time since last use, or hormone dosage and on such study subject characteristics as age and the presence of disease risk factors. The size and quality of the studies relating OC use to a disease are often quite variable. In particular the biases that may affect the results of case-control and cohort studies are rather different. Hence the tables given below provide separate summary relative risk estimates from cohort and from case-control studies. The logarithm of a summary relative risk estimates is calculated as a variance weighted linear combination of the logarithms of the relative risk estimates from two or more specific studies. The reciprocal of the asymptotic variance of this linear combination is then given by the sum of the reciprocals of the log-relative risk variances from the individual studies, thereby allowing approximate 95% confidence intervals to be calculated. Further detail on these "meta-analyses" procedures is given in the previously mentioned review article of Prentice and Thomas.

B. Cerebrovascular Disease

As shown in Table I, the three cohort studies listed above give a summary cerebrovascular disease relative risk estimate of 2.9, with an approximate confidence interval of (2.0, 4.1), for current use of OCs versus never-use. This analysis was based on reports in the early 1980s and hence primarily reflects the risk associated with the higher dose pills in use in the 1960s and 1970s. Most of these cerebrovascular disease events were classified as subarachnoid

hemorrhage, for which the cohort study summary relative risk estimate is 2.0, or nonhemorrhagic (thrombotic, embolic) stroke, for which the relative risk estimate is 3.8.

As shown on the right side of Table I, these same cohort data sources combine to yield evidence of elevated stroke risk among former users of OCs as well, with a relative risk estimate of 1.8 for total cerebrovascular disease, 2.1 for subarachnoid hemorrhage, and 1.9 for nonhemorrhagic stroke. There was no evidence of disagreement among the three cohort studies in respect to the magnitude of the relative risks for either former or current OC use.

Cerebrovascular disease risk has also been studied in relation to OC use in case-control studies. By far the largest such study was carried out by the Collaborative Group for the Study of Stroke in Young Women, which identified 598 nonpregnant women aged 15–44 with various types of cerebrovascular disease and compared their OC use history with that of matched neighborhood and hospital controls. The case-control results for current OC use in Table I arise from this study. These results agree well with the cohort study results, although a somewhat larger estimate of 7.2 is indicated for nonhemorrhagic stroke. This study did not report stroke relative risk estimates for former use of OCs, but two other small case-control studies provide a fatal subarachnoid hemorrhage relative risk estimate of 1.5 associated with former use of OCs, as is shown on the right side of Table I. [*See* STROKE.]

The RCGP study also examined the dependence of the relative risk for total cerebrovascular disease on duration of OC use and time from cessation of OC use. There was a suggestion of a higher relative risk if the duration of current use exceeded 4 years, while there was no evidence that the elevated risk among former users declined with increasing years

TABLE I Summary Relative Risk Estimates and Confidence Intervals (in Parentheses) for Various Cardiovascular Diseases in Relation to Combined OC Use

Disease category (ICD code)	Current OC use		Former OC use	
	Cohort studies	Case-control studies	Cohort studies	Case-control studies
All cerebrovascular disease (430–438)	2.9(2.0,4.1)[a]	2.3(1.6,3.1)	1.8(1.3,2.6)	
Subarachnoid hemorrhage (430)	2.0(1.1,3.6)	2.0(1.2,3.4)	2.1(1.1,3.4)	1.5(1.0,2.1)
Nonhemorrhagic stroke (432–438)	3.8(2.4,6.1)	7.2(3.8,13.7)	1.9(1.2,3.0)	
Myocardial infarction (410)	1.9(1.1,3.3)	2.2(1.6,2.9)	1.1(0.6,1.9)	1.1(0.9,1.3)
Peripheral arterial disease (440–448)	1.6(1.3,2.1)		1.1(0.8,1.5)	
Venous thromboembolism (450–453)	3.2(2.4,4.3)	5.8(4.3,7.9)	1.0(0.6,1.8)	

[a] 95% confidence interval in parentheses.

since cessation of use. However, a study based on the follow-up of 119,000 U.S. nurses from 1976 to 1984, including 282 cerebrovascular disease events, did not detect any relation between stroke risk and former use of OCs. The reasons for the lack of agreement with the longer standing cohort studies previously mentioned are unclear, although hormone dosage, which was not ascertained in the nurses study, may be a contributing factor.

Various studies also examined the dependence of the magnitude of the cerebrovascular relative risk associated with OC use on disease risk factors. The Collaborative Group study found little evidence of a dependence of the relative risk associated with current OC use on cigarette-smoking histories, although the hemorrhagic stroke relative risk estimate was particularly elevated among "regular" cigarette smokers. The RCGP study found similar relative risks for current and former OC use among smokers and nonsmokers. This study also observed somewhat larger stroke relative risks among women aged 35 years or older, compared with that among younger women.

The dependence of the stroke relative risk associated with OC use on hormone dosages is of particular interest in view of the lower dosages in most OC preparations in current use. The RCGP study included experience with three preparations having 50 μg of the estrogen ethinyl estradiol and varying doses (1, 3, or 4 mg) of the progestin norethindrone acetate. Based on 26 cerebrovascular events among never-users and 6, 14, and 5 events at dosages of 1, 3, and 4 mg of norethindrone acetate, the stroke risks were estimated as 2.0(0.8,4.9), 3.8(2.0,7.3), and 6.6(2.5,17.1), respectively, at the three progestin dosages, where approximate 95% confidence intervals are given in parentheses. Although somewhat sparse, these data suggest a strong dependence of stroke relative risks on progestin dosage, at least for this particular progestin. Valuable support for such a dose response has been provided by a 1980 study of reports to the Committee on Safety of Medicine in the United Kingdom, which included reports on 191 occurrences of stroke among current users of the same three preparations. This report also suggested somewhat lower stroke relative risks associated with OC preparations having 30 μg versus 50 μg of ethinyl estradiol. Collectively, however, these data sources do not provide a precise estimate of cerebrovascular disease risks associated with current or former use of low-dose OC preparations. Further observational

study will be required to determine the extent to which the elevations in hemorrhagic and thrombotic stroke risks associated with the older OCs have been obviated by the newer preparations.

Hypertension is the major risk factor for hemorrhagic stroke. In fact, on taking account of random measurement errors in blood pressure measurement one can note that the modest average elevations in blood pressure that have been observed among current OC users provide an explanation for the increased hemorrhagic stroke risks associated with both current and former OC use. Such average elevations for the older OC preparations were in the vicinity of 4–5 mm/Hg systolic and 1–2 mm/Hg diastolic blood pressure. The magnitude of the increase in blood pressure has been shown to vary directly with the progestin dosage, at least for certain OC preparations. [*See* HYPERTENSION]

Blood pressure elevations and consequent structural vascular changes also provide a plausible explanation for the elevated thrombotic stroke risk among former OC users, as well as for a portion of the increase among current users. Other factors, presumably including OC effects on blood coagulability, are likely to be involved in a full explanation of the noteworthy thrombic stroke elevations among current users of the older OCs.

C. Ischemic Heart Disease

The cohort studies mentioned above and a number of case-control studies have been used to examine the relation between current and former OC use and the incidence of ischemic heart disease. For total ischemic heart disease occurrence, reports in the early 1980s from the RCGP and Walnut Creek studies combine to yield a relative risk estimate of 1.4 for current OC use, a value that is just significant at the 0.05 level, and a nonsignificant relative risk estimate of 1.2 for former OC use. When myocardial infarction events were excluded, these relative risk estimates became 1.3 and 1.2, respectively, neither of which was significantly different from zero. However, the myocardial infarction relative risk associated with current OC use is estimated as 1.9 with an approximate 95% confidence interval of (1.1,3.3) based on these two cohort studies, while the corresponding relative risk estimate and confidence interval are 1.1(0.6,1.9) for former OC use, as is shown in Table I. Seven case-control studies of nonfatal myocardial infarction combine to yield a similar summary relative risk estimate of 2.2 for current OC use, with an associated 95% confidence interval of (1.6,2.9). There is no evidence of heterogeneity among the seven studies. A subset of those studies also examined former OC use, giving a summary relative risk estimate of 1.1, identical to that from the cohort studies, with 95% confidence interval (0.9,1.3). Furthermore, the case-control study relative risk estimates for fatal myocardial infarction appear to be quite consistent with those for nonfatal infarction.

In summary, current use of the older OCs is associated with an approximate doubling of myocardial infarction incidence based on a substantial series of studies that generally provide accommodation of such risk factors as cigarette-smoking habits, serum cholesterol concentration, and blood pressure level. Control of these factors is essential because they have evidently played a role in decisions regarding contraceptive choices. However, standard methods of confounding factor control may lend to some underestimation of relative risks associated with OC use, in view of well-documented effects of OCs on blood pressure and on blood lipids and lipoproteins.

The studies summarized above suggest that any increase in myocardial infarction incidence among former OC users must be small. This topic is important, however, in view of the prevalence of this disease and the rapid increase in myocardial infarction incidence rates as women age. Furthermore, the largest of the case-control studies, involving 556 premenopausal myocardial infarction cases admitted to the coronary care units of 155 hospitals in Boston, New York, or Delaware Valley during 1976–1978 and 2,036 control subjects, did report a significant relative risk trend with duration of former OC use. Specifically, relative risk estimates and approximate 95% confidence intervals were 1.0(0.7,1.2), 1.4(1.0,2.0), and 2.3(1.5,3.8) for less than 5, 5–9, and 10 or more years of former OC use, respectively, and the increased risk was evident as long as 5–10 years after discontinuation of use. These findings need to be interpreted with some scepticism in view of Table I and particularly in view of a lack of any association between former OC use and coronary disease in the cohort study of U.S. nurses, which included 485 cases of nonfatal myocardial infarction or fatal coronary heart disease.

Two large case-control studies, reported in 1980 and 1981, examined the dependence of myocardial infarction among current OC users on duration of current use. Although the data cannot be said to be strong, when results from the two studies are combined a significant trend is evident with relative risk estimates and approximate 95% confidence intervals 1.5(0.9,2.5), 2.6(1.6,4.2), and 4.4(2.6,7.4) for less than 5, 5–9, and 10 or more years of current use, respectively.

The RCGP study also examined in a 1982 report the dependence of the relative risk among current OC users on the dose of the progestin norethindrone acetate among preparations having 50 μg ethinyl estradiol. Compared with never-users, the relative risk estimates and 95% confidence intervals of 1, 3, and 4 mg of this progestin are 0.8(0.3,2.0), 1.8(1.0,3.0), and 1.5(0.5,4.9), based on rather limited numbers of cases. Also relative risk estimates for the 3-mg and 4-mg doses compared with the 1-mg dose were 2.2 and 2.2, based on a larger number of reports to the U.K. committee on the Safety of Medicines, and the trend was highly significant. These reports imply a myocardial infarction incidence at 1 mg of norethindrone acetate that is about one-half that at 3 or 4 mg, but they do not provide a precise estimate of relative risk at the 1-mg dose as compared with nonuse.

The RCGP and Oxford cohort studies have remained under active follow-up and have recently reported their myocardial infarction data. Based on a total of only 25 cases in a 1987 report, ever-use of OCs was associated with a twofold increase in myocardial infarction risk in the Oxford cohort, while there was no evidence of increased risk among current users. Experience with preparations having less than 50 μg estrogen was limited in this analysis. A 1989 report from the RCGP study involved 158 myocardial infarction cases. Current OC use was associated with a relative risk estimate of 1.8, which is evidently of borderline significance, while former OC use was associated with a crude relative risk estimate of 1.3 that reduced to 1.0 on adjusting for cigarette smoking and social class. To the extent that the newer low-dose preparations prevail among current users, these results are somewhat disappointing. The authors argue that the excess risk-associated with current OC use is confined to cigarette smokers. Collectively, however, the cohort and case-control studies suggest a relative risk for current OC use that is as large among nonsmokers as smokers. Of course, OC use has a greater impact on absolute myocardial infarction risk among smokers as compared with nonsmokers, because cigarette smoking is a strong risk factor for this disease. Similarly, there does not appear to be a strong dependence of the relative risk associated with OC use on study subject age, based on the collective observational data.

Oral contraceptives have been found to have substantial influences on lipid and lipoprotein metabolism. Serum high-density-lipoprotein (HDL) cholesterol concentration has been found to correlate negatively with progestin dose at a fixed estrogen dose, with 10–20% reductions in HDL cholesterol-associated certain OC preparations. Similarly noteworthy, 10–30% increases in low-density-lipoprotein (LDL) cholesterol concentrations have been shown to follow the use of OCs that are low in estrogen dose and high in progestin dose, whereas preparations that are relatively high in estrogen and low in progestin given rise to elevated HDL concentrations. These results, along with results showing a favorable impact of menopausal estrogens on HDL concentrations, suggest that effects of OCs on cholesterol fractions may explain much of the elevation in myocardial infarction incidence among current OC users. The collective observational data provide little evidence that myocardial infarction risk is affected by former OC use. [*See* CHOLESTEROL.]

D. Peripheral Vascular Disease

As noted in Table I, the RCGP and Walnut Creek cohort study reports in the early 1980s combine to yield a relative risk estimate of 1.6(1.3,2.1) for peripheral arterial disease associated with current OC use. There was little evidence of increased risk among former OC users based on those same data sources.

Venous thromboembolism was one of the earliest adverse effects of OC use to be identified. All three cohort studies reported data on idiopathic venous thromboembolism in the late 1970s. These studies combine to yield a relative risk estimate of 3.2(2.4,4.3) although there is evidence ($p=0.04$) of discrepancy among the estimates, with the RCGP estimate of 2.7(2.0,3.8) being smaller than the other two. There was no evidence of elevated venous thromboembolism risk among former OC users based on data from the RCGP cohort.

The corresponding current OC use relative risk estimate for idiopathic venous thromboembolism was 5.8(4.3,7.9) based on four case-control studies. These studies also combine to yield a smaller current OC relative risk estimate of 2.3(1.7,3.2) for predisposed venous thromboembolism, with predisposition events variously defined as those associated with surgery, severe trauma, or previous thromboembolic events. The relative risks associated with current OC use for pulmonary embolism, the most threatening of the venous thromboembolic events, were similar to those for other such events, and OC relative risks are similar for fatal and nonfatal events.

In summary, idiopathic venous thromboembolism incidence is elevated about fivefold among current users of the older OCs, as compared with a relative risk of about 2.5 for predisposed venous thromboembolic disease. OC use has been found to bring about a reversible increase in the coagulability of blood, involving changes in the platelet, coagulation, and fibrolytic systems, and changes in the structure of veins and arteries. The RCGP data and reports to the British Committee on the Safety of Drugs have been used to examine the dependence of risk on estrogen and progestin dosages. These sources do not allow one to disentangle the roles of the two hormones in a satisfactory manner, although it is reasonable to assume that the higher estrogen dose, if not higher doses of both agents, is associated with increased venous thromboembolism risk. Once again, the relative risk associated with OC use does not appear to depend much on the age or cigarette-smoking habits of the study subject or on the duration of OC use.

III. Risks and Benefits for Cancer

A. Endometrial Cancer

Women treated at the menopause with conjugated estrogens exhibit highly elevated rates of endometrial hyperplasia and of endometrial cancer. There is correspondingly strong evidence that sequential OCs, which exert a net estrogenic effect on the endometrium, also increase endometrial cancer risk. In fact, the summary endometrial cancer relative risk estimate associated with ever-use of sequential OCs is 2.0(1.1,3.8), as is given in Table II, based on three case-control studies reported in the early 1980s.

Estrogens are presumed to increase endometrial cancer risk by enhancing the mitotic activity of endometrial cells, whereas progestins reduce DNA synthesis in the endometrium and lead to endometrial atrophy when given alone. Hence combined OCs, which contain both estrogen and progestin throughout the pill cycle, can be expected to give

TABLE II Summary Relative Risk Estimates and Approximate 95% Confidence Intervals (in Parentheses) for Several Cancers in Relation to OC Use

Disease category	Ever-use of combined OCs	Other relative risk estimates
Endometrial cancer	0.5(0.3,0.7)	Sequential OCs: 2.0(1.1,3.8)
Breast cancer: case-control	1.0(0.9,1.1)	Ever-use before age 25: 1.2(1.1,1.4)
cohort	1.0(0.9,1.1)	
Ovarian cancer	0.6(0.5,0.7)	Long-term use: 0.4(0.3,0.6)
Invasive cervical cancer	1.3(1.2,1.5)	Long-term use: 1.6(1.3,1.9)
Liver cancer (developed countries)	2.6(1.3,5.1)	In developing countries: 0.7(0.4,1.2)
Malignant melanoma: case-control	1.0(0.9,1.2)	Long-term use: 1.3(1.0,1.7)
cohort	1.4(0.8,2.3)	

rise to lower endometrial cancer rates than do sequential OCs, which are now seldom used. That this expectation is correct is noted in Table II. Based on five substantial case-control studies reported in the early- and mid-1980s ever-use of combined OCs is associated with a relative risk of 0.5(0.3,0.7). Hence rather than a twofold increase there is a twofold reduction in endometrial cancer risk associated with combined OC use. These studies suggest somewhat greater endometrial cancer risk reductions with increasing duration of OC use. One of the studies examined risk as a function of time since cessation of use and reported a relative risk of 0.8(0.5,1.3) after 10 or more years since exposure to combined OCs.

B. Breast Cancer

Of the various diseases that may be associated with the use of combined OCs, none have been studied more extensively or intensively than breast cancer. Fully 20 case-control studies involving a total of more than 14,000 breast cancer cases have examined this relation, while the four cohort studies previously mentioned (RCGP, Oxford, Walnut Creek, and Nurses Health Study) have all reported their results on OC use and breast cancer on one or more occasions. As noted in Table II, the case-control studies combine in a meta-analysis to yield a summary relative risk estimate of 1.0 for ever-use of OCs with 95% confidence interval well within (0.9,1.1), whereas the most recent data from the four cohort studies combine to give an identical relative risk estimate of 1.0(0.9,1.1). The cohort study relative risk estimates, however, exhibit somewhat greater heterogeneity than would be expected by chance.

Many of the case-control studies examined breast cancer risk among long-term OC users, defined variously as minimum use from 2 to 15 years. The summary breast cancer relative risk estimate for long-term OC use was 1.0(0.8,1.1), although the individual relative risk estimates again exhibit significant heterogeneity. Similarly, several of the case-control and cohort studies have estimated the breast cancer relative risk a decade or more after the initial OC use. Once again the summary relative risk estimate is close to unity, but the individual estimates show some evidence of heterogeneity.

In spite of some evidence of disagreement among the various studies of OCs and breast cancer, the results mentioned above appear to be quite reassuring. However, the fact that breast cancer risk is enhanced by a number of hormone-dependent events, including early age at menarche, late age at menopause, few or no children, and short or no period of lactation, along with the high incidence of this disease in developed countries where blood hormone concentrations are known to be comparatively high, has caused the OC and breast cancer relation to be the subject of much recent study and controversy. For example, the notion that breast cancer risk may be enhanced whenever OCs increase pertinent hormone concentrations (e.g., in breast tissue) has led to the study of breast cancer risk among women exposed to OCs at young ages, at which time exogenous hormone production may be still increasing, and to the study of breast cancer risk among women exposed to OCs while approaching the menopause, whose endogenous hormone production may have declined. In either situation there is debatable evidence of elevated breast cancer risk among OC users. Such elevations could be masked in the summary relative risk estimates given in Table II, while still being of importance in respect to contraceptive choices at various ages. In respect to exposure to OCs at young ages, six case-control studies of breast cancer before age 45 have reported relative risk estimates associated with ever-use of OCs before age 25. These studies combine to give a significant relative risk estimate of 1.3 with approximate 95% confidence interval (1.1,1.4). However, there is strong evidence ($p<0.001$) of disagreement among the specific studies. Hence, investigation beyond this simple meta-analysis based on published data and possibly additional data collection seems necessary to resolve this issue. A similar relative risk estimate and confidence interval, along with significant evidence of relative risk heterogeneity, arises for the ever-use of OCs before first full-term pregnancy, based on a larger number of studies.

Cellular proliferation in the lobules of the breast is low in the follicular phase of the menstrual cycle and high during the latter half of the luteal phase, immediately after the progesterone and luteal phase estrogen peaks. This suggests exposure to elevated levels or progesterone or perhaps more likely to elevated levels of both estrogen and progesterone may enhance breast cancer risk. Such thinking is consistent with breast cancer relative risks, in the vicinity of 1.3, that have been associated with the use of menopausal estrogens, and with preliminary evidence of larger risk elevations associated with menopausal preparations that include progesterone as well as estrogen. Despite the extensive study that has already taken place, there appears to be a need for further study of OC use and breast cancer, with a particular emphasis on preparation dosages and on pertinent hormone concentrations, for example, in blood or in breast fluid. [*See* BREAST CANCER BIOLOGY.]

C. Ovarian Cancer

Ten case-control studies, reported in 1977–1986, combine to yield a relative risk estimate of 0.6(0.5,0.7) for ever-use of combined OCs in relation to all ovarian malignancies (Table II), of which epithelial cancers predominate. There was no evidence of disagreement among the studies concerning the magnitude of this risk reduction. These studies indicate that protective effect to increase with duration of use with relative risk estimates of about 0.7 for less than 5 years and about 0.4 for 5 or more years of ever-use. Moreover, a protective effect is evidently retained for some years after cessation of

use, with a summary relative risk estimate of 0.5 among former OC users who were 10 or more years from last use. The magnitude of the risk reduction is similar among parous and nulliparous women, arguing against the possibility confounding by infertility. A recent case-control study has noted that a relative risk reduction of a similar magnitude applies to ovarian tumors of low malignancy potential. Another recent case-control study found similar risk reductions for epithelial ovarian tumors among women diagnosed in 1983–1985 as compared with those diagnosed in 1974–1977, perhaps suggesting that this OC benefit is retained by the newer preparations.

Various epidemiologic studies suggest that ovarian cancer risk is decreased by late menarche, pregnancy, and by early menopause, all of which correspond to a reduction in a women's total number of ovulations, as does OC use. It may be that ovulation enhances risk by subjecting the ovarian epithelium to steroid-risk follicular fluid or to elevated levels of pituitary gonadotropins. The fact that menopausal steroids also reduce gonadotropin concentrations in the blood, without a corresponding reduction in ovarian cancer rates, suggests that mechanisms other than gonadotropin reduction may be required to explain the protective effect of OC use on ovarian cancer.

D. Cervical Cancer

Study of the relation of OC use to cervical cancer is hampered by the need to control carefully sexual practices and pap smear screening activities. There have now been six case-control studies that have attempted to control for these as well as other potential confounding factors. These six combine to yield a summary relative risk of 1.3(1.2,1.5) for ever-use of combined OCs (Table II), thereby providing evidence of risk enhancement. Furthermore, the relative risk estimates from the six studies are in reasonable agreement ($p = 0.19$), and there is evidence of increasing relative risk with duration of use, with a summary relative risk of about 1.1 for less than 5 years of use and about 1.6 for more than 5 years. These results are supported by data from three cohort studies. Specifically, in two of these studies, with small number of invasive cervical cancers, all disease events occurred among women who had used OCs, whereas the RCGP study reported an unadjusted relative risk estimate of 2.1(1.1,4.3) for the ever-use of OCs.

There have also been six case-control studies that

have examined OC use in relation to cervical carcinoma *in situ*. These studies yield a summary relative risk estimate of 1.2(1.1,1.4) for ever-use of OCs, and there is evidence of increasing relative risk with longer duration of use. There is, however, evidence of inconsistency ($p < 0.01$) among the study results, possibly because several of those studies made little attempt to control sexual practices. The three cohort study reports on this topic give a somewhat higher relative risk estimate of 1.5(1.1,2.0), again with some evidence ($p=0.06$) of heterogeneity among studies. Cervical squamous dysplasia occurrence has also been studied in relation to combined OC use. Cohort and case-control studies suggest significant relative risks, in the vicinity of 1.4, with evidence of higher relative risk among longer-term OC users. Once again, the individual relative risk estimates are more variable than would be expected on the basis of chance alone, and only limited efforts were made to control confounding.

The studies summarized above provide fairly consistent evidence of association between OC use and squamous cell carcinoma of the cervix. Based on modest sample sizes, these studies also provide arguable evidence of association between such use and adenocarcinoma of the cervix. There has been little attempt to date to investigate mechanisms for such associations. Hormonal mechanisms would appear to merit exploration, as would be consistent with the association of this disease with such reproductive factors as multiparity. However, the strong associations of cervical cancer rates with early age at first intercourse and number of sexual partners leaves open the possibility that the observed associations with OC use could be affected by inadequate confounding factor control. Further studies that include biological measures of exposure to various sexually transmitted viruses would seem to be particularly worthwhile.

E. Liver Cancer

Four case-control studies have examined the association between OC use and hepatocellular carcinoma in developed countries. These four studies combine to give a summary relative risk estimate of 2.6(1.3,5.1) for ever-use of OCs (Table II) and 9.6(4.0,22.8) for long-term use of OCs defined variously as more than 5 or more than 8 years. Although these studies are small and have methodologic limitations, there is considerable consistency among

their results, suggesting that primary hepatocellular carcinoma can be a complication of OC use, particularly long-term OC use.

There have also been two small case-control studies of OC use in relation to hepatic cell adenomas. These benign tumors sometimes result in intraperitoneal or intratumoral bleeding, which may be fatal. These studies indicate that OC users have elevated risk for this tumor and that the relative risk increases dramatically with duration of use. These are, however, rare tumors with the older OCs increasing risk by about three cases per 100,000 users per year.

Fortunately, primary hepatocellular carcinoma among young women is also extremely rare in developed countries, where it has been estimated that about one liver cancer fatality per 100,000 long-term users per year can be expected. There would, however, be some concern if elevated risks persisted for many years beyond cessation of OC use, so that further monitoring and evaluation is indicated.

Of much greater concern is the possibility that relative risks of the magnitudes indicated above would prevail in areas of Asia and Africa in which hepatitis B virus is endemic and liver cancer rates are high. However, a recent sizable case-control study among women in eight generally high-risk countries provide evidence that any elevation of hepatocellular carcinoma rates in the presence of prior hepatitis B virus infection must be small. Specifically, this study gives a liver cancer relative risk estimate of 0.7(0.4,1.2) for ever-use of combined OCs, and there was no evidence of heterogeneity of relative risk estimate among the participating centers in this collaborative study.

F. Malignant Melanoma

Eight cohort studies and three cohort studies have reported malignant melanoma relative risk estimates associated with the ever-use of combined OCs. The case-control studies combine to give a summary relative risk estimate of 1.0(0.9,1.2), while the cohort studies give a summary estimate of 1.4(0.8,2.3). This latter estimate is based on the report of a substantially increased risk in the Walnut Creek cohort, and the three cohort study relative risk estimates are not in agreement ($p = 0.002$). The case-control studies do, however, suggest some modest elevation in risk associated with long-term OC use, defined variously as more than 5 or more than 10 years, with relative risk estimate of

1.3(1.0,1.7). Somewhat larger relative risk estimates were reported among women who were 10 or more years from initial exposure in two of the case-control studies. In summary, these data do not provide convincing evidence that OCs enhance the risk of malignant melanoma but suggest that risk may eventually be elevated among long-term users. [*See* MALIGNANT MELANOMA.]

G. Other Neoplasms

A protective effect of OC use for benign breast disease, including both fibroadenomas and fibrocystic disease, has been well-documented. However, such protection evidently does not convey a reduced breast cancer risk. Reduced rates of retention cysts of the ovary have been observed among the OC users in several studies. Such reduction is not unexpected because OCs inhibit ovulation. Uterine leiomyomas (fibroids) have been variably reported as increased, unchanged, or decreased among OC users in the three long-standing cohort studies. The recent study using the Oxford cohort found that risk consistently decreased with increasing duration of OC use.

A single Australian case-control study reported a reduced risk of colorectal carcinoma among OC users, although none of the three long-standing cohort studies have reported such an association. Initial reports of an elevation in pituitary adenoma rate among women of childbearing age and of an elevation in thyroid cancer among OC users were not confirmed by subsequent studies.

IV. Risks and Benefits for Other Diseases

The two British cohort studies have reported reductions in risk for several menstrual disorders but an increased risk of amenorrhea or scanty menstrual flow among OC users. The amenorrhea can persist for 2 years or more, but no increased risk of permanent infertility has been identified. Reduction in the incidence of iron deficiency anemia has been observed in all three cohort studies.

Elevated incidence of hay fever and allergic rhinitis, with some evidence of a dose response with the estrogen content of the OC preparation, has been reported, as have elevated rates of eczema, conjunctivitis, and diseases of the eyelid. Whether such associations reflect an OC effect on the im-

mune system or whether reporting bias contributes to such associations has not been determined.

All three cohort studies have reported an increased rate of migraine headaches among OC users. There have also been variable reports of increased rates of referral for depression and an increased rate of hospitalization for mental or emotional illness among OC users. Somewhat increased rates of suicide and self-poisoning among OC users have also been suggested by the three long-standing cohort studies. That women with a predisposition for such events more frequently choose OCs as their method of contraception provides a possible explanation for these associations.

A number of other associations with OC use have been reported but have not been conformed by other studies. These include reduction in risk of rheumatoid arthritis and reduction in risk in certain thyroid diseases (euthyroid swelling, thyrotoxicosis, myxoedema).

Certain other conditions, including increased rates of regional enteritis and ulcerative colitis, and small increases in the rates of gastritis, duodenitis, and appendicitis have been reported in two or more of the three cohort studies. Gallbladder disease (cholecystitis and cholelithiasis) has also been related to short-term OC use in several studies. Also OC users have been shown in all three cohort studies to be at increased risk for cervical erosion, vaginal discharge, and pelvic infection. These last associations may simply reflect differing sexual practices between users and nonusers of OCs. However, OC use evidently reduces the risk of pelvic inflammatory disease, perhaps by thickening cervical mucous.

OCs are known to affect carbohydrate metabolism. Evidently estrogen increases glucose levels and suppresses the corresponding insulin response, whereas progestin stimulates insulin production. Hence, OC use may be contraindicated for women with a history of diabetes.

Finally, because OCs provide effective contraception and prevent ovulation, they reduce the risk of various complications of pregnancy and childbirth, including the risk of ectopic pregnancy.

V. Implications for Total Mortality

It is evident that combined OCs convey a range of risks and benefits in respect to human disease. An analysis of total mortality rate provides one method

TABLE III Summary Mortality Relative Risk Estimates and Approximate 95% Confidence Intervals (in Parentheses) for OC Use, Using Data From Three Cohort Studies

Disease category	Ever-use of OCs
Total mortality	1.2(1.0,1.4)
Mortality from circulatory system diseases	2.5(1.6,4.0)
All cancer mortality	1.0(0.8,1.3)
All other mortality	1.1(0.8,1.5)

of bringing together the risks and benefits in a manner that acknowledges aspects of the lethality and incidence rates for the life-threatening diseases. Table III gives summary relative risk estimates for total mortality from the three long-standing cohort studies, based on reports in 1981 that included 249 deaths in the RCGP cohort, 81 deaths in the Oxford cohort, and 138 deaths in the Walnut Creek cohort. The summary total mortality relative risk was 1.2(1.0,1.4) for ever-use versus never-use of the older OCs. The individual relative risk estimates are reasonably consistent ($p = 0.18$) and, overall, provide borderline significant evidence of increased death rates among OC users. Each of the studies controlled for age, parity, cigarette-smoking habits, and aspects of social class or education in these analyses. Twenty-one percent of the deaths were from disease of the circulatory system, for which the three studies combine to give a relative risk estimate of 2.5(1.6,4.0), although there is evidence ($p = 0.01$) of disagreement among the individual estimates. Forty-four percent of the deaths were from cancer, for which the summary relative risk is 1.0(0.8,1.3). Whereas the remaining 35% of the deaths were attributable to other causes, for which the summary relative risk is 1.1(0.8,1.5). Hence, the older OCs evidently increase mortality from cardiovascular diseases with little effect on the mortality rates for all other diseases combined at least within the first decade or so after initiation of use. Additional observation and analysis will be required to determine any longer-term mortality consequences of the older OCs. Importantly, such additional observation will also be necessary to determine the extent to which the newer OCs have reduced or eliminated the elevation in cardiovascular disease mortality risk and the extent to which such OCs have altered the risk–benefit mortality profile for other diseases.

Acknowledgement

This work was supported by grant GM-24472/CA-53996 from the National Institutes of Health.

Bibliography

Brinton, L. A., Tashima, K. T., Lehman, H. F., Levine, R. S., Mallin, K., Savitz, D. A., Stolley, P. D., and Fraumeni, J. F. (1987). Epidemiology of cervical cancer by cell type. *Cancer Res.* **47**, 1706–1711.

Croft, P., and Hannaford, P. C. (1989). Risk factors for acute myocardial infarction in women: Evidence from the Royal College of General Practitioners' oral contraception study. *Br. Med. J.* **289**, 165–168.

Mant, D., Villard-Mackintosh, L., Vessey, M. P., and Yeates, D. (1987). Myocardial infarction and angina pectoris in young women. *J. Epidemiol. Community Health* **41**, 215–219.

Oral contraceptives: Lower-dose pills. (1988). ''Population Reports A7.'' Population Information Program, Center for Communications Programs, The Johns Hopkins University, Baltimore, Maryland.

Prentice, R. L., and Thomas, D. B. (1987). On the epidemiology of oral contraceptives and disease. *Adv. Cancer Res.* **49**, 285–401.

Stampfer, M. J., Willett, W. C., Colditz, G. A., Speizer, F. E., and Hennekens, C. H. (1988). A prospective study of past use of oral contraceptive agents and risk of cardiovascular diseases. *N. Eng. J. Med.* **319**, 1313–1317.

U. K. National Case-Control Study Group. (1989). Oral contraceptive use and breast cancer risk in young women. *Lancet* **i**, 973–982.

The WHO Collaborative Study of Neoplasia and Steroid Contraceptives. (1989). Combined oral contraceptives and liver cancer. *Int. J. Cancer* **43**, 254–259.

Oral Pathology

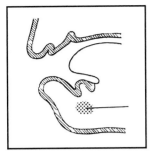

JAMES H. P. MAIN, *University of Toronto*

Glossary

Ameloblast Cell of epithelial origin primarily involved in the formation of dental enamel
Bullae Blister or bleb on a body surface, larger than a vesicle
Cavernous sinus Paired blood sinus lying on the superior surface of the sphenoid bone in the base of the cranial cavity
Dentine Calcified tissue that constitutes the central and larger part of a tooth
Hypoplasia Defective formation or incomplete development of a part
Periodontium Tissues surrounding a tooth including the periodontal membrane and alveolar bone

ORAL PATHOLOGY is the study of diseases of the mouth. Several diseases and abnormalities are unique to the mouth (e.g., dental caries and its sequelae, periodontal disease, and congenital disorders such as cleft lip and palate). In a number of systemic diseases there are specific oral manifestations, and in some cases the symptoms may occur only in the mouth (e.g., in lichen planus). Many other diseases may occur in the mouth as they may do elsewhere in the body (e.g., tumors). The study of all these conditions comprises oral pathology.

I. Developmental Abnormalities

A. Cleft Lip and Palate

These congenital anomalies occur in about one in 1,000 births with racial variations. Types include (1) cleft lip, (2) cleft palate; and (3) cleft lip and palate (unilateral or bilateral). Milder related deformities also occur (e.g., notching of the lip and bifid uvula).

The pathogenesis is believed to be due to lack of growth of mesodermal tissue at around the eighth week of embryonic development with consequent failure of fusion of the processes involved in facial formation. This results in persistence of epithelium, cleft formation, and a deficiency of tissue.

B. Hypoplasia of Enamel

During formation of the enamel, the ameloblasts may be damaged by a number of local or systemic factors of which the following are common:

· Apical infection on a deciduous predecessor
· Trauma
· Exanthematous diseases
· Hypovitaminosis D
· Fluoride in drinking water at concentrations of more than 1 ppm
· Congenital syphilis

These can cause enamel defects in one or many teeth.

C. Amelogenesis Imperfecta

This condition includes a group of hereditary disorders of enamel formation that cause defects in the quantity or quality of the enamel in both deciduous and permanent dentitions. The dentine is normal.

D. Dentinogenesis Imperfecta

In this hereditary disorder, the dentine is defective and the teeth are discolored brown. The enamel is of normal structure but flakes off the dentine readily. The condition may occur on its own or as part of osteogenesis imperfecta.

II. Pathology of Dental Caries

In this common disease, the hard tooth tissues are destroyed, and eventually the soft tissues in the middle of the affected tooth (the dental pulp) become infected by oral bacteria. It is initiated by the action of bacterial products formed in dental plaque. The microbiology and epidemiology are described elsewhere. [*See* DENTAL CARIES.]

A. Caries of the Enamel

The first visible sign of the caries is a white spot on the enamel caused by a loss of translucency. Some parts of the tooth are more susceptible than others, notably pits and fissures and the interproximal surfaces. Histologically the lesion is seen in cross section as a triangular area with the base parallel to the enamel surface, which is itself intact. In the middle of this altered area, "pores" form because of loss of minerals, and it is surrounded by a hypercalcified margin. At this stage remineralization from saliva or medications is possible under certain conditions.

The lesion increases in size until the apex of the triangle reaches the amelo-dentinal junction. At this stage surface cavitation develops and bacterial ingress into the defect.

B. Dentinal Caries

Rapid lateral spread of the caries occurs along the amelo-dentinal junction, undermining the enamel and giving rise to retrograde enamel caries. Bacteria invade the dentinal tubules, and dissolution of the hard tissues occurs with loss of mineral salts. The dental pulp reacts to caries by laying down reparative dentine at the pulpal ends of affected tubules and by producing calcification of the odontoblast processes within the tubules, so sealing them. This reaction is unavailing in the untreated tooth, and in time the rate of carious destruction of the dentine outstrips the rate of production of reparative dentine, resulting in bacterial invasion of the pulp.

III. Sequelae to Dental Caries

Untreated dental caries eventually results in inflammation of the tooth pulp (pulpitis), the commonest cause of toothache. Irritant fillings or tooth cutting operations are other causes. Untreated pulpitis leads to pulpal necrosis and gangrene.

Apical infection results from pulpal gangrene. This is a pyogenic inflammation with mixed oral flora and with pus and granulation tissue formation. Apical infection may remain quiescent for months or years or if the balance of host resistance/organismal virulence is disturbed, it may lead to spreading infection. Commonly the infection spreads buccally, penetrating the alveolar bone, to form a submucosal vestibular abscess. Less often it may spread to a number of potential tissue spaces around the jaws, causing one or more space infections. Infection may also spread into the bone marrow causing osteomyelitis.

The draining lymph nodes (submandibular, submental, and deep cervical) are usually involved. If a hemolytic streptococcus is present in apical infection, it can cause a serious cellulitis of the neck (Ludwig's angina). Rarely apical infections in the maxilla spread to involve veins communicating with the cavernous sinus and cause cavernous sinus thrombophlebitis.

IV. Periodontal Disease

The various forms of periodontal disease are all believed to be caused by a disturbance of the balance between host resistance and noxious products of bacterial dental plaque. Recent studies suggest that the progression of periodontal disease is episodic and site-specific with extended periods of equilibrium between the host reaction and the bacteria. There is ongoing discussion as to whether the bacterial agents are specific or not.

Chronic gingivitis consists of inflammation of the marginal gingivae with no deeper tissue destruction.

Chronic periodontitis is a later development in which there is irregular destruction of the periodontium with the formation of pockets between the teeth and adjacent tissues that are lined by epithelium. Untreated this leads to loosening and ultimately loss of teeth and/or formation of abscesses in the pockets.

Juvenile periodontitis is an adolescent disease in which rapid destruction of the periodontium occurs around the permanent incisors and/or first molars.

Acute necrotizing ulcerative gingivitis (ANUG) is characterized by necrosis of the interdental papillae. The ulceration may spread widely. Large numbers of *Bacillus fusiformis* and *Borrelia vincentii* are present in the slough. Uncommon in healthy people, it is more common and more serious in the malnourished or under stressful conditions [e.g., trenches in warfare (trench mouth)].

V. Impacted Teeth

A tooth is impacted when its eruption is prevented by another tooth or by bone. Impaction is of common occurrence, the third molars being most frequently affected followed by the maxillary canines.

When part of an impacted tooth emerges into the mouth, a mixed infection by oral commensals gains access to the potential space around the crown causing pericoronitis. Infection from this condition may spread as described above.

VI. Specific Infections

A. Herpes Simplex Virus (HSV) Infection

Spread is usually via direct contact with a latent period of 3–5 days. Primary herpetic stomatitis, usually HSV-1, occurs commonly in children and in susceptible adults. The clinical features are of acute painful vesiculo-bullous disease of the oral mucosa with regional lymphadenopathy and mild pyrexia.

The disease resolves spontaneously in 7–10 days in healthy individuals, but the virus remains latent in the cells of the trigeminal ganglion. Recurrent lesions, often induced by minor trauma or sunshine, occur on the skin-vermillion border of the lips (herpes labialis).

Occasional cases of HSV-2 oral infection occur caused by oro-genital contact with the same clinical features. [*See* HERPESVIRUSES.]

B. Herpes Zoster (Shingles)

This disease is caused by the varicella-zoster virus (VZV), a member of the herpes group. Zoster is believed to be a reactivation of latent VZV caused by immunosuppression of various causation and occurs chiefly in older adults.

The clinical features are those of unilateral acute painful vesiculo-bullous disease in the distribution of a sensory nerve lasting 2–4 weeks. Involvement of the second or third division of the trigeminal results in disease affecting the face and mouth. Postherpetic neuralgia occasionally follows resolution.

C. Herpangina

This infection is caused by a coxsackie virus and presents as a vesicular eruption of the soft palate and oropharynx. Constitutional disturbances, lymphadenitis, fever, etc., are usually mild.

D. Human Immunodeficiency Virus

The oral manifestations of human immunodeficiency virus (HIV) infection include:

- Hairy leukoplakia
- Candidiasis
- HIV-associated periodontal disease
- Kaposi's sarcoma
- Lymphoma
- Squamous cell carcinoma

Hairy leukoplakia occurs on the sides of the posterior part of the tongue, usually bilaterally, and is asymptomatic. It appears as white wrinkled lines that may fuse to form plaques. Epstein-Barr virus (EBV) and *Candida* are usually present in the epithelium but not HIV. Presence of this lesion is associated with subsequent development of the acquired immunodeficiency syndrome fairly quickly in a high percentage of patients.

Oral candidiasis is a common finding in HIV-infected persons.

HIV-associated periodontal disease may present as severe ANUG, with heavy infection by the *B. fusiformis* and *B. vincentii,* or as a rapidly progressive chronic periodontitis with alveolar bone destruction.

The tumors listed are those most commonly associated with HIV infection and may present intraorally as elsewhere in the body, Kaposi's sarcoma being common intraorally. [*See* Acquired Immunodeficiency Syndrome (Virology).]

E. Candidiasis

Candida albicans is an oral commensal that becomes pathogenic in immunodepression, general debility, or as a result of chronic local tissue damage. Four types occur:

- Acute pseudomembraneous candidiasis (thrush)
- Chronic hypertrophic candidiasis
- Atrophic candidiasis
- Angular cheilitis (perleche)

The infection will clear with appropriate antibiotic therapy but will recur unless the predisposing factor(s) is eliminated.

F. Actinomycosis

This infection is caused by the *Actinomyces israelii,* often in combination with other bacteria, and occurs in the abdomen, thorax, and cervicofacial region. Infections occur as complications of surgery (e.g., tooth extractions) or secondary to trauma.

The clinical features are those of a persistent swelling around the mandibular angle with draining skin sinuses. The pus contains yellow "sulfur granules" that are colonies of the organism.

G. Syphilis

All stages of syphilis may affect the mouth. In congenital syphilis screwdriver-shaped incisor teeth are seen and mulberry molars. The primary chancre may occur intraorally or on the lip as a painless indurated ulcer teeming with *Treponema pallidum* that lasts for about a month. Secondary syphilis causes a skin rash and "snail track" ulcers on the oral mucosa. Among the many manifestations of tertiary syphilis are glossitis and palatal perforations caused by gumma formation. [*See* Sexually Transmitted Diseases (Public Health).]

H. Tuberculosis

Oral tuberculosis is rare and usually occurs secondary to open pulmonary infection in the form of chronic ulceration, the result presumably of bacterial implantation from infected sputum.

I. Other Fungal Infections

Histoplasmosis, coccidioidomycosis, cryptococcosis, and blastomycosis may also involve the mouth, usually as chronic ulcers and generally secondarily to lung infection.

VII. Cysts

A cyst is a pathological cavity in the tissues lined by epithelium and containing fluid or semifluid material. They are common in the jaws, the epithelium being derived usually from remnants of odontogenic epithelium. Cysts are slow-growing lesions, causing swelling and displacement of teeth. They may become infected.

The commoner types of cysts are

1. Radicular cyst. This arises in a chronic periapical inflammation in which epithelial rests are stimulated to proliferate by the infection. The cyst may remain *in situ* after the tooth has been extracted (residual cyst).
2. Dentigerous cyst. Cysts can arise from the odontogenic epithelium around the crown of an impacted or unerupted tooth, the tooth crown projecting into the lumen.
3. Odontogenic keratocyst. The pathogenesis of this lesion is unknown. They can grow to large size, and "daughter" cysts may bud off the epithelial lining causing a high recurrence rate. They sometimes occur as part of the basal cell naevus syndrome.
4. Inclusion cysts. Epithelium may be enclaved during fusion of the embryonic processes that form the face and jaws. In later life this epithelium may form cysts. The most common type is the incisive canal cyst that lies at the point of fusion of the primitive palate and the two palatal shelves.

VIII. Mucosal Diseases

Lichen planus is a chronic inflammatory condition occurring in adults that may affect skin or mucous membranes or both and that arises because of an abnormality of the immune system. Intraorally it results in hyperkeratosis usually bilaterally and in the form of an interlacing network of fine lines or dots, occasionally as plaques. The adjacent mucosa may be inflamed or ulcerated.

Discoid lupus erythematosus may cause oral lesions that are erythematous or ulcerated with adjacent hyperkeratosis.

Pemphigus vulgaris usually affects the oral mucosa and may start on it. The oral lesions are multiple vesicles that rupture within a few hours, leaving ulcers that may become secondarily infected.

Mucous membrane pemphigoid is a chronic vesiculo-bullous disease usually of the elderly in which the bullae rupture and become secondarily infected. Untreated the ulcers may persist for years.

Erythema multiforme may occur on skin or mucous membranes or on both and is most common in young male adults. It has an acute onset and intraorally results in large serpiginous ulcers, the edges crusted with blood clot and almost always involving the lips. It resolves spontaneously in about 3 weeks and may be an allergic response to herpes simplex or to some other antigen.

Epidermolysis bullosa is an inherited condition that occurs with varying levels of severity and presents in childhood. It affects skin and mucous membranes, the lesions being bullae that form after minor friction so that they frequently form on areas of oral mucosa subject to masticatory stress.

Scleroderma is a generalized disease in which collagen is formed in excessive quantities in the soft tissues and elsewhere. In the mouth it chiefly affects the cheeks, which become stiff and immobile so that oral hygiene measures and dental treatment become difficult resulting in caries and periodontal disease.

IX. Fibro-Osseous Conditions

Fibrous dysplasia is developmental abnormality present in childhood and often affecting the jaws, in which areas of bone are affected by a defect in metabolism that causes resorption of mature bone and its replacement by cellular connective tissue and excessive amounts of immature bone. This results in swelling, deformity, and displacement of teeth. In many cases the condition resolves spontaneously around puberty.

Cherubism is a subtype of fibrous dysplasia, in which only the molar areas of the mandible and, less often, of the maxilla are affected. In early childhood, mild involvement produces a cherubic facies but later is disfiguring.

Cleidocranial dysplasia is a congenital abnormality of bones formed in membrane affecting the skull and clavicles, which are often absent. In this condition the teeth do not erupt.

X. Inflammatory Polyps

Polyps of irritational origin are common in the mouth and consist of varying proportions of granulation and fibrous tissue containing inflammatory cell infiltrates and covered by stratified squamous epithelium that may be hyperkeratotic, hyperplastic, or ulcerated. Many names have been used for these lesions, but all can be classified as fibro-epithelial polyps or, when composed largely of granulation tissue, as pyogenic granulomata. They have minimal neoplastic potential. Generalized fibrous hyperplasia of the gingivae occurs idiopathically and as a result of chronic dilantin therapy.

XI. Tumors

A. Epithelial

1. Papilloma
Small villous exophytic squamous papillomas occur on all parts of the oral mucosa and lips, some of them being verruca vulgaris lesions and others may also be caused by human papilloma virus (HPV) infection. Condyloma acuminatum and focal epithelial hyperplasia also occur because of HPV infection. [*See* PAPILLOMAVIRUSES AND NEOPLASTIC TRANSFORMATION.]

2. Squamous Cell Carcinoma
This is the most common oral malignancy, its incidence varying greatly in different populations and being highest in India where it is associated with the habit of chewing pan (a mixture of tobacco, slaked lime, betel nut and leaf, and other additives). In Western countries, oral cancer accounts for about 4% of all malignancies. It can occur anywhere in the mouth but is most frequent on the sides of the tongue and on the lower lip, where it is associated with actinic damage to the tissues. Excessive consumption of alcohol and tobacco are the only common predisposing factors.

It is predominantly a disease of the elderly. It presents clinically as an area of hyperkeratosis, as a lump, as an ulcer with raised rolled margins, as a velvety red patch, or as a combination of these. Untreated it can cause great local tissue damage and spreads by local invasion and by metastases to the regional lymph nodes in the neck but seldom farther.

3. Mesenchymal and Neurogenic Tumors

Virtually all the benign and malignant tumors of mesenchymal and neurogenic origin may occur in the mouth but with no locally distinctive features.

4. Salivary Tumors

A range of epithelial tumors are found in the major (parotid, submandibular, and sublingual) salivary glands and in the numerous minor glands found all around the mouth. The parotid is by far the most common site, and in it the great majority of tumors are benign. The proportion of malignant tumors is much higher in the submandibular, sublingual, and minor glands.

All benign tumors present clinically as slow-growing masses. Malignant tumors grow more rapidly and may ulcerate. In the parotid they may invade the facial nerve causing unilateral facial paralysis. Metastases usually develop late except for the adenocarcinoma. The adenoid cystic carcinoma is notable for its infiltrative tendencies, often along nerve sheaths.

A simplified list of salivary tumors consists of

· Pleomorphic adenoma	⎤ Benign
· Monomorphic adenoma (various types)	⎟
· Mucoepidermoid tumor	⎤ Low-grade
· Acinic cell tumor	⎦ malignancy
· Adenoid cystic carcinoma	⎤
· Carcinoma ex pleomorphic adenoma	⎟ Malignant
· Adenocarcinoma	⎦

5. Odontogenic Tumors

These tumors are site-specific as they arise from epithelial or mesenchymal or both tissues involved in the process of tooth formation. Histologically, they consist of cells or cell products, including enamel, dentine, and cementum, that are seen in normal tooth formation. They demonstrate a range of behavior from completely benign lesions that are essentially hamartomas to rare metastasizing malig-

nancies. Most are benign. The ameloblastoma, most common of the group, and the odontogenic myxoma are slow-growing infiltrative neoplasms that cause great local destruction and, if arising in the maxilla, may invade the base of the skull and brain. The calcifying epithelial odontogenic tumor contains deposits of amyloid.

Benign

Epithelial tumors
    Ameloblastoma
    Calcifying epithelial odontogenic tumor
    Adenomatoid odontogenic tumor
    Squamous odontogenic tumor
Mixed (epithelial and mesenchymal) tumors
    Odontoma
    Ameloblastic fibroma
    Ameloblastic fibro-odontome
Mesenchymal tumors
    Odontogenic myxoma
    Odontogenic fibroma
    Cementifying fibroma
    Cementoblastoma
Malignant
Epithelial
    Malignant ameloblastoma
    Ameloblastic carcinoma
Mesenchymal
    Ameloblastic fibrosarcoma

Bibliography

Lucas, R. B. (1984). "Pathology of Tumours of the Oral Tissues," 4th ed. Churchill Livingstone, Edinburgh.

Nikiforuk, G. (1985). "Understanding Dental Caries. Vol. 1. Etiology and Mechanisms, Basic and Clinical Aspects." Karger, Basel.

Regezi, J. A., and Sciubba, J. J. (1989). "Oral Pathology. Clinical-Pathologic Correlations." W. B. Saunders, Philadelphia.

Scully, C., and Flint, S. (1989). "Color Atlas of Oral Diseases." J. B. Lippincott, Philadelphia.

Shafer, W. G., Hine, M. K., and Levy, B. M. (1983). "A Textbook of Oral Pathology," 4th ed. W. B. Saunders, Philadelphia.

Wright, B. A., Wright, J. M., and Binnie, W. H. (1988). "Oral Cancer: Clinical and Pathological Considerations." CRC Press, Boca Raton, Florida.

Oxygen Transport to Tissue

I. S. LONGMUIR, *North Carolina State University*

Glossary

Anoxia Total lack of oxygen

Henry's Law When a liquid is equilibrated with a gas, the amount of gas that goes into solution in the liquid is proportional to the partial pressure of that gas in the gas phase. Thus, the concentration of a gas in solution, $C_{O_2} = \alpha\, P_{O_2}$, where α is the Bunsen solubility coefficient and P_{O_2} is the partial pressure of oxygen in the gas phase in equilibrium with the liquid. The activity of a gas in solution is a function of its partial pressure rather than its concentration as is the case with solid solutes.

Hyperoxia and hypoxia Values substantially higher or lower, respectively, than the normoxic values.

Normoxia Partial pressure of oxygen in various parts of healthy human body at rest, breathing air at 760 mm Hg. These values are as follows: $P_{I_{O_2}} = 150$ mm Hg, $P_{A_{O_2}} = 100$ mm Hg, $P_{a_{O_2}} = 96$ mm Hg, $P\bar{v}_{O_2} = 40$ mm Hg.

Partial pressure The partial pressure of a gas in a given volume of a mixture of gases is the pressure that gas would exert if it alone were present in that volume. The sum of the partial pressures of each gas in the mixture is equal to the pressure exerted by the mixture. By convention all pressures in physiology are expressed in millimeters of mercury (mm Hg).

Symbols By convention the following symbols are used: $F_{I_{O_2}}$ is the fraction of oxygen in the inspired air; $P_{I_{O_2}}$ is the partial pressure of oxygen in inspired air; $P_{A_{O_2}}$ is the partial pressure of oxygen in the alveolar air; $P_{a_{O_2}}$ is the partial pressure of oxygen in the arterial blood, i.e., the blood leaving the lung capillaries; $P\bar{v}_{O_2}$ is the mean partial pressure of oxygen in the mixed venous blood, i.e., the blood entering the lung capillaries; $P_{T_{O_2}}$ is the partial pressure of oxygen in the tissues.

ALL HUMAN CELLS require a continuous supply of oxygen, but the urgency with which different cells need oxygen varies. The cells of the retina are the most demanding in this respect. After only a few seconds without oxygen (anoxia) they stop responding to light; the subject "blacks out." Muscle cells, on the other hand, can continue to contract for several minutes in the absence of oxygen. Oxygen is also needed for certain metabolic activities necessary for the maintenance of cell integrity. Cell function can be restored after brief periods of anoxia by restoring the oxygen supply, but if the anoxia is prolonged the cells will die. As little as 4 min of cerebral anoxia causes brain death.

The body's source of oxygen is the ambient air. Some surface cells, such as those of the cornea, can obtain all the oxygen they require directly, but all other cells depend on a complex distribution system. Air is inspired through the nose and mouth and passes down the trachea to the lungs. The main bronchus in each lung branches many times, and the small airways terminate in sacs called the alveoli. Each alveolus is about 0.25 mm in diameter and is surrounded by a mesh of small blood vessels called capillaries. Oxygen diffuses from the air into the capillary blood and reacts with the hemoglobin in the red blood cells. The blood drains into the left heart and is pumped through the aorta and its branches to all the tissues of the body. The smallest branches and the tissue capillaries are vessels about 5.0 μm across and separated from one another by variable distances averaging about 0.1 mm. The oxygen leaves the blood in the tissue capillaries and diffuses through the cell to the sites of utilization.

I. The Oxygen Cascade

The pathway of oxygen from ambient air to the intracellular sites of oxygen utilization is down the "oxygen cascade," so named because at each step there is a fall in partial pressure. The first step is the admission of air to each alveolus. Since air cannot be completely expelled from the lungs, even by forced expiration, the fresh air entering will be mixed with residual air from which some oxygen has been removed. As a result, the average $P_{A_{O_2}}$ is about 50 mm Hg lower than the $P_{I_{O_2}}$. The second step is the diffusion across the alveolar membrane into the blood, where the oxygen binds to hemoglobin. Since the lungs are efficient gas exchangers, the fall in P_{O_2} at this stage is quite small. However the difference between $P_{A_{O_2}}$ and $P_{A_{O_2}}$, the A − a difference, is increased by two other events, shunting and ventilation/perfusion mismatch, as will be explained later. During the third step, the circulation of the blood from lung capillaries to those in the tissue, there is in general a very small fall in P_{O_2} due to oxygen consumption by the blood itself. In certain specialized structures, such as small arterioles invested by venae comites, there can be an appreciable diffusion of oxygen from the arterial blood to the venous blood. The final step of the cascade occurs in the transfer to tissues, where the P_{O_2} at points near the capillary wall may be nearly as high as the $P_{A_{O_2}}$, while at points farthest from open capillaries it may be near zero.

II. Ventilation of the Lung

The lungs consist of a series of branching tubes, which terminate in small air sacs called alveoli. The human lung has 300 million such sacs, each about 0.25 mm in diameter. The sacs are lined with a thin layer of fluid and are therefore essentially bubbles. If the fluid were pure water the hydrostatic pressure inside each bubble would be sufficient to collapse the alveoli. However, this collapsing pressure is opposed by a surface monolayer of detergent (dipalmityllecithin), which efficiently resists surface pressures. Thus, the alveoli do not collapse. At the same time, if the alveoli are made to expand, there comes a point where the surface film is incomplete, and the surface tension of fluid free of monolayer opposes further inflation. Thus by the unique properties of dipalmithyllecithin, also called pulmonary surfactant, the volume range of each alveolus is restricted. Pulmonary surfactant is continuously secreted by cells in the alveolus. If the supply is disrupted by certain toxins, such as paraquat or a $P_{I_{O_2}}$ in excess of 500 mm for a day or more, the alveoli collapse.

If the lungs are removed from the chest cavity, they will collapse, partly because of the small excess of the surface tension of water over the collapse resistance of surfactant and partly because of the large amount of elastic tissue in the solid part of the lung. Normally, the lungs hang in a closed box, the thoracic cavity, which prevents collapse. The volume of the thoracic cavity is increased when the intercostal muscles and the diaphragm contract. This results in the inspiration of air into the lungs. When these muscles relax, the thoracic cavity decreases in volume and air is expired. Virtually all the change in volume of the lungs occurs in the alveoli.

Respiration, the alternative contraction and relaxation of the intercostals and the diaphragm, is controlled by the respiratory centers in the floor of the fourth ventricle of the brain. Although this action is involuntary, it is possible to override it to some extent by voluntary effort. The respiratory center is not an oxygen sensor but responds to changes in arterial blood pH and the partial pressure of carbon dioxide so that the air in the alveoli is kept at a P_{CO_2} of about 40 mm Hg. The lungs have two main functions: the uptake of oxygen and the excretion of carbon dioxide. The latter, although it is an excretion product, is not completely purged from the blood but is kept at a partial pressure of about 40 mm Hg, which keeps the pH of arterial blood at about 7.4, that is, essentially neutral. The effect of this control of carbon dioxide content in the alveolar air is to maintain the $P_{A_{O_2}}$ at about 100 mm Hg when normal air at sea-level barometric pressure is inspired. However, when the $P_{I_{O_2}}$ drops below 150 mm Hg the $P_{A_{O_2}}$ falls below 95 mm. At 70 mm Hg, certain peripheral chemoreceptors such as the carotid bodies are stimulated to override the respiratory center. As a result, the rate and depth of lung ventilation increases. This increases the $P_{a_{O_2}}$. There appears to be a wide variation in the $P_{a_{O_2}}$ at which the override mechanism begins to operate and the degree to which hypoxia stimulates ventilation. The only physiological event that gives rise to a lowered $P_{I_{O_2}}$ is ascent to high altitudes, and subjects with the greatest hypoxic response are best able to ascend to great heights. In addition, living at altitude for some weeks increases the hypoxic drive and enables climbers to ascend further. [*See* RESPIRATORY SYSTEM, PHYSIOLOGY AND BIOCHEMISTRY.]

The air in the lungs is contained in the airways: the main bronchi, the secondary bronchi, the bronchioles, the alveolar ducts, and the alveoli. Oxygen diffuses from the air into the blood only in the latter structures. This air is called the alveolar air and the rest the "dead space" air. It is possible to calculate the volume of the dead space by analyzing an expired breath. At the end of the breath the air will have come from the alveoli only, whereas the whole volume of the breath will contain a mixture of the whole of dead space air plus some alveolar air. Thus, from measurement of the oxygen content of an "end tidal" a sample and of the mixed breath, the dead space air volume can be calculated. From this value the amount of air entering the alveoli per minute can be calculated by subtracting the product of the dead space volume and the breaths per minute from the volume of air inspired per minute. In a normal man at rest the minute alveolar volume ($\dot{V}_A$) is about 5 liters. This volume is increased many times on exercise.

The entire blood output of the right heart is delivered to the capillary meshes surrounding each alveolus. At rest this volume is again about 5 litres per minute ($\dot{Q} = 5$). It is clear that these two fluids, air and blood, need to be delivered equally to all parts of the lungs. In alveoli that are ventilated but not perfused with blood and in those that are perfused but not ventilated there can be no oxygen uptake. Ideally, each alveolus should receive the same volume of both blood and air. Every alveolus in the healthy lung tends to fill with air to about the same extent as result of the surfactant mechanism previously described. The uniform perfusion of the pulmonary circulation is, however, more complex. The pressure in the pulmonary trunk varies between 20 mm Hg when the heart ventricles contract (systole) and 8 mm Hg when they are at rest (diastole). As a result, circulation in capillaries in those parts of the lung that lie above the heart when the subject is erect stops during diastole, whereas the circulation at the base of the lung is continuous. Such a mismatch of ventilation with air and perfusion with blood leads to inefficient gas exchange. There is a mechanism, however, which goes some way to making the distribution of blood in the lung more uniform. It is "oxygen autoregulation." When the oxygen tension in a part of the lung is low, the blood vessels constrict, reducing the blood flow to that region. If a volume of lung is not ventilated for any reason, the pulmonary blood flow ceases altogether. Since the lungs are obliged to accept all the blood ejected from the right ventricle, any reduction in flow in any part of the lungs must result in increased flow in the rest. When the $P_{I_{O_2}}$ falls, as in ascent to high altitudes, all the pulmonary vessels constrict. This results in a rise in pressure in the whole pulmonary circulation and so improves perfusion of the apices of the lung. Thus, the response of the lung blood vessels to hypoxia has the effect of altering the value of $\dot{V}a/\dot{Q}$ toward unity in all parts of the lung.

III. Diffusion across the Alveolar Wall

Since the $P\bar{v}_{O_2}$, the partial pressure of oxygen in the blood entering the capillary nets surrounding each alveolus, is less than that in the alveolar air, oxygen diffuses from air to blood. The rate at which it does so is proportional to the product of the diffusivity of oxygen (D), the area of the alveolar wall (A), and the partial pressure difference between air and blood (ΔP_{O_2}) and inversely proportional to the thickness of the barrier between air and blood (δ). As a result, oxygen flux = $(AD/\delta)\Delta P_{O_2}$. The total alveolar area in humans is greater than 100 m^2, about the size of a tennis court. The thickness of the alveolar wall averages about 1.4 μm. It is clear that the lungs constitute a very efficient gas exchange and in severe exercise can take up several liters of oxygen per minute. At the alveolar level the blood traverses the capillary net in less than a second, about 0.7 sec at rest, and half this during maximal exertion. During this time each 100 ml of blood takes up 5 or more ml of oxygen, and the P_{O_2} rises from about 40 to 95 mm Hg. As the P_{O_2} rises toward that in the alveolar air, the rate of uptake will fall. During expiration no oxygen enters the alveoli, so there will be a fall in P_{aO_2} during this phase of the ventilatory cycle. However, since the volume of air in the alveoli is about 100 times that of the blood in the pulmonary capillaries, this fall will be negligible. During inspiration there will be a rise. These small fluctuations result in a just detectable variation in Pa_{O_2}. Although the Pa_{O_2} in the blood draining each alveolus is very nearly equal to the P_{O_2} in the gas phase in that alveolus, the mean Pa_{O_2} will be a few millimeters of mercury lower than the mean PA_{O_2} because of three factors: (1) the nonuniform distribution of blood and air, (2) shunting, whereby some of the output of the right heart bypasses the alveolar capillaries, and (3) some oxygen consumption by the lung cells. As a consequence, the so called A − a gradient is about 5–6 mm Hg. The diffusing capacity of the lungs can be measured using the rate of oxygen uptake. However, this is difficult because of the uncertainty of the back pressure of oxygen in the blood. One can use another gas, carbon monoxide, which has about the same diffusivity as oxygen. The great merit of this gas is the avidity with which it binds to hemoglobin. As a result, the back pressure problem does not arise and the mass transfer equation can be written $(DA/\delta)/Pa_{CO}$.

In this way the diffusing capacity, DA/δ can be measured from the uptake of carbon monoxide from a single breath.

IV. Transport of Oxygen in the Blood

A. The Equilibrium with Hemoglobin

Oxygen diffuses into the blood plasma first and then passes across the red cell membrane and reacts reversibly with hemoglobin. The red cells contain a high concentration of hemoglobin, about 30% w/v. The reaction of hemoglobin and oxygen is quite fast, and equilibrium is reached while the red cells are in the pulmonary capillaries, a period of less than a second. Each hemoglobin molecule can bind reversibly to four oxygen molecules, which however, are not completely independent of one another. When one molecule of oxygen binds with one site, the shape of the molecule changes, and as a result the affinity of the other three sites for oxygen is increased. Attempts to develop equations relating the amount of oxygen bound to hemoglobin and the P_{O_2} in solution have been only partially successful. However, to understand the role of hemoglobin in oxygen transport it is convenient to consider the oxyhemoglobin dissociation curve. This is a plot of the percentage saturation of the binding sites against the P_{O_2} (Fig. 1, Curves 2, 3, and 4). It will be seen that the curves are S-shaped and that at a P_{O_2} of 95 mm, the normal value for arterial blood, the

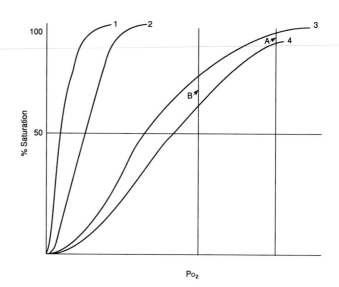

FIGURE 1 Oxyhemoglobin dissociation curves. (1) Hemoglobin monomer, one binding site; (2) hemoglobin solution; (3) red cells—P_{CO_2} that for arterial blood; (4) red cells—P_{CO_2} that for venous blood. The differences between the curves have been exaggerated to clarify the various points made in the text.

hemoglobin is nearly fully saturated. Blood contains about 15 g of hemoglobin per hundred milliliters, and each gram of hemoglobin can bind to 1.3 ml of oxygen. Thus fully saturated blood contains about 20 ml of oxygen bound to hemoglobin. There is also some oxygen in solution, amounting to 0.25 ml at 95 mm Hg. At normal Pa_{O_2} values this is a negligible quantity, but if oxygen is inspired at a pressure of several atmospheres, the amount in physical solution can be sufficient to meet all the tissues' needs for oxygen. However, this partial pressure of oxygen in the blood would be enough to damage the brain and lungs so could not be sustained for more than a brief period. Thus, the major function of the hemoglobin is to raise the oxygen capacity of blood severalfold without increasing the Pa_{O_2} to toxic levels. [*See* HEMOGLOBIN; RED CELL MEMBRANE.]

The shape of the oxyhemoglobin dissociation curve plays an important role in oxygen loading and unloading. As mentioned before, the curve is S-shaped since the effect of the addition of an oxygen molecule is to increase the affinity of the remaining three sites for oxygen. If the hemoglobin molecule is split into four parts, each with one binding site, this effect is lost and the curve becomes a simple hyperbole (Fig. 1, Curve 1). The midpoint of the dissociation curve of blood, the P_{50} is at a P_{O_2} of about 26 mm. A pure solution of hemoglobin has a midpoint of less than 6 mm. With such a low value it can be seen the tissue oxygen tension would have to fall to very low values before a significant amount of oxygen could dissociate. All the hemoglobin in blood is inside the red cells, which also contain $2:3$ bisphosphoglycerate (BPG) in an equimolar concentrate to that of hemoglobin. This molecule binds to deoxygenated hemoglobin, reducing its affinity for oxygen and shifting the P_{50} to a higher value. A number of other substances also shift the P_{50}. The most important ones, physiologically, are CO_2 and pH change. When the blood takes up CO_2 in the tissues, the Pco_2 rises and the pH falls (i.e. acidity increases), shifting the curve to the right (Fig. 1, Curves 3 and 4). In the lungs the reverse changes occur. As can be seen from the figure, the change increases the amount of oxygen taken up in the lungs and the amount given up in the tissues. Thus, the concomitant transport of carbon dioxide increases the efficiency of oxygen transport. A number of congenital and acquired factors influence the dissociation curve. More than a hundred abnormal hemoglobins have been discovered, many of which show reduced efficiency in oxygen transport. The conversion of some hemoglobin to carbonmonoxyhemoglobin or methemoglobin, both present to a small extent in human blood, causes a reduction in P_{50}, which can become significant if either is present in excess of 10%.

B. Blood Perfusion of Tissue

The oxygenated blood drains from the alveolar capillaries through the pulmonary veins into the left side of the heart. From the left ventricle the blood is pumped into the aorta. Arteries branch from this vessel and repeatedly divide into smaller arteries and then arterioles to deliver blood to all the organs and tissues of the body. The smallest vessels are the capillaries, vessels whose walls are one cell thick and have lumen of $5.0~\mu$m, less than the diameter of the red cell. Red cells, therefore, have to be distorted to some extent in order to pass through the capillaries. It is while the red cells are traversing the capillaries that they give up oxygen. They can release oxygen only when the Po_2 in the plasma falls by the diffusion of oxygen into the surrounding tissue where the Po_2 is lower still.

The blood is distributed to the tissue in response to local oxygen need. This distribution is regulated in two ways. In the long term, the size of the arteries appears to be related to the oxygen need of a

particular organ. In the embryo and in healing tissue angiogenesis, the formation of new blood vessels is stimulated by tissue hypoxia. When the oxygen consumption of such tissues as skeletal muscle or the heart is raised by a substantial increase in physical activity, the major vessels in those organs increase in diameter. Because the resistance to blood flow is inversely proportional to the fourth power of the radius, this change is associated with a greatly increased blood flow. The increase in flow due to vessel enlargement occurs in a matter of weeks, but in the short term, on the order of a few seconds, there is another effect, known as "oxygen autoregulation." This term has already been used to describe the reduced blood flow to hypoxic regions of the lung. In the peripheral tissues the response to hypoxia is inverted, and hypoxia increases the blood flow.

All three sets of vessels appear to be involved in the increased tissue perfusion. Arterioles and venules dilate by a relaxation of the smooth muscle in their walls. Capillaries lack this muscle coat; they increase perfusion by recruitment, the opening up of previously closed capillaries. Red cells traverse each capillary at a fairly constant rate. An open capillary has a diameter of about $5.0~\mu$m, and this lumen is further restricted when the endothelial cells which make up the capillary wall swell. Then the red cells can no longer squeeze through, and oxygen delivery from that capillary ceases. Thus, the mechanism of "recruitment" may be the reversal of this reduction by a slight shrinkage of endothelial cells to the point where red cells can flow again. The tissue response to mild hypoxia may be a purely capillary reaction, which reduces the resistance to blood flow, resulting in increased arteriolar flow without any increase in diameter of the latter vessels. In addition, recruitment would also permit more efficient tissue oxygenation because it reduces the intercapillary distances. However, more severe hypoxia results in simultaneous dilatation of both anterioles and venules in addition to capillary recruitment. When the oxygen tension rises, the vessels constrict. Although this constriction is part of the process of "oxygen autoregulation," it is not clear whether or not a separate mechanism is involved.

Although the effects of oxygen tension on tissue perfusion with blood have been known for more than 70 years, the mechanism is still unclear. The most cogent theories are based on the metabolic consequences of oxygen starvation of the oxygen-consuming enzymes. When the cytochrome oxidase system, for lack of oxygen, is unable to produce adenosine triphosphate as fast as it is consumed, adenosine is released. However, arterioles which dilate in hypoxia do so at oxygen tensions above that at which cytochrome oxidase begins to become reduced. Aspirin and similar inhibitors of cyclooxygenase partially inhibit "oxygen autoregulation." This enzyme, which is essential for prostaglandin synthesis, requires a higher P_{O_2} for optimal activity. Its inhibition by lack of oxygen or by drugs may reduce the production of vasoconstrictor molecules. More complex models have been proposed, involving specific "reporter" enzymes whose sole function is in "oxygen autoregulation." It has also been proposed that the vasoconstriction that occurs when the P_{O_2} is high is due to oxygen-free radicals, which are produced in greater amounts in hyperoxia.

C. Transport of Oxygen From Blood to Tissue Site of Utilization

The capillary wall is highly permeable to oxygen, which can diffuse readily through all parts of this structure. Since the cells surrounding the capillary are continuously consuming oxygen, the P_{O_2} in them is lower than in the blood, so oxygen diffuses into the tissue. As the P_{O_2} in the blood falls, oxygen dissociates from oxyhemoglobin as described previously. The first attempt to model the transport of oxygen in tissue was based on two assumptions: Uniform oxygen consumption and uniform passive diffusion. Since oxygen can diffuse for only a limited distance before being used completely, this leads to a picture of bounded cylinders of falling oxygen tension surrounding each capillary. The radius of this cylinder is a function of capillary P_{O_2}, the diffusivity of oxygen in tissue, and the local rate of oxygen consumption. However, the first assumption of uniform oxygen consumption is modified by the finding that the oxygen-consuming enzymes are situated in highly localized structures in the cell (mainly the mitochondria). Attempts to measure the diffusivity of oxygen in cells showed anomalies in intracellular oxygen transport. In red muscle it is clear that myoglobin (a muscle protein) facilitates the transport of oxygen. In all cells oxygen diffuses along the hydrophobic core of membranes ten times as fast as through cytosol. Since most of the respiratory enzymes are localized within organelles associated with membranes, these membranes might con-

stitute an intracellular oxygen transport system. There is some evidence that the endoplasmic reticulum, a double membrane structure, is continuous with both the plasma membrane and the outer mitochondrial membrane. The major respiratory enzyme, cytochrome oxidase, is located in the inner mitochondrial membrane. Some oxidases are situated in the endoplasmic reticulum, and the outer mitochondrial membrane and the rest within peroxysomes attached to the endoplasmic reticulum. Thus, all the oxygen-consuming enzymes are associated with a membrane along which oxygen can diffuse more rapidly than thorough the rest of the cellular constituents. A further complexity has been introduced by the observation that even in the presence of enough oxygen, cells switch off their respiration. It seems that individual cells are either respiring maximally or not at all. Thus, the original cylinder has become greatly distorted both spatially and temporally.

V. Classification of Respiratory Enzymes

Respiratory enzymes can be classified as oxidases, oxygenases, and mixed function oxidases. Oxidases use oxygen as an electron acceptor. Cytochrome oxidase is the terminal enzyme of the electron transport chain in mitochondria and reduces oxygen to water by the addition of four electrons. The principal function of this enzyme system is to generate adenosine triphosphate (ATP), which is the fuel for muscular contraction and many other metabolic events. It is by far the major consumer of oxygen. Other oxidases donate two or one electron to produce hydrogen peroxide or superoxide free radical, respectively, which are utilized for killing foreign organisms.

The oxygenases introduce both atoms of the oxygen molecule into a second molecule. The mixed function oxidases reduce one atom of oxygen to water while incorporating the second atom into another substrate. An alternative classification of these enzymes is on the basis of their affinity for oxygen. The enzyme with the highest affinity is cytochrome oxidase, which can consume oxygen at its maximum rate until only a trace remains. At the other end of the scale are the flavoprotein oxidases, which have much lower affinity. Clearly these enzymes will require much higher oxygen concentrations, but since their major role is synthetic, the limiting factor is either the tissue demand for their products or the availability of substrates on which they act. Other enzymes have an intermediate range of affinities, as demanded by their functions; some of them play a role as oxygen sensors for oxygen autoregulation.

VI. Tissue Oxygen Transport in the Fetus

Tissue oxygen transport in the fetus appears to be qualitatively the same as in the adult but with a striking quantitative difference. The placenta, which is the source of oxygen before birth, is a re-markably inefficient oxygen exchanger compared with the lung. The P_{O_2} in the blood in the umbilical vein in the human fetus is only 29 mm Hg. This figure is further reduced by admixture with fetal venous blood before entering the arterial system. The low fetal Pa_{O_2} is partially compensated for by a low P_{50} in the fetal blood. Nevertheless, the "oxygen cascade" in the fetus has a gradient only one-third that of the adult. This gradient changes dramatically at birth.

Bibliography

Oxygen Transport to Tissue, Vols I–XI. Advances in Experimental Biology and Medicine. Plenum Press New York and London.

Weibel, E. R. (1986). "The Pathway for Oxygen." Harvard University Press, Cambridge and London.

P

Pain

FRAN PORTER, *Washington University School of Medicine*

Glossary

Myelination Process in which a phospholipid sheath forms around nerve fibers. Myelination speeds conduction of the nerve impulse
Nociception Response induced by a painful, injurious stimulus
Noxious stimulus Stimulus that is painful, injurious, or harmful
Psychophysiology Study of the interactions between mental and physiological processes

PAIN REPRESENTS A COMPLEX, psychophysiological event which involves the interaction of sensory, neurochemical, emotional, motivational and cognitive systems. Pain provides important information by conveying the message that "something is wrong." It encourages action to prevent further injury. From an evolutionary point of view, pain is one of the most powerful ways to ensure the survival of an organism in a dangerous or potentially dangerous world.

I. Definition

Pain is typically described along dimensions of quality, location, duration, and intensity yet it is more than the total of these. Perceptions of pain and reactions to it can vary dramatically among individuals and even within the same individual over time, conditions, health, and psychological states. It is, therefore, difficult to precisely quantify pain.

The most widely accepted definition of pain was offered by the International Association for the Study of Pain (IASP) in 1979. This definition states that pain is:

> An unpleasant sensory and emotional experience associated with actual or potential tissue damage, or described in terms of such damage.

This definition emphasizes the importance of the emotional components of the pain experience. It further stresses the individualized nature of pain:

> Pain is always subjective. Each individual learns the application of the word through experiences related to injury in early life.

Thus, exhausting structural or physiologic explanations may not be sufficient to understand the existence, absence or persistence of pain.

II. Physiology of Pain

The physiologic mechanism by which noxious stimuli produce a subjective correlate of pain is far from completely understood. Because pain is fundamentally a subjective experience, there are inherent physiologic limitations to understanding it. However, a series of complex electrical and chemical events between tissue injury and the subjective experience of pain have been identified. These include transduction, transmission and modulation, represented in Fig. 1.

A. Ascending Pain System

Transduction is the process by which noxious stimuli produce electrical activity in specific sensory nerve endings (Fig. 1, left). *Transmission* is the process by which this electrical activity is then transmitted over the peripheral nerve to the spinal cord. The pain message then synapses with cells of origin of the two major ascending pain pathways, the *spinothalamic* and the *reticulothalamic* pathways. These pathways make up a network of relay neurons that ascend from the spinal cord to the brainstem and the thalamus. The message is then transmitted to higher brain centers via reciprocal connections between the thalamus and the frontal and somatosensory cortices. [*See* THALAMUS.]

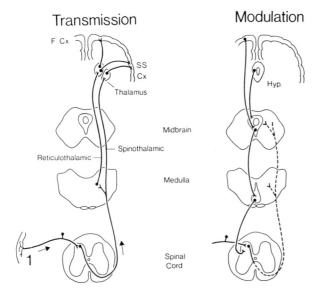

FIGURE 1 Neural pathways of pain transmission and modulation. *Left*, noxious stimuli activate the primary afferent nociceptor's peripheral endings by the process of transduction (1). The neural impulse is then transmitted to the spinal cord where it synapses with the spinothalamic and spinoreticulothalamic pain pathways. The message is then relayed from the thalamus to the frontal cortex and the somatosensory cortices. *Right*, inputs from the frontal cortex and hypothalamus activate cells in the midbrain which control spinal pain transmission cells via cells in the medulla. Reproduced, with permission, from H. L. Fields (1987). ''Pain'' p. 6. McGraw-Hill, New York.

1. Pain Transmission Fibers

There are believed to be two major types of fibers involved in the transduction and transmission of pain signals. These are described by their diameter, impulse conduction rate, extent of myelination, and ability to respond to noxious stimuli. A-delta fibers, the larger (2–5 m) of the two, conduct impulses relatively rapidly (12–80 m/sec), are thinly myelinated, and are specialized to give several types of information. Some respond primarily to mechanical stimulation and probably relay information regarding the site of an injury, while others respond to thermal or chemical stimulation. Still others may be polymodal in responsivity so that they will respond to any type of stimulation (mechanical, thermal or chemical) but only after a certain threshold has been reached. Some A-delta fibers increase their discharge as stimulus intensity is increased so that the greater the stimulation, the more frequent the impulses are fired. Stimulation of A-delta fibers produces a sharp, pricking sensation.

The smaller (0.3–1.0 m) C pain fibers are unmyelinated, conduct impulses relatively slowly (0.4–1.0 m/sec), and respond to noxious stimulation. C fibers are almost all polymodal responders that are activated only by high threshold stimulation. C fiber stimulation is associated with a dull, diffuse, deeply perceived burning sensation.

The receptive field of each nociceptive neuron overlaps with those of other neurons so that every point on the skin surface, for example, lies within the receptive field of two to four neurons. At this peripheral level then, a sophisticated warning system reports that injury is occurring at a specific site via both A-delta and C fibers and provides information about the quality and intensity of the pain via A-delta fibers.

Evidence that both A-delta and C fibers contribute to the perception of pain comes from nerve block studies. Adults report two sensations in response to brief, intense stimuli. These include an early sharp and relatively brief pricking sensation (first pain) and a later dull, prolonged burning sensation (second pain). First pain disappears when myelinated fibers are selectively blocked by pressure, and the second pain is eliminated when unmyelinated fibers are selectively blocked with local anesthesia. It has been proposed that the first pain is elicited by A-delta fiber activity and the second by stimulation of C fibers so that each may contribute uniquely to the quality of the pain sensation.

It has been suggested, however, that this is an

oversimplified view. Most naturally occurring skin stimuli will activate a broad range of receptors, so that the size of the area stimulated, the frequency with which the stimulus is applied and the duration and location of the stimulus will influence the perceived sensation. In addition it is probably incorrect to assume that A-delta and C fibers are solely responsible for pain transmission. When both of these fibers are blocked, activation of A-beta fibers produces a sensation of light pressure or tickling but not specifically pain. Moreover, both A-delta and C fibers can be activated by non-noxious stimuli as well, suggesting that their function is not solely pain related. The possibility remains that there are subcategories of pain fibers, some activated by non-noxious stimuli, others by noxious stimuli, and still others by both noxious and non-noxious stimuli. One could hypothesize that each type would derive from a particular receptor type, but this has yet to be demonstrated.

2. Spinal Cord Projections

Once a pain message reaches the spinal cord, it ascends to the higher brain centers along the laminar organization of the gray matter of the spinal cord. This laminar organization is often used to describe the entry points of the pain fibers and is summarized in Fig. 2. The primary ascending nociceptor input is via A-delta and C fibers. The A-delta thermoreceptors terminate primarily in lamina I and the outermost part of lamina II. C fiber thermoreceptors and mechanoreceptors terminate mainly in lamina II. There are some non-nociceptive A-delta and C primary afferent fibers that also terminate in laminae I and II. The deep part of lamina II and laminae III and IV receive input only from non-nociceptive myelinated primary afferents. The larger diameter myelinated primary afferents (A-alpha and A-beta) fibers that enter the spinal cord in the medial division of the dorsal root terminate in lamina III or deeper. These are primarily low-threshold mechanoreceptors. [See SPINAL CORD.]

B. Descending Pain System

In addition to the ascending pathways, there are descending neural pathways which arise primarily in the midbrain and terminate in the dorsal horn, where they modulate the transmission and perception of pain signals. *Modulation* refers to the regulation of pain transmission neurons by means of selective inhibition at the level of the spinal cord. This descending pathway can be activated by stress or by certain analgesic drugs like morphine. When the pain modulation system is active, noxious stimuli produce less activity in the pain transmission pathway. This, in turn, reduces the perceived intensity of the sensation produced by a noxious stimulus.

The descending pathways are a major component of the brain's own system of analgesia. The close anatomic association of endogenous opioid peptides with the pain modulating networks has provided powerful support for the idea that these peptides actually contribute to analgesia under physiologic conditions. Analgesia can be most reliably activated by noxious stimulation. For example, noxious stimuli that are relatively intense and long lasting can reduce the responses of animals to subsequent noxious stimuli. Such analgesia can be reversed by naloxone, an opioid antagonist, or by lesions of the pain-modulating network.

Analgesia can also be produced by a variety of stressors that are not painful. These include restraint, hypoglycemia, and fear. This analgesic response can also be conditioned, as has been demonstrated in animals that were exposed to a pairing of a light or tone with a noxious stimulus. The precise source and identity of the opioid peptides involved in this analgesic response, however, are not currently known. Nor is it certain whether the anatomy of pain-modulating circuits is the same in humans as in animals. Despite the limitations of current data, it seems probable that this pain-modulating system contributes to the well-described differences in perceived pain among individuals.

III. Pain Theories

Several theories have been proposed to explain the neural mechanisms of pain. An intensity theory first suggested that pain was the result of overstimulation of sensory receptors in any modality. Then a specificity theory proposed that highly specialized modality-specific receptors were responsible for pain. These early theories were replaced by a gate control theory proposed by Melzack and Wall, a revised version of which is illustrated in Fig. 3. This version more completely accounts for both the excitatory–inhibitory and sensory–psychological influences associated with pain than did the previous theories.

According to the gate control theory, a neural mechanism, located in the substantia gelatinosa

| Primary Afferent | Terminal Field | Dorsal Horn Cells | Lamina |

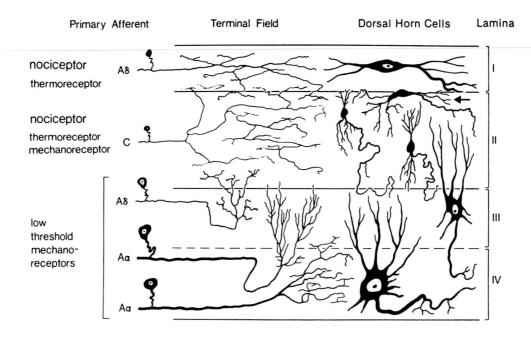

nociceptor Aδ
thermoreceptor

nociceptor C
thermoreceptor
mechanoreceptor

low
threshold
mechano-
receptors

Aδ

Aα

Aα

I

II

III

IV

layer in the dorsal horn, acts as a synaptic gate to increase or decrease the flow of nerve impulses from the receptors via the peripheral nerve fibers to the brain. It suggests that the perceived intensity of pain is the result of a balance of input from myelinated fibers (M), unmyelinated fibers (U), and inhibitory interneurons (I) on pain transmission (T) cells. The activity of the pain transmission cell usually results in the sensation of pain, while the spontaneously active inhibitory interneuron inhibits the same cell, thus reducing perceived pain intensity. The perception of pain also can be inhibited by the excitatory effect of myelinated afferents on the inhibitory interneuron. Unmyelinated nociceptors, in contrast, inhibit the inhibitory interneuron, thus secondarily exciting the pain transmission cell. The unmyelinated primary afferent thus has both direct and indirect excitatory effects on this cell. A balance between these various inputs to that cell produces intermediate levels of pain intensity. The table at the bottom of Fig. 3 shows how perceived pain (pain transmission cell output) is the result of this balance of input from myelinated and unmyelinated primary afferents.

Psychosocial influences play important roles in the gate control system. Cognitive factors and past experiences can dramatically influence reactions to pain. Among pain theories, the gate control theory is the most relevant for understanding the cognitive aspects of pain. Additionally, it provides a theoretical basis for the use of psychological techniques in

FIGURE 2 Summary of major components of the upper laminae of the spinal cord dorsal horn. The primary afferent nociceptor input is via A-delta and C fibers. A-delta nociceptors terminate primarily in laminae I and II. C fiber nociceptors terminate primarily in lamina II. Reproduced, with permission, from H. L. Fields (1987). ''Pain'' p. 49. McGraw-Hill, New York.

pain management as well as sensory, peripheral procedures such as transcutaneous electrical nerve stimulation.

IV. Psychology of Pain

Twentieth century medical science has emphasized the physiologically based sensory qualities of pain. Originally, however, pain was construed more in terms of inherent individual characteristics such as goodness or evil. Ancient civilizations, for example, believed that gods and demons caused pain to punish individuals lacking spiritual goodness. The ancient Greeks and Romans believed that pain was caused by the interaction of earth, fire, air and water with the individual's soul. Aristotle suggested that pain was an emotion, as pervasive as anger, joy or terror, an affect distinct from the classic five senses. This conceptualization would allow for pain to exist in the absence of any tissue injury. Later, Descartes described pain with an exclusive sensory-based model which did not incorporate emotional–psychological components. According to the

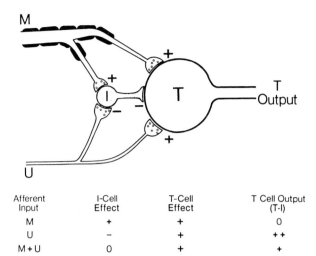

Afferent Input	I-Cell Effect	T-Cell Effect	T Cell Output (T-I)
M	+	+	0
U	−	+	+ +
M + U	0	+	+

FIGURE 3 Revised version of the gate control theory. *Top,* Four neural components of pain: the unmyelinated, nociceptive primary afferent (U), the myelinated non-nociceptive primary afferent (M), the transmission cell (T), whose activity usually results in the sensation of pain, and an inhibitory interneuron (I), which is spontaneously active and whose activity inhibits the T cell, thus reducing perceived pain intensity. *Bottom,* how perceived pain (T cell output) is the result of a balance of input from myelinated (M) and unmyelinated (U) primary afferents. Reproduced, with permission, from H. L. Fields (1987). ''Pain,'' p. 139. McGraw-Hill, New York.

IASP definition, contemporary thought holds that both affective and sensory qualities are components of the pain experience. In fact, painful experiences have been so intimately associated with affective distress that behavior which is lacking this emotional component may be dismissed as not being pain because it lacks face validity.

Pain is an intensely personal experience. The subjective nature of pain makes it difficult to quantify its emotional and psychological components. Recently, three distinct temporal forms of pain have been associated with certain affective states. *Phasic pain* is of short duration and occurs at the onset of injury. Behaviorally, it is characterized by efforts to withdraw from the source of injury accompanied by verbal and nonverbal expressions which are recognized as distress behaviors. *Acute pain* is provoked by tissue damage and includes both phasic pain and a tonic phase, which persists until healing takes place. Acute pain can be associated with fear and anxiety. There is controversy as to whether increased anxiety increases the reaction to painful stimuli. *Chronic pain* persists beyond the period of time required for healing and can be associated with emotional and behavioral disturbances such as depression, irritability, fear, hopelessness, and serious debilitation.

Because pain is a personal experience, there is wide variability in the psychological and emotional reactions to pain. It is clear that the pain experience is potentially affected by childhood experiences, ethnic and cultural variables, socioeconomic factors, genetic predisposition, birth order, gender, and a host of other physical, perceptual, cognitive, and emotional influences. For example, social expectations can strongly influence the way individuals perceive and respond to pain. Thus, wounds may be disregarded by soldiers on the battlefield, athletes on the playing field, and religious zealots during ceremonial rituals, while in other contexts the same tissue damage would be perceived as disruptive and painful. A study contrasting pain endured by wounded soldiers and civilians provided a major challenge to the specificity theory by demonstrating that the context or meaning of pain was a critical variable in the perception of pain. The influence of social expectations has been documented further with respect to cultural—ethnic and gender influences. There are distinct attitudes toward pain and the expression of responses to pain with respect to cultural heritage among Old American, Jewish, Italian, and Irish social groups. Similarly, men tolerated greater degrees of pain than women, differences attributed to gender-influenced social expectations. While most studies have found no difference in pain *threshold* response (the point at which a stimulus is reported to be painful) between men and women, pain *tolerance* (the point at which a stimulus is no longer tolerable) does appear to be influenced by gender-specific expectations. It is important to note that these social influences apply not only to the patient or recipient of pain, but to clinicians as well, whose own experiences and beliefs regarding pain can have significant diagnostic and therapeutic implications for the patient.

Individual personality styles have also been shown to influence pain perception. Much work has been done to identify the pain-prone patient, the individual who, because of personality, appears to be more vulnerable to pain. Other psychological factors such as anxiety and depression have been shown to heighten pain reactions. However, whether pain is ''real'' or ''imaginary,'' or ''physiological'' (organic) or ''psychological,'' to the patient pain is real, even if a doctor can find no physical reason for the pain.

VI. Developmental Aspects of Pain

It has generally been assumed that the ability to perceive and experience pain increases as a child develops; thus, pain is less commonly considered a problem in children than in adults. Traditionally, in fact, it was held that newborns were not even capable of experiencing pain. This belief was founded on several assumptions and the existing but limited knowledge about newborns.

Through the nineteenth century it was held that complete myelination of pain fibers was necessary for the transmission of pain impulses, contributing to the belief that the newborn was protected from pain. We now know that complete myelination is not essential to the transduction and transmission of pain signals. Indeed, the neuroanatomic basis of pain begins to develop very early in fetal life. As early as the 7th week of gestation, sensory receptors begin to develop, and they spread to all cutaneous and sensory surfaces by the 20th week of gestation. Cholinesterase staining studies have further revealed that by 24 weeks thalamocortical synapses between afferent nerves and cortical areas are established. Although these histologic studies have supported the concept that the necessary structural components for pain are present in the fetus, studies to determine their physiologic function have not yet been carried out.

A number of methodological problems is also responsible for the previous assumptions regarding pain in infants and young children. Because infants lack an ability to describe pain to adults, the precise degree to which the infant experiences pain is difficult, if not impossible, to assess. This is problematic and emphasizes the need for indices which are independent of verbal skills.

The absence of discernable responses to presumably painful events has also been interpreted as meaning that infants don't experience pain. Individual differences in response magnitude may reflect differences in developmental stage, clinical conditions, therapies employed for those clinical conditions, endogenous pain-protective mechanisms, or previous experience with pain. Alternatively, the absence of response may simply reflect the inadequacy or inappropriateness of the instrument of assessment.

Finally, the assumption that infants do not remember early postnatal experiences has contributed to the assumption that early pain is neither remembered nor capable of exerting lasting effects on child behavior or development. Although memory for pain has not been systematically studied, it has been shown that the newborn and the young infant can learn and demonstrate memories of sensory events for relatively long time spans. Important animal studies have further demonstrated that even the fetus can remember and respond differentially to sensory stimuli depending on prior experience. A better understanding of how and to what extent early aversive experiences are encoded and affect subsequent behavior is needed.

VII. Assessment of Pain in Children

A. Neonates

The accurate detection of acute pain is a prerequisite for providing maximally effective preventive and palliative efforts. In neonates this is especially challenging because of the absence of self-report. Yet the accurate measurement of pain will facilitate decisions regarding the utility and effectiveness of pain-relieving interventions. Further, pain assessment is essential to furthering our understanding of the nature of pain in infants and children and how it may change throughout the life span.

Scientific investigations of infant pain, however, are appropriately limited by ethical constraints. Infant pain is, therefore, experimentally evaluated predominantly in the context of clinically required or parentally requested invasive hospital procedures. Among healthy full-term infants, these include blood sampling (clinically required) and circumcision procedures (parentally requested). Among premature and sick infants, a broader range of invasive procedures includes blood sampling, lumbar puncture, tracheal intubation and aspiration, peripheral intravenous line placement, chest tube placement, umbilical artery catheterization, and arterial and venous cutdowns. Thus, infants' responses to stimuli that have been found to be unequivocally painful to adults are used as a means of studying pain in preverbal infants. Current assessment techniques are limited to evaluating how infants behave in response to pain (behavioral assessments) and how their bodies react to pain (physiological assessments).

Behavioral assessments include measures of facial expression, body movement, behavioral state, and crying. Darwin was among the first to report that facial expressions were linked with specific emotions and that these were widely recognized, both cross-culturally and across species. Recent research has confirmed that adults can indeed reliably identify emotional states, including pain, in infants only minutes of age using a system to code eye, eyebrow, nose, and mouth positions. Characteristics of the infant and contextual cues, however, contribute to large individual differences in the judgment of the intensity of pain from facial expressions. Facial expressions, therefore, are not standardized indices of infant pain. Shortly after birth infants also demonstrate avoidance of pain by body movements, once thought to be only reflexive and now believed to represent precise and active attempts to withdraw from painful stimuli. Changes in behavioral state after pain also suggest that infants are not merely responding reflexively to pain. Furthermore, striking individual differences in behavioral state after painful unanesthetized hospital procedures, such as circumcision, indicate that even shortly after birth infants have individual styles of coping with pain. Among neonatal responses to pain, however, the cry is the most commonly studied and widely accepted index of pain. Variations in the fundamental frequency (perceived as pitch), duration, harmonic structure and other acoustic characteristics are associated with increases in noxious stimulation. These acoustic variations are recognized by adults, confirming their communicative value.

Although behavioral responses to pain appear to have high face validity and are universally interpreted as meaningful, they are unsuitable as the sole criteria for neonatal pain indices. They are often not applicable to premature or sick infants who, due to their immaturity, clinical condition, or therapeutic program may not demonstrate behavioral responses to pain. Further, behavioral measures have not been standardized and can reflect more the subjective biases of the observer than the infant's pain experience. Physiologic measures, therefore, offer a potentially more objective, quantitative assessment of neonatal pain.

Physiologic measures of pain include heart rate, respiratory rate, transcutaneously measured circulating oxygen and carbon dioxide levels, blood pressure, intracranial pressure, and palmar sweating. Among healthy, full-term infants undergoing noxious stimulation, marked changes in a broad range of these physiologic parameters have been demonstrated, many of which are attenuated with anesthesia. Although many of these physiologic changes

are dramatic, they are often transitory, demonstrating prompt recovery in this population. Sharp physiologic changes have also been widely reported among premature and sick infants in response to noxious stimuli but, more importantly, these changes are not without potential adverse consequences. Bradycardia, cyanosis, hypoxia, increased requirements for oxygenation and ventilation and alterations in circulatory hemodynamics can occur in response to painful stimulation and contribute to the development of clinical conditions which are associated with significant neurodevelopmental morbidity and increased risk of death.

In addition to physiologic measures of pain, hormonal and neurochemical indices are believed to reflect painful events. Changes in serum cortisol, for example, have been documented after routine unanesthetized circumcision procedures in newborns. Endogenously produced opioids such as the endorphins extracted from cerebrospinal fluid also increase after painful disturbances in neonates. It is not clear, however, whether these increased endorphin levels reflect a distinct response or a more generalized response to illness or environmental disturbances.

Currently there is no single, reliable index with which to detect pain in newborns. Although physiologic measures are more readily quantified and less subjective than behavioral parameters, they require further clarification. Distinctions between responses to pain and responses to nonpainful stressors, and the relation between behavioral and physiologic responses need to be systematically investigated.

B. Children

Self-report measures, widely used with adults, may be utilized in children but with attention to several areas of concern. A major problem with children's self-report measures is the bias introduced by the specific demand characteristics of the situation. Children may describe their pain differently to their mothers than to a doctor, for example. Children also may lack the cognitive skills to understand questions regarding pain or to respond in an interpretable fashion to those questions. Finally the open-ended nature of self-report is not amenable to quantification and accurate measurement.

For very young children happy–sad children's faces scales have been shown to be easily understood and inexpensive and appear to have excellent measurement characteristics. Faces depicting different degrees of pain are presented to the child and the child chooses the one that most closely approximates his or her own pain. Children over 7 years of age can also use a "pain thermometer," in which intensity of pain is graphically and numerically represented as a higher degree on the visual thermometer, or other visual analogue scales. The Visual Analogue Scale (VAS) for example, usually consists of a 10 cm line labeled at the ends with "no pain" and "most intense pain imaginable" or similar descriptions. The child appraises his or her pain with respect to where it falls on the scale. The McGill Pain Questionnaire is a verbal assessment scale which is suitable for older adolescents who are able to understand adjectives and can choose which words best describe their pain. This tool is unique among pain rating scales in that it assesses both sensory and emotional responses to pain and, thus, offers some evaluation of the anxiety or apprehension that may accompany pain experiences. The validity and reliability of many of these assessment scales have been well-established. Primarily nonverbal methods have been used to measure the cognitive component of children's pain. These have included asking children to describe the color of their pain or to draw pictures of their pain. Unfortunately, there has been little or no validation of these projective techniques or systematic investigation of their utility in clinical settings.

VIII. Management of Pain in Infants and Children

The management of pain in children is problematic because of the difficulties associated with detecting pain in children, and because only limited research has been done in either area. Questions concerning the underutilization of medication for pain in infants and children have been raised. It is not surprising that newborns and young infants have not been aggressively treated for pain since it has traditionally been held that they were protected from pain owing to their developmental stage. There is clear evidence that infants and young children are undermedicated for pain, both within the hospital setting and by parents, when compared with similarly induced pain in adults. This approach is believed to stem largely from concern regarding the potential toxic or addictive effects of pain medications in children.

More recently (1987), the American Academy of Pediatrics issued a medical policy statement urging clinicians to utilize anesthesia and analgesia to protect infants and young children from unnecessary pain associated primarily with operating room procedures. The risk–benefit ratio of anesthesia in this population, however, has not yet been adequately investigated.

VIII. Testing Pain in Normal Adults

Studies of pain sensation in normal adults began with the assessment of pain thresholds and the goal of duplicating clinical pain sensation to objectively assess the efficacy of analgesia. Mechanical pressure, chemical stimuli, tourniquet pressure, cold, heat, and electrical stimuli were used in these early studies. The methodology and goals of these studies have now evolved to the use of sophisticated psychophysical procedures, behavioral assessments, and advanced physiological techniques to assess the mechanisms of both pain transmission and pain modulation. As part of this process, pain research investigators have attempted to establish standard criteria for experimental pain research stimuli. Gracely's criteria (1984 and 1985) include the following categories:

1. rapid onset;
2. rapid termination;
3. natural;
4. repeatable with minimal temporal effects;
5. objectivity: similar sensitivities in different individuals;
6. excite a restricted group of primary afferents.

It is clear that specific pain-production methods satisfy some but not all criteria. Thus, an ideal stimulus for one experiment may be inappropriate for others.

Similarly, criteria for pain measures have been advanced. Gracely's (1983) characterization of an ideal pain measure includes that it

1. be bias-free;
2. provide immediate information about accuracy and reliability;
3. separate sensory and emotional aspects;
4. assess experimental and clinical pain with the same scale to permit comparisons;
5. provide absolute assessments to allow for intergroup comparisons or comparisons over time.

A variety of verbal (for example, stimulus threshold procedures, visual analogue scales, magnitude estimation) and nonverbal (for example, behavioral, physiological) methods have been extensively used in experimental pain research. These studies have extended beyond demonstration of analgesic efficacy to studies of the mechanisms of pain and analgesia and the use of experimental methods to augment clinical pain report. These studies have also moved from the laboratory to the clinic, providing information about clinical pain syndromes and assisting in the measurement of clinical pain magnitude.

IX. Pain in Older Adults

The elderly represent a group of individuals, much like neonates, who are at risk for pain but who may not report pain. Because of their advancing age, the elderly have an increased incidence of physical illness and associated pain yet they report less pain than their younger counterparts. This phenomenon may be a function of the normal aging process, in which sensitivity to pain stimuli may be decreased. The additional contribution of various diseases including dementia to this failure to report pain has not been documented. However, minimal reporting of pain has been interpreted to mean that old people

do not feel pain as intensely as younger people, yet laboratory studies have shown no age change in pain sensation. These observations have obvious and important implications for both the diagnosis of illness and the clinical management of pain in the elderly. For example, because there is no standardized method of detecting pain or assessing the effectiveness of pain-relieving techniques in elderly individuals, they may be inappropriately medicated for pain. The elderly, therefore represent another group for whom an alternative index of pain is required especially in the context of clinically required medical or care-giving procedures which may induce pain.

Bibliography

Anand, K. J. S., and Hickey, P. R. (1987). Pain and its effects in the human neonate and fetus. *N. Eng. J. Med.* **317,** 1321.
Fields, H. L. (1987). "Pain." McGraw Hill, New York.
McGrath, P. J. and Unruh, A. M. (1987). "Pain in Children and Adolescents." Elsevier, Amsterdam.
Porter, F. (1989). Pain in the newborn. *In* "Clinics in Perinatology," (J. J. Volpe, (ed.) **16,** 549.
Wall, P. D., and Melzack, R. (1989). "Textbook of Pain." 2nd ed. Churchill Livingstone, Edinburgh.

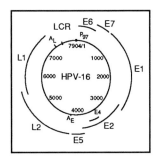

Papillomaviruses and Neoplastic Transformation

KARL MÜNGER AND BRUCE WERNESS, *National Cancer Institute*

LEX M. COWSERT, *ISIS Pharmaceuticals*

WILLIAM C. PHELPS, *Burroughs Wellcome Co.*

Glossary

Colposcope Magnification device used in the gynecological examination of the uterine cervix

Enhancer A DNA element that positively affects gene expression in a position- and orientation-independent manner

Episome Genetic element that exists in an extrachromosomal state

Koilocyte Cervical epithelial cell with marked cytoplasmic clearing; thought to represent a cytopathic effect of human papillomavirus infection

Oncogene Gene, the expression of which leads to cellular transformation in cultures or tumor formation in an animal

Open reading frame Portion of a genome with uninterrupted amino acid coding potential

Tumor suppressor gene Gene, the expression of which is vital for normal cell growth and differentiation. Failure of a cell to produce the functionally active gene product leads to cellular transformation. Such genes are also referred to as "antioncogenes," or "recessive oncogenes."

THE HUMAN PAPILLOMAVIRUSES are a class of small DNA tumor viruses clinically associated with a number of benign and malignant epithelial lesions. These viruses have not yet been propagated in a tissue culture system, and consequently most of the present knowledge on their biology has been obtained from studies with molecularly cloned viral genomes. A number of *in vitro* systems have been developed which allow the genetic dissection of viral functions involved in cellular transformation. A variety of studies have indicated that the human papillomaviruses associated with cervical carcinoma encode two oncogenes: E6 and E7.

The E7 gene encodes a multifunctional protein which is structurally and functionally related to that of another oncogene, the adenovirus E1A proteins. The E7 protein can activate expression of the adenovirus E2 gene by acting on its promoter and is able to cooperate with the *ras* oncogene to fully transform primary baby rat cells. Like the adenovirus E1A proteins, E7 is able to form a stable complex with the retinoblastoma tumor suppressor gene product p105-RB. The E7 gene is sufficient for transformation of established rodent fibroblast cell lines (e.g., NIH3T3 cells).

Studies with human skin epithelial cells (keratinocytes) have shown that the E6 gene is also necessary for transformation, in that combined expression of both the E6 and E7 genes is required to induce keratinocyte immortalization and resistance to signals of terminal differentiation. In a tissue culture system for human keratinocytes which allows the formation of an epidermislike structure *in vitro*, the combined expression of E6 and E7 is sufficient to induce histological abnormalities similar to those observed in severe cervical dysplasia, a precursor of cervical carcinoma.

I. Molecular Characterization and Natural Occurrence of Papillomaviruses

The human papillomavirus (HPV) genome consists of a double-stranded, covalently closed, circular DNA molecule of approximately 8000 base pairs. A schematic physical map of HPV-16 is presented in Fig. 1. All of the viral mRNAs are transcribed from one DNA strand of the papillomavirus genome. The coding strand contains approximately 10 major open reading frames (ORFs). These have been classified as either "early" (E) or "late" (L), based on their position in the papillomavirus genome and their expression pattern in nonproductively versus productively infected cells (Table I). The structural organization of the genome is well conserved among the different papillomaviruses that have been sequenced. A region about 1000 base pairs in length situated between nucleotides 7100 and 100 with no extensive coding potential was originally referred to as the noncoding region. More recently, the term "upstream regulatory region" and "long control region" (LCR) are used, since this region of the genome contains multiple *cis* elements that are critical for the regulation of viral transcription and viral replication.

Papillomavirus genomes are packaged in an icosahedral capsid 55 nm in diameter. They are widespread in nature and have been isolated from a wide range of higher vertebrates, including humans. In all of these species, they generally induce warts and benign papillomas and/or fibroepithelial tumors (i.e., fibropapillomas).

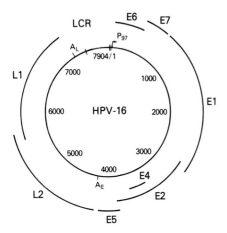

FIGURE 1 The circular genome of HPV-16. The long control region (LCR) contains the only mapped promoter (P_{97}) of HPV-16. E, L, Early and late open reading frames, respectively. A_E, A_L, early and late polyadenylation sites, respectively.

TABLE I Papillomavirus Open Reading Frames and Their Assigned Functions

ORF	Assigned function
E1	Replication (BPV-1)
E2	
Full length	Transcriptional *trans*-activation, DNA binding (BPV-1, HPV-6, and HPV-16); transcriptional repression, DNA binding (HPV-16 and HPV-18)
3' Portion	Transcriptional repression, DNA binding (BPV-1)
E4	Cytoplasmic phosphoprotein in warts (HPV-1)
E5	Transformation, stimulation of DNA synthesis (BPV-1)
E6	Transformation (BPV-1 and HPV-16)
E7	Plasmid copy number control (BPV-1); transformation, transcriptional *trans*-activation (HPV-16 and HPV-18); p105-RB binding (HPV-6, -11, -16, and -18)
L1	Minor capsid protein (BPV-1)
L2	Major capsid protein (HPV-1)

From studies with the cottontail rabbit papillomavirus, it became clear that in the presence of certain chemical compounds (cocarcinogens), some papillomavirus-induced lesions had the potential to progress to malignant carcinomas. This is one of the few animal systems that permit study of the progression of experimentally induced tumors in the natural host organism.

The papillomaviruses are highly host specific and exhibit a high degree of tissue specificity. Regulation of papillomavirus gene expression is thought to be intimately linked to the differentiation program of the epithelial host cells. Differentiation of keratinizing epithelium is a dynamic process in which individual cells move from the most internal layer of the epithelium (the basal layer) to the most superficial layer (the corneum). Based on cellular morphology and biochemical differentiation markers, normally stratified keratinizing epithelium can be divided into four distinct layers: basal, spinous, granular, and cornified.

The basal layer is a single sheet of undifferentiated small cells of uniform morphology. In normal epithelium dividing cells are confined to the basal cell layer. As cells migrate from the basal layer to the cornified layer, they undergo a program of differentiation, resulting in changes in morphology and in the expression pattern of a number of epithelium-specific proteins (e.g., the keratins).

One common feature of papillomavirus infections is the perturbation of the normal differentiation program. In an infected epithelium there is delayed differentiation and disorganization of the granular layer. Mitotically active cells can be observed in suprabasal layers, whereas in normal squamous epithelial cells they are limited to the basal layer. As cells move toward the more superficial layers, the cytoplasm of the infected cells becomes progressively clearer and the cells become larger and more irregularly shaped. It is thought that the virus particles enter through an abrasion in the epithelium and infect the basal cells, where the papillomavirus DNA is maintained at a low copy number as a stable nuclear episome. The full vegetative functions required for the replication and propagation of the papillomaviruses, including the expression of the late structural proteins of the virus particles, are only expressed in the terminally differentiated outermost layers of the epithelium.

So far, the papillomaviruses have not been successfully propagated in the laboratory, and, as a consequence, molecular studies with these viruses have been severely hampered. Only with the introduction of recombinant DNA technology in the late 1970s have detailed genetic studies with these viruses become possible. [See RECOMBINANT DNA TECHNOLOGY IN DISEASE DIAGNOSIS.]

Most of our current knowledge of the molecular biology and genetics of the papillomaviruses has been obtained from studies with bovine papillomavirus type 1 (BPV-1). BPV-1 infects cattle and causes benign fibropapillomas. The cloned genome of BPV-1 DNA readily transforms certain established rodent fibroblast cells *in vitro*. In these transformed cell lines the majority of the BPV-1 DNA is maintained as an autonomously replicating multicopy plasmid with an average copy number of 10–200 per cell nucleus. The viral functions involved in stable plasmid replication in transformed cells could be manifestations of the normal biology of the virus, and therefore, they could represent a model of the latent nonproductive stage of a papillomavirus infection.

II. Clinical Manifestations of Human Papillomavirus Infections

A. General

Warts have been medically recognized since 25 A.D. when Roman encyclopedist Celsus coined the terms "myrmecia," "ficus," and "thymion" to describe common warts. Because most warts spontaneously

regress, early concepts regarding the origin and treatment of warts have been diverse and colorful. That such infections were viral was inferred from experiments by Italian researcher Ciuffo which described the transmission of common human warts by cell-free sterile filtrates of homogenized warts.

In the 1960s virus particles were isolated from such lesions and studied by electron microscopy. It was at first believed that only a single HPV type was responsible for the whole spectrum of warts in humans and that the observed histopathological differences among these lesions were related to the nature of the particular type of epithelium at the site of infection. Only with the application of molecular biological techniques did the heterogeneity of the many HPV types become apparent. More than 60 different HPV types have now been isolated from a wide variety of clinical lesions. The criterion for distinguishing the HPV types is based on DNA hybridization kinetics, since serological reagents are still not widely available to distinguish among them.

B. Benign Skin Warts

Most warts of the skin typically contain HPV types 1, 2, 3, 4, or 10. Common skin warts are most often found in children and young adults. Later in life the incidence of these lesions decreases, presumably due to immunological and physiological changes. Given the ubiquity of skin warts, the reported incidence of malignant progression is remarkably small. Moreover, the sites at which these warts are commonly found (i.e., hands or feet) are often exposed to mechanical trauma and possibly to cocarcinogenic compounds. It therefore seems likely that the potential of malignant progression of an HPV-associated lesion must at least be partially determined by the HPV type.

C. Epidermodysplasia Verruciformis (EV) and Skin Canceer

EV is a rare genetically transmitted disease which is clinically characterized by chronic skin lesions that appear as reddish macules. A variety of HPV types have been associated with EV. With time approximately one-third of these patients develop multiple skin carcinomas; however, only a subset of the EV-associated HPVs, particularly HPV-5 and HPV-8, have been consistently detected in such carcinomas. In general, the carcinomas arise in sun-exposed areas of the skin and thus it is suspected that ultraviolet irradiation could play a cocarcinogenic role with the specific HPV types in their development. [*See* SKIN, EFFECTS OF ULTRAVIOLET RADIATION.]

D. Laryngeal Papillomas

Laryngeal papillomas are benign epithelial tumors of the larynx. A viral etiology for these lesions was suggested by early transmission studies and was confirmed by the observation of virus particles by electron microscopy and more recently by the detection of viral antigens and DNA. Molecular biological studies have demonstrated that most of these lesions contain either HPV-6 or HPV-11. These two HPV types are generally associated with genital condyloma acuminatum (see Section II,E). Spontaneous malignant conversion of laryngeal papillomas is rare; however, they are persistent and tend to expand. It has been observed that therapeutic X-irradiation of juvenile laryngeal papillomas led to an increased conversion to carcinomas 5–40 years after treatment.

E. Genital Papillomas and Genital Cancer

Genital warts are venereally transmitted. Clinically, they can be categorized into two major groups: condyloma acuminatum and flat genital warts.

Molecular studies have demonstrated that greater than 90% of condylomas contain either HPV-6 or HPV-11 DNA. They generally occur on the penis, on the vulva, or in the perianal region. They can spontaneously regress or persist for years. Progression to an invasive carcinoma occurs only at a low frequency.

Unlike other genital warts, those occurring on the uterine cervix usually exhibit a flat, rather than acuminate, morphology; they are usually recognizable only after the application of dilute acetic acid and with the aid of a colposcope. A papillomavirus etiology for cervical dysplasia was suggested in the late 1970s by cytologists who recognized that the cytological changes characteristic of cervical dysplasia observed on a Pap smear were due to HPV infections. This association was important, because previous clinical studies had established that cervical dysplasia [also referred to as cervical intraepithelial neoplasia (CIN)] was a precursor to carcinoma *in situ*, which, in turn, could give rise to invasive squamous epithelial cell carcinoma. HPV types 16 and 18 were cloned directly from cervical carcinoma.

DNA hybridization studies show that more than 70% of the human cervical carcinomas and the de-

rived cell lines harbor either of these HPV types. Another 20% contain additional HPV types (e.g., HPV-31, -33, and -35). The koilocytic cells found in Pap smears and tissue sections, often in association with moderate cervical dysplasia, are now recognized to be manifestations of an HPV infection. Cervical lesions classified by histological criteria as mild to moderate CINs spontaneously regress in about 40% of all cases, while progression to carcinoma *in situ* or invasive cervical carcinoma occurs in a low percentage of cases. Possible cocarcinogens important for progression of these lesions to malignancy may include cigarette smoke.

III. *In Vitro* Transformation Assays for Human Papillomaviruses

A. Transformation of Established Rodent Fibroblasts

Only a limited number of papillomaviruses readily transform established rodent cell lines *in vitro*. In the case of BPV-1, this property has been used to define the viral genes involved in cellular transformation. Similar analyses with the HPVs have been hampered by the inability of many HPV types to induce clear morphological transformation of established rodent cells. Cell transformation was obtained with HPV-16 using a recombinant plasmid encoding the full-length genome of HPV-16, which was transfected into mouse fibroblasts. After 4–6 weeks cell clones exhibiting a transformed morphology were obtained. These are tumorigenic when injected into nude mice. Such *in vitro* transformation assays have permitted a genetic dissection of the transforming potential of HPV-16 and HPV-18, as discussed in Section IV. It is interesting that the genital HPV types clinically associated with a low potential of malignant progression (i.e., HPV-6 and HPV-11; see Section II) also fail to transform these cells *in vitro*.

B. Transformation of Primary Cells

1. Transformation of Primary Rodent Cells

Although HPV-16 and HPV-18 are unable to transform primary rodent cells, they both encode a function that allows the indefinite growth (i.e., immortalization) of these cells. As with other immortalizing oncogenes, they can cooperate with an activated *ras* oncogene to transform primary rat cells.

Only those HPV types that are commonly detected in carcinomas (i.e., HPV-16, -18, -31, and -33) scored positive in this assay, but not those associated with benign lesions (HPV-6 and HPV-11).

2. Transformation of Primary Human Epithelial Cells

Potentially more relevant to human carcinogenesis is the observation that HPV-16 and HPV-18 can immortalize primary human fibroblasts and skin epithelial cells (i.e., keratinocytes). The keratinocyte cell lines derived from these experiments lack the characteristics of transformed cells, but they exhibit altered growth properties and are resistant to terminal differentiation signals (e.g., Ca^{2+} and serum). [See KERATINOCYTE TRANSFORMATION, HUMAN.]

3. Organotypic Culture Systems

Organotypic cultures are obtained by growing primary human keratinocytes on a collagen matrix containing mouse fibroblasts as feeder cells to provide required growth and differentiation factors. The cells are grown as submerged cultures until they form a nearly confluent layer on the collagen matrix and are then lifted and grown at the liquid–air interface, where they undergo terminal differentiation to fully differentiated keratinizing epithelium, made up of several layers imitating regular epidermis (Color Plate 13, top panel).

Primary human foreskin keratinocytes transfected with cloned HPV-16 DNA and grown in organotypic culture show many of the histological features characteristic of naturally occurring HPV lesions, including overcrowding in the parabasal layer as well as some degree of disorganization and vacuolization of the cells in the more superficial layers. The histological appearance is similar to a low-grade CIN. If the transfected keratinocytes are passaged several times and then subjected to organotypic culture, they develop more pronounced histological abnormalities, similar to these of a high-grade CIN (Color Plate 13, bottom panel).

IV. Oncogenic Functions Encoded by HPV-16 and HPV-18

A. Integration and Transcription of HPV-16 in Cervical Carcinoma Cells

In most HPV-associated cervical carcinomas and derived cell lines the majority of the viral DNA is stably inserted (i.e., integrated) in the host DNA.

This is in contrast to precancerous lesions, in which most of the HPV DNA is maintained as autonomously replicating plasmid DNA. The site of integration of the viral genome seems random with respect to the host genome, although integration has been noted in some cases in proximity to cellular protooncogenes. Integration seems to be far more specific with respect to the viral genome. For integration the viral DNA ring must be opened, and this frequently occurs in the E1 or E2 ORFs. Integration is often also accompanied by deletions of the viral genome.

The viral transcripts of several HPV-16 or HPV-18-containing cell lines have been described in some detail. Three major viral mRNA species were detected, all derived from a single promoter located in the viral LCR just upstream from the E6 ORF. These mRNAs include several genes and encode either a full-length E6 gene product or a truncated version (due to internal splicing), termed E6*, as well as the full-length E7 protein. The E6–E7 regions of both HPV-16 and HPV-18 have been implicated in cellular transformation in all of the *in vitro* systems discussed.

As mentioned earlier, the rest of the viral genome is often deleted or disrupted in cervical cancer cells. Of particular interest is the consistent deletion or disruption of the E2 ORF. The HPV-16 and HPV-18 E2 ORF's encode *trans*-acting factors which modulate viral transcription through a conditional enhancer located in the LCR. In HPV-16 and HPV-18 the predominant activity of E2 seems to be to repress transcription. Disruption or deletion of E2 caused by the integration of HPV DNA might, therefore, have profound consequences on the viral transcriptional regulatory circuits. One of the consequences of HPV DNA integration could be a high-level expression of the E6 and E7 genes, which might ultimately result in cellular transformation.

Other ORFs for which little or no mRNA is detected in cervical carcinoma cell lines include E4 and E5. The HPV-16 or HPV-18 E4 ORF has no known function. In BPV-1 the E5 protein encodes a 44-amino-acid oncoprotein which is localized in membranes. The HPV-16 and HPV-18 E5 ORFs may potentially encode similar polypeptides.

B. E6 Protein

Low levels of the full-length E6 proteins have been detected in cervical carcinoma cell lines and in *in vitro* HPV-transformed cell lines by immunoprecipitation. The protein has an apparent molecular weight of 18,000 and is rather basic and cysteine rich. In BPV-1 the E6 protein which is localized in the nucleus and in membranes cooperates with the E5 oncoprotein in the transformation of established rodent cells *in vitro*. It seems to be mainly involved in the induction of growth in semisolid medium and tumorigenicity in nude mice.

The HPV-16 E6 gene could have a similar role. It is required for the full transformation of mouse fibroblasts. In primary human keratinocytes it induces a transiently differentiation-resistant phenotype, and it induces perturbed differentiation in organotypic cultures. The combined expression of E6 and E7 is required for the stable transformation of human epithelial or fibroblastic cells and the induction of the full dysplastic phenotype in organotypic cultures. Comparison of the E6 ORFs of HPV types 6, 11, 16, and 18 has revealed that only HPV-16 and HPV-18, which are associated with a "high" risk for malignant progression, generate E6*, but not HPV-6 and HPV-11, which are associated with a "low" risk. Whether or not E6* polypeptides are synthesized in transformed cells is unknown. It has been hypothesized, however, that only mRNA species containing a spliced E6 ORF would allow for efficient translation of the E7 protein. It is interesting that in all of the *in vitro* transformation assays described earlier, only E6, not E6*, contributes to transformation.

The full-length E6 protein contains four copies of a Cys–X–X–Cys motif. These sequences have been implicated in the binding of Zn^{2+} by the E6 protein and are likely to be important for the structural and biological properties of E6. A number of proteins containing Cys–X–X–Cys sequences are known to be DNA-binding proteins, and their complexes with Zn^{2+} (often referred to as "zinc fingers") might be critical for sequence-specific recognition and binding of the target DNA. E6 polypeptides expressed in bacteria have been reported to exhibit nonspecific DNA binding activity. An E6-dependent enhancer has been mapped in the HPV-18 LCR. The mechanism of E6-mediated transcriptional transactivation is currently not known.

C. E7 Protein

The HPV-16 E7 protein has been immunoprecipitated from cervical carcinoma as a 21,000-molecular weight phosphoprotein. It is more abundant than the E6 protein and is localized to the nucleus by immunofluorescence.

Genetic analyses have demonstrated that the HPV-16 and HPV-18 E7 genes encode the major transforming functions in established rodent fibroblasts. The E7 gene is both necessary and sufficient for the morphological transformation (i.e., "focus" formation) in these cell lines.

The E7 gene is also required for the morphological transformation of primary human cells by HPV-16. This gene alone induces partial differentiation resistance; for the induction of the stably transformed phenotype, however, the expression of the full-length E6 gene is also required.

Similar results were obtained in organotypic culture, in which E7 alone, like E6, induces a perturbed differentiation. In contrast, as mentioned in the previous section, combined high-level expression of E6 and E7 is sufficient to induce the histological changes typical of high-grade CIN lesions (Color Plate 13).

The HPV-16 E7 gene encodes functions similar to those of the adenovirus (Ad) E1A proteins both in the activation of distant genes (i.e., *trans*-activation) and in transformation. Thus, E7 *trans*-activates the Ad E2 promoter and cooperates with an activated *ras* oncogene to fully transform primary baby rat kidney cells. The molecular mechanism of Ad E2 *trans*-activation might be similar, since the two proteins require similar target sequences on the Ad E2 promoter. The HPV-16 E7 protein has a single region between residues 2 and 37 with a striking sequence similarity to portions of the Ad E1A protein (Fig. 2) and with sequences of the large tumor antigens of many small DNA tumor viruses, including simian virus 40 (SV40). This part of the E7 sequence is a fairly well-conserved region in other human papillomaviruses, but not in the BPV-1 E7 gene, which has no Ad E1A-like biological activities.

The carboxy termini of all E7 proteins contain two copies of a Cys–X–X–Cys sequence motif, also present in E6, which have been implicated in the binding of Zn^{2+}. Based on this similarity, it had been proposed that E6 and E7 could have derived from a common ancestor (Fig. 2).

Mutations in the HPV-16 E7 region of similarity to the Ad E1A proteins or in the Cys–X–X–Cys sequence motifs in the carboxy-terminal part of the E7 polypeptide interfere with both cellular transformation and transcriptional *trans*-activation.

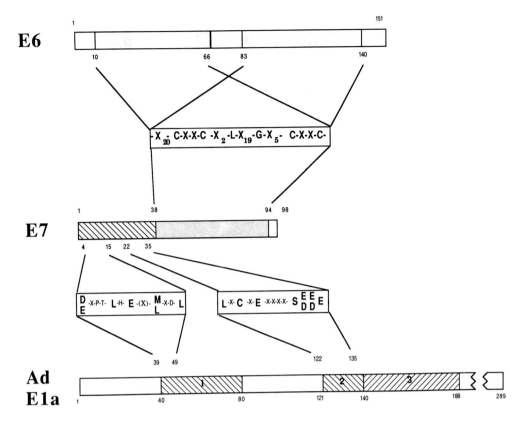

FIGURE 2 Structural similarities of the HPV-16 E6 and E7 genes. The amino-terminal 37 amino acid residues are strikingly similar to portions of conserved domains 1 and 2 of the adenovirus (Ad) E1A protein. In HPV-16 and Ad E1A these sequences are required for cellular transformation properties and contain the sequences necessary for interaction with the retinoblastoma tumor suppressor gene product p105-RB. The carboxy-terminal part of E7 contains two copies of a Cys–X–X–Cys sequence motif which is also conserved in E6. These sequences have been implicated in the binding of Zn^{2+}. Based on these structural similarities, it has been hypothesized that the E6 and E7 genes might have evolved from a common ancestor. The standard single-letter abbreviations for amino acid residues have been used (C, Cys; D, Asp; E, Glu; G, Gly; L, Leu; M, Met; P, Pro; T, Thr; X, any amino acid residue).

V. Interaction of Viral Gene Products with Cellular Proteins

A. Complex of E7 Protein with the Retinoblastoma Tumor Suppressor Gene Product

The protein encoded by the retinoblastoma tumor suppressor gene (*Rb*), p105-RB, can form stable complexes with the proteins of several oncogenes encoded by DNA tumor viruses, including the Ad E1A proteins, the SV40 large tumor antigen, and HPV E7 proteins. The *Rb* locus has been classified as a tumor suppressor gene because its inactivation predisposes to the development of various tumors, most importantly retinoblastoma and osteosarcomas. The exact function(s) of the *Rb* gene is unknown. Association of p105-RB with viral oncoproteins might inactivate it with consequences similar to mutations in *Rb*. The formation of an HPV-16 E7–p105-RB complex can be detected *in vitro* by mixing the HPV E7 proteins with cell extracts containing p105-RB. Immunoprecipitation using an antiserum specific for p105-RB precipitates both this protein and E7. [*See* RETINOBLASTOMA MOLECULAR GENETICS; TUMOR SUPPRESSOR GENES.]

Using such an assay, it was shown that the E7 proteins of HPV-6, -11, -16, and -18 form stable complexes with p105-RB, although the HPV-6 and HPV-11 E7 proteins bind with decreased affinity. The BPV-1 E7 protein did not form a detectable complex. A complex between the HPV-16 E7 protein and p105-RB was also detected *in vivo* in an HPV-16-transformed human keratinocyte cell line.

These data show that the ability of the E7 protein of a given HPV type to form a stable complex with p105-RB *per se* does not correlate with the clinical association of low or high incidence of cancerous progression. Mutational analysis revealed that the sequences on the HPV-16 E7 protein that are necessary for interaction with p105-RB map to only a small region of the sequence similarity between E7, Ad E1A, and SV40 large tumor antigen. Mutations in the HPV-16 E7 protein that abrogate complex formation with p105-RB *in vitro* are also transformation defective; other transformation-defective mutants, however, still associate with p105-RB efficiently. Complex formation with the retinoblastoma tumor suppressor gene product, therefore, seems to be necessary, but not sufficient, for cellular transformation.

VI. Perspectives

The viral and cellular mechanisms by which certain HPV types transform cells *in vitro* could provide insight into the role of these viruses in the multistep process of carcinogenesis. Improved methods to detect HPV-specific DNAs and mRNAs in cells, such as the polymerase chain reaction technique, could be valuable analytical tools for detecting HPV DNAs at an early stage of a viral infection, enabling researchers to study in more detail the various steps in the progression of HPV-induced lesions.

Although the HPV-16 E6/E7 region is sufficient to induce the full dysplastic phenotype characteristic of a high-grade CIN in the organotypic *in vitro* transformation system, additional cellular genetic events are probably necessary for the progression of a benign dysplasia to cancer, because it is observed in only a minority of cases. Studies of somatic cell hybrids formed by the fusion of cervical carcinoma cells with normal human cells implicate mutations in a tumor suppressor gene located on chromosome 11 (therefore, different from the *Rb* locus located on chromosome 13). Moreover, transfer of a normal chromosome 11 into cervical carcinoma cell lines reverses the tumorigenic phenotype. Some studies suggest that this gene might encode a factor(s) directly involved in the transcriptional regulation of HPV gene expression.

As discussed above, formation of a complex of the E7 protein with p105-RB per se might not be sufficient for malignancy. It could, however, be an important step in the induction of cellular prolifera-

tion and/or altered differentiation, which are observed in the majority of the papillomavirus-induced lesions. The amino acid sequences of the HPV-16 E7 protein that are necessary for complex formation with p105-RB are known, and, consequently, it might be possible to develop drugs which specifically interfere with this interaction. It will also be important to investigate whether additional cellular proteins are complexed by the E7 proteins, as found for Ad E1A and the SV40 large tumor antigen. Some of these proteins bind to regions important for cellular transformation, which are conserved in the HPV E7 proteins.

It will also be interesting to find the cellular target(s) for E6, the second oncoprotein encoded by HPV-16. In primary human epithelial cells the biological activities of the E6 and E7 proteins are somewhat overlapping and at the same time complementing. As noted earlier, the E6 and E7 proteins also share some structural features, particularly the Cys–X–X–Cys sequence motifs, and consequently the ability to bind Zn^{2+}. The E6 proteins share no significant amino acid sequence similarity with the Ad E1A proteins and do not form stable complexes with p105-RB. The E6 proteins of HPV-16 and HPV-18, however, can form specific complexes *in vitro* with p53, another tumor suppressor gene product.

Bibliography

deVilliers, E.-M. (1989). Heterogeneity of the human papillomavirus group. *J. Virol.* **63,** 4898–4903.

Dyson, N., Howley, P. M., Münger, K., and Harlow, E. (1989). The human papillomavirus-16 E7 oncoprotein is able to bind to the retinoblastoma gene product. *Science* **243,** 934–937.

Lambert, P. F., Baker, C. C., and Howley, P. M. (1988). The genetics of bovine papillomavirus type 1. *Annu. Rev. Genet.* **22,** 235–258.

Matlashewski, G., Schneider, J., Banks, L., Jones, N., Murray, A., and Crawford, L. (1987). Human papillomavirus type 16 DNA cooperates with activated ras in transforming primary cells. *EMBO J.* **6,** 1741–1746.

McCance, D. J., Kopan, R., Fuchs, E., and Laimins, L. A. (1988). Human papillomavirus type 16 alters human epithelial cell differentiation *in vitro. Proc. Natl. Acad. Sci. U.S.A.* **85,** 7169–7173.

Münger, K., Phelps, W. C., Bubb, V., Howley, P. M., and Schlegel, R. (1989). The E6 and E7 genes of the human papillomavirus type 16 together are necessary and sufficient for transformation of primary human keratinocytes. *J. Virol.* **63,** 4417–4421.

Münger, K., Phelps, W. C., and Howley, P. M. (1990). Human papillomaviruses and neoplastic transformation. In: "Bristol-Myers Cancer Symposia, Vol. **11,**" (R. K. Boutwell and I. L. Riegel, eds) Academic Press, San Diego, 1989, pp. 239–254.

Phelps, W. C., Yee, C. L., Münger, K., and Howley, P. M. (1988). The human papillomavirus type 16 E7 gene encodes transactivation and transformation functions similar to adenovirus E1a. *Cell (Cambridge, Mass.)* **53,** 539–547.

Pirisi, L., Yasumoto, S., Feller, M., Doninger, J., and DiPaolo, J. A. (1987). Transformation of human fibroblasts and keratinocytes with human papillomavirus type 16 DNA. *J. Virol.* **61,** 1061–1066.

Schlegel, R., Phelps, W. C., Zhang, Y.-L., and Barbosa, M. (1988). Quantitative keratinocyte assay detects two biological activities of human papillomavirus DNA and identifies viral types associated with cervical carcinoma. *EMBO J.* **7,** 3181–3187.

Spalholz, B. A., and Howley, P. M. (1989). Papillomavirus–host cell interactions. *In* "Advances in Viral Oncology" (G. Klein, ed.), pp. 27–53. Raven, New York.

Werness, B. A., Levine, A. J., and Howley, P. M. (1990). Association of Human Papillomavirus Types 16 and 18 E6 Proteins with p53. *Science* **248,** 76–79.

Yasumoto, S., Burkhard, A. L., Doninger, J., and DiPaolo, J. (1986). Human papillomavirus type 16-induced malignant transformation of NIH 3T3 cells. *J. Virol.* **57,** 572–577.

Parasitology, Molecular

ALBERT E. BIANCO, *Imperial College of Science, Technology, and Medicine, University of London*

Glossary

Antigenic variation Process of repeated, spontaneous switching in expression of surface antigens in certain protozoan parasites that permits a proportion of the population to evade the host immune response

Fatty acid remodeling Sequence of fatty acid replacements involved in the biosynthesis of a glycosyl phosphatidylinositol membrane anchor for the variant surface glycoprotein of African trypanosomes

Immune evasion Avoidance by parasites of host immune mechanisms, enabling them to establish recurring or persistent infections

Innate host resistance Resistance to parasites mediated by nonimmune factors

Merozoite Daughter cells liberated from schizonts that have the capacity to disperse and invade unparasitized host cells

Micronemes Electron dense, membrane bound vesicles distributed in the cytoplasm in the vicinity of the rhoptries in Apicomplexan protozoa

Parasite Organism living in or on another living organism (or host), from which is obtained most or all essential nutrients, and to which is imparted a degree of damage

Promastigote Flagellated form of certain trypanosomatid protozoa (e.g., *Leishmania*) that occurs in insects and is responsible for transmission of infection

Rhoptries Paired, membrane limited apical organelles found in motile stages of Apicomplexan protozoa ascribed a role in host cell invasion

RNA editing Process involving insertion or deletion of uridine bases to mRNA first identified in mitochondrial transcripts of trypanosomatids

Schizont Stage in the life-cycle of protozoan parasites belonging to the phylum Apicomplexa (e.g., malaria) characterized by the multiplication of organisms through asexual reproduction within a single host cell

MOLECULAR PARISITOLOGY is a broad subject concerned with diverse aspects of parasite biology and the host–parasite relationship. It embraces the study of parasite biosynthesis and biochemistry, molecular genetics, developmental biology, and the interaction of parasites with the host immune response. Included in the term parasite is a diverse array of organisms linked only by the manner in which they associate with another living organism, or host: a relationship broadly defined as one in which the parasite lives in or on the host organism, obtaining essential nutrients and imparting a degree of damage to the host. By convention, viruses, rickettsiae, and bacteria do not qualify as parasites, although it is arguable whether arbitrary exclusion of all of the prokaryotes can be justified. A major consequence of dependency on host metabolism for parasites has been the emergence of degeneracy in the biochemical pathways they can support. On the other hand, their unusual lifestyles have in many instances led to the evolution of elegant and unique biological adaptations. The examples used here to illustrate various aspects of molecular parasitology are drawn from the most familiar protozoan and helminthic parasites of medical importance, in keeping with the theme of an Encyclopedia of Human Biology.

I. Parasite Invasion of Host Cells or Tissues

A. Background

All endoparasites need to gain entry and move to a suitable location within their hosts. This may be achieved passively, as in colonization of the gut lumen following ingestion (e.g., noninvasive *Entamoeba*), or actively, as in migration into blood vessels after penetrating the skin (e.g., larval schistosomes). Commonly, both active and passive elements are involved, as when parasites are transmitted by a hematophagous arthropod and subsequently migrate from the feeding lesion to a location elsewhere in the body (e.g., Plasmodia, filariae). With few exceptions, there is great specificity for the particular cells or tissues that parasites reside in; associated with this has been the evolution of mechanisms that regulate homing behavior (taxes and kineses), host cell recognition, and the process used for invasion. Even parasites that gain entry by simple ingestion and remain in the lumen of the gut may interact with the epithelial lining, and in doing so have evolved elegant mechanisms of adhesion (e.g., *Giardia*) or tissue attachment (e.g., *Trichuris*).

B. Entry into Host Cells

Many protozoan parasites occupy intracellular niches, but few helminths have been able to do this, presumably on account of their size. Exceptions are the infective larvae of *Trichinella* and the early instars of filarial nematodes in arthropods, these being especially small metazoan parasites and occupants of large cells such as those that occur in muscle and invertebrate fat body.

Among the intracellular protozoa, cell–cell interactions are central to the mechanisms of invasion of host cells. This is achieved by one of two principal means, either as an invagination of the host-cell membrane (e.g., *Plasmodium* and *Toxoplasma*) or through activation of host-cell receptors, triggering

an endocytotic response (e.g., *Leishmania*). It is generally accepted in malaria, and is probably applicable to other members of the Apicomplexa (e.g., *Babesia, Theileria, Toxoplasma*) that invasion is a multistep process, involving a sequence of interactions between the parasite and the surface of the host cell. In the case of asexual blood stages of *Plasmodium falciparum*, initial contact with an erythrocyte involves recognition and attachment of the merozoite, followed by reorientation of the apical end to face the host-cell membrane before the rhoptries discharge their contents, and formation of the parasitophorus vacuole begins. Invagination of the membrane necessitates major perturbation of the erythrocyte cytoskeleton, presumed to be mediated by one or more of the products to have been identified in the rhoptries and micronemes, such as a 37 kDa protease that preferentially cleaves spectrin and has some activity against band 4.1. The integral membrane proteins, glycophorins A, B, and C, have been implicated as erythrocyte cell-surface receptors for *P. falciparum* merozoites, while the Duffy blood group determinants (Fy^a, Fy^b) have been ascribed to this role in invasion by *P. vivax* merozoites. However, other receptor–ligand interactions also appear to regulate invasion, and at least one has specificity for infraspecific variants of *P. falciparum*, involving components released from schizonts that bind to both merozoites and erythrocyte membranes.

In contrast with malaria, the hemoflagellate *Leishmania* enters a macrophage following deposition into the host by its sandfly vector. Here, the strategy involves exploiting the natural propensity of the professional phagocyte to take up the parasite, while avoiding subsequent intracellular destruction through activation of the host cell and generation of a respiratory burst. As with malaria, there is controversy about the details of the mechanisms involved, but the available evidence supports the view that two surface molecules of metacyclic promastigotes mediate parasite attachment to the macrophage via complement and mannose–fucose receptors (and possibly others). One of these is a lipophosphoglycan (LPG) that interacts with CR1 (the complement C3b receptor) and CR3 (the iC3b receptor), either directly or via C3 components that associate with LPG on the parasite surface. The other is a 63 kDa glycoprotein (Gp63) that interacts with CR3 either directly or via iC3b, but the molecule also binds via its glycosolated portion to the mannose–fucose receptor (MFR). However, while complement-mediated binding to CR1 or CR3 induces endocytosis without a respiratory burst, attachment to the lectin-binding site of CR3 or MFR triggers macrophage activation. The selection of a particular receptor–ligand interaction is therefore crucial to the outcome of infection, but the factors governing receptor utilization by promastigotes on macrophages *in vivo* have yet to be determined. [*See* LEISHMANIASIS; MALARIA.]

C. Penetration of Host Tissues

This is a property shared by a wide range of helminth species, but is less typical of parasitic protozoa. Migration through host tissues is achieved by mechanical movement, coupled with localized tissue destruction mediated by secreted parasite enzymes. Best characterized of these are the proteolytic enzymes produced by infectious stages of nematodes and trematodes, that are synthesized and secreted by specialized glands that discharge their contents via the mouth or from ducts serving other cephalic structures. Among certain nematode larvae (e.g., *Trichinella* and *Toxocara*), proteases that occur in the secretions of the worms are also associated with the surface of the epicuticle, where they may or may not have functions related to tissue migration.

The path of migration taken by parasites may require them to penetrate a variety of tissues, each composed of a range of macromolecules built into basement membranes, cellular structures, and extracellular matrices. These require degradation by a large battery of specific enzymes, or alternatively by a more limited set of enzymes, each possessing relatively broad substrate specificity. In general it appears that the latter holds true, since the proteases thus far implicated in helminth migratory activity appear to be able to degrade a wide range of the macromolecules that parasites will encounter in their hosts. Moreover, such enzymes may have multiple functions in the life history of the parasite, such as the 37 kDa metalloproteinase of hookworms (*Ancylostoma* spp.), which is used by infective larvae in tissue migration and as an anticoagulant by the gut-dwelling adult worms for feeding on blood.

Tissue penetration involving proteases takes place during invasion of the skin by infective larvae (e.g., *Strongyloides* or hookworm larvae, schistosome cercariae), or following ingestion of the infectious stages which, when activated, traverse the wall of the gut (e.g., *Ascaris*, larval cestodes). It

also occurs during transmission of larvae from mother to fetus across the placenta (e.g., *Toxocara*), and following deposition of larvae into the skin by hematophagus arthopod vectors (e.g., filarial nematodes). Subsequent to the introduction of the parasite, proteases are required for adult worms to migrate through tissues for pairing (e.g., *Loa loa*), or to facilitate dissemination of progeny from the site of production to the site of dispersal in host excreta (schistosome eggs) or by bloodsucking insects (*Onchocerca* microfilariae). In many instances, therefore, the synthesis of proteases is developmentally regulated, which may take the form of continuous expression after initial induction, or more commonly, a discrete period of enzyme production (see Fig. 1).

II. Strategies of Host Modification and Immune Evasion

A. Modifications to the Intracellular Environment

Many intracellular parasites have the ability to radically alter the structure, metabolic activities, and antigenicity of host cells. Such changes do not occur as a simple byproduct of damage (which may also take place), but as a result of specific manipula-

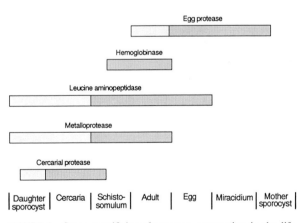

FIGURE 1 Stage specificity of protease expression in the life-cycle of a parasite, represented by the example of the digenetic trematode *Schistosoma mansoni*. In some cases, proteases are expressed in more than one stage, while others are strictly stage specific. Some proteases may be synthesized in one stage (denoted by light shading), then stored and used later in the developmental cycle (denoted by darker shading). [Modified from McKerrow and Doenhoff (1988) *Parasitology Today*, **4**, 334–339. Elsevier Science Publishers Ltd., U.K.]

tion of the parasitized cell. One of the best characterized examples is the parasitized erythrocyte in falciparum malaria, in which changes begin immediately after invasion and progress through schizogony, resulting in a complete transformation in 48 hours. These start with the incorporation of the ring-infected erythrocyte surface antigen (RESA) into the membrane of the newly invaded host cell, the formation of a parasitophorus vacuole (as described in Section I), and the appearance in the cytoplasm of a range of membrane-bounded inclusions, identified as protein-trafficking organelles (e.g., Maurer's clefts). With advancing differentiation of the trophozoite, the erythrocyte loses its natural deformability, and the delineating host membrane develops protrusions that cover the surface, termed knobs. These are associated with several distinct parasite products, one or more of which is believed to mediate adherence to host endothelial cells. This feature of the mature trophozoites and schizonts leads to their sequestration in capillary beds, and is believed to have the crucial role of protecting the distorted erythrocytes from detection and elimination in the spleen. At a metabolic level, the parasitized cell is also dramatically altered, with greatly increased glucose consumption, lactate production, and active uptake of exogenous purines, amino acids, fatty acids, and even iron. There are many further examples among the protozoa of parasites that manipulate the host cell. Among the most interesting are *Theileria* spp., which parasitize bovine lymphocytes and cause cell transformation and lymphoproliferation through what looks tantalizingly like parasite oncogene involvement.

Among the limited examples of intracellular helminths, *Trichinella* infective larvae exhibit the extraordinary capacity to reorganize a fully differentiated muscle cell. There is marked increase in cell size, in the numbers of nuclei, in activation of nucleoli, and hyperplasia of the endoplasmic reticulum. Accompanying this process are metabolic changes, loss of contractile fibers, and eventual formation of a collagenous capsule around the worm, leading to the creation of a totally transformed structure aptly dubbed a nurse cell. It is not clear at what level the parasite mediates redifferentiation, but an intriguing possibility is that secreted products of the stichosome (a specialized glandular tissue) interact directly with host DNA. [*See* TRICHINOSIS.]

B. Modifications to the Extracellular Environment

More typical of metazoan parasites than alteration of an individual host cell is secretion of products with pharmacological properties into the tissues to exert a beneficial, local effect. For example, superoxide dismutase is produced and secreted by most classes of helminths, and is thought to play a role either alone or in combination with other detoxifying enzymes (e.g., catalase, glutathione peroxidase) to defuse the potentially damaging effect of free radicals arising from host cell interactions with the parasite surface. Another product of abundance in the secretions of helminths, in particular in parasitic nematodes, is acetylcholinesterase, which is produced and discharged from specialized glands. While the enzyme undoubtably functions in neurotransmission as in free-living species, there are distinct isomeric forms of acetylcholinesterase that are suggested to have other roles when secreted by parasites into their hosts. These include down-regulation of mucus secretion, of intestinal peristalsis, and of immune effector pathways involving lymphokine production and activation of granulocytes.

Other parasite molecules that may interfere with physiological processes in the host are the enzyme antagonists, which would be most effective at high local concentrations. Serine protease inhibitors have been described from *Ascaris* that may serve to protect the worms against host enzymes in the gut, while *Taenia* metacestodes produce the proteinase inhibitor *taeniaestatin*, which has been demonstrated to be an antagonist of the complement cascade. In fact, taeniaestatin markedly suppresses many host immunological functions, inhibiting complement components C5a, factor D (of the alternative pathway) and C5a-mediated neutrophil chemotaxis and aggregation. These are interesting examples of what may be a widespread phenomenon, but the detection of enzyme inhibitors in parasites should be interpreted with caution, because these may also be used for normal regulatory purposes in cellular metabolism.

C. Mechanisms Used in Host Defense Evasion

Many parasites establish long and persistent infections, either because of the longevity of individuals (e.g., *Trypanosoma cruzi*, *Onchocerca volvulus*) or on account of their ability to produce successive generations in the same host (e.g., African trypanosomes, various roundworms). To do this they must be able to evade rejection by the host immune response, and many and varied are the strategies that have been adopted by parasites.

The common protozoan practice of intracellular localization affords significant protein against immune attack, provided the parasitized cell continues to express Class I MHC antigens that will be recognized as self. In some instances, parasite products appear at the cell surface, so additional mechanisms may be required to avoid the consequences of stimulating a T-lymphocyte response. In many infections, there is suppression of T-lymphocyte functions, which may be confined to responses to parasite antigen, or have a nonspecific component. This is not true only of protozoal infections (e.g., American trypanosomiasis), but is also a recurring theme among the helminthiases (e.g., lymphatic filariasis). A range of parasite products has been detected in serum that block antibody-dependent cytotoxity (ADCC) reactions, or down-regulate interleukin production and proliferation by T cells. Moreover, circulating immune complexes are a notable feature of several parasitic infections and are known to have powerful immunomodulatory effects on cell-mediated immunity.

A chief target of immune attack is the parasite surface, so several mechanisms have evolved to disguise this or protect it against damage from host complement, antibodies, or cells. One of the most celebrated examples is antigenic variation in African trypanosomes, whereby spontaneous switching in surface molecules repeatedly occurs and saves a proportion of the parasites from functional immunity (discussed in greater detail in Section VI,B). A similar phenomenon occurs to a more limited extent in the asexual blood stages of malaria, which also exhibit antigen variability, characterized by infraspecific differences in the epitopes encoded by a wide range of surface and secreted parasite molecules [e.g., the *P. falciparum* erythrocyte membrane protein (PfEMP1), S-antigens)].

An elegant and well-characterized example of antigenic disguise is that exhibited by the trematode, *Schistosoma mansoni*, which soon after infection adsorbs onto its surface, host blood proteins and glycolipids. Alternatively, some parasites synthesize surface molecules that mimic components from the host, such as *Toxocara* larvae, which elaborate products that resemble host blood-group carbohydrates. Even instances of the induction of an immune response may be ineffective, owing to sloughing off of surface antigens (e.g., hookworm larvae), because of cleavage of bound antibody (e.g., *Schistosoma* schistosomules), or simply because it does not lead to destruction of the parasite. Promotion of such ineffectual immune responses may be a further means of defense, hindering functional effector mechanisms as described in the case of *blocking antibodies* in schistosomiasis. Finally, elegant mechanisms have recently been elucidated that allow parasites to survive the binding of host complement to their surfaces, as discussed earlier in the case of *Taenia* metacestodes that produce the inhibitor taeniaestatin, which acts on C5a and the alternative pathway enzyme, factor D. In *Trypanosoma cruzi,* the production of C3 convertase inhibitor by trypomastigotes mimics decay-accelerating factor and limits complement activation on the parasite surface at a step before C3b deposition. In *Leishmania* major, the metacyclic promastigotes resist damage not through interruption of the complement cascade, but instead by abrogating insertion of the lytic C5b-9 complex into the membrane.

III. Molecular Basis of Innate Host Resistance

A. Background

One of the most intriguing, but poorly understood, areas of parasite biology concerns which factors govern host specificity, for both the intermediate and definitive hosts. Usually the host range involves some ecological or behavioral considerations, such as whether parasite and potential host occur in overlapping habitats, or whether host behavior (e.g., dietary preference) does or does not expose it to infection. However, below this layer lies a molecular basis for host susceptibility or refractoriness, one in which resistance is specified by the lack of some essential host factor(s) required by the parasite, or by the presence of a host component(s) deleterious to its survival. Elements of immune effector mechanisms are regarded as distinct from factors governing innate host resistance.

B. Resistance Specified by Deleterious Host Factors

This phenomenon is well illustrated by the African trypanosome, *Trypanosoma brucei brucei,* which has a host range that includes livestock and game animals but does not include humans. The reason for this is that trypomastigotes (blood stages) of *T. b. brucei* are sensitive to a nonimmune factor in human serum, unlike those of *T. b. rhodesiense* and *T. b. gambiense,* which are the causative agents of human trypanosomiasis (or sleeping sickness). This trypanosome lytic factor (TLF) is not present in the serum of individuals with Tangier's disease, a disorder characterized by severe deficiency in the production of circulating high-density lipoprotein (HDL). In normal human serum, the lytic factor partitions with the HDL fraction, and lytic activity has been shown to reside in particles associated with a minor subclass of total HDL. These particles contain at least three unique proteins, together with several apolipoproteins, a proportion of which seem to be crucial to lysis of the trypomastigote in a process that involves uptake of the HDL by endocytosis.

A contrasting example of innate resistance specified by nonimmune host factors in provided by the survival and development of *Onchocerca* larvae in various species of blackflies (Simuliidae). Among closely related blackflies, there may be large variations in susceptibility to *Onchocerca* infection, and host refractoriness correlates with the presence in hemolymph of acellular components toxic to microfilariae. These appear to be lectin-like defense proteins, distinct from the well-documented antibacterial–antibiotic proteins of the Lepidoptera (e.g., attacins, cercropins, lysozyme), and may have the ability to bind the surface of the parasite, which has been shown to carry sugars that change with each stage of development within the vector.

C. Resistance Specified by Missing Host Factors

A good example of this phenomenon is that of erythrocyte phenotypes in relation to susceptibility to *Plasmodium* spp. These include a variety of genetic defects, some of them serious, that have been maintained in human populations in areas endemic for malaria (especially *P. falciparum*). This is because of the protection it confers against plasmodial blood stage infections, which is a reflection of the profound influence these parasites have exerted on

the genetic structure of human populations. Single-gene disorders of erythrocytes associated with protection include sickle cell anemia, alpha and beta thalassemias, ovalocytosis, and deficiencies in the enzymes glucose-6-phosphate dehydrogenase, adenosine triphosphate, pyridoxal kinase, and others. In the case of ovalocytosis, merozoite invasion appears to be blocked by modifications in the composition of the erythrocyte cytoskeleton, while the enzyme deficiencies are thought to impair parasite growth and/or increase sensitivity to oxidant stress. In areas of *P. vivax* transmission, the FyFy genotype is particularly common; this is associated with lack of Duffy blood group determinants which, as discussed earlier, serve as erythrocyte receptors used for invasion by the merozoites (Section I,B).

IV. Molecular Basis of Drug Resistance

A. Background

An aspect of great concern in medical and veterinary parasitology is the development of drug resistance and the consequences this can have on the treatment of individual infections or the implementation of control. Among human infections, the problem is greatest, as it affects protozoal diseases, but drug resistance has also arisen in helminths and has been particularly significant in the management of parasites of livestock. It is perhaps inevitable that the selective pressure exerted by drugs on populations of organisms as resilient as parasites will sooner or later give rise to variants with the ability to resist the effects of the compounds.

B. Genetics of Resistance

Inheritable resistance to drugs may be conferred by a single gene, or involve multiple loci. In theory, single-gene resistance is most likely to arise under conditions of high exposure to the drug, when most parasites are killed. A rare mutation that specifies resistance will immediately confer on a heterozygote a significant advantage (assuming the parasite to be diploid). Once it has arisen, the mutant allele should spread rapidly through the population, provided selection pressure is maintained, and the new phenotype offers sufficient improvement in basic reproductive rate over that of the susceptible population.

Polygenic resistance will be favored at lower exposures to drugs, when a smaller proportion of the parasite population is destroyed. Among the survivors will be existing phenotypes that can tolerate a given dosage of drug, most probably because of the sum effect of several genes governing diverse aspects of their biochemical, physiological, and behavioral makeup. Such resistance may be relatively slow to develop, owing to the requirement for co-selection of several alleles, and because initial selective advantage of a given genotype may not be great. Under these circumstances, different resistant genotypes may arise within geographically separated populations of parasites. In the selection of genes contributing to resistance will be many other alleles that simply segregated fortuitously in the resistance population.

C. Mechanisms of Resistance

These are many and varied and have been most closely studied in the protozoal parasites. However, even in some well-documented examples such as chloroquine resistance in malaria, the study of resistance mechanisms is hindered because of an incomplete understanding of the mode of drug action. The picture is further complicated because a drug-resistant parasite line may simultaneously use several independent mechanisms to promote its survival.

One category of resistance in parasites is manifested by reduced uptake of drug, illustrated by formycin B resistance in *Leishmania tropica,* or chloroquine resistance in rodent and human malarias. Here it appears that resistant organisms switch the system used to import drug into the cell from one involving active transport via ATP-dependent, high-affinity binding sites to one utilizing low-affinity, energy-independent sites. The dynamics of drug efflux from the parasite might be equally important, as proposed for the multidrug resistance phenotype of malignancy in mammalian cells and recently proposed as a factor in defense against chloroquine in *P. falciparum.* Another category of resistance is associated with loss of the target molecule for the drug, again put forward as a mechanism in malarial parasites that become chloroquine resistant. It is postulated that the high-affinity site for chloroquine in these organisms is the breakdown product of hemoglobin, ferriprotoporphyrin IX, and that resistant parasites degrade hemoglobin differently from

susceptibles and export free heme, avoiding the accumulation of ferriprotoporphyrin. Structural modification of the target molecule may also lead to loss in the efficacy of certain compounds, as appears to be the case in nematodes resistant to benzimidazole carbamates (inhibitors of microtubule assembly), which express tubulins that bind the drug less avidly than those synthesized by corresponding wild types.

A relatively common resistance mechanism when drugs operate against a parasite enzyme is compensation by up-regulation in the expression of the target molecule. This is seen in pyrimethamine-resistant malaria in which the drug acts on dihydrofolate reductase. Resistance is associated with gene amplification, increase in available enzyme (possibly through increased rates of transcription), and altered specific activity. Another change that may arise to mitigate the harmful effects of a drug is increased cell concentrations of naturally occurring enzymes involved in detoxification through free radical scavenging. For example, elevated levels of glutathione transferase in the veterinary nematode *Haemonchus contortus* have been linked to strains of the parasite exhibiting cambendazole resistance. Again, this may involve gene amplification, increased rates of transcription, altered stability of the enzyme, or a combination of factors.

V. Novel Molecular Mechanisms Revealed by Investigations of Parasites

A. Background

The evolutionary biology of parasites is governed by many factors, only some of which are shared with their free-living counterparts. The phenomenon of parasitism exerts its own set of selection pressures and these have led to certain adaptations that are unique, or at least are more strongly developed among parasitic species. Indeed, novel molecular interactions, biosynthetic processes and principles of gene regulation have been recently revealed by investigations on parasites that have given important insights into general biology. Here are described three especially interesting examples, all arising from work on African trypanosomes that has consistently led the field with new concepts in molecular parasitology.

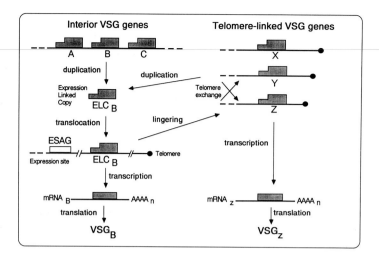

B. Antigenic Variation

The term antigenic variation refers to a process of repeated, spontaneous switching in the expression of surface antigens in certain protozoan species. It is one of the most elegant adaptations for survival among parasitic animals and is certainly the most thoroughly characterized at a molecular level. African trypanosomes, exemplified by *Trypanosoma brucei,* are unrivaled in the art of antigenic variation, but the phenomenon occurs in other parasitic protozoa such as *Plasmodium* spp. and even in some free-living forms, such as the genus *Paramecium.*

In trypanosomiasis, the blood stage trypomastigote lives as an extracellular parasite in its vertebrate host. It rapidly stimulates humoral immune responses and, as a consequence, is subject to attack by antibody-mediated effector mechanisms. However, while the vast majority of trypanosomes of a given generation succumb, a tiny minority with a distinct coat of surface antigens survives to become a subsequent dominant population (homotype) in the blood. The surface of the organism is entirely covered (including its flagellum) with 10^7 molecules composed of a single class of antigen known as the variant surface glycoprotein (VSG), which is a dimeric structure with subunits of 52 to 69 kDa attached to the plasma membrane by a glycosyl phosphatidylinositol (GPI) anchor. Release of the VSG protein may be achieved by the parasite through cleavage with a phosphatidylinositol-specific phospholipase C, and a new VSG molecule may be expressed, antigenically distinct from the first. VSG replacements occur at an estimated rate of 1 to 2×10^{-3} switches per cell per generation and

FIGURE 2 Summary of the DNA rearrangements associated with the sequential expression of individual variant surface glycoprotein (VSG) genes in African trypanosomes. Many basic copy VSG genes are located at interior chromosomal positions as represented by genes A, B, and C. These interior genes are transcribed only after undergoing a duplicative transposition to form an expression-linked copy (ELC) gene in a telomere-linked expression site. Other basic copy VSG genes represented by genes X, Y, and Z are already situated near telomeres. A duplication–transposition event is not essential for their expression, although ELCs of telomere-linked basic copy genes may sometimes occur. Upstream of the expressed telomere-linked VSG genes are expression site associated genes (ESAGs), members of a novel family of isogenes coordinately transcribed with the downstream VSG. ELC genes usually disappear from the genome after the switch to expression of another VSG gene, although they occasionally linger to become nonexpressed, telomere-linked basic copy genes. Expression of telomere-linked genes has also been correlated with exchange of telomeres between chromosomes (telomere exchange) and with displacement of one telomere by a duplicated copy of another telomere region (telomere conversion). However, the precise molecular event that triggers a switch to the transcription of a new ELC or another telomere-linked gene is not known. [Modified from Donelson (1988) In "The Biology of Parasitism. A Molecular and Immunological Approach." MBL Lectures in Biology, Volume 9. pp 371–400. (Englund, P. T. and Sher, A. eds.). Alan R. Liss, Inc., New York.]

are mediated by a complex system of duplicative transpositions of VSG genes (see Fig. 2), of which there are approximately 1000 in the genome. A surprising observation is that all VSG transcripts have a *spliced leader sequence,* a 39 base pair fragment of undetermined function that attaches via *trans*-splicing to the 5′ end of the mature mRNA. It is not known what regulates the order in which VSG genes are expressed, but transcription is rapidly inactivated in the insect-stage procyclics, only to resume in the infectious metacyclics as they prepare to enter the vertebrate host. [*See* TRYPANOSOMIASIS.]

C. RNA Editing

RNA editing is a process involving insertion or deletion of uridine bases to mRNA, first identified in mitochondrial transcripts of trypanosomatids. This phenomenon is undoubtably one of the most significant discoveries to have emerged from molecular studies of trypanosomes and related kinetoplastid parasites (i.e., organisms possessing a modified mitochondrion, or kinetoplast that contains large amounts of circular, mitochondrial DNA). The effect of RNA editing is to modify primary transcripts of what appear to be nonfunctional genes to mature mRNA molecules that have the potential to code for proteins. Sequences identical to edited transcripts do not occur in either the mitochondrion or nuclear DNA, although the mitochondrial DNA contains genes encoding primary transcripts that differ from the edited mRNAs only by mismatches of uridines. Insertion or (less commonly) deletion of uridines within the 5′ region, coding sequence, or 3′ untranslated end, results in frameshifts that correct interrupted opening reading frames, extend coding sequences, and/or introduce in-frame translation initiation and termination codons. Numerous mitochondrial genes in trypanosomatids have been shown to undergo editing of their transcripts, including those coding for cytochome c oxidase subunits, I, II and III, apocytochrome b, and subunits 1, 4, and 5 of NADH dehydrogenase. At present the phenomenon is known from very few organisms other than trypanosomatids, but this is a comparatively recent discovery; it would not be surprising for it to be found to be more widespread in nature.

RNA editing has been proposed as a mechanism for storing genetic information in a compact form, but it also appears to represent a novel form of gene regulation, as the process is under developmental control. Thus, transcripts that are edited at certain stages of the life-cycle are not necessarily modified throughout development, although some RNAs are edited at all points in the life-cycle, indicating that the capability is continuously maintained. Editing of transcripts occurs progressively in a 3′ to 5′ direction, implying that it must be a post-transcriptional process rather than a co-transcriptional one. Directing the position where editing occurs and the number of uridines to be added or deleted are a

newly discovered class of small RNA molecules, aptly termed guide RNAs (gRNA). These are encoded in intergenic regions of the circular, kinetoplast DNA (kDNA) and represent precise complementary sequences within the edited regions or form hybrids just 3′ of the pre-edited region. The search is now on for the enzymes involved in the multistep process of cleavage, transfer of uridylates, and religation within edited RNA molecules.

D. Fatty Acid Remodeling

Glycosylated forms of phosphatidylinositol serve to anchor proteins to plasma membranes in a wide variety of organisms. For example, they attach surface antigens to the free-living protozoan, *Paramecium,* and decay accelerating factor to the surface of erythrocytes in mammals. As described in Section V,B, glycosyl phosphatidylinositol (GPI) is also the linkage of the variant surface glycoprotein (VSG) to the membrane in African trypanosomes. An unusual feature of the lipid anchor is that it contains only myristate, whereas other anchors contain a mixture of fatty acyl or fatty alkyl groups.

Fatty acid remodeling is a sequence of fatty acid replacements involved in the biosynthesis of the mature GPI membrane anchor for the trypanosome VSG. Details of the process were revealed using a cell-free system of biosynthesis developed with isolated membranes of *T. brucei.* It is not clear why myristate is essential for VSG attachment to trypanosomes, but it is only incorporated into late intermediates of GPI synthesis. In the process of GPI addition to newly synthesized VSG molecules, there is evidence for a preformed glycolipid anchor, which has been designated glycolipid A. While myristate is the only fatty acid present in glycolipid A, its biosynthetic precursor glycolipid A′ and earlier intermediates all contain more hydrophobic fatty acids than myristate. In the remodeling of the lipid anchor of these intermediates, one fatty acid is removed from glycerol and replaced with myristate from a myristoyl–coenzyme A donor, and the second is replaced by myristate in an unknown reaction to form glycolipid A. Further work with the cell-free system of GPI biosynthesis should unravel the remaining chemistry of the remodeling process.

The abundance of VSG molecules on individual trypanosomes (10^7 molecules per cell) has been a major asset in elucidating the role of GPI in protein attachment to membranes. It seems likely that investigations on these organisms will continue to be a key contributor of information on the biochemistry and enzymology of GPI anchors in general.

VI. Practical Applications of Parasite Molecules

A. Introduction

This is a vast subject that can only be touched on briefly and will be illustrated with some representative examples of recent advances. There are three broad categories into which most applied research in molecular parasitology falls: the production of reagents for diagnosis–taxonomy, rationale design of drugs, and the development of molecular vaccines.

B. Diagnosis and Taxonomy

Parasitological or clinical criteria are by far the most widely used bases for diagnosis of most parasitic infections. By and large they are inexpensive and reliable, but they tend to be labor-intensive and difficult to standardize. In many situations there is an argument for developing immunodiagnostic assays or tests based on the detection of parasite DNA. Serological procedures are available for a number of infections, including those that cannot be diagnosed by isolating the causative organisms from patients (e.g., toxoplasmosis and toxocariasis).

Two main categories of immunoassays have been designed to detect either parasite antigens (circulating or excreted) or host antibodies. The former have the advantage of picking up active infections only, while the latter are more likely to have greater sensitivity because of the amplifying effect of the immune response. Specificity is a commonly encountered problem in either approach, and makes it relatively easy to diagnose the infection associated with a parasite genus (e.g., schistosomiasis) but more difficult to correctly assign the condition to a particular species (i.e., *S. mansoni* versus *S. haematobium* infection). It is widely recognized that carefully characterized monoclonal antibodies and recombinant polypeptides are needed, and in recent years a number of cloned antigens have given highly encouraging results (e.g., heat-shock protein and cathepsin B-like antigens in schistosomiasis; several low molecular weight antigens in onchocerciasis). MHC restriction of the immune response results in heterogeneity in the production of antibodies, and this may necessitate the use of

cocktails of antigens to gain sensitivity, as in malaria serology. Parasite DNA detection assays are also being developed for such purposes as diagnosing malaria, or for screening donor blood where *Trypanosoma cruzi* is endemic. Detection of infection in vectors, rather than in people, is also important for epidemiological investigations, so DNA probes now pick up African trypanosomes in tsetse flies, *T. cruzi* in reduviid bugs, and malaria sporozoites in mosquitoes. A popular alternative for the detection of *P. falciparum* sporozoites in mosquitoes is to use antipeptide antibodies to the tetrapeptide repeat (NANP or asparagine, alanine, asparagine, proline) of the circumsporozoite protein in a two-site immunoradiometric assay, or its enzyme-linked immunosorbent assay (ELISA) equivalent.

Distinct from the reagents designed to detect infections are probes being developed to differentiate among taxonomically similar organisms. These may be required to identify parasites in a sample from a patient (e.g., Leishmania amastigotes), from an arthropod vector (e.g., filarial larvae in blackflies or mosquitoes), or from areas of contamination in the outside environment (e.g., cestode eggs). The two main tools used here have again been monoclonal antibodies and DNA probes, and a focus of current research is to adapt the assays for use in field laboratories with minimal equipment.

C. Drug Design

The basic strategy is to develop new drugs for the treatment of parasitic infections based on a prior understanding of parasite biochemistry rather than on empirical drug screening. The starting point of such studies is to identify features of parasite metabolism that are unique by virtue of the metabolites involved, or because they are organized into pathways distinct from those of the host. The aspect of metabolism peculiar to the parasite is then exploited in the selection or design of drugs.

Possibly the best example of a general distinction between host and parasite metabolism is the absence of the purine de novo synthetic pathway, which can be seen both in the protozoa and helminths. Salvage pathway networks are therefore an interesting target for chemotherapy, but these are not conserved, and they differ in detail among the various parasitic species. Indeed, a given parasite may utilize multiple pathways of purine salvage, a feature that has frustrated attempts to block the salvage network, together with the problem of a dearth of effective inhibitors of the major salvage enzymes (purine phosphoribosyltransferases, purine nucleoside kinases, phosphotransferases, etc.). A more successful strategy has been to exploit the relaxed substrate specificity of these enzymes by using purine analogs that become toxic when incorporated by the parasite, as in the case of allopurinol (a hypoxanthine analog), which has been used with success against *Leishmania donovani*.

Another common feature of protozoa and helminths is the inability to synthesize long-chain fatty acids or steroids de novo, or to desaturate the fatty acids they have. This is not a specific adaptation to parasitism and occurs in many groups of free-living invertebrates, but it does represent another exploitable distinction in the metabolism of parasites and their hosts. In contrast, several parasitic helminths cannot oxidize lipids because they lack a functional β-oxidation sequence, and here they appear to differ from their free-living counterparts that do possess this function and rely on lipids as substrates for energy metabolism.

In the main it is unprofitable to consider parasite biochemistry so broadly, and most work has focused on specialized phenomena relating to a restricted group of organisms. For example, in kinetoplastid protozoa there is a unique metabolite known as trypanothione, which is involved in reduction of glutathione disulphide and other disulphides by means of nonenzymatic thiol-disulphide exchange. This function replaces glutathione reductase, which kinetoplastids do not possess, and trypanothione-dependent peroxidases remove hydrogen peroxide as substitutes for glutathione peroxidase. Maintenance of trypanothione in the cell as a dithiol [T(SH)$_2$] is dependent on the enzyme trypanothione reductase, and this specialized pathway of polyamine metabolism appears a prime target for chemotherapy as it is vital to the survival of the cell.

D. Vaccination

Commercial vaccines are not yet available for a single human parasitic infection, and only five vaccines have been marketed for veterinary parasites, three against helminths and two against protozoa. All of these have been based on live, attenuated organisms, although the trend in current research is to work toward recombinant or synthetic molecular vaccines.

There is a plethora of studies describing the induction of protection with parasite antigens in laboratory animals, in the main still a long way from serious consideration as potential vaccines. A common denominator of vaccination against parasites is that it is extremely difficult to achieve sterilizing immunity (i.e., when 100% of the organisms in an infection are destroyed), a problem that may be explained largely by the various strategies of immune evasion that these organisms employ (see Section II,C). One of the diseases with the most encouraging progress is cutaneous leishmaniasis, in which human volunteers have been successfully immunized against infection simply by inoculation with live or dead promastigotes. Antigenic fractions and purified molecules (Gp63 and LPG) have also been protective in mice, lending support to the notion that it may be possible to develop a defined leishmanial vaccine for humans.

Far greater attention has been given to the development of a vaccine for *P. falciparum*. The research has been divided among three distinct approaches to vaccination, which include vaccines to prevent infection (directed against the sporozoite), ameliorate pathology (directed against asexual blood stages), or block transmission through mosquitoes (directed against sexual stages). The first of these has been based exclusively on a single molecule, the circumsporozoite protein (CSP), identified to be the target of immunity in human volunteers and animals vaccinated with irradiated sporozoites. Clinical trials with recombinant antigens or synthetic peptides corresponding to regions of a tandemly repeated immunodominant sequence (NANP) induced weak humoral responses and conferred limited protection in a small number of volunteers. This turned attention to strategies that might stimulate T lymphocytes and promote higher antibody titres or cell-mediated immunity, which has led to encouraging developments in vaccine technology, and will ultimately lead to further human trials. Asexual blood-stage parasites are antigenically more complex than sporozoites which has resulted in a less focused program of vaccination. This has involved a relatively large number of molecules from merozoites and the intraerythrocytic parasite forms. Recombinant antigens have been tested in primates but generally have not conferred impressive levels of immunity; nevertheless, it has been claimed that a synthetic peptide vaccine was partially protective when evaluated in a limited trial in humans. Least advanced is a human vaccine based on transmission-blocking immunity, which has the potential to stop parasite development in mosquitoes (and hence transmission), but would not protect those vaccinated directly against the infection. Several key antigens that elicit transmission-blocking immunity have been identified and cloned, and vaccination experiments in animals are now in progress, although human trials have not yet been performed.

A final example is that of the veterinary cestode, *Taenia ovis*. Vaccination of sheep with recombinant antigens has been highly successful in protecting animals against onchospheres (infectious stage) in ovine cysticercosis. With the appropriate adjuvant (saponin), up to 94% protection has been obtained against a controlled parasite challenge. Such a vaccine has the potential to be commercially viable, so it is now undergoing more extensive proving trials. If this becomes, as it well might, the first antiparasitic molecular vaccine in routine use, then it will be of great interest to all molecular parasitologists, as we evaluate its impact on a parasite population placed under highly specific vaccination pressure.

Bibliography

Anderson, R. M., Facer, C. A. and Rollinson, D., eds. (1989). ''Research Developments in the Study of Parasitic Infections.'' Symposia of the British Society for Parasitology and the Linnean Society: Special issue. *Parasitology* **99,** S1–S151. (suppl.), Cambridge University Press, Cambridge, England.

Bryant C., and Behm, C. A., eds. (1989). ''Biochemical Adaptation in Parasites.'' Chapman and Hall Ltd., London and New York.

Englund, P. T., and Sher, A., eds. (1988). ''The Biology of Parasitism. A Molecular and Immunological Approach.'' MBL Lectures in Biology, Volume 9. Alan R. Liss, Inc., New York, N.Y.

McAdam, K. P. W. J., ed. (1989). ''New Strategies in Parasitology.'' Churchill Livingstone, Edinburgh, Scotland.

Trager, W., ed. (1986). ''Living Together. The Biology of Animal Parasitism.'' Plenum Press, New York, N.Y.

Parathyroid Gland and Hormone

PAUL L. MUNSON, *University of North Carolina*

Glossary

Branchial or Pharyngeal Pouch One of a paired set of four embryonic endodermal structures that give rise to the thyroid and parathyroid glands, the thymus, and other important tissues

Osteoblast Bone cell responsible for bone formation

Osteoclast Large, multinucleated cell responsible for bone resorption, both matrix and mineral

Secretory Granules Discrete bundles of granules, enclosed in a membrane, of products of a secretory cell, such as hormones, destined for extrusion from the cell into the extracellular fluid

THE PARATHYROID GLAND is important for good health because it produces, secretes, and regulates the secretion of a hormone, known simply as parathyroid hormone. The major function of the parathyroid gland and parathyroid hormone is to maintain within rather narrow limits a certain concentration of calcium (approximately 10 mg/dl) in the blood plasma that is ideal for efficient neuromuscular function and for the development and maintenance of a healthy skeleton.

I. Parathyroid Gland

A. Anatomy

The parathyroid glands, small tan to reddish-brown structures, usually four in number in the human species, are located in close association with the thyroid gland. Each of the upper two glands, the "external glands," is located on the upper surface of the right or left pole of the thyroid gland. The lower two glands, the "internal glands," are usually located within the lower right and left lobes of the thyroid gland but can also occur behind the esophagus or in the mediastinum. Occasionally, the total number of parathyroid glands in humans is five to eight. Each normal human gland weighs, on the average, approximately 40 mg and measures 4–6 mm in its greatest dimension. In humans, the blood supply for all four glands is from a branch of the inferior thyroid artery, arising from the subclavian artery, and the blood drainage is into the thyroid vein, but there are differences in other species.

B. Histology

The major cells of the parathyroid gland are the "chief cells." These cells, each measuring 6–8 μm in diameter, occur in closely packed sheets. The chief cell is the site of biosynthesis, storage, and secretion of parathyroid hormone. The parathyroid hormone is contained in and secreted from secretory granules in the chief cell. The number of these granules is relatively small compared with other endocrine glands; little parathyroid hormone is stored. Other cells seen on histological sections of the parathyroid gland include fat cells, which constitute 10–50% of the gland in adults, and "oxyphil" cells, less than 5% of the gland, which make their first appearance at puberty and increase in number with age. Neither fat cells nor oxyphil cells have any known function.

C. Embryology

The parathyroid glands originate in the endodermal layers of the third and fourth pairs of branchial pouches. During embryogenesis, they migrate to the thyroid gland.

D. Species Distribution on the Evolutionary Scale

The parathyroid glands occur in all terrestrial vertebrates; they first appear in amphibians and are not present in fish. The glands are ordinarily four in number in most species, as in the human, except for the rat (a widely used species for experiments on parathyroid hormone), in which there are normally only two parathyroid glands, located on the upper poles of the thyroid gland.

II. Parathyroid Hormone; the Polypeptide Chain

Parathyroid hormone is a single-chain polypeptide made up of 84 amino acid residues. It lacks cysteine and substituted amino acid residues but contains two methionines. Neither terminus of the chain is blocked. The molecule does not contain carbohydrates or other nonamino acid cofactors. There is a preponderance of basic amino acid residues; overall, the molecule is basic in character.

The amino acid sequences of parathyroid hormone from several different species—ox, pig, human, and rat—have been determined. They are quite similar but not identical. Fig. 1 shows the structure of the human hormone and the differences in sequence between it and the bovine, porcine, and rat hormones.

III. Biosynthesis of Parathyroid Hormone

The human gene for parathyroid hormone is located on the short arm of chromosome 11. After transcription of the DNA into mRNA in the nucleus of the chief cell, mature mRNA moves into the cytoplasm, and parathyroid hormone is biosynthesized as part of a larger 115 amino acid polypeptide chain known as "preproparathyroid hormone." The synthesis of this larger molecule takes place on polyri-

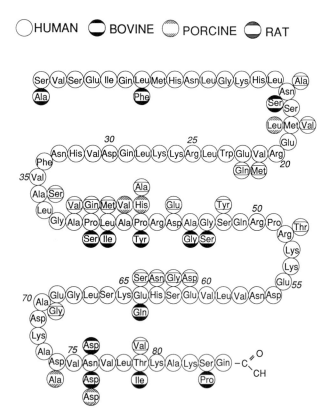

FIGURE 1 Amino acid sequence of human, bovine, porcine, and rat parathyroid hormone. Points of difference between the sequences of the four molecules are indicated by the symbols for species. Reprinted with permission from M. Rosenblatt, H. M. Kronenberg, and J. T. Potts, Jr. (1989). Parathyroid hormone: physiology, chemistry, biosynthesis, secretion, metabolism, and mode of action. In "Endocrinology," 2nd ed. (L. J. DeGroot, ed.), W. B. Saunders Co., Philadelphia, p. 852.

bosomes bound to membranes of the rough endoplasmic reticulum of the chief cell. The next stage consists of two steps. In the first step, 25 amino acids at the amino terminus (known as the signal or leader sequence) are removed by proteolytic enzymes in or near the rough endoplasmic reticulum. The product, "proparathyroid hormone," contains six more amino acids at the amino-terminus than parathyroid hormone. The proparathyroid hormone is then moved to the Golgi apparatus, where it is converted, by another proteolytic enzyme, to parathyroid hormone. Parathyroid hormone (1–84) itself is the major form of the hormone contained in the gland: only 7% of the total is proparathyroid hormone, and there is an even lower percentage of preproparathyroid hormone. After biosynthesis, the parathyroid hormone is transported to secretory granules, from which it is secreted.

Neither preproparathyroid hormone nor proparathyroid hormone is secreted, and neither has any significant biological activity. Their biological significance is confined to their role as biosynthetic precursors of parathyroid hormone.

IV. Terminal Metabolism and Disposition of Parathyroid Hormone

Most of the parathyroid hormone in the plasma is taken up by the liver and kidney. Parathyroid hormone is also bound to receptors on the osteoclasts, but the amount bound, less than 1%, is insignificant compared with liver and kidney.

Proteolytic enzymes of the Kupffer cells in the liver blood capillaries and in other cells at various sites in the kidney split off large carboxy-terminal fragments from parathyroid hormone. Although these fragments are biologically inactive, they may be measured as active intact parathyroid hormone by the usual simple radioimmunoassay methods and thereby confuse interpretation of the results.

Both the intact hormone and the carboxy-terminal fragments are excreted by the normal kidney. The clearance half-time for intact hormone is less than 5 min, but for carboxy-terminal fragments it is at least 20–40 min. In uremia, the half-time of the fragments may be prolonged as much as 100 times. The half-time of intact hormone is prolonged also, but much less so than for the carboxy-terminal fragments because it is taken up and broken down by the liver and the carboxy-terminal fragments are not.

V. Structure–Activity Relationships for Parathyroid Hormone: Antagonists

There is considerable biological cross-reactivity between the species, but the relative potencies of the native hormones from the four different species listed in Fig. 1 differ considerably, depending on what assay method is used for the comparison. In an *in vivo* method (chick hypercalcemia), the ratio of bovine to porcine to human was 1 : 2 : 4. In an *in vitro* method (rat renal adenylate cyclase) the ratio was 9 : 3 : 1. Rat parathyroid hormone (1–34) is considerably more active than all the rest, according to the *in vitro* assay, but the potency value by an *in vivo* assay has not yet been reported. For human biology, the *in vivo* ratios would seem to be more meaningful.

If the methionine residues in parathyroid hormone are subjected to mild oxidation to convert them to methionine sulfoxide, the modified parathyroid hormone loses its ability to inhibit renal reabsorption of phosphate in rats and to stimulate cyclic AMP formation in Japanese quail but retains its activity to produce hypercalcemia and hypocalciuria and to increase production of calcitriol, the hormonal metabolite of vitamin D. A different part of the molecule may be important for each of these activities.

Synthetic polypeptides that contain the amino acid sequence 1-34 of the human and bovine parathyroid hormones are about as active on a molar basis as the entire 1-84 sequence in most assay methods. In the adenylate cyclase assays, *in vitro* and *in vivo*, and the chick *in vivo* assay, the sequence 1-25 was the minimum for retaining any detectable biological activity. Removal of the two amino-terminal amino acids resulted in complete loss of activity. On the other hand, the amino terminus is less important for the cytochemical assay method (described below). Even the synthetic 13-34 human peptide was active in this method.

A substance similar to parathyroid hormone is the "parathyroid hormonelike human hypercalcemic factor of malignancy." (A shorter name now favored is "parathyroid hormonelike peptide.") It has been isolated and characterized from extracts of tumors and found to have homology in structure with the first 13 amino acids of parathyroid hormone. It also shares most of its biological activities with parathyroid hormone. It appears to be responsible for the hypercalcemia of malignancy and is elaborated by several different classes of tumors (squamous, bladder, and ovarian carcinomas), not by the parathyroid gland.

Research toward development of potent antagonists for parathyroid hormone is in progress. A recently synthesized peptide patterned after a partial sequence (7-34) of the parathyroid hormonelike peptide is 6–8 times more potent as an antagonist than the comparable peptide from bovine parathyroid hormone, but it is still not potent enough to be practical as a treatment for hyperparathyroidism.

Along with parathyroid hormone, the parathyroid gland also secretes a larger polypeptide, the so-called "parathyroid secretory protein" (mol wt, 70,000). It is secreted in response to calcium in a manner similar to parathyroid hormone. Parathyroid secretory protein is a glycosylated protein similar or identical to chromogranin A, found in secretory granules of the adrenal medulla. It has recently been renamed "parathyroid chromogranin A." Current evidence suggests that it is a precursor of pancreastatin, which may affect the secretion of parathyroid hormone. (Pancreastatin was discovered as a strong inhibitor of glucose-stimulated insulin release.)

VI. Assay of Parathyroid Hormone in Plasma

The concentration of parathyroid hormone in blood plasma is too low to be measured by any simple bioassay, such as by increase in serum calcium after injection into rats or chicks, but it can be assayed by the cytochemical assay method. This method is based on the determination, in the distal convoluted tubules of the guinea-pig kidney, of the activity of glucose-6-phosphate-dehydrogenase, an enzyme that is activated by intact parathyroid hormone and not by its inactive carboxy-terminal fragments. Unfortunately, the method is too cumbersome and time-consuming for clinical diagnosis, but it has been useful in research. Another limitation of the cytochemical assay method is that it measures the parathyroid hormonelike peptide as well as parathyroid hormone itself.

In preference, radioimmunoassay methods are widely used in clinical diagnosis as well as in research. They can discriminate fairly reliably between a normal person with a rather high but not abnormal concentration of plasma parathyroid hormone and a patient with hyperparathyroidism who

needs treatment. Also, in many cases, a patient with hypercalcemia of malignancy can be discriminated from a patient with hyperparathyroidism.

The actual range of parathyroid hormone concentrations as measured in the plasma of a normal healthy person is rather broad. Depending on the assay used, it varies considerably: a range of 10–55 pg/ml was reported using an immunoradiometric assay and one of 3–29 pg/ml was found using the *in vitro* cytochemical bioassay method. The two methods yielded quite similar values, suggesting that the newer more elaborate and more discriminating radioimmunoassay methods are reasonably reliable for estimating biologically active parathyroid hormone in plasma.

VII. Functions of Parathyroid Hormone and How They Are Performed

A. Calcium Homeostasis

The major function of parathyroid hormone is to raise the concentration of calcium in the blood plasma to an optimum level of about 10 mg/dl and to keep it there. The rapid fall in plasma calcium after removal of the parathyroid glands is illustrated in Fig. 2. In the absence of parathyroid hormone, as shown in the figure, the plasma calcium may fall to as low as 5 mg/dl, a concentration that reflects the apparent equilibrium between bone mineral and body fluids. To avoid an excess of parathyroid hormone that would result in hypercalcemia, the rate of secretion is regulated by negative feedback, as described further below.

The calcium in blood is essentially all in the plasma, where it is divided about equally between protein-bound and ionized calcium. (A small percentage, about 3%, is complexed with organic ions such as citrate.) Only the ionized calcium fraction is physiologically active, but because in most situations the forms of plasma calcium are in rapid equilibrium, any change in total calcium is reflected immediately in a corresponding change in the concentration of ionized calcium. Exceptions to this statement are: In humans with serum protein abnormalities, total serum calcium can change considerably without much change in ionized calcium; and in birds, during the normal egg-laying cycle, the total serum calcium may even double (by binding to phosvitin) without any change in ionized calcium.

Three organ systems affected by parathyroid hor-

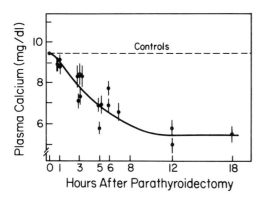

FIGURE 2 Rapid fall in plasma calcium after removal of the parathyroid glands in young male rats. Each point and vertical line represents the mean plus standard error of four to six rats. (Redrawn with permission from data in Fig. 2 in A. H. Tashjian, Jr. (1966), Effects of parathyroidectomy and cautery of the thyroid gland on the plasma calcium level of rats with autotransplanted parathyroid glands. *Endocrinology* **78**, 1144–1153.)

mone contribute to the ability of the hormone to maintain the normal serum calcium concentration: bone, kidney, and intestine.

1. Bone

The mineral phase of bone is predominantly hydroxyapatite, $Ca_{10}(PO_4)_6(OH)_2$. A part of the calcium in bone is adsorbed to the surface of bone in dynamic equilibrium with the calcium in plasma. Parathyroid hormone favors outflow of calcium from bone over inflow into bone and thereby prevents a fall in the plasma calcium when the calcium balance is negative. In the absence of the hormone, inflow is favored, with the result being a lowering of the plasma calcium. The action of parathyroid hormone on a responsive subpopulation of osteoblasts, including those that line bone surfaces, appears to be principally responsible for the minute-to-minute and hour-to-hour regulation of plasma calcium.

A second parathyroid hormone-mediated process that affects plasma calcium is "bone remodeling," in which areas of bone are being "resorbed" (matrix as well as mineral removed) and then replaced by new bone formation. The balance between the two activities of bone formation and bone resorption affects the plasma calcium concentration over the long term, but the changes are too slow to explain the rapid plasma calcium-raising activity of an injection of parathyroid hormone under experimental situations. [*See* BONE REMODELING, HUMAN.]

Bone resorption is performed by the osteoclast, a large, multinucleated bone cell. Although parathy-

roid hormone has a strong effect to increase the number of osteoclasts and their level of activity, no receptors for parathyroid hormone have been found on these cells. On the other hand, osteoblasts (the bone-forming cells) are rich in parathyroid hormone receptors. It has now been shown in *in vitro* experiments that parathyroid hormone stimulates osteoclastic bone resorption only in the presence of osteoblasts; it does not do so in their absence. Parathyroid hormone acts on osteoclasts by stimulating osteoblasts to elaborate an osteoclast-stimulating factor (yet to be identified), and it is in this way that parathyroid hormone effectively but indirectly stimulates bone resorption.

The processes of bone resorption and bone formation are tightly coupled, so that when parathyroid hormone stimulates bone resorption, bone formation occurs also. At normal concentrations of plasma calcium and parathyroid hormone, the two processes tend to be in balance, but during hypocalcemia and at high concentrations of parathyroid hormone the effect on resorption outweighs formation. A separate early effect of small doses of parathyroid hormone to decrease plasma calcium and increase net bone formation has been observed in some experiments in animals. An attempt is being made to exploit this effect of parathyroid hormone for the treatment of osteoporosis.

2. Kidney

Parathyroid hormone also acts on the kidney to promote calcium homeostasis. It does so by (1) reducing the renal excretion of calcium, and (2) stimulating the production of calcitriol, which increases intestinal absorption of calcium and acts on bone to help maintain the plasma calcium concentration.

Most of the calcium in the glomerular filtrate is not excreted in the urine but is reabsorbed in the proximal tubule independently of parathyroid hormone. Nevertheless, the added action of parathyroid hormone to enhance reabsorption of calcium, which is on the thick ascending and granular portions of the distal tubule, is extremely important. The additional calcium reabsorbed under the influence of parathyroid hormone could account for as much as one-fifth to one-third of the total extracellular fluid calcium.

3. Intestine: Calcitriol

The third major organ system by which parathyroid hormone affects calcium metabolism is the small intestine. Most, if not all, of this effect is mediated through the hormonal metabolite of vitamin D, calcitriol (1α,25-dihydroxycholecalciferol), the production of which is increased by parathyroid hormone.

Calcitriol is produced in the kidney by the action of the enzyme, renal 25-hydroxyvitamin D 1α-hydroxylase, on the precursor of calcitriol, calcidiol (25-hydroxycholecalciferol), produced by the liver from cholecalciferol. Cholecalciferol (vitamin D_3) originates in the skin by the action of an ultraviolet part of sunlight (290–315 nm) on 7-dehydrocholesterol. Vitamin D, either vitamin D_2 (ergocalciferol) or vitamin D_3, may also be supplied as dietary supplements in commercial foods (mostly milk). Except for fish with a high fat content there is very little vitamin D in natural foods. Parathyroid hormone affects the production of calcitriol by stimulating the renal 1α-hydroxylase, which results in the supply of an adequate quantity of calcitriol or its analog, 1α,25-dihydroxyergocalciferol, to promote a healthy rate of absorption of calcium by the small intestine. Although all segments of the small intestine—duodenum, jejunum, and ileum—are involved in the absorption of calcium, it is the duodenum that is most sensitive to the effect of calcitriol. [See VITAMIN D.]

B. Effect on Metabolism of Inorganic Phosphate

Parathyroid hormone also has an effect on the concentration of inorganic phosphate in the blood, which, like calcium, is located almost entirely in the plasma. Again, the effect is produced by actions of parathyroid hormone on three organ systems—bone, kidney, and intestine—of which the kidney is predominant.

1. Kidney

About 75% of filtered phosphate is reabsorbed by the proximal tubule, the remainder by the distal tubule and cortical collecting loop. Parathyroid hormone depresses reabsorption of inorganic phosphate in both the proximal and distal renal tubules, thereby increasing the quantity of inorganic phosphate excreted in the urine and decreasing the concentration of inorganic phosphate in the plasma.

2. Bone

The effect of parathyroid hormone to increase outflow of mineral from bone in the short term and during bone resorption in the long term tends to increase the concentration of inorganic phosphate as well as of calcium in the plasma, but the renal effect on phosphate predominates, so that, overall, the result is a decrease in plasma inorganic phosphate, as can be seen in hyperparathyroidism.

3. Intestine

Inorganic phosphate, unlike calcium, is readily absorbed from the small intestine. Nevertheless, calcitriol increases its absorption, and therefore, indirectly, by its effect on calcitriol production, parathyroid hormone favors the absorption of phosphate as well as calcium from the intestine.

4. Importance

The effect of parathyroid hormone on phosphate metabolism is secondary in importance to its effect on calcium metabolism. The supply of phosphate in the food and its absorption from the intestine are ample, unlike the situation for calcium. Therefore, the loss of phosphate due to the action of parathyroid hormone is not usually detrimental. On the other hand, excretion of phosphate protects the body from hyperphosphatemia, which tends to lower plasma calcium by several mechanisms.

C. The Adenylate Cyclase–Cyclic AMP System in the Mechanism of Action of Parathyroid Hormone

Much experimental evidence indicates that the increase in plasma calcium after an injection of parathyroid hormone is mediated by an increase in cyclic AMP. The same is true for the effect of parathyroid hormone on the decrease in reabsorption of inorganic phosphate by the renal tubule. It is thought that after parathyroid hormone binds to its receptor, the receptor interacts with and activates a guanyl nucleotide-binding protein "N," which, in turn, activates adenylate cyclase and increases the hydrolysis of ATP to cyclic AMP. Although the concept that parathyroid hormone acts through cyclic AMP is well supported, there are also indications of an alternative mechanism involving inositol 1,4,5-triphosphate and release of intracellular calcium.

VIII. Regulation of Secretion of Parathyroid Hormone

The most important factor regulating the rate of secretion of parathyroid hormone is the plasma concentration of ionized calcium. (This is unlike the situation in most endocrine cells, in which calcium is required for hormone secretion.) An increase in calcium inhibits secretion of parathyroid hormone, and a decrease "stimulates" it by releasing the gland from inhibition. The interaction between ionized calcium and secretion of parathyroid hormone constitutes a valuable negative feedback system that works to regulate the plasma calcium concentration within narrow limits between 8.8 and 10.5 mg/dl. Figure 3 illustrates changes in the plasma concentration of parathyroid hormone in response to changes in plasma calcium.

The reaction of the feedback system is quite rapid. Observations in experimental animals indicate that the gland responds within 1 min of an induced fall in serum ionized calcium to increase hormone secretion.

The first parathyroid hormone released is that which has been stored in "mature" secretory granules. Later, if the calcium concentration stays low for a long time, with continuous release of parathyroid hormone, a greater proportion of the secreted hormone is newly synthesized hormone. It has been calculated that there is enough parathyroid hormone in the gland to last 7 hr at normal serum calcium concentrations but only 1.5 hr under protracted hypocalcemic conditions.

The fact that calcium and other agents that inhibit parathyroid hormone secretion also inhibit accumulation of cyclic AMP within parathyroid cells suggests that there is an intimate relationship between secretion of parathyroid hormone and the adenyl cyclase–cyclic AMP system. On the other hand, cyclic AMP affects secretion from a preformed hormone pool, while calcium controls secretion of newly synthesized as well as stored parathyroid hormone, suggesting a certain independence of the two factors. We must conclude that, in spite of intensive investigation, the exact mechanisms whereby plasma ionized calcium regulates secretion of parathyroid hormone are not well known.

There are other factors that have been shown to affect the secretion of parathyroid hormone: An increase in the magnesium concentration inhibits the secretion of parathyroid hormone in a manner similar to that of calcium, but the parathyroid is less responsive to magnesium than to calcium. Paradoxically, severe and prolonged hypomagnesemia may lead to hypoparathyroidism. High concentrations of potassium, on the other hand, stimulate secretion of parathyroid hormone. Catecholamines also can increase the secretion of parathyroid hormone, as can various other factors. Nevertheless, the feedback relationship between ionized calcium and parathyroid hormone secretion is the dominating influence on hormone secretion and plasma calcium concentration.

There appears to be a modest circadian rhythm in the rate of secretion of parathyroid hormone, with the rate during the night about twice that in the daytime. In rats, the plasma parathyroid concentration is considerably elevated during lactation. This phenomenon has also been seen in women secreting large amounts of milk, such as mothers nursing twins. Both the circadian changes and the lactation-associated increase in parathyroid hormone secretion appear to be related to factors other than plasma calcium. [*See* Circadian Rhythms and Periodic Processes.]

There is also some evidence that increasing the plasma concentration of calcitriol decreases secretion of parathyroid hormone. The findings suggest some sort of feedback relationship between parathyroid hormone and calcitriol, with parathyroid hormone increasing the production of calcitriol and calcitriol decreasing the secretion of parathyroid hormone. The physiological significance of the interactions of the two feedback systems on each other, if any, has not yet been worked out. In fact, in lactating rats, the concentrations in the plasma of both plasma calcitriol and plasma parathyroid hormone are elevated.

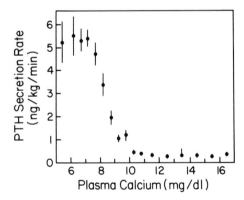

FIGURE 3 Change in plasma concentration of parathyroid hormone in relation to induced changes in plasma calcium concentration. Each point and vertical line represents the mean ± S.E. of repeated measurements in 2–12 calves. (Redrawn with permission from data in Fig. 2 in G. P. Mayer, and J. G. Horst (1978). Sigmoidal relationship between parathyroid hormone secretion rate and plasma calcium concentration in calves. *Endocrinology* **102**, 1036–1042.)

IX. Regulation of the Biosynthesis of Parathyroid Hormone

To supply enough extra parathyroid hormone to reverse prolonged hypocalcemia, hormone production is increased. Evidence suggests that during hypocalcemia, the parathyroid hormone gene is nearly maximally active and that during hypercalcemia it is inhibited. Furthermore, during hypocalcemia most of the biosynthetic product is intact (1-84) nonfragmented parathyroid hormone, whereas during hypercalcemia there is a high percentage of parathyroid hormone fragments due to intraglandular cleavage. Finally, after prolonged hypocalcemia, the number of chief cells is considerably increased. Taken together, this evidence supports the idea that a low concentration of plasma calcium increases net synthesis of parathyroid hormone by stimulating both formation of new chief cells and hormone production within each cell. In contrast, a high concentration of calcium decreases the parathyroid hormone supply. The biosynthesis of parathyroid hormone is, however, poorly understood.

X. Interactions with Other Hormones

A. Calcitriol

In addition to the important interactions of parathyroid hormone with calcitriol that have already been discussed, parathyroid hormone and calcitriol work together in a poorly understood manner to promote the normal growth, development, and maintenance of the skeleton, as well as to increase net outflow of calcium from bone. In vitamin D–deficient animals, a larger amount of parathyroid hormone is needed to produce its usual effect. Excessive treatment with calcitriol or related compounds can, like excessive parathyroid hormone, result in extensive bone demineralization with resulting bone fragility and soft tissue calcification.

B. Calcitonin

A second hormone that interacts with parathyroid hormone is calcitinin, produced in humans and other mammals in the parafollicular or C cells of the thyroid gland.

Adequate doses of calcitonin can antagonize the action of parathyroid hormone by counteracting the effect of parathyroid hormone on outflow of cal-

cium from bone and by inhibiting resorption of bone by osteoclasts. Overall, the effect of calcitonin is to decrease the concentration of calcium in plasma. In spite of expectations, calcitonin has not proved to be a reliable therapeutic agent for the alleviation of hypercalcemia in hyperparathyroidism and other forms of hypercalcemia. On the other hand, calcitonin is at present the treatment of choice for Paget's disease of bone (characterized by increased bone turnover and defective remodeling). In some studies, long-term treatment with calcitonin has been shown to have a beneficial, although small, effect in slowing the progression of osteoporosis in postmenopausal women. In order to avoid the inconvenience of frequent injections of calcitonin, preparations for administration of calcitonin intranasally have been developed and show promise.

C. Glucocorticoids

In a variety of experimental animals (dogs, cats, rats, and mice), it has been found that the hypocalcemia that occurs after parathyroidectomy is greatly reduced by adrenalectomy. According to recent experiments in rats, this effect of adrenalectomy is due to removal of the source of supply of glucocorticoid, which, in rats, is corticosterone. When corticosterone was given to adrenalectomized, parathyroidectomized rats at physiological concentrations, the plasma calcium was reduced to the level after parathyroidectomy alone. Higher doses of corticosterone or the equivalent amounts of hydrocortisone can reduce the plasma calcium concentration even further. In contrast, glucocorticoids did not have a hypocalcemic effect in parathyroid-intact rats, presumably because the secretion rate of parathyroid hormone is increased and thereby counteracts the effect of the glucocorticoids. The mechanism of the glucocorticoid effect and the role glucocorticoids play in normal calcium metabolism remain to be determined.

XI. Deficiency and Excess of Parathyroid Hormone

A. Hypoparathyroidism

Hypoparathyroidism may occur because of congenital absence or deficiency of the parathyroid glands (rare) or, more frequently, because of deliberate or inadvertent parathyroidectomy. Deliberate parathyroidectomy is performed to remove a parathyroid adenoma or parathyroid carcinoma or to relieve secondary hyperparathyroidism (such as may occur as the result of renal failure), but ordinarily not all of the parathyroid tissue is removed. Usually, the remaining gland(s) will increase its function and eventually provide sufficient hormone so that hypoparathyroidism, if it occurs, will be only temporary. Inadvertently, one or more parathyroid glands may be removed during surgical thyroidectomy, resulting in temporary or permanent hypoparathyroidism, depending on how much functional tissue has been left intact.

The immediate result of total parathyroidectomy is a fall in serum calcium, which, if it is severe and not treated adequately, may result in tetany and even death due to respiratory failure (neuromuscular inadequacy). A second immediate result is a decrease in urine inorganic phosphate and hyperphosphatemia. These results are not as serious as the hypocalcemia and are not life threatening. An additional effect that is slower in onset is a fall in the blood concentration of calcitriol (because of diminished production) and the eventual decrease in intestinal absorption of calcium, contributing further to the low level of calcium in the circulation. Over the longer term, the severity of hypoparathyroid hypocalcemia depends on the amount of parathyroid tissue left in the patient, the level of function of this tissue, and the efficacy of supportive treatment.

The immediate treatment of acute hypoparathyroidism is administration of calcium salts intravenously and orally, but this gives only temporary relief. For chronic treatment, parathyroid hormone administration would seem to be appropriate but is not used. Instead, vitamin D_2 or D_3 or a related compound (dihydrotachysterol), or calcitriol is used. They are all active orally, and effective. Initially, the dose is varied, with periodic monitoring of the plasma calcium, toward the objective of maintaining the plasma calcium in the low normal range. However, unexpected hypercalcemia may develop. When this occurs, treatment must be stopped immediately until the serum calcium falls to normal, which takes a variable amount of time, depending on which vitamin D preparation has been used. This time is longer (4–6 weeks) with vitamin D_2 and D_3 because they are slowly released from stores in fat depots. A synthetic compound, closely related in chemical structure to vitamin D, dihydrotachysterol, has a much shorter duration of action (about 2 weeks) and is therefore preferred. Calcitriol is even shorter in its duration of action (2–10 days).

Pseudohypoparathyroidism is the name given to rare types of hypoparathyroidism that are caused not by a deficient supply of parathyroid hormone but by end organ resistance (bone and kidney) to the action of parathyroid hormone. The major symptom is hypocalcemia with all its attendant problems. One result of the hypocalcemia is an increase in the production, secretion, and plasma concentration of parathyroid hormone, which, however, do not correct the hypocalcemia in these patients with end organ resistance to the hormone. Pseudohypoparathyroidism, like hormone-deficient hypoparathyroidism, is effectively treated with a vitamin D preparation.

B. Hyperparathyroidism

Primary hyperparathyroidism may occur because of hyperplasia of the chief cells or development of an adenoma (more rarely, parathyroid carcinoma). The most prominent symptom is hypercalcemia caused by supernormal effects of excessive parathyroid hormone on bone, kidney, and intestine. Important secondary effects may include osteopenia and ectopic calcification in the kidney, aorta, and lung. For reasons unknown, parathyroid hormone secretion by hyperplastic glandular tissue and adenomas is inadequately inhibited by the elevated plasma calcium concentration.

Secondary hyperparathyroidism occurs in association with advanced renal failure and renal osteodystrophy. The latter term includes all the defects in bone and calcium metabolism that appear with the decline in renal function. The increased production and secretion of parathyroid hormone in secondary hyperparathyroidism is assumed to be the result of hypocalcemia, but this has not been well documented. Eventually, during the progression of renal failure, the hypocalcemia is replaced by poorly suppressible hypercalcemia due to the hyperparathyroidism. Factors in renal failure that tend to encourage hypocalcemia are decreased production of calcitriol (with poor intestinal absorption of calcium), hyperphosphatemia, and skeletal resistance to parathyroid hormone.

Primary hyperparathyroidism is treated by surgical removal of the abnormal tissue, whether it is adenoma, carcinoma, or hyperplastic glands. In the latter case, there are two alternative preferred procedures:

1. subtotal parathyroidectomy (removal of three and one-half glands), or
2. removal of all four glands with an immediate autograft of one of the glands into an arm.

Hypocalcemia may be a problem in the immediate period after surgery until the remaining parathyroid tissue has been restored to adequate function and may require treatment similar to that used for hypoparathyroidism from other causes. Development of synthetic antagonists to parathyroid hormone is being attempted but none of them that have been synthesized up to this time is potent enough to be useful in the treatment of hyperparathyroidism.

Nonsurgical approaches to the treatment of secondary hyperparathyroidism, such as combined oral calcitriol and calcium carbonate, are now being used increasingly with considerable success. In cases that require surgery, the recommended procedures are the same as for primary hyperparathyroidism due to hyperplasia.

Bibliography

Aurbach, G. D., Marx, S. J., and Spiegel, A. M. (1985). Parathyroid hormone, calcitonin, and the calciferols. *In* "Williams Textbook of Endocrinology" (J. D. Wilson and D. W. Foster, eds.) 7th ed. W. B. Saunders Co., Philadelphia, pp. 1137–1217.

Cohn, D. V., Kumarasamy, R., and Ramp, W. K. 1986. Intracellular processing and secretion of parathyroid gland proteins. *Vitamins Hormones* **43**, 283–316.

DeLuca, H. F. (1988). The vitamin D story: a collaborative effort of basic science and clinical medicine. *FASEB J.* **2**, 224–236.

Habener, J. F., Rosenblatt, M., and Potts, J. T., Jr. (1984). Parathyroid hormone: biochemical aspects of biosynthesis, secretion, action, and metabolism. *Physiol. Rev.* **64**, 985–1053.

Munson, P. L. (1988). Parathyroid hormone and calcitonin. In "Endocrinology: People and Ideas" (S. M. McCann, ed.). *American Physiological Society,* Bethesda, pp. 239–284.

Raisz, L. G., and Kream, B. E. (1983). Regulation of bone formation. *N. Engl. J. Med.* **309**, 29–35 and 83–89.

Reeve, J., and Zanelli, J. M. (1986). Parathyroid hormone and bone. *Clin. Sci.* **71**, 231–238.

Rosenblatt, M., Kronenberg, H. M., and Potts, J. T., Jr. (1989). Parathyroid hormone: physiology, chemistry, biosynthesis, secretion, metabolism, and mode of action. *In* "Endocrinology" (L. J. DeGroot, ed.) 2nd ed. W. B. Saunders Co., Philadelphia, pp. 848–891.

Parkinson's Disease

EDITH G. McGEER, *University of British Columbia*

Glossary

Blood–brain barrier Membranous wall between blood vessels and brain tissue, which limits the entry of materials into the brain and is, thus, a protective mechanism

Dementia Loss of mental faculties, particularly memory

En cascade Neuronal changes occurring subsequently to, and as a consequence of, the degeneration of other neurons that connect with them

Extrapyramidal system Group of subcortical nuclei, also called the basal ganglia, related to movement control

Free radicals Unstable oxidation products

Inclusion bodies Variety of abnormal swellings visible by histological techniques in degenerating neurons

Neurotransmitter One of a group of specific chemicals used by the nervous system to carry messages from one neuron to another

Peptide Compound made up of a chain (polymer) of amino acids

Peroxides Unstable, highly oxidized compounds

Substantia nigra Major nucleus in the extrapyramidal system

Superoxides Unstable, highly oxidized compounds

PARKINSON'S DISEASE is one of the most common movement disorders, affecting an estimated 2.5% of people over the age of 65 yr, but rarely occurring before 40 yr. It seems to be more common among men than among women. It is an extrapyramidal disorder, with the principal pathology being within the basal ganglia, one of the three systems above the level of the spinal cord, which act to control movement. Pathology in these systems, as well as in the spinal cord or muscles, leads to a variety of diseases with motor symptoms (Fig. 1). The chief symptoms of Parkinson's disease are rigidity, tremor, and difficulty in initiating movement. Ameliorative treatments have been found, based on the major identified neurochemical deficit, but they do not halt the progressive degeneration. The rate of progression can vary greatly from case to case or from time to time in a single case, but the average duration is about 10 yr. The basic cause is unknown, although some cases appear to follow a virus infection, and Parkinson's-like symptoms can be produced in humans or monkeys following exposure to certain toxins. Parkinson's disease is not necessarily accompanied by any mental change, but the tendency to develop dementia is much greater than that in the age-matched spouses of Parkinsonian patients. The occurrence of a familial Parkinsonian–dementia complex among the native population on the island of Guam also suggests that the types of neurons whose losses lead to these two distinct types of symptoms may be vulnerable to similar degenerative processes.

I. General Description

The classical triad of symptoms are rigidity, tremor, and flexion. The rigidity results in a general absence of motor activity and involves all the voluntary muscles. The unnatural immobility of facial muscles leads to the typical Parkinsonian "mask." The tremor affects primarily the fingers and hands, giving rise to a cigarette-rolling movement. It is present at rest but disappears during movement.

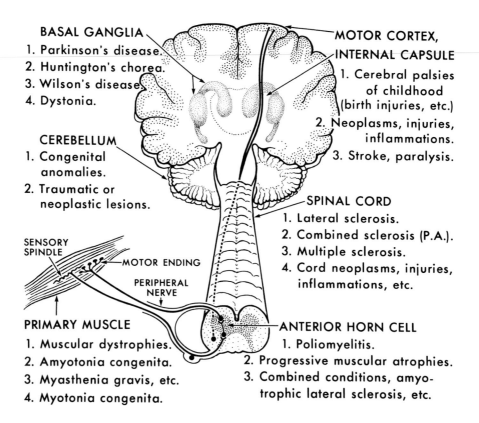

BASAL GANGLIA
1. Parkinson's disease.
2. Huntington's chorea.
3. Wilson's disease.
4. Dystonia.

CEREBELLUM
1. Congenital anomalies.
2. Traumatic or neoplastic lesions.

SENSORY SPINDLE
MOTOR ENDING
PERIPHERAL NERVE

PRIMARY MUSCLE
1. Muscular dystrophies.
2. Amyotonia congenita.
3. Myasthenia gravis, etc.
4. Myotonia congenita.

MOTOR CORTEX, INTERNAL CAPSULE
1. Cerebral palsies of childhood (birth injuries, etc.)
2. Neoplasms, injuries, inflammations.
3. Stroke, paralysis.

SPINAL CORD
1. Lateral sclerosis.
2. Combined sclerosis (P.A.).
3. Multiple sclerosis.
4. Cord neoplasms, injuries, inflammations, etc.

ANTERIOR HORN CELL
1. Poliomyelitis.
2. Progressive muscular atrophies.
3. Combined conditions, amyotrophic lateral sclerosis, etc.

The flexion is of the whole body. The head is flexed on the chest; the body is bowed; and the arms, wrists, and knees are bent. An advanced example of the untreated disease can be diagnosed at a glance but is rarely seen now with the advent of drugs, which commonly relieve all but the final stages. A person with Parkinson's disease has problems in initiating or changing movements; (e.g., getting up from a chair may be a major struggle.) They tend to walk with short, rapid steps and, once started, have difficulty in stopping or changing direction.

II. Central Nervous System Pathology

A. Histological Pathology

The most important and consistent pathological change seen in the brain of Parkinsonian patients is loss of the large, pigmented neurons of the substantia nigra (Fig. 2). In most cases of Parkinson's disease, many other regions of the brain show lesser losses of neurons. However, both the results of drug treatment (Section V) and the appearance of Parkinsonian symptoms in animals, whose only neuronal loss is in the substantia nigra (Sections III and IV), have led to the supposition that the loss of

FIGURE 1 Summary of some diseases that produce motor symptoms and the levels at which they attack the nervous system.

these pigmented neurons is the crucial pathological factor. Perhaps relevant, these pigmented cells are gradually lost during normal human aging (Fig. 4); this loss helps to explain the difficulties of movement seen in aged individuals. [*See* BRAIN.]

A common feature of the histology of Parkinson's disease is the appearance of inclusions called Lewy bodies in the substantia nigral neurons; however, these also appear in various types of degenerating neurons in other neurological diseases.

B. Chemical Pathology

Substantia nigra means black body, and this name was given to the structure because, unlike most of the gray matter of the brain, the substantia nigra of an adult human appears black on gross inspection. The reason is that the large neurons in this nucleus make a neurotransmitter called dopamine, which can be polymerized to melanin, the pigment responsible for the dark coloration of sun-tanned or black skin. Thus, the major chemical defect in Parkinson's disease is believed to be in the dopamine sys-

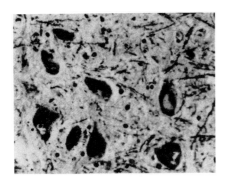
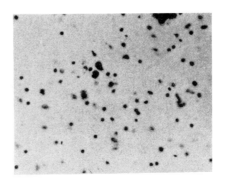

NORMAL   PARKINSONISM

FIGURE 2 Histological changes in the substantia nigra in Parkinson's disease. Note the dramatic loss of the large pigmented neurons.

tems. Dopamine concentrations are significantly decreased all over the brain, but the largest losses are in the caudate and putamen, two other large nuclei of the basal ganglia and the major areas served by the substantia nigra neurons that use dopamine as a neurotransmitter. Neuronal degeneration is not visible in these target areas.

Changes in many other neurotransmitter systems have been reported from postmortem studies on Parkinsonian brains, particularly in the basal ganglia. Some of these may be functional and secondary to changes in the dopamine system or represent degeneration "en cascade." In any case, the relatively minor decreases in the concentrations of these other neurotransmitters do not seem of primary importance to the disease.

III. Theories as to the Cause

A minor proportion (15% or less) of Parkinsonian cases appears to be familial, suggesting that a genetic defect may play some role in the etiology. But the general lack of concordance for Parkinson's disease in identical twins indicates that genetics is not the primary factor in most cases. Parkinson's disease can occur after, and as a result of, an infection of the brain; such "postencephalitic" cases are becoming rarer due to better methods of rapid and effective treatment of the infections. The cause of the majority of cases (so-called idiopathic Parkinsonism) is unknown and may be multiple.

The fundamental problem in this and other neurodegenerative disorders is in discovering why these

particular neurons die. Various theories have been advanced, including viruses or the action of specific neurotoxins. What the toxin or toxins may be remains unknown. It is known that victims of sublethal carbon monoxide poisoning and workers in manganese mines often develop Parkinsonism. One general possibility lies in the vulnerability of these neurons to superoxides, peroxides, or free-radical compounds. A related suggestion is that the melanin that accumulates in such neurons in humans eventually poisons them and that a shortage of antioxidants may favor melanin formation. The finding of a selective destruction of dopaminergic neurons in the substantia nigra in humans and primates treated systemically with N-methyl-4-phenyl-1,2,3,6-tetrahydropyridine (MPTP) has generated great excitement. The neurotoxic potential of this compound was discovered accidentally and tragically when a number of cases of advanced Parkinsonism began appearing in about 1980 among Californians in their late teens or early twenties; a common factor was their use of a street drug that contained MPTP as a minor impurity.

The mechanism by which the destruction of dopamine neurons occurs is not yet completely known, but it involves an oxidation product of MPTP, formed in the brain by the action of an enzyme called monoamine oxidase B (MAO-B), with the oxidation product (MPP^+, Fig. 3) being taken up into dopamine neurons by the specific dopamine uptake mechanisms, which are normally active in recapturing the neurotransmitter released at the synapse. Inhibitors of MAO-B or of the specific uptake process block the toxicity of MPTP. Some believe an environmental toxin similar to MPTP may be responsible for idiopathic Parkinsonism; however, no firm evidence exists. Perhaps the most significant finding so far from MPTP research is that all

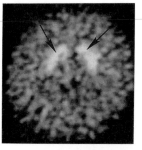

NORMAL   **MPTP-EXPOSED**

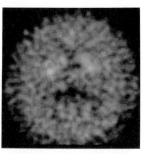

PARKINSON'S DISEASE

FIGURE 3 Conversion of MPTP to MPP+ and selective uptake of the latter by dopamine nerve endings.

FIGURE 5 PET pictures after 6-^{18}F-fluoroDOPA administration to a normal control, an MPTP-exposed, asymptomatic individual, and a Parkinson's disease patient. The light areas indicated by arrows in the normal control correspond to the caudate–putamen, and the lighter these regions, the greater the number of dopamine nerve endings. There is clear loss in the Parkinsonian patient, with a lesser but significant loss in the MPTP-exposed individual. (Photograph courtesy of Dr. D. B. Calne and the UBC/TRIUMF PET program.)

the classical signs of Parkinsonism can occur in humans or monkeys in which dopamine is the only affected neurotransmitter.

Some have argued that idiopathic Parkinsonism may result from an early insult (infectious or toxic) that kills some of the dopamine neurons but that symptoms do not occur until the normal age-related attrition takes the number below a "Parkinsonism threshold" (Fig. 4). A technique to test this hypothesis is now available in positron-emission tomography (PET), which allows study of brain chemistry in humans during life. PET studies with a chemical called 6-^{18}F-fluorodopa can give a picture of the concentration of dopamine nerve endings in the caudate and putamen. Clear decreases can be seen in Parkinsonian patients and even in some MPTP users who do not as yet show clinical symptoms (Fig. 5). Sequential PET studies of such individuals should show whether or not there is a progressive

loss in aging with eventual development of the disease.

IV. Animal Models

Primates treated with MPTP undoubtedly yield the animal model, which most exactly duplicates human Parkinsonism. Treatment of mice, rats, and other small laboratory animals with MPTP is being widely explored, but the effects are generally transient so that an accurate model of dopamine cell loss is not obtained.

Another approach, which has been used for many years in exploring possible therapies for Parkinsonism, is the injection, under anesthesia, of a trace amount of a chemical called 6-hydroxydopamine into rat brain. This close structural relative of dopamine is easily taken up by dopamine neurons and is oxidized to a toxic derivative that selectively de-

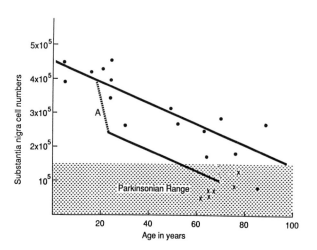

FIGURE 4 Loss of dopamine cells with normal aging (● and solid line) and the hypothesized remote insult plus normal aging (line A) leading to numbers of cells similar to those found in Parkinsonian brains (×).

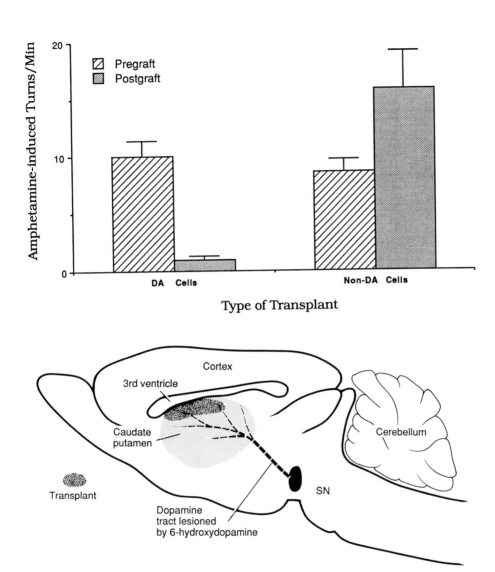

FIGURE 6 A. Bar graph showing typical effect of transplanted fetal substantia nigra cells (dopamine cells) compared with transplanted non-DA cells (sciatic nerve) on rotation in rats with unilateral 6-hydroxydopamine lesions. Rotation was studied immediately before and 4 mo after the transplants. B. Sagittal section of rat brain showing general position of the dopamine projection from the substantia nigra to the caudate–putamen, as well as the position of the transplanted cells in experiments such as those used for 6A.

stroys them. One good example of the use to which this model was put was the demonstration that transplantation of dopamine-producing cells into the brain might be useful in alleviating Parkinsonism. Thus, if 6-hydroxydopamine is injected into only one side of the brain (unilaterally) of adult rats, and the rats are challenged with the dopamine-releasing drug amphetamine, they will rotate because of the unequal release of dopamine in the undamaged and lesioned brain hemispheres. This rotation can be quantitated. The transplantation of substantia nigra cells from fetal rats into such lesioned adults can restore the movement to near normal (Fig. 6A). The transplant is put into the brain just above the caudate (Fig. 6B), so clearly the dopaminergic cells are not involved in the extensive normal circuitry, and yet just getting the dopamine to the caudate–putamen seems to help the symptoms. Transplants into the brain seem much less subject to rejection than transplants into the periphery—a fact that has been used to argue that the brain is "immunologically privileged." The privilege is a limited one, however, and only a fraction of the transplanted cells generally survive. Such transplants have, however, also been reported to "normalize'' the behavior of MPTP-treated monkeys, and these experiments have formed the basis for transplant work in human Parkinsonian patients (Section V).

Most drugs used in the treatment of schizophrenia either block the action of dopamine in brain or deplete its levels. Such drugs can produce Parkinsonian symptoms in both humans and animals, but the symptomatology is generally reversed on removal of the drug. Permanent damage to the dopamine system does not occur, and for this and other reasons animals treated with such drugs are not a good model of human Parkinsonism.

V. Treatment

The treatment of choice for many years in Parkinson's disease has been L-DOPA (also called levadopa). L-DOPA is the normal precursor of dopamine and is effective because it can cross the blood–brain barrier—while dopamine cannot. L-DOPA is readily converted in the brain into dopamine by an enzyme called dopa decarboxylase, which normally occurs in excess and is found in both dopamine neurons and other brain cells. Hence, even when many of the dopamine neurons have been lost, generally enough dopa decarboxylase exists in the brain to effect this conversion. If DOPA is given by itself, much of it is metabolized in the body before it reaches the brain. Hence, it is usually given together with an inhibitor of dopa decarboxylase, so chosen that the inhibitor cannot cross the blood–brain barrier but will block peripheral decarboxylation.

A major problem in the treatment of Parkinsonism with L-DOPA is the occurrence of on/off phenomena, i.e., the L-DOPA may suddenly and transiently lose its effectiveness for no apparent reason. The mechanism is still unknown although changes in receptor sensitivity or in other neurons in the extrapyramidal system have been suggested.

Another treatment approach is the use of synthetic compounds, which are direct dopamine agonists, i.e., they act like dopamine at dopamine receptors. Two types of dopamine receptors are D1 and D2. Action of dopamine at these receptors results in two different types of changes in the chemistry of the membranes and, hence, different modulation of the excitability of the neurons that carry these receptors. D1 sites greatly outnumber

D2 sites, but apparently so far only drugs that act primarily as D2 agonists are effective in Parkinsonism. A well-known example is bromocriptine.

A somewhat different approach is the use of the MAO inhibitor deprenyl. Because dopamine is also metabolized by MAO-B, deprenyl might have some beneficial effect by increasing the life of dopamine in the brain. Such an effect would probably be minor, however, because other brain enzymes also destroy dopamine. The major reason for trying deprenyl is the hypothesis that dopamine cell death in Parkinson's disease may be due to some chemical that, like MPTP, requires oxidation by MAO-B before it becomes toxic. With this hypothesis, deprenyl would not be expected to reverse established symptoms but, rather, might slow the progression of the disease. Such slowing of a disease by a drug is much harder to establish than alleviation of symptoms, but initial reports on deprenyl indicate this approach may be of great benefit.

Another, more debatable approach to treatment of human Parkinsonism is transplants, such as have been shown effective in animal models (Section IV). A major problem is the source of the tissue. For ethical reasons, most centers have used part of the patient's own adrenal medulla. This gland, which lies near the kidneys, normally makes dopamine only as a precursor to noradrenaline and adrenaline and is, therefore, probably not the ideal tissue to use. Only limited success has been reported. Many researchers consider the human experiments premature and liable to cast the whole field of brain transplants into undeserved disrepute, because a good source of tissue has not been established and the risks entailed in broaching the blood–brain barrier, thus possibly allowing infectious agents to enter the brain, may outweigh the potential benefits. An approach that offers some promise for the future in this and other neurodegenerative diseases is to develop tissue culture cell lines expressing the enzyme(s) necessary for synthesis of the desired chemical. Such cell lines in theory can be banked and grown to give an unlimited supply of reproducible cells, which could be treated before implantation with maturation factors that would cause their differentiation into neuronlike cells incapable of cancerous multiplication. One difficulty may be that experiments in animals suggest that the percentage of such cells that survive after transplantation is very low and much less than the percentage of surviving cells in a transplant of fetal tissue. Considerable research along these lines is, however, being done and may provide solutions to this problem.

Bibliography

Calne, D. B., Crippa, D., Trabuchi, M., Comi, G., and Horowski, R. (eds.) (1989). ''Parkinsonism and Aging.'' Raven Press, New York.

Calne, D. B., and McGeer, P. L. (1989). Tissue transplantation for Parkinson's disease. *Can. J. Neurol. Sci.* **15**, 364–365.

Langston, J. W. (1989). Current theories on the cause of Parkinson's disease. *J. Neurol. Neurosurg. Psychiatry,* Special Supplement, 13–17.

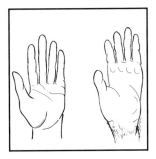

Pathology, Comparative

ROBERT LEADER, *Michigan State University*

JOHN GORHAM, *Washington State University*

I. Background
II. Diseases Induced in Animals
III. Naturally Occurring Conditions
IV. Major Contributions
V. The Impact of Molecular Pathology
VI. Future Developments

Glossary

Antigen Any sort of material such as an organism, toxin, or chemical capable of eliciting a specific immune response

Atherosclerosis Form of arteriosclerosis consisting of deposits of cholesterol-containing plaques in the arteries

Disease Any deviation from normal structure or function of any of the organs of the body

Epitope One specific antigenic site

Genome Complete set of genes, or hereditary substances possessed by an individual, contained in the nucleus of each cell

Germ cell Sperm cells in the male or ova in the female, from which all other body cells are formed, and which carry all hereditary characteristics

Lesion Any measurable pathological deviation in tissue resulting in abnormal appearance or function

Immune system Cellular components, including lymphocytes, macrophages, and granulocytes, along with related biochemical reactions in the body having the function of defending the individual against foreign nonself substances and organisms

Neutralizing antibody Immunoglobulin molecule capable of neutralizing a virus or other infectious agent

Retrovirus Group of viral agents, so-named because they contain the enzyme reverse transcriptase, also known as RNA-dependent DNA polymerase

Somatic cell All cells of the body other than germ cells

PATHOLOGY IS THAT BRANCH of medicine dealing with the essential mechanisms of disease, especially, but not limited to, the structural and functional changes in organs of the body that cause or are caused by disease. Comparative pathology is that branch of pathology that emphasizes comparisons of disease phenomena amongst various species, usually with the ultimate objective of learning more about the diseases of human beings, but at the same time with intrinsic interest in understanding the diseases of all animals including man.

I. Background

Biomedical research and experimental pathology have grown logarithmically over the past 50 years, but comparative medicine goes much further back into history. The study of animals as models for human disease was recorded in Greece as early as the 4th century B.C. by Aristotle, who was funded by Alexander the Great, as later described by Pliny: "Fired by the desire to learn the natures of animals, he trusted the prosecution of this design to Aristotle. For this end he placed at his disposal some thousands of men in every part of Greece, among them hunters, scholars, fishers, park keepers, herdsmen, bee wards as well as keepers of fish ponds and aviaries in order that no creature might escape his notice." One of the most successful examples of the utility of comparative medicine comes from the work of Jenner, who noticed in 1796 that cowpox infection in milkmaids gave them subsequent immunity to smallpox. This eventually led to the total elimination of smallpox, one of the greatest scourges of mankind. This did not happen until more than 180 years after Jenner's first observation, and teaches a lesson in patience as well as the value of keen observation.

II. Diseases Induced in Animals

Comparative pathology can divide easily into two major compartments: studies of naturally occurring diseases of animals and experiments purposefully causing specific conditions in experimental animals. This division is important, both from a practical and an emotional viewpoint. Specific lesions produced in animals are usually intended to mimic as closely as possible some human functional or structural abnormality. Knowledge gained from this process is then extrapolated and applied to the primary condition in the human species with the goal of curing or at least ameliorating the human pain and suffering. Advantage to humans is the main objective, although there may be benefit also to animals other than man.

There has been and continues to be much controversy about the appropriateness and morality of the use of animals in research to gain more knowledge about human disease. This article is a proper forum to address the substantive but not the ethical aspects of these questions. Animal studies have been essential to progress in every field of human biology. Without the use of animals we would not know of the Pavlovian physiological and behavioral responses to sounds and other external stimuli. Paralytic poliomyelitis would remain unconquered. Heart and artery surgery would be impossible. Antibiotics would not be available, and simple infections would still carry the risk of fatality. The recent impact of acquired immune deficiency syndrome on the human population worldwide has been devastating and remains to be solved; but work with a similar disease in rhesus monkeys has given much knowledge about the biology of the virus and its impact on the human immune system.

One need only consider the Nobel Prizes awarded during this century to see the stellar contributions of animals in medical research to the benefit of all animals and human beings. Of the 135 recipients of the Nobel Prize in Physiology or Medicine from 1901 to 1984, most used mammals in their research. One-third were cited for work that involved no warm-blooded animals and an additional 17 were cited for work involving only humans. A wide range of species including vertebrates, invertebrates, cell

cultures, and even higher plants were used as sources of model systems. Careful scrutiny of this series of awards reveals with the greatest clarity the true unity of biological phenomena. Humans are not an isolated species, but relate biologically, emotionally, and behaviorally to all the other passengers on our planet. We all fly together and will either survive or perish together.

III. Naturally Occurring Conditions

Naturally occurring conditions of animals frequently reveal information that directly benefits the species being studied as well as humans. There are many illustrations of this, including the development of canine distemper vaccine many years ago by growing the virus in the membranes of embryonating eggs. Canine distemper is a member of the paramyxovirus group closely related to measles and the successful canine distemper vaccine was later extended by development of a vaccine for measles in children, one of the more important steps in modern control of human infectious diseases. It is not to be claimed that one event led directly to the other, but there was a pattern. The adaptation of a dog virus to growth in eggs, resulting in its attenuation, and subsequent proof that this was an infectious agent capable of stimulating immunity, but without causing disease, was a major achievement. The primary goal was a vaccine that benefited dogs by controlling one of their most devastating infectious diseases; human medicine was later rewarded by the contribution of a significant block of knowledge about the behavior of viruses adapted to growth in nonprimary hosts. The entire chapter of the development of attenuated vaccines for such human diseases as poliomyelitis, measles, and other diseases was aided by the classical work of Haig in canine distemper.

Control of the human immunodeficiency virus (HIV), the cause of the acquired immunodeficiency syndrome (AIDS) is one of the most baffling medical problems facing modern society. The disease was first described and its cause, a retrovirus, determined in the early 1980s. Unfortunately there are currently no vaccines to prevent AIDS, and the treatments that kill the virus in infected cells within the human host have only a partial value. [See ACQUIRED IMMUNODEFICIENCY SYNDROME (VIROLOGY).]

While the AIDS epidemic continues to spread, research workers are investigating animal diseases caused by related retroviruses, which provide convenient and meaningful models for study of host–virus interactions. Most animal species are infected with retroviruses. An understanding of the mechanisms by which these organisms survive in the host and produce chronic disease with highly variable incubation periods is important for our understanding of AIDS and the success of current efforts directed at prophylaxis, treatment, and improved diagnosis of retroviral infections.

Some of the currently known animal retroviral infections that have been intensively studied include maedi/visna of sheep, equine infectious anemia (EIA), simian immunodeficiency, and caprine arthritis-encephalitis. The characteristics of these retroviral diseases are similar to the disease in humans caused by HIV, as shown in Table I. Most of the retroviruses causing this group of diseases are characterized by rapid changes in their antigenicity, which enable them to evade the defense mechanisms of the infected hosts. In the case of equine infectious anemia, the first antigenic change occurs within a few weeks after the first infection and at least six epitope variants can appear within the first year. As in AIDS, and other known retroviral diseases of animals, all horses once exposed remain infected for life. Simian immunodeficiency virus induces suppression of the immune mechanism, which makes it important as an animal model of HIV infection. Caprine arthritis-encephalitis virus is a persistent retrovirus infection of goats and is important to study because viral neutralizing antibodies do not inhibit the progressive course of the disease. It is readily apparent that the interplay between this complex spectrum of viruses and their hosts resembles in many ways the relationship between HIV and the human host, and yet each has its own peculiar biological nature. Much can be learned by studying the entire group.

More recent examples of human and animal benefits from study of animal models can be found in the field of immunology. In the early 1970s veterinarians reported that Arabian foals, which appeared normal at birth, died of a variety of respiratory infections before they reached 5 months of age. It was found that the foals were deficient in both arms of the immune system—antibody (B lymphocytes) and cellular immunity (T lymphocytes). This defect, which is known as *severe combined immunodeficiency* (SCID), is transmitted genetically as an autosomal recessive trait in both humans and horses. Approximately 25% of the Arabian horses in the United States are carriers of this disorder, and have served as a laboratory for further studies of the condition.

Perryman and his coworkers at Washington State University have successfully transplanted bone marrow from a full sibling donor in the treatment of equine human SCID. As bone marrow transplantation is the only means of treating this poorly understood syndrome in children, the work on horses is particularly important. Recently another model of SCID in mice has been discovered. Both animals provide useful models to help us understand this perplexing condition. [*See* BONE MARROW TRANSPLANTATION.]

Animal models have also been intensively studied to gain better understanding of the biochemical nature of hereditary metabolic diseases.

Many important examples of well-recognized animal models, along with analyses of their value in contributing to the solution of human disease, have been published in the scientific literature. The sub-

TABLE I Animal Models of Retroviral Infections

Characteristic of HIV (AIDS)	Most appropriate current animal model (1989)
Immunodeficiency	Feline leukemia
	Simian immunodeficiency virus
Antigenic variation	Maedi-visna in sheep
	Equine infectious anemia
	Caprine arthritis & encephalitis
Encephalitis	Maedi-visna in sheep
Viral persistence	All animal retroviruses

TABLE II Confirmed Animal Models of Human Lysosomal Storage Diseases

Disease	Enzyme defect	Animal species
Pompe Syndrome	Glucosidase	Cow
Mannosidosis	Mannosidase	Cow, Cat, Goat
Gangliosidosis	Gangliosidase	Cat, Calf, Dog
Gaucher	Glucosidase	Dog
Krabbe	Galactosidase	Dog, Mouse
Niemann-Pick	Sphingomyelinase	Dog, Cat
Hurler	Iduronidase	Cat
Maroteaux-Lamy	Arylsulfatase	Cat

(Adapted from Desnick, Robert J., Patterson, Donald F., and Scarpelli, D. G. "Animal Models of Metabolic Diseases." Alan R. Liss Inc., New York.)

jects range from amyloidosis in the mouse to systemic lupus erythematosus in the dog to pattern baldness in the chimpanzee. *Genetically transmitted metabolic defects* in animals can be very useful for investigations of the molecular pathology of the human counterparts. Characterization of the underlying genetic defect and its pathophysiological manifestations allows specific design and evaluation of therapeutic approaches, which could not be applied in clinical trials of the human condition. Furthermore as our knowledge of the splicing of genes becomes more precise, it is possible that studies of these animal models may eventually lead to cures. Such cures would benefit not only human patients but the affected animals as well. Table II outlines examples of some specific metabolic defects that have been well characterized in animals.

IV. Major Contributions

The greatest killer in the populations of Western civilization is *cardiovascular disease,* particularly coronary artery atherosclerosis, which accounts for several hundred thousand deaths in the United States alone each year. Many naturally occurring and induced models in animals have been used in studies of heart disease, including those involving primates, rabbits, dogs, mice, rats, swine, pigeons, chickens, turkeys, and quail. All of these have contributed significantly in their own special way. Recent major advances about the knowledge of the metabolism and heredity of various types of lipids have focused attention on the metabolic and dietary aspects of lipids as risk factors. This has made the studies in primates loom even larger in importance, since these animals are metabolically closer to man than others. Recent work in primates indicates strongly that atherosclerotic plaques produced over many years may be reversible, and may regress significantly under rigorous medical and dietary regimens. This gives hope and encouragement to many millions of patients who have already reached the age where they are most likely to have significant levels of atherosclerosis. A rabbit strain with a genetically transmitted abnormality of lipid metabolism has also proven very important. Each of these models contributes to our knowledge, most because of their similarity to human disease, but often because of specific differences. Differences allow comparisons of definitive steps in mechanisms of

disease, increasing the usefulness of the animal. [*See* ATHEROSCLEROSIS.]

It is important not to overlook the many important contributions of behavioral research to human and animal welfare. The classic example is that of Pavlov (Nobel Prize, 1904), who demonstrated that distinct physiological responses could be elicited in the dog by external stimuli, such as ringing a bell. With the emphasis of modern medicine on *stress,* it is important to be aware of the basic work done in defining the stress syndrome by Selye. Weiss showed that rats subjected to very small electrical shocks developed classical stress manifestations, but were totally protected when warned of the shock by a buzzer. Many other important contributions can be found in research dealing with the processes of learning, and language, particularly in gorillas and chimpanzees; in the rehabilitation of patients with neuromuscular disorders (Nobel Prize, C. S. Sherrington and E. D. Adrian, 1932); and in studies of social behavior patterns (Nobel Prize, K. von Frisch and K. Lorenz, 1973).

Recently considerable attention has been given to the *interdependence of species,* in recognition of the value of companion animals such as cats, dogs, and birds to the well-being of aged and handicapped persons.

Those who work with animal models for human disease often come to realize that there is more than a spatial, mechanical relationship between them and the animals they study. An animal is not a mere test tube. Animals, including mice and rats, and certainly dogs, cats, and primates, react to the attitudes of their keepers; the outcome of experiments can therefore be influenced in a major way by the quality of care given to the animals. Those activists who continuously press for the total elimination of animals from experimental medicine, profess that biomedical laboratories are only a series of cruel torture chambers, where scientists subject animals to the crudest and most useless procedures. Certainly, there are abuses, but they are the exception. There are strict Federal rules for the proper handling of animals, and all laboratories must abide by them in order to receive research support. Federal funding sources have demonstrated their concern by withholding research funds on a number of occasions.

The bond between humans and other animals is firmly rooted in our history from the time of cave dwellers. Many committed scientists are currently devoting careful study to this relationship, and without question, the presence of animals can benefit many patients. Specific examples include favorable responses in blood pressure on the part of hypertensive patients in the presence of pets. This relationship has been especially fruitful in homes for the aged, where the companionship of a cat or dog or bird can make the difference between mere existence and enjoyment of daily life. The relationship between people and animals is a complex one, which cannot be whisked away by wanton accusations of cruelty in laboratory procedures. We are all interdependent. As has been eloquently expressed by Bustad and others, the bond is strong and permanent. If we work to strengthen it, all species will benefit.

V. The Impact of Molecular Pathology

Over the past decade there has been an explosive development of ideas that apply the techniques of molecular biology to pathologic studies in animals. One of the truly exciting areas is the production of *transgenic animals*. These animals will be useful for the study of gene expression and regulation in a wide variety of studies, including pathology, neurobiology, immunology, and in normal as well as abnormal development.

The term *transgenic* was first applied by Gordon and Ruddle in 1981 to animals that have integrated foreign DNA in their somatic or germ cells as a consequence of the introduction of DNA from another individual of the same or different species. In this technique, a fertilized egg is removed from an animal such as a mouse, pig, or cow, and foreign DNA is injected into the male pronucleus. The egg is then implanted into a foster mother. As the fertilized egg develops, the foreign genes are expressed in all new cells produced. Development of this technology in association with DNA splicing was followed by a cascade of many important results. Much attention was given by the lay press to the work of Brinster and Palmiter, who microinjected the gene for rat growth hormone into mouse embryos, resulting in mice that grew much faster and reached weights 70–80% greater than their noninoculation litter mates.

Not long after this discovery, the same idea was applied to the production of domestic animals. Although the results in swine did not give the dramatic success seen in mice, the experiments are continu-

ing. Sheep β-lactoglobulin gene injected into mice caused secretion of this protein in the milk of the mice. In the meantime Chinese workers have successfully introduced rat growth hormone genes into carp and claim that some transgenic fish grow to twice their normal size.

It is now possible to use retroviruses, which can carry genomic material into mammalian cells, as vectors to incorporate desirable genes for such traits as growth, egg production, or disease resistance into recipient animals. The future possibilities for these kinds of experiments to impact on human health are enormous. [*See* RETROVIRUSES AS VECTORS FOR GENE TRANSFER.]

All of these experiments have stimulated stormy discussions among scientists, theologians, ethicists, and politicians. The argument has just opened as this torrent of developments in the manipulation of genes has only begun. It cannot be stopped, indeed will gain momentum in the future, and will be a major future focus of research workers worldwide.

VI. Future Developments

Undoubtedly the application of the emerging knowledge of the transfer of genes will have a major impact on the field of comparative pathology. This is illustrated by a recent paper in *Science,* which is authored by an unlikely consortium of scientists from the U.S. Department of Agriculture Laboratory at Plum Island, the University of California at Davis, and a California biotechnology company, who joined forces to produce a *vaccine by recombinant techniques* similar to those previously used for the production of a smallpox vaccine. This involved insertion into vaccinia virus of specific genes programmed to manufacture immune stimulating factors against a specific disease. The disease was rinderpest, which affects mainly cattle and buffalo, is worldwide in distribution, and causes millions of deaths each year. A modified live virus vaccine grown in tissue culture or in other species is available but is very fragile and often deteriorates in the variable environmental conditions to which it is subjected in the field. The new vaccine provides the safety of a benign agent but with the ability to replicate and be amplified in the inoculated hosts. These advanced techniques resulted in a type of vaccine that may also be effective for immunization against canine distemper and measles, as there is a great deal of cross-immunity amongst the viruses of this closely related family.

A recent article in the *New York Times* looks at the commercial potential for gene-altered or transgenic animals. A wide range of as many as 1000 strains of transgenic mice have already been produced, in addition to 12 strains of transgenic pigs, several breeds of rabbits and fish, at least two breeds of rats, and one transgenic cow with another still under development.

These animals can act as *"molecular factories"* for production of a variety of proteins for medical uses such as immunoglobulins, thyroid, pancreatic, and growth hormones and so forth. But all is not perfect. Only a small proportion of attempts to transplant genes are successful. This weakness is no serious problem in mice, which have a very short generation time, and can be easily bred to large numbers, but poses serious problems in larger species such as cows and pigs. Unquestionably, this area of research will grow tremendously over the next few years, but there will continue to be controversy about its usefulness, and its ethical and religious connotations. Some maintain that we must not tinker with God's work, but the age of the gene is here. The genie has emerged and will not be put back into the bottle.

Bibliography

Bustad, L. K. (1988). Living together: people, animals, environment. A personal historical perspective, *Perspectives in Biology and Medicine* **31**, 2.

Cuthbertson, R. A., and Klintworth, G. K. (1988). Transgenic mice: a gold mine for furthering knowledge in pathology, *Lab. Investigation* **58**, 5.

Gorham, J. R. (1987). Biotechnology and veterinary medicine, *Proceedings of the 91st Annual Meeting of the U.S. Animal Health Association*.

Leader, R. W. (1967). The concept of unity in mechanisms of disease, *The Rockefeller University Review* (Jan.–Feb.).

——— (1967). The kinship of animal and human disease, *Scientific American* **216**, 1.

——— (1969). The Chediak-Higashi anomaly: an evolutionary concept of disease, *NCI Monograph,* no. 32.

Leader, R. W., and Padgett, G. A. (1980). The genesis and validation of animal models, *American Journal of Pathology* **101**, 3s.

Leader, R. W., and Stark, D. (1987). The importance of animals in biomedical research, *Perspectives in Biology and Medicine* **30**, 4.

Narayan, O., Zink, M. C., Huso, D., Sheffer, D., and Crane, S. (1988). Lentiviruses of animals are biological models of the human immunodeficiency viruses, *Microbial Pathogenesis* **5**, 149.

Perryman, L. E., and Magnuson, N. S. (1982). Immunodeficiency disease in animals, *in* "Animal Models of Inherited Metabolic Diseases." Alan R. Liss Inc., New York.

Schmeck, H. (1988). Gene-altered animals enter a commercial era, *New York Times,* Dec. 27, 1988.

Yilma, T., Hsu, D., Jones, L., Owens, S., Grubman, M., Mebus, C., Yamanaka, M., and Dale, B. (1988). Protection of cattle against rinderpest with vaccinia virus recombinants expressing the HA or F gene, *Science* **142**, 1058.

Pathology, Diagnostic

DIANE C. FARHI, MICHAEL E. LAMM, *Case Western Reserve University, and University Hospitals of Cleveland*

Glossary

Anatomic pathology Study of disease by visual and microscopic examination
Autopsy Study of the body after death (postmortem examination)
Clinical pathology (laboratory medicine) Study of disease by laboratory methods
Cytology Microscopic study of cells
Diagnostic pathology Study of disease as applied to the diagnosis and care of patients
DNA Repository of genetic information in cells
Experimental pathology Study of disease in a research laboratory setting
Histology Microscopic study of tissue
Pathology Study of disease

PATHOLOGY IS THE BRANCH OF MEDICINE particularly concerned with the study of disease *per se*, its origins, development, and manifestations. Although its historical orientation has been toward morphological evaluation by naked eye or microscope, pathology today embraces an array of methodologies, including chemical, physical, and immunological. Diagnostic pathology is the application of such techniques along with the pathologist's training and experience to medical diagnosis.

I. Anatomic Pathology

Anatomic pathology is the study of disease through examination of tissues and cells by a variety of methods, ranging from visual inspection of the whole body and its organs to analysis of subcellular constituents. The means of obtaining tissue or cells for study is dictated by the nature and site of disease and the expertise and facilities available; in turn, the means used for obtaining tissue may have an impact on the diagnosis rendered.

The oldest technique for examining tissues is visual inspection of the entire body, body cavities, and organs. This method, used for centuries, was employed as the basis for the first reliable anatomic texts in the fifteenth century and landmark treatises on pathology in the eighteenth and nineteenth centuries. Histology, or the microscopic study of tissues rather than whole organs, began as a discipline in the early nineteenth century, gathering momentum with increasing use and refinement of the microscope and improved methods for preserving and preparing tissues for examination. By the early twentieth century, microscopy became a mainstay of diagnostic techniques in pathology, which it remains today. A variation on microscopy was introduced with the advent of fluorescence technology, which permits the identification of tissue elements through application of fluorescent dyes.

The theory and principle of microscopy as a diagnostic tool were carried further in the mid-twentieth century, when a beam of electrons rather than a beam of light was directed at tissues. Instead of an image formed by transmitted and absorbed light, an image is generated by transmitted and absorbed electrons. The power of this technique, known as electron microscopy, lies in its capacity to resolve images many orders of magnitude smaller than visible by light microscopy. Although not routinely used in diagnostic pathology, the study of cells at this level has contributed much to the understanding of disease processes. [*See* ELECTRON MICROSCOPY.]

To derive the maximum information from any microscopic method, tissues are first stained with any of a variety of reagents. The staining of tissues with biological dyes reached its greatest development in the late nineteenth and early twentieth centuries, at which time the stains now used routinely were formulated. These are based on combinations of naturally occurring and synthetic chemicals with affinities for acidic, basic, and other substances. The cornerstone of diagnostic stains, a combination of hematoxylin and eosin, colors nucleic acids (and therefore nuclei) blue with hematoxylin, and proteins (and therefore cytoplasm) orange-red with eosin. Many other stains are in frequent use, several as direct applications of classical chemistry methods, such as the Prussian blue stain for iron. These stains are employed for the detection of minerals, sugars, pigments, enzymes, and many other substances.

Fluorescent microscopy and electron microscopy also depend on staining techniques for optimum visualization. In the former, a staining reagent may be chemically tagged with a fluorescent marker such as fluorescein, which emits a characteristic wavelength in the visible spectrum when excited by a light of shorter wavelength. The goal of staining in electron microscopy is to render some cellular elements electron-dense, leaving others electron-lucent, in a fashion analogous to light microscopy; this is usually accomplished by staining with heavy metals like osmium, uranium, and lead.

Modern advances in immunology, including the understanding of antigen–antibody interactions, have added a new dimension to the staining of tissues and individual cells. Antibodies can be produced in animals and by cells in tissue culture to a vast and growing array of cellular and extracellular constituents. An antibody directed against the substance of interest is incubated with a slice of tissue or cells in suspension, and the resultant binding of antibody to its target observed. The visualization of this interaction is achieved by tagging the antibody with a fluorescent marker, a heavy metal, or an enzyme that generates a colored product. These techniques, known collectively as immunostaining, have greatly increased our understanding of disease and have been added to the standard diagnostic armamentarium.

The foregoing methods of diagnosis are all applied to the main fields of anatomic pathology: autopsy pathology, surgical pathology, and cytology. These are defined primarily by the way in which tissue is obtained and how it is processed.

Autopsy pathology is the study of the body and its tissues after death, or postmortem. Following medical determination of death and the obtaining of necessary legal permits, the body is examined externally. This may take place at the scene of death in the case of a forensic autopsy (see below) or at a hospital morgue in the case of death at home, in the hospital, or other facility. The body is then examined internally, with care taken not to affect the external appearance, and tissue samples are removed for microscopic study. Specimens may also be removed for study in other diagnostic pathology laboratories (see below). The autopsy is a detailed, comprehensive study of the cause of death and other pathologic processes and results in a report synthesizing the clinical and pathologic findings into a cohesive whole. It is the basis of all anatomic studies and provides opportunities for educating medical students and physicians and for quality control of medical care. In spite of the many advances in clinical medicine, in a significant percentage of cases the autopsy continues to reveal major findings that were unsuspected or misinterpreted during life, even after state-of-the-art diagnostic evaluation. Findings revealed at autopsy may have far-reaching medicolegal implications regarding adequacy of diagnosis and care, aside from the autopsy's importance in forensic pathology (see below).

Autopsy findings are crucial in assessing the natural history of disease and the effects of medical care; in fact, modern medicine as a discipline and as an effective therapeutic endeavor may be said to have originated with the development and acceptance of the autopsy as standard practice. Serious concerns have been raised in recent years regarding the decline of autopsy rates in the United States. If this decline continues, it will undoubtedly prove to be a substantial impediment to progress in understanding human disease.

Surgical pathology is based on examination of tissues removed at surgery. As surgical techniques improved in the nineteenth and twentieth centuries, more specimens were available for pathologic examination antemortem. Interest shifted to detection and diagnosis of disease in the early stages and alterations brought about by treatment. The removal of whole organs or substantial parts of organs for treatment and study has been supplemented or replaced in many cases by tissue biopsy, or removal of a small portion for diagnosis. Improving biopsy techniques have broadened surgical pathology to include some hitherto relatively inaccessible tissues and have permitted repeated tissue studies in difficult or evolving cases. Thus, surgical pathology is a changing area in terms of both new techniques for the study of tissues and new types of specimens being handled. Increasingly sophisticated methods are being brought to bear on ever smaller samples. The questions asked of the surgical pathologist center on the nature of the diagnosis, its distinction from related diagnoses (differential diagnosis), the stage of the natural history at which the disease is seen, and the impact of therapy. Because the patient is alive at the time of pathologic examination, the implications of the pathologist's diagnosis for therapeutic decision-making are great.

An important aspect of surgical pathology involves the communication and interaction of the surgical pathologist and the surgeon while the patient is undergoing an operation. The pathologist may enter the operating room to inspect the abnormality and to render an opinion. Usually a small piece of tissue is given to the pathologist by the surgeon. The pathologist then performs a rapid frozen-section microscopic evaluation, which takes only a few minutes instead of the many hours required for a standard microscopic preparation. Based on the pathologist's frozen-section diagnosis, the surgeon decides on a course of action. The rapidity of frozen-section diagnosis thus allows further surgery or other measures to be taken during the same operation. Without this procedure, definitive surgical treatment might have to be delayed, mandating a separate, second operation. Frozen-section diagnosis also permits other forms of therapy such as antibiotics, anti-cancer agents, or radiation to be instituted immediately for life-threatening infections or malignancies.

Cytology is the microscopic study of individual cells obtained from a patient. It is similar to surgical pathology in the questions asked and implications for clinical care; it differs primarily in the type of specimen submitted for study. Cells are obtained from samples of body fluids and secretions by scraping from external and internal surfaces and by aspiration from internal tissue through a narrow-gauge needle. Cytologic samples are usually collected through less invasive means than surgical pathology specimens and are thus ideally suited for low-cost mass screening studies directed toward the detection of early, potentially curable malignancies and premalignant conditions. A noteworthy example is the contribution of cytopathology to the reduction in morbidity and mortality from carcinoma of the uterine cervix via the Papanicolaou (Pap) test.

II. Clinical Pathology (Laboratory Medicine)

The study of tissues and cells has been augmented in the last century by the study of body fluid components. Organized as a distinct branch of pathology in the 1920s, clinical pathology, or laboratory medicine, is devoted to diagnosis through laboratory methods. This is a rapidly growing area of medicine in which technologic advances have repeatedly revolutionized the scope of laboratory diagnosis. In some areas, subcellular analysis as carried out in anatomic pathology and in clinical pathology merge, and the traditional divisions between these two disciplines disappear. In general, however, the laboratory activities of clinical pathology fall into the following broad diagnostic categories: hematology, blood banking, microbiology, clinical chemistry and toxicology, immunology, and cytogenetics.

Hematology, the study of blood, encompasses the examination of the cellular constituents of blood and blood-forming organs, principally the bone marrow, and the proteins and other substances peculiar to blood, such as the clotting factors. The main activities of a hematology laboratory include quantitative analysis and microscopic evaluation of red and white blood cells and platelets; preparation, staining, and evaluation of aspirated bone marrow samples; tests of coagulation, including quantitative and qualitative evaluation of clotting factors and platelet function tests; detection of abnormal hemoglobins; and specialized tests directed at red and white cell defects, fetal–maternal transfusion, serum viscosity, and others. Many quantitative hematologic tests originally done manually are now performed on automated equipment capable of processing minute amounts of blood rapidly and at low cost; however, skilled microscopic examination remains a very important aspect of this field.

Blood banking, or transfusion medicine, is a related area in that it is concerned with blood, but the emphasis is on blood as a tissue transplant or transfusion. The discipline of blood banking has received major impetus through the needs generated by trauma, surgery, organ transplantation, and treatment of blood disorders. Transfusion of blood may be accompanied by immunologic reactions in the recipient and the transmission of infectious disease. Thus, the activities of the blood bank comprise not only the major public health task of obtaining a continuous supply of fresh blood but assuring its immunologic compatibility and safety when transfused. The blood bank, or transfusion medicine laboratory, collects blood from suitable donors; determines the blood type; tests for infectious agents; separates blood as needed into components such as plasma, platelets, and red blood cells; irradiates or washes components as necessary; tests for donor–recipient compatibility; oversees transfusions; and analyzes adverse transfusion reactions if they occur. Many laboratories also remove blood or specific blood components from patients as a therapeutic maneuver. Improved methods of storing and handling blood are continually being developed to extend the life of this perishable product and permit its more efficient use. In addition, new methods are being rapidly implemented to eliminate transfusion of potentially lethal organisms, such as the human immunodeficiency virus (the agent of acquired immunodeficiency syndrome [AIDS]). [*See* BLOOD SUPPLY TESTING FOR INFECTIOUS DISEASES.]

Microbiology, the study of infectious agents, is one of the oldest branches of clinical pathology. Microbiology laboratories examine a wide variety of specimens harboring diagnostic evidence of infectious disease, including tissues, fluids, secretions, and blood. Originally these laboratories were directed primarily toward the finding and identifying of intact organisms such as bacteria and fungi, but as the spectrum of infectious agents has widened, laboratory techniques have grown to include testing for viral nucleic acids and proteins and immunologic responses, among others. These approaches to the diagnosis of infectious disease are particularly suited to the detection of obscure, latent, or treated infections. The mainstays of diagnosis, however, remain culture in or on suitable media or cells for bacteria, fungi, and viruses and microscopic examination for parasites. Laboratory work

in this area still, to a large degree, relies on skilled microscopic examination. The impact of molecular biology (see below) in microbiology has already been great and is increasing.

Clinical chemistry and toxicology are linked in that these disciplines center on identification and quantification of extracellular molecules in tissues and fluids. The clinical chemistry laboratory measures minerals, fats, proteins, sugars, vitamins, hormones, and a wide variety of other substances in small samples of blood, plasma or serum; in some cases, these tests are made on other specimens, such as urine, cerebrospinal fluid, amniotic fluid, or tissue. Many of these tests are performed in groups designed to identify dysfunction of a particular organ or resolve a specific differential diagnosis. The toxicology laboratory identifies and quantifies therapeutic drugs in body fluids and tissues, as well as alcohol and nonmedically administered narcotics and poisons. Chemical determinations have proliferated dramatically with the advent of highly efficient automated methods and increasingly sensitive analytical techniques, such as radioimmunoassay and gas and liquid chromatography. [*See* GAS CHROMATOGRAPHY, ANALYTICAL; HIGH PERFORMANCE LIQUID CHROMATOGRAPHY; RADIOIMMUNOASSAYS.]

Recent progress in the understanding of the immune system has resulted in the development of clinical immunology laboratories devoted to the identification and monitoring of immunologic disorders. Tests are designed to quantify and assess the function of the major classes of lymphocytes, identify and quantify immunoglobulins and certain other proteins, diagnose cancers of the immune system, identify antibodies of clinical interest, and carry out tissue typing needed for successful transplantation of organs. [*See* LYMPHOCYTES.]

Cytogenetics is that branch of medicine concerned with the detection and analysis of chromosomes, cell elements that contain the DNA governing inherited traits. Laboratories devoted to cytogenetics are often found in departments of pathology, but because much of their work is related to pediatrics and obstetrics, cytogenetics laboratories are sometimes located in these or other departments. The field of cytogenetics has grown substantially in the past few decades and has gained added momentum with the application of molecular biologic techniques capable of identifying individual DNA fragments and even single genes. The cytogenetics laboratory is primarily occupied with growing human cells from various sources in tissue culture, inducing and arresting mitosis, and analyzing the number and configuration of the chromosomes revealed. Chromosomal identification is accomplished by means of sophisticated staining techniques in combination with high-resolution photomicroscopy, and requires highly skilled and experienced personnel. The results have a great impact on prenatal diagnosis, the diagnosis of malignancies and premalignant conditions, and the management of patients undergoing bone marrow transplantation, among others. The increasing use of techniques designed to uncover genetic disorders is rapidly expanding medical options in the diagnosis and management of these diseases; thus, genetic counseling services and the cytogenetics laboratory are very closely allied. [*See* GENETIC COUNSELING.]

III. Forensic Pathology

Forensic pathology, or legal medicine, refers to that branch of pathology concerned with determining the cause of death under unattended, violent, or suspicious circumstances. Under such conditions, the death must be reported to the coroner or medical examiner, who decides if an autopsy should be performed. It is often appropriate to include a toxicological study of body tissues and fluids to determine whether drugs or poisons are present. In a case of traumatic death, the forensic pathologist seeks to ascertain whether death was due to accident, suicide, or homicide. In the case of a bullet wound, at what distance was the gun fired and in which direction? If there were two bullet holes, which was the entrance and which the exit wound, or were there two bullets? Could the wound have been self-inflicted? In the case of a motor vehicle accident, did the driver's preceding death (e.g., from a heart attack) cause the accident, or did the accident cause the driver's death, and if so, was the driver under the influence of drugs or alcohol? When multiple people die in disasters such as earthquakes, explosions, or airplane crashes, or in individual deaths in which the body bears no identification papers, the forensic pathologist endeavors to establish the identity of the deceased.

In situations like those mentioned above, the implications of the pathologist's conclusions extend far beyond determining the cause and circumstances of death from a strictly medical viewpoint.

Important legal and financial considerations may arise that bear on inheritances, insurance coverage, and lawsuits. The forensic pathologist regularly interacts with the police, law courts, and the press. Nor should one minimize the psychological aspects of dealing with bereaved families, for whom the sudden and unexpected death of a family member may induce strong feelings of guilt, especially when a child is involved; it may be comforting indeed if the pathologist can assure the family that the loved one died of natural causes and not as a result of carelessness or inattention. Identifying conditions relevant to genetic counseling is an important part of the pathologist's contribution in cases of stillbirth and deaths of infants.

IV. Co-evolution of Diagnostic Pathology and Clinical Practice

The practice of diagnostic pathology is inextricably linked to the practice of clinical medicine. As advances are made in the clinical diagnosis of certain disorders by epidemiologic evidence, physical examination, or instrumentation, demand grows for new and better laboratory tests to confirm or establish the diagnosis. These tests may range from culture or immunostaining of a newly identified infectious agent to the development of rapid, efficient methods for detecting clinically important enzymes. Conversely, as new technology enters the pathology laboratory from advances in instrumentation or basic biomedical research, the possibility of more rapid and accurate clinical diagnosis increases, spurring interest in diseases that may have been rarely diagnosed, poorly understood, or even unknown. Expertise in clinical medicine and diagnostic pathology grow together, the impetus arising in each depending on the situation. This relationship permits planning of controlled studies, resulting in the careful accumulation of reliable medical knowledge.

V. Pathology as a Bridge between Basic Medical Science and Clinical Practice

Diagnostic pathology is uniquely situated in medicine between clinical practice and basic medical research. The relationship of pathology to basic science is no less important than its position vis-à-vis clinical work; in fact, diagnostic pathology is a crucial bridge between these areas. Through application in pathology laboratories, new techniques developed in academic and industrial research laboratories rapidly gain entry to the diagnostic and therapeutic medical world. In the last few decades, the movement of research tools into diagnostic laboratories has accelerated dramatically; the lag time in this transit is now often measured in months rather than years.

Two examples of this movement are the recent rise of the immunology laboratory, alluded to above, and the current interest in the development of molecular diagnostic laboratories. Clinical immunologists have quickly applied improvements in methods for analyzing proteins to immunoglobulin identification, moving into several generations of immunologic assays. Especially remarkable has been the rapid application of monoclonal antibodies and flow cytometry to diagnostic pathology. Use of such antibodies for cell-surface marker assessment via flow cytometry has changed status from an expensive, cumbersome research procedure to a standard feature of medium to large hospital laboratories; methods have even been developed for transporting specimens in culture media from small hospitals to larger laboratories for these sophisticated tests. The functions of the flow cytometer have grown to include chromosome analysis of malignant cells, and further applications appear limitless.

As a second example, molecular diagnostic laboratories, now at an early stage of development, show great promise. Until recently, this field was confined to the research laboratories in which it was developed. With technical improvements and commercial marketing, analysis for DNA and RNA, including specific segments, is now possible in well-equipped diagnostic laboratories. These procedures include enzymatic digestion of nucleic acids; nucleic acid electrophoresis and hybridization with radioactive or enzyme-labeled nucleic acid probes; nucleic acid electrophoresis and immunostaining; and hybridization of nucleic acids in intact cells, tissue, or tissue digests. Such methods have proved immeasurably useful in protecting transfusion recipients from transmission of the viral agent of AIDS. They have also been used in identifying other transmissible agents, diagnosing malignancies of the immune system, and assessing genetic alterations in tumors. With these techniques, single genes

may be located on individual chromosomes, and the specific genetic make-up of individual human beings may be identified from minute specimens, known as genetic "fingerprinting." These powerful tools hold great implications for the practice of clinical and forensic medicine. In summary, pathology functions as an intellectual bridge between medical research and medical practice. Further, as discussed elsewhere in this article, pathologists must function as a communication bridge between their own discipline and other medical specialists.

VI. Training of a Pathologist

Why might a young physician choose a career in pathology, a field which does not entail the practice of medicine in the usual sense of directly caring for a sick patient? Instead, the pathologist tends to function behind the scene, providing patient-related services indirectly. The pathologist is primarily concerned with diagnosis and has often been referred to as the "doctor's doctor," advising the clinician on the nature of the patient's condition and what needs to be done.

Pathologists are attracted to their profession by its breadth and intellectual challenges, particularly the need to keep one's knowledge current in order to apply new scientific advances to medical problem-solving. Many pathologists enjoy working with analytic instruments and computers. Some are attracted by the opportunity to organize and manage laboratories and the flow of information. Another positive feature is the opportunity to teach in a variety of settings involving medical colleagues, physicians-in-training, medical students, and medical technologists.

After graduation from medical school, the young physician intent on a career in pathology embarks on a period of specialty training to meet the requirements for certification as a pathologist. Training may be confined to either anatomic or clinical pathology (see above), but most pathologists-to-be choose combined training programs, as these best provide the broad range of skills needed for practice in a community hospital. Such programs entail a series of rotations through the different departmental laboratories, followed by more advanced training in the major areas. On the other hand, the opportunity to pursue an aspect of pathology in even greater depth may be especially attractive to an individual planning an academic career in a medical school or teaching hospital. Departments of pathology in such settings have more resources in personnel and facilities, thus offering considerable opportunities for each pathologist to develop and maintain expertise in a subspecialty of diagnostic pathology and/or research. Additional training in some subspecialty area of diagnostic pathology such as cytology, hematopathology, and blood banking is required for subspecialty certification.

The opportunity to pursue research into the causes, nature, and manifestations of disease has always been an attractive part of pathology. Traditionally, research in pathology emphasized morphological evaluation of the changes wrought by different disease processes in cells, tissues, and organs. This aspect continues to be important, but increasingly the pathologist of today uses a wide range of research disciplines available to the modern biomedical researcher, including biochemistry, molecular biology, and immunology to mention a few. There really are no limits to the kind of research a pathologist might do, and the experimental pathologist must be prepared to keep up with new developments in a number of related fields in order to apply them to a research problem. The distinguishing feature of the pathologist's research is not the technique being applied but rather the orientation to disease—its cause, appearance, and detection.

VII. Organization of Pathology in Different Settings

Pathology is practiced differently in various settings. In an academic medical center embracing a medical school and teaching hospital, a department may easily have more than 20 pathologists. Because all the clinical departments in such a hospital are populated by medical specialists and subspecialists, it is incumbent on the pathology department to provide state-of-the-art diagnostic services in a variety of areas. Because no single pathologist or even a small group can be expected to acquire and maintain the enormous depth and variety of expertise needed in a large referral hospital, many diagnostic pathologists, each specializing in a different area, are necessary. For example, the pediatric pathologist primarily interacts with the pediatricians, the neuropathologist with the neurologists and neurosurgeons, and the microbiologist with the clinicians

specializing in infectious diseases. The subspecialty pathologist, by definition, is expected to deal effectively with highly unusual or complicated cases. In contrast to the large teaching hospital, the smaller community hospital is usually staffed with relatively few pathologists, each of whom is likely to be a generalist in anatomic pathology with additional responsibility for one or two aspects of clinical pathology. Although most practice within a hospital setting, some pathologists practice in free-standing private reference laboratories, which provide diagnostic services on specimens obtained from physicians' offices or hospitals.

VIII. The Pathologist and the Practice of Medicine at Large

In a narrow sense, the pathologist functions to perform and interpret diagnostic procedures on specimens obtained from patients; however, more broadly, the pathologist serves as a true clinical consultant very much involved in the care of the individual patient. This aspect of pathology practice is particularly pronounced in smaller hospitals where the clinical services are less highly specialized. The pathologist may be asked to advise whether or not certain tests are appropriate for a given patient's problem, in addition to helping to interpret the results to follow. Besides the role of advisor in the diagnosis and management of individual patients, the pathologist often serves as a general educational resource for the hospital staff. The pathologist is expected to keep abreast of advances in diagnostic medicine and biomedical research and is, thus, in an excellent position to bring new findings and concepts to the attention of clinical colleagues. The educational role of the pathologist is much in evidence when individual cases are discussed in departmental or hospital-wide conferences. In a common format, a case history is presented, followed by a discussion of the clinical and laboratory findings by the clinician and the pathologist. Such conferences provide stimulating interchanges of information and views from which all can learn.

The pathologist is also in a position to monitor the quality of medicine practiced in the hospital at large. Are too many tests being ordered? Are some tests ordered when others would be more useful or when others should be ordered first as a screening procedure? Are too many operations being performed in which pathological examination of removed tissues and organs discloses no abnormality? In short, the pathologist works closely with clinical colleagues and hospital administrators to make sure that laboratory tests are ordered appropriately and performed correctly, promptly, and at the least possible cost.

IX. Conclusion

In conclusion, diagnostic pathology is an intellectually challenging, diverse, and heterogeneous branch of medicine that links basic biomedical research to clinical medical practice. As such, the pathologist must be continually on the alert to adapt emerging medical and scientific knowledge to daily patient-related work.

Bibliography

Henry, J. B. (ed.) (1984). "Clinical Diagnosis and Management by Laboratory Methods," 17th ed. W. B. Saunders, Philadelphia.

Hill, R. B., and Anderson, R. E. (1988). "The Autopsy—Medical Practice and Public Policy." Butterworths, Boston.

Intersociety Committee on Pathology Information. (1989). "Pathology as a Career in Medicine," revised ed. Bethesda, Maryland.

Robbins, S. L., Cotran, R. S., and Kumar, V. (1989). "Pathologic Basis of Disease," 4th ed. W. B. Saunders, Philadelphia.

Rubin, E., and Farber, J. L. (1988). Pathology. J. B. Lippincott, Philadelphia.

Smith, R. D. (1986). Medical education and its impact on the future of pathology. *Arch. Pathol. Lab. Med.* **110,** 296–298.

Smith, R. D., Anderson, R. E., and Benson, E. S. (1985). Manpower needs and supply in academic pathology. *Arch. Pathol. Lab. Med.* **109,** 889–893.

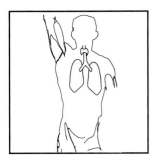

Pathophysiology of the Upper Respiratory Tract

KEVIN T. MORGAN, *Chemical Industry Institute of Toxicology*

DONALD A. LEOPOLD, *State University Hospital–Syracuse*

I. Upper Respiratory Tract Structure and Function
II. Upper Respiratory Tract Defenses
III. Selected Upper Respiratory Diseases
IV. Summary

Glossary

Adduct Chemical addition product

Cribriform plate Thin plate of bone, separating the front of the brain from the olfactory or upper region of the nose, through which the olfactory nerves pass carrying the sense of smell

Dehydrogenase Class of enzymes

Edema Swelling due to fluid accumulation

Histology Microscopic anatomy

Lamina propria Layer of tissue supporting the nasal epithelium that contains many blood vessels, nerves, glands, and other structures

Olfaction Sense of smell

Pathophysiology Study of basic processes responsible for functional disorders of an organism

Reflex Involuntary action in response to a stimulus

Rhinorrhea Free discharge of thin nasal mucus

Serous secretion Clear fluid that moistens membranes

Sinusitis Inflammation of the sinuses

Transudation Passage of serum or other body fluid through a membrane or tissue surface

Viscoelastic Specific physical characteristics that give mucus both fluid and solid properties

MUCH CAN BE LEARNED about biologic processes from studies of the pathophysiology of disease. With the exception of inflammation and infections, injury induced by inhaled air pollutants and sur-gery, and the effects of cardiovascular disease and drugs, the respiratory tract is generally trouble free. However, the upper respiratory tract (nose, pharynx, and larynx) does present an important portal of entry for many pathogenic (disease causing) organisms, including viruses, bacteria, and protozoa. The inspired air is a source of toxic chemicals that can damage the respiratory tract and possibly other organs. Allergens, such as pollen grains or household dust, also enter by this route to play havoc with our upper respiratory tract comfort, in addition to inducing both local and more widespread diseases. Examination of abnormalities associated with diseases and investigation of the mechanisms responsible for these changes can provide insight into the nature of normal body function, in addition to revealing characteristics of many adaptive or defensive responses.

The upper respiratory tract carries out a number of activities that improve our chances of survival in a hostile world. Food can be detected using the sense of smell, which is dependent upon the activity of olfactory structures in the nose. Olfaction can also provide warning of potentially dangerous airborne materials, such as toxic gases. In animals, and probably also in humans, the sense of smell plays a role in sexual activity. The upper respiratory tract, through its air conditioning activity, contributes to general body function by providing protection to the more delicate lining of the lower respiratory tract. These air conditioning functions include warming, cleaning, and humidifying the inspired air. The activity of this region of the respiratory tract is further complicated by its close relationship with the alimentary tract. The pharynx and larynx work together as a mechanical unit to permit passage of food or water into the gastrointestinal tract and transport of vomit in the reverse direction. During swallowing and vomiting it is essential that respiration be only briefly interrupted and that harmful materials such as stomach acid not enter the lower respiratory tract, where they can cause severe damage and even death. The upper respiratory tract can carry out these many tasks only because it also is provided with a number of effective defenses of its own. The wide range of functions carried out by the upper respiratory tract is reflected in its structural, physiological, and biochemical complexity.

I. Upper Respiratory Tract Structure and Function

The upper respiratory tract provides a continuous though complex connecting passage between the ambient air and the lower respiratory tract. In the nasal passages, this airway is tortuous because of the presence of the nasal turbinates and channels which connect to the nasal sinuses. In the pharynx and larynx, the airway is subject to many changes of shape associated with phonation and swallowing. The airway has a moist lining, or mucosa, which exhibits quite specific regional characteristics. The mucosa can be separated histologically and functionally into three major types, each with its own characteristic epithelium: squamous, respiratory, and olfactory. A fourth epithelial type, transitional epithelium, has been described in the nose of non-human primates but not as yet in humans. The pharynx is lined by protective squamous epithelium, consistent with the need for this region to handle air, food, water, and occasionally corrosive stomach acid during vomiting. Much of the larynx is lined by squamous epithelium, though distally; where it connects with the trachea, it is lined by respiratory epithelium. [See RESPIRATORY SYSTEM, ANATOMY.]

Each of these epithelia has different histological, physiological, and biochemical characteristics that influence its susceptibility to specific diseases. Squamous epithelium is much like skin and is not a common site of upper respiratory tract disease. The respiratory epithelium plays a major role in cleaning the inspired air through maintenance of the mucociliary apparatus (see below). Respiratory epithelium represents a common site of upper respiratory tract disease. Finally, the olfactory epithelium, which is confined to a limited region of the upper nasal cavity, is a highly specialized extension of the nervous system that is metabolically quite active and exhibits an array of interesting maladies. The olfactory epithelium is responsible for the sense of smell. Underneath each of these epithelia lie many glands, nerves, and blood vessels which also have regional characteristics. During studies of respiratory disease it is important to be aware of these topographic features of the upper respiratory tract lining and to discern whether changes are localized to specific regions or mucosal types.

Diseases of the upper respiratory tract are frequently induced by inhaled materials, as the nostrils provide a direct portal of entry for many airborne gases or particles. The nose is extremely effective at removing inhaled materials from inspired air. If you blow your nose after exposure to a dusty environment, the material you see deposited in your handkerchief was removed by your nose from the inspired air, demonstrating the ability of this system to protect your lungs. In contrast to adult humans, who can breathe both nasally and oro-nasally, human infants and laboratory rats are obligate nose breathers. This represents an important distinction, as mouth breathing has been shown to permit deeper penetration of gases and aerosols into the lower respiratory tract, presumably as a result of the loss of the effective scrubbing action of the nose. The physicochemical properties of inspired materials play a major role in the effectiveness of the nose to scrub them from the inspired air. Smaller particles (2–5 μm in diameter) and certain gases and vapors (carbon monoxide, acetone, phosgene) readily penetrate this defensive system to enter the lungs. Large particles (over 10 μm in diameter) are almost entirely removed by the nose, as are highly water soluble gases, such as formaldehyde and sulfur dioxide, and possibly also ultrafine particles (<0.2 μm in diameter). It is the removal of these materials, which include potentially infectious agents such as bacteria and viruses, that puts the upper respiratory tract at risk of disease. The upper airways, however, are provided with elaborate defenses.

II. Upper Respiratory Tract Defenses

A. Reflexes

The upper respiratory tract possesses a number of defensive or protective reflex responses. The normal function of these reflexes is to

1. reduce or prevent further entry of invading agents, such as occurs in reflex apnea (cessation of breathing) on exposure to irritant gases

2. remove the offending agent, as in coughing or sneezing
3. carry out adjustments of airway or vascular tone, or changes in quantity and nature of serous or mucus secretions as a component of complex protective responses, including inflammation.

Many of these normal reflexes are exhibited as disease symptoms. The common cold provides many examples of normal reflex defenses, such as coughing, sneezing, increased nasal serous secretions, and changes in vascular perfusion of the nasal mucosa, which are clinically manifested as partial obstruction of the nasal airway. Interestingly, the trigeminal reflex, which causes depression of respiratory rate in laboratory rodents, has been used extensively to assess the toxicity of inhaled irritants and to assess interspecies differences in the amount of potentially toxic airborne materials deposited in the respiratory tract. For researchers studying upper respiratory tract diseases, a wealth of information is to be gained from the way in which these reflexes respond to infectious agents and noxious air pollutants.

B. Mucociliary Clearance

The major routes of inspiratory airflow in the nose are lined by respiratory epithelium that provides an important airway defense system, the mucociliary apparatus. This apparatus plays a major role in cleansing the inspired air of inhaled gases and particulates. The nose filters about 10,000 liters of air per day, and the delicate, but very effective, mucociliary apparatus is readily perturbed by certain air pollutants. Altered mucociliary function can provide a sensitive indicator of early nasal damage.

The mucociliary apparatus derives its name from two of its major components, mucus and cilia. The mucociliary apparatus provides a continual "river" of slimy mucus which is driven over the mucosal surface by the combined action of millions of tiny, hairlike processes, the cilia. The human nose produces about a liter of mucus per day. A mucociliary apparatus is present in both the upper and lower respiratory tracts. In the nose the mucus flows posteriorly toward the nasopharynx, while in the trachea, bronchi, and other lower airways mucus flows upward toward the pharynx. This mucus, with any entrapped materials, is eventually swallowed. In the nose, mucus flow patterns are highly organized. In laboratory animals it has been found that not only mucus flow patterns but also mucus flow rates, cili-

ary length, and beat frequency are highly site specific. Mucus flow rates range from less than 1 mm/min to more than 2 cm/min. Understanding the pathophysiology of the mucociliary apparatus requires knowledge of the cellular and subcellular components of this system.

A wide range of materials have been shown to damage the mucociliary apparatus, including cigarette smoke, toxic gases, and viral and bacterial infections. In the upper respiratory tract, minor or even extensive damage to the mucociliary apparatus may be tolerated with little evident disease. However, in the lower respiratory tract disturbance of this system can be life threatening as a result of obstruction of the small airways, as can occur in patients with chronic bronchitis, a disease associated with cigarette smoking, or cystic fibrosis, a genetically controlled condition that inflicts considerable suffering on affected children and young adults through obstruction of the lower airways with plugs of thick mucus.

Because of the inherent complexity of the system, however, determination of the pathophysiology of these responses is difficult. Defects in mucus flow, and thus reduced clearance efficiency, may result from changes in the viscoelastic properties of mucus, modification of the depth or consistency of the liquid layer (hypophase) that lies beneath the flowing surface blanket of mucus, altered ciliary beat, cellular damage in the supporting epithelium, or disruption of normal function of glands, blood vessels, and nerves in the underlying lamina propria. Futhermore, each of these factors may have a multitude of causes.

It is evident from studies of abnormalities of the mucociliary apparatus that much remains to be learned about this highly effective airway defense system. If material deposited on the nasal surface penetrates the mucociliary apparatus, a number of other defenses are encountered, including both biochemical and immune systems designed to inactivate or detoxify invading agents or organisms. These defenses can contribute to as well as protect from disease processes in the upper respiratory tract.

C. Metabolism of Xenobiotics

Until recently there was little interest in the metabolism of xenobiotics (foreign chemicals) in the upper respiratory tract. However, with the increasing incidence of nasal cancer and other lesions in the nasal passages of laboratory animals used in toxicology studies designed to predict human risk, there has been increased interest in upper airway biochemistry. One important discovery was the finding that nasal tissues, especially the olfactory mucosa, have high concentrations of a number of enzymes that metabolize xenobiotics. These biochemical studies have been extended by the use of histochemistry, a technique that localizes enzymes to specific cell types. A picture of the complex pattern of enzymes that exist in the nose is gradually emerging. However, much remains to be learned about nasal metabolism, especially in humans.

The localization of enzymes that inactivate or activate xenobiotics plays an important role in interpretation of biochemical events that lead to tissue damage and subsequent disease processes. For certain materials, such as formaldehyde and acetaldehyde, metabolism by their respective dehydrogenases results in inactivation through conversion to a less toxic metabolite. It has recently become evident, however, that a number of chemicals are metabolized in the nasal mucosa to more toxic metabolites. These metabolites can produce degenerative changes and in some cases may result in nasal cancer. Examples of the later mode of toxicity include effects of certain nasal carcinogens, including several nitrosamines and phenactin, which induce lesions in the olfactory mucosa. The site specificity of these responses has been attributed to the high levels of cytochromes P-450 in this region of the nose. In fact, recent immunocytochemical studies have demonstrated that the P-450 activity is specifically located in Bowman's glands of the olfactory mucosa and in olfactory epithelial sustentacular cells. Knowledge of enzyme activity and location, and correlation of this information with detailed study of the histopathology of nasal responses to certain chemicals, provides considerable insight into the role of metabolism in nasal disease and disease resistance. [See CYTOCHROME P-450.]

D. Antimicrobial and Immune Defenses

A multitude of host defense mechanisms serve to protect the upper respiratory tract from infection. These defenses include mechanical cleansing, humoral and cell-mediated (immune) responses, and interferon. A number of other substances in nasal secretions may also provide protection against infectious agents. Such substances include lactoferrin, an iron-binding protein with a broad spectrum

antibacterial action. Nasal secretions also contain high levels of lysozyme, an enzyme that kills certain bacteria.

Mechanical cleansing of the nose is carried out by the mucociliary apparatus described above, as well as by sneezing and blowing the nose. However, the efficacy of these activities in protecting against infections remains to be established. Humoral immunity involves both local secretory activity and systemic antibody responses. In fact, considerable amounts of the immunoglobulin, IgA, are present in nasal secretions, while there is much less IgG. IgA in these secretions plays a major role in protection of the nose against both bacterial and viral infections. During viral upper respiratory tract infections there may be considerable transudation of serum proteins, particularly IgG and albumin, which may provide an effective nonspecific inflammatory reaction to infection.

The role of cellular immunity in resisting infections of the upper respiratory tract has received much less attention. As in other tissues, however, infiltration by neutrophils, macrophages, and other cell types represents a consistent feature of nasal inflammation. These cellular responses, which are designed to inactivate invading organisms, represent an important component of nasal inflammation induced by infectious agents and noxious chemicals. Furthermore, the nose acts as an important reservoir for microorganisms that can cause human disease, including *Hemophylus influenzae* (a bacterium), common cold viruses, influenza virus, and respiratory syncytial virus. Evidently, a delicate balance exists between these infectious agents and the nasal defenses, about which much remains to be learned.

III. Selected Upper Respiratory Tract Diseases

There are many upper respiratory tract diseases. A few issues have been selected for discussion to illustrate the value of studies of disease processes for increasing our understanding of human biology. In many instances the nature of the mechanisms responsible for upper respiratory tract diseases are still unknown. However, recent work in laboratory animals has shed considerable light on the nature of these complex processes. In order to discuss potential mechanisms of human diseases, it will be necessary to refer frequently to studies in laboratory animals.

Diseases are often named, classified, and ultimately investigated on the basis of the symptoms they evoke. Diseases of the upper respiratory tract are no exception. Through careful study of the symptoms of a disease process, and identification of the mechanisms responsible for them, a great deal can be learned about normal function and the nature and limits of the body's defense systems. Study of the common cold, for instance, has revealed many aspects of responses of the upper respiratory tract to viral infection, the role of nasal secretions and immune defenses, and resolution of the disease process. It is hoped that such studies will one day lead to the development of effective vaccines. Examination of laboratory animals exposed to toxic chemicals has demonstrated the susceptibility of specific cell types in the olfactory epithelium to each of these chemicals. Enzyme histochemical studies, combined with biochemistry, have revealed the nature of subcellular mechanisms responsible for these highly selective effects of certain chemicals. Furthermore, this work has resulted in increased understanding of the relationships between morphologic and functional changes, and the kinetics of repair and associated cell replication rate. Studies of nasal cancer in laboratory rats have also identified important molecular events, such as the production of DNA-adducts and activation of oncogenes in the nasal passages after exposure to certain carcinogens. This work contributes to understanding normal function through characterization of background levels of adducts and assessment of the role of DNA repair in upper respiratory tract defense. [See REPAIR OF DAMAGED DNA.]

Three areas of research are briefly discussed below—allergic rhinitis, effects of chemical air pollutants, and olfactory dysfunction. Allergic rhinitis was chosen because of its general familiarity. Studies of chemical air pollutants were considered because this is a matter of increasing concern for the public and is an area of active research in both laboratory animals and humans. The issue of olfactory dysfunction was selected because the authors consider this to be of general interest for nonscientists and scientists alike, and it is an area of current research activity for both of the authors.

A. Allergic Rhinitis

Allergic diseases are a major source of illness or disability in people of all ages. Allergic rhinitis presents as a set of symptoms resulting from nasal inflammation induced by allergic or hypersensitivity reactions. Allergic reactions were first described in some of the earliest writings. In fact, the death of one pharaoh was attributed to an allergic reaction to a bee sting in an account provided in Egyptian hieroglyphs. Hippocrates and Galen described allergic reactions in their writings, and they even suggested that they may result from environmental factors. The name "allergy" was given to this condition in 1906 by von Pirquet, who created the term from "allol," the Greek word for "change in the original state." This example demonstrates the common process of disease classification based on a characteristic deviation from the "normal," or healthy, condition. Detailed study of mechanisms of diseases commonly reveals multiple mechanisms inducing a common set of symptoms. Thus, increasing knowledge of mechanisms of disease generally adds to the complexity of this classification process. [See ALLERGY.]

Allergic rhinitis is a common condition which may not be taken seriously by doctors because it is not life threatening. However, a constantly itchy, excessively runny or stuffy nose disturbs sleep, is socially embarrassing, and can clearly impair the quality of life for affected individuals. It may even induce significant changes in patterns of facial growth. A better understanding of this form of rhinitis came after the discovery of immunoglobulin E (IgE) and its role in allergic disease in the mid-sixties. This work has led to considerable research on the mechanisms responsible for allergic conditions. The result of this work has been improved treatment regimens which have recently included the development of immunotherapeutic approaches.

A number of conditions must be differentiated from allergic rhinitis, including viral and bacterial infections, septal deviations, sinusitis, Kartagener's syndrome, vasomotor rhinitis, and even responses to certain medications. Diagnosis is based on exclusion of other diseases, distinction between infectious and noninfectious disease, and separation of allergic from nonallergic patients on the basis of history and a number of allergy tests. Rhinitis is divided into three broad groups—allergic, nonallergic, and infectious. There is some overlap, however, as patients may have two or more of these conditions simultaneously.

Understanding the pathophysiology of allergy requires knowledge of the immune system. The immune system consists of a set of cellular and humoral components that interact with many molecular structures (antigens) in a manner that permits distinction of self from non-self. This process is required for the identification of foreign materials, such as invading bacteria or viruses, and activation of systems designed to eliminate the invaders. This system can also respond inappropriately and lead to hypersensitive (allergic) conditions, such as allergic rhinitis. Interestingly, the science of immunology that examines and defines these complex immune processes, was originally an outgrowth of microbiology, which occupies itself with the study of infectious or disease causing organisms. [See IMMUNE SYSTEM.]

Immunologic studies have demonstrated that hypersensitivity reactions occur as a number of distinct types of disease processes (Types I to IV). Allergic reactions may be very rapid, as in immediate hypersensitivity, such as acute allergies to pollen or household dust. This condition is characterized by sneezing, eye irritation, nasal congestion and obstruction, and rhinorrhea. The mechanisms involved are not entirely known, but they involve IgE, mast cells and basophils, and several physiologically active compounds, or mediators, such as histamine, leukotrienes, and prostaglandins. These mediators induce edema of the nasal mucosa, spasm of smooth muscles, attraction of inflammatory cells, activation of platelets, and increased production of mucus. These complex mechanisms are then expressed as the characteristic symptoms of the disease. In contrast, delayed hypersensitivities, such as contact dermatitis, have a slower onset and involve a completely different set of cellular and tissue reactions. Despite differing mechanisms, acute and delayed hypersensitivities do have common features, including release of histamine and consequent pruritis (itching).

B. Air Pollutants, Toxic Chemicals, and Upper Respiratory Tract Lesion Distribution

It is becoming increasingly evident, especially from inhalation toxicology studies in laboratory animals, that upper respiratory tract lesions induced by noxious chemicals or particulates generally occur in specific locations in the nose or larynx. This is true for lesions on the middle turbinate of nickel workers, formaldehyde-induced lesions in the lateral meatus of rats, and responses to cigarette smoke on the dorsal surface of the epiglottis of rats. Certain materials administered by a noninhalation route (ingestion or injection) have also been found to induce nasal lesions in specific regions of the nose, such as

those produced by phenaceten, and the tobacco-specific nitrosamine, [4-(N-methyl-N-nitrosamino)-1-(3-pyridyl)-1-butanone] NNK.

The site-specific nature of these responses is attributable to (1) regional deposition patterns of inhaled materials, (2) regional tissue susceptibility, or (3) a combination of these factors. Regional deposition may account for the site specificity of lesions as a result of "hot spots" of deposition, resulting in areas which receive a high exposure compared with other regions. This site-specific distribution of deposition is attributable, at least in part, to regional airflow characteristics. For example, the anterior curvature of the middle turbinate is a preferential site of particle deposition in the human nose. This region of the nose is also a frequent site of nasal cancer in nickel workers, presumably as a result of high exposure of this site. Another common site of particle deposition is the region adjacent to the ostium internum, just inside the nasal entrance, where prominent air turbulence is developed, which would be expected to favor deposition. Regional deposition due to airflow characteristics probably also accounts for site specificity of lesions induced by certain toxic gases, such as formaldehyde in rats and rhesus monkeys.

Airflow patterns in the upper respiratory tract are extremely complex, and they can be difficult to study because of the inaccessible nature of these airways, which are encased in bone and cartilage. A number of approaches have been developed to study upper respiratory tract airflow, including the use of airway casts or models. These models, which have been developed in some cases specifically for the purpose of determining airflow patterns for interpretation of toxicology studies have shed considerable light on the nature of nasal airflow.

The major routes of inspiratory nasal airflow pass over regions of the nose that are lined by respiratory epithelium, with smaller flow streams passing to the olfactory region. Differences in airflow patterns between species may account for differences in the distribution of responses to certain toxic materials, such as formaldehyde. Lesions also occur in specific sites in the larynx, though less work has been published on regional deposition patterns in this organ. Common sites of laryngeal lesions which may be attributable to airflow-related deposition patterns include the posterior (dorsal in rats) surface of the epiglottis and the vocal cords. Factors other than airflow may also account for the presence of specific laryngeal lesions, however. Secondary tuberculous laryngitis in humans, for instance, has been attributed to deposition of tubercle bacilli in the larynx as a result of their transport from lungs in the sputum. Study of deposition of materials in the upper airways has led to considerable research on airflow, which in turn has shed light on the important anatomical and physiological features of these complex passageways.

In addition to site-specific deposition, lesions may occur in certain areas of the upper respiratory tract as a result of a regional tissue susceptibility. For instance, inflammation of the sinuses can be exacerbated by poor drainage of inflammatory exudate from these sites. This can result from mucosal swelling in the narrow passageways that normally permit drainage of the sinuses. Activities associated with marked changes in ambient pressure, such as scuba diving and airplane flights, can also initiate sinusitis. Thus the design of sinuses and their system of drainage renders them susceptible to infectious and inflammatory diseases. The more frequent presence of nodules on the vocal cords has been associated with singing or extensive use of the voice. If this association is truly causal in nature, the vibratory function of the vocal cords may account for this apparent susceptibility. Lesions of the olfactory mucosa in rats exposed to a number of chemicals, including nitrosamines, 3-methyl indole, and methyl bromide, have been attributed to site-specific metabolism of these materials, presenting another example of site-specific susceptibility of a tissue resulting in a disease process. In this case, the metabolic processes that may normally play a role in upper respiratory tract defense can induce disease through activation of these materials. The site-specific nature of squamous cell carcinoma of the larynx of tobacco smokers and of the throat of smokers and alcohol drinkers may also provide societally important examples of site specific susceptibility to toxic chemicals.

C. Olfactory Dysfunction

The perception of chemical stimuli is accomplished almost entirely through the first (olfactory) and the fifth (trigeminal) cranial nerves. The olfactory nerve provides information on the quality and intensity of the odorant, while the trigeminal nerve assesses the odorant's pungency. When this system malfunctions, there can either be a change in the intensity of an odorant stimulus (hyperosmia, hyposmia, or anosmia) or a change in its quality (dysosmia). There

is a natural range of olfactory ability that differs among individuals, with women generally having a better sense of smell than men. Olfactory function can also be adversely affected by a number of environmental contaminants, including tobacco smoke. Such variables should be considered when attempting to assess responses to experimental variables in olfactory function studies in human subjects. [See OLFACTORY INFORMATION PROCESSING.]

The first event in the process of olfactory stimulation is the movement of odorous chemical molecules through the narrow, dark passageways of the nose. Any process that obstructs their flow to the olfactory region at the top of the nasal cavity will decrease olfactory ability. Thus, nasal airflow characteristics play an important role in this process. Some materials may naturally fail to reach the olfactory area in sufficient quantities because they are rapidly removed from the inspired air by the more anterior regions of the nose. One group of conditions that adversely influences the sense of smell through impairment of air passage to the dorsal nose are the *obstructing nasal diseases*. These conditions include the common cold, where the accompanying edema of the nasal membrane physically prevents airflow, and thus transport of odorant molecules, to the olfactory receptors. Nasal allergic conditions can also lead to decreased olfactory ability for the same reason—mucosal edema—although this condition generally has a much longer duration. Obstruction from chronic sinusitis, nasal polyps, or nasal tumors can similarly decrease olfactory ability. Medical or surgical therapy is generally useful for people in this category, since their olfactory ability can return if the obstruction is removed. Individuals who have had a laryngectomy and are forced to breathe through a tracheostome in their neck have poor olfactory ability simply because their respiratory airstream no longer flows past the olfactory receptors.

Changes in the nerves of the olfactory system, in the nose or in the brain, account for most other types of olfactory dysfunction. In laboratory animals exposed to gaseous irritants, such as chlorine or dimethylamine gas, there is extensive destruction of olfactory sensory cells. However, very limited information is available on the functional impact of these changes. In the case of methyl bromide, which induces extensive olfactory epithelial degeneration after a single inhalation exposure, olfactory function is also lost. However, the sense of smell returns to these animals during the subsequent recovery process. The ability of the olfactory epithelium to carry out this repair makes it unique as a neural system. In fact, olfactory sensory neurons are one of the few known post-embryonic nerve cells capable of regeneration, as a result of which they have attracted considerable interest by neurobiologists. This is a good example of toxic responses being used for basic research on normal function. Interestingly, in laboratory animals there is frequently information on structural changes in the olfactory epithelium, while olfactory function data are generally not available. In human subjects the reverse situation is usually the case; in humans it is difficult to observe all the stages of an olfactory degenerative process because of ethical difficulties associated with such controlled experiments and the invasive nature of sample collection.

There is an unfortunate group of people who have lost some or all of their olfactory ability after an upper respiratory tract infection. This condition is categorized as *post-infectious olfactory dysfunction*. It is thought that the virus which caused the infection entered the olfactory nerve cells and induced modifications which resulted in their loss of function, or that the infection destroyed the sensory cells completely. This condition can occur after most minor upper respiratory tract infections and usually presents as a failure of olfactory ability to return after the nasal stuffiness has cleared. The loss can be partial or complete, is generally permanent with no known therapy, and no suitable animal model is available to study this condition.

After the odorant molecule has interacted with the primary olfactory neurons, directly or indirectly by currently unknown receptor mechanisms, the chemical information is changed into electrical messages. These messages are produced in the form of action potentials in the sensory neurons. The information is relayed to the olfactory bulb in the brain, which is located on the cranial side of the cribriform plate, through which these nerve fibers pass as many small bundles. The small nerve fibers, as they pass through the cribriform plate are susceptible to mechanical damage during head trauma, especially after occipital injury. *Post-traumatic olfactory loss* tends to be severe and shows little improvement. A curious aspect of this condition is the lack of recovery, even though neurons in the normal individual are constantly regenerating.

Many body functions decline with age, and olfaction is no exception. During the sixth or seventh decades there is a precipitous loss of olfactory func-

tion which is thought to be due to loss of primary neurons. This *age-associated olfactory deficit* can be so severe that the average 80 year old has only half the olfactory ability of a person in his or her thirties. As mentioned above, environmental toxins are another major cause of olfactory dysfunction. Individuals who persistently inhale the vapors of acids, aldehydes, heavy metals, solvents, and other sometimes unknown chemicals can suffer considerable olfactory epithelial damage and loss of their sense of smell. Furthermore, nasal cancer, which may be a consequence of environmental pollutants, can also lead to extensive loss of olfactory function through tissue destruction.

Each of the causes of olfactory dysfunction mentioned above can cause either a decrease or a total loss of olfactory ability. Sometimes, individuals will complain of a change in the quality of their sense of smell. This problem may present as either a distortion of the perception of an odorant, such as a rose smelling like vinegar (parosmia), or the perception of an odor when there is no evidence, in the opinion of adjacent observers, of such an odorant being present (phantosmia). The cause of these problems is all too frequently unknown, but they have been noted to occur in conjunction with many of the previously described causes of olfactory loss. People with these problems are greatly distressed, since much of their life is no longer perceived as pleasant. Such concern is frequently accompanied by loss of weight and possibly clinical depression. However, new therapies are becoming available to relieve these problems. When our olfactory function is normal, we are generally not aware of the pleasure and protection it provides. Dysfunction of our sense of smell, however, will quickly remind us of its importance for our quality of life.

IV. Summary

Much can be learned about normal function from disease processes. Allergic rhinitis represents "inappropriate" responses of normal immune defenses to allergens to yield a disease process characterized by excessive activity of this system as symptomatic hypersensitivity. Studies of the sense of smell have revealed the complex nature of this sensory system, the considerable metabolic capacity of the olfactory mucosa, and the role that metabolism plays in responses of this mucosa to toxic chemicals. Toxicology studies in laboratory animals, designed to assess human risks from exposure to air pollutants, are shedding light on normal upper airway function through the identification of critical physiologic, metabolic, and molecular biologic features of the airway mucosa and its defenses. Research directed at understanding disease processes is continuing to reveal the exquisite design and overwhelming complexity of the upper respiratory tract and is contributing in a significant way to furthering the science of human biology.

Bibliography

Feron, V. J., and Bosland, M. C. (eds.) (1989). "Nasal Carcinogenesis in Rodents: Relevance to Human Health Risk." Pudoc Wageningen, The Netherlands.

Dahl, A. R. (1988). The effect of cytochrome P-450-dependent metabolism and other enzyme activities on olfaction. *In* "Molecular Neurobiology of the Olfactory System, Molecular, Membraneous, and Cytological Studies" (Frank L. Margolis and Thomas V. Getchell, eds.), pp 51–70. Plenum Press, New York.

Korenblat, P. E. and Wedner, H. J., (eds). (1984). "Allergy, Theory and Practice." Harcourt Brace Jovanovich, Orlando.

Mygind, N., and Pipkorn, U., (eds). (1987). "Allergic and Vasomotor Rhinitis: Pathophysiological Aspects." Munksgaard, Copenhagen, Denmark.

Negus, V. (1958). "The Comparative Anatomy and Physiology of the Nose and Paranasal Sinuses." Livingstone Edinburgh and London.

Proctor, D. F. and Andersen, I., (eds). (1982). "The Nose: Upper Airway Physiology and the Atmospheric Environment." Elsevier Biomedical Press, New York.

Reznik, G., and Stinson, S. F., (eds). (1983). " Nasal Tumors in Animals and Man," Vols I, II, and III. CRC Press, Boca Raton, Florida.

Widdicombe, J. G. (1986). The physiology of the nose. *Clinics in Chest Medicine* **7,** 159–170.

Pediatrics

MARY ELLEN AVERY, *Harvard Medical School*

I. History
II. The World's Children
III. The Present Era
IV. Conclusion

Glossary

Antibiotics Substances (e.g., penicillin) derived from molds or bacteria that are used as drugs to combat certain infections caused by other microorganisms

Gene therapy Insertion of a normal gene into an appropriate progenitor (pluripotential) cell, once the gene and its product (a missing enzyme) are isolated, when an individual is born with inborn errors of metabolism, which result from a missing or altered gene that is necessary for the synthesis of an enzyme

Hilar Pertains to hilum of the lung, or lung root, through which nerves and vessels enter the lung; lymph nodes clustered around the major bronchi are called hilar nodes

Neonatology Specialization in the care, development, and diseases of newborn infants

Oral rehydration therapy Oral administration of a solution of sodium chloride and glucose in a 1 : 1 molar ratio with additional potassium chloride to dehydrated individuals; the mixture is balanced to facilitate absorption from the intestine to restore lost liquid; when given slowly (by spoon) to dehydrated infants, it is well tolerated and obviates the need for more expensive (and often unavailable) intravenous medication

PEDIATRICS (Grk. *paies,* child; *iatros,* physician) is the medical specialty dedicated to the maintenance of health, and diagnosis and treatment of diseases of children. Childhood is usually defined as the interval between birth and the completion of adolescence, or through puberty. For children with chronic, handicapping illnesses, a pediatrician may be the primary physician throughout life.

I. History

A. Seventeenth Century

Little evidence indicates that the illnesses of children were the subject of any serious study or the principal concern of physicians before the seventeenth century. For the most part, the care of young children was handled by nurses and mothers. Thomas Sydenham (1624–1689), an English physician, described many of the diseases common in children, such as scarlet fever, measles, and chorea. The most widely read prenatal book in the English language was by Walter Harris, ''Acute Diseases of Infancy,'' published in 1689. It was considered the most popular pediatric book of its era until it was supplanted in 1784 by Michael Underwood's ''A Treatise on the Diseases of Children.''

B. Eighteenth Century

In colonial America, the care of children was provided by mothers, and only sick children were seen by physicians. Hospitals, in the modern sense of the word, did not exist until the late eighteenth century; the delivery of infants was entirely in the hands of midwives, who also served as pediatricians. Although we do not have accurate figures, the mortality among infants in colonial America was staggering. Illnesses rarely seen today, such as diphtheria, cholera, typhoid fever, and smallpox, were among the hazards of childhood. In the absence of understanding the causes of these disorders, let alone the principles of hydration, maintenance, and the use of antibiotics or immunization, treatment of most of the diseases was bloodletting, sweating, or purging.

In the eighteenth century in Salem, Massachusetts, the mortality rate for female children was as high as 313/1,000 livebirths. The average number of children born per family was between 8 and 10, and only an average of 6 survived to adulthood.

In London, attention to the illnesses of children in England was highlighted by the writings of George Armstrong, who opened the first hospital and dispensary for children in 1769.

C. Nineteenth Century

In the nineteenth century, the population of the United States rose from only 5 million at the beginning of the century to approximately 23 million by mid-century. The first pediatric hospital in North America was established in 1855, when the Children's Hospital of Philadelphia opened for "the reception of children suffering from acute diseases and accidents"; this was more than half a century after Hôpital Des Enfants Malades opened in Paris in 1802. The Great Ormond Street Hospital in London was founded in 1852, and The Children's Hospital, Boston, became the second North American children's hospital when it was established in 1869. Vaccination against smallpox was introduced in approximately 1800, when Benjamin Waterhouse vaccinated his 5-yr-old son. Widespread use of vaccination did not occur for another 100 years. In fact, the germ theory of infectious diseases was not widely accepted until the studies of Robert Koch and Louis Pasteur in the late nineteenth century.

Pediatrics as a medical discipline can be dated from the close of the eighteenth to the beginning of the nineteenth century in France. The textbook "Traité des Maladies des Enfants Nouveau-nés et à la Mamelle" was published by Billard in 1828 and is considered to be the first serious text in the field of infant diseases. Contributions of the French included the invention of the stethoscope by Laennec in 1816, as well as the denouncement of the leading intervention of the times—bloodletting. Meanwhile, infant mortality rates in Boston between 1840 and 1845 showed no improvement over those of 1790. As urban centers developed, the death rate for infants actually increased.

Pediatrics did not emerge as a distinct branch of medicine until the second half of the nineteenth century, a period of enormous population growth in the United States. Before then, most hospitals did not admit infants under 2 yr old. By the end of the century, 24 children's hospitals were scattered across the country, and about half of the medical schools had a chair in pediatrics.

The first pediatrician to devote himself full-time to the care of children in the United States was the German-born clinician Abraham Jacobi, who established the first children's clinic at the New York Medical School in 1860. The first full professorship in pediatrics was established by vote of the medical faculty at the Harvard Medical School in 1893 when Thomas Morgan Rotch was appointed full professor. That same year, he presented the first paper on prematurity ever given before the American Pediatric Society, in its fifth year of existence.

The first meeting of the American Pediatric Society was held in September 1889, when Dr. Jacobi delivered his presidential address on "The Relations of Pediatrics to General Medicine." Interestingly, while the first presidents of the Society were (at least part-time) pediatricians, the fourth president was the famous professor of medicine Sir William Osler, who was also greatly interested in diseases of children. During this time (the second half of the nineteenth century), more than one-quarter of the children born in the civilized world died before their fifth birthday, and "summer diarrhea" was the cause of nearly half of these deaths.

D. Twentieth Century

The heralding of modern pediatrics in America dates from publication in 1896 of the textbook "The Diseases of Infancy and Childhood," by L. Emmett Holt, which had a major influence on the development of American pediatrics. In its first edition, this textbook recommended hypodermoclysis, the subcutaneous infusion of saline in the treatment of the dehydration from "cholera infantum," otherwise known as summer diarrhea. This was the first effort to infuse liquids into a dehydrated baby, heralding the more systematic understanding of fluid and electrolyte balance, one of the major achievements of the twentieth century.

A major emphasis in the first part of the twentieth century was infant feeding and the gradual movement toward artificial feeding of infants with cow milk. Before that time, throughout human history, infants had been fed human milk, as experienced indicated it was the only safe and effective way of feeding babies. Once it became possible to pasteurize cow milk and to add carbohydrate to it and dilute it, various formulas were advocated by those who felt that they had embarked on a "scientific" approach to infant-feeding. However, some pediatricians opposed pasteurization for fear it would decrease the nutritional properties of cow milk.

The emergence of biochemical research in pediatrics, initiated by Holt at the Babies Hospital in New York, was given an enormous boost when John Howland and McKim Marriott demonstrated that the acidosis accompanying diarrhea and dehydration was a consequence of bicarbonate loss in the stools; meanwhile Howland assembled a talented group of scientists and clinicians at the Harriet Lane Home of the Johns Hopkins Hospital, many of whom would later head pediatric departments in other medical schools. Among the major advances in the first part of the twentieth century was awareness of the importance of vitamins in the prevention of rickets and scurvy, understanding of the microbial basis of many infectious diseases that regularly weakened and killed children, and the evolution of antibiotics and antisera, which made it possible to treat many of these diseases. Major advances in public health resulted in nearly universal immunization, thus removing many causes of childhood morbidity and mortality, such as diphtheria and tetanus.

II. The World's Children

A. Developing Countries

Pediatrics, as it emerged in Europe and America, has been appreciated around the world and applied in some centers in most every country in the world. However, the tragic fact, as noted by James Grant in the UNICEF publication "The State of the World's Children, 1988," is that one death in every three in the world is the death of a child under 5 yr old. In some developing countries, death rates for children are little different than they were in the nineteenth century in Europe and America. In other words, the progress of recent years has not been extended to more than a fraction of the world's children.

The leading causes of death in poorer countries are diarrheal diseases and malnutrition. Approximately 1 million children die of malaria each year and nearly 2 million die of measles, the latter of which can be prevented by a single vaccination.

Acute respiratory infections, including whooping cough, which can be largely prevented by immunization, also contribute to an estimated 0.6 million deaths. Tetanus, which is also preventable, kills 0.8 million infants a year, and many other diseases are exacerbated in the presence of undernutrition, which is nearly universal in some developing countries.

Significant progress is being made in some areas with the advocacy of oral rehydration therapy and with efforts to achieve universal immunization. Smallpox, once a leading killer of children, no longer exists in the world today because of the eradication of the virus. Possibly, poliomyelitis and measles can also be eradicated, because the viruses that cause them have no animal reservoir, the diseases are readily recognizable, and effective immunization is available. If all susceptibles are immunized throughout the world, the viruses will disappear, and the diseases can be eliminated forever. [*See* INFECTIOUS DISEASE, PEDIATRIC.]

In the United States, infant mortality rates (deaths in the first year of life) were 165/1,000 livebirths in 1900 and 29/1,000 livebirths in 1950 and fell to approximately 10/1,000 livebirths in 1987. In Japan, Scandinavian countries, and some other European countries, infant mortality rates are even lower—<6/1,000 livebirths.

III. The Present Era

A. The Development of Subspecialties

The last half of the twentieth century has seen the development of multiple subspecialties within pediatrics. The largest of these is neonatology (study of the care, development, and diseases of newborn infants), which has contributed to the improved outlook for survival of low birth weight infants in particular (Fig. 1). Other subspecialties in pediatrics include neurology, cardiology and endocrinology (three of the earliest), followed by the study of infectious diseases, genetics, hematology, oncology, nephrology, gynecology, gastroenterology, and pulmonology, to name a few of an ever-expanding list. In addition, behavioral pediatrics and child psychiatry are important subspecialties of pediatrics and psychiatry. The surgical specialties also have their pediatric counterparts in orthopedics, otolaryngology, and ophthalmology.

As subspecialties and further subdivisions of

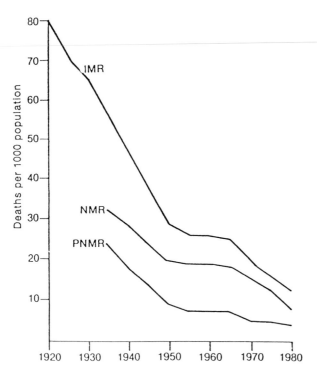

FIGURE 1 Infant mortality rate (IMR) is deaths in the first year of life, neonatal mortality rate (NMR) is deaths in the first 28 days, and postneonatal mortality rate (PNMR) is deaths between one month and one year. The sharp decline in deaths between 1930 and 1950 represents improved standard of living, nutrition, and access to medical care. The decline between 1965 and 1980 parallels the evolution of perinatal intensive care. Between 1980 and 1990 (not shown), the decline in mortality continues, but at a slower rate. Note that about half of all deaths in the first year occur in the first 28 days, and about half of those occur in the first week. The early deaths are mainly caused by complications of prematurity and congenital malformations. (U.S. National data, courtesy of Dr. Marie McCormick.)

them evolve, the need for someone to translate the new knowledge into patient care has emerged, and we now recognize the increasing role of the generalist or the primary care pediatrician to do this.

In the sections that follow, the most common illnesses and injuries that affect children's health are discussed. All of these will present to the generalist but often require the services of the subspecialist. Only by working together can the generalist and subspecialist succeed in achieving the common goal of pediatrics—the health and well-being of the world's children.

B. Diagnosis and Treatment of Diseases

1. Infectious Diseases and Host Defenses

Collectively, infections are the major causes of morbidity in children and contribute substantially to

mortality. Infectious diseases are the consequence of invasion of the body by microorganisms that then induce inflammation. They are classified as viral, bacterial, fungal, protozoal, rickettsial, or parasitic.

The development of effective vaccines against poliomyelitis, measles, diphtheria, pertussis, tetanus, rubella (German measles), and mumps has resulted in the near elimination of these formerly common and sometimes lethal infectious diseases (Table I). In the late 1980s, new vaccines have become available to prevent invasive *Haemophilus* influenza type B infection, which is one of the most common causes of meningitis in infants. A vaccine to prevent varicella infection (chicken pox) is currently being developed. Despite years of sustained research, no vaccine is available for one of the lead-

TABLE I Recommended Schedule for Active Immunization of Normal Infants and Children

Recommended age	Immunization(s)	Comments
2 mo	Diphtheria and tetanus toxoids with pertussis vaccine (DTP) Oral poliovirus vaccine (OPV)	Can be initiated as early as 2 wk old in areas of high endemicity or during epidemics
4 mo	DTP, OPV	Two-month interval desired for OPV to avoid interference from previous dose
6 mo	DTP, OPV	Third dose of OPV is not indicated in the U.S. but is desirable in geographic areas where polio is endemic
15 mo	Measles, mumps, rubella (MMR)	MMR preferred to individual vaccines; tuberculin testing may be done at the same visit
18 mo	DTP, OPV Haemophilus b diphtheria toxoid conjugate vaccine (PRP-D)	
4–6 yr	DTP, OPV	At or before school entry
14–16 yr	Adult tetanus toxoid (Td)	Repeat every 10 yr throughout life

Recommendations of the Committee on Infectious Diseases, 1988, American Academy of Pediatrics, Elk Grove Village, Illinois 60009.

ing causes of death in the world in the 1990s—malaria.

Another monumental achievement has been the development of effective antibiotics, starting with the pioneering observation of Fleming in the discovery of penicillin in 1928 at St. Mary's Hospital, London; shortly thereafter, Waksman at Rutgers discovered streptomycin. Many major accomplishments evolve from here, including the discovery of isoniazid, which accelerated the decline of tuberculosis, for which no effective vaccine exists even now. Although bacteria quickly developed defenses against some of the antibiotics (antibiotic resistance), a number of new compounds were tested and found effective, and the armamentarium of drugs against not only bacteria but some viruses is continually increasing. Despite vaccines and antibiotics, many viral and parasitic illnesses remain lethal at the present time and one in particular, human immunodeficiency virus, which causes acquired immunodeficiency syndrome (AIDS) is in epidemic proportions in certain regions of the world and may become a pandemic if the organism continues to elude efforts to develop an effective vaccine or a drug to combat the infection. [See ANTIBIOTICS.]

The body's defenses against infection include the production of specific antibodies and stimulation of cells that can kill the infecting organism. This complex system is known as the immune system and is orchestrated in part by the central nervous system, lymphoid tissues, and cells derived from precursors in the bone marrow called macrophages. [See IMMUNE SYSTEM.]

2. Clinical Genetics

Surely the greatest scientific discovery of our times (1953) was that of the structure of DNA, the chemical of the genes, which inaugurated the era of molecular biology. Although this new science has provided complete insight into some aspects of growth and differentiation, clinical applications in pediatrics remain relatively few. They are, for the most part, most relevant for prenatal diagnosis of some hereditary diseases. Widely applied examples include sickle-cell anemia, thalassemia, and hemophilia A.

Gene therapy is on the horizon and appears to be possible, but as of 1989 it has not been successful. Some success has been achieved in animals by insertion of a gene in bone marrow stem cells, which can then be reimplanted in the bone marrow.

The virtual explosion of information in molecular biology has brought important new insights into the pathogenesis of genetically determined disorders, of which over 4,500 have been described, each related to a single gene locus. Prenatal diagnosis is feasible by either chromosomal identification or tests on fetal fibroblasts, obtained by amniocenteses before 20 wk of pregnancy.

Many congenital malformations can also be diagnosed prenatally with the advent of ultrasonography. For example, most major malformations of the heart or urinary tract can be diagnosed before birth, so delivery can be planned in a setting where prompt surgical correction is possible.

3. Cardiology

Throughout the twentieth century, but most significantly in the last half of it, surgeons have pioneered in operative correction of congenital malformations of the heart. Initially described in detail by Maude Abbott in the famous Atlas of Congenital Malformations, these cardiac lesions could not be treated until a pioneering surgeon, Robert Gross in Boston, tied the first ductus arteriosis. In addition, another surgical pioneer, Alfred Blalock, with his cardiologist colleague, Helen Taussig, developed in 1945 the diagnosis and treatment of tetralogy of Fallot, a complex malformation that causes cyanotic congenital heart disease. In the 1980s, with the help of cardiopulmonary bypass, complete repair of the most common major malformations has been undertaken operatively, and a number of them recently have been treated through the use of therapeutic catheterization. Cardiac valves that are damaged and constricted can be dilated with balloons, and septal defects can be closed with plastic devices inserted through a cardiac catheter.

New technology has not only provided superior tools for diagnosis, such as modern imaging with computerized scans and magnetic resonance imaging, but has also brought to the bedside efficient, effective, mechanical ventilators and the capacity to monitor continuously the vital signs (temperature, heart rate, respiratory rate, blood pressure, and oxygen saturation of the blood). In pediatric cardiology in particular, ultrasonography has revolutionized noninvasive diagnosis. First the M-mode echo (motion mode) was used for a one-dimensional view of the heart) and then a two-dimensional echo was devised to demonstrate spatial relations. More recently, studies based on the Doppler effect (which detects motion) permitted calculation of flows, and color Doppler adds the capacity to discriminate valvular stenosis and insufficiency as well as intracardiac shunts.

Surgical correction of major lesions, such as transposition of the great vessels, was not possible before cardiopulmonary bypass of blood through an extracorporeal heart–lung machine was achieved. Coupled with deep hypothermia, it is now possible to stop the heart for about 40 min, long enough to enter it and correct major lesions. Cardiac transplantation in children is now performed with constantly improving outcomes.

4. Malignancies

Cancers are the third leading cause of death in children between the ages of 1 and 4 yr, lagging only behind injuries and congenital malformations, and is the second leading cause of death between the ages of 5 and 19 yr. The most common types of tumors by race are shown in Figure 2. Fewer than 10,000 cases are diagnosed each year in the United States. The mortality rate has declined by almost 50% in the past 30 years. The most dramatic improvement has been in the kidney tumor of infants, known as Wilm's tumor, and in certain types of leukemia. [See LEUKEMIA.]

The most common malignancy of childhood, acute leukemia, had been assumed to be always fatal. The remarkable observations of Sidney Farber, Louis Diamond, and others resulted in the effective use of a folic acid antagonist, aminopterin, in producing a remission (although no cures) in children with leukemia. This pioneering observation, published in 1948, opened the era of cancer chemotherapy. A combined approach with surgical excision, radiation, and chemotherapy was applied to many solid tumors and transformed the outlook from uniform fatality to a high probability of recovery.

Bone marrow transplantation is a drastic but sometimes lifesaving treatment for malignant disease and some immunodeficiency states that have failed chemotherapy, surgery, or radiation. Three types of transplantation are being used in the 1980s: autologous transplants involve the patients own cells that have been collected and treated *in vitro;* syngeneic transplants are from one identical twin to another; and allogeneic transplants occur between donor and patient who are not identical twins but have similar histocompatibility antigens. (These antigens are responsible for regulating the immune response, which allows identification of self from nonself; the more nearly identical they are, the less the chance of rejection.)

Human bone marrow transplants were first successful in the 1950s, but only in the mid-1970s were they widely used. Worldwide, by the late 1980s, more than 10,000 bone marrow transplants have been done. [See BONE MARROW TRANSPLANTATION.]

5. Neonatology

Neonatology is the study of events that occur in the first 28 days after birth, which is called the neonatal period. Many conditions are either unique to or most prominent during this critical period when the infant must make an adaptation to extrauterine existence.

The survival of newborn infants, especially those of very low birth weight, is dependent on understanding the course of growth and development *in utero,* some insight into placental function, and measurement of the cardiopulmonary adaptations at the time of birth. Essential to the maintenance of life to all infants is meeting their nutritional requirements. The mother's milk is usually optimal. Modifi-

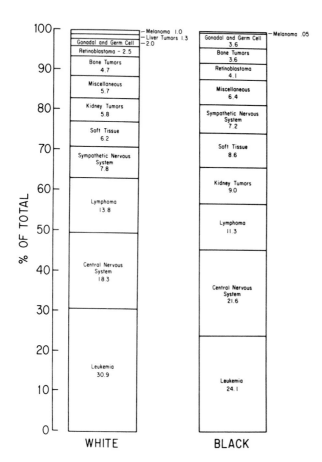

FIGURE 2 Cancer incidence for children ≤15 yr of age by site and race. Data obtained from SEER program 1973–1976. (Reproduced with permission from C. Pratt, 1985, *Pediatr. Clin. North. Am.* **32,** 541.)

cation of cow milk formulas can also support growth if maternal milk for some reason is not available.

Many studies elucidated the main physiologic and biochemical events that occur in the perinatal period. Application of this knowledge made it possible to provide interventions that facilitated the establishment of respiration and its support through mechanical ventilation when the infants are unable to sustain lung function on their own. Essentially, the successful use of ventilators led to the establishment of neonatal intensive care units, where skilled nurses and trained neonatologists can produce the appropriate environmental, mechanical, and nutritional supports for growth after birth, even after birth at 25 wk gestation.

The advances in neonatal care are dependent on advances in technology, such as the miniaturization of equipment that allows measurements of oxygen and carbon dioxide tension (blood gases) on a drop of blood, so that serial evaluation is possible. Microchemical determinations permit assessment of fluid and electrolyte needs as well as caloric requirements. In incubators, temperature can be controlled by servo devices from the infant's skin. Thus, cooling of the skin will trigger more heat, and, conversely, when the infant becomes overheated, the incubator temperature is lowered. This adjustment of the thermal environment is critical for very small infants who lose body heat easily and sometimes are unable to increase metabolism sufficiently to maintain a normal body temperature. The ability to individualize the environment to meet the infant's needs has been a central component of neonatal intensive care.

An aggressive approach to diagnosis and treatment has also greatly improved the outlook for a number of infants. The recognition of the multiple causes of hyperbilirubinemia (jaundice) in the newborn infant and the ability to prevent some and treat others has reduced the risk of one form of cerebral palsy associated with excess bilirubin in the blood during the newborn period.

One of the major causes of respiratory distress in the newborn infant is known as hyaline membrane disease. Advances in understanding this disorder and preventing it have taken place in the past few decades. Infants born before their lung has matured, with respect to the capacity to synthesize and secrete pulmonary surfactants, are predisposed to atelectasis (closure of the terminal airways at end expiration). This situation leads to severe mismatch of ventilation and blood circulation in the lung and often requires ventilatory support. Even with such support, some infants progress from hyaline membrane disease to a chronic lung disease of prematurity, sometimes called bronchopulmonary dysplasia. This latter disorder is presumably the consequence of a combination of injury in an immature lung from the high pressures needed to ventilate the surfactant-deficient lung and, in some instances, the injury produced by the higher inspired oxygen concentration necessary to keep the infant alive in the first days of life.

Other disorders of respiration that can produce major problems are persistent pulmonary hypertension, which can occur in term and post-term infants and, occasionally, those born prematurely. Aspiration of meconium (the material accumulated in the intestinal tract during gestation and passed, usually for the first time, after birth) can produce lethal airway obstruction.

Newborn infants can be born with major malformations, which need to be diagnosed and sometimes treated promptly. These include congenital malformations of the heart, obstruction of the intestine, or deficiencies in the abdominal wall that can be fatal if not treated surgically in the newborn period.

In North America, specialized services for very low birth weight infants, especially pediatric surgical services, cannot be available in all centers where babies are born. Consequently, so-called tertiary care centers, usually associated with major teaching hospitals, have been established, and regionalization of perinatal care is now widely practiced. Where possible, mothers are referred to centers for delivery of the baby when trouble is expected; if the trouble is unexpected and it occurs in a community hospital, transport teams are often available to bring the infant to a neonatal intensive care unit. Transport incubators have been designed so that intensive care can be carried out during the actual period of transport by ambulance or even plane. Secondary care institutions provide individuals trained in the needs of small infants once they have stabilized and no longer require ventilators. Thus, back transport can take place from the tertiary care center to the hospital where the child was born as a step toward discharge home.

6. Pulmonology (Lung Disorders)

The leading causes of admission to most pediatric services in North America in the 1980s are lung disorders in infants and children. In the infant group, hyaline membrane disease, as discussed in neonatology, affects approximately 40,000 infants per year in the United States.

In children in the first and second year of life, bronchiolitis, an inflammation of the small airways, occurs in epidemics usually between October and April in the Northern Hemisphere. It may be associated with respiratory syncytial virus but also occurs with parainfluenza and even adenoviruses. The infants characteristically have overinflated lungs and airway-narrowing that leads to prolonged expiration and wheezes, which are indistinguishable from those in asthma.

In older children, asthma is a major problem and thought to affect 2.4 million children in the United States under the age of 17 yr. Recurrent attacks of airway-narrowing from bronchospasm or edema can be relieved with bronchodilators, and in severe attacks with increased inspired oxygen and corticosteroids. Acute attacks can usually be modified by medications that can be given at home.

Pneumonias caused by either bacteria or viruses are common at any age but are most severe in immunocompromised individuals. Many infants have a relative depression in their gamma globulin (important in defense of bacterial infection) at about 3–5 mo of age. During this period, they are especially vulnerable to pneumonias. Children are also relatively immunocompromised in the wake of any viral infection but notoriously following measles, which allows bacterial pathogens to produce secondary pneumonias and tracheitis.

The ever-increasing population of individuals on chemotherapy for malignant disease usually have some suppression of their immune system, and a major complication of their treatment is pneumonia, often caused by organisms that are not virulent to a normal host.

Most lower airway infectious disease is associated with upper airway infection. Involvement of tonsils and adenoids as well as sinuses is not uncommon as a precipitating event for attacks of asthma and sometimes pneumonia. Sudden infant death syndrome is defined as a quiet death that occurs during sleep, usually without evidence of a struggle, in a circumstance in which no adequate explanation is available after an autopsy. It is reported in 1.5/1,000 livebirths and is more common in blacks and also in males. The age at greatest risk is 2–8 mo, but it has occurred in the first month and rarely after 9 mo. Other findings include an in-

creased risk in preterm infants and a slightly increased risk in siblings of victims. The cause is unknown but is probably multifactorial.

The lung is the organ most commonly involved in tuberculosis, and the form of tuberculosis in childhood is strikingly different from that in adults. The primary lesion can occur at any site in the lung, and the disease is usually not localized to the apices of the lung, as it often is in adults. In children, the so-called primary complex is often asymptomatic, although the enlarged hilar nodes can erode a bronchus and produce endobronchial disease. In the first years of life, a danger exists in the spread of the disease through the bloodstream, so that tuberculous meningitis and osteomyelitis are complications of pulmonary infection. The disease is spread by the airborne route, and, consequently, the lung is always the portal of entry for the tubercle bacillus. A number of antimicrobial drugs, including isonoazid and rifampin, are effective in treating tuberculosis.

Cystic fibrosis is a disorder of the exocrine glands that produces abnormalities in many secretions, including lung mucus. The initial symptoms may be referrable to pancreatic dysfunction in the form of absence of the digestive enzymes, trypsin, and lipases, which results in malabsorption of food. Failure to gain weight and bulky stools may be present. Pulmonary dysfunction may date from early childhood and may be mild or severe. Indeed, the manifestations of the disease vary greatly among individuals. The condition is so strongly associated with elevated concentrations of chloride in the sweat that a positive sweat test is confirmatory of the diagnosis. The inheritance is recessive, and both parents must be carriers of the abnormal genes. The common form of cystic fibrosis is at least related to a gene on the long arm of chromosome 7. Several other genes associated with cystic fibrosis have been identified through linkage studies. Each gene probably specifies a protein associated with epithelial cell chloride channels. [See CYSTIC FIBROSIS, MOLECULAR GENETICS.]

Affected children often have progressive colonization of their lungs with one or another *Pseudomonas* organism, which is often resistant to antibiotics. The predisposition to purulent pulmonary disease is clearly the consequence of some abnormality within the lung, because a few individuals with cystic fibrosis who have had lung transplants from normal individuals have not had recurrence of the disease in the transplanted lung.

Cystic fibrosis is one of the most common hereditary disorders of the white race. It has a prevalence of about 1/2,000 births. The estimated prevalence in the black population is 1/17,000 births, and it is rarest among Asians.

Great strides have been made in prolonging life and improving the quality of life for affected individuals. Formerly considered lethal in childhood, most individuals with cystic fibrosis are now living into their adult years, but as they reach adulthood many of them require repeated hospitalizations for aggressive physical therapy to dislodge their impacted pulmonary secretions and intensive intravenous antibiotic therapy to try to reduce some of the bacterial flora in the lungs. Death is usually associated with pulmonary failure.

7. Gastroenterology

Among the most common symptoms of gastrointestinal disorders in children are vomiting and diarrhea. Diarrheal diseases remain a major cause of morbidity and mortality in many parts of the world and are a continuing reason for hospitalization of infants in the developed world. The cause is most often bacterial or viral contamination of food or water (that can be for a number of different reasons). Although oral rehydration therapy has had a major impact on reduction of deaths from dehydration, some infants sustain water losses through the inflamed intestinal tract to the extent that they are in a state of shock and require intravenous support. The ability to give appropriate concentrations of fluids and electrolytes intravenously has been one of the major advances of the twentieth century. The knowledge of the appropriate concentration of electrolytes and glucose in oral solutions is of lifesaving importance. The World Health Organization and UNICEF have distributed electrolyte and glucose packets, diluteable in water, at very low cost in most developing countries, greatly reducing deaths caused by dehydration that is diarrhea-related. An appropriate oral rehydration solution is usually well tolerated even if infants have been vomiting. It must be given slowly at brief intervals, usually by spoon.

Occasionally, infants do not respond to oral electrolyte solutions, and after a few days attention to caloric intake becomes very important. Intravenous alimentation has been a major advance in this context. Solutions of amino acids, lipids, and glucose can be given intravenously in situations where the intestinal tract must be put to rest to restore the integrity of the mucosa.

Surgical correction of major malformations, such as omphalocele (herniation of the intestinal contents through a defect in the abdominal wall), and relief of intestinal obstructions have been made possible by the ability to maintain the nutritional status of the infant through intravenous alimentation. Among older children, ulcerative colitis and Crohn disease are serious, chronic, inflammatory diseases of the bowel, and they too are occasionally treated by "resting the bowel" and maintaining nutrition with long-term intravenous alimentation.

8. Endocrinology

Endocrinology is the study of the chemical messengers (hormones) that control aspects of growth and metabolism. Pediatric endocrinologists are most often consulted about disorders of growth (too tall, too short) or excessive weight or failure to gain weight. Congenital abnormalities of the reproductive system and external genitalia are other disorders best understood in terms of hormonal imbalance during fetal life. [See ENDOCRINE SYSTEM.]

The endocrine glands of the body include the anterior pituitary, which secretes growth hormone and adrenotropic hormone (ACTH), thyroid-stimulating hormone, gonadotropic hormone, and prolactin. Deficiency of growth hormone results in growth failure in childhood. Recombinant growth hormone produced by genetic engineering is available and effective. Excess growth hormone results in gigantism: after closure of epiphyses of long bones, the excessive growth is in the face, hands, and feet, a condition called acromegaly. The hormones of the posterior pituitary are vasopressin, also called antidiuretic hormone (a deficiency of which will result in polyuria and a condition referred to as diabetes insipidus), and oxytocin (which promotes milk ejection in mothers who breast-feed). The thyroid gland synthesizes thyroxin, which regulates the level of metabolism in all the cells of the body. Deficient thyroxin results in cretinism, which was formerly an important cause of mental and physical retardation. Routine screening of blood of newborn infants for thyroid activity reveals about 1/4,000 infants are deficient. Early treatment is effective. Excessive thyroxin increases metabolism, and weight loss, tachycardia, and heat intolerance are among the symptoms of an overactive thyroid. [See THYROID GLAND AND ITS HORMONES.]

The parathyroid glands synthesize parathyroid hormone, whose principal known target organs are bone and kidney. A lowered concentration of calcium in the blood is the most important factor promoting secretion of parathyroid hormone, which in turn causes increased reabsorption of calcium and decreased reabsorption of phosphate by the renal tubules. The hormone also increases the formation of an active form of vitamin D in the kidney and, hence, facilitates calcium phosphate absorption in the intestinal tract. Deficiency of parathyroid hormone results in low serum calcium levels, which can result in seizures. Loss of calcium from bone results in soft bones and curvature of growing bones, known as rickets. [*See* PARATHYROID GLAND AND HORMONE.]

The adrenal glands consist of a cortex that secretes cortisol, sex hormones (particularly androgens), mineral corticoids, and a medulla that produces catecholamines. Overactivity of the adrenal cortex in fetal life results in masculinization of the female from excessive androgens, a condition known as the adrenogenital syndrome. Gonads, both testes and ovaries, are capable of producing androgens and estrogens; however, under normal circumstances the testes produce predominantly androgens, and the ovaries produce primarily estrogen. The placenta should be viewed as another endocrine organ that secretes many hormones essential for a normal pregnancy. After the first few weeks of fertilization, progesterone becomes the major and indispensable hormone of pregnancy. Human chorionic gonadotrophin is another hormone of pregnancy that circulates in the mother's bloodstream and is the basis of a pregnancy test. [*See* ADRENAL GLAND.]

The pancreas contains islets of cells that produce insulin, glucagon, and somatostatin. Insulin and glucagon regulate circulating glucose levels; insulin deficiency results in diabetes mellitus, insulin excess in hypoglycemia. Diabetes onset in childhood is usually sporadic and may follow a viral illness such as mumps, which can affect pancreatic islet cells. Such individuals become insulin-dependent for life. Only pancreatic transplantation offers a possibility for cure but is not now recommended in children. [*See* INSULIN AND GLUCAGON.]

Metabolism refers to the biochemical processes within cells that maintain their integrity and function. In general, the field of clinical metabolism involves the study of various disorders of chemicals important for life. Most "inborn errors of metabolism" are now understood in terms of their mode of inheritance and their biochemical basis. Some are treated by withholding the substrate, which cannot be metabolized (e.g., a low phenylalanine diet in individuals with phenylketonuria, or no galactose in the diet of an infant who cannot metabolize it, as in galactosemia). Research in the 1990s is focusing on

gene replacement to permit synthesis of the missing gene product (usually an enzyme). As of 1989, no successes have been reported.

Nutrition of infants and children has had much attention through the ages and continues to be a topic of current interest. No serious student of the subject doubts the need for vitamins, minerals, and an appropriate balance of carbohydrate, protein, and fat. A prudent diet for children >2 yr old contains 30–40% of calories in fat, with more emphasis on vegetable and fish protein than on meat protein, to lessen risks of atherosclerosis. Amounts of polyunsaturated fats should about equal the amount of saturated fats. Carbohydrates should supply about half the total calories, and protein should be 10–20% from varied sources. Salt intake should not be excessive. Total calories should be adjusted to achieve optimal weight (neither too thin nor too fat). If a family history of premature death from coronary disease is present, diets more restricted in salt and fat intake are recommended. [*See* NUTRITIONAL QUALITY OF FOODS AND DIETS.]

During the first 15 yr of life, dietary iron intake should be 8–15 mg daily. Iron deficiency results in anemia, which is most common during rapid growth between 6 and 36 mo of age. All children should receive the recommended daily vitamin intake in food or as a supplement. [*See* COPPER, IRON, AND ZINC IN HUMAN METABOLISM.]

9. Nephrology (Kidney Disorders)

The principal problems confronting the pediatric nephrologist relate to congenital malformations of the kidney and urinary tract infections, which may or may not be associated with anatomic abnormalities. Other functions of the kidney that are the concern of nephrologists include disorders of acid–base balance, which are often associated with inborn errors of function of the renal tubule. Hypertension is controlled by the renin–aniogtensin–aldosterone system, as well as extra renal factors such as atrial naturiuretic peptide, vasopressin, and catecholamines. In general, both systolic and diastolic blood pressure rise slowly with age. In children <2 yr old, the upper limit of normal blood pressure is about 110/70; by 10 yr old, it is about 130/90.

The kidney is sometimes the target for immunologic disorders, one of which is glomerulonephritis. This condition, associated with bloody urine, is the result of increased permeability of the capillaries in the glomeruli of the kidneys (and often a reduction of the level of circulating complement C3). Hypertension is common. The condition sometimes is preceded by a pharyngeal or cutaneous infection with group A beta hemolytic streptococci and usually resolves after 1–2 wk.

Another glomerular lesion that leads to increased protein losses in the urine with consequent accumulation of fluid in the body is nephrosis. The condition may be recurrent and chronic, but in about 85% of the cases it is of gradual onset and self-limited. Serum complement C3 is normal in nephrosis, and the condition usually is responsive to corticosteroids. A more serious form, but also fortunately rarer, is the congenital nephrotic syndrome, which may require support of the infant for several months of life to permit adequate growth for kidney transplantation. Renal transplantation is now so successful that virtually any child with chronic renal failure should be considered a candidate regardless of age. Immunosuppression is necessary when the donor kidney is not of identical tissue type.

10. Bone Disorders

Bone disorders of children can be related to inborn errors of metabolism, disturbances in calcium and phosphorus balance, infection, and immunologic disorders, such as rheumatoid arthritis. Adolescents, particularly those who have had excessive stretching of the back ligaments during sports or dancing, for example, may develop curvature of the spine known as scoliosis. Early detection of scoliosis is important, as it is possible to insert a rod that will maintain the spine in a straight position until the vertebrae adequately fuse. Surgery is usually restricted to curves with angles >40–50°. Study of the natural history of scoliosis suggests that curves <40° will not progress further at the end of the growth spurt, whereas those >50° tend to be progressive.

11. Neurology and Behavioral Disorders

In pediatrics, neurology encompasses a wide variety of congenital defects: brain injury associated with asphyxia at birth, infections of the nervous system, and disorders associated with convulsions and coma.

Mental development in children proceeds in a series of successive stages. Within a few hours of a normal birth, an infant fixes on faces and has a generally alert appearance. Enhancement of memory continues during the first months of life. By the end of 1 yr, the infant is formulating speech sounds, which during the second year gradually merge into connected words and sentences. During childhood, development increases to more complex levels of performance, much of which depends on appropriate stimulation and learning experiences.

The period of human development is relatively long compared to most animals and permits time for the brain to acquire and store information and achieve the capacity for reasoning, thus allowing the individual to participate in complex social and technological domains.

In general, the neurologist approaches the patient with the idea of localizing in space and in time the defect that manifests itself as dysfunction of the nervous system and ascertaining its cause. Pediatric neurologists are concerned with developmental delays in motor function, as well as cognitive function, and in speech and language. Infants and children are also subject to some hereditary disorders that lead to progression of neurologic dysfunction by virtue of accumulation of abnormal substances in neurons (Tay-Sachs disease) and demyelinating disorders such as multiple sclerosis and the neurologic lesion of AIDS. In addition, some fixed disorders (i.e., not progressive) include all forms of motor disorders, collectively called cerebral palsy.

Seizures in childhood may be from many different causes. The most common are simple febrile seizures, which are usually generalized and occur most often from 6 mo to 6 yr of age and are, by definition, under 15 min in duration. They are usually single, and full recovery is the rule. More complex seizures require diagnostic investigation to explore the possible causes, such as infections, tumors, malformations, or strokes (vascular occlusions). When the cause cannot be defined, the condition is known as epilepsy. Most of the time, individuals prone to seizures can become symptom-free with appropriate medications. [*See* EPILEPSY; SEIZURE GENERATION, SUBCORTICAL MECHANISMS.]

Neurologists, psychiatrists, and psychologists are concerned with behavioral disorders, some of which are thought to have a biochemical basis. Major depressive disorders can occur in children and adolescents. Young people are also prone to a chronic eating disorder characterized by loss of weight and distortion of body image (anorexia nervosa) or excessive eating followed by induced vomiting (bulimia). [*See* EATING DISORDERS.]

An attention-deficit hyperactivity disorder is characterized by a short attention span, high distractibility, and impulsive behavior. The biologic basis of this relatively common dysfunction in childhood is not known.

A serious pervasive developmental disorder known as infantile autism can lead to a lifetime of impaired ability to communicate verbally with others. These individuals have abnormal motor behavior, emotional lability, and sometimes self-harming behavior. This condition has been associated with mental retardation, although this is not a consistent finding. Approximately one-third of affected individuals can learn to function in sheltered or supervised environments, but two-thirds of affected individuals need constant supervision and support throughout life.

12. Skin Disorders

A wide variety of diseases are manifest in the skin. In general, the most common disorders are simple infectious lesions (such as impetigo), contact dermatitis (such as poison ivy), insect bites, and hypersensitivity lesions (such as hives).

Acne is a common inflammatory condition of skin that appears predominantly over the face and shoulders, about 1–2 yr before the onset of puberty. It results from androgen stimulation of sebaceous (oily) glands. Obstruction of the glands promotes accumulation of secretions (whiteheads) that in time become open and darken as blackheads; when infected they are pimples. Acne may be drug-induced at any age, most commonly by cortisone or similar medications.

A number of more serious congenital lesions are known to involve the skin. These include such diseases as neurofibromatosis (tumors of peripheral nerves that can appear as lumps under the skin) and epidermolysis bullosa, a serious inherited disorder of the epidermis, which leads to peeling of the skin and in its severe form can be fatal.

A group of scaly lesions are also disorders of the epidermis, one of which is psoriasis, which can be present in infancy and is likely to recur; in some instances, it persists throughout life. Another skin disorder is eczema, an inflammatory disorder most common in young children who have remission and exacerbations over a 3–5-yr period. Intense itching and weeping of the lesions are characteristic. A family history of allergic disorders is common.

13. Injuries and Substance Abuse

a. Injuries Each year, more than 22,000 children die in the United States of injury-related causes, most commonly related to motor-vehicle accidents. This statistic makes injuries the leading cause of death beyond the neonatal period. They exceed deaths from cancer, congenital anomalies, and heart disease all together. An estimated 20% of all children will sustain at least one injury that requires medical attention; 17% of all pediatric hospitalizations nationwide are injury-related. The morbidity from injuries is also severe, and they can impose life-long serious disability.

b. Substance Abuse Substance abuse is defined as the intentional introduction of a drug into the body for stimulation and pleasure. Some of the drugs are labelled "recreational," which is unfortunate because most of them, such a tobacco and alcohol, ingested over a period of time can result in addiction as well as injury to the lung (cigarette smoke) and to the brain or liver (alcohol).

More powerful stimulants that have come into wide use for the sake of artificial stimulation include amphetamines, cocaine, and the hallucinogins marijuana, lysergic acid (LSD), and phencyclidine (PSP or angel dust); sedative hypnotics, such as barbiturates and benzodiazepines, especially valium, are also widely used. Among some children, glue- and hydrocarbon-sniffing are practices that can produce very dangerous side effects with injury to the heart, brain, and liver. Cardiac arrythmias and sudden death have been reported. [See COCAINE AND STIMULANT ADDICTION.]

The rising use of these agents among high school students in particular is a cause of major concern. In a 1986 study, more than half of high school seniors reported using marijuana, and about 5% used it daily; cocaine was used by about 13% of high school students. The form of cocaine known as "crack" is particularly dangerous and has been associated with sudden death among its users. [See MARIJUANA AND CANNABINOIDS.]

c. Poisoning Childhood poisonings continue to be an important public health problem. A major effort to respond promptly to poisonings has been undertaken by nationwide effort to establish poison control centers, which in 1985 responded to over 1,400,000 calls. The approach must be prevention, which will require a major coordination of educators to inform school students of the dangers of exposure to poisons, even when they are labeled recreational drugs, and efforts to restrict availability of some of the more dangerous and potentially lethal ones, such as cocaine.

d. Lead Poisoning Lead poisoning remains a significant problem for the fetus *in utero,* as well as infants and children. Lead is ubiquitous in the environment, deriving primarily from automobile exhaust, produced by lead-containing gasoline. Also, paint in older houses being renovated produce a lead-containing dust that can be inhaled, and sometimes the water supply is affected when pipes are soldered with lead. Rarely, acid fruit juices kept in ceramic pitchers with lead paint glazes can leach enough lead into the juice to cause acute and even fatal poisoning. Burning car batteries can cause massive lead intoxication from the inhalation of fumes.

Elevated lead levels in blood have been reported in approximately 12% of black children and 2% of white children in the United States, mostly in those living in urban areas and in old housing with lead-containing plaster and paint. The main source of exposure for adults is the diet, with such products as bone meal (which may be contaminated with lead) and drinking water from lead leached out of pipes that are usually in older homes. Even low levels of exposure before birth are associated with significantly slower achievement of developmental milestones in the first years after birth. Acutely, the symptoms may relate to gastrointestinal problems, abdominal pain, anemia, and neurologic abnormalities.

Treatment requires immediate separation of the affected infant or child from the source of lead and decontamination of the environment. Often, hospitalization of the child is required for adequate chelation therapy to mobilize lead from bone and enhance secretion in the urine.

e. Child Abuse Any form of physical maltreatment of children that inflicts injury, and even neglect, with its consequent emotional deprivation constitutes child abuse. Many state laws mandate reporting even suspicion of injury or neglectful situations so that the situation can be investigated for the protection of the child.

Child abuse and neglect are now recognized as medical conditions with complex aspects that often require a multidisciplinary approach to management. The pediatrician responds to the need to diagnose and treat the injuries and has an obligation to mobilize services of social workers, psychiatrists, and sometimes lawyers to protect the child from recurrent assault.

As the extent of sexual abuse of small children increases, public awareness is increasing as well. In general, forced genital contact or rape are crimes against children and are considered such by the legal system.

f. Burns and Smoke Inhalation Burns are second only to motor-vehicle-related injuries as a cause of mortality and morbidity among children. Sometimes the injury is flame-related and affects the skin; in other circumstances, it involves smoke inhalation, which accounts for nearly half of fire-related deaths. Toxicity of smoke is such that it causes massive pulmonary edema, sometimes after a delay of several hours. Individuals exposed to smoke inhalation must be immediately hospitalized because vigorous respiratory support can be lifesaving. After massive skin burns, loss of water through the skin is a great danger resulting in dehydration, which develops quickly. Rehydration should begin as soon as possible after a burn, using oral electrolyte solution to replace some of the extensive losses of fluid across the burned surface. Intravenous fluids are required with extensive burns. [See THERMAL INJURIES.]

Prevention involves surveillance on many fronts, including the reduction of the exposure of infants to water with a temperature over 50°C, creation of barriers around grates or radiators that may have hot surfaces, and the establishment of smoke detectors and fire extinguishers in all homes. The immediate response to a burn should be the application of cold water to delay the inflammatory response, and, in the event of smoke inhalation, oxygen should be provided to the victim as soon as respiratory distress is evident. Hospitalization is required for all serious burns.

IV. Conclusion

This brief discussion of diagnosis and treatment of diseases, injuries, and psychological disorders is aimed to illustrate some of the problems present in childhood and emphasizes required sophisticated diagnostic and therapeutic interventions. Disorders of other organ systems are often referred to the specialist, and discussion of them can be found in other portions of this Encyclopedia. For example, very few pediatricians treat eye diseases in infancy, although they are the ones that must be alert to detection of them. Likewise, problems of the ears, nose, and throat, other than straightforward infection, are referred to an otolaryngologist.

Pediatricians are child advocates, who point out

the enormous discrepancy between what can be done and what is done for children of the world. Their central role is to maintain the health of the child, diagnose and treat illness, and at all times join forces with parents, educators, and other concerned citizens to work for equal access to preventive and therapeutic services for children everywhere.

Acknowledgment

I am grateful to Dr. Hughes Evans for review of historical sections of this review and to Dr. Lewis First and Dr. Karen McAlmon for helpful discussion of which issues to include and advice on appropriate emphasis.

Bibliography

Antonarakis, S. E. (1989). Diagnosis of genetic disorders at the DNA level. *N. Engl. J. Med.* **320,** 153.

Avery, M. E., and First, L. R. (eds.) (1989). "Pediatric Medicine." Williams & Wilkins, Baltimore.

Cone, T. E. (1979). "History of American Pediatrics." Little, Brown & Company, Boston.

Crick, F. (1988). "What Mad Pursuit: A Personal View of Scientific Discovery." Basic Books, New York.

Grant, J. P. (1988). "The State of the World's Children." Oxford University Press, New York.

Greven, P. J. (1970). "Four Generations, Population, Land and Family in Colonial Andover (Mass.)." Cornell University Press, Ithaca.

Howland, J., and Marriott, W. McK. (1916). Acidosis occurring with diarrhea. *Am. J. Dis. Child.* **11,** 309.

McKusick, V. A. (1988). The morbid anatomy of the human genome. IV. *Medicine* **67,** 159.

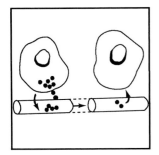

Peptide Hormones of the Gut

TADATAKA YAMADA AND CHUNG OWYANG, *University of Michigan Medical Center*

I. Peptide Hormone Physiology
II. Peptide Hormone Families
III. Clinical Significance of Peptide Hormones

Glossary

Antrum   Distal-most portion of the stomach

Endocrine action   Action that is mediated by transfer of a hormone from the effector cell to the target cell via the circulation

Hormone   Chemical messenger from specific cells in one part of the body that traverses to a nearby cell or to a distant site where it exerts an action on a target cell that recognizes it

Parietal cell   Cell in the stomach lining that is responsible for gastric acid secretion

Peptide   Molecule consisting of a relatively small number of amino acids linked together, amino-terminal end to carboxyl-terminal end; in this chapter, the focus of discussion will be on hormones of the gut that are peptides

Receptors   Sites on a target cell that recognize and bind a specific hormone and initiate the transduction of the binding event to a cellular response

PEPTIDE HORMONES of the gut serve as the chemical messengers that connect cells or organs in the body so that they can perform integrative tasks. The messengers can be delivered via the circulation or the interstitial space. The function of any organ in the gastrointestinal tract is tightly controlled by a variety of neural and hormonal regulators that interact in a temporally appropriate fashion to aid in the process of nutrient assimilation. There exists a diverse array of peptide hormones, many that can be grouped into families that may have arisen from a common ancestor by tandem gene duplication. Peptide hormones have been implicated in the pathogenesis of various gastrointestinal disorders but also hold promise as useful adjuncts in the diagnosis and treatment of other diseases.

I. Peptide Hormone Physiology

A. Introduction

The concept that regulatory substances could be released at one site of the body to control physiological functions at another site arose from the pioneering studies of Bayliss and Starling in 1902, who noted that the duodenum, on acidification, releases a substance that causes the pancreas to secrete bicarbonate-containing juice. This substance, which was known only by its *function* at the time, was called a *hormone* and labeled as *secretin*. Three years later, Edkins described another hormonal activity, labeled as *gastrin,* that mediated the secretion of gastric acid in response to alkalinization of the stomach. During the first half of this century, a number of putative hormonal substances were described by their *functions* in this fashion, although their *structures* were unknown.

In 1959 a dramatic improvement in the detection of these hormonal substances, many of which proved to be small polypeptides, was made possible by the development of the technique of radioimmunoassay by Yalow and Berson. In this technique, the binding of a radioactively labeled antigen (such as a hormone) to an antibody that recognizes it, is competitively inhibited by unlabeled antigen in an assayed sample. As depicted in Fig. 1, the quantity of specific antigen in the assayed sample is a measurable function of the displacement of labeled antigen from the antibody. The development of the radioimmunoassay technique was paralleled by the

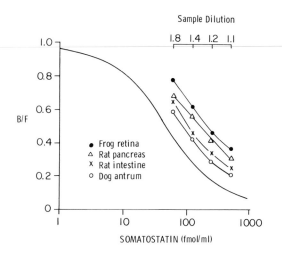

FIGURE 1 Standard curve for radioimmunoassay of somatostatin. The ratio of ^{125}I-Tyr^1-somatostatin bound (B) to antibody or free (F) is a semilogarithmic function of unlabeled somatostatin added to the incubation mixture. Extracts from various organs displace label binding in parallel to the standard curve, and thus the concentration of somatostatin in these samples can be estimated. [From DelValle, J., and Yamada, T. (1990). *Annu. Rev. Med.* **41**, 447–456.]

evolution of advanced peptide purification techniques [e.g., gel filtration (separation of molecules by size), ion exchange chromatography (separation by charge), immunoaffinity chromatography (separation by antigenic properties), reverse-phase high-pressure liquid chromatography (separation by hydrophobicity), and fast protein liquid chromatography (more rapid purification using combinations of the above)]. Moreover, through peptide microsequencing techniques, it became possible to obtain the amino acid sequences of peptides that were purified in only minute amounts. Thus, for example, a peptide such as somatostatin (see below), which was first purified in 1973 from the hypothalami of 500,000 sheep by a team of scientists after several years of struggle, could be purified and structurally analyzed from a single human stomach in 2 weeks by an undergraduate student less than a decade later. The availability of peptide sequences permitted scientists to take an additional step in structural analysis through recombinant DNA technology. By fashioning synthetic oligodeoxynucleotide probes specific for known amino acid sequences (on the basis of the genetic code), it is possible to screen such a "library" of recombinant bacterial or viral clones containing DNA molecules complementary to messenger RNA extracted from a tissue known to contain the peptide hormone of interest. By determining the nucleotide sequence of the recombi-

nant DNA molecule thus identified, one can deduce the amino acid sequence of the precursor molecule that is processed via a variety of steps to form the biologically active peptide. [*See* RADIOIMMUNOASSAYS.]

B. Mechanisms of Peptide Hormone Action

Modern advancements in research technology have reversed the problem faced by scientists studying peptide hormones earlier in this century. Now, instead of physiological functions in search of the hormones responsible for them, we are faced with a host of peptide hormones, the physiological functions of which are unknown. In the broadest sense, these small molecules serve as the means of communication between one part of the body and another, and between distant or adjacent cells. They permit the body to function as an integrated unit rather than as an amalgamation of unrelated parts. The mechanisms by which peptides exert these effects are multiple (Fig. 2). In the classical *endocrine* sense, a peptide hormone is released by cells at one site into the bloodstream and, via the circulation, delivered to a distant target cell possessing receptors that recognize the hormone. Under these circumstances, the physiological event caused by the hormone can be related to its concentration in the circulation. Another mechanism by which peptide hormones can act is called *paracrine* effect. In this case, a peptide is released into the interstitial space and acts on an adjacent cell that has appropriate receptors. *Neurocrine* actions of peptides are specialized forms of paracrine action in which the interstitial space is the synaptic junction between a nerve and its target cell. Under other circumstances, a peptide released locally can directly influence the cell from which it originated. This feedback action is termed *autocrine*. The physiological function of peptide hormones as paracrine, neurocrine, and autocrine effectors is difficult to study because of the inability to sample their concentrations in the interstitial space.[1] Nevertheless, using a variety of means involving pharmacological, histochemical, and biochemical approaches, peptides have been implicated as functioning physiologically

1. This is an unusual use of the term "hormone," which is defined as an agent that is released from one organ or region into the systemic circulation to act on a target in another region. By tradition, peptide effectors are called "hormones" for lack of a better term although they are endocrine peptide, paracrine peptide, neurocrine peptide and autocrine peptide *effectors*.

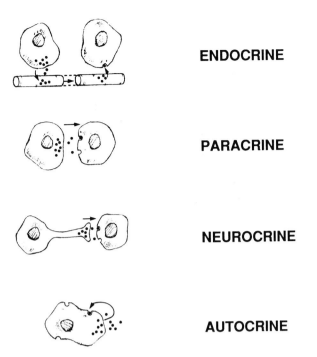

ENDOCRINE

PARACRINE

NEUROCRINE

AUTOCRINE

FIGURE 2 Mechanisms of action of gut peptides. [From DelValle, J., and Yamada, T. (1990). *Annu. Rev. Med.* **41,** 447–456.]

through a combination of these various routes. Some peptides, such as somatostatin, act through all four mechanisms. Others, such as vasoactive intestinal polypeptide (VIP), appear to function only through one (neurocrine in this case).

C. Physiological Functions of Hormones

Most studies of peptide hormone actions must be classified as *pharmacological* as opposed to *physiological,* for the latter implies proximity to the events that occur in real life. To demonstrate the *physiological* effect of a hormone, one must be able to produce an observed *pharmacological* effect of the substance at a target organ by reproducing concentrations of the hormone that the target organ is exposed to under normal or *physiological* circumstances. For example, gastrin is known to be a potent stimulant of gastric acid secretion (see below). When an animal eats a meal containing a substantial amount of protein, the concentration of gastrin in the circulation rises to a certain level, perhaps 50 fmol/ml. This, in turn, is followed by an increase in gastric acid secretion, perhaps to 25 mEq/hr. Most importantly, when the same animal under fasting conditions is given an intravenous infusion of gastrin to produce a circulating concentra-

tion of 50 fmol/ml, a level of gastric acid secretion the same as that observed after a meal is achieved. Moreover, when the gastrin that is released in response to meal ingestion is neutralized in the circulation with antibody to gastrin, the acid secretory response of the stomach is virtually abolished. Such gross *in vivo* studies are often required to ascertain the true physiological role of peptide hormones in integrated systems. Of course, it is possible to circumvent these cumbersome studies if specific receptor antagonists are available. By observing the alterations in physiological functions induced by single applications of a highly selective receptor antagonist for a peptide, it is possible to deduce a functional role for the peptide in question. Unfortunately, such useful antagonists are available for only a few peptides such as opioids or cholecystokinin. Future development of these antagonists will greatly facilitate studies to distinguish between the *physiological* versus *pharmacological* actions of peptide hormones.

D. Gastrointestinal Functions Under Hormonal Control

Virtually every function of the gut is tightly controlled by various hormonal regulators. These functions include exocrine and endocrine secretion, gut motility, intestinal fluid and electrolyte transport, nutrient assimilation, appetite regulation, and even pain perception. Although a description of the various peptide hormones and their actions is detailed below, it is useful to examine their role in controlling one of the major functions of the gut, gastric acid secretion, to gain insight into the intricate interactions between various hormones, and between hormones and other paracrine/neurocrine effectors involved in regulating physiological processes.

The stimulation of acid secretion is initiated by the simple contemplation of food. This so-called *cephalic* phase of gastric acid secretion can be induced by the thought, sight, taste, or smell of food without any of the food entering the stomach. Although under some ordinary conditions the cephalic phase accounts for only a fraction of the acid secretory response to a meal, it can account for as much as two-thirds under some circumstances. The vagus nerve is thought to carry the acid stimulatory signals from the brain to the stomach because vagotomy abolishes the cephalic stimulus. The primary mediator carried by the vagus is thought to be acetylcholine, although gastrin, a known stimulant of acid secretion, is also contained within the nerve. Acetylcholine may stimulate acid secretion by direct action on gastric parietal cells, but indirect actions via stimulation of gastrin secretion from G cells in the stomach or inhibition of the secretion of somatostatin, a hormone known to inhibit acid secretion, may also be important.

Once food enters the stomach, another important series of hormonally mediated events occurs. The proteins contained within a meal, more specifically the breakdown products of proteins (the amino acids and their decarboxylated derivative amines) stimulate gastrin secretion. The stretching of the stomach wall by the ingested meal contributes to gastrin release. As mentioned above, gastrin accounts for the major portion of the acid secretory response to meal ingestion. Gastrin by itself is not an efficacious direct stimulant of gastric parietal cells, which are responsible for acid secretion, but in combination with histamine induces virtually maximal stimulation. Histamine is produced by various cells in the stomach, depending on the animal species, and appears to act as a paracrine effector. Because the intracellular mediator for histamine action (adenylate cyclase/cyclic adenosine monophosphate) is different from that for gastrin (Ca^{2+}/phospholipid dependent protein kinase), the two agents potentiate the actions of each other to produce an effect that exceeds the sum of the individual effects. At the same time that acid secretion is being stimulated by one set of events, another process is set forward to modulate the secretory response by inhibiting it. The ingested meal and the accompanying acid combine to stimulate somatostatin secretion. Somatostatin not only inhibits the secretion of acid by direct action on parietal cells but also acts in conjunction with acid to inhibit the secretion of gastrin.

When the ingested meal enters the small bowel, other events that affect acid secretion occur. Although there appears to be an ill-defined acid stimulatory hormone (termed *enterooxyntin*) released by the presence of nutrients in the intestine, this effect is small and overcome by the greater effects of a series of acid inhibiting intestinal hormones including somatostatin, neurotensin, gastric inhibitory polypeptide (GIP), secretin, and peptide YY (PYY). These inhibitory hormones are all considered to be candidates for the role of "enterogastrone," a hormonal activity described by its function earlier in the century.

The physiological regulation of the stomach

makes intuitive sense from the functional requirements of the digestive process. Acid is needed for the primary digestive enzyme of the stomach (i.e., pepsin), to function optimally. Thus, the initial phase of digestion that occurs in the stomach requires acid. However, when food enters the small intestine, it is further digested by pancreatic enzymes that function optimally in a less acidic environment. Thus, gastric acid secretion must be turned off and the pancreas must be stimulated to secrete enzymes and neutralizing bicarbonate by other hormones such as cholecystokinin (CCK) and secretin which are released by nutrients and acid, respectively, in the intestine. This elegant interplay of hormonal effectors and inhibitors acting in a regionally and temporally integrated fashion applies not only to the stomach but to all organs in the gut and is essential for the normal function of the gastrointestinal tract.

II. Peptide Hormone Families

Structural analysis of peptide hormones has led to the identification of great similarities between groups of them. Because of the similarities within a given family of peptides, its members are presumed to have arisen via tandem duplication of a common ancesteral gene. After duplication, members of a gene family may diverge with respect to structural and functional characteristics. The hormone families consisting of peptides sharing substantial structural similarity are summarized in Table I. A number of peptides that do not fall into a large family group are referred to as orphan peptides.

A. Gastrin/CCK Family

1. Gastrin

The peptide hormone gastrin was postulated to exist as early as 1905 when Edkins first described a substance in the mucosa of the gastric antrum that induced the stomach to secrete acid. For many years thereafter, however, there was controversy over whether this substance was simply histamine or a new substance. This problem was resolved when Gregory and Tracy isolated gastrin from porcine antrum and confirmed its biological potency. The structural feature that is required for gastrin's biological activity is the presence of an amide moiety on the carboxyl-terminal phenylalanine residue. As indicated above, gastrin is released into the cir-

TABLE I Gastrointestinal Peptide Families

Family	Major members	Principal biological actions
Gastrin	Gastrin	↑ acid secretion, ↑ tissue proliferation
	Cholecystokinin (CCK)	↑ pancreatic secretion
		↑ gallbladder contraction
Secretin	Secretin	↑ pancreatic secretion
	Glucagon	↓ sphincter of oddi pressure
		↓ intestinal motility and absorption
	Enteroglucagon	↑ insulin release
	Vasoactive intestinal polypeptide (VIP)	↑ pancreatic and intestinal secretion
	Gastrin inhibitory polypeptide (GIP)	↓ gastric secretion, ↑ intestinal secretion
	Glicentin	↑ hepatic glucose, ↓ acid secretion
	Oxyntomodulin	↑ insulin release, ↓ acid secretion
	Growth hormone releasing factor (GRF)	↑ release of growth hormone
	Peptide histidine isoleucine (PHI)	↑ pancreatic secretion
Pancreatic polypeptide	Pancreatic polypeptide (PP)	↓ pancreatic secretion
	Peptide YY (PYY)	↓ pancreatic secretion
		↓ gallbladder contraction
	Neuropeptide Y (NPY)	↑ vasoconstriction
Opioids	Enkephalin	↓ intestinal transit
	Beta endorphin	↓ intestinal transit
	ACTH	↑ cortisol release
	α-Melanocyte stimulatory hormone (αMSH)	↑ melanin release
	Dynorphin	↓ intestinal transit
Tachykinin-bombesin	Substance P	Contraction gastrointestinal smooth muscle
	Substance K	Contraction gastrointestinal smooth muscle
	Neuromedin K	Contraction gastrointestinal smooth muscle
	Gastrin releasing peptide (GRP)	↑ release gastrin
	Neuromedin B	↓ acid secretion
Orphan peptides	Somatostatin	↓ acid secretion, ↓ pancreatic function
	Neurotensin	↓ acid secretion, ↓ gastric motor activity
	Galanin	↑ plasma glucose, ↓ fundus contraction
	Pancreastatin	↓ islet somatostatin release
	Motilin	↑ motility

culation in response to meal ingestion. There are several discreet phases to this response, including those associated with the sight, thought, taste, or odor of the meal, with gastric distension that results from the presence of food in the stomach, with alkalinization of the gastric lumenal contents by the buffering action of the meal, and with the proteins and their breakdown products in the ingested food. Although gastrin is present in the stomach, it exerts its acid secretory action only after entering the peripheral circulation and then being transported to the basolateral surfaces of the parietal cells. There, specific gastrin receptors of roughly 74,000 daltons in size bind the peptide and transduce the binding signal into a cellular response that appears to be mediated by mobilization of intracellular Ca^{2+} and activation of a Ca^{2+}/phospholipid dependent protein kinase on the cell membrane. Aside from inducing acid secretion, gastrin has another major function to promote growth of gastrointestinal mucosa. This "trophic" action of gastrin is not well-characterized in normal tissues but may have clinical significance in gastrointestinal neoplasms, particularly colon cancer.

2. Cholecystokinin

Cholecystokinin was initially isolated from hog intestine as a peptide hormone 33 amino acids long but has since been found to exist as peptides 58, 39, 33, 22, 8, and 4 amino acids long. It has 100% structural homology with gastrin in the carboxyl-terminal six amino acids, which comprise the biologically active portion of the peptide. It is not surprising, therefore, that CCK can bind to gastrin receptors and exert the same pharmacological effects as gastrin. Normally, however, CCK circulates in the blood in concentrations that are only 5–10% as high as gastrin; thus, under physiological conditions, CCK does not contribute significantly to gastrin's effects. The feature of CCK structure that accounts for its unique function is the presence of a sulfated tyrosine residue 7 amino acids toward the amino-terminal end from the carboxyl-terminal phenylal-anine-amide residue. CCK is produced in discreet endocrine cells scattered throughout the small intestine and in neurons of the cerebral cortex. It is released from the intestine in response to a meal stimulus, but the mechanisms for this release process are not known. It is hypothesized that ingested nutrients, particularly the fats and aromatic amino acids, stimulate the release of a yet unidentified CCK-releasing peptide into the lumen of the small intestine that stimulates CCK release. The pancreatic enzymes released by CCK, in turn, digest the CCK-releasing peptide and, thus, inhibit in a feed-back fashion further stimulation by CCK. The actions of CCK appear to be mediated by two classes of receptors, peripheral-type receptors that require CCK to be sulfated, and central CCK receptors that do not distinguish between sulfated and nonsulfated CCK. The primary "peripheral-type" CCK effects are the stimulation of pancreatic enzyme secretion, the induction of pancreatic growth, and the contraction of the gallbladder. CCK action on the pancreas appears to be mediated by specific receptors that are linked, like gastrin, to mobilization of intracellular Ca^{2+}. The major "neural-type" effects of CCK may be its action on gastrointestinal smooth muscle to delay gastric emptying and enhance intestinal motility. A provocative function that has been ascribed to CCK is that of a satiety signal that causes animals to stop eating.

B. Secretin/Vasoactive Intestinal Polypeptide/Glucagon

1. Secretin

As noted above, secretin was the first hormone for which a biological function was described. It derives its name from its action as a stimulant of pancreatic bicarbonate secretion. It is found in the small intestine, primarily concentrated in the duodenum, as a 27-amino-acid peptide with a carboxyl-terminal amide residue. Secretin is released from the intestine on acidification by the passage of gastric contents into the duodenum. The threshold pH required for its release appears to be 4.5. Nutrients, particularly fats, also can stimulate secretion under experimental conditions, but the physiological significance of this observation is uncertain because no such stimulation is observed in the absence of acid. The major function of secretin is to stimulate the

secretion of bicarbonate from the pancreatic ductu-lar cells, thus accounting for the neutralization of the acid secreted by a meal and emptied into the intestine. Another function of secretin is to inhibit gastric emptying, perhaps thereby protecting the duodenum from exposure to excessive amounts of acid.

2. Vasoactive Intestinal Polypeptide

Vasoactive intestinal polypeptide (VIP) is a peptide with a primary structure similar to that of secretin. It consists of 28-amino-acid residues and derives from a gene that encodes still another related peptide called peptide histidine-isoleucine in animals or (PHI) peptide histidine-methionine in humans (PHM). In contrast to classical peptide hormones such as secretin, VIP is detectable only in neurons of the enteric and peripheral nervous systems. In the gut, VIP neurons are present primarily in the submucous plexus, and their fibers form a rich and dense network within the lamina propria. VIP neurons are innervated by preganglionic cholinergic fibers, thus they appear to function as intermediaries between central nervous system neurons and their target organs in the gut. The major function of VIP appears to be that of an inhibitor of smooth muscle contraction, thus it relaxes the lower esophageal and anal sphincters and promotes both gastric and intestinal relaxation. Another function of VIP, by virtue of its ability to relax vascular smooth muscle, is to enhance gastrointestinal blood flow during the period after ingestion of a meal. VIP may also be important in regulating the hydrated state of intestinal contents via its actions to promote fluid and electrolyte secretion into the lumen of the gut. Under some pathological conditions, the presence of excess VIP may induce diarrhea.

3. Glucagon

Pancreatic glucagon is a 29-amino-acid peptide that also shares a high degree of structural homology with other members of the secretin family. It derives from a larger precursor that may be cleaved to produce a peptide of 69 amino acids known as glicentin or another peptide of 37 amino acids known as enteroglucagon or oxyntomodulin, which is pancreatic glucagon with an extension of eight amino acids at the carboxyl-terminus. The latter two peptides are found in endocrine cells of the intestinal mucosa, whereas the former is found in α-cells in the pancreatic islets of Langerhans. Pancreatic glucagon is considered to be the counterpart to insulin in glucose homeostasis in that it mobilizes glucose from carbohydrate stores in response to hypoglycemia. The specific physiological function of enteroglucagon is not known for certain, but its secretion is stimulated by glucose and digested fats in the intestine. [*See* INSULIN AND GLUCAGON.]

C. Opioids

One of the most profound observations in human biology was the finding that the body makes endogenous peptides that act in the manner of alkaloid opioid analgesics. There are three main classes of these peptides: the enkephalins (methionine or leucine enkephalin) derived from proenkephalin, the dynorphins derived from prodynorphin, and β-endorphin derived from proopiomelanocortin. These peptides have a common structural feature consisting of the pentapeptide Tyr-Gly-Gly-Phe-Met (or Leu). They are widely distributed in the central and peripheral nervous systems, and in the gut, they are found primarily in myenteric plexus neurons that arborize into the circular muscle layers. These peptides act on three classes of receptors labeled mu, delta, and kappa to produce different actions, often with the same physiological effect. For example, mu and delta opioid agonist activity hyperpolarizes myenteric neurons by increasing a K^+ conductance, whereas kappa activity blocks Ca^{2+} channels. Although the mechanisms are different, both actions have the effect of inhibiting neuronal action. The net effect of opioid peptides on gut motility is to slow transit of ingested material through the intestinal lumen. This effect may be exerted either by influencing the activity of other enteric neurons or by direct action on smooth muscle cells. Opioids are also known to have the effect of inhibiting intestinal secretion by diminishing electrolyte transport into and promoting fluid absorption from the intestinal lumen. The combination of delayed transit and decreased secretion produced in the gut by opioids makes them useful in the therapy of diarrheal states.

D. Pancreatic Polypeptide Peptide YY/Neuropeptide Y

1. Pancreatic Polypeptide

Pancreatic polypeptide (PP) is a 36-amino-acid peptide that was initially discovered as an impurity in preparations of insulin. It is localized almost exclusively to the pancreas, where it is found in specific cells in the islets of Langerhans as well as other cells scattered throughout the acinar tissues. PP is released in response to ingestion of nutrients (primarily protein and fats), and this response appears to be mediated by cholinergic neurons because it can be abolished by antagonizing muscarinic cholinergic receptors with atropine. The physiologic function of PP has not been established conclusively, but the peptide appears to be an inhibitor of pancreatic secretion. This action would imply a role for PP in the feed-back regulation of pancreatic secretion stimulated by the enteropancreatic neural reflex. The ingestion of a meal stimulates a cholinergic neural response in the pancreas. The cholinergic neurons appear to act simultaneously both to stimulate pancreatic acinar cells and to stimulate the release of PP, an inhibitor of pancreatic secretion. In this fashion, PP may function to modulate or finely regulate pancreatic secretion.

2. Peptide YY (PYY)

Peptide YY is so named because it is a *p*eptide with a tyrosine (*Y*) residue at the amino-terminus and another tyrosine (*Y*) residue at the carboxyl-terminus. It is one of a series of peptides that have been purified on the basis of their structural properties (specifically, the presence of a carboxyl-terminal amide moiety) without any clues as to their function. Like PP, PYY consists of 36 amino acid residues; however, its localization is very different. It is found mainly in distinct endocrine cells in the small intestine, primarily in the distal portion, and in the proximal colon. PYY is released into the circulation by entry of fat into the small intestine and, unlike release of PP, this response does not appear to be mediated via cholinergic neurons. The main function of PYY appears to be that of an enterogastrone (i.e., an inhibitor of gastric acid secretion derived from the intestine). As noted above, such substances are thought to be important feed-back modulators of the acid secretion induced by a meal. Two other effects of PYY that may be important in processing ingested nutrients is its ability to inhibit pancreatic secretion and delay gastric emptying.

3. Neuropeptide Y

Neuropeptide Y (NPY) derives its name from the fact that it is a *n*euro*p*eptide with amino- and carboxyl-terminal tyrosine (*Y*) residues. It, too, is a 36-amino-acid peptide purified on the basis of its carboxyl-terminal amide moiety. Unlike PP and PYY, however, it is found strictly in neurons of the central and peripheral nervous systems. In the gut, NPY is located in both submucous and myenteric plexus neurons. A portion of the NPY fibers is associated with blood vessels in the gut. These fibers derive from extramural neurons containing norepinephrine. It is not clear how NPY release is stimulated in the gut, but its action appears to produce sustained vasoconstriction and, in so doing, reduces enteric blood flow and motility.

E. Tachykinins

The tachykinin family consists of a number of amphibian, molluscan, and mammalian peptides including substance P, one of the most thoroughly studied of all the gut peptides. These peptides were termed "tachykinins" because of their characteristic rapid onset of action in gut tissues in contradistinction of the "bradykinins," which elicit much slower responses. Occasionally the tachykinins are referred to as "neurokinins." The carboxyl-terminal amino acid sequence-Phe-X-Gly-Leu-Met-NH$_2$ is highly conserved in the members of the tachykinin family. The amino acid at position X defines much of the specificity of a given tachykinin. For example, the aromatic amino acid phenylalanine determines substance P specificity, whereas the aliphatic amino acid valine is characteristic for neurokinin A and B. Tachykinins are present in almost all layers of the intestinal wall and can be synthesized by neurons in the myenteric and submucous ganglia. Substance P can also be found in enterochromaffin cells of the gut. It is of particular note that tachykinins are found in the dorsal root and nodose ganglia. The fact that dorsal root ganglion neurons that innervate the gut may send processes to the skin may explain the referred pain noted in many gastrointestinal disorders. The receptors for the tachykinins have been characterized particularly well because their genes have been cloned and analyzed. They exhibit a structure characteristic of other receptors whose signals are mediated by guanine nucleotide binding proteins in that they traverse the cell membrane seven times with extracellular domains containing sites for addition of carbohydrates and intracellular domains containing sites for addition of phosphate residues. The primary effects of tachykinins appear to be stimulatory on gastrointestinal motility via excitatory neurons regulating peristalsis, although the physiological significance of these effects are not confirmed. Other potentially important effects of tachykinins include regulation of blood flow, intestinal secretion, and pancreatic secretion.

F. Orphan Peptides

1. Somatostatin

Somatostatin received its name from its initially described ability to inhibit pituitary growth hormone secretion. It was isolated from the hypothalamus almost by accident by investigators seeking to find the stimulant of growth hormone. The peptide exists in two primary molecular forms of 14 and 28 amino acids in length, respectively. Somatostatin is found in a wide variety of tissues in the body including the central, peripheral, and enteric nervous systems and the gastroenteropancreatic systems. In the gut, it is found in intrinsic and extrinsic neurons of the submucosal and deep muscular layers as well as in specialized D cells in the mucosal lining of virtually the entire gastrointestinal tract. The release of somatostatin from the gut in response to a meal may result from a multiplicity of factors. Nutrients may act directly to stimulate D cells that communicate with the lumenal contents in the gastric antrum. Other factors (e.g., postprandial gastric acid and gastrin secretion) may also contribute to somatostatin secretion. The major function of somatostatin seems to be to act as a natural brake for the body. It is characterized by virtually uniform inhibitory actions in the gut, on exocrine secretion, endocrine secretion, motility, fluid and electrolyte transport, blood flow, and tissue growth and proliferation. The mechanisms by which somatostatin exerts these effects on the target cell are uncertain but may involve inhibitory actions at multiple sites including the cell's receptors for other neurohumoral messengers, the intracellular mechanisms that transduce the receptor signal to a cellular action, or even the terminal action of the cell itself (e.g., exocytosis). Given its widespread inhibitory effects, it is not surprising that somatostatin has potential usefulness in treating a variety of gastrointestinal disorders resulting from too much activity of one sort or another (e.g., diarrhea).

2. Motilin

Motilin is a 22-amino-acid peptide found in discreet endocrine cells in the small intestine. It derives its name from its actions on gut motility. Unlike other peptide hormones of the gut, motilin concentrations in the blood decrease in response to a meal in seemingly paradoxical fashion. Instead, during the interdigestive period there are regular periodic increases in plasma motilin levels, which coincide with acceleration of gastric emptying and initiation of a wave of intestinal contraction that courses through the entire gut at regular intervals of 60–120 minutes. These contractions, called the migrating motor complexes, are known as the "housekeepers" of the gut that function to clean it out from time to time. Ingestion of a meal prevents the normal cyclic increases in plasma motilin and, thereby, the migrating motor complexes. The mechanism by which motilin exerts its effects has yet to be elucidated, but experimental studies suggest that the neural circuits responsible for programming the migrating motor complexes reside in the enteric nervous system and involve cholinergic pathways. Motilin presumably acts on enteric neurons as well as on gastric smooth muscle to initiate the intestinal housekeeping process.

3. Galanin

Galanin, a 29-amino-acid peptide, is another of the gut hormones purified on the basis of its carboxyl-terminal amide moiety. Its name derives from the presence of a glycine residue at its amino-terminus and an alanine residue at its carboxyl-terminus. In the gut it is found exclusively in enteric neurons concentrated mainly in the small intestine. Like somatostatin, galanin exerts primarily inhibitory action. For example, it appears to be an important regular of glucose homeostasis by virtue of its inhibitory action on insulin secretion. Its function in the gut may result from its inhibitory actions on motility.

4. Neurotensin

As its name suggests, neurotensin was initially isolated on the basis of its ability to influence systemic blood pressure (i.e., it induces hypotension). It is a 13-amino-acid peptide that is related to two other peptides that share some structural homology at the carboxyl-terminus, neuromodulin N and xenopsin. Neurotensin in the gut is primarily found in endocrine-type cells in the ileal mucosa, with smaller amounts being present in the rest of the gut. Small amounts are also found in enteric neurons. The peptide is most potently released by ingested fat. Although the physiologic function of neurotensin has not been determined, it may play a role in a variety of gastrointestinal events activated by fat in the lumen including inhibition of acid secretion, en-

hancement of intestinal fluid secretion, stimulation of pancreatic output, and increase in blood flow.

III. Clinical Significance of Peptide Hormones

A. Disease Pathogenesis

Peptide hormones of the gut have been implicated in the pathogenesis of a wide variety of human disorders including peptic ulcer disease, hypergastrinemia syndromes, irritable bowel syndrome, gallstones, diarrheal disorders, reflux esophagitis, Hirschprung's disease, and even morbid obesity. The most striking correlation between peptide hormones and gastrointestinal disease is found in the many syndromes associated with hormone-secreting tumors of the pancreas. These interesting "accidents" of nature have provided clues as to the physiological function of some peptide hormones. Because the cells that produce hormones in the pancreas are thought to have an embryological origin common to other neuroendocrine tissues, the pancreatic endocrine neoplasms almost always involve more than a single hormone and often involve other neuroendocrine organs. Nevertheless, the specific syndromes that arise from these tumors generally can be ascribed to the effects of a single hormone.

The most common of the pancreatic endocrine neoplasm syndromes was described by Zollinger and Ellison in 1955 when they noted the association between hypersecretion of gastric acid, refractory peptic ulcer disease, and nonbeta cell tumors of the islets of Langerhans in the pancreas. Later the substance responsible for the bulk of clinical findings was identified as gastrin. One interesting effect of the uncontrolled acid hypersection induced in this disorder by gastrin is that many of the patients have diarrhea largely because of the inability of pancreatic enzymes to digest ingested nutrients in the presence of an acidic environment in the intestine. Gastric acid secretion has long been known to be a physiological action of gastrin, but another curious manifestation of Zollinger-Ellison syndrome, gastric mucosal hyperplasia, contributed to the discovery that gastrin has another important function as a promotor of tissue growth.

A syndrome complex of severe watery diarrhea, low serum potassium concentrations, and reduction in gastric acid secretion in association with a pancreatic islet cell tumor was described in 1958 by

Verner and Morrison. The diarrhea was so severe that the disorder was called "pancreatic cholera" by some. Initially, this syndrome was thought to be a variant of Zollinger-Ellison syndrome; however, the offending agent was later identified to be VIP. A cardinal feature of the disease is the presence of diarrhea in the absence of nutrient ingestion, implying that it results from excess intestinal secretion rather than nutrient maldigestion or malabsorption. Like cholera toxin, VIP induces the activation of adenylate cyclase and subsequent production of cyclic adenosine monophosphate in intestinal epithelial cells. This induces the secretion of water and electrolytes into the intestinal lumen. In some patients, the diarrhea can exceed 8 L/day; thus severe dehydration and internal electrolyte disturbances may result.

Another hormone, somatostatin, was implicated in the pathogenesis of a clinical syndrome in a somewhat circuitous fashion. Although the hormone was described in 1973, no clinical abnormality had been associated with its overproduction until 1977 when a diabetic patient with modest diarrhea underwent surgery for gallstones and was found to have an incidental pancreatic tumor. On histochemical analysis of the tumor, it was noted to be abundantly occupied by cells containing somatostatin. It was only in retrospect that the symptoms of the patient could be correlated to the actions of the hormone. The ability of somatostatin to inhibit gallbladder contraction led to stasis of bile and gallstone formation, the inhibition of digestive enzyme secretion from the pancreas by somatostatin led to

nutrient maldigestion and diarrhea, and the inhibitory action of somatostatin on insulin secretion led to diabetes. Subsequently, a host of other patients with somatostatin-secreting tumors of the pancreas have been identified.

Aside from these disorders, there are numerous other syndromes associated with overproduction of insulin, glucagon, substance P, neurotensin, adrenocorticotrophic hormone (ACTH), and PP among others. The features of the various syndromes are summarized in Table II. In some patients, the features of one or more of these syndromes may be combined in a clinical entity known as multiple endocrine neoplasia Type I. The vast majority of patients with this syndrome have pancreatic endocrine neoplasms producing numerous peptide hormones in the pancreas; however, most of these patients also have hormone-secreting tumors in their pituitary and parathyroid glands. The disease is generally transmitted in an autosomal dominant pattern; thus the relatives of all patients with peptide-secreting endocrine neoplasms should be screened carefully for signs of endocrine abnormalities.

B. Disease Diagnosis

Peptide hormones, because of their many actions, are potentially useful as aids in diagnosis and treatment of a number of pathological conditions. Obviously, the measurement of peptide hormones in the plasma is an essential element of diagnosing endocrine neoplasms. Some endocrine neoplasms, be-

TABLE II Pancreatic Endocrine Neoplasm Syndromes

Peptide hormone	Regulatory action	Clinical manifestations
Gastrin	Acid secretion Cell proliferation	Gastric acid hypersecretion Peptic ulcer disease Diarrhea
Somatostatin	Widespread inhibitory actions	Gallstones Diabetes Diarrhea
Vasoactive mediated polypeptide (VIP)	Intestinal fluid and electrolyte secretion Vasomotor actions	Watery diarrhea Electrolyte abnormalities Flushing
Insulin	Decreased blood sugar Anabolic state	Hypoglycemia Behavioral disturbances
Glucagon	Increased blood sugar Catabolic state	Skin rash Diabetes Anemia
Adrenocorticotrophic hormone (ACTH)	Cortisol secretion	Moon-like face Muscle wasting Weakness

cause of their pancreatic ductular origin, can be induced to secrete hormones in response to stimulation with secretin, a hormone that generally induces bicarbonate secretion from pancreatic ducts. This "secretin test" is particularly useful in diagnosing patients with Zollinger-Ellison syndrome whose serum gastrin levels are only marginally elevated. Secretin and CCK infusions are also useful in evaluating patients with disorders of pancreatic exocrine function. Because these two peptides induce bicarbonate and enzyme secretion from the pancreas, respectively, by infusing the peptides and measuring the maximal secretory responses of the pancreas, it is possible to quantify the relative degree of organ dysfunction. CCK's other primary action to stimulate gallbladder contraction can also be used diagnostically to evaluate patients with chronic abdominal pain that may originate from abnormalities in the motility of the biliary tree. Other hormones that have effects on the motility of the intestine are useful adjuncts in radiological and endoscopic evaluation of the gastrointestinal tract. Both glucagon and somatostatin in appropriate doses can relax the smooth muscles in the intestine. This effect "quiets" the movements of the intestine sufficiently to take more detailed x-rays of the gut and also to permit cannulation of the biliary tree where it terminates in the duodenum so that it can be injected with radiological contrast dye for x-ray evaluation (i.e., endoscopic retrograde cholangiopancreatography for ERCP).

C. Disease Therapy

The application of hormones in therapy of pathological states has been restricted to only a few special circumstances. Obviously, the opioids have proved to be useful in treatment of pain. Opioids are also contained within many standard remedies for diarrhea. These substances are effective by virtue of their ability to inhibit intestinal motility as well as secretion of fluid and electrolytes. Erythomycin derivatives have been found, quite by accident, to substitute for motilin in inducing intestinal smooth muscle contraction. This property has been explored as a potential means for treatment of patients with intestinal dysmotility.

In the past, native peptide hormones have not been useful as therapeutic agents because they are small, easily degraded molecules and, thus, disappear quickly after being administered. Recently, several peptide analogues with greatly prolonged

half-lives have been synthesized. Thus, for example, somatostatin analogues may have a plasma half-life of 2 hours or more, whereas the natural compound disappears within minutes. One new somatostatin analogue, octreotide, has been used effectively in treating patients with endocrine neoplasms. In a sense, somatostatin analogues are ideal for treatment of these disorders because the peptide has the capability of inhibiting both the secretion of peptides from tumors as well as the effects of the secreted peptides at their target organs. Because of somatostatin's widespread inhibitory actions, the octreotide analogue has potential usefulness in the treatment of a variety of other disorders such as diarrhea, pancreatitis, motility disorders, and gastrointestinal bleeding.

Other applications of peptide hormone biology to clinical conditions involve the use of specific peptide receptor antagonists. Naloxone, an opioid receptor antagonist, has proved to be a useful antidote for narcotic overdoses. More recently, antagonists for CCK have been developed as potentially useful agents in treating disorders of intestinal motility or appetite regulation. Gastrin receptor antagonists with potential applications in treatment of peptic ulcer disease are being developed.

Given the multitude of effects that peptide hormones have in the body, the development of additional stable analogues of peptide hormones or selective peptide receptor antagonists holds great promise for treatment of many other human disease states.

Bibliography

DelValle, J., and Yamada, T. (1989). Secretory tumors of the pancreas. In "Gastrointestinal Diseases," 4th ed. (M. H. Slesinger and J. S. Fordtran, eds.). W. B. Saunders, Philadelphia.

Makhlouf, G. M. (1989). "Neural and Endocrine Biology, Vol. 2: Handbook of Physiology, Sec. 6: The Gastrointestinal System." American Physiological Society, Bethesda, Maryland.

Walsh, J. H. (1987). Gastrointestinal hormones. In "Physiology of the Gastrointestinal Tract," 2nd ed. (L. R. Johnson, ed.). Raven Press, New York.

Wolfe, M. M., and Jensen, R. T. (1987). Zollinger-Ellison syndrome: Current concepts in diagnosis and management. N. Engl. J. Med. 317, 1200.

Yamada, T. (1987). Local regulatory actions of gut peptides. In "Physiology of the Gastrointestinal Tract," 2nd ed. (L. R. Johnson, ed.) Raven Press, New York.

H-Tyr-D-Ala-Gly-N-Me-Phe-Gly-ol
H-D-Phe-Cys-Tyr-D-Trp-Orn-Thr-Cys-Thr-NH₂
H-Tyr-D-Pen-Gly-Phe-D-Pen-OH
H-D-Phe-Cys-Phe-D-Trp-Lys-Thr-Cys-Thr-ol
H-Sar-Arg-Val-Tyr-Val-His-Pro-Ala-OH
pGlu-His-Trp-Ser-Tyr-D-Ser (t-Bu)-Leu-Arg-Pro
H-D-Arg-Arg-Pro-Hyp-Gly-Thi-Ser-D-Phe-Thi-Arg
Ac-Ser-Tyr-Ser-Nle-Glu-His-D-Phe-Arg-Trp-Gly

Peptides

TOMI K. SAWYER AND CLARK W. SMITH,
The Upjohn Company

I. Naturally Occurring Peptides
II. Biotechnology and Biomedical Research

Glossary

Anabolic hormone Chemical messenger (e.g., peptide, steroid, biogenic amine) which stimulates directly or indirectly biochemical changes of intermediary metabolism which result in the biosynthesis of macromolecules (e.g., protein, polysaccharide, lipid, nucleic acid)

Hyperglycemic action Effect producing abnormally high concentrations of glucose in the blood

Heterodimeric peptide Peptide (or protein) having two nonidentical subunits (e.g., α and β) which interact usually in a noncovalent manner to produce the biologically active form of the molecule

Juxtaglomerular cells Epithelial cells of the afferent arteriole located in the kidney and proximate to the point where the arteriole abruptly branches into a network of capillaries

Nociceptic Pain sensory responsive

Second messengers Chemical messengers (e.g., cyclic AMP, calcium, inositol triphosphate) which are produced intracellularly in response to extracellular "first" messengers (e.g., peptide, biogenic amine) acting through cell membrane receptors and transmembrane signalling mechanisms

Secretagogue Chemical messenger (e.g., peptide, steroid, biogenic amine) which stimulates the release (or secretion) of a specific cellular product

Secretory granules Cytoplasmic storage vesicles derived from the Golgi complex; contain secretory products of a cell

Trophic growth effect Growth stimulatory effect which may be induced directly or indirectly by trophic chemical messengers at their target tissues (e.g., muscle, bone)

Signal sequence Amino-terminal sequence of amino acids (primarily hydrophobic) which is requisite for transport of newly synthesized cytosolic peptides (or proteins) into the lumen of the endoplasmic reticulum where the signal sequence is then enzymatically removed

Tyrosine autophosphorylation Protein (e.g., receptor, enzyme) self-phosphorylation which may be induced by hormonal and/or allosteric mechanisms and results in functional modification due to the formation of phosphate ester linkage on the amino acid side-chain group of Tyr

PEPTIDES ARE chemical messengers that exist in all forms of life and serve as hormones, neurotransmitters/neuromodulators, and growth factors, regulating the cardiovascular, gastrointestinal, reproductive, and immune systems in higher organisms, including humans. All peptides are composed of amino acids (i.e., residues) covalently linked by amide (CONH) bonds formed between an α-carboxyl (CO₂H) group of one residue and an α-amino (NH₂) group of another residue. The specific ordered-sequence (Fig. 1) primary structure of a peptide determines both its three-dimensional properties (i.e., secondary structure, or conformation, and tertiary structure) and its biological properties. Specific examples of such properties for a few selected peptides are described in Section II.

Peptide hormones (e.g., adrenocorticotropin and insulin) normally exist in very low concentrations (10^{-6} to $10^{-12} M$) in bodily fluids or tissues, in which they persist for short times (1–30 minutes), because they are rapidly inactivated by proteolytic enzymes and/or excreted (i.e., cleared) from the body. Their transitory actions are suitable for continuous regulation (i.e., homeostasis). Another common trademark of peptides is their specificity at target cells

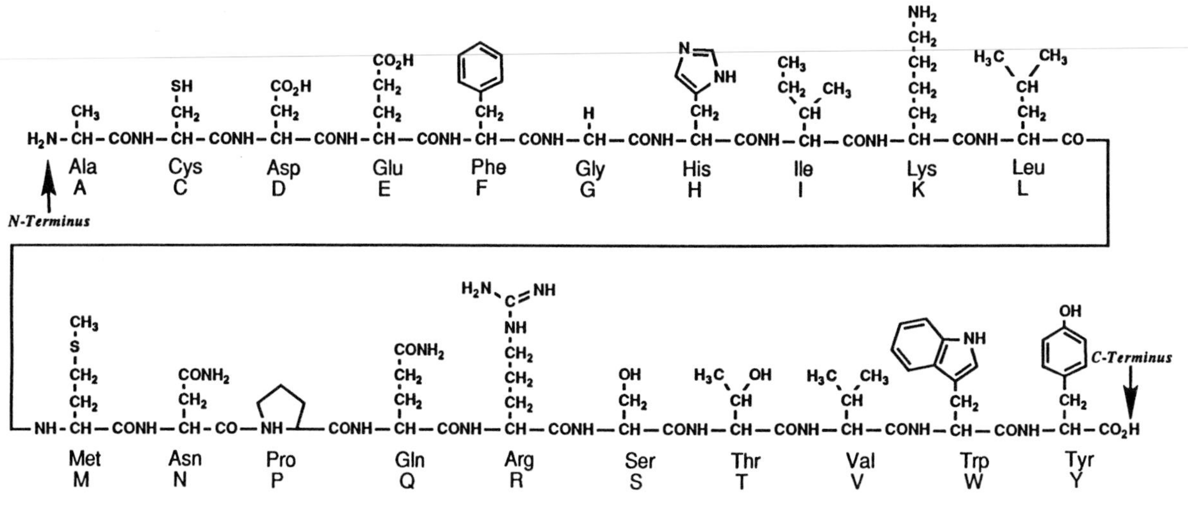

FIGURE 1 A hypothetical peptide composed of 20 different amino acids. The three-letter and one-letter abbreviations for each residue are given. N-Terminus, Amino terminus; C-Terminus, carboxy terminus.

(e.g., secretory or neuronal), which depends on the existence of receptors, or proteins typically found on the cell surface, each capable of recognizing and binding a particular peptide, forming with it biochemically active complexes. These peptide–receptor complexes are reversible, and dissociation ultimately results in an unstimulated state of the target tissue. However, the cellular response, once initiated, can proceed independently of the initial events occurring at the cell surface, until it is terminated by other signals or deactivation processes. As a result of these chemical and biological characteristics, research on peptides relies on many technological tools, such as (1) radioimmunoassay, to detect and quantitate peptides in various biological tissues or fluids; (2) radioreceptor assay, to determine peptide–receptor interactions and explore the mechanism of action of peptides; and (3) artificial synthesis, either direct or indirect, through DNA recombination technology, to prepare the quantities of peptides required for *in vitro* or *in vivo* studies.

I. Naturally Occurring Peptides

A. Biosynthesis and Metabolism

Peptide hormones are synthesized, like all proteins, on ribosomes by translating specific mRNAs, which reproduce the sequences of the genes present in the DNA of chromosomes. Most peptides are made up of biologically inert and higher-molecular-weight precursor molecules (i.e., prepropeptides), which are later split, or processed, enzymatically. The "pre" portion of the precursor usually comprises

an amino-terminal extension of 15–30 amino acids to the sequence of the main peptide, called the signal sequence. This sequence directs the transport of the peptide to the endoplasmic reticulum and is removed as the peptide is transported into its cavity.

The remaining peptide can still be an inactive precursor, a propeptide, which can be processed in various compartments (e.g., secretory granules or the blood). The peptide can be modified by sulfation, glycosylation, amidation, or acetylation, and, if it is to be secreted, it is packaged into secretory vesicles within the Golgi apparatus. Peptide neurotransmitters/neuromodulators are similarly biosynthesized and stored in secretory vesicles within neurons until they are released at synapses.

Degradation and inactivation of peptides can occur at a variety of physiological sites (e.g., the liver, kidneys, lungs, and brain) through the action of proteolytic enzymes (exopeptidases or endopeptidases) that recognize and cleave the bonds between certain amino acid residues.

B. Mechanisms of Action

The high specificity of peptides for their respective target tissues is achieved by release close to the site of action and rapid degradation, as well as by high specificity for their receptors. Receptors are cell surface macromolecules that bind specific peptides,

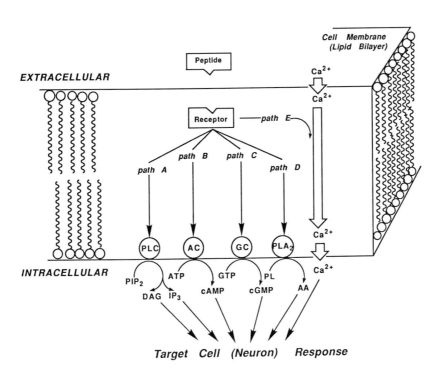

FIGURE 2 Potential mechanisms of action of peptide hormones (see text for discussion). PLC, Phospholipase C; AC, adenylate cyclase; GC, guanylate cyclase; PLA$_2$, phospholipase A$_2$; PIP$_2$, phosphoinositol bisphosphate; DAG, diacylglycerol; IP$_3$, inositol triphosphate; PL, phospholipid; AA, arachidonic acid.

but with high affinity and noncovalently, according to their unique amino acid sequence and resultant three-dimensional conformation and topography. Peptide receptor subtypes have also been identified and pharmacologically defined (e.g., μ, δ, and κ opioid receptors for the endogenous opioid peptides: endorphin, enkephalin, and dynorphin, respectively). The mechanisms of action of the majority of peptides are based on physiological regulation of peptide receptor number and various forms of information transfer (i.e., signal transduction).

As shown in Fig. 2, signal transduction from the peptide–receptor complex to the cellular genes can occur by transmembrane modulation of certain enzymes, such as adenylate cyclase (which elevates cAMP), guanylate cyclase (which elevates cGMP), calmodulin (which mobilizes Ca^{2+}), and/or enzymes that affect the metabolism of arachidonic acid (via prostaglandin synthesis) or phosphatidylinositol (via inositol triphosphate and diacylglycerol synthesis). Several key intracellular signals (i.e., secondary messengers) can mediate the biological roles—either stimulatory or inhibitory—of a particular peptide. In many cases GTP-binding pro-

teins localized at the cell membrane can act as intermediaries of signal transduction between the peptide–receptor complex and the enzymes mentioned.

C. Peptide Biological Systems

A myriad of biological activities have been described for the more than 200 peptides that have been isolated and characterized in humans or other vertebrate species. Only a few peptides (Table I) are described here to exemplify some of their more fundamental biological properties. An important theme of this section is that many of the classical peptide hormones, which are chemical messengers of the endocrine system, have now been found within the brain, the gastrointestinal tract, and many other tissues. Thus, as an extension of their hormonal properties, which were first described, several of these peptides manifest neural, metabolic, growth, cardiovascular, gastrointestinal, reproductive, and immune regulatory functions (Table I).

1. Adrenocorticotropic Hormone and Melanocyte-Stimulating Hormone

Adrenocorticotropic hormone (ACTH) and melanocyte-stimulating hormone (MSH) are structurally similar peptide hormones derived primarily from the pituitary gland. Proopiomelanocortin is a precursor protein (31,000 molecular weight) for both ACTH and several molecular variants of MSH

TABLE I Some Known Biologically Active Peptides

Peptide	Amino acids[a]	Major biological actions[b]
Adrenocorticotropic hormone	39	Adrenal steroidogenesis increased
Angiotensin II	8	Vasoconstriction increased, aldosterone secretion increased, dipsogenia (i.e., thirst) increased
Bradykinin	9	Smooth muscle contraction increased, vasodilation increased, inflammation algesia increased
Calcitonin	32	Blood Ca^{2+} decreased
Cholecystokinin 8	8	Pancreatic enzyme and electrolyte secretion increased, gallbladder contraction increased, satiety increased, secretion of insulin and glucagon increased
Corticotropin-releasing factor	41	Corticotropin secretion increased
Dynorphin	17	Sedative analgesia increased, feeding behavior increased
β-Endorphin	31	Opiatelike activity increased, central analgesia increased, respiratory depression increased, euphoria increased
Leu-Enkephalin	5	Spinal analgesia increased, emotional effects increased
Epidermal growth factor	53	Epidermal growth and keratinization increased, growth of corneal epithelium increased
Follicle-stimulating hormone	220	Spermatogenesis in males increased, ovarian follicle growth and estradiol synthesis in females increased
Gastrin 17	17	Pepsin, gastric acid–HCl, and pancreatic enzyme secretion increased; stomach, gallbladder, and intestine smooth muscle contraction increased
Glucagon	29	Blood glucose increased, gluconeogenesis increased, glycogenolysis increased
Gonadotropin-releasing hormone	10	Luteinizing hormone and follicle-stimulating hormone secretion increased
Insulin	51	Blood glucose decreased, protein synthesis and lipolysis increased
Luteinizing hormone	210	Testicular androgen synthesis in males increased, ovarian estradiol and progesterone syntheses in females increased
α-Melanotropin	13	Melanin biosynthesis and secretion increased
Motilin	22	Gastrointestinal villous motility increased
Nerve growth factor	118	Sympathetic neurite development increased
Oxytocin	9	Milk secretion increased, uterine contraction increased
Parathyroid hormone	84	Blood calcium increased
Prolactin	199	Milk synthesis increased, corpus luteum progesterone synthesis increased
Secretin	27	Pancreatic secretion of H_2O, bicarbonate, electrolyte, and protein increased
Somatomedins	50–80	Peripheral nervous system growth and development increased
Somatostatin 14	14	Somatotropin, thyrotropin, parathyroid hormone, calcitonin, renin, and gastric acid secretion increased
Somatotropin	191	Hepatic somatomedin synthesis increased
Somatotropin-releasing hormone	44	Somatotropin secretion increased
Substance P	11	Peripheral nervous system pain transmission increased, central nervous system pain transmission decreased
α_1-Thymosin	28	Lymphocyte proliferation and differentiation increased
Thyrotropin	201	Thyroid hormone T_3 and T_4 synthesis and secretion increased
Thyrotropin-releasing hormone	3	Thyrotropin and prolactin secretion increased
Vasoactive intestinal polypeptide	28	Vasodilation increased, bronchodilation increased, glucagon and insulin secretion increased, glycogenolysis increased, lipolysis increased
Vasopressin	9	Renal H_2O absorption increased, vasoconstriction increased

[a] Prevalent form of the peptide in humans.
[b] Major actions (some peptides affect multiple biological actions) as related to either an increased/stimulated or decreased/inhibited physiological response.

ACTH H-Ser-Tyr-Ser-Met-Glu-His-Phe-Arg-Trp-Gly-Lys-Pro-Val-Gly-Lys-Lys-Arg-Arg-Pro-Val-Lys-Val-Tyr-Pro-Asn-Gly-Ala-Glu-Ser-Ala┐
└Glu-Ala-Phe-Pro-Leu-Glu-Phe-OH

ANG II H-Asp-Arg-Val-Tyr-Ile-His-Pro-Phe-OH

Bradykinin H-Lys-Arg-Pro-Pro-Gly-Phe-Ser-Pro-Phe-Arg-OH

CCK 33 H-Lys-Ala-Pro-Ser-Gly-Arg-Val-Ser-Met-Ile-Lys-Asn-Leu-Gln-Ser-Leu-Asp-Pro-Ser-His-Arg-Ile-Ser-Asp-Arg-Asp-Tyr(SO$_3$H)-Met-Gly-Trp-Met-Asp-Phe-NH$_2$

Enkephalin-Met[5] H-Tyr-Gly-Gly-Phe-Met-OH

β-Endorphin H-Tyr-Gly-Gly-Phe-Met-Thr-Ser-Glu-Lys-Ser-Glu-Thr-Pro-Leu-Val-Thr-Leu-Phe-Lys-Asn-Ala-Ile-Val-Lys-Asn-Ala-His-Lys-Lys-Gly-Gln-OH

Dynorphin H-Tyr-Gly-Gly-Phe-Leu-Arg-Arg-Ile-Arg-Pro-Lys-Leu-Lys-Trp-Asp-Asn-Gln-OH

Gastrin 17 (pyro)Glu-Gly-Pro-Trp-Leu-Glu-Glu-Glu-Glu-Glu-Ala-Tyr(SO$_3$H)-Gly-Trp-Met-Asp-Phe-NH$_2$

Glucagon H-His-Ser-Gln-Gly-Thr-Phe-Thr-Ser-Asp-Tyr-Ser-Lys-Tyr-Leu-Asp-Ser-Arg-Arg-Ala-Gln-Asp-Phe-Val-Gln-Trp-Leu-Met-Asp-Thr-OH

GnRH (pyro)Glu-His-Trp-Ser-Tyr-Gly-Leu-Arg-Pro-Gly-NH$_2$

α-MSH Ac-Ser-Tyr-Ser-Met-Glu-His-Phe-Arg-Trp-Gly-Lys-Pro-Val-NH$_2$

Somatostatin 14 H-Ala-Gly-Cys-Lys-Asn-Phe-Phe-Trp-Lys-Thr-Phe-Thr-Ser-Cys-OH

FIGURE 3 Primary structures of selected peptides of human origin (see text for discussion). ACTH, Adrenocorticotropic hormone; ANG, angiotensin; CCK, cholecystokinin; GnRH, gonadotropin-releasing hormone; α-MSH, α-melanocyte-stimulating hormone; ANG, angiotensin.

(i.e., α-MSH, β-MSH, and γ-MSH). α-MSH is identical to the first 13-amino acid sequence of ACTH, but, as a result of posttranslational modifications, is N-acetylated and C-amidated (Fig. 3). The biogenetic regulation and enzymatic processing of proopiomelanocortin have been intensively studied and are a classical example of prepropeptide hormone biology. [*See* PITUITARY.]

ACTH secretion from the pituitary gland is stimulated by the hypothalamic peptide, corticotropin-releasing factor. ACTH activates the adrenal gland to secrete the corticosteroids (i.e., aldosterone and cortisol), which, in turn regulate the metabolism of carbohydrates, lipids, and proteins as well as Na$^+$ resorption at various target tissues. ACTH is also active on melanocytes (MSH-like stimulation of melanin synthesis/secretion) and adipose tissue (lipolysis). [*See* ADRENAL GLAND.]

The release of MSH from the pituitary gland is controlled by mechanisms that are more poorly defined, but may include the inhibition of secretion by hypothalamic catecholamines, which, if interrupted, results in MSH release. α-MSH activates melanin pigmentation in most vertebrate species,

and it is significantly more potent (i.e., more than 100-fold) than ACTH in this effect. The structural similarity between MSH and ACTH results in some overlap of biological properties. Other examples of activities attributable to both MSH and ACTH include certain central nervous system–behavioral effects (i.e., arousal, attention, memory retention, and learning) and antipyretic or pyretic thermoregulation. Interestingly, the melanotropic activity of MSH is uniquely enhanced by heat–alkali treatment, which results in chemical racemization (i.e., the stereochemical conversion of an L- to a D-amino acid) of specific residues within its sequence. Racemized MSH effects highly potent and sustained (prolonged) melanotropic activity *in vitro* and *in vivo*. [*See* CATECHOLAMINES AND BEHAVIOR.]

ACTH and MSH appear to act through a common molecular mechanism—the stimulation of adenylate cyclase—which then produces cAMP (the secondary messenger), increasing its concentration within the target cell.

2. Angiotensin and Bradykinin

The regulation of blood pressure involves the biological activities of several peptide and nonpeptide factors. Angiotensinogen-derived peptides, more specifically angiotensins II (ANG II) and III (ANG III), are physiologically active in increasing blood pressure (i.e., hypertensive). Both ANG II (Fig. 3) and ANG III are small linear peptides ultimately generated from ANG, a biologically inactive precursor glycoprotein secreted from the liver. ANG is processed to give another biologically inactive precursor, ANG I, which is a decapeptide corresponding to the amino-terminal sequence of ANG. Conversion of ANG to ANG I is specifically effected by circulating renin, an enzyme secreted from the juxtaglomerular cells of the kidney in response to hypovolemia (i.e., a reduction in blood volume), an increase in the sodium concentration in the blood, and/or a central nervous system stimulus (catecholamine related). ANG I is converted to ANG II by the angiotensin-converting enzyme, in lung tissue. Finally, removal of the first amino acid at the amino terminus of ANG II by aminopeptidase yields ANG III.

ANG II is an extremely potent vasoconstrictor on arterioles and effects an immediate elevation of blood pressure. Long-range effects of ANG II and/or ANG III include the stimulation of aldosterone release from the adrenal cortex, which then effects Na^+ retention and a fluid volume increase (i.e., antidiuresis). All of the components of the ANG–renin–angiotensin-converting enzyme system have been identified in the brain, and two biological actions that are known in this regard include the stimulation of drinking behavior and the stimulation of pituitary vasopressin (i.e., antidiuretic hormone) secretion.

ANG II acts via specific receptors, but the mechanism is not known in detail. On endothelial (i.e., blood vessel) tissue the signaling pathway apparently involves the regulation of a Ca^{2+} channel, rather than the activation of adenylate cyclase. The resultant changes in cytosolic Ca^{2+} levels seem to provide the molecular basis for muscular contraction in the blood vessel wall.

Kinins are another family of peptides which, in contrast to the ANGs, are physiologically important in effecting vasodilation (i.e., hypotensive), among many other activities. In response to localized tissue injury, the kinins are derived from plasma protein precursors, called kininogens, via the enzymatic processing activities of plasma and/or tissue kallikreins. Specifically, bradykinin (Fig. 3) and kallidin are the biologically active kinins, and, in contrast to the ANGs, are inactivated by the angiotensin-converting enzyme (or kininase II). A bradykininlike peptidergic system also exists in the brain. [*See* KININS: CHEMISTRY, BIOLOGY, AND FUNCTIONS.]

Bradykinin contracts venous smooth muscle to provide increased blood flow and vascular permea-

bility to damaged tissue. Included in this biological action are the activation of sensory pain fibers and stimulation of the synthesis of both prostacyclin and endothelium-derived relaxing factor. As a mediator of inflammatory pain transmission, bradykinin has recently been shown to act on neurons that respond to pain.

Bradykinin receptors apparently are pharmacologically distinct in different target tissues. Receptor-mediated signal transduction for bradykinin appears to include phosphatidylinositol hydrolyis, guanylate cyclase activation (i.e., increased cGMP), and arachidonic acid metabolism via phospholipase C and/or phospholipase A_2 mechanisms. It is believed that each signal transduction pathway is associated with a different receptor.

3. Cholecystokinin and Gastrin

As exemplified above for ACTH and MSH and for the opioid peptides, chemical similarity exists for two families of gastrointestinal peptides, namely, cholecystokinin and gastrin (Fig. 3) and secretin-related peptides (see Section I,A,7). In the cases of both cholecystokinin (CCK) and gastrin, many molecular variants of each peptides have been identified (e.g., CCK 39, CCK 33, CCK 12, CCK 8, gastrin 34, and gastrin 17). Nevertheless, usually only one of these molecular forms constitutes the main form stored in the secretory granules of the cell in which the peptide is synthesized. The carboxy-terminal pentapeptide sequences of the CCK and gastrin peptides are identical. Both CCK and gastrin peptides could possess a tyrosine residue which is sulfated on the side chain, but the sites at which this amino acid exists in the peptides are different. The difference might play a major role in effecting their different biological functions and specificities. Recently, the common carboxy-terminal tetrapeptide sequence of CCK or gastrin (i.e., CCK 4 or gastrin 4) has been discovered in the pituitary gland and the hypothalamus. [*See* HYPOTHALAMUS.]

CCK 33 is secreted from the duodenal mucosa when food is present (particularly amino acids, HCl, and certain free fatty acids) and causes contraction and emptying of the gallbladder, as well as stimulation of the release of pancreatic enzyme. Other known physiological roles of the CCK peptides include gastric emptying and potentiation of secretin-induced pancreatic bicarbonate secretion. CCK peptides might also regulate the growth of the exocrine pancreas; however, in contrast to the gastrins, CCK peptides does not exhibit a trophic growth effect on the gastrointestinal mucosa. Recently, it has been proposed that CCK 8 could function as a satiety hormone, and within the central nervous system CCK might possess neurotransmitter/neuromodulator (e.g., dopaminergic) activities.

The mechanism of action of CCK is initiated by binding to its receptors, which differ according to target tissue. Recent studies show that brain and pancreatic CCK receptors are structurally and functionally distinct. CCK binding to its pancreatic receptors effects Ca^{2+} outflux and increases cytosolic cGMP. Concomitant increased metabolism of phosphatidylinositol has also been determined.

Gastrin is produced in the antral mucosa of the stomach and in the G cells of the duodenal mucosa. Similar to CCK, food is the primary physiological stimulus of gastrin secretion, and its release is further regulated by the autonomic nervous system and, possibly, by other peptides, somatostatin, and vasoactive intestinal polypeptide.

Gastrin is the most potent known secretagogue of gastric acid. It also stimulates pepsinogen secretion (perhaps indirectly via HCl release), relaxes the pyloric sphincter, stimulates pancreatic enzyme secretion, and increases intestinal motility.

4. Cytokines

Only recently, the biological responses of the immune system have been shown to be governed by soluble peptides, known as cytokines, which regulate the proliferative and differentiative responses of the respective immune lineage cells. [*See* CYTOKINES IN THE IMMUNE RESPONSE.]

These peptides have multiple biological functions, formerly attributed to separate factors. For example, the 17,500 molecular weight interleukin 1 (IL-1) possesses the biological activities of lymphocyte-activating factor, endogenous pyrogen, leukocytic endogenous mediator, mononuclear cell factor, catabolite, osteoclast-activating factor, and hemopoietin 1. Two forms of IL-1 (α and β) have been identified. Both forms are unique in that they are synthesized as precursors (31,000 molecular weight) that do not have signal cleavage sequences, despite the fact that IL-1 is found extracellularly (see Section I,A). Although IL-1α and IL-1β have only 26% amino acid homology, they interact with the same receptor. The pair serves as a prime example of how different ligands with identical receptor interactions can be made through the assembly of similar overall three-dimensional structure using different building blocks.

IL-1 facilitates the release of another cytokine, IL-2 (14,700 molecular weight), from T lymphocytes, formerly known as T cell growth factor, killer cell helper factor, or thymocyte mitogenic factor. In turn, IL-2 stimulates the release of another cytokine, γ-interferon, which is believed to act as a powerful differentiation signal, inducing macrophage killer activity, cytotoxic T cell differentiation, and B cell immunoglobulin secretion. This cascade of cytokines is only one example of the broader observation that, during activation of cellular or humoral immunity, some cytokines are dependent on the action of others that precede them in a precise sequence of production and site of action. Thus, not only do the cytokines regulate the biological response of cells in a primary manner, they further control the secondary response of the cells to secrete other cytokines needed to expand or support the differentiation of antigen-specific cells. Furthermore, it is now known that many biological systems, in addition to the immune system, are modified by the actions of secondary cytokines synthesized and secreted in response to an initiating cytokine. [*See* INTERFERONS; INTERLEUKIN-2 AND THE IL-2 RECEPTOR.]

5. Enkephalin, Endorphin, and Dynorphin

The endogenous opioid peptides were discovered after the elucidation of receptors within the central nervous system which were pharmacologically characterized as receptors for opiates (i.e., morphine, an agonist; naloxone, an antagonist). Because there was no evidence to suggest that morphinelike alkaloids occur naturally in the animal species, it was proposed that there might exist an endogenous opiate-like substance which would serve as natural ligands at these receptors. Several opioid peptides were subsequently discovered and now include endorphin, enkephalin, and dynorphin (Fig. 3). Several molecular variants of each of these peptides exist (e.g., Met- and Leu-enkephalin), and each peptide is derived from a prepropeptide. Proopiomelanocortin is a precursor for endorphin. The biological sources of the opioid peptides include the brain and the pituitary gland have also been localized to peripheral neural nerves of the gastrointestinal tract and other tissues. Discrete neuronal localization of endorphins, determined by immunocytochemistry, has shown that high concentrations are found in brain regions involved with pain transmission, respiration, motor activity, pituitary hormone secretion, and mood.

The primary role of the opioid peptides is analgesic, as a result of enkephalin-containing spinal cord neurons connected to peripheral sensory nerve cells that convey pain information. These primary afferent neurons utilize the neurotransmitter peptide, substance P, to transmit the pain stimuli, and the release of substance P is directly suppressed by enkephalin, or opiate agonist. Interestingly, acupuncture-induced analgesia is believed to occur via indirect stimulation of such endorphin receptors.

The opioid receptors have been classified into three groups by both pharmacological and biochemical methods: μ, δ, and κ. These receptors appear to mediate such actions as central analgesia (i.e., via β-endorphin), emotional behavior and spinal analgesia (i.e., via Leu-enkephalin), and sedative analgesia and appetite stimulation (i.e., via dynorphin). The principle receptor-mediated mechanism of action for the opioid peptides is by the inhibition of adenylate cyclase, and the biochemical basis of tolerance could be related to decreased receptor signaling, and that of withdrawal symptoms is connected to overproduction of cAMP, due to a compensatory mechanism of opioid-induced chronic suppression of adenylate cyclase activities.

6. Gonadoliberin

Hypothalamic gonadoliberin, or gonadotropin-releasing hormone [GnRH, or luteinizing hormone (LH)/follicle-stimulating hormone (FSH) RH] (Fig. 3), is a decapeptide that stimulates the secretion of both gonadotropins, lutropin (i.e., LH) and follitropin (i.e., FSH), from the pituitary gland. The action of GnRH on FSH release is selectively inhibited, in turn, by the antisecretagogue known as inhibin. Gonadal steroids regulate the action of GnRH by negative feedback inhibition of LH/FSH. The biosynthesis of GnRH is modulated by the neurotransmitter γ-aminobutyric acid. Extrahypothalamic GnRH also exists and regulates the biological activity of the testes, ovary, and placenta and also exhibits central neurotransmitter effects related to sexual behavior. In addition, such GnRH is involved in tumor growth (e.g., prostate cancer).

The mechanism of action of GnRH is apparently mediated primarily via specific receptors that stimulate adenylate cyclase. As a result of subsequent increased intracellular Ca^{2+}, gonadotropin secretory granules release their contents to their extracellular space. GnRH is released in a pulsatile manner, and receptor down-regulation (i.e., decreased activity) is observed if exogenous GnRH is administered chronically, with resulting tissue refractoriness and, in the case of females, ovulation inhibition.

7. Insulin and Glucagon

Insulin is a heterodimeric peptide consisting of an A chain (21 amino acids) and a B chain (30 amino acids), which are covalently linked by two disulfide links between cysteine residues. Proinsulin contains an interconnecting sequence of 23 amino acids between the amino terminus of the A chain and the carboxy terminus of the B chain. The chemical complexity of insulin has prevented its total synthesis by standard chemical methods. Therefore, insulin research and insulin therapy (i.e., in diabetes mellitus) have relied primarily on insulins isolated from various species or on the biosynthetic production of recombinant insulin. The three-dimensional structure of insulin, determined by X-ray crystallography, suggests that a specific conformation of the peptide backbone and the spatial proximity of amino acids derived from both the A and B chains are required for biological activities.

Insulin is secreted from the β cells of the pancreatic islets in response to an elevated blood glucose concentration. The major physiological role of insulin is to lower blood glucose (i.e., hypoglycemic action) by enhancing the cellular uptake of glucose and stimulating the subsequent conversion of glucose to glycogen, protein, or fat in liver, muscle, and adipose tissues, respectively. Insulin stimulates the active transport of glucose, amino acids, K^+, Na^+, and PO_4^{-2} into its target tissues. Insulin secretion is inhibited by normal levels of blood glucose, as well as by somatostatin derived from pancreatic D cells.

The mechanism of action of insulin is complex. The insulin receptor (450,000 molecular weight) is a membrane glycoprotein with an $\alpha_2-\beta_2$ subunit structure, linked by disulfide bonds. Insulin binds to the α subunits, causing the β subunits to undergo tyrosine autophosphorylation and to become an active receptor kinase. It is important to note that, in contrast to other peptide hormones which affect intracellular protein phosphorylations via, for example, cAMP as a secondary messenger, insulin causes a direct phosphorylation event. This biological mechanism is also observed for growth factors. Insulin-bound receptors rapidly become clustered; invaginate, forming coated pits, lined with clathrin molecules at the cytoplasmic side; and, finally, internalize by vesicular endocytosis. The immediate intracellular events associated with insulin action could include modification of membrane protease activity, activation of phosphokinase C or phospholipase C, and/or Ca^{2+} mobilization.

Glucagon (Fig. 3) is derived primarily from the α cells of the pancreas. It is chemically similar to other peptide hormones (e.g., secretin, vasoactive intestinal peptide, and gastric inhibitory peptide). Glucagon affects glucose homeostasis in a manner opposite that relative to insulin. It produces hyperglycemic action by directly stimulating hepatic glucose release by glycogenolysis (i.e., the conversion of glycogen to glucose) and gluconeogenesis (i.e., the conversion of amino acids and glycerol to glucose). The gluconeogenic action of glucagon is particularly important in the physiological events associated with prolonged fasting and exercise, whereas its glycogenolytic activities are rapid and short term to elevate blood glucose levels. Glucagon is lipolytic, thereby liberating free fatty acids and glycerol.

The mechanism of action of glucagon at its target tissues is initiated by receptor-specific binding, followed by transmembrane signaling to adenylate cyclase and an elevation of cytosolic cAMP. The glycogenolytic action is mediated by cAMP-dependent intracellular conversion of inactive phosphorylase *b* to active phosphorylase *a*. The glucagon receptor (63,000 molecular weight) is a dimeric protein, with binding sites for both glucagon and a guanine nucleotide triphosphate-binding signal-transducing protein. [*See* INSULIN AND GLUCAGON.]

8. Somatotropin and Somatostatin

Somatotropin, or growth hormone (GH), is primarily produced by the pituitary gland, and its secretion is regulated by hypothalamic somatotropin-releasing factor and somatotropin release-inhibiting factor (i.e., somatostatin). Human GH is chemically similar to prolactin: It consists of 191 amino acids and has two intramolecular disulfide bonds.

GH is an anabolic hormone (i.e., it promotes the accumulation of reserves) in terms of its direct or indirect (via somatomedins) effects on metabolism, cellular proliferation, and growth. It stimulates amino acid uptake into muscle protein and stimulates the extracellular deposition of collagen. GH stimulation of hepatocytes effects glycogenolysis (i.e., a hyperglycemic response, as described for glucagon) and secretion of somatomedins, which, in turn, stimulate cellular growth in many target tissues.

The mechanism of action of GH is not well understood. It apparently binds to specific receptors on its target tissues, but the subsequent steps of intracellular signaling are complex. The cloning of cDNA encoding GH receptors has shown a structural similarity between cell surface GH receptors and circulating (i.e., serum-derived) GH-binding proteins. It has been proposed that the GH receptor is a transmembrane protein with an extracellular domain and a cytosolic domain, connected by a short intramembrane sequence, and that signal transduction is direct (as opposed to, for example, signaling mechanisms based on adenylate cyclase).

Somatostatin is perhaps the most studied and important peptide inhibitory factor known. It contains 14 amino acids and is cyclic, by virtue of an intramolecular disulfide bridge between two cysteine residues (Fig. 3). Somatostatin secretion from the hypothalamus functions primarily to inhibit pituitary GH release. Somatostatin is widely distributed in the central nervous system and is known to potentiate L-dopa and affect sedation and hypothermia. Pancreatic somatostatin inhibits the production of insulin and glucagon and attenuates gastric acid secretion.

Two types of somatostatin receptor have been characterized in pituitary, brain, and pancreatic tissues. Binding of somatostatin to its receptors is believed to cause inhibition of adenylate cyclase activity, which is stimulated by the agonists (e.g., somatotropin-releasing factor). However, other signaling mechanisms (e.g., Ca^{2+} related) might also be pertinent.

II. Biotechnology and Biomedical Research

A. Recombinant Peptides

It is no coincidence that our progress in understanding the biochemical properties of certain peptides corresponds with advances in modern molecular biology and the ability to clone and express the genes in prokaryotic or eukaryotic cell lines. Some of the peptide hormones are present in such small quantities that their isolation and complete identification by protein sequencing techniques are extremely laborious. The advent of recombinant DNA technology has permitted peptide sequence determination through use of the corresponding cDNA. Moreover, the expression of the cloned genes has yielded peptides in sufficient quantities to characterize and explore their biological activities. For example, genes of about 15 human cytokines have been cloned. They include three colony-stimulating factors (macrophage, granulocyte, and granulocyte–macrophage), interferons α and γ, interleukin-1 (α and β) through -6, lymphotoxin, tumor necrosis factor, and transforming growth factor β. The molecular weight of these cytokines (i.e., 17,000–30,000) spans the gray zone of an arbitrary definition of where peptide ends and protein begins, and the smallest cytokines reach the absolute size limits (i.e., about 150 residues) for chemical synthesis. Thus, all peptides can be obtained by either chemical synthesis or recombinant DNA technology. Chemical synthesis is important for the lower-molecular-weight peptides, which are usually unstable to the proteases present in bacterial expression systems.

The list of biologically important human peptides produced by recombinant DNA technology continues to expand. Human growth hormone (i.e., somatotropin), important for treating dwarfism of hypopituitary origin (Table II), was first obtained only from cadaver brains. It is now readily available through bacterial culture. As mentioned earlier (Section I,C,7), the chemical complexity of insulin makes its synthesis impractical, so replacement therapy for diabetes uses porcine insulin isolated from the pig pancreas. However, porcine insulin does not have the same sequence as human insulin, and the risk of immune response is higher than with the human sequence. Again, recombinant DNA technology now allows the production of human insulin. The A and B chains of recombinant human insulin are produced separately, not together, as in natural proinsulin. Extensive research is required to discover conditions under which biologically active insulin could be obtained with its two disulfide bonds correctly formed.

Recombinant DNA technology has been important not only for the production of high-molecular-weight peptides, but even more for conducting research into the mechanism of peptide–receptor interactions. It is now possible to systematically alter peptide sequences by site-directed mutagenesis of the gene DNA and study structure–activity relationships.

TABLE II Abnormal Conditions of Excessive or Deficient Physiological Levels of Naturally Occurring Peptides

Peptide	Relationship	Symptom (Disease)
Adrenocorticotropic hormone	Deficiency	Abnormal carbohydrate metabolism (Addison's disease)
	Excess	Protein catabolism increased (Cushing's disease)
Gonadotropin-releasing hormone	Deficiency	Secondary hypogonadism
	Excess	Secondary hypergonadism
Insulin	Deficiency	Hyperglycemia and glucosuria (diabetes mellitus or insulin-dependent diabetes, if resulting from cytotoxic autoantibodies to β-cells)
Parathyroid hormone	Excess	Abnormal blood Ca^{2+} increased, hyperparathyroidism
Somatomedin	Deficiency	Abnormal growth decreased (Larin-type dwarfism)
Somatotropin	Deficiency	Abnormal growth decreased (hypopituitary dwarfism)
	Excess	Abnormal growth increased (giantism in children)
	Excess	Abnormal growth increased (acromegaly in adults)
Thyrotropin	Excess	Abnormal metabolism increased (Graves' disease)
Vasopressin	Deficiency	Hypovolemia–dehydration increased (pituitary-type diabetes insipidus)

B. Peptides Therapeutics

For their important physiological actions, peptides are at the basis of several human pathophysiological (disease) states (Table II). Replacement therapy using naturally occurring, recombinant, or synthetic peptides is frequently used to offset abnormally low levels of the endogenous peptide (e.g., diabetes mellitus, Addison's disease, and dwarfism). Medicinal chemistry and/or biotechnology strategies to design more effective (i.e., in terms of biological potency, metabolic stability, duration of action, and oral absorption) peptide-based drugs have become a major challenge and effort for the pharmaceutical industry.

TABLE III Some Known Synthetic Peptides Demonstrating Noteworthy Biological Properties[a]

Synthetic Peptide	Structure–Activity Relationships
H-Tyr-D-Ala-Gly-N-Me-Phe-Gly-ol	Potent μ-selective opioid agonist
H-D-Phe-Cys-Tyr-D-Trp-Orn-Thr-Cys-Thr-NH$_2$	Potent μ-selective opioid antagonist
H-Tyr-D-Pen-Gly-Phe-D-Pen-OH	Potent δ-selective opioid agonist
H-D-Phe-Cys-Phe-D-Trp-Lys-Thr-Cys-Thr-ol	Potent somatostatin agonist
H-Sar-Arg-Val-Tyr-Val-His-Pro-Ala-OH	Potent ANG II antagonist
pGlu-His-Trp-Ser-Tyr-D-Ser(t-Bu)-Leu-Arg-Pro-NHEt	Potent GnRH agonist
H-D-Arg-Arg-Pro-Hyp-Gly-Thi-Ser-D-Phe-Thi-Arg-OH	Potent bradykinin antagonist
Ac-Ser-Tyr-Ser-Nle-Glu-His-D-Phe-Arg-Trp-Gly-Lys-Pro-Val-NH$_2$	Potent MSH agonist

[a] N-Me-Phe, N^{α}-methyl-L-phenylalanine; Gly-ol, glycinol; Orn, L-ornithine; Pen, L-penicillamine; Sar, sarcosine; D-Ser(t-Bu), D-serine; Pro-NHEt, L-proline ethylamide; Hyp, 4-hydroxy-L-proline; Thi, L-2-thienylalanine; Nle, L-norleucine.

Synthetic peptide derivatives might simply contain D-amino acid substitutions or side chain or backbone amide modifications, or might be changed with respect to their three-dimensional properties (i.e., conformation and structural flexibility) via the use of intramolecular bridging (i.e., cyclization). Such synthetic peptide analogs might act as superagonists or competitive antagonists. In addition, some exogenous (i.e., not found in the host animal) naturally occurring peptides could possess some of these chemical or pharmacological properties when given *in vivo*. Peptide antagonists are useful for examining the precise physiological roles of peptides, and perhaps as drugs to block the undesired effects of inappropriate (or excessive) levels of an endogenous peptide. A few selected peptide derivatives relevant to this discussion are shown in Table III. Finally, the search for nonpeptide compounds that bind to endogenous peptide receptors (e.g., morphine at the opioid receptor) will undoubtedly advance our understanding of peptide biology and, possibly, provide new leads in the design, discovery, and development of peptide-related drugs.

Bibliography

Farrar, W. L., Ferris, D. K., and Harel-Bellan, A. (1989). The molecular basis of immune cytokine action. *CRC Crit. Rev. Ther. Drug Carrier Syst.* **5**, 229.

Hadley, M. (1984). "Endocrinology." Prentice-Hall, Englewood Cliffs, New Jersey.

Nogrady, T. (1988). "Medicinal Chemistry: A Biochemical Approach," 2nd ed. Oxford Univ. Press, New York.

Wallis, M., Howell, S. L., and Taylor, K. W. (1985). "The Biochemistry of the Polypeptide Hormones." Wiley, New York.

Perception

H. BOUMA, D. G. BOUWHUIS, *Institute for Perception Research/IPO in Eindhoven*

E. P. KÖSTER, J. H. A. KROEZE, *Utrecht University*

E. J. M. EIJKMAN, *Nijmegen University*

F. L. ENGEL, J. 'T HART, A. J. M. HOUTSMA, F. L. VAN NES, J. A. J. ROUFS, *Institute for Perception Research/IPO in Eindhoven*

Glossary

Attention Selective perception of certain objects or events out of many possible ones

Hearing Perception mediated by certain acoustical vibration impinging upon the ear

Kinesthetic perception Perception mediated by mechanical and thermal stimulation outside or inside the body

Perception Interface between the outside world and the human brain, as regards the intake of sensory data, the processing into sensory attributes and the recognition as objects or events

Smell Perception mediated by certain volatile chemical substances impinging upon the sensory surfaces of the nose

Taste Perception mediated by certain chemical substances impinging upon the tongue

Vision Perception mediated by certain electromagnetic radiation impinging upon the eye

HUMAN PERCEPTION is taken here as the physico-chemical interface between the outside world and the human brain, plus the sensory and cognitive processing of the data transforming these into information. Each sense organ is generally activated by its specialized type of stimulation and gives rise to its specific set of sensory attributes. Senses that take in stimulation from distant sources are hearing, through certain acoustical vibration, vision, through certain electromagnetic radiation, and smell, by a number of volatile chemical substances. Sense organs that react to stimulation from close sources are taste, by certain chemical substances in the mouth, and the skin senses (and kinesthetic senses), from mechanical stimulation at the outside or inside of the body. Within each of the senses, the sensory attributes and their dynamic variations are subjected to processes of recognition, i.e., interpreted within a framework of evolved experience and recent context. An important special case is the recognition of spoken language (speech) through the ear or written language (reading) through the eye, where the language elements are symbolic carriers of information between humans. In the overwhelmingly rich sensory stimulation to which humans are exposed, powerful perceptual selection processes make the chaotic input amenable. All perceptual functions are far from constant because they are in active development over life from one's earliest to final moment.

I. General

A. Transduction

Transducers are located in the sense organs, which, after initial filtering, map the physical data to neural data. So in the eye, imaging is an optical filtering process and the rod and cone receptors of the retina are the transducers proper. In the ear, the ear canal, the chain of middle ear bones, and basilar membrane provide filtering, and the hair cells act as transducers. When physical parameters are outside the range of filters and transducers, no perception results, even if perceptual terms are commonly used (infrared, ultrasound).

B. Sensory Attributes

After transducers, neural processing takes over. This is quite complex and leads to (1) largely or partly automatic reflexes in muscles and glands and (2) impressions and interpretations about the outside world (perception). The two categories are interdependent, and active perception, in which movements of the observer occur, may be different from passive perception. Sensory attributes are coupled to physical stimuli and reflect direct processing, which may be of a complex nature; examples in vision are brightness and color as coupled to spatial luminance distribution and wavelength composition; examples in hearing are loudness and sound direction as dependent on physical spectral intensities and differences in intensity and time of the signals for the two ears, respectively. For the sense organs, with the possible exception of smell, the number of sensory attributes is relatively small. Constant stimulation need not produce constant sensation (adaptation). The field of study that relates sensory attributes to physical stimuli is called psychophysics.

C. Recognition

Another type of processing is the interpretation of the sensory attributes of one or more senses by mapping them onto internal concepts of the outer world that have developed earlier. The central term here is recognition, because learning has been involved, perceptual experience is relevant, and interpretation may well differ among observers. The activated concepts imply meaning to the observer, which relates to networks of relevant associations. Examples are recognition of faces and of musical sounds. Associations may sometimes be described in physical terms, i.e., in terms of physical stimuli, which would produce similar activations. The recognition and associations form a basis for further cognitive evaluation and for intended movements, which bring about changes in the physical world. Natural language is concerned with the manipula-

tion of symbolic information. The symbols themselves, speech or print, are coupled to meaning by language conventions rather than by physical attributes. Units are called "words," and their ordering rules are described by the syntax. Natural language permits an incredibly rich gamut of expressions from a similarly rich gamut of inner worlds. Subtlety and ambiguity lie close together. Speech perception and reading are the two perceptual components.

D. Selection

Perception is usually quite robust and stable in the presence of wide variations of physical stimuli (perceptual constancies). This should perhaps be attributed to the types of processing involved and to priorities in processing (selection). Priorities in processing are dependent both on sensory attributes coupled to external physical stimuli (conspicuity) and on internal factors relating to the search of specific information by the subject (directed attention). As the relevant distribution of sensory attributes becomes more complex, priorities in processing are more pertinent. Conspicuity and directed attention may be in mutual conflict; however, the rule seems to be that the two processes are cooperative in a sense that the desired information is coded in terms of its most conspicuous attributes.

Robustness and stability of perception are prerequisites for integrating the perceiving subject in each of the many physical environments, thus preventing perceptual chaos. The basis for this seems to be that the subjects act on an internal mode of the physical world and incoming data is interpreted in terms of shifts and changes in this internal model. Components of such internal models reside, as we say, in long-term memory and are revived as required in the internal model of the world "here and now." Language data are coded as internal speech in a short-term buffer. It holds some five to nine words or word combinations. Several other memory components are short term as well, but properties of the above language type are best known.

E. Development

Because it is brought about by physical stimuli, perception depends on the physical environment. However, the great amount of learning involved in perception makes perceptual processes themselves also very much dependent on the world around us.

In this sense, human perception bears the marks of the environment that we live in. Much of the present environment has been brought about by technology and consequently both our internal concepts as built up in the perceptual process and the perceptual processes themselves are geared to this technological world.

II. Hearing

A. Transduction

The human hearing organ consists of an acoustical–mechanical part comprising outer, middle, and inner ear, and an electrochemical part, which includes a variety of neural structures that run from the 8th (auditory) nerve to the temporal lobe of the brain's cortex. The outer ear (pinna, ear canal) mainly transmits sound waves from the outside air to the eardrum. In the middle ear, three small bones (malleus, incus, stapes) with attached ligaments and muscles connect outer ear (eardrum) and inner ear (oval window). The lever action of these bones and the area reduction of eardrum to oval window form a mechanical transformer between the low impedance of air in the ear canal and the much higher impedance of the fluid inside the inner ear. Automatic stiffening of the middle ear muscles when the ear is exposed to intense sound (middle ear reflex) causes sound attenuation, which protects the inner ear. The inner ear (cochlea) is a rolled-up tube partitioned almost over its entire length (35 mm) by the basilar membrane. This membrane assumes different resonance patterns, dependent on the vibration frequency of the stapes. Thousands of (inner) hair cells, situated along this membrane and innervated by fibers of the 8th nerve, convert the mechanical motion of the membrane into electrical pulses, which propagate along 8th-nerve fibers to other nuclei of the auditory nervous system. The actual processing of those nerve pulses, which results in the sensation of hearing, may occur at many neural levels anywhere between the termination of the 8th nerve (cochlear nucleus) and the auditory cortex. [See COCHLEAR CHEMICAL NEUROTRANSMISSION; EARS AND HEARING.]

The sensitivity range of the human ear is on the one hand limited by the hearing threshold, which is shown in Fig. 1. On the other hand, our hearing range is limited by the discomfort or pain threshold, which is typically around 130 dB, independent of

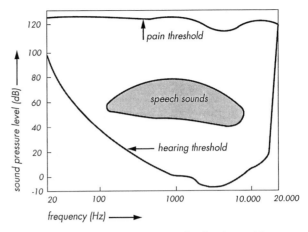

FIGURE 1 Region of sound pressure level and sound frequency where normal hearing occurs.

frequency. For long-duration sound exposures, a hearing-damage threshold can be as low as 90 dB.

B. Auditory Attributes

Relationships between perceptual, subjective attributes of sound and its physically measurable, objective characteristics are well-established. Perceived loudness of a tone is a power function of its sound pressure (exponent 0.6). Perceived pitch has a simple monotonic relationship with frequency (tone height) but also a cyclic one (tone chroma). This causes pitches of successive octaves to sound similar. Perceived timber depends primarily on the spectral profile of a sound but also on temporal features such as attack and decay. Perceived location of sound depends systematically on interaural arrival-time differences (low frequencies) and interaural intensity differences (high frequencies). There are also many cross-relationships. Loudness and timber both depend on frequency, and the pitch of a tone can change when only its intensity changes.

C. Recognition

What we hear is not only determined by the absolute or differential sensitivity of our ears but also by our memory capacity for sound. Roughly speaking, we can barely distinguish tones that differ by 1 dB in intensity (i.e., 12% in sound pressure) or 0.2% in frequency. Integrating this over the entire intensity and frequency ranges of our ear, we find that we can distinguish about 340,000 different tones. However, if we have to recognize or identify every one of those tones, we end up making mistakes. In fact, the largest number of different tones that we can reliably identify is about seven. Identification performance is primarily limited by memory capacity and not so much by differential sensitivity of our ears. Memory capacity for sound can grow with training. During infancy and childhood, we learn to recognize and handle many different speech sounds of the language (categorical perception). Trained musicians, especially those who have absolute pitch, can recognize considerably more than seven notes.

D. Hearing Impairments

Hearing impairments are first of all those associated with upward threshold shifts. They can be broadly divided into conductive and sensori-neural hearing losses. The former are broadband and exhibit simultaneous upward shifts of hearing and pain thresholds. The latter are often frequency-specific and generally show a mere increase in hearing threshold without an equivalent rise in pain threshold. Reduction of dynamic range by a sensori-neural disorder often causes an abnormal growth of perceived loudness with sound intensity, called recruitment. Some other hearing disorders, not necessarily connected with hearing loss, are tinnitus (hearing a constant tone when there is no sound) and monaural diplacusis (hearing several tones or noise when the sound is a pure tone). Binaural diplacusis (hearing slightly different pitches in each ear when they receive the same tone) is a common phenomenon but can become excessive in pathological cases. Noise-induced temporary threshold shifts are temporary and recover with time. When, however, exposure levels are too high (gun shots) or the ears have been exposed for many years to sound levels of 85 dB or higher (factory noise), permanent threshold shifts may result.

Current technology has given us virtually unlimited control over sounds we can produce (computer music) and over means of encoding or decoding sound (telephone communication). Often, one can save communication costs or sound storage space by simply not encoding those portions of a sound that our ears would not perceive anyway (perceptual entropy). For the hearing-impaired it is sometimes useful to process the sound of a hearing aid so that it will match the person's residual hearing. Finally, for those who suffer from total bilateral cochlear hearing loss, various cochlear electrode implant techniques are available through which the 8th nerve can be stimulated by electrical transformations of sound, resulting in sensations that resemble hearing.

III. Speech

Speech is the sound produced by the vocal mechanism of humans as a carrier of meaningful messages that are coded according to the rules of language.

A. Production

The energy is provided by a stream of air from the lungs; the various speech organs convert it into a source sound and shape it into the many different speech sounds. For most speech sounds, the source consists of vocal cords vibration, which gives rise to a quasi-periodic sound with many overtones. Varying the repetition frequency results in changes in perceived pitch.

B. Vowels and Consonants

The source sound is shaped in the vocal tract. This consists of the pharyngeal and oral cavities (and the nasal cavity, which is only relevant for nasal sounds, such as the first and last sounds in "man"). The vocal tract has a flexible form, mainly determined by the movability of the tongue body and tip. At many different places in the oral cavity, a reduction of its cross-sectional area can be made, which gives rise to as many differences in timber, noticeable in, for example, the 12 vowels of English in the words "heat," "hit," "head," "hat," "hard," "hot," "hod," "hook," "hoot," "hut," "herd," and "the".

If somewhere in the vocal tract a sufficiently narrow constriction is made, or a closure followed by a release, the air stream from the lungs may become turbulent. This causes noise and, in these cases, the sound source is situated in the vocal tract itself. Examples are "s" and "t." Again, the position of the constriction determines the nature of the speech sound (e.g., "k" [back] and "t" [front]).

The articulatory characteristics mentioned thus far are reflected in acoustic properties: Figure 2a and b shows the waveforms of the vowels in "heat" and "hoot". The energy distribution as a function of frequency (the spectrum) appears to be specific

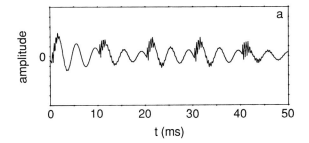

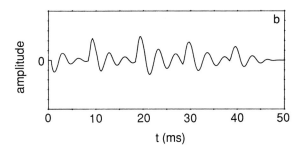

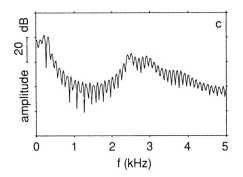

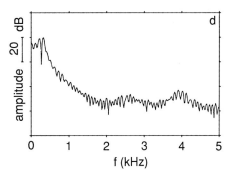

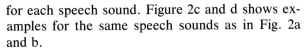

FIGURE 2 (a) Waveform of the vowel in "heat." (b) Waveform of the vowel in "hoot." (c) Spectrum of the vowel in "heat." (d) Spectrum of the vowel in "hoot."

for each speech sound. Figure 2c and d shows examples for the same speech sounds as in Fig. 2a and b.

Many of the acoustic features give rise to distinct auditory attributes, such as hissing sounds ("s") vs. sounds with pitch ("m"), high ("heat") vs. low ("hoot") timber, short ("t") vs. long ("s") sounds. But this does not imply that normal speech perception is based on the identification of all the auditory attributes for each of the successively incoming speech sounds. Indeed, isolated speech sounds can only be recognized with about 70% certainty.

C. Recognition

Cognitive attributes must also be taken into account: We should consider that the listener largely bases his or her interpretation on stored knowledge. An appeal to cognition can take place at different levels of abstraction. For instance, it has become apparent that in a word recognition task, the number of perceived syllables together with the location

of the word stress strongly reduces the number of candidate words. Nevertheless, only the low level of speech sound recognition is decisive in distinguishing between, for example, "elegant," "element," and "elephant." On the other hand, if the word is presented in a syntactically correct sentence, its status as either a noun or an adjective may be revealed, and this will help to separate the first word from the other two (e.g., the context "That's an . . . solution" only allows "elegant").

Clearly, in normal discourse the contextual information is of great help in these situations. But a truly complicating factor is that in fluent speech there are practically no pauses between words (like spacings in print). Our impression is that we can nevertheless process the continuous stream of speech as if it were segmented into words. This can hardly be understood without the assumption that a constant, intensive appeal is made to an internally represented dictionary. Thus, speech perception is a very complicated activity, in which knowledge of the language plays an important part, on levels ranging from phonology (sound structure) to semantics (meaning). There is an enormous gap between the abundant information available in the speech signal and the restricted amount of it the listener can cope with. Therefore, speech perception must be considered to be a highly selective process. [*See* LANGUAGE.]

D. Speech Technology

Speech perception can be studied by means of measuring the acceptability of artificial speech. An interesting by-product of this method is that, in the end, it may lead to reliable acoustic specifications for speech synthesis-by-rule. This is, in fact, one of the objectives of speech technology, which furthermore is concerned with—parsimoniously coded—speech transmission, with automatic speech recognition and with speaker identification. The products of these efforts are beginning to find their way in a world that becomes increasingly technical: Computer systems that respond to spoken commands or queries, in artificial speech rather than in orthographic form on a screen, are already one of the possibilities. But also communicatively disabled persons may benefit from these facilities: a reading machine for the visually impaired, a keyboard-to-speech device for vocally handicapped people. Mention may be made of semi-automatic devices as well, such as the intonable electrolarynx for laryngectomees or those who lack the use of their voice temporarily.

IV. Vision

A. The Eye and Visual Pathways

The eye consists of a bulblike transparent body (vitreous body) that is enveloped by the sclera. At the front side of the eye bulb, a clear window enables light to enter the eye, through a hole in the iris diaphragm (pupil) and fall onto the retina. The optics of the eye (cornea and lens) create an image of the outside world on the receptor mosaic within the retina. There are two kinds of photoreceptors: rods and cones, with rods operating at night and cones at daylight. They transduce the light distribution into a stream of nervous activity, which is modified at different retinal cell layers and guided into the optic tract. Eventually, the information stream of the left half of the retinas reaches the left part of the visual cortex of the brain, the right retinas feeding the right part of the cortex, both passing a cross-over (optic chiasm) and a neural nucleus (lateral geniculate nucleus). [See EYE, ANATOMY; VISION (PHYSIOLOGY); VISUAL SYSTEM.]

B. Physical Stimuli

Light, emitted by a natural or artificial light source, is reflected by object surfaces and projected onto the retina. It enables a human being to obtain information from the outside world, to orient themself and to act accordingly. The relevant physical magnitude of the reflected light is luminance, which can be characterized as light density. The illumination pattern at the retina caused by the properties of the visible surround is proportional to its luminance pattern. Spectral components of this light stimulate three kinds of cone differently, which causes color vision. [See RETINA.]

C. Visual Attributes

Variations in luminance and spectral content evoke the visual attributes brightness, brightness contrast, sharpness, and color. Details can be detected by differences in brightness and color. Visual acuity characterizes the ability to see small details. Generally, the visual attributes are not trivially linked to the physical stimuli. The luminance distribution evokes, for instance, an internal image of the outside world that is largely determined by object properties such as surface reflectance, whereas the light source properties and observer position are relatively unimportant. Other visual attributes that also mainly reflect object properties are depth, movement, and texture.

D. Cognitive Aspects

It is commonly acknowledged that the main goal of vision is to derive three-dimensional object shape from the information contained in the retinal luminance distribution. In other words, the visual system aims at reconstructing the outside world from the light density variations.

Visual attributes such as brightness, color, movement, and depth can be regarded as lower-level interpretations that form the input to higher-level interpretations. These involve the grouping of related items within the visual field to larger wholes, and exchanges with memory items. For instance, features of body shape, color of hair, and movement enable us to identify an individual.

In visual recognition, in search and in reading, the eyes normally move over the text in irregular jumps (saccades), each eye pause giving rise to recognition of objects in the fixation area. Integration across eye saccades occurs smoothly in an unknown way.

E. Visual Disorders

Defects may occur anywhere in the visual pathways. Optical defects may result in a retinal image that is not optimal. Lens errors, for instance, may evoke blurred imaging on the retina, while light straying at particles in the optic media diminishes contrast. Neural defects may already occur at lower processing levels, such as color blindness. A typical example of malfunctioning at higher levels is word blindness (dyslexia).

F. Vision in a Technical World

Luminance distributions of pictures displayed on film and television are intended to evoke the same perceptual sensations as the corresponding luminance distributions that occur in real scenes. This does not imply that they should be identical. In fact, the displayed images are often far more parsimonious than the real ones. Generally, it is sufficient if the differences cannot be seen. A typical example is the apparent continuity of a television image, which in fact is a line-type image that is periodically refreshed.

V. Reading

Reading concerns the visual perception of language signs. The signs are coupled to meaning by convention, either directly as in ideographs or by alphabetic or syllabic code (characters), which relates to speech sounds. [*See* READING PROCESSES AND COMPREHENSION.]

A. Eye Movement Control

Generally, the process of reading is preceded by searching for the desired part of a text on a page. During search, the eyes skim over the page, their motion being guided by certain text attributes. Just as in other static visual situations, the eyes move stepwise so that in a series of fixations of about 250 msec the successive characters can be imaged on the central part of both retinas. Saccades between fixations are 8 ± 4 letter spaces in alphabetic languages. Other saccades are necessary from the end of a line or column to the beginning of the next one, and back to previous parts of the text. [*See* EYE MOVEMENTS.]

B. Word Recognition

During fixation pauses, information is extracted from the visual reading field, which measures 10–20 letter positions of alphabetic text or 1–3 ideographs.

Both character recognition and word contour recognition contribute to this process. Luminance contrast and discriminability between characters of similar configuration are important.

C. Temporal Integration

Time is a crucial factor in reading: If the reading speed is too low, it takes too long to absorb the full content of a sentence and, therefore, its context, thus preventing integration. There is a minimum reading rate of 20 words per minute. Reading rate is the result of a complex interplay between necessary recognition time and the prevention of backward masking, saccade length, and accruing comprehension.

D. Reading and Language

A reader's knowledge of the language interacts with all components of reading. Visual patterns are decoded as characters and words, in turn representing sounds of speech and semantic units. Characters in a prescribed order and orientation represent words of the language, with a different sound pattern, and a specific grammatical function and meaning. Fluent reading requires extensive use of redundancy of the printed text at graphical, orthographic, lexical, syntactic, and semantic levels.

E. Reading Disorders

In our culture, which depends so much on the printed text, a reading disorder is a considerable handicap. Reading disorders may be due to insufficient intellectual development or a visual impairment. A specific reading disorder—dyslexia—occurs in about 6% of the male population. Its origin is now sought in the chain of cognitive processes following vision. [*See* DYSLEXIA–DYSGRAPHIA.]

F. Reading in a Technical World

An increasing part of what people read now is presented with electronic means on electronic displays. Reading proceeds basically in the same way for paper-based as for screen-based texts, but negative effects may result from display properties. The presentation of bright text on a dark background, for instance, together with the presence of a reflecting glass sheet in front of the screen, may hamper reading if the surrounding luminance is high. With proper care, many such negative effects may be avoided. Also sharpness, color, contrast, and character configuration are relevant.

VI. Smell

A. Transduction

The olfactory epithelia, two pigmented areas of 2–4 cm^2 in the olfactory clefts on both sides of the nose, consist of receptor cells, sustentacular cells, and Bowman's glands, which secrete the watery olfactory mucus (composition unknown) in which the odor-receptive cilia (80) sprouting from the protruding knobs of the bipolar receptor cells are bathed. The axons of these cells (neurons that are replaced by new ones from a basal cell layer during life) reach the olfactory bulb, where they form synapses. In this layered structure, many interconnections are found. The mitral cells (secondary neurons) form the olfactory tract to central parts of the brain. Granular cells receive afferent and centrifugal input and exert inhibitive influences on other cell types in the bulb. Odorous molecules can reach the receptors via ortho-nasal (sniffing) or retro-nasal (exhalation of vapors from the mouth) stimulation. The nature of the receptive mechanism is unknown, but receptor proteins are involved. On average, the olfactory receptor responds to 30% of the odors presented, but different receptors respond to different sets of odors and correlations between the sensitivities to odors are low (average $r = \pm 0.30$), suggesting a fair degree of independence and specificity. [See OLFACTORY INFORMATION PROCESSING.]

Human olfactory sensitivity strongly varies among odors and among individuals. Adaptation (reduction of sensitivity during stimulation) is strong and recovery after adaptation slow. Odors often suppress each other in mixtures.

B. Chemical Stimuli

The physicochemical properties determining the odor of a molecule are unknown. Volatility is a prerequisite, and water and lipid solubility, molecular shape, and functional groups are important. Enantiomers can have distinguishable odors.

C. Sensory and Cognitive Attributes

Odors warn (fire, cadavers), convey pleasure (food), and pay a role in sex (perfume, body odors).

Odorous quality is quite varied: Many thousands of odors can be distinguished, but not verbally labelled. Odor memory is predominantly episodic. Odors have (inborn) or acquire strong affective values. Pleasant odors become unpleasant when strong.

D. Olfactory Disorders

Specific anosmia (i.e., insensitivity to one odor or a group of closely related odors in otherwise normally sensitive persons) is not exceptional. General anosmia (complete insensitivity), hyposmia (reduced sensitivity), parosmia (disturbed perception of the nature of the stimulus), and kakosmia (perceiving all odors as putrid) can be caused by head traumata, viral infections, hormone or neurotransmitter deficiencies, or obstructions in the nasal pathways. Recovery prospects depend on the nature of the cause. Parosmia can occur as a transient stage in the regrowth of olfactory receptors after destruction of the epithelium. Olfactory hallucinations are sometimes caused by brain tumors.

E. Odors in a Technical World

Industry, intensive agriculture, waste treatment, and traffic frequently cause malodors. Direct scaling methods for odor nuisance have been developed. Malodors may cause social problems, sleeplessness, and nausea.

VII. Taste

A. Transduction

Taste is a part of complex oral sensations, to which touch, pain, and temperature may also contribute. Transductors are the taste buds in the lingual papilla or in the soft palate and epiglottis. They consist of 50–150 elongated epithelial cells with microvilli projecting outward into a "taste pore" in the epithelial layer. Taste cells on the anterior tongue are innervated by the chorda tympani (CT), those on the tongue's edges by the CT and glossopharyngeal nerve (GN), and taste buds on the back of the tongue by the GN only. The vagal nerve (VN) innervates the epiglottis. Gustatory nerve targets include the neocortex, required for associations such as the retention of learned taste aversions, and the limbic system, serving hedonic aspects. Stimulus intensity, reflected by neural response magnitude, predicts sensation strength, whereas qualities (sweet, salty, sour, bitter) are associated with four nerve fiber groups. Most stimuli activate fibers of all groups, leading to distinct across-fiber patterns. In suprathreshold mixtures, mutual suppression may occur. Weak stimuli sometimes show mixture-enhancement. With prolonged stimulation, taste declines and water may take on a different aftertaste (adaptation). Recovery occurs after stimulus cessation. Cross-adaptation between different taste substances mainly occurs within and not between taste qualities. Only small regional sensitivity differences between qualities exist on the tongue. Bitterness is stronger on the back of the tongue than on the tip; however, on the tip, bitterness recognition is superior. [*See* TONGUE AND TASTE.]

B. Taste Stimuli and Sensory Attributes

Sweet is initiated by binding to complementary proteineous sites. In addition to sugars, many sweeteners are known. One is the protein thaumatin, equi-sweet to a 3,000-fold sucrose weight. Aspartame, equi-sweet to a 200-fold sucrose weight, is widely used in foods and drinks. Sensitivity to bitter-tasting urea is bimodally distributed in the population, suggesting a genetically controlled receptor substance. Acids taste sour. Many salts taste salty, but they often are predominantly bitter. [*See* SWEETNESS, CHEMISTRY.]

C. Taste Disorders

Taste complaints, in most cases, can be ascribed to misperceived olfactory disturbances. Except for slow decline in the elderly, taste rarely causes complaints. The decline of taste in the elderly is not uniform across compounds, so the receptor composition on the tongue may not be uniformly altered. Localized loss or change of taste may be associated with destruction of tongue tissue, nerve damage (e.g., a tumor), or central pathology.

D. Taste Technology

Taste substances are used in many products. Often sugars are replaced by noncaloric or noncariogenic sweeteners. Low-sodium diets and low-energy drinks require carefully balanced substitutes so as to maintain their hedonic value. Industrial and consumer taste panels assist in psychophysical assessment of most foods and beverages.

VIII. Skin Senses

A. Transduction

The skin senses (or somatic senses) subserve the perceptions of touch, pressure, warmth, cold, pain, and also electric current. Their main purpose is to explore form and roughness of surfaces, to search for comfort with respect to temperature, to avoid painful contacts that could do harm, and to inform about movement and position of the body and the limbs. In the evolutionary process, they are probably among the oldest senses whereby animals tried to probe their environment. In two different layers of the skin, the dermis and the epidermis, a manifold of nervous structures can be discovered, which support the skin sensitivity. Their supposedly specialized functions are still a matter of debate. In the dermis, there are rather large so-called Pacinian corpuscles. They are internally shaped like an onion and are almost certainly subserving the sense of touch. Also in the dermis, fluid-filled compartments containing collagen fibrils with nerve endings are found, called Ruffini organs. In the epidermis, one finds Meissner's corpuscles, consisting of laminar cells interwoven with nerve endings, and also disc-like Merkel's corpuscles. Some of these corpuscles are believed to function in a network of touch and pressure-sensitive transducers. There is also an abundance of free nerve endings in the skin, the function of which is most difficult to establish. Somehow, their responses elicit different sensations of warmth, cold, and pain, which are very well discernible. Finally, in the hairy skin one finds hair follicle nerve end organs responding to the bending of hairs. [*See* PROPRIOCEPTORS AND PROPRIOCEPTION; SKIN AND TOUCH.]

The large number (of the order of 10^6) of nerve fibers ending in the skin layers are distributed very unevenly over the body surface. The sparse distribution on the back in contrast to the dense distribution on the fingertips, mouth, and tongue reflects the different sensitivities in these areas.

One form of perception, located deeper in the body, is kinesthesis, the sense of movement and of posture; this will not be treated here.

B. Sensitivity to Physical Stimuli

Because the sensitivity over the skin area varies considerably, only approximate indications can be given here. Sensations of the sense of touch (also often referred to as the sense of vibration) requires an amplitude of indentation of the skin of >0.1–10 μm in the frequency range of 20–1,000 Hz. A feeling of pressure can be elicited by placing weights of >10–100 mg on the skin. Warmth can be felt by temperature rises of >0.1–0.2°C over a sufficiently large area within a sufficiently short time interval. Cold sensations begin to arise at a smaller temperature drop (−0.05°C) under approximately similar conditions. Pain can be evoked under a variety of conditions: by damaging the skin mechanically, by applying an aggressive chemical to the skin surface, by exposing the skin to temperatures of >45°C, or by sending an electric current in the mA range through the skin.

C. Sensory Attributes

A conspicuous property of the skin senses is the fact that, except for pain, they all show a marked decrease in perceived "strength of sensation" when a stimulus remains constant over time. For instance, the pressure of a coin placed on the skin is felt for only a short time. Plunging in cold or warm water gives initially an overwhelming sensation of cold or warmth, but this sensation wears off rather quickly. The sense of vibration escapes the prevailing adaptation by the nature of its stimulus, which is a rapid change of indentation of the skin. A further important aspect of skin sensation is the fact that its magnitude increases as the stimulus area is enlarged up to an upper bound. This is called the summation area. The summation area of touch, pressure, pain, cold, and warmth gets larger in this order. The summation area is also approximately inversely related to the density of nerve fibers and is therefore smallest for the tongue and the fingertips. Finally, the summation area is related to the so-called two-point threshold, which is the minimum distance of two pointlike stimuli that are just perceived separately.

D. Cognitive Attributes

In exploring the environment, touch among the skin senses is dominant. It is amazing how many different textures of solid surfaces can be recognized. Even the nature of fluids may be felt by moving the hand in the fluid.

The touch sense also can have an important signalling function in interpersonal relationships as the way of touching one another can have an emotional content. The senses of warmth and cold do not seem to carry more specific conscious information than is related to comfort. The alarm function of the sense of pain is evident; many fast retraction reflexes depend on this sense. Intractable pain causes patients suffering from this pathological condition to experience an unbearable pain in the skin (often in the face) for which no adequate external reason can be found.

E. Technical Use of the Skin Senses

Of human's sensory systems, the skin senses play a restricted role in communication. It is not until the visual or the auditive sense is impaired or overloaded that the tactile sense is called upon for extra communication. In the past, many ingenious devices employing the sense of touch have been put to use for the visually handicapped. The temperature senses are not commonly thought of as useful for communication purposes owing to their relatively slower response, their poorer localizability, and their fast adaptation. Attempts have been made to enlarge the information capacity of the human operator by recruiting the skin senses in a so-called "skin vision" system, whereby information from the outside world is "projected" onto the skin. The interpretation of such "projections" will have to be learned. Up until the present, only blind people have been applying such projection systems with limited success.

IX. Perceptual Selection

A. Reduction of Complex Data

A drastic reduction occurs in the amount of perceptual data impinging on the observer. This occurs through body, head, and eye movements and through sensory and cognitively controlled selective processes.

B. Object and Subject Factors

We classify the factors that control the selective processes into object and subject factors. The object factors are of external origin and relate to properties of the stimuli (conspicuity). The subject factors are of internal cognitive origin and derive from the subject's motives, desires, and expectations (directed attention). The two factors are also named involuntary and voluntary determinants of selective attention, respectively.

Voluntary selection may be precategorical (i.e., directed at expected stimulus location) or postcategorical (i.e., depending on stimulus recognition). Benefit primarily exists in the neglect of nonattended stimuli. Also, some direct benefit of selective attention can be demonstrated. An example of precategorical voluntary attention is the neglect of messages to one ear when the two ears receive different speech messages. The "cocktail party effect" is related in fact to both precategorical selection (sound direction) and postcategorical selection (speech content). Visually, only certain attributes can be effectively selected among other attributes present. Selection also occurs among signals coming simultaneously from different senses. So pain may be sometimes reduced by shifting attention toward other perceptual information.

Figure 3a demonstrates how the conspicuity of a target (center) depends on differences with neighboring distractors. Figure 3b indicates that a high conspicuity corresponds to a large visual field in which the stimulus will be detected.

C. Selection in a Technical World

Research on auditory attention was initiated in the early 1950s through a need to understand the performance difficulties faced by air traffic controllers and pilots who were required to respond quickly and accurately to a range of visual and auditory inputs. However, how to catch and hold attention is also a very practical problem for the advertiser, the road sign designer, the newspaper make-up editor, and, nowadays, the designer of multimodal human–computer interfaces.

X. Development of Perceptual Functions

When during a person's lifetime unchanged stimuli gradually give rise to different sensations and cognitive percepts, we speak of perceptual development. Changes in perception are brought about by maturation, sensory development, perceptual learning, and physiological changes, that, ultimately, may also lead to deterioration of perceptual functions.

A. Infancy

In infants, the components of the visual system mature in the order in which visual information is pro-

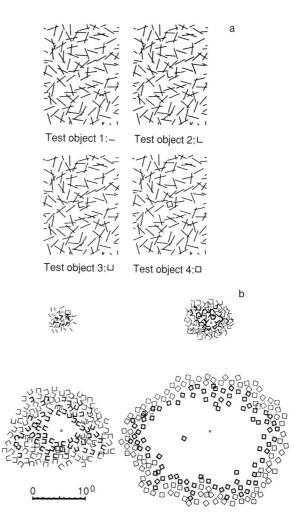

FIGURE 3 (a) Differential conspicuity as a result of target-background differences for test objects 1–4. In each case, the target object is in the center. (b) Areas around the fixation point in which targets 1–4 can be detected in a single glance in the same background as in Figure 3a. Target objects in bold were detected; the others were not discovered.

cessed. With the retina maturing first and the cortex last, elementary stimulus variables can be handled better than complex patterns in infancy. A corollary of this is that during the first 2 mo both foveas are not consistently aligned for binocular fixation. A sensory limit on blur detection amounts to ±1.4 diopters, while convergence may vary in a 3° range in 1-mo-old infants. Stereoacuity develops rapidly from 3 mo of age onward, and near space perception after 4–5 mo is similar to that of adults, but visual space seems to be confined to some 1.5 m.

Although newborns might be color-deficient, as the lateral geniculate nucleus and the prestriate cortex are not yet mature then, no evidence indicates this. At 4 mo, infants can perceive hues and do so categorically, in much the same fashion as normal trichromatic adults. By then, myelinization of the optic radiations is complete.

Visual acuity is moderate at birth; at <1 mo of age resolution is just <1°, but at 2 mo acuity is good for distances at least between 30 cm and 1.5 m. At 6 mo, acuity is about 5″ of arc, while at 1 yr it is comparable with that of adults (at least 1″ of arc). Shape perception shows a gradual development too. At 1 mo, infants prefer to look at a face-like pattern rather than a plain geometrical one, but they scan details only at 2 mo. Whether or not infants up to 6 mo of age have the same kind of size and shape constancy as adults is unknown; stimulus invariance seems to involve much perceptual learning and not to be a direct consequence of maturation.

Hearing, coupled with localization, can be shown to exist at birth but is more sensitive to complex sounds than to pure tones. Recognition of the mother's voice may happen early but is probably fully evolved at 1 or 2 mo. Ample evidence indicates that 1–4-mo-olds can easily distinguish between many different speech sounds and categorize them.

One-year-old infants can recognize objects by touch alone, but little is known about earlier touch perception, which is strongly linked to unconditioned reflexes.

A surprising ability, present at 3 wk, is to imitate adults' facial gestures. Full control of facial gestures also provides evidence on the ability to recognize smells (6 wk) or tastes. Apparently, preference for sweets is not learned or modified by experience.

B. Childhood

While young children can distinguish detailed visual features quite well, at 5 yr of age they are still unable to grasp fully the relation between the whole and the parts. Only well after 10 yr of age are they able to complete an embedded figure test at adult speed. They are also less sensitive to figure orientation. Size constancy is much better than that of infants but is restricted largely to distances of 3 m. When growing up, children are increasingly better able to confine their attention to specific information bearing parts of the stimulus pattern as a whole and use their attention strategically for perceptual tasks.

C. Learning

The simplest kinds of learning consist of habituation and sensitization. Habituation occurs when repeated presentations of a stimulus evoke increasingly lower sensation levels. Sensitization, on the contrary, increases sensation level after sufficient stimulus repetition but, in general, requires cognitive effort. Both concepts are to be distinguished from adaptation, which may encompass both effects. Adaptation is operative in a short time range (minutes or less), and its effects are completely reversible. Habituation and sensitization, however, have enduring effects, up to weeks or months or even longer, and may not be reversible at all. Practice is a common means to induce sensitization and is effective in a large range of psychophysical tasks. [*See* LEARNING AND MEMORY.]

The human perceptual world derives a great deal from extensive experience with the objective physical environment (e.g., resulting in the perception of the world as upright). Perceptual learning is said to occur when basic stimulus features are interpreted on a higher abstract level to represent a different unit. This procedure may be repeated many times, whereby the stimulus is residing over a hierarchy of ever more detailed subordinate feature categories. Thus, abstraction becomes a means for perceptual efficiency when dealing with masses of environmental stimulation. Conversely, attention is increasingly employed to select wanted information and to cancel or suppress irrelevant perceptual information. Reading is a good example of cumulative and abstract perceptual learning; from visual features to characters, symbols or words, phrases, sentences, and thematic and semantic units. The ability to recognize words routinely takes years of practice; at 10 yr of age, normal readers are some 150 msec slower than adults to name shortly presented words and about 10% less accurate. Text redundancy (e.g., residing at the orthographic, lexical, and syntactic level) can only be employed after extensive practice; both sensitization and perceptual learning as well as selective attention must take place.

Most stimulus patterns, for reasons of efficiency, tend to be perceived as instances of categories; the latter are obtained by continuous exposure and learning and depend heavily on human perceptual experience. Perceptual categories form the framework of the recognition process.

D. Physiological Changes

One of the best-documented cases of physiological change is that of decreased accommodation range of the eye, known as presbyopia. Due to continuous growth of fibrous tissue within the lens capsule, it becomes less elastic and transparent. Whereas the accommodation range at 8 yr of age may span up to 19 diopters, it decreases almost linearly until the age of 50 yr, after which it stabilizes at a range of 1 diopter on average. The presence of fibers and particles within the lens and the eyeball fluid will scatter incoming light and cause glare, resulting in reduced subjective contrast and acuity. Yellowing of the lens causes a slight reduction of sensitivity to green, blue, and violet for older people. In addition, they have more difficulty in ignoring irrelevant visual stimulation, needing more time to process information.

A common aging form in hearing is presbyacusis involving a progressive hearing loss for high frequencies. This goes together with a decrease in speech intelligibility from the 30th year onwards.

During a person's lifetime, changes occur in the anatomical structure, the neural pathways, and physiological functioning of all sensory systems; in general, this implies reduced sensitivity and performance levels. In many cases, however, sensory sensitivity of older people is hardly or not at all impaired compared with that of younger ones; older people generally employ a stricter criterion for perception and recognition, which means their sensitivity only seems affected, but it is actually equally effective. This appears also to be the case for touch sensitivity and pain. Little is known about age-related changes in taste and smell.

Bibliography

Aslin, R. N., Alberts, J. R., and Peterson, M. R. (eds.) (1981). "Development of Perception," Vol. 2. Academic Press, New York.

Bartoshuk, L. M. (1988). Taste. *In* "Stevens' Handbook of Experimental Psychology" (R. C. Atkinson, R. J. Herrnstein, and G. Lindzey, eds.), 2nd ed., Vol. 1, Ch. 9, "Perception and Motivation" (R. D. Luce), pp. 461–499. John Wiley & Sons, New York.

Boff, K. R., Kaufman, L., and Thomas, J. P. (1986). "Handbook of Perception and Human Performance." John Wiley and Sons, New York.

Borden, G. J., and Harris, K. S. (1984). "Speech Science Primer; Physiology, Acoustics and Perception of Speech," 2nd ed. Williams & Wilkins, Baltimore.

Charness, N. (ed.) (1985). "Aging and Human Performance." John Wiley & Sons, Chichester.

Singer, H., and Ruddell, R. B. (eds.) (1985). "Theoretical Models and Processes of Reading," 3rd ed. Lawrence Erlbaum Associates, International Reading Association, Newark, Delaware.

Thomas, E. F., and Silver, W. L. (eds.) (1987). "Neurobiology of Taste and Smell." John Wiley & Sons, New York.

Tinker, M. A. (1964). "Legibility of Print." The Iowa State University Press, Ames.

Perception, Food and Nutrition

RICHARD D. MATTES and MORLEY R. KARE, *Monell Chemical Senses Center*

Glossary

Cephalic phase response Vagally mediated preabsorptive physiologic responses to sensory stimulation

Chemical senses Taste and smell

Flavor Amalgam of input from the auditory, visual, somesthetic, olfactory, and gustatory systems

Gustatory Pertaining to taste

Hedonic Pertaining to pleasure

Olfactory Pertaining to smell

Satiety Decreased hunger

Somesthetic Pertaining to touch, temperature, or pain

Trigeminal Fifth cranial nerve, which conveys information about touch, temperature, pain, and chemical irritation

AMONG THE NUMEROUS FACTORS regulating ingestive behavior are the flavors of foods. The perception of these flavors is determined by the interaction of the sensory properties of a given food and the sensory capabilities of the particular individual. Since these sensory parameters vary between and within both products and individuals over time, characterization of the role of sensory factors in food consumption has proven elusive. Nevertheless, evidence that sensory perception and food intake are related is available from clinical populations with disorders of food intake or sensory function, as well as from healthy individuals. In addition to an influence on food selection, sensory

stimulation (particularly chemosensory stimulation), may alter the utilization of nutrients derived from ingested foods. Finally, the use of these nutrients can, in turn, alter sensory function since sensory systems are comprised of tissues with nutrient requirements.

I. Sensory Measures

Sensory measurement as it relates to foods is complicated by the fact that there are two levels at which assessment is required. The first concerns the sensory properties of foods; the second, the sensory capacities of individuals. Sensory properties of foods consist of attributes such as texture, color, and odor as determined by each item's chemical composition. Whether these characteristics are detected and how they are interpreted are determined by the function of each individual's sensory systems. The interaction between the two determines the perceived flavor of foods. For example, humans are polymorphic with respect to their sensitivity to phenylthiocarbamide (PTC) and related compounds. Some individuals (approximately 70% of the caucasian U.S. population) find low concentrations of such compounds in the food supply to have a bitter taste, whereas others fail to perceive any bitterness from the same foods. This may influence the acceptability of these items since bitterness is generally a negative taste attribute.

Study of the sensory properties of foods has generally fallen into the domain of food science. This knowledge is used for such applications as new product development, product matching, product improvement, quality control, and product ratings. A variety of tests are used for each of these purposes. Difference and sensitivity tests are designed to determine whether particular product characteristics are discernible. Judgments about whether a stimulus is present in a given context (e.g., sucrose in ketchup) or what the quality of a stimulus may be (salty for sodium chloride in soup) are the two most common sensitivity measures. They are termed detection and recognition thresholds, respectively, and are only statistical concepts. They merely represent the level of stimulus (e.g., spice) required so that the number of correct responses (e.g., present or not present) by observers in a defined testing situation will exceed chance performance with a preset level of probability. The difficulty of the task and, as a consequence, the absolute threshold value will be determined by testing conditions.

Descriptive sensory tests are used to identify and quantify product attributes. For example, judgments may be sought for the viscosity of an array of foods with varying gelatin levels. Once again, the absolute values of responses will be determined by the testing paradigm. For example, viscosity ratings for a particular food following sampling of a large array of either less viscous or more viscous samples will probably be higher in the first instance than in the second.

The third class of tests, affective tests, provide insights on the palatability and acceptability of products. Sensory responses may be obtained for levels of specific product attributes (e.g., sourness of lemonade) or for the product as a whole. While the focus is on the product, wide intra- and interindividual variability in hedonic responses necessitates the use of a well-defined panel of judges tested under highly controlled conditions. An example of an intrasubject factor is the level of hunger or satiety of the respondent. Having subjects rate a rich dessert item when hungry or full will likely lead to discrepant responses. Differing previous experiences with foods can result in discrepant hedonic ratings between subjects. While overtly bitter foods are generally regarded as distasteful in Western cultures where exposure to such items is limited, they are considered more acceptable to groups whose customary diet includes bitter items.

Study designs for product evaluation generally involve obtaining sensory evaluations of foods or beverages with varying properties from observers with carefully controlled or at least well-defined characteristics (e.g., age, sex, educational level, economic status). This is particularly true of sensitivity and descriptive type tests. Instrumental analyses are also used, but in many instances, the sensitivity of human sensory systems cannot be duplicated. Trained panelists are commonly used for these pur-

poses. A trained panelist is an individual who, through practice and experience, is very familiar with a product or line of products and can reliably evaluate each of its attributes. It should be emphasized that these individuals are not selected on the basis of some innate supersensory capacity. Instead, they have learned to use their sensory capacities more fully. In contrast, untrained panelists are individuals with no particular expertise with a product. Often they are potential consumers of a product and are used primarily for tests of product acceptability.

Interest in the sensory capacities of humans has been particularly strong in the field of psychology as well as the basic physical sciences (e.g., physiology, chemistry, anatomy). Measurements are again generally made in the three domains of sensitivity, description, and affect. However, because of the difference in orientation (i.e., focus on the respondent instead of the stimulus), testing procedures often vary. Test stimuli are commonly held constant and individual variation in responses to these stimuli are the measure of interest.

Individual differences in sensory function may be attributable to both innate physiologic as well impinging environmental or external factors. An example of one innate physiologic influence is an individual's genetic constitution, which determines sensitivity to the odors of different compounds. An undetermined number of individuals in the population have specific anosmias, the inability to perceive the odor of low concentrations of certain compounds (e.g., androstenone, the odorant attracting swine to truffles). Since the perceived odor of a food is comprised of a combination or profile of the volatile compounds it releases, the absence of one constituent part will alter the overall quality of the food's odor.

External influences on sensory function include such factors as state of hunger, health status, and smoking habits. As these factors are in a constant state of flux, their influence and, as a consequence, the sensory function of an individual can vary widely over time. For example, nasal congestion may reduce olfactory (but not gustatory) sensitivity, smoking diminishes taste sensitivity acutely, and being hungry can enhance the pleasantness of tastes and smells.

Preference is the sensory attribute most strongly associated with food selection and ingestion. Preference responses are obviously based upon discernment of product characteristics, but measures of sensitivity and intensity hold little predictive power with respect to ingestive behavior. They appear to be translated into an integrated hedonic message that is used to evaluate the acceptability of the item. Given the complexity of information used to derive such a judgment, it is not surprising that measurement of preferences is similarly complex. Responses to questions about the preferred frequency of ingestion of a food, preferred level of an attribute in the food, or preference for foods with a given profile of attributes can, and typically do, provide different impressions. For example, an affectively neutral food (e.g., bread, milk, butter) may be ingested frequently while a highly preferred item (e.g., pumpkin pie, lobster, caviar) may be ingested only on special occasions. A food with a high level of a given component (e.g., salt on potato chips) may be highly preferred, but other items with high levels of the same component can be disliked (e.g., anchovies). Only when multiple dimensions of taste preferences are assessed can predictions be made about consummatory behavior.

II. Sensory Influences on Food Intake

Human sensory systems evolved, in part, to aid in the identification, procurement, and ingestion of healthful foodstuffs. Each system contributes different information, but because the chemical senses serve as the final checkpoint before ingestion, research has focused primarily on these senses.

Whether taste is comprised of four primary qualities (sweet, sour, salty, bitter), which in different combinations yield all gustatory sensations, or whether there are multiple unique tastes remains controversial. Nevertheless, most researchers support the four-quality view and it has been proposed that sensitivity to each of the qualities has conveyed an adaptive protective benefit. Sweet taste may have aided in the identification of energy sources since many carbohydrates are sweet. Salt sensation could have helped ensure intake of adequate electrolytes as many possess a salty note. Sour sensitivity could have been used to avoid substances at pH extremes and it was useful to recognize and avoid bitter substances since a relatively high proportion of such compounds are toxic. Many Japanese feel that the taste of monosodium glutamate (*umami*) constitutes a fifth basic taste and could signal the presence of protein sources. [*See* TONGUE AND TASTE.]

The prevailing view with regard to olfaction is that there are no olfactory primaries. This sensory system is more like the immune system in its ability to recognize a seemingly endless array of unique compounds. While teleological arguments have been made with respect to "gustatory wisdom" and food intake, no such speculation has been offered for the olfactory system. It should be noted that the oral cavity also receives trigeminal innervation (adding mouth feel, temperature, and pain). This system probably mediates the perception of fats in the diet as well as acids, bases, astringent compounds, and the textural properties of foods. [*See* OLFACTORY INFORMATION PROCESSING; PERCEPTION OF COMPLEX SMELLS.]

Due, in part, to the currently greater interest in the role of other physiological systems (e.g., neural, endocrine) in the control of ingestive behavior, the contribution of sensory input under present living conditions is not well characterized. Perhaps the most compelling evidence stems from assessments of individuals with nutrient-related health disorders. There are case reports of individuals with salt-wasting pathologies craving salt. Protein-calorie malnourished children have also shown a greater preference for stimuli containing an amino acid or protein source than nutritionally replete peers. There are few additional examples of such gustatory wisdom although subtle sensory-based adjustments may occur without notice since the potential problems are mitigated. In contrast, the historical failure of individuals with scurvy, beriberi, and pellagra to correct their problem with available foods argues against a functional role of gustatory wisdom. Moreover, appropriate dietary responses to many chronic diseases such as diabetes, hypertension, obesity, and dental caries are not only absent, but efforts to promote more healthful diets are often met with strong opposition.

Food cravings and food aversions, which are often based upon the sensory qualities of foods, also hold nutritional implications. For example, cravings expressed by pregnant women, bulimic patients, and depressed patients treated with tricyclic antidepressants have generally involved sweet items. Such cravings can influence nutritional status if the sought-after foods are ingested to the exclusion of items supplying needed nutrients, or if the craved items contain compounds that alter the bioavailability of nutrients provided by other foods. An estimated one-third to two-thirds of the population have formed a food aversion at some point in life.

These aversions are marked decreases in the acceptability of particular foods after their ingestion has been temporarily paired with illness, such as nausea or emesis. Reports indicate that the taste or smell of the targeted food(s) are sufficient to elicit revulsion responses. [*See* Dietary Cravings and Aversions during Pregnancy.]

There are approximately two million Americans suffering from taste and/or smell disorders. These range from slight diminutions of sensitivity to complete loss of function. Persistent tastes or smells and distorted sensations are also reported. If the chemical senses influence ingestive behavior, such patients would be expected to display dietary alterations. Research has generally failed to reveal any dietary abnormalities other than complaints about the loss of enjoyment of food. However, recent work indicates that compensatory dietary responses vary between individuals and may be offsetting within this patient population. Some patients report increasing intake in an attempt to achieve a sought-after sensory experience or to mask an unpleasant sensation, whereas others reduce consumption in frustration with the lack of enjoyment or because food elicits an unpleasant sensation. When patients are divided according to the type of response they adopt, clinically meaningful changes in food intake and body weight are observed.

Diverse experimental evidence supports a view that sensory factors influence ingestive behavior in healthy individuals. Studies focusing on individual meals have shown that the palatability of foods influences eating behaviors such as rate of chewing and number of chews per bite. It is inferred that such behaviors are related to actual food intake. Other work has shown that variety in a meal promotes greater total intake relative to when a single food is available. When chemosensory receptors are bypassed by delivering food directly into the stomach via a tube (intragastric feedings), patients often express dissatisfaction due to the lack of oral stimulation. Such patients can maintain adequate caloric intake via this intragastric feeding regimen. However, when oral intake is permitted, intragastric feeding does not induce a compensatory reduction in voluntary oral intake to offset the extra calories derived from the intragastric feed. Thus, the desire for oral stimulation may override short-term satiety cues.

Common experience supports a role for sensory factors in the selection and ingestion of foods in a longer time frame. The sensory attributes of diets

from different cultures are widely discrepant and are resistant to change as evidenced by the reluctance of immigrants to modify their dietary practices. Efforts to supplement the diets of developing nations with various nutrients by introducing them into staple foods have frequently been unsuccessful due to some change they impart in a sensory property of the food. For example, efforts to add iron to sugar have not succeeded, in part, because the sugar becomes discolored, rendering it unacceptable to consumers. The resources and effort expended by food companies on the sensory appeal of products provides further support. Many food producers suffered large financial losses when marketing more healthful "natural" or modified (e.g., low-sodium, low-fat) products with compromised sensory attributes.

There are numerous mechanisms by which sensory factors can influence dietary behavior. First, it has been argued that humans possess certain innate sensory preferences. The most commonly cited example is for the sweet taste. This quality is able to promote fetal drinking and increase sucking responses in neonates. Changes in mimetic reflexes as well as certain autonomic responses (e.g., heart rate) have also been recorded. In adults, there is evidence that 40–50% of calories are derived from primarily sweet-tasting foods. Preference for the salty taste may also be innately determined, but shows a maturational lag. It is manifest at approximately 6 months of age. Adults consume approximately 40% of calories from primarily salty-tasting foods. [*See* Salt Preference in Humans.]

Food intake may also be negatively influenced by sensory qualities. Deviations in any sensory attribute (e.g., color) will often lead to rejection of a product with which an individual is familiar. In addition, some sensory properties are simply aversive. Calories derived from primarily sour and bitter foods constitute only about 8% and 2% of total energy intake. Thus, it can be argued that if foods possess these qualities, they are either avoided or adulterated to mask these taste qualities.

A third mechanism stems from the view that humans have an innate preference for sensory variety. One clear example of this mechanism is the greater preference rating given novel foods relative to staple foods by long-term residents of refugee camps. Monotonous diets have also been advocated as a means to achieve weight reduction. The frustration with this approach is so great, however, that compliance with such regimens is universally poor.

One final mechanism (though this list is not intended to be exhaustive) involves the role of sensory cues as predictors of the metabolic consequences of foods. Through experience, the satiety value of foods is learned and portion sizes are adjusted accordingly. Sensory cues, which have consistently been associated with each food, supply needed information. In studies where the energy values of meals have been surreptitiously manipulated, 25–50% of study participants were found to behave as though they were attending to the previously learned sensory cues rather than the actual metabolic implication of the meal. That is, they ingested less of a meal previously associated with a higher energy content and more of a meal previously associated with a low energy content. Humans are able to maintain adequate levels of energy intake in the absence of sensory input. However, when sensory information is available, it may be prepotent at least in some individuals over the short-term.

III. Sensory Influences on Nutrient Utilization

Sensory stimulation in the head and neck region exerts nutritional effects beyond its role in food selection via cephalic phase responses. These salivary, gastric, pancreatic exocrine, and pancreatic endocrine responses to the thought, sight, sound, odor, feel, or taste of foods appear to prime the body to better absorb and utilize ingested nutrients. While the existence of such reflexes has been documented in humans, their physiological role remains uncertain. This is due in part to the fact that they are short-lived (i.e., lasting several minutes), tend to be small relative to the comparable postingestive response and are difficult to elicit reliably under experimental conditions. Nevertheless, evidence that postingestive responses are often shorter in duration and lower in magnitude in the absence of any cephalic phase component supports a view that they serve to optimize food digestion. Generally, the more components of the normal ingestive process one recruits, the larger will be the cephalic phase response. Thus, as one progresses from simply thinking of food to combining this with the sight and then the smell followed by the taste of the item and finally swallowing, the magnitude of the response increases. Evidence of the efficacy of gusta-

tory stimulation at each site of a cephalic response is substantially stronger than that for a visual priming effect.

In some, but not all, studies the mere sight of food elicits a cephalic phase release of saliva. The response is strongest for items that evoke the largest release of saliva when ingested. Olfactory stimulation is a sufficient condition, as odors alone can trigger a release of saliva. Gustatory stimulation, particularly with acidic solutions, elicits the largest cephalic phase salivary response. Whole mouth salivary flow rates can rise from basal levels of approximately 0.3 ml/min to well over 1 ml/min. The nature of the response may be modified by an individual's state of hunger and the palatability of the stimulus, though the extent to which these factors may influence the response is not well established. [See SALIVARY GLANDS AND SALIVA.]

To the extent that the cephalic phase salivary response influences the release of saliva during food ingestion, it may exert an impact on digestion via three mechanisms. First, the passage of a food through the gastrointestinal tract is facilitated by the addition of fluid and glycoproteins derived from saliva. Second, saliva contains digestive enzymes that can initiate the breakdown of starch and lipids. The importance of starch digestion in the oral cavity is typically minimal since amylase is rapidly inactivated in the acidic environment of the stomach. Lipid digestion can be more important under certain conditions such as feeding of the preterm infant. Finally, there is limited evidence that saliva entry into the stomach inhibits gastric emptying and may thereby influence hunger sensation and food digestion.

Thought, sight, smell, and taste are each singly capable of eliciting a release of gastric acid. Cognitive, visual, and olfactory stimulation can elicit a release that averages between one-quarter and two-thirds of that noted following mastication, but not swallowing of food. Gustatory stimulation is the strongest cephalic phase trigger. Responses are also directly related to stimulus palatability. Stimulation with an unpleasant or tasteless stimulus can have no effect on gastric acid release whereas provision of a preferred food will result in a marked release of acid. An exaggerated response has been reported in duodenal ulcer patients.

A nutritional impact of the cephalic phase gastric response can be presumed due to the magnitude of the response and importance of gastric digestive functions. The release of gastric acid may serve to activate preenzymes (e.g., pepsinogen), which then promote the formation of numerous digestive enzymes. The acidic secretion as well as gastrin may also enhance pancreatic secretion.

While the passage of gastric contents into the duodenum elicits a release of pancreatic secretions, an independent effect of cephalic stimulation has also been documented. Responses are observed in individuals who are achlorhydric (i.e., absence of hydrochloric acid in gastric juice) or who have undergone procedures to block the passage of gut contents into the duodenum. Combined visual, olfactory, and trigeminal stimulation is sufficient for a pancreatic exocrine response, but taste is more potent. Sham feeding (ingestion of food coupled with diversion of the food out of the body via a gastric fistula) has been associated with elevations of trypsin, lipase, and chymotrypsin of 35–66%, 16–50% and 25–60%, respectively. The total volume of the pancreatic secretion is enhanced, but there appears to be a particularly marked effect on release of digestive enzymes. Thus, an especially enzyme-rich secretion is produced.

The cephalic phase contribution to pancreatic exocrine responses to food ingestion may not only serve to augment the release of digestive enzymes, but of satiety hormones as well. Cholecystokinin is reported to possess satiety properties and animal studies indicate this effect is enhanced by oropharyngeal stimulation. The effects of cephalic stimulation on the release of the many presumptive pancreatic satiety hormones has yet to be studied.

The most extensively studied cephalic phase response is that of insulin. The thought, appearance, and odor of food have led to elevations of plasma insulin, though only a subset of test subjects display such a response in any given study. The poor reliability of the response may be due to its fragility or variations in experimental conditions. For example, an exaggerated response has been reported in obese individuals, a delayed response onset time has been noted in adolescents relative to adults, and greater responses have been observed when more palatable stimuli are used. A cephalic phase release of pancreatic polypeptide has also been described.

The role of insulin in the control of hunger and satiety has not been established. To the extent that this hormone does exert an impact on hunger and satiety, the cephalic phase response may contribute to these sensations and their modification by eating. Pancreatic polypeptide is believed to inhibit gastric

and pancreatic exocrine secretion and to relax the gall bladder and may thereby influence digestion.

Direct evidence of these presumed dietary effects of cephalic phase responses is not available. However, suggestive observations have been reported. In three independent studies of preterm infants, those provided with oral stimulation during nasogastric feeds displayed enhanced growth efficiency. That is, greater weight gain was achieved on a comparable level of energy intake. The stimulus in each case was a rubber pacifier. Similar results have been noted in infants provided with nonoral tactile/kinesthetic stimulation so that mediation by cephalic phase responses cannot be assumed. Second, patients receiving total parenteral nutrition and thus receiving no oral stimulation have been reported to require unexpectedly high levels of energy to maintain body weight. However, recent studies also raise questions about this observation. Under conditions of normal health and food intake, cephalic phase responses may serve more as the trigger or primer for responses that occur as food passes along the gastrointestinal tract than as an effective digestive response itself.

IV. Nutritional Influences on Sensory Function

All sensory systems are comprised of metabolically active neural and supporting tissues. Normal functioning of these tissues require the provision of appropriate levels of nutrients. Deficiencies or excesses of nutrients or food constituents can result in a derangement of sensory function. Examples of such effects are blindness due to vitmain A deficiency, deafness due to excessive alcohol or quinine ingestion, burning mouth syndrome due to B vitamin deficiency, and loss of taste due to zinc deficiency.

The mechanisms by which nutritional disorders can alter sensory function are multiple as exemplified by the impact of nutrient deficiencies on the sense of taste. In order for a stimulus to be perceived, it must gain access to its appropriate receptor. In vitamin A deficiency, keratinatious material plugs taste pores and impedes the access of tastants to receptors in this area. Second, once a stimulus reaches a receptor, the interaction must result in the generation of a signal to identify the presence and nature of the stimulus. The transduction processes for the sense of taste remain largely unknown. Whether specific nutrients such as zinc are involved is still a matter of speculation. In the case of vision, however, the role of vitamin A in the transduction process is well documented. Third, sensory receptors must be supported by other types of tissues. Nutrient deficiencies or excesses can result in the death of these supporting tissues. Deficiencies of numerous nutrients lead to atrophy of taste papilla that contain the taste cells. Fourth, once an effective stimulus-receptor interaction occurs, information about the presence of this stimulus must be relayed to the central nervous system for decoding. The function of neural tissue that conveys this information is again dependent upon the organisms nutritional status. Deficiencies (e.g., pyridoxine) or excesses of nutrients or food contaminants (e.g., lead) can lead to neuropathies and taste changes. Finally, incoming sensory information must be processed centrally. Central processing disorders such as Wernicke-Korsakoff syndrome, which results from alcoholism, or schizophrenia, which may in some cases also have nutritional antecedents, have also been associated with abnormal responses to taste and smell stimuli and with olfactory and gustatory hallucinations.

Normal variations in nutritional status (i.e., shifts from states of mild depletion to repletion) can also alter sensory judgments. However, in this instance evidence indicates that the changes are confined to affective responses with reports of alterations in sensitivity being attributable to response bias. Food may be less appealing following a meal, but the functionality of the gustatory and olfactory systems is not impaired. The influence of marginal nutrient deficiencies on sensory function remains an area of considerable speculation, but of little research.

In summary, food selection and perhaps nutrient utilization are influenced by the interplay between the sensory properties of foods and the sensory capabilities of consumers. Mechanisms by which sensory information may exert its impact on these two processes have been proposed although the importance of each remains incompletely characterized. In any case, through an impact on food intake and digestion, sensory factors influence an individual's nutritional status and the status of the sensory systems themselves. Any disruption in this reciprocal relationship between sensory function and nutrition will likely lead to a compromised status of each.

Bibliography

Amerine, M. A., Pangborn, R. M., and Roessler, E. B., eds. (1965). "Principles of Sensory Evaluation of Food." Academic Press, New York.

Carterette, E. C., and Friedman, M. P., eds. (1974). "Handbook of Perception: Psychophysical Judgment and Measurement." Academic Press, New York.

Kare, M. R., and Brand, J. G. (1986). "Interaction of the Chemical Senses with Nutrition." Academic Press, New York.

Kare, M. R., and Maller, O., eds. (1977). "The Chemical Senses and Nutrition." Academic Press, New York.

Kramer, A., and Szczesniak, A. S., eds. (1973). "Texture Measurements of Foods." D. Reidel Publishing Co., Dordrecht, Netherlands.

Mattes, R. D. (1987). Sensory influences on food intake and utilization in humans, *Human Nutrition: Applied Nutrition* **41A,** 77–95.

Meiselman, H. L., and Rivlin, R. S., eds. (1986). "Clinical Measurement of Taste and Smell." Macmillan, New York.

Schiffman, S. S. (1983). Taste and smell in disease: part I, *New England Journal of Medicine* **308,** 1275–79.

Schiffman, S. S. (1983). Taste and smell in disease: part II, *New England Journal of Medicine* **308,** 1337–43.

Sensory Evaluation Division of the Institute of Food Technologists (1981). Sensory evaluation guide for testing food and beverage products, *Food Technology* (November), 50–59.

Solms, J., Booth, D. A., Pangborn, R. M., and Raunhardt, O., eds. (1987). "Food Acceptance and Nutrition." Academic Press, New York.

Perception of Complex Smells

DAVID G. LAING, *CSIRO Sensory Research Centre, Australia*

Glossary

Odor The smell emitted by a substance

Odorant A substance that has an odor

Odor suppression A reduction in the perceived strength of an odor by another odor

Olfaction The sense of smell

Olfactometer An instrument for measuring the sensitivity of the sense of smell

Olfactory bulb A bulbous anterior projection of the brain in which the olfactory nerves terminate

Olfactory cortex A subcortical region of the brain dealing with olfaction

A complex smell can arise from a single odorant or from mixtures of odorants. A single odorant may have a number of odorous features that can give the impression of a mixture of odorants. For example, the odorant amyl valerate is described as fruity, fragrant, sweet, sour, sweaty, putrid, and sickening. However, since it is generally accepted that complex smells consist of a number of odorants, this definition will be used here. The major characteristics of complex smells are their perceived strength, commonly called perceived intensity; quality or type of smell; and pleasantness. Measurement of odor intensity can be achieved by having people

rate this attribute on a number scale. Such a scale might be structured so that zero equals no odor and 100 equals extremely strong. This is but one example of the many scaling procedures used. Odor pleasantness is measured using similar rating methods. Measuring odor quality is more difficult because quality is dependent on semantics and experience. The descriptor putrid may mean rotten meat to one person and feces to another, yet the two odors are quite different unpleasant smells. Standardizing descriptions of odors is a major problem in olfaction and complicates measurements of what we smell in odor mixtures. Another way to measure odor quality is to evaluate odor similarities between pairs of substances.

I. Introduction

We rarely smell a single odor. Almost all the odors we encounter at home from cooking, the bathroom, or the garden, or in the workplace from cars, perfumed passersby, sewage treatment plants, and factories, are complex mixtures of odorants. Often they contain dozens, sometimes hundreds of odorous constituents. However, despite our familiarity with a particular complex smell such as freshly cut grass or coffee, we tend not to analyze the mixture into its many odorous components. Rather we use object terms such as *cut grass* and *coffee* to describe what we smell.

In contrast to the nonanalytic behavior of the average human toward complex smells, perfumers and flavorists spend their working hours learning and practicing how to construct and dissect complex smells. These specialists learn by experience which odors can be mixed to produce new smells or mimics of natural mixtures. Often odorants from natural sources are used. Spearmint oil is primarily responsible for the characteristic smell of spearmint chewing gum, and rose oil is a common ingredient of floral perfumes. Many odorants are synthesized by chemists in the laboratory, often because of the high cost or low availability of natural materials. These synthetic odorants are used in many flavor and perfume formulations and include synthetic fish, whisky, and meat flavors, and musk perfumes. Blending odorants to produce specific fragrances and flavors is considered by the professionals to be an art and is likened to the production of a symphony by an orchestra; all the notes and sounds blend in harmony to produce the desired effect.

Unlike fragrances and food flavors, complex smells from factories, sewage and waste treatment plants, feces, and rotting food are usually unpleasant. Their unpleasantness often arises from high concentrations of individual odorants that do not blend well with each other. Hydrogen sulphide or rotten egg gas is a common and characteristic constituent of sewage; although many other odorous substances are present, often its concentration is high enough so that it alone will be the cause of annoyance and complaint. Similarly, individual unpleasant odorants are the cause of food taints. A common food contaminant is the musty substance 2,4,6-trichloroanisole, which is produced by the action of bacteria on the byproducts of chlorinated sterilants used in water supplies and food manufacturing plants.

Thus complex smells can be natural, synthetic, pleasant, or unpleasant. The primary sensation detected by the nose may be due to stimulation by a wide array of odorants blended to produce an aroma characteristic of an object, or it may arise from a single odorant that can be readily distinguished from the weaker odors of many minor components.

II. Instrumental Analysis of Complex Smells

Complex smells from flowers, cooking, and factories often contain many odorous substances. A significant number of these odorants will be present at very low concentrations below the level that the nose can detect, while others will be substantially above the detection level. Humans appear to have a limited capacity for distinguishing the constituents of complex smells, so the task of identification has fallen to instruments. The two major instruments

used in odor analysis are the gas chromatograph and the mass spectrometer. The former is used to separate the constituents of mixtures; the latter, to identify them. With the gas chromatograph, a sample of a complex smell is injected into a stream of inert gas such as nitrogen or helium, which is passed through a narrow column containing a thin film of liquid material, the stationary phase, to which the constituents of the smell adhere with different affinities. Different odorants partition at different rates between the stationary phase and the gas phase, so they gradually move along the column at different speeds to a detector. The amount of substance is registered as a peak on a recorder. Many odor laboratories fit a sniffing tube to the outlet of the chromatograph so that each constituent can be sniffed as it emerges (Fig. 1). Each peak on the recorder can therefore be labeled with a smell description. The complete set of peaks and descriptions from a complex mixture is called an aromagram (Fig. 2). Using this combined gas chromato-

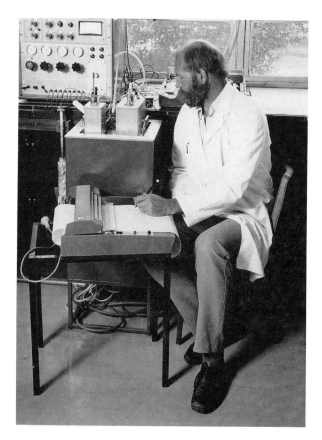

FIGURE 1 Sniffing the constituents of an odor mixture as they emerge from the outlet of a gas chromatograph.

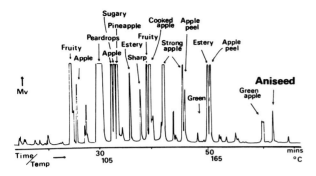

FIGURE 2 An aromagram, showing the many constituents of a complex mixture of odors from apples and their descriptions. Column: 150 m × 0.76 mm Carbowax 20 M; programmed from 65°–210°C at 2–4°C/min. (From Williams and Knee, (1977) with permission.)

graph–sniffing technique, it is possible to determine the types of individual smells in a mixture. Identification of the individual smells is achieved by passing the separated components directly into a mass spectrometer. Within this instrument, the odor molecules are bombarded with a strong electrical charge, causing them to fragment into charged particles called ions. These particles, which are characteristic for each odorant, are guided by a magnetic field to a detector, their location on the detector being dependent on the mass of the ion. A computer, containing large banks of information about how different molecules fragment, is often linked to the spectrometer and can give immediate information on the likely identity of each constituent. Thus by means of the nose, gas chromatograph, and mass spectrometer, it is possible to identify many of the constituents of complex smells. However, these instruments have shortcomings. High temperatures are used in gas chromatographs, and this can cause decomposition of mixture constituents. Also some powerful odorants cannot be detected by the instruments at the levels in which they are present in a mixture. The final array of compounds identified, therefore, may not always represent all the constituents of a mixture. [*See* GAS CHROMOTOGRAPHY, ANALYTICAL.]

Nevertheless, perfumers and flavorists make considerable use of the information from these instruments. By combining their experience with such information, they can reconstruct or create mixtures with great similarity to the natural counterpart and produce the fragrances and flavors on the market today.

III. Olfactometers

Researchers studying the perception of complex mixtures in a university laboratory or odorous air pollutants in the suburbs use an instrument called an olfactometer to prepare and present single- or multiple-odorant stimuli to subjects (Fig. 3). Despite the very large number of olfactometers having different designs, similar principles in various combinations are utilized for stimulus preparation and control. In a typical modern olfactometer, a stream of nitrogen is passed over each of the individual odorants stored in glass tubes. The vapor from the odorants saturates each nitrogen stream which is then separately diluted to desired odor intensity levels by streams of pure air. These air streams containing different odorants are then mixed before delivery to an outlet where they are sniffed by a subject. Mixing of odorants may be controlled manually or by a computer, and the responses of subjects recorded on scoresheets or by the computer. Provided the amounts of odorants presented at the outlet are not too low, a gas chromatograph can be used to measure the final concentrations of the individual components. Olfactometers are constructed from materials that exhibit low odor adsorbence and include glass, Teflon, Plexiglas and stainless steel.

IV. Ability of Humans to Detect and Identify the Constituents of Mixtures

Humans can discriminate among literally thousands of odors when the odorants are not presented mixed to the nose. Indeed it is difficult to find two odors that cannot be discriminated from each other. This impressive ability, however, does not appear to be retained when mixtures of odors are sniffed. Presentation of stimuli, consisting of between one and five odorants, to panels of untrained or trained laypersons, and to a group of professional perfumers and flavorists, showed that few people can identify more than three components in a mixture (Fig. 4). Furthermore, with mixtures of four or five odorants, most subjects only chose three odorants, regardless of whether they were correct or incorrect, suggesting that some of the odorants had vanished by blending or fusing with the other odorants. These results are supported by the finding that mixtures have a small range of perceived complexity. When

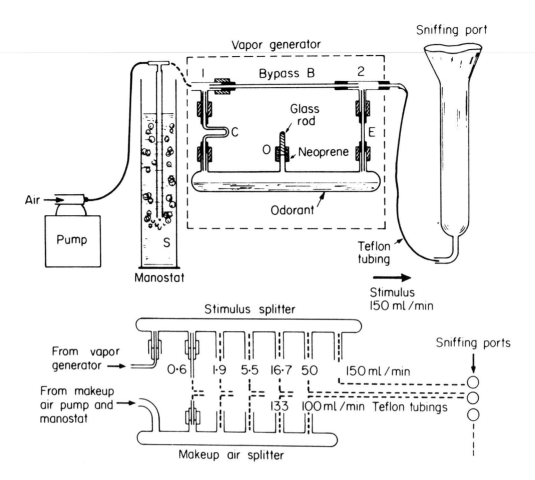

FIGURE 3 Diagram of an olfactometer, an instrument used to produce precise quantities of an odorant to the noses of subjects. [From Dravnieks (1975) with permission.]

subjects were asked to rate the perceived complexity of stimuli consisting of between one and five odorants, only a small increase in ratings was recorded as the number of constituents increased. In some instances, complexity actually decreased with the addition of another odorant, suggesting that fusion or blending of odorants had occurred. It appears, therefore, that the sense of smell has a low capacity for resolving mixtures into their components, in contrast to its ability to discriminate single odors from each other. [*See* OLFACTORY INFORMATION PROCESSING.]

Studies with very simple two-odor mixtures have shown that the ability to discriminate an odor is enhanced if the odor is familiar or particularly unpleasant. However, the effects of familiarity and pleasantness on the discrimination and identification of odorants in more complex mixtures remain to be determined.

Overall, the information available indicates that it is very difficult for humans to perceive and identify more than three odorants in a mixture. This sug-

gests that the sense of smell is synthetic rather than analytic. That is, it appears that odors may blend perceptually with others and lose their identity, or perhaps produce a new odorous sensation. The latter effect has yet to be documented in psychophysical studies. If the sense of smell is analytic then, as with tones in audition, we would expect to be able to distinguish all the components of a mixture, provided the concentration of each were above the recognition threshold. The low resolution of the constituents of mixtures indicates that the perception of smells resembles the perception of colors that fuse to form new colors. Red and yellow form a homogeneous orange, while blue and yellow fuse to produce green. Clearly the analogy with color is not absolute, but until we understand more fully why odorants disappear when presented in mixtures, it is perhaps the closest one at present.

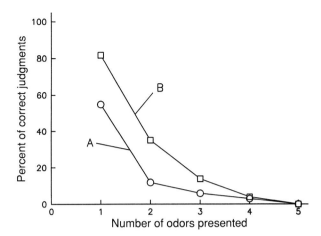

FIGURE 4 Percentage of judgments correctly identifying the components of stimuli consisting of one to five odorants. Function A indicates the percentage of times that the correct identification was selected. Function B shows the percentage of times that the correct selection was made, but others were also (incorrectly) selected. [From Laing and Francis, (1989) with permission.]

V. Odor Interactions

The fact that humans cannot smell all the components in a mixture can in part be explained in terms of the general effects observed when odors are mixed. It has been demonstrated, for example, that when two odors are mixed, the perceived strength of one or both odors may be reduced. The degree of reduction, which is called odor masking or suppression, is dependent on the perceived intensity (before mixing) of the individual components and the types of odorants. The spearmint-smelling odorant carvone, for example, strongly reduces the perceived intensity of citrus-smelling limonene, or the vinegar smell of propionic acid. In contrast, the latter two odors have little or no effect on the perception of carvone unless they are present at a very high intensity level. On the other hand, the odorants benzaldehyde (almond) and eugenol (cloves) suppress each other equally when both are present at similar (unmixed) intensities, but when they are mixed at unequal intensity levels, the higher-intensity component will have a disproportionate suppressing effect on the other.

These examples demonstrate that odor suppression can reduce the chance of perceiving a constituent of a mixture, and provide an indication that the

types of odorants in a mixture and/or their intensities influence how many constituents are perceived in a mixture.

Although little objective evidence from human studies has been reported, enhancement of the perception of one odorant by another is also reported to occur. This effect, called synergism, currently appears to be more anecdotal than factual, although electrophysiological studies of the responses of receptor cells in some aquatic species clearly demonstrate it can occur. Whether the effect seen at the receptor level can be measured in the whole organism, as is required in psychophysical tests, has yet to be determined.

The substantial apparent loss of information on odor constituents in complex mixtures is also reflected in measurements of the total perceived intensity of mixtures. When the vapors of two odors are mixed, a number of outcomes are possible, but in many instances the main effect is that the intensities of the individual odorants are not simply additive. The total perceived intensity of a binary mixture may smell (1) as strong as the sum of the perceived intensities of the unmixed components, exemplifying complete addition; (2) more intense than the sum of its components, exemplifying hyperaddition or synergism, which occurs rarely; or (3) less intense than the sum of its components, showing partial addition, which is the most common outcome. Thus the sense of smell compresses information on intensity, with the total perceived intensity of a mixture rarely exceeding the sum of the intensities of the individual (unmixed) components. Interestingly, it has been reported that, as a general rule, when up to three odors are mixed, their intensities are at least partially additive, but when more than three are mixed, there is little or no increase in the total intensity of the mixture above the level recorded for ternary mixtures.

It appears that an additive process dominates the outcome for two- and three-component mixtures, while an interactive process dominates four- and five-component mixtures. This finding, along with those describing how many odorants can be perceived in a mixture, strongly suggests a significant effect in information processing occurs in the olfactory system when more than three odors are present in a mixture. Part of this loss of information may be due to individual odorants suppressing others; however, sufficient evidence now exists to indicate that a more profound mechanism is responsible.

VI. Predicting How Mixtures Smell

Currently there is no satisfactory method for predicting which constituents of a mixture will be perceived. A variety of methods have been reported, but none has been found to be widely applicable. Predictions based on the quantity of each constituent in a mixture and their detection or recognition threshold, or the perceived intensities (unmixed) of individual constituents, have only been partially successful, primarily because no method takes account of the perceptual interactions such as odor suppression that occur between odorants. Accordingly, none has been universally accepted as a means of predicting what we will smell in a mixture.

Greater success has been achieved in predicting the overall intensity of simple mixtures. The general observation by Zwaardmaker in the late nineteenth century, subsequently confirmed, was that when two odors are mixed, the perceived intensity of the mixture is always less than the sum of the perceived intensities of the component odors. This indicated that the sense of smell compresses information about odor intensity. In 1973 the vector model of odor interaction proposed that binary odor mixtures should be treated with the same rule as that used to add vectors, like a parallelogram of forces (Fig. 5). Under these conditions a unique number, cos α, characterizes a pair of odorants whatever their concentration and levels of intensity. The concept of an angle characterizing a given pair of odorants had previously been applied to describe odor similarity. As shown in Fig. 5, in the proposed vector model of odor interaction, two odors in a mixture are represented by the vectors i and j and the lengths of these

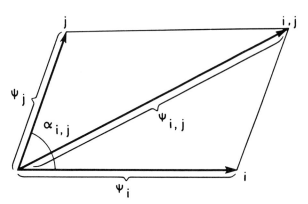

FIGURE 5 The vector model of perceptual odor interaction (see text for explanation). (From Berglund, Berglund, Lindvall, and Svensson. [Copyright 1973 , by the American Psychological Association. Reprinted with permission. *J. Exp. Psychol.* **100**, 29–35.]

R_i, R_j, denote their odor intensities. The length of the resultant vector R_{ij} represents the odor intensity of the mixture ij, as perceived by an observer. The angle α_{ij} between any two vectors i and j is constant for each pair of odorants, is independent of concentration, and is assumed to represent a perceptual relationship between odors. The mathematical formulation of the vector model for binary mixtures is

$$R_{ij} = (R_i^2 + R_j^2 + 2 R_iR_j\cos \alpha_{ij})^{1/2}.$$

It was proposed that the angle α_{ij} be used as an index of the common perceptual qualitative content between two odors. The value of α for all odors studied has been between 102° and 115°. Although satisfactory correlations between predicted and experimental values for α have been obtained for a number of binary mixtures, the accuracy of the predictions falls off significantly with four- and five-component mixtures. It is possible that this decrease in accuracy is related to the substantial loss of information or nonadditivity of intensity that has been noted in mixtures containing more than three components. Also, although the predictions for binary mixtures are satisfactory, the value of α does not appear to be related to the similarity of the qualities of the two odors as originally proposed. Values of between 102° and 115° have been reported for α, but there appears to be no trend across reported values to indicate that there is a relationship between the qualitative similarity of odorants in a mixture and the angle α. For example, values of 102° and 115° for α have been reported for mixtures of the dissimilar substances pyridine-hydrogen sulphide and carvone-eugenol respectively.

Another shortcoming of the vector model is that it assumes the effects of components are reciprocal. When odor suppression occurs, for example, it assumes that two odors will suppress each other equally. This does not usually occur. Also, the model cannot handle the occurrence of synergism when the perceived intensity of a mixture is increased beyond that expected for complete addition of the intensities of the individual (unmixed) components.

Following suggestions that cos α reflects not only sensory interactions, but may also indicate the direct influence of the psychophysical power law, more sophisticated models were developed in the 1980s that incorporate features of the vector model and include terms that reflect the influence of the power law on interactions. These new developments in mathematical models await more vigorous testing with a wide variety of odorants than has so far occurred with binary mixtures, as well as with more complex mixtures.

According to these authors, the first applications of this model seem to allow a discrimination between the olfactory interactions that occur at an integrated level of the brain and those that occur at the level of olfactory receptors. Another conclusion could be the prediction of a synergistic (suppressing) effect of odorants eliciting low (high) values of the power law exponent for the type of interaction occurring at the level of olfactory receptors.

Research into air pollution would benefit greatly from the development of an appropriate model for predicting intensity of mixtures. Being able to estimate the strength of mixtures and of individual odors, in a plume at different distances from an odor source such as a factory or sewage treatment plant, would provide a basis for siting at locations that would cause least public nuisance.

VII. Mechanisms of Mixture Perception

Little is known about how odor interactions such as suppression and synergism occur, or whether these events are of peripheral or central origin. If suppression occurs in the nose, the mechanism is likely to involve competition for receptor sites, or inhibition of cell responses, if the suppressing odorant occupies sites elsewhere on the cell membrane that can affect activation of the cell. On the other hand, if interaction originates in the brain, it is more likely to result from inhibition of secondary or tertiary neurons by the suppressing odorant.

Electrophysiological studies of receptor cell responses of insects and crustaceans, e.g., the lobster, have demonstrated that odor suppression can occur at the periphery. Interestingly, studies with the lobster showed that when presented alone, some odorants do not always have a stimulatory effect on some cells, but when mixed with other stimulating chemicals, they inhibit the responses of the cells to these latter substances. For example, the responses of some cells stimulated by the amino acid taurine, but not stimulated by glycine, were eliminated when a mixture of the two substances was presented. Whether glycine acted as an antagonist and blocked sites normally occupied by taurine, or acted allosterically by occupation of a nearby site on the membrane and affected the binding of

taurine, is not known, but either mechanism is possible.

Evidence supporting a peripheral mechanism for odor suppression in mammals was also obtained from a comparative study of the responses of humans and rats to binary odor mixtures. Rats were presented with a mixture of limonene and propionic acid in a ratio in which humans could only smell limonene, and the responses of neurons in the rat olfactory bulb that receive direct input from the receptor cells were measured. Little activity occurred in the bulbar neurons normally activated by the particular concentration of the acid.

Suppression of mixture components at central locations in the olfactory system, particularly in the olfactory lobe, has also been reported in insects and aquatic animals, but no comparable studies have been reported in mammals. A tentative model has been proposed to account for the central suppression observed in these species. The similarities between the basic anatomical arrangement of cells and their projections, at least between the periphery and the olfactory bulb (the deutocerebrum in these species), could be relevant to mechanisms in mammals. The model suggests that excitatory receptor cells activate second-order inhibitory interneurons, which in turn inhibit responses of second-order neurons in these species. The second-order neurons in mammals are mitral cells, whose activity is known to be modulated, indeed inhibited by two types of interneurons, granule cells and periglomerular cells, thus providing a parallel across species.

Another approach to the peripheral versus central question has been to compare the effects when two odorants are delivered simultaneously but separately to each nostril (dichorhinic stimulation), or simultaneously to both nostrils, as occurs during natural sniffing (birhinal stimulation). With the former, no interaction can occur between two odorants at the receptor cells, because the nostrils are essentially isolated from each other. Any interaction must be of central origin. The results of such studies indicate that suppression of one odor by another can occur with dichorhinic stimulation, but the suppression is weaker than that found with birhinal stimulation. Of significance, however, is that in a number of instances when strong suppression of one odorant occurred during birhinal stimulation, little or no suppression occurred with dichorhinic stimulation. For example, substantial suppression occurred with particular mixtures of the odorants limonene and propionic acid, limonene and pinene, and pinene and propionic acid, but it was not evident with the corresponding dichorhinic mixtures. Strong suppression of an odor was witnessed with birhinal stimulation when little or no suppression occurred during dichorhinic stimulation, so it appears that the mechanisms underlying the suppression observed with the two methods of stimulation are different. Whether the suppression observed with the two methods has any relationship has yet to be determined.

It has also been proposed that the extent to which two odorants affect perception of each other is an indication of the extent to which receptor cells have sites on their surface membrane common to both odorants. For example, if odorant A has little or no effect on the perception of odorant B, it can be assumed that cells activated by B have no sites for A. On the other hand, if B strongly suppresses A, then cells responsive to A may have sites available for occupation by B. These sites need not be at the same location and could be located elsewhere on the membrane. Occupation of sites elsewhere could trigger intracellular processes that inhibit activation of the cell by the suppressed odorant. To resolve this question, both biochemical and physiological studies of the responses of cell membranes to odor mixtures are required.

The loss of information about the constituents of mixtures that occurs when more than three or four odors are sniffed simultaneously suggests operation of a fundamental mechanism that restricts the amount of information processed by this sense. This loss of information, however, can be accounted for from our present knowledge of the anatomy and physiology of the olfactory system. Although not proven, the following is a current explanation for the restricted processing capability of the sense.

When an odorant is sniffed, a unique pattern of cells is activated in the nose and olfactory bulb. Different odorants produce different patterns. The patterns in the nose indicate that there are cells in specific regions of the receptor area that have a special sensitivity and/or selectivity for a particular odor. These peripheral patterns are reflected in patterns of activated cells in the olfactory bulb, because there is a moderate orderly projection of the axons of receptor cells to cells in the bulb. Cells in the dorsal region of the nose, for example, largely project to cells in the dorsal regions of the bulb.

However, this order is lost when axons exit the bulb, as projections from the bulb to the olfactory cortex exhibit a very low degree of order. A small region of the cortex samples a broad area in the bulb, and small areas in the bulb project to large regions of the cortex. This arrangement of bulbar–cortical nerve projections suggests that processing of information at the cortex could involve significant convergence of signal input, as axons from scattered bulbar sites converge on a target cell in the cortex. The unique patterns of neural activity that characterize different odors at the periphery and the bulb are likely to be lost with convergence of the bulbar axons at the cortex, as they produce a representation of a complex odor that provides little information about the odorants in the stimulus mixture.

Despite the loss of order of projections at the cortex, it is clear that some constituents of a mixture are discriminated. Physiological studies of the time taken by odorants to stimulate receptor cells indicate that these times vary markedly between odorants and for different concentrations of an odorant. Time differences of several hundred milliseconds have been reported. If these differences between signal inputs from different odorants are maintained at the olfactory cortex, then it is possible that these temporal differences allow at least some odorants to be discriminated. Indeed the odorants identified may represent the odorants with the fastest and slowest stimulation times. In other words, the odorants identified are likely to be those whose input to the olfactory cortex are widely separated in time. Identification of the constituents of mixtures may therefore be limited by the convergence of neural input at the olfactory cortex but be aided by temporal separation of neural input from each odorant.

VIII. Summary

Methods for studying the perception of complex smells include instrumental methods for separating and identifying the constituents of mixtures, and delivering controlled quantities of odorants to subjects in olfactory tests. In contrast to these precise methods, analysis of mixture constituents by the nose appears to be rudimentary. The sense of smell seems to have a significant shortcoming in its information-processing capacity, the evidence available indicating it is limited in its ability to resolve more than three or four components in multicomponent mixtures. This limit in capacity to resolve the constituents of odor mixtures may also be related to the nonaddition of the perceived intensity of constituents in mixtures containing more than three odors. However, mixture phenomena such as odor suppression and synergism are being reported in electrophysiological studies of neurons at peripheral and central levels of the olfactory system, and it may be possible in the not too distant future to explain the responses of humans to odor mixtures in terms of physiological mechanisms.

Bibliography

Laing, D. G., Cain, W. S., McBride, R. L., and Ache, B. W., eds. (1989). ''Perception of Complex Smells and Tastes.'' Academic Press, Sydney, Australia.

Laing, D. G., and Francis, G. W. (1989). The capacity of humans to identify odors in mixtures. *Physiol. Behav.* **46,** 809–814.

Laffort, P., Etcheto, M., Patte, F., and Marfaing, P. (1989). Implications of power law exponent in synergy and inhibition of olfactory mixtures. *Chemical Senses* **14,** 11–23.

Rabin, M. D. (1988). Experience facilitates olfactory quality discrimination. *Perception & Psychophysics* **44,** 532–540.

Erickson, R. P. (1985). Grouping in the chemical senses. *Chemical Senses* **10,** 333–340.

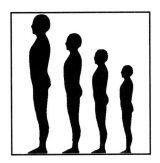

Personality

NATHAN BRODY, *Wesleyan University*

Glossary

Behavioral inhibition system System influenced by septal and hippocampal regions of the brain that inhibits action
Correlation Statistical measure of the relationship among a set of paired scores; it varies from +1.00 to −1.00. Zero correlation indicates lack of relationship; positive or negative correlations measure the magnitude of a positive or negative relationship among the scores
Extraversion Personality dimension related to the tendency to be alone or with other people
Factor analysis Statistical procedures used to determine the dimensionality of a correlation matrix
Neuroticism Personality dimension related to adjustment and the tendency to develop mild psychopathologies
Trait Tendency of an individual to exhibit consistent behavior in diverse situations
Within-family environment Variations in the environment encountered by individuals reared in the same family

PERSONALITY PSYCHOLOGY may be defined as the study of individual differences among people. There are approximately 30,000 words in English that in one way or another refer to differences among individuals. It is possible to use this extensive linguistic resource to develop a scientifically viable descriptive system for individual differences in personality through the use of a statistical procedure called factor analysis.

I. Description of Personality

A. Factor Analysis

The use of factor analysis can be explained by reference to a hypothetical study that is assumed to be representative of a large class of such investigations. Assume that a group of individuals is presented with a relatively comprehensive set of terms that may be used to describe themselves. The terms are arrayed in the form of scales anchored at each end by terms that are opposite in meaning e.g. neurotic and normal. Each bipolar scale permits the individual to rate himself or herself with a number reflecting position on the scale. If there are 100 scales, each person's self description on these scales may be exhaustively described by his or her 100 ratings. If there are 100 subjects in this study an exhaustive description of the data set would contain 10,000 numbers. It is possible to develop a more economical description of the data set. It is entirely likely that there are significant redundancies among the set of personality descriptors selected for investigation. For example, the description "neurotic" is usually assigned the description "poorly adjusted" rather than "well adjusted." The redundancies among all possible pairs of descriptive scales may be noted by obtaining a correlation for all possible pairs or scales indicating the degree of relationship between each pair. There are 4,950 such correlations in our hypothetical sample. It is likely that there will be subsets of scales that exhibit relatively high correlations among the members of the set but relatively low correlations with other scales. The set of clustered scales may be said to measure a common hypothetical factor. Factor analysis may be used to discover and define the underlying dimensions or number of factors needed to summarize and account for the redundant relationships that are obtained among the set of correlations.

In countless investigations of what is assumed to be a relatively exhaustive set of personality descriptors it has been found that the matrix of correlations may usually be explained by assuming that there are a relatively limited number of factors present—typically varying from three to six. Among the best-defined and constantly recurring factors is *extraversion*—a tendency to prefer to be with people or to be alone. The opposite or low end of the factor is called *introversion*. Other factors that are constantly found include *neuroticism* (sometimes called *adjustment*), and *impulsivity,* a factor that includes such descriptive scales as persistence and conscientiousness. Among other factors that have been found are *culture, agreeableness,* and *aggression.* The exact subset of factors needed to define self-report ratings of personality is still a matter of dispute. There is virtually unanimous agreement that the subset of factors must include extraversion, neuroticism, and impulsivity. There is also some agreement, if not unanimous accord, that the set of factors is relatively limited, and that something like five or six dimensions are relatively exhaustive.

A conceptually analogous experiment may be repeated using characterizations made by someone who is acquainted with the ratee, instead of self-report ratings of one's personality. Such ratings may be factor analyzed and it is usually found that the dimensional structure of these ratings corresponds to the dimensional structure of the self-report ratings. Moreover, factor scores derived from self-report measures of personality tend to correlate with factor scores derived from personality ratings made by individuals who are acquainted with the person. Thus, one's self-description of personality tends to be in partial agreement with the description of personality obtained from one's acquaintances.

B. Trait Theory

The discovery that descriptions of personality may be summarized in terms of a limited number of dimensions is, strictly speaking, a discovery about the language used to describe personality. It is possible to argue that the dimensions of personality derived from factor analysis do in fact provide us with an adequate descriptive account of personality. Each of the factorially derived dimensions of personality may be assumed to be a measure of a personality trait. A trait is defined as a consistency in behavior exhibited by an individual in diverse settings. Historically, there have been two views

about trait descriptions of personality. One group of psychologists holds that the descriptions are linguistic fictions that do not in point of fact describe consistencies in the social behavior of individuals. Another group argues that the descriptions are accurate reflections of consistent behavior patterns. The former group argues that correlations between trait measures derived from self-reports or ratings and actual measures of behavior are quite low; they also claim that individuals tend to exhibit characteristically different behaviors in different social settings. Thus, an individual might be aggressive in job settings involving subordinates, and nonaggressive and meek in the home with his or her spouse. Such an individual's characteristic behaviors would not be adequately summarized by assigning a score on a dimension of aggressiveness to the individual. These psychologists tend to argue that personality may be optimally understood in terms of the idiosyncratic pattern of responses of each individual to the characteristic social situations that he or she encounters.

It is increasingly clear that the debate between trait theorists and their critics is subject to empirical resolution. The resolution may be accomplished by reference to the use of aggregated measures of behavior. Any single observation of behavior is likely to be unreliable as an index of the characteristic behavior of an individual. Consider an example. To predict punctuality from trait ratings of conscientiousness, a psychologist might observe whether or not individuals for whom conscientiousness ratings had been obtained arrive at class on time. If the correlation between the ratings and the behavioral index of being on time were low, the psychologist might conclude that there is little relationship between the trait rating and this particular index of social behavior. The conclusion may, however, be in error because a single observation of the punctuality of an individual is not likely to be a reliable index of the characteristic behavior of that individual. A person who is characteristically punctual might be late on a single occasion due to unusual and unrepeatable circumstances. [*See* BEHAVIOR MEASUREMENT IN PSYCHOBIOLOGICAL RESEARCH.]

The use of aggregated measures of behavior in which individuals are observed in the same settings on several occasions is likely to increase the reliability of the measurements and result in more accurate predictions of behavior from trait ratings. It is generally conceded that individuals are quite consistent in their behavior in the same social situation.

Individuals tend to be temporally consistent in their behavior in the same setting. They are also consistent in their characteristic trait ratings. Longitudinal studies of adult personality with decade or longer time lags between measurements find that individuals rate themselves in a similar way on different occasions.

Critics of trait theory generally concede that individuals are temporally consistent in their behavior. They have, however, argued that this does not provide support for trait theory. The critical issue for them is the extent to which individuals are consistent in their behavior in different settings. Consider again our punctuality example. Critics would argue that aggregated measures of punctuality measured by observing time of arrival at a class would not necessarily predict aggregated measures of time of arrival at a party. Therefore they would argue that personality is temporally consistent and situationally inconsistent. Trait theorists can again rescue their position by the use of aggregated measures. Assume that aggregated measures of behavior in different situations that are assumed to be indicative of a trait are obtained for a group of individuals. It is possible to randomly divide the set of measures into two subsets. An aggregated score for each of the subsets can be obtained for each individual. Each score represents the average level of behavior reflecting a particular trait exhibited in a particular subset of situations. The pairs of scores will have a substantial positive correlation indicating that an individual's tendency to behave in a manner that is reflective of a particular trait in one set of situations will predict with substantial accuracy the individual's tendency to exhibit behavior indicative of that trait in a different set of situations. This analysis indicates that the dispute between trait theorists and their critics is not really substantive but rather reflects a preference for different levels of analysis. Trait theorists are interested in the broadest level of aggregated consistency in behavior that is exhibited by an individual. They would want to know what an individual's characteristic behavior is in a variety of situations. Their critics are more likely to argue that two individuals with the same aggregated trait score are likely to exhibit characteristically different patterns of behavior in different situations. They, therefore, prefer to describe personality in terms of an analysis that emphasizes the pattern of differential responsiveness of individuals to different situations.

Although trait descriptions may not be entirely accurate descriptions of the fine-grained dimensions of an individual's social behavior, there is some evidence that they do relate to important social outcomes. The utility of a trait descriptive analysis of personality can be found in longitudinal studies in which trait measures are predictive of measures of behavior obtained years later. Consider some examples. Ratings of aggressiveness based on peer ratings of a child's classmates in elementary school have been found to predict criminality and indices of antisocial behavior obtained in young adulthood. Measures of temperament that may be precursors of adult traits can also be predictive of social behavior. For example, in one study ratings of difficult temperament in infants during the first year of life were predictive of reading problems and behavioral difficulties in adjusting to elementary schools. Trait ratings of neuroticism and impulsivity based on the aggregated ratings of five acquaintances at the start of a marriage have been found to be more predictive of marital unhappiness and divorce than social background measures of the partners or knowledge of life crises. In particular, if the male partner in a prospective marriage was described as nonneurotic and nonimpulsive and the female partner was described as nonneurotic, the marriage was more likely to be described as successful, and was less likely to lead to divorce than if the male partner was described as neurotic and impulsive and the female partner was described as neurotic. Individuals who rate themselves as impulsive or who are rated as impulsive by their acquaintances are likely to have academic difficulties in college even if they have high scores on tests of academic ability. These and many other findings indicate that trait ratings predict important social outcomes. These findings suggest that trait ratings capture characteristic differences in the behavior of individuals that lead, over time, to different social outcomes.

II. Behavioral Genetics of Personality

A. Twin Studies and Family Studies

Why do individuals differ in traits? It is a truism to assert that heredity and environment jointly determine individual differences in trait ratings. Behavior-genetic methods may be used to provide a more precise understanding of the nature of genetic and environmental influences on traits. Behavior-genetic analyses are variants and elaborations of two fundamental research designs introduced by Galton in the latter part of the nineteenth century—the twin method and the family method. The twin method capitalizes on an experiment of nature. There are two kinds of twins—monozygotic (MZ) and dyzogotic (DZ). The former derive from the splitting of a single fertilized egg and are, as a result, genetically identical. The latter derive from the fertilization of two separate eggs and are as genetically similar as siblings. The family method in its more refined versions involves comparisons between children in adoptive families and biological families. The resemblance between adopted children and their biological and adoptive parents provides evidence about the relative importance of genetic influences and differences attributable to variations in the environment associated with being reared in different families. In recent years there has been an increase in both the number and sophistication of studies of the behavior genetics of personality traits. There are, however, major lacunae in the existing literature. There are, for example, virtually no studies of personality traits based on ratings by acquaintances for adults. Most of the studies are based on self-report trait measures. Nevertheless, a large body of evidence supports several tentative generalizations about the role of heredity and environment in the development of individual differences in traits. Twin studies indicate that MZ twins reared in the same home exhibit correlations of about .5 on self-report measures of personality traits. If trait scores were determined solely by genotypes, the correlation would be 1.00. This implies that personality traits are influenced by the environment. Since the correlation of .5 is based on MZ twins reared in the same home, this environmental influence must be attributable to differences experienced by MZ twins reared in the same home. These are called within-family environmental influences.

Twin studies almost invariably find that the correlation among same-sex DZ twins is less than that for MZ twins. Depending on the trait being measured, the correlations are generally .25 or lower. The differences in the magnitude of the correlations of MZ and DZ twin pairs is usually interpretable as a measure of the influence of genetic factors on the trait. Since MZ twin pairs are genetically identical, and DZ twin pairs are not, it can be argued that the greater similarity in genetic characteristics of MZ twins leads to greater similarity on the phenotypic characteristic. Note that this conclusion assumes

that MZ and DZ twin pairs experience environments of equal variability within the pair. Although there is little persuasive reason to doubt this assumption, we shall directly examine its validity when we consider studies of separated twins reared in different families. A crude measure of the degree of genetic influence on a trait, ignoring several complexities, may be obtained by doubling the difference between within-pair MZ and DZ correlations for the trait. This analysis suggests that approximately 50% of the phenotypic variance on personality traits is attributable to genetic influences.

In behavior-genetic analyses it is traditional to partition the total phenotypic variability in a trait into component sources of variance that may be thought of as the characteristic influences on trait variation. Our analyses of twin studies of personality have established that personality traits are influenced both by genotypes and by variations in the within-family environment. A quantitative analysis of these two sources of variance shows that they account for all of the phenotypic variability of personality traits. This analysis leads to a somewhat counterintuitive result—personality traits are not influenced by variations attributable to being reared in different homes. The family that rears a child does not, on this analysis, influence personality trait scores. This conclusion is also supported by studies of twins reared in different families.

Three modern studies have supplied data on personality resemblance for twins reared in separate homes. While the samples are relatively small, the results indicate that MZ twins reared in separate families tend to be as similar to each other as MZ twins reared in the same family. These studies support the conclusion that there are no between-family environmental influences on personality traits.

Twin studies also provide information about the nature of genetic influences on different personality traits. It is possible to distinguish between additive and nonadditive genetic influences. Nonadditive genetic influences tend to decrease the phenotypic similarity of individuals who are not genetically identical. Nonadditive influences are attributable to dominance and to epistasis—the interactive influence of genes. In the limiting case nonadditive influences can lead to zero phenotypic resemblance among individuals who are not genetically identical. Evidence of relatively high correlations for MZ twin pairs combined with relatively low correlations for same-sex DZ twin pairs, and low correlations between parents and children, provide support for the existence of nonadditive genetic influences. Extraversion appears to be a trait that is subject to nonadditive genetic influence. Twin studies indicate that the correlation among same-sex DZ twin pairs for this trait is less than half the value of the MZ correlation. Since DZ twins have a genetic correlation of at least .5, a phenotypic correlation for DZ twin pairs of less than half that of the corresponding MZ pairs suggests that there are nonadditive genetic influences on extraversion.

Family studies provide data that is in some respects congruent with the results of twin studies. Correlations between parents and children and between biological siblings reared in the same family for self-report measures are relatively low—generally below .3. In adopted families the correlations between adopted parents and their adopted children and between biologically unrelated siblings reared in the same family are close to zero. These results imply that the relatively low degree of resemblance between parents and children and between siblings reared together in biological families are primarily attributable to the genetic relationship among family members. The absence of significant relationships between adopted parents and their adopted children, as well as between biologically unrelated children reared in the same family, provides additional support for the assertion that personality traits are not influenced by variations attributable to being reared in different families.

Genetic and Environmental Influences

Although genetic analyses provide us with some insights into the sources of individual differences in personality traits, they do not provide a clear understanding of the ways in which genetic influences and within-family environmental influences combine to determine a person's personality traits. At this stage of our knowledge we can only speculate about the influence of the environment on personality. Traditionally, personality theorists tended to believe that personality was determined by socialization practices associated with variations in the environment of different families. The personality of one's parents, the atmosphere of the home, and the ways in which children were fed, toilet trained and disciplined, were among the influences stressed by psychologists as critical for the development of personality. Behavior-genetic research indicates that all of these influences can be eliminated as determinants of individual differences in personality traits.

We can only speculate about possible within-family environmental influences. Among the variables that might be important are prenatal influences, birth traumas, and illnesses. There may be important variations in the actual or perceived socialization experience encountered by children in the same family. One child may be favored or treated more leniently than another. It should also be noted that the term "within-family environmental influence" is a misnomer, because it encompasses events that occur outside of the context of the family. These might include the influence of friends, lovers, books, religious experiences, and a huge array of potential environmental events that occur outside the home. We have virtually no knowledge about the potential influence of any of the above types of within-family environmental events on individual differences in personality traits.

Our knowledge of the way in which genotypes influence personality is equally speculative. One account suggests that genotypic influences develop from passive to active over time. Passive influences refer to genotypically influenced variations in response to events encountered by an individual. Such variations might lead individuals to be treated differently and to experience different environments. In the most active influence an individual might learn to select environmental niches that are compatible with his or her genotypically influenced personality characteristics. Consider an example. A child might be born with a genetic disposition to be fearful. Such a child is likely to find the world threatening. His or her peers, observing the child's fearfulness, may become threatening and select the child as a victim. In this example the genotypes would influence the environmental events encountered by the child. The child might eventually learn to structure his or her environment to avoid threats and fear-inducing events. This attitude in turn might influence such things as choice of friends, profession, and recreational activities. This example shows that we cannot understand the impact of the environment on individuals without taking into account the differences in the characteristics that individuals bring to the environment.

III. Biological Basis of Personality

A. Eysenck's theory

Evidence that personality traits have a genetic basis suggests that traits must have a biological basis.

That is, genes exert their influence on personality by influencing the structure and functioning of the nervous system. Among the more influential accounts of the biological basis of personality are those advanced by Hans Eysenck and Jeffrey Gray. Eysenck's theory was substantially revised in 1968 and it is that version that shall be considered here. Eysenck's theory provides a biological account of extraversion and neuroticism. We shall consider his theory of extraversion because it has received more research support. Eysenck assumes that individual differences in extraversion are related to individual differences in the arousability of a diffuse cortically arousing ascending reticular system in the brain. Introverts relative to extraverts would have a hyperarousable reticular system. Eysenck, borrowing a Pavlovian concept, assumes that there is a limit to cortical arousal. At this upper threshold of arousal, transmarginal inhibition occurs and is likely to reduce or prevent further cortical arousal. Since introverts have hyperaroused nervous systems they are likely to encounter transmarginal inhibitions at lower levels of stimulation than extraverts. These assumptions imply that introverts are likely to be more aroused than extraverts when they encounter stimuli that are low in intensity and extraverts are likely to be more aroused than introverts when they encounter stimuli that are high in intensity.

A number of studies using the galvanic skin response—a psychophysiological index of arousal—have provided support for Eysenck's theory. Thus, in studies in which individuals were exposed to tones of different loudness, introverts were found to have more intense galvanic skin responses than extraverts when loudness was low, and less intense responses when the loudness was high. In another study, in which stimulation was varied by manipulating caffeine dosage, introverts were found to exhibit more intense levels of galvanic skin response than extraverts when they were exposed to placebos or low- or medium-dose levels of caffeine. At high-dose levels of caffeine, extraverts had slightly more intense galvanic skin responses than introverts.

Extraverts and introverts, when exposed to stimuli that differ in arousal potential, not only exhibit different levels of psychophysiological arousal, but are also likely to respond at different levels of efficiency. For example, in a study in which auditory thresholds for the detection of sounds were measured under different conditions of light intensity, extraverts had slightly higher auditory thresholds at low levels of light intensity and lower auditory thresholds at high levels. As light intensity increased, extraverts tend to improve their performance (i.e., lower their thresholds) whereas introverts tended to decline in performance. Another experiment employed a vigilance task in which individuals were required to maintain vigilant attention and detect the occurrence of a randomly occurring visual stimulus. It was found that introverts were more vigilant and alert than extraverts when performing the task under relatively low noise levels whereas extraverts were more vigilant and alert under conditions of relatively high noise levels. There is also evidence that introverts prefer situations that are lower in stimulation and arousal potential than extraverts.

This brief review of research indicates that there is some evidence supporting Eysenck's theory of extraversion with respect to psychophysiologial indices, performance measures, and preferences for different situations. How do these differences relate to the fundamental definition of extraversion in terms of a preference to be with other people or to be alone? There is research suggesting that the mere presence of other individuals leads to the development of a hyperaroused state in the human organism. The higher level of arousal induced in introverts may be excessive, leading to inefficient performance. Parties and gatherings of large numbers of individuals are likely to be noisy and high in arousal potential, and therefore are likely to be more aversive to introverts than to extraverts. Thus, on this analysis, differences in the arousability of the nervous system in different situations may be used to explain the characteristic social behaviors of individuals who differ in extraversion.

B. Gray's Theory

Jeffrey Gray has proposed a biological theory of anxiety. The potential to experience anxiety is assumed to be related to neuroticism and extraversion. Individuals who are high on neuroticism and low on extraversion (i.e., introverts) are assumed to be subject to the development of anxiety. Gray assumes that anxiety is related to the hyperarousal of a behavioral inhibition system that is related to activity in the septo-hippocampal regions of the brain. Antianxiety drugs such as the benzodiazepines (commercially available as Librium and Valium), barbiturates, and alcohol are all assumed to decrease the activity of the behavioral inhibition system. Similarly, lesions in the septo-hippocampal regions of the brain are also assumed to decrease the activity of the behavioral inhibition system. [*See* HIPPOCAMPAL FORMATION.]

The behavioral inhibition system is activated by threats of punishment and loss of reward as well as by certain innate triggers. The system, as is implied by its name, is assumed to inhibit motor action, alert the organism, and induce a state of hyperattentional arousal that leads the organism to scan the environment for potential danger or potential aversive events. The system is not assumed to be directly related to learning or performance. Rather, its sole function is to stop action in order to permit an appraisal of potential dangers in the environment. Gray assumes that introverted neurotics, who have hyperarousable behavior inhibition systems, are more responsive than stable extraverts to potential negative effects in the environment.

Gray has attempted to extend his theory to an explanation of several neurotic conditions. He assumes that introverted neurotics are susceptible to anxiety attacks, phobias, and obsessive-compulsive behaviors. Each of these forms of neurotic behavior is related to the activity of the behavioral inhibition system. Anxiety states are defined as the phenomenological representation of the arousal of the behavioral inhibition system. Gray assumes that most phobias are triggered by a genetically programmed tendency for potentially dangerous stimuli to elicit activation of the behavioral inhibition system. He indicates that the stimuli that elicit phobias appear to be the same in most cultures. Introverted neurotics, by virtue of their hyperarousable behavioral inhibition system, are likely to respond with inhibitory responses and withdrawal when encountering stimuli that trigger potential phobic responses. Obsessive-compulsive behaviors are also assumed to be a consequence of arousal of the behavioral inhibition system. These responses are also assumed to be cross-culturally similar and to serve a common function—a tendency to focus attention on potentially dangerous components of the environment. Obsessive and compulsive thoughts and actions are assumed to reflect a preoccupation with potential dangers in the environment. Compulsions may be understood as a tendency to remove environmental contaminations. Obsessive thoughts may be understood as deriving from a heightened attention to potential dangers attributable to the arousal of the behavioral inhibition system.

It should be noted that Gray's theory of personality differs from Eysenck's in two principal respects. First, Gray places affect rather than generalized cortical arousal at the center of his analysis of differences between introverts and extraverts. Second, Gray assumes that the most important dimensions of personality are defined by the combination of extraversion and neuroticism. He contrasts neurotic introverts with stable extraverts and neurotic extraverts with stable introverts. Eysenck tends to treat neuroticism and extraversion as independent dimensions of personality.

IV. Conclusion

This overview of personality has presented a conception of personality based on trait theory. This theory assumes that there is a limited number of basic personality dimensions that is influenced by genotypes, which in turn are assumed to influence the structure and function of the nervous system. It should be noted that there are a number of other approaches to understanding personality that do not assign a central role to biological concepts.

Bibliography

Brody, N. (1988). "Personality in Search of Individuality." Academic Press, San Diego.

Carver, C. S., and Scheier, M. F. (1988). "Perspectives on Personality." Allyn & Bacon, Boston.

Eysenck, H. G., and Eysenck, M. W. (1985). "Personality and Individual Differences: A Natural Science Approach." Plenum, New York.

Gray, J. A. (1982). "The Neuropsychology of Anxiety: An Enquiry into the Functions of the Septo-Hippocampal System." Oxford University Press, New York.

Plomin, R., DeFries, J. C., and McLearn, G. E. (1980). "Behavioral Genetics: A Primer." Freeman, San Francisco.

Personality Disorders (Psychiatry)

DEBORAH B. MARIN, *Cornell University Medical College*

ALLEN J. FRANCES, *Cornell University Medical College*

THOMAS A. WIDIGER, *University of Kentucky*

Glossary

Cognition Term which reflects a person's quality of perceiving, recognizing, judging, reasoning, and imagining

Affective instability Rapid fluctuations of mood that last several hours to a few days

Serotonin Specific chemical agent found both in the central nervous system and peripheral tissues

Electroencephalogram Recording of electric potentials of the brain that is derived from leads placed on the scalp

DSM-III-R "The Diagnostic and Statistical Manual of Mental Disorders" (third edition, revised) (DSM-III-R) is the currently used system to diagnose personality disorders

PERSONALITY IS an individual's constellation of predictable and enduring patterns of response, behavior, and cognition that occurs in everyday life. Everyone has a personality, features of which may become apparent very early in life. A personality disorder exists when an individual's personality traits are inflexible and maladaptive and impair social and occupational functioning. These disorders are chronic behavior disturbances with an early and insidious onset that appear by late adolescence or early adulthood. To varying degrees, personality disorders impact on all facets of personality and include cognition, mood, behavior, and interpersonal style in a pervasive manner. Recent research has focused on determining the most appropriate ways to define personality disorders as well as elucidating their development, treatment, and impact on other illnesses.

I. Historical Background

Personality disorders develop from the same factors that contribute to the development of normal personality. Biological, social, and psychological factors all contribute to varying degrees to different personality traits and disorders. Genotypic variability is adaptive from an evolutionary perspective for the species, but can at times be maladaptive for the unfortunate individuals who receive the maladaptive genetic variants. There are data to support genetic contributions to at least some personality disorders. Personality is also quite responsive to the environment, and maladaptive personality traits will at times be due to pathologic familial, social, and/or economic factors. For example, it can be adaptive to be submissive and dependent in one's family of origin but not after one leaves home. Given the amount of time spent with and the importance of the family of origin, this behavior pattern will be slow to respond to the adult change in one's environment. Some personality traits will be adaptive in one situation but not in others (e.g., aggressive competitiveness can contribute to success in a career but not in a marriage). It is then inaccurate to say that personality disorders are simply maladaptive, for they will often represent extreme or atypical variants of adaptive genetic variation and/or behavior patterns that are adaptive in one situation or time but not in others.

Personality has been a major focus of theory and research for psychology since its founding as a science in the late nineteenth century. Psychologists have been attempting to identify and measure the basic dimensions of personality, which is not easy given that there are at least 2700 different trait terms in the English language. Some have suggested that only two basic dimensions of control (dominance versus submission) and affiliation (love versus hate) are necessary to describe all variation in personality style, while others have distinguished between 810 different character types.

II. Epidemiology

The prevalence of personality disorders has been assessed in inpatient and outpatient psychiatric, medically ill, and community samples. These varied populations as well as differing assessment techniques have yielded a wide range of rates. Most epidemiologic studies have been conducted on psychiatric inpatients who have additional psychiatric diagnoses. In these samples, rates as high as 67% have been reported, with many patients having two or more personality diagnoses. Personality disorders are common in psychiatric outpatients, with rates ranging from 12 to 100%. In the medical setting, outpatients who are at risk for human immunodeficiency virus (HIV) infection have a 10% rate of personality disorders. There are no systematic studies using DSM-III-R criteria that assess all personality disorders in the general population. Earlier studies demonstrate that up to 30% of community samples have maladaptive personality traits. The overall rates of personality disorders range from 5 to 10% in the community. Antisocial personality disorder, the most rigorously studied personality disorder, has a prevalence of about 3% in the community.

III. Diagnosis

Personality disorders differ quantitatively, not qualitatively, from normality since they frequently encompass behaviors which occur in diminished number or intensity in the "normal" population. The distinction between health and illness also varies among cultures which have different societal norms.

Personality traits represent dispositions to react to the environment across many situations and at different points in time. "State" conditions reflect a person's condition at a given point in time. Many psychiatric conditions, for example depression, can significantly alter a person's interpersonal style. Upon remission from such a state condition, the individual's personality may be quite different from that observed when ill. It is crucial to determine if a patient's behavior represents a change from his or her baseline or is an enduring personality style. [*See* DEPRESSION.]

When determining if a behavior is maladaptive enough to be consistent with a personality disorder, attention must also be turned to the situation in which the behavior occurs. An observed behavior may be in response to the demands of such a situation or social role instead of a personality style. Behaviors should not be designated as personality traits if they occur only in response to a certain situation or role. In a catastrophic situation, all differences in behavior may be eliminated among individuals. Yet in everyday life, people definitely exhibit different reactions to a given situation.

DSM-III-R uses a multiaxial approach to assess mental disorders and highest level of functioning of an individual. Axis I diagnoses include clinical syndromes like schizophrenia, depression, substance abuse, and anxiety disorders. Axis II includes the personality disorders. Diagnoses on both axes are made if a person has coexisting conditions. There are 11 personality disorders which are grouped into 3 clusters.

The odd or eccentric cluster includes schizotypal, schizoid, and paranoid personality disorders. Schizotypal personality disorder encompasses a combination of odd or peculiar thought, speech, behavior, and perception. Content of thought may include paranoid ideation and eccentric beliefs that are not consistent with cultural norms. Speech is often digressive, vague, or inappropriately abstract, with concepts being expressed in an unclear or odd manner. Behavior and appearance are often odd and eccentric. Such a person is usually withdrawn, experiences social anxiety, and may have brief breaks with reality. Unlike a schizophrenic patient, who has frank hallucinations or delusions, schizotypal patients experience subtle distortions of their environment.

An individual with schizoid personality disorder tends to be a loner who does not form relationships or responds to others in a meaningful manner. A person with this disorder neither seeks nor enjoys close relationships with either friends or family. Such a person frequently appears aloof, indifferent, and not responsive to praise, criticism, or any feelings experienced by others.

The patient with paranoid personality disorder has the pervasive and unwarranted tendency to interpret the actions of others as being purposefully demeaning or threatening. Without justification, such a person will question the loyalty of friends or peers. He or she will be easily slighted, tend to react with excessive anger, and bear grudges for a long time. Paranoid beliefs may result in hypervigilence, secretiveness, hostility, and anxiety. It may be difficult to distinguish this disorder from schizotypal personality since mistrust of others and social anxiety are shared by both diagnoses.

The dramatic cluster includes histrionic, borderline, narcissistic, and antisocial personality disorders. Patients with these disorders frequently are flamboyant, impulsive, and seeking of social contact. An individual with one of these disorders will frequently have traits, if not the diagnosis, of another disorder in this group.

The person with histrionic personality disorder tends to be attention seeking, self-dramatizing, and excessively gregarious. Typically, there is excessive concern with physical appearance and a constant need for reassurance, approval, or praise from others. An individual with this disorder usually is very self-centered and becomes inordinately angry when their demands are not met. Such an individual often behaves in a seductive, manipulative, exhibitionistic, and shallow manner.

Narcissistic personality disorder describes a person who is grandiose, entitled, exploitive, arrogant, and preoccupied with fame, wealth, and achievement. There exists a grandiose sense of self-importance and a tendency to exaggerate their accomplishments and talents. Yet, self esteem is often very fragile and is reflected in hypersensitivity to the evaluation by others. Such a person lacks empathy and has shallow relationships.

The antisocial patient lacks empathy, social responsibility, and guilt. Such a person displays a consistent pattern of behavior which disregards conventional limitations imposed by society. A veneer of charm may mask an inherent lack of concern for the rights and mistreatment of others. There frequently exists impulsivity, as seen with frequent physical fights and abusive behavior, in conjunction with a lack of appropriate responses to apparent consequences of one's actions.

Borderline personality disorder encompasses a behavior pattern of intense and chaotic relationships with fluctuating and extreme attitudes toward others. A pervasive identity disturbance is reflected in uncertainty about several life issues, including self-image, sexual orientation, long-term goals, types of friends, or values. These patients are affectively unstable and experience marked, yet brief, mood shifts from baseline mood to anxiety, irritability, or depression. These individuals often engage in impulsive and self-destructive behaviors which include activities like substance abuse, casual sex, shoplifting, and binge eating. Repeated suicidal threats or behavior are common in more severe forms of this disorder. This diagnosis frequently overlaps with histrionic and schizotypal personality disorders.

The avoidant or anxious cluster includes dependent, obsessive–compulsive, avoidant, and passive–aggressive personality disorders. Dependent personality disorder encompasses a pattern of excessive reliance on others which is reflected in the tendency to permit others to make decisions, feel helpless when alone, to subjugate one's needs to others, and to tolerate mistreatment. Such an individual requires excessive advice and reassurance and cannot function when assertiveness is required. He or she will not uncommonly live with a domineering, overprotective, and infantalizing person.

The avoidant patient is inhibited, introverted, and anxious. Additional features include low self esteem, hypersensitivity to rejection, apprehension, social awkwardness, and fear of being embarrassed. The criteria regarding social withdrawal overlap with schizoid personality disorder. However, the schizoid patient is indifferent to others while the avoidant person desires relationships yet is too shy to obtain them. The avoidant and dependent personality disorders share characteristics of interpersonal insecurity with the desire for relationships and low self esteem.

The person with obsessive–compulsive personality disorder tends to be perfectionistic, constricted, and excessively disciplined. Behavior is rigid, formal, and intellectualizing. There often exists a preoccupation with trivial details or procedures that hinders the individual from having a broad view of a situation. Decision making is very difficult and often postponed or avoided to the point of interfering with task completion. An individual with this disorder resists authority of others while unreasonably insisting that others agree with their views.

Passive–aggressive personality disorder is reflected in a pattern of passive resistance to authority and responsibility. Indirect resistance, that often includes behaviors like procrastination or tardiness, results in persistent social and occupational ineffectiveness. Such a person will also dawdle and intentionally be inefficient when asked to do a task that he or she does not want to do. Associated symptoms include complaining, irritability, whining, discontent, and lack of receptiveness to useful suggestions. Anger is usually expressed through resistant and negativistic actions.

DSM-III-R permits the listing of personality traits if an individual does not have all criteria necessary for a personality disorder as well as the diagnosis of multiple axis II disorders. These provisions reflect the fact that all individuals have certain personality traits and that patients with a personality disorder frequently possess more than one axis II diagnosis.

IV. Etiological Models

Different theories exist that attempt to explain the development of personality and personality disorders. Much of the supporting evidence for these hypotheses is theoretical because it is often very difficult to test them in controlled experimental situations. At this time, there is no definitive model that has been proved to be more correct than the others. Earlier writings emphasized either environment or biology as causative factors. In contrast, more recent literature proposes that personality development depends on the interaction of an individual's constitution and experiences within and without the home.

A. Psychoanalytic Theory

Psychoanalytic theory proposes that an infant is born with instinctual attachments and sexual and aggressive drives. Personality development is a process in which the child gains increasing control of his instinctual impulses by learning and developing identifications with important caretakers. A successful effort for adaptation results in normal behavioral tendencies. If a child's needs are too intense to be satisfied adequately or if expected needs are not satisfied because of lack of response from the environment, the resulting experiences may serve as a trauma with resultant ''fixation'' points which impedes subsequent maturation. Disturbances in development at particular development phases will result in specific characteristic personal-

ity formations. The quality of the interaction between the individual and their environment results from unconscious conflicts which influence behavior. [See PSYCHOANALYTIC THEORY.]

B. Social Learning Theory

Social learning theory proposes that social experiences, particularly those occurring early in one's development, are the major determinants of behavior and personality. This theory places less emphasis on one's constitution and more on the environment. Specific social experiences result in an individual's behavior, cognition, and affect. For learning theorists, attention is directed to careful identification of conditions that determine how behavior is learned and unlearned. These early experiences have great impact in early life and are more difficult to modify than those experiences occurring later in life. The developing child learns not only from exposure to and instruction by his parents, but also from events observed in their environment. For example, styles of interpersonal communication are obtained from observation of everyday events. Excessive, insufficient, erratic, or inconsistent methods used by parents to modulate a child's behaviors will result in maladaptive personality patterns. For example, the child will become socially avoidant and withdrawn if he or she cannot meet excessive parental demands and harassments.

C. Temperament

Very young infants already differ from one another in biological functions, reactivity to stimuli, adaptability to change, characteristic moods, distractibility, and persistence. These innate endowments constitute each person's temperament, which interacts with the environment in ways that accentuate initial behaviors. An individual's temperament appears to endure during the course of development. A good match between a child's temperament and environment fosters a healthier personality development.

Genetic studies are very useful ways to determine the relative contribution of environment and heredity to personality. The literature investigating the genetic influences on personality has used different methods and patient populations, including twins reared together and apart, and first degree relatives of patients with personality disorders. Family studies suggest that schizotypal and paranoid personality disorder patients have a familial association with schizophrenia. Borderline personality disorder appears to run in families and to have an association with mood disorders, alcoholism, and substance abuse.

Twin studies have determined that both environment and genetics contribute to the development of personality traits. The genetic influence appears to account for approximately 50% of measured personality traits. Most of the environmental contribution is due to the individual specific, rather than a common family, environment.

D. Biology

Several advances in biological techniques have helped to elucidate the biological substrates of personality. Many of these techniques were initially used to investigate the biological correlates of depression and schizophrenia. Patients with personality disorders and maladaptive extremes of certain traits differ from control groups in a number of biological parameters. Several studies have shown an inverse correlation between central serotonin levels and impulsivity and physical aggression. These behaviors underlie both certain personality dimensions as well as personality disorders, particularly borderline and antisocial personality disorders. Borderline and antisocial patients exhibit abnormal brain functioning as measured by the electroencephalogram. Schizotypal patients demonstrate abnormal results in tests measuring information processing. Neuroendocrine abnormalities have been noted in patients with borderline personality disorder. [*See* DEPRESSION: NEUROTRANSMITTERS AND RECEPTORS.]

Although most of these studies are being used for research purposes at present, biologic tests such as those described above may aid in personality evaluation in the future and lend credence to the role of biological substrates in personality characteristics.

V. Personality Disorders and Other Diagnoses

By definition, the presence of a personality disorder impairs one's ability to adjust to stress. Therefore, it is to be expected that personality disorders will impact upon both psychiatric and medical conditions. Personality pathology influences the predisposition, presentation, course, and treatment response of several psychiatric illnesses, including

schizophrenia, mood, anxiety, eating, and substance abuse disorders. The exact nature of the relationship between personality disorders and the major syndromes can be conceptualized in different ways. Certain personality traits may (1) predispose, (2) modify, (3) represent a complication of, (4) represent an attenuated form of, or (5) coexist independently with specific axis I disorders.

Borderline, histrionic, dependent, and avoidant personality disorders frequently occur in patients with major depression. Conversely, individuals with borderline, histrionic, dependent, avoidant, antisocial, schizotypal, obsessive–compulsive, and passive–aggressive diagnoses have an increased incidence of mood disturbances. The presence of character pathology in depressives may be associated with a different clinical presentation than that seen in patients with major depression alone. Specifically, the presence of borderline character disorder is associated with increased incidence of anxiety, substance abuse, as well as attempted and completed suicide. The presence of personality pathology in general has been correlated with earlier onset and poorer treatment outcome of depression.

It is difficult to ascertain if certain personality disorders predispose to or represent a complication of schizophrenia because of this illness's chronic course and not uncommon long prodrome. Schizotypal and paranoid personality disorders have a familial association with schizophrenia. Schizotypal personality disorder is the most frequently diagnosed personality disorder in schizophrenics.

Borderline, antisocial, and histrionic patients often exhibit impulsive behavior, as evidenced by their increased use of alcohol and other psychoactive substances. In the histrionic or borderline patient, substance abuse may represent an attempt to alleviate the affective instability. For the antisocial patient, alcoholism may result from or initiate an antisocial lifestyle. Complications of substance abuse in such patients include increased impulsivity and self-destructiveness, particularly attempted and completed suicide.

Individuals with avoidant personality disorder may develop simple phobias, generalized anxiety disorder, or panic disorder. Social hypersensitivity and perception of threat predispose schizotypal and paranoid patients to develop anxiety disorders.

Debate exists regarding the relationship between personality and the onset and course of medical illness. Nonetheless, the occurrence of serious medical illness interacts with and may augment the personality traits possessed by any patient. Medical illness represents a particularly difficult event for patients with personality disorders because of their impaired ability to deal with stress. A patient's personality traits also affect the quality of the doctor–patient relationship, acceptance of treatment, and outcome. Assessment of a patient's personality optimizes the patients coping with a medical condition. The compulsive patient will do best with a physician who invites him or her to be an active participant in the treatment plan. The dependent patient may best benefit from a relationship in which the physician makes the decisions.

VI. Treatments

The different treatments of personality disorders include psychodynamic, supportive, interpersonal, behavioral, cognitive, and pharmacologic therapies.

There is very little empirical basis to the choice of a particular treatment option. Interpretation of results is also difficult since these patients frequently have more than one personality diagnosis and coexisting axis I conditions.

Unlike the other therapies, which are more target symptom oriented, psychodynamic psychotherapy focuses on changing the patient's personality structure. Therefore, this therapy usually strives to decrease the inflexible quality of the maladaptive traits and increase self-awareness of behaviors in order to improve interpersonal functioning. This treatment also strives to reduce the automatic quality of behavioral responses that result from unconscious conflicts. Patients with dependent, obsessive–compulsive, avoidant, histrionic, and passive–aggressive personality disorders are most likely to benefit from this form of therapy. [*See* PSYCHOANALYTIC PSYCHOTHERAPY.]

Supportive psychotherapy attempts to aid the patient's coping skills without changing his or her character structure. Supportive treatment can help the patient through periods of medical, interpersonal, or occupational stresses. For instance, admiring the narcissistic patient, providing information to the obsessive–compulsive patient, or maintaining distance from the schizoid patient can obtain these goals.

Interpersonal psychotherapy attempts to improve the patient's skills in establishing new relationships and solving problems in existing relationships. Such an approach depends on the fact that personality styles often elicit complementary responses in others. Such an approach may be particularly suitable for personality disorders since these conditions often represent maladaptive interpersonal styles. In this treatment, the therapist assumes an interpersonal style that encourages more adaptive and flexible functioning in the patient in order to diminish the usual pattern of mutually debilitating relations.

Behavior therapy promotes the patient's involvement in more adaptive and pleasurable activities. Behavior techniques involving assertiveness training and graded exposures for social anxiety may be useful for dependent and avoidant patients, respectively. The schizoid patient may benefit from shaping of his or her social behavior.

Cognitive therapy focuses on diminishing central, irrational assumptions causing patients' beliefs and behaviors. Such a therapeutic approach may be particularly useful for personality disorders which frequently include specific cognitive styles. The borderline patient may benefit from addressing the exaggerated attitudes in order to develop more realistic perceptions of others. Cognitive techniques that focus on assumptions of threat and inadequacy may benefit the avoidant and dependent patient. The obsessive–compulsive patient may benefit by addressing his or her irrationally rigid, severe beliefs and moral standards.

Pharmacotherapy may be useful when it focuses on such features as affective dysregulation in borderline personality disorder, perceptual disturbances of schizotypal personality disorder, or anxiety or avoidant personality disorder. Low-dose neuroleptics have been found to be useful for anxiety and cognitive disturbances of schizotypal and borderline personality disorders. Antidepressant regimens may be very useful for mood disorders frequently seen in patients with axis II disorders.

VII. Conclusions

Everyone has a personality that imparts a particular and somewhat predictable quality to their relationships, responses to the environment, and cognitive style. A personality disorder exists when a specific constellation of personality traits causes pervasive and enduring maladaptive behaviors across many situations. These disorders rarely occur in isolation and impact on both other psychiatric diagnoses and medical illness. A complex interaction between the individual and their environment results in these conditions. A growing literature is elucidating the underpinnings of these conditions and treatment options.

Bibliography

American Psychiatric Association (1987). "Diagnostic and Statistical Manual of Mental Disorders," 3rd ed., revised. American Psychiatric Association, Washington D.C.

Fisher, F. D., and Leigh, H. (1986). Models of the doctor-patient relationship. *In* "Psychiatry" (R. Michels, J. O. Cavenar, H. K. H. Brodie, A. M. Cooper, S. B. Guze, L. L. Judd, and G. L. Klerman, eds.). Lippincott, Philadelphia.

Frances, A. J., and Widiger, T. (1986). The classification of personality disorders: An overview of problems and solutions. *Psych. Assoc. Annu. Rev.* **5**, 240–258.

Liebowitz, M. R., Stone, M. H., and Turkat, I. D. (1986). Treatment of personality disorders. *Am. Psych. Assoc. Annu. Rev.* **5**, 356–393.

Merikangas, K. R., and Weissman, M. W. (1986). Epidemiology of DSM-III axis II personality disorders. *Am. Psych. Assoc. Annu. Rev.* **5**, 258–277.

Marin, D. B., Frances, A. J., and Widiger, T. A. (1990). Personality disorders. *In* "Clinical Psychiatry for Medical Students" (Alan Stoudemire, ed.), in press. Lippincott, Philadelphia.

Siever, L. J., Klar, H., and Coccaro, E. (1985). Psychobiologic substrates of personality. *In* "Biologic Response Styles" (Howard Klar and Larry Siever, eds.). American Psychiatric Association, Washington, D.C.

Widiger, T. A., and Hyler, S. E. (1986). Axis I/II interactions. *In* "Psychiatry" (R. Michels, J. O. Cavenar, H. K. H. Brodie, A. M. Cooper, S. B. Guze, L. L. Judd, and G. L. Klerman, eds.). Lippincott, Philadelphia.

Person Perception

JOHN N. BASSILI, *University of Toronto*

Glossary

Actor–observer effect The tendency to attribute other people's behaviors to their personalities while attributing our own to features of the situation

Associative networks The metaphor used to characterize the mental representation of information in terms of *nodes* and *links*

Attributions The inferences people make to explain behavior

Construct accessibility The state of readiness of our social knowledge structures for processing incoming information

Discounting principle The notion that confidence in the role of a cause in producing an effect decreases if other plausible causes are also present

Display rules Rules about the appropriateness of emotional expressions in particular situations

Implicit personality theory The informal assumptions people make about relationships between character traits

Impression management The processes people use to control self-relevant information

Information integration The notion that impressions require the integration of information about others and the models of how this integration is effected

Memory-based judgments Judgment processes that require the recall of information stored in memory

On-line judgments Judgment processes that occur during the encoding of information

RESEARCH ON PERSON PERCEPTION explores how we form impressions of people and how we use these impressions to make judgments about them. At times, impressions are subjective, involving assumptions based on the way people look or on hearsay information. At other times, impressions involve scrupulous efforts at objectivity, such as when a jury has to decide the guilt or innocence of an accused. However impressions are arrived at, they play a central role in social functioning, because they represent an important aspect of people's understanding of their social world.

I. The Challenge of Person Perception

The aim of the perceiver is usually to form an accurate impression of another person. Two elements of person perception pose a challenge in reaching this goal. The first is that unlike inanimate objects, people care about the impressions others form of them and take an active role in controlling the information about themselves that others are permitted to see. This process is known as *impression management*. To form accurate impressions of others, the perceiver must see through the impression management efforts of others, a task that is not always easy or successful. The second challenge for the perceiver is created by the indirect nature of person perception. The elements of the impressions we form of others include character traits, intentions, interests, abilities, moods, and attitudes. None of these properties can be perceived directly. There is no look, sound, or feel to the trait of kindness, the intention to help, or the state of empathy. These properties have to be inferred from outward manifestations such as actions, demeanors, verbalizations, and expressions.

II. Surface Information: Appearances, Expressions, and Demeanors

A. Assumptions Based on Appearance: The Case of Physical Attractiveness

Appearance often provides the first information we have about another person. A person's sex, height, weight, and age, as well as the way he or she is dressed and groomed are readily apparent. This information can lead the perceiver to make strong assumptions about the person.

Several studies have demonstrated that we make positive assumptions about physically attractive people. In one study, subjects were given pictures of physically attractive people, some who were average in attractiveness, or some who were unattractive. The subject then rated the kindness, sensitivity, interestedness, strength, poise, modesty, sociability, and sexual warmth of the person depicted in the picture. Averaging across these characteristics, it was found that attractive persons were rated highest and unattractive persons, lowest. This bias in favor of attractive people extended to ratings of assumed occupational status, marital competence, and happiness.

The tendency to generalize from one positive characteristic, such as attractiveness, to the whole of the person is called the *halo effect*. Early research on the effects of physical attractiveness on person perception uncovered so many characteristics favoring attractive people that some researchers concluded that attractiveness surrounded the person with a positive aura. It was found that physically attractive people are more likely to have their written work evaluated positively, to be recommended for hiring following a job interview, to receive·lighter sentences from mock juries, and to be seen as effective psychological counselors. The bias was found to affect the perception of men and women equally and to extend to the perception of children. Adults, for example, judge transgressions committed by attractive children less severely than the same acts committed by unattractive children. Together, these findings suggest that in the eyes of the perceiver, what is beautiful is good.

An accumulation of research findings on the effects of attractiveness on perception, however, sug-

gests that the notion that what is beautiful is good is too general and undifferentiated. With the many positively biased assumptions made about attractive people, there are negative assumptions and evaluative characteristics on which no bias is shown. Attractive people are dealt with harshly by mock juries if it appears that they exploited their looks in the commission of their crime, as in a swindle or con game. Moreover, studies analyzing ratings of personality characteristics have found that physical attractiveness has its impact primarily on ratings of sociability, social competence, and popularity. The impact of physical attractiveness on strength, adjustment, and intelligence is only moderate, and it is negligible on integrity and concern for others.

The heterogeneity of perceptions of attractive people suggests that the halo effect metaphor is inappropriate. Instead, the assumptions we make about attractive people stem from implicit personality theories. These are informal theories we hold about the kinds of personal characteristics that tend to go together in people. In the same way that we assume intellectuals are introverted and idealistic, we also assume that attractive people are sociable and popular. These assumptions are fairly specific and reflect the links that exist in our implicit personality theories between attractiveness and other trait attributes.

The assumptions we make about physically attractive people are not necessarily inaccurate. *Self-fulfilling prophecies* describes the tendency to act on assumptions in such a way as to elicit confirmatory behavior from other people. It has been shown, for example, that college men who phoned a college woman they assumed to be attractive exhibited more socially skilled verbal behavior and elicited more socially skilled verbal behavior from her than when they assumed that she was unattractive. It is possible that the positive attention attractive people receive increases their self-confidence and provides them with opportunities to hone their social skills.

B. The Perception of Emotions and the Detection of Lies

Facial expressions are particularly rich in their potential for expressing emotions. Consider the faces shown in the top panel of Fig. 1. Even if these pictures were not labeled, you would have no difficulty identifying each of the emotions expressed. Research has demonstrated that people are highly accurate at recognizing the posed expressions of happiness, sadness, fear, surprise, anger, and disgust. It is not difficult to understand why this is the case. The expressions shown in Fig. 1 are very rich in information. Each expression contains a number of distinctive elements involving the shape of facial features (e.g., the smiling mouth in happiness) and the presence of wrinkles on the forehead in fear.

In addition to static information in these pictures, research has shown that dynamic motion information is also distinctive in each of the expressions and provides further cues for their identification. The white dots shown in the bottom panel of Fig. 1 are randomly placed on blackened faces. When stationary, these dots do not reveal much about the surface on which they are placed. Observers cannot even guess that they are placed on faces. This indicates that all feature information in normal photographs is eliminated in the blackened faces. Yet, when videotapes of these faces are shown to viewers, they can immediately tell that they are looking at faces and are remarkably accurate at recognizing each of the six emotions. Static and dynamic facial information together provide rich cues for the recognition of emotions.

In his theory of evolution, Darwin argued that there was great survival value to the accurate signaling of emotional states. Darwin proposed that because there is a link between the signals used by subhuman primates and those used by humans, these signals are universal. Research on the universality of emotional expressions has produced findings consistent with Darwin's position. In fact, the six emotional expressions shown in Fig. 1 are recognized much the same way across cultures. The possibility that this consistency is caused by exposure to western facial expressions is ruled out by a study conducted in a remote part of New Guinea with people who had no contact with western culture. Brief stories describing an emotion-inducing situation were read to the New Guineans, who were asked to select from a number of pictures the one that depicted the emotion. On average, the choices made by members of this group agreed with those of western subjects in over 80% of the cases. Only in the case of fear and surprise was the confusion rate high in the New Guinean group, a finding that suggests surprising situations may generally have a fearful component in this culture.

The impressive recognition rates of expressions such as those shown in Fig. 1 should not be taken as evidence that people are always accurate at perceiving the emotional states of others. Matching six expressions with six corresponding emotion labels is not nearly so challenging as identifying emotions in everyday life, for several reasons. First, the emotional expressions used in research, eliciting impressive recognition rates, are usually posed by highly trained actors. Second, these expressions are pure, in that they convey a single emotion rather than complex blends (such as fearful surprise in reaction to a disgusting stimulus). Third, and most important, emotional expressions are often influenced by learned *display rules* about the appropriateness of facial expressions in various situations.

The effect of display rules on emotional expressions was demonstrated in research with bowlers. The facial expressions of bowlers were observed to discover whether smiling resulted from happiness with a successful performance or from social engagement with partners. If a bowler is happy following a good roll, then the frequency of smiles should be higher following a strike or a spare than following a poor roll. In fact, the association between performance and smiling was weak. Smiling was found to depend primarily on whether the bowler was engaged in social contact with his or her companions.

Perhaps the most instructive situation with respect to the accurate identification of emotions involves the detection of deceit in situations in which the liar is anxious about being apprehended. The criminal under investigation and the smuggler questioned by a customs officer usually experience fear. Yet the facial expression presented in such situations seldom resembles the expression of fear shown in Fig. 1. Under such circumstances, people mask the fear they actually feel and often simulate other feelings such as relaxation or joviality.

How successful is the falsification of emotional expressions? In a realistic and well-controlled study, passengers about to board a plane were given fake contraband to smuggle past a customs inspector. A randomly selected control group of passengers were given no contraband. The "smugglers" and the "innocent" participants were videotaped during the interview with the customs inspectors. These interviews were later shown to undergraduate students, who coded the behavior of the travelers and also tried to pick out the smugglers among them.

The study yielded several interesting results. The verbal and nonverbal behavior of the smugglers did not differ systematically from those of the innocent travelers. This indicated that smugglers were suc-

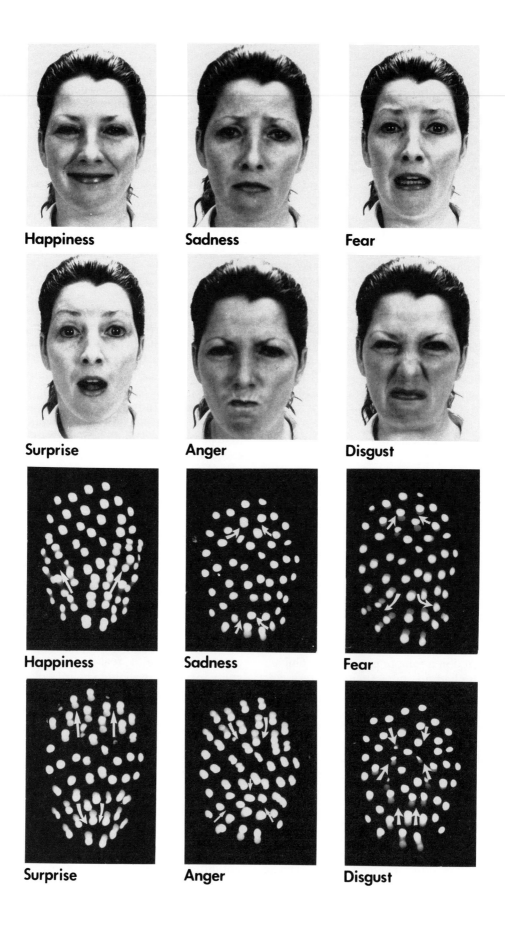

Happiness

Sadness

Fear

Surprise

Anger

Disgust

Happiness

Sadness

Fear

Surprise

Anger

Disgust

cessful in masking the cues that would have given away their guilt. Customs inspectors were no more likely to search smugglers than innocent travelers, and undergraduate students who later viewed the videotapes were also unable to pick them out. Despite this lack of accuracy, customs inspectors and undergraduate observers agreed on who they thought was smuggling contraband. Both groups relied on the same cues to decide who was guilty. Specifically, travelers who paused before giving their answers, who gave short answers, who avoided eye contact with the customs inspector, and who shifted their bodies a lot were deemed to be hiding something, even though these cues were not reliably connected with actual guilt.

This study illustrates that people can mask their true feelings with success, and that perceivers face a serious challenge in trying to see through the mask. In fairness, the passengers who served as subjects in this study did not experience the level of anxiety a real smuggler would experience. The subjects, therefore, may have had an easier time controlling their self-presentations. It is likely that intense emotions tend to leak, despite efforts at masking them. Research suggests that the lower body is more likely than the face to carry cues of leakage. A cautious assumption about accuracy in the recognition of emotions in general and of deceit in particular is that people differ in their skills at falsifying the information they present to others and perceivers differ in their skills at seeing through falsification.

II. Understanding the Causes of Behavior

The cues we gather from appearances, expressions, and demeanors form only a small part of the information we use in person perception. When we go beyond the snap judgments we form on the basis of these cues, we usually pay increasing attention to a

FIGURE 1 Posed expressions of six basic emotions. The bottom panel shows time exposures of blackened faces covered with white dots. The camera shutter was opened just before the beginning of an expression and was closed just after the emotion reached its apex. [The bottom panel is from Bassili, J. B. (1979). "Emotion recognition: The role of facial movement and the relative importance of upper and lower areas of the face." *J. Pers. Soc. Psychol.*, **37**, 2049–2058. American Psychological Association, reprinted by permission of the publisher.]

person's actions. The inferences we make to explain behavior are called *attributions*. In 1958 Fritz Heider introduced the field of *attribution theory* by arguing that to function successfully in the social environment, a person has to relate the actions of others to their personal characteristics and to characteristics of the situation. It matters, for example, whether a boss attributes an employee's lateness to exceptional traffic conditions or to the employee's lack of commitment to the job, or whether a husband attributes his wife's frequent trips to a city to the presence of a large business account or to that of a lover.

A. The Covariation Principle

At the heart of attribution theory is the notion that people tend to attribute effects to causes with which they covary. The method is analogous to John Stuart Mill's method of difference, namely that the cause of an event resides in the conditions present when the event occurs and absent when it does not.

In 1967 Harold Kelley formulated an important model of attribution based on the principle of covariation. According to this approach, there are three causes for social behavior: the *person* who performed the behavior, the *entity* to which the person responded, and the *circumstances* that surrounded the behavior. Consider the case of Paul, who laughs uproariously at a comedian at an office party. An observer could attribute Paul's behavior to something about him (a giddy personality), something about the entity (the comedian being funny), and/or something about the circumstances (the festive mood of the party).

Kelley suggested that perceivers examine three types of information in their efforts to explain behavior: *distinctiveness* (does the person respond to other entities in the same way?), *consensus* (do most other people respond to this entity in the same way?), and *consistency* (does the person respond to this entity in the same way in different circumstances?). In our example, Paul's behavior would be low in distinctiveness if he responds with uproarious laughter to other comedians, whereas it would be high in distinctiveness if he does not easily laugh at other comedians. Consensus would be low if few other people laughed at this comedian, whereas it would be high if most other people laughed at this comedian. Finally, consistency would be low if Paul did not laugh at the comedian at other times (such as when the comedian appeared on television or at a nightclub), and it would be high if Paul laughed at the comedian at other times.

To make an attribution, people look at the pattern of information formed by distinctiveness, consensus, and consistency. Three patterns are particularly important. First, high distinctiveness, high consensus, and high consistency tend to yield an entity attribution. That is, if Paul laughs at few other comedians, if most other people laugh at this comedian, and if Paul laughs at this comedian every time he sees him, then the perceiver will attribute Paul's laughter at the office party to the talent of the comedian. Second, low distinctiveness, low consensus, and high consistency tend to yield a person attribution. That is, if Paul laughs at all comedians, if most other people do not laugh at this comedian, and if Paul laughs at this comedian every time he sees him, then the perceiver will attribute Paul's laughter to something about Paul (his giddiness). Third, high distinctiveness, low consensus, and low consistency tend to yield a circumstance attribution. That is, if Paul laughs at few other comedians, if few other people laugh at this comedian, and if Paul does not laugh at this comedian on other occasions, then the perceiver will attribute Paul's laughter to something about the particular circumstances at the office party.

It is instructive to relate the information contained in distinctiveness, consensus, and consistency to the covariation principle. Distinctiveness represents variation in the entity (the comedian), consensus represents variation in the person (Paul), and consistency represents variation in the circumstances (the office party). According to the covariation principle, effects are attributed to the causes with which they covary. High distinctiveness indicates that the laughter disappears when the comedian is different, thus contributing to the impressions that it is something about the comedian that caused Paul's laughter. Low consensus indicates that the laughter disappears when the person is different, thus contributing to the impression that it is something about Paul that caused his laughter. Low consistency indicates that the laughter disappears when the circumstances are changed, thus contributing to the impression that it is something about the office party that caused Paul's laughter.

The covariation principle of attribution, as it is embodied in Kelley's model, has received strong empirical support. A number of studies have presented subjects with scenarios describing a person's reaction to an entity in a particular situation and have manipulated distinctiveness, consensus, and consistency information. In general, subjects' attributions to each of the three causes have been consistent with the predictions of the model.

B. Correspondent Inference

Kelley's model provides a comprehensive analysis of causal inference. There is one class of attribution, however, that is particularly important in person perception—attributions to people's personality characteristics. The theory of correspondent inference, postulated by Edward Jones and his colleagues, focuses on these attributions. Correspondent inference refers to the perceived congruence between a behavior and an underlying personality characteristic. Suppose you saw a person behaving in a friendly manner. Under what conditions would you infer that the person has a friendly disposition?

The first step in making correspondent inferences is to infer whether the behavior was intended. Under most circumstances, we do not learn much about other people from behaviors that they do not intend. The case of accidental behavior is particularly relevant here. An intended shove may lead to the inference of an aggressive disposition, but a shove resulting from a loss of balance normally does not. There are exceptions to this rule. Traits such as clumsiness, forgetfulness, and insensitivity result in behavior that is largely unintended. Thus, while intention is necessary to the inference of most dispositions, there are cases in which the disposition itself implies that the person does not have control over the behavior.

Having established the presence or absence of intention, the perceiver next attempts to determine what trait caused the behavior. A common way for the perceiver to make trait inferences is to focus on the effects of behavior. This is illustrated best in situations where a person chooses between paths of action that have different effects. Imagine a person contemplating the purchase of one of two cars. Both cars are known for their performance and reliability, but one has attractive, conservative lines, while the other is bedecked with wings and scoops that give it an eye-catching appearance. The person decides to purchase the showy car. According to correspondent inference theory, a perceiver would infer that the person likes getting attention. This is because the attention-getting property of the car is the only effect that is not common to the two alternatives. The fact that the chosen car has good performance and reliability is not particularly informative because the other car also has these characteristics. The theory of correspondent inference also predicts that the perceiver will make more confident inferences when there are few rather than many effects that are not common to the alternatives. If, in our example, the showy car was also cheaper, smaller, and more economical than the conservative car, then the perceiver could not have attributed its purchase specifically to its eye-catching quality.

Predictions of the model have received good empirical support. Subjects in some studies read descriptions of situations where actors made choices between several alternatives. In general, the greater the number of effects that were not common to the alternatives, the less confident subjects were of their attributions.

C. The Discounting Principle and the Correspondence Bias

The idea that confidence in the role of a particular cause in producing an effect decreases as other plausible causes are also present has been embodied in a general principle called *discounting*. The principle applies to situations in which a behavior may have been caused by a personality disposition or by any number of causes external to the person. Imagine Peter being courteous to Paul. This behavior may have been caused by Peter's trait of courtesy or by something else. For example, courtesy may have been normative in that interaction, or Peter may have been ingratiating himself to Paul in preparation for asking of him a large favor, or perhaps Peter's wife implored him to show courtesy to her friend Paul. According to the discounting principle, the more the perceiver is aware of alternative plausible causes, the less confident he or she will be that any one of the causes was responsible for a behavior.

The discounting principle was illustrated in a series of experiments on the perception of other people's attitudes. Subjects in these experiments read essays and were asked to judge the attitudes of the authors. The contents of the essays, as well as the conditions under which they were purportedly written, were varied. Some essays were favorable to Fidel Castro, the Cuban leader, and the rest were unfavorable to him. Moreover, some subjects were told that the position taken by the author was freely chosen, but others were told that the author was assigned the position by the experimenter.

Subjects who thought that the authors had chosen the direction of their essays inferred that the opinions expressed by the authors reflected their true attitudes. Discounting, however, occurred in cases where subjects thought that the position taken by the essay was assigned by the experimenter. From the point of view of the subject, there are two plausible causes for the opinions expressed in such cases: the authors' attitudes, and the demands made by the experimenter. As a result, subjects were less likely to infer that the authors' personal attitudes were consistent with those expressed in the essays in these conditions.

Something intriguing emerged in these experiments on the perception of attitudes. Despite the fact that subjects in these experiments were told explicitly that the positions presented in the essays had been assigned to the authors, they still inferred that the written positions reflected the authors' true attitudes. In other words, while discounting occurred in these experiments, its effect was far from complete. This phenomenon has become known as the *correspondence bias*. The bias refers to people's tendency to make attributions that give too much weight to personal dispositions, and not enough weight to situational forces. So pervasive has this bias proven in attribution research that it has also become known as *the fundamental attribution error*.

D. Salience and the Actor–Observer Effect

Fritz Heider's pioneering work in attribution anticipated the correspondence bias. According to Heider, other people's behavior is usually so salient to us that we tend to focus on it rather than on the situation. Taken a step further, this idea suggests that any salient stimulus is likely to be seen as playing an exaggerated causal role.

Several experiments have supported this notion. For example, if the attention of the perceiver is focused on one member of a group during a conversation, that person will be perceived as playing a more central causal role in directing the conversation than other members of the group. Perceivers have also been found to attribute more causality to a person who stood out by being the only member of a visible minority in the group, or even by being dressed in a way that was different from the rest of the group. These effects are not limited to situations in which the perceiver has little information relevant to causality other than salience.

There is one important exception to the correspondence bias. It involves situations in which people make attributions about their own behavior rather than that of others. Instead of attributing their own behavior internally, people tend to attribute it to external causes. For example, students who are asked why they chose their major tend to give answers that describe properties of the field (e.g., computer science is an important technological field). When asked about their best friend's choice of majors, however, they give dispositional reasons (he has always been a computer whiz). This asymmetry in attribution for a person's own behavior versus that of others is called the actor–observer effect.

The effects of salience on causal attributions can help explain the actor–observer effect. As observers, we tend to focus on behavior. As actors, however, we do not focus on our own behavior. Instead, our attention is directed at the features of the environment to which we are reacting. What is salient to actors and to observers is very different, and this is why they tend to make attributions to different causal factors.

E. Construct Accessibility and the Interpretation of Behavior

Salience is not entirely a property of the information available to a perceiver. Another important determinant of how people attend to and interpret social information is the level of activation of their social knowledge structures. Abstract knowledge structures that are relevant to the interpretation of information about people are known as *trait constructs*. Research has shown that trait constructs become more *accessible* immediately after their use. This phenomenon is known as priming. As a result of priming, a trait construct is more likely to be used to interpret new information and to affect relevant judgments. To illustrate the effect of construct accessibility on the interpretation of social behavior, consider having lunch with a colleague at work. During lunch you talk about shy people. A little later you are joined by an employee who has only been with the firm a few days. He asks if he can join you, sits quietly, and eats his lunch without making much eye contact. According to the notion of construct accessibility, you will be more likely to attend to these apparently shy behaviors and more

likely to interpret them as reflecting a shy disposition after having been engaged in conversation about shyness than if your construct of shyness had not been primed by the prior conversation. This is particularly true with ambiguous behaviors. If the behaviors were obviously shy, then you would perceive them as such whether or not your construct for shyness were primed. If the behaviors were not at all shy, then the construct of shyness would not be relevant to the behaviors and would not be likely to be invoked in interpreting them.

The level of activation of social constructs is affected not only by priming. Each of us tends to pay attention to particular aspects of social behavior more than to others. Some of us are particularly attuned to the competitiveness of behavior; others, to its friendliness. Chronic differences in levels of construct activation account for these differences. Together, priming and chronic effects play an important role in the interpretation of ambiguous social information.

F. Self-Serving Biases in Attribution

Classical theories of attribution focus on the rationality of the perceiver. The covariation and the discounting principle, as well as the theory of correspondent inference, all paint a picture of a quasi-scientific perceiver eager to determine the causes of behavior rationally. The impact of salience, a nonrational basis for causal inference, on attribution demonstrates that there are limits to the metaphor of the scientific perceiver. In addition, research has revealed other biases demonstrating that perceivers sometimes make defensive rather than rational attributions.

A well-documented bias in attribution is the tendency to take personal responsibility for successes while attributing failures externally. There are two motivational reasons for this. First, such attributions serve to protect and enhance the way people see themselves. In one study, subjects were led to believe that they had failed at a task and were then induced to attribute their failure either to the difficulty of the task or to their own lack of ability. Subjects were more upset and had lower self-esteem when they attributed failure to lack of ability rather than to the difficulty of the task. Second, these attributions serve an impression-management function by bolstering the image the person presents to others. People who expect to fail at a task tend to exert less effort than people who expect to succeed, presumably because this provides them with an excuse for their poor performance that doesn't reflect on their own ability.

Another well-documented bias involves the attribution of responsibility for misfortunes such as accidents and crimes. In such situations people tend to blame the victim, even with little evidence. This finding suggests that people make attributions that promote their belief in the fairness of the world. There is an important motivational reason for maintaining a belief in a just world. If the world is just and victims of mishaps deserve their misfortune, then such misfortunes will not happen to blameless people like us. This interpretation is supported by data that show that our propensity for blaming the victim decreases when we feel similar to him or her. If a person who is similar to us deserves an awful calamity, then perhaps the same could happen to us!

The self-serving biases documented by attribution researchers serve to make a broader point about person perception. Motivation often plays an important role in our interpretation of information about people. Our sentiments about others, about ourselves, and our particular goals in social interactions influence the way we perceive events around us.

G. When Are Attributions Made?

Heider conceived of attributions as a necessary step in understanding social reality. He believed that the motivation to engage in attribution is natural and spontaneous. Recently a number of approaches have been taken to test this assumption. One approach has been to analyze naturally occurring responses to social events to identify the factors conducive to attributional thinking.

A particularly interesting example of this approach is a study of commentary given by members of winning and losing teams following baseball and football games. The explanations in newspaper articles for the outcome of a number of games were coded for attributional content. This coding revealed a tendency to attribute wins internally and losses externally. More to the point, the sheer number of attributions was greater following an unexpected outcome than following an expected one. An unexpected outcome in this context involves a loss by the favored team to the underdog. One reason for attributions, therefore, is to make sense of unex-

pected events. Other research involving the coding of responses has demonstrated that people are also more likely to make attributions following failure than following success at a task, possibly because such outcomes also tend to be unexpected.

Trait attributions are particularly important in person perception. Recent studies have explored spontaneous attribution of traits to others. These studies present subjects with sentences of the form: "The secretary solved the mystery halfway through the book," and then use indirect memory techniques to test whether subjects made the trait attribution suggested by the sentence (e.g., the secretary is clever). The results of these studies suggest that trait attributions are made spontaneously with only moderate frequency. In general, people have to be intent on forming impressions of others before attributing traits to them.

IV. Organizing and Using Our Mental Representations of Others

A. Impression Formation

In his pioneering work on how people form impressions of others, Solomon Asch presented subjects with lists of traits that described other persons and asked them about the impressions they had formed. Asch was particularly interested in the capacity to take disparate, and even contradictory information about others, and to make the information fit in a way that revealed the fabric of others' characters. To illustrate, consider a person who is both brilliant and silly. A perceiver may resolve the apparent contradiction in this description by thinking of a person who is brilliant intellectually but silly with practical day-to-day chores.

According to Asch, people can form coherent impressions of others, because social information can usually be given a number of interpretations. Brilliance can be interpreted as meaning great depth of intelligence or dazzling astuteness. Silliness can be interpreted as shallowness or mindlessness. The perceiver can successfully juxtapose the trait of brilliance and that of silliness because each of these traits has an interpretation compatible with an interpretation of the other (depth of intelligence and mindlessness).

Several phenomena of person perception derive from this theoretical stance. *Central traits* play a particularly important role in guiding the interpretation of other traits in an impression. Consider the following two descriptions of another person: (1) intelligent, skillful, industrious, warm, determined, practical, cautious; (2) intelligent, skillful, industrious, cold, determined, practical, cautious. The impressions engendered by these two descriptions differ radically despite the fact that the descriptions differ by only a single trait (warm versus cold). The traits warm and cold are central in this context because they influence the interpretation of each of the other traits. Warm and cold are the only traits in the list that convey information about the interpersonal motives of the target person. These traits allow the perceiver to read different meanings, that are primarily relevant to talents and abilities, into each of the other elements of the description.

Other phenomena compatible with this approach are the *primacy effect* and the *context effect*. The primacy effect is the tendency for early information to have more impact on impressions than later information. According to Asch, early information sets an expectation about the person. This expectation becomes relevant to interpreting subsequent information. The context effect is the tendency for a trait to be perceived more positively in the context of positive traits and more negatively in the context of negative traits than it is perceived in a neutral context. This results from the impact of context on the interpretation of the target trait.

B. Information Integration

The notion that our impressions of others represent a complex interpretive process has not been accepted universally. The best-articulated alternative to this approach has been presented by Norman Anderson, who noted that evaluations constitute the most powerful element of impression formation. Anderson proposed an averaging model of evaluations in impression formation. This is the model in its simplest form. Suppose a personnel manager meets a job candidate who appears intelligent and efficient but who has poor social skills. These traits differ on how positive and negative they are, and this is what the personnel manager focuses on. Assume that on a scale ranging from -10 to $+10$, intelligence is a $+9$; efficiency, a $+8$; and poor social skills, a -5. According to the averaging model, the personnel manager will form an overall impression of the job candidate that represents the average of the three traits, that is, $9 + 8 - 5 / 3 = 4$.

The averaging model is noteworthy for its simplicity and its openness to precise experimental testing. Norman Anderson and his colleagues have conducted an extensive research program to test the model and have garnered impressive evidence for it. One particularly interesting test involves the comparison of the averaging model with its obvious counterpart, the additive model. According to the additive model, an impression is the sum rather than the average of the scale values of the component traits. Thus, in our example, the overall evaluation of the job candidate would be $9 + 8 - 5 = 12$. The fact that 12 is larger than 4 is not important, because these numbers are meaningful only in the context of each model. However, the comparison between the two models becomes quite revealing if we assume that the candidate is also athletic, and that this quality has a scale value of $+2$. If we incorporate this value into the impression according to the averaging model, the overall evaluation becomes $9 + 8 + 2 - 5 / 4 = 3.5$. On the other hand, if we incorporate this value into the impression according to the additive model, the overall evaluation becomes $9 + 8 + 2 - 5 = 14$. Averaging has resulted in a less positive impression, and adding has resulted in a more positive one. Clearly, the two models make very different predictions about impression formation. The research that has tested these predictions has generally favored the averaging model. The practical lesson for us is that if we have succeeded in creating a positive impression with our best traits, we should resist the temptation to cram our self-presentation with mildly positive information about ourselves.

C. Associative Networks and the Structure of our Impressions

With the advent of the information-processing revolution in psychology, researchers have become interested in specifying the exact nature of the mental representations formed in person perception. Reid Hastie as well as Thomas Srull and Robert Wyer proposed that these representations are composed of idea *nodes* and of associative *links*. Figure 2 shows such an associative network. At the top level of the structure is a node identifying features of the person such as his or her proper name. At the second level are concepts that describe the person and that organize more specific information about him or her. Trait concepts are assumed to play a domi-

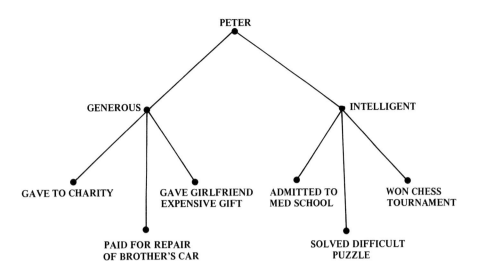

FIGURE 2 An associative network structure representing trait and behavior information. [Reprinted with permission from Bassili, J. N., ed. (1989). "On-Line Cognition in Person Perception." Lawrence Erlbaum Associates, Hillsdale, New Jersey.]

nant role at this level, because they describe others and organize information about them. At the bottom level are specific facts about the person, such as behaviors.

Most of the evidence for these structures comes from research on memory for information about others. Such research is based on the notion that the retrieval of information from an associative network involves movement from one node to another along existing links. For example, in recalling information about Peter, the retrieval process begins at the top node and then descends either toward the generous node or the intelligent node. For simplicity we will assume that the links to the two nodes are of equal strength and that the selection between them is random. Suppose that the search proceeded down to the generous node. The information contained in that node would be recalled (i.e., Peter is generous). The retrieval process would then continue randomly to one of the three behavioral nodes subsumed under the trait generous. Upon retrieving that information, the process would return to the higher node and select the next link randomly from the four emanating from that node. It is possible for the retrieval process to return to the Peter node, or even for it to retrace its steps to the behavioral node just retrieved. In the latter case, the process would return to the generous node and select a pathway again. The retrieval process would stop upon reaching a point where already recalled information is encountered repeatedly.

D. Evidence for Trait-Behavior Associations in Impressions

One of the most important assumptions in the network structure we have been discussing is that behaviors are organized in terms of the trait concepts they exemplify. This assumption has been supported by research showing that when we recall the behaviors of a person about whom we have formed an impression, we tend to recall behaviors that exemplify a particular trait as a group. In recalling information about Peter, a subject is likely to report the generous behaviors and the intelligent behaviors in clusters.

The fact that clustering results from links with trait concepts is supported by a number of other findings. There is little clustering by trait category when the information about a person is learned without intention to form an impression. As we saw earlier, trait concepts are not normally applied to the interpretation of behavior under such conditions. It is interesting that with a drop in clustering, these conditions also result in much poorer recall than conditions in which information is learned with the intention of forming an impression. Research has shown that subjects are more likely to recall a behavior exemplifying a trait if they had also recalled the trait than if they had not. In other words, trait concepts provide powerful organizing elements in our memory structures and act as midway cueing stations for the recall of others' behaviors.

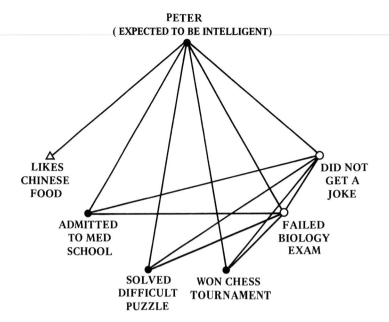

FIGURE 3 Example of associations formed between behaviors that are consistent with (black circles), inconsistent with (white circles), or irrelevant to (triangle) an expectation.

E. Memory for Behaviors that Confirm or Disconfirm our Expectations

So far we have considered associations between behaviors and trait concepts without considering direct associations between behaviors. Interbehavior associations have been explored in one of the most extensive research programs in person memory. This research is based on the phenomenon that people remember behaviors inconsistent with their expectations about another person better than behaviors that fit their expectations. This is not to say that we do not remember consistent behaviors well. Such behaviors are remembered much better than behaviors irrelevant to an expectation. Still, the question arises as to why inconsistent behaviors are recalled best.

The answer is that inconsistent behaviors are surprising and puzzling, and this makes us think about them in relation to other behaviors of the person. According to the associative approach, the act of thinking about two behaviors in relation to each other creates an association between them. The result of such thinking is that inconsistent behaviors are associated more with other behaviors than with consistent behaviors. This is illustrated in Fig. 3. Applying the search procedures discussed earlier to this network, it becomes apparent that nodes representing inconsistent behaviors are more likely to be reached than nodes representing consistent behaviors. The latter nodes, in turn, are more likely to be reached than nodes representing irrelevant behaviors.

F. On-Line Versus Memory-Based Evaluations

Earlier we saw that evaluations are of primary importance in impression formation. The mental representations we have of others provide the basis for the evaluative judgments we make about them. There are two situations in which this can happen: those in which we evaluate people immediately as we acquire information about them, and those in which our evaluations are based on information we recall from memory. In other words, the former type of evaluations are performed *on-line*, whereas the latter are *memory-based*. As it turns out, there are important differences between the cognitive representations that are used in on-line versus memory-based evaluations.

On-line evaluations are by far the more common. Suppose that having just been introduced to Peter, you learn that he was admitted to medical school, that he gave to a charity, etc. Two things will happen as you acquire each piece of information about Peter. The information will have some impact on your general liking for him, and the information will be stored in memory in a network such as the one shown above. The important thing to note here is that your liking for Peter will be based on an evaluative representation, updated simultaneously with the appearance of each new piece of information.

This representation is independent of the associative network in which the behavioral information is stored.

Suppose that a few weeks later, a mutual acquaintance asks you about your feelings toward Peter. Research suggests that your response will rely on the latest update of your general evaluation of Peter rather than on your memory of the information that led to your evaluation. You may answer that you like Peter very much, without remembering why you feel this way. In fact, research has often demonstrated that there is little connection between feelings and the informational basis for these feelings. With the passage of time, the recall of information deteriorates and is independent of the evaluative representation of a person.

There is one situation in which the relationship between evaluations and memory for relevant information is good. This is when the evaluations are memory-based. Suppose that you had no interest in forming an impression of Peter when you learned behavioral information about him. In this case, your mental representation would consist of a simple list of behaviors linked with Peter. Upon having to answer your friend's query a few weeks later, you would have to determine your feelings on the basis of information you could retrieve from memory. What is forgotten will have no impact on your evaluation. Naturally, what you feel toward Peter and what you remember of him will be closely connected in this situation.

of attribution that we infer the intentions and character traits of others. Because behaviors are often more salient to us than the situational context within which these behaviors occur, our attributions are usually biased toward the person. The opposite is true of attributions about our own behavior. The accessibility of our social constructs also plays a role in determining the salience of information and its interpretation. Construct accessibility is determined by recent prior activation and by chronic predisposition. Attributions are generally motivated by a desire to form an impression about a person or to understand why an unexpected event occurred.

The information we acquire about others is organized into meaningful impressions. We are able to form coherent impressions of others despite the disparate information we have about them because social information can usually be interpreted in a number of ways. Our overall evaluations of others are often based on the average value of the information we have about them. Models of how we store information about others assume that our mental representations consist of nodes and of associations between them. These models have been successful in explaining why we tend to remember behaviors that violate our expectations better than behaviors that confirm them. They have also been instrumental in delineating the differences between evaluations formed during acquisition of information about a person and evaluations based on memory.

V. Summary

The processes through which we get to know people are collectively known as person perception. These processes can be triggered by superficial information such as appearance or nonverbal behaviors. We often form preliminary impressions about people on the basis of such information. These impressions can be elaborate because of assumptions we make on the basis of implicit personality theories. One example of this is the assumption that physically attractive people have better social skills and are generally more outgoing than physically unattractive people.

When we observe others' behavior, we are often motivated to understand its causes. We usually do so by attributing the behavior to properties of the person or of the situation. It is through this process

Bibliography

Bassili, J. N., ed. (1989). "On-line Cognition in Person Perception." Lawrence Erlbaum Associates, Hillsdale, New Jersey.

Ekman, P. (1985). "Telling Lies: Cues to Deceit in the Marketplace, Politics, and Marriage." W. W. Norton Co., New York, N.Y.

Fiske, S. T., and Taylor, S. E. (1984). "Social Cognition." Addison-Wesley, Reading, Massachusetts.

Higgins, E. T., and Bargh, J. A. (1987). Social cognition and social perception. *In* "Annual Review of Psychology," (M. R. Rosenzwieg and L. W. Porter, eds.). Annual Reviews, Palo Alto, California.

Ross, M., and Fletcher, G. J. O. (1985). Attribution and social perception. *In* "Handbook of Social Psychology." (G. Lindzey and E. Aronson, eds.). Random House, New York, N.Y.

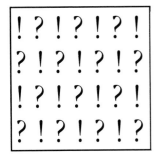

Persuasion

JOHN T. CACIOPPO, RICHARD E. PETTY, GARY G.
BERNTSON, *Ohio State University*

Glossary

Attitude General evaluative response or disposition to respond
Influence Effect of events and others on behavior
Persuasion Power of persons to alter attitudes and behavior through information

ALL ORGANISMS HAVE BIOLOGICAL mechanisms for approaching, acquiring, or ingesting certain classes of stimuli and withdrawing from, avoiding, or rejecting others. For simple organisms, the stimuli that potentiate approach or withdrawal, the form of the response, and the mediating mechanisms are relatively fixed. For more complex organisms, such as humans, multiple mechanisms contribute to approach and withdrawal tendencies. This potentiation can manifest consciously (cognitively, emotionally) as well as behaviorally and can be stored in memory in the form of attitudes; both the eliciting stimulus and the form of the response are subject to generalization and modification. Persuasion refers to the power of persons to alter attitudes and behavior through information. Attitude and behavior

change resulting from a communication constitute persuasion regardless of the communicator's intent or the recipient's awareness that an attitude has been changed. Not included under the rubric of persuasion are changes in knowledge or skill (i.e, education) or changes in behavior that require another's surveillance or sanctions (i.e., compliance). Persuasion, therefore, represents a form of self- and social-control that does not rely on coercion. In addition, innate and relatively inflexible predilections to approach or withdraw (such as reflexes or fixed action patterns) and irreversible changes in parameters of approach or withdrawal (such as diminished response vigor due to aging) may be related to attitudes but are not considered instances of persuasion. Traditionally, studies of the antecedents of persuasion have focused on characteristics of the source of a recommendation and on rational or emotional forms of argumentation linking a particular recommendation to a person's beliefs and values; however, any information that changes an individual's predilection to react to a class of persons, objects, or issues in a consistently positive or negative fashion could be included under the rubric of persuasion. Studies of the consequences of persuasion have tended to focus on changes in attitudes and cognition and the persistence of these changes, but the physiological and behavioral effects have also been investigated.

I. Attitudes and Persuasion

People's perceptions of events and people in their world are organized in part in terms of long-term evaluative responses to stimuli. This organization is reflected in the structure of language; people's conceptual organization of motivation, emotions, and moods; facial expressions of emotion; and everyday behaviors. The evaluative (attitudinal) categorization of stimuli has even been found to emerge in some circumstances prior to an individual recognizing the eliciting stimulus.

Attitudes, in turn, influence what people perceive and feel about their world and can have direct and indirect effects on behavior across a wide range of situations. The direct effect of attitudes on behavior represents the tendency for people to approach, acquire, support, protect, and promote liked, in contrast to disliked, objects, persons, and issues. Although additional psychological operations may intervene between attitudes and behavior, such as accessing the relevant attitude and formulating a behavioral intention, these operations primarily pertain to the response side of the information-processing sequence (e.g., response execution). Promotional campaigns for products in stores, for instance, are designed simply to mobilize people to try the products. Finally, the causal relationship between attitudes and behavior is reciprocal rather than unidirectional. The feedback from actions or inactions toward some target constitutes a powerful source of information that can shape subsequent beliefs and attitudes about the target. Product trials, for instance, can provide compelling reasons for individuals for purchasing, or not purchasing, particular products again. [*See* ATTITUDE AND ATTITUDE CHANGE.]

The indirect effect of attitudes on behavior stems from their influence on individuals' selective attention to, interpretation of, and recollection of people and events in their world and, subsequently, on their behavior and on the behavior of others toward them. In addition to the direct behavioral effects of attitudes, therefore, attitudes can also shape an individual's experience and representation of the world. If information favoring both sides of an issue is presented to groups with opposing attitudes, the discrepancy between the groups' attitudes might intuitively be expected to diminish somewhat; however, the opposite result has been found to be the case. Individuals often accept the evidence that supports their initial attitude and are critical of the evidence contradicting their initial attitude. The result is that both groups find reasons to strengthen their initial attitude, and disagreement between the two groups can actually be heightened by information favoring both positions.

In addition to having a directive function that channels activity into specific types of responses toward certain stimuli, some people also view attitudes as having the dynamic function of energizing people to act. Thus, individuals who come to feel strongly about another person, topic, or issue may not only channel their thoughts and behavior toward the target accordingly when given the opportunity but may also be excited to create opportunities to act in a positive or negative fashion toward the target.

To summarize, attitudes are central to people's conceptual organization of their world and to the organization of their behavior. Because both attitudes and behavior are multiply determined, the correspondence between attitudes and behavior tends to be modest but significant. The correspondence between attitudes and behavior can be enhanced by (1) general rather than specific measures of attitudes and behaviors, (2) direct experience with the attitude target, (3) prior knowledge and thought about the target, (4) a highly accessible attitude toward the target, (5) weak social norms governing behavior toward the target, and (6) personal control over one's behavior toward the target. By influencing attitudes and factors that moderate attitude–behavior correspondence, persuasion can potentiate a broad class of positive or negative behaviors toward a person, object, or issue.

II. Two Routes to Persuasion

A. Background

The resolution of conflicts and the mobilization of effort to serve the goals of a collective are cornerstones of civilization. The history of mankind reveals physical force and intimidation—not persuasive skill—to be the key mode of achieving political, social, and economic control. William McGuire has noted that persuasion has played a central role in social control in only four epochs: in Athens from 427 to 338 B.C. (during which time Plato and Aristotle considered the processes underlying persuasion), in Rome from approximately 150 to 43 B.C. (during which time Cicero wrote about oration and persuasion), in Europe from approximately 1470 to 1572 (during the Italian Renaissance), and in the present period of the mass media, which began to take form in the eighteenth century. The scientific study of what and how factors affect persuasion gained momentum during World War II, when the mass media played an important role in recruiting and indoctrinating troops, maintaining the morale of the Allied forces and residents, and assaulting the morale of the Axis troops.

This early research was organized by the question "who said what to whom, how and with what effect?"—that is, in terms of the external stimulus factors of source (e.g., expertise, trustworthiness), message (e.g., one-sided, two-sided), recipient (e.g., sex, intelligence), and modality (e.g., print, auditory) factors. Two assumptions underlying much of this early work were that each of these factors had general and independent effects on persuasion, and a close correspondence existed between attitude change and behavior change across situations. Both assumptions proved to be oversimplifications. After accumulating a vast quantity of data and a large number of theories, surprisingly little agreement existed concerning if, when, and how the traditional source, message, recipient, and modality variables affected persuasion. Existing literature supported the view that nearly every independent variable studied increased persuasion in some situations, had no effect in others, and decreased persuasion in still other contexts. This diversity of results was even apparent for variables that on the surface, at least, appeared to be quite simple. For example, although it might seem reasonable to propose that by associating a message with an expert source agreement could be increased (e.g., see Aristotle's "Rhetoric"), experimental research suggested that expertise effects were considerably more complicated. Sometimes expert sources had the expected effects, sometimes no effects, and sometimes reverse effects.

B. The Central and Peripheral Routes to Persuasion

Empirical and theoretical advances over the past decade have led to a more comprehensive framework for organizing, categorizing, and understanding the basic processes underlying these diverse effects. Specifically, the many different empirical findings and theories in the field have been viewed as emphasizing one of two relatively distinct routes to persuasion. The first is attitude change that occurs as a result of a person's careful and thoughtful consideration of the merits of the information presented in support of an advocacy (central route); the second is that occurring as a result of some simple cue in the persuasion context (e.g., an attractive source) that induces change without necessitating scrutiny of the merits of issue-relevant information (peripheral route). This model of the psychological operations underlying persuasion,

which is depicted in Fig. 1, highlights that attitudes are multiply determined and that attitudes whose verbal expression is similar may have different antecedents and consequences. For instance, the issue-relevant thinking that characterizes the central route to persuasion can result in the integration of new arguments, or one's personal translations of them, into one's underlying belief structure for the attitude object. In addition, by scrutinizing the strengths and weaknesses of a recommendation, the information and the consequent attitude is rendered more coherent, accessible, and cross-situational. Attitudes formed through the central route, therefore, are relatively persistent, resistant to counterpersuasion, and predictive of intentional behavior.

If people were to try to adopt only those attitude positions about which they had thought carefully, they either would be unable to venture into novel situations or would be unable to respond to the myriad stimuli to which they are exposed each day. This is true even though individuals might be motivated generally to hold correct attitudes. The numerous stimuli to which individuals must respond daily, coupled with individuals' limited time and cognitive resources, make it adaptive to also be capable of using cognitively less demanding shortcuts (e.g., simple cues, habits, rules-of-thumbs) to guide attitudinal reactions in some situations. Although such a mechanism for attitude change (peripheral route) can guide responses to a wide variety of stimuli while minimizing the demands on individuals' limited cognitive resources, the resultant attitudes and behavior are based on information that is only superficially or peripherally related to the actual merits of the position. Hence, some responses potentiated by this generally adaptive mechanism may be unreasonable and maladaptive. These maladaptive features of attitudes derived through the peripheral route are diminished somewhat by their relatively short persistence, susceptibility to change, and weak influence on behavior.

C. The Elaboration Likelihood Continuum

The model outlined in Fig. 1 has provided a general framework for understanding how a variety of factors, such as speed of speech and source credibility, can increase, decrease, or have no effect on attitude change. If the central route is followed, the perceived cogency of the message arguments and factors that may bias argument processing (e.g., prior knowledge, initial opinion) are predicted to be

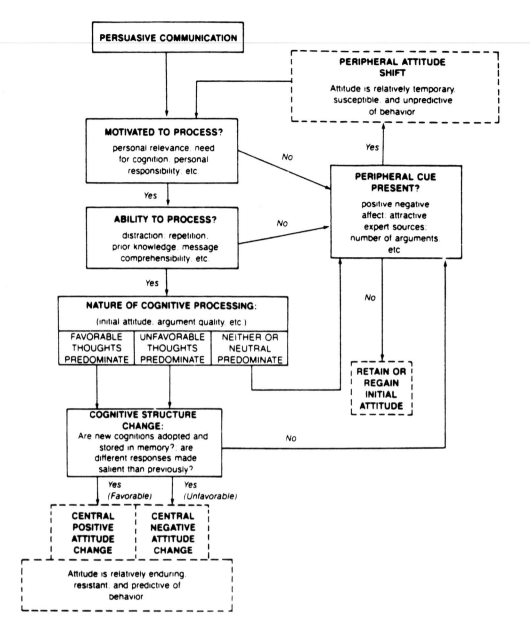

FIGURE 1 Elaboration likelihood model of persuasion. This figure depicts the two anchoring endpoints on the elaboration likelihood continuum: the central and peripheral routes to persuasion. (From R. E. Petty and J. T. Cacioppo, 1986, "Communication and Persuasion: Central and Peripheral Routes to Attitude Change," Springer-Verlag, New York.)

important determinants of the individual's acceptance or rejection of the recommendation, and factors in the persuasion setting that might serve as peripheral cues are relatively unimportant determinants of attitudes. If, on the other hand, the peripheral route is followed, the strength of the message arguments and factors that bias argument processing becomes less important and peripheral cues become more important determinants of attitudes, i.e., a trade-off occurs between the central and the peripheral route to persuasion.

Importantly, the conditions that lead to influence through the central versus the peripheral route have

also been specified. For instance, many attitudes and decisions are either perceived to be personally inconsequential or involve matters about which people are uninformed. In these situations, people may still want to be correct in their attitudes and actions, but they are not willing or able to think a great deal about the arguments for or against a particular position. Peripheral cues provide a means of maximizing the likelihood that one's position is correct while minimizing the cognitive requirements for achieving this position.

Implicit in the central route, on the other hand, is that people must relate the incoming message arguments to their prior knowledge in such a way as to evaluate the cogency and scope of the arguments—that is, they expend cognitive effort to examine the information they perceive to be relevant to the central merits of the advocacy. When conditions foster people's motivation and ability to engage in this issue-relevant thinking, the elaboration likelihood is said to be high. This means that people are likely to attend to the appeal, attempt to access relevant information from both external and internal sources, and scrutinize or make inferences about the message arguments in light of any other pertinent information available. Consequently, they draw conclusions about the merits of the arguments for the recommendation based on their analyses and derive an overall evaluation of, or attitude toward, the recommendation. Thus, the central and the peripheral routes to persuasion can be viewed as anchors on a continuum ranging from minimal to extensive message elaboration or issue-relevant thinking, and issue-relevant thinking may be relatively objective or biased. Factors governing an individual's motivation and ability to scrutinize the truthfulness of various attitude positions determine whether the central or the peripheral route operates.

Motivational variables are those that propel and guide people's information processing and give it purposive character. A number of variables have been found to affect a person's motivation to elaborate on the content of a message. These include (1) task variables such as the personal relevance of the recommendation, (2) individual difference variables such as need for cognition, and (3) contextual variables such as the number of sources advocating a position. These kinds of variables act on a directive, goal-oriented component, which might be termed intention, and a nondirective, energizing component, which might be termed effort or exertion.

Intention is not sufficient for motivation, for instance, because one can want to think about a message or issue but not exert the necessary effort to move from intention to thought and action. If both intention and effort are present, then motivation to think about the advocacy may exist, but message elaboration may still be low because the individual does not have the ability to scrutinize the message arguments. A number of variables can affect an individual's ability to engage in message elaboration, including task variables such as message comprehensibility, individual difference variables such as intelligence, and contextual variables such as distraction and message repetition. Contextual variables that affect a person's ability to elaborate cognitively on issue-relevant argumentation can also be characterized as factors affecting a person's opportunity to process the message arguments.

Experiments have demonstrated that if task, individual, and contextual variables in the influence setting combine to promote motivation and ability to process, then the arguments presented in support of a change in attitudes or behavior are thought about carefully. If the person generates predominantly favorable thoughts toward the message, then the likelihood of acceptance is enhanced; if the person generates predominantly unfavorable thoughts (e.g., counterarguments), then the likelihood of resistance, or boomerang (attitude change opposite to the direction advocated), is enhanced. The nature of this elaboration (i.e., whether favorable or unfavorable issue-relevant thinking) is determined by whether the motivational and ability factors combine to yield relatively objective or relatively biased information processing and by the nature of the message arguments. If elaboration likelihood is low, however, then the nature of the issue-relevant thinking is less important, and peripheral cues become more important determinants of attitude change (see Fig. 1).

A number of recent experiments have explored ways to stimulate or impair thinking about the message arguments in a persuasive appeal. Distraction, for instance, can interfere with a person's scrutiny of the arguments in a message and thereby alter persuasive impact. In an illustrative experiment on distraction and persuasion, students listened to a persuasive message over headphones while monitoring in which of the four quadrants of a screen a visual image was projected (a distractor task). In the low distraction condition, images were presented once every 15 sec, whereas in the high distraction condition images were presented once every 5 sec. Importantly, neither rate of presentation was so fast as to interfere with the students' comprehension of the simultaneously presented persuasive message, but the students' argument elaboration was much more disrupted in the high than low distraction condition. The results revealed that the students were less persuaded with distraction when the arguments were strong, but more persuaded with distraction when the arguments were weak.

Numerous task, contextual, and individual difference variables have been identified that enhance or impair argument elaboration by affecting a person's motivation or ability. Moderate levels of repetition of a complicated message can provide individuals with additional opportunities to think about the arguments and, thereby, enhance argument processing. Messages worded to underscore the self-relevance of the arguments enhance individuals' motivation to think about the arguments. Being singly responsible rather than one of many assigned to evaluate the recommendation can induce more issue-relevant thinking, as individuals are unable to diffuse their responsibility for determining the veracity of the recommendation.

III. Argument Elaboration Versus Peripheral Cues as Determinants of Persuasion

The hypothesis that there is a tradeoff between argument scrutiny and peripheral cues as determinants of a person's susceptibility or resistance to persuasion has also been supported by recent research. In an illustrative study, two kinds of persuasion contexts were established: one in which the likelihood of relatively objective argument elaboration was high, and one in which the elaboration likelihood was low. This was accomplished by varying the personal relevance of the recommendation: students were exposed to an editorial favoring the institution of senior comprehensive exams at their university, but some students were led to believe these comprehensive exams would be instituted next year (high elaboration likelihood), whereas others were led to believe the exams would be instituted in 10 years (low elaboration likelihood).

To investigate the extent to which students' argument scrutiny determined attitudes, half of the students heard eight cogent message arguments favoring comprehensive exams, and the remaining students heard eight specious message arguments favoring the exams. Finally, to examine the extent to which peripheral cues were important determinants of attitudes, half of the students were told the recommendation they would hear was based on a

report prepared by a local high school class (low expertise), whereas half were told the tape was based on a report prepared by the Carnegie Commission on Higher Education (high expertise). Following the presentation of the message, students rated their attitudes concerning comprehensive exams and completed ancillary measures. Results indicated that argument quality was the most important determinant of the students' attitudes toward comprehensive exams when they believed that the recommendation was consequential for them personally, but that the status or expertise of the source was the most important determinant of the students' attitudes when they believed that the recommendation would not affect them personally. These results held even though comprehension of the message arguments, and judgments of the expertise of the source, were equal across the experimental groups.

IV. Objective Versus Biased Argument Processing

Message processing in persuasion research was traditionally thought to imply objective processing. This, too, was an oversimplification. When an individual is motivated to scrutinize arguments for a position, there are no assurances that the information processing will be objective or rational. Objective argument processing means that a person is trying to seek the truth wherever that may lead. When a variable enhances argument scrutiny in a relatively objective manner, the strengths of cogent arguments and the flaws in specious arguments become more apparent. Conversely, when a variable reduces argument scrutiny in a relatively objective fashion, the strengths of cogent arguments and the flaws of specious arguments become less apparent. Objective processing, therefore, has much in common with the concept of "bottom-up" processing in cognitive psychology because elaboration is postulated to be relatively impartial and guided by data (in this case, message arguments).

In contrast, biased argument processing means that there is an asymmetry in the activation thresholds for eliciting favorable or unfavorable thoughts about the advocacy. Consequently, the encoding, interpretation, and recall of the message arguments are distorted to make it more likely that one side will be supported over another. Biased processing has more in common with "top-down" than "bottom-up'' information processing, because the interpretation and elaboration of the arguments is governed by existing cognitive structures, such as a relevant knowledge or attitude schema, which guide processing in a manner favoring the maintenance or strengthening of the original schema. Research on factors such as the role of initial attitudes described in Section I has demonstrated that people are sometimes motivated and able to augment even specious arguments to arrive at a more cogent line of reasoning for their desired position.

V. Persuasion Variables Have Multiple and Interactive Effects

Another reason the processes underlying persuasion have appeared enigmatic is that some variables may increase argument processing at one level of the factor, but may actually bias or decrease argument processing at a different level of that factor. For instance, repeating a long or complicated persuasive message can provide individuals with additional opportunities to think about the message arguments and, therefore, enhance relatively objective argument scrutiny. Excessive exposures to a persuasive message can become tedious, however, and can actually motivate a person to reject the recommendation. Hence, the same stimulus factor—message repetition—has quite different effects on issue-relevant thinking as the amount of this factor increases.

Factors previously thought to have simple effects on information processing and persuasion have also been found to have quite different effects depending on the presence or absence of other factors. For instance, presenting a persuasive message on a non-involving issue in rhetorical rather than declarative form can increase an individual's propensity to think about the message arguments. When the recommendation is already personally involving, however, the insertion of rhetorical questions in the message arguments can actually interfere with the individual's ongoing idiosyncratic argument scrutiny.

In sum, the introduction of new factors (e.g., arguments presented in rhetorical rather than declarative form) can have striking but explicable effects on people's cognitive processes and attitudes. Current models of persuasion are now able to account for rather complicated patterns of data even though the intervening processes are fairly straightforward.

VI. Biological Aspects of Attitudes and Persuasion

Despite early conceptualizations of attitudes as postural orientations and neural predispositions to respond, programmatic research on the biological mechanisms underlying attitudes and persuasion is still fairly recent. This gap in theory and research on persuasion is due in part to the methodological approaches championed by the pioneers in the field, the interest by early theorists in applying persuasion research to address social problems (e.g., wartime propaganda), and the relative ease for governments and institutions to manipulate environmental rather than biological factors to achieve social control. Occasionally, study of the biological aspects of attitudes and persuasion has been dismissed because attitudes are said to be acquired through experience, as if environmental influences were antithetical to biological mechanisms. More recently, attention is being paid to the physiological manifestations by which the elementary psychological operations underlying persuasion can be indexed, and to the physiological mechanisms through which attitudes and persuasion are accomplished.

A. Genetic Factors

The achievement of strong behavioral proclivities in animals through selective breeding raises the possibility that some attitudes, and the manner in which individuals respond to information pertinent to attitude change, may be partially determined genetically. The existing data, though sparse, support both possibilities. Most studies bearing on the genetic contributions to attitudes and persuasion are based on the similarities observed between monozygotic twins reared apart from an early age. These studies have documented genetic contributions to individual differences in general intelligence, positive and negative affective dispositions, interests, general social attitudes (e.g., liberalism–conservativism), and job satisfaction. Although some of the individual differences shaped by genetic factors may influence attitudes directly, most of these dispositional factors (e.g., intelligence, affective disposition, interests, values) would likely influence attitudes and persuasion by increasing objective or biased message processing or by affecting what constitutes a compelling argument or peripheral cue to a particular individual.

The contribution of genetic factors tends to be

modest, so that environmental factors are also major determinants of attitudinal reactions. Perhaps a more important finding to emerge from research on behavioral genetics is that genetic and environmental factors are not as separable as once thought. Environmental factors can inhibit or trigger the expression of genetic influences, and genetic factors can lead individuals to seek and remain in certain environments. Monozygotic twins reared apart, for instance, have been found not only to express similar levels of satisfaction with their jobs, but to hold jobs that are similar in terms of complexity, motor skills, and physical demands. Possible mechanisms of heritability for this finding range from affective disposition to cognitive and physical capacity.

B. *In Utero* Factors

Based on the billions of dollars spent annually each year on advertising, it has been estimated that the average person in the United States has the potential to be exposed to over 1,400 persuasive appeals per day. Even if only a small fraction of these appeals are effective, this deluge of appeals suggests that an individual's attitudes are under nearly constant challenge. As noted above, attitudes based on little prior knowledge are particularly susceptible to change. One of the more surprising findings is that repeated, unreinforced exposures to a novel or unfamiliar stimulus result in a positive attitude toward the stimulus; i.e., repeated exposure to a novel stimulus that results in neither reward nor punishment breeds preference for this stimulus over a similar stimulus to which an individual has not been exposed. This mere exposure effect has been demonstrated using stimuli as diverse as nonsense words, ideographs, polygons, and faces, and the mere exposure effect is enhanced by factors such as a heterogeneous exposure sequence, a moderate number of presentations of the target stimulus (e.g., <100), brief exposure durations (e.g., <5 sec), and a delay between the stimulus presentations and attitude measurement.

Attitude change due to information emanating from the environment has also been documented *in utero* and appears to be a variation on the mere exposure effect. In an illustrative study, pregnant women recited a speech passage aloud each day during their last six weeks of pregnancy. Their newborns were tested within a few days following birth to determine whether the sounds of the recited passage were more reinforcing (i.e., preferred) than the

sounds of a novel passage. A matched group of newborns who had not been exposed previously to either passage were also tested to determine whether one passage was simply more likeable. Results revealed that the passage to which the fetuses had been exposed during the third trimester was preferred over the comparison passage, whereas the matched group of newborns exhibited no preference for one over the other passage. *In utero* recordings indicate that the auditory frequencies to which fetuses are exposed range between approximately 125 and 1,000 Hz. The fundamental frequency of the speech of women, but not men, tends to fall within this frequency range. As would be expected, therefore, this "mere exposure" effect is not found for fathers' voices, but instead a preference for paternal voices over less familiar male voices develops postnatally once paternal voices have become perceptually salient among other male voices.

C. Autonomic Factors

The activity of the autonomic branch of the peripheral nervous system has been of interest in studies of attitudes and persuasion because it was thought to reflect, if not represent, the emotional substrate of attitudes and attitude change. Reports of the autonomic differentiation of attitudes and emotions appear occasionally, but the autonomic responses found to differentiate hedonic states have tended to differ across studies, as factors such as the implications of the attitude for an individual's action varies. A more robust finding to emerge from psychophysiological research is that autonomic activity varies as a function of the intensity of the emotion (regardless of valence), cognitive effort, and behavioral effort aroused by an attitude object or persuasive appeal. In an illustrative study, students indicated their agreement or disagreement with 20 controversial statements. Several weeks later students were tested individually while a measure of autonomic activation—palmar skin resistance—was recorded. The 20 statements were again read to each student, but in addition students were told that a fictitious majority of their peers held a similar or dissimilar attitude. Students were then asked to restate their original attitude toward each statement, at which time skin resistance was measured. The major finding was that autonomic activity was higher the greater the discrepancy between the attitudes of the student and the fictitious majority except when students were absolutely confident of their original attitude.

General autonomic arousal, as might be achieved by exercise or by watching a sexually explicit film, has also been found to enhance an individual's normal affective reaction to an unrelated but evocative stimulus presented several minutes after the initial evocative stimulus. The specific mechanism responsible for this effect is not entirely clear, although the effect appears to be limited to instances in which (1) the second stimulus itself elicits a clear and dominant affective response and (2) the second stimulus is presented while individuals' autonomic activity is still elevated from the initial evocative stimulus but the individuals no longer feel aroused. More recent studies that have examined the effect of exercise-induced arousal on argument scrutiny and persuasion have found that peripheral cues (e.g., celebrity status) have a greater influence on attitudes at high than moderate levels of arousal, whereas the quality of the message arguments have a greater influence at moderate than high levels of arousal.

D. Somatic Factors

Attitudes were defined initially in terms of a postural orientation and the resulting disposition for an ensuing action. Subsequent theory and experimentation has focused on the latter dynamic component evoked by a stimulus. Somatic factors have been of interest in these inquiries for two reasons: (1) measures of skeletomuscular activity have provided information regarding mental and behavioral responses to attitude stimuli, and (2) skeletomuscular activity has been found to influence attitudes and persuasion.

As noted above, the initial studies of bodily responses and attitudes focused on autonomic activity. Over the past two decades, however, somatic responses to emotionally evocative stimuli have been found to be highly differentiated. Research on facial efference (measured using electromyography) and observable facial actions has revealed that (1) individuals perform at better-than-chance levels when categorizing facial expressions of happiness, sadness, fear, anger, disgust, surprise, and contempt; (2) the inductions of what individuals report as being positive and negative emotional states are associated with distinctive patterns of facial actions (emotional expressions); (3) distinctive expressions of emotion are displayed by neonates and the blind

as well as sighted adults; (4) cultural influences can, but do not necessarily, alter these expressions significantly; (5) the variability in emotional expressions that can be observed across individuals and cultures is attributable to factors such as differences in which emotion, sequence of emotion, or blend of emotions are evoked and to cultural prescriptions regarding the display of emotions; and (6) although many subtle emotional processes or events are not accompanied by visually perceptible expressive facial actions, the valence and intensity of these emotions are accompanied by distinctive patterns of facial efference which are measurable using surface electromyography. Heightened facial electromyographic activity can also be recorded over the perioral (speech muscle) regions during silent language processing, and this has provided a method of examining gross differences in message processing in persuasion.

Manipulations of somatic activity such as body posture and facial expression also appear capable of modulating affective reactions, biasing message processing, and influencing persuasion. Early research, for instance, suggested that people have difficulty feeling a particular emotion (e.g., joy) when they are posed in a contrasting stance (e.g., anger). In a more recent study, students were led to believe that they were participating in a study on the comfort and sound quality of stereo headphones when listeners were engaged in movement (e.g., dancing, jogging). Some students were told that they should move their heads up and down (vertical movements condition) about once per second to test the headphones, whereas others were told to move their heads from side to side. A final group of students heard no specific statements about head movements. Head movements were chosen because of their strong association with agreeing and disagreeing responses in a wide variety of cultures. Students heard musical selections and either an editorial in favor of raising tuition at their university or one in favor of reducing tuition. Following the broadcast, the students answered questions, including what they thought tuition should be. Results revealed that students who heard the editorial favoring an increase in tuition supported a higher level of tuition than did students who heard the editorial favoring a reduction in tuition. More interestingly, this effect was modulated by head movements. Vertical head movements (as if nodding in agreement) led to the most attitude change to both editorials, and horizontal head movements (as if shaking in disagreement) led to the least attitude change to the editorials.

Somatic manipulations have been less effective in inducing affective responses, but a growing number of studies suggests that somatic events can modulate affect and persuasion. This finding has stimulated research on the mechanisms underlying this effect and underlying attitudes and persuasion generally.

VII. Central Nervous System Substrates of Dispositions to Respond

From a biological perspective, attitudes and persuasion are ultimately products of the operation of the nervous system. In this regard, the features of the conceptual central and peripheral routes to persuasion evidence striking parallels with functional levels of organization in the nervous system. Although both the central and peripheral routes to persuasion can involve the highest levels of the central nervous system (CNS) (e.g., the cerebral cortex), it is probable that the elaboration likelihood continuum has its ultimate origin in fundamental ontogenetic and phylogenetic trends in CNS development. Historical and recent findings from the neurosciences support this view.

The highest levels of the CNS show the greatest expansion and elaboration through both the development of the individual (ontogeny) and that of the species (phylogeny), and serve to differentiate the adult human from the infant, and from other animals. In contrast, lower levels of the CNS (e.g., the spinal cord) evidence a more common, primordial organization throughout ontogeny and across phylogeny. Basic approach–withdrawal dispositions, however, are intrinsic to all levels of CNS organization, as documented by both experimental studies in animals and clinical findings in humans with spinal cord injuries. Cord transections isolate spinal networks from higher neural influences, leaving the lower regions of the body (e.g., the legs and trunk) under the exclusive control of spinal mechanisms. In spite of this loss of higher neural controls, the spinal cord is intrinsically capable of supporting reflexive limb-withdrawal to a noxious stimulus. This response is mediated, in part, by a relatively simple three-neuron circuit. The simplicity of this basic reflex circuit, however, does not imply an immutability of spinal networks. Just as attitudinal dispositions are subject to change, so are spinal

dispositions. For instance, spinal networks can learn to withdraw from innocuous stimuli which, although not painful themselves, come to predict the occurrence of pain stimuli through Pavlovian conditioning.

The isolated spinal cord can also evidence primitive approach responses, as indicated by basic genital reflexes to tactile stimulation (erection, pelvic thrusting, and ejaculation). Thus, basic approach–withdrawal behaviors can be seen not only in the simplest of organisms, but at the lowest level of organization in the mammalian NS. Moreover, these basic approach–withdrawal dispositions do not differ from attitudinal dispositions in modifiability—spinal reflexes are clearly subject to learned modification with experience. At the same time, no one would mistake the primitive responses of the isolated spinal cord for the richness of the reactions of an intact organism to aversive or sexual contexts. The fundamental distinctions between these classes of reaction are two: (1) response variability and persistence and (2) the level of contextual control.

The first striking difference between the spinal organism (i.e., an organism with a lesion separating the spinal cord from the brain) and an organism with an intact connection between the brain and spinal cord is the repertoire of responses to, for example, an aversive stimulus. Although a spinal organism may show limb-withdrawal, the intact organism also exhibits more global escape and avoidance responses, aggression, vocalization, or instrumental responses that serve to eliminate or diminish the aversive stimulus. A related difference is apparent in the persistence of behavioral responses. Although the spinal withdrawal is highly stimulus-bound, the intact organism may evidence behavioral activation, agitation, and escape attempts that persist well after the pain stimulus is withdrawn; i.e., the response of the intact organism is less dependent on the immediate sensory environment. Indeed, the stress that frequently characterizes human existence in contemporary societies is seldom directly related to pain stimuli.

This latter feature anticipates the second major distinction in the approach–withdrawal responses of spinal and intact organisms. That difference is in the complexity of contextual controls over behavior. The approach–withdrawal responses of the spinal organism appear to be sensitive only to relatively simple dimensions of a stimulus, such as its modality, intensity, or body location. In contrast, the aversive or sexual reactions of an intact organism frequently depend on highly complex relational features of the social–environmental context. In the intact organism, sexual arousal may not require direct tactile stimulation but may be manifest in the presence of a specific individual or by the thought of a specific individual, in an appropriate social context.

Studies of decerebrate organisms further illustrate the increased behavioral flexibility and the expansion of relational contextual controls over behavior that result from higher-order neural organizations. The decerebrate organism can display highly complex orofacial consummatory responses (orientation, chewing, swallowing) to palatable items placed within the mouth and vigorous rejection responses to nonpalatable items. These approach–withdrawal behaviors parallel the basic capacities of the spinal preparation, but evidence a degree of complexity and integration beyond that characteristic of spinal systems. Moreover, these complex reflexive responses are influenced not only by the immediate sensory features of the stimulus (palatability) but are also sensitive to internal motivational conditions (e.g., food deprivation, metabolic need). Thus, while mechanisms for reflex ingestive responses may be relatively hard-wired, their ultimate expression is further controlled by an additional class of internal stimuli.

Furthermore, although decerebrates are sensitive to metabolic needs and are capable of competent ingestive (i.e., approach–withdrawal) responses in the presence of a suitable goal object, they still fail to adequately regulate food intake or maintain body weight in typical environments. What appears to be lacking in these organisms is the ability to anticipate metabolic need, to evidence goal-searching behavior in the absence of an immediate stimulus for ingestion, or to respond to the normal contextual controls over food intake (e.g., social convention, passage of day, consideration of caloric need). These controls require a contextual representation which transcends simple dimensions of environmental stimuli (e.g., presence of food, level of food deprivation), and entails relational aspects among stimuli (e.g., passage of time together with the presence of food, or hunger together with the memory of the location of food). This latter class of contextual controls, entailing relational features among stimuli, or transcendent representations of the environment, liberates an organism from the immediate dictates of the sensory environment and confers

what has been interpreted as deliberative or goal-directed action.

The progression of increased behavioral flexibility and the expansion of relational contextual controls over behavior constitute hallmarks of higher-level neural organizations. These higher-level systems appears to be organized in a hierarchical fashion, and both extend the sensory processing of lower systems and expand on the motor repertoire and flexibility of lower mechanisms. Moreover, this hierarchical organization permits multiple levels of analysis and control over behavioral processes. In response to a pain stimulus, for example, lower-level processing may predominate initially, resulting in a rather stereotyped but highly adaptive, short-latency limb-withdrawal. The significant advantage of lower-level processing is that, although somewhat inflexible, it is highly efficient and places minimal burdens on higher-level processing substrates. Indeed, for the initial protective response, elaborate processing of the stimulus is not necessary, and in fact may be maladaptive. Lower-level processing, however, does not preclude further analysis at higher perceptual levels. In the case of a pain stimulus, this further analysis may be manifest in subsequent emotional reactions (e.g., fear, anxiety), which may motivate subsequent behavior (e.g., avoidance, aggression).

This pattern of multiple-level analysis and control confers significant advantages. By their nature, higher-level organizations must integrate information from varied modalities and sources and exert control over diverse aspects of behavior. This convergence of sensory information, the need for integration with prior memories, and the divergence of output control can create a processing bottleneck, which taxes the information-processing capacity of neural networks. Consequently, these higher-level systems may require active attentional focus and may have limited capacity for multiple concurrent activities. Those stimuli or conditions that do not effectively compete for attentional resources may be subject to only lower-level processing or to fairly elementary processing by higher-level neural networks. An important question, and an actively researched area, relates to the determinants of which stimuli are selected for further processing. This is also a fundamental question in the area of attitudes and persuasion, as it addresses the distinction between peripheral and central routes to persuasion and their underlying mechanisms.

Parallels can be noted in the distinctions between

central and peripheral routes to persuasion and the continuum between hierarchical and lower-level information processing. The peripheral processing route, lower-level neural processing, and elementary information processing within a higher level of neuraxis are characterized by (1) minimal cognitive elaboration, (2) limited flexibility, (3) stimulus-dependency or relative lack of persistence, and (4) reliance on rather simple, nonrelational features of the stimulus context. In contrast, the central route to persuasion as well as higher-level neural processing are featured by (1) elaborate, integrative analysis of multiple stimuli, (2) maximal flexibility, (3) persistence and resilience, and (4) reliance on complex relational features of the stimulus context (i.e., abstractions). In both neural processing and persuasion, analysis may frequently take place across multiple levels or within a level to varying degrees, or may shift between one level or another, depending on the context and competing demands.

Basic knowledge of neural organization, therefore, may offer insights into the processes of attitude formation and persuasion. The processes underlying the neural control of approach–withdrawal reactions, emotions, and knowledge are yielding to current investigations in the social sciences and neurosciences. As a result, the neural systems and neurochemical mechanisms underlying the actions of antianxiety, antidepressant, and antipsychotic agents are beginning to be understood. With rapidly developing neuroimaging techniques, views are available of aspects of brain function in conscious individuals during a variety of cognitive activities. These techniques, together with more basic studies of cellular processes, information transformations among neurons, and behavioral analyses of attitudes and persuasion, should provide the tools for a truly interdisciplinary study of social psychological phenomena such as attitudes and persuasion.

Bibliography

Arvey, R. D., Bouchard, T. J., Jr., Segal, N. L., and Abraham, L. M. (1989). Job satisfaction: Environmental and genetic components. *J. Appl. Psychol.* **74,** 187–192.

Bornstein, R. F. (1989). Exposure and affect: Overview and meta-analysis of research, 1968–1987, *Psychol. Bull.* **106,** 265–289.

DeCasper, A. J., and Spence, M. J. (1986). Prenatal maternal speech influences newborns' perception of speech sounds. *Infant Behav. Develop.* **9,**133–150.

McGuire, W. (1985). Attitudes and attitude change. *In* "Handbook of Social Psychology" (G. Lindzey and E. Aronson, eds.). Random House, New York.

Petty, R. E., and Cacioppo, J. T. (1986). "Communication and Persuasion: Central and Peripheral Routes to Attitude Change." Springer-Verlag, New York.

Pratkanis, A. R., Breckler, S. J., and Greenwald, A. G. (eds.) (1989). "Attitude Structure and Function." Erlbaum, Hillsdale, New Jersey.

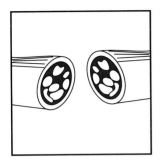

Phagocytes

SUZANNE GARTNER, *Henry M. Jackson Foundation Research Laboratory*

Glossary

Complement A group of serum proteins activated in a cascading fashion following interaction with antibody–antigen complexes or certain bacterial components

Cytokine A molecule secreted by a cell that can mediate and regulate the functions of that cell or those of other cells

Fc fragment A portion of an immunoglobulin molecule generated by papain digestion. It is composed of two heavy chains and has binding sites for activation of complement, but no binding sites for antigen

Granuloma A small nodular lesion associated with chronic inflammation and characterized by the presence of numerous macrophages, which can fuse to form multinucleated giant cells

Major histocompatibility complex (MHC) Four closely linked loci on chromosome 6 (in humans) that code for strong transplantation antigens. These antigens are present on the surface of all cells except mature erythrocytes. The MHC is also called the human leukocyte antigen (HLA) complex

Opsonization An enhancement of phagocytosis mediated by the coating of the antigen with specific antibody

Phagocytosis The process whereby phagocytes ingest and degrade material

Phagolysosome An intracellular vesicle in phagocytes created by the fusion of a primary lysosome with a membrane-bound vacuole containing engulfed material (a phagosome); phagolysosomes are also called secondary lysosomes

Receptor A protein structure, located on the surface of a cell, that serves as a specific recognition and binding site for other molecules

HUMAN PHAGOCYTES are bone-marrow–derived cells of two categories, polymorphonuclear leukocytes and mononuclear phagocytes, whose specialized function is the engulfment and digestion of particulate matter. Together these groups of cells serve as a first line of defense against invading microbes. Mononuclear phagocytes also perform a number of other diverse functions. As scavengers, they continuously remove and degrade debris and aging cells from the tissues. As participants in the immune response, they process and present antigen, act as cytotoxic effector cells and produce and secrete a number of immunoregulatory molecules. Phagocytes also play a role in certain disease conditions, most notably chronic diseases and diseases in which macrophages serve as host cells for the pathogen.

I. Definition, Description and Origin

Phagocytes are defined functionally by their ability to engulf matter such as particulate material or other cells from their environment. This ability is characteristic of many species within the plant and animal kingdoms and is a major means by which unicellular organisms obtain nutrition. It has been suggested that this ability also played a role in the evolutionary process from unicellular to multicellular organisms. For example, it is conceivable that mitochondria, the energy-generating organelles present in higher cells, arose from ingested bacteria. In humans, essentially only two kinds of cells, polymorphonuclear leukocytes and mononuclear phagocytes, have the capacity to phagocytize. Both kinds of cells originate from a common precursor in the bone marrow and are present in blood and tissues, and both play key roles in defending the body from invading potential pathogens.

A. Polymorphonuclear Leukocytes

Polymorphonuclear leukocytes (commonly called *polys*), as the name suggests, are characterized by the irregularity of their nuclear structure. This category includes two kinds of phagocytic cells, the neutrophil and the eosinophil, along with the non-phagocytic basophil.

1. The Eosinophil

This cell is named for its most striking feature, eosinophilia, the bright red color its numerous large granules take on as a consequence of being stained with the acid eosin. Ultrastructurally these membrane-bound granules are ovoid, and each contains a single, irregular, electron-dense crystalloid core. The core is composed of a poorly characterized insoluble basic protein, while the matrix of the granule contains a variety of hydrolytic enzymes including histaminase, arylsulfatase-B, and phospholipase-D, as well as a peroxidase distinct from that present in neutrophils. They account for 1 to 6% of the white cells in circulating blood, but more than 99% of their total numbers are resident in the bone marrow and other tissues. Predominant sites include the skin, lungs, gastrointestinal tract, and vagina—tissues frequently exposed to the external environment. The halflife in blood of an eosinophil has been estimated to be approximately 5 hours. Once in the tissues, the eosinophil does not reenter the circulation except under certain pathological conditions. Like neutrophils, they are mobile and respond to chemotactic factors. Eosinophils are highly phagocytic for antigen–antibody complexes but generally less phagocytic than neutrophils. They can ingest bacteria and fungi. However, in spite of the fact that they produce more peroxide than neutrophils, they are less effective in killing bacteria, presumably because they lack lysozyme and phagocytin, and also because they cannot utilize the peroxidase-H_2O_2-Cl pathway.

Two primary functions have been ascribed thus far to the eosinophil, a role in the termination of allergic reactions and a role in host defense against parasites. Type I (anaphylactic) hypersensitivity is a systemic or local, rapidly developing immunologic reaction resulting from the release of potent vasoactive amines from basophils and mast cells. This release is triggered by the binding of a previously seen antigen to specific IgE molecules attached to the basophils and mast cells. Also released is eosinophil chemotactic factor of anaphylaxis (ECF-A), which attracts eosinophils to the area. The eosinophils undergo degranulation, which results in the release of the granule enzymes, enzymes that degrade those released by the basophils and mast cells. The final result is the termination of the anaphylactic reaction. In parasite-associated granulomas, as many as 50% of the cells present are eosinophils. Antibody-dependent, eosinophil-mediated destruction of the parasite is thought to be accomplished as follows: Antiparasite IgG and IgE antibodies bind to specific receptors for the Fc part of antibodies present on the eosinophil, which leads to degranulation of the cell. The released basic proteins and peroxidase then act to kill the target parasite. The activation of the eosinophil by cytokines has also been suggested to have a role in this process.

2. The Neutrophil

Neutrophils (so called because they do not stain with acidic dyes like eosinophils, nor basic dyes like basophils) are the most abundant of the leukocytes in blood, where they usually constitute 55–65% of the total population. The majority of neutrophils, however, reside in the bone marrow where they serve as a rapidly mobilizable defense reserve. Their total lifespan is thought to be about 8 days, approximately 10 hours in the blood, 1 to 2 days in the tissues, and the remainder in the marrow. The mature neutrophil possesses a highly lobulated nucleus, usually consisting of five lobes joined together by thin strands of nuclear material. A more immature form of the neutrophil with a bilobular nucleus can also be seen in blood. An increase in the numbers of these so-called bands in blood is indicative of their recruitment from the marrow, in response to injury or infection.

The cytoplasm of neutrophils contains azurophilic granules, designated primary granules, as well as numerous smaller granules, designated specific granules. The azurophilic granules are large, spherical, and electron-dense and are actually lysosomes. They contain a number of enzymes such as acid hydrolases and neutral proteases (collagenase, elastase, cathepsin) as well as potent antibacterial enzymes such as lysozyme, myeloperoxidase, and D-amino-oxidase. The specific granules are small, more numerous, and irregular in shape and density. They contain the bactericidal enzymes lysozyme and lactoferrin, as well as alkaline phosphatase, whose function is not really known.

Neutrophils play a primary role in the acute inflammatory response. At the site of injury or infection, they phagocytize and destroy invading microorganisms and cell debris. Neutrophils also destroy microbes and degrade debris by releasing their enzymes directly into their environment. They have little capacity for protein synthesis, and hence are incapable of any prolonged, continuous function. Neutrophils, along with mononuclear phagocytes and K cells, can also participate in antibody-dependent, cell-mediated cytotoxicity (ADCC). This is a process that has been observed *in vitro* whereby an IgG-coated target cell is lysed on contact with the neutrophil. It involves the binding of the Fc receptor on the neutrophil to the Fc portion of the exposed IgG molecule. Low concentrations of antibody are sufficient, phagocytosis is not involved, and the phenomenon is complement independent. [*See* NEUTROPHILS.]

B. Mononuclear Phagocytes

The mononuclear phagocyte system, previously termed the reticuloendothelial system (RES), includes the monocytoid components of the bone marrow, the monocytes of the peripheral blood, and the tissue macrophages. These cells are specialized in phagocytosis and intracellular digestion. Unlike neutrophils, mononuclear phagocytes are general scavengers, in that they continually engulf and degrade aging and decaying cells from the tissues. They also perform a number of other functions. In particular, macrophages function as mediators of the immune response. Many tissue macrophages have specialized names. For example, Kupffer cells are macrophages in the liver, and microglial cells are presumed to be a kind of brain macrophage. Tissue macrophages are thought to develop from blood monocytes following their migration from blood into the tissues. [*See* MACROPHAGES.]

In conventional fixed and stained preparations, monocytes are comparatively large, and possess a large, eccentrically located nucleus, which is often indented or horseshoe-shaped. They have small azurophilic granules scattered throughout their cytoplasm. These granules are primary lysosomes containing a number of enzymes functional at acid pH including peroxidase, acid phosphatase and aryl-sulfatase, and acid nucleases. (Secondary lysosomes present in macrophages are vacuolar structures that are the sites of current or past digestion.) Monocytes compose 2–10% of the blood leukocytes. Their stores in the bone marrow are small, and their transit time in the blood is estimated to be about 32 hours. As tissue macrophages, they appear to live at least several weeks, and perhaps several months. Unlike neutrophils, monocytes are capable of continuous lysosomal activity and regeneration, and they can utilize the aerobic, in addition to the anaerobic, metabolic pathway.

The microscopic appearance of the macrophage varies considerably depending on the activation state of the cell and its tissue microenvironment. Resting cells appear rounder, and small, active ones are larger and possess several cytoplasmic projections. Like monocytes, quiescent macrophages have many primary lysosomes. During phagocytosis, lysosomes fuse with phagosomes (vacuoles containing the ingested material) creating phagolysosomes. As a consequence, the number of lysosomes decreases in the actively phagocytic macrophage. However, because lysosomes can be regenerated in these cells, their declining numbers can be later increased. Phagolysosomes are the sites of enzymatic digestion of the phagocytized material. Residual matter may remain in the cell indefinitely or be released into the surrounding tissue by exocytosis. Mature macrophages often contain a number of residual bodies—storehouses of incompletely digested material. These features are responsible for the name macrophage which is derived from the Latin meaning large eater, and contribute to the scavenger appearance of the macrophage, irregular with respect to shape and inclusions. The tremendous degradative ability of the macrophage is essential for the disposition of metabolic breakdown products, a prime example being the ingestion and breakdown of aged red blood cells by spleen macrophages.

II. Phagocytosis

Phagocytosis, also called endocytosis, is defined as the process of internalization and degradation of extracellular material. Essentially three distinct steps—recognition, engulfment, and degradation—characterize the phenomenon. Recognition is mediated through specific receptors located on the surface of the phagocyte. Some receptors can bind the foreign material directly. For example, human phagocytes have receptors (mannose–fucose receptors) that directly bind to sugars present on the walls of yeast cells. More frequently, however, phagocytes use receptor for the Fc part of antibodies and for proteins of the complement system, which are present in blood. In these cases the foreign material (antigen) is first bound by immunoglobulin (IgG) or immunoglobulin and the C3 component of complement. The antigen is then indirectly bound to the phagocyte either by the binding of the Fc portion of the antigen-complexed immunoglobulin molecule to the Fc receptor on the phagocyte, or by the binding of complement-coated, antigen–antibody complexes to the phagocyte via the complement receptor. The distribution of the receptor–ligand complexes on the surface of the phagocyte plays a role in the efficiency of phagocytosis. Clustered antibody or complement molecules are much more effective. Thus erythrocytes coated with clusters rather than monomers of IgG are phagocytized more avidly.

It should be mentioned that most cells (one exception being the mature erythrocyte) are capable of ingesting material from their environment. This is a receptor-mediated process called pinocytosis, and it differs from phagocytosis in three principal ways: (1) pinocytosis, but not phagocytosis, can occur at temperatures below 18°C; (2) phagocytosis, but not pinocytosis, requires the participation of actin filaments within the cell; and (3) particles taken up via pinocytosis are generally less than 1 micron in diameter, whereas phagocytized particles can be considerably larger (e.g., other cells). The target material for pinocytosis is soluble molecules, whereas the target material for phagocytosis is usually particulate matter. Also, pinocytosed material is taken up in special clathrin-coated vesicles.

Certain features of the receptor–ligand interaction merit further attention. The binding of the targeted material to the receptor of the phagocyte initiates a series of events. However, binding alone is not sufficient, at least in some cases. For example, quiescent macrophages can bind but not phagocytize complement-coated particles. It has been shown, however, that the addition of T-cell lymphokines or other activators of macrophages (e.g., phorbol myristate acetate) can drive the process of phagocytosis to completion. Another important feature is that the phagocytosis receptors, the Fc, complement, and oligosaccharide receptors, all function independent of each other, although Fc and complement receptors can work cooperatively in the phagocytosis of a particle. Furthermore, the binding of a ligand to one kind of receptor does not appear to signal the other kinds of receptors.

Following binding of the target particle to the phagocyte, the membrane of the cell begins to advance, ultimately enclosing the particle within a membrane-bound vesicle. This process is accomplished via the polymerization and depolymerization of actin filaments located in the cytoplasm beneath the membrane. The movement of the cell membrane leads to additional ligand–receptor interactions which, in turn, lead to further advancement of the membrane. This process has been likened to the closing of a zipper. A single membrane-to-membrane fusion event closes the phagosome. The phagosome may then move deeper into the cytoplasm of the cell. Engulfment can be a relatively rapid process. A neutrophil can ingest a chain of 10 to 15 bacteria within a few minutes, and a macrophage can ingest even more. Both neutrophils and macrophages contain cytoplasmic stores of membrane components; it is presumed that they are used to replace membrane consumed during phagocytosis. Indeed, recycling of membrane components has been documented for phagocytosing macrophages. Obviously phagocytosis is an energy-requiring process. Macrophages and neutrophils have considerable stores of glycogen and obtain most of their energy from glycolysis. Macrophages also have large stores of creatine phosphate, which they tap during phagocytosis and other times of high energy demand.

The initial event of degradation is the acidification of the phagosome, which results in the dissociation of the bound material from its receptor. Acidification is a consequence of the pumping of protons into the phagosome and is triggered when the particle becomes bound to its membrane receptor. The dissociated receptor is often recycled back to the plasma membrane. The acidified phagosome then fuses with a lysosome whose potent degradative enzymes function optimally at pH 4.0. Enzymatic digestion of the engulfed material follows. Two categories of bactericidal mechanisms have been identified, one oxygen dependent and the other oxygen independent. The term respiratory burst refers to the increase in oxygen uptake by macrophages and neutrophils during phagocytosis. This process is not actually activated by the ingestion of material, nor is it a necessary part of phagocytosis. However, much of the oxygen is reduced to superoxide anion, which presumably is toxic for the ingested microorganisms. The other major oxygen-dependent pathway is the H_2O_2-myeloperoxidase-halide system. Activation of an oxidase for reduced pyridine nucleotide during phagocytosis leads to the liberation of H_2O_2 within the phagolysosome. Although the details of the process are not known, it is clear that the bactericidal effect of H_2O_2 is tremendously enhanced in the presence of the enzyme myeloperoxidase plus a halide ion, chloride ion being the most physiologically relevant. This mechanism appears to be effective against viruses and fungi as well.

Two oxygen-independent bactericidal mechanisms have been proposed: (1) the inhibition of bacterial growth by acid pH, and (2) the direct digestion of cell wall and membranes of the bacteria via specific enzymes. The pH of the phagolysosome continues to remain low (at approximately 4.0) as a consequence of the production of hydrogen ion, which is derived from the action of carbonic anhydrase or the production of lactic acid. Few bacteria can proliferate at this pH; many can be directly killed by the lactic acid. In addition, the enzyme lysozyme can digest, to at least some degree, the bacterial cell walls of all bacteria but especially those of gram-positive cocci, by hydrolyzing the muramic acid-N-acetyl glucosamine bonds. In the neutrophil, this digestion can be followed by the action of the arginine-rich cationic protein, phagocytin, which serves to lyse bacterial membranes. Under certain *in vitro* circumstances, neutrophils can be induced to release their lysosomal enzymes into their environment. The *in vivo* relevance of this phenomenon is unclear, but possibly it plays a role in certain pathological situations such as immune complex–associated glomerulonephritis. Not surprising is the fact that some pathogens have developed ways to circumvent their destruction by modifying their host and adapting themselves so that they can survive and even proliferate within macrophages. Examples include the tuberculosis bacillus and certain retroviruses. Such pathogens employ a number of approaches to accomplish this, and clearly this phenomenon represents a major mechanism of disease generation.

A final point regarding phagocytosis is the movement of phagocytes to a site of inflammation, which is mediated by chemotactic factors, a process called chemotaxis. Neutrophils, eosinophils, and macrophages are all susceptible to chemotaxis, although they do not all respond to the same attractants. The response is a sensitive one, in that these cells can recognize a gradient of chemoattractant as small as 1% across the dimension of the cell. The two most important chemoattractants for neutrophils are bacterial products and the components of the complement system. Eosinophils also respond to complement (C5 fragments), and at sites of anaphylactic allergy, they respond to ECF-A released from IgE-sensitized basophils or mast cells when these cells contact the specific antigen. Chemotactic factors for monocytes include the C3 and C5 fragments of complement, various bacterial factors, denatured proteins arising from dead cells, and uncharacterized factors from neutrophils and lymphocytes. Genetic or acquired deficiencies in chemotaxis are associated with an increase in susceptibility to microbial infection.

III. Macrophage Surface Receptors

Cells of the mononuclear phagocyte lineage display a number of important receptor proteins on their surfaces. Those involved in phagocytosis, the mannose–fucose receptor and the Fc and complement receptors, were mentioned earlier, but the latter two require further discussion. In addition, mononuclear phagocytes possess receptors for certain growth factors and cytokines, most importantly receptors for (1) interferons alpha, beta, and gamma; (2) interleukin-1; (3) tumor necrosis factor; and (4) colony-stimulating factor. Macrophages also produce and secrete many of these factors.

Human macrophages have three distinct Fc receptors. FcRI can bind both monomeric and polymeric IgG with high affinity, and its expression can be increased by interferon gamma. FcRI manifests a IgG subclass preference IgG1 > IgG3 > IgG4 > IgG2. The two other Fc receptors, FcRII and FcRIII, bind IgG with a lower affinity. The FcRII receptors are also found on other types of cells, including B lymphocytes. Several receptors for the different complement fragments have been detected on the surface of human macrophages. The best characterized are the CRI and CR3 receptors, but also present are the CR4, C5a, C1q and Factor H receptors. CR1, previously named the immune adherence receptor, is present on a number of different kinds of cells (neutrophils also), and is the receptor for the C3b component of complement. CR3 is present on mononuclear phagocytes and granulocytes (some lymphocytes also), and binds to the iC3b component of complement. The CR1 and CR3 receptors participate not only in phagocytosis but also play a role in antibody-dependent cellular cytotoxicity (ADCC). The C5a receptor on the macrophage membrane plays a key role in arachidonic acid metabolism by mediating the release of arachidonate derivatives. The lipid arachidonic acid is a major constituent of the macrophage membrane, which is released into the cytoplasm of the cell following appropriate signaling. The arachidonic acid is then metabolized to form a family of compounds called eicosanoids. These compounds are potent, short-lived, short-range hormones. Representatives are the prostaglandins, leukotrienes and thromboxanes. [*See* LYMPHOCYTES.]

By a process of differentiation and maturation called hematopoiesis, cells of the blood derive from a single precursor—the pluripotent stem cell—that resides in the bone marrow. This pluripotent cell gives rise to a number of other progenitor cells, each of which becomes committed to only one of the specific differentiation pathways. Assays have been developed that permit the clonal growth of the various progenitor cells in soft agar. Specific factors, referred to as colony stimulation factors (CSFs), are required to drive the clonal proliferation of the cells and also appear to regulate hematopoiesis *in vivo*. Hematopoietic cells of varying degrees of differentiation display receptors for these factors, which are produced by several kinds of cells located in various sites throughout the body.

Macrophage CSF (M-CSF, CSF-1) is produced by macrophages and fibroblasts. Its receptor is present only on macrophages. Granulocyte–macrophage CSF (GM-CSF) is produced by macrophages, bone marrow stromal cells, and activated lymphocytes; its receptor is present on hematopoietic stem cells and myeloid cells. Granulocyte CSF (G-CSF) is produced only by macrophages and stromal cells. Its receptor is present on both granulocytes and a progenitor cell which can give rise to either macrophages or granulocytes. Finally, interleukin-3 (multi-CSF) is produced by activated lymphocytes, and its receptor is present on a progenitor cell capable of giving rise to granulocytes, mast cells, and lymphoid cells. In all cases, the binding of these regulatory molecules to their appropriate target cells induces cell maturation. Of particular interest is that the CSF-1 receptor, a tyrosine kinase, is identical to the product of the proto-oncogene c-*fms*. This receptor is present on monocytes and macrophages at all stages of differentiation, with the most mature cells bearing the greatest number of receptors.

The interleukin-1 (IL-1) receptor is an 80 kilodalton glycoprotein with a high affinity for its ligand. Both IL-1 alpha and IL-1 beta bind with high affinity and compete with each other for binding sites. The tumor necrosis factor (TNF) receptor is a protein with molecular weight in the range of 60 to 80 kDa. Like IL-1, it exhibits a high affinity for its ligand. Similarly, the receptors for interferon alpha and beta are high-affinity protein receptors with molecular weights in the 100 to 120 kDa range. Although many of the postbinding events regarding these regulatory molecules have not been delineated, it is thought that the receptor-factor complexes are endocytosed.

IV. Macrophage-Derived Cytokines

As alluded to earlier, macrophages produce a number of soluble factors with a variety of functions. Those to be noted here include interleukin-1, interleukin-6, interferons alpha and beta, and tumor necrosis factor. These mediators act locally, have multiple activities, and are synthesized *de novo* following stimulation of the cell. Although macrophages are a major source of these factors, other kinds of cells can also produce them. Certain features of these cytokines are shown in Table I. Two additional cytokines produced by macrophages have been identified. One, monocyte-derived neutrophil chemotactic factor (MDNCF), is produced in response to endotoxin, tumor necrosis factor, and interleukin-1, and promotes degranulation and activation of neutrophils as well as chemotaxis. The other, designated IP10, is 42% homologous with MDNCF and is produced by macrophages following interferon gamma stimulation. Its functions await elucidation. [*See* CYTOKINES IN THE IMMUNE RESPONSE.]

Macrophages also produce and secrete a considerable number of other other molecules. An important example is apolipoprotein E, which participates in the transport of lipids.

TABLE I Macrophage-derived Cytokines

Characteristic	Interleukin 1 (IL-1α, IL-1β)	Interleukin 6 (IL-6)	Tumor necrosis factor (TNF)	Interferon (IFNα, IFNβ)
Gene:	Chromosome 2	Chromosome 7	Chromosome 6	Chromosome 9
Protein: (mature form)	159aa[a] 153aa	190aa (glycosylated)	157aa	IFNα I 165–166aa II 172aa (nonglycosyl.) IFNβ 166aa (glycosylated)
Receptor:	High-affinity	High-affinity	High-affinity	High-affinity
Effector function:	Many kinds of cells respond by means of several different mechanisms.			Inhibition of cell & viral growth; enhances accessory cell activity of macrophages
	↑ T- & B-cell growth; regulates production of cytokines by other cells	↑ Ig secretion from B lymphoblasts; ↑ growth of thymocytes and some T-cell clones	Lyses some tumor cells; ↑ IL-2R on T cells; activates macrophages; activates neutrophils	
In vivo effects:	Primary mediator of systemic inflammatory reaction (e.g., fever, neutrophilia)	Participates in acute phase inflammatory reaction	Cachexia, regression of some tumors; mediator of acute-phase inflammatory reaction	Modulation of antibody production; graft rejection; antiviral effect

[a] A.a. = amino acids

V. Macrophage Activation and Cytotoxicity

A. Activation

Activated macrophages are characterized by morphologic alterations as well as by changes in the secretory status of certain products. Activated macrophages become larger and their cytoplasm more ruffled, and their overall protein content increases, along with increases in the production of the lysosomal enzymes. Their surface receptor repertoire changes (e.g., the number of Fc receptors increases), and as a consequence of continuous *endocytic sampling* of their surroundings, they turn over the equivalent of their entire membrane approximately every 30 minutes. They also have an enhanced killing capacity for both endocytosed bacteria and tumor cells. The activation state can be induced by either an immunologic or a nonimmunologic interaction. In the former case, the macrophage is activated as a consequence of its interaction with a sensitized T lymphocyte, while in the later case, the activator is either a bacterial component or another soluble mediator. Unfortunately, controversy exists with regard to lymphocyte-derived macrophage activating factors. Many immunologists believe that the previously described macrophage-activation factor (MAF) is indistinguishable from gamma interferon. However, most

would agree that both granulocyte-macrophage colony stimulating factor and interleukin-4 (previously called B cell stimulatory factor 1) are potent inducers of macrophage cytotoxicity.

Clearly T lymphocytes play a primary role in the activation of macrophages at sites of inflammation. The microbes or tumor cells provide antigens, which stimulate the proliferation of lymphokine-secreting, sensitized T cells. Interactions with these T cells activate the macrophages. However, the capacity of macrophages to be activated by the lymphokine-producing T cells is limited. Likewise, the cytotoxic activity of these macrophages is short-lived; it peaks within 8 to 12 hours following lymphokine stimulation and is totally gone within 24 hours. These processes are controlled by means of a negative feedback mechanism, presumably because strict regulation serves to depress the activation response as soon as it is no longer required.

B. Cytotoxic Effector Function

Many of the known mechanisms whereby macrophages kill microbes were mentioned earlier. With regard to tumor cell killing, tumor necrosis factor can be directly toxic to tumor cells. Also, it is thought that cytolytic proteases can act synergistically with reactive oxygen intermediates to kill tumor cells. However, some tumor-targeted killing is oxygen independent, as immunologically activated macrophages have been shown to kill tumor cells under anaerobic conditions. The killing of tumor cells by activated macrophages appears to require intimate contact between the two cells, because soluble toxic factors cannot be demonstrated. Although activated but not nonactivated or inflammatory macrophages can bind selectively to a number of different tumor cells, no receptor associated with this binding has been identified.

C. The Delayed-Type Hypersensitivity Response

The delayed-type hypersensitivity response (also known as Type IV hypersensitivity) is initiated by antigen-sensitized T lymphocytes at the site of antigen deposition. A classic example is the tuberculin skin reaction, which is characterized by reddening and a hardening of the tissue at the site of injection. The response peaks about 24 to 72 hours following injection and then slowly resolves. An important feature of the reaction, which accounts for the hardening of the tissue, is the deposition of fibrin at the site. Although sensitized T cells are the initiators of the response, their numbers account for only approximately 10% of the cells in the lesion. The predominant cell in the reaction is the monocyte–macrophage. The hypersensitivity response is amplified by the sensitized T-cell–driven recruitment of monocyte–macrophages and additional unsensitized T lymphocytes, the recruitment being lymphokine mediated. The ultimate goal is the rapid accumulation of a population of nonspecific, cytotoxic macrophages. These effector macrophages can destroy a wide variety of targets including bacteria, viruses, fungi, and some parasites and tumor cells. The response is governed by a number of complex interactions between the T cell and the macrophage, and between their secreted lymphokines.

VI. Antigen Presentation

A primary event in the initiation of an immune response is the proper presentation of the antigen to lymphocytes. The cells capable of this function are called antigen presenting cells (APC). They are the mononuclear phagocytes, Langerhans' cells of the skin, follicular dendritic cells of lymph nodes, and B lymphocytes. Proper recognition requires that the antigen be codisplayed with molecules of the Class II major histocompatability complex (MHC class II). The CD4+ helper T lymphocytes recognize antigen in the context of MHC class II, while the CD8+ cytotoxic T cells recognize antigen in the context of MHC class I molecules. Obviously, appropriate expression of MHC class II molecules on macrophages is crucial for optimal antigen presentation. The expression of the MHC molecules is both positively and negatively regulated in many different ways. For example, interferon gamma produced by activated T cells or by tumor necrosis factor–activated natural killer cells can act to increase MHC class II expression on macrophages. In contrast, prostaglandins can decrease the expression. [*See* CD8 AND CD4: STRUCTURE, FUNCTION, AND MOLECULAR BIOLOGY.]

It is clear that macrophages in some way process antigen, but the precise steps involved have not been elucidated. Unlike the B-cell response, the T-cell proliferative response to antigen presented by APC can be induced by denatured or fragmented proteins. Macrophage processing of antigen involves internalization of the material followed by intravesicular chemical processing at low pH. (These vesicles are endocytic vacuoles that have been acidified but have not fused with primary lysosomes, so they lack hydrolases.) Approximately 30 to 60 minutes after its native form is internalized, fragments of the processed antigen appear as *immunogenic determinant* on the surface of the macrophage. How the determinant reaches the cell surface is unclear, but evidence from the murine system, indicating that the mouse equivalent of MHC class II proteins can bind peptides, has led to the hypothesis that these proteins act as transporters of the immunogenic peptides.

VII. Phagocytes in Inflammation

Acute inflammation is a short-lived response to tissue injury and is characterized by the emigration of leukocytes, particularly neutrophils, to the site of the injury, and by the exudation of fluid and plasma proteins. Chronic inflammation is a more diverse process and usually of longer duration. Chronic inflammatory lesions are characterized by the presence of lymphocytes and macrophages and by the proliferation of fibroblasts and small blood vessels (angiogenesis). These are dynamic processes, and phagocytes are key participants. [*See* INFLAMMATION.]

Neutrophils and monocyte–macrophages are primarily responsible for the destruction of invading microbes present in acute inflammatory lesions. A few additional points concerning their emigration should be mentioned. A phenomenon called margination, in which leukocytes moving within the bloodstream orient themselves along the periphery of the flow, has been observed. This occurs more frequently when the flow slows or becomes stagnant, as at sites of inflammation. Ultimately, many leukocytes stick to the endothelial cell surface. Although the mechanisms responsible for the sticking have not been thoroughly delineated, chemotactic factors are thought to play a role. Following adhesion, the leukocytes migrate along the endothelium until they locate a hole, usually the junction between neighboring endothelial cells. The leukocyte then moves itself through the hole, traverses the basement membrane, and ultimately journeys into the tissue.

The predominant cell type within an acute inflammatory lesion is a function of the nature of the initiating stimulus and the age of the lesion. Within the first 6 to 24 hours, neutrophils are more frequently represented. After that time, monocytes begin to predominate. The production of a monocyte chemoattractant by neutrophils and the longer life-span of the monocyte–macrophage are thought to account for this situation. Some examples of the nature of the stimulus are the following: (1) acute *Pseudomonas* infections are characterized by a preponderance of neutrophils for up to 48 hours; (2) eosinophils predominate in certain hypersensitivity reactions; and (3) in some viral infections, lymphocytes are the first emigrating cells.

The accumulation and persistance of macrophages in chronic inflammatory lesions occur as a result of the continued recruitment of monocytes from the blood, the local proliferation of macrophages, and the prolonged survival and immobilization of macrophages. These cells continue to attempt to destroy and/or digest the targeted material. If the material is too large to be engulfed or resists degradation, the macrophages may form a granuloma. Granulomas are collections of inflammatory cells, mostly macrophages, rimmed by lymphocytes. They are usually 1–2 mm in diameter. Ultimately the macrophages can fuse to one another to form a multinucleated foreign body giant cell. (Foreign body giant cells in which the nuclei are located on the periphery of the cell are called Langhans' giant cells.) Classic examples of diseases associated with chronic inflammation include tuberculosis and leprosy (both caused by mycobacteria), schistosomiasis, and cat-scratch fever. Also, chronic inflammation resulting from the presence of large, toxic, indigestible material such as asbestos fibers is involved in the development of certain lung diseases. Finally, macrophages may play another role in chronic inflammation. Macrophage-derived factors have been implicated in the angiogenesis and fibroblast proliferation characteristic of chronic lesions.

Bibliography

Dean, R. T., and Jessup, W., eds. (1986). "Mononuclear Phagocytes: Physiology and Pathology." Elsevier, Amsterdam, The Netherlands. (available from Elsevier, New York)

Gallin, J. I., Goldstein, I. M., and Snyderman, R., eds. (1988). "Inflammation: Basic Principles and Clinical Correlates." Raven Press, New York, N.Y.

Paul, William, ed. (1989). "Fundamental Immunology," 2nd edition. Raven Press, New York, N.Y.

Powandi, M., Oppenheim, J. J., Kluger, M. J., and Dinarello, C. A., eds. (1988). "Monokines and Other Nonlymphocytic Cytokines." Alan R. Liss, New York, N.Y.

van Furth, R., ed. (1985). "Mononuclear Phagocytes—Characteristics, Physiology and Function." Nijhoff, Amsterdam, The Netherlands. (available from Kluwer Academic, Norwell, Massachusetts)

pH and Carbohydrate Metabolism

ROBERT D. COHEN, *University of London*

I. Basic Mechanisms
II. Physiological Interactions of Carbohydrate and Acid–Base Metabolism
III. Concluding Remarks

Glossary

Acidosis Lowering of pH in a body fluid compartment (e.g., extracellular, intracellular); acidosis may be "respiratory," because of CO_2 retention, or "metabolic," because of other causes

Gluconeogenesis Synthesis of new glucose from precursors (e.g., amino acids) and intermediates of carbohydrate catabolism (e.g., lactate, pyruvate, 2-oxoglutarate)

Glycolysis Conversion of monosaccharides (e.g., glucose and fructose to pyruvate and lactate)

pH Negative logarithm (base 10) of the hydrogen ion (proton) activity. Thus if hydrogen activity is 10^{-7} molar, pH is 7, which is neutral. Lower pH is acid, higher is basic

ANIMALS POSSESS powerful mechanisms for maintaining the pH of extra- and intracellular fluids within relatively narrow ranges. This is because hydrogen ion activity is a determinant of the structural conformation of proteins (e.g., enzymes) and the state of ionization of weak acids and bases. Changes in these properties influence fluxes through biological membranes and through metabolic pathways. Accordingly, there are important effects of pH changes on several major pathways of carbohydrate metabolism, including the conversion of glucose and fructose to pyruvate and lactate, and subsequent oxidation of pyruvate to carbon dioxide and water; these catabolic pathways are major sources of energy production. In addition, the ana-

bolic conversion of precursors such as lactate and pyruvate and certain amino acids to glucose (gluconeogenesis) is also substantially influenced by pH. The situation is complicated by the fact that these reaction pathways are often H^+ (proton) producing or consuming, and their rates are therefore themselves determinants of pH, both in the cells where they occur and elsewhere. These interactions between pH and carbohydrate metabolism are considered here, beginning with an outline of basic mechanisms, followed by an attempt at physiological integration.

I. Basic Mechanisms

A. Effect of pH on Individual Reactions of Carbohydrate Metabolism

Both catabolic (degradation) and anabolic (synthetic) pathways of carbohydrate metabolism are characterized by bottlenecks, i.e., nonequilibrium (rate-controlling) steps that are markedly affected by pH in the cellular compartment in which they take place. For the glycolytic pathway, which is located entirely within the cell cytosol, the main bottleneck is the conversion of fructose 6-phosphate to fructose 1,6-bisphosphate, catalyzed by the enzyme phosphofructokinase (PFK) (Fig. 1). PFK is markedly inhibited by lowering of cytosol pH probably by increasing the inhibitory action of ATP on the enzyme. The gluconeogenic pathway from 3-carbon precursors such as lactate and pyruvate occurs mainly in the liver and kidney and has several potentially rate-controlling steps. Those most obviously affected by pH are the conversion of pyruvate to oxaloacetate by pyruvate carboxylase, which takes place in mitochondria, and the subsequent conversion of oxaloacetate to phosphoenolpyruvate by the enzyme phosphoenolpyruvate carboxy-

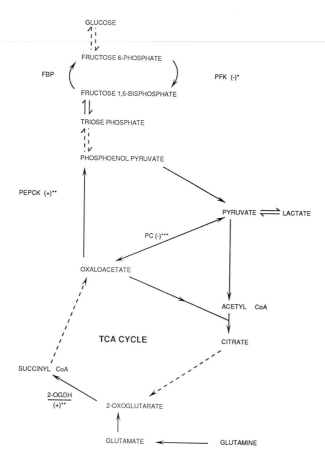

FIGURE 1 Composite diagram of main reactions of carbohydrate metabolism indicating (*underlined*) the catalytic steps at which physiologically relevant effects of pH change have been clearly demonstrated. Glycolysis (glucose → pyruvate) occurs in all tissues. Gluconeogenesis (pyruvate → glucose, or glycogen) is confined to liver, kidney, and, under some circumstances, carbon skeletal muscle. The (TCA) tricarboxylic acid cycle occurs in all tissues demonstrating oxidative metabolism. (+) or (−) indicates overall stimulatory or inhibitory effects of lowering pH. *Asterisks* indicate those tissues in which these effects may be important: *most tissues; **kidney; ***liver. PFK, phosphofructokinase I; PC, pyruvate carboxylase; PEPCK, phosphoenolpyruvate carboxykinase; FBP, fructose 1,6-bisphosphatase; 2-OGDH-2-oxoglutarate dehydrogenase. The *interrupted arrows* represent transitions involving multiple enzyme steps. See text for further explanation.

kinase (PEPCK). The latter reaction is entirely cytosolic in some species but, in others including humans, takes place both in the cytosol and in the mitochondria.

There are quite different physiologically relevant modes of action of pH on these two enzymes of gluconeogenesis. In the case of hepatic pyruvate carboxylase *in vitro,* a fall of pH within the physiological range inhibits the obligatory allosteric acti-

vation by acetyl coenzyme A. In the case of (renal) PEPCK, lowering the blood pH of the animal (i.e., induction of acidosis) induces its synthesis, with consequent acceleration of gluconeogenesis.

A further mode of pH action on enzymes is exemplified by the marked activation of renal 2-oxoglutarate dehydrogenase observed in acid conditions both *in vitro* and *in vivo;* this effect is achieved by increase in affinity of the enzyme for its substrate. 2-oxoglutarate dehydrogenase catalyses the conversion of 2-oxoglutarate to succinyl coenzyme A in the tricarboxylic acid cycle; it is also concerned in the pathway of glutamine metabolism, and its role in this respect appears to be an important regulator of the renal adaptation to acidosis (see next section).

A final important effect of pH changes is in the posture of "redox" reactions. Using the cytosolic conversion of lactate to pyruvate as an example, this reaction is represented by the equation

$$Lactate + NAD^+ \rightarrow Pyruvate + NADH + H^+$$

in which NAD^+ is reduced to NADH. This reaction, catalyzed by lactate dehydrogenase, is close to equilibrium under most circumstances and the law of mass action may therefore be applied:

$$\frac{[Lactate]}{[Pyruvate]} = \frac{K[NADH]\,[H^+]}{[NAD^+]}$$

A change in the left-hand side (usually denoted as L/P) may therefore be due to either or both a change in $[NADH]/[NAD^+]$ or in $[H^+]$. Thus a fall in cytosolic pH of 0.3 units (corresponding to a twofold change of $[H^+]$) would result in a doubling of L/P without change in the redox status as determined by $[NADH]/[NAD^+]$. Similar considerations apply to other redox reactions (e.g., those catalyzed by 3-hydroxybutyrate dehydrogenase and by glutamate dehydrogenase, both of which are intramitochondrial).

In all the above examples the effect is substantial within the range of pH that might be expected in the relevant organs (and cell compartments within those organs) under physiological and pathophysiological conditions.

B. Effect of pH on the Supply of Substrate to the Reactions of Carbohydrate Metabolism

1. Membrane Transport Effects

pH has major effects on the transport of substrates across cell membranes. The most well-characterized example in carbohydrate metabolism is that of lactate, in relation both to its exit from cells in organs producing lactate (e.g., skeletal muscle and erythrocytes) and to its entry into cells in organs responsible for lactate disposal (e.g., the liver and kidneys). Because of its considerable physiological and clinical importance, the lactate example will be discussed in some detail.

Lactate (and almost certainly pyruvate) traverses cell membranes by at least two mechanisms, passive nonionic diffusion not mediated by a carrier and a transporter-mediated process, both of which are pH-dependent. Nonionic diffusion is a very general mode of distribution of weak acids and bases across biological membranes; the molecular species actually traversing the membrane is the uncharged form, which is much more soluble in the membrane lipid bilayer than is the charged form. For a weak acid, the distribution of the total acid (i.e., dissociated plus undissociated) between two compartments (1) and (2) (separated by a lipid membrane), in which the hydrogen ion activity is $[H^+]_1$ and $[H^+]_2$, respectively, is given by

$$\frac{R_1}{R_2} = \frac{[H^+]_2 \, ([H^+]_1 + K)}{[H^+]_1 \, ([H^+]_2 + K)} \qquad [1]$$

where R is the total concentration of acid as defined above, K is the acid dissociation constant, and the subscripts refer to the two compartments. In deriving this equation, it is assumed that the permeability to the charged moiety is negligible compared with that of the undissociated acid, that transit of the undissociated acid across the membrane is by passive diffusion alone, and that equilibrium has been reached. The main implication of Eq. 1 is that weak acids will tend to be concentrated in the more alkaline of the two compartments, and that if $[H^+]_1$ and $[H^+]_2 \ll K$ then $R_1/R_2 = [H^+]_2/[H^+]_1$. Under these conditions at equilibrium, a gradient of 1 pH unit between the two compartments results in a 10-fold concentration difference of the total weak acid. Departures from the theoretical distribution may result from (a) nonequilibrium conditions (e.g., removal of the weak acid by metabolism after traversing the membrane), or (b) significant permeability of the ionized form, in which case the electrical potential across the membrane becomes an additional determinant of R_1/R_2. It should be noted that for some weak acids such as lactate (with pK_a 3.8), the concentration of the undissociated moiety is several orders of magnitude smaller than that of the lactate ion in the physiological range of pH. The nonionic

diffusion mechanism is under some circumstances the principal mode of transit across cell membranes because the low concentration of the undissociated form is more than compensated by the much greater ability to cross the cell membrane. It is important to appreciate that this mechanism is always kinetically first order (i.e., it does not saturate) in contrast to the transporter mechanism described in the next paragraph.

The transporter mechanism for entry of lactate (ionized) is semispecific, certain other monocarboxylates, notably pyruvate, also being transported. In the majority of mammalian species examined, it is largely stereospecific for the natural L-isomer of lactate. It is widely distributed but has been best characterized in erythrocytes and in hepatocytes. The erythrocyte lactate transporter has been reconstituted in active form into artificial liposomes. In hepatocytes, the lactate transporter activity is markedly increased in the starved and diabetic states as opposed to the fed state. Transport is notably pH-dependent, a downhill proton gradient across the membrane being associated with high transporter activity and vice versa. The mechanism of the pH dependency is at least partly explained by the transporter acting as a "symport" for both lactate ions and protons. Although the stoichiometry for lactate and H^+ appears to be 1:1 in some cells, in others (e.g., the hepatocyte) the ratio of lactate/H^+ transported is in excess of unity. Another possibility is that the pH effect is mediated by proton-directed conformational changes in the transporter. The distinct pyruvate transporter of the inner mitochondrial membrane is similarly pH-dependent. [*See* CELL MEMBRANE TRANSPORT.]

2. Supply of Substrate to Organs of Carbohydrate Metabolism

Complete metabolism of glucose to CO_2 and water requires oxygen. Oxygen lack diverts pyruvate to lactate, instead of the tricarboxylic acid cycle, to reconvert NADH to NAD^+ and thus allow anaerobic glycolysis and some ATP production to continue. Oxygen is also required on the synthetic side (i.e., gluconeogenesis), which has a particularly heavy requirement for ATP.

pH affects oxygen delivery to the various organs in several ways. The first is through a complex series of changes in the affinity of hemoglobin for oxygen. Acute falls in blood pH produce an immediate change resulting in the more ready release of oxygen from hemoglobin to the tissues. However, in more chronic acidosis the affinity returns toward its original value, because of a pH-induced decline in the erythrocyte content of 2,3-diphosphoglycerate (2,3 DPG), which binds to hemoglobin, causing the decreased affinity. 2,3-DPG synthesis is unique to erythrocytes and takes place in a "side-loop" of the glycolytic pathway arising as an alternative route of 1,3 diphosphoglycerate metabolism. A fall in erythrocyte pH lowers 2,3-DPG by a complex mechanism.

The second major effect of pH on substrate supply for carbohydrate metabolism is on blood flow through the organs concerned. Acidosis has markedly negative effects on myocardial contractility, leading to diminished myocardial output and consequent fall in individual organ blood supply. This has the potential for increasing anaerobic glycolysis and the production of lactic acid in peripheral organs, and of diminishing lactate disposal in organs such as the liver. Lactate uptake by the perfused rat liver falls slowly at first with diminishing blood flow but declines rapidly when flow is below 25–30% of normal, and the liver may switch to lactic acid production. A third important effect of lowered pH on carbohydrate metabolism is on the entry of glucose into cells, which in many cases is dependent on the presence of insulin. It has long been apparent that in diabetic ketoacidosis there is a marked resistance to the blood glucose lowering effect of insulin. At least part of this phenomenon is related to an acidosis-induced decrease in the number of insulin receptors in insulin-sensitive cells, with consequent impairment of insulin-dependent glucose entry. [*See* INSULIN AND GLUCAGON.]

3. Proton Production and Consumption During the Metabolism of Carbohydrate

When an electrically neutral carbohydrate such as glucose is fully converted to neutral products such as carbon dioxide and water, there is no net production or consumption of protons. In contrast, if the metabolic reaction gives rise to charged anionic products such as lactate or pyruvate, protons are released in equivalent amounts.

$$C_6H_{12}O \rightarrow 2CH_3\,CHOHCOO^- + 2H^+$$
$$\text{glucose} \qquad \text{lactate}$$

Similarly, if as in gluconeogenesis, anionic precursors are converted to glucose, protons are consumed in the overall reaction.

$$2CH_3\,CHOHCOO^- + 2H^+ \rightarrow C_6H_{12}O_6$$

There is therefore *in the steady state* a close stoichiometry of proton production or consumption as indicated in the above equations. Thus in glycolysis, the production of two protons would be expected for each molecule of glucose converted to lactate. That this stoichiometry is disrupted in non-steady state conditions can be illustrated by writing down the equations of the individual reactions of glycolysis, having full regard to the ambient pH, the acid dissociation constants of phosphate, and the adenine nucleotides ATP and ADP and the charge on the reactants. Summation of these equations indicates that the process of glycolysis of one glucose molecule results in the production of two lactate ions, the conversion of two molecules of each of ADP and inorganic phosphate to two molecules of ATP, and an amount of protons considerably less than the two expected. However, this situation does not represent a steady state, because, for instance, the ATP concentration has been increased. If one then takes into account the proton production accompanying the full use of that additional ATP (for energy-driven reactions, e.g., muscle contraction), so that the true steady state is maintained, it is simple to show the expected stoichiometry of two protons produced per one molecule of glucose converted to lactate. [*See* ADENOSINE TRIPHOSPHATE (ATP).]

These proton producing and consuming reactions alter the pH at the site where they occur and thus modify the rates of those reactions by feed-back. Intracellular acidification in skeletal muscle has been clearly demonstrated in skeletal muscle during anaerobic exercise in humans using noninvasive ^{31}P-magnetic resonance spectroscopy. In isolated rat liver perfused with medium containing lactate as the sole substrate, the hepatic venous effluent is always slightly more alkaline than the inflowing perfusate, despite a higher partial pressure of carbon dioxide (P_{CO_2}) than in the portal vein. This could in theory be due to entry of the lactate solely as the undissociated acid, protons being thus continuously abstracted from the perfusate. However, this cannot be more than a part of the process, because it has been shown that hepatic intracellular pH rises with increasing lactate uptake; the simplest explanation is that some fraction of the lactate entering the hepatocyte does so in the ionized form, and intracellular protons are consumed during its subsequent conversion to glucose. There is evidence that the lactate transporter transfers lactate at least partly in the ionized form.

II. Physiological Interactions of Carbohydrate and Acid–Base Metabolism

In normal resting adult humans, about 1,300 mmol lactate enters the circulation per day, principally from skeletal muscle, skin, gut, brain, and erythrocytes. The similar quantity of protons produced with this lactate titrates tissue and blood bicarbonate. In health, the bicarbonate deficit thus created is corrected when lactate is taken up and converted to neutral products (e.g., glucose, or CO_2 and water), principally by the liver, but also by the kidney and heart and under some circumstances by skeletal muscle itself. The process of regeneration of bicarbonate may be represented as follows:

$$2CH_3 \, CHOHCOO^- + 2H^+ \rightarrow C_6H_{12}O_6$$

$$2H_2O \rightarrow 2H^{+\nearrow} + 2OH^-$$

$$2OH^- + 2CO_2 \rightarrow 2HCO_3^-$$

As indicated above, the first and probably all these reactions take place within the liver cell, and virtually all the regenerated bicarbonate is then exported to the circulation to restore the deficit. Lactate oxidation to CO_2 and water has the same overall effect and is more prominent in kidney, heart, and skeletal muscle than in the liver. Even in resting humans, the proton turnover associated in this way with lactate production and removal forms a major fraction of proton turnover through the circulation.

This cycle of lactate production and disposal, with its accompanying proton turnover, is affected by pH at the many stages detailed in the first part of this account. The marked drop in intracellular pH (pH_i) in skeletal muscle accompanying glycolysis has been suggested as a mechanism of fatigue, related to feed-back inhibition of glycolysis at the PFK step by low pH (see Fig. 2). The declining force during fatigue is negatively and linearly related to pH_i. Furthermore, exercise endurance in humans is diminished by induction of acidosis with ammonium chloride and enhanced by administration of sodium bicarbonate. Measurements of intermediate metabolism of glycolysis in skeletal muscle during these maneuvers is consistent with an effect at the PFK step. Yet if this mechanism were the principal explanation of fatigue, a fall in muscle ATP would be expected. In fact the fall during fatigue is relatively slight, and some additional factor such as compartmentation of ATP would have to be

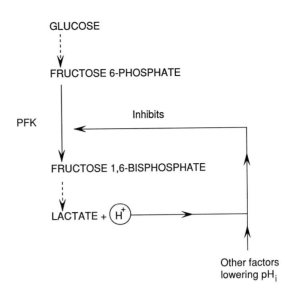

FIGURE 2 Skeleton of glycolysis, demonstrating negative feedback of generated hydrogen ions on the phosphofructokinase (PFK) step. Other factors that lower cell pH (pH_i) have the same effect.

present. The matter thus remains unresolved. It has been suggested that the inhibition of PFK by lowered pH has been evolved to protect the animal against the likely fatal acidosis which would be caused by totally unbridled glycolysis. [*See* GLYCO-LYSIS.]

A further example of a pH effect on glycolysis probably exerted at the PFK step is seen during respiratory alkalosis, in which both extra- and intracellular pH has been elevated by lowering of the partial pressure of CO_2 due to increased pulmonary excretion of CO_2. Glycolysis appears to be stimulated in many tissues, including erythrocytes, and blood lactate is thereby increased. The extent of this effect is, however, variable, depending *inter alia* on the response of lactate removal mechanism. In normal humans spontaneous hyperventilation, or mechanical ventilation under anesthesia, produces only small increments in blood lactate; in contrast, more substantial elevations are seen in mechanically hyperventilated dogs. The production of lactic acid in response to low P_{CO_2} goes some way toward correcting the alkalosis.

Lactate produced by glycolysis and entering the circulation is removed principally by the liver; any tendency to mild lactic acidosis is counteracted by increased hepatic lactate uptake, due to both the elevation of blood lactate concentration and the increased pH gradient across the hepatocyte plasma membrane, which stimulates both the transporter and diffusion mechanism of lactate entry. At normal or slightly elevated blood concentrations of lactate (i.e., between 0.5 and 1.5 mmol/L), the transporter pathway seems to be the dominant entry mechanism (Fig. 3). Consistent with these mechanisms is the observation that hepatic lactate disposal increases in humans undertaking moderate exercise, despite a 50% fall in hepatic blood flow. Observations both in humans and in animals suggest that hepatic lactate uptake increases with blood concentration up to a level of 3–5 mM and then remains constant despite further elevation. At 3–5 mM the lactate transporter is already saturated, and the bulk of entry must be via the nonionic diffusion pathway, which cannot saturate. The plateau effect above 3–5 mM must therefore be due to saturation of one of the nonequilibrium reactions of gluconeogenesis. The role of the liver in correcting the very

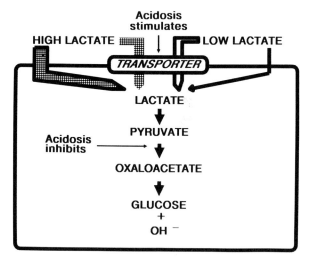

FIGURE 3 Possible model of effects of acidosis on hepatic disposal of lactate. A schematic hepatocyte is depicted with a lactate transporter at the perisinusoidal pole. The top right-hand corner illustrates events at normal or only minimal raised blood lactate concentration. A minor fraction of lactate entry takes place through the nontransporter diffusion pathway. The major fraction is transporter-mediated. Both pathways are stimulated by an increased proton gradient downhill from the outside to the inside of the cell, with an increase in overall lactate disposal. The top left-hand corner depicts events occurring at high blood lactate concentrations (>2 mmol/L). The transporter is saturated, so no further increase can take place via this pathway. The main entry is through the diffusion pathway. Although this may potentially increase, as shown, the point of rate limitation has shifted to the pyruvate → oxaloacetate step of gluconeogenesis, which is inhibited by acidosis, so overall lactate disposal falls.

marked degree of lactic acidosis (blood lactate up to 20–25 mmol/L), which may be achieved during severe exercise, is therefore limited, and other mechanisms of the correction seen on cessation of exercise must be involved. A striking observation is that the rate of decline of lactic acidosis after cessation of maximal exercise is greater if the subject persists with submaximal exercise than if he or she rests completely. This phenomenon is due to the use of lactate as a fuel of oxidation in submaximally exercising skeletal muscle. This cannot happen in maximum exercise when the demand for oxygen outstrips the supply and anaerobic glycolysis therefore produces rather than consumes lactate.

Severe lactic acidosis also occurs in circulatory insufficiency (shock), due to both increased production of lactate and protons because of tissue hypoxia and major impairment of lactate removal. Observations in perfused rat liver have demonstrated that provided perfusate lactate levels are substantially elevated, thereby saturating the transporter, acidosis itself is a powerful inhibitor of lactate disposal via gluconeogenesis. This contrasts with the stimulatory effect of acidosis seen at lower lactate levels, presumably related to stimulation of the transporter. At the higher levels of lactate, the site of the inhibitory action of acidosis on gluconeogenesis is at a step between pyruvate and oxaloacetate (Fig. 3). The degree of inhibition is closely related to hepatic pH_i whether this be achieved by raising PCO_2 ("respiratory acidosis") or lowering bicarbonate ("metabolic acidosis"). There are several stages in the pyruvate → oxalocetate step, which are possible sites of the inhibitory effect of low pH; an obvious but unproved possibility is the effect of low pH on the allosteric activation of pyruvate carboxylase, referred to earlier.

The pathological significance of this inhibitory effect is considerable. In lactic acidosis the prime homeostatic requirement is to dispose of lactate at an accelerated rate so that bicarbonate may be rapidly regenerated. This inhibitory effect of acidosis on gluconeogenesis is precisely the opposite of what is required and, in fact, exacerbates the lactic acidosis. Several types of lactic acidosis have a particularly fulminating course, and it has been suggested that this may in part be due to a vicious circle set up by acidotic inhibition of gluconeogenesis from lactate. In clinical shock, the prospect for survival becomes increasingly poor when blood lactate concentration exceeds 5 mmol/L; the mechanism described above may be one component in this poor prognosis. The antidiabetic biguanide phenformin was responsible for a large number of fatalities because of fulminating lactic acidosis in the 1970s, before this drug was abandoned. Phenformin has multiple effects on biological membranes because of its insertion into them and consequent disturbances of surface charge. One of its actions is to inhibit the lactate transporter; it may be speculated that for this reason the homeostatic defenses against mild lactic acidosis are removed, allowing lactic acidosis to develop to the degree in which the inhibitory effect on gluconeogenesis is evident. Unlike in recovery from exercise, there is no substantial alternative disposal route for lactate hydrogen ions in either shock or phenformin toxicity.

The kidney has a smaller role in lactate disposal than the liver, but one that animal studies suggest could be significant. The kidneys use lactate both as a fuel of oxidation and for conversion into glucose. In normal rats, they account for about 20% of the disposal of an intravenously injected sodium lactate load. If rats are made progressively acidotic by feeding ammonium chloride, the rate of extrarenal removal of lactate falls. However, this is partially counteracted by an increased rate of renal lactate disposal, so that in severely acidotic rats the kidneys may account for up to 40% of removal of the lactate load. Whether this effect is seen in humans is unknown. It seems likely to be due to the induction of PEPCK as described in Section I. This effect does not occur in liver, thus accounting for the difference in behavior of the liver and kidneys toward lactate in acidosis. It should be noted that lactate loss in the urine is insignificant until blood lactate reaches 8–10 mmol/L and urinary excretion of buffered protons is small compared with proton removal by the liver during lactate metabolism.

The complex interactions of pH and carbohydrate metabolism are well-illustrated in diabetic ketoacidosis. In this condition there is gross overproduction of glucose by the liver, accounting for the major part of the hyperglycemia. The substrates for glucose formation include both amino acids and 3-carbon precursors such as lactate and pyruvate. There may be severe extracellular acidosis, usually attributed to the production of 3-hydroxybutyrate and acetoacetic acids ("ketone bodies") by the liver, and it is at first sight anomalous that diabetic ketoacidosis is only rather infrequently accompanied by severe lactic acidosis; the latter would be expected because, *inter alia,* of the acidotic inhibition of gluconeogenesis from lactate described

above. Some recent observations provide a possible solution to this paradox. Firstly it is not possible to inhibit gluconeogenesis from lactate by perfusing livers taken from diabetic ketoacidotic rats with acidotic medium, whereas, as indicated earlier, this inhibition is readily demonstrated in livers taken from normal rats. Surprisingly, the intracellular pH in perfused livers from diabetic ketoacidotic rats is higher than that in the "normal" perfused livers, although both sets of livers are perfused with similarly acidotic media. Secondly, if rats are made progressively acidotic by inducing diabetic ketoacidosis, by ammonium chloride ingestion, or by hydrochloric acid infusion, hepatic cell pH (pH_i) measurements *in vivo* by ^{31}P-nuclear magnetic resonance spectroscopy indicate that although hepatic pH_i falls in ammonium chloride and hydrochloric acid acidosis, there is almost no effect of progressive blood acidosis on hepatic pH_i in the diabetic ketoacidotic livers. It may be that this unique protection of hepatic pH_i in livers from diabetic ketoacidotic animals permits normal or increased gluconeogenesis from lactate to proceed despite severe blood acidosis. The basis of this protection is speculative but may be due to stimulation of gluconeogenesis from lactate by the ketogenic state, with consequent increased generation of bicarbonate. This bicarbonate could be sufficient to neutralize protons produced during ketogenesis and thus to prevent a fall in hepatic pH_i.

The acid–base consequences of fructose administration have been of considerable interest because this monosaccharide was included as a component of intravenous feeding solutions for clinical use. It was noted that subjects to whom these solutions were administered at a substantial rate developed lactic acidosis. Unlike glucose, fructose is rapidly converted in part to lactate by the liver. It has been shown in both humans and animals that this is due to brisk phosphorylation of fructose to fructose 1-phosphate by fructose 1-kinase; this reaction requires ATP, which is rapidly depleted in liver cells during fructose infusion. Studies in rat perfused livers in which hepatic cytosol pH has been followed either by ^{31}P-magnetic resonance spectroscopy or by intracellular pH-sensitive microelectrodes have shown striking falls during fructose infusion. This fall in cytosol pH is initially partially due to the H^+ released in the fructokinase reaction during the phase of fall in ATP concentration; later, low cytosol pH is maintained by the substantial quantities of lactate and protons released during fructolysis.

Fructose has now been abandoned as an intravenous feeding agent. It should be noted that the ordinary absorption of fructose from the hydrolysis of dietary sucrose does not proceed at rates sufficient to produce the above effects.

The kidney has classically been regarded as having a major role in acid–base homeostasis. Although this view has recently been subjected to substantial reinterpretation, it is undisputed that acidosis results in a substantial increase in the rate of urinary excretion of ammonium ions. [*See* pH Homeostasis.]

This has probably been wrongly interpreted as representing the excretion of protons buffered as NH_4^+ but is more correctly regarded as diverting nitrogen to urinary excretion as NH_4^+ rather than to urea synthesis; urea synthesis incurs the undesirable penalty of generating two protons for every molecule of urea synthesized in the liver, so some switch from urea synthesis to NH_4^+ exertion in acidosis would be advantageous for acid–base homeostasis. The mechanism of increased ammoniagenesis is now fairly well understood and involves well-defined interactions of pH, carbohydrate, and amino acid metabolism. NH_4^+ is derived from the renal hydrolysis of glutamine, followed by oxidative deamination of the resulting glutamate to 2-oxoglutarate, two molecules of NH_4^+ being generated. The latter reaction, catalyzed by glutamate dehydrogenase, is at near equilibrium, and it is necessary to remove the products to allow the reaction to proceed. NH_4^+ is removed by nonionic diffusion into the acidified urine. 2-Oxoglutarate is removed by gluconeogenesis, the flux-controlling enzyme in this process being PEPCK, which, as indicated above, is increased in quantity by chronic acidosis. Inhibition of this enzyme prevents the increase in NH_4^+ production seen in acidosis. In addition, 2-oxoglutarate dehydrogenase is stimulated by acute acidosis; this effect diminishes renal 2-oxoglutarate concentration even when PEPCK is inhibited.

III. Concluding Remarks

Cells possess three main categories of defense against tendencies to alter their internal pH. These are (1) physicochemical buffering, (2) movements of protons (or other relevant ions, e.g., OH^-, HCO_3^-) across the plasma membrane, and (3) alteration in the internal production or consumption of protons. This account has been substantially concerned with the last of these three mechanisms and has reviewed the manner in which changes in flux through the pathways of carbohydrate metabolism may alter proton production or consumption. It has been shown that these changes are sometimes homeostatic and at others antihomeostatic for the maintenance or correction of acid–base status. Furthermore, they are frequently themselves caused by pH alteration.

Such emphasis as has been given has been in consideration of perturbations in the acidotic direction. The reason is that in terms strictly of carbohydrate metabolism, this is the direction in which most of the major disturbances occur—exercise, shock, and ketoacidosis being prime examples. However, there is a view that the threat of disturbance in the alkalotic direction has been much underestimated, particularly in biological and evolutionary terms rather than from the clinical perspective. But because this threat derives from the metabolism of the carbon skeletons of amino acids, it has not been considered here. The reader is referred to the appended bibliography for source material on this topic.

Bibliography

Atkinson, D. E., and Bourke, E. (1984). The role of ureagenesis in pH homeostasis. *Trends Biochem. Sci.* **9,** 297–300.

Cohen, R. D. (1990). *In* "The Metabolic and Molecular Basis of Acquired Disease" (R. D. Cohen, B. Lewis, K. G. M. M. Alberti, and A. M. Denman, eds.). Bailliere Tindall & Cox, London.

Cohen, R. D., and Guder, W. G. (1988). Carbohydrate metabolism and pH. *In* "pH Homeostasis" (D. Häussinger, ed.). Academic Press, London.

Madias, N. E. (1986). Lactic Acidosis. *Kidney Int.* **29,** 752–774.

Iles, R. A., Stevens, A. N., and Griffiths, J. R. (1982). NMR studies of metabolites in living tissue. *Prog. Nucl. Magnetic Resonance Spectroscopy* **15,** 49–200.

Pharmacogenetics

BERT N. LA DU, JR., *University of Michigan*

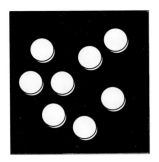

Glossary

Ecogenetics Essentially, the same meaning as pharmacogenetics but used in reference to chemicals of the environment, rather than chemicals that are used as therapeutic agents

Polymorphic enzymes Multiple forms of an enzyme, differing slightly in their structure and catalytic activities

PHARMACOGENETICS is the study in humans and animals of genetically transmitted traits that confer exaggerated or unusual responses to therapeutic drugs and to chemicals of our environment. These genetic traits may affect pharmacokinetics, defined as the processes by which drugs are absorbed from their site of administration, distributed throughout the body, metabolized, and eliminated from the body. They can also alter the pharmacodynamics (i.e., the tissue responsiveness to the drug at the organ or cellular level) because of some structural alterations in receptor proteins at the sites inside the cell or the cellular membranes, where critical interactions take place between drug molecules and receptors.

The objectives of pharmacogenetic investigations are listed in Table I. Apparently, pharmacogenetics is a very pragmatic specialty that seeks to understand the basis of unusual reactions to drugs and to improve the effectiveness and safety of drugs by avoiding these adverse reactions.

I. Examples

About 60 different pharmacogenetic conditions are known, and some representative examples are briefly characterized in Table II. They range from the striking differences among people in their ability to taste phenylthiourea (PTC), to the serious, life-threatening hereditary condition called malignant hyperthermia with muscular rigidity. The former represents an inherited difference in the bitter taste threshold for PTC, which has not yet been explained at the molecular level but appears to represent no significant clinical vulnerability; the latter is a very serious and, fortunately, rare condition, which is most likely to occur during surgery in young, healthy adults who have often been given two commonly used drugs during anesthesia: halothane and succinylcholine. Until quite recently, nearly half of the subjects who developed a pronounced rise in body temperature and a generalized rigidity of skeletal muscles died from the resulting heart or kidney failure. The unusual susceptibility is inherited as a dominant trait, so many family members of affected people are also at risk, even though the condition is rare. A pharmacological test, based on the contractile response of a skeletal muscle biopsy preparation to caffeine, has been used successfully to identify susceptible relatives. As more is learned about the genetic basis of this condition, developing simpler diagnostic tests should be possible.

II. Underlying Causes

Many pharmacogenetic conditions are relatively rare, and the majority are of the pharmacokinetic

TABLE I Objectives of Pharmacogenetic Studies

1. Identify the genetic trait associated with the pharmaco-genetic condition.
2. Determine the mechanism by which the trait causes the unusual response to the drug, or drugs.
3. Determine the mode of inheritance of the trait (dominant or recessive; autosomal or sex-linked).
4. Estimate the gene frequencies of the trait in the general population and selected ethnic and geographic areas.
5. Evaluate the clinical importance of the trait.
6. Develop a simple test to detect the trait, without exposure to the drug, if a diagnostic test is needed.

type, characterized by a marked reduction in the rate of drug metabolism. Slow metabolizers with a deficiency of one of the drug metabolism enzymes are likely to develop a much higher drug concentration than normal in the blood and tissues. If repeated, standard doses of the drug are given to such people, the drug may accumulate and reach toxic concentrations, unless appropriate adjustments are made in the dose and dosing interval to compensate for the individual's reduced rate of metabolism. [*See* PHARMACOKINETICS.]

New pharmacogenetic conditions have been dis-

TABLE II Some Examples of Pharmacogenetic Conditions

Defective protein or polymorphism	Susceptible phenotype	Drugs or foreign compound	Disorder	Incidence
N-acetyltransferase	Slow acetylator	Isoniazid	Polyneuritis	60% Caucasian; 10–20% Orientals
Serum cholinesterase	Atypical and other variants	Succinylcholine	Prolonged paralysis	1 : 2,000 Caucasians
Serum paraoxonase polymorphism	Slow metabolizer	Paraoxon	Pesticide poisoning	50% Caucasians
Debrisoquine hydroxylation deficiency	Slow metabolizers	Debrisoquine, sparteine	Specific P-450 deficiency, overdose toxicity	6–9% Caucasians
Mephenytoin hydroxylation deficiency	Slow metabolism	Mephenytoin, mephobarbital	Specific P-450 deficiency, overdose toxicity	5% Caucasian; 23% Japanese
Aldehyde dehydrogenase deficiency	Poor metabolizer	Ethanol	ALDH isozyme I deficiency, overdose toxicity	30–50% Orientals
Phenyloin hydroxylation deficiency	Slow metabolizers	Phenytoin	Specific P-450 deficiency, overdose toxicity	Very rare
Glucose-6-phosphate-dehydrogenase deficiency	Deficient red blood cells	Primaquine, many drugs	Specific enzymatic defect, hemolysis	Frequent in American Blacks and in tropical countries
Malignant hyperthermia	Defective binding of calcium in muscle membrane	Halothane, succinylcholine	Hyperthermia, hyperrigidity (often fatal)	About 0.005% in Caucasians
Warfarin resistance	Resistant phenotype; requires 20 times as much drug as normal	Dicoumarol-type anticoagulants	Receptor defect; increased affinity for vitamin K	Very rare
Porphobilinogen-synthetase inducibility	Abnormal inducibility of enzyme	Barbiturates	Drug induces δ-amino-levulinate dehydratase, porphyria	Rare
Hypoxanthine-guanine phosphoribosyl transferase	Enzymatic deficiency	Allopurinol	Drug relatively ineffective for treating gout in patients with this deficiency	Rare
Catechol-o-methyl transferase	Enzymatic deficiency	Isoproterenol, L-dopa	Uncertain	Rare
Unknown	Bitter taste in "tasters"	Phenylthiourea	Marked variation in taste threshold	Nontasters, 30% Caucasians

covered primarily by investigating patients who have shown an exaggerated response to a drug or experienced some unusual adverse reaction to a drug. The investigation may establish that the patient is unusually slow in metabolizing the drug, or the patient transformed the drug to a different pattern of drug metabolites, some of which produce toxic reactions. If close relatives of the patient can also be tested without undue risk, the slow metabolism may be duplicated or the usual metabolic pattern may be obtained in these family members. Another family member with the reduced rate of metabolism would increase the likelihood that genetic factors are responsible for the unusual findings. The first patient showing an unusual resistance to a coumarin anticoagulant drug required 20 times the usual dose of the drug to obtain the expected anticoagulant effect. By remarkable coincidence, this patient had an identical twin brother, who also received a coumarin-type anticoagulant within a few weeks of his brother's treatment. He displayed exactly the same unusual type of drug resistance. A survey of additional family members revealed that several others were also resistant to the drug, and resistance was inherited as a dominant trait.

Most pharmacogenetic conditions are completely unsuspected. The affected people have no indication of their unusual genetic trait and show no disturbance in their metabolism of sugars, proteins, fats, or other ordinary chemical compounds. From this observation, it is clear that pharmacogenetic traits selectively involve different metabolic systems and a different set of enzymes from those concerned with the metabolism of endogenous chemicals that arise during the course of normal intermediary metabolism.

An interesting biological problem is posed by these distinctions in metabolic systems. What is the nature of the special enzymatic systems utilized by mammalian species to inactivate and eliminate drugs and other foreign chemicals, and how did these drug metabolism enzyme systems evolve? We know that there are two major pathways by which drug molecules are transformed in the body; (1) direct attack by special enzymes such as the P-450 enzymes of the liver endoplasmic reticulum, which catalyze drug oxidations, and reductions to produce primary changes in the drug molecule; and (2) the conjugation reactions, catalyzed by transferase enzymes that utilize endogenous donor compounds such as activated sulfate or glucuronate. These conjugation reactions generally yield products of the drug that are inactive pharmacologically because important functional groups are blocked in the course of forming the glucuronides, acetylated derivatives, or sulfate conjugates (Fig. 1).

Surprisingly, these inactivating systems for drug molecules show extremely high variability from animal species to species, and even within the human population, from one person to another. Half-lives of drugs (the time required for half of the drug to be removed from the body) can vary as much as 10–20-fold. This wide variation explains why a standard dose of a drug will be too high for some people but too low for many others, even though it is calculated to be the right amount for the "average" person. Individualization of the drug dose is often necessary to obtain optimal therapeutic results in most patients.

The other important observation is that whether a person is a fast or slow metabolizer of a particular drug, he or she will persist in his or her individual rate, if rechallenged, with little change over many years. The logical conclusion to be drawn from these two findings is that there is considerable genetic heterogeneity in the drug metabolism enzyme systems, and these characteristics are set for each individual primarily by genetic rather than environmental factors. It is not unexpected, then, that a large number of genetic traits in humans have been identified that are responsible for the wide range of activities of the drug metabolism enzymatic systems. Some of these genetic traits cause the individuals carrying them to react in very unusual ways,

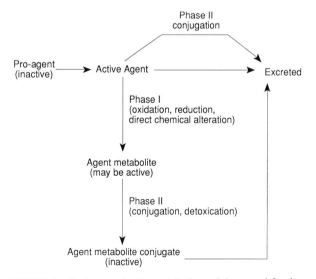

FIGURE 1 Basic steps in the metabolism of drugs and foreign chemicals.

even when the drug is given in the generally recommended dose.

Pharmacogenetic studies attempt to identify the biochemical basis for the conditions and determine which enzymatic system is altered. Some pharmacogenetic conditions can now be diagnosed by a simple blood test, and the drug sensitivity can be accurately predicted without exposing the subject to the drug. For example, several of the serum cholinesterase variants that cause an exaggerated response to the muscle-relaxant agent succinylcholine can be accurately diagnosed by a simple *in vitro* test with serum (or plasma). The sample is incubated with its substrate benzoylcholine and certain inhibitors, dibucaine and fluoride, to determine the status of the enzyme in question, cholinesterase. A number of the cholinesterase variants have been analyzed at the DNA level and shown to represent point mutations or frameshift mutations that affect the amount or quality of the enzymatic structure. Although the succinylcholine sensitivity condition is rare, it is inherited as a simple Mendelian trait. Close relatives (siblings, parents, and children) are much more likely to share the same type of drug sensitivity than nonrelatives. Accordingly, these "high risk" individuals may want to have their blood typed before receiving the drug to find if they would also be sensitive to the drug.

The isoniazid acetylation polymorphism can be used to illustrate a topic of current interest in pharmacogenetic investigation. About half the population in this country are slow and half rapid, so the slow acetylation phenotype, determined by a double dose of a recessive gene, is not rare, as is so often the situation with genetic disorders. Also, the consequence of being either a rapid or a slow metabolizer is probably related directly to the level of N-acetyltransferase activity of the liver. Unfortunately, this is a difficult organ to evaluate by direct tests, but information about the clinical importance of the acetylation polymorphism can also be obtained indirectly. For example, it can be determined whether certain adverse drug reactions or complications occur with greater frequency in people who are rapid or slow acetylator phenotypes. Up until the last few years, however, it was necessary to administer to people test drugs, such as isoniazid, to identify their status, by measuring the proportion of the free drug and the acetylated drug in a urine sample.

The *in vivo* typing method has limited the number of studies that could be done, but an alternative method has recently been developed that permits acetylator phenotyping by measuring the ratios of certain urinary metabolites of caffeine after ingestion of a test dose of the drug, or after drinking a cup of regular coffee. No doubt, the next advance will be a diagnostic test that can be carried out on DNA samples from blood leukocytes when a person's acetylator genotype can be deduced from specific DNA sequences. As it becomes practical to study a larger number of subjects, there will be better opportunities to find correlations between rapid or slow acetylator phenotype and adverse reactions to isoniazid and other drugs. Slow acetylators of isoniazid seem to be more likely to have severe adverse reactions to several other drugs, develop lupus erythematosis either spontaneously or as an adverse drug reaction, and seem to have bladder tumors more often after chronic exposure to some industrial chemicals.

Because many more people are fast or slow acetylators than the number who develop the above-mentioned complications, additional genetic and environmental factors must contribute to the development of the toxic reaction or protect against its development. A more complete study requires that additional analyses be undertaken to evaluate what other genetic markers and environmental components occur regularly in the subgroup that did develop the toxic drug reactions.

Liver enzymes such as the *P*-450 cytochrome systems, flavin-monooxygenases, xanthine oxidase, and several alcohol and aldehyde dehydrogenases that catalyze various types of drug oxidation reactions are particularly important in drug metabolism. Of these, the cytochrome *P*-450 systems have received particular attention in recent years, and they are the target of much current pharmacogenetic interest because of their unique roles in so many types of drug metabolism. Some genetic variants or deficiencies of specific *P*-450s are included in Table II, but many more types are suspected as being associated with slow metabolism or with adverse reactions from toxic metabolites of other drugs and environmental chemicals. Again, it has been very difficult to obtain enough human liver tissue to characterize the individual *P*-450 proteins and identify the variations in their structure that result from genetic mutations. [*See* CYTOCHROME *P*-450.]

II. Future Directions

Drug metabolism and toxicity studies have been done with rats and mice of defined, highly inbred genetic strains, and with congenic animals produced by selective breeding, to carry particular genetic traits on a specified genetic background. These pharmacogenetic experiments in the animals have identified certain forms of *P*-450 that seem to be of particular importance in the metabolism of selected types of drugs and foreign chemicals such as the hydroxylation of polycyclic hydrocarbons, the hydroxylation of debrisoquine and sparteine, the hydroxylation of mephenytoin, and the oxidation of ethanol. Cytochrome *P*-450 enzymes in human liver seem to be analogous with similar special substrate preferences. A number of the genes coding for the *P*-450 proteins have been cloned, and some genetic defects that modify the structure of the proteins or affect the rate of synthesis of the enzyme have been identified. These advances utilizing the methods of molecular biology have revolutionized the study and diagnosis of molecular disorders caused by genetic defects, and these methods also allow investigators to assess the status of a number of the drug metabolism enzymes, including the liver *P*-450 enzymes, by analyzing the person's blood cell genomic DNA. The new DNA analytic and diagnostic methods are overcoming the major obstacles in human pharmacogenetic studies of the past: problems of safety in studying drug effects in patients, if these studies may provoke toxic reactions, and the difficulty in obtaining for detailed study the drug metabolism enzymes that are present primarily in liver or other tissues that are very difficult to sample directly. These new techniques will result in rapid advance of pharmacogenetic investigations toward the objectives listed in Table I, particularly in the identification of new traits that cause or contribute to adverse drug effects, and the development of simple diagnostic tests to identify susceptible people.

Only a few pharmacogenetic examples have been discovered that represent pharmacodynamic alterations in tissue-intrinsic sensitivity to a drug. Resistance to coumarin-type anticoagulants is one example. These anticoagulants decrease the rate of synthesis of several liver proteins required in blood clotting. The resistant individuals show a normal response to such drugs, but it requires about 20 times as much drug than a normal person. Because these resistant people also have an exaggerated response to the antidotal effect of vitamin K, it is believed that essential receptors in the liver have been altered because of some genetic mutation that has increased the binding affinity for the vitamin and decreased it for the anticoagulant drug. This example serves as a prototype, and we can anticipate that there will be many more pharmacogenetic conditions of this type. Analyses are currently being made using the techniques of molecular biology of many specific drug receptor proteins, and genetics variants can be expected.

Bibliography

Calabrese, E. J. (1984). "Ecogenetics: Genetic Variation in Susceptibility to Environmental Agents." Wiley-Interscience, New York.
Kalow, W., Goedde, H. W., and Agarwal, D. P. (1986). "Ethnic Differences in Reactions to Drugs and Xenobiotics." Alan R. Liss, New York.
Omenn, G. S., and Gelboin, H. V. (1984). "Genetic Variability in Response to Chemical Exposure." Banbury Report #16, Cold Spring Harbor Laboratory, New York.
Vogel, F., and Motulsky, A. G. (1986). "Human Genetics. Problems and Approaches," 2nd ed. Springer, New York.

Pharmacokinetics

BRUCE CAMPBELL, *Servier Research and Development Limited, Slough, England*

Glossary

Bioavailability   The amount of orally administered drug that becomes available to the systemic circulation following formulation dissolution, absorption, and hepatic first pass, all processes that may reduce the total or absolute bioavailability

Clearance   A proportionality constant, measured in units of flow ($L.h^{-1}$), that relates the rate of elimination to the measured concentration. It provides a measure of the extraction or removal of a drug from a given volume of blood as it passes through a particular eliminating organ (e.g., liver, kidney, or whole body) at equilibrium over a given time

Drug Metabolism   This is the mechanistic study of how the body chemically modifies a foreign chemical (xenobiotic) to produce metabolites, in its attempt to remove the compound from the body. Often, but not always, this detoxifies the compound by increasing polarity to enable more rapid renal excretion, but metabolites can be more active or toxic than the parent drug.

Eoogenetics   Essentially the same as pharmacogenetics, but used in reference to chemicals of the environment, rather than chemicals used as therapeutic agents

Halflife   The time taken, measured in hours, for the drug plasma (blood) concentration to fall by one-half

Pharmacodynamics   This is the study of the action of a drug in terms of beneficial (pharmacology) or detrimental effects (toxicology) through its interaction with enzymes, ion transport, or receptors, and can be thought of as what the drug does to the body

Pharmacogenetics   This is the study of the inherited variation in drug response; in the context of this chapter, it is the genetic variation in pharmacokinetics or metabolism

Polymorphic enzymes   Multiple forms of an enzyme, differing slightly in their structure and catalytic activities

Steady state   The blood equilibrium plateau or plasma level reached following repeated administration of a drug when the amount absorbed (rate in) equals the amount eliminated (rate out), normally taken to occur 3–5 halflives after dosing commences

Therapeutic window   The range of drug levels between a minimal effective level (*MEL*) when desirable activity starts, and a minimal toxic level (*MTL*) when toxicity or unacceptable side effects occur

Volume of distribution   A proportionality constant measured as the volume (L) that a drug distributes in the body at equilibrium, relating the drug concentration in plasma (blood) to the amount of drug in the tissues

PHARMACOKINETICS comes from the Greek words *pharmakon* (drug or poison) and *kinesis* (movement) and is the study of the time course of movement of a chemical within the body. Pharmacokinetics, also called *biodisposition*, can be used to examine the fate of any natural or synthetic chemical, drug, pesticide, industrial food additive, protein, etc., in any animal species or plant. However, most recent studies have been undertaken with drugs in animals and man, so for convenience, pharmacokinetics will be discussed in terms of this association. It is measured by the analysis of the drug or its metabolite in body fluids, tissues, excreta, or expired air at various times after administration. In simple terms, it describes what the body does to the drug and examines *absorption, distribution, metabolism,* and *elimination* (ADME), normally in mathematical terms reducing complex physiological processes into relatively simple equations. Parameters derived from such equations are specific for the compound and study conditions (age, disease, race, smoking), but may also be used predictively in association with pharmacodynamics (see VIII) or toxicity to characterize the likelihood of therapeutic or adverse effects of compounds when the conditions are changed.

I. History and Introduction

Although many scientists, since the advent of modern medicines, have described the time course of drug action, it is generally accepted that it was Theorell, a Swedish physiologist, who in 1937 first proposed physiologically based kinetic models. He divided the body into discrete compartments, each with its own volume, representing the blood, a drug depot, intra- and extracellular fluid, kidney elimination, and an organ for degradation. Using rates of exchange from each compartment, he was able to describe the movement of drugs within the body making certain simple assumptions and using differential rate equations. However, it was not for a further 16 years that the actual word *pharmacokinetics* was first coined by F. H. Dost in his book *Der Blutspregel-Kinetic der Konzentration Sabläufe in der Krieslauffussigkeit.* A further 9 years later, the first symposium was held on the subject in Borstel, Germany. Over the subsequent years up until 1970, there was a growing awareness of the importance of this subject, and the fundamental concepts and mathematical equations were derived and tested by various researchers including Wagner, Gillette, Nelson, Levy, Gibaldi, and Riegelman in the United States, and Kruger-Thiemer, Rowland, and Dettli in Europe. Meanwhile, Williams in London was painstakingly compiling, by review and research, the basis for the understanding of drug metabolism. Unfortunately, analytical methods were

relatively crude and unable to measure less than μg.ml^{-1} concentrations, and were often nonspecific, measuring both unchanged drug and metabolites. The full use of the newly derived kinetic formula, therefore, had to wait a further 10 years for the development of specific analytical techniques, capable of routinely measuring nanogram (10^{-9} grams) or lower quantities in body fluids.

The large amounts of data generated by such work led to another problem, that of computer power to store and analyze the results. Although some programs were available in the early 1970s (e.g., NONLIN, AUTOAN), these needed relatively expensive hardware and some expertise. In the 1980s, there has been a proliferation of simpler programs that can run on less expensive microcomputers and can be used by expert and student alike. These include ELSMOS, SIPHAR, ELSFIT, MK model, PC NONLIN, BIOPAK, BIOV and others.

Although the youngest of the disciplines in drug development, pharmacokinetics has now become one of the most important. It alone provides a synthetic framework for all the other disciplines of *drug discovery, pharmacology, toxicology, biopharmacy, formulation development* and *clinical studies.* For example, it can provide a greater insight into which animals to choose for preclinical studies, what dosage changes are needed for aged or diseased patient, the choice of coadministered drug and food, the frequency of drug administration required, what level of compliance is achieved, why some patients do not respond or have side effects, etc. Lack of pharmacokinetic information is also the single most important reason why new drugs are not passed for marketing by the U.K. regulatory authorities-Committee on Safety of Medicines (CSM). This pivotal position for pharmacokinetics has lead to a plethora of papers and journals, and many drug-related reports now make some reference to this ubiquitous science.

II. Analytical Methods

Over the last 20 years, there have been dramatic changes in analytical techniques for the measurement of chemicals in biological fluids, with sensitivities improving by more than a millionfold. This has been achieved both by a greater efficiency of extraction and by improvement in the detectors of chromatographic systems.

TABLE I Major Body Fluids Used for Study of Pharmacokinetics

Plasma	Urine	Expired air	Cerebrospinal fluid
Red cells	Bile	Tears	Saliva
Hair	Feces	Milk	Sperm
Serum	Sweat		Vaginal fluid

A. Extraction

Before analysis, the compound is extracted from its biological matrix (Table I), the contaminating impurities removed, and the solution finally concentrated. Extraction can be achieved by either organic solvents not miscible with water, or column chromatography.

(1) Solvents (ether, chloroform, hexane, etc.) are shaken with the aqueous body fluid adjusted to a pH that maximizes the formulation of a lipid-soluble form of the drug. The latter preferentially dissolves in the organic layer, leaving the biological contaminants in the aqueous layer.

(2) For column extraction, the biological fluid is poured into a glass or plastic column containing a solid phase separator, which traps the drug by either ion exchange or lipophilicity (see III.B.2), allowing unwanted contaminants to pass through and be washed away. The drug is ultimately released by use of mild acids, bases, or solvents, and the extract is concentrated before analysis.

B. Paper and Thin-Layer Chromatography

Separation and semiquantitation of compounds can be achieved by applying the concentrate onto paper (paper chromatography [PC]) or a thin layer of silica cellulose or alumina attached to glass plates (thin-layer chromatography [TLC]), and allowing a solvent mixture to run up the paper or layer. The differential affinity of the drug for the cellulose and solvent allows separation of the components. Visualization of individual spots or bands is achieved by placing them under ultraviolet light or a light that causes them to fluoresce, or by spraying with coloring reagents and comparing the relative distance run from the origin (RF value) to known standards. Quantitation (10^{-4}–10^{-7} g.ml^{-1}) is achieved directly by (1) visual comparison with standards; (2) cutting the paper and solubilizing the drug; (3) *densitometry,* using a measure of the amount of light absorbed when a particular wavelength is passed through the drug spot; or (4) a *linear analyzer,* a direct measure of radioactivity of any radiolabeled compounds.

C. Gas Liquid Chromatography

This technique, developed in the 1960s, revolutionized pharmacokinetics, because gas liquid chromatography (GLC) could separate and accurately measure compounds down to 10^{-9} g.ml^{-1} simply and reproducibly. The concentrated mixture is injected into a long, coiled glass column (2 mm × 2 m) containing a coated solid support held in a temperature-programmable oven and purged by a low-pressure gas system (nitrogen or helium). The drug, metabolites, solvent, etc., are volatilized at an optimal temperature; the rate at which they pass through the column is proportional to their binding to the coating on the solid phase and to a lesser extent their molecular weight. The coating and temperature are variable, so optimal separations can be achieved. The method has been improved by using very thin but longer (0.5 mm × 30 m) capillary columns, allowing small amounts to be quantified with better peak resolution. The separated compounds are measured by various types of detectors: (1) nonspecific flame ionization detector (FID); (2) nitrogen specific; (3) phosphorous specific; (4) electron capture (EC); and (5) mass spectrometry (MS). The amount detected is compared to a known amount of an added internal standard using calibration curves. The GLC technique, however, has the drawback of not being able to analyze temperature-labile molecules and those with large molecular weights.

D. High Performance (Pressure) Liquid Chromatography

The most commonly used analytical technique, high performance liquid chromatography (HPLC) has now taken over from GLC as it combines sensitivity (10^{-9} g.ml^{-1}), stability, and a greater adaptability to automation. It is an analogous technique to GLC with the concentrated extract applied to a stainless steel column packed with a solid phase, but used at room temperature. The compounds are forced through under high pressure (2000 psi) with either a simple single-solvent (isocratic) or a multiple-solvent (gradient) mixture, and the compounds either differentially bind to the solid phase (normal phase: NP-HPLC) or more commonly, differently solubilize in a gradient elution (reverse phase: RP-

HPLC). Altering the amount of polar to nonpolar constituents in the solvent allows the compounds to move through the column at different speeds to achieve separation. Detection devices are (1) ultraviolet absorbance; (2) fluorescence, normal and laser; (3) radioactivity; (4) electrochemical; and (5) mass spectrometry. [*See* HIGH PERFORMANCE LIQUID CHROMATOGRAPHY.]

E. Radioimmunoassay and Radioreceptor Assay

1. Radioimmunoassay

These biological analytical techniques developed in recent years have provided a breakthrough in the routine and rapid measurement of drugs and hormones at low levels in the body. They were used initially in 1960 by Yalow and Berson in New York to measure insulin and by Ekins in London to measure thyroxine. A competition assay method can be outlined as follows and divided into *antibody production, synthesis of radiolabel,* and *assay:* (1) the drug (or hapten) to be analyzed, unless of high molecular weight (>5000 Da) is reacted with a protein (normally albumin) to produce an *immunogen* capable of being antigenic; (2) the immunogen (~100 μg) is injected intradermally in an oily *adjuvant*, e.g., Freunds to enhance the immune response, normally into guinea pigs, rabbits, sheep, etc., at monthly intervals for periods up to 6 months. This stimulates the production of *antibodies (immunoglobulins)* specific to the injected immunogen and harvested from the animal for use; (3) in a competitive radioimmunoassay (RIA), the original drug is radiolabeled with either I^{125} or tritium to obtain a label of high specific activity and this is added to the antibody at a concentration for all the radioactivity to be bound; (4) when an unknown amount of the drug to be analyzed is added to a solution containing the radioactive drug–antibody complex, an amount of radioactive drug is displaced proportional to the added cold drug; (5) the remaining radioactive drug–antibody complex is removed by charcoal, centrifugation, or second antibody precipitation. The supernatant-free radioactive drug is counted by scintillation spectrophotometry for tritiated compounds and by gamma counting for I^{125}. Newer techniques of *scintillation proximity assay* obviate the need for the removal of the drug–antibody complex. By constructing calibration curves with known quantities of the cold drug, the amount of displaced radioactive ligand proportional to the concentration of the drug in the sample relates

to the concentration of the unlabeled drug in the sample. [*See* RADIOIMMUNOASSAYS.]

RIA is rapid (300 samples/day), sensitive (<10–100 pg.ml^{-1}), and simple (little expertise required); automation is easy, and is cheap for many samples; sample volumes are low (100–300 μl); and no extraction is necessary.

Among its disadvantages are a long development time (up to 6 months); initial development is expensive, particularly for few samples; and specificity may be poor with crossreactivity to metabolites or compounds of similar structures.

Another type of RIA is the *solid phase assay,* which may be either direct or indirect. In the direct assay, the antigen is bound to the plastic surface of the wells in microtiter plates. Iodine^{125}-labeled, antigen-specific antibodies are then applied in solution for approximately an hour at 37°C. After this, the plates are washed, and the amount of bound radioactivity is counted in a gamma counter. For greater sensitivity, the indirect assay may be more appropriate. In this case, the antigen is bound to the plate as before and a nonradiolabeled, antigen-specific antibody is applied. After incubation and washing, a second radiolabeled antibody which recognizes the first antibody as an antigen is added. After incubation and washing, the plate is counted as before. This has the advantage of magnifying the amount of radiolabel that will be attached (albeit indirectly) to the original antigen.

A variation on this procedure is ELISA (enzyme-linked immunosorbent assay). It is similar in principle to the solid phase RIA with the exception that an enzyme (for example, alkaline phosphatase) is coupled to the antibody rather than a radiolabel. Subsequent incubation of the antibody-bound enzyme with its substrate produces a colored product in proportion to the concentration of the original antigen. Such a reaction may be simply quantified by colorimetric analysis.

2. Radioreceptor Assay

This technique is in many ways similar to RIA and was first described for the assay of ACTH by Lefkowitz in 1970 using homogenates of adrenal cortical tissue. In radioreceptor assay (RRA), the binding of the radioactive ligand is, in contrast to RIA, to a specific receptor, normally at which the drug or hormone may elicit its pharmacological activity; this is not always the case. The receptors are prepared directly from tissues of lung, brain, kidney, platelets, etc., and may be crude homogenates or purified and solubilized preparations. All the advantages of RIA are applicable to RRA, with the additional advantage that only the active biological component is measured. Thus, where a racemate is administered, only the active eutomer that has a high specificity for the active receptor will be measured, and therefore RRA can be thought of as a biological activity measurement. It is a technique relatively more quickly developed than RIA. However, the disadvantages include lack of specificity, since an active metabolite or other similarly active drug will be measured. Multiple receptor binding, e.g., β blockers to $\beta 1$ and $\beta 2$ receptors, makes interpretation difficult. Frequent production of receptors with each batch need to be tested and characterized before use. Because of these disadvantages, neither RIA nor RRA has been used extensively for pharmacokinetic analysis, mainly because of lack of specificity of analysis.

F. Metabolite Identification

Mass spectrometry allows for the specific identification of an unknown compound by identifying the component parts or masses once the molecule has been fractured by *electron impact, chemical ionization,* or *fast atom bombardment.* Other techniques for metabolite identification include *nuclear magnetic resonance* (NMR), in which the drug is placed within a magnetic flow and bombarded with radio frequencies that resonate the odd-numbered nucleons of ^1H ^{15}N, inducing a transition energy state within the molecule, and which can interact with protons in neighboring atoms. The resultant spectra are a finger print for the molecule under investigation. *Infrared spectroscopy* is also used for metabolite identification.

III. Physicochemistry and Membrane Transport

For any drug to produce an activity, it should normally be absorbed, pass through various tissues and body fluids and finally interact with its site of action. To do this, it must be able to dissolve in aqueous fluids, and pass through the various lipoprotein membranes in the body. This is known as *drug transport* and the rates and extent of these processes are dependent on the physicochemical properties of the drug and the part of the body involved.

A. Membrane Transport

Membranes throughout the body are a lipoprotein sandwich composed of an inner, mainly lipid core, and an outside surface layer of protein. This is owing to the configuration of the phospholipid with the *hydrophobic* or *lipophilic* (water hating or *lipid*) portion of the molecule facing inward while the *hydrophilic* (water liking or polar) layer faces the outside, more aqueous, environment. In addition, there are both narrow aqueous-filled channels and larger aqueous pores crossing the membrane. Chemicals can be transported by two basic mechanisms: (1) *passive diffusion* and (2) *active transport.* [*See* MEMBRANES, BIOLOGICAL.]

Passive diffusion is the movement of a molecule down a concentration gradient until equilibrium occurs. It is the main transporting mechanism of most xenobiotic chemicals.

The rate of penetration of such molecules by diffusion is dependent on the *permeability* (P) and *surface area* (S) and the concentration difference $(C_{n1} - C_{n2})$ of the molecules on either side of the membrane according to *Fick's Law* of diffusion.

$$\text{Rate of penetration} = (C_{n1} - C_{n2}) \cdot (S) \cdot (P). \quad (1)$$

Although the concentration gradient is the driving force for molecular transport under normal conditions, the permeability determines and controls the rate and extent of transport. Of the four major determinants of permeability, *membrane composition and thickness, lipid solubility,* and *ionization* of the molecule, the latter two are by far the most important physicochemical factors influencing both drug transport and binding. *Blood flow* is another important consideration for transport. If a drug can easily pass through membranes, then there is no barrier for its diffusion into tissues. Under these circumstances, the rate of uptake can be reduced if blood flow or perfusion is slow, producing a *perfusion rate limitation* to transport. On the contrary, if a drug cannot easily diffuse into the tissue, blood flow changes will have little effect, and the drug has a *diffusion rate limitation.*

Active transport normally carries water-soluble endogenous compounds like glucose, amino acids, etc., across and up a concentration gradient through water channels. Unlike passive transport, this process is compound specific, usually *carrier mediated* or *enzyme linked,* and is not normally associated with drugs except some that are secreted in the bile or by the kidneys.

B. Physicochemistry

1. Ionization

Most drugs are either *weak acids* (A) or *weak bases* (B) and therefore exist in solution, either as their polar ionized form (acid/B(A$^-$; base H$^+$) or in lipid-soluble unionized form (acid HA; or base B) depending on the pH of the solution of body fluid. When the pH equals the pKa, the ionization constant of the compound, concentrations of both are equal (equations 2 and 3):

$$\text{Unionized} \qquad \text{Ionized}$$

$$[HA] + [H_2O] = [H_3O^+] + [A^-] \text{ (Acid)} \qquad (2)$$

$$[B] + [H_2O] = [BH^+] + [OH^-] \text{ (Base)} \qquad (3)$$

but when the pH becomes more acidic, the equilibrium shifts to more unionized for acids or less unionized for bases, and vice versa when the pH is basic. It is probable that only the unionized lipid form is transported, so both the pKa of drug and the pH are important for movement across the membrane.

2. Lipid Solubility

Lipid solubility is a measure of the extent that a drug can dissolve in a water-immiscible organic solvent and can be measured by equilibrating a drug between an aqueous buffer of controlled pH (7·4) and an organic solvent, normally octanol. The ratio of drug concentrations in the solvent and aqueous phase is called the distribution coefficient (D), but because this is dependent on the pKa of the drug and pH of the solution. The real distribution or partition coefficient (P) of the unionized drug is given by equations 4 and 5, depending on whether it is an acid or base.

$$\text{Log P (acid)} + \log D + \log (1 + 10^{pH-pKa}) \qquad (4)$$

$$\text{Log P (base)} = \log D + \log (1 + 10^{pKa-pH}) \qquad (5)$$

C. Quantitative Structure Kinetic Relationships

Certain physicochemical properties of a drug can be important determinants of its kinetics and metabolism and have been used to describe *quantitative structure pharmacokinetic relations (QSPR)* in an attempt to predict certain kinetic parameters.

1. Diffusion

As early as 1897, Overton and Meyer, suggested that drugs and chemicals cross lipid membranes by dissolving in the lipid layer, and that the rate and extent of this diffusion were linearly related to the lipophilicity (log P) of the unionized form, because only this would be transported.

$$\log \text{Diffusion} = b \log P + a \qquad (6)$$

Although for some systems, this relationship may be true, e.g., the greater the lipophilicity, the faster and more extensive the transport, it has become apparent that it is unable to explain the observation that at values of high log P, transport becomes slower. More recently, Hansch and others have defined a parabolic relationship (equation 7) between diffusion and lipophilicity, suggesting that there is a maximal value of log P above which a very lipid drug stays associated with the lipoprotein membrane rather than passing through.

$$\log \text{Diffusion} = -a \log P - b \, (\log)^2 + C \qquad (7)$$

Using this relationship, it is possible to design drugs according to whether inclusion or exclusion to a particular organ or system, e.g., CNS, is required. It is of interest that the optimal value of log P is approximately 2 (octanol: buffer pH 7·4) regardless of whether transport is across the skin, stomach, intestine, blood–brain barrier, kidney tubule, or even plant leaves.

2. Lipoprotein Binding

Nonspecific protein and tissue binding is also directly related to lipophilicity, but unlike diffusion, the relationship is not parabolic but linear across most values of log P. However, at log P values greater than 2, the relative tissue binding becomes greater than the plasma protein binding, and distribution volume throughout the body becomes disproportionately greater. Thus a drug with a high lipid solubility can be predicted to have (a) high protein binding and may displace similarly bound drugs; (b) high distribution volume leading to low blood levels, long halflife, infrequent dosing, slow buildup to steady state and (c) low renal clearance due to low filtration (high protein binding), and high tubular reabsorption leading to high hepatic clearance and first-pass metabolism and circulatory metabolites.

3. pH Partition Hypothesis

Although the rate of transport is dependent on log P, the relative amount of drug on one side of a membrane compared to the other is governed by the relative ratio of ionized to unionized drug on both sides, because it is the unionized form that is transported.

This can be calculated from the pH difference of tissue (pH_t) and plasma (pH_p) and the pKa of the drug according to the Henderson–Hasselbalch equations 8 and 9:

$$\text{WEAK ACID} \quad \frac{\text{Conc in tissue}}{\text{Conc in plasma}} = \frac{1 + 10^{pH_t-pKa}}{1 + 10^{pH_p-pKa}} \quad (8)$$

$$\text{WEAK BASE} \quad \frac{\text{Conc in tissue}}{\text{Conc in plasma}} = \frac{1 + 10^{pKa-pH_t}}{1 + 10^{pKa-pH_p}} \quad (9)$$

Thus, weak bases will concentrate in more acidic solutions compared to plasma ($pH_p 7\cdot4$) e.g., stomach pH 1–2, red cell pH 7·2, milk pH 7·2, urine pH 6–7, while the opposite is true for weak acids. Unfortunately, few kineticists examine the physicochemical aspects of a drug, but much can be learned and predicted about the drug's behavior in the body from this information.

IV. Biodisposition

A. Absorption

The process by which a drug moves from its site of administration into the body or to where it is actually measured is called *absorption*. Normally this is following oral administration but can also describe many other routes, in fact, any route where the drug must cross membranes before entering the systemic circulation (Fig. 1). Before a drug is absorbed from any part of the *gastrointestinal (GI) tract,* it must be in solution so that the unionized part can be transported by passive diffusion. In a few cases, highly ionized compounds like the quarternary ammonium drugs are slowly and incompletely absorbed. The important factors that control absorption are formulation, solubility, pKa and pH, intestinal surface area, other factors.

1. Formulation

Formulations are developed to control the delivery of the drug into the body. Thus solutions, syrups, suspensions, soft gelatin capsules and sublingual lozenges are rapid delivery systems with early peak blood levels (15 min–1 hour) while compressed tablets and hard gelatin capsules are slower delivery systems yielding peak levels of drugs of 1 to 4 hours, depending on the rate of dissolution of the formulation. *Dissolution* is a process by which a formulation disintegrates into granules, deaggregates into particles and dissolves into solution. It can be controlled and slowed down by *drug delivery*

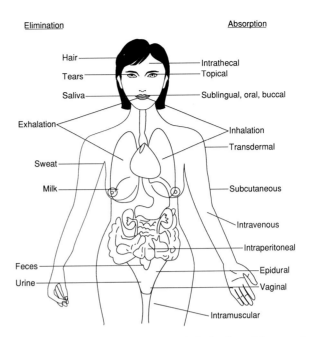

FIGURE 1 Routes of absorption and elimination of drugs and chemicals from the human body.

systems which include (a) enteric coating, (b) sustained release, and (c) osmotic pumps. Enteric coating is used to protect an acid-labile drug from stomach degradation in order to be released and absorbed lower in the GI tract. It can also protect against drug-induced irritation. Sustained release slows the release of a drug to achieve constant or controlled levels so that one dose can be administered less frequently. This is particularly useful for rapidly eliminated drugs. Osmotic pump systems achieve the same effect by allowing water to pass through a semipermeable membrane and slowly pump the drug out as a solution at a constant rate.

2. Solubility

Although the lipid-soluble form passes through membranes for GI-tract absorption, the drug must initially be in aqueous solution. Thus, most drugs, unless they have an inherent water solubility are administered as their salts, hydrochloride or sulphate for weak bases, sodium or potassium for weak acid.

3. pKa

The rate of absorption from the gastrointestinal tract, buccal cavity, vagina, or skin, etc., will to some degree be dependent on the pKa of the drug and pH of the body fluid. Once into solution, the ease of membrane transport is proportional to the amount of unionized drug in equilibrium and its lipid solubility. Thus, in the acid pH (1–2) of the stomach, weak acids are unionized and some absorption can take place while weak bases remain ionized and are not absorbed. However, few drugs, are absorbed to any great extent in the stomach.

4. Surface Area

The stomach is a physiological vat with a relatively smooth surface, secreting acids and enzymes to degrade food. The combination of a small surface area, 1 m^2, and low blood flow 150 ml.min^{-1}, when compared with the intestine with its highly convoluted microvilli that increase the surface area and blood flow to 200 m^2 and 1000 ml.min^{-1} respectively, means that most drugs are absorbed from the intestine regardless of their pKa, although the rates may be different.

5. Other Factors

A number of other factors influence drug absorption including (a) food may have little effect for some drugs, but for others, it can enhance or inhibit absorption by altering dissolution and transport or delaying gastric emptying; (b) fluids will normally enhance the rate of absorption; (c) supine posture or no physical activity can slow down absorption owing to delayed gastric emptying and reduced splanchnic (GI tract) blood flow. Increased activity will increase absorption; (d) coadministration of drugs may enhance or inhibit total absorption because of their effects on gastric motility, blood flow, and hepatic metabolism; (e) diseases such as cancer, Crohn's disease, and cardiac failure can also reduce absorption.

B. Bioavailability

When drugs are administered, they are conveniently packaged in formulations such as tablets, capsules, suppositories, sprays, patches, etc., to aid in the absorption of the product. Bioavailability measures the factors that influence the rate and extent of arrival of a drug at its site of action, but in practical terms, it is defined as the fraction (F) of the dose reaching the systemic circulation as unchanged (not metabolized) drug following administration by any route. *Absolute bioavailability* compares the amount of drug available from the test formulation with an IV dose, while *relative bioavailability* compares the availability of two dif-

ferent formulations that may be used clinically, and if comparable, can be said to be bioequivalent.

Different methods of measurement, design, and statistical analysis are used. Various kinetic parameters are used to measure bioavailability in terms of rate (a) peak concentrations (C_{max}) and extent, (b) time to C_{max}(T_{max}), and (c) infinite **area under** the plasma (blood) time **curve** (AUC) after single and repeated dosing. Experiments are undertaken on a sufficiently large number of healthy volunteers (n = 12–50) so that the data will provide an 80% probability of detecting a 20% difference in these parameters, (80/20 power rule) particularly AUC, as the measure of extent of absorption. Various study designs are used including (a) *crossover design,* in which each subject takes two formulations to obtain a within-subject comparison, or (b) a *latin-square design* for more than 2 to 4 formulations, or (c) *incomplete block* for more than 4 formulations, in which subjects do not take all the formulations. Statistical analysis of the parameters is under constant discussion but includes (a) the standard *power approach* or hypothesis testing using analysis of variance (ANOVA) to test the null hypothesis for no difference; (b) *confidence intervals* (symmetrical and nonsymmetrical) about a mean difference, which provides a more meaningful basis for clinical decisions; (c) *repeated measure* of ANOVA for all individual plasma levels, or lastly; (d) a new but perhaps the most relevant, *Bayesian* approach using predefined or determined kinetic or clinical objectives.

In general, formulation differences in effect are likely to be more important for drugs with a narrow therapeutic window, with a steep slope of the kinetic dynamic relationship. There is also a growing awareness that clinical end points can and should also form the basis of more meaningful comparative bioavailability testing whenever possible.

C. Distribution

Once a drug is absorbed into plasma, it is rapidly circulated around the body by the blood once every 2–3 minutes. The process by which a drug is reversibly transferred from one part of the body to another is called distribution and is usually calculated from the site of measurement (plasma) and expressed as a dilution volume. A drug will pass through membranes to and from various body fluids, subsequently binding to tissue macromolecules. The extent of distribution for any drug is largely dependent on its lipid solubility and its relative binding to blood and tissue protein. Thus, a drug may be highly bound to plasma protein but still distribute extensively into the rest of the body if its affinity to tissues is greater.

1. Volume of Distribution

Various kinetic distribution volumes can be measured (see Glossary and Definitions) from plasma or blood data with perhaps the most frequently used being the apparent volume of distribution at equilibrium (V). Others include the initial dilution volume (V_i) that a drug would occupy when first dosed, the steady state volume (V_{ss}) when all tissues are in equilibrium with plasma but assuming no elimination, and V_u, the volume of free unbound drug. Although V_i may be similar to the plasma volume after an IV dosing, the volumes have no physiological volume. Values at equilibrium often are much larger than a *body volume*. *All volumes* of distribution can be thought of as only dilution values or mathematical balancing terms.

2. Blood Protein Binding

Most drugs bind to blood components to some degree. Although all binding will affect kinetics, it becomes more important when it exceeds 90%. In the red cell, the major proteins for binding are hemoglobin and the enzyme carbonic anhydrase, while in plasma, the two main proteins are albumin, to which many acidic and neutral drugs bind, and α-glycoproteins to which basic drugs tend to bind. Some binding can also occur to globulins and lipoproteins. Protein binding can be measured *in vitro* by incubating the drug (C) with plasma or specific protein and then separating and measuring the free drug (Cu) from that bound (C–Cu). Percentage binding (PB) is given by

$$PB = \frac{C - Cu}{C} \cdot 100. \qquad (10)$$

Separation of Cu is achieved by (a) *equilibrium dialysis* using passive diffusion into a protein-free buffer (pH 7·4) through a semipermeable membrane (SPM); (b) *ultrafiltration,* which forces the aqueous solution of free drug through a semipermeable filter; (c) gel filtration, which separates protein-bound drug from free drug by molecular weight sieving; (d) *electrophoresis,* which achieves separation by differential migration in an electrical field.

In practice, however, the percentage unbound (PB-100) or fraction unbound (fu) is clinically most useful. A change from 90 to 95% bound may seem little but actually reduces fu by 50%. Binding can be reduced in disease states because of low albumin levels (hepatic and cardiac failure, nephrotic syndrome), while binding to α-glycoprotein (a stress protein) is increased owing to increased levels in inflammatory arthritis, Crohn's disease, myocardial infarction, burns, etc. It has long been thought that changes in protein binding would alter the therapeutic activity of the drug, as more free drug would be available at the site of action. With the exception of high clearance drugs administered intravenously, this is not the case, as more drug is at the same time available for elimination and at steady state; although the total levels may alter, the free active drug levels remain constant if there is no change in free clearance. Theoretically, it can be argued that steady-state levels (Cu_{ss}) are dependent on free clearance (CL_u) and not volume; if CL_u does not alter, then neither will Cu_{ss}. Protein binding is therefore of less importance than once considered.

D. Elimination

If the body was unable to eliminate drugs, one dose would last a life time. Nonspecific mechanisms must therefore exist to remove foreign substances from the body. There are many routes possible (Fig. 1) but the most important are excretion, metabolism, biliary, exhalation, and lactation. The kinetic parameters associated with elimination are clearance, halflife, and mean residence time.

a. Clearance is a proportionality factor relating the rate of drug elimination (k_{el}) to plasma drug concentration (C).

$$k_{el} = CL \cdot C. \qquad (11)$$

and the total body clearance (CL) is the sum of the individual clearances from each eliminating organ. Thus, clearance can also be thought of physiologically, as the loss or extraction of drug by that organ as blood passes through. It is therefore related to the blood flow (Q) and the degree of extraction (E) for the particular drug.

$$CL = Q \cdot E. \qquad (12)$$

For highly extracted or hepatically metabolized drugs where $E \simeq 1$, the clearance approaches hepatic blood flow (Q_H) (1500 ml.min^{-1}) (see equation 12), and elimination is only limited by perfusion through the organ and not by events within the liver

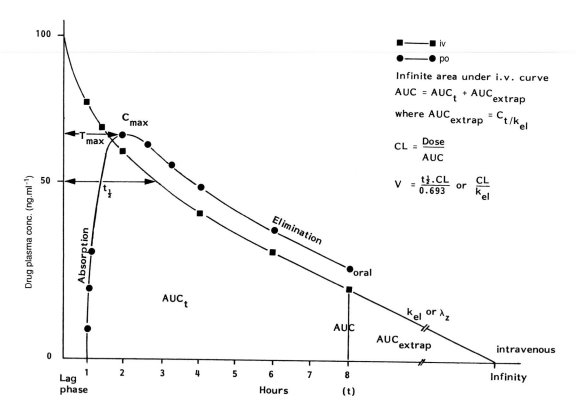

FIGURE 2 Plasma concentrations of a drug following single oral and intravenous administration showing derivation of some kinetic parameters. The area under the plasma time curve (AUC$_t$) can be measured by summation of each area under sequential time passes.

(perfusion-rate limited). These drugs have high first-pass metabolism, low systemic bioavailability, and the clearance can be markedly changed by alterations in portal blood flow (disease, food, concomitant drugs).

In contrast, for low-extraction drugs, $E < 0.3$, extraction is not controlled by blood flow but by other processes: (a) enzymatic reactions; (b) biliary excretion; (c) the degree of protein binding within blood; or (d) diffusion into tissue (diffusion-rate limited).

Clearance is important to kinetics, as estimates of availability or the amount entering the body (F) can be made from intravenous data assuming only hepatic metabolism

$$F = 1 - \frac{CL}{Q_H}. \qquad (13)$$

In addition to the total amount absorbed (F), the most important determinant of average steady state drug levels (C$_{ss}$.av) after repeated dosing, is clearance and in particular, it is the free drug clearances which are likely to affect activity as shown in Equation 14.

$$C_{ss}.av = \frac{F.D.\tau.}{CL}. \qquad (14)$$

where F.D. is the absorbed dose administered over a certain dosing interval, τ. Clearance (CL) can be simply measured (equation 15) by calculating the area under the infinite plasma concentration time curve (AUC) (Fig. 2) after intravenous administration. It cannot be measured after oral dosing unless the extent of absorption is known.

$$CL = \frac{Dose}{AUC_{iv}}. \qquad (15)$$

b. Half-Life The time required for a drug to decline to half its concentration has long been considered to be an important kinetic parameter, perhaps because it can be easily measured (Fig. 2). It fact, it is a hybrid constant dependent on clearance and the distribution volume (V)

$$t\tfrac{1}{2} = \frac{0.693 . CL}{V} \qquad (16)$$

and a long half life may be due to a high distribution volume or a slow clearance. Similarly, both volume and clearance could alter without changing the half life. Although a poor indicator of elimination, it gives information on (a) time of drug in the body; (b) time to reach steady state on continual drug administration ($\approx$4-5 $\times$ $t_{\frac{1}{2}}$); and (c) time for dosing interval, approximately every half life. Drugs may have several half lives, often dependent on different rates of elimination from different volumes, but the last or terminal half life ($t_{\frac{1}{2}z}$) is the most important and used for most kinetic calculations. Half life is related to the elimination constant k_{el} (λ_2) according to equation 17.

$$t_{\frac{1}{2}} = \frac{0.693}{k_{el}}. \qquad (17)$$

c. Mean Residence Times (MRT) These are taken from the statistical moments analysis of chemical engineers to overcome the problems of multiple half lives, and recirculation with secondary peaks, by giving an average time that the drug has spent in the body, sometimes expressed as turnover or sojourn times.

1. Renal Excretion

The major route of elimination of the more polar drugs and metabolites is by renal excretion.

$$\text{Rate of urinary excretion} = CL_R \cdot C \qquad (18)$$

where CL_R = renal clearance and C = plasma drug concentration. The appearance of the drug in urine is the net result of (a) filtration, (b) secretion, and (c) reabsorption.

Rate of excretion = Rate of filtration
+ Rate of secretion − Rate of reabsorption (19)

Filtration, a passive process of the kidney nephron, occurs across the glomerular membranes in the Bowman's capsule. Filtration rate at any instant is dependent on plasma concentration; protein binding (only free drug Cu is filtered); molecular weight (exclusion above 1–2,000 Da); and glomerular filtration rate of plasma water (GFR). In healthy individuals, GFR is approximately 125 ml.min^{-1} measured by inulin or creatinine, compounds that are not protein bound, secreted, or reabsorbed. For a drug with a filtration rate based on unbound drug different from this value, other processes must be occurring.

$$\text{Rate of filtration} = GFR \cdot Cu. \qquad (20)$$

Secretion, an active process, transports drug from blood into the proximal tubule of the nephron if renal clearance is greater than fu.GFR. There are separate mechanisms for acids and bases, but each lacks specificity, which can lead to dangerous drug interactions of coadministered drugs showing the same secretory mechanisms, e.g., quinidine and cimetidine.

$$CL_R = fu.GFR + \left[\frac{\text{secretion rate} - \text{reabsorption rate}}{C} \right]. \qquad (21)$$

Reabsorption is a passive process that transports the drug back into the circulating blood from urine all along the nephron, starting at the proximal tubule where water is absorbed. It occurs to some extent for many drugs and is dependent on urine flow; urinary pH; pKa; and lipid solubility. Urine flow is important for drugs that are extensively reabsorbed ($CL_R < 20$ ml.min^{-1}). At high flow rates, because the system is never in equilibrium, the urine will wash the drug through the nephron without allowing adequate time for diffusion. The rate of reabsorption is dependent on the concentration gradient and on the ability of the unionized drug to diffuse through tubular membranes. The extent of reabsorption is greater for high-lipid-soluble drugs. According to the Henderson–Hasselbach equations 8 and 9, it is dependent on the pKa and the pH difference between urine and plasma. Urinary pH can fluctuate between $5\cdot0$ to $7\cdot5$ (average $6\cdot3$), depending on diurnal variation, exercise, food, drugs, etc., so the pKa of the drug can play an important role in renal excretion. Thus excretion of a weak base (pKa 10) can vary from 1% of the dose in alkaline to 80% in acid urine. In general, highly polar drugs are not reabsorbed; strong bases are not reabsorbed when the pKa is >12 nor for acids with pKa <2 irrespective of urinary pH, as they exist mostly in the ionized form. Urinary pH fluctuations have a pronounced effect on reabsorption for lipid-weak bases with pKa between $7\cdot0$ (high reabsorption) and 11 (low), while for weak acids, the pKa needs to be between $3\cdot0$ (low) and $7\cdot5$ (high).

2. Metabolism

Many drugs are lipid soluble with some degree of protein binding with a subsequent reabsorption from the kidney tubules and a reduced renal filtration, so the body needs to remove the xenobiotic by biotransformation to more water-soluble compounds, which will be excreted in the kidneys. This biotransformation primarily occurs in the liver, the first point of entry of a drug into the body after ingestion, although it can also occur to a much lesser extent in the kidneys, lungs, blood (enzymic hydrolysis), as well as the gut wall and intestinal contents (microflora). When a drug is metabolized during absorption through the liver, it is called first-pass hepatic metabolism and is important for drugs with high clearances. When hepatic clearance is high (>1000 ml.min^{-1}), this is generally undesirable, because of (a) less available drug; (b) extensive production of active or toxic metabolites during the absorption process; (c) larger intersubject variation in blood levels; (d) drug interactions; and (e) greater variation in disease.

Metabolism as an enzyme process is relatively nonspecific and can deal potentially with any xenobiotic using two basic mechanisms, phase I and phase II. Phase I reactions, mostly oxidation and reduction, use mixed-function oxidase enzymes located on the endoplasmic reticulum in the microsomal fraction of hepatic cells and occurs by binding to cytochrome P-450. There are at least 15 subtypes involved in different phase I reactions for different groups of compounds.

Phase II reactions normally involve the formation of conjugation or addition products occurring with a diverse group of transferase enzymes, UDP-glucuronyl, sulphur, methyl, acetyl, and glutathione-S, transferases. Unlike phase I metabolites, phase II compounds are polar and normally inactive. An individual's metabolic status is dependent on many factors including genetic, race, gender, food, and disease states. Coadministration of drugs, chemicals, or even brussel sprouts, may increase the number of microsomal enzymes causing induction, or decrease the number, inhibition, leading to lower or higher clearances of metabolized drugs respectively. Some animals may be deficient in certain pathways for example, the dog (acetylation), the cat (glucinoridation), and the guinea pig (mercapturic acid formation). Species differences in metabolism can be compared using *in vitro* culture techniques from isolated hepatic tissue: (a) microsomal (phase I only); (b) hepatic slices; (c) hepatocytes; or (d) isolated specific enzymes (e.g., UDP-glucuronyl transferase), and are especially useful for choosing the best species for experimental studies to compare with man.

3. Biliary Excretion

Biliary excretion is a complex and poorly understood process in which drugs are cleared from circulation by being actively secreted against a concentration gradient from hepatocytes into bile. It is a universal but relatively slow process. Biliary flow (BF) is only 0.5 ml.min^{-1}, and it is thought to be a safety overflow valve for compounds not easily metabolized or renally excreted.

$$\text{Biliary clearance} = \text{BF} \cdot \left(\frac{C_{bile}}{C}\right). \qquad (20)$$

where C_{bile} = drug concentration in bile, and C = plasma drug concentration.

There are separate mechanisms for biliary secretion of acids, bases, and unionized compounds, with two main controlling factors, molecular weight and polarity. The structure of the drug is also important because the process is active, but little is known about this mechanism.

a. Molecular Weight Molecular weight is a major determinant as many drugs or metabolites with a MW between 250 and several thousand are excreted in the bile. However, the lower-end threshold is species dependent, so that for large animals, man and monkey, secretion will not occur below approximately 500 Da, while biliary excretion for small animals like rat and mouse occurs above 250 Da.

b. Polarity In addition to high MW, high polarity will encourage biliary excretion. Phase II conjugated metabolite, e.g., glucuronides, which have both an increase in the MW of 176 Da, and an increase in polarity, are preferentially eliminated via the bile.

Once excreted into the bile, the drug passes down the biliary tract into the gallbladder (absent in rats), the contents of which (bile acids) are emptied into the small intestine during meals to aid the digestive process. The drug may be directly absorbed again or the conjugate may be hydrolyzed by gut flora back to the unconjugated drug (aglycone) which may also be reabsorbed. This process can continue in a cycle: absorption, biliary excretion, absorption, etc., and is known as enterohepatic circulation. Under extreme conditions, it can cause secondary *absorption* peak oscillations as small amounts of drug escape the cycle and enter the systemic circulation. Similarly, escape may occur down the gastrointestinal tract causing fecal excretion but because enterohepatic circulation can occur after IV administration, high fecal drug levels do not always equate with poor absorption.

4. Lactation

Although this is not a major route of elimination, excretion into milk is important because during weaning, neonates may be exposed to unacceptable large doses of the drug via ingestion of the mother's milk. This is of particular concern if additionally, neonatal clearance is reduced. The amount of drug in milk is, according to the Henderson-Hasselbach equation, dependent on (a) pH of milk; (b) pKa; (c) lipid and protein binding in milk and plasma; and (d) time of lactation. The pH partition hypothesis enables calculation of the free drug ratio in milk and plasma, and as milk pH is slightly more acid $(7 \cdot 2)$ than that of plasma, basic drugs will be concentrated in milk. However, the protein content of milk is relatively low, and therefore the degree of uptake is particularly related to the lipid solubility of the drug. However, the binding to milk fat and protein is dependent on the time of lactation, as early milk (colostrum) is low in lipid, but subsequently increases during the first month postpartum. The infant dose from milk can be calculated from maternal steady-state levels (C_{SS}), the ratio of the areas under the milk/plasma time curves over the same period $(AUC_M/AUC$ and the volume of milk ingested (V_{milk}), on average about 150 ml.kg^{-1}.day^{-1},

$$\text{Infant dose} = C_{ss} \cdot \left(\frac{AUC_M}{AUC}\right) \cdot V_{milk}. \quad (23)$$

E. Pharmacogenetics

All of drug biodisposition and response is, to some degree, controlled by inheritance, but the differences between individuals are normally diffuse and indistinguishable. When specific feature differences of genetic origin are measurable (blue or brown eyes), this is known as *genetic polymorphism*. In drug metabolism, genetic polymorphism can be detected in population studies by polymodal frequencies of rates of metabolism; clearance; or steady-state drug levels showing two peaks, one for slow and the other for fast eliminators. There are perhaps three well-recognized metabolic polymorphisms: (a) *acetylation*, due to differences in the enzyme n-acetyl-transferase. The prevalence of slow acetylators is Egyptians 91%, North Americans 60%, British 40%, Eskimos 10%, for drugs like isoniazid,

procainamide, hydralazine and many sulphonamides where levels can differ more than 4-fold in the patient population receiving the same dosage regimen of drug leading to side effects and toxicity or ineffective therapy; (b) *hydroxylation*, due to deficiency of certain cytochromes of P-450. Prevalence in Caucasians is 8%, but it is higher for Orientals, for drugs like debrisoquin, nortriptyline, phenformin, perhexiline and metoprolol; and (c) *hydrolysis*, due to differences in plasma cholinesterase, occurring in 1 of 2500 in the general population, in particular, for succinylcholine. [*See* PHARMACOGENETICS.]

Because of the potential side effects associated with slow metabolism, attempts have been made to identify or phenotype patients before therapy by administering single or multiple markers. However, the clinical value of this approach has yet to be proven.

F. Repeat-Dose Kinetics

When a drug is repeatedly dosed, plasma levels will build up or accumulate to steady-state levels (C_{ss}) when the rate of drug entering the body equals the rate of elimination (Fig. 3). The time taken to reach steady state is usually between 3 and 5 half lives, but may be longer if small but significant deep compartments with longer unrecognized terminal elimination rates are present. For drugs with short half lives, much less than the dosing interval, steady state is reached after one dose. As stated previously, the magnitude of these plateau levels is dependent on only clearance and dosage, and not on body weight. However, the volume of distribution will influence the magnitude of the maximum and minimum steady state levels $(C_{ss}$ maximum and C_{ss} minimum) and half-life (Fig. 3). Thus, obese subjects with a potentially larger volume will reach C_{ss} more slowly and with longer half lives; the oscillations between maximal and minimal levels will be smaller at a fixed dosing interval. The infinite area under the plasma time curve (AUC) is theoretically equal to the area under the plasma time curve during a dosing interval at steady state. If this is not found, then the kinetics of the drug may have changed (*inhibition or induction*) during repeated dosing with a corresponding change in k_{el} at the end of dosing, assuming no change in distribution. Similarly, the accumulation of the drug can be calculated by comparing the AUC during the dosing interval after a single dose and at steady state.

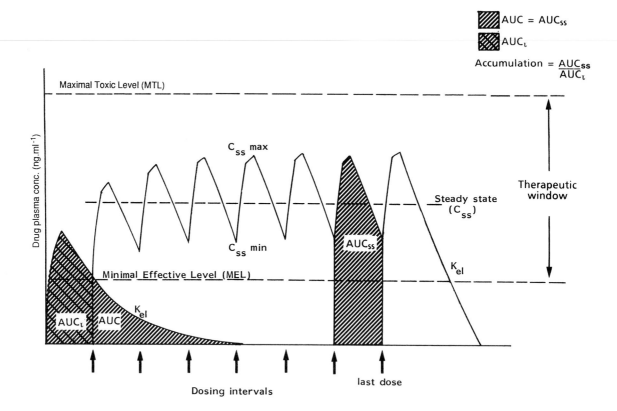

FIGURE 3 The buildup of plasma levels to steady state within an ideal therapeutic window during repeat dosing of a drug.

G. Stereoselectivity

Most drugs have one or more carbon atoms, each one of which has four bonds. If all the chemical substituents associated with these bonds are different, then the carbon atom can be said to be *asymmetric,* and the molecule will have a *chiral* center. Chirality comes from the Greek word ($\chi\varepsilon\iota\rho$) meaning a hand. A molecule with an asymmetric carbon atom can exist as two separate but spatially different (stereo) and noninterchangeable compounds, and just like our own hands, are nonsuperimposable mirror images of each other but with the same elementary chemical composition. These compounds are called *enantiomers* (Greek $\varepsilon\nu\alpha\nu\tau\iota o$ opposite shape), and there may be two or more existing for one formula, depending on the number (n) of asymmetric carbon atoms, according to the formula:

$$\text{Number of Isomeric pairs} = 2^n. \qquad (24)$$

If isomerism occurs at only one asymmetric atom in a molecule with more than one center of asymmetry, then the result is a stereoisomer, not a mirror image, and is called a *diastereoisomer,* which unlike enantiomers, have different physicochemical properties. Enantiomers are said to be optically active; when they are in solution, they bend polarized light either to the right [dextro (Greek) rotary] or left [laevo (Greek) rotary] and are respectively symbolized by the prefix D- or (+) and L- or (−). Confusion can arise by the additional use of prefixes R- (recto [Lat] = right) and s- [sinister (Lat) = left], which are designations of an absolute configuration. In nature, dominance of one or other of the isomers is due to a preferential stability of one form (e.g., D-glucose L-amino acids), but chemical synthesis usually produces an equal mixture of both, called a *racemate* (prefix rac-). Approximately 25% of all drugs have asymmetry, and more than 80% are administered as the racemate of two or more separate compounds. Because the body can perceive these spatial differences stereoselectively, where one enantiomer has a predominance over the other, there is now increasing awareness that those isomers may be stereospecific with different pharmacology, toxicology, pharmacokinetic and metabolic profiles.

In order to show stereoselective differences in kinetics, it is necessary to separate and measure the

TABLE II Common Metabolic Pathways

Phase I	Phase II
Oxidation	Conjugation
Deamination	Glucuronide
Dealkylation	Peptide
Epoxidation	Sulphate
Hydroxylation	Glutathione
Sulphoxidation	Methylation
Reduction	Acetylation
Alcohol dehydrogenase	Acylations
Azoreduction	Mercapturic acid
Nitro reduction	
Dehalogenation	
Hydrolysis	
Amide hydrolysis	
Ester hydrolysis	
Aromatization	

isomers. This can be achieved by GLC or HPLC using recent progress in chiral chromatographic discrimination using (a) chiral derivitization by reagents, (b) chiral elements in HPLC, or (c) chiral stationary phases. Although relatively few chiral drugs have been investigated (<10%), it has been shown that only those processes in which the drug binds to a *receptor* or a specific site-recognition protein, will there be stereoselectivity in kinetics. Thus, physical processes of passive diffusion or absorption, distribution, and renal elimination (filtration and reabsorption) do not normally show selectivity, while active processes such as renal and biliary secretion and metabolism do show enantiomeric differences. Protein binding is also stereoselective but the difference in free-drug concentration rarely exceeds two-fold, and volumes of distribution are rarely different.

The most important aspect of stereoselective kinetics is the differences in the rates and sometimes routes of metabolism, which have been shown to occur for all the pathways given in Table II. With the limited available knowledge, there do not appear to be any ground rules, and some species are different from others; thus deamination is faster for the D-isomers of phenylethylamines in man but not in rats, while ring hydroxylation is faster for L-isomers in man but not other animals. For some of the non-steroidal anti-inflammatory drugs, there is metabolic *enantiomeric conversion* from the inactive R-isomer to the active s-isomer. The combination of different isomeric clearances leading to different steady-state levels of compounds with sometimes

markedly different activities does lead to problems in interpreting kinetic–dynamic interaction and the use of therapeutic drug monitoring when racemates are administered. It is now accepted that isomers, when administered together as the racemate, should be thought of as two or more separate compounds, which should be analyzed separately and the pharmacokinetic analysis undertaken accordingly. Calculation of kinetic parameters from combined levels can lead to conclusions that are not simply an average of the two isomers.

V. Pharmacokinetic Analysis

This is the most difficult and complex aspect of pharmacokinetics, and it is not possible within the scope of this chapter to go into great detail. If there is further interest, the books of Rowlands and Tozer (1989), or Gibaldi and Perrier (1982) will provide a more complete description. The aim of pharmacokinetic analysis is to convert large quantities of drug-level data, normally from plasma, into a small number of parameters that can be used predictably to define the biodisposition of the drug or chemical under investigation. Although there are three types of such data analysis: direct; empirical; and theoretical. In practice, their use overlaps depending on the analytical objectives.

A. Direct Analysis

For this approach, the parameters are taken directly from the data either with no further manipulation, e.g., for C_{max}, T_{max}, etc., or by relatively simple empirical computation of the data; e.g., the area under the concentration time curve to time t (AUC_t) is the sum of

$$AUC_t = \frac{(t_1 - 0) \cdot (C_0 + C_1)}{2}$$
$$+ \frac{(t_2 - t_1) \cdot (C_1 + C_2)}{2}$$
$$+ \ldots + \frac{(t_t - t_{t-1}) \cdot (C_{t-1} + C_t)}{2}. \quad (25)$$

where C_0, C_1, and C_t and plasma drug concentrations from zero time up until any time ι.

B. Empirical Analysis

This *model-independent* approach makes no assumptions about the physiological processes in-

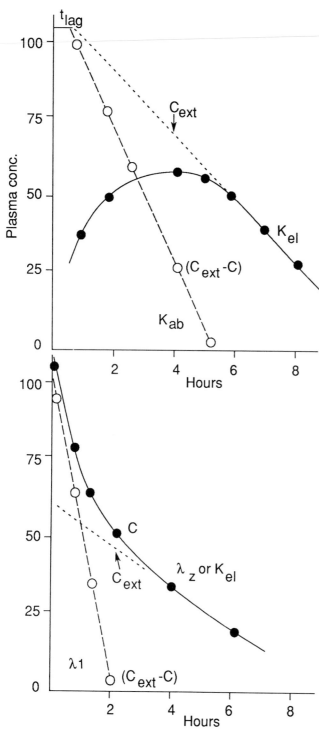

FIGURE 4 The method of residuals is used to graphically calculate (a) absorption rate (K_{ab}) and (b) initial elimination rate exponent (λ_1). The terminal log linear elimination is back extrapolated (----) and any drug concentrations along with line are called C_{ext}. The observed concentration C is subtracted from C_{ext} at the same time and plotting these residuals (C_{ext}-C) against this time will give a straight line (— —) with slopes (first-order rates only) equal to (a) absorption or (b) initial elimination rates.

volved. For most drugs, the concentration-time profiles after an IV injection can be described by a polyexponential equation

$$C_t = C_1 e^{-\lambda_1 \cdot t} + C_2 e^{-\lambda_2 \cdot t} + \cdots + C_n e^{-\lambda_n \cdot t} \quad (26)$$

where C is the drug concentration and λ the exponential coefficient for each in phase of the curve. In practice, n is rarely more than 3, because phases greater than this cannot easily be separated. When the data are fitted by these equations, then the same kinetic parameters as derived from direct analysis can be calculated, e.g.,

$$AUC = \frac{C_1}{\lambda_1} + \frac{C_2}{\lambda_2} \cdots \frac{C_N}{\lambda_N}. \quad (27)$$

In general, estimates of the parameters by empirical analysis are better than those by direct analysis, particularly half-life, and computed maximal blood levels. Empirical analysis also describes mathematically the complete curve even at time points when no sampling has been made, so it also has the advantage of enabling better prediction or simulation of blood levels after single or repeated administration, and can be combining with simultaneous pharmacodynamic modeling. Unfortunately, empirical analysis has the disadvantage that it cannot be used to predict the effect on pharmacokinetics of physiological pathological changes such as renal or hepatic disease.

C. Theoretical Analysis

This *model-dependent* approach assumes that some underlying process is contributing to the observed changes in body-fluid levels. There are 3 different types of these models: compartment, physiological, and systems.

1. Compartment

The representation of pharmacokinetics as if the body consists of a number of discrete compartments has been an extensively used model. Although these compartments do not need to have any physiological identification, they are generally considered to correspond approximately to (a) central: blood and rapidly perfused tissues (lung, heart, kidneys, etc.), (b) tissue: more slowly perfused tissue (skin, muscle, etc.), (c) deep: very slowly perfused tissue (fat, bone, etc.), and (d) elimination (urine, feces, etc.) (Fig. 5).

The concepts behind this modeling are simple. The drug is assumed to go in and out of the com-

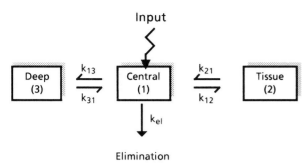

FIGURE 5 Three-compartment model.

partments at rates directly proportional to the amount in each, but which may be different for each compartment. Unfortunately, although the concept is easy, the calculation of the rate constants of exchange (k) and the amount of drug in each compartment at any one time is not, particularly for multiple compartments. Thus, for a simple one-compartment IV bolus model, the rate constant for elimination k_{el} can be determined by integrating the differential equations in Equation 28

$$dA/dt = -k_{el}.A, \text{ and; } A = Dose_{e^{-k_{el}t}}$$
$$dE/dt = +k_{el}.A. \qquad (28)$$

where A is the amount of drug in compartment 1.

However, as the number of compartments and differential equations increases, the arithmetical calculation becomes of such complexity that results of the required numerical integration are not always reliable. The main advantage of this modeling is that once solved, the equations allow computation of the amount of drug in any compartment at any time after dosing and are conceptually easy to understand, particularly by the nonpharmacokineticist. However, in recent years, such compartmentilization has come under criticism because it uses an abstract *black box* approach with little real anatomical basis, is prone to error, and can be overinterpreted, particularly if the model is incorrect.

2. Physiological

Although based on compartmental concepts, physiologically based modeling attempts to overcome some of their disadvantages by defining the kinetic processes in terms of actual blood flow, protein binding, membrane permeability, real tissue volumes, and metabolic clearances. Such an approach leads to a more realistic and understandable handling of kinetic data and is of potentially greater clinical significance, particularly when kinetic pre-

dictions of the effects of disease or age are needed. In addition, utilizing the comparative physiology in animals, such models allow a better basis for comparing species differences in biodisposition. Despite the obvious advantages of such modeling, the technique has not gained widespread acceptance for the following reasons: (a) difficulty of measurement of the necessary physiological variables, particularly in humans; (b) models tend to be overambitious while body-fluid analysis is insufficient for such overcomplexity; (c) statistical validity of these models is difficult to check. Because of this, together with the inherent variability for such parameters, the final results can sometimes be rendered meaningless.

Despite these drawbacks, some success has been achieved with anti-cancer agents and anesthetics, and the future will see more investigators attempting to overcome the problems by using simpler and more verifiable models.

3. Systems Dynamics

This, the newest of pharmacokinetic approaches, gets away from the highly structured and restrictive compartmental models previously described. It uses, instead, a mathematical approach to quantify general properties of physiological systems without the necessity to describe preconceived kinetic models. It therefore uses fewer but more verifiable assumptions and parameters. This model-independent procedure considers drug transport as a stochastic or random behavior of drug molecules in the body and uses mathematical operations such as *superposition* for multiple-dose predictions, *deconvolution* for curve fitting and disposition, and *decomposition* for absorption, distribution, and elimination. In the main, it assumes linear processes to be operational, but nonlinearity may also be modeled. Perhaps its greatest advantage will come from the use of systems analysis in pharmacodynamics, as it can be theoretically based on the current understanding of the mechanisms of action involved. Systems analysis, however, does have some disadvantages, the main being that this more generalized approach does not provide all the kinetic parameters of conventional approaches, and that the mathematics at the moment is unfamiliar to many scientists.

4. Summary

Kinetic models should (1) simply explain the biodisposition of a drug in terms of a small number of parameters from biofluid analysis; (2) be able to extrapolate or predict other situations; (3) relate to pharmacodynamic processes. However, kinetic modeling is only as good as the data obtained and the assumptions used. No one model is the only answer, and sometimes the model can be wrong, even though the data fit is adequate. The following questions should always be asked: Is the model too complex or too specific to answer the objective of the study, based on the quantity or quality of the data, and are other models more useful alone or in combination?

VI. Population Kinetics

An important use of kinetics is the prediction of dosage changes for certain patients groups: elderly, obese, renal or hepatic failure, etc. This can be achieved by measurement of kinetic parameters in a large population where individual estimates are statistically combined to produce overall population parameters that can be simply related to certain demographic details (age, weight, renal clearance, etc.). The errors of these estimates can also provide confidence in the accuracy of these population parameters. Three approaches can be used to relate response (steady-state drug levels) to the covariate (time): naive averaging of data (NAD); two-stage; mixed-effect model (MEM) or population approach; and nonparametric maximum likelihood (NPML).

A. Naive Averaging of Data (NAD)

This approach averages all the plasma levels from all individuals studied at each observational time and analyzes the data for kinetics as if they came from one individual. This method is commonly used, but should be avoided because it smooths out the data, providing no information about the individual or the error, and often the apparent population estimates are incorrect. The naive pool method (NPD) is a more general variant of the NAD, fitting all the data in one step. It is used for initial population estimates, but suffers from the same problems.

B. Standard Two-Stage Method (STS)

This allows, in the first stage, for an evaluation of kinetics from each individual, and in the second stage, combines these to produce mean population parameters and relates them to various pathological

factors to show the influence, say, of body weight, on steady-state levels. There are several variations and improvements to the two-stage approach including global two-stage (GTS), iterative two-stage (ITS), and generalized iterative two-stage (GITS). They use either the initial estimates of population kinetics to feed back and improve individual values to obtain better population results (Bayesian approach), or more improved algorithms with more appropriate data weighting to compute the basic kinetics.

These procedures described above suffer in that they use traditional designed studies comparing data from specific illness groups with those from healthy volunteers. The disadvantages include (1) all population groups cannot be studied; (2) large amount of controlled data needed from small groups (n = 10–20); (3) sparse data cannot be used; (4) highly controlled hospitalized studies needed; (5) data may be polarized using extreme differences, healthy or severe renal failure patients, for example.

C. Population Kinetics

To overcome these problems and to obtain kinetic parameters more representative for all types of patients, Sheiner and Beal have introduced the concept of population kinetics. This is based on the principle of extended least squares and is computed on a program called NONMEM (nonlinear mixed-effect model). Instead of using a small number of the individuals to define the population, the emphasis is reversed so that a small number of random, but well-defined, concentrations from a large number of individuals are used to make up a population from which individual kinetics can be defined. The advantages of this approach are that (1) it can use sparse data, only 2 or 3 points per individual at any time, but well documented; (2) data can come from any study where drug levels are measured; (3) data from specific studies, e.g., smoking, can be added at any time to improve estimates; (4) it provides good average population estimates and are obtained; (5) population standard deviations and estimate of residual error are all obtained, providing statistical confidence in data. This analysis can be extended to important pathophysiological variables related to disease states. Thus, it is possible to say that clearance (CL) of a particular drug and therefore steady-state levels are related to body weight (W), age (Ag) and renal clearance (CL_R) as shown in equation 26:

$$CL = (W \cdot 0.2) \cdot \left(\frac{CL_R}{70}\right)^{-0.7} \cdot \left(\frac{Ag}{65}\right)^{-0.5} \cdot \quad (29)$$

The disadvantages are that (1) the software is not easy to use, and few people are trained to use this approach; (2) the randomness of study design is a difficult concept to accept; (3) the applicability and acceptability of the data generated have not yet been fully tested, either clinically or with regulatory authorities. However, this population approach linked with Bayesian feedback, incremental learning, and improvement of estimates can, however, be used for all aspects of drug development including pharmacodynamics, toxicokinetics, etc.

D. Nonparametric Likelihood Estimation

Nonparametric likelihood estimation is a new approach to population kinetics and gives estimates of the shapes of the distribution curves from sparse data providing information on symmetry, modality,

degree of tail, etc. It will probably be used as an exploratory technique before use of parametric population analyses.

VII. Effect of Age and Disease

The most important kinetic changes that may occur in disease, smoking, pregnancy, etc., compared to those in healthy volunteers, are summarized in Table III. The extent of the changes, if they occur at all, is very much dependent on the drug and the individual.

A. Elderly

Although many physiological changes occur in elderly patients, the most important is the reduction in renal blood flow (50% of normal at 65), which has important consequences for renally cleared drugs. Other changes include reduction in cardiac index (distribution and elimination), hepatic efficiency

TABLE III Major Changes in Kinetic Parameters Which Can Occur in Age and Disease Compared with Healthy Volunteers

	Elderly age>65	Young age<5	Hepatic	Renal CL_CR<50	Cardiac failure	Pregnancy	Burns	Intestinal	Smoking
Absorption (F)	↓	↓	↑ (L)	↓ ↔	↓	↑	↑ ↓	↓	↔
Peak levels (C_{max})	↑ ↔	↑	↑	↑ ↓ ↔	↓	↔	↓	↓	↔
Fraction unbound (f_u)	↑ ↓ (G A)	↑	↑	↓ ↑ (G A)	↓ (G) (myocardial infarction)	↑	↑ ↓ (G)	↑ ↓ (A G) (Crohn's)	↓ (G)
Distribution (V)	↑ ↓ (L B)	↑	↓	↑ (P)	↓ ↑ (L P)	↑	↑	↓ ↔ (G)	↔
Half life ($t_{\frac{1}{2}}$)	↑	↑	↑	↑ (P)	↑	↔	↑ ↓	↑ ↔ (G)	↔
Hepatic clearance (CL_H)	↓	↓	↓	↓ ↔	↓	↑	↓ ↑ (H L)	↔	↑ ↔
Renal clearance (CL_R)	↓	↑ (N)	↔	↓ (P)	↓	↑	↑ ↓	↔	↔
Steady state levels (C_{ss})	↑	↑	↑ (L)	↑ (P)	↑	↓ ↔	↓ ↑	↓	↑ ↔
Dosage changes	↓	↓	↓ (L)	↓ (P)	↓	↔ ↑	↔	↑	↓ ↔

L - Lipid soluble drugs	P - Polar drugs	---- - for some drugs	↑ - major change
G - α-glycoprotein binding	A - Albumin binding	↑ - minor changes	↔ - no change
N - Neonates	H - High clearance drug		

(metabolism), splanchnic blood flow (absorption), breathing capacity (elimination of anesthetics), and protein binding (distribution). There are large individual differences, and age alone is not a good index of functional deterioration.

B. Neonates

The newborn has an increased gastric pH, blood flow, circulating endogenous fetal compounds (proteins, bilirubin), but decreased gastric emptying, intestinal blood flow, hepatic metabolism (drug dependent), α-glycoprotein levels, and particularly renal clearance (large intersubject variation). However, within the first few weeks and months after birth, maturation can dramatically alter physiological processes, and it is very difficult to predict dosage changes, as each drug and each subject behave differently.

C. Renal

Renal failure can occur directly or indirectly as a consequence of hypertension, diabetes, infection, etc., and can be estimated by creatinine clearance. The main change is a reduced renal clearance, which substantially reduces the clearance for those drugs primarily eliminated in the urine. Several nomograms are available relating renal clearance to dosage changes for such drugs. It should also be remembered that although the parent drug may not be renally eliminated, the more polar metabolites will be, and if active or toxic, then dosage changes may be necessary. The acidosis caused by retention of uric acid may reduce absorption, and alters tissue distribution and protein binding. Hepatic clearance can also be indirectly reduced by circulating inhibitors.

D. Hepatic

Hepatic disease can include cirrhosis, acute and chronic, hepatitis, and obstructive jaundice. Unlike renal disease, there are few reliable markers for hepatic function, and simple correlations with kinetic parameters are difficult. The major pathological changes are a reduction of metabolism, hepatic blood flow, and albumin concentrations, and an increase in portacaval shunting. Hepatic clearance can be substantially reduced ($\approx 50\%$) for highly extracted drugs attributable to reduced blood flow and bypass shunting, while for poorly extracted drugs, clearance may be reduced, increased, or remain the same.

E. Cardiac Failure

Cardiac disease can include cardiac failure, hypovolemic shock, and myocardial infarction, but the main physiological change is a reduced blood perfusion to all organs, resulting in reductions in rates of absorption, distribution, and elimination. Although in some cases these factors cancel each other out, the net overall effect is longer half-lives and higher steady-state levels for many drugs.

F. Gastrointestinal Tract

These disorders can include cancer, Crohn's disease, irritable bowel, achlorhydria, and in general, they significantly reduce the rate and extent of absorption of drugs. The stress protein, α-glycoprotein, is increased, which may reduce the volume of distribution of certain basic drugs.

G. Pregnancy

There are many physiological changes associated with pregnancy. These will differ according to fetal growth and hormonal balance. There are reductions in gastric activity, intestinal motility, and albumin, while there are increases in tidal volume, cardiac output, blood flow and volume, fat deposition, edema, volume (fetus, placenta, breasts, etc.) and particularly renal clearance (50%), but hepatic clearance remains unchanged. For most drugs, steady-state levels do not change, but increased dosage is needed for some drugs that are extensively renally cleared.

VIII. Kinetic–Dynamic Relationships

Unfortunately, the relative ease of measuring drug concentrations in blood compared to objective dynamic measurements has led medicinal scientists to assume that blood levels are synonymous or simply related to activity. Often this is not the case, and large changes in levels may produce only small changes in effect or vice versa. It is therefore necessary to relate or model drug levels with activity in order to maximize the use of kinetics.

A. Receptor Binding

Clark, in the 1930s, first proposed from chemical reactions that for a drug to illicit an activity, it must *interact* in some way with a *receptor substance* and that this could be expressed mathematically by the law of mass action, thus

$$Drug + Receptor \rightleftharpoons$$
$$Drug - Receptor \rightarrow effect. \quad (30)$$

Rearrangement produces the Michaelis–Menten type equation, which relates the effect to the maximal effect (E_{max}), the dissociation constant (K_D) and the free-drug concentration at the receptor site (D_u).

$$Effect = \frac{E_{max} \cdot D_u}{K_D + D_u}. \quad (31)$$

This produces a sigmoidal curve when effect is plotted against drug concentration with a small effect at low levels and greater response linear portion, which then attains a maximal effect when all the receptors are filled (Table IV).

Unfortunately for most drugs, it is not possible to sample at the site of action, and the assumption must be made that the free drug in blood (Cu) is readily accessible to the receptor. Thus equation 29 becomes

$$\begin{array}{cc} Blood & Receptor \\ C \rightleftharpoons C_u \rightleftharpoons D_u \rightleftharpoons DR \rightarrow effect. \end{array} \quad (32)$$

B. Dynamic Models

Assuming that the system is in equilibrium and ignoring the kinetic control of C, it is thus possible to relate drug levels in blood to the measured effect, using different models. A number have been used, but those used most frequently will be discussed in terms of their appropriateness and how they have been derived (Table IV).

TABLE IV Summary of Various Kinetic-Dynamic Models With Their Derivation and Limitations

Type of Model	Equation	Comment
Fixed effect	$E = E_{max}$ if $C \geqq C_{threshold}$	Trigger or all or nothing effect past a certain plasma level
Linear	$E = SC + const$	Only describes part A-B of response
Log linear	$E = S \log C + const$	Describes 20-80% of response, B-C
E_{max}	$E = \dfrac{E_{max} \cdot C}{EC_{50} + C}$	Describes whole curve A-D
Sigmoidal E_{max}	$E = \dfrac{E_{max} \cdot C^{\gamma}}{EC_{50}^{\gamma} + C^{\gamma}}$	Describes whole unit including slope function A-D

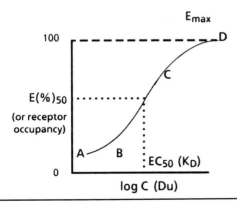

E is the calculated effect, S the slope of a linear regression, C is the measured plasma (blood) concentration, E_{MAX} the maximum effect achievable, EC_{50} the concentration at which 50% of the E_{max} occurs, the Hill coefficient or slope function (sigmoidicity) and constant is the effect intercept when C = 0.

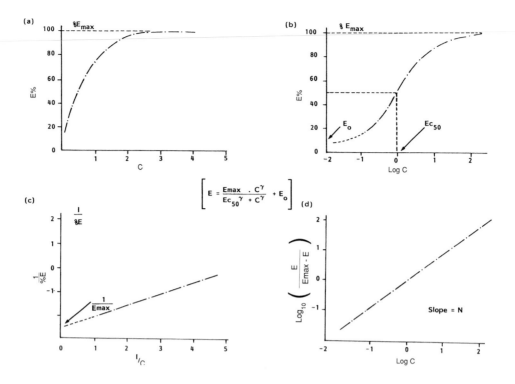

$$E = \frac{E_{max} \cdot C^{\gamma}}{Ec_{50}^{\gamma} + C^{\gamma}} + E_o$$

FIGURE 6 Graphical calculations of parameters from sigmoidal E_{max} kinetic–dynamic curves.

1. The *fixed effect relationship* is an all-or-nothing effect above or below a certain drug concentration with no progressive change in activity with changes of levels. This may occur for drugs with sharp dose–response curves, but may be an artifact, a consequence of combining data from different individuals (naive pool) (see VII.G).

2. The *linear relationship,* in which effect is plotted directly against concentration, has been used extensively in early kinetic–dynamic modeling, but suffers in that it cannot describe a maximal effect. It is often derived from (a) insufficient data points; (b) mixing individuals; and (c) no data points above the EC_{50}, (the drug concentration giving 50% maximal effect, E_{max}).

3. The *log linear relationship* in which effect is plotted against the log plasma concentration, describes only the central 20–80% of the complete curve.

4. The E_{max} relationship, however, describes the concentration–effect curve over the full range of activity, while the *sigmoidal E_{max} relationship,* the most complete of the relationships, uses in addition the concept of steepness of the curve by using the Hill coefficient (Y). This need not be an integer with values commonly ranging from $0 \cdot 5$ to $2 \cdot 0$, and although it has no physiological meaning, it can be clinically important. When the slope is steep (Y > 1) there is a rapid change in activity for a small change in drug levels around the EC_{50} (*narrow therapeutic window*), and it is difficult to titrate a desired effect. Conversely, with a shallow slope (Y < 1), the activity occurs over a wide range of drug concentrations, and there is a built-in safety mar-

gin owing to a *wide therapeutic window* depending on the balance of efficacy and toxicity. Note that although Y may be different, E_{max} and EC_{50} can remain the same. This type of relationship, when γ is small, can produce apparent paradoxes in which the kinetic halflife may be only 2 hours, but the activity can continue for more than 24 hours.

In general, whatever the apparent shape of the curve, the sigmoidal E_{max} model should be attempted initially before using simpler functions.

C. Parameter Measurement

For simple linear relationships, regression analysis is sufficient, but for the more complex equations, nonlinear least squares regression analysis is required. This can be achieved with various software programs like SAS, NONLIN, and ELSMOS, but parameter estimation can be achieved graphically by rearrangement of equation 33 and use of the plots shown in Fig. 6.

$$E = \frac{E_{max} \cdot C^{Y}}{EC_{50}^{Y} + C^{Y}} + Eo \qquad (33)$$

where Eo is the baseline effect.

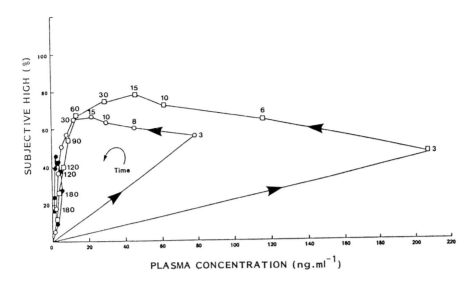

FIGURE 7 Anticlockwise hysteresis of Δ-9-tetrahydrocannabinol when plasma concentrations are related to a subjective high, after different routes of administration. (Chiang *et al.*, 1984). □, IV; ○, smoking, ●, PO.

D. Kinetic–Dynamic Modeling

It has been assumed so far that the drug in blood is in instantaneous equilibrium with the drug at the receptor site, but for most drugs, this is not true, and the pharmacokinetic changes may have to be considered. To overcome this, both the dynamic and kinetic modeling should be done simultaneously. The resultant equations are complex and may compound any errors inherent in either system. In addition, it is necessary to give the active site a size or compartment. To overcome this problem, researchers have used a so-called effect compartment model, a minimalistic conceptualization in that the active site is assumed to have a negligible volume. Various software programs are available to undertake this analysis, including ELSMOS, SIPHAR, and NONLIN. One of the problems that these programs can overcome is hysteresis.

E. Hysteresis (Anticlockwise Hysteresis)

The word comes from the Greek word Υστερησις, meaning to come late. In this context, it has been coined for the delay in obtaining maximal activity compared with time of maximal plasma drug levels. It is often caused by the time needed for a drug to be transported from the systemic circulation to the site of action. This is graphically shown when plasma levels are plotted against activity and the points are joined up in time order to produce an anticlockwise directional curve (Fig. 7).

Thus, for an IV administration with initial high plasma levels, but nothing in the brain, there is an increasing effect with time for equivalent drug levels. Inhalation produces less hysteresis, while the slower oral absorption shows none, as there is more time for equilibrium between plasma and brain. Note that after 15 minutes, all routes of administration show the same kinetic–dynamic curve, and thus for modeling, only those latter time points should be used. Removal of hysteresis can also be achieved mathematically by using a *link-effect model*, which collapses the hysteresis and provides estimates of the rate of uptake into the active site.

Hysteresis can also occur with (1) *active agonist metabolites,* which produce maximal blood levels later than the parent drug. This can occur for many drugs, particularly those with which first-pass metabolism is important; (2) *delayed or cascade activity.* For some drugs, e.g., antihypertensives and antidepressants, there is a delay of several weeks before maximal activity is seen. Similarly, the anticoagulant action of warfarin takes several days owing to the cascade activity of the various clotting factors in blood; or (3) *sensitization,* in which the activity increases with time because of the action of the drug itself, e.g., an increased number of receptors (up-regulation).

F. Proteresis (Clockwise Hysteresis)

This phenomenon can be graphically seen as a decreasing effect with time for equivalent drug concentrations. It occurs when there is (1) *tolerance,* due to a reduction in the number of receptors (down-regulation) or an endogenous transmitter, and occurs with addictive drugs but also with the nitrate vasodilators and antibiotics; (2) *active but antagonistic metabolites* that can inhibit the action of the administered drug; (3) *feedback regulation,* in which a physiological mechanism, or enzymatic, biochemical, or nerve function, comes into play to counteract and dampen the action of the drug with time; and (4) *learning* owing to insufficient time for subject familiarization with psychometric or behavioral tests for drugs that depress the CNS.

G. Individual Differences

Frequently, investigators use data from all the subjects together (naive pool method) or use mean data to obtain kinetic–dynamic relationships. This incorrectly assumes that all individuals respond similarly to the drug, and can lead to incorrect results. Whenever possible, relationships should be obtained from individual data sets, and then means or medians should be calculated from the derived subject parameter values.

IX. Therapeutic Drug Monitoring

For many drugs, there are large intersubject variations in blood levels, for certain drugs up to 40-fold, attributable to race, age, food, disease, coadministered drugs, gender, etc., so it has long been the goal of pharmacokinetics to be able to tailor a particular dosage for a particular patient so that steady-state levels would be within the therapeutic window or targeted concentration range. The measurement of drug levels to provide this prediction, either as single-dose kinetics or more routinely at steady state, is called therapeutic drug monitoring (TDM) and is being used in hospitals for drugs with a narrow therapeutic window like phenytoin, digoxin, theophylline, lithium, gentamicin and warfarin. For this service to be of benefit, several assumptions have been made: (a) steady-state levels can be predicted from a few blood levels; (b) all subjects produce a similar response from comparable drug levels; (c) the procedure is cost effective. Steady-state levels can now be relatively well predicted, using

initial population kinetic values and Bayesian feedback methods to impose the fit for the individual patient from only three or four drug levels; however, most up-to-date evidence now shows that few drugs have a drug-level response curve which is the same between or within individuals. Even for those drugs with which TDM is extensively used, data do not support clear definitions of a therapeutic window, often because of the lack of carefully controlled studies. Because of the lack of reliability of TDM, the cost–benefit ratio of such expensive and time-consuming practices is now being questioned by the payers of health care. Many clinicians still use clinical end-points to decide on dosage changes, with drug levels considered only as backup information.

X. Future

The importance of pharmacokinetics in drug development and understanding drug action cannot be overemphasized. As a science, it has matured and come a long way from the highly mathematical derivations that few nonkineticists understood, to more physiological interpretations of the parameters that provide comprehensible rationales for drug usage. There is an increasing awareness that the basis for kinetics has been laid down, and few fundamental changes are necessary except for fine tuning. However, the future will see (a) improvements in analytical methods as lower doses of more specifically acting drugs are used, and a better separation of asymmetric molecules; (b) use of noninvasive techniques such as NMR, PET (position emission tomography), and bioelectrical impedance, which can measure drugs within the body by scanning and not by taking samples, allowing human distribution studies to be undertaken; (c) greater awareness of chronopharmacokinetics, in which the kinetics parameters will change depending on when the drug is administered; (d) greater understanding of metabolic differences between animals, and between individuals differing by race, environment, way of life, and genetic polymorphism; and (e) better indices of disease states, such as hepatic, renal, respiration, cardiac, so that drug dosage can be adjusted accordingly. There is now the realization that kinetics is only a probe; in isolation, without relating it to dynamic activity, it is of little interest.

More emphasis will be put on dynamic modeling, using the newer population approaches, systems analyses, and Bayesian feedback. Even the newest of mathematical theories, deterministic chaos, is now being used in kinetic dynamic modeling to more fully cope with the complexities of the human body. These techniques will form the basis for prediction kinetics, replacing the more classical kinetic methods and naive therapeutic or compliance monitoring, and lead to a greater understanding of the disease processes themselves.

TABLE V Symbols and Derivations

Symbol	Definition	Derivation
Ae (Au)	Total amount of unchanged drug excreted into the urine [mg]	Data
AUC (AUC$_\infty$)	Area under the plasma drug concentration time curve to infinity [ng.h.ml^{-1}]	(i) AUC$_t$ + C$_t$/k$_{el}$ (ii) $\dfrac{D}{V.k_{el}}$
AUC$_t$	Area under the plasma drug concentration time curve to the last measured time point, (t) [ng.h.ml^{-1}]	$(t_1 - t_0).(C_0 + C_1)/2 +$ $(t_2 - t_1).(C_1 + C_2)/2 + \ldots$ $(t_t - t_{t-1}).(C_t + C_{t-1})/2$
AUMC	Infinite area under the first moment versus time curve [ng.h^2.ml^{-1}]	$C_1/\lambda_1^2 + C_2/\lambda_2^2 + \ldots + C_n/\lambda_n^2$
C (Cp)	Total concentration of drug in the plasma [ng.ml^{-1}]	Data
C$_b$	Total concentration of drug in blood (ng.ml^{-1})	Data
C$_0$	The initial or back extrapolated to zero time, plasma concentration [ng.ml^{-1}]	Data
CL (Clp)	Total plasma clearance of drug [ml.min^{-1}]	(i) $\dfrac{F.D}{AUC}$ (ii) k$_{el}$.V
CL$_B$	Total blood clearance (ml.min^{-1})	$\dfrac{F \cdot D}{AUC \text{ (blood)}}$
CL$_{CR}$	Renal clearance of creatinine [ml.min^{-1}]	$\dfrac{\text{Urinary excretion rate}}{\text{Plasma concentration}}$
CL$_H$	Hepatic clearance [ml.min^{-1}]	(i) E$_H$. Q$_H$ (ii) CL − C$_R$
CL$_{int}$	Intrinsic clearance of drug from eliminating organ [ml.min^{-1}]	(i) $\dfrac{CL_H}{Cu}$ (ii) $\dfrac{CL_R}{Cu}$
CL$_R$	Renal clearance of drug [ml.min^{-1}]	$\dfrac{Ae}{AUC}$
C$_{max,ss}$	The maximum concentration at steady state [ng.ml^{-1}]	C$_{max}$ + (C$_{min}$. e^{-kel} . t$_{max}$)
C$_{min}$	Minimum plasma concentration reaching during a dosing interval [ng.ml^{-1}]	Data
C$_{min,ss}$	The minimum trough drug level at steady state [ng.ml^{-1}]	(i) Data (ii) $\dfrac{C_{min}}{1 - e^{-kel}}$ '(i.v.)
C$_{av,ss}$ (C$_{ss}$)	Average concentration at steady state during repeated dosing [ng.ml^{-1}]	$\dfrac{AUC}{\iota}$
C$_t$	Plasma concentration at last measured time point	Data
Cu	Unbound concentration in plasma [ng.ml^{-1}]	Data
D	Dose administered [mg]	Data
D$_L$	Loading dose to achieve desired steady state levels [mg]	V . C$_{ss}$
D$_M$	Fixed maintenance dose to keep steady state levels constant [mg]	D$_L$. (1 − e$^{kel.\iota}$)
E	Effect (unit of effect or %)	$\dfrac{E_{max} . C^Y}{EC_{50}^Y + C^Y}$
EC$_{50}$ (E$_{50}$)	Concentration of drug producing 50% of maximal effect [ng.ml^{-1}]	Data
E$_H$	Hepatic extraction ratio (no units)	$\dfrac{CL}{Q_H}$
E$_{max}$	Maximum effect (unit of effect or 100%)	Data
E$_0$	Baseline effect (unit of effect or %)	Data
ER (E)	Extraction ratio (no units)	$\dfrac{CL}{Q}$

(Table continues)

TABLE V *(continued)*

Symbol	Definition	Derivation
F	Fraction of drug entering the body intact or absolute availability (no units)	$\dfrac{AUCp.o..Di.v.}{AUCi.v..Dp.o.}$
f_e	Fraction of absorbed drug extracted unchanged into the urine (no units)	(i) $\dfrac{F \cdot A_e}{D}$ (ii) $\dfrac{CL_R}{CL}$
F_{max}	Theoretical maximum oral availability for a drug assuming first order first pass hepatic metabolism	(i) $1 - ER$ (ii) $1 - (CL_H/Q_H)$
fu_p fu_b	Fraction unbound of total drug concentrations in plasma/blood (no units)	$\dfrac{Cu}{C}$ or $\dfrac{Cu}{C_b}$
GFR	Glomerular filtration rate [ml.min^{-1}]	$\approx CL_{CR}$ normally 120 ml.min^{-1}
k_{12}	Transfer rate constant from central (1) to peripheral (2) compartment [h^{-1}]	from integration (1) $\dfrac{dC_1}{dt} = -k_{12}.C_1 + k_{21}.C_2$
k_{21}	Transfer rate constant from peripheral (2) to central (1) compartment [h^{-1}]	$\dfrac{dC_2}{dt} = k_{12}.C_1 - k_{21}.C_2$
k_a	Absorption rate constant [h^{-1}]	Data
K_D	Dissociation constant of a drug from its receptor [ng.ml^{-1}]	Data
k_e	Urinary excretion rate constant [h^{-1}]	$C_R.C$
k_{el}	Elimination rate constant [h^{-1}]	$\dfrac{0.693}{t\frac{1}{2}}$
MRT	Mean residence time within the body [h]	$\dfrac{AUMC}{AUC}$
Q_H (HBF)	Hepatic blood flow (hepatic artery + portal vein) [ml.min^{-1}]	On average = 1050 + 300 ml.min^{-1}
Q_R (RBF)	Renal blood flow [ml.min^{-1}]	On average = 1100 ml.min^{-1}
Rac (Ra)	Accumulation ratio (no units)	(i) $\dfrac{AUC}{AUC_t}$ (ii) $\dfrac{1}{1 - e^{-kel_t}}$
S	Slope of linear models of kinetic-dynamic relationships	Data
t	Time after drug administration [h]	Data
t_{lag}	Lag time [h]	Data
T_{max}	Time of maximal plasma concentration during a dosing interval [h]	Data
$t\frac{1}{2}$ ($t\frac{1}{2}\beta$)	Half life of the terminal slope [h]	(i) $\dfrac{0.693}{\lambda_z}$ (ii) $\dfrac{0.693 \cdot V}{CL}$
V (Vd or Vd$_{area}$)	Apparent volume of distribution derived from total plasma concentration after distribution equilibrium has been achieved [l]	(i) $\dfrac{D}{AUC \cdot \lambda_z}$; (ii) $\dfrac{t\frac{1}{2} \cdot Cl}{0.693}$
V_1 (V_0)	Initial volume of distribution at zero time [l]	$\dfrac{D}{C_o}$ (for iv bolus)
V_p; V_T	Plasma volume; tissue volume	Data
V_{ss}	Apparent volume of distribution under steady state conditions derived from total plasma concentrations [l]	MRT . CL
Y (N)	Slope factor or Hill Coefficient in concentration–response relationship (no units)	Data
λ_i	Exponent of the ith exponential term of a polyexponential equation [h^{-1}]	Data
λ_1, λ_2	Exponential coefficients [h^{-1}]	Data
λ_z	Terminal exponential coefficient	Data
ι	Dosing interval [h]	Data

[a] e.g., for two compartment model.

Because of the mathematical nature of pharmacokinetics, symbols are frequently used to simplify equations. This list is not exhaustive but defines those that are most frequently used together with alternatives ( ) and frequently used units [ ] and derivation. Data means that the parameter can be obtained directly without computation.

(Note: To convert ml.min^{-1} to L.h^{-1} divide by 16.6)

Acknowledgments

I would like to thank Bob Ings and Clive Bowman for the comments, criticisms, and necessary cutting, Kerrie Campbell for her artistic acumen, and Julia Chatten for her perseverence in producing this paper.

Bibliography

Campbell, D. B. (1990). The use of kinetic-dynamic interactions in the evaluation of drugs. *Psychopharmacology*, **100**, 433–450.

Gibson, G. G., and Skett, P. (1986). "Introduction to Drug Metabolism." Chapman and Hall, London, New York.

Rowland, M., and Tozer, T. N. (1989). "Clinical Pharmacokinetics, Concepts and Application, 2nd Ed. Lea & Febiger, Philadelphia and London.

Benet, L. Z., Massond, M. S., and Gambertoglio, J. G. (1984). "Pharmacokinetic Basis for Drug Treatment." Raven Press, New York, N.Y.

Gibaldi, M., and Perrier, D. (1982). "Pharmacokinetics." Marcel Dekker, New York, N.Y.

Phenylketonuria, Molecular Genetics

RANDY C. EISENSMITH and SAVIO L. C. WOO, *Howard Hughes Medical Institute, Baylor College of Medicine.*

Glossary

ASO Allele-specific oligonucleotide. An oligonucleotide used as a hybridization probe to identify a specific point mutation present in an individual's genomic DNA

cDNA Complementary DNA. A DNA molecule prepared from messenger RNA using the enzyme reverse transcriptase

Guthrie test A semiquantitative bacterial inhibition assay for the determination of serum phenylalanine levels in newborns. The Guthrie test requires only a small volume of blood, is specific, inexpensive, and well suited to handle large numbers of samples

Haplotype The genetic constitution of an individual with respect to one member of a pair of allelic genes

PCR Polymerase chain reaction. The amplification of specific regions of DNA through the repeated annealing and subsequent extension of oligonucleotide primers by the heat-stable enzyme Taq polymerase

RFLP Restriction fragment–length polymorphism. Benign nucleotide substitutions within the genomic DNA of individuals that have no phenotypic manifestations; they can be detected only by restriction endonucleases

Classic phenylketonuria (PKU) is an autosomal recessive genetic disorder caused by a deficiency of hepatic phenylalanine hydroxylase (PAH). The disorder is characterized by an accumulation of phenylalanine in the serum, resulting in hyperphenylalaninemia (HPA) and associated abnormalities in aromatic amino acid metabolism. Untreated PKU patients develop severe postnatal brain damage and irreversible mental retardation. PKU is among the most common inborn errors of amino acid metabolism in man, with an average incidence of approximately 1 in 10,000 Caucasian births.

I. Biochemical Basis of Classic Phenylketonuria

Classic phenylketonuria was first reported in 1934 by Folling, who observed the presence of increased levels of phenylpyruvate in the urine of a group of mentally retarded patients. A year after the discovery of this disease, Penrose confirmed Folling's earlier suspicion that PKU is an autosomally transmitted recessive genetic disorder. The biochemical defect present in PKU patients was established by Jervis in 1947, when he observed that the administration of phenylalanine produced a prompt elevation in serum tyrosine levels in normal individuals, but not in individuals with PKU. From these studies, he concluded that PKU patients lack the ability to convert phenylalanine to tyrosine. This conclusion was supported by the subsequent demonstration of the conversion of phenylalanine to tyrosine in postmortem liver samples from normal individuals, but not in those from PKU patients.

In studies inspired by the prior success of dietary management in the treatment of galactosemia, several investigators including Louis Woolf, Horst Bickel, and others began the administration of low-phenylalanine diets to young PKU patients in the early 1950s. This dietary treatment led to the reduction of serum phenylalanine and urinary phenylpyruvate levels in these children. In addition, some improvement in behavioral performance has been reported in PKU patients receiving dietary therapy.

As these initial findings were extended in additional studies, it became apparent that there is an inverse relationship between the age of onset of dietary therapy and the ultimate IQ level attained by treated patients. This observation provided a strong incentive for the implementation of neonatal screening programs for PKU. With the development of the Guthrie test, such neonatal screening programs became feasible. From the collective results of mass screening programs in Western countries, the incidence of PKU has been estimated at approximately 1 in 10,000. This value corresponds to a carrier frequency of approximately 1 in 50 for this autosomally transmitted recessive disorder.

II. Phenylalanine Hydroxylating System in Man

When Jervis first demonstrated that liver samples from PKU patients were unable to convert phenylalanine to tyrosine, relatively little was known about the enzyme or enzymes responsible for this process. This situation improved dramatically throughout the 1960s and the early 1970s, in large part owing to the work performed in the laboratory of Seymour Kaufman. The results of this work are summarized in Fig. 1.

Phenylalanine hydroxylase (PAH), a mixed-func-

tion oxygenase, catalyzes the hydroxylation of the essential amino acid L-phenylalanine to L-tyrosine. This enzyme, expressed exclusively in the liver in man, utilizes L-phenylalanine and molecular oxygen as substrates, and requires the cofactor L-erythrotetrahydrobiopterin (BH_4) as an electron donor. Although the cofactor is oxidized to the dihydro- form during each reaction cycle, adequate cofactor levels are maintained in the liver through the reduction of dihydrobiopterin to tetrahydrobiopterin by the enzyme quinonoid dihydrobiopterin reductase (QDPR).

Most cases of PKU are associated with a deficiency of PAH activity (classical PKU). However, deficits in QDPR activity or other enzymes involved in the biosynthesis and metabolism of the cofactor BH_4 can also produce elevations in serum phenylalanine levels that lead to PKU. Although these atypical forms of PKU represent only about 1 to 2% of all PKU cases, the existence of atypical PKU complicates somewhat the diagnosis of PKU patients. Furthermore, patients with atypical PKU require significantly different regimens of treatment and counseling, and may face significantly different prognoses than patients with PAH deficiencies. The characterization of this multienzyme phenylalanine hydroxylating system marked the end of the classical era of PKU research.

III. The Molecular Basis of Phenylketonuria

A. Molecular Cloning of Human PAH and Determination of Its Primary Structure

The molecular era of PKU research began with the cloning of the rat PAH cDNA in the Woo laboratory. First, PAH mRNA was purified from rat liver by polysome immunoprecipitation and used to prepare a rat PAH cDNA. The authenticity of this clone was established by hybrid-selected translation and confirmed by matching the nucleotide sequence with the partial amino acid sequence of the purified rat enzyme. Second, the rat PAH cDNA clone was used as a specific hybridization probe to isolate several human PAH cDNA clones from a human liver cDNA library. The longest of these clones contained an open reading frame that encoded a protein comprising 451 amino acids. The predicted amino acid sequence of human PAH, as deduced from the nucleotide sequence of this clone, was shown to be

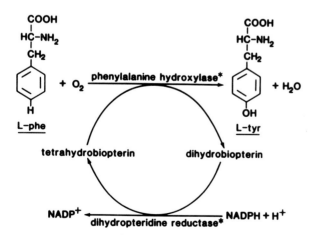

FIGURE 1 The phenylalanine hydroxylating system in man. The asterisks denote gene products implicated in the most common genetic disorders of phenylalanine metabolism.

over 90% homologous to the amino acid sequence of rat PAH.

Native human PAH initially appeared to be a polymeric enzyme comprising multiple subunits, so it was not clear whether human PAH is a heteropolymer or a homopolymer. This issue was critical to the understanding of the genetics of PKU. If the enzyme is a heteropolymer, multiple genetic loci may be expected. To resolve this issue, an expression vector was produced that contained the human PAH cDNA driven by an appropriate promoter. Introduction of this expression vector into cultured mammalian cells resulted in the expression of PAH mRNA, the production of immunoreactive PAH protein, and the presence of pterin-dependent enzymatic activity similar to authentic human PAH. These gene-transfer experiments demonstrated that human PAH is a single-gene product, permitting the use of the human PAH cDNA as a probe to perform molecular analysis of the PAH gene and the PKU locus in man.

B. Molecular Structure of the Human PAH Gene and Its Chromosomal Location

Chromosomal assignments for human genetic loci can be made by using cloned genes in molecular hybridization studies to probe genomic DNA isolated from human–rodent cell hybrids that contain different combinations of human chromosomes. A panel of such hybrid cell lines was analyzed by Southern hybridization using the human PAH cDNA clone as a hybridization probe. These studies indicated that the human PAH locus is on chromosome 12. This result was confirmed in subsequent experiments using both deletion chromosome mapping and *in situ* hybridization. Detailed examination of the results of the *in situ* hybridization experiments further localized the human PAH loci to the 12q22–q24.1 region of chromosome 12. The PAH cDNA had previously been shown to contain all of the genetic information necessary for expression of PAH, which is the enzyme deficient in PKU, so the PKU locus in man must also lie within this region. [*See* CHROMOSOMES.]

The structural organization of the PAH gene was subsequently established by Southern analysis of normal human genomic DNA using the human PAH cDNA as a hybridization probe. Preliminary observations suggested that the chromosomal PAH gene is over 65 kilobases (kb) long and contains multiple intervening sequences. Because of the large size of

this gene, a human genomic DNA library was constructed using cosmid vectors, which permitted the analysis of relatively large fragments. Screening of this cosmid library with the human PAH cDNA resulted in the isolation of four cosmids, which contain overlapping PAH genomic sequences. Detailed analysis of these four cosmid clones indicated that the human PAH gene is approximately 90 kb in length, contains at least 13 exons, with intron sizes ranging from 1 to 23 kb, and encodes a mature messenger RNA of approximately 2.4 kb.

C. Restriction Fragment–Length Polymorphisms and Prenatal Diagnosis of Phenylketonuria

An examination of the PAH locus in normal and PKU individuals by Southern analysis, using the human PAH cDNA as a hybridization probe, reveals the presence of several restriction fragment–length polymorphisms (RFLPs) in or near the human PAH gene. These RFLPs are the result of benign nucleotide substitutions within the genomic DNA of individuals with no phenotypic manifestations, and the RFLP can be detected only be restriction endonucleases. The relationship of eight RFLP sites to the 13 exons that constitute the complete coding region of the 90 kb human PAH gene are shown in Fig. 2. The four most common of the 50 or so haplotypes identified thus far are shown at the bottom of this figure. [*See* DNA MARKERS AS DIAGNOSTIC TOOLS.]

Because these RFLPs are tightly linked to the

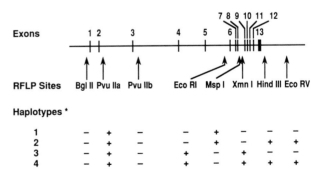

FIGURE 2 The relationship between restriction fragment-length polymorphic sites and the 13 exons that constitute the complete coding region of the human PAH gene (top). The four most common PAH haplotypes among Caucasians in the Northern European population (bottom). The pluses and minuses denote the presence or absence of a particular-sized fragment following digestion of genomic DNA with a given restriction endonuclease.

* A total of 50 haplotypes have been documented as of December 31, 1988.

PAH gene, they can be used to follow the transmission of normal or mutant chromosomes in PKU families by comparison of the hybridization patterns of parental and proband DNA samples following digestion with one or more restriction endonucleases. Figure 3 shows the first practical application of RFLP haplotype analysis in the detection of mutant PAH chromosomes in a PKU family. In this figure, genomic DNA from the father, the mother, the proband, and an unborn child were isolated, digested with the restriction endonuclease Hind III, and analyzed by Southern hybridization using the human PAH cDNA as a probe. Both parents are heterozygous for the Hind III polymorphism, as evidenced by the presence of both the 4.2 and 4.0 kb bands, while the proband contains only the 4.0 kb band. Thus, the mutant PAH gene is associated with the 4.0 kb allele in this family. This type of analysis was informative in this family, so fetal DNA was obtained and similarly analyzed. The fetus was also homozygous for the 4.0 kb fragment, thus permitting a prenatal diagnosis of PKU. This diagnosis was confirmed shortly after birth.

Although the application of haplotype analysis permitted the development of prenatal diagnosis of PKU, this technique is not without limitations. Most importantly, prenatal diagnosis using RFLP analysis can be provided only to those families with a prior incidence of PKU. Unfortunately, the vast majority of new cases of PKU (over 95%) are the result of random mating events, and thus are undetectable by traditional haplotype analysis. To overcome these limitations, the specific molecular lesions responsible for PKU must be isolated and characterized.

D. The Molecular Lesions Associated with the Four Most Prevalent Mutant PAH Haplotypes among Northern Europeans

Haplotype analysis of a number of PKU families from many European countries indicated that two PAH haplotypes (haplotypes 2 and 3) are much more prevalent among mutant chromosomes than among normal chromosomes. Close associations between RFLP haplotypes and specific mutations had previously been observed in other genetic disorders such as β-thalassemia, so the relationship between PAH haplotypes 2 and 3 and PKU chromosomes was examined by direct molecular analysis.

1. Molecular Analysis of PAH Haplotype 2 and 3 Mutant Chromosomes

The PAH genes from PKU patients bearing PAH haplotype 2 or 3 chromosomes were isolated by molecular cloning, and sequence analysis was performed. The results of these analyses from an individual homozygous for mutant haplotype 2 revealed the presence of a C-to-T transition in exon 12, which resulted in the substitution of the amino acid tryptophan for arginine at amino acid residue 408 of the PAH protein. Similar analyses from an individual homozygous for mutant haplotype 3 revealed the presence of a G-to-A transition at the consensus splice–donor site at the exon 12–intron 12 boundary region.

Expression analysis of these two mutations was performed to verify that these missense mutations were in fact the cause of PKU in these individuals. With regard to the mutation present on the mutant haplotype 2 chromosome, oligonucleotide-directed mutagenesis was performed to create an expression vector that contained the mutant PAH cDNA. The normal and mutant constructs were then introduced into cultured mammalian cells and levels of PAH mRNA, immunoreactive PAH protein, and PAH enzyme activity were assayed. In these experiments, both the normal and mutant constructs produced similar levels of PAH mRNA, but the mutant construct failed to produce PAH immunoreactivity or PAH enzyme activity. Thus, the missense mutation associated with mutant haplotype 2 chromosomes is indeed a PKU mutation. Similar results were obtained in mutagenesis and expression studies of the splicing mutation using an expression vec-

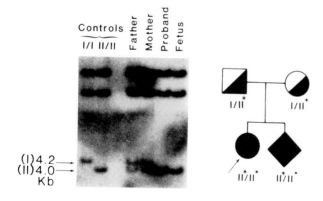

FIGURE 3 The application of RFLP haplotype analysis for the detection of mutant PAH alleles in a family with a history of phenylketonuria (see text for details).

tor that contained a mini-PAH gene, consisting of the entire PAH cDNA interrupted at the appropriate point by intron 12.

2. Linkage Disequilibrium between Specific Mutations and PAH Haplotype 2 and 3 Mutant Chromosomes

Characterization of the molecular lesions associated with mutant haplotype 2 and 3 chromosomes permitted the direct detection of these mutations in genomic DNA by allele-specific oligonucleotide (ASO) hybridization analysis. Such analyses have demonstrated that among Northern Europeans, nearly all haplotype 2 mutant chromosomes bear the codon 408 missense mutation and nearly all haplotype 3 mutant chromosomes bear the splicing mutation. Furthermore, these studies demonstrated that the codon 408 mutation is not present on the vast majority of nonhaplotype 2 mutant chromosomes in most populations and is never present on normal chromosomes, suggesting that this mutation is both responsible for PKU and distinct from other mutations present on other mutant chromosomes. ASO hybridization analysis has shown similar results for the splicing mutation and haplotype 3 mutant chromosomes. The fact that these mutations are not present among most other mutant chromosomes indicates that other mutant chromosomes must bear novel PKU mutations, further confirming the heterogeneous nature of this disorder at the molecular level.

These data suggest that the association observed between haplotype 2 mutant alleles and the codon 408 missense mutation, or between haplotype 3 mutant alleles and the splicing mutation, may be the result of a recent mutational events on normal haplotype 2 or normal haplotype 3 chromosomes, which were then distributed throughout European populations before there had been sufficient time for transfer of these mutant fragments to chromosomes of other haplotypes by crossover or gene conversion events. Thus, these mutations are in linkage disequilibrium with regard to their particular chromosomal haplotype.

Because of this linkage disequilibrium between specific mutations and RFLP haplotypes in the PAH gene, it is possible, within specific populations, to determine the genotype of PKU patients with regard to the mutant chromosome they carry. Thus, PKU patients who are homozygous for mutant haplotype 2 or 3 most likely bear the same mutation on both chromosomes, while those individuals containing both mutations are most likely compound heterozygotes bearing both mutant chromosomes. Likewise, because of the tight linkage disequilibrium observed between mutation and haplotype for the codon 408 mutation or the splicing mutation, patients who are heterozygous for mutant haplotype 2 or 3 with any of the other haplotypes are also likely to be compound heterozygotes. By this type of analysis, it is estimated that about 75% of all PKU patients in the Northern European population are genetic compounds.

3. Molecuar Analysis of PAH Haplotype 1 and 4 Mutant Chromosomes and Linkage Disequilibrium between Specific Mutations and PAH Haplotype 1 and 4 Chromosomes

Detailed sequence and hybridization analysis of mutant haplotype 1 chromosomes have thus far revealed the presence of at least two missense mutations associated with haplotype 1 PKU chromosomes of Northern Europeans. Two different G-to-A transitions have been observed in exon 7 of the PAH gene. One results in the substitution of glutamine for arginine at residue 261 of the PAH protein (the R261Q mutation), and the second results in the substitution of lysine for glutamate at residue 280 of the PAH protein (the E280K mutation). Likewise, similar analyses performed on mutant haplotype 4 chromosomes have revealed at least two mutations associated with haplotype 4 PKU chromosomes of Northern Europeans. A G-to-A transition in exon 5 of the PAH gene results in the substitution of glutamine for arginine at residue 158 of the PAH protein (R158Q), while a C-to-T transition in exon 7 results in the conversion of an arginine codon at position 243 to a termination codon (R243X). The results of population-screening studies for these mutations demonstrated that not all of the mutant chromosomes of a given haplotype contain the same molecular lesion, a finding in contrast to the initial reports on mutant haplotypes 2 and 3, where nearly all of the mutant alleles of a given haplotype examined contained the same molecular lesion. These results indicate that mutant haplotype 1 and 4 chromosomes bear multiple, independent mutations, at least one of which is in linkage disequilibrium with respect to its haplotype background. The fact that the mutations associated with haplotypes 1 and 4 are exclusive of chromosomes of other haplotypes, but not inclusive of all chromosomes of a given haplotype, may reflect the predominance of these two haplotypes among normal PAH alleles.

IV. Population Genetics and Carrier Screening of Phenylketonuria

A. Population Genetics of Phenylketonuria

Because PAH haplotypes 2 and 3 are relatively rare among normal chromosomes in various European populations, it was somewhat puzzling to observe that these two mutant chromosomes account for about half of all PKU chromosomes in Europe. In general, several different mechanisms may account for the relatively high frequency of PKU chromosomes in man. First, these mutations may have occurred initially on rare chromosomal backgrounds which were then expanded in the general population, the *founder and drift hypothesis*. However, the probability of two such events both having occurred on rare chromosomal backgrounds is relatively low. Alternatively, these mutations might have occurred in populations where haplotype 2 and 3 chromosomes were better represented, and were then maintained at a relatively high frequency by some sort of selective advantage, the *heterozygote selection hypothesis*. As yet, the nature of this heterozygote advantage, if present, remains unclear.

In addition to these two mechanisms, hypermutability in the PAH locus could also account for the relatively high frequency of PKU in certain populations. Although recent evidence demonstrates that approximately half of the known PKU mutations involve CpG dinucleotide mutational "hot spots," this fact alone cannot explain the prevalence of PKU. Many genes associated with less prevalent genetic disorders contain similar numbers of CpG dinucleotides. Furthermore, if recurrent mutation were prevalent at these dinucleotide sequences, these mutations should be linked to a number of different haplotypes based on the relative frequency of these haplotypes among normal individuals. Although recurrent mutation has been observed in the PAH gene, it is a relatively rare event most often observed in only a few chromosomes. Finally, the relative lack of deletion mutations in the PAH gene further argues against the *hypermutability hypothesis*.

Having analyzed multiple polymorphic protein markers in various African, European, and Asian populations, Cavalli-Sforza and colleagues hypothesized that there was a major vectorial human migration from the Middle East toward Northern Europe, which occurred about 10,000 to 20,000 years ago. They theorized that this migration was the result of the advent of farming technology in the founding population, which gave them an advantage over the native early Europeans. Based on this hypothesis, it might be suggested that the specific mutations associated with PKU chromosomes occurred or were already present in this or other founding populations, and were spread throughout the European continent by migration some 10,000 to 20,000 years ago. Thus, the relatively high frequency of PKU is probably attributable to an expansion of several predominant mutant alleles in this population rather than to a high rate of mutations of the human PAH gene. [*See* POPULATION GENETICS.]

B. Carrier Screening of Phenylketonuria by Direct Mutation Analysis

Although the percentage of PAH mutations detectable by ASO hybridization is still rather low (approximately 50% of the U.S. population), it is in fact high enough for ASO hybridization–based, carrier-screening programs to exert a significant influence on the incidence of PKU. Table 1 illustrates the theoretical reductions in the incidence of PKU in the U.S. population through the implementation of carrier-screening programs at various levels of accuracy. In the absence of screening, the incidence of PKU among Caucasians results in the birth of 400 affected individuals in the United States annually. However, if 50% of all carriers of mutant PAH alleles could be detected through carrier-screening programs, and if as a result of such screening, matings between carriers failed to produce offspring with PKU, roughly 300 cases of PKU could be eliminated annually in the United

TABLE 1 Prevention of PKU by Carrier Screening

Screening accuracy	Carrier frequency	PKU frequency	PKU incidence[a]
0%	(1/50)^2 × 1/4 = 1/10,000		400
50%	(1/50 × 50%)^2 × 1/4 = 1/40,000		100
75%	(1/50 × 25%)^2 × 1/4 = 1/160,000		25
90%	(1/50 × 10%)^2 × 1/4 = 1/1,000,000		4

[a] Based on 4×10^6 annual births in the United States.

States. As more PKU mutations are characterized, and the accuracy level of carrier screening is increased, these theoretical reductions become even more dramatic. With the characterization of 90% or more of the PAH mutations in the general population, the frequency of PKU could theoretically be reduced to 1 in 10^6, corresponding to only few new cases of PKU each year in the United States.

The application of ASO hybridization analysis to the detection of known PAH mutations is limited somewhat by the amount of genomic DNA required to produce a detectable hybridization signal. However, this limitation can be easily overcome through the application of the polymerase chain reaction (PCR), a method that permits the amplification of specific regions of genomic DNA. An example of the combination of PCR amplification of genomic DNA and ASO hybridization of the amplified product in the detection of the mutations associated with mutant PAH haplotypes 2 and 3 is shown in Fig. 4. This methodology permits the unequivocal determination of the respective genotype of normal individuals, affected individuals, and individuals who are carriers for these two PKU mutations. Furthermore, this methodology has now been refined to the degree that only a single drop of blood or a single strand of hair can be used as the source of genomic DNA. This procedure can also be fully automated, permitting the rapid, simultaneous analysis of many samples. With these techniques, and with the characterization of additional PAH mutations, carrier detection of PKU in the general population will soon be technologically feasible.

Before carrier screening of PKU and similar recessive genetic disorders can be implemented in a given population, however, serious ethical consideration must be given to the potential medical and social effects of such tests. First and foremost, the test should be voluntary, and the results of testing must be kept confidential. They are to be used solely by individuals to make informed reproductive decisions. Second, the implications of a negative result in a screening test must be addressed. The accuracy of carrier detection by gene screening can never be 100% because of the presence of novel mutations, so a negative result does not totally preclude the possibility of producing an affected offspring. Consequently, the public must also be educated about the relative limitations of such testing procedures and how screening can reduce but not completely eliminate the incidence of genetic disorders such as PKU. Obviously, the extent of the

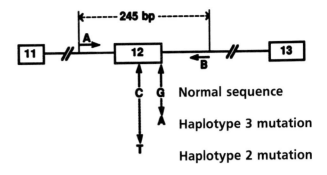

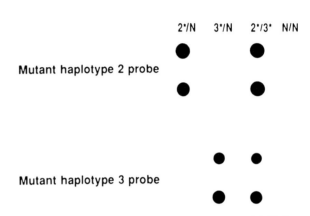

FIGURE 4 The application of PCR amplification and ASO hybridization techniques for the detection of mutant PAH alleles in two individuals bearing one mutant haplotype 2 allele and one normal allele (2*/N), from two individuals bearing one mutant haplotype 3 allele and one normal allele (3*/N), from two individuals who are compound heterozygotes for mutant haplotype 2 and mutant haplotype 3 (2*/3*), or from two normal individuals (N/N). The sepcific segment of genomic DNA from the region of the human PAH gene containing the mutation sites associated with mutant PAH haplotypes 2 and 3 was amplified as shown in the schematic diagram at the top of this figure. The amplified material was then blotted onto nitrocellulose membrane and hybridized with an oligonucleotide specific for either the haplotype 2 mutation or the haplotype 3 mutation.

reduction in the incidence of a genetic disorder like PKU is ultimately dependent upon the percentage of mutant genes that can be reliably detected in a given population, which is in turn determined by studying the individuals who constitute these populations.

Acknowledgment

This work was supported in part by NIH grant HD-17711 to S. L. C. Woo, who is also an Investigator with the Howard Hughes Medical Institute.

Bibliography

DiLella, A. G., Marvit, J. and Woo, S. L. C. (1987). The molecular genetics of phenylketonuria. *In* ''Amino Acids in Health and Disease: New Perspectives.'' (S. Kaufman, ed.) UCLA Symposia on Molecular and Cellular Biology, New Series **55**. Alan R. Liss, Inc., New York, N.Y.

Güttler, F. (1980). Hyperphenylalaninemia: Diagnosis and classification of the various types of phenylalanine hydroxylase deficiency in childhood. *Acta Paediatr. Scand. Suppl.* **280,** 7.

Kaufman, S. (1976). Phenylketonuria and its variants, *Adv. Neurochem.* **2,** 1.

Scriver, C. R., and Clow, C. L. (1980). Phenylketonuria and other phenylalanine hydroxylation mutants in man. *Annu. Rev. Genet.* **14,** 179.

Scriver, C. R., Kaufman, S., and Woo, S. L. C. (1988). Mendelian hyperphenylalaninemia. *Annu. Rev. Genet.* **22,** 301.

Scriver, C. R., Kaufman, S., and Woo, S. L. C. (1989). The hyperphenylalaninemias. *In* ''The Metabolic Basis of Inherited Disease, Volume II,'' 6th ed. (C. R. Scriver, A. L. Beaudet, W. S. Sly and D. Valle, eds.). McGraw-Hill, Inc., New York, N.Y.

Woo, S. L. C. (1989). Molecular basis and population genetics of phenylketonuria. *Biochemistry* **28,** 1.

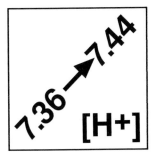

pH Homeostasis

DIETER HÄUSSINGER, *Medizinische Universitätsklinik, Freiburg, Federal Republic of Germany*

I. Intracellular pH Homeostasis
II. Extracellular pH Homeostasis

Glossary

Acids, bases Most commonly, an acid is a proton donor, and a base is a proton acceptor

Buffer Solution of a conjugate acid–base pair capable of ameliorating the effect of added acid or base on pH in that solution

Buffer strength In its differential form, defined as the ratio $\dfrac{dB}{dpH}$, with dpH reflecting the increase of pH following addition of a small amount of base (dB); the highest buffer strength of a buffer system is found when the pH of the solution equals the pK value of the buffering conjugate acid–base pair

Conjugate acid–base pair Acids dissociate into a proton and their conjugate base, whereas proton-binding to a base leads to the formation of its conjugate acid

pH Initially defined as the negative decadic logarithm of the free proton concentration; the modern pH scale is only a relative measure of the proton's chemical potential; thus, the inverse antilog of pH represents the proton concentration in a first approximation. The term "free proton concentration" is convenient and widely used; however, physico-chemically the free H^+ has a short lifetime and is present predominantly as the hydronium ion:

$$H^+ + H_2O \rightleftharpoons H_3O^+$$

BIOLOGICAL COMPOUNDS SUCH AS PROTEINS have multiple acid or base groups; the degree of their protonation is critical for their function. The different intra- and subcellular compartments exhibit a characteristic and remarkably constant proton concentration, i.e., pH, within narrow limits, although numerous reactions taking place in these compartments will consume or produce protons. Such compartmental homeostasis of pH is achieved by a variety of H^+ or HCO_3^- pumps and exchangers, which are located in the bordering membranes. These transport proteins not only stabilize the compartmental pH but also can build up differences of proton concentrations at the two sides of a biological membrane, generating electrochemical proton gradients. The pH in the various compartments differs, and the functions of the proton concentration and the transmembrane proton concentration gradients are variable. For example, the low lysosomal pH allows enzymatic hydrolysis, and pH gradients are used in endosomes for the sorting of receptor–ligand complexes and serve as the driving force for the transport of biogenic amines in chromaffin granules. The so-called proton motive force, i.e., the electrochemical proton gradient across the inner mitochondrial membrane, drives adenosine triphosphate (ATP) synthesis. On the other hand, the free proton concentration is a physiologically important regulator of biological processes. Small, controlled changes of the cytoplasmic pH under the influence of growth factors and hormones may be important in normal and malignant growth. Mechanisms responsible for the maintenance of physiological proton concentrations in cellular and subcellular compartments (intracellular pH homeostasis) depend on the proton concentration in the external environment. For mammals, control of the proton concentration in the extracellular fluid (extracellular pH homeostasis) has become systemic, i.e., affects the whole body. Specialized cell types have evolved for the regulation of extracellular systemic pH by the interplay of different organs, mainly the lung, liver, and kidney. Thus, homeostasis of pH evidently does not mean an absolute constancy of pH but can be defined as the integration of events allowing the free proton concentration in the various compartments to play its role as a physiological coordinator of biological processes.

I. Intracellular pH Homeostasis

A. General Considerations

Because cells consist of different compartments with different free proton concentrations, the term intracellular pH (pH$_i$) describes more or less an average pH of many compartments. Accordingly, pH$_i$ depends on the technique employed for its determination and represents an experimentally determined cellular pH, which will be an average of the pH of those compartments reached by the respective pH$_i$-determining technique. In many cases, pH$_i$ will come close to the cytosolic pH; then intra- and extracellular pH differences will approximate the pH gradient across the plasma membrane. All techniques for pH$_i$ assessment yielded transmembrane pH gradients far below those predicted theoretically by the Nernst equation, indicating that the distribution of H$^+$ across the plasma membrane is not in electrochemical equilibrium.

$$V_m = R \cdot T \cdot F^{-1} \ln([H^+]_e/[H^+]_i),$$
$$\text{(Nernst equation)}$$

where V_m is the voltage difference across the membrane in volt, R is the gas constant, T is the absolute temperature, F is Faraday's constant, and subscripts $_i$ and $_e$ refer to intracellular and extracellular, respectively. At 37°C, this equation becomes, in terms of pH,

$$V_m = 0.061(pH_i - pH_e).$$

Thus, with a membrane potential of −60 mV (inside negative) and pH$_e$ of 7.4, the electrochemical equilibrium would predict a pH$_i$ of 6.4, whereas actually pH$_i$ values of about 7.0 are measured. Thus, cells must have mechanisms for H$^+$ extrusion, which keep the pH$_i$ at values above equilibrium. Many of these pH$_i$-controlling systems are biochemically or functionally characterized; they underly a complex regulation that provides the basis for a control and coordination of biological processes by the proton concentration.

B. Methods for Assessment of pH$_i$ and Subcellular pH

1. Weak Acid or Base Distribution

This technique for determination of pH$_i$ relies on the permeability of biological membranes for the uncharged, but not for the charged, species of a metabolically inert weak acid–base pair. After equilibration between the intra- and extracellular space, the concentrations of the ideally freely diffusible, uncharged species are equal in both compartments, whereas the concentrations of the charged corresponding acid or base are determined by the proton concentrations on either side of the membrane. For the weak acid HA, which dissociates to A$^-$ + H$^+$, the distribution of A$^-$ between the intra- and extracellular space is inversely related to that of H$^+$:

$$[H^+]_i/[H^+]_e = [A^-]_e/[A^-]_i.$$

Employing the weak acid HA as an indicator (i.e., at a concentration sufficiently low to avoid significant pH$_i$ poisoning), pH$_i$ can be calculated from pH$_e$, the pK value of the indicator system HA/A$^-$, and the concentrations of dissociated and undissociated indicator [HA + A$^-$] in the intra- and extracellular space:

$$pH_i = pK + \log[(10^{pH_e - pK} + 1)$$
$$\times ([A^-]_i + [HA]_i)([A^-]_e + [HA]_e)^{-1}]$$

The most commonly used pH$_i$ indicator is the weak acid 5,5-dimethyloxazolidine-2,4-dione (DMO). DMO is frequently employed as a ^{14}C-labeled compound, and the distribution of radioactivity between the extracellular space and tissue allows determination of pH$_i$ after correction for intra- and extracellular water spaces. The technique is comparatively easy to handle; however, repetitive or continuous determinations of pH$_i$ are not possible. pH$_i$ values determined by means of weak acid–base indicators will represent some average of the subcellular compartments among which the indicator is distributed. Weak acid–base pairs can also be used to determine the pH in subcellular organelles such as mitochondria, isolated by suitable cell subfractionation techniques.

2. Fluorescent Dyes

This technique employs an indicator whose spectrofluorometric properties are a function of the environmental pH. This indicator is loaded into the cells, which are excited by light of a suitable wavelength, while monitoring pH-dependent absorbance or fluorescence at another wavelength. Because the spectrofluorometric characteristics of a dye are different in extra- and intracellular environments, the system requires calibration. Indicator compounds frequently used are fluorescein derivatives. Biscarboxyethylfluorescein (BCECF), which is not electrically charged, rapidly permeates the plasma membrane and is trapped inside the cells after hydrolysis by intracellular esterases, yielding the active pH-sensitive compound carboxyfluoresceine. This technique allows continuous pH_i registration and is applicable to single and/or small cells and isolated organelles. Time-dependent bleaching and leakage of the dye out of the cells are the drawbacks.

3. pH-Sensitive Microelectrodes

The difference of electrical potential between a pH-sensitive (glass or liquid ion exchange) microelectrode and a reference microelectrode with very small tips (<0.2 μm), both introduced into a single cell, is used to measure pH_i. The electrodes can be placed separately into adjacent, electrically coupled cells or can be combined to one double-barreled microelectrode. pH_i values determined by this technique most closely approach the cytosolic pH and allow the continuous measurement of pH_i. This technique is limited to cells large enough to sustain microelectrode puncture.

4. Nuclear Magnetic Resonance Spectroscopy

Because the nuclear magnetic resonance (NMR) frequency of endogenous inorganic phosphate is dependent on the $H_2PO_4^-/HPO_4^{--}$ ratio (so-called chemical shift), the inorganic phosphate peak of the ^{31}P-NMR spectrum can be used to determine pH_i. "Pure" inorganic phosphate signals from the intracellular compartment can be obtained. How much subcellular compartments can be distinguished is not fully resolved and depends on the tissue: whereas in liver ^{31}P-NMR probably gives whole cell, rather than cytosolic pH, in heart muscle it largely reflects the cytosolic pH. Besides ^{31}P-NMR, ^{1}H-NMR and ^{19}F-NMR also have been employed to study pH_i, using the pH-dependent ^{1}H chemical shift on histidine and the pH-dependent distance of the two ^{19}F peaks of difluoromethylalanine. A major advantage of the NMR technique is that the relationship between pH_i and metabolic processes can be studied *in vivo*. [*See* MAGNETIC RESONANCE IMAGING.]

C. Mechanisms and Regulation of Proton Translocation Across Biological Membranes

In principle, two mechanisms for proton transport across biological membranes can be distinguished: electroneutral exchange mechanisms (Na^+/H^+ antiport, HCO_3^-/Cl^- exchange, K^+/H^+ ATPase) and electrogenic proton pumps (H^+ ATPases). These transporters are responsible for the maintenance of pH in the respective compartments and the generation of electrochemical proton gradients. They are discussed in the chapter on Ion Pumps. There is, in addition nonionic diffusion. Many weak acids and bases rapidly equilibrate across the membrane by nonionic diffusion of their uncharged species. In the case of weak acids, such as many monocarboxylic acids, diffusion into the cell is accompanied by an import of protons. Another example is CO_2, which rapidly diffuses across membranes into cells, where it is hydrated to H_2CO_3, which rapidly dissociates to H^+ and HCO_3^-, leading to a fall of pH_i. When uncharged species of weak bases such as NH_3 enter the cell, a rise of pH_i occurs due to protonation of NH_3 yielding NH_4^+. [*See* ION PUMPS.]

D. How Cells Respond to an Intracellular Acid Load

When the interior of a cell is subjected to an acid load, pH_i initially drops and recovers with time to the initial pH_i value. Several mechanisms contribute to this pH_i homeostatic response: physicochemical buffering, adaptations of cellular metabolism, stimulation of proton extrusion out of and bicarbonate import into the cell, and probably, to a minor extent, a concentrative shift of protons from the cytosol into subcellular organelles.

Physicochemical buffering is the most rapid response to an intracellular acid load and represents proton-binding to various intracellular bases according the mass action law with formation of the corresponding acids. Physicochemical buffering depends on the buffer strength of the cell interior (normally about 70 mmol/liter) and determines the maximal deviation of pH_i following an intracellular acid or alkali load. Part of the intracellular acid load is also removed from the cytosol by H^+ uptake into various organelles (H^+ ATPases of the vacuolar system). Simultaneously, H^+ extrusion from other organelles such as mitochondria decreases. Such net proton shifts into subcellular organelles represent another buffering mechanism with respect to cytosolic pH (organellar buffering); however, its quantitative importance is yet unclear. Buffering mechanisms ameliorate pH changes but can neither prevent the pH_i change nor return pH_i to its original level. Return to normality is brought about by H^+ extrusion from the cell via Na^+/H^+ exchange and/ or HCO_3^- import into the cell via Cl^-/HCO_3^- exchange mechanisms in the plasma membrane.

These transporters are activated when pH_i falls. During this process, H^+ previously bound to intracellular buffers is released again according to the mass action law; the entire acid load is finally extruded until pH_i and the cellular buffering mechanisms are returned to their initial state.

As long as pH_i is below its normal value, alterations of cellular metabolism accompany the pH_i recovery phase. They can be seen as metabolic adaptations to the need of proton removal during an intracellular acidic challenge; the opposite happens in an alkaline challenge. Because protons are formed or consumed in a variety of metabolic reactions, an inhibition of H^+ formation and/or a stimulation of H^+ consumption provides a pH_i homeostatic response. One example is the inhibition of H^+ producing glycolysis in acidosis due to a 90% inhibition of phosphofructokinase activity when the pH_i falls from 7.2 to 7.1. Further, lowering of the steady-state concentrations of organic mono-, di-, and tricarboxylates consumes H^+: their tissue levels drop following a decrease of pH_i. Metabolic adaptations for pH_i homeostasis may exhibit organ-specificity: in the liver, a decrease of pH_i inhibits urea synthesis, a liver-specific pathway consuming HCO_3^- (or producing H^+), which plays a major role in whole-body pH homeostasis (see also Section II.C).

E. Biological Significance of pH_i and Subcellular pH Homeostasis

The free proton concentration in subcellular compartments is a physiologically important regulator of a variety of biological processes. pH values of different subcellular compartments are given in Table I. Only a few examples are given below.

1. The pH_i Signaling Hypothesis

Activation of Na^+/H^+ exchange under the influence of a variety of mitogens and growth factors

TABLE I pH Values in Different Compartments

Compartment	pH
Extracellular	7.36–7.44
Cytosol	6.9–7.2
Mitochondria	7.4–7.7
Lysosomes	4.5–5.4
Endosomes	about 5
Secretory vesicles	
Parotid gland	6.8
Pancreatic islet cells	5.0

was found to increase cytosolic pH from its normal value of 6.9–7.2 by up to 0.4 pH units. This persistent rise in pH_i was seen as an early signal, which triggers proliferation and cell growth. Indeed, inhibitors of Na^+/H^+ exchange blocked the proliferative response to many growth factors. Furthermore, mutant fibroblasts, lacking Na^+/H^+ exchange activity, which do not show an increase of pH_i following mitogen treatment, have an impaired activation of DNA synthesis. However, in many cases a proliferative response to mitogens and growth factors does not seem necessarily dependent on Na^+/H^+ exchange activity or a rise in pH_i; this is especially true if bicarbonate is present in the experimental systems. Thus, a rise in pH_i has probably only a permissive role for the initiation of cell growth and differentiation in conjunction with other important signals. Stimulus-activation of Na^+/H^+ antiport under these conditions may not only trigger but also act to counter an increased intracellular proton formation accompanying metabolic hyperactivity during proliferative responses. Elevations of pH_i may also play a role in malignant transformation of cells: tumor-promoting phorbol esters stimulate Na^+/H^+ exchange probably via activation of protein kinase C, growth factor-like substances are produced by a number of transformed cell lines, and the oncogene product of the simian sarcoma virus shows marked homology with platelet-derived growth factor. Some evidence indicates that activation of Na^+/H^+ exchange is involved in tissue regeneration and organ hypertrophy.

A rise in pH_i following hormonal activation of Na^+/H^+ antiport will stimulate phosphofructokinase and could, in part, explain the stimulation of glycolysis by Ca^{++}-mobilizing hormones such as alpha-adrenergic agonists or vasopressin. [*See* GLYCOLYSIS.]

In neutrophils, the action of chemotactic peptides is blocked by inhibition of the Na^+/H^+ exchanger. Furthermore, manipulation of pH_i in neutrophils modulates their chemotactic responsiveness. The stimulus-activated production of partially reduced reactive oxygen species by neutrophils and macrophages, such as O_2^- or O_2^{--}, is accompanied by a considerable intracellular acidification, which is counteracted by Na^+/H^+ exchange. Also, some evidence supports a regulatory role of the Na^+/H^+ exchanger on the activation of O_2^--producing membrane-bound reduced nicotinamide adenine dinucleotide phosphate oxidase in neutrophils. [*See* NEUTROPHILS.]

2. pH_i and Membrane Potential

In a variety of tissues, a rise of pH_i (intracellular alkalosis) leads to a hyperpolarization of the cell membrane potential due to an increase of K^+ conductance by opening pH-sensitive, Ca^{++}-activated K^+ channels. Conversely, intracellular acidosis leads to a depolarization. The effects of pH_i on membrane potential in turn affect Na^+-coupled metabolite transport into cells, as well as electrogenic bicarbonate transport. Thus, in the basolateral membrane of the proximal kidney tubule depolarization leads to an intracellular alkalinization, and hyperpolarization acidifies the cell interior. Intracellular acidification affects the electrical coupling between epithelial cells through gap junctions, causing uncoupling due to pH sensitivity of gap-junction channels.

3. pH_i and Regulation of Cell Volume

When cells are exposed to anisotonic media, osmotic water fluxes lead to cell swelling or shrinkage. Subsequent activation of volume-regulatory mechanisms restore the initial cell volume within minutes. Regulatory volume decrease following osmotic cell swelling is due to an increased K^+ conductance in most cell types. Regulatory volume increase following cell shrinkage in hypertonic media is amiloride-sensitive in many cell types such as rat liver cells, dog erythrocytes, or human lymphocytes. Parallel activation of Na^+/H^+ and of Cl^-/HCO_3^- exchange in these cells mediates regulatory volume increase resulting in a net gain of Na^+, Cl^-, and H_2O inside the cell.

4. Regulation of Intracellular Membrane Flow

The vacuolar system comprises the endoplasmic reticulum, the Golgi apparatus, endosomes, lysosomes, secretory vacuoles, and other membrane-bordered structures participating in transport and processing of soluble or membrane-bound macromolecules. This system is dynamic and presents a constant flow of membrane among the different structures. Many components of the vacuolar system exhibit an acidic internal pH, created by H^+ pumps, which is an important prerequisite for proper functioning of the vacuolar system.

The low intralysosomal pH provides not only optimal conditions for the activity of lysosomal hydrolases but may also serve as signal for the activation of several lysosomal proteases. For example, cathepsin D is transported to the lysosome as a catalytically inactive precursor but is autocatalytically activated by the low intralysosomal pH through proteolytic cleavage. Endosomes represent an intermediary, prelysosomal compartment involved in the transport of internalized insoluble or soluble material or receptor–ligand complexes from the plasma membrane to the lysosomes. In these vesicles, sorting of receptor–ligand complexes takes place under low pH, which favors receptor–ligand dissociation; this determines the fate of the receptor and the ligand (either recycling or lysosomal degradation). The acidic endosomal pH, however, is only one signal for sorting and routing; other factors include Ca^{++} and ligand valency. The low endosomal pH also plays a role in targeting lysosomal proteins from the Golgi apparatus to the lysosomes. Newly synthesized acid hydrolases bear phosphomannosyl groups that bind to the membrane-bound mannose-6-phosphate receptors at neutral pH but rapidly dissociate at acidic pH. Such dissociation occurs in the acidic compartment of uncoupling of receptors and ligands (CURL); the mannose-6-phosphate receptor recycles to the Golgi apparatus, whereas the hydrolases reach lysosomes. Acidotropic agents, such as chloroquin or methylamine, perturb the intravesicular acidic pH and divert precursors of lysosomal enzymes to a secretory pathway out of the cell. Also, some evidence indicates that acidic pH is involved in the accumulation of proteins to be secreted in secretory vesicles. [*See* GOLGI APPARATUS.]

II. Extracellular pH Homeostasis

A. General Considerations

The pH in various intracellular compartments critically depends on the pH in the cellular environment, i.e., the extracellular fluid (pH_e) in higher organisms. This is because intra- and extracellular pHs are linked by several processes that transport protons or bicarbonate either actively (H^+ pumps and Na^+/H^+ and HCO_3^-/Cl^- exchange) or passively (nonionic diffusion, passive H^+, and HCO_3^- movements due to electrochemical gradients). By these mechanisms, changes in pH_e will affect pH_i in the same direction.

The bicarbonate buffer is of special importance for homeostasis of pH_e. In the extracellular space, the bicarbonate buffer is in chemical equilibrium with other buffer systems (collectively referred to as nonbicarbonate buffers, such as phosphate and proteins). According to the Henderson–Hasselbalch equation, the pH of a bicarbonate buffer is determined by the ratio of concentration of bicarbonate to that of physically dissolved CO_2, which is the product of the physical solubility of CO_2 and the partial CO_2 pressure:

$$pH_e = pK' + \log([HCO_3^-]_e/[CO_2]_e).$$

In organisms, the concentrations of HCO_3^- and CO_2 can be effectively regulated by excretion or metabolic disposal. The open system maintaining a steady state of $[CO_2]$ substantially increases operational buffer capacity. A physiological pH_e of about 7.4 is maintained by a coordinate interplay of specialized cells, mainly in lung, kidney, and liver. These cells act to stabilize not only the extracellular $[HCO_3^-]/[CO_2]$ ratio (and accordingly pH_e) but also the absolute concentrations of CO_2 and HCO_3^- in the extracellular space, despite the fact that CO_2 and HCO_3^- are continuously generated during the respiratory oxidation of foodstuff; thus, constancy of pH_e is largely brought about by finely tuned mechanisms disposing of CO_2 and HCO_3^- from the organism at the same rate as they are generated. Carbonic anhydrases accelerate CO_2/HCO_3^- equilibria in the various compartments and guarantee sufficiently rapid transfer of CO_2 between tissues and extracellular fluid on the one hand and the extracellular fluid and the external environment on the other.

B. CO_2/HCO_3^- Equilibria and Transport Between Tissues: The Role of Carbonic Anhydrases

The following reactions are involved in the equilibrium of CO_2 and the HCO_3^- system:

$$CO_2 + H_2O \rightleftharpoons H_2CO_3 \qquad \text{(reaction I)}$$
$$H_2CO_3 \rightleftharpoons H^+ + HCO_3^- \qquad \text{(reaction II)}$$

Reaction II is very fast and reaches chemical equilibrium almost instantaneously, whereas reaction I is slow: at 37°C and a physiological pH of 7.4, the uncatalyzed hydration–dehydration reactions will be half complete within 3.6 sec after a step concentration change of one of the reaction components. Because the blood transit time in capillaries (where gas exchange takes place) is <1 sec, carbonic anhydrase (CA) activities are required to accelerate formation of HCO_3^- from CO_2 in peripheral capillaries and of CO_2 from HCO_3^- in lung capillaries.

We know five different CA isoenzymes (CA I–V) (Table II), which facilitate CO_2/HCO_3^- equilibration and transport across membranes or supply HCO_3^- for biosynthetic processes. They are distinguished with respect to their sensitivity to sulfonamide inhibition, serological characteristics, and tissue distribution. Some examples on the physiological roles of CA will be given. CO_2 produced in peripheral tissues rapidly diffuses to the blood and equilibrates with the plasma and the red cell interior. High CA activities in the erythrocytes guarantee rapid equilibration of the CO_2/HCO_3^- system inside them. The accompanying intracellular acidification produces an allosteric change of hemoglobin, which favors tissue oxygenation (Bohr effect). Electroneutral HCO_3^-/Cl^- exchange mediates exit of HCO_3^- to the plasma. Whereas inside the erythrocytes, CO_2/HCO_3^- equilibria are reached during capillary transit in the presence of CA, this does not necessarily apply for the plasma CO_2/HCO_3^- system, especially in tissues lacking extracellular membrane-bound CA, such as liver. Membrane-bound CA present in capillaries of skeletal muscle accelerates CO_2/HCO_3^- equilibration in plasma, so that equilibrium is reached already in end-capillary blood. The processes occurring during gas exchange among blood, plasma, and muscle tissue are a mirror image of those taking place during the gas exchange in alveolar capillaries. In the lung, besides a membrane-bound CA on the luminal surface of lung capillaries, an intracellular CA has been described that participates in alveolar CO_2 exchange by facilitating CO_2 diffusion.

C. Organ-specific Contributions to pH_e Homeostasis

pH homeostasis in the extracellular space is brought about by a complex interplay between specialized organs, mainly lung, liver, and kidney. Their coordinate action adjusts pH_e close to 7.4 by regulating the extracellular CO_2 and HCO_3^- concentrations. These organs have sizeable excretory or metabolic pathways for irreversible elimination of CO_2 or HCO_3^- (or of constituents of other extracellular buffers in chemical equilibrium with the HCO_3^-/CO_2 system). These pathways, in turn, are controlled by pH_e, $[CO_2]_e$ and $[HCO_3^-]_e$, so that there is a feedback control circuit between pH_e and CO_2/HCO_3^- elimination.

TABLE II Carbonic Anhydrases (CA)

Isoenzyme	Localization	Tissue	K_i (acetazolamide)
CA I	cytoplasmic	lung, erythrocytes, brain, colon, cornea	2×10^{-7} M
CA II	cytoplasmic or membrane-bound	lung, erythrocytes, white skeletal muscle, liver, kidney, bone, brain, stomach, small intestine, lung	10^{-8} M
CA III	cytoplasmic	red skeletal muscle, liver, erythrocytes	3×10^{-4} M
CA IV	membrane-bound	skeletal muscle, kidney, lung, brain	10^{-6}–10^{-7} M
CA V	mitochondrial	liver, kidney	10^{-7} M

Clearly, any organ can influence pH_e by either release or uptake of acids or bases into or from the extracellular space. These processes, however, must be viewed as an integral part of overall metabolic fluxes. Their potential effects on pH_e are usually balanced by proton fluxes in opposite directions. One example is gastric acid secretion, which transiently renders the blood somewhat alkaline (alkaline tide), until it is compensated by pancreatic HCO_3^- secretion. Similarly, during severe exercise, lactic acid production by the muscle may result in transient extracellular acidosis owing to accumulation of lactate plus H^+ in the extracellular space but spontaneously resumes following removal of lactate via gluconeogenesis or oxidation, i.e., processes consuming protons. Normally, the steady-state concentrations of acidic or basic metabolic intermediates in the extracellular space remain constant, because flux rates through reactions generating protons are matched by proton-consuming reactions of equal magnitude (biochemical steady state). Under pathological conditions only, imbalances of these metabolic proton fluxes give rise to acid–base disturbances (for example, when the oxidative pathway from glucose to CO_2 is interrupted during hypoxia, whereas glycolytic pyruvic and lactic acid formation continues). Although normally intermediate metabolite concentrations remain remarkably stable despite wide variations of flux through the whole metabolic pathway, the amount of acidic or basic end products formed during the metabolism of carbohydrate, fat, and protein is variable and provides a continuous threat to acid–base homeostasis. In man, these end products include CO_2 (12–20 moles/day), HCO_3^- (about 1 mol/day), NH_4^+ (about 1 mol/day), H_2SO_4 (10–20 mmoles/day), and $H_2PO_4^-$ (10–20 mmoles/day). Elimination of these metabolic end products at the same rate as they are generated to achieve a biochemical steady state is the hallmark of pH_e homeostasis. Interestingly, major pH homeostatic mechanisms are frequently linked to other vital functions, explaining at least in part a limitation of pH homeostatic responses. For example, respiration links CO_2 elimination to oxygen uptake. Under extreme conditions the need for sufficient oxygen uptake may conflict with hyperventilatory CO_2 exhalation. Similarly, bicarbonate homeostasis is closely linked to ammonium homeostasis, and under certain conditions ammonium toxicity can only be avoided for the price of acid–base disturbances.

1. Role of the Lungs

The lungs are the major organ for irreversible elimination of CO_2; this process underlies a complex feedback control by the arterial pH and PCO_2. Pulmonary CO_2 excretion is regulated by alveolar ventilation ($\dot{V}_A$), which may rise in humans from 5 liters/min at rest to 50 liters/min during heavy exercise. The relationship between $\dot{V}_A$, the arterial partial pressure of CO_2 ($PaCO_2$), that in inspired air ($PiCO_2$) and metabolic CO_2 production $\dot{M}_{CO_2}$ is described by the following equation:

$$PaCO_2 - PiCO_2 = \dot{M}_{CO_2}(\dot{V}_A \cdot \beta)^{-1},$$

where β is the capacitance coefficient ($\beta = 0.0517$ mmol/1 torr). In resting humans, $PiCO_2$ is close to 0, $\dot{M}_{CO_2} = 10$ mmol/min, and $\dot{V}_A = 5$ liters/min. Accordingly, a normal arterial $PCO_2 = 40$ torr results. Increases in $\dot{M}_{CO_2}$ or $PiCO_2$ lead to an increase of $PaCO_2$ and a decrease in pH_e when $\dot{V}_A$ remains unchanged. Changes of arterial pH and

PCO$_2$ are sensed by central (medulla oblongata) or peripheral (carotid and aortic bodies) chemoreceptors, resulting in an increased impulse rate of afferent nerve fibers from the chemoreceptors to the respiratory centers in the brain stem. This signal stimulates efferent output of motoneurons to respiratory muscles and increases alveolar ventilation until arterial PCO$_2$ levels and pH are returned to normal. Thus, a control circuit with negative feedback loop exists for stabilization of arterial PCO$_2$. The adequate stimulus of the chemoreceptors appears to be the pH rather than the PCO$_2$; in addition, arterial (peripheral) chemoreceptors are sensitive to changes in the arterial oxygen concentration. The pH-sensitivity explains why PaCO$_2$ levels become subnormal when blood bicarbonate concentration is lowered, stabilizing the pH$_e$ close to 7.4. This situation exists, for example, during severe exercise, when plasma bicarbonate decreases due to net lactic acid production by the muscle and the arterial PCO$_2$ is adjusted by ventilation to values below 40 torr. Good evidence indicates that ventilation control by signals from chemoreceptors is supplemented by afferent impulses from muscle or joint mechanoreceptors and from collaterals from cortical motor centers. [*See* RESPIRATORY SYSTEM, PHYSIOLOGY AND BIOCHEMISTRY.]

2. Role of the Kidneys

The mechanisms by which renal function contributes to homeostasis of pH$_e$ include the reabsorption of filtered bicarbonate from the tubular fluid, net excretion of protons or bicarbonate, which can be assessed by titration of the final urine back to the pH of the extracellular fluid and participation in the coordinate regulation of bicarbonate and ammonium homeostasis in an interorgan team effort with the liver.

The glomerular filtrate contains bicarbonate at a concentration equal to that in plasma, i.e., normally about 25 mM. Reabsorption of filtered bicarbonate largely occurs in the proximal tubule involving luminal Na$^+$/H$^+$ exchange, and ATP-driven H$^+$ secretion, membrane-bound luminal and cytosolic CAs, and a basolateral Na$^+$ coupled HCO$_3^-$ transport. *In vitro* proximal tubular H$^+$ secretion underlies complex control by hormones (parathyroid and thyroid hormones, glucocorticoids) and the acid–base status. In acidosis, proximal tubular bicarbonate reabsorption is increased by activation of Na$^+$/H$^+$ exchange and insertion into the luminal membrane of ATP-driven H$^+$ pumps usually stored in endocy-

totic vesicles. The distal tubule largely determines the final urinary pH, by increasing acidity through an electrogenic H$^+$ ATPase located in the luminal membrane. Proton secretion into the lumen is accompanied by intracellular formation of HCO$_3^-$, which leaves the cell across the basolateral membrane via a Cl$^-$/HCO$_3^-$ exchange. The cortical collecting duct can exhibit bicarbonate absorption or secretion depending on the acid–base status. Alkalosis stimulates bicarbonate secretion, whereas acidosis stimulates bicarbonate reabsorption. Two types of intercalated cells are involved in this process. Type A cells absorb HCO$_3^-$ and contain an apical H$^+$ ATPase and a basolateral Cl$^-$/HCO$_3^-$ exchanger. An opposite polarity of these transport systems in type B cells allows bicarbonate secretion. Upon an acute decrease of pH in the cell, new H$^+$ pumps are inserted into the apical cell membrane of type A cells by migration from internal membranes via an exocytotic mechanism; whether or not an increase of cell pH induces endocytotic removal of the H$^+$ pumps is yet unclear. Chronic adaptation to acidosis includes an increase of H$^+$ secreting type A cells. Some evidence indicates that acid–base status may regulate the interconversion of type A and B cells, and both cell types may represent a different functional state of the same cell. As a pH homeostatic response, interconversion of type A and B cells could then change from bicarbonate secretion to absorption or vice versa.

The pH in urine normally varies between 5 and 8, and the net amount of protons excreted or retained by the kidney can be estimated by adding measured amounts of acid or alkali to the urine (titration) until its pH reaches the same value as plasma pH. This titratable acid excretion is normally about 30–50 mmol/day and largely reflects protons bound to urinary buffers such as phosphate. This is the direct net effect of the kidneys on pH$_e$ homeostasis; This seems a small contribution, especially when comparison is made with CO$_2$ excretion by the lungs (10–20 mol/day) or with HCO$_3^-$ disposal during hepatic urea synthesis (about 1 mol/day; vide infra). When bicarbonate is infused, however, urinary bicarbonate excretion can provide a spillover for excess bicarbonate, which cannot be eliminated metabolically (see below). Above a certain threshold concentration of bicarbonate in plasma, further increases result in increased urinary bicarbonate excretion proportional to the amount of filtered bicarbonate. This bicarbonate threshold concentration varies with the PCO$_2$; its increase in respiratory acidosis (increased steady-state PCO_2 in chronic lung disease to maintain normal daily CO_2 exhalation) allows an increase of the plasma $[HCO_3^-]$ to maintain normal pH.

There has been some confusion on the role of the kidneys in pH homeostasis regarding their role in ammonium excretion (so-called renal ammoniagenesis). In physiology, NH_4^+ excretion into urine was viewed as net proton excretion (so-called nontitratable acid). Although NH_4^+ is an acid according to the Bronstedt–Lowry definition, it is a very weak acid (pK = 9.1), which at physiological pH values dissociates protons in only minimal amounts. Thus, NH_4^+ movements from plasma to urine are *per se* without a relevant effect on plasma pH. The role of renal ammoniagenesis, therefore, required reinterpretation on the basis of a close link of ammonium and bicarbonate homeostasis in the body.

3. Coordinate Regulation of Bicarbonate and Ammonium Homeostasis by Liver and Kidneys

Whereas CO_2 and H_2O are the only end products of fat and carbohydrate metabolism, the complete oxidation of proteins or bipolar amino acids at a physiological pH also yields HCO_3^- and NH_4^+ as well as small amounts of H_2SO_4 (derived from sulfur-containing amino acids). Due to the chemical structure of an "average" amino acid, HCO_3^- and NH_4^+ are formed in almost equimolar amounts, i.e., about 1 mol/day, each, in a person oxidizing 100 g protein/day. The major route for elimination of this metabolically generated HCO_3^- is urea synthesis. This liver-specific pathway can be viewed chemically as an irreversible, energy-driven neutralization of the strong base HCO_3^- (pK = 6.1 for the CO_2/HCO_3^- system) by the weak acid NH_4^+ (pK = 9.1 for the NH_4^+/NH_3 system):

$$2HCO_3^- + 2NH_4^+ \longrightarrow H_2NCONH_2 + CO_2 + 3H_2O$$

energy

Accordingly, urea synthesis consumes HCO_3^- and NH_4^+ in the same stoichiometry as these compounds arise during protein breakdown. The daily average excretion of 30 g urea in humans corresponds to an irreversible bicarbonate elimination of about 1 mol/day. Hepatic urea synthesis (and accordingly the rate of irreversible HCO_3^- consumption in the liver) reveals a complex and sensitive feedback control by the extracellular pH, $[HCO_3^-]$ and $[CO_2]$. Acid–base control of urea synthesis occurs at the level of mitochondrion: the formation of NH_4^+ and HCO_3^- inside mitochondria by glutaminase and CA V is under pH control and determines the rate of synthesis of carbamoylphosphate, a major rate-limiting step of the urea cycle. Thus, when the urea cycle is inhibited in acidosis, the sparing of bicarbonate is a direct pH–homeostatic response of the liver. A sophisticated structural and functional organization of the pathways of urea synthesis and glutamine metabolism in the liver acinus allows modulation of the urea cycle according to the requirements of acid–base homeostasis without threat of hyperammonemia. This is because in the liver acinus ammonia-rich blood from the viscera first gets into contact with cells capable of urea synthesis (but not glutamine synthesis). These cells take up as much NH_4^+ as is needed for acid–base-controlled HCO_3^- disposal via the urea cycle. NH_4^+ that is not utilized by ureogenic cells is washed to a more downstream, small hepatocyte population, which is located just before the venous outflow and is devoid of urea cycle enzymes. These cells take up NH_4^+ with high affinity and synthesize glutamine. Thus, an increase in hepatic glutamine formation maintains nontoxic ammonium levels when urea synthesis is switched off in acidosis to spare bicarbonate: hepatic NH_4^+ detoxication is shifted from urea to glutamine synthesis. The final elimination of surplus NH_4^+ is provided by renal ammoniagenesis. In this process, glutaminolysis in the kidney liberates NH_4^+, which is excreted into the urine. Thus, renal ammoniagenesis provides a spillover for elimination of NH_4^+, without concomitant HCO_3^- consumption, and may be viewed primarily as an ammonium homeostatic mechanism. Similar to hepatic urea synthesis, renal ammoniagenesis is sensitively regulated by the acid–base status: renal NH_4^+ excretion is stimulated in acidosis in parallel with inhibition of bicarbonate-consuming urea synthesis in the liver. Thus, liver and kidney act as a team for maintaining ammonium and bicarbonate homeostasis. These relationships are schematically depicted in Figure 1: a pH-controlled shift of NH_4^+ elimination from urea synthesis to urinary excretion maintains ammonium homeostasis in the whole body and provides the basis for an undisturbed feedback circuit between acid–base status and hepatic HCO_3^- consumption. Failure of the renal ammoniagenesis indirectly leads to metabolic acidosis: the pH-sensitive control of urea synthesis is overridden by the

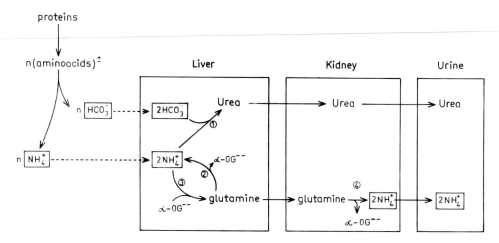

FIGURE 1 Coordinate regulation of HCO_3^- and NH_4^+ homeostasis. NH_4^+ and HCO_3^- are produced in almost equivalent amounts during protein breakdown. Whereas urea synthesis irreversibly consumes HCO_3^- and NH_4^+ in equal proportions, HCO_3^- is spared when urea synthesis is switched off in acidosis (bicarbonate homeostatic response of the liver). Under these conditions, NH_4^+ homeostasis is maintained by NH_4^+ excretion into urine (ammonium homeostatic response of the kidney). A feedback control loop between bicarbonate-consuming urea synthesis and the extracellular pH, $[CO_2]$ and $[HCO_3^-]$, adjusts hepatic HCO_3^- consumption to the acid–base status. Numbers in circles refer to major points of flux control by the extracellular acid–base status. In metabolic acidosis, flux through the urea cycle (reaction 1) and hepatic glutaminase (reaction 2) is decreased, whereas flux through hepatic glutamine synthetase (reaction 3) and renal glutaminase (reaction 4) is increased. (From Häussinger et al., 1986, *Biochem. J.* **236,** 261–265.)

need to eliminate potentially toxic NH_4^+, resulting in excessive stimulation of urea synthesis and concomitant HCO_3^- consumption. Under normal conditions, production of NH_4^+ during protein oxidation exceeds HCO_3^- generation by about 40 mmol/day due to neutralization of H_2SO_4, which is simultaneously formed from sulfur containing amino acids. This explains the normal urinary NH_4^+ excretion of about 40 mmol/day, which can rise up to 400 mmol/day in severe metabolic acidosis, when the bicarbonate-consuming urea synthesis is strongly inhibited.

D. Disturbances of pH_e Homeostasis and Compensation

Disturbances of pH_e homeostasis occur when acid or alkali loads (or losses) overwhelm mechanisms of pH_e homeostasis or when pH_e homeostatic mechanisms are impaired due to pulmonary, hepatic, or renal dysfunction. pH_e disturbances are classified as acidosis (plasma pH <7.36) and alkalosis (plasma pH >7.44), both of which can be of metabolic, respiratory, or mixed respiratory–metabolic origin. In pure metabolic pH derangements, the primary disturbance is a fall or rise of plasma $[HCO_3^-]$. Respiratory compensation occurs by adjusting the P_{CO_2} to values restoring a near-physiological pH_e of 7.4, according to the Henderson–Hasselbalch equation. In pure respiratory pH derangements, the primary disturbance is a deviation of the arterial P_{CO_2} from its normal value (38–42 mmHg), which is secondarily compensated by lowering or increasing the plasma $[HCO_3^-]$ (metabolic compensation). Respiratory disturbances can be compensated metabolically, and, vice versa, metabolic disturbances are compensated through respiration, with restoration

of the pH_e to almost normal values; the compensation, however, will cause abnormalities of plasma $[HCO_3^-]$ and $[CO_2]$. Their restoration to the normal $[HCO_3^-]$ value of 25 mmol/liter is only achieved when the underlying primary disturbance is corrected. Compensation is achieved by an inhibition of urea synthesis (sparing of HCO_3^-) together with an increased renal ammoniagenesis (ammonium homeostatic response) and is quantitatively less important via increased net H^+ excretion by the kidneys (titratable acid). When compensation by these mechanisms fails, persistent severe metabolic acidosis results in compensatory bone resorption, which supplies base equivalents (carbonate, phosphate).

E. Functional Significance of pH_e Homeostasis

Apart from its relevance for pH_i homeostasis, homeostasis of pH_e is crucial for a variety of events occurring in the extracellular space, as shown by the following examples. pH_e affects hormone-binding to cell-surface receptors, explaining relative insulin resistance and a decreased catecholamine activity in acidosis. Carrier-mediated lactate uptake into hepatocytes is stimulated in acidosis, whereas amino acid uptake decreases. In many organs, a fall in pH_e leads to vasodilatation. An acidic pH_e in capillary blood decreases the affinity of hemoglobin for oxygen, facilitating oxygen delivery to peripheral tissues.

Bibliography

Astrup, P., and Severinghaus, J. W. (1986). "The History of Blood Gases, Acids and Bases." Munksgaard, Copenhagen.

Atkinson, D. E., and Camien, N. M. (1982). The role of urea synthesis in the removal of metabolic bicarbonate and the regulation of blood pH. *Curr. Top. Cell. Reg.* **21**, 261–302.

Grinstein, S. (ed.) (1989). "Na^+/H^+ Exchange." CRC Press, Oxford.

Häussinger, D. (ed.) (1988). "pH Homeostasis." Academic Press, London.

Mellman, I., Fuchs, R., and Helenius, A. (1986). Acidification of the endocytic and exocytic pathways. *Ann. Rev. Biochem.* **55**, 663–700.

Tashian, R. E., and Hewett-Emmett, D. (eds.) (1984). Biology and chemistry of the carbonic anhydrases. *Ann. N.Y. Acad. Sci.* **429**

Phospholipid Metabolism

DENNIS E. VANCE, *University of Alberta*

Glossary

Diacylglycerol An intermediate in phospholipid biosynthesis and an activator of protein kinase C
Phosphatidylcholine Quantitatively, the most important phospholipid found in human cells
Phosphatidylethanolamine Quantitatively, the second most important phospholipid found in human cells
Phospholipases Enzymes that degrade phospholipids
Phospholipid transfer protein Transfers phospholipids between cellular membranes

PHOSPHOLIPIDS ARE BIOLOGICAL COMPOUNDS that contain phosphate and have both hydrophobic and hydrophilic components. Phospholipids are apparently critical for life, because all living organisms and cells have phospholipids in their membranes. Phospholipids provide the basic structure and the permeability barriers of cellular membranes. The central role of phospholipids in biology is underscored by the nearly complete lack of genetic alterations in their metabolism in humans and other animals, suggesting that such alterations are lethal. The metabolism of phospholipids includes the pathways for synthesis (anabolism) and degradation (catabolism). In addition, the pathways and mechanisms by which phospholipids move among the organelles of cells is an important aspect of phospholipid metabolism. How cells regulate metabolism of phospholipids is also discussed. Phospholipids are also a major source of cellular second messengers in the transduction of signals from the cell membrane (examples are diacylglycerols, inositol-trisphosphate, and platelet-activating factor). How these second messengers arise from phospholipids and how the cell regulates the metabolism of these compounds are also considered. The structures of phospholipids and their roles in membranes has been considered in the chapter on Lipids.

I. Phospholipid Synthesis in Procaryotes

Although the emphasis of this chapter is on metabolism of phospholipids in animal systems, it is instructive to consider briefly the procaryotes, chiefly *Escherichia coli*. The metabolic pathways of phospholipids in this organism have been clearly defined, and many of the enzymes involved in biosynthesis have been purified. In *E. coli*, phospholipid biosynthesis can be considered to begin with the synthesis of fatty acids from acetyl-ACP and malonyl-ACP, as diagrammed in Fig. 1. Almost exclusively, the purpose of fatty acid synthesis in this bacterium is for phospholipid synthesis. The major fatty acid derivatives formed are palmitoyl-ACP, a saturated fatty acid with 16 carbons in a thioester linkage to acyl carrier protein (ACP), and vaccenoyl-ACP, an ACP derivative of a monounsaturated fatty acid with 18 carbons (Fig. 1). These fatty acyl-ACP derivatives react sequentially with glycerol-3-phosphate (G-3-P) to yield phosphatidic acid, which is present as a trace component in *E. coli* and immediately reacts with cytidine triphosphate (CTP) to give cytidine diphosphate (CDP)-diacylglycerol. The pathways branch (Fig. 1) to yield either phosphatidylethanolamine (PE; the major phospholipid

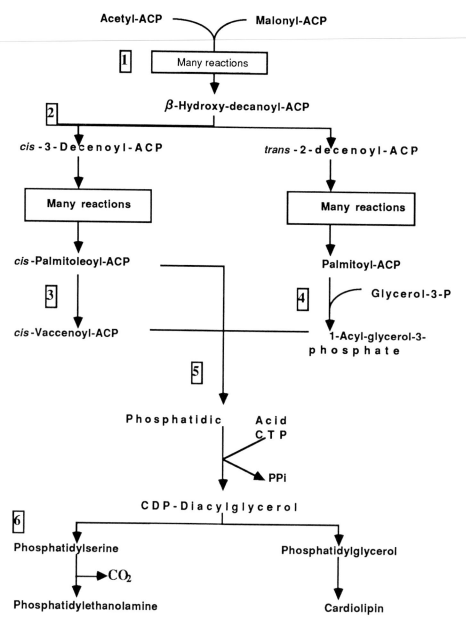

FIGURE 1 Biosynthetic pathway for phospholipids in *E. coli*. The synthesis of fatty acids begins with the condensation of acetyl-ACP and malonyl-ACP (1). Many additional reactions occur until a branch point is reached (2). At this juncture the cell commits itself to make either a saturated or monounsaturated fatty acyl-ACP derivative. The first unsaturated fatty acyl-ACP formed is *cis*-palmitoleoyl-ACP, and this is elongated to the 18-carbon monounsaturated fatty acyl derivative *cis*-vaccenoyl-ACP (3). The synthesis of the phospholipids begins with the acylation of glycerol-3-P with a 16-carbon acyl group, palmitoyl-ACP (4). Usually, an unsaturated fatty acid is inserted into the 2 position to give phosphatidic acid (5). This compound reacts with CTP to yield CDP-diacylglycerol, which is then converted to the major phospholipids in *E. coli* (6). ACP, acyl carrier protein; CDP, cytidine diphosphate; CTP, cytidine triphosphate; PPi, inorganic pyrophosphate.

in *E. coli* comprising 75–85% of the total) or phosphatidylglycerol and cardiolipin (which account for the other 15–25% of the phospholipids in the organism). The synthesis of the acyl-ACP derivatives takes place in the cytosol, whereas the acylation of G-3-P and subsequent reactions occurs at the inner surface of the inner membrane of *E. coli*.

Regulation of phospholipid synthesis in *E. coli* apparently occurs at an early but unknown step of fatty acid synthesis. This seems reasonable because the fatty acids in this organism (in contrast to mammals) are primarily used for phospholipid synthesis. Such regulation is also energy-efficient, as 94% of

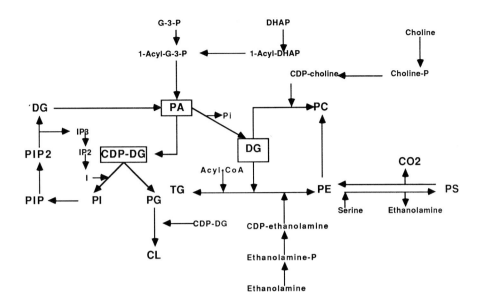

FIGURE 2 Outline for the biosynthesis of the major phospholipids in human cells. The conversion of PE to PC appears to be quantitatively significant only in the liver. CDP-DG, cytidine-diphosphodiacylglycerol; CL, cardiolipin; DG, diacylglycerol; DHAP, dihydroxyacetone phosphate; G-3-P, glycerol-3-phosphate; I, inositol; IP2, inositol-diphosphate; IP3, inositol-triphosphate; PA, phosphatidic acid; PC, phosphatidylcholine; PE, phosphatidylethanolamine; PG, phosphatidylglycerol; Pi, inorganic phosphate; PI, phosphatidylinositol; PIP, phosphatidylinositol-4-phosphate; PIP2, phosphatidylinositol-4,5-bisphosphate; PS, phosphatidylserine; TG, triacylglycerol.

the energy consumed in phospholipid synthesis is used for the manufacture of the fatty acyl components. The growth temperature of the organism determines the proportion of unsaturated fatty acids made and incorporated into phospholipids by modulating the synthesis of cis-3-decenoyl-ACP (Fig. 1). How the cell decides whether to partition CDP-diacylglycerol toward PE or phosphatidylglycerol and cardiolipin is not known. The commitment to make one of the phospholipids is apparently not based on the amounts of enzymes, as overexpression of phosphatidylserine synthase by 10-fold had no effect on the amount of phosphatidylserine and PE made. Similarly, an increase in the amount of phosphatidylglycerolphosphate synthase did not preferentially direct CDP-diacylglycerol toward phosphatidylglycerol or cardiolipin.

In the presence of an appropriate culture medium, bacteria continue to grow and divide; hence, phospholipid turnover (i.e., degradation and resynthesis) apparently is not required. Thus, the major phospholipid of E. coli, PE, is relatively stable and is not degraded or metabolized further. In contrast, the phospholipids of animal cells are in a constant state of synthesis and turnover.

II. Phospholipid Synthesis in Animals

A. Metabolic Pathways

The biosynthesis of phospholipids is far more complex in animal cells than in E. coli. Six major phospholipid species are found in animal cells compared with only three in the bacterium. The fatty acid substituents are much more varied, particularly in that animal cells have polyunsaturated fatty acids, which are generally absent in bacteria. In addition, in animal cells, phospholipid biosynthesis occurs on several different organelles.

The general scheme for phospholipid synthesis in animal cells is presented in Fig. 2. In animal cells, the biosynthesis of fatty acids is not directly linked with phospholipid synthesis as in E. coli. Although fatty acids are required for phospholipid synthesis, fatty acids are also a major source of energy and are stored in triacylglycerols for future energy requirements. As in E. coli, phospholipid synthesis begins with the acylation (attachment of a fatty acid) of G-3-P, usually with a saturated fatty acyl-CoA. A second acylation, usually with an unsaturated fatty acyl-CoA, yields phosphatidic acid. An alternative pathway beginning with dihydroxyacetone-phosphate (an intermediate in glycolysis) is equally important for the biosynthesis of phosphatidic acid

(Fig. 2). Phosphatidic acid is at a branch point in metabolism leading to either CDP-diacylglycerol or diacylglycerol.

As in *E. coli,* CDP-diacylglycerol is the precursor of phosphatidylglycerol and cardiolipin (Fig. 2). One difference is that in animal cells, cardiolipin arises from the condensation of phosphatidylglycerol with CDP-diacylglycerol, whereas in *E. coli,* two molecules of phosphatidylglycerol condense to form cardiolipin and glycerol. The other phospholipid made from CDP-diacylglycerol is phosphatidylinositol (Fig. 2). This lipid is further metabolized to phosphatidylinositol-4,5-P_2 and is discussed later (Section V, Functional Aspects of Phospholipids). Unlike *E. coli,* CDP-diacylglycerol is not a precursor of phosphatidylserine or PE in animal cells.

The major metabolic fates for diacylglycerol are the synthesis of phosphatidylcholine (PC), PE, and triacylglycerol (Fig. 2). The synthesis of triacylglycerol is important for the storage of energy in liver, adipose, and other cells but will not be discussed further in this chapter. PC is formed via the transfer of phosphocholine from CDP-choline to diacylglycerol by an enzyme that occurs on the endoplasmic reticulum and Golgi membranes and possibly at lower concentrations on other cellular membranes. CDP-choline is made from choline via phosphocholine as shown in Fig. 2. Choline is a dietary requirement for humans and other animals. Choline deficiency results in reduced levels of PC in liver, serum, and other tissues with an accumulation of triacylglycerol in liver. Such a deficiency is largely a laboratory phenomenon due to the ubiquitous occurrence of PC, the major source of dietary choline, in animal and vegetable sources in our daily diet. The major mechanism in animals for the biosynthesis of choline is via the conversion of PE to PC (and the subsequent catabolism of PC to choline) via the addition of three methyl groups by a single methyltransferase enzyme found principally on the endoplasmic reticulum. In animals, PE methyltransferase is recovered in significant amounts only from the liver. Approximately 30% of PC made in the liver has been estimated to derive from methylation of PE.

Quantitatively, PE is the second major phospholipid found in animals. Ethanolamine is phosphorylated by a cytosolic kinase, which is also responsible for the phosphorylation of choline. The other reactions of PE biosynthesis are depicted in Fig. 2 and seem to be catalyzed by different enzymes

from those used to make PC. PE can also be made from the decarboxylation of phosphatidylserine in mitochondria. The relative contributions of the two major pathways for PE biosynthesis are not established.

The biosynthetic pathway for phosphatidylserine in animals is quite different from that in *E. coli.* Phosphatidylserine, which comprises 5–10% of animal cell membranes, is made via an exchange of serine for ethanolamine from PE on the endoplasmic reticulum as shown in Fig. 2. This appears to be the sole pathway for the biosynthesis of ethanolamine in nature.

The fatty acid substituents of PC, PE, and other phospholipids can be remodeled by a deacylation–reacylation pathway (Fig. 3). This seems to be an important mechanism for the enrichment of polyunsaturated fatty acids into phospholipids, particularly for the introduction of arachidonic acid (the precursor of prostaglandins and other eicosanoids). In the type II cells that line lung alveoli, this pathway is used for the enrichment of a saturated fatty acid (palmitic acid) into PC. The resulting dipalmitoyl-PC is a major component of lung surfactant secreted by these type II cells. The surfactant provides a surface tension, which prevents lung collapse when a person breathes out air. In premature infants with respiratory distress syndrome, a defect appears in the secretion of lung surfactant.

Another major phospholipid found in animal cells is sphingomyelin. PC is the donor of phosphocholine for the synthesis of sphingomyelin (Fig. 4). This reaction is believed to occur on either Golgi and/or plasma membranes of cells. The plasma membrane

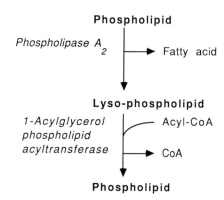

FIGURE 3 Deacylation–reacylation of a phospholipid occurs in most human cells and provides a mechanism for the cell to alter the fatty acid composition of a phospholipid. This is a major pathway for the introduction of arachidonic acid into phospholipids.

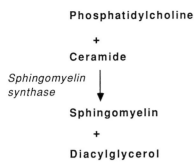

Phosphatidylcholine

+

Ceramide

Sphingomyelin synthase

Sphingomyelin

+

Diacylglycerol

FIGURE 4 Biosynthesis of sphingomyelin occurs via the transfer of phosphocholine from phosphatidylcholine to ceramide. Ceramide is a sphingolipid, which has a different structure but a similar shape to diacylglycerol.

is enriched in sphingomyelin compared with other membranes of the cell. As its name suggests, sphingomyelin is abundant in the myelin membrane.

B. Regulation

The understanding of how animal cells regulate the rate of phospholipid synthesis is at a relatively primordial level compared with that of other lipid metabolic pathways, such as fatty acid or cholesterol biosynthesis. One major obstacle has been the relatively modest variations in the rate of phospholipid synthesis as a result of different perturbations, such as fasting. Secondly, most of the enzymes have not been purified or only recently have been obtained in pure form, as these enzymes are usually tightly integrated into cellular membranes. A third difficulty is that many of the substrates are water-insoluble.

The regulation of PC biosynthesis is best understood. Control of this pathway appears to be centered on the enzyme (CTP–phosphocholine cytidylyltransferase), which converts phosphocholine to CDP-choline (Fig. 2). The cytidylyltransferase is a dimer of identical subunits that is recovered from the cytosol in an inactive form and from cellular membranes (endoplasmic reticulum and Golgi), where the enzyme is activated by phospholipids. Considerable evidence indicates that the rate of PC biosynthesis in animal cells is governed by the reversible translocation of the cytidylyltransferase between the cytosol and cellular membranes. Current studies are directed toward elucidation of the mechanism, which regulates the subcellular distribution of the cytidylyltransferase. One possibility is that the supply of fatty acids may be important. When

fatty acids (e.g., 0.3 mM) are added to the medium of cultured animal cells, an induction of PC biosynthesis is coincident with a movement of cytidylyltransferase from cytosol to cellular membranes; when fatty acids are removed from the cell medium, the rate of PC synthesis returns to control levels, and, correspondingly, the cytidylyltransferase moves from membranes back to the cytosol. Only one study shows such a mechanism operates in an intact animal; thus, more experiments are required to validate the physiological significance of fatty acid-mediated translocation of the cytidylyltransferase.

A second possible mechanism for control of PC biosynthesis is by cyclic adenosine monophosphate (cAMP)-dependent phosphorylation of the cytidylyltransferase. *In vitro* the pure enzyme is a substrate of this kinase, which reduces the affinity of the cytidylyltransferase for membranes. Removal of the phosphate group from the enzyme with a protein phosphatase causes an increased binding of the enzyme to membranes where it is activated. In agreement with this result, addition of cAMP analogues to liver cells in culture causes an inhibition of PC synthesis and a concomitant loss of cytidylyltransferase from membranes into cytosol. Currently, no evidence exists for or against the involvement of protein phosphorylation in regulation of cytidylyltransferase in a physiologically relevant model system.

A third, and perhaps more important, mechanism for control of PC biosynthesis has recently been discovered. When liver cells are deprived of choline and methionine, the rate of PC biosynthesis is decreased. As a result, the concentration of PC in the cell decreases by 10–20%. Under these conditions, the amount of cytidylyltransferase bound to cellular membranes doubles. When the cells are supplemented with choline or methionine, PC biosynthesis rapidly increases, and the level of PC in the cells rises. As the concentration of PC approaches normal levels, the need for increased PC biosynthesis is relaxed, and the excess cytidylyltransferase on the membranes is released into the cytosol. Thus, there appears to be feedback regulation by PC on the amount of cytidylyltransferase bound to membranes and, as a consequence, the rate of PC biosynthesis.

In summary, PC biosynthesis via the choline pathway is usually regulated by the reversible translocation of the rate-determining enzyme CTP–phosphocholine cytidylyltransferase between cytosol

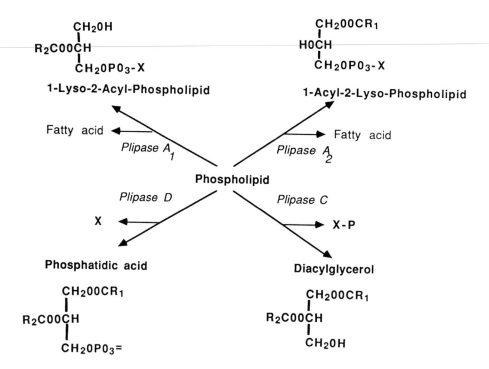

FIGURE 5 Four major phospholipase reactions found in human cells. X represents the head group of the phospholipid, such as choline or inositol, and X-P represents the phosphorylated head group. Plipase is an abbreviation for phospholipase. The designation Lyso indicates the absence of a fatty acyl group at the indicated position on the phospholipid.

and membranes. The physiologically relevant mechanisms involved in translocation of this enzyme still need to be defined.

As stated previously, PC can also be made in liver by the methylation of PE. The PE methyltransferase appears to be constitutively expressed in liver and present at very low levels in nonhepatic tissues. Studies in intact animals provide evidence that the supply of methionine, the precursor of S-adenosylmethionine (the methyl donor), regulates the rate of PE methylation. Similarly, in cultured cells the supply of PE has also been demonstrated to alter the rate of methylation. Thus, biosynthesis of PC via this pathway seems to be regulated by the supply of substrates.

III. Phospholipid Catabolism— Phospholipases

A. Background

Phospholipases, which are present in virtually all cells, are enzymes that degrade phospholipids, as shown in Fig. 5. The products are further catabolized, re-incorporated into phospholipids, or serve as second messengers, as discussed in Section V. These enzymes can be classified into functional categories—digestive and regulatory. The digestive phospholipases have been studied most extensively, because they are proteins that are usually

soluble in aqueous buffers. This category includes phospholipases, isolated from pancreas, that are secreted into the intestinal lumen and digest dietary phospholpids. Snake venom is another important source for phospholipases. A third source of digestive phospholipases is the lysosomes of cells, which degrade phospholipids that enter the cell via phagocytosis. The regulatory phospholipases are also discussed in Section V.

B. Phospholipases A_2

This class of phospholipases cleaves the fatty acyl group from the 2 position from phospholipids. The pancreatic phospholipase A_2 is secreted from the pancreas into the intestine as a proenzyme, where it is activated by proteolytic cleavage to yield an enzyme with a molecular weight of approximately 14,000. This processing allows the enzyme to bind to phospholipid aggregates and catalyze the hydrolysis of the fatty acid from the 2 position of the phospholipid, as shown in Fig. 5. The binding and catalytic sites appear to be at different positions on the phospholipase. Phospholipase A_2 purified from cobra venom is homologous to the pancreatic enzyme but is active as a dimer of two identical subunits, whereas the pancreatic enzyme acts as a monomer. Both enzymes have an absolute requirement for calcium.

A phospholipase A_2 similar in size to the pancreatic and venom enzymes has been purified from liver mitochondria. This enzyme is hydrophobic and is recovered from the inner and outer mitochondrial membranes. The function and regulation of this enzyme are unknown. Possibly, it is important in the deacylation–reacylation cycle to modulate the acyl composition of mitochondrial phospholipids.

C. Phospholipases A_1, C, and D

Phospholipases A_1 are widely distributed in nature, usually have a broad specificity for phospholipid substrates, and often catalyze the hydrolysis of lysophospholipids. A soluble enzyme has been studied from rat liver lysosomes that has optimal activity at pH 4.0 and does not require calcium. Two phospholipases A_1 have been purified from *E. coli.* Curiously, bacterial mutants defective in either or both of these enzymes have normal growth characteristics and phospholipid turnover. Hence, the function of these enzymes in *E. coli* remains uncertain.

Phospholipases C have also been most intensely studied in bacteria, particularly *Bacillus cereus;* two have been isolated from this organism. One enzyme with broad specificity has a requirement for zinc. A second does not require zinc but, curiously, is specific for phosphatidylinositol, which is not a phospholipid found in bacteria. The mammalian phospholipases C that have been studied are also generally specific for phosphatidylinositol and related inositol-containing phospholipids and require calcium. However, one described lysosomal phospholipase C does not need calcium and has a pH optimum of 4.5.

Plants were originally identified as the only source of phospholipases D; however, this enzyme activity has been recently described in several animal cells and tissues.

IV. Transfer of Phospholipids in Cells

In mammalian cells, the bulk of the phospholipids are made on the cytosolic face of the endoplasmic reticulum yet are distributed on every membrane in the cell. The phospholipid composition of membranes within cells differs, and there is an asymmetry in the distribution of phospholipids between the two leaflets of some, if not all, membranes. How the specific phospholipid composition of individual membranes is established and maintained is not known.

The first question is how do the phospholipids move from the site of synthesis on the cytosolic face to the inner leaflet of the endoplasmic reticulum bilayer? In pure phospholipid vesicles, phospholipids do not spontaneously cross the lipid bilayer. Because phospholipids do cross the bilayer in biological membranes, a protein catalyst of lipid transbilayer movement (flip-flop) might be involved. There are now several reports of proteins called flippases, which accelerate the movement of phospholipids from one side of the endoplasmic reticulum to the other. However, definite proof of their biological function is still lacking. [*See* Cell Membrane Transport.]

The mechanism of the movement of phospholipids from their sites of synthesis on the endoplasmic reticulum to other cellular membranes is better understood than the transbilayer movement of phospholipids. Phospholipids are poorly soluble in an aqueous environment; thus, solubilization and diffusion seems an unlikely mechanism for movement of these lipids between membranes. A likely possibility is the transfer of phospholipids by binding to transport proteins. Proteins that catalyze the exchange or transfer of phospholipids between membranes *in vitro* have been isolated and studied intensely. However, what function, if any, these proteins perform *in vivo* remains a moot question over 20 years after their discovery. Possibly, these proteins are only involved in the exchange of phospholipids between membranes and not in the net transfer from the endoplasmic reticulum to a target membrane. Another likely mechanism for lipid transport is via vesicles that bud from the endoplasmic reticulum and transport lipids and proteins to Golgi and other organelles. These vesicles would contain phospholipids and fuse with target membranes. Thus, in addition to the transport of proteins, these vesicles would transport phospholipids. A third possible transport mechanism is continuity of the membranes between adjacent organelles and movement of phospholipids between organelles by diffusion within the bilayer. Future studies should establish which of these possible mechanisms function in eucaryotic cells and the quantitative contribution of the various transport processes.

V. Functional Aspects of Phospholipids

A. Sources of Second Messengers

It has been known for many decades that the precursor of the eicosanoids, arachidonic acid, was present at the sn-2 position of phospholipids. This fatty acid is probably released by the action of a phospholipase A_2, and is rapidly converted into prostaglandins, thromboxanes, leukotrienes, or one of the hydroxy-eicosanoic acids. Eicosanoids are 20-carbon derivatives of fatty acids that have hormonelike actions near their site of synthesis. Among many different functions, they have been shown to be important in the modulation of adenylate cyclase, facilitation of ion transport, platelet aggregation, bronchoconstriction, increase of vascular permeability in skin, and chemotaxis of polymorphonuclear leukocytes.

Only in the last decade, scientists have recognized the importance of other species of phospholipids as storage forms of second messengers, which are active in various biological processes. Foremost is the lipid phosphatidylinositol-4,5-bisphosphate, the metabolism of which is shown in Fig. 2. This lipid is present in small quantities (<1% of the total cellular phospholipids) in the plasma membrane of many different types of cells. When a hormone or other agonist binds to a receptor on the membrane, a specific phospholipase C is activated and degrades the lipid to inositol-1,4,5-triphosphate and diacylglycerol, both of which are second messengers. The inositol derivative binds to the endoplasmic reticulum and causes the release of calcium into the cytosol of the cell, where it activates many different cellular processes. For example, the binding of the hormone vasopressin to liver cells results in the breakdown of phosphatidylinositol-4,5-bisphosphate to inositol-1,4,5-trisphosphate. This molecule mobilizes calcium, which activates the enzyme phosphorylase, which in turn degrades glycogen. The other product, diacylglycerol, activates an enzyme called protein kinase C, which phosphorylates a large variety of different proteins involved in stimulus–response couplings. Thus, protein kinase C has been implicated in catecholamine secretion, insulin release, steroid synthesis, arachidonic acid release, and many other cellular responses. Considerable evidence indicates that guanine triphosphate-binding proteins act as a bridge between the receptor and the activation of phospholipase C.

Other evidence indicates that PC can be degraded in response to vasopressin with the release of di-acylglycerol and phosphocholine. This might be a mechanism for amplification of the activation of protein kinase C without the additional effect on calcium release.

A surprise discovery in 1979 was platelet-activating factor (1-alkyl-2-acetyl-sn-glycerol-3-phosphocholine). This compound aggregates platelets at a concentration of 10^{-11} M and induces an antihypertensive response when 60 ng are injected intravenously into rats. A variety of blood cells, healthy tissues, and tumor cells synthesize and degrade platelet-activating factor. It is made in cells stimulated by agents such as the calcium ionophore A23187. The biosynthesis involves the acetylation at the sn-2 position of 1-alkyl-2-lysoglycerophosphocholine, which arises from 1-alkyl-2-acylglycerophosphocholine via the action of a phospholipase A_2. Alternatively, platelet-activating factor can be made by the transfer of phosphocholine to 1-alkyl-2-acetyl-glycerol. [See PLATELET-ACTIVATING FACTOR PAF-ACETHER.]

B. Anchoring of Proteins to Membranes

In recent years, some proteins have been discovered to link covalently to phosphatidylinositol in the

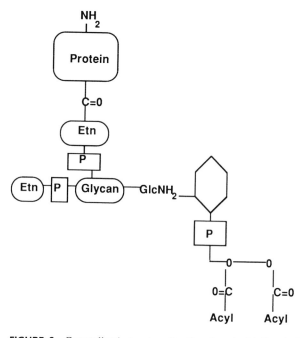

FIGURE 6 Generalized structure for the phosphatidylinositol–carbohydrate–ethanolamine–phosphate structure, which anchors certain proteins to the plasma membrane of a few types of cells. The remaining part of the structure represents phosphatidylinositol. Etn, ethanolamine; $GlcNH_2$, glucosamine; Glycan, an oligosaccharide; P, phosphate.

plasma membrane. In this linkage, the COOH-terminal amino acid is bonded in an amide linkage to the amino function of ethanolamine phosphate, which is linked via a series of carbohydrates (terminating in glucosamine) to the inositol moiety of phosphatidylinositol (Fig. 6). Among the proteins found in such a linkage are cell-surface hydrolases (e.g., alkaline phosphatase), protozoa coat proteins, proteins on the surface of lymphocytes, and proteins involved in the adhesion of cells to other cells. The protein attached to the carbohydrate can be released from the membrane by phospholipase C cleavage of the phosphatidylinositol moiety. It is not known if this degradation is simply a normal event in membrane recycling or if the cleavage might be important in regulation of interactions at the cell surface. Presumably, this particular linkage of proteins to cell membranes is somewhat advantageous and will be elucidated in future studies.

VI. Conclusion

Ten years ago, few scientists would have predicted the large number of functions that are now ascribed to phospholipids. Because the structure in the phospholipids is so diverse, we can only wonder what lies ahead in the next decade.

Bibliography

Ferguson, M. A. J., and Williams, A. F. (1988). Cell-surface anchoring of proteins via glycosyl-phosphatidylinositol structures. *Ann. Rev. Biochem.* **57,** 285.

Kennedy, E. P. (1986). The biosynthesis of phospholipids. *In* "Lipids and Membranes; Past, Present and Future" (J. A. F. Op den Kamp, B. Roelofsen, and K. W. A. Wirtz, eds.). Elsevier Science Publishers B.V., Amsterdam.

Vance, D. E. (ed.) (1989). "Phosphatidylcholine Metabolism." CRC Press, Boca Raton, Florida.

Vance, D. E., and Vance, J. E. (eds.) (1985). "Biochemistry of Lipids and Membranes." Benjamin-Cummings Publishing Co., Redwood City, California.

Waite, M. (1987). "The Phospholipases." Plenum Press, New York.

Phosphorylation of Microtubule Protein

JESÚS AVILA, JAVIER DÍAZ-NIDO, *Centro de Biología Molecular (CSIC-UAM), Madrid*

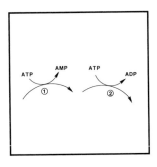

Glossary

Microtubule protein Protein components of the microtubule fraction obtained from cytosol through cycles of polymerization–depolymerization (assembly–disassembly); contains tubulin and microtubule-associated proteins

Neuronal plasticity Ability of nerve cells (neurons) to alter their morphology (especially their dendritic spine pattern) and physiology (especially their extent of response to other neurons depending on their immediate environment, which consists of hormones, trophic factors, stimulatory and inhibitory neurotransmitters, and neuromodulators)

Phosphoprotein phosphatase Enzyme that hydrolyzes monophosphoric esters on proteins, with release of inorganic phosphate

Phosphorylation Metabolic process of introducing a phosphate group into an organic molecule; protein phosphorylation usually refers to the formation of monophosphoric esters with hydroxy amino acids present in proteins

Posttranslational modification Any chemical modification of a protein that occurs after the completion of the polypeptidal chain; it comprises both the irreversible splitting off of certain peptide fragments and the reversible modifications of various amino acids by addition or removal of various chemical moieties (phosphate, adenosine phosphate-ribose, acetate, fatty acid, sugars, sulfate, etc.)

Protein kinase Enzyme that catalyzes the transfer of a phosphate group from a high-energy donor, usually adenosine triphosphate, to a protein to form a phosphoprotein

PROTEIN PHOSPHORYLATION is a chemical modification consisting of the formation of monophosphoric esters, mainly at hydroxy amino acids (serine, threonine, and tyrosine), which usually results in changes in the properties of the modified proteins. Both the major component of microtubules, tubulin, and the microtubule-associated proteins (MAPs) appear to be modified in this way. As microtubule functions within cells depend on the ability of microtubules to polymerize (assemble) and depolymerize (disassemble) and to interact with other cell structures, the modification of microtubule proteins by phosphorylation affect important aspects of cell physiology and pathology that depend on these properties.

I. Introduction

Microtubules are long, slender polymers built of the protein tubulin, present in the cytoplasm of all eukaryotic cells, forming part of the cytoskeleton, the system of various filament types, that spans the cell. Microtubules seem to perform a multitude of essential roles within the cells, including the generation and maintenance of cell shape, the intracellular transport and sorting of cytoplasmic organelles, the assembly of the mitotic (and meiotic) spindle at cell division, the subsequent segregation of chromosomes, and the formation of microtubule arrays involved in cell motility, such as cilia and flagella. This variety of functions is based on the ability of microtubule protein to polymerize and depolymerize reversibly and to interact with other subcellular structures.

Protein phosphorylation, in the same way as other posttranslational modifications of proteins, has been found to control many metabolic processes through the modification of some key enzymes, and it is thought also to regulate the ability of certain structural proteins to associate with other proteins. According to this view, phosphorylation of microtubule protein might provide a mechanism for regulating the assembly and disassembly of microtubules, as well as their interaction with other structures, thus controlling the cellular functions in which microtubules are involved. [See MICROTUBULES.]

Although they are found in all eukaryotic cells, microtubules are especially abundant in neurons, where they constitute the inner scaffolding of their appendages (neurites), axons, and dendrites and also serve as tracks for organelle transport between cell bodies and neurite endings. For this reason, brain extracts have been the primary source of microtubule protein through cycles of *in vitro* polymerization–depolymerization at appropriate temperatures. These preparations contain tubulin as the major component, in addition to a group of proteins referred to as MAPs (see Table I). MAPs are thought to promote microtubule assembly, stabilize

the resulting microtubules, and interact with other cytoskeletal elements and cell organelles, thus being largely responsible for the functional diversity of microtubules.

Brain microtubule protein preparations also contain very minor components, among which a variety of enzymes involved in modifications of microtubule proteins have been the best characterized. These enzymes are not considered MAPs, as they usually bind to MAPs rather than to tubulin. Phosphorylation is the main modification affecting both tubulin and MAPs, and, consequently, it is one of the most important putative regulatory mechanisms of microtubule functions. Thus, a great deal of attention has been paid to the protein kinases found in brain microtubule protein preparations. These include a cyclic adenosine monophosphate (cAMP)-dependent protein kinase, also called protein kinase A (PKA), a type II calcium–calmodulin-dependent protein kinase, and a casein kinase II-related enzyme. Moreover, the presence of other MAP kinases not yet characterized is indicated. Some of them might correspond to the calcium–phospholipid-dependent protein kinase (protein kinase C [PKC]) and/or tyrosine-specific protein kinases, which have been found to phosphorylate some microtubule components (mainly MAPs).

The study of microtubule protein phosphorylation is carried out through two complementary approaches. One is the *in vitro* approach; it analyzes which microtubule components are phosphory-

TABLE 1 Brain Microtubule Proteins

Protein	M_r (SDS-PAGE)	Main localization	Phosphorylation characteristics
MAP-1A	350K	Neurons (axon/dendrite), glia	By casein kinase II
MAP-1B[a]	325K	Neurons (axon/dendrite), glia	By casein kinase II
MAP-1C	300K	Neurons (axon)	
MAP-2(A, B)	270K	Neurons (dendrite)	By protein kinase A, Ca/CaM kinase, and protein kinase C
MAP-4	210K	Glia	
MAP-3	180K	Neurons, glia	
Kinesin	120K	Neurons, glia	
Chartins	70–80K		In response to NGF
MAP-2C	70K	Immature neurons (axon/dendrite)	By protein kinase A, Ca/CaM kinase, and protein kinase C
Tau	50–65K	Neurons (axon)	By Ca/CaM kinase (mode I), protein kinase C (mode II), casein kinase II, and protein kinase A
α-tubulin	52K	Ubiquitous	By Ca/CaM kinase II
β-tubulin	50K	Ubiquitous	By Ca/CaM kinase II
			The minor neural isoform by casein kinase II

[a] MAP-1B is also referred to as MAP-1(x), MAP-1.2, and MAP-5.

lated, at which residues (amino acids) on the molecules, which protein kinases are implicated, and which consequences these phosphorylation reactions have on the functional properties of the modified molecules. Using this approach, brain tubulin and MAPs have been determined to be differentially phosphorylated by distinct protein kinases in *in vitro* assays. The second approach attempts to correlate the phosphorylation of certain microtubule components with cellular events in which cytoskeletal changes take place and is mainly based on the results obtained from *in vivo* phospholabeling experiments. This *in vivo* approach allows the determination of the physiological relevance of the phosphorylation reactions observed *in vitro*.

II. *In Vitro* Phosphorylation of Brain Microtubule Protein

We will consider the various enzymes capable of carrying out the phosphorylation of microtubule proteins.

Cyclic AMP-dependent protein kinase (cAMPdpk) is a tetramer consisting of two regulatory (R) subunits (M_r 49,000–55,000) joined by a disulfide bond and two noncovalently bound catalytic (C) subunits (M_r 40,000). The complete enzyme (R_2C_2) is inactive, and when cAMP binds to the regulatory subunits, it dissociates, releasing the active catalytic subunits.

There are two cAMP-dependent protein kinases, referred to as type I and type II, which differ in their regulatory subunits: R_I (M_r 49,000) or R_{II} (M_r 52,000–55,000). Both enzymes are present in the cytosolic, membrane, and nuclear fractions of brain homogenates; the type II kinase is largely localized to neurons, whereas the type I enzyme is predominantly present in non-neural cells. Interestingly, 30–35% of total cAMP-dependent protein kinase found in brain cytosol is associated with microtubule protein. This localization may be due to the association of the R_{II} subunit to MAP-2, which is present preferentially in neuronal cell bodies and dendrites. MAP-2 can be cleaved by limited proteolysis into a carboxy-terminal tubulin-binding domain (M_r 35,000–40,000) and a large amino-terminal domain (M_r 240,000), which normally projects out of the microtubule and binds the R_{II} subunit.

The preferred substrates for type II cAMP-dependent protein kinase in brain microtubules are MAP-2 and tau proteins. Up to 13 residues (serine

and threonine) of the MAP-2 molecule can be phosphorylated *in vitro* by the enzyme.

Type II calcium–calmodulin-dependent protein kinase (CaM-kinase II, Ca/CamdpK), also referred to as the multifunctional Ca^{2+}–calmodulin-dependent protein kinase, is a large and complex enzyme with a molecular weight of approximately 600,000, consisting of α (M_r 50,000), β (M_r 60,000), and β' (M_r 58,000) subunits. The α and β subunits are catalytic and bind calmodulin but are structurally different and encoded by distinct genes. The complete enzyme is apparently made up of 12 subunits, primarily varying proportions of α and β subunits. This gives rise to various isoenzymes ranging in molecular mass from 450,000 to 700,000, which are distributed differentially in distinct brain regions at different developmental stages. For example, isoenzymes with a high proportion of α subunits are predominantly expressed in the mature forebrain, whereas those with a high proportion of β subunits are expressed early in development and predominate in the cerebellum. [*See* ISOENZYMES.]

Type II Ca–calmodulin-dependent protein kinase is probably the most abundant kinase in brain, constituting 0.5–1% of brain protein. It is highly concentrated in the telencephalon, especially within the hippocampus, where it accounts for nearly 2% of total protein. The enzymes with a high proportion of α subunit are enriched in postsynaptic densities within mature dendrites. A significant fraction of the enzyme is also found associated with brain microtubules and neurofilaments. The proteins to which the kinase is bound have not yet been characterized. Although a complex of tubulin and CaM-kinase II has been isolated from brain cytosol, whether the kinase is directly bound to tubulin or associated with some MAP present in the complex is not known.

Purified CaM-kinase II requires Ca^{2+} and calmodulin for initial enzyme activity. Upon activation by Ca^{2+}–calmodulin, the kinase phosphorylates itself and acquires a calcium-independent phosphotransferase activity for a time interval longer than the duration of the activating calcium signal. These properties make this enzyme a candidate as a "memory" molecule.

The preferred substrates for CaM-kinase II in brain microtubules are MAP-2, tau, and tubulin. It phosphorylates MAP-2 *in vitro* in at least eight residues, which are probably different from those that are phosphorylated by cAMPdpK.

CaM-kinase II also phosphorylates tau proteins,

probably inducing a conformational change, because the electrophoretic mobility of tau is decreased, and its antigenicity is changed. This type of tau phosphorylation is known as mode I.

CaM-kinase II phosphorylates both α- and β-tubulin subunits on serine and threonine residues at the tubulin carboxy-terminal region, which differ from those phosphorylated by other protein kinases (such as casein kinase II).

Casein kinase II is a ubiquitous protein kinase that preferentially phosphorylates serine and threonine residues surrounded by acidic amino acids, such as those found in casein and phosvitin. It differs from other serine–threonine protein kinases, including cAMPdpK, CaM-kinase II, and PKC, which mainly phosphorylate residues close to basic amino acids. Casein kinase II consists of two catalytic (α) subunits (M_r 43,000) and two regulatory (β) subunits (M_r 24,000). The mechanism regulating the activity of this protein kinase remains to be established. Some evidence suggests that it consists of phosphorylation–dephosphorylation by other protein kinases. Casein kinase II is found in both the nucleus and the cytoplasm of most eukaryotic cells. Within the cytoplasm, it appears to be partly bound to microtubules, probably to high molecular weight MAPs, mainly MAP-1 polypeptides. It is also associated with neurofilaments.

The preferred substrates for casein kinase II in brain microtubules are MAP-1 and tau polypeptides and a minor neuronal-specific β-tubulin isotype. In contrast, MAP-2, which is the preferred substrate for most protein kinases, is barely phosphorylated by it. MAP-1 polypeptides (MAP-1A and MAP-1B) are phosphorylated in a high number of residues. These polypeptides are rather specific substrates for this kinase, for they are hardly phosphorylated by other protein kinases (including cAMPdpK, CaM-kinase II, and PKC).

Tubulin treated with alkaline phosphatase to remove endogenously bound phosphate is also a major target for casein kinase II, which phosphorylated primarily a single serine residue (serine number 444) located within the carboxy-terminus of a minor neuronal-specific β-tubulin isoform.

Calcium–phospholipid-dependent protein kinase, also referred to as PKC, is an abundant enzyme present in both the soluble and particulate fractions from brain homogenates. Seven subspecies, coded for by distinct genes, have been identified. The enzyme is a single monomer (M_r approximately 80,000) consisting of an amino-terminal regulatory domain and a carboxy-terminal protein kinase domain. The kinase domain can be obtained as a soluble catalytic fragment of M_r 51,000 upon limited proteolysis with the Ca^{2+}-dependent protease calpain. Whereas the complete enzyme requires the presence of cofactors such as calcium, diacylglycerol, phosphatidyl serine, arachidonic acid or phorbol esters for activity, the free proteolytic fragment is fully active in the absence of these molecules.

The preferred substrates for PKC in brain microtubule protein preparations are MAP-2 and tau proteins. MAP-2 is phosphorylated in both the tubulin-binding and the projection domains. Ten residues phosphorylated in the projection domain cannot be phosphorylated by other protein kinases, including cAMPdpK and CaM-kinase II. Tau proteins are phosphorylated without changes in electrophoretic mobility (mode II of tau phosphorylation). Tau phosphorylation is partly inhibited in the presence of calcium-binding proteins such as calmodulin and S100b, which bind to tau in the presence of calcium.

Other novel serine–threonine protein kinases that phosphorylate microtubule proteins *in vitro* have been described. For example, a MAP kinase, which is not activated by cyclic nucleotides, calmodulin, or phospholipids and is inhibited by calcium ions, phosphorylates tau, altering its electrophoretic mobility (mode I). The phosphorylation of tau by this enzyme is enhanced in the presence of tubulin, which is not phosphorylated by it.

Another novel serine–threonine protein kinase, called MAP-2 kinase, isolated from insulin-stimulated adipocytes, phosphorylates MAP-2 *in vitro*. *In vivo*, however, its main target is S6 kinase, a protein kinase phosphorylating ribosomal S6 protein. A similar protein kinase has been detected in human embryonic fibroblasts stimulated by various mitogenic factors.

Finally, some tyrosine-specific protein kinases may phosphorylate some microtubule components *in vitro*. Thus, the insulin as well as the erythrocyte growth factor (EGF) receptors, which display such activity, phosphorylate both MAP-2 and tubulin. Of particular interest is the phosphorylation of α-tubulin by these kinases, as one of the modified residues is the C-terminal tyrosine, which is susceptible to removal by specific carboxypeptidases. Tubulin phosphorylated at its C-terminal tyrosine does not assemble into microtubules.

III. Functional Effects of Microtubule Protein Phosphorylation

Microtubule protein phosphorylation may regulate microtubule function through modulation of the ability of microtubule proteins to assemble and disassemble and to interact with other cytoskeletal elements and cell organelles. As a first approach to test for this possibility, the effect of *in vitro* phosphorylation of microtubule components on the assembly and interaction properties of microtubules has been analyzed.

Phosphorylation of the minor neuronal-specific β-tubulin isotype by casein kinase II does not affect its ability to assemble *in vitro*. Furthermore, polymerized tubulin is a better substrate for casein kinase II than unpolymerized tubulin.

In contrast, tubulin phosphorylation by CaM-kinase II notably decreases its ability to polymerize and bind to MAPs. Phosphorylation of C-terminal tyrosine in α-tubulin by the insulin receptor and other tyrosine-specific protein kinase also prevents tubulin polymerization.

Ca^{2+}–calmodulin-dependent phosphorylation of tubulin enhances its association with phospholipid membrane vesicles. This observation is important in view of the fact that tubulin is also associated with various membrane organelles: the proportions present in the microtubule, soluble, and membrane pools may be regulated by Ca^{2+}–calmodulin-dependent phosphorylation.

Different protein kinases phosphorylate different residues at the carboxy terminus of tubulin, a domain involved in the regulation of polymerization and in the interaction with MAPs. As distinct tubulin isoforms have characteristic carboxy-terminal sequences, their differential phosphorylation by various kinases might be a mechanism to regulate microtubule assembly and interaction with MAPs.

The extensive *in vitro* phosphorylation of MAP-2 or tau with cAMP-dependent protein kinase, CaM-kinase II, PKC, and the EGF receptor-associated kinase decreases their binding to tubulin, inhibiting microtubule assembly and causing their disassembly. Brain protein phosphatases 1, 2A, and 2B (which is designated calcineurin) dephosphorylate MAP-2 and tau previously phosphorylated by CaM-kinase II and the catalytic subunit of cAMP-kinase, restoring their tubulin-binding abilities. Thus, microtubule disassembly–assembly might be regulated by the phosphorylation–dephosphorylation of

MAP-2 and tau by their protein kinases and phosphatases. Extensive cAMP-dependent phosphorylation of MAP-2 or tau also prevents their ability to cross-link actin microfilaments.

A large number of phosphorylation sites on MAP-2 are located not on the tubulin-binding domain but on the projection domain; different residues are phosphorylated by the various kinases. The phosphorylation of this domain may affect the interaction of MAP-2 with cytoskeletal elements other than microtubules, including neurofilaments.

In contrast, the *in vitro* phosphorylation of both MAP-1A and MAP-1B by casein kinase II does not inhibit their ability to copolymerize with microtubule protein.

IV. *In Vivo* Phosphorylation of Brain Microtubule Protein

The *in vitro* studies do not provide indications of the physiological relevance of the observed modifications because some proteins are phosphorylated *in vitro* by certain protein kinases that do not do so *in vivo*. For instance, the main *in vivo* substrate for the MAP-2 kinase found in insulin-treated adipocytes seems to be not MAP-2 (which is neuronal-specific) but a S6 kinase (referred to as S6 kinase II).

In vivo studies of phosphorylation events have been correlated with cytoskeletal changes in neuronal morphogenesis and plasticity. Neuronal morphogenesis involves the extension of processes (axon and dendrites) and the establishment of functional synaptic contacts. These processes are driven by an increased assembly and stabilization of microtubules. Neuronal plasticity involves changes in the dendritic branching of neurons in response to the signals they receive, and it may require transient events of microtubule disassembly. [*See* PLASTICITY, NERVOUS SYSTEM.]

An experimental approach to study the role of phosphorylation involves the analysis of microtubule proteins in brain at different developmental stages following an intracranial injection of labeled phosphate into living animals. Another approach uses brain slices or neuronal cultures in which the variation in the phosphorylation state of microtubule proteins can be correlated with the cytoskeletal changes.

The major microtubule protein components phosphorylated in rat brain upon intracranial injection of labeled phosphate are the high molecular weight MAPs MAP-1A, MAP-1B, and MAP-2. A slight labeling of tau proteins and of β-tubulin can also be detected. MAP-1B, MAP-2, tau, and β-tubulin are phosphorylated in adult rat brain and in that of 5-day-old rat pups, when most neurons are extending their processes; whereas MAP-1A phosphorylation is observed only after neuronal maturation takes place, in adult rat brain. Both MAP-1B and a minor neuronal-specific β-tubulin are phosphorylated in cultures of embryonal carcinoma, pheochromocytoma, and neuroblastoma when the cells extend neurites. In primary cultures of rat brain neurons, MAP-1B, MAP-2, and β-tubulin are phosphorylated.

The analysis of the phosphorylated proteins reveals that both MAP-1A and MAP-1B are phosphorylated in serine and threonine residues; both contain numerous phosphorylated residues. MAP-1B is the major microtubule-associated phosphoprotein in cultured neuronal cells and in immature rat brain, whereas the expression and phosphorylation of MAP-1A is notably enhanced in adult rat brain where microtubules achieve a high density and stability. Phosphorylated MAP-1A and MAP-1B can efficiently coassemble with microtubule protein.

MAP-2 is phosphorylated *in vivo* in serine, threonine, and tyrosine residues. Up to 46 esterified phosphates are present per MAP-2 molecule. This highly phosphorylated MAP-2 does not copolymerize with microtubules. After the phosphate content of MAP-2 decreases to 10–16 phosphate molecules owing to phosphatase action, MAP-2 can coassemble with microtubules. The residual phosphates present may partly correspond to those introduced by PKC on the projection domain of the MAP-2 molecule. From comparison of *in vitro* and *in vivo* phosphorylated proteins it appears highly plausible that different protein kinases, including cAMP-dependent kinase, PKC, CaM-kinase II, and tyrosine-specific protein kinases, participate in the *in vivo* phosphorylation of MAP-2. Phosphorylation in the tubulin-binding domain of MAP-2 may favor microtubule disassembly, and that on its projection domain may affect its interaction with other cytoskeletal elements.

The phosphorylation of MAP-2 may be important in neuronal plasticity. In fact, MAP-2 is notably enriched on dendritic microtubules. Phosphorylation regulates the role of MAP-2 in microtubule assembly, and its ability to cross-link actin. MAP-2 phos-

phorylation may give rise to a relatively uncross-linked, malleable cytoskeleton, which would allow cell shape changes. Moreover, MAP-2 phosphorylation may occur in response to the increased intra-neuronal levels of cAMP, Ca^{2+}, or diacylglycerol occurring at postsynaptic sites upon neurotransmitter binding, providing a pathway for neurotransmitter-induced neuronal cytoarchitectural modifications (see below).

Tau proteins are phosphorylated *in vivo* in both developing and adult brain. Tau purified from brain is phosphorylated by the mode I and mode II mechanisms. Immunocytochemical studies show that mode I phosphorylated tau may be largely restricted to cell bodies and dendrites and can be visualized after phosphatase treatment of brain sections. If mode I phosphorylation of tau is due to the action of CaM-kinase II, as suggested by *in vitro* studies, tau phosphorylation within dendrites may occur in response to increased Ca^{2+} levels and could result in microtubule disassembly. Interestingly, highly phosphorylated tau is associated with the paired helical filaments (PHFs) found in Alzheimer's disease patient brains (see below). [*See* ALZHEIMER'S DISEASE.]

As mentioned above, tubulin is also phosphorylated *in vivo*. β-tubulin contains bound phosphate at its carboxy-terminal domain when purified from brain. The modified residue is serine 444 (see Fig. 1) of the minor neuronal-specific β-tubulin, which is phosphorylated when neuronal cells are induced to extend axonlike processes. This phosphorylation reaction is catalyzed by a casein kinase II-related enzyme and takes place mainly on assembled tubulin. Cell fractionation studies indicate that phosphorylated tubulin is largely restricted to the cytoskeletal fraction and is nearly absent from the soluble fraction, suggesting that it is associated with stable axonal microtubules (see below).

Finally, three proteins of M_r 69,000, 72,000, and 80,000, referred to as chartins, have been found phosphorylated in some cultured cells of neuronal origin treated with neurotrophic factors. Highly phosphorylated chartins are bound to assembled

FIGURE 1 Sequence containing the phosphorylated residue on β-tubulin. This sequence is present at the carboxy terminus of the minor neural β isoform.

microtubules, and their phosphorylation correlates well with neurite outgrowth (see below).

V. Possible Roles of Microtubule Protein Phosphorylation in Neuronal Cells

Microtubules are necessary for the extension and maintenance of the neuronal processes that endow neurons with their characteristic shapes and also serve as tracks for bidirectional transport of various organelles between cell bodies and process tips. Phosphorylation of microtubule components may modulate the functionality of microtubules during neuronal morphogenesis, plasticity, and intracellular organelle transport.

A. Role of Microtubule Protein Phosphorylation in Neuronal Morphogenesis

The generation of mature neurons from undifferentiated neuroblasts involve five fundamental events, in which the microtubule cytoskeleton plays a prominent role: (1) axonal sprouting, (2) dendritic extension and arborization, (3) axonal maturation, (4) dendritic maturation, and (5) synaptogenesis.

Both axon and dendrites grow in length through specialized terminal, motile appendages called growth cones, which guide the extending neurite to its target. Growth cone motility requires the presence of actin-myosin and also a paucity of assembled microtubules. The growth cone cytoplasm contains a large amount of unpolymerized tubulin, perhaps due to some modifications, including phosphorylation. Both α- and β-tubulin can be phosphorylated in vitro in cytoskeletal fractions prepared from growth cones of neonatal rat forebrains; 9–13 phosphate groups are incorporated per molecule of tubulin dimer. The modified residues include tyrosine and serine, suggesting the involvement of neuronal tyrosine-specific protein kinases, similar to those of growth factor receptors or oncogene products. In addition, calcium-dependent protein kinase may be also involved as nerve growth cones have a high concentration of Ca^{2+}.

The progressing growth cone leaves behind a slender, stable neurite filled with microtubules. This conversion requires a net microtubule assembly, which must also occur during the late maturation of neurites.

The major microtubule components that are phosphorylated in parallel to the increase in microtubule assembly, which accompanies neurite extension and maturation, are MAP-1B, the minor neuronal-specific β-tubulin and chartins.

MAP-1B phosphorylation is not affected by drugs that inhibit microtubule assembly and block neurite outgrowth, suggesting that MAP-1B phosphorylation is a step prior to microtubule assembly. It is possible that MAP-1B phosphorylation favors microtubule nucleation, promoting the appearance of arrays of bundled microtubules inside the developing neurites. This possibility is supported by the fact that monoclonal antibodies that recognize phosphorylated epitopes on MAP-1 polypeptides also react with microtubule-organizing centers in undifferentiated neuroblasts and non-neural cells. The phosphorylation of MAP-1B may also be required to assemble a stable microtubule scaffolding within developing and mature axons.

The minor neuronal-specific β-tubulin isoform is phosphorylated with embryonal carcinoma, pheochromocytoma, and neuroblastoma cells are induced to extend axonlike processes. Tubulin phosphorylation is abolished by microtubule-depolymerizing drugs and stimulated by taxol, which promotes tubulin assembly, suggesting that phosphorylation takes place on assembled tubulin. Phosphorylated tubulin is mainly associated with the cytoskeletal fraction, indicating that β-tubulin phosphorylation might modify and stabilize preformed microtubules inside neurites. The fact that the casein kinase II-related enzyme responsible for β-tubulin phosphorylation acts preferentially on assembled tubulin supports this hypothesis.

Chartins also become phosphorylated during neurite outgrowth in pheochromocytoma (PC12) cells induced to differentiate by the addition of nerve growth factor (NGF) to the culture medium. Phosphorylation is inhibited by forskolin, which elevates intracellular cAMP levels. Forskolin enhances the initial NGF-induced sprouting of short cytoplasmic extensions, while it suppresses the subsequent growth of long axonlike processes. This suggests that elevated cAMP levels may inhibit neurite extension by interfering with NGF-dependent chartin phosphorylation. The fact that microtubule-depolymerizing drugs abolish chartin phosphorylation indicates that the phosphorylation of microtubule-bound chartins may favor the stabilization, elongation, or bundling of preformed microtubules or their interaction with other cytoskeletal elements, thereby facilitating the growth of long axonlike processes. [See NERVE GROWTH FACTOR.]

B. Role of Microtubule Protein Phosphorylation in Neuronal Plasticity

The developing visual system of cats has been a very useful model in the study of the biochemical basis for neuronal plasticity, as there is a critical period during which visual experience permanently molds some aspects of visual function.

The cAMP-stimulatable phosphorylation of MAP-2 within dendrites of visual cortex neurons is strongly dependent on visual experience. Under normal rearing conditions, *in vitro* MAP-2 phosphorylation in visual cortex homogenates increases during development. It is diminished after rearing in the dark but this decrease can be reversed by a short exposure to light. Furthermore, the effect of visual stimulation is specific for the visual cortex. The decreased *in vitro* cAMP-stimulated MAP-2 phosphorylation after rearing in the dark probably indicates a more phosphorylated state of MAP-2 *in vivo* as compared to MAP-2 in normally reared animals. The enhanced *in vitro* MAP-2 phosphorylation observed after exposure to light would then indicate a dephosphorylation of cAMP-stimulated sites in MAP-2 *in vivo*. [See VISUAL SYSTEM.]

Visual plasticity is abolished after perfusing the visual cortex with 6-hydroxydopamine, a potent neurotoxin specific for neurons that use norepinephrine as transmitter, and is restored after perfusion with norepinephrine or cAMP derivatives. Thus, visual plasticity seems to depend on the norepinephrine stimulation of receptors in visual cortex neurons, which results in increased intracellular cAMP levels. Cyclic AMP may facilitate visual cortex plasticity through MAP-2 phosphorylation, producing a transient release of the cytoskeletal scaffolding within dendrites. This could permit changes in dendritic shape and, consequently, in the number of synapses. Analogously, the more dephosphorylated state of MAP-2 within the light-stimulated neuronal pathways suggests a stabilization of the dendrite scaffolding. Thus, MAP-2 phosphorylation may perform a general role in neuronal plasticity, allowing rapid cytoskeletal rearrangements within dendrites in response to changes in the synaptic input that such dendrites receive.

It appears plausible that the phosphorylation of microtubule proteins, mainly of MAP-2 and tau pro-

teins, may also participate in the molecular changes underlying other examples of neuronal plasticity, including those that lead to long-term potentiation (LTP) in hippocampal synapses, as the induction and maintenance of LTP depend on the activation of PKC and/or CaM-kinase II and may be correlated with structural changes in the modified synapses.

C. Role of Microtubule Protein Phosphorylation in the Intraneuronal Transport of Organelles

Microtubules serve as tracks for the conveyance of organelles within cells, with the cooperation of some microtubule-interacting ATPases. The best characterized ATPases are kinesin and MAP-1C, which promote movements in opposite directions. Little is known about the phosphorylation features of these molecules.

However, some evidence from the study of fish melanophores suggests that protein phosphorylation–dephosphorylation may underlie the regulation of bidirectional transport of pigment granules along microtubules. Movement away from the cell center requires cAMP is accompanied by the phosphorylation of a M_r 57,000 polypeptide, and is prevented in the presence of the protein inhibitor of the cAMP-dependent protein kinase. Furthermore, phosphatase inhibitors block both the movement of pigment granules toward the cell center and the concomitant dephosphorylation of the M_r 57,000 polypeptide. These findings correlate well with the fact that pigment aggregation and dispersion are regulated by hormones that alter intracellular cAMP levels. However, the identity of the M_r 57,000 protein and its role in pigment granule transport on proteins are still unknown.

D. Possible Implications of Microtubule Protein Phosphorylation in Neurological Disorders

As microtubule protein phosphorylation may be critically involved in the regulation of nerve cell shape maintenance and in intraneuronal transport mechanisms, its alterations would lead to changes in the organization of microtubules, their associated proteins, and their interactions with other cell organelles, giving rise to a variety of neuropathies.

It has been suggested that certain neurotoxins, such as aluminum salts, 2,5-hexanedione, acrylamide, and iminodipropionitrile, may alter the pattern of phosphorylation of neurofilaments and microtubule proteins. Intrathecal or intracranial injections of aluminum in rabbits produces a severe neurodegenerative syndrome, characterized by the intraneuronal accumulation of hyperphosphorylated neurofilament proteins. In addition, chronic oral administration of aluminum to rats increases the phosphorylation of MAP-2 and the M_r 200,000 neurofilament subunit in the brain stem and cerebral cortex. This increased phosphorylation may constitute the first step in the deposition of hyperphosphorylated cytoskeletal proteins in aberrant structures within neurons. It has also been shown that aluminum interacts with several highly phosphorylated cytoskeletal proteins including MAPs and neurofilaments, causing their irreversible aggregation.

Accumulations of abnormally phosphorylated cytoskeletal proteins are found in several neurodegenerative diseases, including Alzheimer's disease, Pick's disease, amyotrophic lateral sclerosis, and Parkinson's disease. Of these diseases, senile dementia of Alzheimer's type (SDAT) has been the most thoroughly studied, as it is the most frequent neurodegenerative disorder.

SDAT is characterized by the degeneration of neurons in various brain centers. One of the major histopathological hallmarks is the presence of neurofibrillary tangles (NFTs), intraneuronal inclusions stained by silver methods. NFTs are composed of large numbers of pairs of approximately 10-nm helically entwined filaments referred to as paired helical filaments, constituted mainly by highly phosphorylated tau proteins. Highly phosphorylated tau is seen also in the cytoplasm of some neurons containing delicate fibrillary inclusions, which may be precursors of NFTs. Thus the accumulation of hyperphosphorylated tau appears to be one of the earliest cytoskeletal changes in the process of NFT formation. In support of this view is the finding of two hyperphosphorylated tau proteins with abnormal M_r of 64,000 and 69,000, respectively, in Alzheimer's patient brain cortices. These proteins are never detected in brains from healthy subjects, suggesting that they are specific markers for the early events of NFT formation.

VI. Phosphorylation of Microtubule Proteins in Non-neural Cells

Microtubule proteins in non-neural cells have not been characterized with great detail because of the relative paucity of microtubules. An exception are platelets, in which microtubules play an essential role in maintaining cell shape. Phosphorylation of α- and β-tubulin correlates well with the changes in cell shape that accompany platelet aggregation, but the underlying mechanism is unknown.

Microtubule protein phosphorylation may be also implicated in the regulation of cell division in all eukaryotic cells. The onset of mitosis is characterized by disassembly of the interphase microtubule network and assembly of the mitotic spindle. The phosphorylation of microtubule components is possibly important, because a cascade of phosphorylation events seems to be responsible for the entry into mitosis. Protein kinases are possibly involved in cell-cycle control. For example, the activity of casein kinase II is enhanced by mitogens and varies throughout the cell cycle. In addition, an essential component of an inducer of mitosis, the M phase promoting factor, is a M_r 34,000 protein kinase homologous to the cdc 2 gene product of the fission yeast *Schizosaccharomyces pombe*. This cdc 2 kinase is present in all eukaryotic cells tested, including yeast, starfish, clam, frog, rat, mouse, and human. During the cell cycle of cultured human cells (HeLa), the abundance of this protein kinase varies little, but its activity varies more than 50-fold and is maximal during mitotic metaphase. Interestingly, this kinase is localized in the nucleus during interphase and becomes cytoplasmic during mitosis, with a fraction tightly associated with centrosomes (mitotic poles). This distribution is quite similar to that described for some proteins immunologically related to brain MAPs. Furthermore, a monoclonal antibody recognizing a phosphorylated epitope present in several cytoskeletal proteins including brain MAP-1 reacts strongly with the centrosomes in dividing cells, with maximal staining at prometaphase, when the microtubule-nucleating ability of centrosomes is highest. The blocking of this phosphorylated epitope by the addition of the specific antibody, or its modification by the removal of phosphate after alkaline phosphatase treatment leads to the inhibition of the nucleation of microtubules from mitotic centrosomes. Thus, it has been suggested that the phosphorylation of some centrosomal components, including some MAPs, may initiate the assembly of mitotic spindle microtubules. Supporting this view is the fact that the addition of cdc 2 kinase to cell-free extracts from interphase frog eggs induces a change in microtubule dynamics similar to that observed *in vivo* during the interphase–metaphase transition. Cdc 2 kinase may directly phosphorylate some spindle components and also activate a cascade of protein kinases (including tyrosine-specific protein kinases, MAP-2 kinase, and S6 kinase II) that in turn may phosphorylate cytoskeletal proteins. [*See* Mitosis.]

The phosphorylation of specific spindle components may also participate in the chromatid separation. During anaphase, the microtubules attached to the chromatids shorten as the chromatids move toward the poles. A M_r 62,000 protein, which is a component of isolated sea urchin egg mitotic spindles, is a substrate for an endogenous calcium–calmodulin-dependent protein kinase also associated with the mitotic spindle. The phosphorylation of this protein seems to result in the depolymerization of microtubules in isolated mitotic spindles. Thus, the phosphorylation of this protein may trigger microtubule disassembly, initiating anaphase. This picture is supported by the metaphase arrest observed in dividing sea urchin eggs microinjected with antibodies to the M_r 62,000 protein.

Finally, protein phosphatases are involved in the completion of mitosis. This has been shown by studying the temperature-sensitive cell-cycle mutation bim G11 in the fungus *Aspergillus nidulans,* which results in a block of anaphase separation. The corresponding wild-type gene encodes a putative protein phosphatase with strong homology to mammalian type 1 protein phosphatase. During the exit from mitosis, the bim G protein dephosphorylates the mitotic pole phosphoproteins, which have been phosphorylated during the entry into mitosis (mainly at prometaphase).

Bibliography

Aoki, C., and Siekevitz, P. (1988). Plasticity in brain development. *Sci. American* **259,** 34–42.
Díaz-Nido, J., Serrano, L., Méndez, E., and Avila, J. (1988). A casein kinase II related activity is involved in phosphorylation of microtubule associated protein MAP-1B during neuroblastoma cell differentiation. *J. Cell Biol.* **106,** 2057–2065.
Gard, D. L., and Kirschner, M. (1985). A polymer-dependent increase in phosphorylation of β-tubulin accompanies differentiation of a mouse neuroblastoma cell line. *J. Cell Biol.* **100,** 764–774.
Kennedy, M. B. (1988). Synaptic memory molecules. *Nature* **335,** 770–772.
Maller, J. L. (1990). Xenopus oocytes and the biochemistry of cell division. *Biochemistry* **29,** 3157–3166.
Matus, A. (1988). Microtubule associated proteins: Their role in determining neuronal morphology. *Ann. Rev. Neurosci.* **11,** 29–44.
Serrano, L., Díaz-Nido, J., Wandosell, F., and Avila, J. (1987). Tubulin phosphorylation by casein kinase II is similar to that found in vivo. *J. Cell Biol.* **105,** 1731–1739.
Verde, F., Labbé, J. C., Doreé, M., and Karsenti, E. (1990). Regulation of microtubule dynamics by cdc 2 protein kinase in cell-free extracts of Xenopus eggs. *Nature* **343,** 233–238.

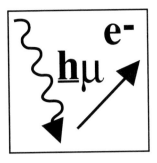

Photoelectric Effect and Photoelectron Microscopy

O. HAYES GRIFFITH, KAREN K. HEDBERG,
University of Oregon

Glossary

Fluorescence microscopy Type of optical microscopy in which the image is formed by light emitted by the specimen (e.g., fluorescence) when it is illuminated by an exciting light; the microscope is operated in the dark-field mode with a blocking filter so that only the fluorescent light, and not the exciting light, reaches the ocular; in combination with a fluorescent dye-labeled antibody to label-specific sites, this is called immunofluorescence microscopy

Ionization potential Minimum voltage (usually reported in units of energy as electron volts) required to remove an electron completely from a molecule (or atom or ion); the ionization energy is the potential times the charge of the electron

Photoelectric effect Ejection of electrons from the surface of a solid or a liquid, or from a gas when it is illuminated by light

Photoelectron imaging Any form of imaging in which the source of information is the distribution of points from which electrons are ejected from the specimen by the action of ultraviolet light (i.e., photoelectric effect); the highest resolution technique of photoelectron imaging is presently photoelectron microscopy, also known as photoemission electron microscopy

Photoelectron microscopy Type of microscopy in which a specimen is exposed to light and the resulting emitted electrons are used to form an image of the exposed surface; the electrons are first accelerated and then focused by means of an electron lens system similar to that of a transmission electron microscope; in physics, photoelectron microscopy is often referred to as photoemission electron microscopy, the highest resolution technique of photoelectron imaging

Photoelectron spectroscopy Analysis of the kinetic energies of electrons ejected from gases, liquids, or solids by electromagnetic radiation, usually ultraviolet light, X-rays, or synchrotron radiation; the ionization potentials identified from the photoelectron spectra are used to determine the electronic structure of molecules and surfaces

Work function Corresponds to the threshold of photoemission (the onset of which is generally not sharp) in plots of the electron emission from a conductive metal surface versus wavelength of incident light; it is determined from a fit of the experimental curve with an analytical equation (e.g., Fowler equation)

PHOTOELECTRON MICROSCOPY IS THE electron optical analogue of fluorescence microscopy. Light is focused on a specimen, and the emitted electrons are accelerated and imaged by means of an electron optical system. The enlarged image is captured on film and yields a photograph of the exposed surface of the specimen. Contrast is provided by intrinsic differences in the ionization potentials of cell components, by extrinsic labeling with photoemissive markers, and by surface topography. Photoelectron microscopy can be used to study a variety of biological specimens, including well-spread cells in culture, cell organelles, cytoskeletal structures, membranes, and DNA.

I. The Photoelectric Effect

The basic concept of the photoelectric effect is illustrated in Figure 1. Historically, the photoelectric effect has played important roles in establishing electrons as subparticles of atoms and in developing the quantum theory of radiation (e.g., the concept that despite its wave nature, light nevertheless has some properties akin to those of particles). Einstein received the Nobel Prize in 1921 for his seminal 1905 paper that introduced the concept of units of light energy called photons, or quanta. Any material will exhibit a photoelectric effect if the incident light quanta are of sufficiently high energy (e.g., ultraviolet [UV] light) to eject electrons. All three states of matter—solid surfaces, liquids, and gases—can exhibit a photoelectric effect.

Several types of information are available by the photoelectric effect. These include (1) the number of electrons released per incident photon (this number, which is usually much less than one, is also called the efficiency of photoemission, or the photoelectron quantum yield), (2) the kinetic energy of the emitted electrons, (3) the angular distribution of emitted electrons, and (4) for solids, the positions from which the electrons leave the surface. Different instruments developed over the years utilize each of these sources of information.

Today, the photoelectric effect is used in many ways. Photocells and photomultipliers take advantage of the fact that light is converted into an electron current (e.g., Fig. 1). Photocathodes are surfaces specially prepared to have a high efficiency for emitting electrons. The best photocathodes generally contain the element cesium (e.g., the cesium–antimony or silver–oxygen–cesium photocathodes). Most optical spectrophotometers in chemistry and medical laboratories contain a photocell to detect changes in transmitted light by recording changes in the electric current. Image intensifiers and television cameras contain photocathodes to convert the light image into an electron image, which can then be amplified and scanned electronically.

Primary processes of photosynthesis are closely related to the photoelectric effect. The chief difference is that in the (external) photoelectric effect the electrons are removed essentially to infinity, whereas in the light reactions of photosynthesis the photoejected electrons are trapped near the reaction centers.

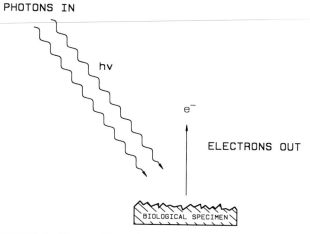

FIGURE 1 Diagram illustrating the photoelectric effect. Light (hν) striking a surface will cause the ejection of electrons providing that the energy of the photons is sufficiently great. For most materials, UV light is required.

Photoelectron spectrometers are instruments that shine a monochromatic beam of light on a specimen and measure the kinetic energy, number, and sometimes the angular distribution of emitted photoelectrons. Gases, in particular, exhibit very sharp and well-resolved photoelectron spectra, from which information about the electronic structure of the molecule is derived. UV light removes only the more loosely bound or valence electrons, including those which are involved in chemical reactions. Hence, UV photoelectron spectroscopy is used to study the molecular orbitals of atoms and molecules. If the energy source is a beam of X-rays, the technique is called X-ray photoelectron spectroscopy (XPS), or electron spectroscopy for chemical analysis. X-ray photons have sufficient energy to remove core electrons as well as valence electrons, and XPS is therefore applicable for chemical analysis (e.g., detecting elements and oxidation states).

II. The Photoelectron Microscope

In photoelectron microscopy, instead of collecting the total current (as in a photocell) or analyzing the kinetic energies (as in a photoelectron spectrometer), the positions at which the electrons leave the specimen are preserved by means of a series of electron lenses. Historically, simple versions of photoelectron microscopes were built in Germany as early as 1932 for the study of metals. The first use

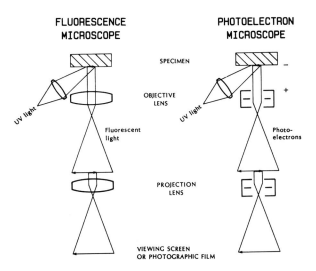

FIGURE 2 Simplified diagrams comparing a fluorescence microscope (left) with a photoelectron microscope (right). Both microscopes rely on light as the source of excitation, and both magnify the image in the same general way. The fluorescence microscope uses glass optical lenses, and the photoelectron microscope uses electron lenses.

of photoelectron microscopy to image organic and biological surfaces was in 1972. This remains a relatively small field compared to other forms of electron microscopy. [*See* ELECTRON MICROSCOPY.]

A strong analogy exists between photoelectron microscopy and the widely used technique fluorescence microscopy. The essential features of these two methods are compared in Figure 2. Fluorescence involves the absorption of light by the specimen followed by emission of light at longer wavelengths. A glass objective lens focused on the specimen collects a portion of this fluorescent light and forms an enlarged image of the pattern of emission. The image is further enlarged by an ocular (projector lens), resulting in an image of the biological specimen in which bright areas correspond to regions rich in fluorescence emission. The idea in photoelectron microscopy is to retain the contrast inherent in an emission experiment, i.e., imaging bright (strongly photoemissive) objects against a dark background, while greatly improving the resolution compared with that of optical methods. UV radiation is used to eject electrons, and the electrons are accelerated and focused by electron objective, intermediate, and projector lenses. Thus, photoelectron microscopy is the electron optical analogue of fluorescence microscopy.

III. Advantages and Limitations

Photoelectron microscopy is still a developing technique and instrumentation is not yet widely available. The resolution of current photoelectron microscopes is between 5 and 10 nm. The main advantage of photoelectron microscopy is the source of contrast. Small differences in the bonding of the outermost electrons can cause significant contrast between molecules that would be otherwise difficult to detect, e.g., hemes, chlorophylls, and aromatic hydrocarbons. A second advantage is the extremely high sensitivity to surface relief as encountered in studies of fine fibers, membrane proteins, cytoskeletal elements, and DNA. A third advantage is lower specimen damage compared with electron beam methods. The main limitation is that, due to the extreme sensitivity to topography, the specimens must be relatively flat. Another limitation is caused by the fact that light travels unimpeded through water but electrons do not. It follows that, for example, fluorescence microscopy is performed on wet and often living specimens, whereas the specimens for photoelectron microscopy must be frozen or dehydrated. This is a limitation of essentially all electron microscopes. Thus, a frequent strategy is to use a combination of optical microscopy and electron microscopy. Photoelectron microscopy can be viewed as one specialized type of electron microscopy. It differs from the better known types such as transmission electron microscopy and scanning electron microscopy in that there is no electron gun to provide the electrons. The specimens are the source of electrons in photoelectron microscopy.

IV. Selected Applications

Photoelectron microscopy, like scanning electron microscopy, is a technique for studying the exposed surfaces of specimens; therefore, the study of cell surfaces is an appropriate area of application of photoelectron microscopy. An example is a study in which photoelectron microscopy was used to examine the effects of phorbol ester tumor promoters on cell-surface fibronectin. Fibronectin is a large glycoprotein that plays a role in adhesion, migration, and differentiation. It binds to the surfaces of fibroblastic cells (cells of mesenchymal origin, generally destined to be connective tissue) at specific transmem-

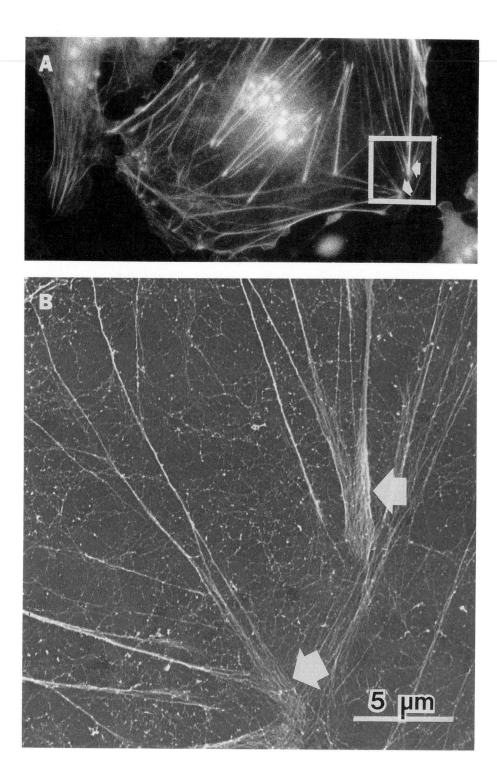

brane receptors. To distinguish it from other proteins on the cell surface, fibronectin on human fibroblasts was labeled by an indirect antibody method with 6 nm diameter colloidal gold as the marker (e.g., immunophotoelectron microscopy).

FIGURE 3 Comparison of fluorescence (A) and photoelectron (B) images of the cytoskeleton of a mouse fibroblast (Swiss 3T3 cell line) cultured on a sample mount. The box in (A) indicates the boundaries of the region shown in the higher magnification photoelectron micrograph (B). The surface membranes of the cells were removed with neutral detergent, and then the cells

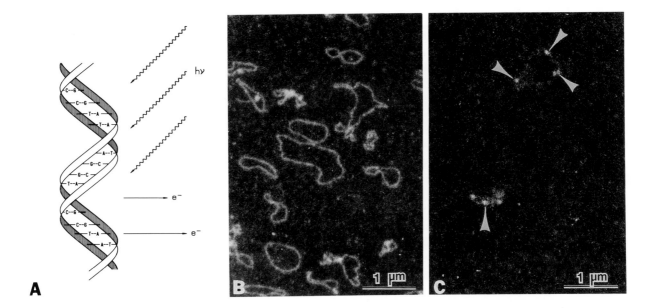

FIGURE 4 Photoelectron imaging of DNA. A. Diagram illustrating photoemission of DNA. B. Photoelectron images of DNA plasmids (pBR322) prepared by the cytochrome *c* spreading technique. C. *Escherichia coli* RNA polymerase protein complexes (arrows) bound to naked DNA plasmids (no cytochrome *c*). (Adapted, with permission, from Griffith, Habliston, Birrell, Skoczylas, and Schabtach, 1990, "Biopolymers," in press.)

Colloidal gold is much more photoemissive than cell-surface components, and the small gold spheres are readily recognized as patterns of bright dots against the darker cell-surface background. The contrast provided by this label can be boosted by silver enhancement, which provides a coating of metallic silver on the gold particles. This approach

were selectively labeled for fluorescence visualization of filamentous actin by means of rhodamine-conjugated phalloidin. The bright streaks in (A) indicate the presence of actin microfilament bundles crisscrossing the main cell body. The bright spots in the center of (A) are nonspecific fluorescence from the nuclear region of the cell. The fluorescent label seen in (A) does not also serve as a photoelectron label. The major source of contrast in the photoelectron image (B) is the topography of the exposed cellular structures. This topographical contrast allows the actin microfilament bundles present in the boxed region of (A) to be easily identifiable in (B). Many of the microfilament bundles in the fluorescence micrograph (A) appear to have blunt, slightly bulbous ends. In the photoelectron micrograph (B), these ends are resolved to be splayed meshworks of filaments terminating on the lower cell surface (arrows). These regions, known as adhesion plaques or focal contacts, are believed to contain transmembrane proteins, which link the actin cytoskeleton inside the cell to adhesive proteins on the substrate beneath the cell.

made it possible to observe the change in patterns and the release of cell-surface fibronectin induced by the tumor promoter. The images of cell surfaces are well resolved because photoelectron emission from within or beneath the cell does not reach the sample surface and, therefore, does not contribute to the image. For this kind of study, cells are generally unstained and uncoated. Even without labeling, the cell surfaces are visible by the topographic contrast inherent in photoelectron microscopy. Thus, photoelectron microscopy can be used to simultaneously image both labeling patterns and the topography of the cell surface.

Components inside cells may be studied after removal of the cell membrane with detergents or by mechanical shearing. The cytoskeleton of cultured cells, including microtubules, intermediate filaments, and actin-containing filament bundles, are within the useful imaging range of photoelectron microscopy. These photoelectron images are especially detailed and informative. One representative application is the study of effects of activators of protein kinase C on the actin cytoskeleton of cultured cells. Colloidal gold, or silver-enhanced colloidal gold, is the most frequently used marker although much smaller (e.g., molecular) markers are possible. Double-labeling, with one fluorescent marker (e.g., a dye) and one photoemissive marker, can also be used so that both fluorescence and photoelectron images can be obtained from the same specimen. Another variation is to use a fluorescence marker, but no photoemissive marker, and to compare the fluorescence (labeled) image with the photoelectron (unlabeled) image of the same region. In this case, the photoelectron image is primarily due to the topography of the cytoskeletal elements. An example is shown in Figure 3, with a fluorescence image (upper half) and an enlarged photoelectron image of the boxed area (lower half); note the much higher resolution of the photoelectron image.

As a final example, recently nucleic acids have been imaged by photoelectron microscopy. The photoemission of DNA is illustrated in Figure 4A. This experiment is reminiscent of autoradiography, where electrons are released by radioactive decay. However, in photoelectron microscopy the DNA is not radiolabeled, and the electrons are released by the action of UV light. This makes a higher resolution possible because the number of electrons released is larger and their kinetic energies are much smaller than with autoradiography. A photoelectron image of pBR322 DNA plasmids spread by the standard cytochrome *c* method used in transmission electron microscopy is shown in Figure 4B. The cytochrome *c* coats the DNA so the effective diameter is probably 7–10 nm, but still much less than that of conventional electron microscopy preparations, which utilize metal coating or staining. A photoelectron image of naked DNA plasmids (no cytochrome *c*) with RNA polymerase complexes bound to them is shown in Figure 4C. The DNA is only 2 nm in diameter. The significance of photoelectron imaging of DNA lies in the information content. The photoelectron image is formed by valence electrons, which are sensitive to molecular structure. This technique may prove useful in studies of protein–DNA interactions and small molecule (carcinogen)–DNA interactions. A rapid method of physical mapping of chromosomes (but not base sequencing at present instrument resolution) is another possibility if the differences in ionization potentials of the bases provide sufficient contrast to produce a modulation of brightness along the image of the DNA duplex.

Bibliography

Eland, J. H. D. (1984). "Photoelectron Spectroscopy." Butterworths, London.

Ghosh, P. K. (1984). "Introduction to Photoelectron Spectroscopy." Wiley Interscience, John Wiley & Sons, New York.

Griffith, O. H., and Rempfer, G. F. (1987). Photoelectron imaging: Photoelectron microscopy and related techniques. *In* "Advances in Optical and Electron Microscopy," Vol. 10 (R. Barer and V. E. Cosslett, eds.), pp. 269–337. Academic Press, London.

Griffith, O. H., and Rempfer, G. F. (1988). Photoelectron imaging and photoelectron labeling. *Ultramicroscopy* **24,** 299.

Habliston, D. L., Birrell, G. B., Hedberg, K. K., and Griffith, O. H. (1986). Early phorbol ester induced release of cell surface fibronectin: Direct observation by photoelectron microscopy. *European J. Cell Biol.* **41,** 222.